npn Bipolar Junction Transistor (BJT)

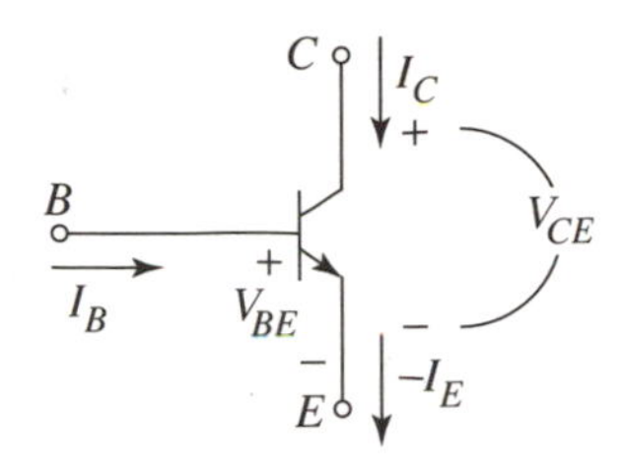

Constant-Current (Forward Active) Model

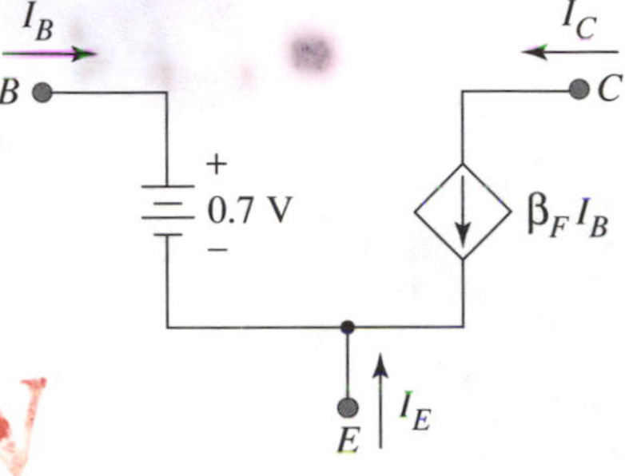

Ebers-Moll Equations

$$I_C = \alpha_F I_{ES}(e^{V_{BE}/V_{th}} - 1) - I_{CS}(e^{V_{BC}/V_{th}} - 1)$$

$$I_E = -I_{ES}(e^{V_{BE}/V_{th}} - 1) + \alpha_R I_{CS}(e^{V_{BC}/V_{th}} - 1)$$

Reciprocity $\alpha_F I_{ES} = \alpha_R I_{CS} = I_S$

Small-Signal Model

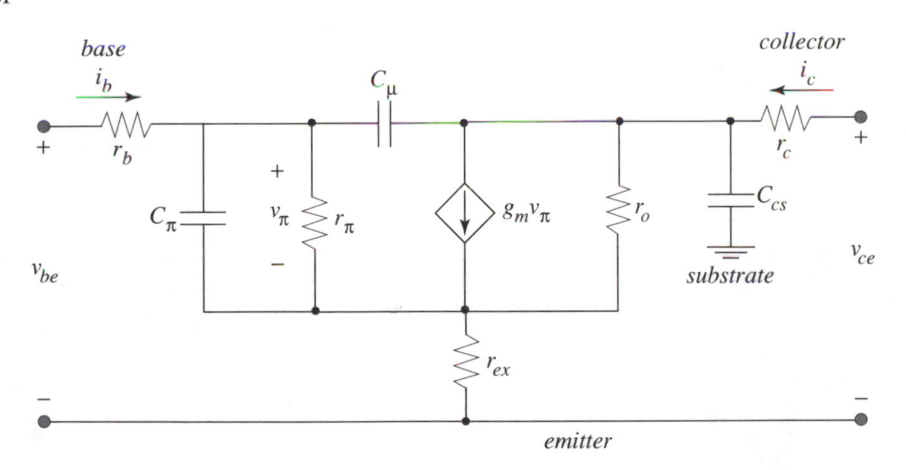

$$g_m = \frac{I_C}{V_{th}} = \frac{q}{kT}I_C \qquad C_\pi = g_m \tau_F + \sqrt{2}C_{jEo}$$

$$r_\pi = \frac{\beta_o V_{th}}{I_C} = \frac{\beta_o}{g_m} \qquad C_\mu = \frac{C_{\mu o}}{\sqrt{1 + V_{CB}/\phi_{Bc}}}$$

$$r_o = \frac{V_{An}}{I_C} \qquad C_{cs} = \frac{C_{cso}}{\sqrt{1 + V_{CS}/\phi_{Bs}}}$$

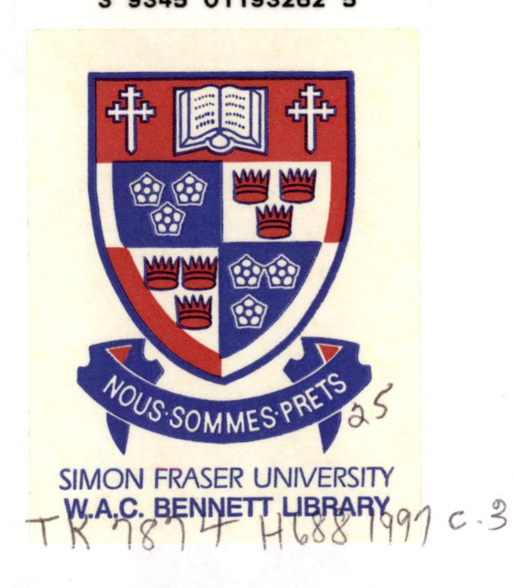

MICROELECTRONICS

AN INTEGRATED APPROACH

Roger T. Howe

Department of Electrical Engineering and Computer Sciences
University of California at Berkeley
Berkeley, California

Charles G. Sodini

Department of Electrical Engineering and Computer Science
Massachusetts Institute of Technology
Cambridge, Massachusetts

Prentice Hall Electronic and VLSI Series
Charles G. Sodini, Series Editor

Prentice Hall, Upper Saddle River, New Jersey 07458

Library of Congress Cataloging-in-Publication Data

Howe, Roger Thomas.
Microelectronics : an integrated approach / Roger T. Howe, Charles G. Sodini.
p. cm.
Includes bibliographical references and index.
ISBN 0-13-588518-3
1. Microelectronics. 2. Solid state electronics. I. Sodini, Charles G. II. Title.
TK7874.H688 1997
621.3815--dc20 96-17867
CIP

Acquisitions editor: Eric Svendsen
Production service: TKM Productions
Editorial/production supervision: Sharyn Vitrano
Copy editor: TKM Productions
Interior design: Lisa Delgado
Cover design: Joseph Sengotta
Art director: Amy Rosen
Managing editor: Bayani Mendoza DeLeon
Editor-in-chief: Marcia Horton
Director of production and manufacturing: David W. Riccardi
Manufacturing buyer: Julia Meehan
Editorial assistant: Kathryn Cassino

Printed in the United States of America

10 9 8 7 6 5 4 3 2

ISBN 0-13-588518-3

Prentice-Hall International (UK) Limited, London
Prentice-Hall of Australia Pty. Limited, Sydney
Prentice-Hall Canada Inc., Toronto
Prentice-Hall Hispanoamericana, S.A., Mexico
Prentice-Hall of India Private Limited, New Delhi
Prentice-Hall of Japan, Inc., Tokyo
Simon & Schuster Asia Pte. Ltd., Singapore
Editora Prentice-Hall do Brasil, Ltda., Rio de Janeiro

To our families,

Jenny

Nathan, Julie, Debbie and Greg

R. T. H.

Anne

Maria and John

C.G.S.

Contents

chapter 5 DIGITAL CIRCUITS USING MOS TRANSISTORS 259

chapter 6 THE pn JUNCTION DIODE 331

chapter 7 THE BIPOLAR JUNCTION TRANSISTOR 391

chapter 8 SINGLE-STAGE BIPOLAR/MOS TRANSISTOR AMPLIFIERS 461

chapter 11 DIFFERENTIAL AMPLIFIERS 685

chapter 12 FEEDBACK AND OPERATIONAL AMPLIFIERS 751

Preface

Our Vision

Electronics in the Electrical Engineering Curriculum

Analog and digital circuit design and electronic devices are one of the core course sequences in the undergraduate electrical engineering curriculum. As a gateway to the exciting and rapidly growing field of microelectronics, it is essential that these courses be taught well, both for future specialists and for those concentrating in other areas of electrical engineering or computer science. Students demand to learn material that is as close to modern applications as possible in order to stay motivated. Faculty are eager to introduce new concepts that previously have been taught in senior and even graduate-level courses into these core courses. At the same time, the courses must provide a solid foundation in the fundamentals of electronics rather than a sampling of the most popular current topics. Finally, circuit design, as its name implies, is a subject that should be well-suited for honing students' skills as engineering designers. All of these factors have forced a re-examination of the electronics sequence and opened the door to fresh approaches that are better adapted to the needs of students in the late 1990s.

A New Approach

This text offers an approach to analog/digital electronics and basic device physics that provides an answer to the changing demands on the undergraduate electronics courses. We have described these subjects in the context of modern silicon integrated circuit technology. This focus on ICs provides an efficient way to learn the central concepts of device physics and analog/digital circuits. It also motivates students by offering them many opportunities for early exposure to up-to-date applications.

In developing this text, we have paid careful attention to limiting the amount of material to only those concepts that are necessary for building a solid foundation in integrated devices and circuits. We have targeted an audience consisting of students who are embarking on a lifetime career in microelectronics *and* those students who are completing their study of this field. Our criterion in selecting topics was to avoid overwhelming students with extraneous information, while providing them with an appreciation for this exciting field. We made a concerted effort to partition material between the core electronics sequence and senior-level advanced courses in device physics, analog IC design, and digital VLSI design.

A Flexible Organization

This text is extremely flexible and has been class-tested in different "flavors" of core electronics courses. We feel that a large cross section of schools can make use of this text in a gateway course or course sequence. The device physics material is covered in multiple passes of increasing depth. The analog circuit design chapters are divided into CMOS-only and Bipolar/CMOS modules. As a result, the professor can set the precise device/circuit balance point to tailor a course to the specific needs of his or her students.

For example, a one-semester, device physics course could be taught using Chapters 1–4, 6, and 7 including both the first and second pass material. A one-semester circuits-only course could be taught using Chapters 1, and selected sections of 4 and 7 to describe the MOSFET and bipolar transistor; Chapter 5 for digital circuits; and Chapters 8–11 for analog circuits. By only considering CMOS circuits, the course could include one of the capstone chapters—Chapter 12 on op amp design or Chapter 13 on memory circuits. The book can also be used in a one-semester course that covers both circuits and device physics by using only the first pass of the device physics material and omitting Chapters 12 and 13. Our courses at UC Berkeley and MIT follow this syllabus.

The book can also be used in a two-semester core electronics sequence that covers all chapters. Today, many schools offer a one-semester device physics course followed by a two-semester circuit design course. By judicious elimination of non-essential material, this text allows undergraduate electronics to be taught in two semesters instead of three. As a result, microelectronics is more accessible to those not planning to concentrate in this area. It is our hope and intent that this text communicates the excitement and bright future of microelectronics to a broad cross section of students.

Why Use This Text?

This text is designed to increase the effectiveness of core electronics courses. Some of its features are:

1. An integrated circuit context throughout for both devices and circuits. Examples and problems use modern device parameters so that students are calibrated in today's IC technologies.
2. Analog and digital (including memory) circuit design are treated in a modular form in a single text. Students are able to learn the analog nature of digital circuits such as static RAM memory cells and sense amplifiers.
3. Device physics is dealt with in multiple passes that increase in complexity. Professors can incorporate varying amounts of this material depending on the needs of their students, without loss of continuity.
4. A basic introduction to IC technology. Cross sections and layout help to put abstract device and circuit concepts into real physical structures. Students gain an appreciation for the variations in device and circuit parameters.
5. Real-world design examples. Students are walked through the design process starting with simple approximations used in initial hand design and ending with refinement using SPICE simulations.
6. Capstone chapters for both analog and digital circuit design tie together concepts taught throughout the text. Chapter 12 treats the design of integrated CMOS and BiCMOS operational amplifiers. Chapter 13 discusses the design of MOS static and dynamic memory circuits.
7. Exercises, problems, and design problems of varying complexity are included in Chapters 2–13. Through the worked examples and design examples in the text, the student is exposed to the solution methods of IC engineers.

Supplements

Answers to even-numbered exercises are available from the book's web site:
http://www.prenhall.com/howe

Several supplements are also available from the book's web site, which will make it easier to use this innovative text. Lecture material from several versions of the course can be downloaded. In addition, teaching tips by Professor Mark Law of the University of Florida on using the text will be available. Comments and questions on the text can be submitted through an on-line form, which we hope will bring about greater interaction between the students and us.

A one-semester laboratory manual is available free of charge from the web site, which is used in a UC Berkeley junior-level electronics laboratory. The lab is based on a set of Microlinear, Inc. BiCMOS tile array "lab chips" that allow undergraduate students, for the first time, to measure modern integrated devices and analog and digital circuit building blocks. This set of chips is available from Electronic School Supply, Inc., 3130 Skyway Drive, Suite 108, Santa Maria, CA 93455, USA, (800) 726-0084.

A solutions manual is available free of charge to instructors adopting the text, with worked-out answers to all exercises, problems, and design problems.

Acknowledgments

There were several important contributors that helped us in the early stages of the development of this text and also through their thoughtful reviews during its creation. We are grateful for the helpful comments and suggestions by the following groups: Professors Supriyo Bandyopadhyay at the University of Notre Dame, Giorgio Casinovi at Georgia Institute of Technology, Bradley Clymer at the Ohio State University, Terri Fiez at Washington State Univeristy, Steven Garverick at Case Western Reserve University, Ted Higman at the Universtiy of Minnesota, Ross Holstrom at the Univeristy of Massachusetts at Lowell, Robert Krueger at the University of Wisconsin at Milwaukee, Stephen Long at the University of California at Santa Barbara, Randy Moss at the University of Missouri at Rolla, Dr. Andrew Robinson, formerly of the University of Michigan, and now at Advanced Technology Laboratories, and Dr. Laurence Walker of Digital Equipment Corporation.

The material in this book has been greatly influenced by our colleagues at UC Berkeley and MIT. In particular we want to thank Professors Bernhard Boser, Paul Gray, Richard Muller, Bill Oldham, and Donald Pederson at UC Berkeley and Professors Tayo Akinwande, Anantha Chandrakasen, James Chung, Jesus del Alamo, Clifton Fonstad, Qing Hu, Harry Lee, Rafael Reif, and Martin Schlecht at MIT.

Many of our students provided detailed comments and criticism. These include: Archana Goyal, Gani Jusuf, Cuong Pham, Wayne Yeung, and James Young at UCB and Tracy Adams, Steven Decker, Iliana Fujimori, Jeffrey Gealow, Gary Hall, Donald Hitko, Michael Lim, Joseph Lutsky, Daniel Reif, and Ching-Chun Wang at MIT. Wayne Yeung helped in developing examples for the device physics chapters. We would like to especially thank Raja Jindal at MIT who spent countless hours in helping to prepare and edit several chapters and Frank Cheung at UCB who tirelessly checked examples and prepared the solutions manual for the end-of-chapter problems. We would also like to thank Ms. Patricia Varley at MIT for her huge effort in manuscript word processing.

We are grateful to our editors at Prentice Hall for their efforts in bringing this book through the process of idea to reality. Alan Apt was instrumental in getting this text off to a good start and Sondra Chavez did a superb job on the developmental editing as well as helping us through the maze of manuscript preparation. Eric Svendsen and Marcia Horton provided valuable guidance near the end of this project. We would like to thank Mr. Ralph Pescatore at TKM for the final production and Mr. Scott Smith at Academy Art Works for final figure preparation.

Roger T. Howe

Charles G. Sodini

PRENTICE HALL SERIES IN ELECTRONICS AND VLSI

Charles G. Sodini-*Massachusetts Institute of Technology*
Series Editor

Prentice Hall has formed a partnership with Charles G. Sodini to enhance the field of electrical engineering. With the inception of this series, Prentice Hall affirms its commitment to publish innovative books at the graduate and undergraduate level in Circuit and Network Analysis, Electronics, Analog Electronics, Digital Electronics, VLSI Technology, VLSI Design, CAD Tools, and Device Physics. This text is the second of this series' offerings, which includes the recently published *Digital Integrated Circuits: A Design Perspective* by Jan Rabaey, and will soon feature a new offering in Device Physics. Prentice Hall is committed to publishing groundbreaking works in these fields, and establish this series' importance to the field of Electrical Engineering.

chapter 1

Introduction to Microelectronics

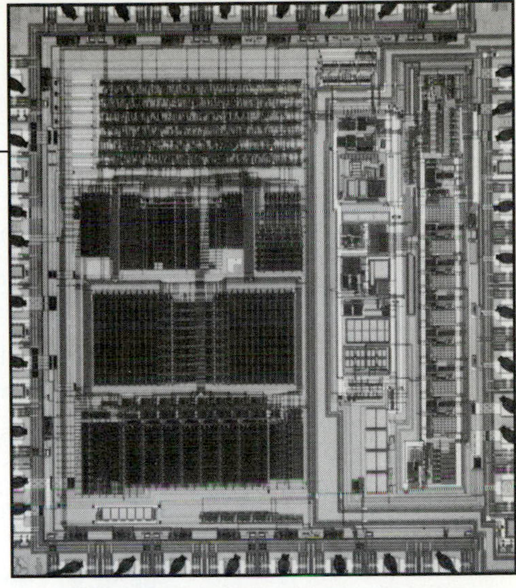

The AD7721 is a 16-bit Sigma-Delta Analog-to-Digital converter. This mixed-signal chip includes a 7th order analog modulator and digital filters to remove high frequency quantization noise. (Courtesy of Analog Devices, Inc.)

With the development of the integrated circuit the microelectronics industry is undoubtedly the most influential industry to appear in our society. Its impact on almost every person in the world exceeds that of any other industry since the beginning of the Industrial Revolution. The reasons for its success are:

- Exponential growth of the number of functions on a single integrated circuit,
- Exponential reduction in the cost per function,
- Exponential growth in sales (economic importance) for approximately forty years.

This growth has led to unprecedented quality at lower prices for consumer electronics such as televisions, cellular telephones, Video Cassette Recorders (VCR), stereo systems, etc. The computational power available to the individual has increased to the point that it has changed the way we think about problem solving. Communication technology including wired and wireless networks are leading us into the information age. These technological advances will continue to change the way we live.

The innovation responsible for these impressive results is the integration of electronic circuit components fabricated in silicon integrated circuit technology. In this chapter, we will broadly describe some of the basic concepts required to understand analog and digital circuit design. The connection between integrated circuit technology and circuit design is the device physics that governs the operation of basic components—primarily transistors, capacitors, and resistors.

Chapter Objectives

- Provide real-world motivation for learning and understanding microelectronics.
- Understand the basic function of the primary digital building block, the *inverter,* and gain an appreciation for the practical implementation issues.
- Review linear circuits including Kirchhoff's current and voltage laws, controlled sources, and the ideal operational amplifier.
- Use examples to introduce the concepts of bias, differential measurement, and charge-based circuits.
- Introduce semiconductor memory design as an example of digital/analog device technology interrelationships.
- Provide a roadmap for the subjects that will be introduced throughout this text.

1.1 INTRODUCTION

This section outlines a visual representation of the impressive increasing complexity of microelectronics over the last fifty years. Figure 1.1 shows the first transistor invented at AT&T Bell Laboratories in 1947.

Figure 1.2 shows a logic gate using transistor-transistor logic capable of performing the NAND logic function. The integrated circuit technology used to produce the transistors in Fig. 1.2 was developed in the early 1960s. Another impressive invention surfaced by the end of that decade—the first "industry standard" operational amplifier called the *μA709 op amp* as shown in Fig. 1.3.

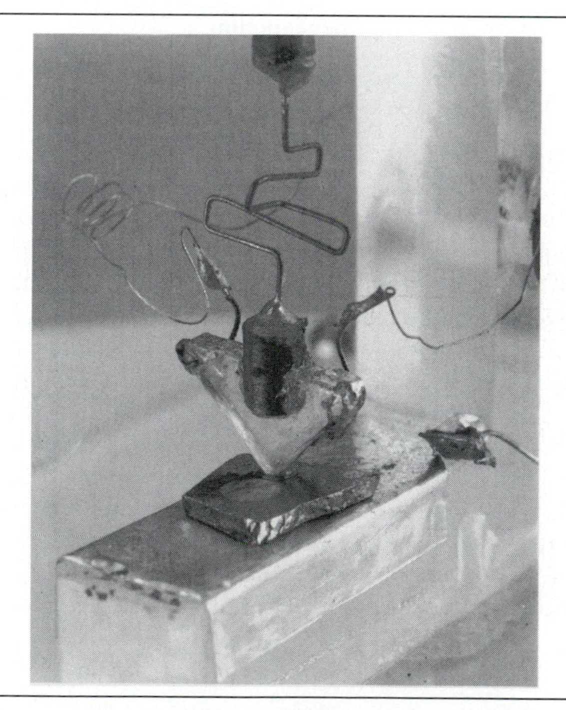

Figure 1.1 First transistor invented at Bell Laboratories, 1947. This point contact transistor used the same basic operation as modern, bipolar junction transistors. (Property of AT&T Archives. Reprinted with permission of AT&T.)

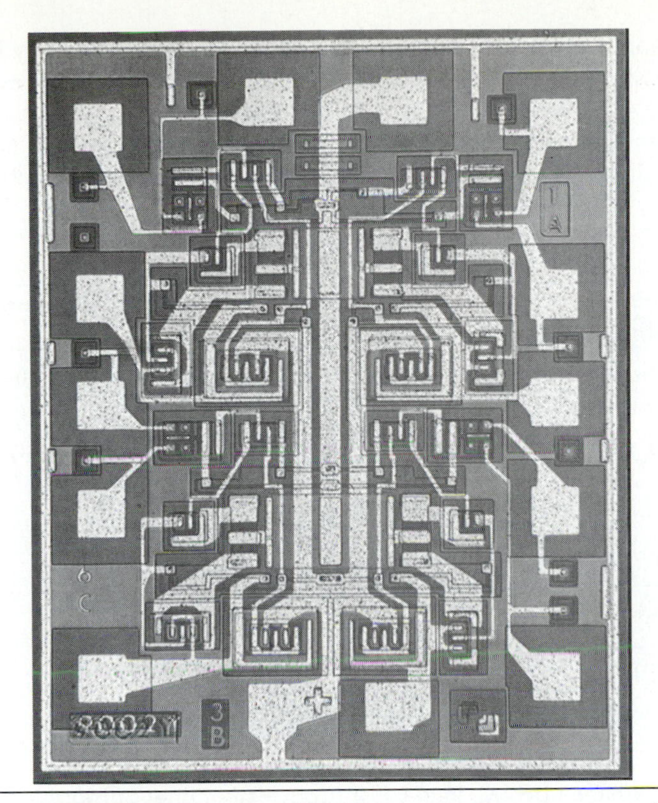

➤ **Figure 1.2** A quad two-input NAND gate using transistor-transistor logic. (Courtesy of National Semiconductor Corp.)

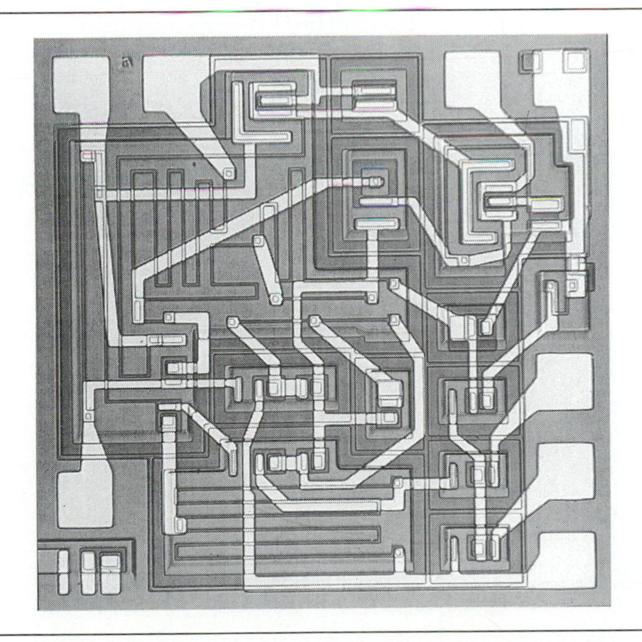

➤ **Figure 1.3** The first general purpose operational amplifier μA709. (Courtesy of National Semiconductor Corp.)

Semiconductor memory for computational systems began in the early 1970s. One of the most successful early designs at standardizing **dynamic random access memory, (DRAM)** was the 4096-bit DRAM shown in Fig. 1.4.

Although some microprocessors predated it, the architecture that has had the most impact in today's personal computer systems was the Intel 8086 microprocessor shown in Fig. 1.5.

➤ **Figure 1.4** A 4096 bit DRAM, Intel 2104. (Courtesy of Intel Corporation.)

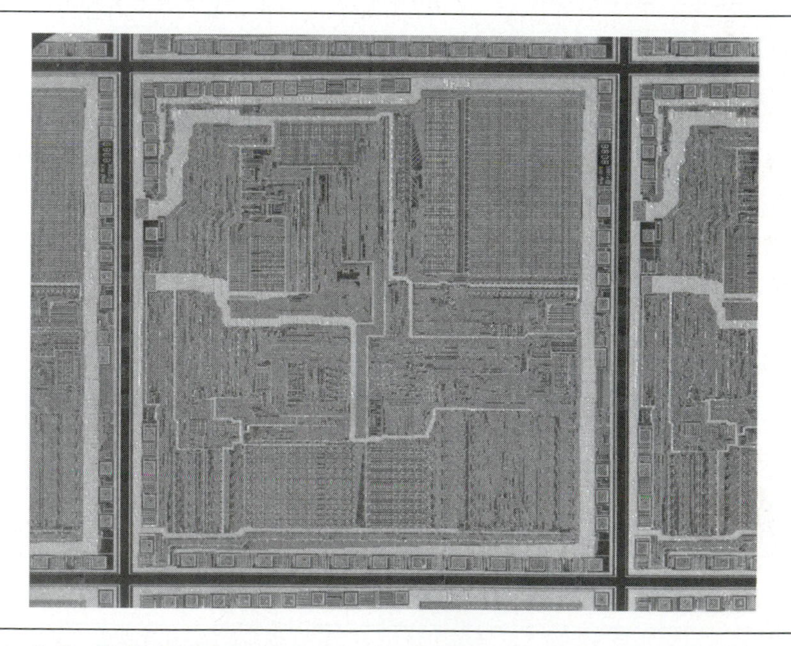

➤ **Figure 1.5** A 16 bit microprocessor, Intel 8086. (Courtesy of Intel Corporation.)

Continued integration of more complex functions on integrated circuits led to a family of **digital signal processors (DSP)** found in many consumer electronic systems. DSP chips are responsible for the wide variety of image and sound capabilities found in today's entertainment systems. The TMS320 shown in Fig. 1.6. was developed at Texas Instruments and was one of the first integrated DSP architectures.

More recently, the combination of digital signal processing and communication circuits offer communication between computational systems. A fax modem chip set is shown in Fig. 1.7. The mixed signal chips include both analog and digital circuits,

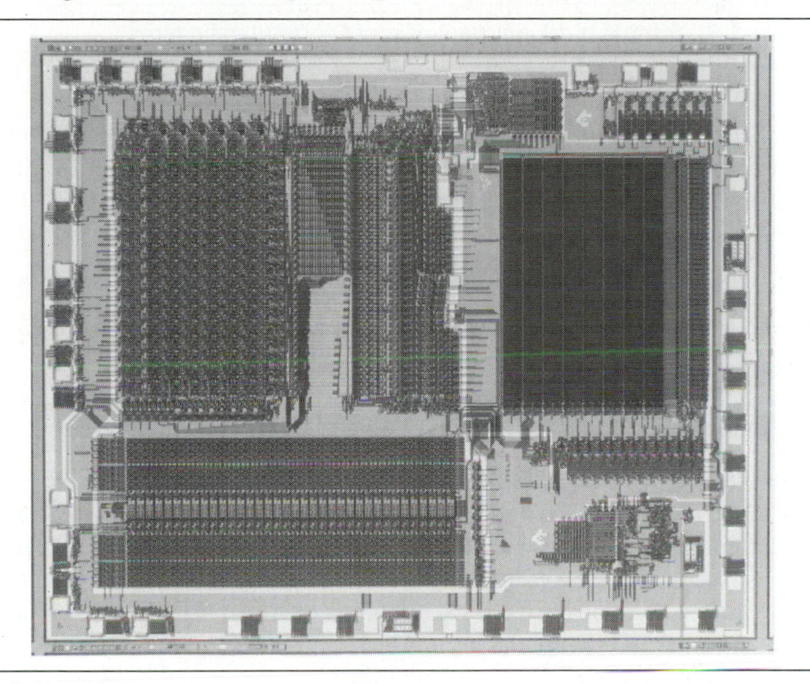

➤ **Figure 1.6** A digital signal processor, Texas Instruments TMS320. (Courtesy of Texas Instruments Corporation.)

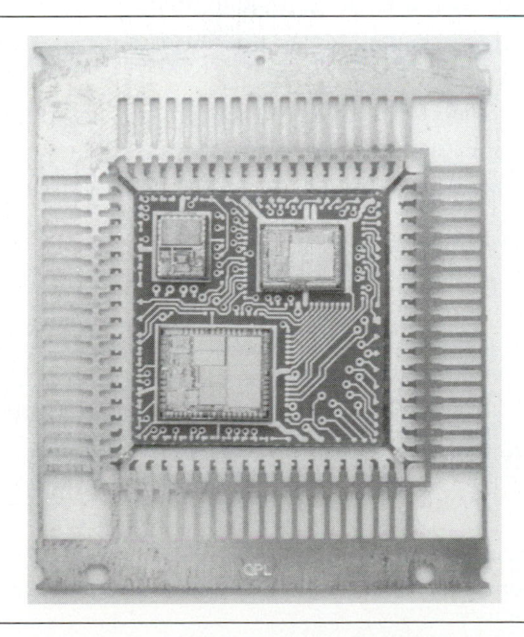

➤ **Figure 1.7** A fax modem chip set demonstrating mixed signal (analog and digital) integration, Rockwell RC 144DPL. (Courtesy of Rockwell Corporation.)

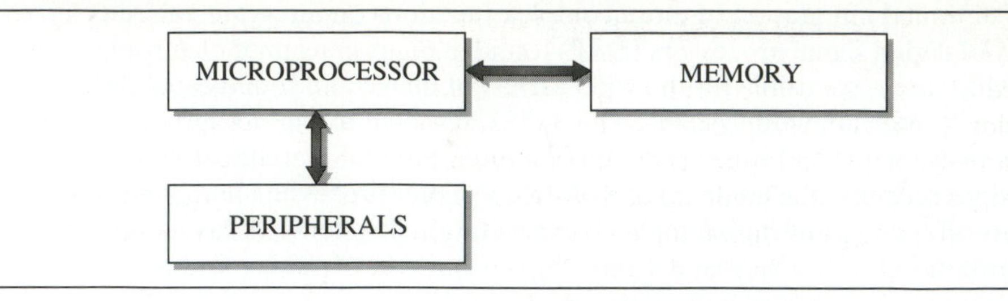

➤ **Figure 1.8** Simple block diagram of most computer systems.

making it a good example of the integration of disciplines required for modern microelectronic design.

Fig. 1.8 shows a simple block diagram of most computer systems. Most computer systems are composed of:

Processing unit—usually based on a microprocessor chip,

Memory unit—based on a variety of semiconductor memory and higher density magnetic memory such as hard disks,

Peripherals—interface the computational system to the outside, real world including printers, displays, keyboards etc.

Microelectronic systems have become more important as the interface between the real world and digital computational world is bridged. For example, compact disc audio is not possible unless the analog signal produced by the voice or musical instrument is amplified and then converted from the analog to the digital domain. Figure 1.9 shows a block diagram of a typical microelectronic interface system that converts physical signals to digital signals.

In this chapter, we will:

1. Describe the fundamental circuit used in digital circuits—namely the inverter.
2. Describe the conversion of light, temperature, and acceleration to an analog signal represented as a voltage.
3. Use these examples to introduce the concepts of bias, differential measurement, and charge-based circuits.
4. Provide a review of linear circuits including the ideal op amp.
5. Introduce semiconductor memory.

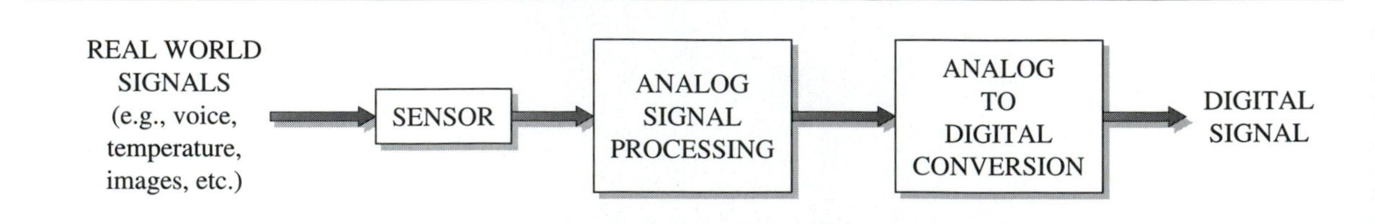

➤ **Figure 1.9** Typical microelectronics interface system: physical-to-digital conversion.

➤ 1.2 THE DIGITAL INVERTER

Figure 1.10 is a die photograph of the Intel Pentium™ microprocessor. The major functional blocks of this processor are also shown on the die photo. If we take a specific architectural block such as the integer execution unit and enlarge the die photo, we would find that it is made up of thousands of digital logic gates. We will study the analysis and design of digital logic gates fabricated with metal oxide semiconductor (MOS) transistors in Chapter 5. In this section, we will describe the basic function of the most primitive digital logic gate—the **inverter**.

The **ideal inverter** takes as its input the logic state **0** or **1** and returns the opposite state. One way to build the ideal inverter is to use voltage to represent the states **0** and **1.** We will define a logic **0** when the voltage is less than 0 V and a logic **1** when the voltage is greater than 0 V. To build an ideal inverter, we need an **ideal switch** that has the following properties.

1. If a logic **1** is input to the switch, the switch is considered closed with 0 Ω on-resistance.
2. If a logic **0** is input to the switch, the switch is considered open with an on-resistance that approaches infinity.
3. The time to change the state of the switch should be as small as possible.

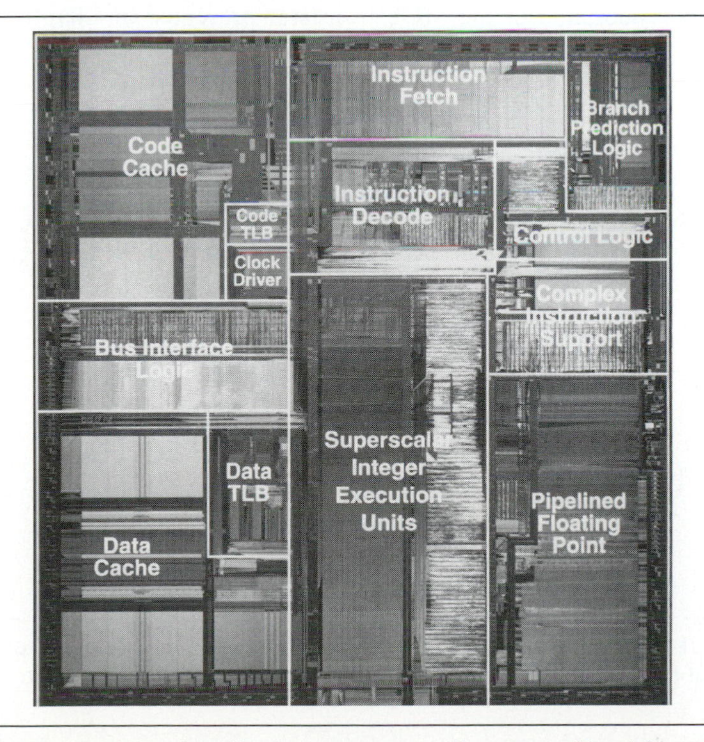

➤ **Figure 1.10** Intel Pentium™ microprocessor. (Courtesy of Intel Corporation.)

We define a second ideal switch called $\overline{\textbf{switch}}$, which is open when a logic **1** is input and closed when a logic **0** is input. An ideal inverter formed by these ideal switches is shown in Fig. 1.11.

We have provided a positive and negative power supply. If the input is a logic **1**, the bottom switch will be closed, the top switch will be open, and the output voltage will become V^- representing a logic **0**. If we input the opposite state, the top switch will be closed, the bottom switch will be open, changing the output voltage to V^+, which represents a logic **1**.

In practical inverters it is more convenient to use a single power supply, for example 5 V. If a logic **1** and logic **0** are redefined as having a transition point at one-half the power supply value, we can still achieve ideal inverter operation. The switch diagram is shown in Fig. 1.12(a). The corresponding voltage transfer function between the input voltage and output voltage is shown in Fig. 1.12(b). We see that when a logic **0** is input, represented by an input voltage less than $V_{DD}/2$, the output voltage will be at V_{DD}. If the input voltage is greater than $V_{DD}/2$, the bottom switch in the switch diagram will be closed, the top switch will be open, and the output voltage will go to 0 V representing a logic **0**. Figure 1.12(c) shows an actual voltage transfer characteristic for typical inverters found in modern microprocessor technology. Notice that the edges are more rounded and the transition is not vertical, but the overall appearance of the voltage transfer function approximates the ideal characteristic.

It is important to understand the basic operation of the transistors before learning how to design the transistors and circuits to yield a good approximation to the ideal voltage transfer characteristic. This operation is governed by basic device physics; hence, we begin this text with the basic device physics concepts necessary to understand transistor operation. Chapter 3 will discuss the electrostatics required to understand the operation of the MOS transistor presented in Chapter 4. The key

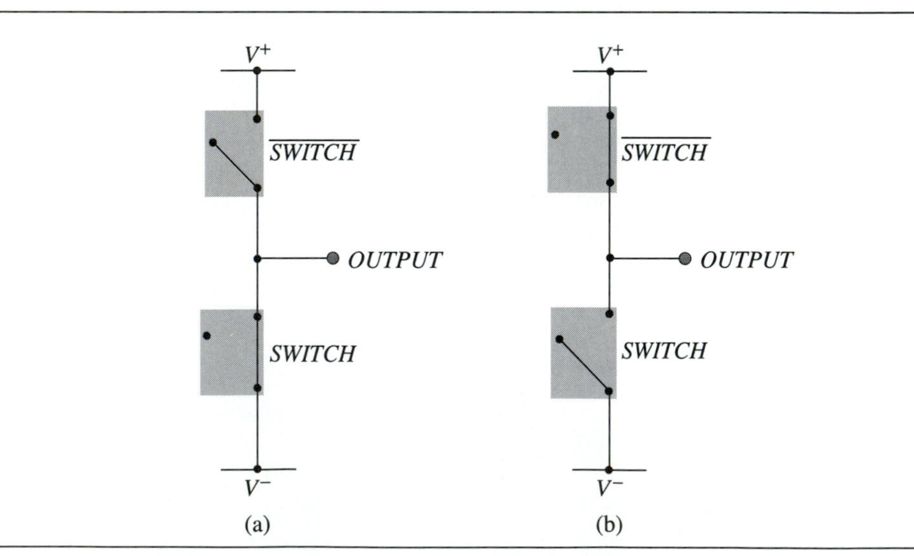

➤ **Figure 1.11** Ideal inverter using ideal switches and dual power supplies. (a) Logic **1** input and a logic **0** output. (b) Logic **0** input and a logic **1** output.

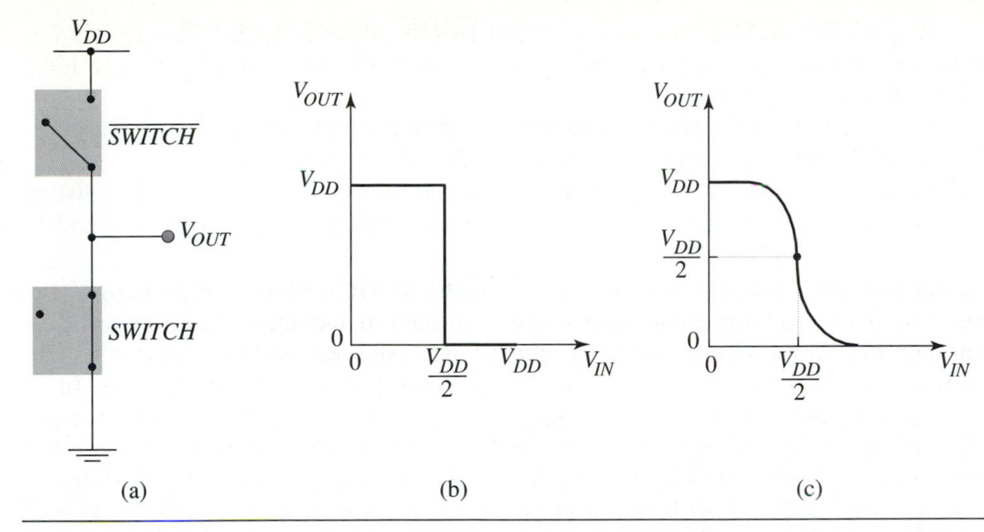

➤ **Figure 1.12** Single power supply inverter. (a) Inverter formed by ideal switches. $V_{IN} > V_{DD}/2$ (b) Ideal voltage transfer function. (c) Actual voltage transfer function.

point is that a basic knowledge of transistor operation is required in order to learn how to design circuits that approach ideal operation.

1.2.1 Practical Inverter-Dynamic Operation

A load capacitance C_L is attached to the output in a practical inverter. This capacitance models the gates that the inverter is driving. Most of the time required to change the state of the output is the time it takes to charge and discharge the load capacitance. In general, the time to open and close the switches is small compared to charging and discharging the load capacitance. With this assumption in mind, we will describe the time it takes to change the state of the output from a logic **1** to a logic **0** using an idealized current-voltage transistor characteristic found in Fig. 1.13. We will learn about MOS and bipolar transistor *I-V* characteristics in Chapters 4 and 7, respectively.

The current-voltage characteristic for the transistor approximates the ideal switch. If the input to the transistor is a logic **0**, there is no current through the transistor for all voltages across it, which means it is an open switch. When the input to

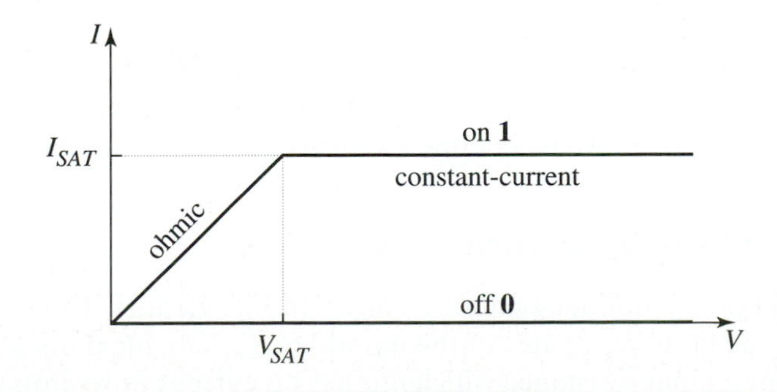

➤ **Figure 1.13** Idealized transistor *I-V* characteristic.

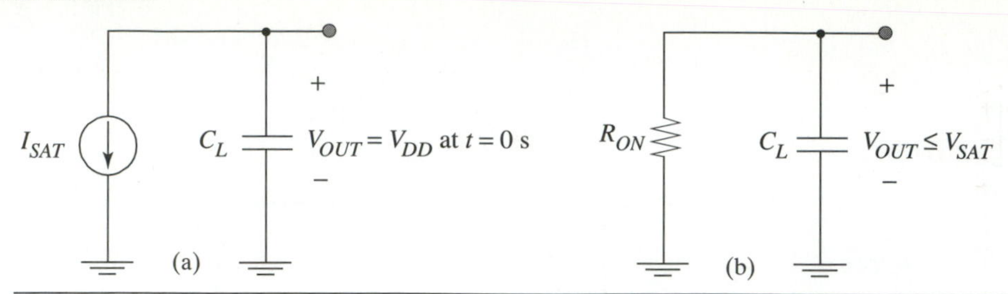

Figure 1.14 Model to demonstrate inverter transient operation switching V_{OUT} from V_{DD} to 0 V. (a) Current source of value I_{SAT} discharging C_L when $V_{SAT} \leq V_{OUT} \leq V_{DD}$. (b) Resistor of value $R_{ON} = V_{SAT}/I_{SAT}$ discharging C_L when $V_{OUT} < V_{SAT}$.

the switch is a logic **1**, the current-voltage characteristic has two regions of operation. For low voltages below V_{SAT}, the current is proportional to the voltage, which is similar to the *I-V* characteristic of a resistor. If the voltage across the switch is greater than V_{SAT}, the current saturates at a value called I_{SAT}. Therefore, we have two regions of operation—the ohmic region and constant-current region.

Figure 1.14 shows half the inverter to describe its transient operation. Assume the output voltage is equal to V_{DD} at $t = 0^-$. If a logic **1**, $V_{IN} = V_{DD}$ is input to the switch at $t = 0$, the output voltage will begin to fall as capacitor C_L is discharged. If we assume V_{DD} is greater than V_{SAT} at $t = 0^+$, the current through the switch will be equal to I_{SAT} and the time varying output voltage is given by

$$V_{OUT}(t) = V_{DD} - \frac{I_{SAT}t}{C_L} \text{ for } V_{OUT} \geq V_{SAT} \tag{1.1}$$

When the output voltage becomes less than V_{SAT}, the switch will enter its ohmic region. The time to discharge the capacitor will be governed by the *RC* time constant where the capacitance is C_L, and the resistance of the switch is given by

$$R_{ON} = V_{SAT}/I_{SAT} \text{ for } V_{OUT} \geq V_{SAT} \tag{1.2}$$

To increase the speed of this transient operation, we want the transistor to have a low on-resistance and a large saturation current. In addition, we would like the load capacitance to be small. In Chapter 4, we will learn enough about the operation of the MOS transistor to understand how to achieve these goals.

1.3 MICROELECTRONIC SENSING SYSTEMS

This section will demonstrate how to use an operational amplifier to convert physical quantities such as light, temperature, and acceleration into a voltage signal. Before beginning this description, we will define an ideal operational amplifier and describe its function.

1.3.1 The Ideal Operational Amplifier

An ideal operational amplifier symbol is shown in Fig. 1.15(a). The input voltage is applied across the input resistance of the amplifier R_{in}. The **ideal operational amplifier** has an infinite input resistance which implies no current flows into the amplifier. The input voltage is amplified by a voltage controlled voltage source that has a gain which is a function of signal frequency. The voltage gain $a(j\omega)$ approaches infinity for

(a) (b)

➤ **Figure 1.15** Operational amplifier (a) Symbol (b) Ideal linear circuit model ($R_{in} \rightarrow \infty$; $R_{out} \rightarrow 0\ \Omega$).

all input signal frequencies in the ideal case. In other words, the ideal operational amplifier is capable of amplifying with an infinite voltage gain any voltage signal at any frequency. In practice the gain and phase response of an operational amplifier varies with frequency. You will learn more about these responses in Chapter 10 and how to design an operational amplifier in Chapter 12. Finally, the ideal operational amplifier has an output resistance, R_{out} equal to zero so the voltage generated by the voltage controlled voltage source appears directly across the load resistor R_L. A circuit model for the ideal operational amplifier is shown in Fig. 1.15(b).

The goal in operational amplifier design is to continue to improve the device and circuit design to approximate the ideal op amp specification. Since many operational amplifier designs use both MOS and bipolar transistors, we will discuss the device physics that governs bipolar transistor operation in Chapters 6 and 7 before discussing the basic circuit building blocks required to design an operational amplifier in Chapters 8 and 9. Chapter 10 discusses the frequency response of these circuit blocks. Chapter 11 explains the differential amplifier that is required at the input of an operational amplifier. Chapter 12 presents the concepts of feedback and operational amplifier design.

1.3.2 Light-to-Voltage Conversion

When light is shined on a reverse-biased pn junction diode, current flows in the diode that is proportional to the intensity of the light. Figure 1.16 shows a circuit with a reverse-biased pn junction diode and a corresponding model representing the signal current that is proportional to light and a source resistance R_S.

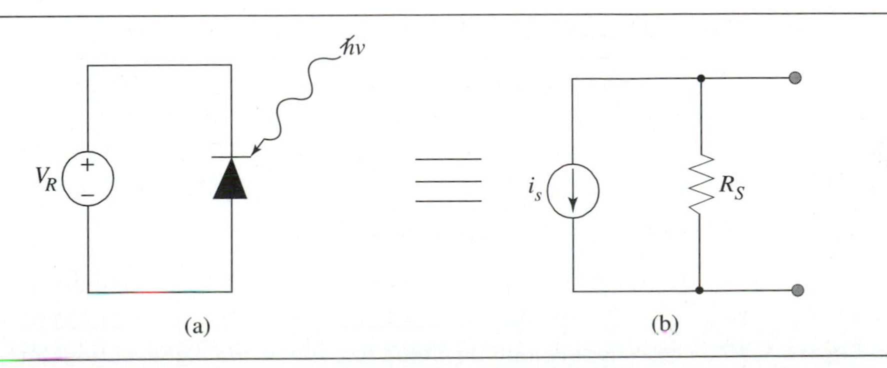

➤ **Figure 1.16** Light shining on a reverse-biased pn junction diode. (a) Circuit. (b) Model with signal current source i_s proportional to light intensity and source resistance R_S.

For the purposes of this discussion, assume the light is proportional to the signal current and the source resistance is infinite. If we feed this signal current into an operational amplifier with a feedback resistor R_F connected between the output and negative input of the operational amplifier (Fig.1.17(a)), the output voltage will be proportional to the signal current.

To see why this happens, consider applying the circuit model for the ideal operational amplifier with the added feedback resistor as shown in Fig. 1.17(b). We can analyze this circuit by writing an equation for the input voltage given by

$$v_{in} = i_s R_F - a v_{in} \tag{1.3}$$

Rearranging, we find

$$i_s = \frac{(1 + a) v_{in}}{R_F} \tag{1.4}$$

We can also write an equation relating the input voltage to the output voltage as

$$v_{in} = v_{out} / a \tag{1.5}$$

Substituting Eq. (1.5) into Eq. (1.4) and recognizing that the gain of the operational amplifier is much greater than 1, we can solve for the transfer function between the output voltage and input signal current given by

$$\frac{v_{out}}{i_s} = \left(\frac{a}{1 + a}\right) R_F \approx R_F \tag{1.6}$$

An easier way to solve for this transfer function is to recognize that the input to the operational amplifier is a **virtual ground,** which is equivalent to saying that the input voltage and input current are equal to zero. The input current is equal to zero by the definition of an ideal operational amplifier. Why is the input voltage $v_{in} = 0$ V? The gain of the ideal op amp is infinite, while its output is bounded. We can write KCL at the negative input node of the operational amplifier in Fig. 1.17(a) by recognizing that the current through the feedback resistor is simply v_{out}/R_F. The current

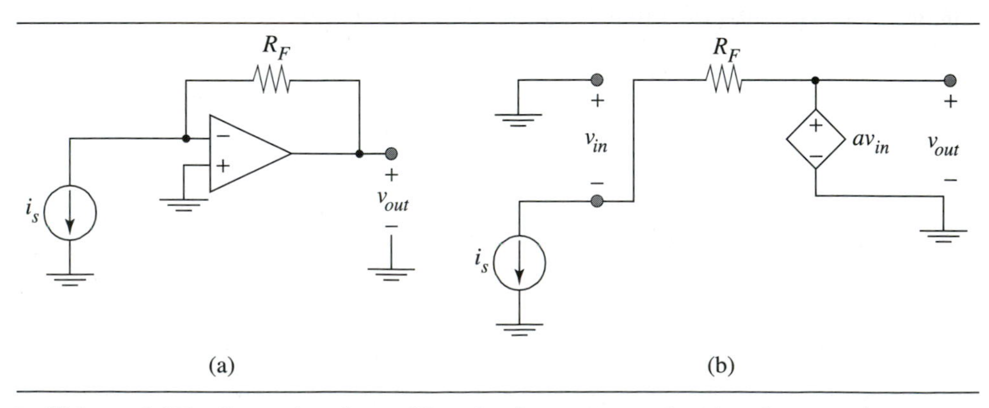

Figure 1.17 Operational amplifier circuit to convert the signal current into a voltage. (a) Circuit diagram. (b) Linear circuit model.

from the signal source must be equal to that current. By equating these two, we find the same result as given in Eq. (1.6).

The pn junction diode has a current-voltage relationship given by

$$i_D = I_o(e^{qv_D/kT} - 1) \tag{1.7}$$

We will learn the device physics necessary to derive this equation in Chapter 6. If the voltage across the diode, v_D is positive, we say that the diode is forward-biased and the current exponentially increases with the voltage. A reverse-biased diode has a negative voltage across the diode. If it is much less than kT/q, which is approximately 26 mV at room temperature, the exponential term is small. Hence, the pn junction current is given by

$$i_D = -I_o \quad for \quad v_D \ll kT/q \tag{1.8}$$

I_o is proportional to the intensity of the light; hence, the signal current through the reverse-biased diode is proportional to the light intensity.

The ideal operational amplifier can be used to provide this reverse bias by applying a negative voltage $-V_R$ at the positive input of the ideal op amp as shown in Fig. 1.18. This circuit can be analyzed by applying the principle of the virtual ground and writing a KCL equation at the negative input node. The voltage across the input of the op amp is 0 V, which implies that the voltage at the negative input of the op amp $v_- = -V_R$. We start by finding the current through the feedback resistor i_{R_F} and equate it to the signal current in the pn junction diode

$$i_s = \frac{v_{OUT} - (-V_R)}{R_F} \tag{1.9}$$

Rearranging this equation, we find

$$v_{OUT} = R_F i_s - V_R \tag{1.10}$$

We see that the output voltage is proportional to the signal current, but has an additional term $-V_R$ that is an offset voltage. In other words, if there is no light and the signal current is equal to zero, we still have an output voltage equal to $-V_R$. We can

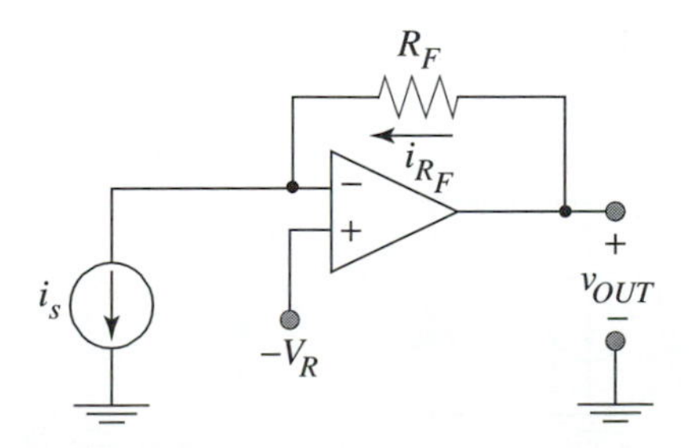

➤ **Figure 1.18** Operational amplifier with $-V_R$ applied to the positive input to reverse bias the pn junction diode.

remove this offset voltage by biasing the operational amplifier with a bias current that is equal to V_R/R_F as shown in Fig. 1.19.

This bias current precisely cancels the current through the feedback resistor that was caused by applying $-V_R$ at the positive input of the op amp. Biasing amplifiers so that the output is in the middle of the operating range (in this case, 0 V) is a common analog circuit design technique.

1.3.3 Temperature-to-Voltage Conversion

Many resistors in integrated circuit technology vary with temperature. If the temperature range is small enough, this variation can be approximated as a linear function so the change in resistance is proportional to temperature. Another principle in analog circuit design is that given a non-linear function between two variables, one can approximate it as a linear function if the change is small enough. Figure 1.20 is a graph that demonstrates this concept.

The temperature dependence of the resistor across this small range $T_1 < T < T_2$ is given by

$$R_x = R_o + \Delta R = R_o + \alpha(T - T_o) \tag{1.11}$$

We can place this temperature dependent resistor in an operational amplifier circuit in the non-inverting configuration shown in Fig. 1.21.

Using the virtual ground concept, we know V_R appears across R_2, yielding a current through that resistor. In addition, the output voltage causes a current to flow into the series combination of the temperature dependent resistor R_x and the temperature independent resistor R_2. An equation relating these currents is

$$\frac{V_R}{R_2} = \frac{v_{OUT}}{R_x + R_2} \tag{1.12}$$

which can be rewritten as

$$v_{OUT} = \left(1 + \frac{R_x}{R_2}\right)V_R \tag{1.13}$$

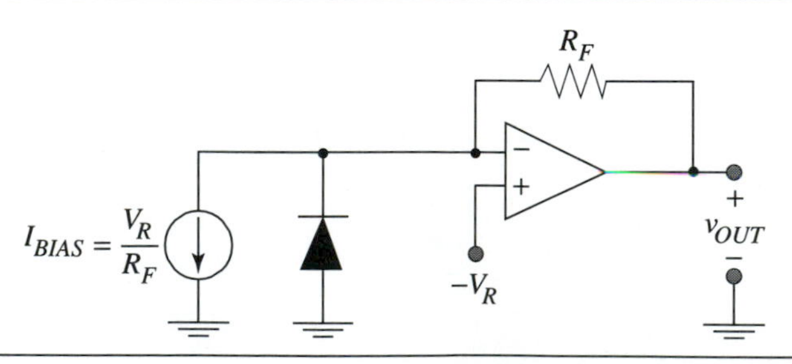

Figure 1.19 Operational amplifier circuit with bias current to remove output offset voltage.

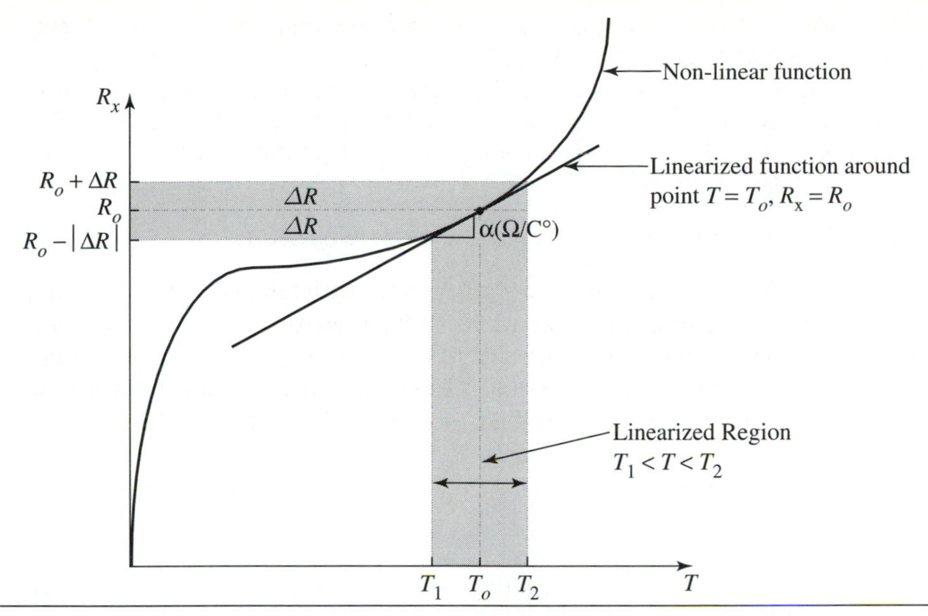

➤ **Figure 1.20** Graphical demonstration of linearizing a function across a small range.

Substituting Eq. (1.11) yields

$$v_{OUT} = \left(1 + \frac{R_o}{R_2}\right)V_R + \frac{\alpha}{R_2}(T - T_o)\,V_R \tag{1.14}$$

When the temperature being measured is equal to T_0, we find the output voltage still has a non-zero value or an **offset voltage**.

Another technique pervasive in analog integrated circuit design is to duplicate the measurement circuit with elements that do not change with the measured signal. In other words, we build the same circuit found in Fig. 1.21, but substitute a temperature independent resistor for R_x, that has a value equal to R_o at T=T_o. Building matched elements that have similar values is not difficult using integrated circuit technology and is a common analog circuit design strategy. The differential measurement circuit is shown in Fig. 1.22.

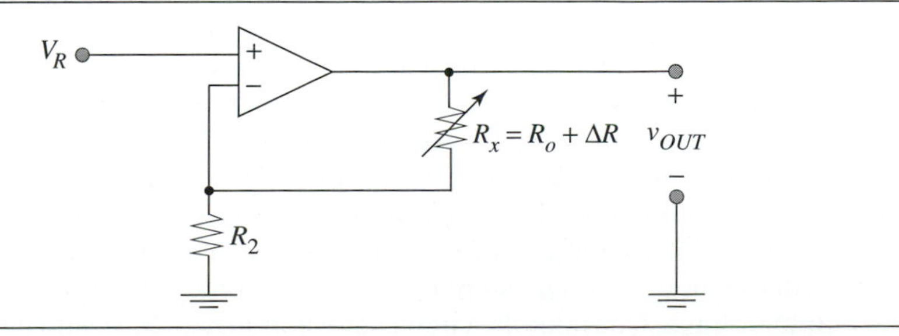

➤ **Figure 1.21** Temperature dependent resistor R_x placed in an operational amplifier in the non-inverting configuration.

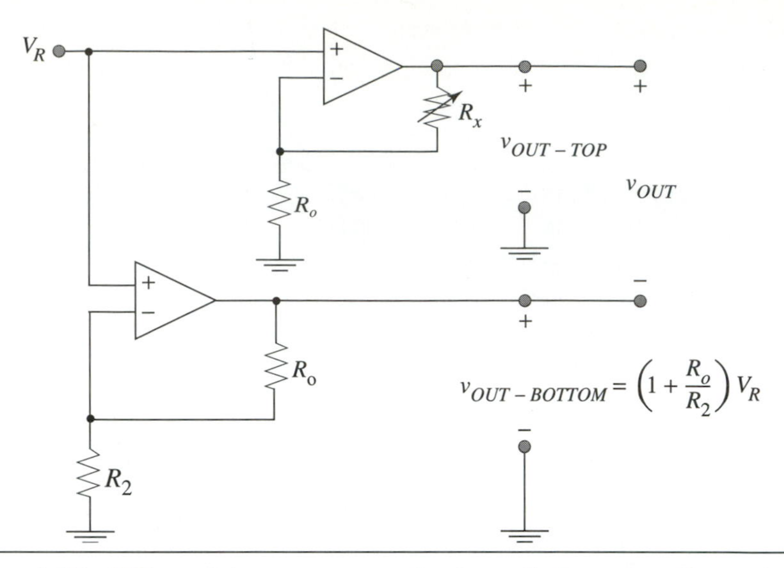

➤ **Figure 1.22** Differential measurement circuit to eliminate the offset voltage. If R_o and R_x are the same value for the top and bottom circuit at $T = T_o$, then $v_{OUT} = 0$ V when $T = T_o$.

The output voltage of interest is the difference between the output voltage of the top op amp and the bottom one. It should be noted that the output voltage of the bottom op amp is precisely the voltage that must be subtracted from the output voltage of the top op amp to remove the offset voltage.

Many of the details presented in the above discussion will be covered in-depth in this text. It is assumed that the student is familiar with circuits involving ideal operational amplifiers.

1.3.4 Acceleration-to-Voltage Conversion

Recently, silicon integrated circuit technology has become a platform for micromechanical devices as well as electronic devices. This technological advance has led to the integration of micromechanical sensors to act as a transducer between physical and electronic signals. A large number of micromechanical devices have been reported in the literature, several having become commercially viable. An example of a commercial product which integrates both micromechanical and electronic devices is shown in Fig. 1.23. The ADXL-50 from Analog Devices is an integrated accelerometer that detects automobile collisions for airbag deployment.

One approach to converting a displacement into an electrical signal is by measuring the change in capacitance. Fig. 1.24(a) shows a conceptual representation of a suspended mass made of low resistivity silicon between two conducting plates. The mass moves when this device is accelerated and the capacitance between nodes 1 and 2 changes. If the motion is small enough, we can linearize the capacitance vs. acceleration curve around the point $a = a_o$, $C_{1-2} = C_o$, as shown in Fig. 1.24(b).

We will use an operational amplifier circuit to convert the change in capacitance to an output voltage. Consider the circuit shown in Fig. 1.25(a) with clock waveforms ϕ_1 and ϕ_2 shown in Fig. 1.25(b). The operation begins with ϕ_1 high, implying that switches S_1 and S_2 are closed and switch S_3 is open. At this time capacitor C_F, which is

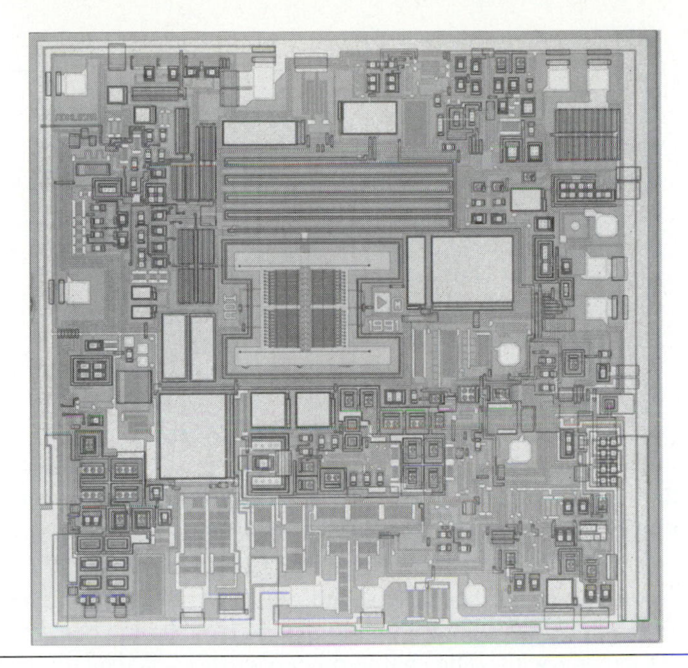

➤ **Figure 1.23** Accelerometer demonstrating the integration of micromechanical and electronic devices on a single integrated circuit, Analog Devices ADXL-50. (Courtesy of Analog Devices Inc.)

tied between the output and negative input of the op amp, is discharged and the output voltage v_{OUT} is equal to 0 V since the positive input of the operational amplifier is grounded. At the same time, the top plate of the variable capacitor C_{1-2} is charged to a voltage V_R. During this time, the charge on this sensing capacitor is given by

$$Q_{C_{1-2}} = (C_o + \Delta C)V_R \tag{1.15}$$

On the next phase when ϕ_2 is high and ϕ_1 is low, switches S_1 and S_2 are open and switch S_3 is closed. During this time, the charge that had been stored on capaci-

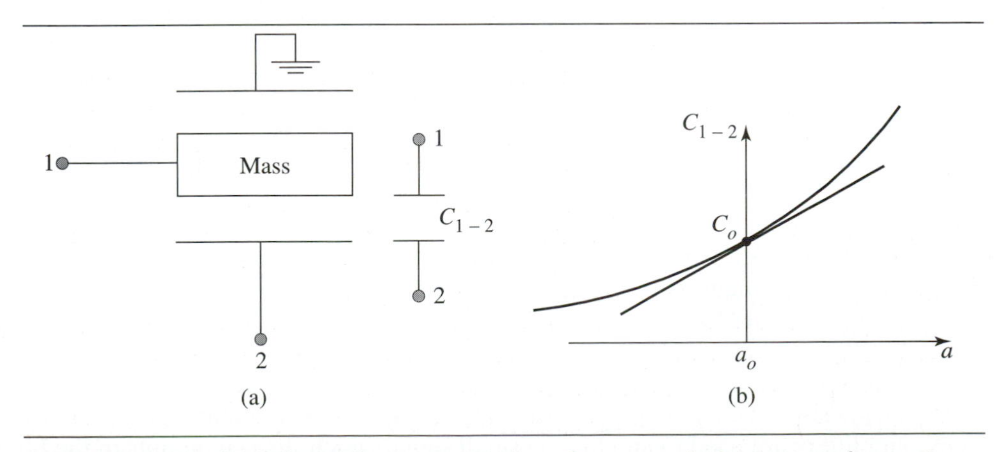

➤ **Figure 1.24** (a) Schematic representation of a suspended mass accelerometer. (b) Capacitance vs. acceleration curve.

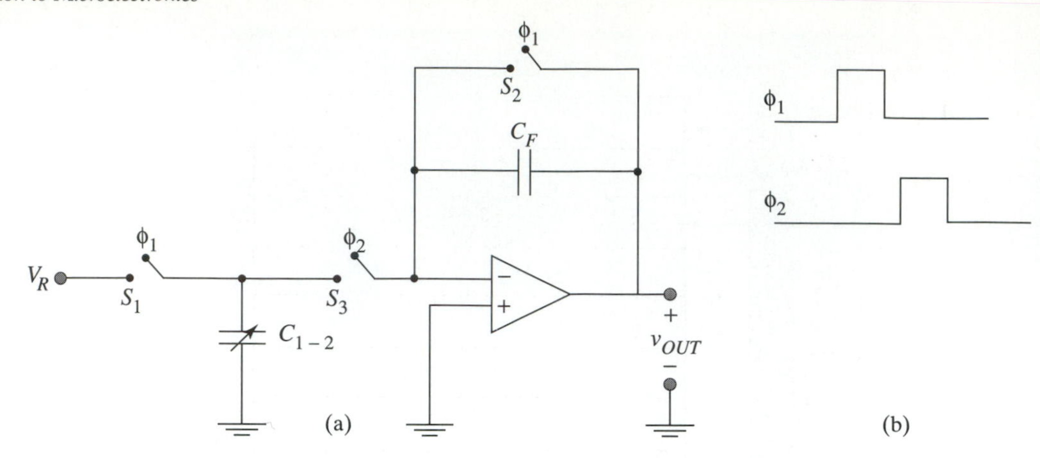

➤ Figure 1.25 Operational amplifier circuit to convert change in capacitance to an output voltage. (a) Circuit diagram. (b) Clock waveforms.

tor $C_{1\text{-}2}$ discharges to ground because the negative input of the op amp is at virtual ground. However, since the op amp can have no input current, charge can only flow to the feedback capacitor C_F. The charge on capacitor C_F is exactly the same as was stored on the sensing capacitor $C_{1\text{-}2}$. The output voltage of the op amp is now given by Q_{C_F}/C_F since the left side of the feedback capacitor is at virtual ground. The output voltage is

$$v_{OUT} = -\frac{Q_{C_F}}{C_F} = -\left(\frac{C_o + \Delta C}{C_F}\right)V_R \tag{1.16}$$

Notice that the output voltage is proportional to the change in capacitance due to the acceleration plus an offset term. As in the previous cases, differential techniques could be applied to remove this offset voltage.

In this section, we described several electronic circuits which convert physical quantities to an electrical output voltage. An ideal operational amplifier was used to perform this conversion. These examples are intended to be a review of the ideal operational amplifier and an introduction to the concepts of bias and differential measurements.

➤ 1.4 MEMORIES

Memory circuits are an excellent example of the interrelationship of process technology, device physics, digital and analog circuit design. A block diagram of a typical memory is shown in Fig. 1.26.

The memory cells store one bit of information and are arranged in a two-dimensional array. Each memory cell is accessed to either be read or written by a control signal called the **word line.** The data being written or read from the memory are connected to the cell via the **bit line** that is perpendicular to the word line. On the left side of the array, addresses are input to a row decoder to select the specific word line that chooses a row of memory cells to be read or written. At the bottom of the array, sense amplifiers are used to amplify the small signal which the cell outputs to the bit line, to a full logic level during a read operation. During a write operation, these sense amplifiers act as drivers to write the information back into the memory cell. Finally,

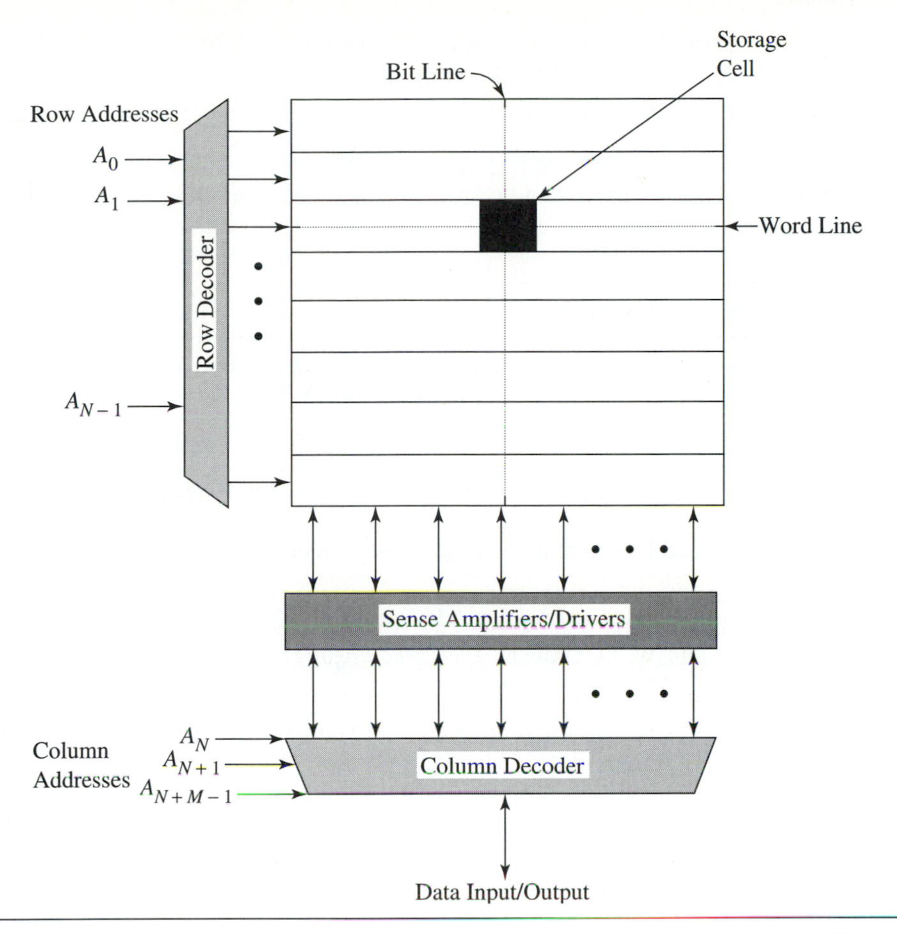

➤ **Figure 1.26** Block diagram of a typical semiconductor memory.

the column decoder is used to select the appropriate vertical line so data from a single cell can be read or written. A more complete description of semiconductor memory operation is given in Chapter 13.

Semiconductor memory design requires extremely dense circuits in order to achieve maximum memory capacity. Innovation in process technology and a detailed understanding of device physics is required to continue to shrink the memory cell size. The sense amplifiers use differential circuits, as well as feedback, to amplify the small signal from the memory cell. The decoders use specially designed inverters that are precharged to improve their density and power dissipation. Chapter 13 which discusses semiconductor memories is not only an introduction to this exciting area, but also a review of the concepts presented throughout the text.

➤ SUMMARY

This chapter offered a brief overview of the type of material that will be presented throughout the text. In addition to the examples given, we reviewed many of the basic concepts required to study microelectronics. The concepts we expect the student has seen in previous courses are listed below. A review of these concepts is given at the beginning of the appropriate chapters.

- Chapter 1—Basic linear circuit techniques using Kirchhoff's current and voltage laws, controlled sources, and the ideal op amp
- Chapter 3—Basic electrostatics
- Chapter 5—Basic Boolean algebra and logic functions
- Chapter 10—Use of phasors and complex variables for sinusoidal steady-state circuit analysis

FURTHER READING

There are several introductory texts that cover the concepts needed as a prerequisite to this text. Some excellent examples are listed below.

1. R. A. DeCarlo and P. M. Lin, *Linear Circuit Analysis Vols. 1 & 2*, Prentice Hall, 1995.
2. J. D. Irwin and D. V. Kerns, Jr., *Introduction to Electrical Engineering*, Prentice Hall, 1995.
3. W. H. Hayt, Jr. and J. E. Kemmerly, *Engineering Circuit Analysis*, 5th Ed., Mc Graw Hill, 1993.
4. C. R. Paul, *Analysis of Linear Circuits*, McGraw Hill, 1989.
5. S. E. Schwarz and W. G. Oldham, *Electrical Engineering,* 2nd Ed., Oxford University Press, 1993.

chapter 2

Semiconductor Physics and IC Technology

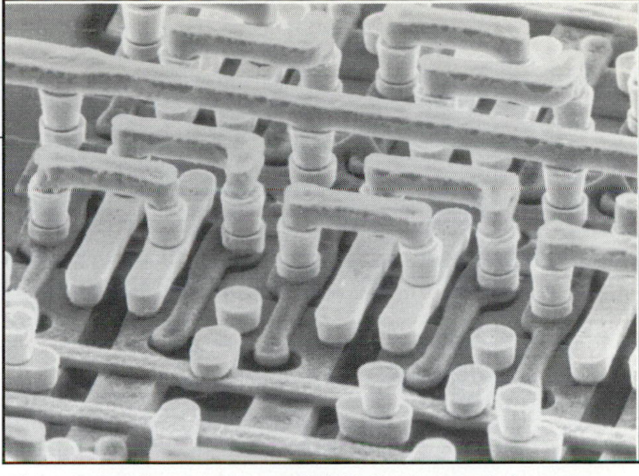

This scanning electron micrograph (SEM) photograph shows the complexity of modern integrated circuit interconnect technology. The interlevel dielectrics have been etched away to show the metallization patterns. (Courtesy of International Business Machines Corporation. Unauthorized use not permitted.)

Microelectronic engineers need a working knowledge of the physical processes underlying the behavior of electronic devices. This chapter provides a short introduction to semiconductor physics, with the goal of developing the basic concepts and terminology needed to model electronic devices described in Chapters 3, 4, 6, and 7. In addition a basic familiarity with IC fabrication is essential to understand the device structures in later chapters. This chapter includes a short overview of the essential features of this powerful technology.

Chapter Objectives

- The bond model for silicon—the electron and hole as mobile charges.
- The control of electron and hole concentrations through doping.
- Drift and diffusion of electrons and holes.
- IC processes and the connection between mask layout and device cross section.
- Sheet resistance and the integrated-circuit resistor.

2.1 PURE SEMICONDUCTORS

To begin our discussion of semiconductor physics, we consider the dominant semiconductor used in modern microelectronics—crystalline silicon. Silicon is in Group IV of the periodic table, with atomic number 14. Recalling basic atomic chemistry, its 14 electrons occupy the first three energy levels. The first three orbitals, 1*s*, 2*s*, and 2*p*, are filled by 10 electrons, leaving four electrons to occupy the 3*s* and 3*p* orbitals. In order to minimize the overall energy, the 3*s* and 3*p* orbitals hybridize to form four, tetrahedral 3*sp* orbitals. The electronic configuration of silicon is

$$(1s)^2\ (2s)^2\ (2p)^6\ (3sp)^4$$

Each of the four outer hybrid orbitals have one electron and are capable of forming a bond with a neighboring atom. These four electrons that participate in bonding are called **valence electrons**. The inset on the left shows the electronic configuration of silicon.

The crystalline form of silicon has the diamond crystal structure. Each silicon ion in the crystal has four nearest neighbors that are aligned with the tetrahedral 3*sp* hybrid orbitals. Each ion contributes an electron to fill the hybrid orbital. Figure 2.1 shows the unit cell of silicon, with the tetrahedral bonds from one atom highlighted. The edge of the unit cell is $a_o = 5.43$ Å.

How many atoms should be assigned to the unit cell? The eight corner atoms are shared among the eight cells touching the corner and therefore, contribute one atom to the unit cell. The six atoms on the faces of the cube are shared with adjacent cells and thus, contribute three atoms. Finally, there are four atoms that lie entirely within the cube, for a total of eight atoms in the unit cell.
The atomic density of crystalline silicon, in atoms per cubic centimeter, is

$$N_{Si} = 8 \cdot n_{cell} = 8 \cdot \frac{1}{a_o^3} = \frac{8}{(5.43 \times 10^{-8}\text{cm})^3} = 5.00 \times 10^{22}\text{cm}^{-3} \tag{2.1}$$

where n_{cell} is the number of unit cells per cm^3. Since each atom has four tetrahedral bonds and contributes one electron to each bond, the concentration of valence electrons is $4\,N_{Si} = 2 \times 10^{23}\ \text{cm}^{-3}$.

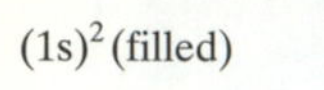
$(1s)^2$ (filled)

$(2s)^2$ (filled)

$(2p)^6$ (filled)

$(3sp)^4$

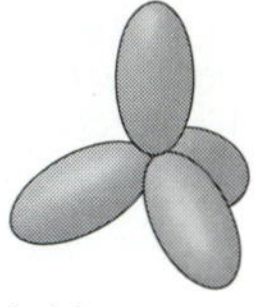

4 tetrahedral hybrid orbitals, half-filled with 4 valence electrons

electronic orbitals in silicon

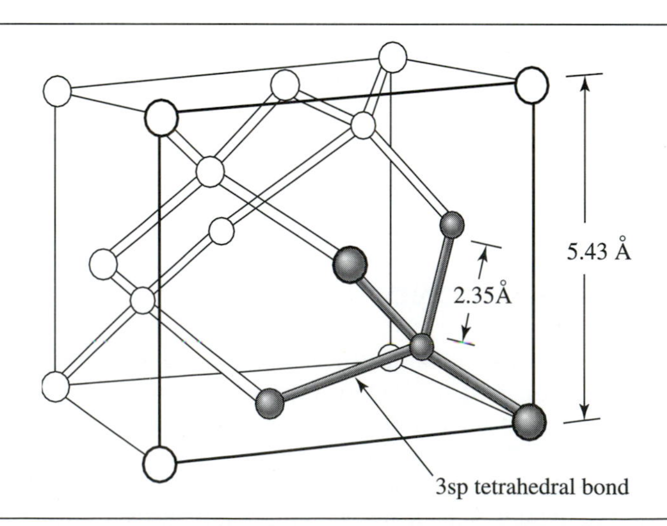

➤ **Figure 2.1** Unit cell of crystalline silicon (After R. S. Muller and T.I. Kamins, *Device Electronics for Integrated Circuits,* 2nd ed., Wiley, 1986.)

2.1.1 Electrons, Holes, and the Bond Model

We now investigate the electronic properties of the bonded network of silicon atoms in Fig. 2.1, which we refer to as the silicon **lattice**. In order to sketch the silicon lattice conveniently, we use the two-dimensional collapsed view shown in Fig. 2.2. The portion of the lattice shown in Fig. 2.2 is in the interior, or **bulk**, of the silicon crystal away from any surfaces. Each valence electron in a bond is symbolized by a connecting line, which means that two lines are used to represent each tetrahedral bond between silicon atoms. Note that this convention differs from the one used in chemistry, where a single line represents both electrons (in a single bond).

We refer to this abstract picture of the silicon crystal as the **bond model.** The silicon ions (nucleus and 10 inner electrons) each have a charge of ($+4\ q$), where $q = 1.6 \times 10^{-19}$ C is the magnitude of the charge of an electron. The four valence electrons per ion in the surrounding four bonds contribute a charge of ($-4\ q$), making the crystal electrically neutral. In Fig. 2.2, all of the bonds are shown to be complete, which is true only for a perfect crystal at a temperature of absolute zero (0 degrees Kelvin).

An **intrinsic semiconductor** is one in which there are no impurities. A perfect crystal of intrinsic silicon at absolute zero has all of its valence electrons tied up in bonds since this is the lowest energy configuration. Since there are no "loose" or "free" electrons to carry current, intrinsic silicon is an insulator at absolute zero.

At temperatures above absolute zero, lattice vibrations occur that can impart enough energy to the valence electrons to break a very small percentage of the bonds. Each broken bond results in a mobile electron that can move around the crystal. This process is known as **thermal generation**. Note that the electron is still confined to the

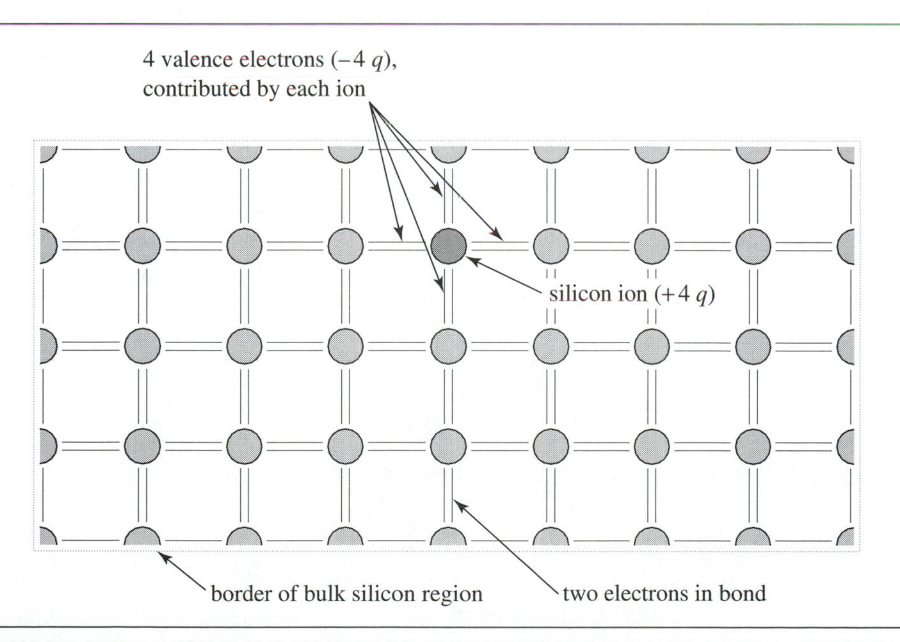

➤ **Figure 2.2** Collapsed two-dimensional silicon lattice at absolute zero, with all bonds complete. This portion of the crystal is in the interior, or bulk, away from surfaces. It is electrically neutral since the four valence electrons balance the net charge on the silicon ion.

crystal since much more energy is required for it to escape into the surrounding ambient.

Figure 2.3 shows the silicon lattice after one bond has been broken due to thermal generation. The mobile electron has a negative charge $-q = -1.6 \times 10^{-19}$ C and leaves behind an incomplete bond that takes on a charge of $+q$. At room temperature, the electron moves quite rapidly and undergoes frequent collisions as suggested by the trajectory sketched in Fig. 2.3.

Since electrons have particle-like behavior in a vacuum, it seems plausible that the mobile electron resulting from the broken bond is capable of moving through the crystal lattice. For example, a cathode-ray tube used in a computer monitor works by steering electron trajectories with electric fields. What is surprising is that the positive charge arising from the broken bond is also mobile with similar properties to the electron, except that it has an equal and opposite charge. Figure 2.3 depicts that the positive charge is also free to move around the lattice. This mobile positive charge is called a **hole**.

2.1.2 Limitations of the Bond Model

The bond model is useful in visualizing the creation of electron-hole pairs. However, it has some shortcomings. Electrons and holes in crystals have both particle-like and wave-like behavior that can only be understood from the perspective of quantum mechanics. The bond model necessarily depicts both the electron and the hole as localized in space, which is not correct.

Fortunately, it is not necessary to take a solid-state physics course prior to studying semiconductor devices and circuits. This text considers that electrons and holes can effectively be considered as *classical* particles that obey Newton's Laws and have unit negative and positive charge. Quantum mechanics has a negligible effect on the

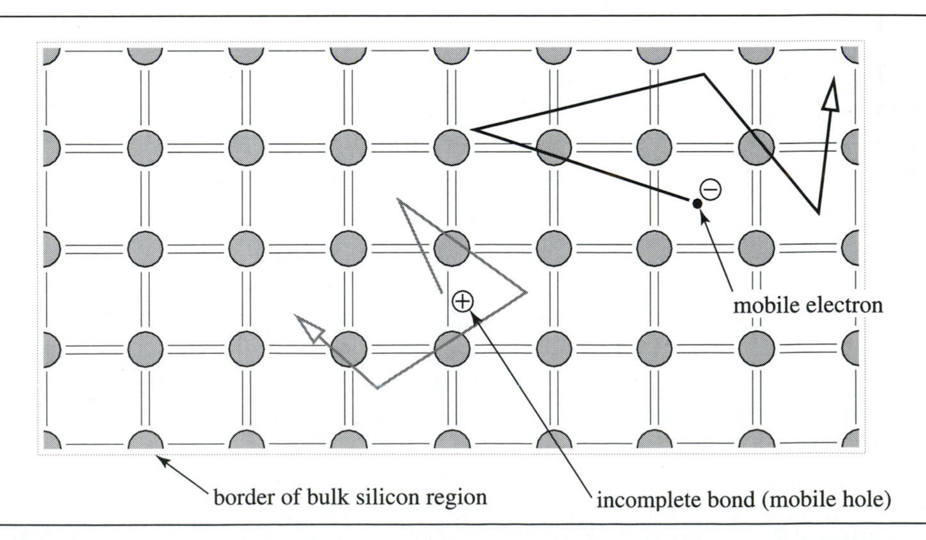

➤ **Figure 2.3** Electron and hole resulting from breaking one bond in intrinsic silicon.

accuracy of models for microelectronic devices that have dimensions of 0.1 μm or larger.[1]

2.2 GENERATION, RECOMBINATION, AND THERMAL EQUILIBRIUM

We have seen that mobile electrons and holes can be created in pure, also known as **intrinsic**, silicon by the breaking of bonds due to thermal generation. The thermal generation rate $G_{th}(T)$, with units of $cm^{-3}s^{-1}$, increases exponentially with temperature. Another type of generation process is **optical generation**, with rate G_{op}. Optical generation occurs when a bond absorbs a photon with sufficient energy to break it. The minimum energy for bond-breaking in silicon is around 1.1 eV, corresponding to a wavelength of around 1 μm (in the near-infrared region of the optical spectrum.) In contrast to thermal generation, the optical generation rate is essentially independent of temperature. The total generation rate G is the sum of the thermal and optical rates

$$G = G_{th}(T) + G_{op} \tag{2.2}$$

If mobile electrons and holes are continuously generated, a balancing process is necessary that continuously removes them. **Recombination** is the elimination of an electron and hole from the crystal by completing one of the broken bonds. Recombination processes are classified according to what happens to the energy released when the bond forms.

Thermal recombination—the energy goes to vibrations in the crystal lattice. This is equivalent to saying the silicon crystal is heated.

Optical recombination—the energy goes to photon emission. Optical recombination is of negligible importance in silicon. In other semiconductors such as gallium arsenide, optical recombination is a very significant component of the total recombination rate. In these materials, light emission due to optical recombination is the basic principle behind light-emitting diodes and laser diodes.

For recombination to occur, both an electron and a hole are necessary. The recombination rate will be small if there are few electrons despite an abundance of holes or vice versa. Therefore, it is reasonable that the recombination rate is proportional to the *product* of the electron and hole concentrations

$$R = k(n \cdot p) \tag{2.3}$$

where n is the electron concentration (cm^{-3}), p is the hole concentration (cm^{-3}), and k is the recombination rate constant (cm^3s^{-1}). In contrast, the generation rate G is independent of the electron and hole concentrations since the number of broken bonds is a tiny fraction of the total number of bonds in the crystal.

If the environment of the silicon crystal (optical radiation flux, electric field intensity, temperature, etc.) is *constant,* then the electron and hole concentrations must eventually reach stable values that are set by the balance between the recombi-

steady-state

1. Current research is exploring **nanoelectronic** devices with critical dimensions on the order of 10 nm or less. Quantum effects are crucial to understanding their basic operation.

nation and generation rates. We define this situation as **steady-state**. The total recombination rate and total generation rates must be equal for steady-state conditions

$$R = k(n \cdot p) = G = G_{th}(T) + G_{op} \tag{2.4}$$

Given the particular environmental conditions that determine $G_{th}(T)$ and G_{op}, the product of the electron and hole concentrations is

$$n \cdot p = \frac{G_{th}(T) + G_{op}}{k} \tag{2.5}$$

thermal equilibrium

An idealized, special case of the steady-state condition called **thermal equilibrium** is of great importance in semiconductor physics. Thermal equilibrium is defined as the prolonged absence of external energy sources. The silicon sample is completely isolated from optical or electrical sources (e.g., power supplies and light sources) by placing it in a "black box" at a constant temperature. Some optical generation occurs in addition to thermal generation, due to black-body radiation at the sample temperature. For the case of thermal equilibrium, Eq. (2.5) reduces to

$$n_o \cdot p_o = \frac{G^o_{th}(T) + G^o_{op}(T)}{k_o} \tag{2.6}$$

where all quantities are subscripted or superscripted with "o" to denote thermal equilibrium. The right-hand side of this equation is a function of temperature only (the single variable in thermal equilibrium). Defining this function as $n_i^2(T)$, we can re-express Eq. (2.6) as the ***mass-action law***

mass-action law

$$\boxed{n_o \cdot p_o = n_i^2(T)} \tag{2.7}$$

This important result relates the product of the thermal equilibrium electron and hole concentrations to a constant that depends only on temperature. The unusual name for this equation is borrowed from chemistry, since Eq. (2.7) is analogous to an equilibrium equation for a chemical reaction. For example, the equilibrium between hydrogen ions (concentration $[H^+]$) and hydroxyl ions (concentration $[OH^-]$) reacting to form water is expressed as

$$[H^+] \cdot [OH^-] = K_{eq}(T) \tag{2.8}$$

where $K_{eq}(T)$ is the equilibrium constant.

In pure silicon in thermal equilibrium, the electron and hole concentrations are equal since the electrons and holes are created in pairs when the bonds are broken. Therefore, it follows from substitution in Eq. (2.7)

$$n_o \cdot p_o = n_o^2 = p_o^2 = n_i^2(T) \tag{2.9}$$

and consequently, the electron and hole concentrations in pure silicon are both equal to the **intrinsic concentration** $n_i(T)$ in thermal equilibrium. Before proceeding further, it is worthwhile to consider the order of magnitude of the electron and hole concentrations in pure silicon at room temperature

intrinsic concentration

$$n_i \cong 10^{10}\ \text{cm}^{-3} \text{at } 300\ \text{K} \tag{2.10}$$

Given that the number of bonds is $4 \times N_{Si} = 2 \times 10^{23}\ \text{cm}^{-3}$, at room temperature only an extremely small fraction (1 in 20 trillion) of the bonds are broken. As temperature increases, the intrinsic concentration increases exponentially—approximately doubling for every 10 °C rise over room temperature.

2.3 DOPING

Silicon would be of little use if the hole and electron concentrations were both equal to $10^{10}\ \text{cm}^{-3}$. Fortunately, the hole and electron concentrations can be controlled by adding minute quantities of specific elements to the silicon crystal. There are a variety of ways to incorporate these impurities into the silicon lattice using a process called **doping**. We will first consider the case of **donors**, which are elements in Group V of the periodic table.

2.3.1 Donors

The bond model is useful for visualizing how donor atoms control the concentration of electrons in silicon. Typical donors for silicon are phosphorus (P), arsenic (As), and antimony (Sb), all of which are in Group V of the periodic table. Figure 2.4 shows the bond model of silicon with the addition of an arsenic atom on a substitutional lattice site. Since arsenic has five valence electrons, it is able to form bonds with all four neighboring silicon atoms. The fifth valence electron is left after the bonds are

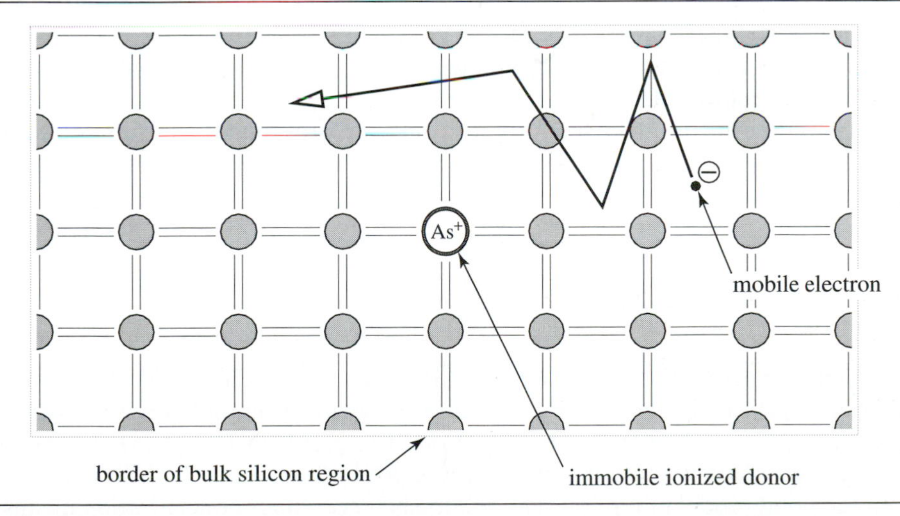

Figure 2.4 Bond model with arsenic incorporated into a substitutional site in a bulk region of the silicon crystal. Due to thermal excitation, arsenic loses (donates) its fifth valence electron as a mobile electron resulting in an immobile positive charge.

formed and is very loosely bound to the arsenic atom. For temperatures $T \gtrsim 100$ K = −173° C, there is enough thermal energy for this fifth electron to leave and move through the lattice, leaving behind an immobile, positively charged arsenic ion.

The charges in the bulk silicon region in Fig. 2.4 are the key to quantifying the electron and hole concentrations for arsenic-doped silicon. The region was electrically neutral before the arsenic was incorporated into the lattice. After the arsenic is added and ionizes, which occurs for $T \gtrsim 100$ K = −173° C, the region is still neutral since the negative electron charge is balanced by the positive charge on the arsenic ion. Adding up all the charge densities (units: C / cm^3) with their proper signs, we require that the net charge density ρ is zero:

$$\rho = -qn_0 + qp_o + qN_d = q(p_o + N_d - n_o) = 0 \tag{2.11}$$

The concentration of dopants is extremely low and their presence in the lattice does not significantly alter the mass-action law in thermal equilibrium. Solving Eq. (2.7) for the hole concentration and substituting into Eq. (2.11), we find that

$$\frac{n_i^2}{n_o} + N_d - n_o = 0 \tag{2.12}$$

By multiplying both sides by the electron concentration n_o and rearranging, the result is a quadratic equation

$$n_o^2 - N_d n_o - n_i^2 = 0 \tag{2.13}$$

Using the quadratic formula, we find that the electron concentration in silicon doped with donors is

$$n_o = \frac{N_d + \sqrt{N_d^2 + 4n_i^2}}{2} = \frac{N_d}{2} + \frac{N_d}{2}\sqrt{1 + \frac{4n_i^2}{N_d^2}} \tag{2.14}$$

where we have taken the positive root since the electron concentration must be positive. The hole concentration p_o can be found from the electron concentration using the mass-action law in Eq. (2.7).

In practice, the donor concentration is much greater than the intrinsic concentration $N_d \gg n_i$. The square root in Eq. (2.14) is approximately one and the electron concentration n_o is

n-type electron concentration

$$\boxed{n_o \cong \frac{N_d}{2} + \frac{N_d}{2} = N_d} \tag{2.15}$$

Silicon doped with $N_d \gg n_i$ is called **extrinsic**. The electron concentration n_o in extrinsic silicon is determined by the donor concentration and is not a function of temperature-dependent thermal generation. The result in Eq. (2.15) can be easily remembered since it corresponds to "one electron per donor," which can be seen from the bond model in Fig. 2.4. The relatively large donor concentration means the thermally generated electrons can be neglected, as quantified by Eq. (2.14).

The thermal equilibrium hole concentration in silicon doped with donors can be found from the mass-action law

n-type hole concentration

$$p_o = \frac{n_i^2}{n_o} \cong \frac{n_i^2}{N_d} \qquad p_o = \frac{10^{20}}{N_d}\ \text{cm}^{-3} \qquad (T = 300\text{K}) \tag{2.16}$$

For example, for a donor concentration of $N_d = 10^{14}$ cm^{-3}, the hole concentration at room temperature is only $p_o = 10^6$ cm^{-3}—a factor of 100 million lower than the electron concentration! By adding donor impurities to the silicon sample, the mass-action law in Eq. (2.7) depresses the hole concentration. Few holes remain in equilibrium since recombination with the large number of electrons removes most of the them. Silicon doped with donors is referred to as **n-type**, in which electrons are the **majority carriers** (by far the most numerous) and holes are the **minority carriers**.

The lower limit of useful donor concentration is set by the requirement that $N_d \gg n_i$ in Eq. (2.14), so $N_{d_{min}} \cong 10^{12}$ cm^{-3}. On the opposite extreme, there is a maximum amount of substitutional impurities that the silicon lattice can absorb without forming a new material phase with a different crystal structure and electronic properties. This maximum varies among the donor elements, but a typical value is around $N_{d_{max}} \cong 10^{19}$ cm^{-3}. This corresponds to about 1 in 5000 lattice sites occupied by a donor since the atomic density of silicon is $N_{Si} = 5 \times 10^{22}$ cm^{-3}.

➤ EXAMPLE 2.1 Electron and Hole Concentrations in n-type Silicon

Find the concentration of holes and electrons for a bulk silicon sample in thermal equilibrium at room temperature that has a donor concentration $N_d = 10^{14}$ cm^{-3}.

SOLUTION

Using Eq. (2.14), the electron concentration is

$$n_o = \frac{10^{14}}{2} + \frac{10^{14}}{2}\sqrt{1 + \frac{4 \cdot 10^{20}}{10^{28}}} = 1.00000001 \times 10^{14} \quad \text{cm}^{-3}$$

The hole concentration p_o can be found from the mass-action law in Eq. (2.7) since the sample is in thermal equilibrium

$$p_o = \frac{n_i^2}{n_o} = \frac{(10^{10})^2}{1.00000001 \times 10^{14}} = 9.9999999 \times 10^5\ \text{cm}^{-3}$$

Verify that the bulk silicon region is electrically neutral by adding the charge densities of the holes, ionized donors, and electrons

$$\rho = q\,(p_o + N_d - n_o) = q\,(9.9999999 \times 10^5 + 1 \times 10^{14} - 1.00000001 \times 10^{14}) = 0$$

Note that since N_d exceeds n_i by four orders of magnitude, we could have used Eq. (2.15) to find $n_o \cong N_d = 10^{14}$ cm^{-3}, and then used the mass-action law to find the hole concentration to be $p_o \cong 10^6$ cm^{-3}.

2.3.2 Acceptors

A substitutional impurity from Group III of the periodic table that has only three valence electrons is called an **acceptor**. The usual acceptor atom for silicon is boron (B). Figure 2.5 is the bond model's picture of doping by acceptors.

Since the Group III boron atom only has three valence electrons, it can complete only three of the four bonds with its silicon neighbors. However, an electron in an adjacent bond requires only a small amount thermal energy to jump over and complete the fourth bond. This process of "accepting" an electron accounts for the name of this class of dopants. The resulting hole is free to move around the crystal, whereas the immobile acceptor takes on a negative charge due to the extra electron as shown in Fig. 2.5.

Given that the acceptor concentration is N_a (cm^{-3}), the equilibrium hole and electron concentrations can be found using the same approach used for donors. The bulk silicon region is electrically neutral and the hole, electron, and ionized acceptor charge densities must therefore sum to zero:

$$\rho = -qn_o + qp_o - qN_a = q(p_o - n_o - N_a) = 0 \tag{2.17}$$

The electron concentration can be eliminated by substituting $n_o = n_i^2 / p_o$ from the mass-action law in Eq. (2.7)

$$p_o - \frac{n_i^2}{p_o} - N_a = 0 \tag{2.18}$$

Using the quadratic formula to solve this equation yields an expression for the hole concentration in silicon doped with acceptors

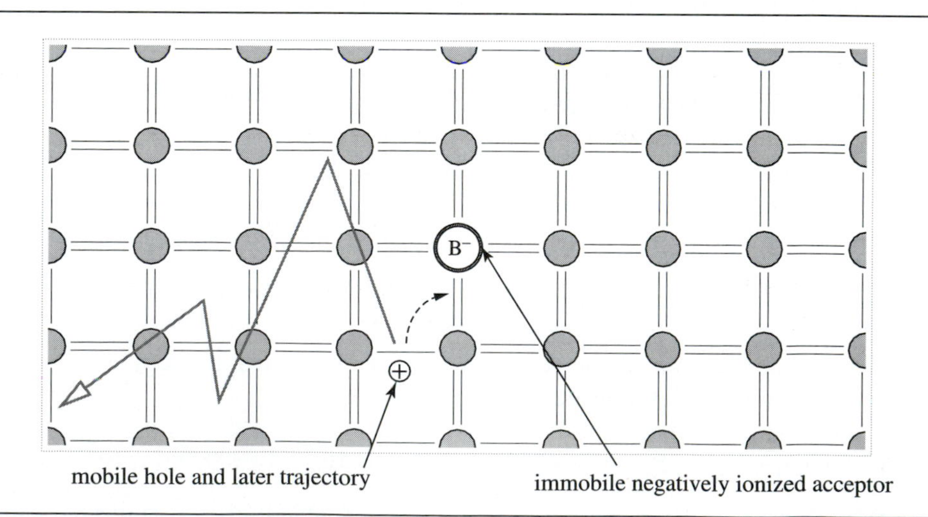

➤ **Figure 2.5** Bond model showing boron on a substitutional lattice site in a bulk region of the silicon crystal. A hole is created when the incomplete fourth bond is filled by acceptance of an electron from a neighboring bond.

$$p_o = \frac{N_a}{2} + \frac{N_a}{2}\sqrt{1 + \frac{4n_i^2}{N_a^2}} \qquad (2.19)$$

For the typical case where N_a is orders of magnitude larger than n_i, the intrinsic hole concentration, the square root term is approximately one and the hole concentration is

$$\boxed{p_o \cong N_a} \qquad (2.20)$$

p-type hole concentration

By applying the mass-action law, we find that the equilibrium electron concentration at room temperature is given by

$$\boxed{n_o = \frac{n_i^2}{p_o} \cong \frac{n_i^2}{N_a} = \frac{10^{20}}{N_a}\ \text{cm}^{-3} \text{ at 300 K}} \qquad (2.21)$$

p-type electron concentration

Silicon doped with acceptors is called **p-type**, in which holes are the majority carriers and electrons are the minority carriers. P-type silicon is **complementary** to n-type silicon, in which the relationship between holes and electrons is reversed.

Due to the mass-action law, doping with $N_a = 10^{15}\text{cm}^{-3}$ of boron results in a hole concentration $p_o = 10^{15}\text{cm}^{-3}$ and electron concentration $n_o = 10^5\ \text{cm}^{-3}$. As with donors, the minimum useful acceptor concentration to have extrinsic, p-type silicon is determined by the intrinsic concentration n_i. At room temperature $N_{a_{min}} \approx 10^{12}\ \text{cm}^{-3}$ for $N_a \gg n_i$. The maximum concentration of substitutional boron acceptors in silicon is on the order of $N_{a_{max}} \approx 10^{19}\ \text{cm}^{-3}$ since a boron-silicon compound begins to precipitate at higher concentrations.

2.3.3 Donors and Acceptors: Compensation

Silicon is often doped with both donors *and* acceptors. To find the equilibrium electron and hole concentrations, first consider the bond model in Fig. 2.6. Using the same reasoning used for the donor-only and acceptor-only cases, it can be seen that charge neutrality applies to the bulk silicon region in Fig. 2.6. The charge densities from electrons, holes, ionized donors, and ionized acceptors must therefore sum to zero

$$\rho = -qn_o + qp_o + qN_d - qN_a = q(p_o - n_o + N_d - N_a) = 0 \qquad (2.22)$$

When $N_d > N_a$, we can substitute for the hole concentration in thermal equilibrium $p_o = n_i^2 / n_o$ from the mass-action law and find

$$n_o^2 - (N_d - N_a)\,n_o - n_i^2 = 0 \qquad (2.23)$$

This quadratic equation is identical to Eq. (2.13) with the **net** donor concentration $N_d - N_a$ replacing the donor concentration. The solution can be found by substitution into Eq. (2.14)

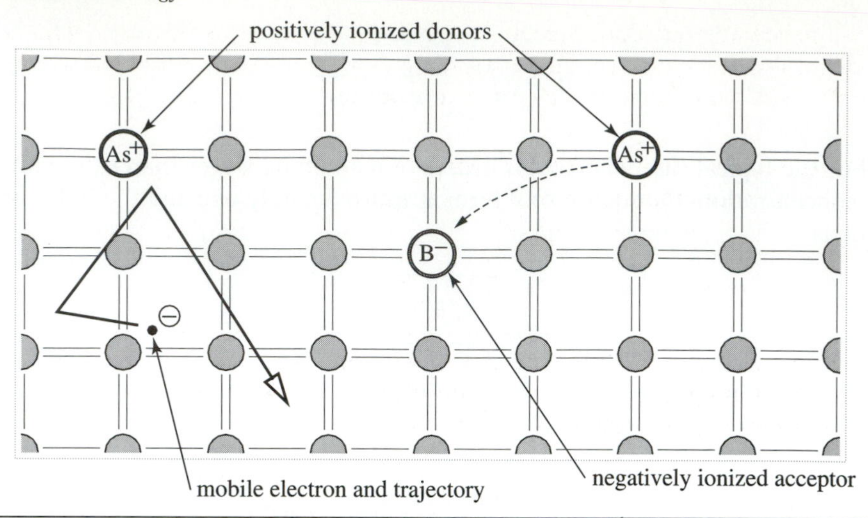

➤ **Figure 2.6** Bond model showing a bulk section of silicon that is doped with both arsenic (donors) and boron (acceptors) with $N_d > N_a$. Donated electrons ionize the acceptors by completing their half-filled bonds, with the excess $N_d - N_a$ becoming mobile electrons. This process is called **compensation.**

$$n_o = \frac{(N_d - N_a)}{2} + \frac{(N_d - N_a)}{2}\sqrt{1 + \frac{4n_i^2}{(N_d - N_a)^2}} \tag{2.24}$$

where we assume that the donor concentration is greater than the acceptor concentration $N_d > N_a$. In practice, the excess of donors over acceptors $N_d - N_a$ will usually greatly exceed the intrinsic concentration since separately $N_d \gg n_i$ and $N_a \gg n_i$. For example, if the donor concentration exceeds the acceptor concentration by only 5% at a typical doping level of about 10^{15} cm^{-3}, then

$$N_d - N_a = 1.05 \times 10^{15}\text{cm}^{-3} - 1.0 \times 10^{15}\text{cm}^{-3} = 5 \times 10^{13}\text{cm}^{-3} \gg n_i \tag{2.25}$$

Assuming $N_d - N_a \gg n_i$, we find that

compensated n-type electron concentration

$$\boxed{n_o \cong N_d - N_a \qquad (N_d > N_a)} \tag{2.26}$$

The physical interpretation of Eq. (2.26) is that electrons donated by the donors first ionize all the acceptors, with only the net donor concentration in excess of the acceptor concentration contributing mobile electrons to the crystal. This process is called **compensation.**

Since the sample is in thermal equilibrium, the minority hole concentration can be found by the mass-action law

compensated n-type hole concentration

$$\boxed{p_o \cong \frac{n_i^2}{N_d - N_a} \qquad (N_d > N_a)} \tag{2.27}$$

At first glance, the fact that there are negligible holes in a sample doped with acceptors seems counter-intuitive. However, the mass-action law depresses the hole concentration since the (majority) electron concentration far exceeds n_i.

For the case where the acceptor concentration is greater than the donor concentration, similar reasoning shows that the sample is p-type with

compensated p-type carrier concentration

$$p_o \cong N_a - N_d \qquad n_o \cong \frac{n_i^2}{N_a - N_d} \qquad (N_a > N_d) \tag{2.28}$$

In summary, a silicon crystal that is doped with one type of impurities may be compensated by doping with an excess of impurities of the opposite type. In section 2.5, we will see that compensation is a cornerstone of integrated device fabrication.

➤ EXAMPLE 2.2 Electron and Hole Concentrations for Compensated Silicon

a) Find the thermal equilibrium electron and hole concentrations at room temperature for a bulk silicon region doped with donors at $N_d = 10^{14}\text{cm}^{-3}$ and with acceptors at $N_a = 5 \times 10^{13}\text{cm}^{-3}$.

SOLUTION

The sample is n-type since the donor concentration exceeds the acceptor concentration. Since the net donor concentration $N_d - N_a = 10^{14}\ \text{cm}^{-3} - 5 \times 10^{13}\ \text{cm}^{-3} = 5 \times 10^{13}\ \text{cm}^{-3} \gg n_i$, we can equate the electron concentration to the net donor concentration (Eq. 2.26)

$$n_o \cong N_d - N_a = 5 \times 10^{13}\ \text{cm}^{-3}$$

The mass-action law in Eq. (2.7) determines that the minority hole concentration is

$$p_o \cong \frac{n_i^2}{N_d - N_a} = 2 \times 10^6\ \text{cm}^{-3}$$

To verify that the region is electrically neutral, we add the contributions of the positive holes and ionized donors with the negative electrons and negative ionized acceptors

$$\rho = q(p_o + N_d - n_o - N_a) = q(2 \times 10^6 + 10^{14} - 5 \times 10^{13} - 5 \times 10^{13}) \approx 0$$

Note: The charge density would have been exactly zero if the exact expression for the compensated electron concentration in Eq. (2.24) had been used.

(b) Now consider a bulk n-type silicon sample with $N_d = 10^{14}\ \text{cm}^{-3}$. We would like to add acceptors to make the hole concentration $p_o = 5 \times 10^{13}\ \text{cm}^{-3}$. What acceptor doping concentration N_a is needed? What is the electron concentration?

SOLUTION

The desired hole concentration is much greater than the intrinsic concentration; therefore, the compensated sample is p-type. Using Eq. (2.28), we can solve for the acceptor concentration

$$p_o \cong N_a - N_d \quad \rightarrow \quad N_a \cong p_o + N_d = 5 \times 10^{13}\,\text{cm}^{-3} + 10^{14}\,\text{cm}^{-3} = 1.5 \times 10^{14}\,\text{cm}^{-3}$$

The electron concentration follows from the mass-action law

$$n_o = \frac{n_i^2}{p_o} = \frac{10^{20}}{5 \times 10^{13}} = 2 \times 10^6\,\text{cm}^{-3}$$

2.4 CARRIER TRANSPORT

At room temperature, the holes and electrons in silicon are in agitated, random motion. Their average velocity is called the **thermal velocity** v_{th}, and is about 10^7 cm/s—only a factor of 3000 less than the speed of light! Holes and electrons collide frequently by scattering off ionized impurities, crystal defects, surfaces, and each other. A typical value for the average **collision time** τ_c at room temperature in bulk silicon is about 1×10^{-13} s = 0.1 picoseconds. In between collisions, carriers on average will travel a distance

$$\lambda = v_{th}\,\tau_c = 10\text{ nm} = 0.01\ \mu\text{m} \tag{2.29}$$

where λ is defined as the **mean free path**. State-of-the-art silicon microelectronic devices have dimensions that are at least 0.1 μm, but are typically much larger. Therefore, holes and electrons undergo many collisions while traversing a device structure. In the next section, this important fact will allow us to find simple equations that describe the average transport of charge carriers due to electric fields or to concentration gradients.

2.4.1 Drift Velocity

In thermal equilibrium, the electrons and holes in a region of the silicon lattice have no net motion. In other words, the average position of all the electrons (or separately, all of the holes) does not change with time. Figure 2.7(a) illustrates this situation for electrons. The random trajectories of three electrons along the x axis are shown in Fig. 2.7(a) for a time interval $\Delta t = 7\,\tau_c$, during which seven collisions occur. To show clearly the electron motion, the x-axis trajectories are spread out and all trajectories are shown originating at $x = x_i$. The average change in position after the interval Δt is

$$\overline{\Delta x} = \frac{(x_{f,1} - x_i) + (x_{f,2} - x_i) + (x_{f,3} - x_i)}{3} \approx 0 \tag{2.30}$$

using the approximate values for the position changes in Fig. 2.7(a). Of course, the average position change for electrons in a region of the silicon lattice will be taken over a huge number of electrons. Since the trajectories are random, $\overline{\Delta x}$ will be zero after any time interval $\Delta t \gg \tau_c$.

The situation with an applied electric field E is different, as shown in Fig. 2.7(b). It is important to first be sure of the sign for the electric field and other directional variables. By convention, the electric field is positive when it points in the same direction as the reference direction for the x axis. Thus, the electric field E in Fig. 2.7(b) is positive. The electrons are subject to an electrostatic force

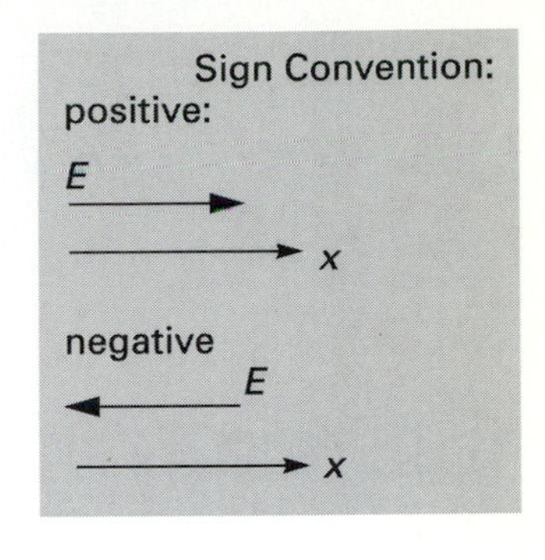

$$F_{el} = -qE \tag{2.31}$$

that points in the $-x$ direction. During their "free flight" in between collisions, the electrons with mass m_n experience a constant acceleration $a = F_{el} / m_n = -qE/m_n$ that deflects their trajectories toward the $-x$ direction as shown in Fig. 2.7(b).

Averaging the position changes over the interval Δt, we find

$$\overline{\Delta x} = \frac{(x_{f,1} - x_i) + (x_{f,2} - x_i) + (x_{f,3} - x_i)}{3} < 0 \tag{2.32}$$

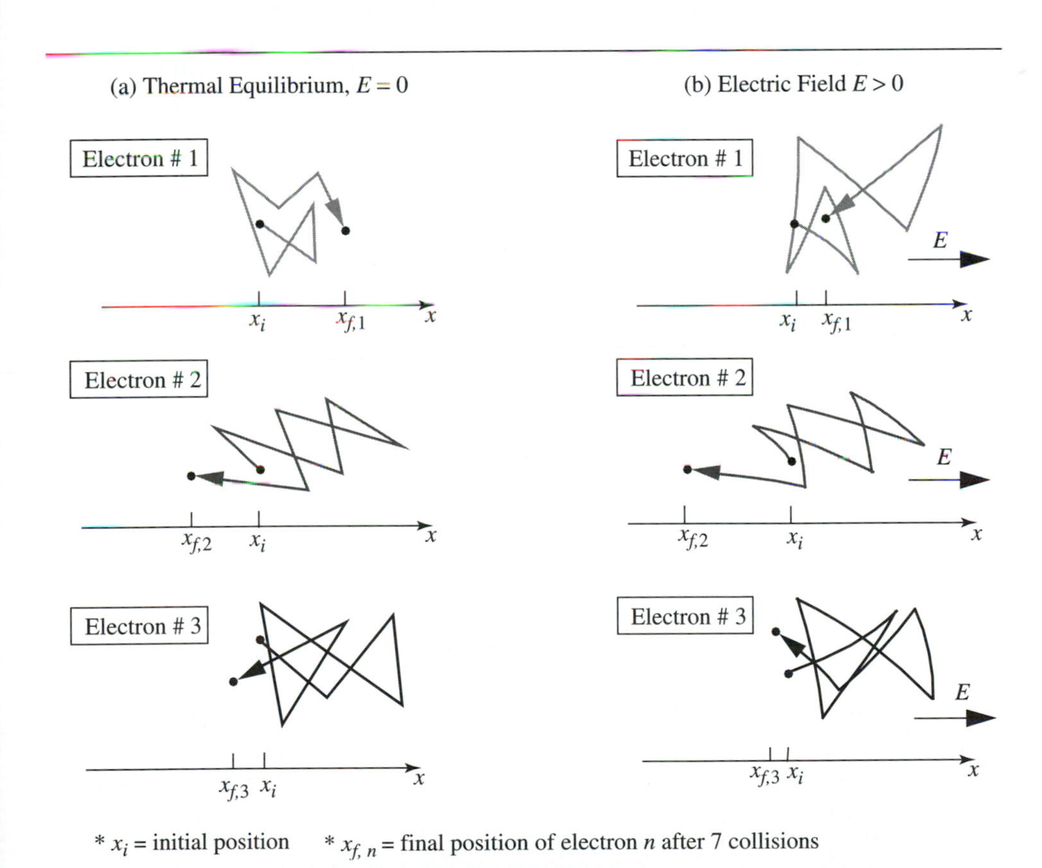

➤ Figure 2.7 (a) Trajectories for three electrons in thermal equilibrium (zero applied electric field) for seven collisions. (b) Trajectories for the same three electrons for an electric field $E > 0$. All of the trajectories are drawn with the same initial position x_i.

using the three sample trajectories in Fig. 2.7(b). Dividing the average position change by the time interval, we define the **drift velocity** of the electrons as

$$v_{dn} = \overline{\frac{\Delta x}{\Delta t}} \tag{2.33}$$

which is negative for $E > 0$. Note that the drift velocity is the average net velocity for a huge number of electrons in a region of the silicon lattice. Experimentally, the drift velocity v_{dn} is found to be proportional to the electric field

electron drift velocity

$$\boxed{v_{dn} = -\mu_n E} \tag{2.34}$$

where μ_n is the **electron mobility** with units of $cm^2/(Vs)$.

The electron mobility is a function of both the temperature and total doping concentration. Mobility decreases with increasing temperature since thermal agitation causes lattice vibrations to scatter electrons. Ionized dopants also scatter electrons and, therefore, reduce the collision time. This means the mobility is a decreasing function of $N_d + N_a$. Figure 2.8 gives experimental measurements of electron mobility for $N_d + N_a = 10^{13}$ to 10^{20} cm^{-3} at room temperature. It is useful to have a rough number for electron mobility in silicon for approximate calculations. Substituting a typical total dopant concentration of 2.5×10^{16} cm^{-3}, Fig. 2.8 gives an electron mobility of $\mu_n \approx 1000$ $cm^2/(Vs)$.

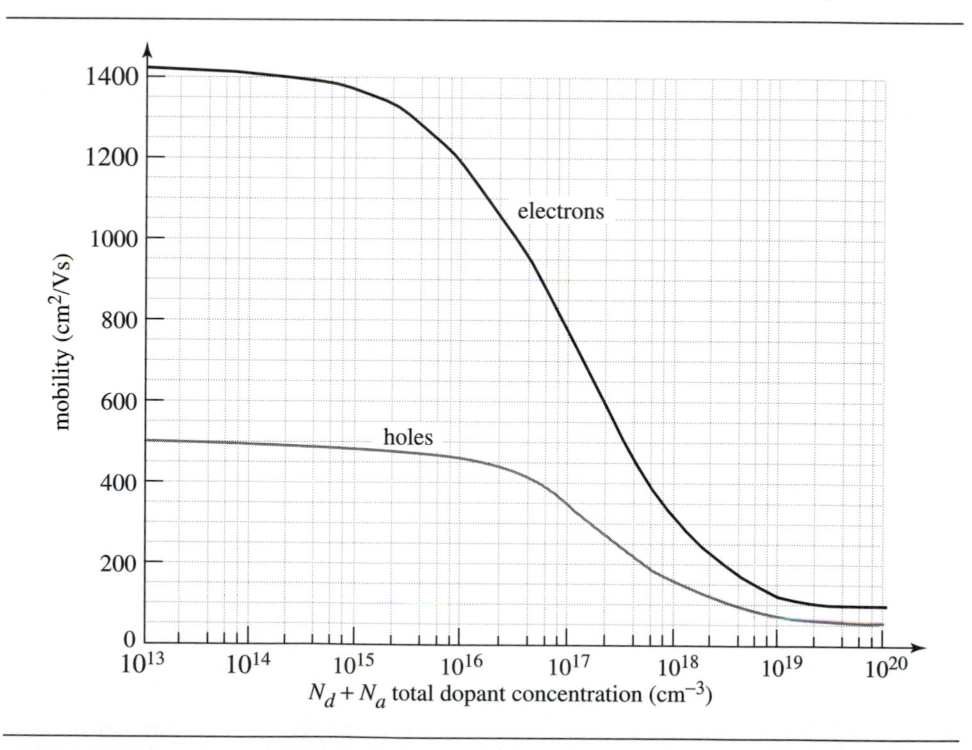

➤ **Figure 2.8** Linear-log plot of electron and hole mobilities at room temperature, as functions of the total doping concentration $N_d + N_a$. After R. S. Muller and T. I. Kamins, *Device Electronics for Integrated Circuits,* 2nd ed., Wiley, 1986.

Holes also drift in an electric field—the holes drifting *with* the field on average, because the positive charge causes them to accelerate in the same direction as E between collisions. The drift velocity for holes is also proportional to the electric field

$$\boxed{v_{dp} = \mu_p E} \tag{2.35}$$

hole drift velocity

where μ_p is the **hole mobility**. In silicon, hole mobilities are lower than electron mobilities. They range from 100–500 $cm^2/(Vs)$ at room temperature as shown in Fig. 2.8. For a typical dopant concentration of 2.5×10^{16} cm^{-3}, the "round number" for hole mobility is $\mu_p \approx 400$ $cm^2/(Vs)$.

The linear relationship between drift velocity and electric field breaks down when the electric field is high enough that the drift velocity approaches $v_{sat} = 10^7$cm/s. Figure 2.9 shows $|v_d|$ as a function of electric field for electrons and holes and illustrates that both carriers have a maximum drift velocity of about 10^7 cm/s. This phenomenon is known as **velocity saturation**.

Advances in integrated-circuit technology are enabling ever-smaller electronic devices, a trend that is increasing the internal electric fields. Velocity saturation is an important phenomenon for accurate modeling of VLSI device characteristics. In this book, we will assume that the linear relationships between electric field and drift velocity in Eqs. (2.34) and (2.35) are valid unless velocity saturation is specifically mentioned.

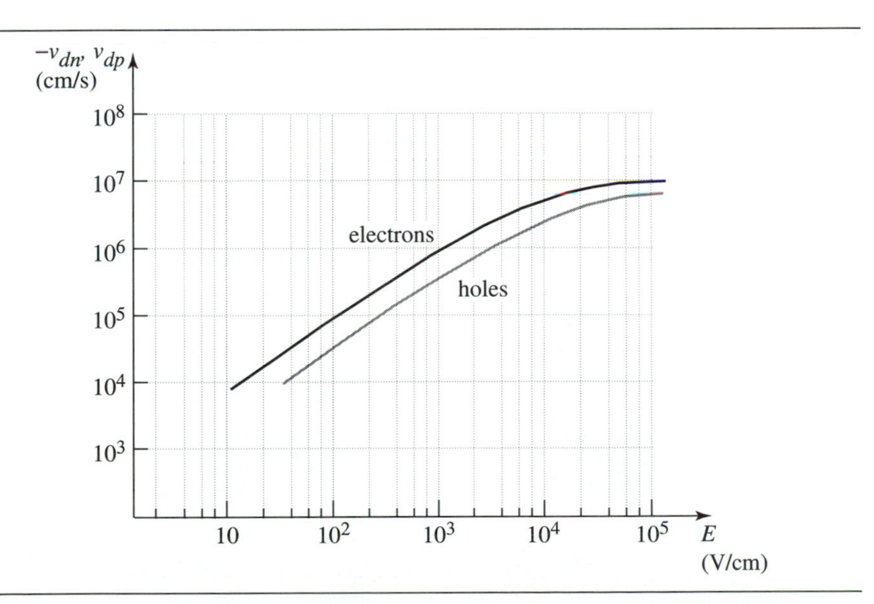

➤ **Figure 2.9** Log-log plot of the drift velocity as a function of electric field for electrons and holes, showing the linear region where the velocity is proportional to the field and velocity saturation at high fields. After R. S. Muller and T. I. Kamins, *Device Electronics for Integrated Circuits,* 2nd ed., Wiley, 1986.

2.4.2 Drift Current Density

Drift of charged particles results in the transport of charge through the silicon lattice. We quantify the transport of charge by the **current density** J, which is defined as the rate at which charge passes through a reference plane in the $+x$ direction per unit area. The units for current density are C/(s cm^2) or A/cm^2. The sign convention for current density is illustrated in Fig. 2.10. J is positive when positive charge passes in the $+x$ direction. Since the charge carriers can be positive (holes) or negative (electrons), there are two ways that J can be positive.

1. Holes can pass through the reference plane in the $+x$ direction.
2. Electrons can pass through the reference plane in the $-x$ direction.

Note that either case results in a net transport of positive charge through the reference plane in the $+x$ direction making the current density J positive.

The hole current density due to drift in an electric field is expressed as J_p^{dr}— subscript p designates holes and superscript dr, drift. This can be expressed in terms of the electric field by directly applying the definition of current density, as shown in Fig. 2.11(a).

Over a time interval Δt, holes in the volume $\Delta V = A(v_{dp}\,\Delta t)$ will drift across the reference plane in the $+x$ direction. The number of holes ΔN_p that cross the plane in the interval Δt is given by the product of the concentration p of holes and the volume ΔV:

$$\Delta N_p = p\Delta V \tag{2.36}$$

The hole charge drifting across the reference plane ΔQ_p is the product of ΔN_p and the charge per hole q. According to the definition of current density

$$J_p^{dr} = \frac{\Delta Q_p}{A\Delta t} = \frac{qp\Delta V}{A\Delta t} = \frac{qpAv_{dp}\Delta t}{A\Delta t} = qpv_{dp} \tag{2.37}$$

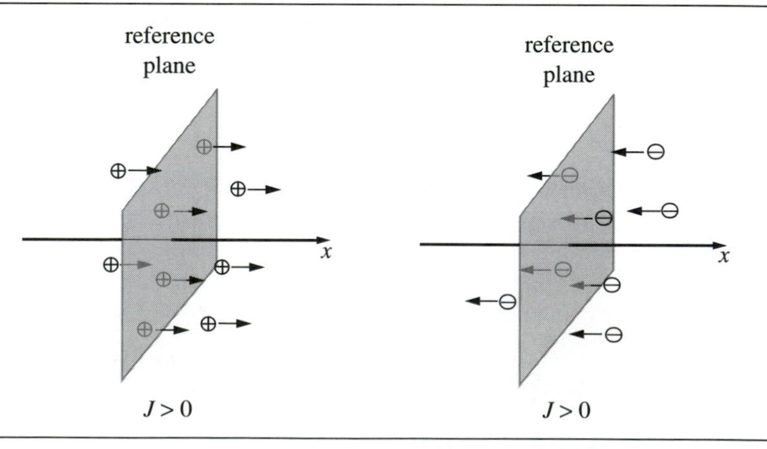

➤ **Figure 2.10** Sign convention for current density J. Both cases represent positive values of J since there is a net transport of positive charge in the $+x$ direction.

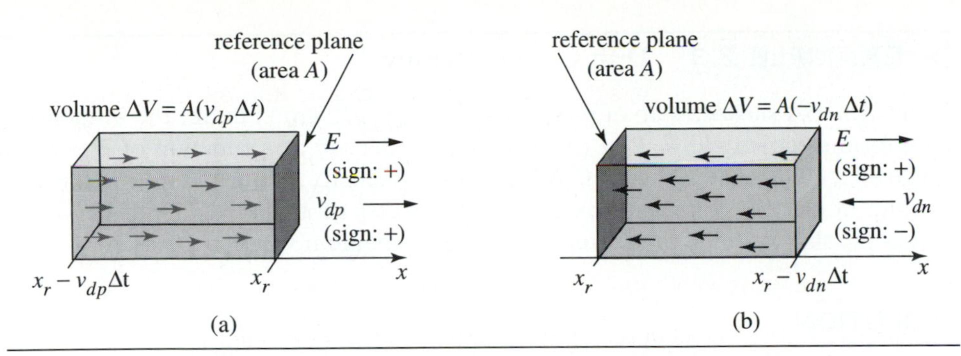

➤ **Figure 2.11** (a) The definition of current density applied to drifting holes, and (b) to drifting electrons. The reference plane is located at $x = x_r$ for this example. For an electric field pointed in the $+x$ direction, holes drift through the reference plane from left to right, whereas electrons drift from left to right. The current density $J^{dr} > 0$ for both cases.

Substituting for the drift velocity from Eq. (2.35), we find that

$$\boxed{J_p^{dr} = qp\mu_p E \text{ for } (v_{dp} \ll v_{sat})} \tag{2.38}$$

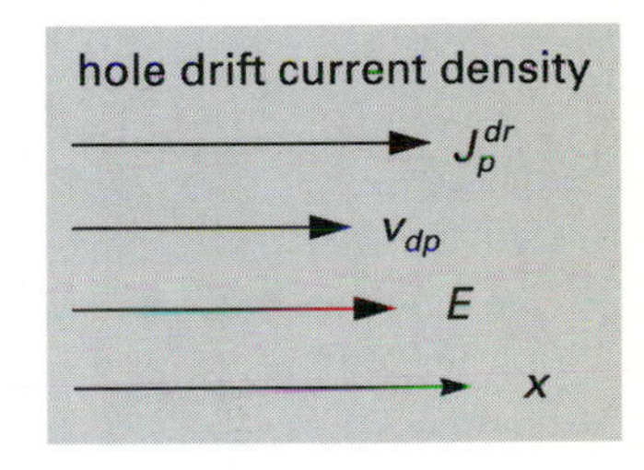

For electrons, the drift velocity is opposite in direction to the electric field as shown in Fig. 2.11(b). On average, electrons in a volume $\Delta V = A(-v_{dn}\,\Delta t)$ drift across the reference plane in the time interval Δt. The number of electrons crossing the plane is therefore $\Delta N_n = n\Delta V$, and the electron charge is equal to

$$\Delta Q_n = (-q)\,\Delta N_n = -qnA\,(-v_{dn}\Delta t) \tag{2.39}$$

The electrons drift across the reference plane in the $-x$ direction as shown in Fig. 2.11(b). Therefore, the drift current density is given by

$$J_n^{dr} = -\left[\frac{\Delta Q_n}{A\Delta t}\right] \tag{2.40}$$

Substituting the electron charge from Eq. (2.39) and the drift velocity from Eq. (2.34), the electron drift current density is given by

$$J_n^{dr} = -\left[\frac{(qnAv_{dn}\Delta t)}{A\Delta t}\right] = -qnv_{dn} = -qn\,(-\mu_n E\,) = qn\mu_n E \tag{2.41}$$

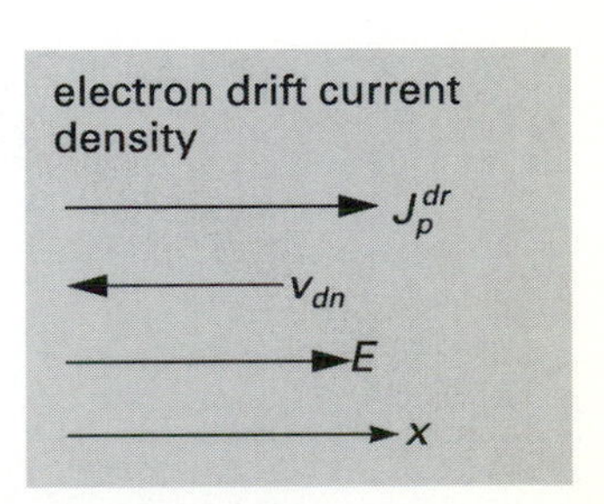

Therefore, the electron drift current points in the same direction as the electric field. Restating Eq. (2.41)

$$\boxed{J_n^{dr} = qn\mu_n E, \text{ for } |v_{dn}| \ll v_{sat}}, \tag{2.42}$$

➤ EXAMPLE 2.3 Drift Current Density

Figure Ex2.3A shows a slab of p-type silicon ($N_a = 10^{16}$ cm^{-3}) under the influence of a uniform electric field E. Plot the drift current density as a function of E over the range -5×10^5 V/cm $< E < 5 \times 10^5$ V/cm. Note: Velocity saturation occurs for holes drifting in electric fields with magnitudes in excess of 10^5 V/cm. Find the voltage across the slab at which the current reaches its maximum value.

SOLUTION

For small electric fields, we can use Eq. (2.38) to find the relationship between E and the drift current density

$$J_p^{dr} = qp\mu_p E = qN_a \mu_p E \qquad v_{dp} \ll v_{sat}$$

where we have substituted $p = N_a$ for the hole concentration. Even though the holes are drifting with the electric field and the slab is not in thermal equilibrium, their concentration remains approximately equal to the acceptor concentration. As we will see later in this chapter, metal coatings on the left and right faces of the silicon slab can supply and remove the drifting carriers.

From Fig. 2.8, the hole mobility is $\mu_p = 425$ cm^2/(Vs) and the low-field drift current density is $J_p^{dr} = (0.68)E$ (units: J: A/cm^2, E: V/cm).

According to Fig. 2.9, this linear relationship is valid until the hole drift velocity saturates, which occurs at electric fields greater than $E_{crit} \approx 10^5$ V/cm = 100 kV/cm. Since we don't have a model for velocity saturation, we limit the drift current density at $J_{p,max}^{dr} = (0.68) \cdot E_{crit} = 68$ kA/cm^2 as shown in Fig. Ex2.3B

Substituting for the thickness of the slab, the voltage needed to saturate the drifting holes is $E_{crit} \times 1\ \mu$m = (10^5V/cm) (10^{-4} cm) = 10 V. The maximum current is given by the product of $J_{p,max}^{dr}$ and the area of the slab: I_{max} = (68 kA/cm^2) ($5 \times 5 \times 10^{-8}$ cm^2) = 17 mA.

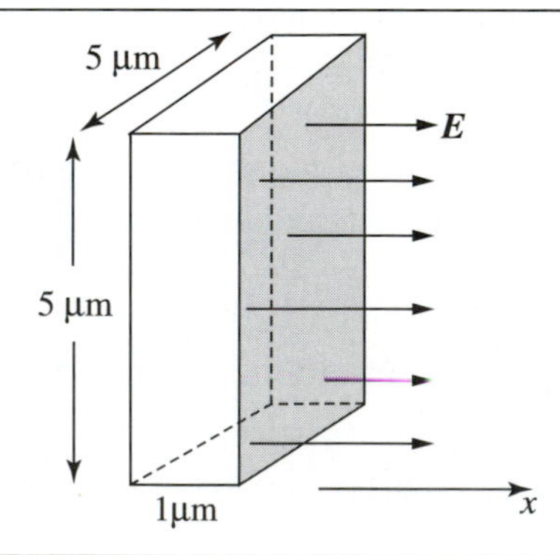

➤ **Figure Ex2.3A** P-type silicon slab with uniform applied electric field E.

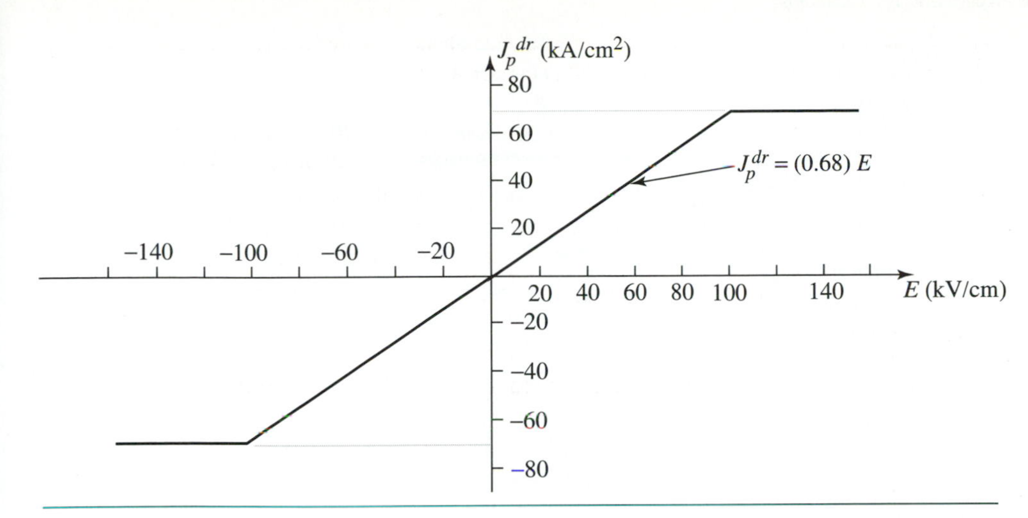

➤ **Figure Ex2.3B** Drift current density through slab as a function of the applied electric field.

2.4.3 Diffusion Current Density

Diffusion is the process where gradients in concentration cause a net transport of particles. For example, suppose an ammonia bottle is opened in one corner of a room in which the air is at room temperature. Even if the air in the room is perfectly still, the random motions (with frequent collisions) of the air and ammonia molecules enable the transport of ammonia throughout the room. The net transport will be from regions of high concentration in the corner with the open bottle, to regions of low concentration in the opposite corner. If the bottle is capped, the concentration of ammonia will equalize throughout the room after a period of time. Diffusion reduces and eventually eliminates all gradients in ammonia concentration throughout the room.

Since electrons and holes also undergo random motion with frequent collisions, gradients in the electron or hole concentrations result in diffusion transport. In order to model diffusion, we consider holes having a concentration $p(x)$, which decreases with x as shown in Fig. 2.12. On the average, holes crossing the reference plane at $x = x_r$ started within one mean free path λ of x_r. The diffusion current density J_p^{diff} of holes across the reference plane has its origin in the difference in the number of holes in the two regions of volume $\Delta V = A\lambda$ to the left and right of the reference plane. Only half of the holes starting in either region will cross the reference plane in the time interval τ_c, while the other half are moving away from the reference plane. Therefore, the hole diffusion current density at $x = x_r$ is given by

$$J_p^{diff}(x = x_r) = q\left[\frac{\frac{1}{2}p\,(x = x_r - \lambda)\,A\lambda - \frac{1}{2}p\,(x = x_r + \lambda)\,A\lambda}{A\tau_c}\right] \tag{2.43}$$

where $A\lambda$ is the volume. Note that holes crossing from the right to the left, opposite to the $+x$ direction, enter into Eq. (2.43) with a negative sign.

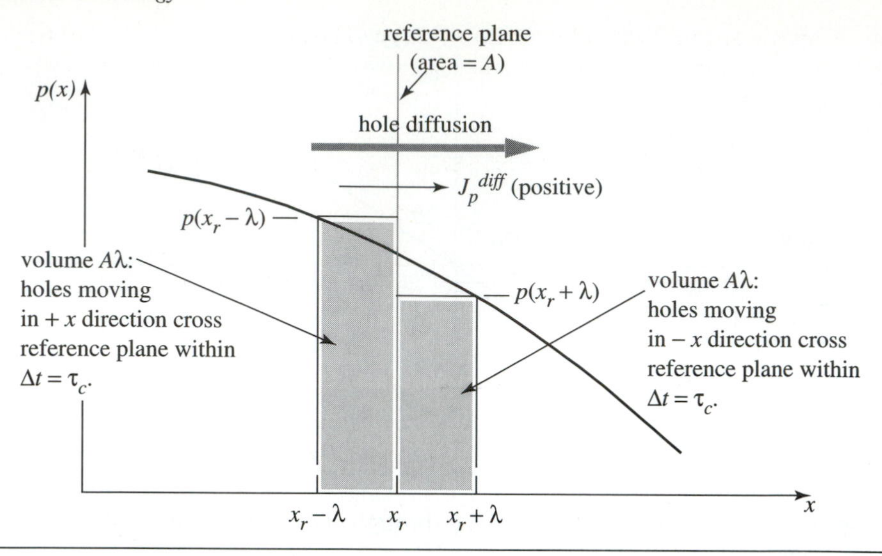

➤ **Figure 2.12** One-dimensional picture for quantifying the hole diffusion current density as a function of the gradient in hole concentration.

For Eq. (2.43) to be a good approximation, the mean free path (about 0.01 μm at room temperature) must be much smaller than the distance over which $p(x)$ is varying. This requirement is usually satisfied for microelectronic device structures. Note that we have highly exaggerated the gradient in the hole concentration in Fig. 2.12 to make the difference in the hole population across the reference plane clearly visible. Using first-order Taylor expansions for $p(x = x_r + \lambda)$ and $p(x = x_r - \lambda)$ and simplifying, we find that

$$J_p^{diff}(x = x_r) = \frac{q\lambda}{\tau_c}\left(\frac{\left(p(x_r) - \left.\frac{dp}{dx}\right|_{x_r}\lambda\right) - \left(p(x_r) + \left.\frac{dp}{dx}\right|_{x_r}\lambda\right)}{2}\right) \tag{2.44}$$

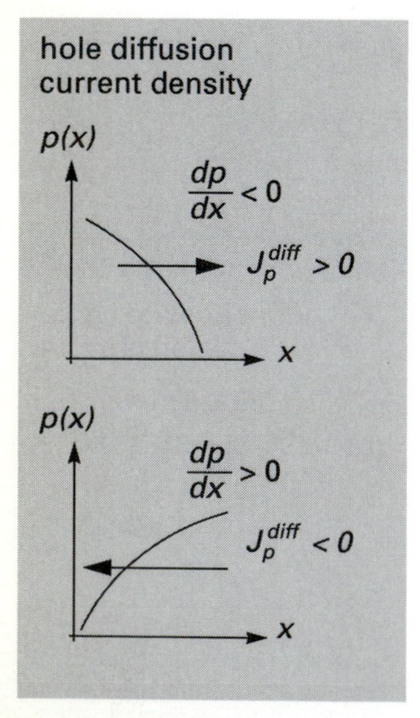

Canceling the hole concentration and collecting terms, Eq. (2.44) reduces to

$$J_p^{diff}(x = x_r) = -q\left(\frac{\lambda^2}{\tau_c}\right)\left.\frac{dp}{dx}\right|_{x = x_r} = -qD_p\left.\frac{dp}{dx}\right|_{x = x_r} \tag{2.45}$$

In Eq. (2.45), the term λ^2/τ_c is identified as the **hole diffusion coefficient** D_p, with units of cm^2/s. The diffusion current density for holes is

$$\boxed{J_p^{diff} = -qD_p\frac{dp}{dx}} \tag{2.46}$$

The proportionality of the diffusion current density to the concentration gradient is known as **Fick's Law** and is a fundamental result for diffusion processes. The negative sign in Eq. (2.46) reflects the fact that particles diffuse from regions of high concentration toward regions of low concentration. The concentration gradient dp/dx in Fig. 2.12 is negative, so the holes diffuse in the $+x$ direction and the diffusion current density is positive.

For the case of the electron concentration gradient shown in Fig. 2.13, we find the net transport of charge across the reference plane in the $+x$ direction as

$$J_n^{diff}(x = x_r) = (-q)\left[\frac{\frac{1}{2}n(x = x_r - \lambda)A\lambda - \frac{1}{2}n(x = x_r + \lambda)A\lambda}{A\tau_c}\right] \tag{2.47}$$

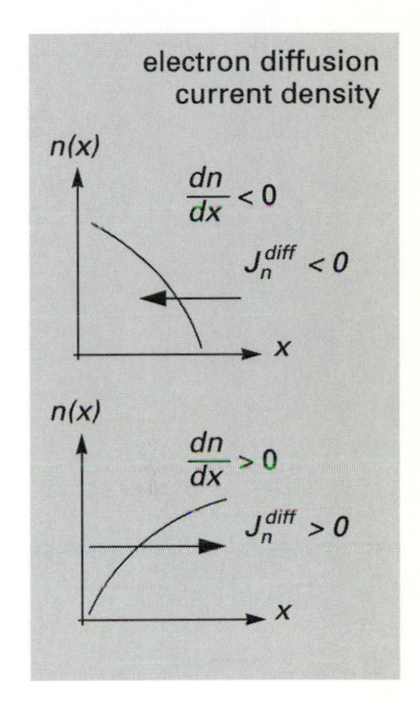

This is identical in form to Eq. (2.43) for holes, except we have accounted for the negative charge on the electrons crossing the reference plane. Again, use first-order Taylor expansions for the concentrations at $x_r - \lambda$ and $x_r + \lambda$

$$J_n^{diff}(x = x_r) = \frac{(-q)\lambda}{\tau_c}\left(\frac{\left(n(x_r) - \left.\frac{dn}{dx}\right|_{x_r}\lambda\right) - \left(n(x_r) + \left.\frac{dn}{dx}\right|_{x_r}\lambda\right)}{2}\right) \tag{2.48}$$

Collecting terms, the electron diffusion current density can be expressed by

$$J_n^{diff}(x = x_r) = q\left(\frac{\lambda^2}{\tau_c}\right)\left.\frac{dn}{dx}\right|_{x = x_r} = qD_n\left.\frac{dn}{dx}\right|_{x = x_r} \tag{2.49}$$

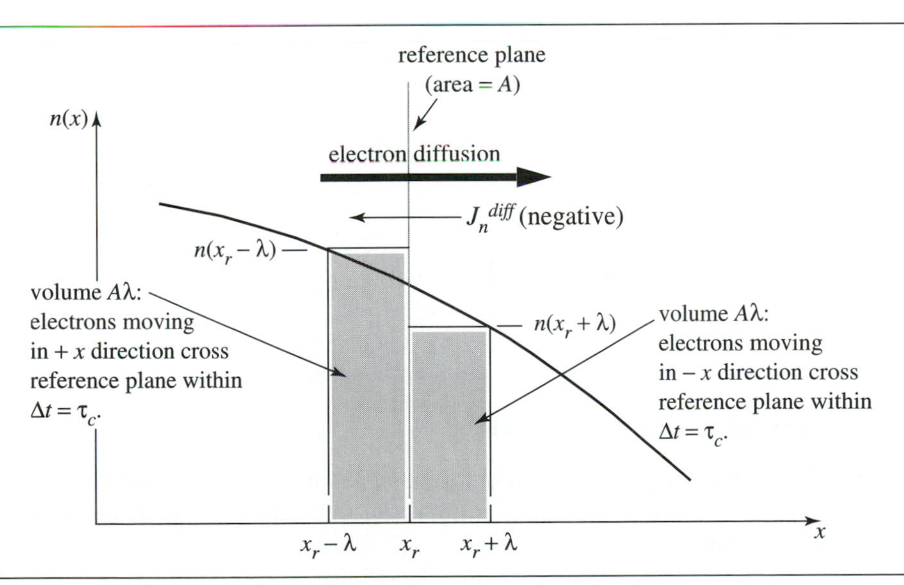

➤ **Figure 2.13** One-dimensional picture for quantifying the electron diffusion current density as a function of the gradient in electron concentration. Note that the electrons diffuse left-to-right (in the $+x$ direction), but that J_n^{diff} points in the $-x$ direction and is negative.

where the **electron diffusion coefficient** is identified as D_n.The negative gradient in electron concentration ($dn/dx < 0$) shown in Fig. 2.12 means the electron diffusion current density is negative (i.e., it points in the $-x$ direction). The physical reason behind this result is that the electrons diffuse from regions of high concentration to regions of low concentration in the $+x$ direction in Fig. 2.13. However, each electron carries a charge $-q$, which results in J_n^{diff} pointing in the $-x$ direction.

The diffusion coefficients for holes and electrons are functions of doping concentration and temperature since these are determined by the mean free path and collision time. The mobility μ and diffusion coefficient D were derived from physical models based on large numbers of carriers undergoing random thermal motion with frequent collisions. In a solid-state physics course, it is shown that the two constants are proportional and their ratio is equal to

$$\frac{D}{\mu} = \frac{kT}{q} \tag{2.50}$$

where $k = 1.38 \times 10^{-23}$ J/K is Boltzmann's Constant. This important result is called the **Einstein Relation**—named after Albert Einstein, who was active in solid-state physics in addition to his other interests. Eq. (2.50) holds separately for holes and for electrons

Einstein Relation

$$\frac{D_p}{\mu_p} = \frac{kT}{q} \quad \text{and} \quad \frac{D_n}{\mu_n} = \frac{kT}{q} \tag{2.51}$$

One implication of the Einstein Relation is that only the mobility need be measured. The diffusion coefficient can be found directly from Eq. (2.51). Using room-temperature values for the electron and hole mobilities in silicon (1000 and 400 cm²/(Vs)), typical round numbers for the electron and hole diffusion coefficients are

$$D_n = 25 \text{ cm}^2/\text{s and } D_p = 10 \text{ cm}^2/\text{s} \tag{2.52}$$

The **thermal voltage** is defined as the quantity kT/q:

thermal voltage

$$V_{th} = \frac{kT}{q} \tag{2.53}$$

The thermal voltage has units of Volts and will appear in many device physics equations. It ranges from 25 mV in a cool room in which the temperature is 17 °C or 62 °F to 26 mV in a somewhat warm room at 28.5 °C or 83 °F. In this book, we will choose either $V_{th} = 25$ mV or 26 mV—whichever results in "round numbers."

2.4.4 Total Current Density

As we have seen in the previous two sections, electrons and holes drift in an electric field and diffuse due to concentration gradients. The total current density J is the sum of all four components

$$J = J_n + J_p = J_n^{dr} + J_n^{diff} + J_p^{dr} + J_p^{diff}$$
$$J = qn\mu_n E + qD_n\frac{dn}{dx} + qp\mu_p E - qD_p\frac{dp}{dx} \tag{2.54}$$

The total electron and total hole current densities J_n and J_p are important results, therefore they are given separately

electron and hole total current densities

$$J_n = J_n^{dr} + J_n^{diff} = qn\mu_n E + qD_n \frac{dn}{dx} \quad (2.55)$$

$$J_p = J_p^{dr} + J_p^{diff} = qp\mu_p E - qD_p \frac{dp}{dx} \quad (2.56)$$

In most cases, we will be able to simplify Eqs. (2.55) and (2.56) for a particular device structure and focus on the dominant components of the current densities.

➤ EXAMPLE 2.4 Diffusion Current Density

Figure Ex2.4 shows a 1 μm-thick slab of p-type silicon having $N_a = 10^{17}$ cm^{-3}. By continuously injecting minority electrons on the left side at $x = 0$ and extracting them from the right side at $x = 1$ μm, we have set up a linear gradient in the electron concentration $n(x)$. What is the electron diffusion current density in the slab?

SOLUTION

On the linear plot of $n(x)$, we note that the electron concentration at $x = 1$ μm, 10^3 cm^{-3}, is indistinguishable from zero. For a total doping concentration of 10^{17}cm^{-3}, Fig. 2.8 indicates that the *electron* mobility is $\mu_n = 750$ cm^2/(Vs). Substituting into the Einstein Relation from Eq. (2.51), we find that the electron diffusion coefficient is

$$D_n = V_{th} \cdot \mu_n = (25 \times 10^{-3}\text{V}) \cdot (750\text{cm}^2/\text{Vs}) = 18.75\ \text{cm}^2/\text{s}$$

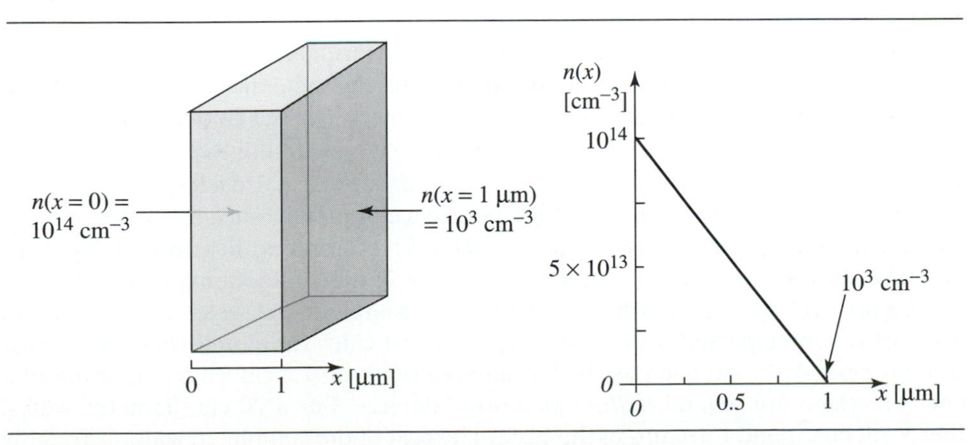

➤ Figure Ex2.4 Electron concentration through a slab of p-type silicon.

We apply Eq. (2.49) to calculate the electron the diffusion current density in the silicon slab

$$J_n^{diff} = qD_n \frac{dn}{dx} = (1.6 \times 10^{-19})\,(18.75)\left(\frac{10^3 - 10^{14}}{10^{-4}}\right) = -3\ \text{A/cm}^2$$

Note that the current density is negative since the electrons are diffusing from left to right and carry a negative charge. The magnitude of the diffusion current density in this example is also small in comparison with typical majority carrier drift current densities. A voltage of less than 1 mV between the left and right faces of the slab is all that is needed to cause a drift current of majority holes of 3 A/cm^2.

2.5 SILICON INTEGRATED CIRCUIT TECHNOLOGY

The focus of this text is silicon **integrated** devices and circuits, in which an electronic functional block, subsystem, or entire system is implemented on a single integrated circuit (IC). A basic appreciation of the manufacturing processes used to make ICs will be helpful in later chapters since it will give you an appreciation for the dimensional scales and physical structures of integrated devices. As you will see, the origin of the important "parasitic" circuit elements in ICs is obvious from the device structures.

This section provides a basic introduction to **IC fabrication technology**. This fascinating discipline draws from materials science and engineering, chemical engineering, and applied physics, as well as electrical engineering. By many measures, silicon IC technology is the crowning technological achievement of the late 20th century. Few areas of life have been unaffected by the penetration of silicon ICs such as microprocessors, microcontrollers, and modems. Since the fundamental physical limits of these products are at least two decades away, the early 21st century is certain to see further advances in the core fabrication processes that will have great impact on computers and other electronic systems. In addition, "spin-offs" from IC technology are catalyzing new fields such as micro electromechanical systems (MEMS). Silicon MEMS have display, sensing, or actuating functions. They may incorporate 10^6 micromechanical elements on the silicon chip along with a complex electronic interface. In summary, although the first silicon IC dates to around 1960, the end of the IC revolution is by no means in sight.

Figure 2.14 shows the fundamental unit of IC manufacturing—a silicon **wafer**. The typical wafer diameter is 20 cm in the mid-1990's. The wafer is a nearly perfect crystal that has been polished to be extremely flat. Wafers of this diameter cost about US$50.00 in large quantities. A group of around 24 wafers, called a **lot**, undergoes the same sequence of fabrication steps together. In a complex IC such as a microprocessor or dynamic random-access memory (DRAM), the process has around 200 steps and takes about three weeks to complete for a typical manufacturing schedule.

Figure 2.14 shows that when the process is complete, the wafer is tiled with an array of ICs interspersed with a few diagnostic test chips for monitoring the fabrication process. A typical complex IC has an area of about 1.5 cm × 1.5 cm, or about 2 cm^2, on which are several *million* electronic devices. For a 20 cm-diameter wafer, there will be around 150 chips of this area. The cost of the completed wafer is roughly US$2,000, which is a function of the fabrication process. How many of the chips work

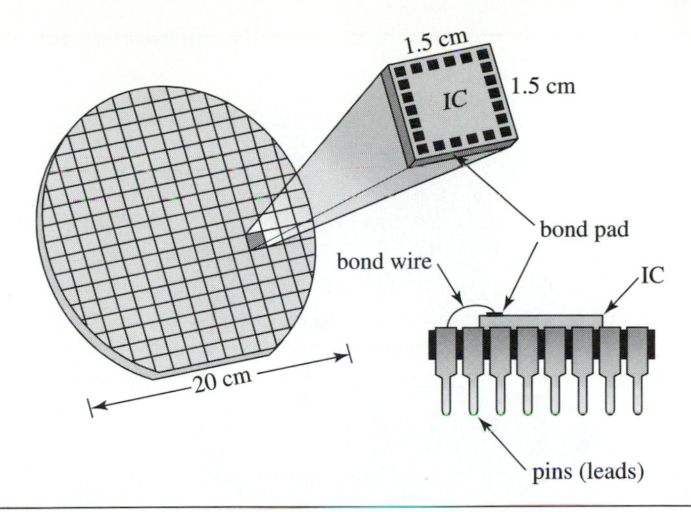

➤ **Figure 2.14** Silicon wafer and inset showing a single IC chip (die), with bond pads, and a side view of a dual in-line (DIP) package, shown with the lid removed and only a single bond wire. Wafer diameter and chip area are typical for mid-1990's large-volume, complex ICs.

(the **yield**) will affect the price at which the manufacturer sells each chip. The yield is a function of the quality control in the manufacturing plant, as well as the design margins built into the process and circuit. For example, a high-performance IC may have dimensions that approach the technological limits with the result that the yield is low. However, the high performance may justify a very high price and make the process profitable despite the low yield. As a particular IC design matures in production, the yield can be predicted to improve to around 90% or higher, which allows the selling price to drop steadily. The IC industry is extremely competitive. Companies aggressively track the **yield curve** of most products by dropping prices as yields increase. This maximizes volume and makes it more difficult for other companies to enter into direct competition on the product.

General-use ICs are packaged in plastic dual in-line (DIP) packages as shown in Fig. 2.14. In recent years, there has been great progress in economical packaging of ICs into high-density **multichip modules** for use in applications such as lap-top computers and cellular phones. Instead of the DIP shown in Fig. 2.14, chips are often encapsulated in low-profile surface mount packages or mounted directly on the circuit board using a "flip-chip" technique. Electrical connections between the board and inverted chip are made through electroplated solder bumps, rather than through bond wires, allowing for many more interconnects per chip.

2.5.1 Photolithography and Pattern Transfer

The essential process in IC fabrication is the imaging, alignment, and transfer of complex patterns onto the silicon wafer. A modern fabrication process uses around 15–20 separate patterning steps to define the transistors, diodes, and several levels of electrical interconnections (i.e., thin-film wires) that make up the IC. A **wafer stepper** is used to transfer patterns from an optical plate called a **mask**, to a photosensitive film coating the top surface of the wafer called **photoresist**. The process shown schematically in Fig. 2.15 depicts the exposure of a single die to the image of the mask.

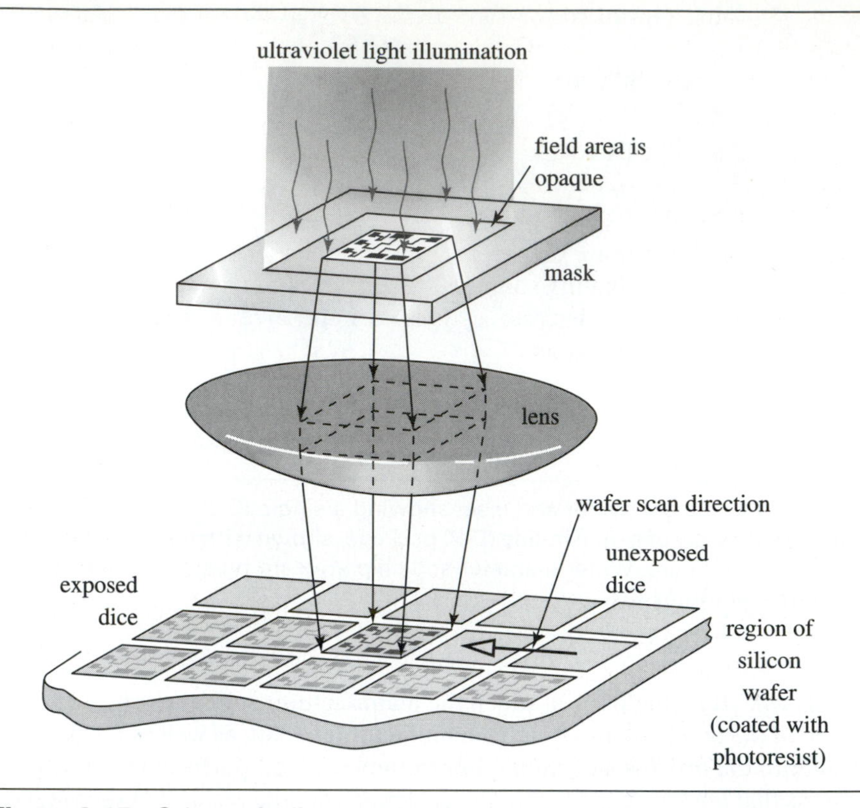

➤ **Figure 2.15** Schematic diagram showing imaging of the mask pattern on one die in a wafer stepper. The mask pattern is shown in gray on exposed dice. Unexposed dice are shown in outline only. In reality, the image in the photoresist is not visible until after development.

After completion of the exposure, the wafer stepper indexes to the next die, automatically aligns to special features previously etched on it, and repeats the exposure. Since the image of the mask pattern must be distortion-free and the error allowed in the alignment is less than 0.1 μm, it is not surprising that a high-volume wafer stepper costs well over US$2 million in the mid-1990s.

Figure 2.16 shows the step-by-step process of how the transfer from the mask to the wafer is accomplished through the photolithography and pattern transfer process. Figure 2.16 also introduces the customary way of visualizing three-dimensional IC structures. The **layout** consists of the series of masks used to create the IC structure. Lines across the layout denote the locations of **cross sections** through the silicon wafer. Typically, only the top few μm of the wafer are needed to show the device structure. In order to draw the cross sections, we must know the sequence of masking and fabrication steps. This sequence is called the **fabrication process**, or **process** for short. In analogy with food preparation, the fabrication process is sometimes called the **recipe**.

The single-mask photolithography and pattern transfer process in Fig. 2.16 involves five steps.

1. A 500 nm-thick film of silicon dioxide, SiO_2, is grown on the silicon wafer. In order to grow the thermal SiO_2 film, the wafer is placed in a furnace at an elevated temperature. Silicon surfaces react with oxygen to form a very high quality insulating film.
2. The wafer is uniformly coated with a 1 μm-thick film of photoresist by applying a mixture of photoresist and solvent, spinning the wafer at several thousand rpm for 30 seconds, and then baking to drive off the solvent.
3. Areas of this film that are exposed to ultraviolet light (UV) during the exposure step dissolve in an alkaline solution—a process known as **development**. Note carefully the differences in cross sections *A-A* and *B-B* after development of the mask image in the photoresist.

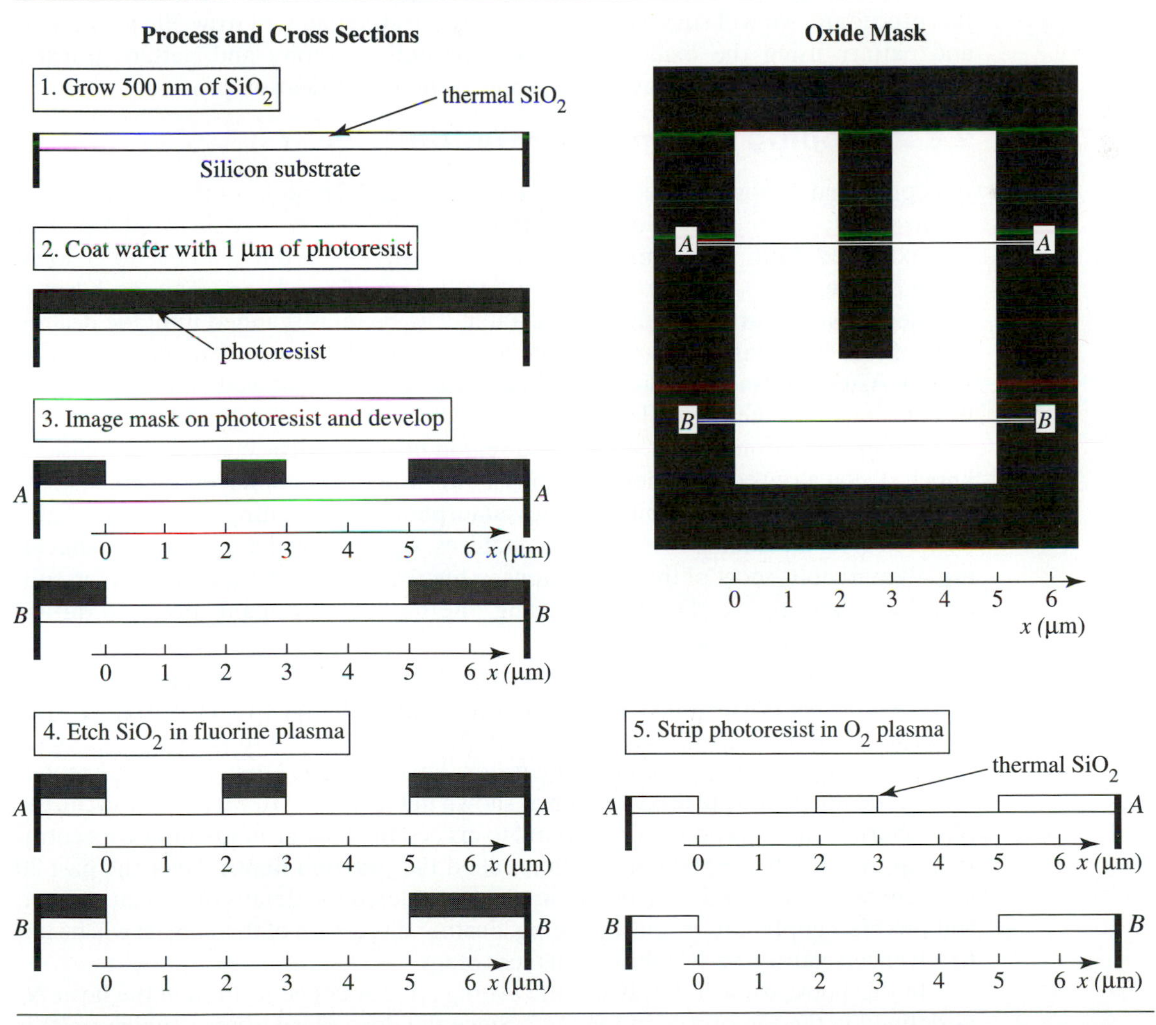

➤ Figure 2.16 Cross sections for transferring the oxide mask pattern to a thermal oxide film. A shorthand description of this process is "grow 500 nm of thermal SiO_2 and pattern using the oxide mask."

4. After baking the photoresist to drive off residual solvents, the wafer is immersed in a vacuum chamber containing a fluorine **plasma**. A typical example of a plasma is the glow discharge in a fluorescent light. The plasma reacts with SiO_2 and etches it much faster than it etches the photoresist film. Once the silicon dioxide is etched away from the exposed areas, the underlying silicon is also etched, but very slowly. This enables the etch to be terminated before significant silicon erosion occurs.
5. The final step is to remove the photoresist in an oxygen plasma. This plasma aggressively attacks organic materials, such as photoresist, but does not etch silicon or silicon dioxide. Comparing the cross sections for steps 3 and 5 in Fig. 2.16, we see that the mask pattern has been successfully transferred to the silicon dioxide film.

As we mentioned, an IC process involves many photolithography and pattern transfer steps on deposited films. In order to avoid repeating steps 1–5 each time we use this procedure, we will summarize Fig. 2.16 by the statement: "Grow 500 nm of SiO_2 and pattern using the oxide mask." The photolithography and pattern transfer sequence can be thought of as a subroutine of the IC fabrication process.

2.5.2 Doping by Ion Implantation

Acceptors and donors can be incorporated into selected regions of the top surface of a silicon wafer by means of **ion implantation**. Ions of boron, arsenic, or phosphorus are extracted from a plasma, accelerated to energies from 20 keV to 3 MeV (eV = electron volt), and formed into a tightly aligned ion beam. The ion beam is scanned over the surface of the silicon wafer. The beam is scanned until the desired **dose** of ions is implanted into the wafer. The ion dose has units of ions per unit area (cm^{-2}). After entering the silicon, the bombarding ions are slowed and then stopped by collisions with the silicon lattice.

Ion implantation creates a damaged region near the surface, since the collisions knock silicon atoms out of their positions in the lattice. The damaged region loses its crystal structure entirely and becomes **amorphous**. By heating the silicon wafer above 900° C, the damaged region recrystallizes, a process which is known as **annealing**. Remarkably, most of the dopant ions in the recrystallized region end up on substitutional lattice sites. During annealing, the implanted dopant ions also diffuse further into the silicon wafer.

Figure 2.17 illustrates how ion implantation can be used to dope specific regions of the wafer. We start with the patterned oxide film from Fig. 2.16, grown on a p-type wafer with a background donor concentration $N_a = 10^{15}$ cm^{-3}. A dose of $Q_d = 10^{14}$ cm^{-2} phosphorous ions is implanted into the wafer. Their energy is such that the phosphorus ions do not penetrate the SiO_2 film as shown in Fig. 2.17. After a thermal cycle, the phosphorus concentration $N_d(x)$ is found to exceed the background donor concentration up to a depth $x_j = 500$ nm, which is called the **junction depth**. Over the past 20 years, extensive research has investigated and modeled the details of dopant concentrations after implantation and annealing. Further discussion of this subject will be left to advanced courses on IC fabrication technology.

In this book, we will simply use the average donor concentration in the layer N_d as defined in the ion profile in Fig. 2.17. Since the dose Q_d of donors (units: cm^{-2}) is located in a layer of depth x_j after annealing, the average concentration is

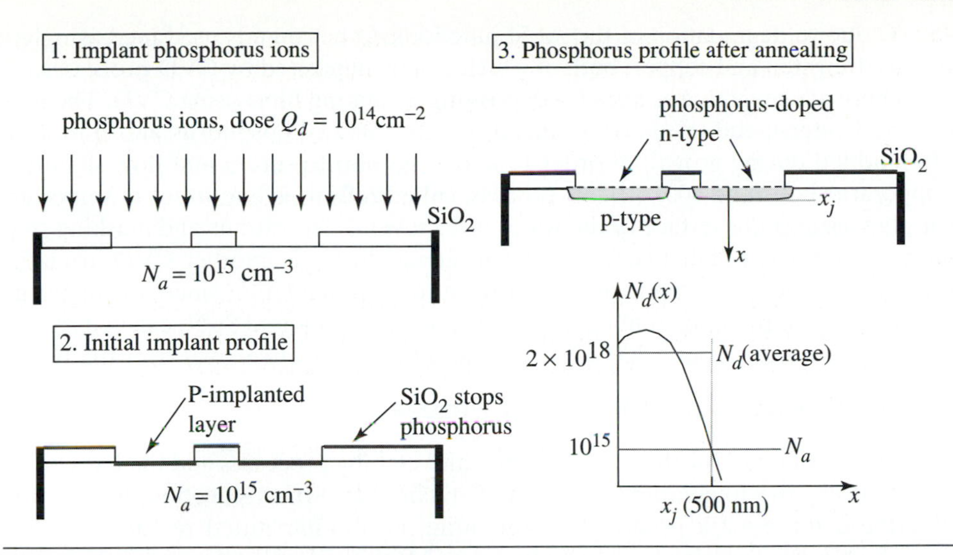

➤ **Figure 2.17** Phosphorus implantation into the patterned oxide film shown in Fig. 2.16. The average concentration of phosphorus is defined in the plot of $N_d(x)$ after annealing.

$$N_d = \frac{Q_d}{x_j} \tag{2.57}$$

For the case shown in Fig. 2.17, $N_d = (10^{14}\text{ cm}^{-2})/(5 \times 10^{-5}\text{ cm}) = 2 \times 10^{18}\text{ cm}^{-3}$.

2.5.3 Deposited Conducting and Insulating Films

In order to fabricate integrated circuits, we need conducting films that can be patterned into the microscopic wiring called **interconnects**. The interconnects enable electrical signals to be propagated around the IC. In addition, insulating films are needed to enable crossovers between two or more levels of interconnects. In the mid-1990s, three to five levels of interconnect are used for efficient packing and for minimum interconnect lengths in high-density digital ICs. As we will see in Chapter 5, short interconnects are important for achieving high switching speed.

Our first conductor is **polycrystalline silicon** or **polysilicon**. This material is an aggregate of crystallites or grains of single crystal silicon. Thin films of polysilicon are deposited on the surface of a silicon wafer at temperatures around 600 °C in a process known as **chemical vapor deposition** (CVD). Polysilicon is usually deposited with a maximum doping level of phosphorus. As we will see in the next section, polysilicon is a mediocre conductor; however, it has the virtues of being thermally stable and free of contaminants.

Aluminum is the most commonly used conducting film for interconnects. It is deposited by **sputtering**, a process in which material is deposited on the wafer from a nearby target that is bombarded by ions from a plasma discharge. One drawback of aluminum interconnects is that they cannot tolerate temperatures higher than about

450 °C due to degradation of the Al-Si junctions. Other metals used increasingly in ICs are tungsten and copper, both of which can be deposited by CVD processes.

There are several processes for depositing insulating films using CVD. The most common inter-metal insulator is silicon dioxide, SiO_2. Phosphorus and boron are often added during growth in order to allow the film to soften and flow slightly at temperatures around 900° C. This process, called **reflow**, is important in smoothing (or planarizing) the surface of the wafer after several depositions and masking steps have created too rough a surface. Silicon nitride, Si_3N_4, is another CVD insulating film that is able to mask a high-temperature thermal oxidation since it is impermeable to oxygen. In plasma-enhanced CVD processes, SiO_2 and Si_3N_4 can be grown at low temperatures (around 300° C) for use as a final protective layer over the IC.

2.5.4 Layouts and Process Flows

You are now familiar with the lithography and etching processes and the basic materials that are the foundation of silicon IC technology. In this section, we will first describe a three-mask process for fabricating an ion-implanted resistor and then describe a more complex process to make an MOS field-effect transistor. You need not understand anything at this point about how these structures function as electronic devices. In following these sample processes, focus on the connection between the layout and the evolving device cross sections.

Figure 2.18(a) depicts the layout of the resistor, as it would be viewed in a computer-aided design (CAD) program. At first glance, you would expect that these patterns would correspond exactly to the actual masks used for photolithography. In fact, a mask that is nearly opaque (referred to as **dark field** mask) will obscure the other masks. Even with the use of color and a variety of "fill" and "border" designs, a nearly opaque mask will make it difficult to see features on the other masks in a layout. Fortunately, there is a simple solution: we display the negative, or complement, of the dark field mask in depicting the layout. **Clear field** masks have few opaque regions and are displayed directly. Of course, we must keep a careful record of which are clear field and which are dark field, so that the correct masks can be made. Many wafer lots have been ruined due to miscommunication on this minor, but critically important point.

In the resistor layout shown in Fig. 2.18(a), the oxide mask is identified as a dark field mask. Since the negative of the oxide mask is shown in the layout, the rectangle is clear on the actual mask. You should note carefully from cross sections *A-A* and *B-B* after step 1 that the oxide has been etched from the region of the wafer exposed by the clear rectangle. If this mask had been clear field, the oxide would have been removed everywhere *except* the rectangle, which would have been dark on the physical mask.

We now describe the process sequence in Fig. 2.18(b). The first step is to grow 500 nm of thermal oxide, which is done by heating the wafer in a furnace with oxygen flowing through it. This film is patterned using the oxide mask and the photolithography and etching process shown in Fig. 2.16. The wafer is then implanted with a dose of phosphorus ions that is sufficiently heavy to compensate the background acceptor concentration in the p-type silicon wafer. The thermal SiO_2 is thick enough to stop the phosphorus ions from penetrating into the underlying silicon substrate and is, therefore, effective at masking the implant.

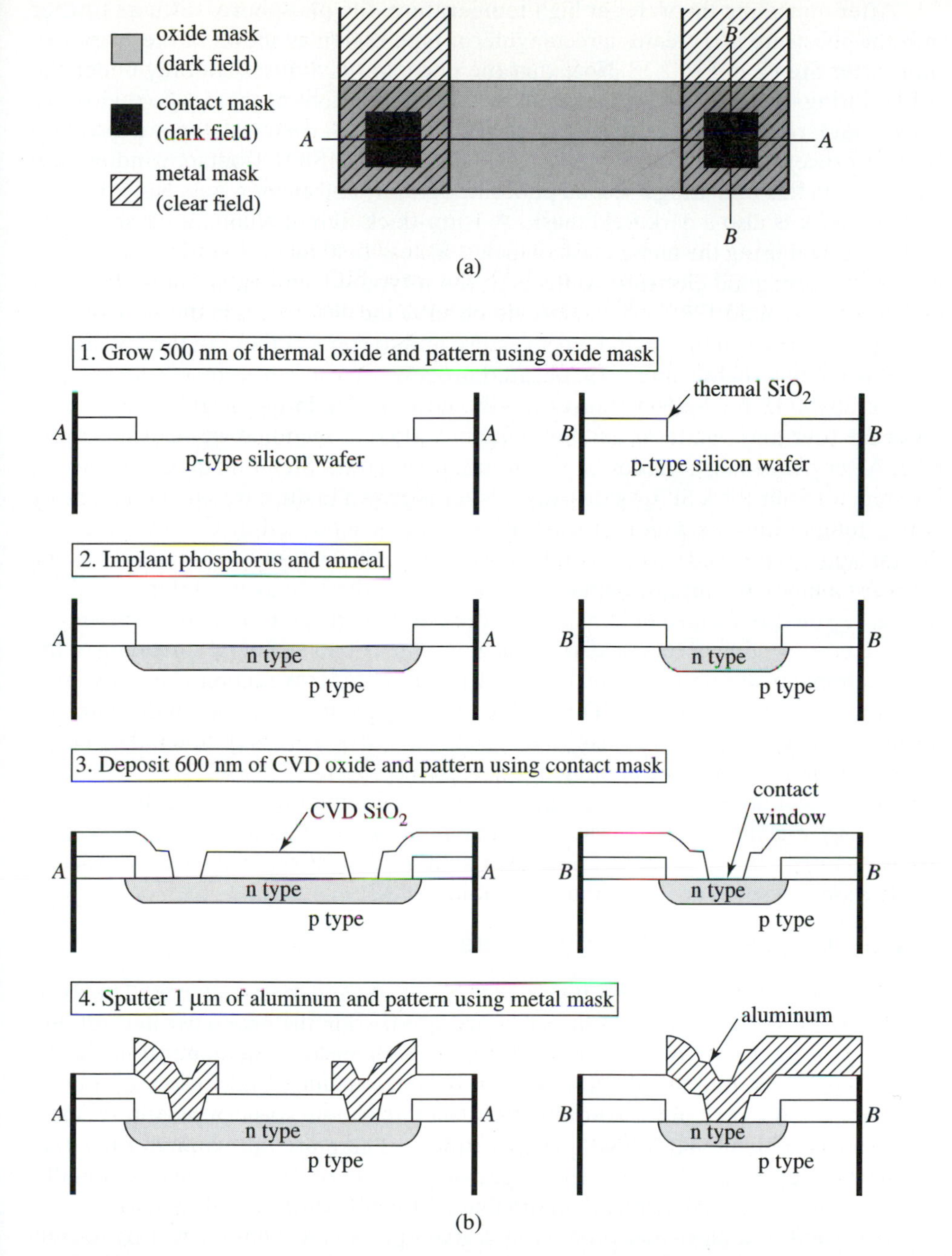

➤ **Figure 2.18** (a) Layout and (b) process flow with cross sections *A-A* and *B-B* for fabrication of an ion-implanted resistor.

After annealing the wafer at high temperature, the phosphorus diffuses further into the substrate and creates a rectangular n-type region as shown in the cross sections after Step 2 in Fig. 2.18. Note that the phosphorus diffuses laterally under the oxide during the anneal, with the result that the dimensions of the n-type region are somewhat larger than the rectangle on the oxide mask. A chemical-vapor deposition (CVD) process is then used to deposit 600 nm of CVD SiO_2. Contact windows are patterned in this film using a second photolithography and etching step. Note that the contact mask is also a dark field mask. A 1 μm-thick film of aluminum is sputtered and patterned using the metal mask, which is a clear field mask. In order for the aluminum to form good electrical contacts to the n-type silicon region, the wafer must be heated to around 425 °C. It is helpful to observe the differences in the *A-A* and the *B-B* cross sections in the completed structure and to connect these to the layout.

With only a slightly more complicated process, we can fabricate a "starter" version of an MOS field-effect transistor—the core device in digital ICs. Figure 2.19 gives the four mask patterns and the sequence of cross sectional views of the structure. After patterning a window in the initial 500 nm thick SiO_2 film in the first masking step, a 15 nm-thick SiO_2 "gate oxide" layer is grown in the bare silicon regions by a high-temperature oxidation process. Polysilicon is deposited by CVD with n-type doping and is patterned in the second masking step. The polysilicon film serves as the mask for a high-dose arsenic implant that penetrates the gate oxide. After annealing, n-type regions are formed on each side of the polysilicon "gate" that are seen only in the *B-B* cross section perpendicular to the gate. There is some lateral spreading of the arsenic implant during the anneal as seen in the *B-B* cross section. The next step involves the deposition of a CVD SiO_2 film, etching contact windows to the gate and to the two n-type regions adjacent to it. Finally, the MOSFET is completed by deposition and patterning of the aluminum interconnect layer.

An IC technology is often specified by λ, its minimum feature size. In an MOS technology, λ is the etched polysilicon gate length as shown in the final cross section in Fig. 2.19(b). The trend has been to ever-smaller dimensions, with state-of-the-art production processes at $\lambda < 0.5$ μm in the mid-1990s.

2.5.5 Geometric Design Rules

In Fig. 2.19(a), the contact nests inside both the oxide and metal patterns. The overlaps between patterns on different masks are specified in the **geometric design rules** for the process. These rules allow for systematic changes in the dimensions of the device structures, due to lithography, diffusion, or etching. They also incorporate knowledge of the random variations in these factors and the misalignment tolerances of the mask making and lithography equipment. There are "proximity" rules that specify the separation between the edges of patterns on different masks. Finally, there are rules that define the minimum line and space widths for each mask.

The yield of a particular product in a given process technology will be heavily influenced by the margins built into the design rules. With an IC containing 10^6 MOSFETs, each with at least three contacts, carefully chosen design rules for the contact mask nesting are essential for achieving a balance between high yield and a reasonable die size. Systematic experimentation and modeling of statistical variations are required to optimize the design rules for a particular technology and product family.

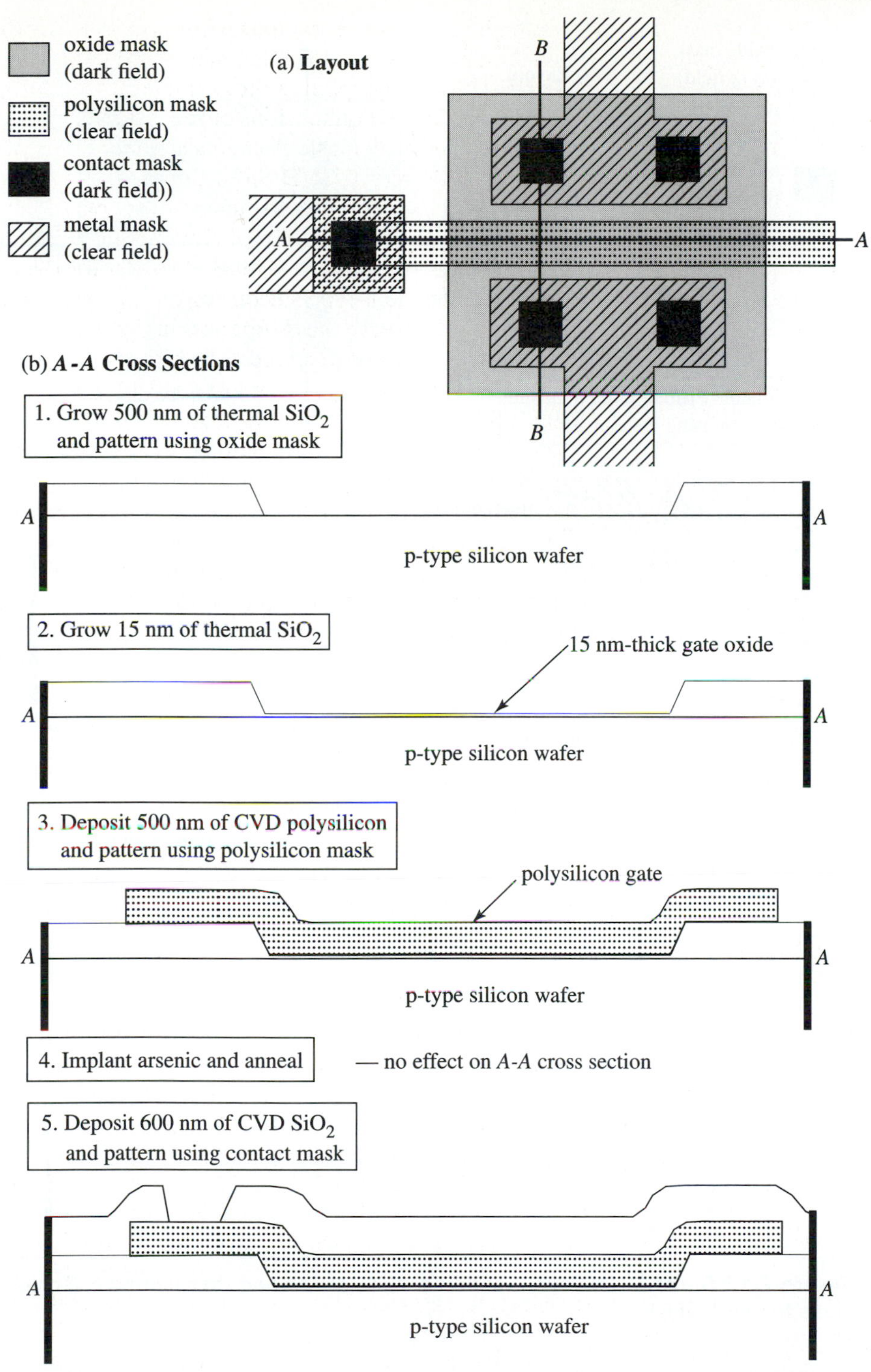

➤ **Figure 2.19** (a) Layout and (b) *A-A* cross sections for a simple MOSFET

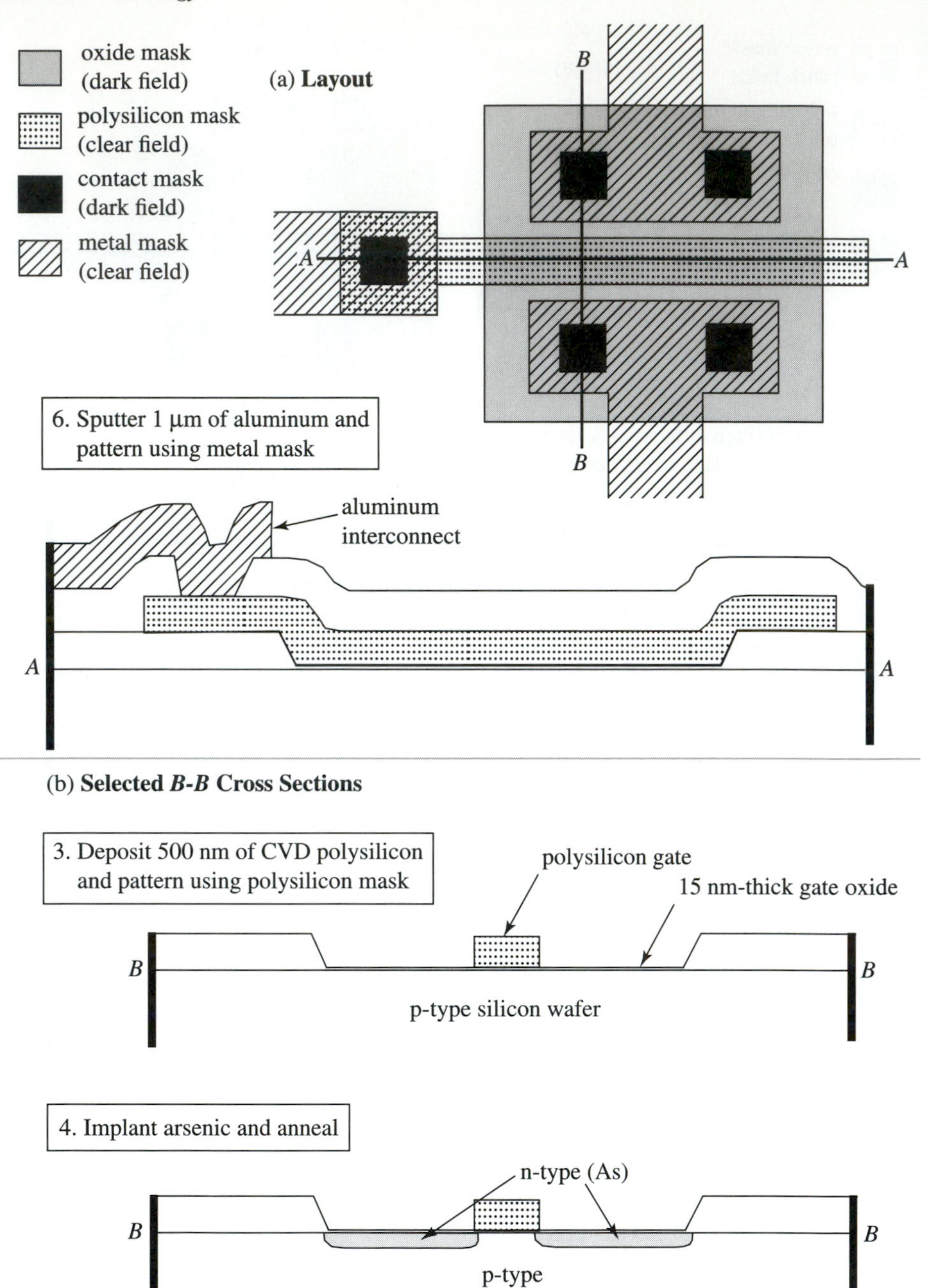

➤ Figure 2.19 (cont.) (a) Layout and final *A-A* section and (b) selected *B-B* cross sections for a MOSFET.

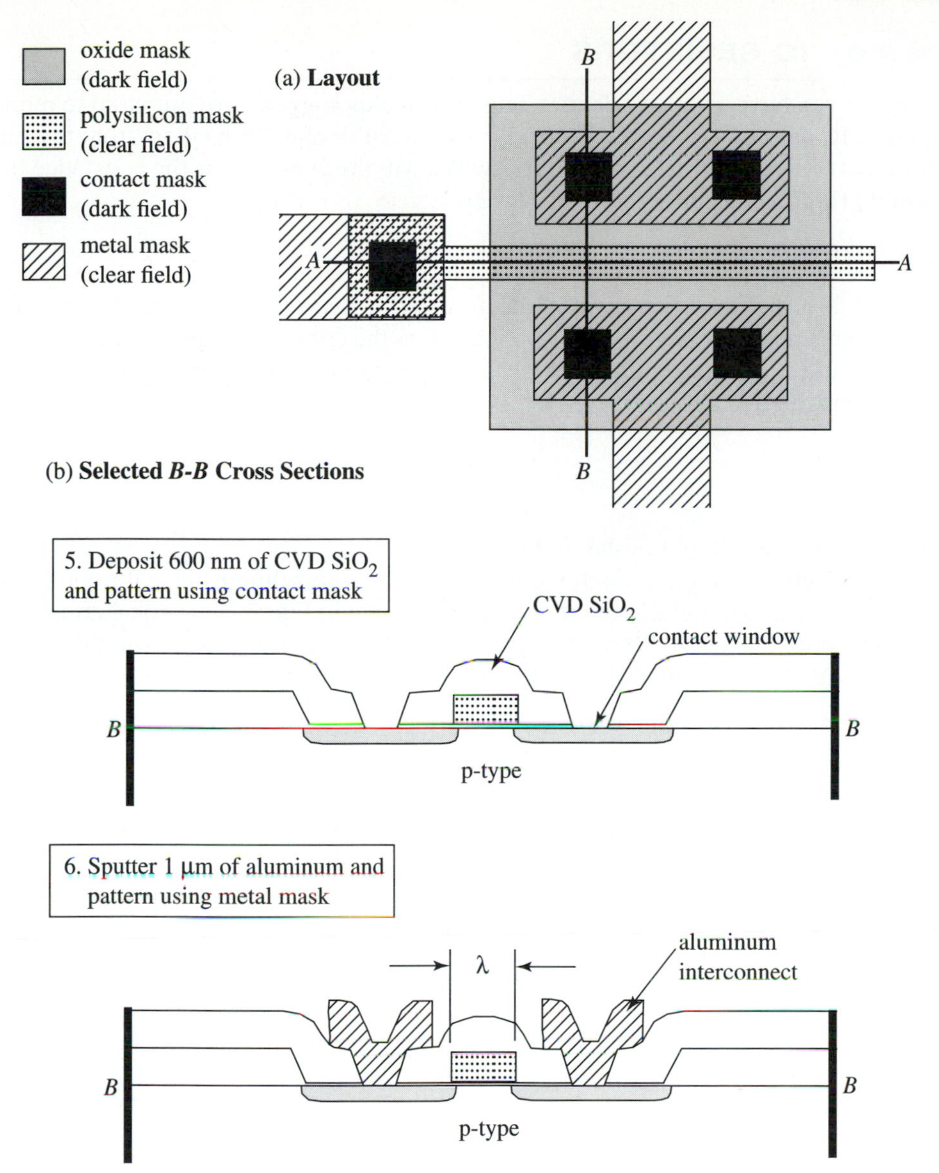

➤ **Figure 2.19 (cont.)** (a) Layout and (b) selected *B-B* cross sections for the simple MOSFET.

➤ 2.6 IC RESISTORS

Now that you have a basic understanding of IC technology, we are prepared to model our first IC device—a silicon resistor. First we must derive Ohm's Law from the fundamental concept of drift current density. We will then re-express the resistance in a form that will be more useful for IC design.

2.6.1 Ohm's Law

Figure 2.20 is a schematic view of a slab of n-type silicon with two sides having metallized contacts. A DC voltage V is applied across the contacts. What is the magnitude of the resulting current I? Our first observation is that there is an electric field E in the silicon with a magnitude given by

$$E = \frac{V}{L} \tag{2.58}$$

where L is the length of the slab in Fig. 2.20.

The electric field points in the direction of the x coordinate since it points from the contact with the higher potential (the nearer one in Fig. 2.20) to the contact with the lower potential. This electric field will cause electrons and holes in the silicon slab to drift, with the resulting drift current density being given by

$$J = J_n^{dr} + J_p^{dr} = qn\mu_n E + qp\mu_p E = (qn\mu_n + qp\mu_p)\left(\frac{V}{L}\right) \tag{2.59}$$

in which we have assumed that the electric field is low enough that velocity saturation is not an issue. There is no concentration gradient along the slab, since the mobile carriers are supplied and removed by the metal contacts. As a result, diffusion is ignored in Eq. (2.59). Since the current I [units: A] is the product of the current density J [units: A/cm^2] and the cross sectional area of the slab $A = W\,t$, we can find an equation relating current and voltage

$$I = J \cdot A = (qn\mu_n + qp\mu_p)\left(\frac{V}{L}\right)(Wt) = \left[(qn\mu_n + qp\mu_p)\left(\frac{Wt}{L}\right)\right] \cdot V \tag{2.60}$$

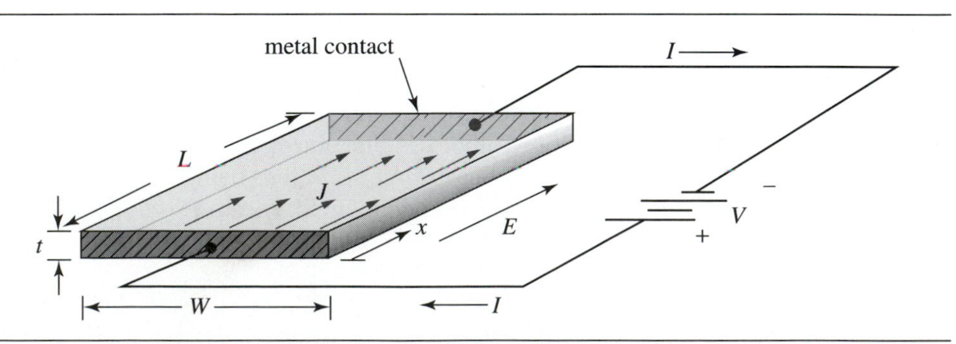

➤ **Figure 2.20** n-type silicon slab with length L, width W, and thickness t. A voltage V is applied across two metal contacts at opposite ends.

Ohm's Law states that $I = V/R$ and therefore, we can identify the resistance R of the slab from Eq. (2.60)

$$R = \frac{1}{(qn\mu_n + qp\mu_p)\left(\frac{Wt}{L}\right)} = \left(\frac{1}{(qn\mu_n + qp\mu_p)}\right)\left(\frac{L}{Wt}\right) \tag{2.61}$$

The first term on the right-hand side of Eq. (2.61) is defined as the **resistivity**, ρ [units: Ω cm]:

$$\rho = \frac{1}{(qn\mu_n + qp\mu_p)} \tag{2.62}$$

Substituting for the resistivity, Eq. (2.61) reduces to a simple expression

$$R = \rho\left(\frac{L}{Wt}\right) \tag{2.63}$$

In order to evaluate the resistivity, we must substitute the electron and hole concentrations n and p. The slab in Fig. 2.20 was assumed to be n-type with a donor concentration of N_d. Although the slab is not in thermal equilibrium, the carrier concentrations are not significantly perturbed from their equilibrium values by their constant drift across the slab. Substituting $n = N_d$ and $p = n_i^2/N_d$ for the n-type slab in Eq. (2.62), the resistivity ρ_n of n-type silicon is given by

n-type resistivity

$$\rho_n = \left(qN_d\mu_n + \frac{qn_i^2\mu_p}{N_d}\right)^{-1} \cong \frac{1}{qN_d\mu_n} \tag{2.64}$$

In other words, we can neglect the contribution of minority carriers in the drift current density since their concentration is orders of magnitude smaller than the majority carriers. For a p-type silicon region, holes are the majority carriers and the resistivity ρ_p is

p-type resistivity

$$\rho_p = \left(q\frac{n_i^2}{N_a}\mu_n + qN_a\mu_p\right)^{-1} \cong \frac{1}{qN_a\mu_p} \tag{2.65}$$

For doping concentrations of 10^{13} cm^{-3} to 10^{19} cm^{-3}, the resistivity of silicon ranges from about 500 Ω–cm to 5 mΩ–cm.

2.6.2 Sheet Resistance

Integrated-circuit resistors are fabricated from regions of one doping type in a substrate of opposite type. Figure 2.21 shows the layout and cross section of an n-type, ion-implanted IC resistor, that is fabricated using the process illustrated in Fig. 2.18. Metal interconnects make electrical contact at each end of the implanted region through contact windows etched through the deposited oxide film. The rectangular central portion of the implanted region has length L, width W, and thickness t. The design rules require that the ends of the n-type region are widened to allow room for the contact windows.

If a voltage is applied across the contacts to the resistor, the resulting electric field will cause a lateral electron drift current density as shown in Fig. 2.21(b). As we will see in Chapter 3, if we apply positive voltages to both contacts with respect to the substrate, there is negligible current across the n-p junction between the n-type region and the p-type substrate. Hence, the n-type region is electrically isolated from the substrate. This simple structure hints at some of the difficulties in integrated circuit design, in that a p-n junction is the unwanted by-product of making an ion-implanted resistor. We will revisit the IC resistor in Example 3.6 in Chapter 3 and develop a more accurate model for it, after we have studied the reverse biased p-n junction.

From the previous section, the resistance of the rectangular region is given by

$$R = \rho_n\left(\frac{L}{Wt}\right) = \left(\frac{1}{qN_d\mu_n}\right)\left(\frac{L}{Wt}\right) \tag{2.66}$$

In IC design, it is very convenient to re-express Eq. (2.66) in a form that separates those parameters determined by the fabrication process (mobility μ_n, donor concentration N_d, and thickness t of the implanted layer) from those that are determined by

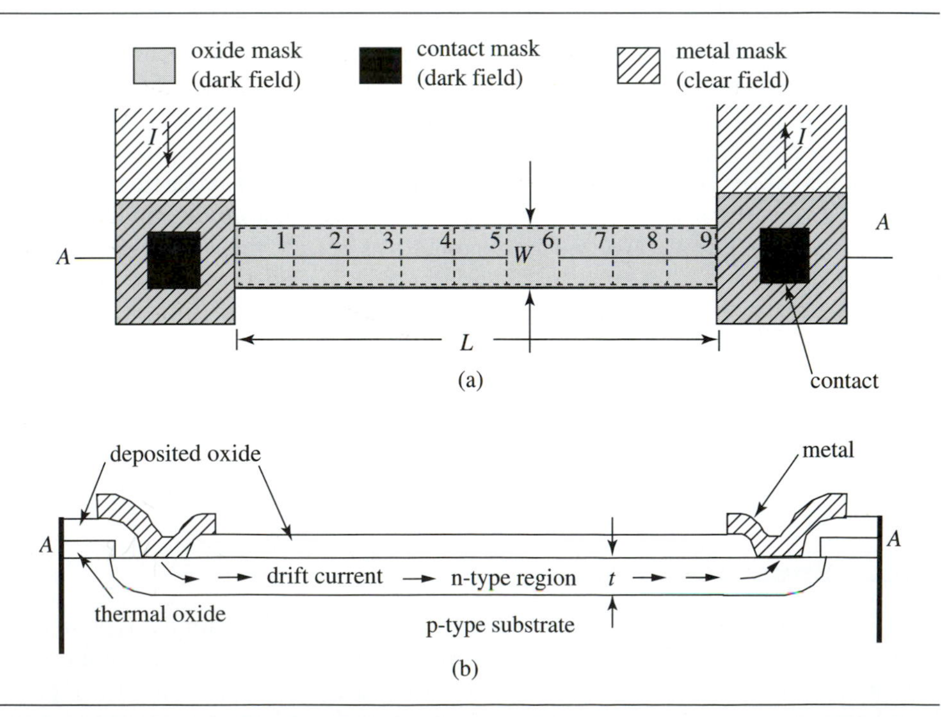

➤ **Figure 2.21** (a) Layout (b) cross section of an n-type resistor having $L/W = 9$ that is fabricated using the process sequence described in Fig. 2.18. Arrows show direction of electron drift current density for the case of a higher voltage on the left-hand contact.

the layout (length L and width W of the resistor). Collecting terms in Eq. (2.66) accordingly

sheet resistance

$$R = \left(\frac{1}{qN_d\mu_n t}\right)\left(\frac{L}{W}\right) = R_\square\left(\frac{L}{W}\right) = R_\square N_\square \tag{2.67}$$

in which we have defined the **sheet resistance** $R_\square = (q\,\mu_n\,N_d\,t)^{-1}$. The ratio of length to width is also defined as $N_\square = (L/W)$, the number of "squares." The units for sheet resistance are by convention $\Omega/\square$, pronounced "ohms per square." Of course, the number of "squares" is dimensionless. The layout in Fig. 2.21(a) illustrates the "square" concept by dividing the central portion into 9 segments of length W. In practical cases, the length L is often an integer multiple of the width W and $N_\square$ will be an integer.

For average doping levels of 10^{15} cm^{-3} to 10^{19} cm^{-3} and a typical layer thickness of $t = 1$ μm, the sheet resistance ranges from 50 kΩ /□ to 10 Ω /□. For other IC materials, heavily phosphorus-doped (n^+) gate polysilicon films are typically 500 nm thick and have $R_\square$ approximately 20 Ω /□. Aluminum interconnections are 1 μm in thickness and have a sheet resistance of around 30 mΩ /□—far lower than for the polysilicon or implanted silicon layers.

Why is sheet resistance such a useful parameter to the IC designer? The answer is that the parameters of the fabrication process such as the resistor thickness, doping concentration, and mobility are all "givens" for an IC designer. The layout shown in Fig. 2.21(a) is all that is under the IC designer's control in a particular process technology.

Thus far, we have neglected the contributions of the contact regions at the ends of the resistor to the total resistance. The details of how the current spreads out from the metal-silicon contact determine the resistance for a particular contact layout. We will neglect any additional resistance due to the metal-silicon junction not being an ideal ohmic contact. For the "dogbone" style contact shown in Fig. 2.22(a), each contact region adds 0.65 squares. It is convenient to express the additional resistances in units of squares, so that they can be counted into the total $N_\square$. In Fig. 2.22(b), the effect of the 2-D distribution of current density at a right angle bend is that the "corner square" contributes only 0.56 squares to the total $N_\square$. Accurate accounting for corners is important, since large-value resistors are typically folded in order to fit compactly into the layout. Example 2.5 illustrates the concept of sheet resistance for resistor layout.

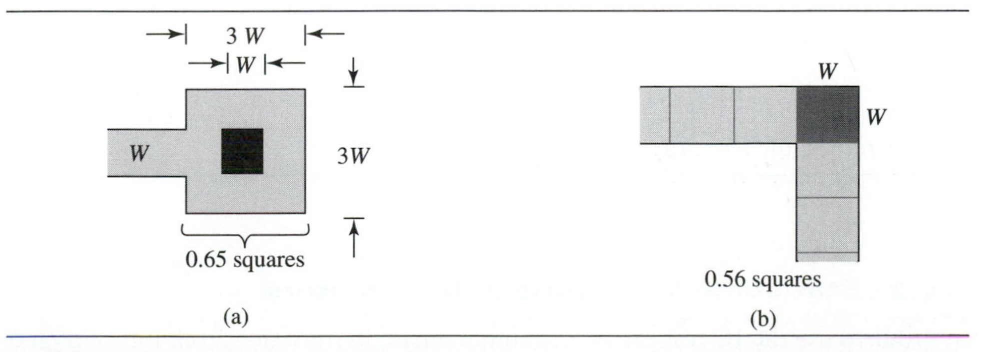

➤ **Figure 2.22** Effective fractional square for (a) a "dogbone" contact area and (b) a corner. After R. C. Jaeger, *Introduction to Microelectronic Fabrication*, Addison-Wesley, 1988.

➤ EXAMPLE 2.5 IC Resistor Layout

(a) Given that a dose $Q_d = 10^{12}$ cm^{-2} of arsenic is implanted into p-type silicon, with a junction depth of $x_j = 400$ nm, estimate the sheet resistance of this n-type layer.

SOLUTION

The average arsenic concentration in the implanted layer is the thickness of the layer (in cm) divided into the dose (per cm^2)

$$N_d = \frac{Q_d}{x_j} = \frac{1 \times 10^{12}}{4 \times 10^{-5}} = 2.5 \times 10^{16}\ \text{cm}^{-3}$$

From Fig. 2.8, the electron mobility for the average concentration is $\mu_n = 1000$ cm^2/Vs. Substituting into the definition of sheet resistance in Eq. (2.67), we find that

$$R_{\square} = \frac{1}{q\mu_n N_d x_j} = \frac{1}{(1.6 \times 10^{-19})\ (1000)\ (2.5 \times 10^{16})\ (4 \times 10^{-5})} \cong 6\ \text{k}\Omega/\square$$

In letting $x_j = t$ in finding $R_{\square}$, we assume there is no penetration of the n-type region by the charged layer that forms at the p-n junction. This approximation is reasonable so long as the background acceptor concentration in the substrate, N_a, is much less than the donor concentration N_d in the implanted layer. We will quantify this approximation in Chapter 3. We have also made this same assumption in letting the electron concentration $n = N_d$, rather than $N_d - N_a$. For a typical substrate doping, these approximations lead to small errors.

(b) For this n-type layer, find the length L for a straight resistor (no corners) of value $R = 56$ kΩ. The minimum width $W = 2.5$ μm and the two contact areas have the layout shown in Fig. 2.22(a)

SOLUTION

First, the total number of squares needed is

$$N_{\square} = R/R_{\square} = 56\text{k}\Omega\ /\ (6\ \text{k}\Omega/\square) = 9.3\ \square$$

Subtracting off the contributions of the contacts, the rectangular portion of the resistor has

$$(L/W) = N_{\square} - 2\ (0.65) = 9.3 - 1.3 = 8$$

Therefore, the length $L = 8 \times 2.5\ \mu\text{m} = 20\ \mu\text{m}$.

2.6.3 Statistical Variations in IC Resistors

In order to use the IC resistor as a circuit element, its variation from lot to lot and over temperature must be understood. We will introduce a simple way to quantify the effect of manufacturing variations on the resistance, but will leave the subject of temperature effects to a more advanced course. This section is applicable to other device structures since the sources of variation are fundamental to IC manufacturing.

From Eq. (2.66), the resistance is the product of the sheet resistance and the number of squares, each of which contains parameters or dimensions that vary from chip to chip:

$$R = \left(\frac{1}{qN_d\mu_n t}\right)\left(\frac{L}{W}\right) \tag{2.68}$$

Since a background in probability and statistics is *not* assumed, the variation of a particular quantity, such as the doping concentration N_d, will be expressed simply as

$$N_d = \overline{N_d}(1 \pm \varepsilon_{N_d}), \tag{2.69}$$

where $\overline{N_d}$ is the average doping and ε_{N_d} is defined as the **normalized uncertainty**. It is convenient to use the normalized uncertainty rather than the absolute uncertainty in the particular quantity. We can determine the range of variation by multiplying the average by the normalized uncertainty. For example, a doping concentration defined by an average of $\overline{N_d} = 10^{17}\ \text{cm}^{-3}$ and a normalized uncertainty of $\varepsilon_{N_d} = 0.025$ means that

$$N_d = (10^{17}\ \text{cm}^{-3})\ (1 \pm 0.025) = 10^{17}\ \text{cm}^{-3} \pm 2.5 \times 10^{15}\ \text{cm}^{-3} \tag{2.70}$$

Returning to Eq. (2.68), we substitute for each term that is subject to manufacturing variations:

$$R = \left(\frac{1}{q\overline{N_d}(1+\varepsilon_{N_d})\,\overline{\mu_n}(1+\varepsilon_{\mu_n})\,\bar{t}(1+\varepsilon_t)}\right)\left(\frac{\bar{L}(1+\varepsilon_L)}{\overline{W}(1+\varepsilon_W)}\right) \tag{2.71}$$

It is reasonable that the average resistance should be

$$\bar{R} = \left(\frac{1}{q\overline{N_d}\,\overline{\mu_n}\,\bar{t}}\right)\left(\frac{\bar{L}}{\overline{W}}\right) \tag{2.72}$$

However, how do we predict the normalized uncertainty ε_R given the normalized uncertainties in the parameters and dimensions? Applying an important result from statistics, the normalized uncertainty in the resistance is the square root of the sum of the squares of the normalized uncertainties of each variable

$$\varepsilon_R = \sqrt{\varepsilon_{N_d}^2 + \varepsilon_{\mu_n}^2 + \varepsilon_t^2 + \varepsilon_L^2 + \varepsilon_W^2}\,. \tag{2.73}$$

This result is valid when the variation of one variable is independent of those of the other variables and when all variables follow the Gaussian, or normal, probability distribution. In practice, Eq. (2.73) is a good first-cut estimate for the case where the normalized uncertainties are relatively small (< 0.1). This "statistical" approach is convenient and we will use it to estimate the effects of process and layout variations—even though we are not going to verify the underlying assumptions.

We now turn to the origin of the variations in the terms in Eq. (2.71). The ion implanter will largely determine the run-to-run control on the doping concentration. Mobility variations are due to random local variations in crystal defects. Note that there is also coupling between the mobility and the doping concentration, as shown in Fig. 2.8. This coupling will be ignored in our first-cut estimate of uncertainty in Eq. (2.73). The thickness t of the resistor is subject to variations in the time and local wafer temperature during the furnace annealing process.

The lateral dimensions of the resistor are affected by random linewidth variations from the pattern transfer processes, as shown in Figure 2.23. Defining the uncertainty on one edge as $\delta/2$, the width and length of the resistor are

$$W = \overline{W} \pm \frac{\delta}{2} \pm \frac{\delta}{2} = \overline{W} \pm \delta \text{ and } L = \bar{L} \pm \frac{\delta}{2} \pm \frac{\delta}{2} = \bar{L} \pm \delta, \tag{2.74}$$

where we have added the uncertainties from each edge. The width and length can be rewritten in the form

$$W = \overline{W}\left(1 \pm \frac{\delta}{\overline{W}}\right) = \overline{W}(1 \pm \varepsilon_W) \text{ and} \tag{2.75}$$

$$L = \bar{L}\left(1 \pm \frac{\delta}{\bar{L}}\right) = \bar{L}(1 \pm \varepsilon_L) . \tag{2.76}$$

From Eqs. (2.75) and (2.76), the normalized uncertainties in the width and length are inversely proportional to their average values:

$$\varepsilon_W = \delta / \overline{W} \text{ and } \varepsilon_L = \delta / \bar{L} . \tag{2.77}$$

Therefore, the designer has control over the normalized uncertainties in width and length and can make trade-offs between the resistor area (the product of width and length) and its normalized uncertainty. It is common for the length to greatly

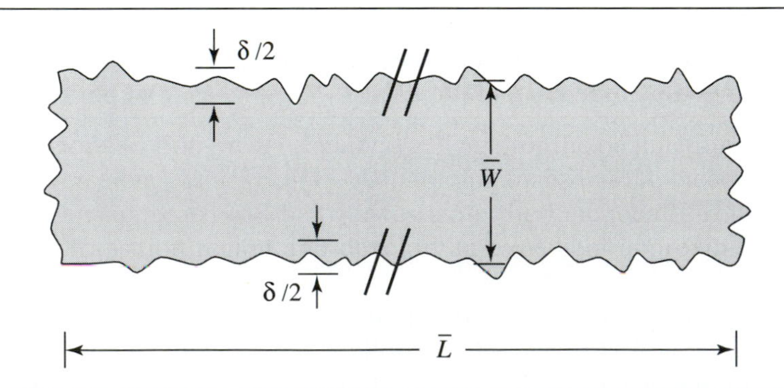

➤ **Figure 2.23** Highly magnified top view of the resistor showing the uncertainty $\delta/2$ in the edges due to random variations in the photolithography and etching processes. Note that the sketch only shows the ends of the relatively long and narrow resistor.

exceed the width, in which case the normalized uncertainty in the length becomes negligibly small, according to Eq. (2.77)

What are the magnitudes of the process uncertainties that enter into Eq. (2.73)? To answer this question, we must first specify the collection of samples used to evaluate the average and variation. The variation is smallest for structures that are nearby on the same chip, since they are subjected to nearly identical processes. If an entire wafer is considered, then the spread in parameters and dimensions increases. The uncertainties increase further when a wafer lot is evaluated and are largest when measurements are made on samples from many lots. Finally, the IC process equipment and the organization and discipline of the workforce are major factors in setting the baseline levels of process uncertainties.

The statistics on process variations are highly proprietary for each manufacturer, needless to say. However, some rough numbers are useful for gaining an appreciation for the limits of IC technology. Nominally identical, adjacent ion-implanted resistors can have a normalized uncertainty as low as 10^{-3}. With careful attention to the layout, resistor ratios can have a similar level of precision. A further benefit of adjacent resistors is that they are at nearly the same temperature. As a result, their resistances track with changing ambient temperature. When the set of resistors is expanded to include samples from a wafer, a lot, or a series of lots, the normalized uncertainty increases by a factor of 10 to 100—not including the significant variation over temperature. As a result, the precision is too poor to allow a designer to use an IC resistor for circuits where its absolute value must be tightly controlled. The absence of precision integrated resistors has been a major motivator of new approaches to circuit design.

➤ EXAMPLE 2.6 IC Resistor Process Variations

An ion-implanted resistor with the cross section shown in Fig. 2.21 is designed to have a resistance $R = 20\text{ k}\Omega$. The process parameters and their normalized uncertainties over a series of lots are: $N_d = (4 \times 10^{17}\text{ cm}^{-3})(1 \pm 0.04)$, $\mu_n = (450\text{ cm}^2/\text{Vs})(1 \pm 0.02)$, $t = 550\text{ nm}(1 \pm 0.035)$.

(a) Find the average sheet resistance and its normalized uncertainty.

SOLUTION

Substituting the average values for the parameters into the definition of sheet resistance in Eq. (2.67), we find that

$$\overline{R}_{\Box} = \frac{1}{(1.6 \times 10^{-19}\text{ C})(4 \times 10^{17}\text{ cm}^{-3})(450\text{ cm}^2/\text{Vs})(0.55 \times 10^{-4}\text{ cm})} = 630\ \Omega/\Box.$$

The normalized uncertainty in the sheet resistance is given by

$$\varepsilon_{R_{\Box}} = \sqrt{\varepsilon_{N_d}^2 + \varepsilon_{\mu_n}^2 + \varepsilon_t^2} = \sqrt{(0.04)^2 + (0.02)^2 + (0.035)^2} = 0.06$$

(b) Over several wafer lots, the linewidth edge variation is $\delta = 0.15\ \mu\text{m}$. For nominal resistor widths of $W = 2\ \mu\text{m}$ and $W = 4\ \mu\text{m}$, find the normalized uncertainty for the resistor.

SOLUTION

The number of squares required is:

$$N_{\square} = \frac{R}{R_{\square}} = \frac{20 \text{ k}\Omega}{630 \ \Omega/\text{sq.}} = 31.75$$

Due to the large number of squares, the uncertainty in the length can be neglected. According to Eq. (2.77), the normalized uncertainty in the resistor width is

$$\varepsilon_W = \frac{0.15 \ \mu\text{m}}{W} = 0.075 \ (W = 2 \ \mu\text{m}) \text{ and } \varepsilon_W = 0.038 \ (W = 4 \ \mu\text{m})$$

Therefore, the normalized uncertainties of the resistor for the two widths are

$$\varepsilon_R = \sqrt{\varepsilon_{R_{\square}}^2 + \varepsilon_W^2} = \sqrt{(0.06)^2 + (0.075)^2} = 0.096 \ (W = 2 \ \mu\text{m}) \text{ and}$$

$$\varepsilon_R = \sqrt{\varepsilon_{R_{\square}}^2 + \varepsilon_W^2} = \sqrt{(0.06)^2 + (0.038)^2} = 0.0625 \ (W = 4 \ \mu\text{m})$$

The nominally 20 kΩ IC resistor has an absolute uncertainty of nearly 2 kΩ for the narrow width layout and about 1.25 kΩ for the wider layout. The penalty for the wider layout is an increase in area by a factor of 4 since the length must also be doubled to keep the number of squares at 31.75.

For large resistor values, it is desirable to fold the resistor in order to make a more compact IC structure. We conclude the chapter with a design example which explores the geometry involved in designing for a square layout.

DESIGN EXAMPLE 2.7 Folded IC Resistor Layout

Given an n-type layer with a sheet resistance $R_{\square} = 1 \text{ k}\Omega/\square$, we want to lay out a folded resistor with $R = 75 \text{ k}\Omega$. The geometric design rules for the n-type region specify that the minimum width is $W_s = 2 \ \mu\text{m}$ and that the minimum gap between n-type regions is $W_g = 3 \ \mu\text{m}$, as defined in the sample layout in Fig. Ex2.7. The contact regions each contribute 0.8□. In order to make the resistor easy to pack into the IC layout, we want to fold the resistor.

What is the segment length L_s (defined in the sample layout) that results in a nearly square resistor layout? How much area does the resistor require? You may neglect the portions of the contact areas outside the folded segments in finding the resistor area.

SOLUTION

Assuming that the resistor consists of N_s segments, the total number of squares $N_{\square}$ is

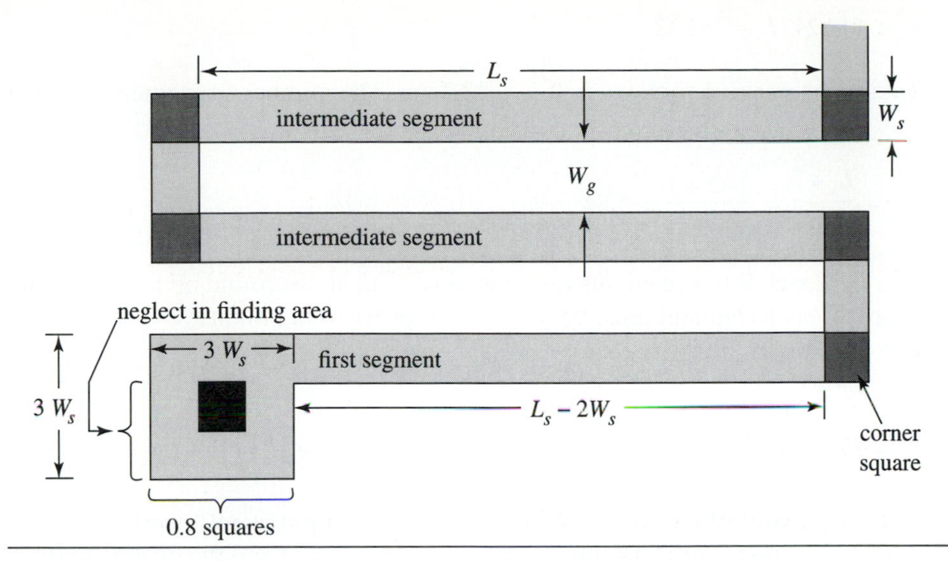

➤ **Figure Ex2.7** Layout of a folded IC resistor with offset contacts.

$$N_\square = 2\,(0.8) + 2\left(\frac{L_s - 2W_s}{W_s}\right) + (N_s - 2)\left(\frac{L_s}{W_s}\right) + (N_s - 1)\left(2\,(0.56) + \frac{W_g}{W_s}\right)$$

in which the first two terms are for the first and last segments (with the contacts), the third term is for the intermediate segments of length L_s, and the final term is for the folds (with the two corner squares contributing 0.56 squares each). Using the minimum width $W_s = 2\ \mu\text{m}$ and the minimum gap $W_g = 3\ \mu\text{m}$, we find

$$N_\square = 1.6 + L_s - 4 + \left(\frac{(N_s - 2)}{2}\right)L_s + (N_s - 1)\ (2.62) \ = \ \left(\frac{L_s}{2} + 2.62\right)N_s - 5.02$$

In order to achieve a square layout, Fig. Ex2.7 indicates that

$$L_s + 2W_s \ = \ (N_s - 1)\ (W_s + W_g) + W_s \ = \ (N_s - 1)\ (5\ \mu\text{m}) + 2\ \mu\text{m}$$

in which the left hand side is the total length of one segment and the right hand side is the length of the side with the folds. Solving for the number of segments, we find that

$$N_s \ = \ \frac{L_s + 2W_s + W_g}{W_s + W_g} \ = \ \frac{L_s + 7}{5}$$

The total number of squares needed is $N_\square = R/R_\square = 75\ \text{k}\Omega\ /\ (1\ \text{k}\Omega/\square) = 75$. Substituting for N_s in the expression for the total number of squares, we find a quadratic equation for the segment length

$$75 \ = \ \left[\frac{L_s}{2} + 2.62\right]\left(\frac{L_s + 7}{5}\right) - 5.02\text{, which simplifies to}$$

$$L_s^2 + (12.24)\,L_s - 763.32 = 0$$

Taking the positive root, we find $L_s = 22.2$ µm. The number of segments needed is

$$N_s = \frac{L_s + 7}{5} = \frac{22.2 + 7}{5} = 5.84$$

One approach is to use an integer number of folds so we round up to $N_s = 6$. The length of each segment must be modified in order to maintain $N_{\square} = 75$ using the new number of folds

$$N_{\square} = 75 = \left(\frac{L_s}{2} + 2.62\right)N_s - 5.02 = \left(\frac{L_s}{2} + 2.62\right)6 - 5.02 \rightarrow L_s = 21.4\ \mu\text{m}$$

The final layout of the folded resistor with $L_s = 21.4$ µm is not a perfect square, since the segments are 25.4 µm long (counting the folds) and the edge with the 6 folds is 27 µm long. An alternative to forcing N_s to be an integer would be to adjust the lengths of the first or the last segments, in order to make the overall layout closer to a square. In practice, the integer number of folds is close enough to a square layout. The area of the resistor is

$$A = (L_s + 2W_s)\,[\,(N_s - 1)\,(W_s + W_g) + W_s] = (21.4 + 4)\,(5 \cdot 5 + 2) = 686\ \mu\text{m}^2$$

➤ SUMMARY

This chapter has provided a foundation in two subjects that are essential for integrated device and circuit modeling: the physics of holes and electrons in silicon and the technology of IC fabrication. Key concepts that you should understand are:

- The bond model and its description of an electron and a hole in silicon.
- The concept of thermal equilibrium and the mass-action law.
- Hole and electron concentrations for silicon with donors, with acceptors, or with both donors and acceptors.
- Drift velocity and drift current density.
- Diffusive transport and diffusion current density.
- The total electron and hole current densities.
- Lithography and pattern transfer.
- Masked ion implantation of impurities.
- Cross sections through device structures from the layout and the sequence of fabrication steps (process).
- Resistivity and sheet resistance.
- Ion-implanted IC resistors and their layout, including equivalent squares for contact regions and corners.

- Normalized uncertainty and its calculation for a function of several variables, given their uncertainties.

FURTHER READING

Semiconductor Physics

1. C. G. Fonstad, *Microelectronic Devices and Circuits*. McGraw-Hill, 1994, Chapters 1 and 2. About the same level.
2. R. F. Pierret, *Semiconductor Fundamentals*, 2nd Ed. Modular Series on Solid State Devices, Vol. I, G. W. Neudeck and R. F. Pierret, eds., Addison-Wesley, 1988, Chapters 1 and 3. Somewhat more advanced.

IC Technology

1. W. Maly, *Atlas of VLSI Technologies*, Addison-Wesley, 1986. Several step-by-step process sequences provide a good intuitive understanding of the connection between layout, process, and device cross sections.
2. R. C. Jaeger, *Introduction to Microelectronic Fabrication*, Modular Series on Solid State Devices, Vol. V, G. W. Neudeck and R. F. Pierret, eds., Addison-Wesley, 1988, Chapters 1 and 2. Early chapters provide a good overview of IC fabrication.

PROBLEMS

Note: Assume that the sample temperature is 27 °C , unless otherwise noted in the problem statement.

EXERCISES

E2.1 A silicon wafer is doped with donors at a concentration of $N_d = 10^{15}$ cm^{-3}.

(a) What is the electron concentration n_o (cm^{-3}) at room temperature?

(b) What is the hole concentration p_o (cm^{-3}) at room temperature?

E2.2 A silicon wafer is doped with acceptors at a concentration of $N_a = 10^{14}$ cm^{-3}.

(a) What is the electron concentration n_o (cm^{-3}) at room temperature?

(b) What is the hole concentration p_o (cm^{-3}) at room temperature?

E2.3 Given that a silicon wafer is doped with acceptors with $N_a = 10^{14}$ cm^{-3}. We add a donor concentration of $N_d = 7.5 \times 10^{13}$ cm^3 to a particular region.

(a) What type is the region (n or p)?

(b) What is the electron concentration n_o (cm^{-3}) in the region?

(c) What is the hole concentration p_o (cm^{-3}) in the region?

E2.4 Given that a silicon wafer is doped with donors with $N_d = 5 \times 10^{17}$ cm^{-3}. We add an acceptor concentration of $N_a = 5.5 \times 10^{17}$ cm^3 to a particular region.

(a) Which carrier is the majority carrier?

(b) What is the hole concentration p_o (cm^{-3}) in the region?

(c) What is the electron concentration n_o (cm^{-3}) in the region?

E2.5 For a wafer doped with acceptors at $N_a = 10^{14}$ cm^{-3}, what concentration of donors must be added in order to reduce the hole concentration by 90%?

E2.6 In a subatomic particle detector, extremely low doped silicon substrates are needed. What donor concentration is required in order to have an electron concentration $n_o = 8\,n_i$? You will need to account for thermal generation for this problem.

E2.7 In modern VLSI processes, the dimensions of device structures are approaching 1 μm^3. For a donor concentration $N_d = 5 \times 10^{14}$ cm^{-3}, find the number of electrons and the number of holes in this volume. *Comment*: your answer will show one reason why there is a tendency for doping concentrations to increase as device dimensions are scaled into the submicron region.

E2.8 In a silicon region with donor concentration $N_a = 10^{16}$ cm^{-3}, we apply an electric field in the $+x$ direction with a magnitude of 10^3 V/cm.

(a) What is the electron drift velocity (magnitude and sign)?

(b) What is the electron drift current density (magnitude and sign)?

(c) How long is required for an electron to drift 1 μm, on average?

(d) How many collisions occur while it is drifting? You can assume that the collision time for the drifting electron is $\tau_c = 0.1$ ps.

E2.9 Repeat E2.8 with a donor concentration of $N_d = 10^{15}$ cm^{-3}.

E2.10 In a silicon region with acceptor concentration $N_a = 10^{18}$ cm^{-3}, we apply an electric field in the $+x$ direction with a magnitude of 2×10^3 V/cm.

(a) What is the hole drift velocity (magnitude and sign)?

(b) What is the hole drift current density (magnitude and sign)?

(c) How long is required for a hole to drift 1 μm, on average?

(d) How many collisions occur while it is drifting? You can assume that the collision time for the drifting hole is $\tau_c = 0.05$ ps.

E2.11 A silicon region is doped with acceptors at $N_a = 10^{15}$ cm^{-3} and with donors at $N_d = 8 \times 10^{14}$ cm^{-3}.

(a) What type is the silicon (n or p)?

(b) What electric field must be applied for the magnitude of the drift velocity of the majority carriers to be 4×10^6 cm/s?

(c) What is the drift current density of the majority carriers for the case of the electric field in part (b)?

E2.12 By optical means, we set up a minority hole concentration gradient across a 2 μm-wide silicon region that is given by:

$$p(x) = (10^{18}\,\text{cm}^{-4})\,x$$

where x is the coordinate in the direction of the concentration gradient. The donor concentration in the region is $N_d = 10^{16}$ cm^{-3}.

(a) Find the hole diffusion current density in the region.

(b) How long does a hole take to diffuse across the 2 μm-wide region?

E2.13 The minority hole concentration gradient across a 2.5 μm-wide silicon region is given by:

$$p(x) = (5 \times 10^{18}\,\text{cm}^{-4})\,(2.5\,\mu\text{m} - x)$$

where x is the coordinate in the direction of the concentration gradient. The donor concentration in the region is $N_d = 5 \times 10^{17}\ \text{cm}^{-3}$.

(a) Find the hole diffusion current density in the region.

(b) How long does a hole take to diffuse across the 2.5 μm-wide region?

E2.14 In a silicon sample, we establish a minority electron concentration gradient over the region $x \geq 0$ that is given by

$$n(x) = (10^{15}\text{cm}^{-3})\, e^{-x/(2\ \mu\text{m})}$$

The acceptor concentration is $N_a = 10^{17}\ \text{cm}^{-3}$.

(a) Find the magnitude and sign of the electron diffusion current density at $x = 0$.

(b) Plot the electron diffusion current density over the interval $0 < x < 10$ μm.

E2.15 The gradient in minority hole concentration in a silicon sample over the region $x \geq 0$ is given by

$$p(x) = (10^{13}\text{cm}^{-3})\,(1 - e^{-x/(1\ \mu\text{m})})$$

The donor concentration is $N_d = 5 \times 10^{15}\ \text{cm}^{-3}$.

(a) What is the hole concentration at $x = 0$?

(b) What is the hole diffusion current density at $x = 0$? Provide an explanation for the apparent discrepancy between parts (a) and (b).

(c) Plot the hole diffusion current density over the interval $0 \leq x \leq 5$ μm.

E2.16 By continuous improvements in fabrication technology, three generations of a microprocessor IC have areas that are given in the table below. The cost of processing a 20 cm-diameter wafer increases with each generation, due to the need for more expensive process equipment.

Technology	Chip Area	Wafer Cost
1st Generation	$2 \times 2\ \text{cm}^2$	$1750
2nd Generation	$1.5 \times 1.5\ \text{cm}^2$	$2000
3rd Generation	$1 \times 1\ \text{cm}^2$	$2250

The yield of good chips for each generation is 75%.

(a) Find the number of good chips on a 20-cm wafer for each generation. Do not consider the complication of dice that intersect the edge of the wafer—an estimate of the chip count per wafer is adequate.

(b) What is the manufacturing cost per good microprocessor chip for each generation?

E2.17 For an implant dose $Q_a = 10^{18}\ \text{cm}^{-2}$ of acceptors and a post-anneal thickness of $t = 0.5$ μm, what is the acceptor concentration in the layer?

E2.18 Given a silicon wafer with an acceptor concentration of $N_a = 10^{16}$ cm^{-3}. We would like to use ion implantation to dope a surface region so that the average acceptor concentration is $N_a = 10^{17}$ cm^{-3}. If the post-anneal thickness of the region is $t = 0.1$ μm, what dose is needed?

E2.19 Starting with a silicon wafer with a donor concentration of $N_d = 10^{17}$ cm^{-3}, we would like to use ion implantation to convert a surface region to p-type, with an average acceptor concentration $N_a = 3 \times 10^{17}$ cm^{-3}. What implant dose is needed, if the post-anneal thickness of the implanted region is 0.25 μm.

E2.20 The five-step process for a two-layer structure made from silicon dioxide and polysilicon is as follows:

1. Grow 0.5 μm of silicon dioxide (SiO_2)
2. Pattern using the oxide mask.
3. Grow 500 Å of SiO_2.
4. Deposit 1 μm of polysilicon.
5. Pattern using the polysilicon mask.

The CAD layout for this structure is shown in Fig. E2.20.

(a) Sketch the cross section *A-A* from the process sequence, assuming that the oxide mask is dark field and the polysilicon mask is clear field.

(b) Repeat part (a), with the oxide mask clear field and the polysilicon mask dark field.

(c) Repeat part (a), with the oxide mask dark field and the polysilicon mask dark field.

(d) Repeat part (a), with the oxide mask clear field and the polysilicon mask clear field.

E2.21 The layout in Fig. E2.21 is for the ion-implanted resistor process outlined in Fig. 2.18.

(a) Sketch the cross section *A-A* from the layout in Fig. E2.21.

(b) Sketch the cross section *B-B* from the layout in Fig. E2.21.

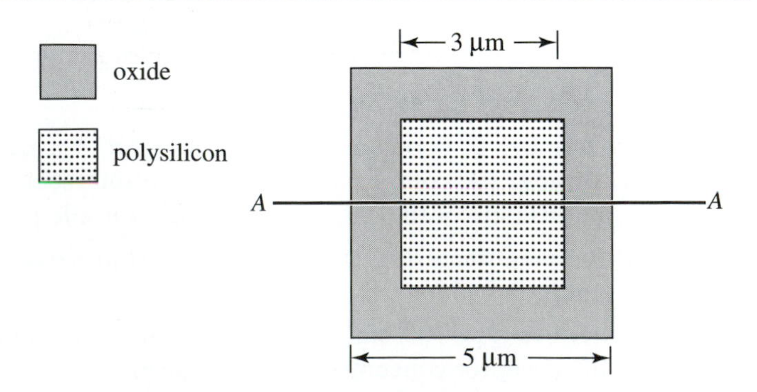

➤ Figure E2.20

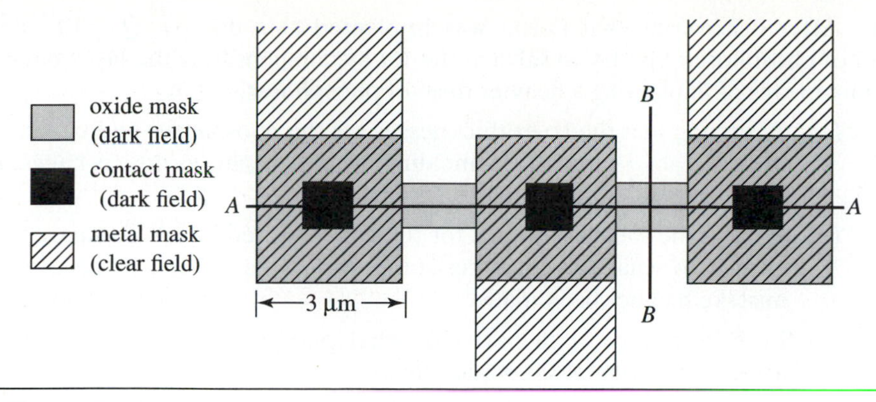

➤ **Figure E2.21**

E2.22 The layout in Fig. E.2.22 is for the simple MOS transistor process described in Fig. 2.19 (b).

(a) Sketch the cross section along *A-A*.
(b) Sketch the cross section along *B-B*.
(c) Sketch the cross section along *C-C*.
(d) Sketch the cross section along *D-D*.

E2.23 The parameters for an ion-implanted region are: arsenic dose $Q_d = 10^{12}\ \text{cm}^{-2}$, thickness after anneal $t = 1.1\ \mu\text{m}$. The background doping of the p-type substrate is $10^{14}\ \text{cm}^{-2}$.

(a) What is the average arsenic concentration in the ion-implanted layer?
(b) What is the sheet resistance in units of Ω/□.

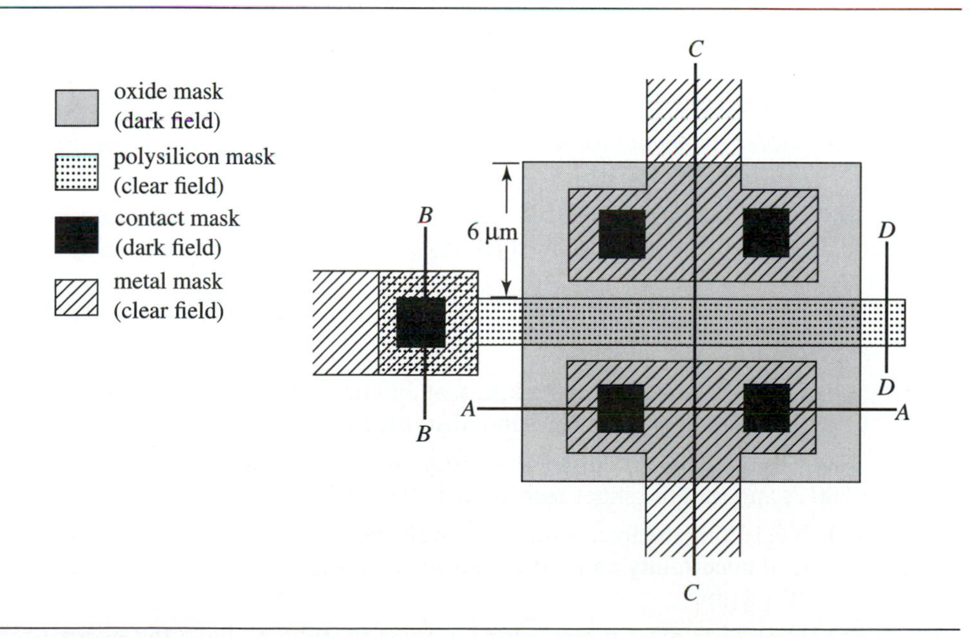

➤ **Figure E2.22**

E2.24 After discovering that boron was implanted at a dose of $Q_a = 10^{11}$ cm^{-2} instead of phosphorus, you try to salvage the lot by compensating the layer back to n-type by means of implanting a heavier dose of phosphorus.

(a) Assuming that the layer thicknesses for the phosphorus and boron are the same at $t = 1$ μm after annealing, what phosphorus dose is necessary to achieve the desired electron concentration?

(b) What is the sheet resistance for the compensated layer? Compare your answer to what the sheet resistance would have been if the original mistake had not been made.

(c) From your answer in part (b), what phosphorus dose is needed to achieve the original sheet resistance?

E2.25 Given a 1 μm-thick implanted layer and a maximum average dopant concentration of 1×10^{20} cm^{-3} in this layer.

(a) What implant dose is needed to reach the maximum concentration? Neglect the background doping in the substrate.

(b) For an n-type layer, what is the minimum sheet resistance? Note that the mobility degrades with increasing doping concentration, as shown in Fig. 2.8.

(c) For a p-type layer, what is the minimum sheet resistance? Note that the mobility degrades with increasing doping concentration, as shown in Fig. 2.8.

E2.26 Given a sheet resistance of 100 Ω/□ in a 1 μm-thick, n-type implanted layer.

(a) What is the average donor concentration in the layer? Neglect the background doping in the substrate.

(b) Sketch the layout for a 2.5 kΩ resistor, using the "dogbone" style contacts shown in Fig. 2.22(a).

E2.27 Given a sheet resistance of 250 Ω/□ in a 1 μm-thick, p-type implanted layer.

(a) What is the average acceptor concentration in the layer? Neglect the background doping in the substrate.

(b) Sketch the layout for a 2 kΩ resistor, using the "dogbone" style contacts shown in Fig. 2.22(a).

E2.28 Given that the normalized uncertainties are: $\varepsilon_{N_d} = 0.025$ in the donor concentration and $\varepsilon_t = 0.02$ in the layer thickness. The uncertainty in the mobility can be neglected.

(a) For an average donor concentration of 10^{16} cm^{-3}, find the variation in N_d.

(b) For an average layer thickness of 500 nm, find the average sheet resistance and its normalized and absolute uncertainties.

E2.29 Given that the uncertainty in the lateral dimension of an ion-implanted resistor is $\delta = 500$ Å and that the sheet resistance is 100 Ω/□.

(a) Neglecting the uncertainty in the sheet resistance, what are the normalized uncertainty and the variation in a 5 kΩ resistor for a resistor width $W = 2$ μm.

(b) How wide must the resistor be made in order to have the normalized uncertainty less than 0.01 (1%)?

E2.30 For a doped polysilicon film, the sheet resistance is 25 Ω/□ for a 5000 Å-thick film. If the thickness of the film is controlled to ±2.5% over the wafer, what is the variation in the resistance of a 20 kΩ resistor?

PROBLEMS

P2.1 A sample of silicon is doped with phosphorus with a concentration of 10^{15} cm^{-3}.

(a) What is the average volume occupied by an electron in this sample, in μm^3?

(b) What is the average volume occupied by a holes in this sample, in μm^3?

P2.2 A region of a silicon wafer is doped with three impurities:

arsenic: 10^{16} cm^{-3}, boron: 1.15×10^{16} cm^{-3}, phosphorus: 2.5×10^{15} cm^{-3}.

Find the electron and hole concentrations n_o and p_o.

P2.3 A sample of silicon is doped with $N_d = 1.1 \times 10^{13}$ cm^{-3} and $N_a = 1 \times 10^{13}$ cm^{-3}.

(a) Find the electron and hole concentrations n_o and p_o at room temperature.

(b) Show that the charge density $\rho = 0$ (exactly).

P2.4 The intrinsic concentration in silicon n_i has a strong temperature dependence, as shown by these measurements over the temperature range for automotive electronic applications:

T (° C)	n (cm^{-3})
–55	2.24×10^6
0	1.42×10^9
70	2.82×10^{11}
125	5.06×10^{12}

A course in semiconductor physics shows that the functional dependence of the intrinsic concentration is

$$n_i = A \cdot T^{3/2} e^{-6608.7/T},$$

where T is in Kelvin.

(a) Find the constant A by fitting the function to the measurements above.

(b) How rapidly the intrinsic concentration varies near room temperature is important for device modeling. Find the temperature increase (in degrees Centigrade) needed to double n_i and the temperature decrease needed to halve n_i.

P2.5 A sample of silicon is doped with $N_d = 10^{13}$ cm^{-3}. As the temperature increases, the contribution of thermal generation to the electron and hole concentrations eventually becomes significant. Using the result from P2.4(a) for the intrinsic concentration as a function of temperature,

(a) Plot $n_o(T)$ over the temperature range $-55°\text{ C} < T < 125°\text{ C}$ on a log-linear plot.

(b) What is the maximum temperature for which $p_o/n_o < 10^{-3}$?

P2.6 A sample of silicon is doped with $N_a = 2 \times 10^{14}$ cm^{-3} and $N_d = 9 \times 10^{13}$ cm^{-3}. Using the result from P2.4(a) for the intrinsic concentration as a function of temperature,

(a) Plot $p_o(T)$ over the temperature range $-55°$ C $< T < 125°$ C on a log-linear plot.

(b) What is the maximum temperature for which $n_o/p_o < 5 \times 10^{-4}$?

P2.7 Figure P2.7 is a top view of a silicon region which has metal electrical contacts on two sides, across which a voltage $V_A = 2$ V is applied. The silicon layer is doped with arsenic at a concentration of 10^{13} cm^{-3} and is 2 μm thick and you can assume room temperature. You should use the appropriate mobility for the given dopant concentration.

(a) Find the value of the electric field E_x in the silicon.

(b) Find the value of the drift velocity of electrons and of holes in the silicon. How long does is take an electron, on average, to drift across the silicon region.

(c) Find the drift current densities for electrons and for holes. What is the percentage error in the total current I if the minority carrier drift current density is neglected?

(d) What is the value of the resistance R in ohms?

P2.8 If the arsenic concentration is increased to 10^{18}cm^{-3}, repeat P2.7.

P2.9 Plot the maximum current I_{max} across the silicon region in Fig. P2.7 as a function of doping concentration from $N_d = 10^{13}$ cm^{-3} to $N_d = 10^{18}$ cm^{-3}, using a log-log scale. You can assume that the saturation velocity of electrons is $v_{sat} = 10^7$ cm/s and should include the dependence of mobility on doping concentration. Plot the applied voltage $V_{A,max}$ at which I_{max} is reached as a function of doping concentration.

P2.10 Figure P2.10 is a top view of a 1 μm-thick silicon region with variable acceptor concentration, across which a voltage V_A is applied through metal electrical contacts. Assume that the hole saturation velocity is $v_{sat} = 10^7$ cm/s.

(a) For an applied voltage $V_A = 1$ V, plot the electric field $E(x)$. *Hint*: you will find it helpful to consider the silicon region as three series resistors.

(b) What is the maximum current I_{max} through this structure?

(c) What voltage V_{max} must be applied to reach this maximum current?

(d) Plot the electric field $E(x)$ corresponding to V_{max}.

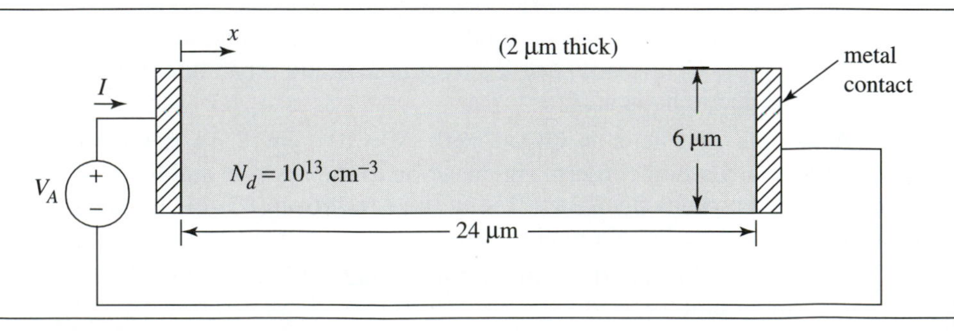

➤ **Figure P2.7**

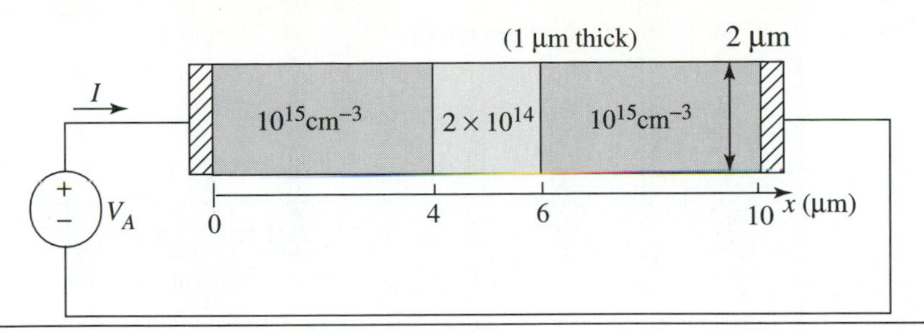

➤ **Figure P2.10**

P2.11 Figure P2.11 is a top view of a 1 μm-thick silicon region with variable donor concentration. A voltage V_A is applied through metal electrical contacts.

(a) For an applied voltage $V_A = 1$ V, what is the current *I?* You may find it helpful to view the silicon region as two resistors in parallel.

(b) Plot the current as a function of voltage over the range $-20\text{ V} < V_A < 20$ V. You can assume that the saturation velocity of electrons $v_{sat} = 10^7$ cm/s and should include the variation of mobility with donor concentration.

P2.12 The top view of a 1 μm-thick silicon region with a varying width is shown in Fig. P2.12. The donor concentration is $N_d = 10^{17}$ cm^{-3} throughout. You can assume that 25% of the applied voltage is dropped across the "neck," as long as the saturation velocity is not reached in any part of the region.

(a) Find the ratio of the current density in the wide region to that in the narrow region, for the case where $V_A = 2$ V.

(b) *Sketch* the electric field $E(x)$ along the center line of the structure, for the case when $V_A = 2$ V. The numerical value in the regions of constant width should be accurate. It will be helpful to view the silicon region as three resistors in series, with the voltage drop across the "neck" region given as 25% of V_A

(c) What is the maximum current I_{max} through this structure? What is the applied voltage when I_{max} is reached?

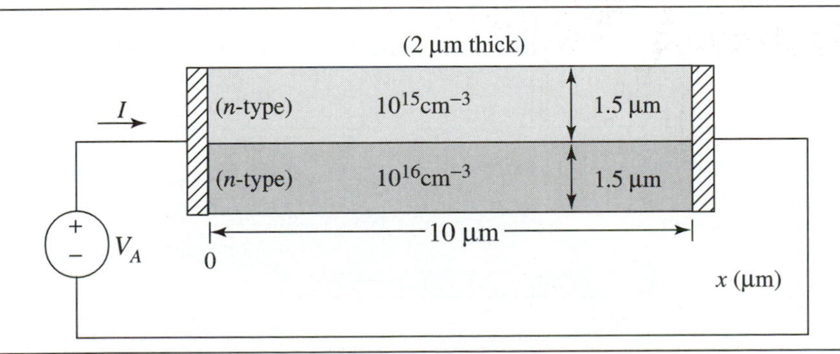

➤ **Figure P2.11**

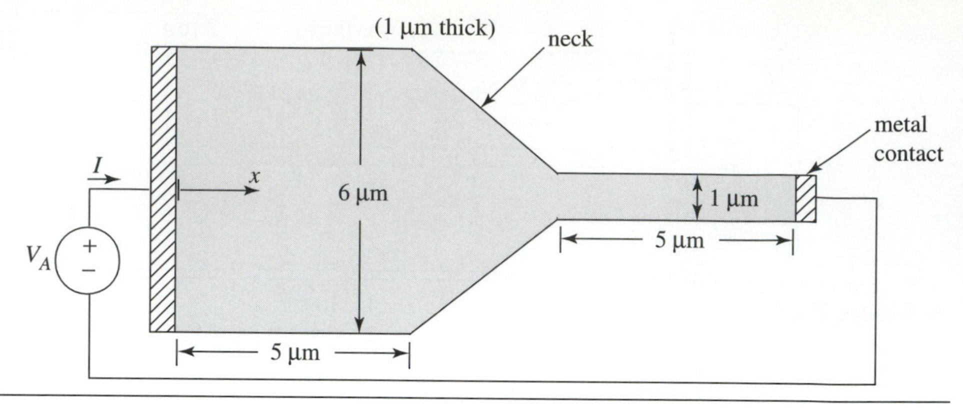

➤ **Figure P2.12**

P2.13 In the slab of silicon region in Fig. P2.13, the minority carrier concentration is maintained at its thermal equilibrium value at the bottom surface ($x = 0.3$ μm). The silicon is doped with acceptors at $N_a = 10^{16}$ cm^{-3}.

(a) If the minority carrier concentration at the top surface is 1/10 the majority carrier concentration and we assume a linear gradient from top to bottom, what is the diffusion current density in the slab?

(b) How long does it take for a minority carrier to diffuse across the slab, on average?

P2.14 When a gradient in minority carrier concentration is set up in a semiconductor region, an identical gradient in the majority carrier concentration is quickly established, in order to keep the silicon region electrically neutral. This situation is shown in the carrier concentration distributions in Fig. P2.14.

(a) Find the minority electron diffusion current density.

(b) Find the majority hole diffusion current density.

(c) Since the silicon region is open-circuited, the total current density must be zero. Find the required drift current density.

(d) Assuming that the contribution of the minority carriers to the drift current is negligible (which it is), what electric field is required in the sample to generate the drift current density needed from part (c).

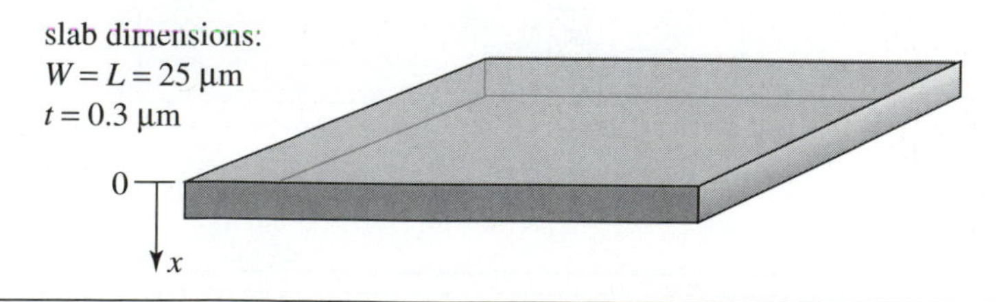

➤ **Figure P2.13**

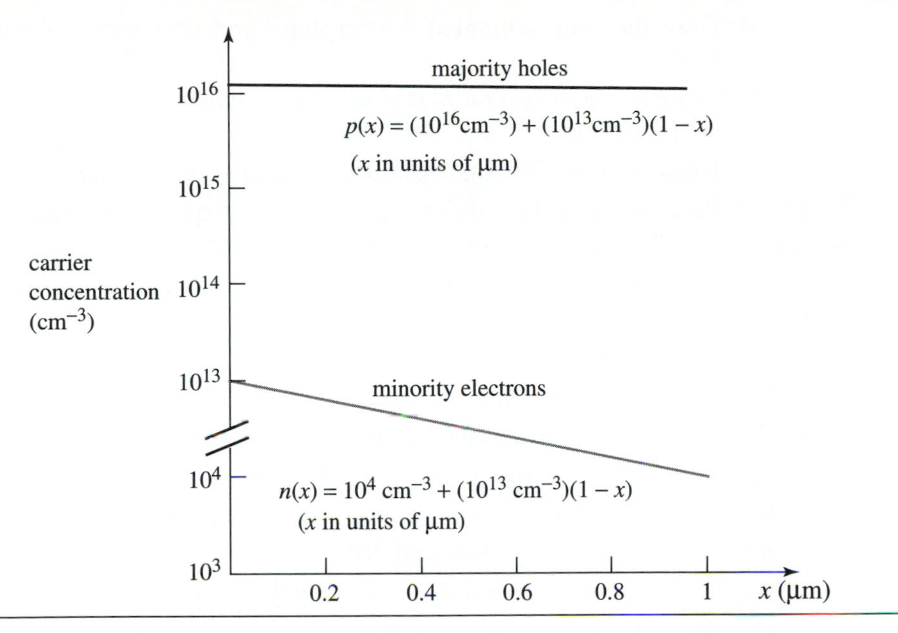

➤ **Figure P2.14**

P2.15 The process sequence and layout for making a simple polysilicon micromechanical bridge are listed below and in Fig. P2.15. Using IC-based processes, we can make structures for sensing non-electrical inputs, such as acceleration.

1. Deposit 2 μm of CVD SiO_2.
2. Pattern with the anchor mask.
3. Implant phosphorus and anneal.
3. Deposit 2 μm of phosphorus-doped polysilicon and anneal.
4. Pattern with the poly mask.
5. Etch for 4 minutes in hydrofluoric acid (SiO_2 etch rate: 1 μm per minute); rinse wafer in water and dry.

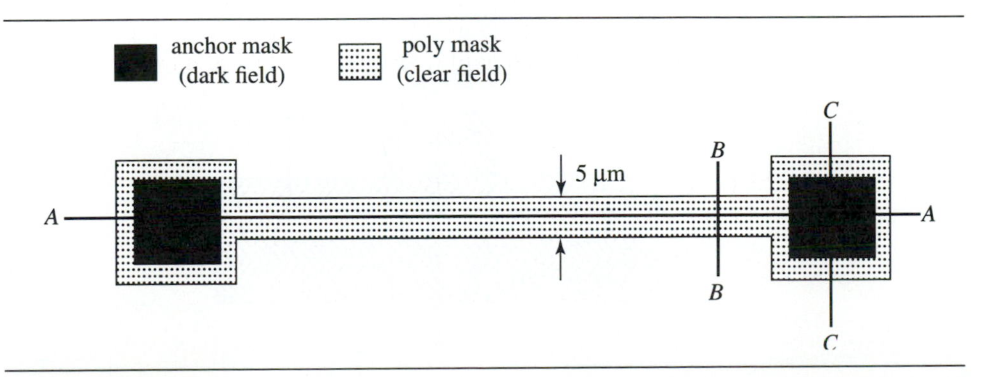

➤ **Figure P2.15**

(a) Draw the cross sections *A-A* after step 4 and after completion of the process.

(b) Draw the cross section *B-B* after step 5 is half completed (2 minutes of HF etching.)

(c) Draw the cross section *C-C* after completion of the process.

P2.16 This problem considers the fabrication of a two-level micromechanical structure, which is accomplished by extending the process in P2.15.

Steps 1-4: same as those in P2.15, with the masks renamed as *anchor1* and *poly1*

5. Deposit 1 μm of SiO_2.
6. Pattern the oxide with the anchor2 mask.
7. Deposit 1 μm of phosphorus-doped polysilicon and anneal.
8. Pattern the second polysilicon layer with the poly2 mask.
9. Etch for 12 minutes in hydrofluoric acid (SiO_2 etch rate: 1 μm per minute); rinse wafer in water and dry.
   (a) Draw the cross section *A-A* after step 8.
   (b) Draw the cross section *B-B* after completion of the process.
   (c) Draw the cross section *C-C* after completion of the process.
   (d) Draw the cross section *D-D* after completion of the process.

P2.17 Texturing the silicon surface is a way to increase the surface area of a capacitor; the following process and layout are one way to achieve this goal.

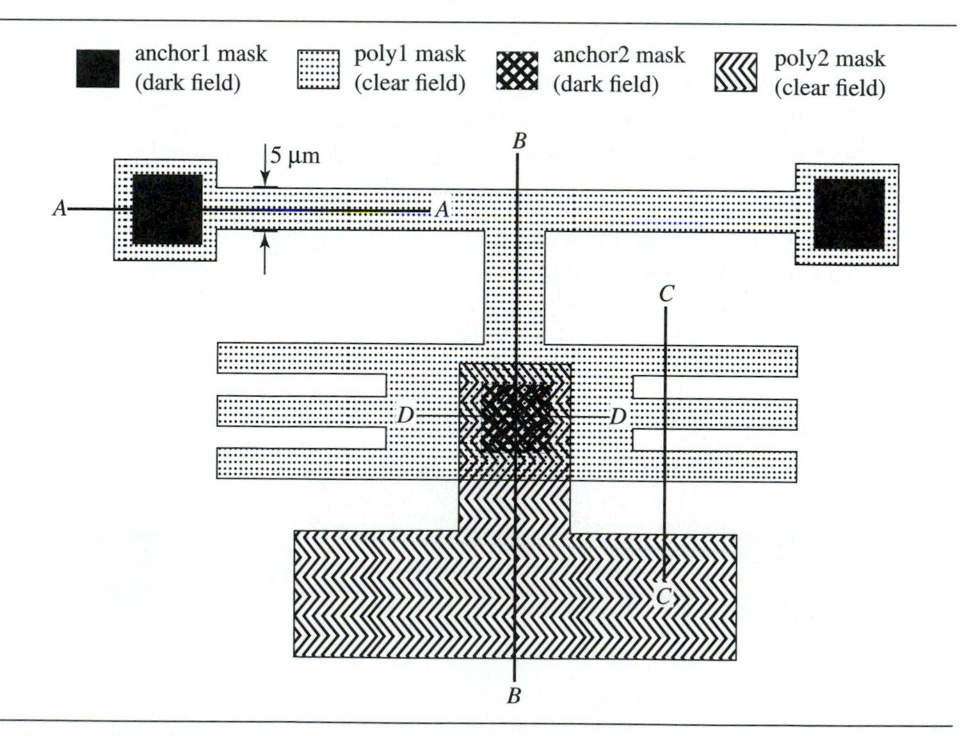

➤ Figure P2.16

1. Etch 5 μm deep into the silicon substrate using the trench mask*.
2. After stripping the photoresist and cleaning the wafer, grow 500 Å of thermal SiO_2.
3. Deposit 7500 Å of phosphorus-doped polysilicon.
4. Pattern using the polysilicon mask.

(a) Draw the cross section *A-A* after step 2.

(b) Draw the cross section *A-A* after step 3.

(c) Draw the cross section *B-B* after completion of the process.

P2.18 A simple double metallization process is as follows; the mask layout is shown in Fig. P2.18.

1. Grow 0.5 μm of thermal SiO_2.
2. Deposit 0.5 μm of phosphorus doped polysilicon and pattern using polysilicon mask.
3. Deposit 0.5 μm of CVD SiO_2 and pattern using contact mask.
4. Deposit 1 μm of sputtered Al and pattern using metal1 mask.
5. Deposit 0.5 μm of CVD SiO_2 and pattern using via mask.
6. Deposit 1 μm of sputtered Al and pattern using metal2 mask.

(a) Accurately sketch the cross section *A-A*, assuming that the deposited CVD films conformally coat the surface.

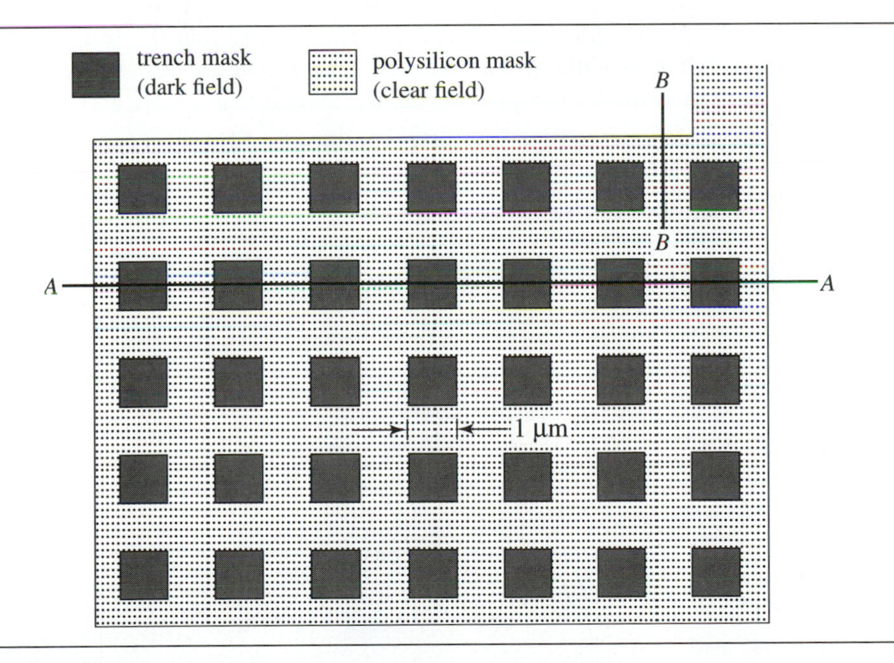

➤ Figure P2.17

*A reactive-ion etch is used for the trench etch into the silicon substrate, which is optimized to yield vertical walls in silicon with negligible mask undercutting.

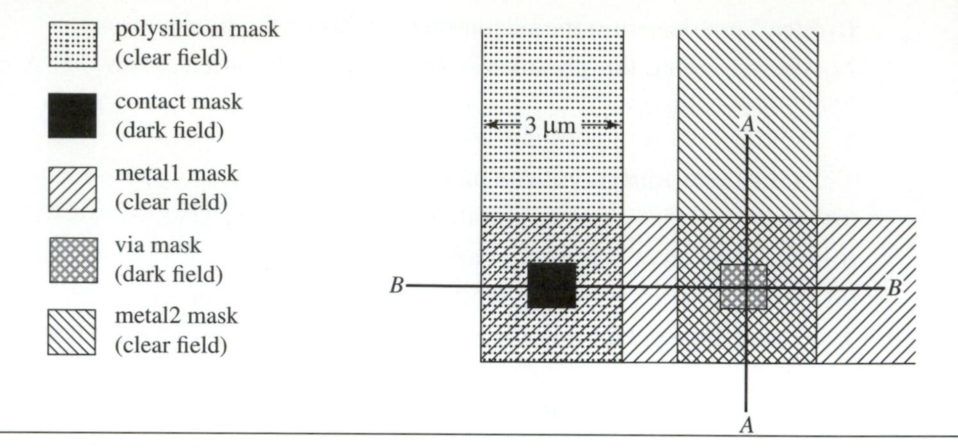

➤ **Figure P2.18**

(b) Accurately sketch the cross section *B-B*, assuming that the deposited CVD films conformally coat the surface.

P2.19 Using the double metal process from P2.18, consider the layout in Fig. P2.19.

(a) Accurately sketch the cross section *A-A*, assuming that the deposited CVD films conformally coat the surface.

(b) Accurately sketch the cross section *B-B*, assuming that the deposited CVD films conformally coat the surface. Note that the rough topography will make lithography difficult. The geometric design rules can prevent this situation by prohibiting the stacking of metal1 and metal2.

P2.20 The mask layout in Fig. 2.20 uses the simple MOSFET process in Fig. 2.19(b).

(a) Sketch the cross section along *A-A*.

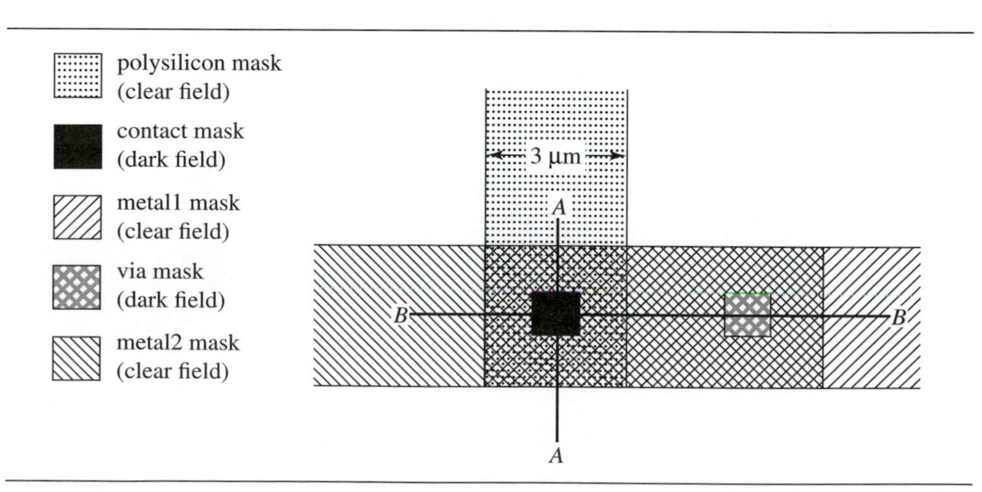

➤ **Figure P2.19**

(b) Sketch the cross section along *B-B*.

(c) Sketch the cross section along *C-C*.

P2.21 The resistor whose layout is shown in Fig. P2.21 is a test structure on an integrated circuit.

The measured resistance is $R = 22.3\ \text{k}\Omega$. What is the sheet resistance of the ion-implanted layer in ohms per square? You can assume that the contact regions each contribute 0.65 squares.

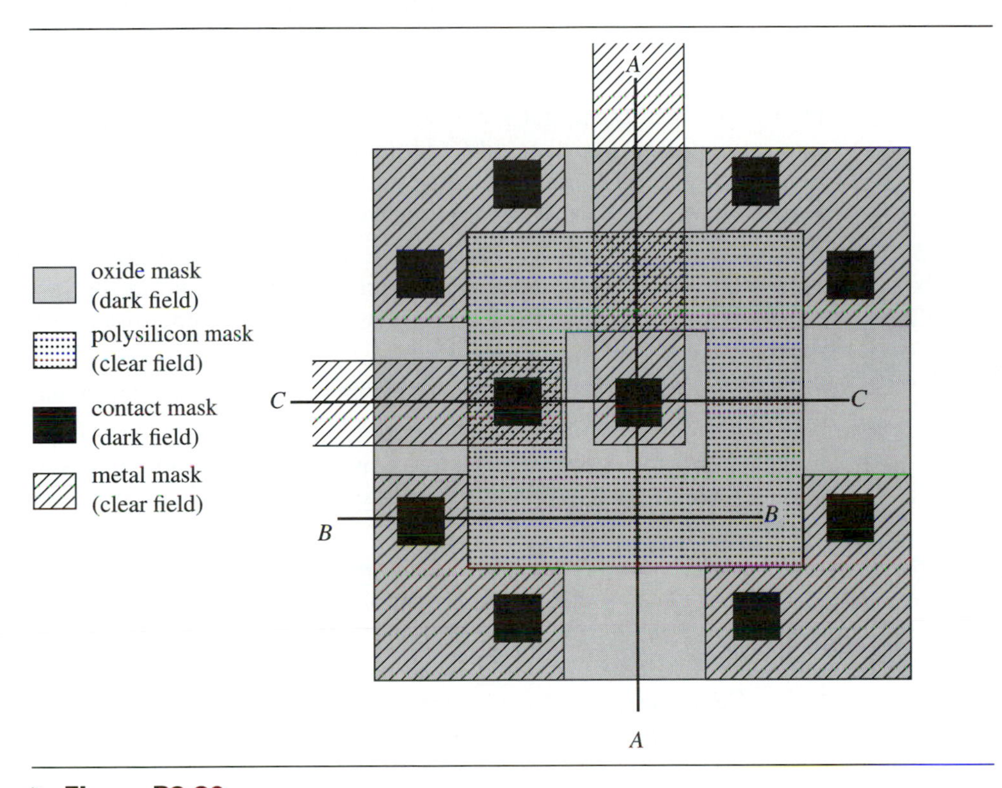

➤ **Figure P2.20**

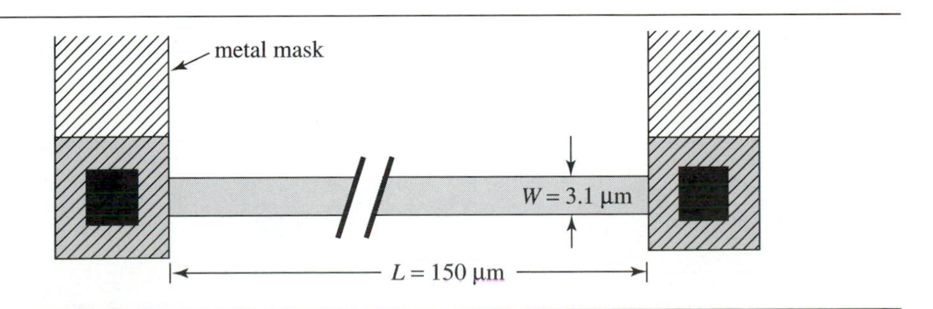

➤ **Figure P2.21**

P2.22 The two ion-implanted resistors in Fig. P2.22 have a different style of contact region compared to that shown in Fig. 2.22. The longer one has a measured resistance of $R_1 = 395\ \Omega$, with the shorter one measuring $R_2 = 270\ \Omega$.

From these measurements and the dimensions, what is the sheet resistance of this layer and the effective number of squares for this type of contact region?

P2.23 Given a p-type ion-implanted resistor with an average acceptor concentration of $N_{a1} = 10^{16}\ \text{cm}^{-3}$ over the depth $0 < x < 0.5\ \mu\text{m}$ and $N_{a2} = 7.5 \times 10^{15}\ \text{cm}^{-3}$ over the depth $0.5\ \mu\text{m} < x < 1.25\ \mu\text{m}$. Find the sheet resistance in $\Omega/\square$.

P2.24 Given an n-type ion-implanted layer with thickness $t = 1\ \mu\text{m}$ and average doping concentration $N_d = 10^{17}\ \text{cm}^{-3}$.

(a) What is the sheet resistance?

(b) In order to meet the target sheet resistance of $750\ \Omega/\square$, we can perform a second implant anneal to add additional dopants to a layer of thickness $0.25\ \mu\text{m}$. Neglecting the change in the overall layer thickness after annealing, what dose and dopant type (n or p) are needed?

P2.25 The normalized uncertainty in the average acceptor concentration in an implanted layer is $\varepsilon_{N_a} = 0.025$. The uncertainties in the mobility and in the layer thickness ($\bar{t} = 1.4\ \mu\text{m}$) are negligible for this problem. Finally, the dimensions of the resistor pattern have an uncertainty of $\delta = 400\ \text{Å}$.

(a) For a 120 square resistor with $\overline{N_a} = 10^{17}\ \text{cm}^{-3}$ and $\overline{W} = 2.5\ \mu\text{m}$, what is the average resistance and its uncertainty?

(b) If we increase the doping by a factor of two, what is the width necessary in order to keep the average resistance the same? What is the new uncertainty?

P2.26 It is an attractive idea, at first glance, to compensate a silicon region so that $N_d = N_a$ and $n_o = p_o = n_i$. Assume that the wafer is doped with acceptors at precisely $N_a = 10^{16}\ \text{cm}^{-3}$ and that we can introduce donors uniformly into a 1 μm-thick surface region with a normalized uncertainty of 0.005.

(a) If we wish to guarantee that the surface region stays p-type, what donor concentration should we add? Note that the normalized uncertainty is 0.005.

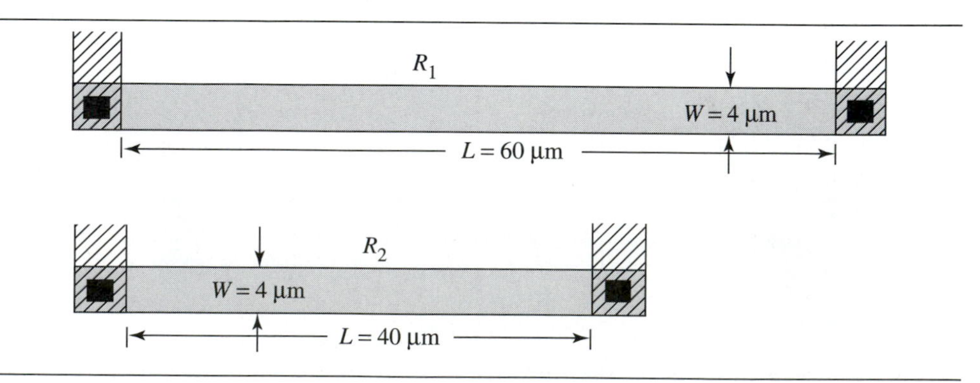

➤ Figure P2.22

(b) What is the range of sheet resistances for the surface region, given the doping in part (a)?

Design Problems

D2.1 There are often geometric constraints in implementing resistors on an integrated circuit. Figure D2.1 defines the outline of the area in which the 160 kΩ resistor must fit, given the following geometric design rules. The locations for the "dogbone" style contacts are set by the other circuit elements are cannot be moved. The oxide mask pattern can abut the edges of the outline in Fig. D2.1.

Geometric design rules:

minimum width of oxide mask pattern = minimum resistor width= $W = 2\ \mu\text{m}$

minimum spacing between oxide patterns = $W_g = 2.5\ \mu\text{m}$

Resistor layer characteristics:

doping concentration $N_d = \overline{N}_d(1 \pm \varepsilon_{N_d}) = 10^{17}\ \text{cm}^{-3}(1 \pm 0.025)$

thickness $t = \bar{t}(1 \pm \varepsilon_t) = 0.6\ \mu\text{m}\ (1 \pm 0.02)$

variation in resistor lateral dimension: $\delta = 650$ Å.

(a) Lay out the 160 kΩ resistor to fit inside the outline in Fig. D2.1.

(b) What is the uncertainty in the resistance?

(c) Modify your layout in (a) to minimize the uncertainty, while keeping within the boundaries of the outline. By what factor can you reduce the uncertainty?

D2.2 In some cases, the IC designer aims for the maximum resistance possible in a particular area. In this problem, you will lay out a polysilicon resistor in an area of 100 μm × 120 μm. The contacts should use the "dogbone" style in Fig. 2.22(a) and are

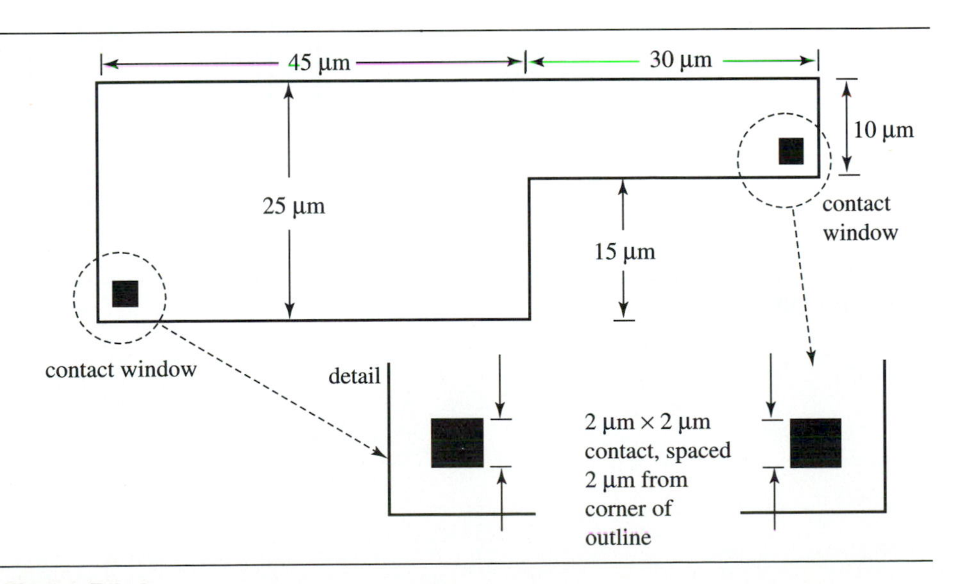

➤ **Figure D2.1**

to be located as shown on Fig. D2.2. The polysilicon resistor can abut the edges of the outline. The geometric design rules and parameters for the polysilicon film (lot-to-lot uncertainties) are:

Geometric design rules:

minimum polysilicon line width $W = 1\ \mu\text{m}$

minimum spacing between polysilicon lines = $W_g = 1\ \mu\text{m}$

Polysilicon film parameters:

resistivity $\rho = \bar{\rho}(1 \pm \varepsilon_\rho) = (1\text{m}\Omega - \text{cm})(1 \pm 0.05)$

thickness $t = \bar{t}(1 \pm \varepsilon_t) = 0.5\ \mu\text{m}\ (1 \pm 0.05)$

variation in polysilicon line: $\delta = 500\ \text{Å}$.

(a) Design the polysilicon resistor layout for the maximum number of squares in this area, given the design rules and fixed location of the contacts.

(b) What is the maximum resistance and its variation?

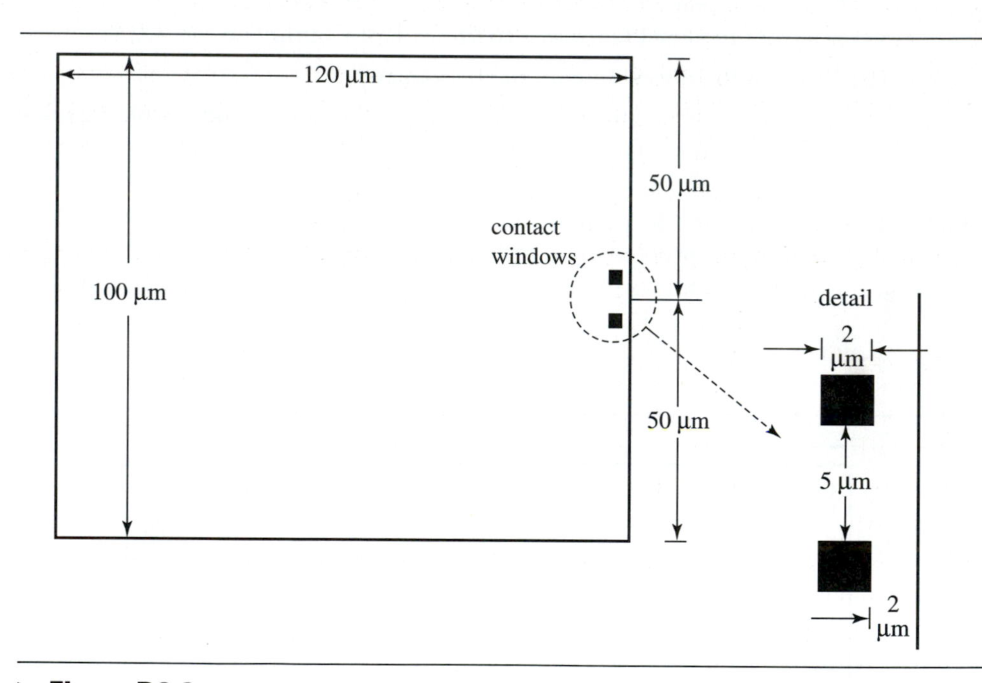

➤ **Figure D2.2**

chapter 3

PN Junction and MOS Electrostatics

SEM of trenches used to form extremely dense MOS capacitors used in state-of-the-art dynamic random access memories. (Courtesy of International Business Machines Corporation. Unauthorized use not permitted.)

With a basic understanding of semiconductor physics and IC technology from Chapter 2, we are ready to tackle two important device structures: the pn junction and the metal-oxide-semiconductor (MOS) capacitor. A good command of electrostatics is needed to find the electrostatic potential, electric field, and charge density in these structures. The electrostatics of the MOS capacitor is the foundation for understanding the MOS field effect transistor in Chapter 4. The electrostatics of the pn junction plays a similar role for the pn diode in Chapter 6 and the bipolar transistor in Chapter 7.

Chapter Objectives

- The interrelationships between charge density, electric field, and potential.
- The boundary conditions for electric field and electrostatic potential.
- The connection between potential and carrier concentration in thermal equilibrium.
- The depletion approximation and its application in simplifying the analysis of the pn junction and MOS capacitor.
- The qualitative spatial variation of charge density, electric field, and potential through the pn junction and MOS capacitor.
- Quantitative models for the depletion widths, electric fields, and potential for both structures.
- The definition of capacitance and its application to the reverse-biased diode and the MOS capacitor.

3.1 APPLIED ELECTROSTATICS

In introductory physics courses, an overview of the classical theory of electromagnetism is presented. These courses focus on the underlying assumptions, the three dimensional mathematical structure, and the inherent beauty of this remarkably powerful theory. Problem solving is usually limited to highly symmetrical geometries where the various surface integrals can be evaluated exactly. As a result, students often master the mathematics of electromagnetism, but fail to gain an intuitive grasp of the subject. When presented with a simple problem in one dimension, students are often unsure which way the electric field points, let alone grasp the details of the potential variation!

In order to understand semiconductor devices, we must develop a practical, *working knowledge* of electrostatics. Semiconductor device structures have complicated boundaries that make it difficult, if not impossible, to find closed-form analytical solutions in two or three dimensions. Therefore, we will use the one-dimensional electrostatic equations to develop insight and a basic understanding of device operation. For detailed modeling, device physicists solve the three-dimensional equations numerically using simulation programs.

3.1.1 Gauss's Law

We will first collect the basic electrostatic equations that relate charge density, electric field, and electrostatic potential. Gauss's Law connects the electric field E [units: volts per centimeter, V/cm] with the charge density ρ [units: Coulombs per cubic centimeter, C/cm^3]. Its one dimensional, differential form is

Gauss's Law in differential form

$$\frac{dE}{dx} = \frac{\rho}{\varepsilon} \tag{3.1}$$

where ε is the electric permittivity [units: Farads per centimeter, F/cm]. If the electric field is known, we can differentiate it (i.e., find its slope) and use Eq. (3.1) to plot the charge density. In many cases, we will know the charge density ρ and need to find the electric field. Integrating Eq. (3.1) along an interval $x_a \le x' \le x$ we find the integral form of Gauss's Law:

Gauss's Law in integral form

$$\int_{x_a}^{x} d[\varepsilon E(x')] = \varepsilon E(x) - \varepsilon_a E(x_a) = \int_{x_a}^{x} \rho(x')dx' = Q(x) \tag{3.2}$$

where $Q(x)$ is the charge in the interval [units: C/cm^2]. The possibility of a change in permittivity due to a material interface in the interval has been accounted for by keeping the permittivity together with the field in the integral. This result is useful since we can often conclude from the distribution of charge in semiconductor device structures that $E(x_a) = 0$ for a particular boundary x_a.

3.1.2 Potential and Poisson's Equation

The electrostatic potential $\phi(x)$ is defined with respect to its value at a point x_o as the integral of the negative of the electric field E from x_o to x:

definition of electrostatic potential

$$\phi(x) - \phi(x_o) = \int_{x_o}^{x} -E(x)dx \tag{3.3}$$

Eq. (3.3) allows us to calculate the potential *difference* between two points given the electric field. To find a numerical value for $\phi(x)$, it is necessary to select an arbitrary reference point x_o where by definition $\phi(x_o) = 0$. We will consider the question of selecting references for $\phi(x)$ in semiconductors later in this chapter.

Differentiation of Eq. (3.3) allows us to find the electric field given the potential

$$E(x) = -\frac{d\phi(x)}{dx} \tag{3.4}$$

The potential can be related to the electric field, and vice versa, through Eqs. (3.3) and (3.4). Since the derivative of the electric field is proportional to the charge density through the differential form of Gauss's Law (Eq. (3.1)), differentiation of Eq. (3.4) and substitution leads to a second-order differential equation for the potential:

Poisson's Equation

$$\frac{d^2\phi(x)}{dx^2} = -\frac{dE(x)}{dx} = -\frac{\rho(x)}{\varepsilon} \tag{3.5}$$

The result, called Poisson's Equation, can be useful since it directly links the potential with the charge distribution.

3.1.3 Intuition and Sign Conventions

Even though Eqs. (3.1), (3.4), and (3.5) are ordinary differential equations with constant coefficients, the mathematical details of solving them can get in the way of finding the answer. For instance, a sign error will cause the electric field to point in the wrong direction. In this section we will introduce some techniques for sketching the answers *before* solving the equations.

As an example to motivate these techniques, Figure 3.1 sketches the charge density in a material with permittivity ε. Positive charges are located on the top (negative x) and negative charges are on the bottom (positive x). The *total* charge in the region is zero—which is always true for electronic device structures, as we will discover later in this chapter. From Eq. (3.2), the electric field $E(-2\ \mu m)$ at $x = -2\ \mu m$ and the electric field $E(3\ \mu m)$ at $x = 3\ \mu m$ are equal

$$E(3\mu m) - E(-2\mu m) = \int_{-2\mu m}^{3\mu m} \frac{\rho(x)}{\varepsilon} dx = 0 \tag{3.6}$$

In device physics, we are interested in the internal electric field caused by the charge distribution inside the device. For our purposes, any external electric field at the boundaries of the charged region $E(-2\ \mu m) = E(3\ \mu m)$ will be ignored and, for convenience, set to zero.

In this text, the x axis will be drawn horizontally with the positive direction to the right. The physical orientation of the x axis will be specified on the device cross section as in the top of Fig. 3.1. The charge distribution can be integrated graphically to find the electric field, using the integral form of Gauss's Law

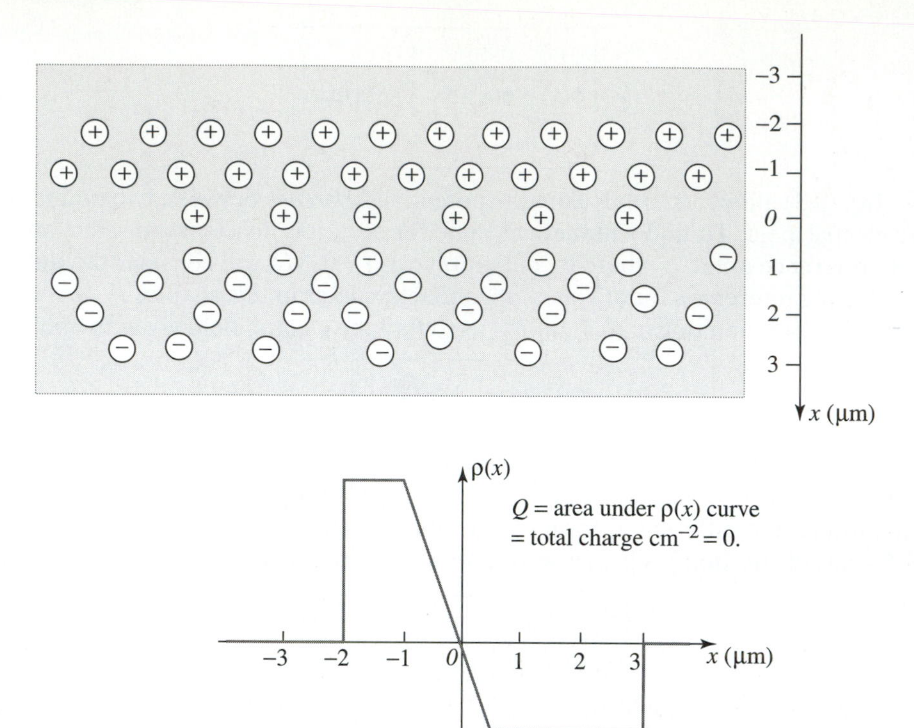

➤ **Figure 3.1** Charge distribution and corresponding charge density $\rho(x)$. The area under the curve is the total charge Q (per cm^2) and is zero for this charge distribution—and for *all* charge distributions in silicon devices in this text.

$$E(x) = E(-2\mu m) + \int_{-2\mu m}^{x} \frac{\rho(x')}{\varepsilon} dx' = \int_{-2\mu m}^{x} \frac{\rho(x')}{\varepsilon} dx' \tag{3.7}$$

since $E(-2\ \mu\text{m}) = 0$. Figure 3.2(a) shows the electric field.

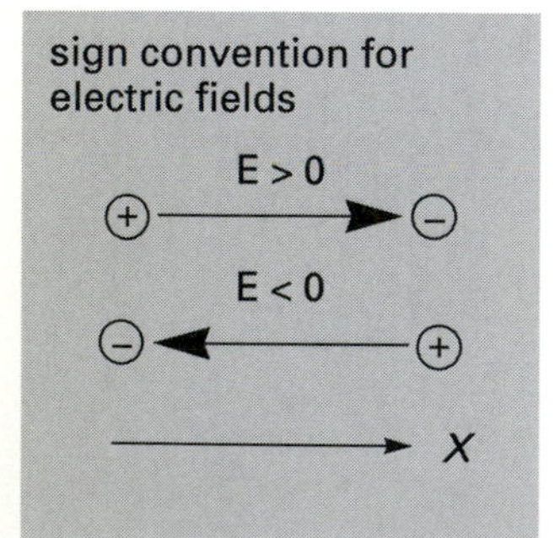

The electric field varies linearly with x for regions with constant charge density and varies quadratically in the interval $-1\ \mu\text{m} < x < 0.5\ \mu\text{m}$, where the charge density varies linearly. The electric field in Fig. 3.2(a) reaches its maximum value which is equal to

$$E_{max} = \int_{-2\mu m}^{0} \frac{\rho(x')}{\varepsilon} dx' \tag{3.8}$$

The electric field is positive over the entire charged region, meaning that it points in the same direction as the x axis. Note that $E(x)$ is positive in the interval $0 < x < 3\ \mu\text{m}$,

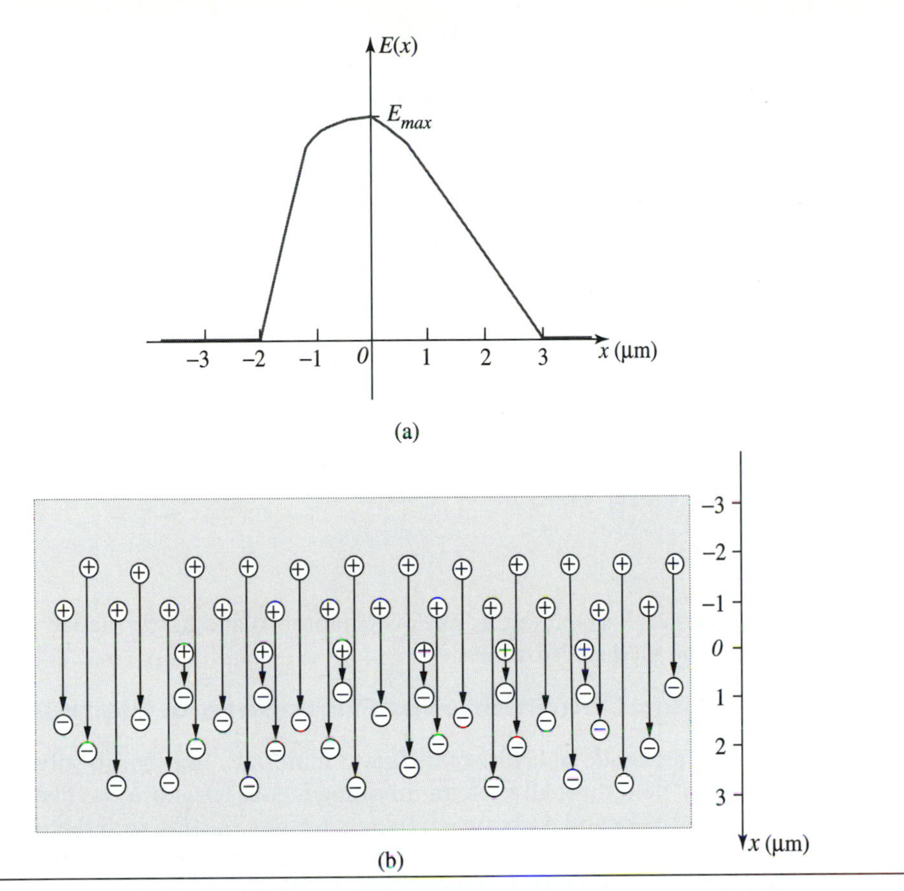

➤ **Figure 3.2** (a) Electric field and (b) qualitative picture of electric field lines connecting the positive and negative charges for the charge distribution in Fig. 3.1.

where the charge density $\rho(x)$ is negative. In order to clarify the connection between charge density and electric field, Figure 3.2(b) shows electric field lines connecting the positive and negative charges. From this figure, we can see that the electric field's magnitude increases (i.e., there are more field lines) as we move from $x = -2$ μm to the origin, after which it decreases as the field lines terminate on the negative charges.

From the electric field $E(x)$, the electrostatic potential $\phi(x)$ can be found by integration from Eq. (3.3). Since we have not yet considered a potential reference in silicon, we will set the potential for $x < -2$ μm to an arbitrary potential ϕ_o in Fig. 3.3. Since the electric field is zero outside the charged region, the potential in this region is constant. To check the sketch of $\phi(x)$, the potential at $x = -2$ μm must be higher than at $x = 3$ μm since the electric field points from regions of high potential (positive charge) to regions of low potential (negative charge)

In summary, the basic features of the electric field and potential can be found from the charge density without excessive use of formal mathematics. In order to understand electronic devices, it is very important to *know* which way the electric field points or whether an answer for the potential "makes sense." There will be

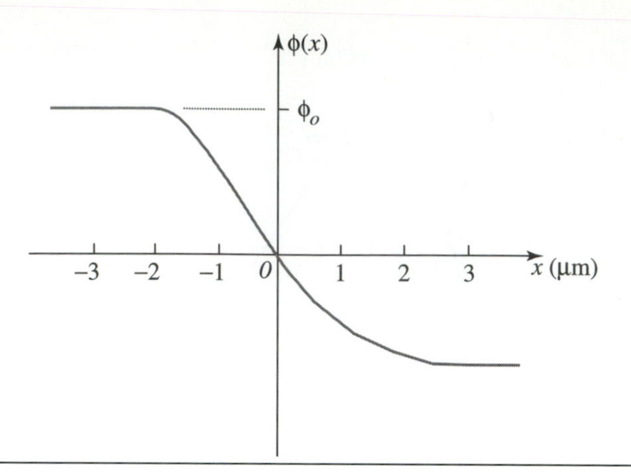

➤ **Figure 3.3** Electrostatic potential for the charge distribution in Fig. 3.1. The reference potential is ϕ_o at $x < -2$ μm.

many opportunities in this chapter, as well as Chapters 4 and 6, to practice the skills needed to master applied electrostatics.

3.1.4 Boundary Conditions for Potential and Electric Field

Electronic devices are made of layers of different materials. In order to solve for the potential or electric field through such an inhomogeneous structure, we need conditions for ϕ and E at the boundary between two materials. Starting with the potential, we observe that an abrupt jump in ϕ would lead to an infinite electric field at the boundary according to Eq. (3.4). Since infinite electric fields are not possible (they would tear the material apart), we conclude that $\phi(x)$ must be continuous

continuity of potential at a boundary

$$\boxed{\phi(x_b^-) = \phi(x_b^+)} \tag{3.9}$$

where the boundary is located at $x = x_b$.

On the other hand, the electric field usually jumps at a boundary. Figure 3.4(a) shows the case of a boundary between materials 1 and 2, where we have located the boundary at $x = 0$ for convenience. The electric field is indicated at two points located a very small distance Δ on either side of the boundary.

We can use the integral form to Gauss's Law to relate the electric field at $x = -\Delta$ to that at $x = +\Delta$

$$\int_{-\Delta}^{+\Delta} d\varepsilon E(x) = \varepsilon_2 E(x{=}\Delta) - \varepsilon_1 E(x = -\Delta) = \int_{-\Delta}^{+\Delta} \rho(x)dx = 0 \tag{3.10}$$

The permittivity is left inside the integral on the left hand side since it varies with position through the interval $-\Delta \le x \le \Delta$ The boundary condition is found by letting $\Delta \to 0$

electric field jump for charge-free boundary

$$\boxed{E(0^+) = \left(\frac{\varepsilon_1}{\varepsilon_2}\right) E(0^-)} \tag{3.11}$$

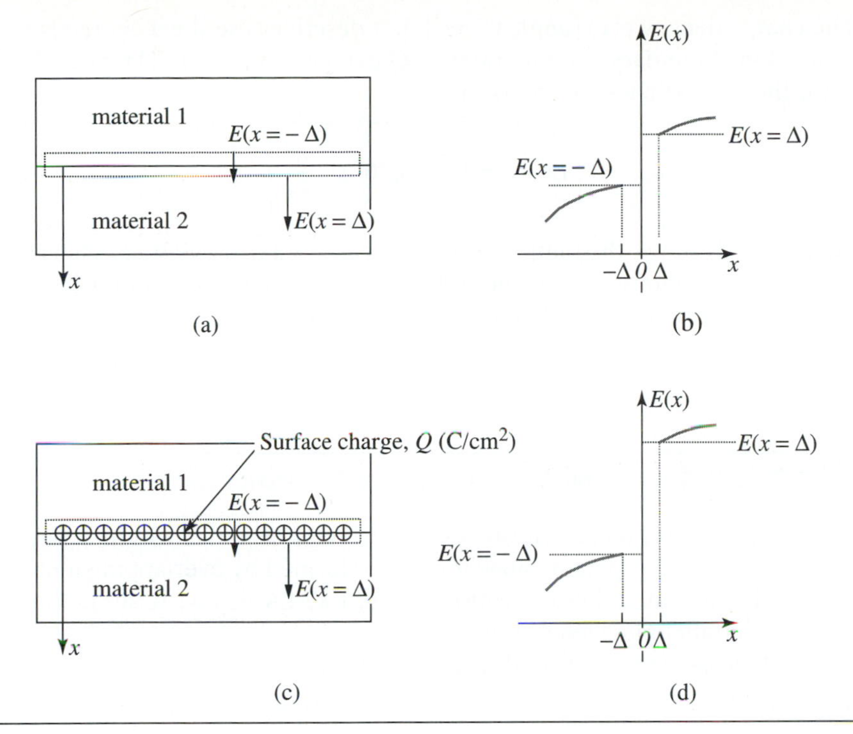

Figure 3.4 (a) Boundary between materials 1 and 2 with permittivities $\varepsilon_1 > \varepsilon_2$ and (b) resulting jump in the electric field. (c) Boundary, with a surface charge Q located at the interface and (d) resulting jump in the electric field.

in which $E(0^+)$ is the electric field in material 2 at the boundary and $E(0^-)$ is the electric field in material 1 at the boundary. Figure 3.4(b) shows the jump in the electric field with Δ not quite zero, for the case where $\varepsilon_1 > \varepsilon_2$. When $\Delta \to 0$, the plot of electric field becomes discontinuous at the boundary.

A common interface in microelectronic devices is between silicon ($\varepsilon_s = 11.7\ \varepsilon_o$ where $\varepsilon_o = 8.85 \times 10^{-14}$ F/cm is the permittivity of vacuum) and SiO_2 (silicon dioxide) which has $\varepsilon_{ox} = 3.9\ \varepsilon_o$. For this case, the electric field in the silicon $E_s(0^+)$ is related to the electric field in the oxide $E_{ox}(0^-)$ by

$$E_s(0^+) = \left[\frac{3.9}{11.7}\right] E_{ox}(0^-) = \frac{E_{ox}(0^-)}{3.0} \tag{3.12}$$

The electric field therefore drops by a factor of three from the SiO_2 side to the silicon side of the boundary.

In some cases of practical importance, there can be a sheet of charge Q [units: C/cm^2] at the boundary as shown in Fig. 3.4(c). The presence of this charge sheet changes the integral in Eq. (3.10)

$$\int_{-\Delta}^{+\Delta} d\varepsilon E(x) = \varepsilon_2 E(x{=}\Delta) - \varepsilon_1 E(x{=}-\Delta) = \int_{-\Delta}^{+\Delta} \rho(x)dx = Q \tag{3.13}$$

The charge density $\rho(x)$ [units: C/cm^3] that describes the sheet charge is a "delta function" at the boundary—it integrates to Q even when $\Delta \rightarrow 0$. The boundary condition for the case with a sheet charge is

electric field jump for charged boundary

$$E(0^+) = \left(\frac{\varepsilon_1}{\varepsilon_2}\right)E(0^-) + \frac{Q}{\varepsilon_2} \tag{3.14}$$

Figure 3.4(d) shows the jump in E for the case of a positive sheet charge Q and $\varepsilon_1 > \varepsilon_2$. The discontinuity in E is increased by the presence of the positive charge sheet, in comparison to the uncharged boundary in Fig. 3.4(a)-(b).

EXAMPLE 3.1 Metal-Metal Capacitor Electrostatics

In many IC processes, there are two or more levels of metallization separated by deposited oxide films. A planar capacitor can be formed by overlapping metal layers as shown in the layout and cross section in Fig. Ex3.1A below. Assume that a one-dimensional, parallel-plate analysis is adequate.

(a) Find the capacitance C in fF/(µm)^2, where 1 fF = 10^{-15} F.

SOLUTION

By definition, capacitance is the derivative of charge with respect to voltage $C = dQ/dV$; in this case, C is constant. We are interested in finding the capacitance per unit area, so the charge Q is also interpreted as per unit area. When a positive voltage V is applied to the capacitor, a positive surface charge density $+Q$ accumulates on the top plate of the capacitor, which is mirrored by an equal and opposite negative charge surface charge density $-Q$ on the bottom plate.

The close-up view of the gap between the plates in Fig. Ex3.1B shows an interior region of the capacitor with all the electric field lines pointing in the $+x$ direction. In this one-dimensional analysis, we will neglect the 2-D effects at the perimeter of the capacitor and assume that all the field lines are parallel to the x axis

As shown in the close-up view of the capacitor, the electric field is constant in the oxide since there is no charge present (see Eq. (3.1), Gauss's Law). Since the electric field is constant and the voltage drop across the gap is V, it follows that $E = V/t_d$. The boundary condition in Eq. (3.14) connects the electric field to the charge density Q [C/cm^2] on the plates

$$E = \left(\frac{\varepsilon_o}{\varepsilon_{ox}}\right)E_{metal} + \frac{Q}{\varepsilon_{ox}} = \frac{Q}{\varepsilon_{ox}}$$

where ε_{ox} is the permittivity of the deposited SiO_2 and $E_{metal} = 0$ (no electric field inside metals). We can substitute $E = V/t_d$ and find the capacitance per unit area

$$\frac{V}{t_d} = \frac{Q}{\varepsilon_{ox}} \rightarrow C = \frac{dQ}{dV} = \frac{\varepsilon_{ox}}{t_d}$$

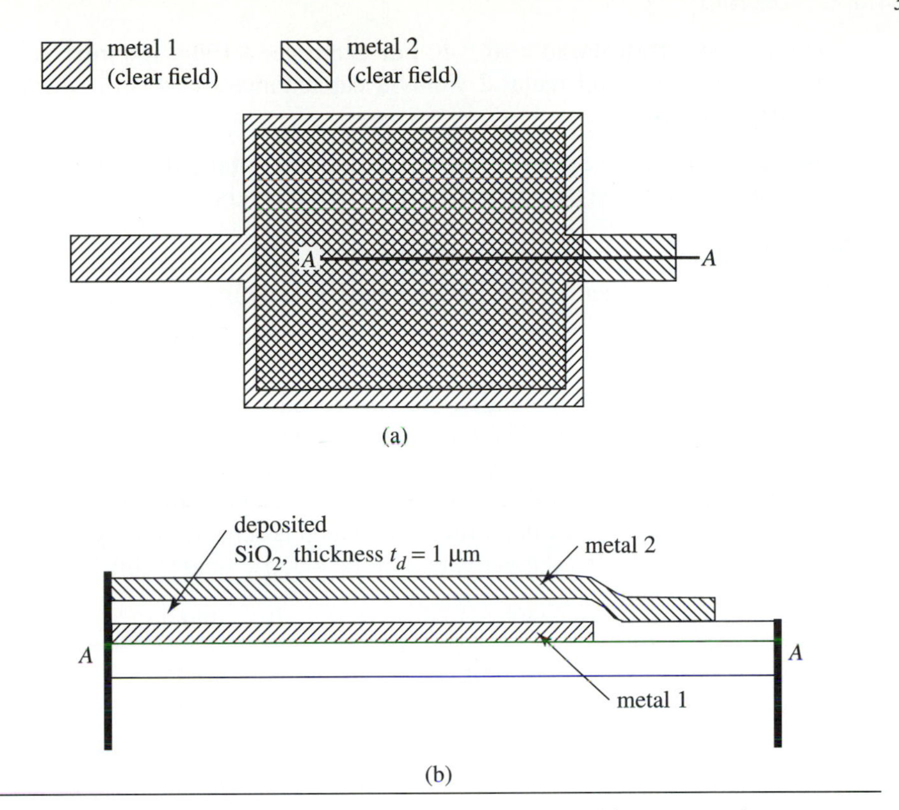

➤ **Figure Ex3.1A** Metal-metal IC capacitor: a) layout and (b) cross section.

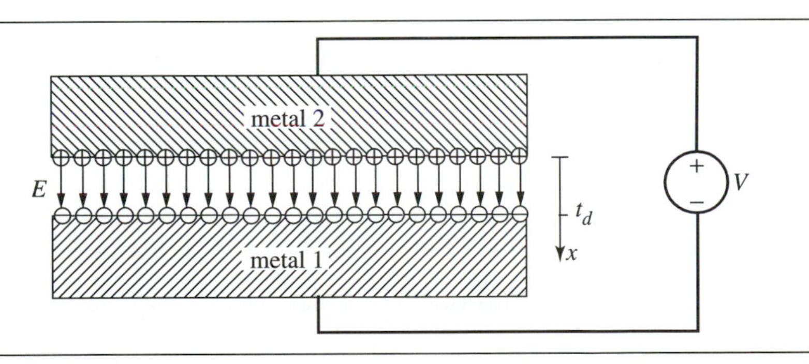

➤ **Figure Ex3.1B** Close-up of metal-metal capacitor with applied voltage, showing surface charges on top and bottom metal plates and electric field lines.

Substituting $\varepsilon_{ox} = 3.9\ \varepsilon_o = 3.45 \times 10^{-13}$ F/cm and $t_d = 1\ \mu m = 1 \times 10^{-4}$ cm for the deposited oxide film, we find that

$$C = 3.45 \times 10^{-9} \mathrm{F/cm}^2 = 34.5\ \mathrm{aF/\mu m^2}$$

where 1 aF = 1 attofarad = 10^{-18} F. For example, a 100 × 100 μm² overlap area between metal 1 and metal 2 yields a capacitance of 345 fF, where 1 fF = 1 femtofarad = 10^{-15} F.

(b) Sketch the charge density, electric field, and potential when a voltage $V = 1$ V is applied between the top and bottom metal layers.

SOLUTION

For an applied voltage of $V = 1$ V, the electric field in the gap is

$$E = (1\ \text{V})/(1 \times 10^{-4}\ \text{cm}) = 10\ \text{kV/cm}$$

From the definition of potential, the electrostatic potential $\phi(x)$ has a constant slope equal to –10 kV/cm. The bottom electrode is taken as the zero reference in the plot of $\phi(x)$ below. Finally, the charge density ρ [C/cm³] is zero everywhere except on the inside surfaces of the metal plates. The mathematical function for the surface charge Q is a delta function, which is represented by "spikes" at the two surfaces. Note that the derivative of the electric field is consistent with these spikes at $x = 0$ and $x = t_d$.

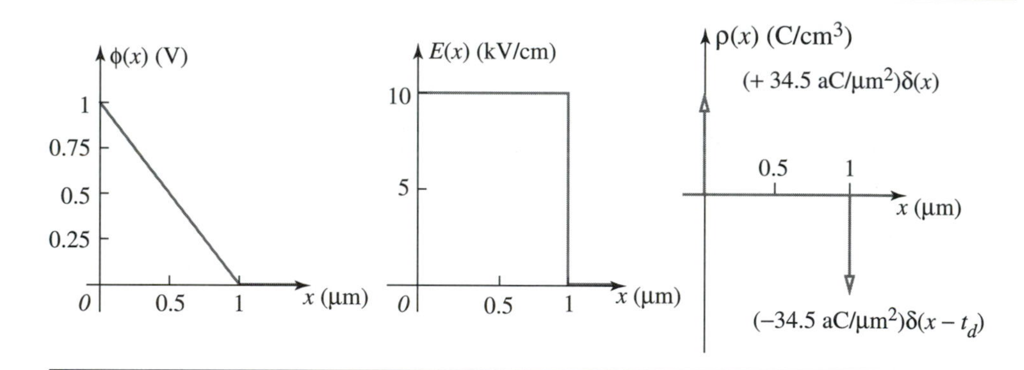

➤ **Figure Ex3.1C** Potential, electric field, and charge density for the capacitor with $t_d = 1$ μm.

EXAMPLE 3.2 Field and Potential from a Distributed Charge Density

(a) Given the following charge distribution, sketch the electric field $E(x)$ and find the numerical value of $E(0)$.

SOLUTION

We should first be sure which way the electric field points.The electric field points from positive charges to negative charges—in the opposite direction as the x axis. Therefore, the electric field is negative. A sketch of the charge distribution with the electric field lines is shown in Fig. Ex3.2B, which indicates that the electric field is constant between the charged layers for $-1\ \mu\text{m} < x < 1\ \mu\text{m}$, and zero outside the charged layers for $x < -2\ \mu\text{m}$ or $x > 2\ \mu\text{m}$. Our basic intuition from the parallel-plate capacitor in Ex. 3.1 (derived from Gauss's Law) also applies to this "smeared" charge distribution.

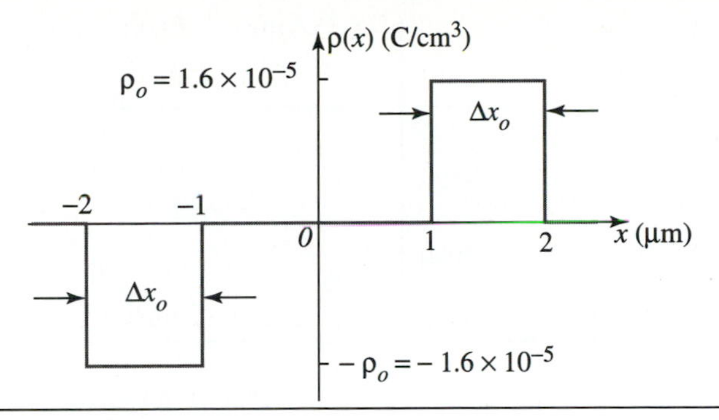

➤ **Figure Ex3.2A** Arbitrary charge distribution in silicon.

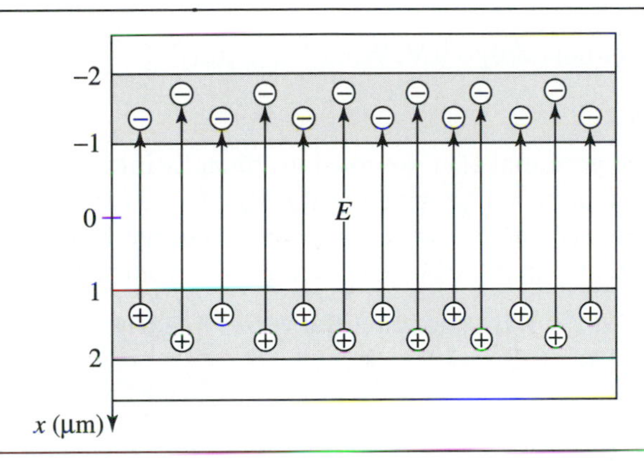

➤ **Figure Ex3.2B** Sketch of electric field lines pointing from positive to negative charges.

We now want to find the field in between the charged layers. Rather than integrate the charge density, we can use the integral form of Gauss's Law to relate the total charge $+Q$ or $-Q$ [C/cm^2] in one of the charged regions to the electric field

$$E(x) - E(-2\mu m) = \frac{-Q}{\varepsilon_s} = \frac{-\rho_o \Delta x_o}{\varepsilon_s} = \frac{-1.6 \times 10^{-9}}{1.036 \times 10^{-12}} = -1.55 \times 10^3 \text{ V/cm}$$

where the boundary at x is anywhere between the charged layers: $-1\ \mu\text{m} < x < 1\ \mu\text{m}$. Since the field at the other boundary at $x = -2\ \mu\text{m}$ is zero, the electric field between the charged layers is $E' = -1550$ V/cm. Within the charged layers, the charge density is constant and the electric field must vary linearly with x according to the differential form of Gauss's Law. This knowledge, combined with the electric field being zero at the boundaries of the charged region at $x = -2\ \mu\text{m}$ and $x = 2\ \mu\text{m}$, enables us to sketch the trapezoidal shape of $E(x)$ in Fig. Ex3.2C. By "turning the crank" and integrating $\rho(x)/\varepsilon_s$, we could have directly calculated the functional form of $E(x)$ assuming we didn't make algebraic errors along the way. By reasoning through the implications of Gauss's Law and the shape of the charge distribution, we are certain we have the correct answer before doing the math.

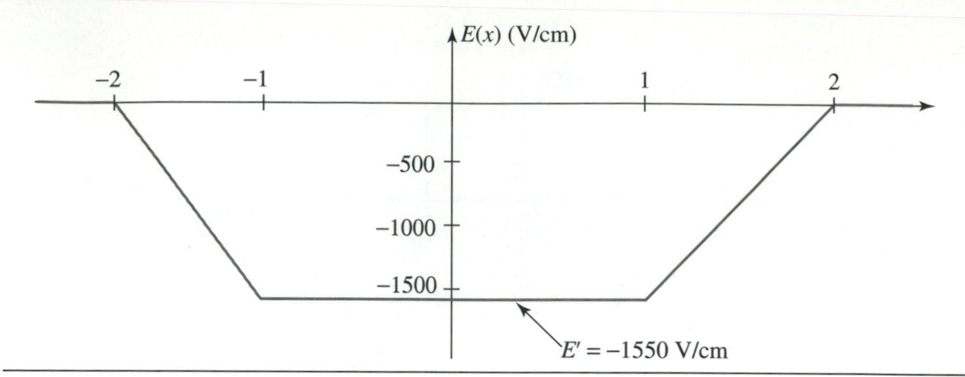

➤ **Figure Ex3.2C** Plot of electric field $E(x)$ for the charge distribution in this example.

(**b**) Sketch $\phi(x)$ assuming $\phi(0) = 0$ V. What is $\phi(2\ \mu\text{m})$?

SOLUTION

In order to plot the potential $\phi(x)$, we must graphically integrate $-E(x)$. Starting from $x = 0$, at which point we are given that $\phi(0) = 0$, we can integrate in the $+x$ direction. We know $\phi(2\ \mu\text{m}) > 0$ since the positive charge is on the right side of the plot of $\rho(x)$ and also that the electric field points from right to left.

Over the interval $0 < x < 1\ \mu\text{m}$, the potential varies linearly since the field is constant. The potential at $x = 1\ \mu\text{m}$ is given by

$$\phi(1\mu\text{m}) = \phi(0) + \int_0^{1\mu m} -E(x')\,dx' = (1550\ \text{V/cm})(10^{-4}\text{cm}) = 155\ \text{mV}$$

Over the interval $1\ \mu\text{m} < x < 2\ \mu\text{m}$, the potential varies quadratically since the field is a linear function. The potential change from $x = 1\ \mu\text{m}$ to $x = 2\ \mu\text{m}$ is given by the area under $-E(x)$

$$\phi(2\mu\text{m}) - \phi(1\mu\text{m}) = \left(\frac{1}{2}\right)(10^{-4}\text{cm})\ (1550\text{V/cm}) = 77.5\text{mV}$$

Therefore, the potential at $x = 2\ \mu\text{m}$ is $\phi(2\ \mu\text{m}) = 155 + 77.5\ \text{mV} = 232.5\ \text{mV}$. Since the electric field is symmetric about $x = 0$, the potential for $x < 0$ is $\phi(-x) = -\phi(x)$. The potential $\phi(x)$ is plotted below. If the function for $\phi(x)$ is needed, then the negative of $E(x)$ can be integrated and the solutions in each interval patched together using the continuity boundary condition.

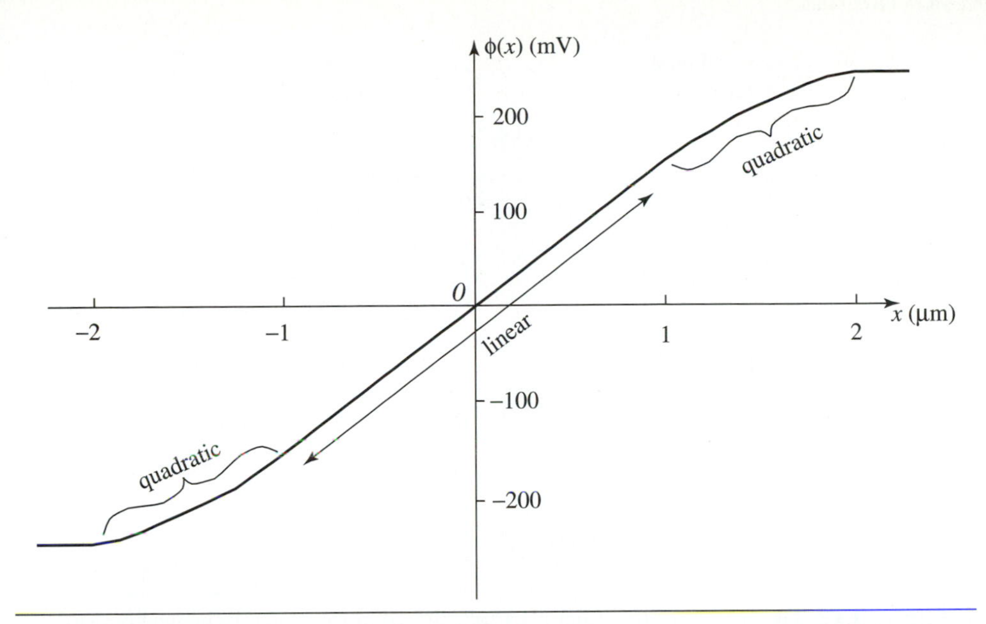

➤ **Figure Ex3.2D** Plot of the potential for the distributed charge density in Example 3.2

EXAMPLE 3.3 Metal-Oxide-Semiconductor (MOS) Capacitor

Another type of common IC capacitor is the Metal-Oxide-Semiconductor (MOS) capacitor. It is formed by a metal layer, an oxide layer, and a layer of silicon. A single mask is needed to create the basic sandwich as shown in the layout and cross section in Fig. Ex3.3A.

(a) Given the qualitative picture of the charge distribution and electric field lines shown in Fig. Ex3.3B, plot the charge density $\rho(x)$. Find an equation relating Q_G and X_d.

SOLUTION

First, we need to define the coordinate system. The origin is located at the interface between the oxide and p-type silicon. The oxide and metal extend into the $-x$ region and the silicon extends into the $+x$ region. The charge on the metal gate is located at the bottom surface as a sheet charge density, with units C/cm^2. On the plot of charge density in C/cm^3 in Fig. Ex3.3C, the gate charge is plotted as a delta function $Q_G\delta(x + t_{ox})$ that is located at the metal/oxide interface at $x = -t_{ox}$.

The total charge (gate charge + charge in the silicon substrate) is zero; therefore, the total charge in the silicon per unit area must be equal and opposite to the charge on the gate

$$Q_G = -(\rho_o X_d) = -\rho_o X_d \rightarrow X_d = Q_G/(-\rho_o)$$

(b) Find and sketch the electric field $E(x)$ and find an expression for the field in the oxide, E_{ox}, in terms of ρ_o, X_d, and ε_{ox}.

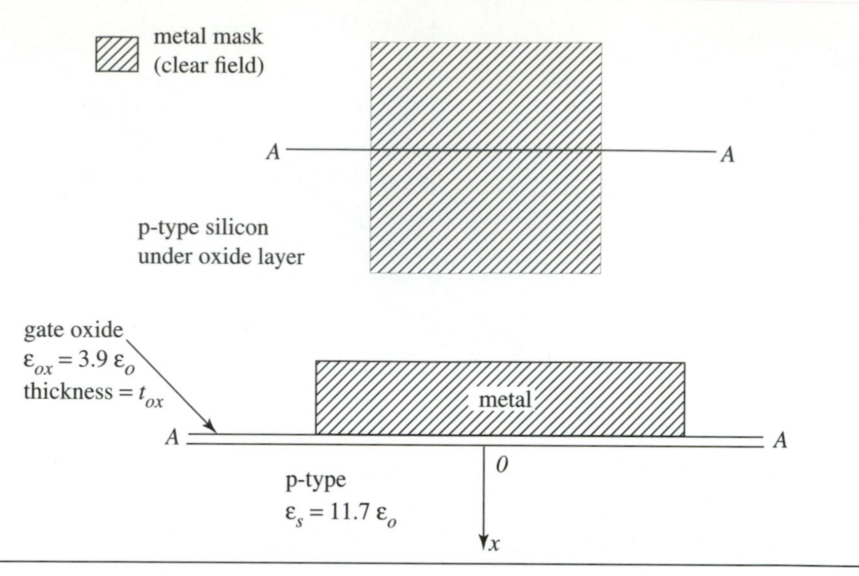

➤ **Figure Ex3.3A** Metal-oxide-silicon capacitor: (a) layout and (b) cross section.

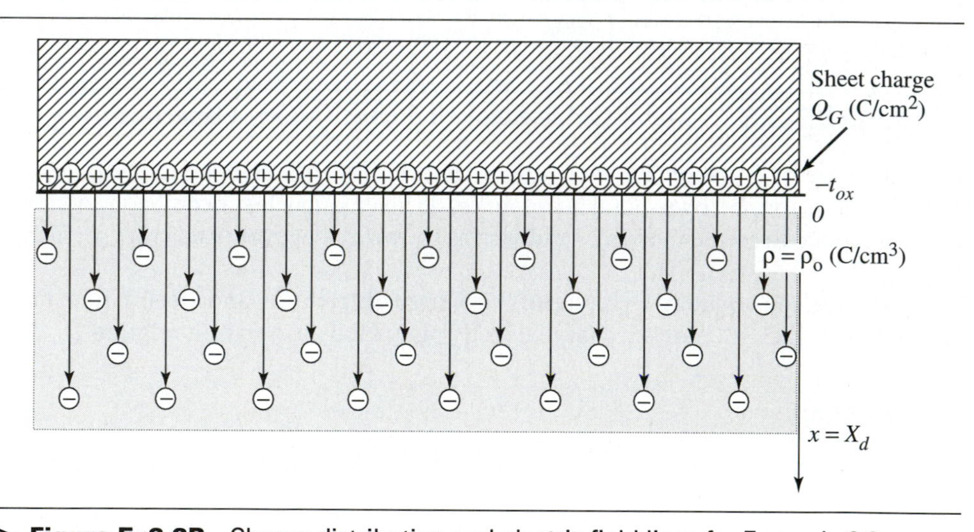

➤ **Figure Ex3.3B** Charge distribution and electric field lines for Example 3.3.

SOLUTION

The electric field points from positive charges to negative charges in the $+x$ direction, and is, therefore, positive. The electric field is confined to the region $-t_{ox} < x < X_d$ because the total charge in this interval is zero. In the oxide $\rho(x) = 0$ and the differential form of Gauss's Law

$$\frac{dE}{dx} = \frac{\rho(x)}{\varepsilon_{ox}} = 0\text{, for } -t_{ox} < x < 0$$

indicates that the field $E(x) = E_{ox}$ is constant for $-t_{ox} < x < 0$. In the charged region in the silicon, the charge density is constant at $\rho(x) = \rho_o$ and, therefore, the

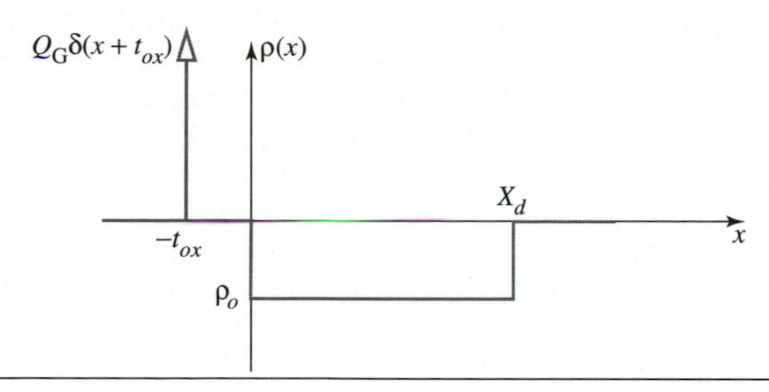

➤ Figure Ex3.3C Charge density $\rho(x)$ corresponding to the qualitative charge distribution in Fig. Ex3.3B.

electric field is a linear function of x there. Since the field is known to be zero at $x = X_d$, we only need the value of the field at $x = 0$ to complete the sketch. The boundary condition at the oxide/silicon interface is

$$\varepsilon_{ox} E_{ox} = \varepsilon_s E(x = 0^+) \rightarrow E(x = 0^+) = (\varepsilon_{ox}/\varepsilon_s) E_{ox} = E_{ox}/3$$

Gauss's Law in integral form is a convenient way to find the value of the electric field at the boundary of a charged region. Using Eq. (3.2) over the interval $0 \le x \le X_d$, we find

$$\varepsilon_s E(X_d) - \varepsilon_{ox} E_{ox} = \int_0^{X_d} \rho_o dx = \rho_o X_d$$

Since $E(X_d) = 0$, the oxide field is

$$E_{ox} = \frac{-\rho_o X_d}{\varepsilon_{ox}}$$

Note that there is the possibility of sign errors in applying Eq. (3.2); however, we are not depending on the math to tell us the sign since we already *know* $E_{ox} > 0$. The electric field $E(x)$ is plotted in Fig. Ex3.3D.

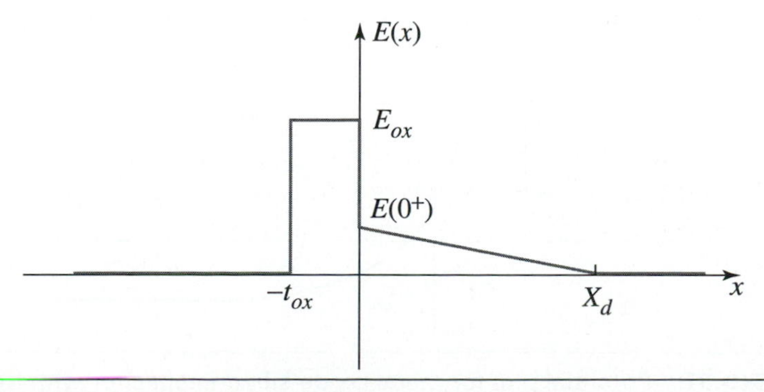

➤ Figure Ex3.3D Sketch of electric field for metal-oxide-silicon capacitor with the charge distribution in Fig. Ex3.3B.

(c) Sketch the potential $\phi(x)$ through the structure given that the potential of the metal gate is $\phi(-t_{ox}) = \phi_m$. Find an expression for the surface potential $\phi_s = \phi(0)$ in terms of ϕ_m, ρ_o, X_d, ε_{ox}, and t_{ox}. In addition, find an expression for the potential drop across the charged region in the silicon in terms of ρ_o, X_d, and ε_s.

SOLUTION

As we learned in Example 3.2, there is no need to integrate $-E(x)$ in order to sketch the potential. From the definition of potential and the sketch of $E(x)$ above, the potential is a linear function in the oxide and a quadratic function of x in the charged region of the silicon substrate. Poisson's Equation is useful for finding the concavity of the potential in the charged region

$$\frac{d^2\phi(x)}{dx^2} = -\frac{\rho_o}{\varepsilon_s} > 0$$

Therefore, the potential is "concave up" in the charged region. Finally, we also know that the potential is continuous at the boundary $x = 0$ between the oxide and silicon. We can piece together the linear and quadratic solutions and sketch the potential between the metal and underlying substrate ($x > X_d$), the potential of which is defined as ϕ_{sub}. Figure Ex3.3E is a plot of the potential through the MOS capacitor.

On the sketch of $\phi(x)$ below, the total drop has been divided into two parts, the drop across the oxide V_{ox} and the drop across the charged region in the silicon V_B.

$$\phi_m - \phi_{sub} = V_{ox} + V_B$$

Note that the polarities of V_{ox} and V_B are defined on the potential plot in Fig. Ex3.3E to avoid confusion with signs. The drop across the oxide can be found by integrating the oxide field from $x = -t_{ox}$ to $x = 0$.

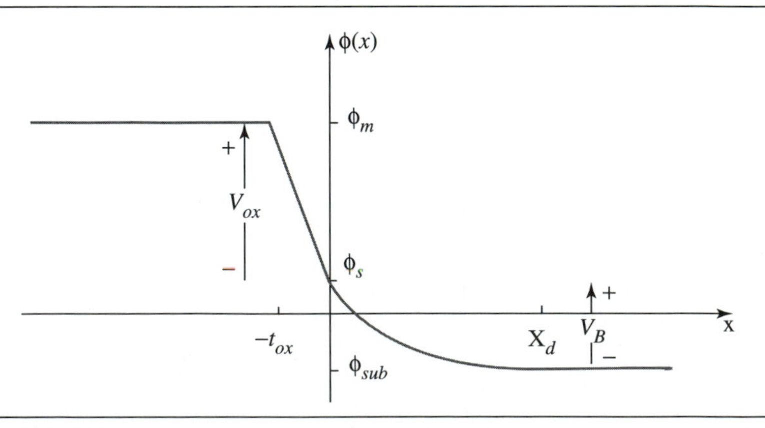

➤ Figure Ex3.3E Potential plot for metal-oxide-silicon capacitor with the charge distribution in Fig. Ex3.3B.

$$\phi_s - \phi_m = \int_{-t_{ox}}^{0} (-E_{ox})\, dx = -E_{ox}(0 - (-t_{ox})) = -V_{ox}$$

and therefore, $V_{ox} = E_{ox}t_{ox}$. The surface potential can be found in terms of the given parameters by substituting for the oxide field

$$\phi_s = \phi_m - E_{ox}t_{ox} = \phi_m - \frac{(-\rho_o)X_d t_{ox}}{\varepsilon_{ox}}$$

This expression can be simplified by defining the charge Q_B [C/cm^2] in the substrate as $Q_B = \rho_o X_d = -Q_G$, and by recognizing that the oxide capacitance per unit area is $C_{ox} = \varepsilon_{ox}/t_{ox}$. The surface potential and drop across the oxide are related to these charges by

$$\phi_s = \phi_m + \frac{Q_B}{C_{ox}} \text{ and } V_{ox} = -\frac{Q_B}{C_{ox}} = \frac{Q_G}{C_{ox}}$$

The last result is not unexpected since the voltage drop V_{ox} must be proportional to the charge $\pm Q_G$ stored on each side of the insulating oxide.
Finally, Poisson's Equation is used to find the drop V_B across the charged region

$$\frac{d^2\phi(x)}{dx^2} = -\frac{\rho_o}{\varepsilon_s} \text{ for } 0 < x < X_d$$

This ordinary differential equation can be solved for $\phi(x)$. The drop across the charged region is

$$\phi(0) - \phi(X_d) = V_B = \frac{1}{2}\left(-\frac{\rho_o}{\varepsilon_s}\right)X_d^2 = \left(-\frac{\rho_o X_d}{2\varepsilon_s}\right)X_d = \frac{-Q_B}{2\varepsilon_s}X_d$$

Note that we have quantified the essential features of the electric field and the potential in the MOS structure without depending on formal mathematical analysis.

3.2 CARRIER CONCENTRATION AND POTENTIAL IN THERMAL EQUILIBRIUM

In order to find complete solutions for the electrostatic potential in silicon, we must define a potential reference. Although the choice is arbitrary, we will find it very convenient to select a reference that is related to the thermal equilibrium carrier concentration.

Recall from Chapter 2 that thermal equilibrium is equivalent to the absence of *any* stimulus to the device—zero applied voltage, no external light source, etc. In this state, the current density J_o must be zero. Perhaps less obvious is that the electron *and* hole current densities J_{no} and J_{po} must *each* be equal to zero in thermal equilibrium. The physical reasoning is that the populations of electrons and the holes are each in equilibrium and, therefore, must have zero current densities. For example, $J_{no} \neq 0$ implies that electrons will pile up with time at one end of a structure in thermal equilibrium, which is hardly reasonable.

Setting the electron current density equal to zero in thermal equilibrium:

$$0 = qn_o\mu_n E_o + qD_n \frac{dn_o}{dx} \tag{3.15}$$

Solving for the gradient and using Einstein's Relation (Eq. 2.51) to eliminate the ratio of the mobility to the diffusion coefficient, we find that

$$\frac{dn_o}{dx} = \left(-\frac{\mu_n}{D_n}\right)n_o E_o = \left(-\frac{q}{kT}\right)n_o\left(-\frac{d\phi_o}{dx}\right) \tag{3.16}$$

where the negative gradient of the potential ϕ_o has been substituted for the electric field. Solving for the differential potential $d\phi_o$, we find

$$d\phi_o = \left(\frac{kT}{q}\right)\frac{dn_o}{n_o} = V_{th}\frac{dn_o}{n_o} \tag{3.17}$$

in which we have substituted the thermal voltage $V_{th} = kT/q$.

This differential equation can be solved by direct integration from a reference point x_o to an arbitrary point x

$$\phi_o(x) - \phi_o(x_o) = V_{th}\ln\left(\frac{n_o(x)}{n_o(x_o)}\right) \tag{3.18}$$

To find a numerical value for $\phi_o(x)$, we need to select a reference for the potential. By convention, the reference for potential is chosen to be the point where the electron concentration is the intrinsic concentration

potential reference in thermal equilibrium

$$\phi_o(x_o) = 0 \text{ when } n_o(x_o) = n_i \tag{3.19}$$

Substituting this reference into Eq. (3.18) defines the potential in terms of the electron concentration in thermal equilibrium

$$\phi_o(x) = V_{th}\ln\left(\frac{n_o(x)}{n_i}\right) \tag{3.20}$$

Note that n-type silicon has a positive electrostatic potential since $n_o > n_i$ when electrons are the majority carriers. In order words, a region of n-type silicon is *relatively* positive, when compared with an intrinsic region.

By inverting this equation, we can express the electron concentration in terms of the electrostatic potential

$$n_o(x) = n_i e^{\phi_o(x)/V_{th}} \tag{3.21}$$

A parallel derivation for hole concentration leads to the following results

$$\phi_o(x) = -V_{th}\ln\left(\frac{p_o(x)}{n_i}\right) \qquad p_o(x) = n_i e^{-\phi_o(x)/V_{th}} \tag{3.22}$$

so that p-type silicon ($p_o > n_i$) has a negative electrostatic potential, with respect to intrinsic silicon.

3.2.1 The 60mV Rule

In order to get an easy-to-remember rule, we replace the natural logarithm with the base 10 logarithm in Eq. (3.20) and evaluate the expression at a "warm" room temperature (V_{th} = 26 mV):

60mV rule

$$\boxed{\phi_o(x) = (26\text{mV})\ln(10)\log\left(\frac{n_o(x)}{10^{10}}\right) = (60\text{mV})\log\left(\frac{n_o(x)}{10^{10}}\right)} . \tag{3.23}$$

This **60 mV rule** can be evaluated without a calculator for carrier concentrations that are equal to 10^{14}cm^{-3}, 10^{15}cm^{-3}, etc. For example, $n_o = 10^{16}$ cm^{-3}corresponds to the potential

$$\phi_o = (60\text{ mV})\log(10^{16}/10^{10}) = (60\text{ mV})(6) = 360\text{ mV} \tag{3.24}$$

We can also express the potential as a function of the hole concentration in thermal equilibrium by substituting the mass-action law $n_o = n_i^2/p_o$ into Eq. (3.23):

$$\phi_o(x) = -(60\text{mV})\log\left(\frac{p_o(x)}{10^{10}}\right) \tag{3.25}$$

For a hole concentration of $p_o = 10^{16}$ cm^{-3}, the potential is ϕ_o= –360 mV. The carrier concentrations can range from the intrinsic concentration to about 10^{19} cm^{-3} for electrons or for holes, which corresponds to a potential range in thermal equilibrium of $-540\text{ mV} \le \phi_o \le 540\text{ mV}$. The potential of silicon that is extremely heavily doped with donors (called **n^+ silicon**) is $\phi_{n+} = 550$ mV. For **p^+ silicon,** which is heavily doped with acceptors, the potential is $\phi_{p+} = -550$ mV. Figure 3.5 shows the 60 mV rule in graphical form.

n^+ and p^+ silicon

The derivation of Eq. (3.23) only required that the electron and hole currents are zero throughout the sample. In addition to thermal equilibrium, this requirement is also satisfied in capacitors since the insulator prevents carrier transport through the structure.

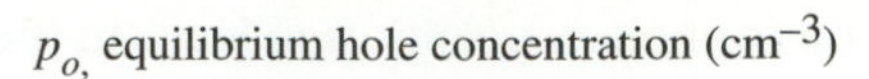

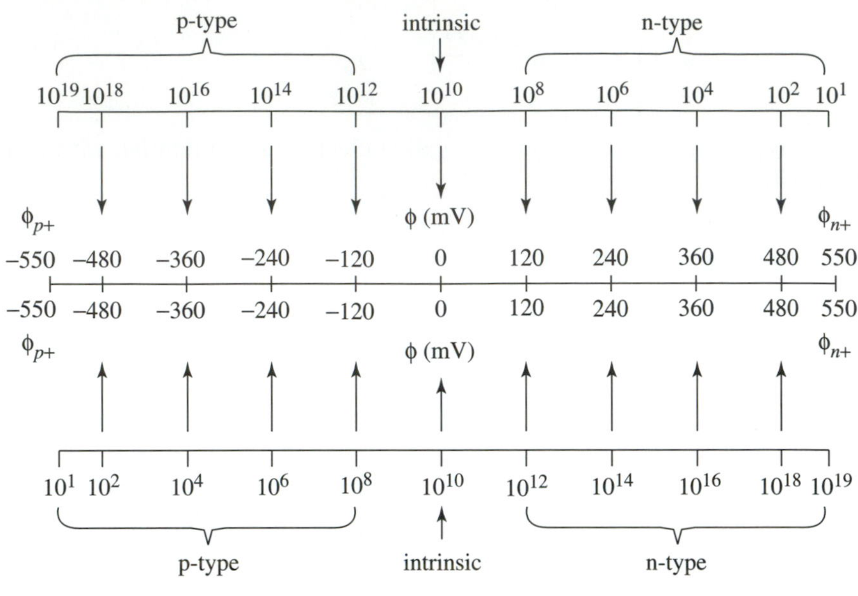

Figure 3.5 60 mV rule relating electrostatic potential in thermal equilibrium to hole or electron concentration at room temperature.

3.3 THE pn JUNCTION: A FIRST PASS

A **pn junction** consists of a p-type silicon region immediately adjacent to an n-type silicon region. As we have seen in Chapter 2, pn junctions are built into nearly all IC device structures—either intentionally or as by-products of the fabrication process. Figure 3.6(a) shows the cross section of a pn juction. The p and n regions are contacted by metal interconnections on the top surface of the chip which are used to apply a voltage across the junction. For now, we will ignore the underlying p-type substrate that forms another junction with the n side of the pn junction. The substrate is electrically contacted and set at a voltage (often ground) to ensure that adjacent device structures have minimum electrical interaction.

The essential features of the pn junction can be understood from the one-dimensional simplification shown in Fig. 3.6(b). For mathematical simplicity, the doping concentrations N_a on the p-side and N_d on the n-side are assumed to be constant. Furthermore, it is assumed that N_a and N_d already incorporate any doping compensation effects and thus represent the net concentrations of acceptors and donors.

How do we develop a qualitative picture of the electrostatics of the pn junction? As a start, we consider the case of thermal equilibrium. We know that the hole and electron current densities $J_{po} = 0$ and $J_{no} = 0$ throughout the structure. Therefore, it follows that the hole drift current density must be equal in magnitude and opposite in direction to the hole diffusion current density throughout the pn junction in thermal equilibrium.

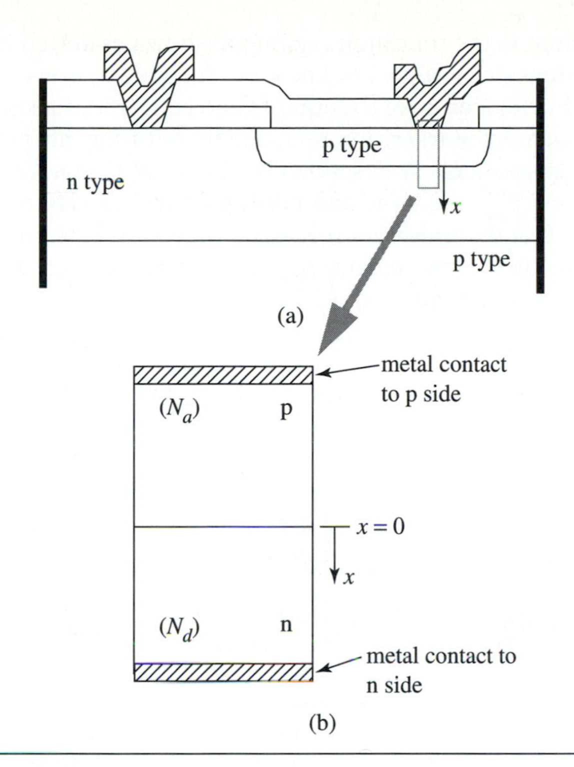

➤ **Figure 3.6** (a) Cross section of a typical pn junction fabricated in an IC process, (b) one-dimensional pn junction consisting of boxed region in (a) with metal contact to n-side added at the bottom. The metallurgical junction (boundary between p-type and n-type regions) is located at $x = 0$.

$$J_{po} = J_{po}^{dr} + J_{po}^{diff} = qp_o\mu_p E_o - qD_p\frac{dp_o}{dx} = 0 \rightarrow qp_o\mu_p E_o = qD_p\frac{dp_o}{dx} \tag{3.26}$$

Similarly, the electron drift and diffusion current densities must cancel

$$J_{no} = J_{no}^{dr} + J_{no}^{diff} = qn_o\mu_n E_o + qD_n\frac{dn_o}{dx} = 0 \rightarrow qn_o\mu_n E_o = -qD_n\frac{dn_o}{dx} \tag{3.27}$$

A rough sketch of the hole concentration $p_o(x)$ is helpful in developing insights into the implications of Eq. (3.26) for the pn junction. Far enough from the metallurgical junction at $x = 0$, the hole concentration on the p-side of the junction must be $p_o(x) = N_a$. In effect, we are asserting that the p-type region can be considered as bulk silicon when it is far enough from the junction since the effects of the pn junction are confined to a region near $x = 0$. We shall quantify the width of the "transition region" when the junction is quantitatively analyzed later in the chapter. However, as a preview of that analysis, the order of magnitude of a typical transition width is 1 μm. On the n-side of the junction, the hole concentration far enough from the junction is $p_o(x) = n_i^2/n_o = n_i^2/N_d$–the hole concentration in bulk n-type silicon. In Fig. 3.7(a), the hole concentration $p_o(x)$ is sketched through the pn junction in thermal equilibrium. The transition region is confined to a region of undetermined width $-x_{po} < x < x_{no}$, where $-x_{po}$ is its border on the p-side and x_{no} is its border on the n-side of the junction in thermal equilibrium. Using similar reasoning, we can sketch the electron concentration $n_o(x)$ in thermal equilibrium through the pn junction.

The concentrations in the transition region are shown as dotted lines connecting the bulk concentrations since we do not know the functional forms.

The most obvious feature of the equilibrium concentrations $p_o(x)$ and $n_o(x)$ is the large gradients across the transition region. The diffusion current densities are proportional to these gradients as described in Eqs. (3.26) and (3.27). Figure 3.7 shows that both J_{po}^{diff} and J_{no}^{diff} are large and positive in the transition region. In order for $J_{po} = 0$ and $J_{no} = 0$ in the transition region, as they must in thermal equilibrium, it is necessary that there be balancing drift current densities $J_{po}^{dr} = -J_{po}^{diff}$ and $J_{no}^{dr} = -J_{no}^{diff}$. Substituting into Eq. (3.26), the required electric field E_o in thermal equilibrium is

$$E_o = \frac{qD_p \frac{dp_o}{dx}}{qp_o\mu_p} = \frac{-qD_n \frac{dn_o}{dx}}{qn_o\mu_n} \tag{3.28}$$

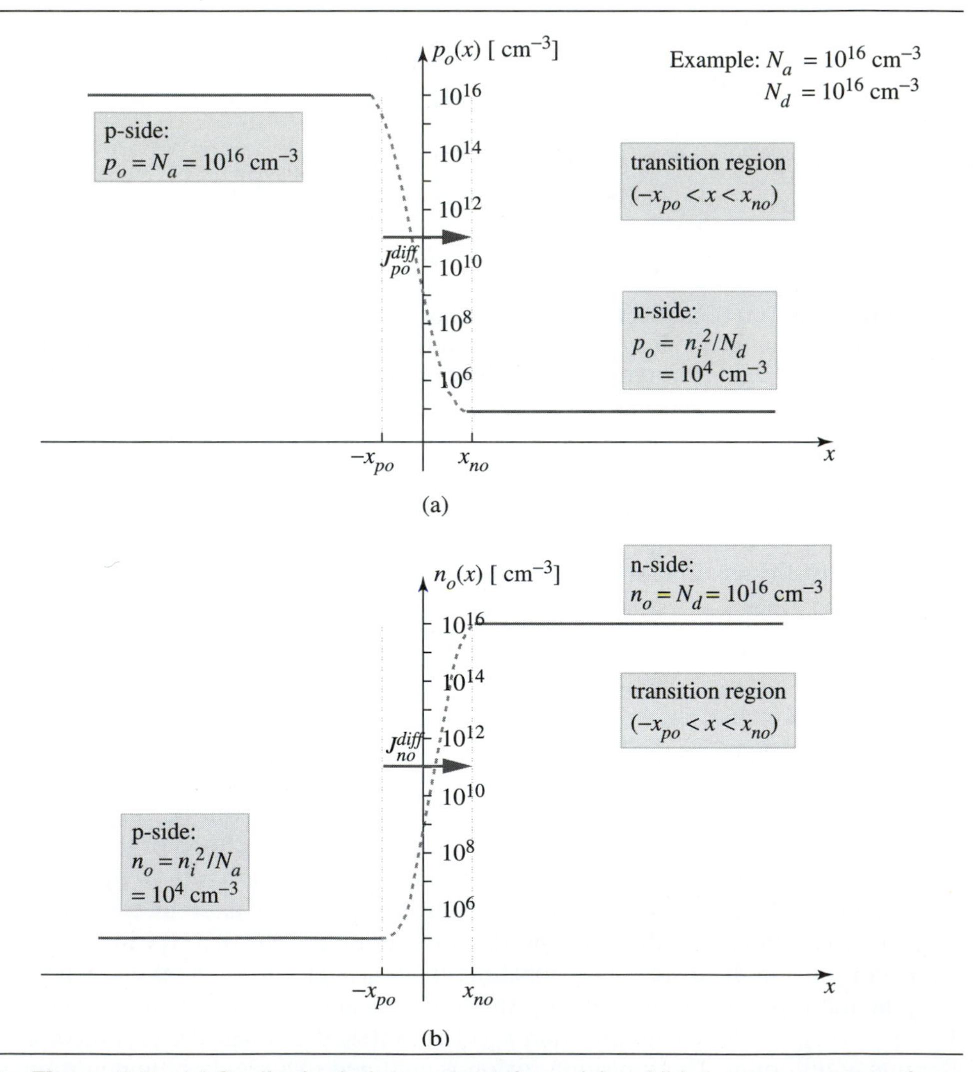

➤ **Figure 3.7** (a) Qualitative hole concentration $p_o(x)$ and (b) electron concentration $n_o(x)$ in thermal equilibrium. The doping concentrations are 10^{16} cm^{-3} for this example. The equilibrium hole and electron diffusion current densities are both large and positive in the transition region because of the large concentration gradients.

Since we do not know the functional forms for $p_o(x)$ or for $n_o(x)$, we cannot quantify the electric field in the transition region. However, we do know from Eq. (3.28) and from Fig. 3.7 that the electric field in the transition region must be *large* in magnitude and must point in the $-x$ direction so its sign is *negative*. In contrast, outside the transition region p_o and n_o are constants and we conclude that the diffusion current densities and the electric field are zero.

How can a large electric field exist in the transition region? From our understanding of electrostatics, there must be a region of negative and positive charge near $x = 0$. There are four charged species in silicon, two mobile (holes and electrons) and two fixed (ionized donors and ionized acceptors). The charge density in thermal equilibrium $\rho_o(x)$ is given by

$$\rho_o(x) = q(p_o - n_o + N_d - N_a) \tag{3.29}$$

In the transition region on the p-side of the junction over the interval $-x_{po} \le x \le 0$, there are very few electrons (see Fig. 3.7(b)) and no donors, so Eq. (3.29) simplifies to

$$\rho_o(x) \cong q(p_o - N_a)\ ,\ \text{for } -x_{po} \le x \le 0 \tag{3.30}$$

As shown in Fig. 3.7(a), the thermal equilibrium hole concentration $p_o \ll N_a$ in the interval $-x_{po} \le x \le 0$, which leads to a *negative* charge density

$$\rho_o(x) < 0\ ,\ \text{for } -x_{po} \le x \le 0 \tag{3.31}$$

On the n-side of the transition region in the interval $0 \le x \le x_{no}$, there are few holes (see Fig. 3.7(a)) and no acceptors, therefore Eq. (3.29) simplifies to

$$\rho_o(x) \cong q(-n_o + N_d)\ ,\ \text{for } 0 \le x \le x_{no} \tag{3.32}$$

The electron concentration in thermal equilibrium $n_o(x) < N_d$ in this interval (see Fig. 3.7(b)), making the charge density there *positive*

$$\rho_o(x) > 0\ ,\ \text{for } 0 \le x \le x_{no} \tag{3.33}$$

From the sketches of $p_o(x)$ and $n_o(x)$ in Fig. 3.7 and Eqs. (3.30) and (3.32), we can develop the rough sketch of the thermal equilibrium charge density $\rho_o(x)$ in Fig. 3.8(a). Since the doping is symmetrical in Fig. 3.7 ($N_a = N_d = 10^{16}\ \text{cm}^{-3}$), the transition region is equally split around $x = 0$ in order to maintain charge neutrality. Given the rapid change in concentrations in Fig. 3.7 (note the logarithmic scale), the slopes in $\rho_o(x)$ at $x = -x_{po}$ and $x = x_{no}$ should be greater than depicted in Fig. 3.8(a). Inside the transition region, the mobile carrier concentrations are negligible for finding the charge density. In effect, the mobile carriers have been depleted from the transition region. We will refer to the high-field region at the junction as the **depletion region**.

The sketches for the electric field $E_o(x)$ and the potential $\phi_o(x)$ in thermal equilibrium are given in Figs. 3.8(b) and 3.8(c). The field is found by graphical integration and is approximately triangular in shape. In the regions away from the junction, the potential is given by the 60 mV rule, as shown in Fig. 3.8(c). The potential on the p-side is defined as ϕ_p and is –360 mV for this diode, since $N_a = 10^{16}$ cm^{-3}. On the n-side, the potential is ϕ_n = 360 mV since $N_d = 10^{16}$ cm^{-3}. The **built-in potential** ϕ_B is defined as

$$\phi_B = \phi_n - \phi_p = 360 \text{ mV} - (-360 \text{ mV}) = 720 \text{ mV}. \tag{3.34}$$

The sign of the built-in potential is consistent with the direction of the electric field in Fig. 3.7(b) and the fact that the charge density is positive on the n-side of the

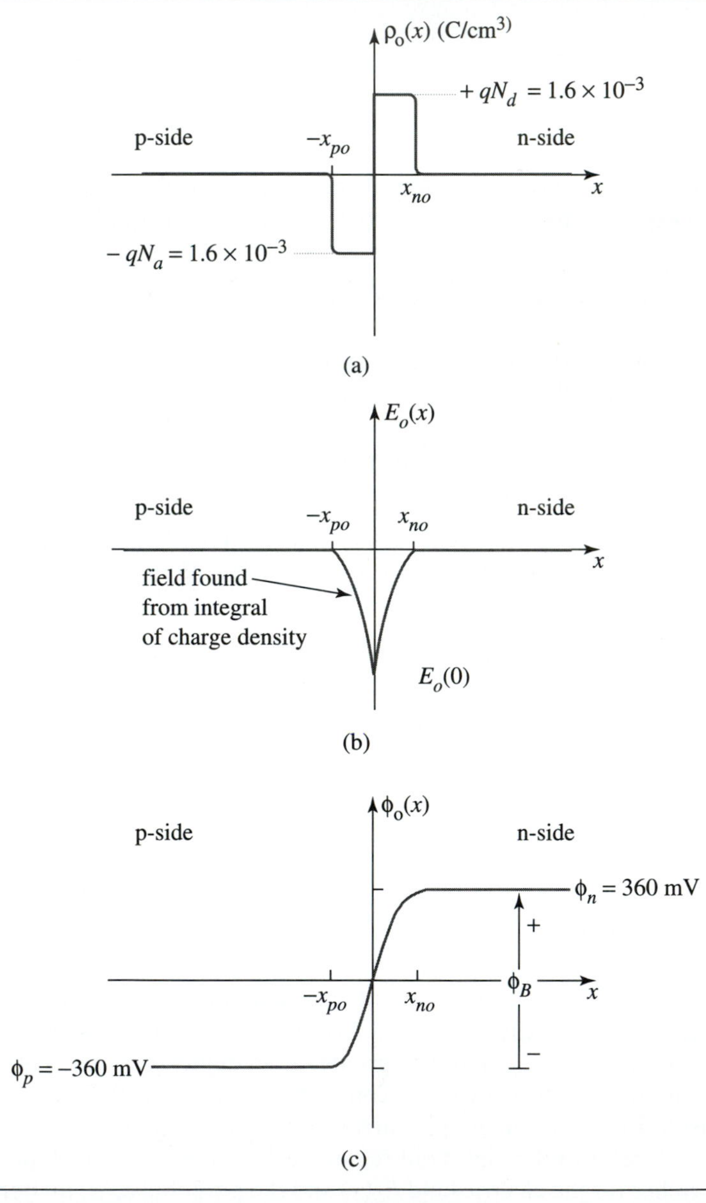

➤ **Figure 3.8** Qualitative electrostatics of the pn junction in thermal equilibrium: (a) charge density, (b) electric field, and (c) electrostatic potential. The doping is $N_a = N_d = 10^{16}$ cm^{-3}, as given in Fig. 3.7, which leads to $x_{no} = x_{po}$.

transition region. One way of rationalizing the potential barrier at the junction is that it prevents holes from flooding from the p-side (where they are majority carriers) to the n-side (where they are minority carriers.) Of course, the potential barrier also confines the majority electrons to the n-side of the transition region.

As an electrical device structure, the pn junction has an interesting feature: it has a stored charge with no applied voltage, due to the built-in potential barrier. The charge stored per unit area on the p-side is called the **depletion charge**. In thermal equilibrium, the depletion charge Q_{Jo} [units: C/cm^2] is

$$Q_{Jo} = -qN_a x_{po} \tag{3.35}$$

The sign of the depletion charge stored on the p-side of junction is negative, which is consistent with the sign of the potential drop from the p-side to the n-side of the junction in thermal equilibrium:

$$\phi_p - \phi_n = -\phi_B < 0 . \tag{3.36}$$

Having developed a qualitative picture of the pn junction in thermal equilibrium, we now extend our understanding to the case of an applied voltage. In this chapter, we are considering only the electrostatics of the junction and will only discuss the reverse-biased situation. The reference polarity for the applied voltage V_D is chosen so that a forward bias is positive, as shown in Fig. 3.9(a) and (b), so for reverse bias $V_D < 0$.

In Fig. 3.9(a), we have defined the potential increase from the p to the n side as the **barrier potential** ϕ_j. As we know from basic electronics, the current through the reverse-biased diode is negligible. Since there can be no resistive potential drops across the p and n bulk regions, the applied voltage must be dropped across the depletion region. Accounting for the presence of the built-in potential ϕ_B, the barrier potential under an applied bias is

$$\phi_j = \phi_B - V_D . \tag{3.37}$$

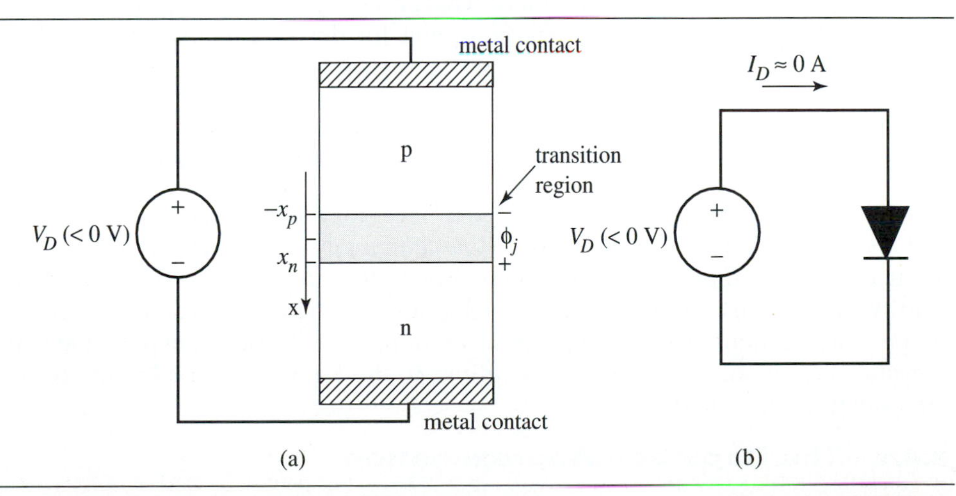

Figure 3.9 (a) pn junction with negative applied voltage V_D (reverse bias) and (b) corresponding circuit schematic.

In thermal equilibrium, $V_D = 0$ and Eq. (3.37) correctly indicates that the barrier potential must equal the built-in potential ϕ_B. Under reverse bias with $V_D < 0$, the barrier at the junction increases, which is consistent with the polarities of V_D and ϕ_j in Fig. 3.9(a).

We have *not* derived Eq. (3.37) from first principles, but rather have established that it correctly accounts for the presence of a built-in barrier at the junction in thermal equilibrium. In Section 3.5, we will investigate in detail how the internal barrier ϕ_j is offset from the externally applied voltage V_D.

What is the effect of the increase in potential barrier on the charge distribution and electric field? More charge is needed on each side of the junction in order to generate enough electric field to yield the larger potential barrier. Since the mobile carriers have been depleted from the depletion region, the charge density is constant at $-qN_a$ on the p-side and $+qN_d$ on the n-side of the junction. In order to get more charge, the width of the depletion region must increase. In Fig. 3.10(a), the edges of the depletion region are shown to have doubled over their thermal-equilibrium values, for the particular reverse bias voltage.

In Fig. 3.10(b), the magnitude of the peak electric field is double that in thermal equilibrium, since the charge on each side of the junction has doubled. Finally, the potential plot in Fig. 3.10(c) shows that the barrier potential under reverse bias has increased over its value in thermal equilibrium: $\phi_j = \phi_B - V_D > \phi_B$. For convenience in plotting $\phi(x)$, we have kept the n-side of the junction fixed at its equilibrium potential and applied the negative diode voltage to the p-side. We can visualize the reverse bias voltage $V_D < 0$ as pulling down the potential on the p-side of the junction, as shown in Fig. 3.10(c).

The effect of reverse bias is to increase the magnitude of the depletion charge q_J, as can be seen by comparing Fig. 3.10(a) with Fig. 3.8(a). As we will derive in Section 3.5, the functional dependence of the depletion charge on the applied bias is:

$$Q_J(V_D) = Q_{Jo}\sqrt{1-(V_D/\phi_B)} = -qN_a x_{po}\sqrt{1-(V_D/\phi_B)}\ , \tag{3.38}$$

where Q_{Jo} (< 0) is the depletion charge in thermal equilibrium. In summary, the pn junction under reverse bias is a non-linear charge-storage element. Its charge-voltage characteristic in Eq. (3.38) is the starting point for deriving equivalent circuit elements in Section 3.6.

3.4 THE pn JUNCTION IN THERMAL EQUILIBRIUM

From the previous section, we have gained a "feel" for the pn junction in equilibrium and under reverse bias, which will be valuable in analyzing the electrostatics of the pn junction. Our approach will be to investigate the functional form of the charge density in the transition region and develop an expedient approximation that will simplify the mathematics. We will then solve for the electric field and potential in the transition region and connect these solutions to the known potential in the regions far away from the junction.

3.4.1 The Depletion Approximation

With confidence that we have a correct picture of the electrostatics in Fig. 3.8, we now make a very important approximation that will greatly simplify the mathematics.

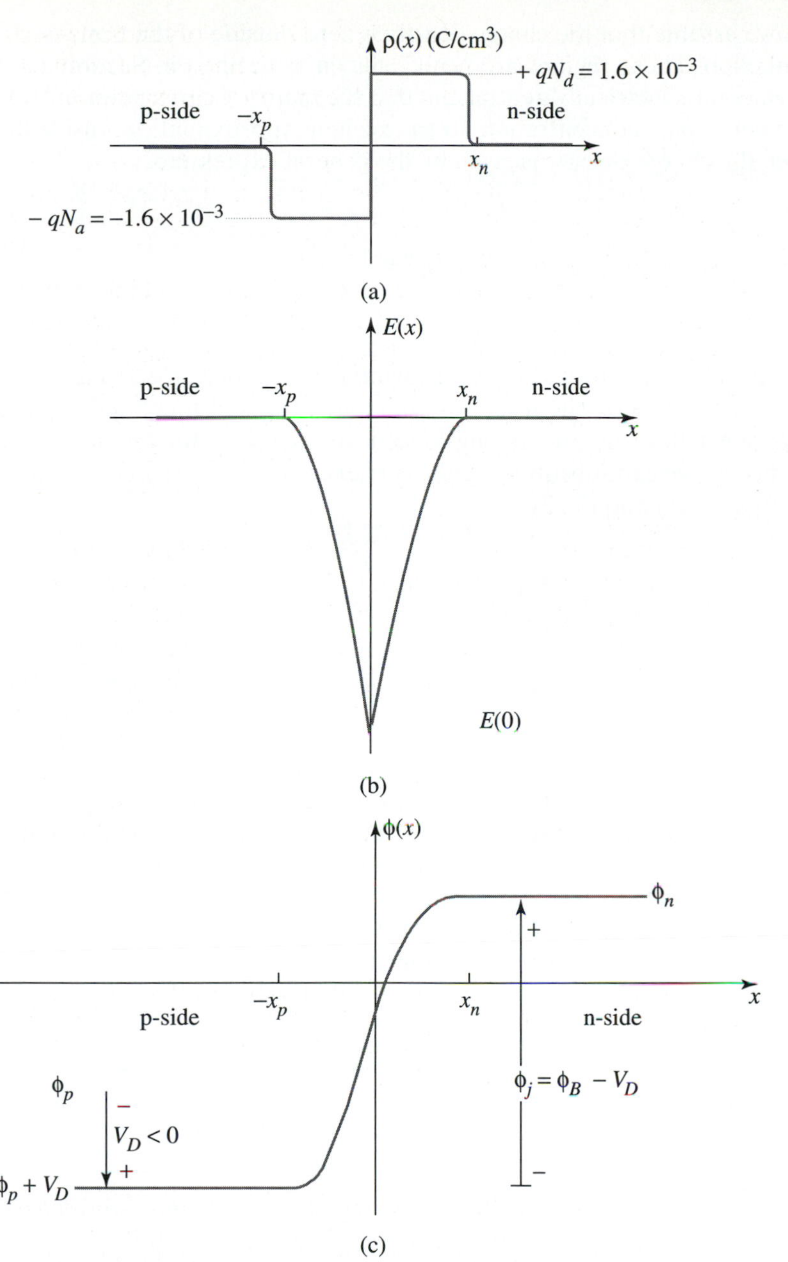

➤ Figure 3.10 Qualitative electrostatics of the pn junction under reverse bias $V_D < 0$ which is sufficient to double the width of the depletion region: (a) charge density, (b) electric field, and (c) electrostatic potential. The doping is $N_a = N_d = 10^{16}$ cm^{-3}, as in Fig. 3.9.

First, we assume that the charge density is zero outside of the transition region near the junction. These regions are "bulk" silicon as defined in Section 2.3. Recall that bulk silicon is electrically neutral and that the majority carrier concentration is equal to the net doping concentration, to an excellent approximation. Inside the transition region, the charge density is given by the general expression

$$\rho_o(x) = q\,(p_o - n_o + N_d - N_a) \cong \begin{cases} q\,(p_o - N_a) & (-x_{po} \le x \le 0) \\ q\,(N_d - n_o) & (0 \le x \le x_{no}) \end{cases} \tag{3.39}$$

where the approximations involve neglecting the minority electrons on the p-side and the minority holes on the n side—both many orders of magnitude less the majority carrier concentrations and doping concentrations. Since the pn junction is in thermal equilibrium, we can substitute for the carrier concentrations in terms of the potential from Eqs. (3.21) and (3.22)

$$\rho_o(x) \cong \begin{cases} q\left(n_i e^{-\phi_o(x)/V_{th}} - N_a\right) & (-x_{po} \le x \le 0) \\ q\left(N_d - n_i e^{\phi_o(x)/V_{th}}\right) & (0 \le x \le x_{no}) \end{cases} \tag{3.40}$$

The exponential functions of the potential present a daunting challenge in solving for $\phi_o(x)$, which leads us to explore whether there is a way to make a further simplification of Eq. (3.40). On the p-side, the hole concentration falls off exponentially with distance for $x > -x_{po}$ as shown qualitatively in Fig. 3.7(a). On the n-side, the electron concentration is similarly negligible for points well inside the transition region. For points very near the edges of the transition region, the majority carrier concentrations are close to the doping concentrations and cannot be neglected. We propose a simplifying approximation to Eq. (3.40) that will be referred to as the **depletion approximation**.

depletion approximation

$$\rho_o(x) \approx \begin{cases} -qN_a & (-x_{po} \le x \le 0) \\ qN_d & (0 \le x \le x_{no}) \end{cases} \quad \text{and} \quad \rho_o(x) = \begin{cases} 0 & (x \le -x_{po}) \\ 0 & (x_{no} \le x) \end{cases} \tag{3.41}$$

In effect, we consider the transition region to be *completely depleted* of mobile holes and electrons. The depletion approximation is slightly inaccurate near the edges of the depletion region since it overestimates the charge density there by neglecting the majority carriers. In Fig. 3.8(a), the charge density $\rho_o(x)$ is shown to taper off at the edges of the junction, which is what Eq. (3.40) predicts and what actually occurs. The depletion approximation in Eq. (3.41) gives a reasonably accurate solution for the depletion region width.

What is the motivation for making this approximation? The answer is simple–the mathematics of solving Poisson's Equation using Eq. (3.40) in the depletion region is too messy. By making the depletion approximation and dividing the pn junction into a depletion region, sandwiched between bulk p and n regions, we can develop relatively simple solutions for the potential, field, and depletion region width. These solutions will be useful for gaining an appreciation of the essential physics of pn junctions.

Several device simulators are available for numerically solving the two- or three-dimensional versions of Poisson's Equation in semiconductors should we desire to study a particular device structure with more accuracy.

3.4.2 Quantitative Analysis Using the Depletion Approximation

In Eq. (3.41), we have an approximate expression for the charge density $\rho_o(x)$ in thermal equilibrium. We now integrate the charge density twice to find first the electric field $E_o\,(x)$ and then the electrostatic potential $\phi_o(x)$. Gauss's Law relates the charge density to the derivative of the electric field

$$\frac{dE_o}{dx} = \frac{\rho_o(x)}{\varepsilon_s} \tag{3.42}$$

where $\varepsilon_s = 11.7\,\varepsilon_o$ is the electric permittivity of silicon. On the p-side of the depletion region $-x_{po} < x < 0$, we can integrate Eq. (3.42) and find that

$$E_o(x) = \int_{-x_{po}}^{x} \frac{\rho_o(x')}{\varepsilon_s} dx' + E_o(x = -x_{po}) = \int_{-x_{po}}^{x} \frac{-qN_a}{\varepsilon_s} dx' + 0 = -\frac{qN_a}{\varepsilon_s}(x - (-x_{po})) \tag{3.43}$$

where we have used the fact that the electric field is zero at the edge of the depletion region in thermal equilibrium since there is no charge density in the bulk p region. We can integrate the field and find an expression for the electrostatic potential on the p-side, $-x_{po} < x < 0$

$$\phi_o(x) = \int_{-x_{po}}^{x} -E_o(x')\,dx' + \phi_o(x = -x_{po}) = \frac{qN_a}{2\varepsilon_s}(x + x_{po})^2 + \phi_p \tag{3.44}$$

where ϕ_p is the potential of the p-type bulk p region in thermal equilibrium. According to Eq. (3.22), this potential is

$$\phi_p = -V_{th} \ln\left(\frac{N_a}{n_i}\right) \tag{3.45}$$

Using exactly the same approach, we can integrate the charge density in Eq. (3.41) starting from the n-side of the depletion region $0 < x < x_{no}$

$$E_o\,(x = x_{no}) = E_o\,(x) + \int_{x}^{x_{no}} \frac{\rho_o(x')}{\varepsilon_s} dx' = E_o\,(x) + \int_{x}^{x_{no}} \frac{qN_d}{\varepsilon_s} dx' = E_o\,(x) + \frac{qN_d}{\varepsilon_s}(x_{no} - x) \tag{3.46}$$

in which we have been careful to keep the limits of integration consistent with the positive x direction. Since the electric field $E_o(x = x_{no}) = 0$ at the edge of the depletion region, we can solve Eq. (3.46) for $E_o(x)$ on the n-side of the depletion region $0 < x < x_{no}$

$$E_o(x) = -\frac{qN_d}{\varepsilon_s}(x_{no} - x) \tag{3.47}$$

The potential for $0 \le x \le x_n$ can be found by direct integration of the electric field

$$\phi_o(x = x_{no}) = \phi_o(x) + \int_x^{x_{no}} -E(x')dx' = \phi_o(x) + \int_x^{x_{no}} \frac{qN_d}{\varepsilon_s}(x_{no} - x')\,dx' \tag{3.48}$$

Solving for $\phi_o(x)$ and evaluating the integral, we find

$$\phi_o(x) = \phi_n - \frac{qN_d}{2\varepsilon_s}(x_{no} - x)^2 \qquad (0 \le x \le x_n) \tag{3.49}$$

where ϕ_n is the potential of the bulk n region from Eq. (3.20)

$$\phi_n = V_{th} \ln\left(\frac{N_d}{n_i}\right) \tag{3.50}$$

The two solutions for the electric field must connect at $x = 0$, the metallurgical junction, since there is no sheet of charge at this surface (see the discussion on boundary conditions for the electric field in Section 3.1). By substituting $x = 0$ into Eqs. (3.43) and (3.47), we find that

$$-\frac{qN_a}{\varepsilon_s}(0 + x_{po}) = -\frac{qN_d}{\varepsilon_s}(x_{no} - 0) \tag{3.51}$$

Simplifying, we find

$$N_a x_{po} = N_d x_{no} \tag{3.52}$$

This result states that the negative charge (per unit area) on the p-side of the depletion region $-qN_a x_{po}$, is equal in magnitude and opposite in sign to the positive charge on the n-side: $+qN_d x_{no}$. Equation (3.52) is hardly surprising since we are safe in assuming that the total charge in IC device structures is zero, as discussed in Section 3.1.

In order to find expressions for the edges of the depletion region $-x_{po}$ and x_{no}, we first use Eq. (3.52) to substitute for one of them into the two solutions for $\phi_o(x)$, Eqs. (3.44) and (3.49). We then require that they match at the metallurgical junction $x = 0$

$$\frac{qN_a}{2\varepsilon_s}(0 + x_{po})^2 + \phi_p = \phi_n - \frac{qN_d}{2\varepsilon_s}(x_{no} - 0)^2 = \phi_n - \frac{qN_d}{2\varepsilon_s}\left(\frac{N_a}{N_d}x_{po}\right)^2 \tag{3.53}$$

Solving for x_{po} and substituting the built-in potential $\phi_B = \phi_n - \phi_p$, we find that

$$x_{po} = \sqrt{\left(\frac{2\varepsilon_s \phi_B}{qN_a}\right)\left(\frac{N_d}{N_d + N_a}\right)} \tag{3.54}$$

The width on the n-side can now be found directly from Eq. (3.52)

$$x_{no} = \sqrt{\left(\frac{2\varepsilon_s \phi_B}{qN_d}\right)\left(\frac{N_a}{N_d + N_a}\right)} \tag{3.55}$$

The sum of Eqs. (3.54) and (3.55) yields X_{do}, which is the width of the depletion region in thermal equilibrium

$$\boxed{X_{do} = x_{no} + x_{po} = \sqrt{\left(\frac{2\varepsilon_s \phi_B}{q}\right)\left(\frac{1}{N_a} + \frac{1}{N_d}\right)}} \tag{3.56}$$

equilibrium depletion width

Figure 3.11 summarizes the electrostatics of the pn junction in thermal equilibrium using the depletion approximation.

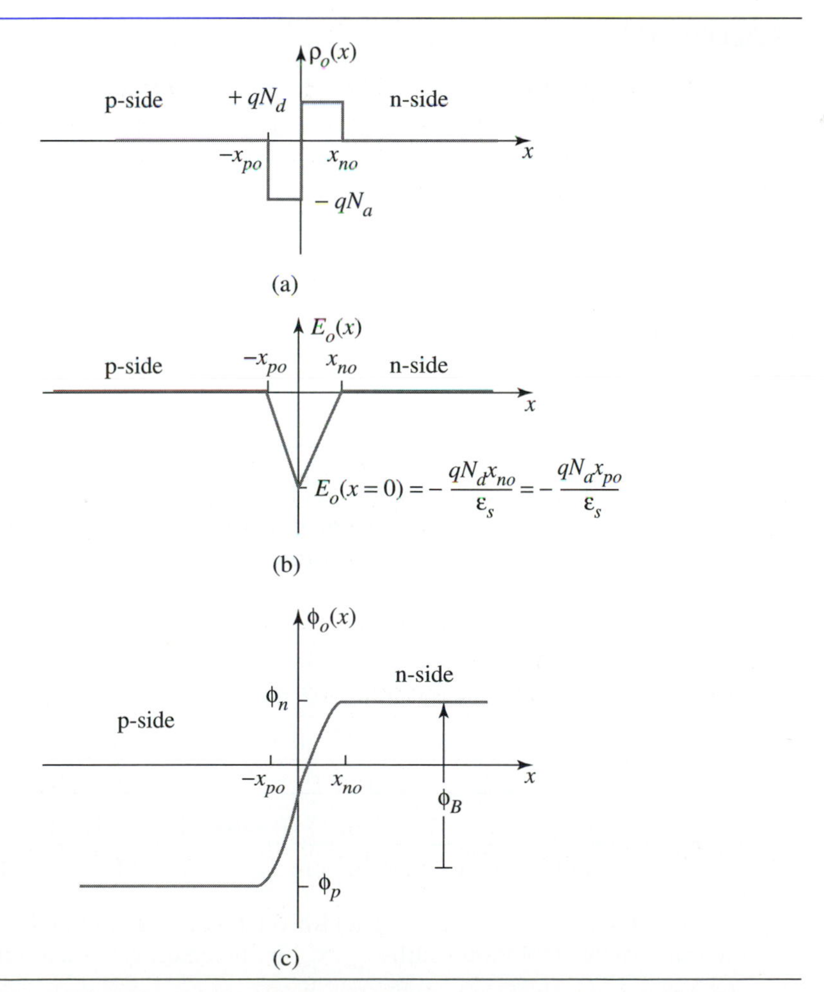

➤ **Figure 3.11** (a) Charge density, (b) electric field, and (c) potential in thermal equilibrium based on the depletion approximation.

EXAMPLE 3.4 Depletion Regions in Thermal Equilibrium

In this example, we will investigate the electrostatics of three pn junctions with progressively more asymmetrical doping. Our goal is to plot $\rho_o(x)$, $E_o(x)$, and $\phi_o(x)$ for junctions 1–3 in thermal equilibrium. The p-side is located on the negative side of the x axis as in Fig. 3.5.

Junction	N_a [cm^{-3}]	N_d [cm^{-3}]
1	5×10^{16}	5×10^{16}
2	5×10^{16}	2×10^{17}
3	5×10^{16}	5×10^{17}

SOLUTION

We start by applying the Eqs. (3.23 and 3.25) to find the potentials on the p and n sides and the built-in voltage.

Junction	ϕ_p [mV]	ϕ_n [mV]	ϕ_B [mV]
1	−402	402	804
2	−402	438	840
3	−402	462	864

Note that the built-in voltage is a weak function of the n-side doping concentration, increasing less than 10% between junctions #1 and #3 that have an order-of-magnitude difference in N_d. We substitute into Eqs. (3.54) and (3.55) to find the edges of the depletion region on the p-side and n-side. The results for junction 2 are

$$x_{po} = \sqrt{\left(\frac{(2)\,(11.7)\,(8.85 \times 10^{-14}\,\mathrm{Fcm^{-1}})\,(0.840\mathrm{V})}{(1.6 \times 10^{-19}\mathrm{C})\,(5 \times 10^{16}\mathrm{cm^{-3}})}\right)\left(\frac{2 \times 10^{17}}{2 \times 10^{17} + 5 \times 10^{16}}\right)} = 134\,\mathrm{nm}$$

$$x_{no} = \sqrt{\left(\frac{(2)\,(11.7)\,(8.85 \times 10^{-14}\,\mathrm{Fcm^{-1}})\,(0.840\mathrm{V})}{(1.6 \times 10^{-19}\mathrm{C})\,(2 \times 10^{17}\mathrm{cm^{-3}})}\right)\left(\frac{5 \times 10^{16}}{2 \times 10^{17} + 5 \times 10^{16}}\right)} = 33\,\mathrm{nm}$$

In the plot of the charge density $\rho(x)$ for the three cases in Fig. Ex3.4A, note that the ratios of the depletion widths x_{po}/x_{no} are inversely related to the doping ratios N_a/N_d.

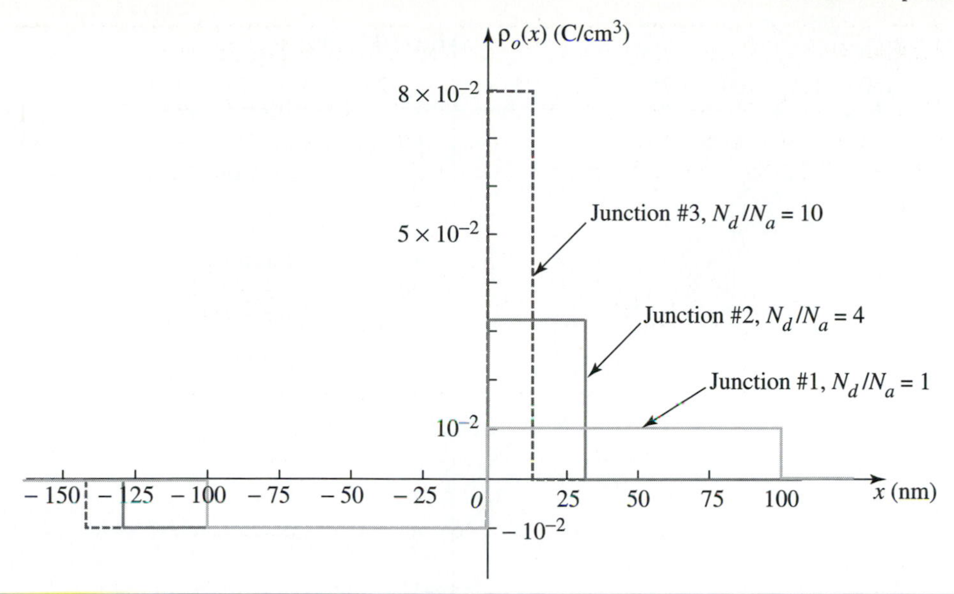

➤ **Figure Ex3.4A** Charge density in thermal equilibrium for the three junctions.

The numerical results for the depletion widths are summarized in this table.

Junction	x_{po} [nm]	x_{no} [nm]
1	102	102
2	134	33
3	143	14.3

To plot the electric field in equilibrium, we need only find its value at $x = 0$ from Eq. (3.47) since we know the shape of $E(x)$ from Fig. 3.11(b). For junction 2, the calculation is

$$E_o(0) = -\frac{qN_d x_{no}}{\varepsilon_s} = \frac{(1.6 \times 10^{-19}\text{C})\ (2 \times 10^{17}\text{cm}^{-3})\ (3.3 \times 10^{-6}\ \text{cm})}{(11.7)\ (8.85 \times 10^{-14}\text{Fcm}^{-1})} = -102\ \text{kV/cm}$$

The peak electric field weakly depends on the doping ratio of the junction when the concentration on one side is fixed. The electric field is plotted in Fig. Ex3.4B, with the numerical results for the peak electric field summarized in the following table.

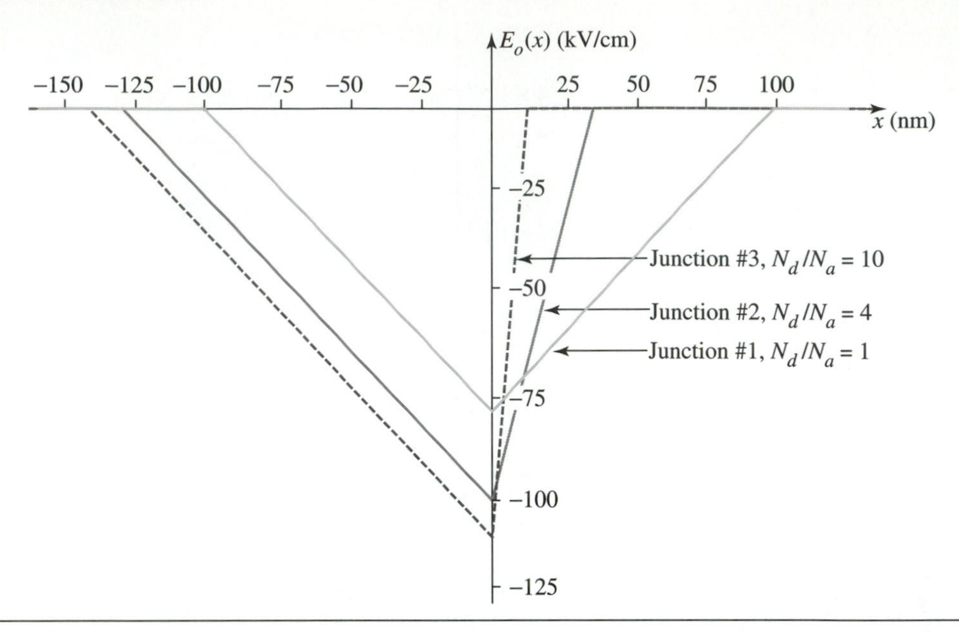

➤ **Figure Ex3.4B** Equilibrium electric field $E_o(x)$ for the three pn junctions in Example 3.4.

Junction	$E_o(0)$ [kV/cm]
1	−79
2	−102
3	−110.5

Poisson's Equation tells us that if the charge distribution is constant, then the potential will behave quadratically as shown in Eqs. (3.44) and (3.49). Evaluating Eq. (3.44) for junction 2, we find that the potential at $x = 0$ is

$$\phi_o(0) = \frac{(1.6 \times 10^{-19}\text{C})\,(5 \times 10^{16}\text{cm}^{-3})\,(1.34 \times 10^{-5}\text{cm})^2}{2\,(11.7)\,(8.85 \times 10^{-14}\text{F/cm})} - 402\text{mV} = 289\ \text{mV}$$

In summary, only a factor of 10 difference in doping is needed for the depletion region to be mostly located on the lightly doped side. Many IC pn junctions have higher doping ratios and can be considered "one-sided" for practical purposes.

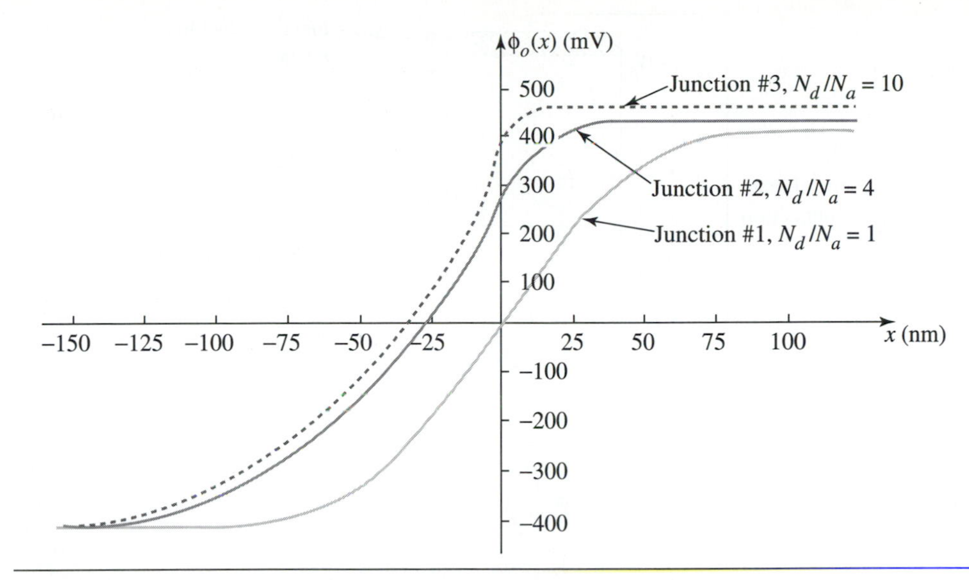

➤ **Figure Ex3.4C** Equilibrium potential $\phi_o(x)$ for the three pn junctions having different doping ratios.

3.4.3 Metal-Semiconductor Contact Potentials

The fact that a pn junction in thermal equilibrium has a built-in potential barrier of approximately 1 V is initially puzzling. For example, the sum of the potential drops around the short-circuited pn junction in Fig. 3.12(a) must be zero since there are no voltage or current sources in the loop and the diode is not illuminated. How then is the built-in potential barrier cancelled around the loop? The metal and bulk silicon regions have constant potentials in equilibrium. Inspection of Fig. 3.12(a) shows the only possibility is that built-in potential drops occur at the metal contacts to the p and n bulk regions. The extended thermal equilibrium potential plot in Fig. 3.12(b) represents these potential changes qualitatively as abrupt jumps at the metal-silicon contacts. We will leave the analysis of the potential variation at the contacts to more advanced courses. In this text, we will ignore that there is a non-zero width to these potential changes.

These contact potentials occur between any dissimilar pair of materials because of the difference in the potential energy of the conduction electrons across the junction. By properly engineering the metal-silicon contact, it is possible to approach an **ohmic contact**. The essential property of an ohmic contact is that it presents no barrier to hole or electron flow in either direction. These specially processed metal-silicon junctions behave like batteries, in that their potential drops are not functions of the forward or reverse current through them. The abrupt changes in potential ϕ_{pm} and ϕ_{mn} shown in Fig. 3.12(b) indicate qualitatively the narrow region of potential change that occurs in ohmic metal-silicon contacts. Note that the metal is *not* at zero potential in Fig. 3.12(b) since the potential reference for the plot is the *internal* silicon reference. The point where $\phi(x) = 0$ occurs when the electron and hole concentrations equal the intrinsic concentration, which is located inside the depletion region on the n-side of the metallurgical junction, from Fig. 3.12(b).

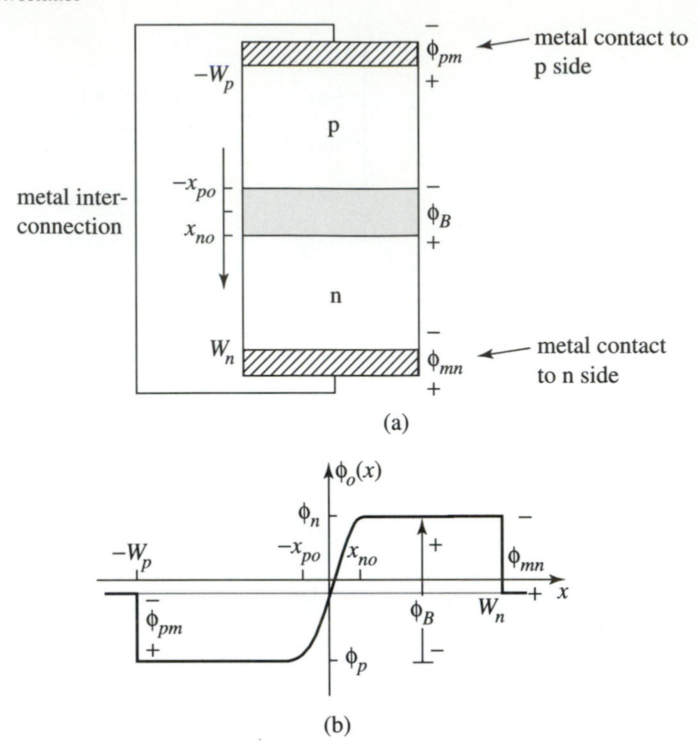

➤ **Figure 3.12** (a) pn diode including ohmic metal-silicon contacts, (b) extended plot of potential in thermal equilibrium including the metal interconnections.

By summing the potential drops around the loop in Fig. 3.12(a), Kirchhoff's voltage law (KVL) around the loop yields

$$0 = \phi_{pm} + \phi_B + \phi_{mn} \tag{3.57}$$

Solving for the sum of the contact potentials, we find that

$$\phi_B = -(\phi_{pm} + \phi_{mn}) \tag{3.58}$$

At first glance, it seems strange that the potential drops at the contacts would have to add up to cancel the built-in potential of the junction. Were this not the case, KVL would imply that current would flow in thermal equilibrium when a resistor is connected across the metal contacts to the p and n regions. This conclusion violates our basic understanding of the meaning of equilibrium; namely, that the net current must be zero for this state.

➤ 3.5 THE pn JUNCTION UNDER REVERSE BIAS

Given our quantitative model for the pn junction in thermal equilibrium, we are in position to analyze how the electrostatics are changed by application of a reverse bias voltage. In Fig. 3.13, we have connected a battery $V_D < 0$ V across the pn junction.

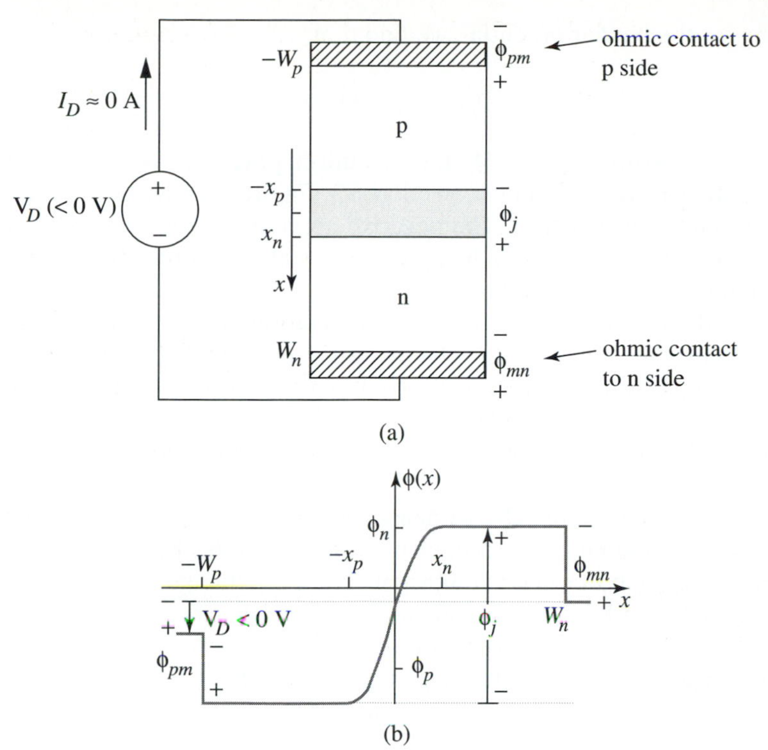

➤ **Figure 3.13** (a) pn junction with negative applied voltage V_D (reverse bias) and (b) plot of the potential across the structure under reverse bias including the ohmic contacts. The "o" subscript has been removed from all variables since the structure is no longer in thermal equilibrium.

Since V_D is negative, it lowers the potential of the p-side of the junction with respect to that of the n-side, which we will consider fixed at ϕ_n. (Alternatively, we could consider the p-side fixed and then $V_D < 0$ V would raise the n-side potential. Recall from Section 3.1 that only potential differences matter.) The effect of the reverse bias is to increase the potential barrier ϕ_j between the p and n bulk regions as shown in Fig. 3.13(b).

Since we are no longer in equilibrium, we must consider whether or not there is a significant current I_D through the pn junction. Raising the potential barrier between the p-side and n-side of the junction is *not* likely to result in a large net carrier transport. In fact, reverse-biased diodes are considered open circuits to a first approximation. We will use this fact and assume $I_D \approx 0$ A for the reverse-biased diode in this chapter. In Chapter 6 we will verify this assumption in a detailed study of the diode under forward and reverse bias.

The potential barrier ϕ_j across the depletion region under reverse bias can be found by applying Kirchhoff's voltage law. Note that the metal-silicon contact potentials at the ohmic contacts are unaffected by the applied voltage. The sum of the potential drops is given by

$$-V_D - \phi_{pm} - \phi_j - \phi_{mn} = 0 \tag{3.59}$$

Solving for the barrier potential, we find that

$$\phi_j = (-\phi_{pm} - \phi_{mn}) - V_D = \phi_B - V_D \tag{3.60}$$

where we have substituted Eq. (3.58) for the built-in potential. We asserted this result for ϕ_j in our first-pass treatment in Section 3.3. Figure 3.13(b) shows the potential variation through the structure. The negative applied voltage increases the barrier height and separates the potential levels of the metal interconnections, which is consistent with our concept of a voltage source.

Now that the barrier height is known as a function of the applied voltage V_D, we can proceed with the analysis of the electrostatics under reverse bias. After applying the depletion approximation, we must solve for the electric field and the potential in the depletion region, assuming it is confined to the interval $-x_p(V_D) < x < x_n(V_D)$. The boundaries of the depletion region are functions of the applied bias since they will change with the barrier height ϕ_j.

Rather than integrate the charge density $\rho(x)$ over the depletion region, we can avoid needless mathematics by recognizing that the form of $\rho(x)$ is *identical* to the charge density $\rho_o(x)$ in thermal equilibrium. The only change is that the boundaries are now $-x_p$ and x_n and the barrier height is ϕ_j. Therefore, the previous solutions for $\phi_o(x)$ and $E_o(x)$ can be adapted to the reverse bias case by simply substituting $\phi_j = \phi_B - V_D$ for ϕ_B. The edges of the depletion region under reverse bias ($V_D < 0$) are given in Eqs. (3.61) and (3.62), and their sum yields the total width in Eq. (3.63)

depletion widths under reverse bias

$$x_p(V_D) = \sqrt{\left(\frac{2\varepsilon_s(\phi_B - V_D)}{qN_a}\right)\left(\frac{N_d}{N_d + N_a}\right)} = x_{po}\sqrt{1 - (V_D/\phi_B)} \tag{3.61}$$

$$x_n(V_D) = \sqrt{\left(\frac{2\varepsilon_s(\phi_B - V_D)}{qN_d}\right)\left(\frac{N_a}{N_d + N_a}\right)} = x_{no}\sqrt{1 - (V_D/\phi_B)} \tag{3.62}$$

$$X_d(V_D) = \sqrt{\left(\frac{2\varepsilon_s(\phi_B - V_D)}{q}\right)\left(\frac{1}{N_a} + \frac{1}{N_d}\right)} = X_{do}\sqrt{1 - (V_D/\phi_B)} \tag{3.63}$$

where the thermal equilibrium values x_{po}, x_{no}, and X_{do} are defined in Eqs. (3.54)–(3.56). Figure 3.14 shows the effects on the depletion width, electric field, and potential of applying voltages of $V_D = -2.4$ V and $V_D = -6.4$ V across a pn junction with a thermal equilibrium barrier height of $\phi_B = 0.8$ V. These values of reverse bias will double and triple the widths of the depletion region over the thermal equilibrium width since

$$\sqrt{1 - (V_D/\phi_B)} = \sqrt{1 - ((-2.4)/0.8)} = \sqrt{4} = 2 \text{ and} \tag{3.64}$$

$$\sqrt{1 - (V_D/\phi_B)} = \sqrt{1 - ((-6.4)/0.8)} = \sqrt{9} = 3 \tag{3.65}$$

We conclude this section with a numerical example.

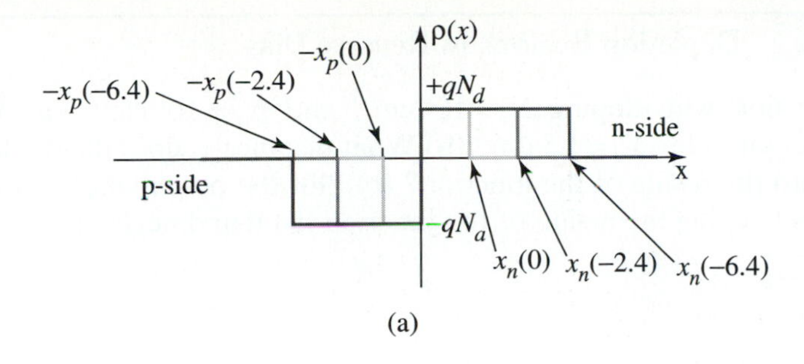

(a)

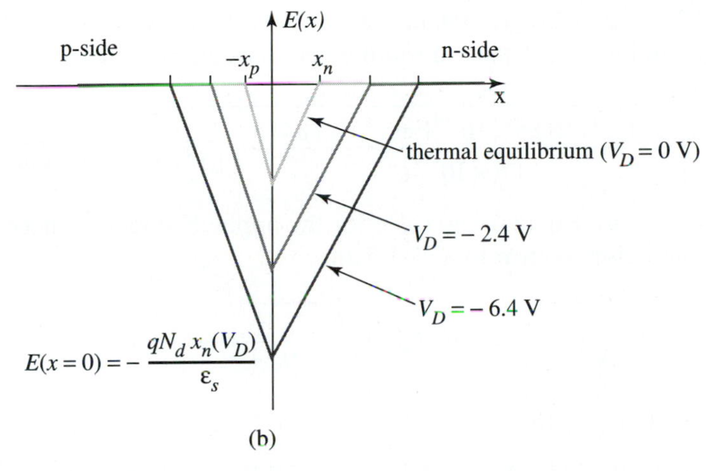

(b)

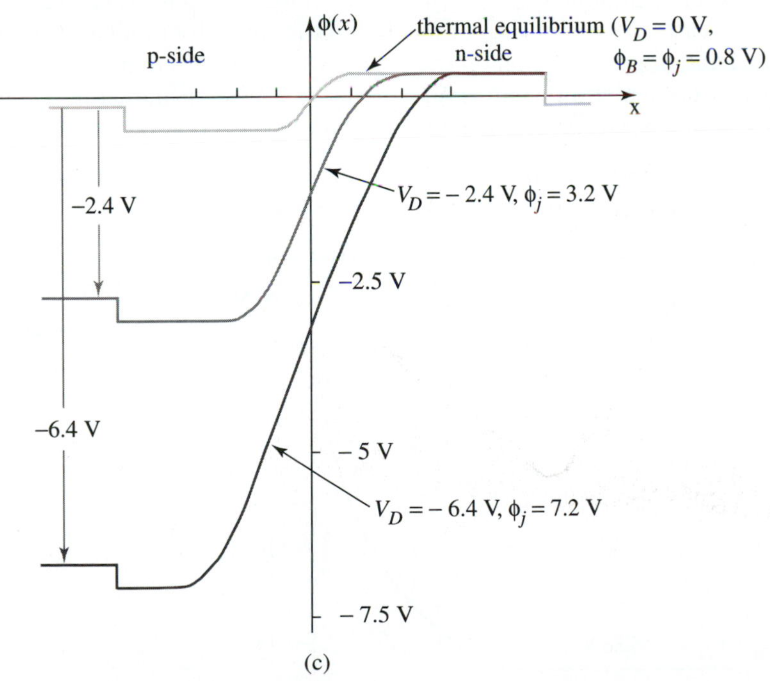

(c)

➤ **Figure 3.14** pn junction under reverse biases $V_D = -2.4$ V and $V_D = -6.4$ V. (a) charge, (b) electric field, and (c) potential, according to the depletion approximation.

EXAMPLE 3.5 Depletion Regions in Reverse Bias

Consider a pn junction with doping $N_a = 10^{16}\text{cm}^{-3}$ and $N_d = 10^{18}\text{cm}^{-3}$. **(a)** What applied voltage is required for $X_d = 1.5\ \mu\text{m}$? **(b)** What distance x_n does the depletion region penetrate into the n-side of the junction? **(c)** Plot the potential $\phi(x)$ for this applied reverse bias keeping the n-side of the junction fixed and neglecting contact potentials.

SOLUTIONS

(a) In thermal equilibrium, the potential in the bulk regions can be found using Eqs. (3.23) and (3.25): $\phi_n = 480$ mV, $\phi_p = -360$ mV, and $\phi_B = \phi_n - \phi_p = 840$ mV. The thermal equilibrium depletion width is found using Eq. (3.56)

$$X_{do} = \sqrt{\frac{2\,(11.7)\ (8.85\times10^{-14}\text{F cm}^{-1})\ (0.84V)}{1.6\times10^{-19}C}\left(\frac{1}{10^{18}\text{cm}^{-3}}+\frac{1}{10^{16}\text{cm}^{-3}}\right)} = 0.33\mu\text{m}$$

We can now use Eq. (3.63) to solve for the applied voltage V_D necessary for the depletion region to grow to $X_d = 1.5\ \mu\text{m}$

$$X_d(V_D) = X_{do}\sqrt{1-\frac{V_D}{\phi_B}} = (0.33\mu m)\sqrt{1-\frac{V_D}{0.84\text{V}}} = 1.5\,\mu\text{m}$$

which yields $V_D = -16.5$ V.

(b) In order to find the width on the n-side of the junction, we can substitute into Eq. (3.62) after calculating x_{no} or recall that the charge per unit area on the n-side is equal and opposite to that on the p-side, which implies that $x_n N_d = x_p N_a$. From (a), the total depletion width is $X_d = x_n + x_p = 1.5\ \mu\text{m}$. Eliminating x_p and noting that $N_d = 100\ N_a$, it follows that $x_n = x_p/100 \cong X_d/100 = 15$ nm.

(c) In plotting the potential across the junction, the n-side remains at $\phi_n = 0.48$ V while the p-side potential drops to $\phi_p + V_D = -16.86$ V.

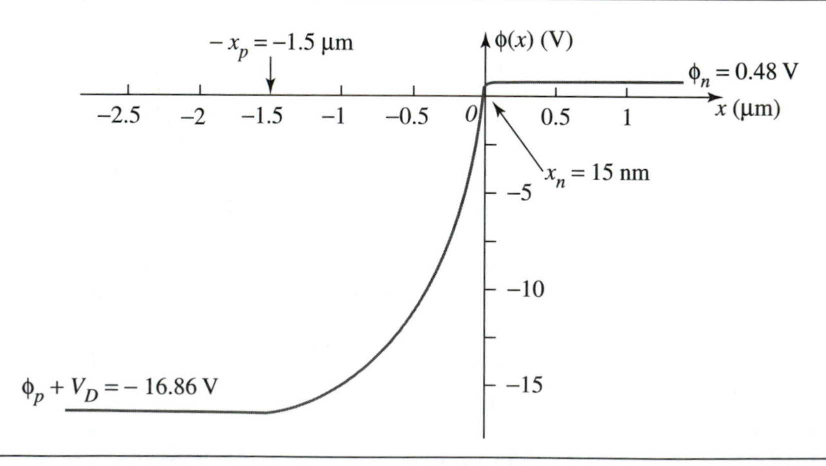

➤ Figure Ex3.5 Potential plot for Example 3.5 pn junction under applied bias of $V_D = -16.5$ V.

➤ 3.6 DEPLETION CAPACITANCE

The pn junction under reverse bias is a two-terminal charge-storage element. It is essential to connect the electrostatics of the reverse-biased pn junction with its circuit models. We will explain modeling of this device structure from several perspectives since the pn junction is universal in ICs and also, since this is our first exposure to circuit modeling.

As we have seen from Sections 3.3–3.5, the charge on the p-side of the depletion region is a function of the applied voltage. Repeating Eq. (3.38), the depletion charge varies with the applied voltage as

$$Q_J(V_D) = -qN_a x_{po}\sqrt{1-(V_D/\phi_B)} = -qN_a x_p(V_D),\ V_D \le 0, \tag{3.66}$$

with the right-hand side coming from Eq. (3.61). Unlike a conventional capacitor, the depletion charge does not change linearly with applied voltage.

To proceed further, we must become more precise in defining symbols for electrical quantities. The convention in this text is that an upper-case symbol with an upper-case subscript (e.g., Q_J or V_D) will refer to a DC or average quantity. A lower-case symbol with an upper-case subscript (e.g., q_J or v_D) will indicate the total quantity, including its time variation.

We have not considered the possibility that the applied voltage or the depletion charge may be functions of time in our analysis thus far. We will model the time-varying depletion charge by simply substituting q_J and v_D into Eq. (3.66). This "quasi-static" approach is an excellent approximation over the range of frequencies for microelectronics. Our starting point for developing circuit models for the reverse-biased pn junction is

$$q_J(v_D) = -qN_a x_{po}\sqrt{1-(v_D/\phi_B)},\ v_D \le 0 \tag{3.67}$$

Figure 3.15 is a plot of Eq. (3.67) for negative applied voltages.We have normalized q_J by its magnitude in thermal equilibrium $qN_a x_{po}$ and v_D by the built-in potential ϕ_B.

A numerical circuit simulator such as SPICE incorporates Eq. (3.67) into its model for a reverse-biased pn junction with an abrupt change in doping concentration at the metallurgical junction. For hand analysis, however, it is quite burdensome to have nonlinear models in the node or loop equations. Therefore, we will develop an approach to linearizing the charge-storage characteristics of the pn junction that will be valid under conditions that often occur in practical cases. We will derive an equivalent capacitor that describes the incremental charge storage for small deviations from a specified DC voltage. This technique, termed **small-signal modeling**, is a powerful approach to nonlinear circuit elements and will be used later for transistors.

small-signal modeling

In order to further develop this technique, we introduce the special case where the total depletion charge q_J can be written as the sum of its DC (average) value Q_J and an incremental (i.e., very small) component that is considered time-varying $q_j = q_j(t)$

$$q_J = Q_J + q_j,\ \text{where } |q_j| \ll |Q_J| \tag{3.68}$$

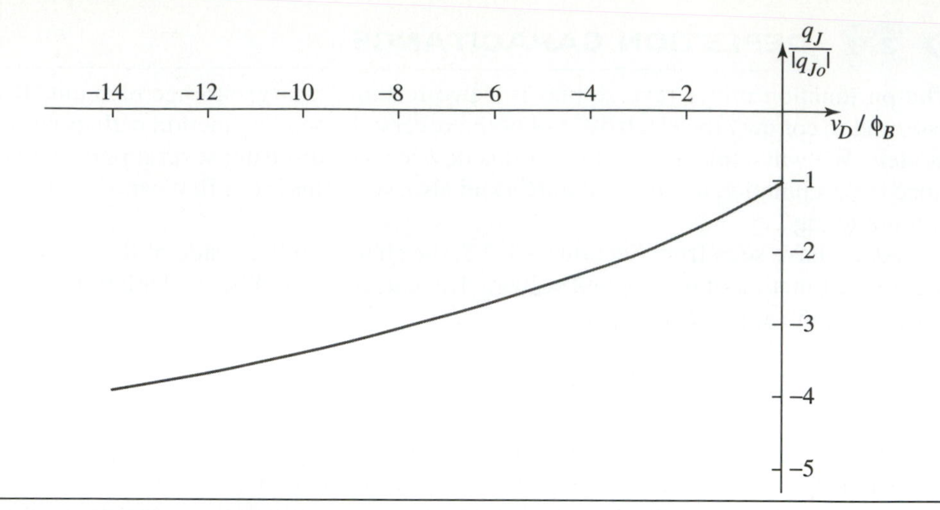

➤ **Figure 3.15** Depletion charge q_J, normalized to the magnitude of the equilibrium charge $|q_{Jo}| = qN_a x_{po}$, as a function of the applied voltage v_D, normalized to ϕ_B, the built-in potential barrier.

For this case, we can also express the applied voltage v_D as the sum of its DC value V_D, and a much smaller time-varying component $v_d = v_d(t)$

$$v_D = V_D + v_d \text{ where } |v_d| \ll |V_d| \tag{3.69}$$

The lower-case variables with lower-case subscripts in these equations are reserved for "small signals" which meet the test of being very small compared with the average value of the signal. It is helpful to interpret Eqs. (3.68) and (3.69) on a graph of the depletion charge before pursuing the mathematical analysis further.

3.6.1 Graphical Interpretation

Figure 3.16 shows the depletion charge as a function of the applied voltage with the function $q_J(v_D)$ evaluated at V_D and at $v_D = V_D + v_d$. For illustrative purposes, we have assumed that $v_d \ll |V_D|$ is a constant, positive voltage. From Eq. (3.68), the small-signal depletion charge q_j is given by

$$q_j = q_J - Q_J = q_J(v_D) - q_J(V_D) \tag{3.70}$$

in which we have substituted for the DC depletion charge $Q_J = q_J(V_D)$. In Fig. 3.16, the small-signal depletion charge is the product of the small-signal voltage and the slope of the function $q_J(v_D)$, evaluated at $v_D = V_D$

$$q_j = (slope) \times v_d \tag{3.71}$$

The slope has units of $(\text{C/cm}^2)/\text{V} = \text{F/cm}^2$ and is defined as the **depletion capacitance** $C_j(V_D)$. As shown in Fig. 3.16, the slope of $q_J(v_D)$ decreases with increasing reverse bias on the junction. We will consider the physical interpretation of this fact after developing an equation for the depletion capacitance.

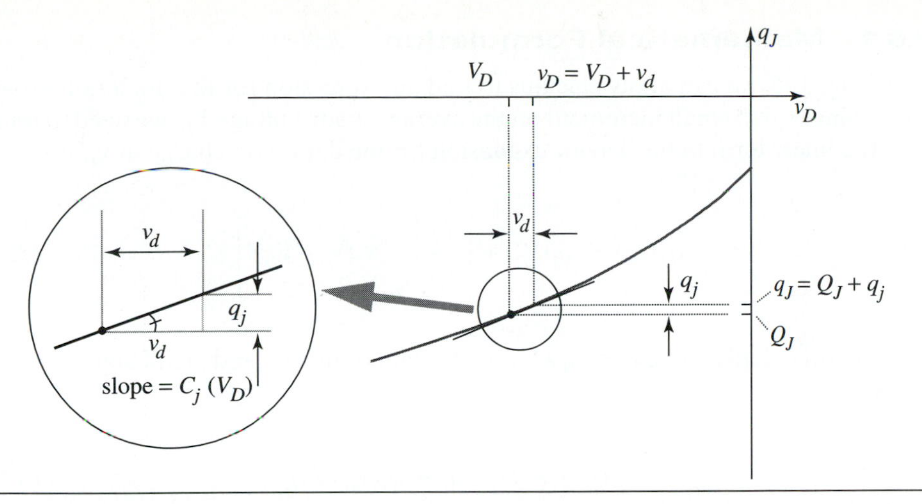

➤ **Figure 3.16** Graphical interpretation of the depletion capacitance as the slope of the depletion-charge vs. applied voltage curve evaluated at the point $q_J(V_D) = Q_j$.

The concept of small-signal modeling can be viewed from a circuit perspective as well. In Fig. 3.17(a), the total circuit quantities are broken into their DC and small-signal components. In many cases, we are concerned only with the small-signal behavior of the device structure. Therefore, it is convenient to work with the small-signal equivalent circuit for the reversed-biased pn junction shown in Fig. 3.17(b). Needless to say, this circuit describes only the response of the reverse-biased junction to small changes in the applied voltage. In order to find the numerical value for the capacitance from the slope of $q_J(v_D)$, we must first know the DC bias voltage V_D.

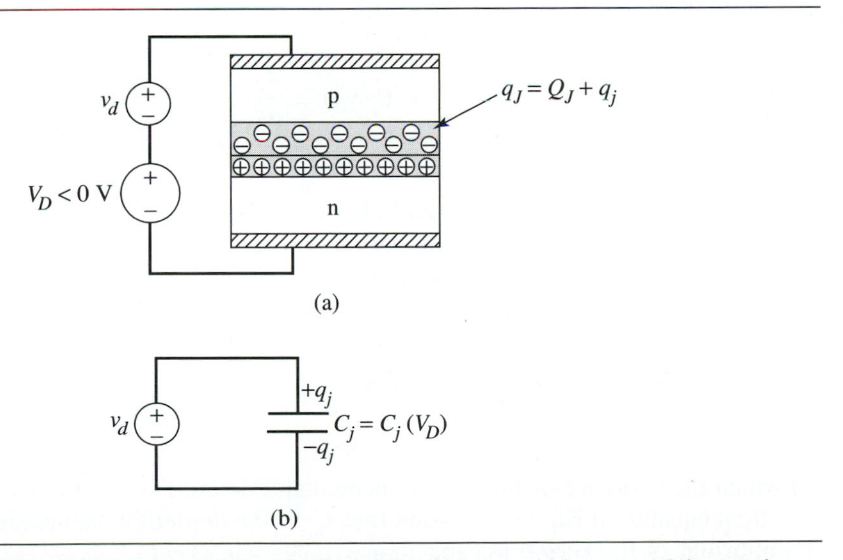

➤ **Figure 3.17** (a) Circuit model with the total voltage $v_D = V_D + v_D$ and the total charge q_J broken into their DC and small-signal components and (b) small-signal equivalent circuit for the reverse-biased pn junction.

3.6.2 Mathematical Formulation

From Fig. 3.16, we can apply calculus to find an expression for the depletion capacitance. Since v_d is a small increment on the average diode voltage V_D, we need to retain only the linear term in the Taylor expansion for the depletion charge at v_D

$$q_J(v_D) = q_J(V_D) + \left[\left.\frac{dq_J}{dv_D}\right|_{V_D} \times (v_D - V_D)\right] \tag{3.72}$$

The total depletion charge $q_J(V_D)$ is the sum of its DC and small-signal components

$$q_J(v_D) = Q_J + q_j = q_J(V_D) + q_j \tag{3.73}$$

Substitution of Eq. (3.73) and the small-signal voltage $v_d = v_D - V_D$ into the Taylor expansion results in cancellation of the DC depletion charge

$$q_J(V_D) + q_j = q_J(V_D) + \left.\frac{dq_J}{dv_D}\right|_{V_D} \times v_d \text{ and} \tag{3.74}$$

$$q_j = \left.\frac{dq_J}{dv_D}\right|_{V_D} \times v_d \tag{3.75}$$

We identify the derivative in Eq. (3.75) with the slope in Eq. (3.71) and therefore, the depletion capacitance is defined as

depletion capacitance

$$\boxed{C_j = C_j(V_D) = \left.\frac{dq_J}{dv_D}\right|_{V_D}} \tag{3.76}$$

in which the second equality is a reminder that C_j is a function of the DC bias voltage.

In order to evaluate the functional form of the depletion capacitance, we substitute the expression for the depletion charge as a function of the diode voltage from Eq. (3.67) into the definition

$$C_j = \frac{d}{dv_D}\left\{-qN_a x_{po}\sqrt{1 - v_D/\phi_B}\right\}\Big|_{V_D} = \frac{qN_a x_{po}}{2\phi_B\sqrt{1 - V_D/\phi_B}} = \frac{C_{jo}}{\sqrt{1 - V_D/\phi_B}} \tag{3.77}$$

in which the various constants are collected and defined as C_{jo}. Closer inspection of the final equality in Eq. (3.77) shows that C_{jo} is the depletion capacitance in thermal equilibrium, or the **zero-bias** capacitance, $C_j(V_D = 0\text{ V}) = C_{jo}$.

There should be a simple physical interpretation for C_{jo}, so we substitute from the definition of x_{po} in Eq. (3.54) to find

$$C_{jo} = \frac{qN_a x_{po}}{2\phi_B} = \frac{qN_a}{2\phi_B}\sqrt{\left(\frac{2\varepsilon_s\phi_B}{qN_a}\right)\left(\frac{N_d}{N_d+N_a}\right)} = \sqrt{\frac{q\varepsilon_s}{2\phi_B}\left(\frac{N_aN_d}{N_d+N_a}\right)} \quad (3.78)$$

The right-hand equality has many common terms with X_{do}, the depletion width in thermal equilibrium (Eq. (3.56)). Factoring out the permittivity of silicon, Eq. (3.78) simplifies to

$$C_{jo} = \varepsilon_s\sqrt{\frac{q}{2\varepsilon_s\phi_B}\left(\frac{1}{N_a}+\frac{1}{N_d}\right)^{-1}} = \frac{\varepsilon_s}{X_{do}} \quad (3.79)$$

zero-bias depletion capacitance

As a final expression for the depletion capacitance that will be useful in the next section, we substitute for C_{jo} in Eq. (3.77)

$$C_j(V_D) = \frac{C_{jo}}{\sqrt{1-V_D/\phi_B}} = \frac{\varepsilon_s}{X_{do}\sqrt{1-V_D/\phi_B}} = \frac{\varepsilon_s}{X_d(V_D)} \quad (3.80)$$

depletion capacitance for reverse bias

where we have substituted for the depletion width under reverse bias in Eq. (3.63).

The normalized depletion capacitance C_j/C_{jo} is plotted in Fig. 3.18. The capacitance decreases with increasing reverse bias, which is consistent with its interpretation as the slope of the depletion charge function plotted in Fig. 3.15. Since a typical equilibrium barrier height $\phi_B \approx 1$ V, the normalized voltage is roughly the same as the DC bias.

3.6.3 Physical Interpretation

Our final perspective on the depletion capacitance will be to develop some insight into the physical meaning of what appears to be a purely mathematical concept. An important clue is provided by Eq. (3.80), which is identical to the capacitance of a par-

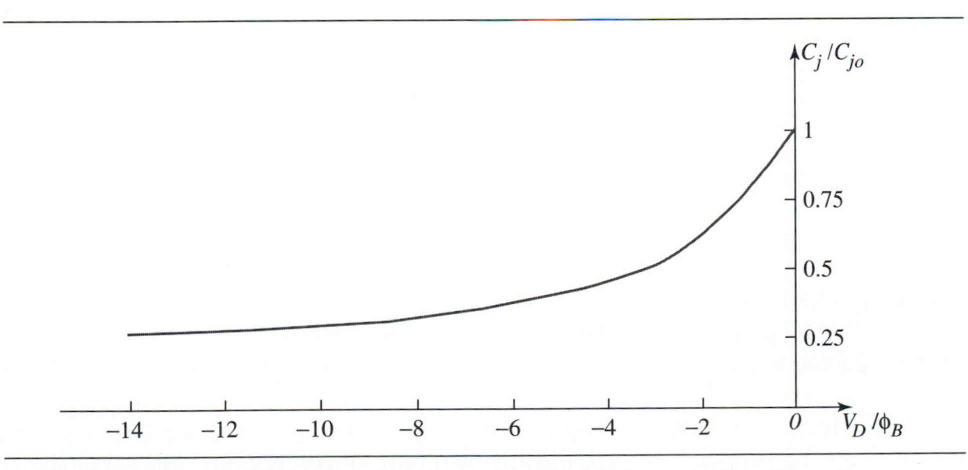

➤ **Figure 3.18** Depletion capacitance, normalized to its zero-bias (thermal equilibrium) value, as a function of the applied DC voltage, normalized to the barrier height ϕ_B in thermal equilibrium.

allel-plate capacitor with its electrodes separated by the depletion width $X_d(V_D)$ at the particular DC voltage. What charges are separated by X_d? From Eq. (3.76), the depletion capacitance is the ratio of the small-signal depletion charge q_j to the small-signal voltage v_d

$$C_j = \frac{q_j}{v_d} = \frac{\varepsilon_s}{X_d} \tag{3.81}$$

which indicates that the small-signal charge layers $\pm q_j$ are separated by X_d.

In Fig. 3.19, the charge density in the depletion region is plotted, first for the case where the diode voltage is $V_D < 0$ V in (a), and then where a positive small-signal voltage is added to give $v_D = V_D + v_d$ in (a). The depletion region narrows slightly in the second case since the positive v_d means that $v_D > V_D$. The depletion charge q_J for (b) is less negative than is Q_J in (a) since the depletion region is narrower for v_D.

We graphically subtract (a) from (b) to find the incremental charge density that accounts for the small-signal depletion charge q_j since by definition $q_j = q_J - Q_J$. The result is plotted in Fig. 3.19(c), in which the change in the charge density due to the addition of v_d is confined to the edges of the depletion region. In the limit of $v_d \rightarrow 0$, the small-signal charges $\pm q_j$ become sheets that are separated by a gap width of X_d, which explains the physical origin of the parallel-plate formula in Eq. (3.81).

Although we have derived Eq. (3.81) for the case of an abrupt pn junction with constant doping, it is in fact a general result for any pn junction. The functional dependence of the depletion width on the reverse bias will depend on the details of the particular dopant distributions and consequently, the dependence of C_j on the reverse bias may differ slightly from that in Fig. 3.18.

3.6.4 SPICE Models for IC pn Junctions

SPICE is a powerful circuit simulator that is commonly used to simulate integrated circuits. In SPICE, the depletion capacitance of a pn junction under reverse bias is modeled by the following equation

$$C_j = \frac{CJO}{(1 - V_D / VJ)^{M}} \tag{3.82}$$

in which the SPICE symbols are $CJO = C_{jo}$, the zero-bias (equilibrium) capacitance, $VJ = \phi_B$, the built-in barrier voltage, and M is the grading coefficient of the junction. From Eq. (3.80), $M = 0.5$ for the abrupt pn junction that we have modeled in this chapter. For IC pn junctions, M is usually curve-fit to measurements of $C_j\,(V_D)$ or is extracted from a numerical device simulator. Typical values for M range from 0.3 to 0.7.

The effect of changing the width of the depletion region is that a displacement current must be supplied through the adjacent bulk silicon regions. The small-signal junction capacitance incorporates this effect into the circuit model, which is illustrated by the following example. This example revisits the ion-implanted resistor and adds charge-storage at the pn junction to the small-signal model.

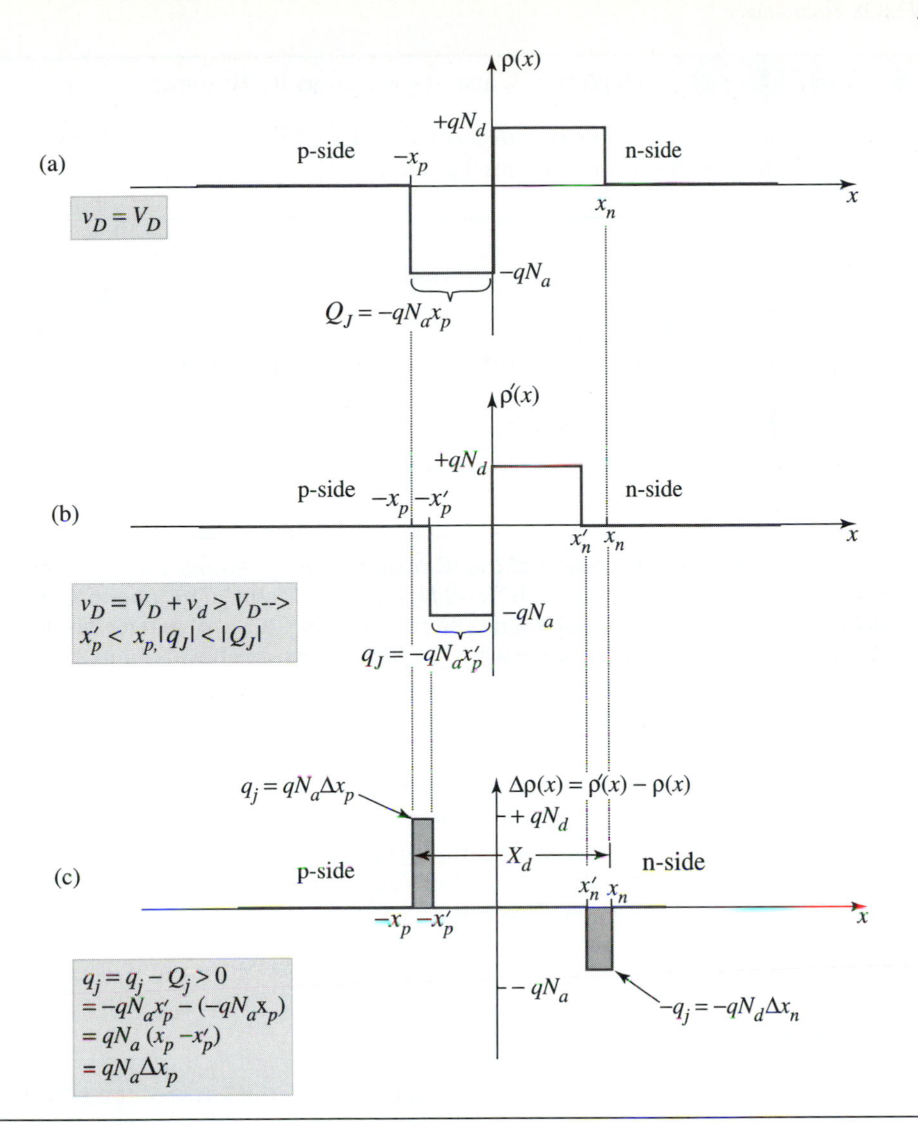

➤ **Figure 3.19** (a) Charge density $\rho(x)$ in depletion region for a reverse bias of $V_D < 0$, (b) charge density $\rho'(x)$ in depletion region for a perturbed reverse bias $v_D = V_D + v_d$ with $v_d > 0$, and (c) difference $\Delta\rho(x)$ between (b) and (a) showing incremental depletion charge $\pm q_j$ separated by approximately the depletion width X_d. Note that the magnitude of v_d is exaggerated in order to clarify its effect on the depletion region width.

EXAMPLE 3.6 Depletion Capacitance of an IC Resistor

(a) For the ion-implanted resistor described in Ex. 2.5 and shown in cross section below, find the depletion width x_{no} for $V_D = 0$ V.

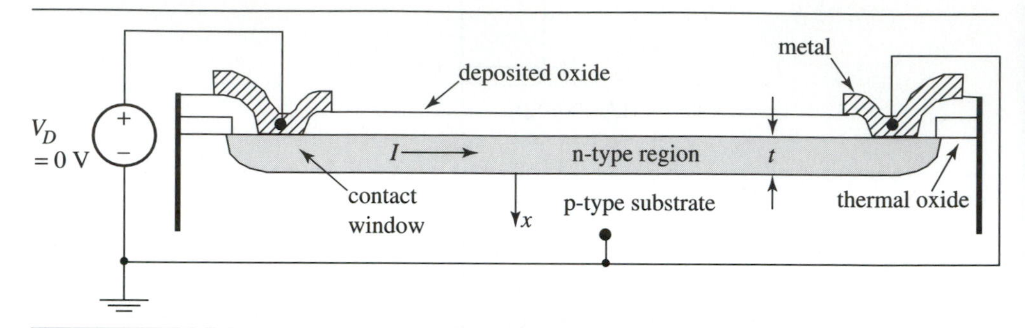

➤ **Figure Ex3.6A** Cross section of the IC resistor from Example 2.5. The doping concentrations are $N_d = 2.5 \times 10^{16}$ cm^{-3} and $N_a = 2.5 \times 10^{15}$ cm^{-3}. The junction depth is $t = 0.4$ μm = 400 nm, the width of the n-type region is $W = 2.5$ μm, and its length is $L = 20$ μm. The contact regions are each $3W \times 3W$ in area, as shown in Fig.2.22(a).

SOLUTION

Since N_a is a factor of 10 less than N_d, most of the depletion region will reside on the p-side of the junction. In order to use Eq. (3.55) to find the depletion edge in thermal equilibrium, we first find the potentials of each side under zero bias using Eqs. (3.23) and (3.25): $\phi_n = 384$ mV, $\phi_p = -324$ mV. The depletion edge on the n-side is then $x_{no} = 58$ nm from Eq. (3.55).

(b) Recalculate the sheet resistance by correcting the thickness of the n-type region for the depletion at the junction.

SOLUTION

There are no carriers in the depletion region, so the thickness of the n-type bulk region t' is

$$t' = t - x_{no} = 400\text{ nm} - 58\text{ nm} = 342\text{ nm}$$

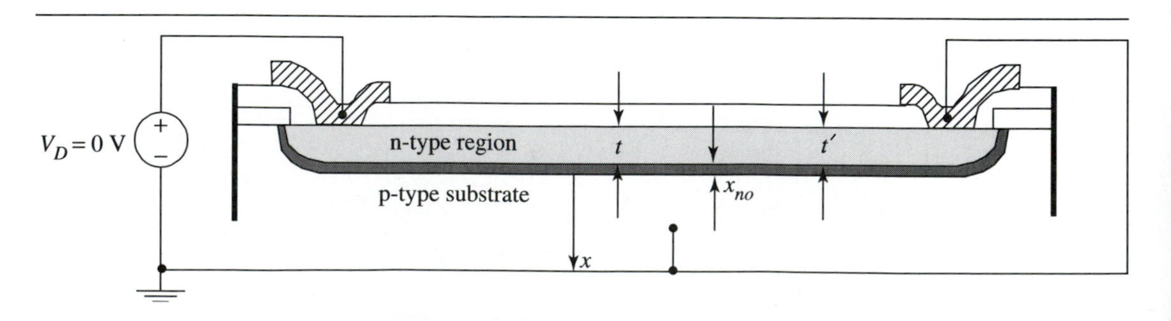

➤ **Figure Ex3.6B** IC resistor showing the junction depth t and the narrowing of the n-type region due to depletion width x_{no} on the n-side of the junction.

The sheet resistance increases due to the encroachment of the depletion region into the n-type region as shown in Fig. Ex3.6B.

$$R_{\square} = \frac{1}{qN_d\mu_n t'} = \frac{1}{qN_d\mu_n(t - x_{no})}$$

$$R_{\square} = \frac{1}{(1.6\times10^{-19}\,\mathrm{C})(1050\,\mathrm{cm^2V^{-1}s^{-1}})(2.5\times10^{16}\,\mathrm{cm^{-3}})(400\,\mathrm{nm} - 58\,\mathrm{nm})} = 6.9\mathrm{k}\Omega/\square$$

represents a 15% increase over the value in Ex. 2.5 where the depletion region was neglected. The resistance is $R = (6.9\ \mathrm{k}\Omega/\square)(9.3\square) = 64.2\ \mathrm{k}\Omega$.

(c) Calculate the depletion capacitance C_R in fF between the n region and the p-type substrate assuming the DC voltage $V_D = 0$ V.

SOLUTION

The area of the ion-implanted n-type region from the layout (including the two contact areas) in Example 2.5 is

$$A_R = L\,W + 2(3W)^2 = 162.5\ \mu\mathrm{m}^2 = 1.625\times10^{-6}\,\mathrm{cm},^2$$

where the area of each contact region $(3W)^2$ is included.

The depletion capacitance (per unit area) for $V_D = 0$ is

$$C_{jo} = \frac{\varepsilon_s}{X_{do}} = \frac{(11.7)\,(8.85\times10^{-14}\,\mathrm{F\,cm^{-1}})}{0.635\mu\mathrm{m}} = 16.3\times10^{-9}\ \mathrm{F/cm^2}$$

in which we used $X_{do} = x_{no} + x_{po} = x_{no} + (N_d/N_a)\,x_{no} = 11\,x_{no} = 635$ nm. Multiplying by the area A_R, the capacitance is

$$C_R = C_{jo}A_R = (16.3\times10^{-9}\mathrm{F/cm^2})\ (16.25\times10^{-6}\mathrm{cm^2}) = 26.5\ \mathrm{fF}$$

If a DC reverse bias is applied across the pn junction and the depletion region widens, then the resistance will increase, due to further reduction of the bulk n region thickness, and the capacitance will decrease, due to the wider depletion region.

(d) Draw a simple small-signal model of the ion-implanted resistor for the case where $V_D = 0$ V.

SOLUTION

The ion-implanted resistor with depletion capacitance is best modeled as a series of N segments with resistance R/N and capacitance C_R/N. For simplicity, we lump the resistance and capacitance into single elements in the circuit below in Fig. Ex3.6C.

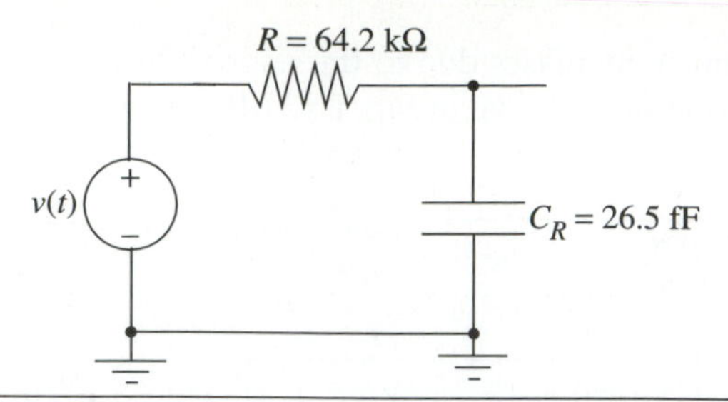

➤ Figure Ex3.6C Small-signal circuit for the ion-implanted resistor in Example 2.5 with zero DC bias between the n region and p-type substrate. The resistance includes the effect of encroachment by the depletion region. The capacitance is the depletion capacitance of the entire n region.

➤ 3.7 THE MOS CAPACITOR: A FIRST PASS

We now turn our attention to the more complex, but far more important metal-oxide-semiconductor (MOS) capacitor. Figure 3.20 shows a cross section of the three-layer sandwich in which a heavily doped, n-type polycrystalline silicon layer serves as the "metal" layer. In practice, polysilicon is the most common material used for the "metal" in integrated MOS structures, having supplanted aluminum and other metal films since the 1970s. The oxide film is a thermally grown layer of SiO_2. The silicon substrate could be either p-type or n-type—we will first concentrate on the former. Metal interconnections to the polysilicon gate and to the p-type bulk silicon are shown schematically in Fig. 3.20.

3.7.1 MOS Capacitor in Thermal Equilibrium

As we did with the pn junction, we will start our investigation of MOS electrostatics with the structure in thermal equilibrium. Since the oxide is a near-perfect insulator, we short together the gate and bulk metal layers to enable the charge exchange needed to reach equilibrium as indicated in Fig. 3.20. For the pn junction, $V_D = 0$ V also was a necessary condition for equilibrium.

The boundary condition on the potential in the silicon, far from the oxide, is given by Eq. (3.25)

$$\phi_p = -V_{th} \ln\left(\frac{N_a}{n_i}\right) \tag{3.83}$$

For a typical doping concentration of 10^{17} cm^{-3}, the potential of the p-type bulk region far from the surface is $\phi_p = -420$ mV. The n^+ designation on the polysilicon gate indicates that it is very heavily doped with donors, to the point that its potential can be taken as the maximum possible for silicon in thermal equilibrium

$$\phi_{n+} = 550\,\text{mV} \tag{3.84}$$

Based on our intuitive understandin\g of electrostatics, we can understand the qualitative electrostatics in the MOS structure in thermal equilibrium. The sign of the potential drop (gate positive, p-type silicon negative) means that the electric field E_o

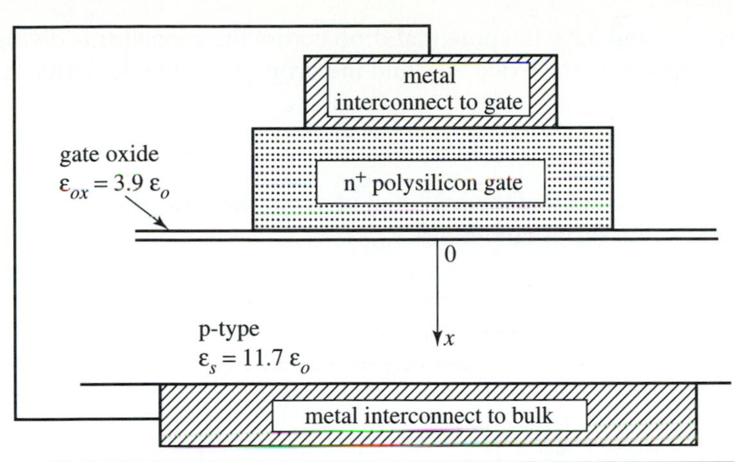

Figure 3.20 MOS capacitor with p-type substrate. Gate and bulk metal contacts are shorted together for thermal equilibrium.

is positive (points from gate to bulk, in the $+x$ direction as defined in Fig. 3.20). Therefore, a positive charge must be present on the bottom of the polysilicon gate and there must be a balancing negative charge in the p-type silicon underlying the gate oxide. The gate oxide itself will be considered a charge-free perfect insulator in this text. In more advanced courses, the effect of oxide charge on the MOS structure will be investigated in detail.

As we saw with the pn junction in Section 3.2, negative charge density in p-type silicon can occur if a depletion region forms under the gate oxide. Figure 3.21 shows a sketch of the charges in the MOS capacitor with p-type substrate, in the state of thermal equilibrium. Since the gate is highly conductive n^+ polysilicon, the gate charge Q_{Go} (units: C/cm^2) is depicted as a sheet charge located on its bottom surface. An intuitive picture of why the depletion region forms is to view the positive charge on the gate as repelling mobile holes in the silicon under the gate oxide, leaving behind the immobile, negatively charged acceptor ions to a depth X_{do}. The gate charge Q_{Go} is equal and opposite to the bulk charge Q_{Bo} in the silicon substrate

$$Q_{Go} = -Q_{Bo} = -qN_aX_{do} \tag{3.85}$$

Just as we found for the pn junction, there is charge stored on the MOS capacitor in thermal equilibrium. It is helpful to sketch the potential variation through the MOS structure in equilibrium. Example 3.3 analyzed the electrostatics of the MOS capacitor with the charge distribution in Fig. 3.21. Figure 3.22 is a sketch of the potential $\phi(x)$, which is based on the analysis in Ex. 3.3.

The potential drop across the MOS structure is the difference between the potentials of the n^+ polysilicon gate and the p-type substrate. For the plot in Fig. 3.22, we have used a substrate doping concentration of $N_a = 10^{17}\ cm^{-3}$ and the total drop is:

$$\phi_{n+} - \phi_p = 550\ mV - (-420\ mV) = 970\ mV \tag{3.86}$$

In Figs. 3.21 and 3.22, the potential drop across the capacitor is divided into the sum of the drop across the oxide $V_{ox,o}$ and the drop across the depletion region V_{Bo}

$$\phi_{n+} - \phi_p = V_{ox,o} + V_{Bo} \tag{3.87}$$

Finally, we have defined the potential at $x = 0$ (the SiO_2-silicon interface) as the thermal equilibrium **surface potential**, ϕ_{so}.

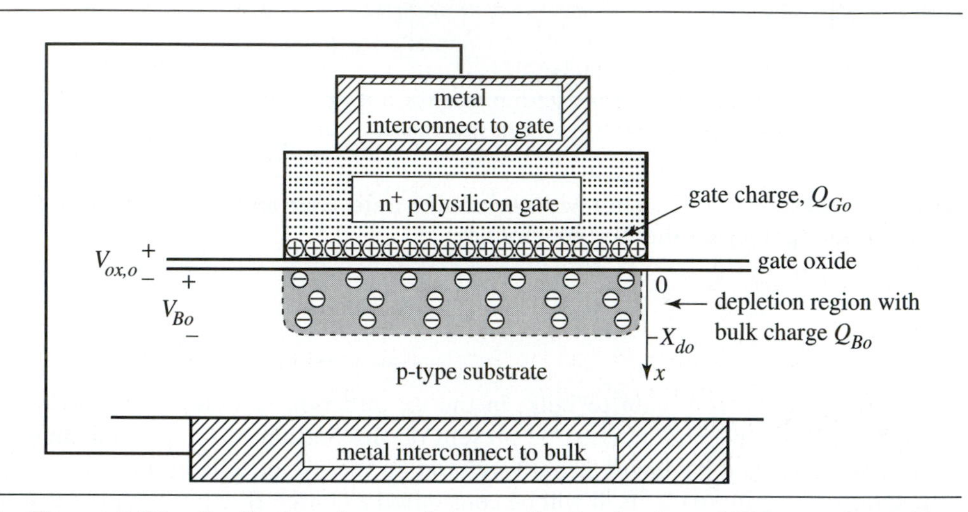

➤ **Figure 3.21** Qualitative picture of charge distribution in an MOS capacitor with p-type substrate in thermal equilibrium.

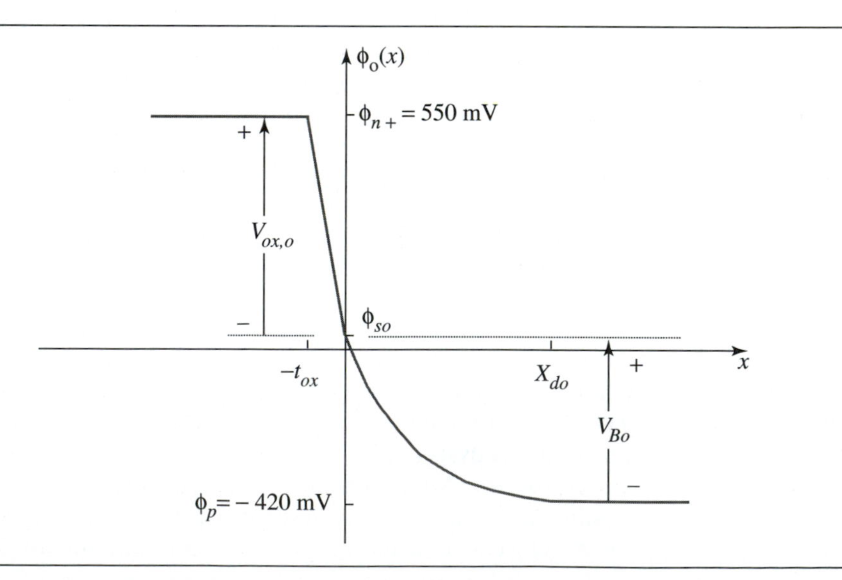

➤ **Figure 3.22** Potential plot for MOS capacitor with p-type substrate in thermal equilibrium. The substrate doping concentration is $N_a = 10^{17}$ cm^{-3}.

3.7.2 MOS Capacitor under Applied Bias

The effect on the gate charge of an applied voltage V_{GB} is rather complicated. The easiest place to start is by applying a voltage that is equal and opposite to the "built-in" voltage across the MOS capacitor in thermal equilibrium. We define this special voltage as the **flatband voltage**, V_{FB}:

flatband voltage for p-type substrate

$$\boxed{V_{FB} = -(\phi_{n+} - \phi_p)} \qquad (3.88)$$

For a substrate acceptor concentration of 10^{17} cm^{-3}, the flatband voltage is $V_{FB} = -970$ mV. In plotting the potential variation under applied bias, we will keep the substrate fixed and move the gate. With $V_{GB} = V_{FB} = -970$ mV, the n$^+$ polysilicon gate's potential is pulled down to 550 mV – 970 mV = –420 mV, as shown in Fig. 3.23 (a). Therefore, there is no internal potential drop across the MOS capacitor when $V_{GB} = V_{FB}$ and as a result, the gate charge is zero:

$$Q_G(V_{GB} = V_{FB}) = 0 \qquad (3.89)$$

If we make the applied voltage more negative than the flatband voltage, $V_{GB} < V_{FB}$, then the gate potential drops below that of the substrate. The gate charge therefore becomes negative, as shown in Fig. 3.23(b), and the nature of the charge in the substrate changes. The identity of the positive substrate charge is not hard to determine, since there are plenty of positively charged holes in the p-type silicon that can accumulate under the gate oxide, as shown in Fig. 3.23(b), to cancel the negative gate charge. The MOS capacitor is said to be in **accumulation** for $V_{GB} < V_{FB}$. The potential and charge density in accumulation are shown in Fig. 3.23(b). These plots are identical to those for the parallel-plate capacitor analyzed in Ex. 3.1. Therefore, the gate charge in accumulation is linearly related to the difference between the applied voltage and the flatband voltage:

gate charge in accumulation for p-type substrate

$$\boxed{Q_G = C_{ox}(V_{GB} - V_{FB}) \text{ for } V_{GB} \le V_{FB}} \qquad (3.90)$$

where $C_{ox} = \varepsilon_{ox}/t_{ox}$ is the capacitance per unit area of the gate oxide.

We now apply voltages $V_{GB} > V_{FB}$. Since $V_{GB} = 0 \text{ V} > V_{FB}$, we already have examined this case when we considered the MOS capacitor in thermal equilibrium. The gate charge is positive and increases with increasing applied bias

$$Q_G(V_{GB}) = -Q_B = -(-qN_aX_d) = qN_aX_d(V_{GB}) \text{ (depletion)} \qquad (3.91)$$

where $X_d(V_{GB})$ indicates the dependence of depletion width on V_{GB}. In order to quantify $X_d(V_{GB})$, we need to find the dependence of the surface potential on the gate-bulk voltage V_{GB}, which we will do in the next section.

Figure 3.24(a) shows the electrostatics of the MOS capacitor biased in **depletion**, with $V_{GB} = 250$ mV. For this bias voltage, the electrostatic analysis in the next section shows that the surface potential is $\phi_s = 185$ mV. The surface potential corresponds to silicon that is slightly n-type—opposite to the p-type substrate. Even though the MOS capacitor is under applied bias, $J_n = 0$ and $J_p = 0$ through the structure due to

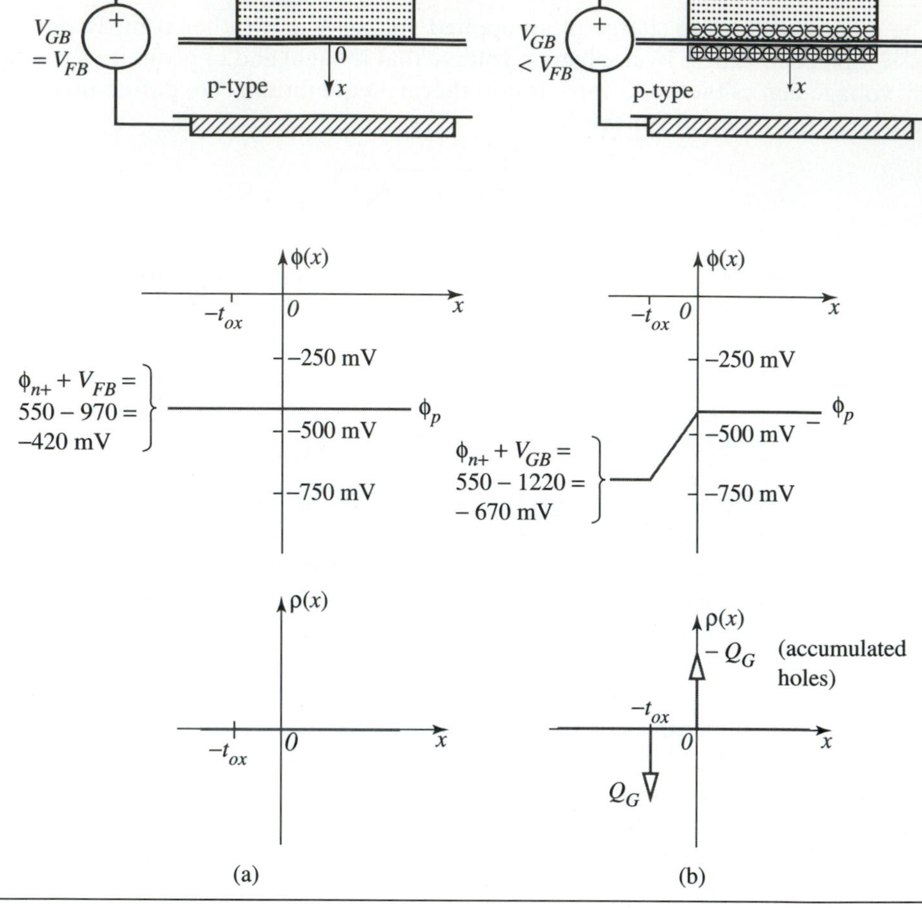

➤ **Figure 3.23** MOS capacitor with p-type substrate under an applied bias. (a) $V_{GB} = V_{FB}$ (flatband) and (b) $V_{GB} = -1.22$ V $< V_{FB}$ (accumulation.) The potential and charge densities are plotted, for the case where the substrate is doped at $N_a = 10^{17}$ cm^{-3}.

the SiO_2 insulating film. Therefore, the carrier concentration is linked to the electrostatic potential by the analysis in Section 3.3.1. Note that this argument is not valid for the reverse-biased pn junction since there is a current, albeit small, through the structure when $V_D < 0$ V.

According to Eq. (3.21), the surface electron concentration at $x = 0$ is

$$n_s = n_i \cdot 10^{\phi_s/(60\text{ mV})} = 10^{10} \cdot 10^{185/60}\text{ cm}^{-3} = 1.2 \times 10^{13}\text{ cm}^{-3} \quad (3.92)$$

This concentration of electrons is negligible compared to the ionized acceptor concentration in the depletion region. As a result, the additional negative charge at the SiO_2–silicon interface from the electrons at $x = 0$ does not appear in the charge density plot in Fig. 3.24(a).

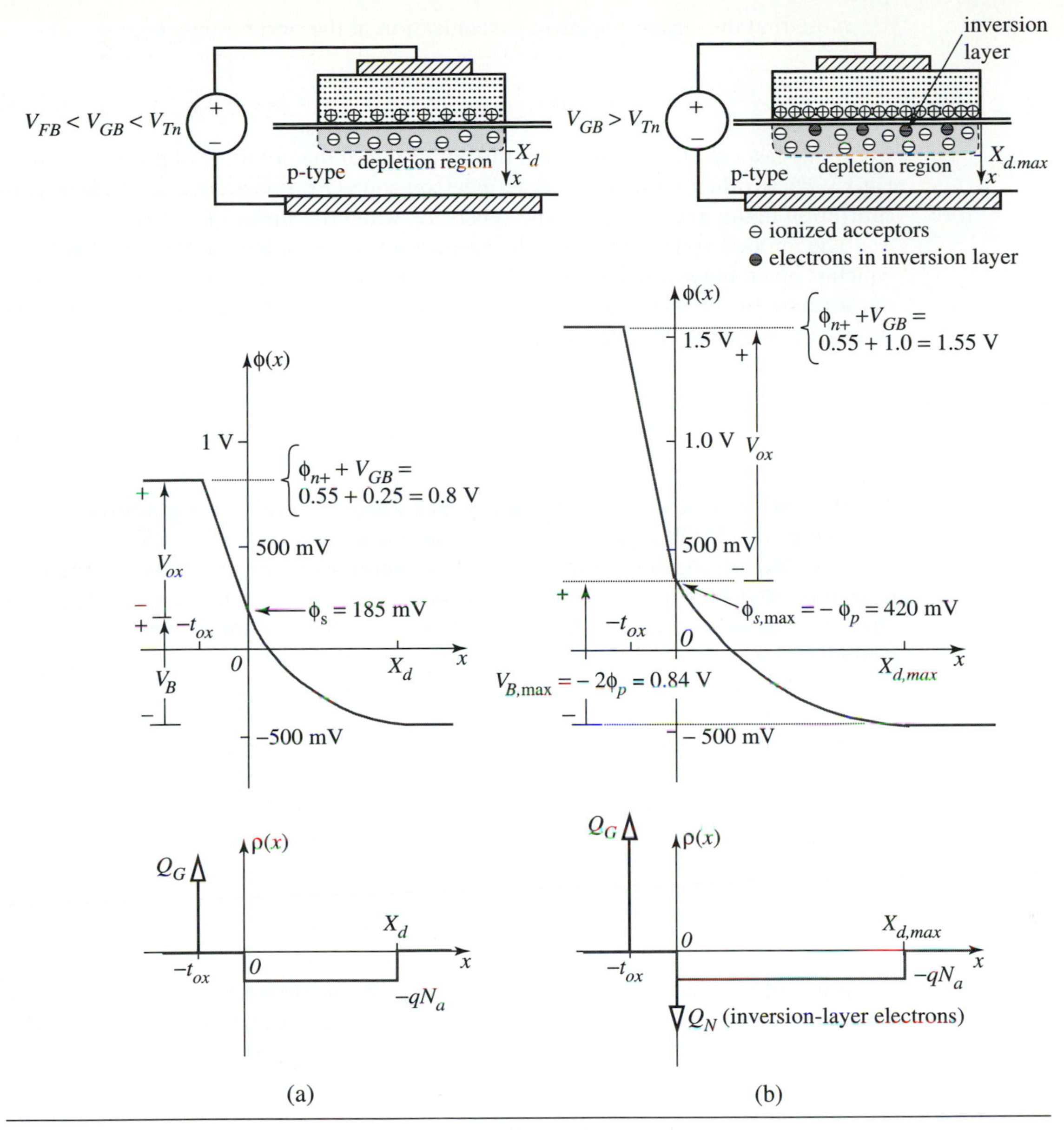

➤ **Figure 3.24** MOS capacitor with p-type substrate under an applied bias. (a) $V_{FB} < V_{GB} = 250$ mV and $V_{GB} < V_{Tn}$ (depletion) and (b) $V_{GB} = 1.0\text{ V} > V_{Tn}$ (inversion). The potential and charge densities are plotted, for the case where the substrate is doped at $N_a = 10^{17}\text{ cm}^{-3}$.

If the gate-bulk voltage V_{GB} is increased further, the surface potential will continue to rise and eventually, there will be a significant electron concentration at the surface. For a first-order analysis of the MOS capacitor, we consider that the surface has **inverted** when the surface potential becomes equal and opposite to the potential of the substrate:

$$\phi_s = -\phi_p \tag{3.93}$$

Note that the surface electron concentration at the onset of inversion is

$$n_s = n_i e^{\phi_s/V_{th}} = n_i e^{-\phi_p/V_{th}} = N_a, \tag{3.94}$$

where the last equality follows from the definition of the potential of p-type silicon in Eq. (3.22). In other words, the surface electron concentration is equal to the hole concentration in the undepleted p-type substrate, when the surface has just inverted.

The applied voltage at which the surface inverts is called the **threshold voltage**, which is given the symbol V_{Tn} for a MOS capacitor on a p-type substrate. Note that the **n** refers to the n-type surface after inversion—not to the p-type substrate. The threshold voltage is the sum of three terms

threshold voltage for p-type substrate

$$\boxed{V_{Tn} = V_{FB} - 2\phi_p + \frac{1}{C_{ox}}\sqrt{2q\varepsilon_s N_a(-2\phi_p)}} \tag{3.95}$$

The derivation of Eq. (3.95) will be left to Section 3.8. A typical range for the threshold voltage of a MOS capacitor on an n-type substrate is $V_{Tn} = 0.5 - 1$ V.

The electron concentration at the surface increases exponentially with increasing ϕ_s as indicated by Eq. (3.94). After inversion occurs, further increases in V_{GB} only slightly increase the surface potential. Note that the maximum possible surface potential is $\phi_{s\,(max)} = 550$ mV, which is only 130 mV higher than the value of $-\phi_p = -(-420$ mV), for the case where $N_a = 10^{17}$ cm^{-3}. In order to simplify the electrostatics of the MOS capacitor, we will assume that the surface potential stops increasing after the onset of inversion for $V_{GB} > V_{Tn}$ and remains "pinned" at $-\phi_p$, which we will consider its maximum.

Figure 3.24(b) shows the electrostatics for a MOS capacitor biased with $V_{GB} > V_{Tn}$. The surface potential is $\phi_s = -\phi_p$, which means that the bulk voltage V_B (the drop across the depletion region) has reached its maximum value

$$V_B = V_{B,\max} = \phi_{s,\max} - \phi_p = -\phi_p - \phi_p = -2\phi_p \tag{3.96}$$

Much of the electrostatics in inversion is easily quantified, since the bulk voltage is a known quantity for $V_{GB} \geq V_{Tn}$. For example, the depletion width and bulk charge are constant at their maximum values $X_{d,\max}$ and $Q_{B,\max}$ in inversion. The oxide voltage is

$$V_{ox} = (V_{GB} + \phi_{n+}) - \phi_s = V_{GB} + \phi_{n+} + \phi_p \text{ for } V_{GB} \geq V_{Tn} \tag{3.97}$$

We now consider the important question of how the gate charge varies with applied bias in inversion. The gate charge balances the bulk charge $Q_{B,\max}$ in the depletion region and the electron charge Q_N in the surface inversion layer

$$Q_G(V_{GB}) = -(Q_N + Q_{B,\max}) \text{ for } V_{GB} \geq V_{Tn} \tag{3.98}$$

For $V_{GB} = V_{Tn}$, we consider that the electron charge $Q_N \approx 0$ since the MOS capacitor is at the onset of inversion, which implies that

$$Q_G(V_{Tn}) \approx -Q_{B,\max} \tag{3.99}$$

Taking equalities in these approximations, we find that the additional gate charge from increasing the bias voltage above threshold is

$$Q_G(V_{GB}) - Q_G(V_{Tn}) = -Q_N \quad (3.100)$$

In other words, the additional gate charge is balanced by the inversion-layer electron charge. The added gate charge and the inversion-layer electrons are separated by the gate oxide, from which we conclude that the parallel-plate capacitance applies

$$Q_G(V_{GB}) - Q_G(V_{Tn}) = C_{ox}(V_{GB} - V_{Tn}) \text{ for } V_{GB} \geq V_{Tn} \quad (3.101)$$

From Eqs. (3.99) and (3.101), the gate charge in inversion is

gate charge in inversion for p-type substrate

$$\boxed{Q_G(V_{GB}) = C_{ox}(V_{GB} - V_{Tn}) - Q_{B,\max}} \quad (3.102)$$

A qualitative picture of the functional dependence of the gate charge on the applied bias for a MOS capacitor with p-type substrate is given in Fig. 3.25, for the case of a substrate doping of $N_a = 10^{17}$ cm^{-3} and a 150 Å-thick oxide. From Eq. (3.88), the flatband voltage is $V_{FB} = -0.97$ V and from Eq. (3.95), the threshold voltage is $V_{Tn} = 0.6$ V. The gate charge varies linearly in accumulation and in inversion, since the additional charge in the substrate due to a change in the gate-bulk voltage is stored at the SiO_2 silicon surface.The slope in these regions is the oxide capacitance per unit area, C_{ox}. In depletion, the substrate charge is smeared over the depletion region. Therefore, Q_G has a non-linear dependence on V_{GB} which has been sketched qualitatively in Fig. 3.25.

In order to quantify the charge storage in the MOS capacitor, we must solve for the charge, field, and potential in the three regions of operation. Prior to tackling the

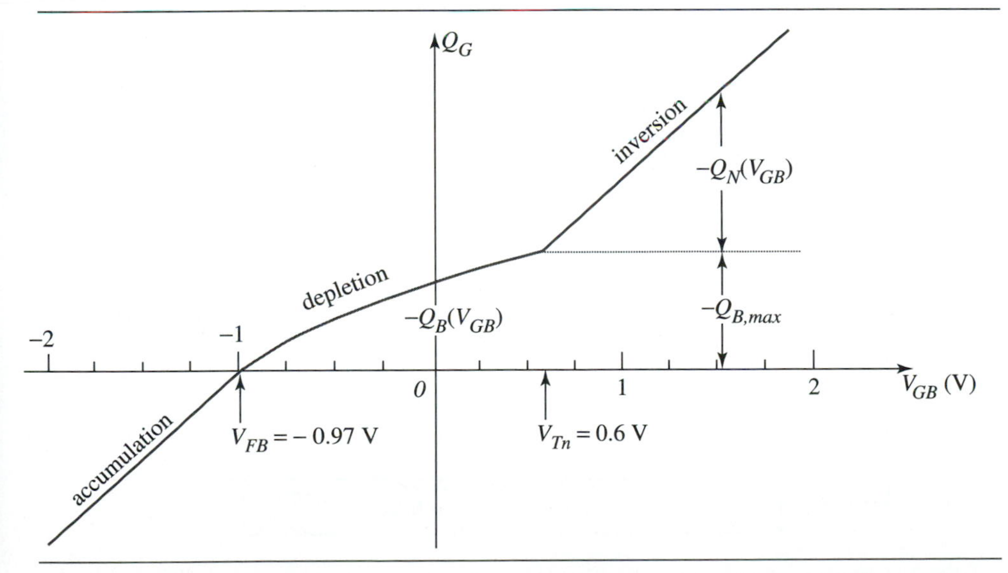

➤ **Figure 3.25** Gate charge as a function of gate-bulk voltage for an MOS capacitor with a 150 Å-thick gate oxide and a substrate doping $N_a = 10^{17}$ cm^{-3}.

rather involved algebra, it is helpful to summarize the qualitative behavior of the MOS capacitor on an n-type substrate.

3.7.3 MOS Capacitor on an n-type Substrate

Figure 3.26 shows a MOS capacitor with an n^+ polysilicon gate which is fabricated with an n-type substrate. We will use a typical substrate doping of $N_d = 10^{17}$ cm^{-3}, as we did in the previous section. Our qualitative understanding of the MOS capacitor on a p-type substrate is a good foundation for understanding this "complementary" structure.

In thermal equilibrium, the potential drop from the n+ polysilicon gate to the n-type substrate for a typical doping concentration $N_d = 10^{17}$ cm^{-3} is:

$$\phi_{n+} - \phi_n = 550 \text{ mV} - 420 \text{ mV} = 130 \text{ mV} \quad (3.103)$$

Therefore, the electric field in the oxide is positive (in the same direction as x) in Fig. 3.26 and we conclude that the charge on the gate is positive and that the charge in the n-type substrate is negative. The negative charge consists of accumulated electrons at the SiO_2 silicon interface under the gate. We can visualize the positive charge on the gate attracting the electrons and causing them to accumulate under the gate. In contrast to the situation with a p-type substrate, the MOS capacitor on an n-type substrate is in accumulation when in thermal equilibrium.

Following our procedure from the previous section, we can null the internal voltage drop by applying the flatband voltage $V_{GB} = V_{FB}$

flatband voltage for n-type substrate

$$V_{FB} = -(\phi_{n+} - \phi_n) = -130 \text{ mV} \quad (3.104)$$

For applied voltages more positive than the flatband voltage, the electric field is positive and the MOS capacitor is in accumulation. As we have seen, thermal equilibrium with $V_{GB} = 0 \text{ V} > V_{FB}$ falls into this region. The gate charge in the accumulation region is

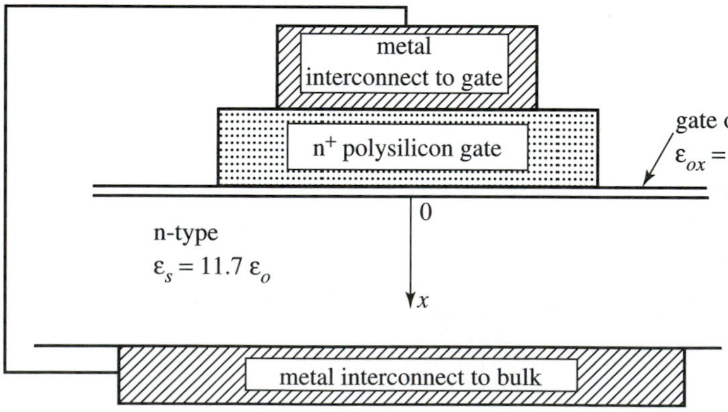

➤ **Figure 3.26** MOS capacitor with n-type substrate. $V_{GB} = 0$ for investigating the capacitor in thermal equilibrium.

$$\boxed{Q_G = C_{ox}(V_{GB} - V_{FB}) \text{ for } V_{GB} \geq V_{FB}} \tag{3.105}$$

gate charge in accumulation for n-type substrate

For applied voltages more negative than the flatband voltage, the electric field reverses and a positive charge is needed in the silicon substrate. Since the substrate is n-type, a depletion region forms under the gate that consists of positively ionized donors. The gate charge for depletion is negative for this case

$$Q_G(V_{GB}) = -Q_B = -qN_d X_d(V_{GB}) \tag{3.106}$$

As the gate-bulk voltage becomes more negative, the surface potential is pulled down and the surface hole concentration p_s increases. The gate-bulk voltage when the surface potential is equal and opposite to that of the n-type substrate ($\phi_s = -\phi_n$) is defined as the threshold voltage, which is given by

$$\boxed{V_{Tp} = V_{FB} - 2\phi_n - \frac{1}{C_{ox}}\sqrt{2q\varepsilon_s N_d(2\phi_n)}} \tag{3.107}$$

threshold voltage for n-type substrate

The signs of the potential drops across the depletion region and across the oxide are negative for the case of an n-type substrate. For voltages more negative than the threshold voltage, the MOS capacitor is inverted and the gate charge balances holes in the surface inversion layer and the maximum depletion charge:

$$\boxed{Q_G = -Q_P - Q_{B,\,\max} = C_{ox}(V_{GB} - V_{Tp}) - Q_{B,\,\max} \text{ for } V_{GB} \leq V_{Tp}} \tag{3.108}$$

gate charge in inversion for n-type substrate

The gate charge as a function of the applied bias voltage is sketched in Fig. 3.27.

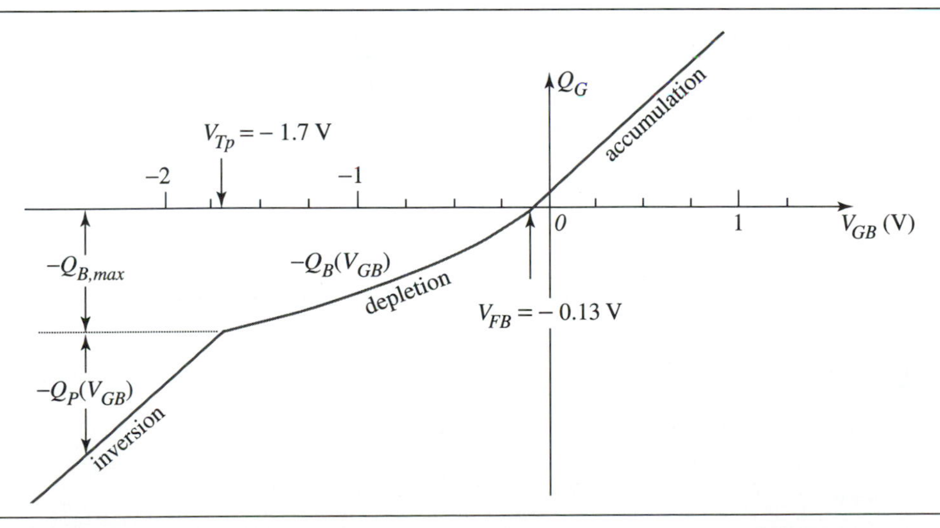

➤ **Figure 3.27** Gate charge as a function of gate-bulk voltage for an MOS capacitor on an n-type substrate with doping $N_d = 10^{17}$ cm^{-3} and a 150 Å-thick gate oxide.

EXAMPLE 3.7 **MOS Electrostatics in Accumulation and in Inversion**

In this example, we consider a MOS capacitor with $t_{ox} = 200$ Å on a p-type substrate with acceptor concentration $N_a = 5 \times 10^{16}$ cm^{-3}.

(a) Find the electric field in the oxide and the charge per unit area in the silicon substrate (units C/cm^2) for $V_{GB} = -2.5$ V.

SOLUTION

The first step is to determine whether the capacitor is in the accumulation, depletion, or inversion regions. The flatband voltage for an MOS capacitor on a p-type substrate with these parameters is given by Eq. (3.88)

$$V_{FB} = -(\phi_{n+} - \phi_p) = -(550 - (-402)) = -0.95\text{V}$$

and the threshold voltage is found from Eq. (3.95)

$$V_{Tn} = -0.95\text{V} - 2(-0.40) + \frac{\sqrt{2(1.6 \times 10^{19})(1.04 \times 10^{-12})(5 \times 10^{16})(-2)(-0.40)}}{(3.45 \times 10^{-13})/(2 \times 10^{-6})}$$

$$V_{Tn} = -0.95\text{V} - 2(-0.40) + 0.67 = 0.52 \text{ V}$$

In evaluating the final term, we have substituted the permittivities of silicon and SiO_2

$$\varepsilon_s = 1.04 \times 10^{-12} \text{ F/cm and } \varepsilon_{ox} = 3.45 \times 10^{-13} \text{ F/cm}.$$

The applied bias is $V_{GB} = -2.5 \text{ V} < V_{FB} = -0.95$ V and we conclude that the p-type substrate is accumulated and the electrostatics of Fig. 3.23(b) apply. We can find the electric field from the information on the potential plot in accumulation:

$$E_{ox} = \frac{V_{ox}}{t_{ox}} = \frac{(V_{GB} + \phi_{n+}) - \phi_p}{t_{ox}} = \frac{-2.5 + 0.55 - (-0.40)}{2 \times 10^{-6}} = -7.8 \times 10^5 \text{ V/cm}$$

The charge in the substrate consists of accumulated holes. It is opposite in sign to the gate charge, which is given by Eq. (3.90)

$$Q_P = -Q_G = -C_{ox}(V_{GB} - V_{FB}) = -1.72 \times 10^{-7}(-2.5 - (-0.95)) = 2.67 \times 10^{-7} \text{ C/cm}^2$$

(b) Find the numerical value of the depletion width and depletion charge when the capacitor is biased in the inversion region.

SOLUTION

In inversion, the potential drop across the depletion region is

$$V_{B,\max} = \phi_s - \phi_p = -\phi_p - \phi_p = -2\phi_p = -2(-0.40) = 0.8 \text{ V}.$$

Adapting the results of Ex. 3.3, the relationship between the depletion region width $X_{d,\max}$ and the potential drop $V_{B,\max}$ is

$$V_{B,\max} = \frac{1}{2}\left(\frac{qN_a}{\varepsilon_s}\right)X_{d,\max}^2$$

Solving for the depletion region width, we find that

$$X_{d,\max} = \sqrt{\frac{2\varepsilon_s V_{B,\max}}{qN_a}} = \sqrt{\frac{2\,(1.035\times10^{-12})\,(0.8)}{(1.6\times10^{-19})\,(5\times10^{16})}} = 1440\ \text{Å}.$$

The bulk charge in the depletion region is

$$Q_{B,\max} = -qN_aX_{d,\max} = -(1.6\times10^{-19})(5\times10^{16})(1.44\times10^{-5}) = -1.15\times10^{-7}\ \text{C/cm}^2.$$

(c) Find the electric field in the oxide and the inversion-layer electron charge, for $V_{GB} = 2.5$ V.

SOLUTION

We know that the MOS capacitor is inverted, since $V_{GB} = 2.5\ \text{V} > V_{Tn} = 0.52$ V. The electrostatics in Fig. 3.24(b) apply for this region. From the potential plot, we can evaluate the electric field in the oxide:

$$E_{ox} = \frac{V_{ox}}{t_{ox}} = \frac{(V_{GB}+\phi_{n+}) - (-\phi_p)}{t_{ox}} = \frac{2.5+0.55-0.40}{2\times10^{-6}} = 1.32\times10^6\ \text{V/cm}.$$

Electrons in the inversion layer are one component of the total charge in the silicon substrate. Using Gauss's Law, we can relate the electric field in the oxide to the substrate charge:

$$E_{ox} = \frac{-(Q_{B,\max}+Q_N)}{\varepsilon_{ox}}$$

Solving for the inversion-layer charge, we find that

$$Q_N = -\varepsilon_{ox}E_{ox} - Q_{B,\max} = -4.55\times10^{-7} - (-1.15\times10^{-7}) = -3.4\times10^{-7}\ \text{C/cm}^2$$

We could have found the electron charge directly from Eq. (3.101)

$$Q_N = -C_{ox}(V_{GB}-V_{Tn}) = -\left(\frac{3.45\times10^{-13}}{2\times10^{-6}}\right)(2.5-0.52) = -3.4\times10^{-7}\ \text{C/cm}^2$$

(d) Electric fields with magnitudes greater than $E_{ox,\max} = 5\times10^6$ V/cm will cause irreversible damage to the gate oxide. Find the permissible range of gate-bulk voltages.

SOLUTION

In accumulation, the electric field in the oxide is

$$E_{ox} = \frac{V_{ox}}{t_{ox}} = \frac{(V_{GB} + \phi_{n+}) - \phi_p}{t_{ox}}$$

Solving for the most negative gate-bulk voltage using $E_{ox} = -5 \times 10^6$ V/cm, we find that

$$V_{GB,\min} = E_{ox}t_{ox} - \phi_{n+} + \phi_p = (-5 \times 10^6)(2 \times 10^{-6}) - 0.55 + (-0.4) = -10.95 \text{ V}$$

In inversion, the electric field in the oxide is

$$E_{ox} = \frac{V_{ox}}{t_{ox}} = \frac{(V_{GB} + \phi_{n+}) - (-\phi_p)}{t_{ox}}$$

Substituting the maximum positive field $E_{ox} = 5 \times 10^6$ V/cm, we find that

$$V_{GB,\max} = E_{ox}t_{ox} - \phi_{n+} - \phi_p = (5 \times 10^6)(2 \times 10^{-6}) - 0.55 - 0.4 = 9.05 \text{ V}$$

In this example, we have found numerical values for several properties of the MOS capacitor when biased in accumulation and in inversion. A clear understanding of the MOS structure and its regions of operation, combined with one-dimensional electrostatics, can prove adequate for solving practical problems—without the need for complicated analysis.

➤ 3.8 THE ELECTROSTATICS OF THE MOS CAPACITOR

In the previous section, the basic features of the MOS capacitor were outlined without deriving the results from a thorough analysis of the electrostatics. In this section, the MOS capacitor is considered more systematically, beginning with the MOS capacitor on a p-type substrate in thermal equilibrium. We will rely on the previous section to provide the context for the electrostatics.

3.8.1 MOS Electrostatics in Thermal Equilibrium

Figure 3.28 shows the MOS capacitor with $V_{GB} = 0$ V and indicates the locations and symbols for the internal potential drops. In particular, the contact potentials ϕ_{mn+} between the metal and n^+ polysilicon gate, and ϕ_{pm} between the p-type bulk and its metal contact are included in this section. These battery-like potential drops were described in Section 3.4.3 in the analysis of the pn junction under reverse bias. We define the potential drop across the oxide is defined as $V_{ox,o}$ and the drop across the depletion region as V_{Bo}, where the subscript **o** indicates thermal equilibrium.

The charge density $\rho_o(x)$ through the MOS capacitor on a p-type substrate is sketched in Fig. 3.29(a), for the case of thermal equilibrium. We would like to quantify the width of the depletion region X_{do} as a function of the material properties and oxide thickness. The bulk charge Q_{Bo} in the depletion region must be equal in magnitude and opposite in sign to the gate charge

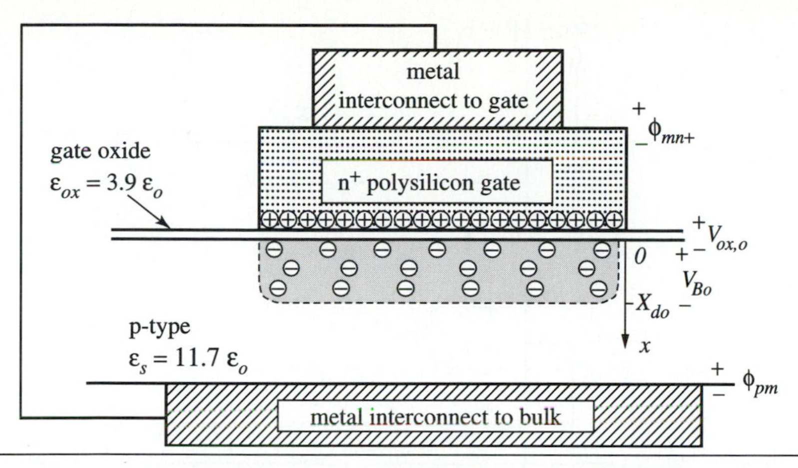

Figure 3.28 MOS capacitor with p-type substrate, with the charge distribution taken from Fig. 3.21. Gate and bulk metal contacts shorted together for the thermal equilibrium analysis.

$$Q_{Bo} = -qN_a X_{do} = -Q_{Go} \tag{3.109}$$

where N_a is the doping concentration in the bulk. As in Section 3.7, numerical values will be calculated for a substrate acceptor concentration of $N_a = 10^{17}$ cm^{-3} and an oxide thickness $t_{ox} = 150$ Å. The charge density in Fig. 3.29(a) is the same as in Ex. 3.3. Following the same steps as in that analysis, we graphically integrate $\rho_o(x)$ and find that the electric field in the gate oxide is constant and that $E_o(x)$ varies linearly in the depletion region as shown in Fig. 3.29(b). The boundary condition on electric field

$$\varepsilon_{ox} E_o(x = 0^-) = \varepsilon_s E_o(x = 0^+) \tag{3.110}$$

is used to connect the electric field solutions across the oxide/silicon interface. The electric field drops a factor of three across $x = 0$ since the permittivity of silicon is a factor of three larger than that of oxide.

The first step in plotting the potential is to apply Kirchhoff's voltage law to the MOS capacitor in thermal equilibrium. From Fig. 3.28,

$$-\phi_{mn^+} - V_{ox,o} - V_{Bo} - \phi_{pm} = 0 \tag{3.111}$$

Solving for the internal drop from the n$^+$ polysilicon gate to the p-type bulk, we find that

$$V_{ox,o} + V_{Bo} = -(\phi_{mn} + \phi_{pm}) \,. \tag{3.112}$$

For the MOS capacitor, the internal potential drop is known to be the difference in the thermal equilibrium potentials of the n$^+$ gate and the p-type bulk:

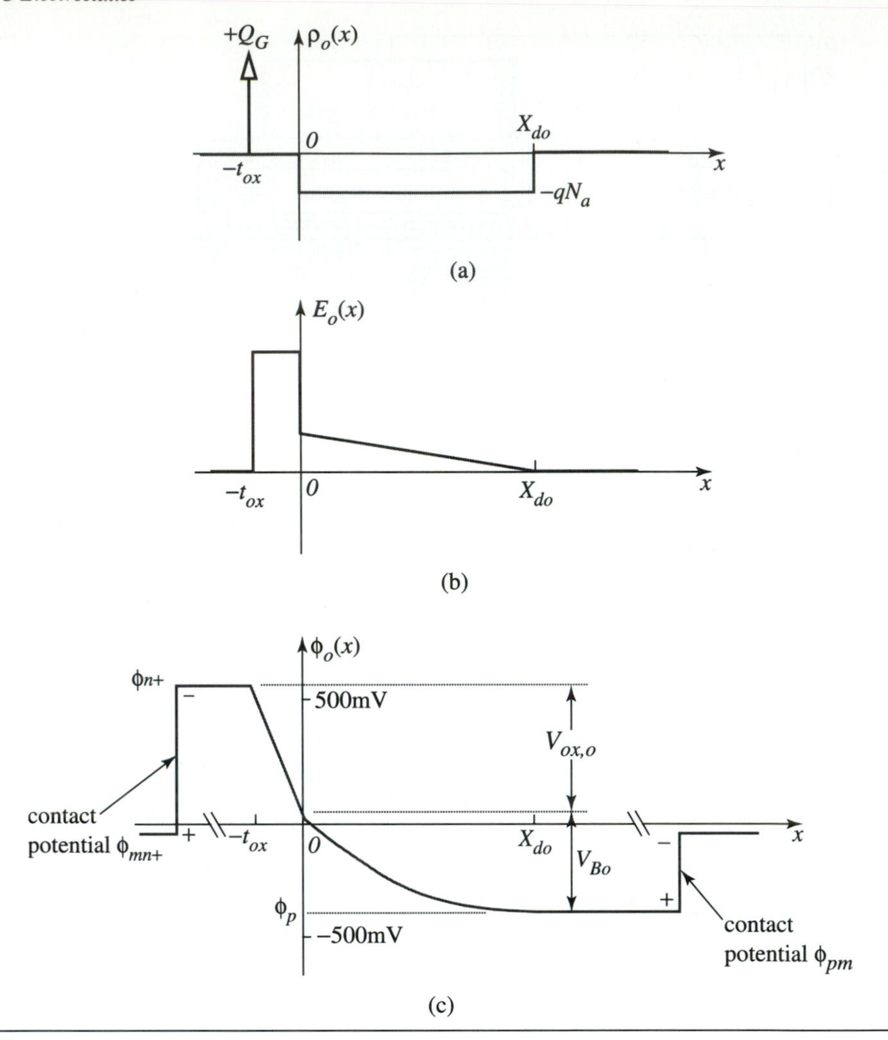

➤ Figure 3.29 MOS capacitor in thermal equilibrium: (a) charge density $\rho_o(x)$, (b) electric field $E_o(x)$, and (c) electrostatic potential $\phi_o(x)$. The potential jumps at the metal contacts to the n^+ polysilicon gate and to the p-type bulk are both negative quantities in (c).

$$V_{ox,o} + V_{Bo} = \phi_{n+} - \phi_p \tag{3.113}$$

By equating Eqs. (3.113) and (3.112), the sum of the contact potentials is equal to the difference in the potentials of the gate and the bulk

$$\phi_{n+} - \phi_p = -(\phi_{mn+} + \phi_{pm}) \tag{3.114}$$

which is similar to Eq. (3.58) for the pn junction in thermal equilibrium.

In order to plot the potential $\phi_o(x)$ in thermal equilibrium, we graphically integrate the negative of the electric field across the depletion region and the oxide. The potential varies linearly in the oxide since E_o is constant there, and quadratically in the depletion region. As a sign check, we verify that the potential of the gate comes out higher than that of the underlying bulk silicon. The contact potentials ϕ_{mn+} between the metal and n$^+$ polysilicon gate, and ϕ_{pm} between the p-type bulk and its metal contact are shown schematically in Fig. 3.29(c) as abrupt jumps. For the case of thermal equilibrium, note that the gate and bulk metal contacts are at the same potential in Fig. 3.29(c).

In order to quantify the potential variation $\phi_o(x)$ and the width of the depletion region X_{do}, it is necessary to wade through considerable algebra. This is an unfortunate feature of MOS device physics, as will be seen in this chapter and Chapter 4. Nevertheless, it is worthwhile to work out the detailed solutions for $E_o(x)$ and $\phi_o(x)$.

To start, Gauss's Law in integral form connects the electric field just inside the silicon to the depletion charge Q_{Bo} [C/cm^2]

$$E_o(0^+) = \frac{-Q_{Bo}}{\varepsilon_s} = \frac{qN_aX_{do}}{\varepsilon_s} \tag{3.115}$$

Since there is no charge layer at the oxide/silicon interface, the boundary condition in Eq. (3.11) connects the silicon field to the oxide field on the other side of the interface

$$E_o(0^-) = \frac{\varepsilon_s}{\varepsilon_{ox}}\left(\frac{qN_aX_{do}}{\varepsilon_s}\right) = \frac{qN_aX_{do}}{\varepsilon_{ox}} = E_{ox} \tag{3.116}$$

From the definition of potential, it follows that the potential in the oxide where $-t_{ox} \le x \le 0$ is given by

$$\phi_o(x) - \phi_{n+} = \int_{-t_{ox}}^{x} -E_{ox}dx = -E_{ox}(x+t_{ox}) = -\frac{qN_aX_{do}}{\varepsilon_{ox}}(x+t_{ox}) \tag{3.117}$$

It is not difficult to make a sign error, but we can check our algebra since we know the potential decreases through the oxide. The potential at the oxide/silicon interface can be found by evaluating Eq. (3.117) at the origin

$$\phi_o(0) = \phi_{n+} - \frac{qN_aX_{do}t_{ox}}{\varepsilon_{ox}} \tag{3.118}$$

This result can be expressed in terms of $V_{ox,o}$ which is the drop across the oxide

$$V_{ox,o} = \phi_{n+} - \phi_o(0) = \frac{qN_aX_{do}}{C_{ox}} = \frac{Q_{Go}}{C_{ox}} \tag{3.119}$$

where we have substituted the oxide capacitance $C_{ox} = \varepsilon_{ox}/t_{ox}$ [F/cm^2]. In Eq. (3.119), we have also substituted the equality between the charge (per unit area) on the gate Q_{Go} and the negative of the charge (per unit area) in the silicon $-Q_{Bo} = qN_aX_{do}$.

We now turn our attention to the depletion region in the silicon substrate. From the integral form of Gauss's Law, the electric field $E_o(x)$ in the charged region $(0 \le x \le X_{do})$ is given by

$$\varepsilon_s E_o(x) - \varepsilon_s E_o(0^+) = \int_{0^+}^{x} -qN_a dx = -qN_a x \tag{3.120}$$

Solving for $E_o(x)$ and substituting Eq. (3.115) for $E_o(0^+)$, the electric field at the interface, we find the linear equation sketched in Fig. 3.29(b)

$$E_o(x) = \frac{qN_a(X_{do} - x)}{\varepsilon_s} \tag{3.121}$$

We then integrate the electric field to find $\phi_o(x)$ in the charged region $0 \le x \le X_{do}$

$$\phi_o(x) - \phi_o(0) = \int_{0^+}^{x} -E_o(x)dx = \left(-\frac{qN_a}{\varepsilon_s}\right)\left(X_{do}x - \frac{x^2}{2}\right) \tag{3.122}$$

Substituting $\phi_o(0)$ from Eq. (3.118), the potential in the charged region is given by

$$\phi_o(x) = \left(\phi_{n+} - \frac{qN_a X_{do}}{C_{ox}}\right) - \frac{qN_a}{\varepsilon_s}\left(X_{do}x - \frac{x^2}{2}\right) \tag{3.123}$$

Since the potential at the bottom of the charged region is $\phi_o(X_{do}) = \phi_p$, we can solve for the thermal equilibrium built-in potential across the polysilicon-oxide-silicon sandwich

$$\phi_{n+} - \phi_p = \frac{qN_a X_{do}}{C_{ox}} + \frac{qN_a X_{do}^2}{2\varepsilon_s} \tag{3.124}$$

The first term in on the right-hand side of Eq. (3.124) is $V_{ox,o}$, the equilibrium drop across the oxide (see Eq. (3.119)), therefore the second term is the potential drop across the charged region V_{Bo}, as can also be shown from Eq. (3.122). This result is represented graphically in Fig. 3.29(c).

The width of the depletion region can be found by solving for X_{do}

$$X_{do} = t_{ox}\left(\frac{\varepsilon_s}{\varepsilon_{ox}}\right)\left(\sqrt{1 + \frac{2C_{ox}^2(\phi_{n+} - \phi_p)}{q\varepsilon_s N_a}} - 1\right) \tag{3.125}$$

Evaluating Eq. (3.123) at $x = 0$, we find that the thermal equilibrium surface potential is

$$\phi_{so} = \phi_o(x{=}0) = \phi_{n+} - \frac{qN_a X_{do}}{C_{ox}} \tag{3.126}$$

where X_{do} is given in Eq. (3.125).

EXAMPLE 3.8 MOS Capacitor in Thermal Equilibrium

It is always helpful to substitute typical dimensions and material properties into complicated expressions such as Eqs. (3.125) and (3.126) in order to develop a feel for the magnitudes of the quantities. For a typical gate oxide thickness of $t_{ox} = 15$ nm and a typical bulk doping of $N_a = 10^{17}$ cm^{-3}, we find that the depletion layer is

$$X_{do} = 76 \text{ nm}$$

and that the surface potential at the oxide/silicon interface is

$$\phi_{so} = \phi(x = 0) = 550 \text{ mV} - 529 \text{ mV} = 21 \text{ mV}$$

For this particular case, the potential drop across the oxide $V_{ox,o} = 529$ mV is slightly larger than the drop across the depletion region $V_{Bo} = 21 - (-420) = 441$ mV, with the depletion region width equal to about five times the oxide thickness.

3.8.2 MOS Electrostatics under Applied Bias

From the numerical results in Ex. 3.8, we note that the surface potential for this example corresponds to silicon that is nearly intrinsic. In other words, the "built-in" voltage drop across the MOS capacitor, $\phi_{n+} - \phi_p$, has set up electric fields that have affected the electrical properties of the silicon underlying the gate oxide. This phenomenon, termed the **field effect**, is controllable through an applied voltage on the gate. We will see in the next chapter that it forms the basic principle of MOS transistors.

Having done a thorough analysis of the electrostatics of the MOS capacitor in thermal equilibrium, we now turn to the case where an external bias V_{GB} is applied between the metal interconnections to the polysilicon gate and p-type bulk silicon. Figure 3.30 shows the MOS capacitor with a voltage source V_{GB} connected from the

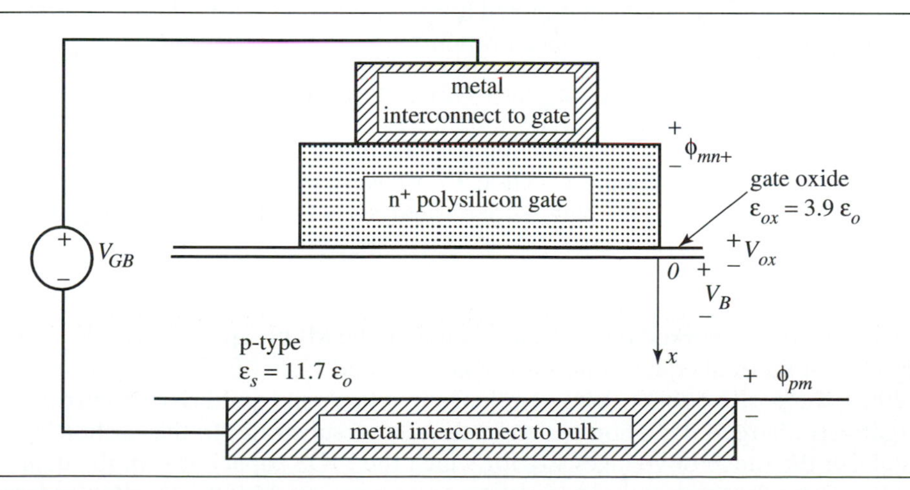

➤ **Figure 3.30** MOS capacitor on a p-type substrate for one-dimensional analysis of charge distribution ρ(x), electric field E(x), and potential φ(x) for the case of an applied bias V_{GB}.

gate to the bulk. We would like to investigate the effect of V_{GB} on the potential $\phi(x)$ through the structure. For convenience, the metal interconnection to the bulk will be kept fixed in plotting the change in potential under applied bias. As we have seen in Section 3.6, the electrostatics of the MOS capacitor are very different depending on the sign and magnitude of the applied voltage. We now quantify the charge, electric field, and potential for the MOS capacitor for the different regions of applied bias.

FLATBAND

A good starting point is to apply a gate voltage that is *opposite* to the built-in internal potential drop $\phi_{n+} - \phi_p$, which we define as the flatband voltage V_{FB}. For example, the built-in drop is 970 mV for the case where the substrate doping is $N_a = 10^{17}$ cm^{-3}. Applying $V_{GB} = V_{FB} = -970$ mV shifts the gate interconnect lower by 970 mV with respect to the bulk interconnect. The contact potential ϕ_{mn+} is unchanged by the applied bias, so the polysilicon gate potential is lowered to

$$\phi_{n+} - V_{FB} = 550\,\text{mV} - 970\,\text{mV} = -420\,\text{mV} \tag{3.127}$$

The bulk metal contact is considered fixed under applied bias and its contact potential ϕ_{pm} is also unchanged, so the p-type bulk remains at $\phi_p = -420$ mV. Therefore, there is no potential drop between the n^+ polysilicon and p-type bulk in flatband. The potential is therefore constant at $\phi(x) = -420$ mV throughout the MOS structure when biased at flatband. There is no electric field or charge density in the MOS capacitor for this case either, as shown in Fig. 3.31.

ACCUMULATION

The electrostatics of the MOS capacitor are very different for voltages $V_{GB} < V_{FB}$ compared with $V_{GB} > V_{FB}$. In the former case, the potential of the n^+ polysilicon gate is pulled less than that of the p-type bulk, leading to a negative charge on the gate and a positive charge in the silicon underneath the gate oxide. As we have seen in Section 3.7, the p-type substrate has a high concentration of mobile holes that accumulate at the SiO_2–silicon interface due to attraction by the negative gate charge. An excess of holes over the acceptor concentration results in a net positive charge at the silicon surface. The surface potential is pulled lower, due to the surface hole concentration p_s exceeding the bulk doping concentration:

$$\phi_s = -V_{th}\ln(p_s/n_i) < -V_{th}\ln(N_a/n_i) \tag{3.128}$$

However, the logarithmic function is weak and it is a reasonable approximation that $\phi_s \approx \phi_p$ in accumulation. In Section 3.7, we were able to quantify the electrostatics in accumulation since the parallel-plate analysis in Ex. 3.1 could be adapted to this case.

DEPLETION

We have already worked through an example of the MOS capacitor with $V_{GB} > V_{FB}$ in the case of thermal equilibrium since $V_{GB} = 0$ V and $V_{FB} = -0.97$ V. As shown in Fig. 3.29(a), the positive sheet charge on the gate in thermal equilibrium is mirrored in a negatively charged depletion region in the silicon substrate. In this section, we will establish the range of voltages V_{GB} for which the MOS capacitor is in the depletion region. Since the form of the charge density for a depleted MOS capacitor is identical to that in thermal equilibrium, the results of the equilibrium electrostatic analysis can be adapted and applied here.

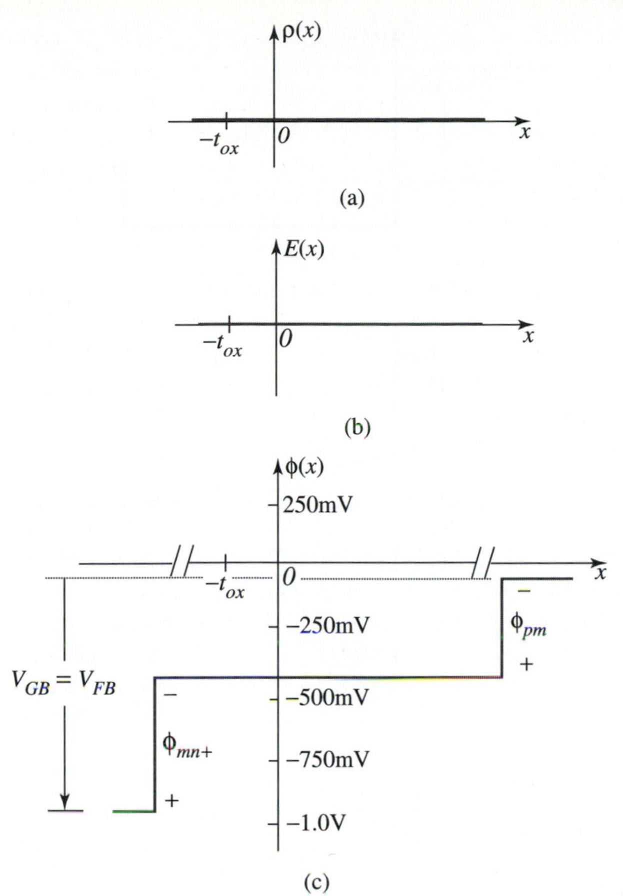

➤ **Figure 3.31** MOS capacitor on a p-type substrate biased in flatband with $V_{GB} = V_{FB} = -0.97$ V. (a) Charge density, (b) electric field, and (c) potential. The arrow on the potential plot graphically indicates that the flatband voltage pulls the gate metallization lower relative to the bulk metallization.

Figure 3.32 shows the electrostatics of a MOS capacitor with $V_{GB} = 250\text{mV}$ and $V_{FB} = -970$ mV. Later in this section, we will justify that the capacitor is depleted for this particular gate-bulk bias. The only difference from equilibrium is that the internal voltage drop across the oxide and depletion region is now

$$V_{GB} - V_{FB} = V_{ox} + V_B = 250 - (-970) = 1220\,\text{mV} = 1.22\text{ V} \tag{3.129}$$

rather than $-V_{FB} = \phi_{n+} - \phi_p = 0.97$ V. Making the substitution $V_{GB} - V_{FB}$ for $-V_{FB}$ in the equilibrium depletion width in Eq. (3.125) and $V_{GB} + \phi_{n+}$ for ϕ_{n+} in the equilibrium surface potential in (3.126), we find that

$$X_d(V_{GB}) = t_{ox}\left(\frac{\varepsilon_s}{\varepsilon_{ox}}\right)\left(\sqrt{1 + \frac{2C_{ox}^2 (V_{GB} - V_{FB})}{q\varepsilon_s N_a}} - 1\right) \tag{3.130}$$

$$\phi_s(V_{GB}) = (V_{GB} + \phi_{n+}) - \frac{qN_a X_d(V_{GB})}{C_{ox}} \tag{3.131}$$

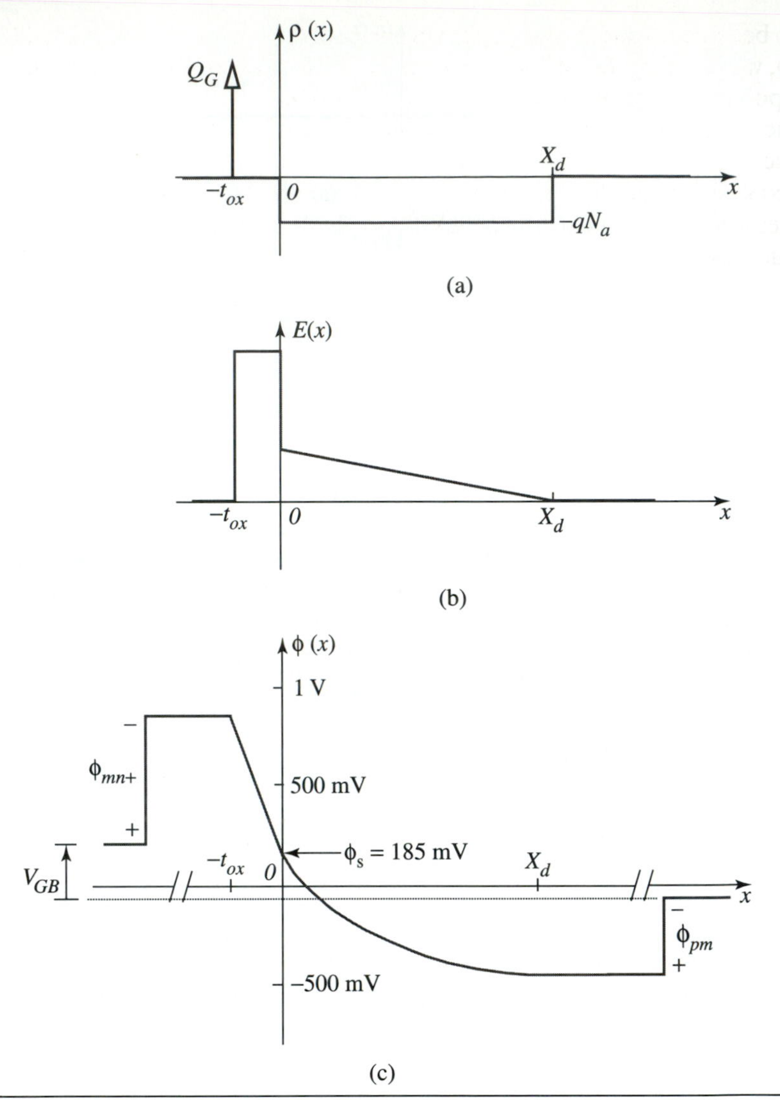

➤ **Figure 3.32** MOS capacitor on a p-type substrate biased in depletion with an applied bias V_{GB} = 250 mV for V_{FB} = –970 mV: (a) charge density, (b) electric field, and (c) potential.

For V_{GB} = 250 mV, the depletion region width increases to X_d = 88.5 nm and the surface potential is ϕ_s = (250 mV + 550 mV) – 615 mV = 185 mV.

We discussed the implications of the surface becoming n-type in Section 3.7. Here, we will quantify the onset of inversion and derive the expression for the threshold voltage.

The Threshold Voltage

As we increase the applied bias, the surface potential rises according to Eq. (3.131). At some point the assumptions underlying the depletion charge distribution in Fig.

3.32(a) become invalid. For example, if we substitute $V_{GB} = 1$ V into Eqs. (3.130) and (3.131), we find that the surface potential $\phi_s = 740$ mV, which is greater than the maximum potential possible in silicon ($\phi_{n+} = 550$ mV).

The electron concentration increases exponentially as the surface potential increases, according to Eq. (3.92), to the point where it eventually becomes the dominant component of the negative charge in the silicon substrate. After the surface becomes strongly n-type, the charge density in the silicon substrate must be modified to include the electron contribution

$$\rho(x) = -q(N_a + n(x)) = -q(N_a + n_i e^{\phi(x)/V_{th}}) \text{ for } 0 \le x \le X_d \qquad (3.132)$$

Substitution of Eq. (3.132) into Poisson's Equation results in a nonlinear differential equation for the potential $\phi(x)$. The solution will be left for an advanced device physics course. As usual, our goal is to find an approximation that allows us to capture the essential results in relatively simple equations.

The exponential increase in the electron concentration with increasing surface potential turns out to be the key to making a reasonable approximation to the charge density. We are able to set a critical surface potential ϕ_s' below which $n(x)$ can be neglected in Eq. (3.132) and the MOS capacitor considered depleted. The **onset of inversion** is defined to be when the surface potential is equal and opposite to the potential of the bulk p-type substrate.

$$\boxed{\phi_s' = -\phi_p} \qquad (3.133)$$

onset of inversion for p-type substrate

In simplifying the charge density in Eq. (3.132), we will neglect the surface electron concentration for surface potentials less than the critical surface potential at the onset of inversion

$$n_s \approx 0 \text{ for } \phi_s \le \phi_s' = -\phi_p \qquad (3.134)$$

Note that the surface electron concentration when $\phi_s' = -\phi_p$ is

$$n_s = n_i e^{\phi_s'/V_{th}} = n_i e^{-\phi_p/V_{th}} = N_a = p(x > X_d) \qquad (3.135)$$

In other words, Eq. (3.135) states that at the critical surface potential, the electron concentration is equal to the acceptor concentration, or that the surface is "as much n-type as the bulk is p-type." Although the electron concentration $n_s = N_a$ at this point, we will nevertheless neglect n_s in solving the electrostatics for $\phi_s' = -\phi_p$ according to the approximation in Eq. (3.134). This approach seems likely to lead to large errors, but the very rapid (exponential) change in n_s with surface potential makes the results reasonably accurate and very useful for understanding the essentials of MOS electrostatics.

The applied gate bias V_{GB} at which the onset of inversion occurs is obviously a very important quantity for the MOS capacitor. We define it as the **threshold voltage** V_{Tn}[1] for capacitors on p-type substrates. In order to find an expression for V_{Tn}, we can

1. Not to be confused with the thermal voltage, $V_{th} = kT/q$.

directly apply our understanding of the electrostatics of the depleted MOS capacitor. According to Eq. (3.134), $n_s \approx 0$ for $\phi_s \leq -\phi_p$ and so the charge density from Fig. 3.32(a) remains valid. We can solve for V_{Tn} by substituting $\phi_s = -\phi_p$ into Eqs. (3.130) and (3.131). However, there is little insight gained. Instead, we will begin from the plot of potential at the onset of inversion in Fig. 3.33.

At the onset of inversion, the potential drop V_B' across the depletion region is a known quantity

$$V_B' = \phi_s' - \phi_p = -\phi_p - \phi_p = -2\phi_p \tag{3.136}$$

The charge density in the depletion region is $\rho(x) = -qN_a$ and can be integrated twice using Poisson's Equation, as we did in Ex. 3.3. At the onset of inversion, the depletion width has increased to its maximum value which we call $X_{d,\max}$. Integration of the constant charge density in the depletion region leads to

$$V_B' = \left(\frac{qN_a}{2\varepsilon_s}\right)X_{d,\max}^2 = -2\phi_p \tag{3.137}$$

Solving for $X_{d,\max}$ we find it is related to the drop across the depletion region by

$$X_{d,\max} = \sqrt{2\frac{\varepsilon_s}{qN_a}(-2\phi_p)} \tag{3.138}$$

The charge in the depletion region at the onset of inversion $Q_B' = Q_{B,\max}$ is the product of its charge density and the depletion width

$$Q_{B,\max} = -qN_a X_{d,\max} = -\sqrt{2q\varepsilon_s N_a(-2\phi_p)} \tag{3.139}$$

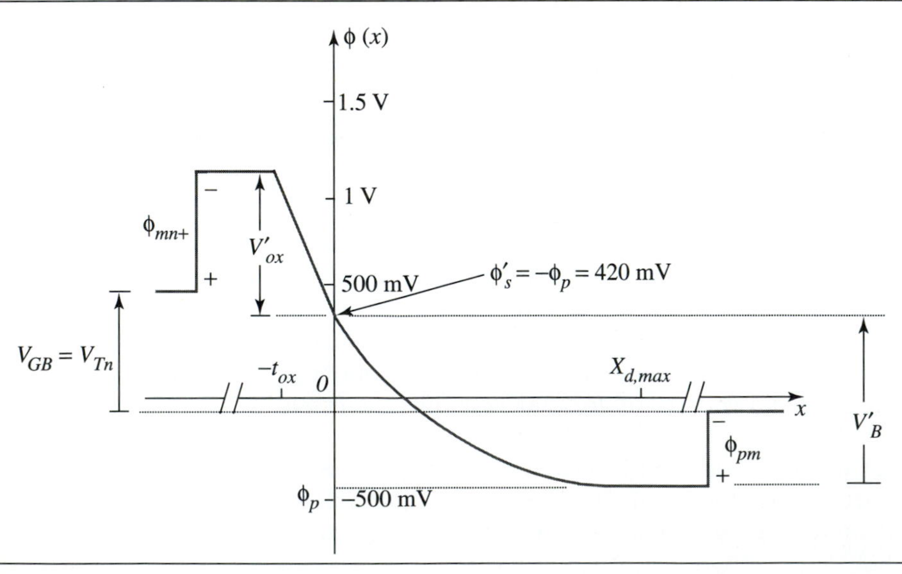

Figure 3.33 Potential for gate biased at the threshold voltage $V_{GB} = V_{Tn}$, the onset of inversion. The surface potential is equal and opposite to the bulk potential.

Gauss's integral law can be applied to find the electric field in the oxide, which when multiplied by the oxide thickness yields V_{ox}', the potential drop across the oxide in Fig. 3.33

$$V_{ox}' = E_{ox}' t_{ox} = \frac{-Q_{B,\max} t_{ox}}{\varepsilon_{ox}} = \left(\frac{t_{ox}}{\varepsilon_{ox}}\right)\sqrt{2q\varepsilon_s N_a(-2\phi_p)} \qquad (3.140)$$

The term in parentheses in Eq. (3.140) is the inverse of the oxide capacitance C_{ox}. Since the internal voltage drop in the MOS capacitor is the sum of the drops across the depletion region and across the oxide as shown in Fig. 3.33

$$V_{GB} - V_{FB} = V_{Tn} - V_{FB} = V_B' + V_{ox}' \qquad (3.141)$$

Substituting from Eqs. (3.137) and (3.140), we can solve for the threshold voltage on a p-type substrate, which was given previously in Eq. (3.95) in Section 3.7

$$V_{Tn} = V_{FB} + V_B' + V_{ox}' = V_{FB} - 2\phi_p + \frac{1}{C_{ox}}\sqrt{2q\varepsilon_s N_a(-2\phi_p)} \qquad (3.142)$$

This expression hardly seems like a "neat" final result. However, V_{Tn} is the sum of three simple terms. First, the flatband voltage is the built-in offset across the MOS structure. The second term is the drop across the depletion region at the onset of inversion. The final term is the magnitude of the depletion charge at inversion divided by the oxide capacitance, which turns out to be the potential drop across the gate oxide. For the typical case where t_{ox} = 15 nm and $N_a = 10^{17}$ cm^{-3}, the threshold voltage from Eq. (3.142) is V_{Tn} = 600 mV.

INVERSION

What happens when the applied gate-bulk voltage V_{GB} is increased further? From Eq. (3.135), small changes in the surface potential lead to large increases in the surface electron concentration. As we know from Eq. (3.23), a 60 mV increase in ϕ_s over ϕ_s' corresponds to an order-of-magnitude increase in n_s. Therefore, we conclude that the surface potential only slowly increases as V_{GB} is raised, since the electron charge will increase linearly with V_{GB}. This insight allows us to greatly simplify the electrostatics in inversion, by making the **delta-depletion approximation**

delta-depletion approximation

$$\boxed{\phi_s \approx \phi_s' = -\phi_p \text{ for } V_{GB} \geq V_{Tn}} \qquad (3.143)$$

In other words, we consider the surface potential **pinned** at $-\phi_p$ in inversion. The approximation that ϕ_s is pinned means that we cannot use Eq. (3.21) to find the electron concentration at the surface in the inversion region. However, we will be able to find the inversion-layer electron charge Q_N [C/cm^2] by other means.

Now that the boundary between depletion and inversion has been established using the delta-depletion approximation, we can analyze the electrostatics of the inverted MOS capacitor. Our primary goal is to find the relationship between the electron charge in the inversion layer and the applied voltage V_{GB}.

Figure 3.34(a) shows the charge density for the case where V_{GB} = 1.0 V when V_{Tn} = 600 mV. Electrons at the SiO_2–silicon surface constitute a sheet charge Q_N

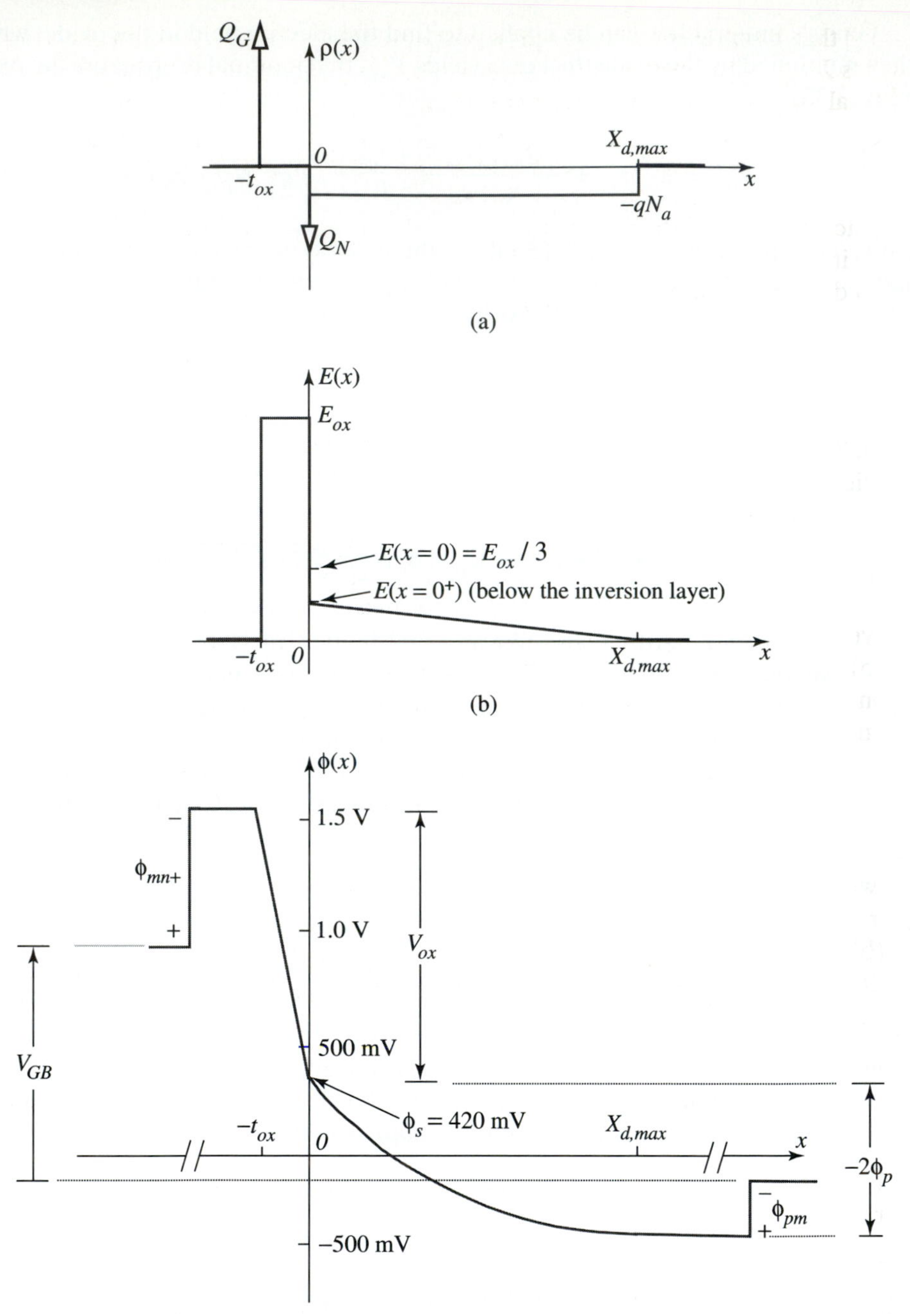

➤ Figure 3.34 MOS capacitor on a p-type substrate biased in inversion with $V_{GB} = 1$ V, $V_{Tn} = 600$ mV: (a) charge density, (b) electric field, and (c) potential.

[C/cm^2] that is represented by the delta function in Fig. 3.34(a). The depletion region width is $X_{d,\max}$ after inversion because the drop across it remains fixed at $-2\phi_p$. The potential drop across the oxide increases since $V_{GB} > V_{Tn}$ and

$$V_{GB} - V_{FB} = V_{ox} - 2\phi_p \tag{3.144}$$

Once again, the integral form of Gauss's Law is useful for connecting V_{ox} to the charge in the silicon that consists of the sum of the electron charge Q_N, and the (maximum) depletion charge $Q_B' = Q_{B,\max}$

$$V_{ox} = E_{ox} t_{ox} = \left(\frac{t_{ox}}{\varepsilon_{ox}}\right)(-Q_{B,\max} - Q_N) \tag{3.145}$$

In Fig. 3.34(b), the electric field drops a factor of three from the oxide to just inside the silicon at $x = 0$

$$E(0) = \frac{-Q_N - Q_{B,\max}}{\varepsilon_s} = E_{ox}\left(\frac{\varepsilon_{ox}}{\varepsilon_s}\right) = \frac{E_{ox}}{3} \tag{3.146}$$

In inversion, there is a second discontinuity in the electric field, as shown in Fig. 3.34(b). The electrons in the inversion layer form a charge sheet located at the SiO_2–silicon interface. Therefore, the electric field just below the inversion layer at $x = 0^+$ does not include the contribution of the inversion charge

$$E(0^+) = -\frac{Q_{B,\max}}{\varepsilon_s} = E(0) - \frac{(-Q_N)}{\varepsilon_s} \tag{3.147}$$

The width of the inversion layer is determined in an advanced course to be on the order of 50 Å, so the drop in electric field is not quite as abrupt as shown in Fig. 3.34(b).

Our goal is to express the inversion-layer charge Q_N as a function of the applied voltage, so we substitute Eq. (3.145) into Eq. (3.144)

$$V_{GB} - V_{FB} = V_{ox} - 2\phi_p = \frac{(-Q_{B,\max} - Q_N)}{C_{ox}} - 2\phi_p \tag{3.148}$$

in which we have also substituted $C_{ox} = \varepsilon_{ox}/t_{ox}$. Solving Eq. (3.148) for the inversion-layer charge, we find that

inversion-layer electron charge for p-type substrate

$$\boxed{Q_N = -C_{ox}\left(V_{GB} - V_{FB} + 2\phi_p - \frac{-Q_{B,\max}}{C_{ox}}\right) = -C_{ox}(V_{GB} - V_{Tn}) \text{ for } V_{GB} \geq V_{Tn}} \tag{3.149}$$

where we have identified the threshold voltage from Eq. (3.142). Note that there is *no* inversion layer charge unless the gate voltage is greater than the threshold voltage. This result is central to the model of the MOS field-effect transistor that will be developed in Chapter 4.

The results of this section for the MOS capacitor on a p-type substrate can be summarized by writing the gate charge as a function of the applied gate-bulk bias in accumulation, depletion, and inversion.

Gate charge for:
accumulation
depletion
inversion

$$Q_G = C_{ox}(V_{GB} - V_{FB}) \text{ for } V_{GB} \leq V_{FB} \quad (3.150)$$

$$Q_G = -Q_B(V_{GB}) = \frac{q\varepsilon_s N_a}{C_{ox}}\left(\sqrt{1 + \frac{2C_{ox}^2(V_{GB} - V_{FB})}{q\varepsilon_s N_a}} - 1\right)(V_{FB} \leq V_{GB} \leq V_{Tn}) \quad (3.151)$$

$$Q_G = C_{ox}(V_{GB} - V_{Tn}) + \frac{q\varepsilon_s N_a}{C_{ox}}\left(\sqrt{1 + \frac{2C_{ox}^2(V_{Tn} - V_{FB})}{q\varepsilon_s N_a}} - 1\right) \text{ for } V_{GB} \geq V_{Tn} \quad (3.152)$$

The depletion charge in Eq. (3.151) is found from Eq. (3.130) which gives the depletion width as a function of bias $X_d(V_{GB})$. The results in Eqs. (3.150)–(3.152) quantify the sketch of Q_G versus V_{GB} in Fig. 3.25 and will be useful for finding the MOS capacitance in the next section.

We conclude this section with an example that adapts the quantitative analysis of the MOS capacitor to the case of an n-type substrate.

EXAMPLE 3.9 MOS Electrostatics with an N-type Substrate

Given a MOS capacitor on an n-type substrate with an n$^+$ poly gate (ϕ_{n+}= 550mV), a donor concentration $N_d = 10^{16}$ cm^{-3}, and an oxide thickness t_{ox} = 500 Å. The contact potentials have values ϕ_{mn+} = −400 mV and ϕ_{nm}= 210 mV.

(a) Find the flatband voltage V_{FB} .

SOLUTION

The flatband voltage is equal and opposite to the built-in voltage in the MOS structure

$$V_{FB} = -(\phi_{n+} - \phi_n) = -(550 \text{ mV} - 360 \text{ mV}) = -190 \text{ mV}$$

(b) Plot the potential $\phi_o(x)$ in thermal equilibrium

SOLUTION

The MOS capacitor is accumulated in thermal equilibrium, since accumulated electrons under the gate balance the positive gate charge. We can sketch the potential since the substrate charge is a delta function at the SiO_2 silicon interface with value $Q_{No} = -Q_{Go}$, as shown in Fig. Ex3.9A.

(c) Find the threshold voltage V_{Tp}.

SOLUTION

In order to invert the surface, we must first apply V_{FB} to reach flatband. To deplete the substrate, we must apply $V_{GB} < V_{FB}$ to repel the mobile electrons from the surface and leave the positively charged ionized donors. At the onset of inversion, the surface potential will be lowered to the point where it is equal and opposite to that of the n-type bulk: $\phi_s' = -\phi_n$. Adding the flatband voltage, the

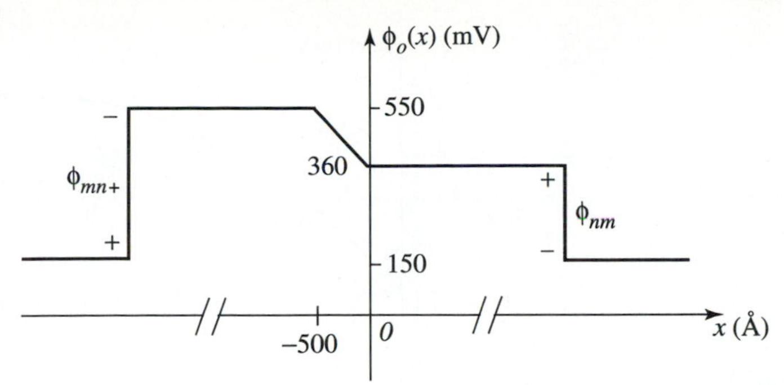

➤ Figure Ex3.9A Potential of the MOS structure with n-type substrate in thermal equilibrium.

voltage drop across the depletion region $V_B{}' = V_{B,\max} = -2\phi_n$ and the voltage drop across the oxide V'_{ox} the threshold voltage V_{Tp} is

$$V_{Tp} = V_{FB} - 2\phi_n + V'_{ox} = V_{FB} - 2\phi_n - \frac{Q_{B,\max}}{C_{ox}}$$

Note that the drop across the oxide is negative since the substrate has a positive depletion charge $Q_{B,\max} > 0$ and the gate charge is negative. Substituting for the maximum depletion charge $Q_{B,\max}$, we find

$$V_{Tp} = V_{FB} - 2\phi_n - \frac{\sqrt{2q\varepsilon_s N_d(2\phi_n)}}{C_{ox}} = -0.19 - 0.72 - \frac{4.88 \times 10^{-8}\,\mathrm{C/cm^2}}{6.9 \times 10^{-8}\,\mathrm{F/cm}} = -1.62\ \mathrm{V}$$

(d) Plot the potential $\phi(x)$ for $V_{GB} = V_{Tp}$.

SOLUTION

When $V_{GB} = V_{Tp} = -1.62$ V, the n$^+$ polysilicon gate has a potential of $0.55 - 1.62 = -1.07$ V. The potential in the n-type bulk remains fixed at $\phi_n = 0.36$ V with the surface potential at the onset of inversion $\phi_s = -\phi_n = -0.36$ V. The potential varies quadratically in the depletion region, which has a width

$$X_{d,\max} = \sqrt{\frac{2\varepsilon_s(2\phi_n)}{qN_a}} = \sqrt{\frac{2 \cdot 11.7 \cdot 8.85 \times 10^{-14} \cdot 0.72}{1.6 \times 10^{-19} \cdot 10^{16}}} = 0.31\ \mu\mathrm{m}$$

and linearly in the oxide as shown in Fig. Ex3.9B.

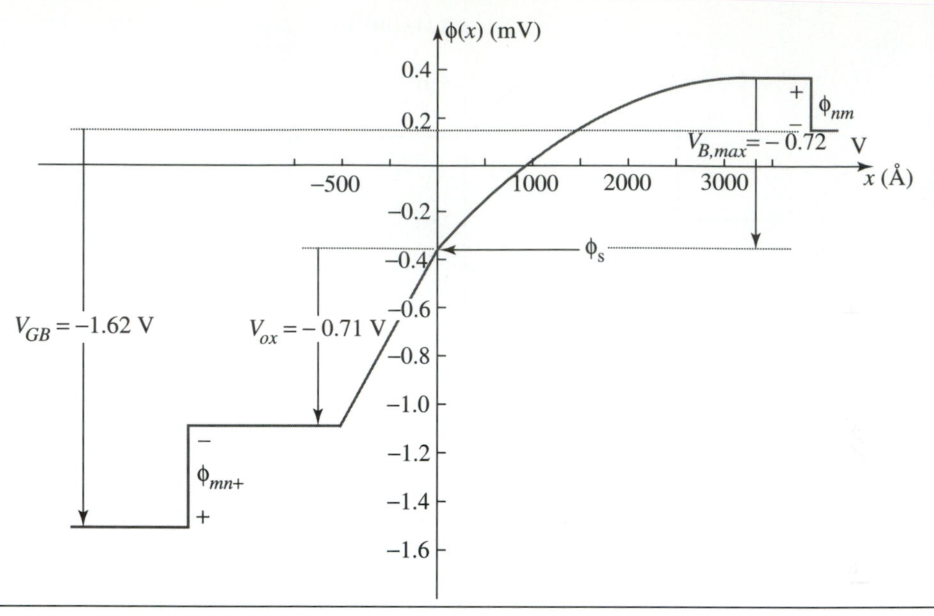

➤ **Figure Ex3.9B** Potential of MOS structure with n-type substrate for the case where $V_{GB} = V_{Tp}$.

➤ 3.9 CAPACITANCE OF THE MOS STRUCTURE

In this section, we will investigate the small-signal modeling of the MOS structure. Our approach will parallel Section 3.6 for the pn junction under reverse bias. We begin by viewing the capacitance as a graphical derivative of the charge-voltage characteristics of the capacitor sketched in Fig. 3.25 (for p-type substrate) and Fig. 3.27 (for n-type substrate). By differentiating the expressions for gate charge as a function of gate-bulk bias, we can find mathematical results for the MOS capacitance. Finally, we can gain insight into the physics of the MOS capacitor by determining where the small-signal increment in gate charge is mirrored in the underlying silicon substrate, for the three regions of gate-bulk bias.

3.9.1 Graphical Interpretation

Compared to the pn junction, the MOS structure has a much more complicated charge-storage characteristic. For case of a p-type substrate, we can apply the definition of the capacitance as the slope of $q_G(V_{GB})$ to Fig. 3.25 (repeated for convenience in Fig. 3.35(a)) to derive the plot of $C(V_{GB})$ in Fig. 3.35(b). We have used symbols for the total variables (DC + small-signal) in Fig. 3.35(a), since we are assuming that our analysis in Section 3.7 and Section 3.8 can be extended to time-varying signals.

By definition, the capacitance of the MOS structure is the slope of the $q_G(v_{GB})$ curve in Fig. 3.35, (a) evaluated at a particular DC bias voltage V_{GB}. In accumulation and inversion, the gate charge has a term that is proportional to v_{GB}, with a slope equal to C_{ox}. Therefore, the capacitance is independent of the DC bias and is equal

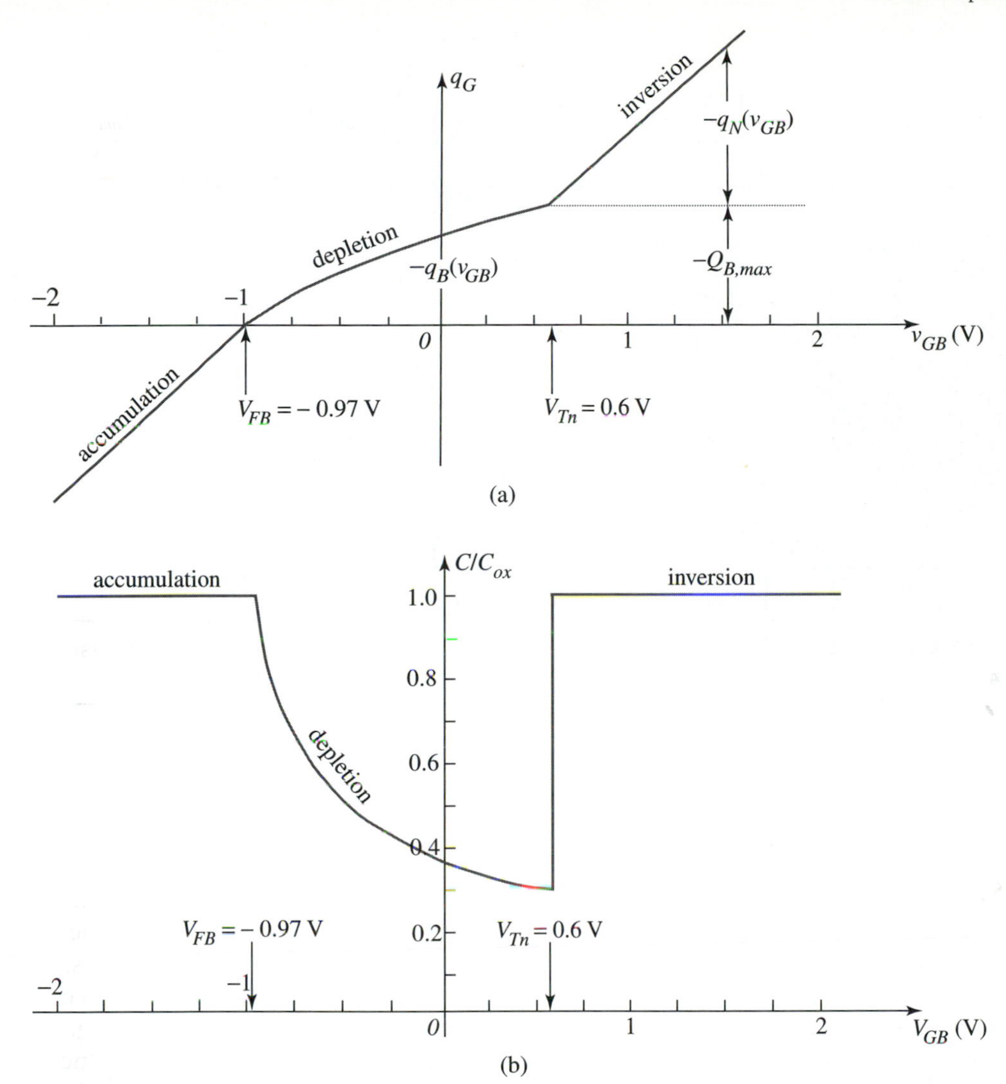

➤ **Figure 3.35** (a) Gate charge as a function of gate-bulk voltage for an MOS capacitor on a p-type substrate with a 150 Å-thick gate oxide and a substrate doping $N_a = 10^{17}$ cm^{-3} and (b) capacitance as a function of gate-bulk voltage, found by graphically differentiating (a).

to the oxide capacitance $C_{ox} = \varepsilon_{ox}/t_{ox}$. In Fig. 3.35(b), the capacitance is plotted normalized to C_{ox}. Later in this section, we will explain the reason for the variation of the MOS capacitance with gate bias in the depletion region. The abrupt jump in capacitance at the threshold voltage is a convenient way to measure V_{Tn}.

For an MOS capacitor on an n-type substrate, the gate charge versus gate-bulk voltage is plotted in Fig. 3.27, which is repeated as Fig. 3.36(a). Graphical differentiation yields the capacitance as a function of the DC gate-bulk voltage V_{GB} in Fig. 3.36(b). Note that substrate type can be determined from the capacitance plots in Figs. 3.35(b) and 3.36(b).

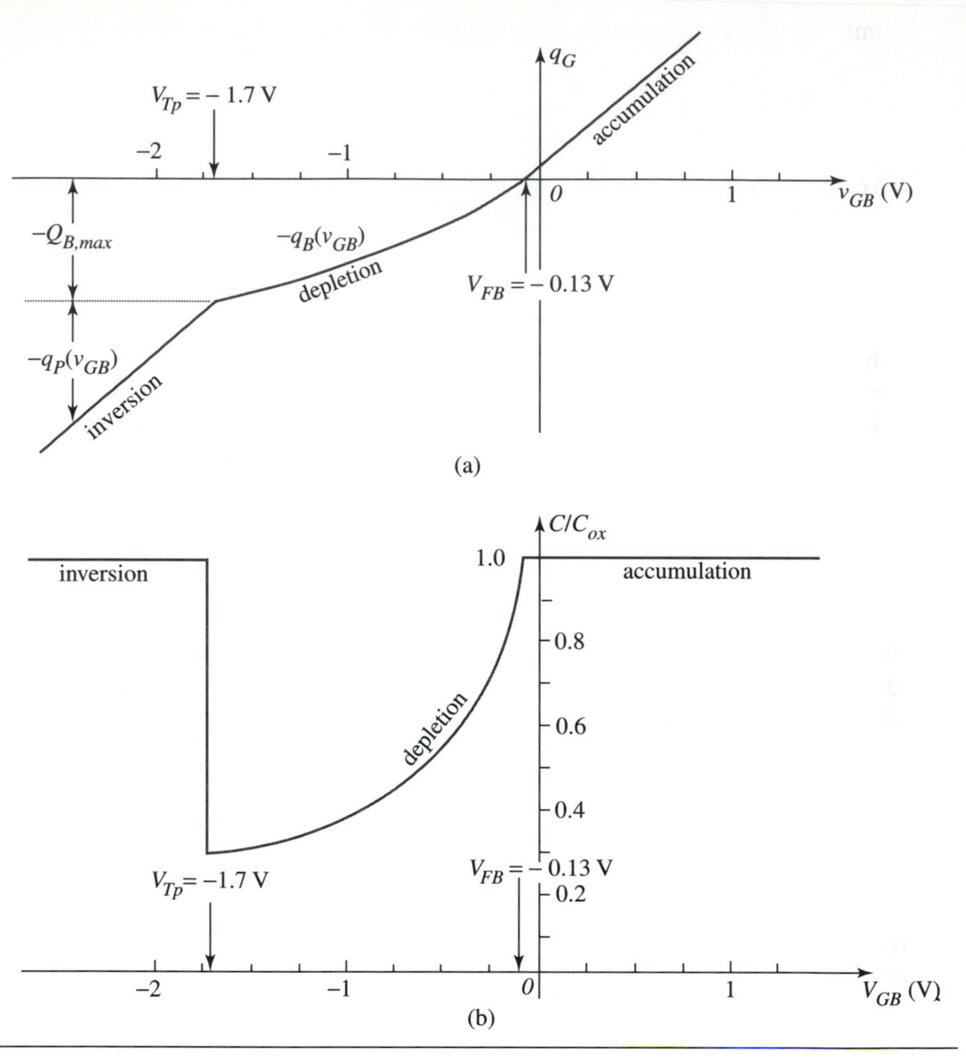

➤ **Figure 3.36** (a) Gate charge as a function of gate-bulk voltage for an MOS capacitor on an n-type substrate, with a 150 Å-thick gate oxide and a substrate doping $N_d = 10^{17}$ cm^{-3} and (b) capacitance as a function of gate-bulk voltage, found by graphically differentiating (a).

3.9.2 Mathematical Formulation

The expressions for gate charge as a function of gate-bulk bias in Eqs. (3.150)–(3.152) can be differentiated to find the capacitance in accumulation, depletion, and inversion. The capacitance of the metal-oxide-semiconductor structure is defined as

$$C = \left.\frac{dq_G}{dv_{GB}}\right|_{V_{GB}} \tag{3.153}$$

In accumulation, the MOS capacitance for a p-type substrate is

$$C = \frac{d}{dv_{GB}}[C_{ox}(v_{GB} - V_{FB})]\Big|_{V_{GB} \le V_{FB}} = C_{ox} \text{ for } V_{GB} \le V_{FB} \tag{3.154}$$

For the case of inversion, the MOS capacitance for a p-type substrate is the same

$$C = \frac{d}{dv_{GB}}[(C_{ox}(v_{GB} - V_{Tn}) - Q_{B,\max})]\Bigg|_{V_{GB} \ge V_{Tn}} = C_{ox} \tag{3.155}$$

since the bulk charge is fixed at $Q_{B,\max}$ in inversion. The situation in depletion requires differentiation of Eq. (3.151), which is valid for gate-bulk biases $V_{FB} \le V_{GB} \le V_{Tn}$

$$C = \frac{dq_G}{dv_{GB}}\Bigg|_{V_{FB} \le V_{GB} \le V_{Tn}} = \frac{C_{ox}}{\sqrt{1 + \dfrac{2C_{ox}^2(V_{GB} - V_{FB})}{q\varepsilon_s N_a}}} \tag{3.156}$$

This result allows the capacitance versus gate-bulk bias to be quantified in the depletion region. The minimum capacitance occurs just prior to inversion, which we can evaluate by substituting $V_{GB} = V_{Tn}$ into Eq. (3.156)

$$C(V_{GB} = V_{Tn}^-) = \frac{C_{ox}}{\sqrt{1 + \dfrac{2C_{ox}^2(V_{Tn} - V_{FB})}{q\varepsilon_s N_a}}} \tag{3.157}$$

In order to interpret the meaning Eq. (3.156), we will express it in terms of the depletion width $X_d(V_{GB})$ from Eq. (3.130), which we repeat here for convenience

$$X_d(V_{GB}) = \left(\frac{\varepsilon_s}{C_{ox}}\right)\left(\sqrt{1 + \frac{2C_{ox}^2(V_{GB} - V_{FB})}{q\varepsilon_s N_a}} - 1\right) \tag{3.158}$$

After some algebra, the capacitance in the depletion region can be written as

$$C = \frac{C_{ox}}{\left(\dfrac{C_{ox}}{\varepsilon_s}\right)X_d(V_{GB}) + 1} = \frac{C_{ox}\left(\dfrac{\varepsilon_s}{X_d(V_{GB})}\right)}{C_{ox} + \left(\dfrac{\varepsilon_s}{X_d(V_{GB})}\right)} \text{ for } V_{FB} \le V_{GB} \le V_{Tn} \tag{3.159}$$

From Section 3.6, we can identify the capacitance C_b of the depletion region as

$$C_b = \frac{\varepsilon_s}{X_d(V_{GB})} \tag{3.160}$$

which enables Eq. (3.159) to be rewritten as the capacitances C_{ox} and C_b in series

$$C = \frac{C_{ox} C_b}{C_{ox} + C_b} \text{ for } V_{FB} \leq V_{GB} \leq V_{Tn} \tag{3.161}$$

Later in this section, we will see why the oxide and depletion capacitances are in series combination when the MOS capacitor is biased in depletion.

EXAMPLE 3.10 MOS Capacitance-Voltage Curves

Shown below is a measured "C-V" curve for an MOS capacitor. The gate is n^+ polysilicon with ϕ_{n+}= 550 mV.

(a) From the normalized capacitance-voltage characteristic, determine the substrate type and doping level.

SOLUTION

To make sense of the *C-V* curve, we must determine whether point *A* or *C* corresponds to flatband. Going from point *A* to point *B*, the capacitance decreases with increasing voltage, which corresponds to a widening depletion region. Also, the transition from *B* to *C* is much more abrupt than the transition from *A* to *B*. These observations are consistent with *A* being the flatband condition and *C* being inversion. From the characteristic, we estimate that

$$V_{FB} = -1\text{V} \qquad V_T = 1\text{V}$$

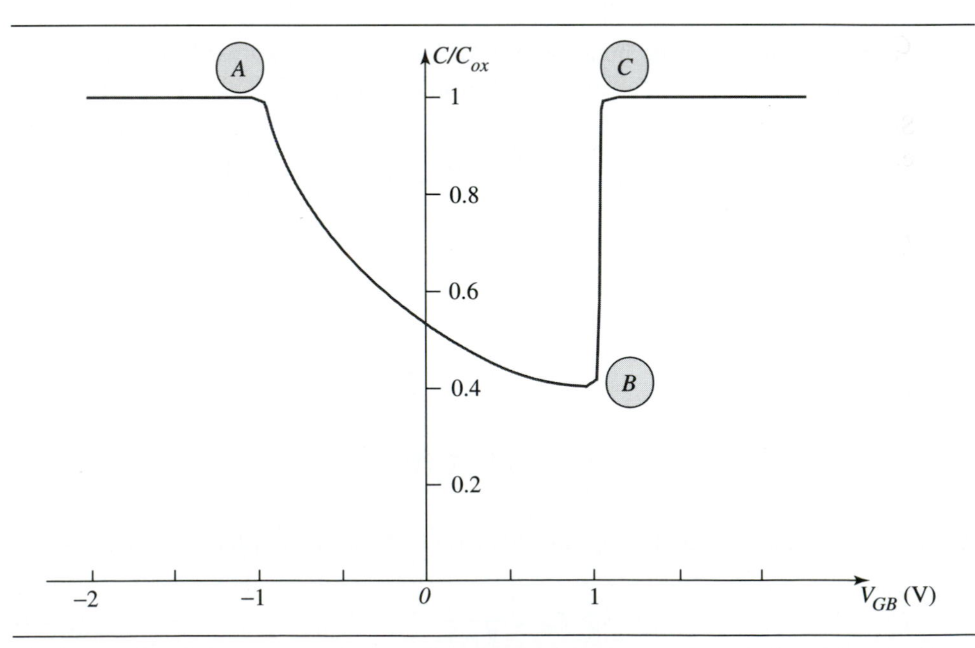

➤ **Figure Ex3.10** Measured capacitance-voltage curve for an MOS structure with the capacitance normalized to its maximum value—the oxide capacitance.

Using the definition of the flatband voltage

$$V_{FB} = -1\text{V} = -(0.55\text{V} - \phi_{sub}) \rightarrow \phi_{sub} = -450\,\text{mV}$$

where ϕ_{sub} is the substrate potential. Since ϕ_{sub} is negative, the substrate is p-type. To find the acceptor doping level in the substrate, we apply the 60mV rule.

$$\phi_p = (-60\text{mV})\log\left(\frac{N_a}{1\times 10^{10}\text{cm}^{-3}}\right) = -450\,\text{mV} \rightarrow N_a = 3.16\times 10^{17}\,\text{cm}^{-3}$$

(b) Determine the oxide thickness t_{ox}.

SOLUTION

To find the oxide thickness t_{ox}, we must first find the oxide capacitance. Since the capacitance has been normalized, it is not possible to use the measurements in accumulation or in inversion to find C_{ox} directly. Instead, we use the depletion region curve where the capacitance is the series combination of C_{ox} and the capacitance of the depletion region C_b. From Eq. (3.157), the minimum normalized capacitance is given by

$$C_{\min}/C_{ox} = C(V_{GB} = V_{Tn}^{-})/C_{ox} = \frac{1}{\sqrt{1 + \dfrac{2C_{ox}^2(V_{Tn} - V_{FB})}{q\varepsilon_s N_a}}}$$

Solving for the oxide capacitance, we find that

$$C_{ox} = \frac{\varepsilon_{ox}}{t_{ox}} = \sqrt{\frac{q\varepsilon_s N_a}{2(V_{Tn} - V_{FB})}\left[\left(\frac{C_{ox}}{C_{\min}}\right)^2 - 1\right]}$$

Substituting $C_{\min}/C_{ox} = 0.4$ and the other parameters from part (a) into the expression for the oxide thickness, we find that

$$t_{ox} = \frac{\varepsilon_{ox}}{\sqrt{\dfrac{q\varepsilon_s N_a}{2(V_{Tn} - V_{FB})}\left[\left(\dfrac{C_{ox}}{C_{\min}}\right)^2 - 1\right]}}$$

$$t_{ox} = \frac{(3.9)(8.85\times 10^{-14})}{\sqrt{\dfrac{1.6\times 10^{19}(1.035\times 10^{-12})(3.16\times 10^{17})}{2(1-(-1))}\left(\left(\dfrac{1}{0.4}\right)^2 - 1\right)}} = 132\,\text{Å}$$

An alternative approach is to use the threshold voltage to calculate C_{ox} from Eq. (3.95). Using this approach, we find that $t_{ox} = 125$ Å which is about 5% lower. Uncertainties in reading the threshold voltage from the *C-V* curve and limitations of the delta-depletion approximation account for the discrepancy.

(c) From your results from part (a) and (b), find the gate charge Q_G for $V_{GB} = -3$ V and for $V_{GB} = +3$ V

SOLUTION

For $V_{GB} = -3\text{ V} < V_{FB}$, the MOS capacitor is in accumulation. The gate charge is

$$Q_G = C_{ox}(V_{GB} - V_{FB}) = 2.55 \times 10^{-7}\text{F/cm}^2(-3\text{V} - (-1\text{V})) = -5.1 \times 10^{-7}\text{ C/cm}^2$$

For $V_{GB} = 3\text{ V} > V_{Tn}$, the MOS capacitor is in inversion. The gate charge is equal and opposite to the sum of the magnitudes of the inversion electron charge and the magnitude of the maximum depletion charge. The latter is found from Eq. (3.140) and Eq. (3.142)

$$Q_G = C_{ox}(V_{GB} - V_{Tn}) - Q_{B,\max} = 5.1 \times 10^{-7} + 2.8 \times 10^{-7}\text{ C/cm}^2 = 7.9 \times 10^{-7}\text{ C/cm}^2$$

3.9.3 Physical Interpretation

In order to make physical sense of this rather complicated capacitance-voltage curve, we will investigate where the increment in gate charge q_g is mirrored for each of the three different regions of operation: accumulation, depletion, and inversion. Starting with a MOS capacitor biased in accumulation, Fig. 3.37 illustrates the charge density before and after a positive increment q_g is added to the DC gate charge Q_G. The increment in gate charge is due to a positive increment in the gate voltage v_{gb}. In Fig. 3.37(a), we note that $Q_G < 0$ for accumulation. The charge increment on the gate is mirrored in the accumulated holes under the gate oxide as shown in the small-signal charge density plotted in Fig. 3.37(c). The incremental charge distribution is identical to that of the parallel-plate capacitor in Ex. 3.1. Therefore, the capacitance of the MOS structure in accumulation is C_{ox}, the oxide capacitance.

Figure 3.38 shows the more complicated situation for a MOS capacitor biased in depletion. In this case, the gate charge is positive and the gate-bulk bias voltage is in the range $V_{FB} \le V_{GB} \le V_{Tn}$. The gate charge increment is mirrored at the bottom of the depletion region as shown in Fig. 3.38(c). The incremental charges $+q_g$ and $-q_g$ are separated by an oxide layer of thickness t_{ox} with capacitance $C_{ox} = \varepsilon_{ox}/t_{ox}$, and a depletion region of thickness $X_d(V_{GB})$ with capacitance $C_b = \varepsilon_s/X_d(V_{GB})$. The equivalent circuit is the series combination of C_{ox} and C_b as shown in Fig. 3.38(c). Therefore, the MOS capacitance when biased in depletion is

$$C = \frac{C_{ox}C_b}{C_{ox} + C_b} \text{ for } V_{FB} \le V_{GB} \le V_{Tn}. \tag{3.162}$$

which explains the physical basis for Eq. (3.161).

For an inverted MOS capacitor, an increment in gate voltage is mirrored in an increase electron charge in the inversion layer as shown in Fig. 3.39. The depletion charge is fixed and does not vary with gate voltage as is reflected in the constant depletion width $X_{d,\max}$ in inversion. As a result, the capacitance in inversion is again the oxide capacitance

$$C = C_{ox} \text{ for } V_{GB} \geq V_{Tn} \tag{3.163}$$

Our analysis of the MOS structure has not considered the time variation of voltages or charges. In fact, our use of the logarithmic relationship between potential and carrier concentration for finding the carrier concentrations in the silicon substrate means we have implicitly assumed that sufficient time has elapsed for transient currents to die out. Measurement of the capacitance-voltage curves shown in this chapter requires use of a slow ramp of the bias voltage V_{GB} and a low-frequency AC

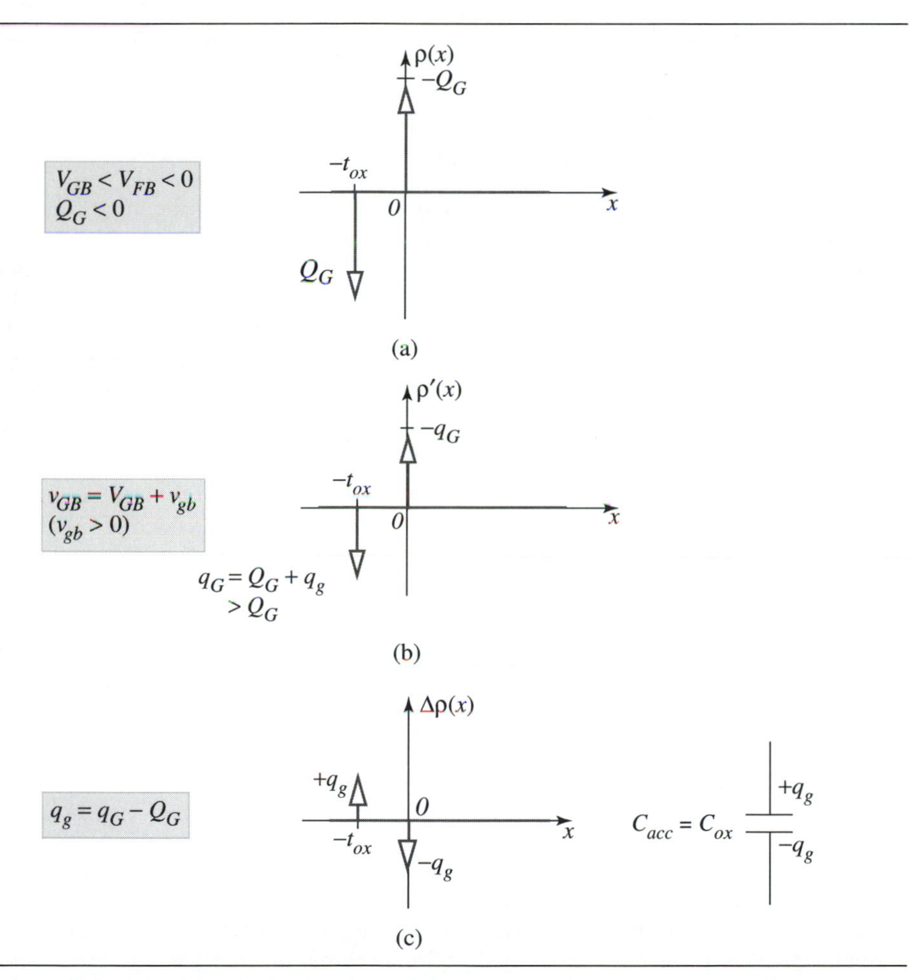

➤ **Figure 3.37** Charge density $\rho(x)$ for a MOS structure on a p-type substrate in accumulation: (a) bias voltage $V_{GB} < V_{FB}$, (b) perturbed charge density $\rho'(x)$ for $v_{GB} = V_{GB} + v_{gb}$, where the incremental voltage $v_{gb} > 0$, and (c) incremental charge density $\Delta\rho(x) = \rho'(x) - \rho(x)$ showing that the incremental charges $\pm q_g$ are separated by the gate oxide. The small-signal model in accumulation is the oxide capacitance.

incremental voltage $v_{gb}(t)$. We will leave further discussion of the various ways to measure MOS *C-V* curves to more advanced texts in device physics or IC fabrication technology.

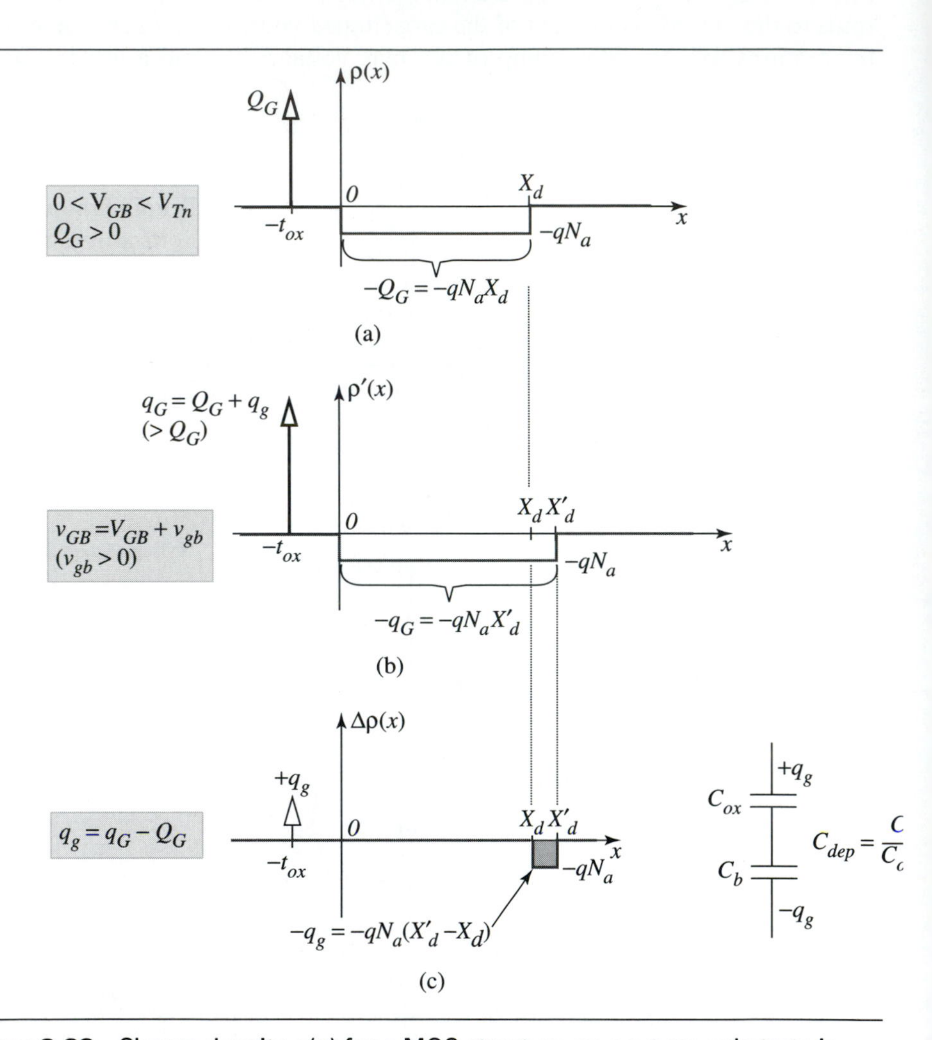

➤ **Figure 3.38** Charge density $\rho(x)$ for a MOS structure on a p-type substrate in depletion: (a) bias voltage $V_{FB} < V_{GB} < V_{Tn}$, (b) perturbed charge density $\rho'(x)$ for $v_{GB} = V_{GB} + v_{gb}$, where the incremental voltage $v_{gb} > 0$, and (c) incremental charge density $\Delta\rho(x) = \rho'(x) - \rho(x)$ showing that the incremental charges $\pm q_g$ are separated by the gate oxide and the depletion region. The small-signal model in depletion is two capacitors in series.

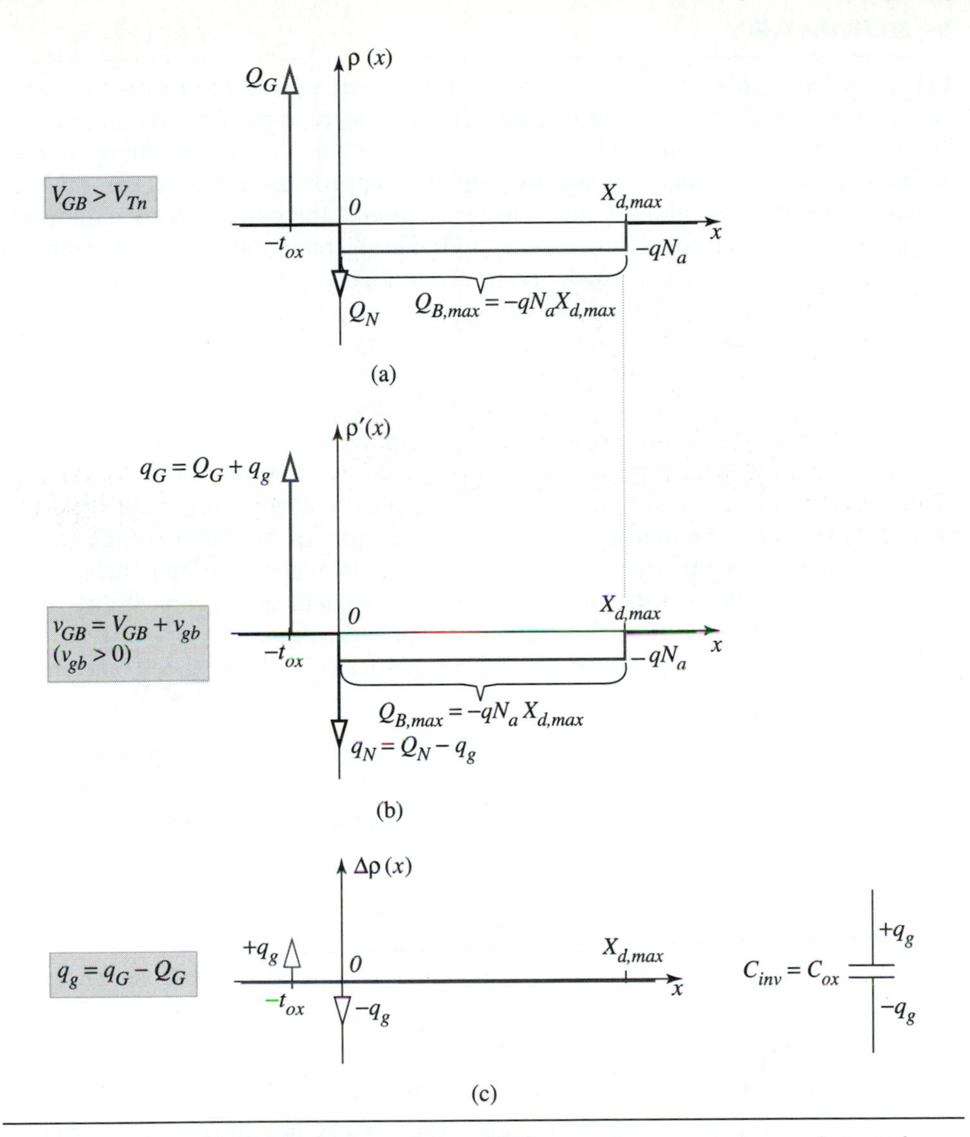

➤ **Figure 3.39** Charge density $\rho(x)$ for a MOS capacitor on a p-type substrate in inversion: (a) bias voltage $V_{GB} > V_{Tn}$, (b) perturbed charge density $\rho'(x)$ for $v_{GB} = V_{GB} + v_{gb}$, where the incremental voltage $v_{gb} > 0$, and (c) incremental charge density $\Delta\rho(x) = \rho'(x) - \rho(x)$ showing that the incremental gate charge is mirrored by an increment in the electron inversion charge. Therefore, the small-signal model in the inversion region is again the oxide capacitance.

➤ SUMMARY

This chapter began with a review of basic electrostatic theory and its application to one-dimensional device structures. These concepts were applied to the pn junction, first in thermal equilibrium and then under a reverse bias $V_D < 0$. The charge density in this structure was modeled using the depletion approximation, in which the electron and hole concentrations in the transition region at the junction were assumed to be negligible. In order to pin down the potential in silicon in thermal equilibrium, we selected a potential reference such that intrinsic silicon is at zero volts. The relationship of potential to the logarithm of carrier concentration followed from this reference, with n-type silicon having a positive potential and p-type silicon a negative potential. The very important concept of small-signal modeling was introduced by finding the capacitance of the reverse-biased pn junction.

The MOS capacitor was the second important device structure analyzed in this chapter. Its nonlinear charge-storage characteristics were described by finding the charge stored on the gate as a function of the applied voltage using the delta-depletion approximation. An unusual and very useful feature of the MOS structure is the possibility of inverting the type of the substrate, which means creating a surface layer that is of opposite type to the substrate. Finally, the small-signal capacitance of the MOS capacitor was found as a function of the applied voltage.

The following important concepts and techniques should have been mastered in Chapter 3.

- Interrelationships between charge density, electric field, and electrostatic potential in one dimension from Gauss's Law and the definition of potential.
- One-dimensional boundary conditions on electric field and potential.
- Techniques for graphically sketching the charge density, electric field, and potential without doing the formal mathematics.
- Equilibrium in the pn junction as a balance of drift and diffusion current densities.
- The logarithmic relationship between potential and electron and hole concentration in thermal equilibrium.
- Charge density in pn junctions using the depletion approximation.
- Electrostatics of the pn junction under reverse bias.
- Contact potentials and the use of Kirchhoff's voltage law to show the current is zero in a short-circuited pn junction in equilibrium.
- Capacitance of the pn junction and its interpretation from the small-signal model of charge storage.
- The MOS structure and its electrostatics in thermal equilibrium.
- Accumulation, depletion, and inversion of the MOS capacitor and the corresponding charge densities
- The delta-depletion approximation and the quantitative analysis of MOS electrostatics
- Capacitance of the MOS structure in accumulation, depletion, and inversion.

FURTHER READING

Reverse-Biased pn Junction

1. G. N. Neudeck, *The PN Junction Diode*, Modular Series on Solid State Devices, 2nd ed. Vol. II, G. W. Neudeck and R. F. Pierret, eds., Addison-Wesley, 1989, Chapter 2. About the same level.
2. R. S. Muller and T. I. Kamins, *Device Electronics for Integrated Circuits*, 2nd ed., Wiley, 1986, Chapter 4. More advanced discussion of the depletion approximation; includes Fermi level concept.

MOS Capacitor

1. C. G. Fonstad, *Microelectronic Devices and Circuits*. McGraw-Hill, 1994, Chapter 9. About the same level; includes a discussion of the effects of oxide charge.
2. R. F. Pierret, *Field Effect Devices*, Modular Series on Solid State Devices, Vol. IV, G. W. Neudeck and R. F. Pierret, eds., Addison-Wesley, 1983, Chapters 2–4. More advanced discussion, including the concept of the Fermi level.

PROBLEMS

EXERCISES

Permittivities: crystalline silicon: $\varepsilon_s = 1.035 \times 10^{-12}$ F/cm
silicon dioxide (SiO_2): $\varepsilon_{ox} = 3.45 \times 10^{-13}$ F/cm
silicon nitride (Si_3N_4): $\varepsilon_n = 8.63 \times 10^{-13}$ F/cm

E3.1 The charge distribution in a sample of silicon is given in Fig. E3.1.

(a) Determine the sign of the electric field at $x = -750$ nm and $x = 250$ nm.

(b) Determine the numerical value of the electric field at $x = -250$ nm.

(c) Plot the electric field as a function of x.

E3.2 The charge distribution in an IC structure is shown in Fig. E3.2.

(a) Find the numerical value of the surface charges Q so that the total charge is zero in the structure. Note that the units of Q are C/cm^2.

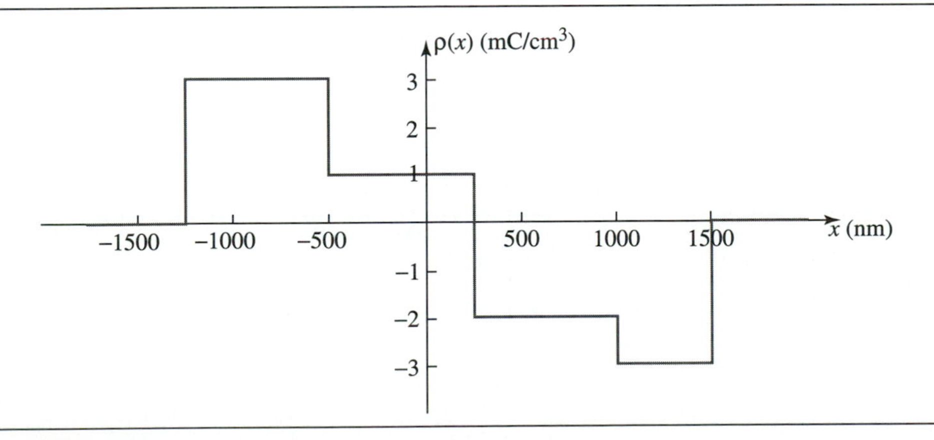

Figure E3.1

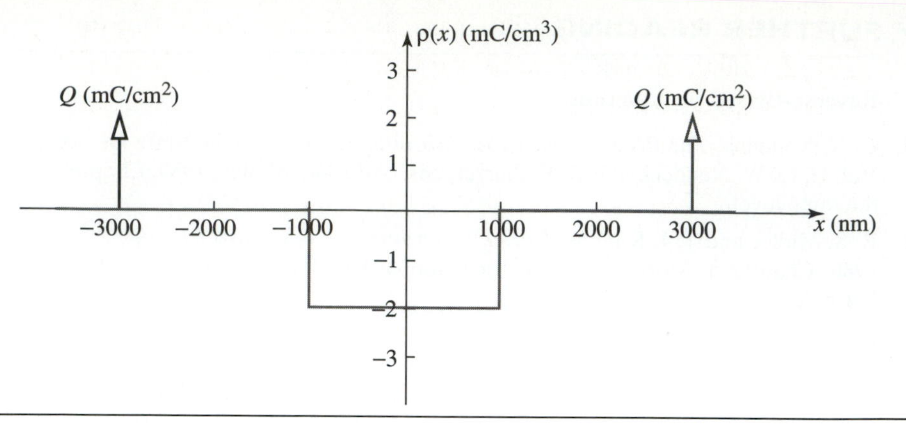

➤ **Figure E3.2**

(b) Plot the electric field $E(x)$ through the structure.

(c) Sketch the potential $\phi(x)$, given that the potential $\phi(\pm 3000) = 0$.

E3.3 The charge distribution in a silicon region is plotted in Fig. E3.3.

(a) Find the width Δ so that the silicon region is electrically neutral. The boundary at $x = -300$ nm is fixed.

(b) Find the numerical value of the electric field at $x = -250$ nm and at $x = 150$ nm

(c) Plot the electric field $E(x)$

(d) If the potential $\phi(-300 \text{ nm}) = -500$ mV, find the potential at $x = 350$ nm.

E3.4 The charge distribution in a silicon region is given in Fig. E3.4. There is no need to find $E(x)$ to solve (a)–(c).

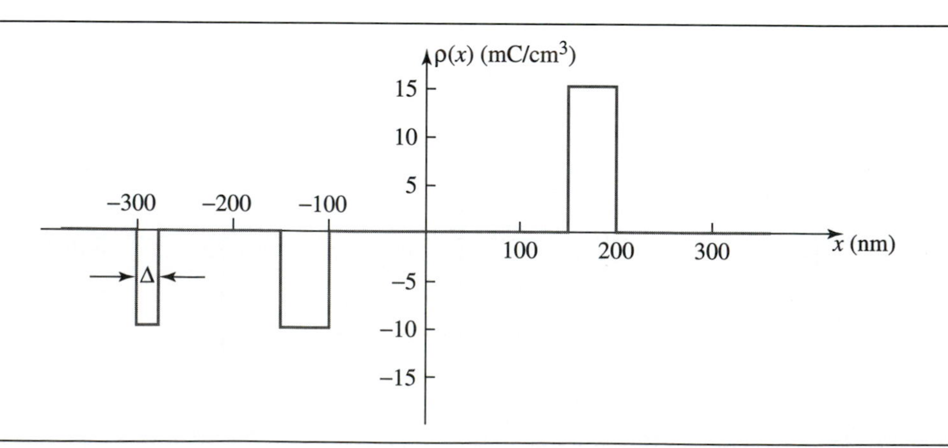

➤ **Figure E3.3**

(a) Find the numerical value of the electric field E at $x = -1.5$ μm.
(b) Find the numerical value of the electric field E at $x = 0$.
(c) Find the numerical value of the electric field E at $x = 1.5$ μm.

E3.5 The electric field in a silicon structure is given in Fig. E3.5.
(a) Plot the charge density $\rho(x)$.

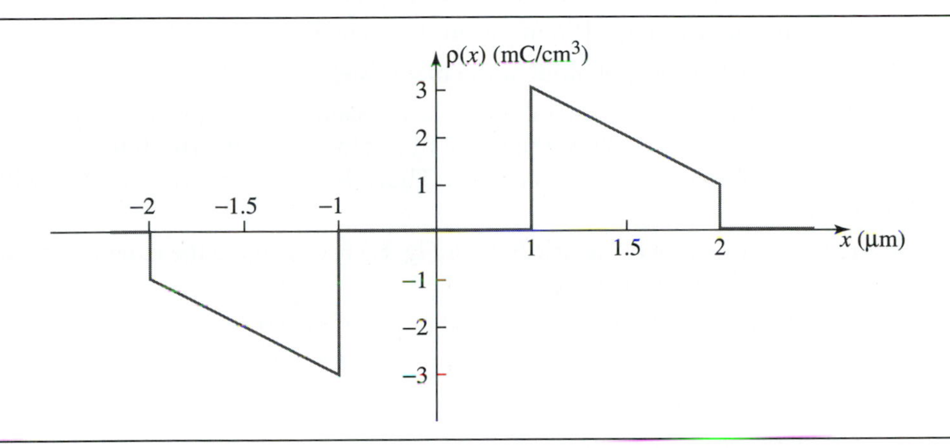

➤ Figure E3.4

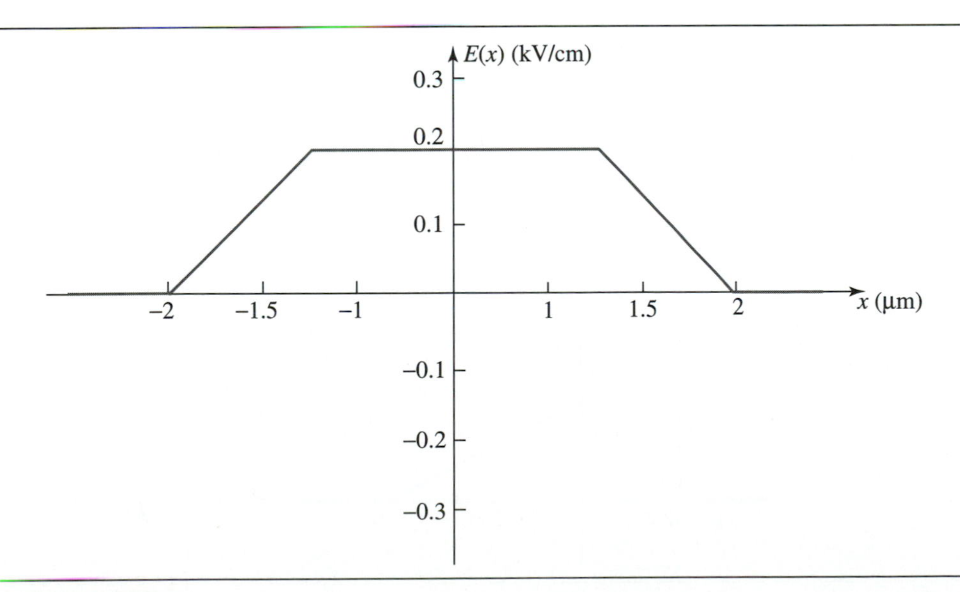

➤ Figure E3.5

(b) If the potential $\phi(-2\ \mu m) = -500$ mV, what is the potential at $x = 0$? What is the potential at $x = 2\ \mu m$. Note that you don't need to solve for the potential function in order to answer this question.

(c) Sketch the potential $\phi(x)$, given that the potential $\phi(-2\ \mu m) = -500$ mV.

E3.6 In the plot of the electric field $E(x)$ in Fig. E3.6, the materials consist of a metal film ($x < t_i$), an insulator with unknown permittivity ($t_i < x < 0$) and silicon ($x \geq 0$).

(a) Given that there is no sheet charge located at $x = 0$, what is the numerical value of the permittivity of the insulating film?

(b) Sketch the charge density $\rho(x)$ through the structure.

(c) How much charge [C/cm^2] is on the metal film at $x = t_i$?

(d) Given that the potential of the metal is $\phi_m = 350$ mV, sketch $\phi(x)$.

E3.7 The electric field in a metal-SiO_2-silicon sandwich is shown in Fig. E3.7. The metal is located at $x < -100$ Å, the oxide film is located between $-100\ \text{Å} < x < 0$, and the silicon substrate ranges over $x > 0$. There is a sheet charge at $x = 0$ with units of C/cm^2.

(a) From the plot of electric field in Fig. E3.6, determine the magnitude and sign of the sheet charge at $x = 0$.

(b) What is the charge on the metal film at $x = -100$ Å?

(c) Sketch the potential $\phi(x)$, given that $\phi(-100\ \text{Å}) = -850$ mV.

E3.8 The electric field $E(x)$ in a silicon region is given in Fig. E3.8.

(a) Plot the charge density $\rho(x)$.

(b) Given that the potential $\phi(0) = 0$, sketch the potential $\phi(x)$.

E3.9 The potential $E(x)$ through a silicon sample is graphed in Fig. E3.9.

(a) Sketch the potential $\phi(x)$, given that $\phi(0) = 0.5$ V.

(b) Sketch the charge density $\rho(x)$.

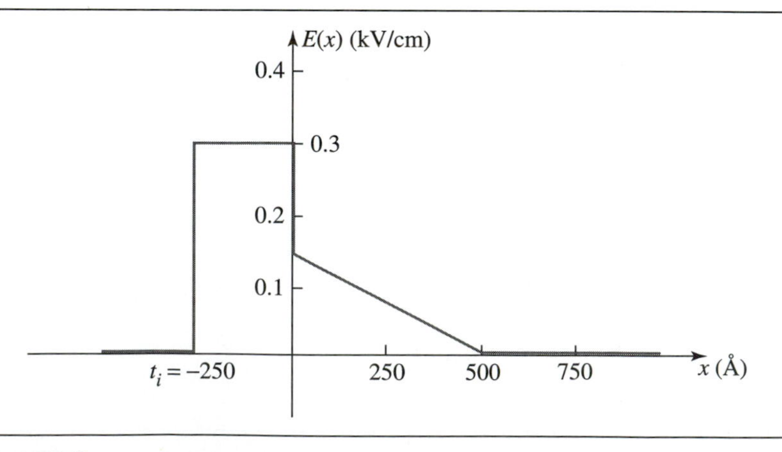

➤ **Figure E3.6**

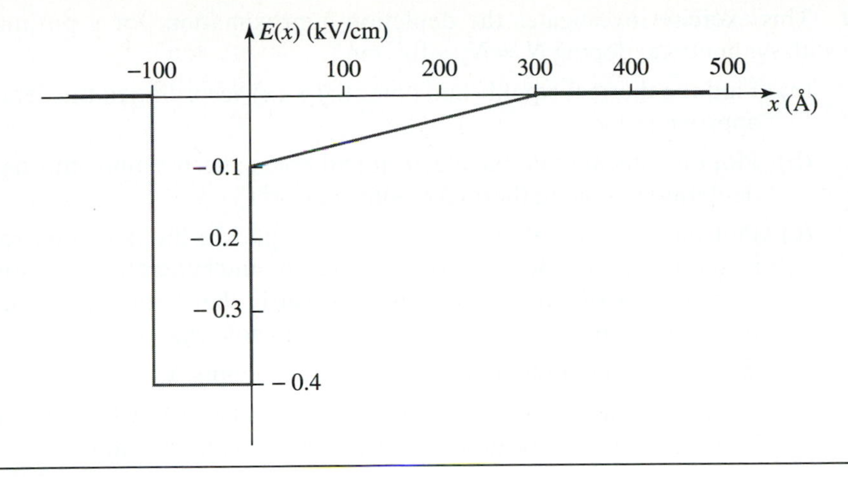

➤ Figure E3.7

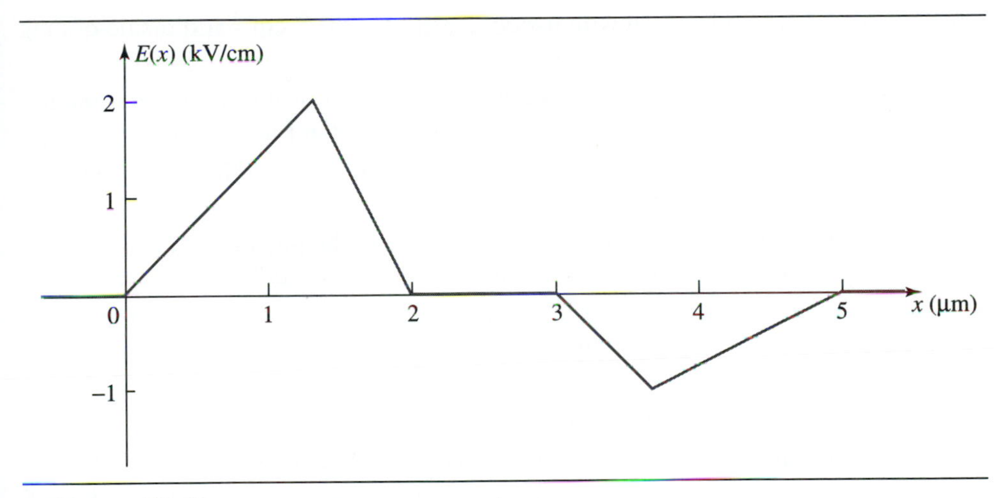

➤ Figure E3.8

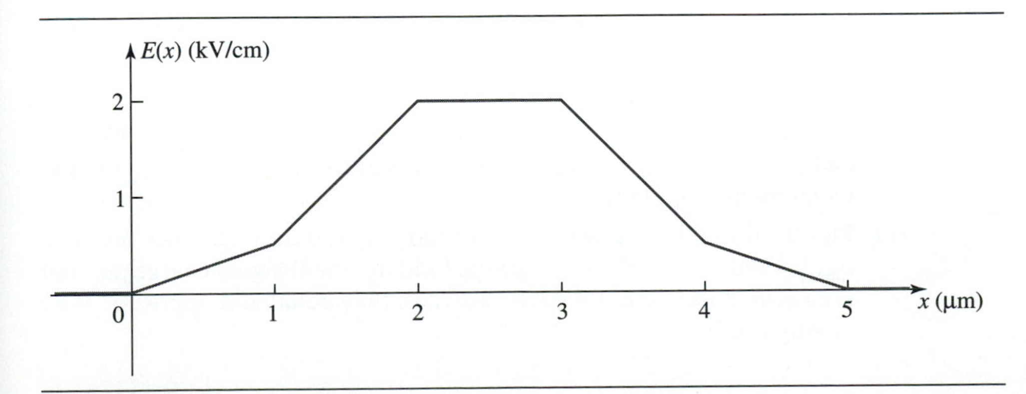

➤ Figure E3.9

E3.10 This exercise investigates the depletion approximation, for a pn junction diode with symmetrical doping $N_a = N_d = 10^{15}$ cm^{-3}.

(a) Find the thermal equilibrium potential $\phi_o(x)$ according to the depletion approximation.

(b) Plot on a linear scale the electron and hole concentrations through the depletion region, in thermal equilibrium.

(c) Plot on a linear scale the charge density $\rho(x)$ in the depletion region, including the contributions of the electrons and holes from your answer in part (b). Estimate the percentage error in the charge on one side of the junction, if we neglect the electron and hole charge?

E3.11 Consider a pn junction diode with symmetrical doping $N_a = N_d = 10^{17}$ cm^{-3}.

(a) What is the magnitude of the electric field $|E_o(x = 0)|$ (where the metallurgical junction is located at $x = 0$) in thermal equilibrium?

(b) For what value of bias voltage V_D is the electric field at $x = 0$ increased by a factor of 3?

(c) Plot $|E(0)|$ as a function of V_D for $-7.5 \text{ V} < V_D < 0$.

E3.12 Consider a pn junction with p-side doping $N_a = 10^{19}$ cm^{-3} and n-side doping $N_d = 10^{17}$ cm^{-3}.

(a) Sketch the potential in thermal equilibrium, using the approximation that the junction is "one-sided" as discussed in Ex. 3.5.

(b) What is the electric field $|E_o(x = 0)|$ (where the metallurgical junction is located at $x = 0$) in thermal equilibrium?

(c) If the maximum electric field allowable in the junction is $E_{\max} = 5 \times 10^5$ V/cm what is the maximum reverse bias permitted?

E3.13 Low doped silicon is used in particle detectors, in order to obtain a wide depletion region. Consider a pn junction with p-side doping $N_a = 5 \times 10^{12}$ cm^{-3} and n-side doping $N_d = 10^{17}$ cm^{-3}.

(a) What is the thermal equilibrium depletion region width?

(b) What is the reverse bias needed to increase the depletion width to 50 μm?

(c) If the maximum electric field allowable in the junction is $E_{\max} = 5 \times 10^5$ V/cm what is the maximum depletion width? What is the bias voltage corresponding to the maximum depletion width?

E3.14 We would like to investigate the charge distribution in a pn junction with doping levels $N_a = 10^{16}$ cm^{-3} and $N_d = 5 \times 10^{16}$ cm^{-3}.

(a) For $V_D = -2.5$ V, plot the charge density $\rho(x)$.

(b) We add to V_D a time-varying voltage $v_d(t) = (250 \text{ mV})\cos(\omega t)$. Plot the charge density when $v_D(t) = V_D + v_d(t)$ is at its maximum ($\rho_{\max}(x)$) and its minimum ($\rho_{\min}(x)$).

(c) Plot the difference between $\rho_{\max}(x)$ and $\rho(x)$ and the difference between $\rho_{\min}(x)$ and $\rho(x)$ on the same graph. Find the small-signal charge per unit area q_j in units C/cm^2 for each case. Are they equal and opposite? Why or why not?

(d) Find the depletion capacitance $C(V_D)$ for $V_D = -2.5$ V and the small-signal depletion charge q_j, assuming that $v_d(t)$ can be considered a small signal. Compare with your answer to (c).

E3.15 In order to evaluate the junction diode as a circuit element, we would like to know the capacitance per unit area and the tuning range.

(a) Plot the capacitance per unit area in thermal equilibrium versus the p-side doping concentration from $N_a = 10^{13}$ cm^{-3} to 10^{19} cm^{-3}. Assume that the n-side doping concentration is $N_d = 10^{19}$ cm^{-3}.

(b) If the maximum electric field allowable in the junction is $E_{max} = 5 \times 10^5$ V/cm, plot the maximum reverse bias $V_{D,max}$ on the capacitor as a function of the p-side doping concentration.

(c) From your results in part (b), plot the ratio of $C(0)/C(V_{D,max})$ as a function of p-side doping concentration.

E3.16 We are interested in using a reverse-biased pn junction as a small-signal capacitor. The doping levels are $N_a = 10^{16}$ cm^{-3} and $N_d = 10^{18}$ cm^{-3}. We would like to have a capacitor with $C = 500$ fF ±150 fF.

(a) What is the capacitance per unit area [units: fF/μm^2] for $V_D = 0$ V.

(b) What is the area needed for the capacitor and the voltage range needed for tuning? Note that $V_D = 0$ V must correspond to $C = 650$ fF, the maximum capacitance.

E3.17 An MOS capacitor is fabricated on a p-type silicon substrate with doping concentration $N_a = 10^{16}$ cm^{-3}, using an n$^+$ polysilicon gate.

(a) What is the flatband voltage, V_{FB}?

(b) What is the surface potential in inversion?

(c) What is the maximum width of the depletion region?

(d) For $t_{ox} = 200$ Å, what is the threshold voltage V_{Tn}?

E3.18 An MOS capacitor is fabricated on an n-type silicon substrate with doping concentration $N_d = 5 \times 10^{15}$ cm^{-3}, using an n$^+$ polysilicon gate.

(a) What is the flatband voltage, V_{FB}?

(b) What is the surface potential in inversion?

(c) What is the maximum width of the depletion region?

(d) For $t_{ox} = 150$ Å, what is the threshold voltage V_{Tp}?

E3.19 In an IC containing MOS structures, the n$^+$ polysilicon gate is often used as an interconnect layer. In the cross section in Fig. E3.19, the polysilicon layer forms a parasitic MOS capacitor over the "field" region in between devices. The oxide thickness $t_{ox1} = 200$ Å and the substrate doping is $N_a = 5 \times 10^{16}$ cm^{-3}.

(a) Find the threshold voltage V_{Tn1} of the MOS capacitor formed over the thin oxide of thickness t_{ox1}.

(b) What thickness t_{ox2} of field oxide is required if the threshold voltage V_{Tn2} of this MOS capacitor must be a factor of 5 higher than V_{Tn1}?

(c) If the substrate doping under the field regions is increased by a factor of 10 and the doping in the thin-oxide region kept at 5×10^{16} cm^{-3}, what is the thickness t_{ox2} such that $V_{Tn2} = 5\ V_{Tn1}$?

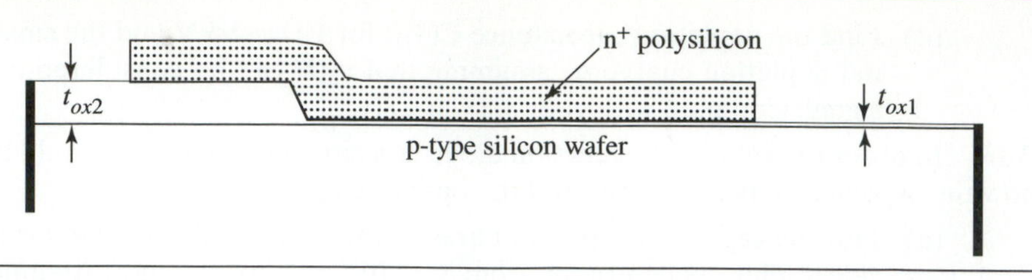

➤ **Figure E3.19**

E3.20 It is occasionally useful to design a MOS capacitor with a threshold voltage $V_T = 0$ V. The oxide thickness is $t_{ox} = 250$ Å and we use an n$^+$ polysilicon gate.

(a) What p-type substrate doping concentration is required for $V_{Tn} = 0$ V? An iterative approach may be helpful, since the substrate potential ϕ_p is a weak function of the doping concentration.

(b) Sketch the potential $\phi_o(x)$ in thermal equilibrium.

E3.21 This exercise concerns an MOS capacitor with an n$^+$ polysilicon gate, a 500 Å-thick gate oxide, and a p-type substrate with acceptor concentration $N_a = 7.5 \times 10^{15}$ cm^{-3}.

(a) If the gate charge is $Q_G = -1 \times 10^{-7}$ C/cm^2, sketch the charge density through the MOS capacitor.

(b) For the gate charge in part (a), what is the gate-bulk bias V_{GB}?

(c) If the gate charge is $Q_G = +2.5 \times 10^{-7}$ C/cm^2, sketch the charge density through the MOS capacitor.

(d) For the gate charge in part (c), what is the inversion charge Q_N in C/cm^2?

(e) For the gate charge in part (c), what is the gate-bulk bias V_{GB}?

E3.22 For this exercise, the MOS capacitor has an n$^+$ polysilicon gate, a 450 Å-thick gate oxide, and an n-type substrate with donor concentration $N_d = 7.5 \times 10^{16}$ cm^{-3}.

(a) If the gate charge is $Q_G = 2 \times 10^{-7}$ C/cm^2, sketch the charge density through the MOS capacitor.

(b) For the gate charge in part (a), what is the gate-bulk bias V_{GB}?

(c) If the gate charge is $Q_G = -5 \times 10^{-7}$ C/cm^2, sketch the charge density through the MOS capacitor.

(d) For the gate charge in part (c), what is the inversion charge Q_P in C/cm^2?

E3.23 The MOS capacitor is sometimes used for storing charge. Consider an MOS capacitor with n$^+$ polysilicon gate, $t_{ox} = 100$ Å, and a p-type silicon substrate with an acceptor concentration $N_a = 10^{17}$ cm^{-3}.

(a) What is the threshold voltage, V_{Tn}?

(b) For a capacitor with area 1 μm × 5 μm and a bias voltage $V_{GB} = 2.9$ V, what is the inversion layer electron charge, Q_N in C? What is the number of electrons stored in the capacitor?

E3.24 The capacitance-voltage curve for an MOS capacitor with an n$^+$ polysilicon gate is given in Fig. E3.24.

(a) What is the flatband voltage, V_{FB}? What is the threshold voltage, V_T?

(b) What is the oxide thickness, t_{ox}?

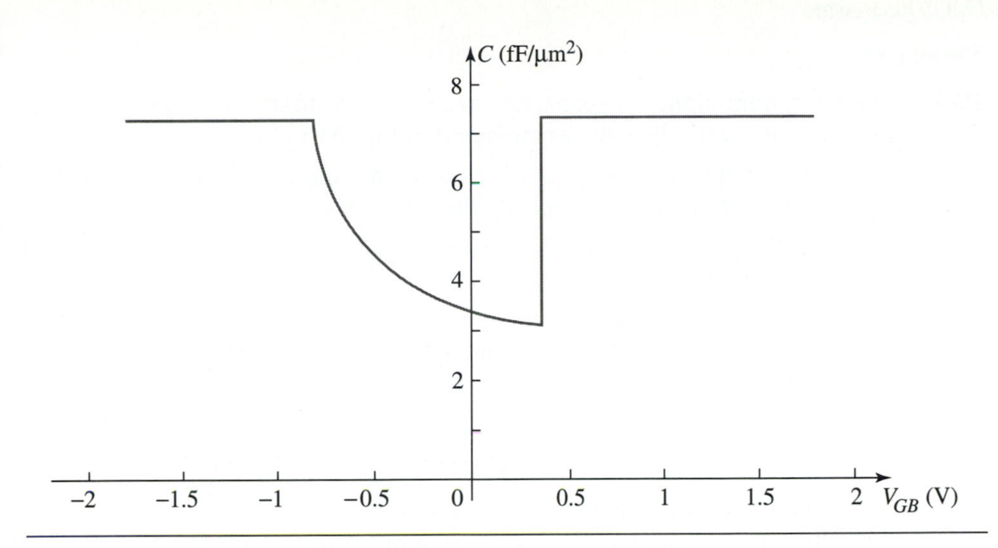

➤ **Figure E3.24**

(c) What is the substrate type?

(d) What is the substrate doping?

E3.25 The capacitance-voltage curve for an MOS capacitor with an n^+ polysilicon gate is given in Fig. E3.25.

(a) What is the flatband voltage, V_{FB}? What is the threshold voltage, V_T?

(b) What is the oxide thickness, t_{ox}?

(c) What is the substrate type?

(d) What is the substrate doping?

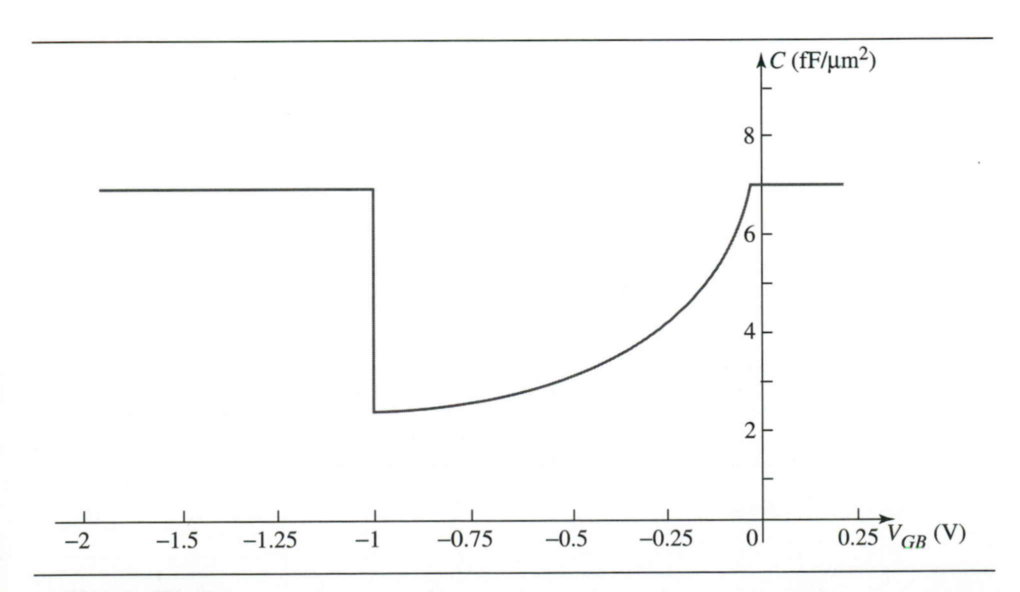

➤ **Figure E3.25**

PROBLEMS

P3.1 An aluminum-aluminum capacitor has a composite dielectric of Si_3N_4 (silicon nitride) and SiO_2 with thicknesses indicated in Fig. P3.1.

(a) For an applied voltage $V_{21} = 1$ V between the metal layers, plot the electric field $E(x)$ in the composite dielectric ($0 < x < 500$ Å).

(b) Plot the potential $\phi(x)$, with the bottom plate (metal 2) being considered the reference ($\phi(500\ \text{Å}) = 0$).

(c) What is the magnitude and sign of the sheet charge on the top plate (metal 2), for $V_{21} = 1$ V? A circuit approach to this part might be helpful.

P3.2 A region of silicon has the charge density shown in Fig. P3.2.

(a) What is the sheet charge Q [units: C/cm^2] needed to balance the negative charge density over the interval 1 μm $< x <$ 2.5 μm.

(b) Find the electric field at $x = 0.5$ μm.

(c) Find the potential at $x = 0$, given that $\phi(2.5\ \mu\text{m}) = -0.4$ V.

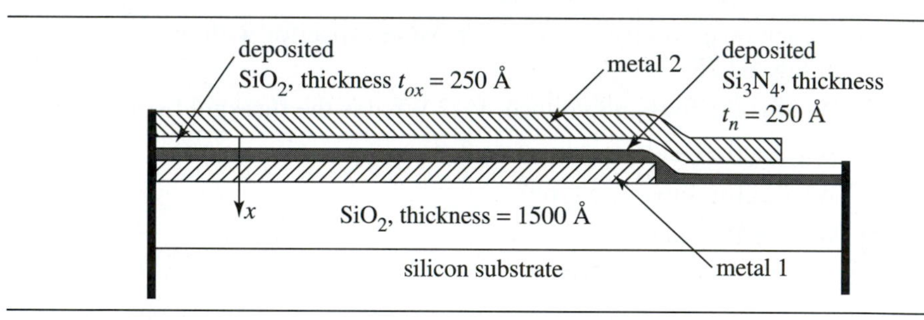

➤ **Figure P3.1**

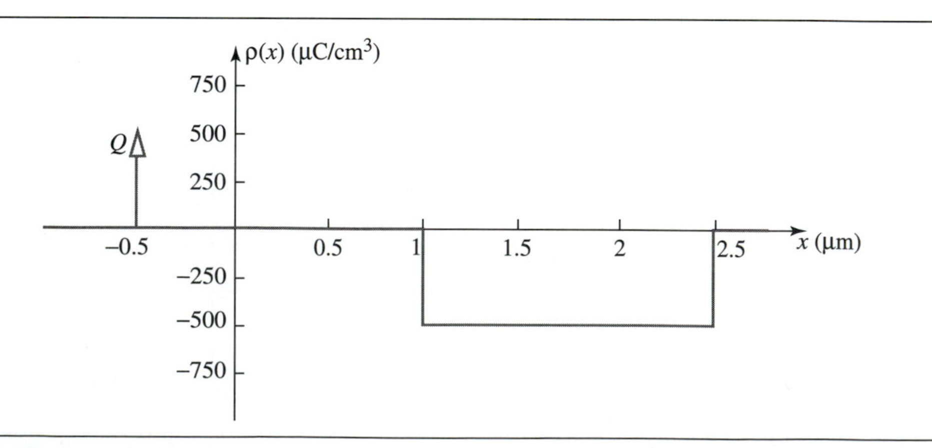

➤ **Figure P3.2**

P3.3 A region of silicon has the charge density given in Fig. P3.3. The sheet charge $Q = -2.5 \times 10^{-7}$ C/cm^2.

(a) Find the sheet charges Q_a at $x = 0$ and $x = 1.75$ μm needed to have the region of silicon be electrically neutral.

(b) Sketch the electric field $E(x)$ through the structure.

(c) Sketch the potential $\phi(x)$ given that $\phi(-1\ \mu m) = 0$.

P3.4 In the aluminum-aluminum capacitor shown in cross section in Fig. P3.1, we now add a sheet charge $Q_I = 5 \times 10^{10}$ C/cm^2 at the interface between the SiO_2 and Si_3N_4 at $x = 250$ Å.

(a) For $V_1 = V_2 = 0$, plot the electric field $E(x)$ through dielectric.

(b) Repeat part (a) for $V_1 = 1$ V and $V_2 = 0$ V.

(c) Plot the potential $\phi(x)$ for part (b).

P3.5 In Fig. P3.5, there is a material interface at $x = 0$ at which a sheet charge is also located, with value $Q = -2.5$ nC/cm^2.

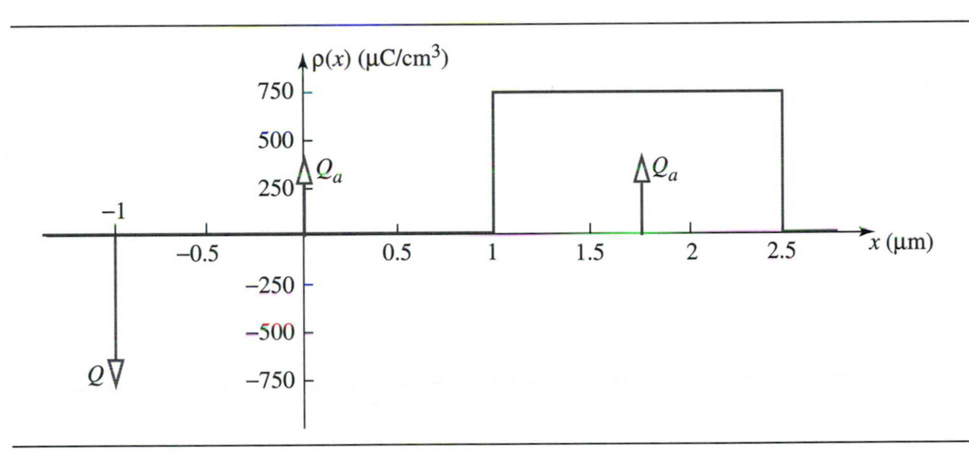

➤ **Figure P3.3**

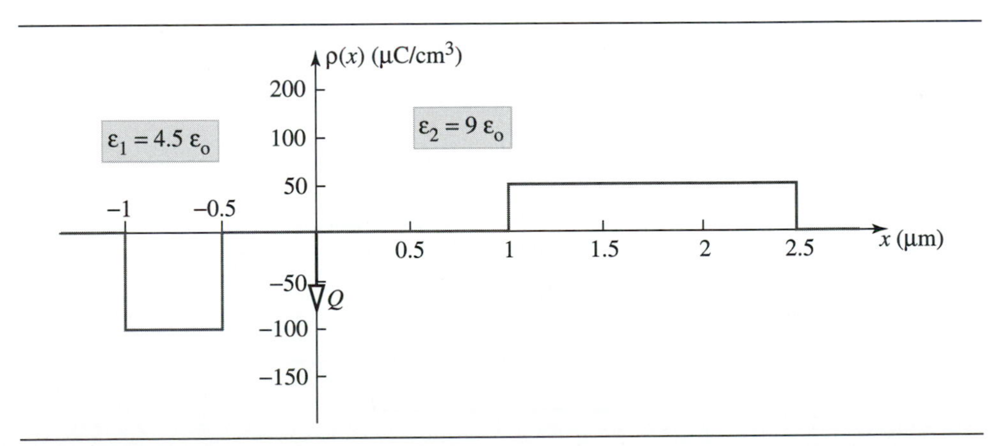

➤ **Figure P3.5**

(a) What is the relationship between $E(0^-)$ and $E(0^+)$?

(b) Sketch the electric field through the structure.

P3.6 The silicon structure in Fig. P3.6 consists of two pn junctions that are separated by 1 μm. The n-type to metal contact potential $\phi_{nm} = 400$ mV $= \phi_n - \phi_m = -\phi_{mn}$.

(a) Find the depletion widths for each junction in thermal equilibrium.

(b) Plot the electric field $E_o(x)$ through the structure in thermal equilibrium.

(c) Plot the potential $\phi_o(x)$ in thermal equilibrium, including the contact potentials.

P3.7 The pn junction in Fig. P3.7 has a stepped doping concentration on each side of the metallurgical junction.

(a) Find the depletion width X_{do} in thermal equilibrium.

(b) If the n and p sides of the junction are each depleted by 1 μm, sketch the charge density, the electric field, and the potential.

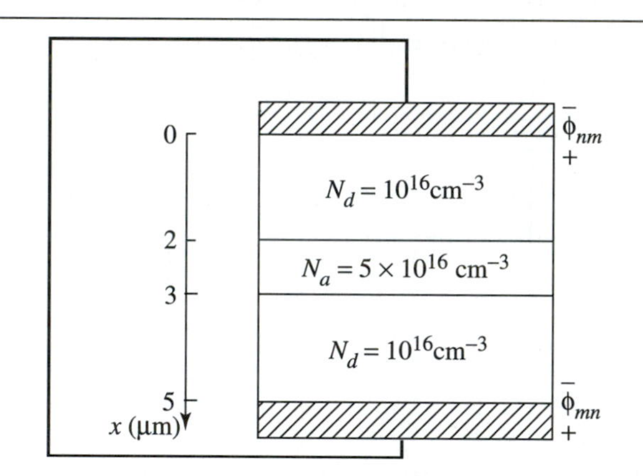

➤ Figure P3.6

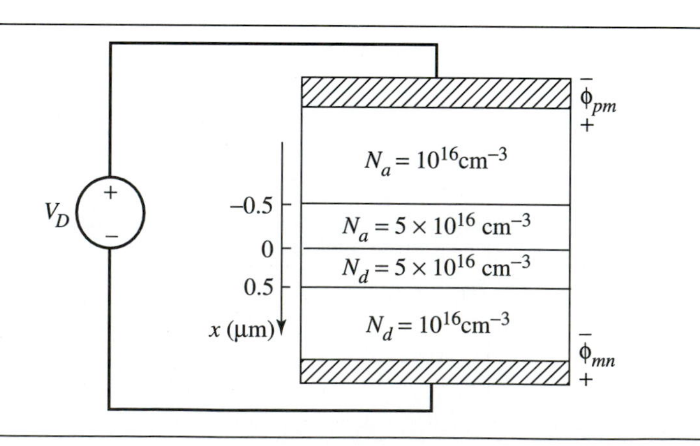

➤ Figure P3.7

(c) Find the applied voltage V_D (< 0 V) that corresponds to the depletion width in part (b).

P3.8 The pn junction diode in Fig. P3.8 has a stepped doping concentration.

(a) Find the depletion width X_{do} in thermal equilibrium.

(b) If the n and p sides of the junction are depleted to ± 1.0 μm, sketch the charge density, the electric field, and the potential.

(c) Find the applied voltage V_D (< 0 V) that corresponds to the depletion width in part (b).

P3.9 Figure P3.9 shows a p-i-n diode. The *i* stands for intrinsic silicon, which in practice means silicon that is compensated with $N_a = N_d$.

(a) Sketch the charge density, electric field, and potential in thermal equilibrium.

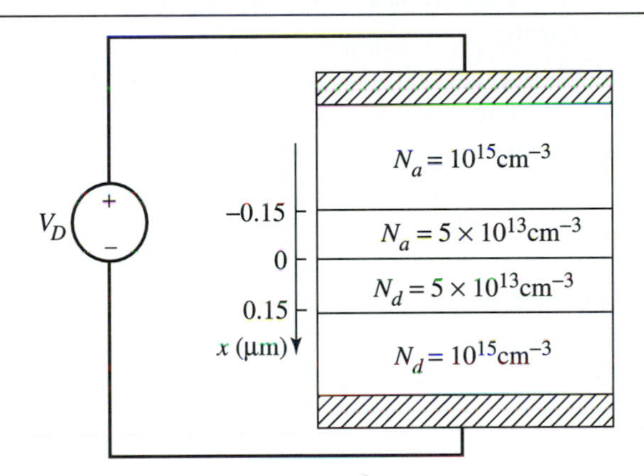

➤ **Figure P3.8**

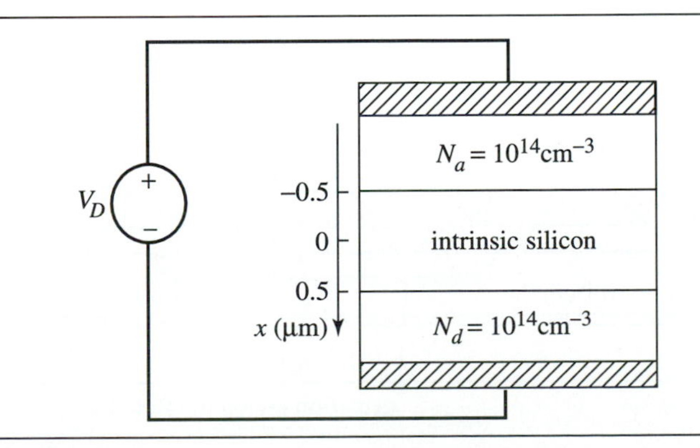

➤ **Figure P3.9**

(b) For an applied bias $V_D = -3$ V, find the widths of the depletion regions in the p and the n silicon.

P3.10 The IC structure shown in cross section in Fig. P3.10 is part of a voltage-controlled resistor.

(a) What is the thickness t of the undepleted n-type layer for $V_C = 0$. The distance between the metallurgical junctions between the n-type layer and the p$^+$ layers is $t_o = 1$ μm.

(b) For what value of V_C does $t = 0$? This condition is called pinch-off.

(c) Sketch the potential $\phi(x)$ at pinch-off.

P3.11 This problem continues the voltage-controlled resistor from Fig. P3.10.

(a) Plot the sheet resistance of the n-type layer as a function of V_C.

(b) Plot the junction capacitance per unit area (fF/μm^2) of the n-type layer as a function of V_C.

P3.12 Sketch the depletion capacitance of the pn junction in Fig. P3.8 as a function of the applied bias V_D. The capacitance at any breakpoints (where the slope of $C(V_D)$ changes) should be accurate.

P3.13 Sketch the depletion capacitance of the pn junction in Fig. P3.9 as a function of the applied bias V_D. The capacitance at any breakpoints (where the slope of $C(V_D)$ changes) should be accurate.

P3.14 The analysis of the electrostatics of the pn junction remains valid for small forward biases $V_D < 250$ mV. What is the percentage increase in junction capacitance per unit area over that of thermal equilibrium is found for the voltage-controlled resistor in Fig. P3.10 if we use $V_C = -250$ mV?

P3.15 The fabrication process used to make the voltage-controlled resistor in Fig. P3.10 has the following normalized uncertainties for each junction depth x_j and each doping concentration N.

$$x_j = \bar{x}_j(1 \pm 0.02) \text{ and } N = \bar{N}(1 \pm 0.04)$$

(a) Find the uncertainty in the pinch-off voltage, assuming that all of the contributions are independent.

(b) Find the uncertainty in the zero-bias junction capacitance per unit area.

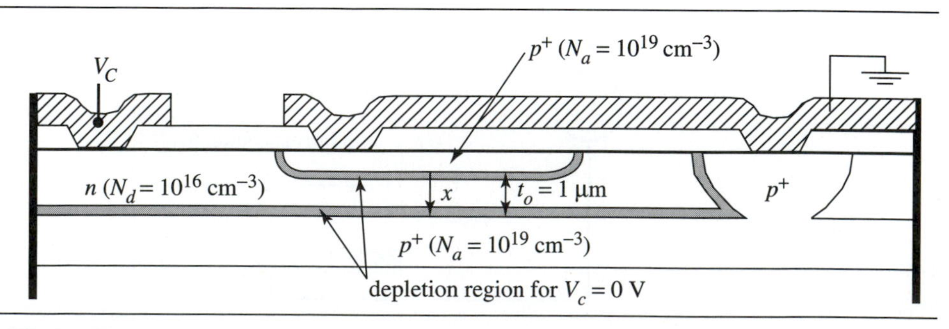

➤ **Figure P3.10**

P3.16 Figure P3.16 is the cross section of an ion-implanted region. In this problem, we will develop a very simple model for the extra capacitance contributed by the edges (or sidewalls) of the region. The bottom of the region has area $W \times W$ and the edges are modeled as half-circles in cross section, with a radius equal to the junction depth x_j. The doping concentrations for the bottom and sidewalls are approximated by:

$$\text{bottom: } N_a = 10^{17}\ \text{cm}^{-3},\ N_d = 10^{16}\ \text{cm}^{-3}$$

$$\text{sidewall: } N_a = 5 \times 10^{17}\ \text{cm}^{-3},\ N_d = 10^{17}\ \text{cm}^{-3}$$

(a) Find the sidewall capacitance per unit edge length in thermal equilibrium by applying the one-dimensional depletion model to the semi-circular edge of the region. Use $x_j = 0.75\ \mu\text{m}$.

(b) Plot the ratio of the sidewall capacitance to the total junction capacitance in thermal equilibrium, as a function of the lateral dimension W over the range $W = 2\ \mu\text{m} - 100\ \mu\text{m}$. Use $x_j = 0.75\ \mu\text{m}$.

P3.17 Heavily boron-doped polysilicon (referred to as p^+ polysilicon) is sometimes used as a gate material. Its thermal equilibrium potential is $\phi_{p+} = -550$ mV. The gate oxide is $t_{ox} = 200$ Å and the substrate doping is $N_a = 7.5 \times 10^{16}\ \text{cm}^{-3}$ for this problem.

(a) Find the flatband voltage V_{FB}. What state is the capacitor in for the case of thermal equilibrium?

(b) Find the threshold voltage V_{Tn}.

(c) Sketch the charge density, electric field, and potential through the MOS capacitor in thermal equilibrium.

(d) Sketch the charge density, electric field, and potential through the MOS capacitor for $V_{GB} = V_{Tn}$.

P3.18 Using the p^+ polysilicon gate introduced in P3.17, we fabricate an MOS capacitor with a gate oxide is $t_{ox} = 250$ Å and a substrate doping is $N_d = 5 \times 10^{16}\ \text{cm}^{-3}$ of

(a) Find the flatband voltage V_{FB}. What state is the capacitor in for the case of thermal equilibrium?

(b) Find the threshold voltage V_{Tp}.

(c) Sketch the charge density, electric field, and potential through the MOS capacitor in thermal equilibrium.

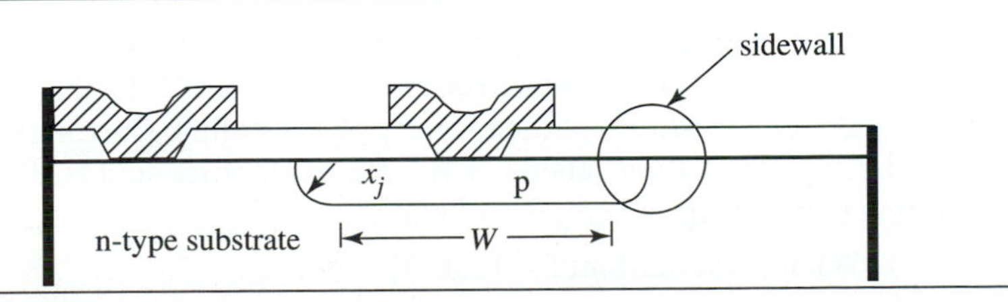

➤ **Figure P3.16**

(d) Sketch the charge density, electric field, and potential through the MOS capacitor for $V_{GB} = V_{Tp}$.

P3.19 A composite dielectric is sometimes useful in an MOS capacitor. For this problem, the gate insulator consists of 50 Å of oxide (on the substrate), over which is 450 Å of *vacuum*. The gate is n$^+$ polysilicon and the substrate is doped p-type with $N_a = 5 \times 10^{15}$ cm^{-3}.

(a) What is the threshold voltage?

(b) Sketch the electric field through the MOS capacitor for $V_{GB} = -2.5$ V.

(c) Sketch the electric field through the MOS capacitor for $V_{GB} = +2.5$ V.

(d) For $V_{GB} = 2.5$ V, what is the change in the gate charge if the vacuum gap is reduced by 10%? A pressure transducer has been made using this principle.

(e) What is the electrostatic force f_E per unit area on the polysilicon gate for a gate-bulk bias $V_{GB} = 2.5$ V? This effect must be accounted for in designing a pressure sensor.

P3.20 This problem considers the effect of fixed charge at the SiO_2–silicon interface on the parameters of an MOS capacitor. For a MOS capacitor with n$^+$ polysilicon gate, $t_{ox} = 150$ Å, and $N_a = 10^{17}$ cm^{-3}, we have $Q_F = +2.5 \times 10^{-8}$ C/cm^2 located at $x = 0$.

(a) Sketch the charge density in thermal equilibrium for this MOS capacitor. What is the shift in the flatband voltage V_{FB} due to the fixed charge?

(b) What is the shift in the threshold voltage V_{Tn} due to the fixed charge?

(c) Sketch normalized *C-V* curve with and without the fixed charge.

P3.21 This problem considers the effect of fixed charge at the SiO_2–silicon interface on the parameters of an MOS capacitor. For a MOS capacitor with n$^+$ polysilicon gate, $t_{ox} = 150$ Å, and $N_d = 10^{17}$ cm^{-3}, we have $Q_F = +2.5 \times 10^{-8}$ C/cm^2 located at $x = 0$.

(a) Sketch the charge density in thermal equilibrium for this MOS capacitor. What is the shift in the flatband voltage V_{FB} due to the fixed charge?

(b) What is the shift in the threshold voltage V_{Tp} due to the fixed charge?

(c) Sketch normalized *C-V* curve with and without the fixed charge.

P3.22 The control of the threshold voltage is critical for VLSI CMOS processes. In this problem, we explore the implications for the control of gate oxide thickness. We would like the threshold voltage to be $V_{Tn} = 500$ mV ± 50 mV in a particular process.

(a) Find the required average gate oxide thickness is $\overline{t_{ox}}$ in Å.

(b) Assuming that the substrate doping is $N_a = 10^{17}$ cm^{-3} and is perfectly controlled, what is the allowable variation in the gate oxide thickness?

(c) Now consider that $N_a = 10^{17}$ cm$^{-3} \pm 2 \times 10^{15}$ cm^{-3}. How much does the substrate doping variation add to the variation in V_{Tn}?

P3.23 Due to a processing error, the n$^+$ polysilicon gate of an MOS capacitor is merely n-type with a doping concentration $N_d = 10^{17}$ cm^{-3}. In this problem, assume you can treat the electrostatics of the n-type polycrystalline silicon gate exactly as you would treat n-type crystalline silicon. The gate oxide thickness is 200 Å and the substrate doping concentration is $N_a = 10^{17}$ cm^{-3}.

(a) What is the flatband voltage V_{FB}?

(b) For $V_{GB} = 2$V, what is the width of the depletion region in the n-type gate polysilicon? Sketch the electric field and the potential.

(c) Sketch the normalized *C-V* curve for this capacitor.

P3.24 In this problem, the substrate is doped at $N_a = 5 \times 10^{16}$ cm^{-3} for $0 < x < 500$ Å and $N_a = 5 \times 10^{17}$ cm^{-3} for $x > 500$ Å. The gate oxide is 250 Å thick and the gate is n$^+$ polysilicon.

(a) Sketch the charge density in thermal equilibrium.

(b) What is the flatband voltage V_{FB}?

(c) What is the threshold voltage V_{Tn}?

P3.25 Due to an error in implantation, the substrate doping for an MOS capacitor is $N_a = 10^{16}$ cm^{-3} rather than 5×10^{16} cm^{-3}. The gate oxide is 250 Å thick and the gate is n$^+$ polysilicon. We would like to remedy this error by doing another boron implant.

(a) What are the minimum dose and depth of the implant required to obtain the desired threshold voltage?

(b) Sketch the normalized *C-V* curve before and after the corrective implant.

P3.26 Sketch the normalized *C-V* curve for the MOS capacitor with the stepped substrate doping in P3.24.

P3.28 The layout and cross section of a MOS structure is shown in Fig. P3.28.

(a) What are the threshold voltages of the thin oxide and thick oxide portions of this MOS capacitor?

(b) For $V_{GB} = 5$ V, what is the ratio of the inversion layer charge stored on the thick-oxide to the thin-oxide portions?

(c) Sketch the capacitance (in fF) versus V_{GB} for this structure.

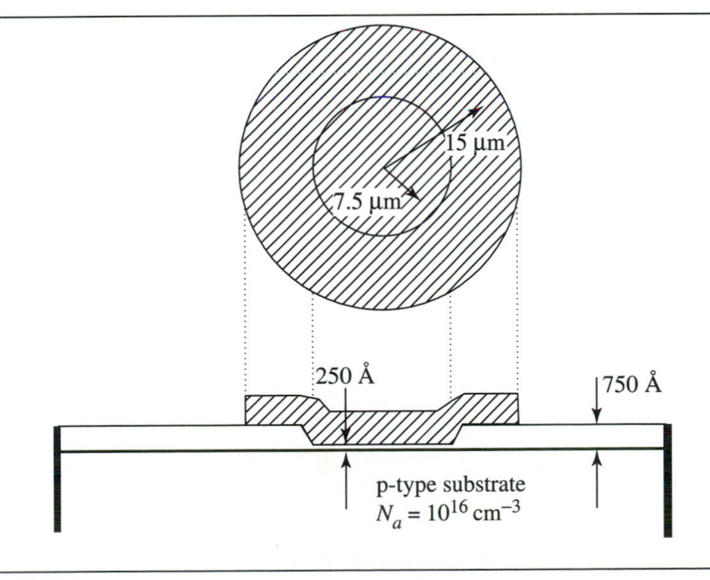

➤ **Figure P3.28**

DESIGN PROBLEMS

D3.1 The junction capacitance versus voltage curve is given below for a one-sided pn-junction. Determine the area and the (nonuniform) doping concentration that are required to have this capacitance curve. Note the breakpoints at $V_D = -4$ V and at $V_D = -8$ V.

D3.2 The requirement for a particular CMOS process is that the threshold voltage variation is $V_{Tn} = 0.7 \pm 0.1$ V. The oxide thickness is controlled to $t_{ox} = \overline{t_{ox}}(1 \pm 0.1)$ and the doping concentration is controlled to $N_a = \overline{N_a}(1 \pm 0.02)$. The possible range for the oxide thickness is limited to $100\text{Å} \le t_{ox} \le 500$ Å and the doping concentration is limited to $10^{15} \le N_a \le 10^{17}$ cm^{-3}.

(a) Choose an average doping concentration and an average oxide thickness that yield $\overline{V_{Tn}} = 0.7$ V.

(b) Find the variation in the threshold voltage, given the uncertainties in the oxide thickness and the doping concentration. Iterate to find a solution for $\overline{t_{ox}}$ and $\overline{N_a}$ that satisfy the 100 mV variation in the threshold voltage.

(c) Rather than use the statistical model for the uncertainties, it is more conservative to use worst-case analysis in which the signs of the variations in t_{ox} and in N_a are chosen to largest magnitude shift in V_{Tn}. Repeat (b) using this approach and compare your answers with the statistical model.

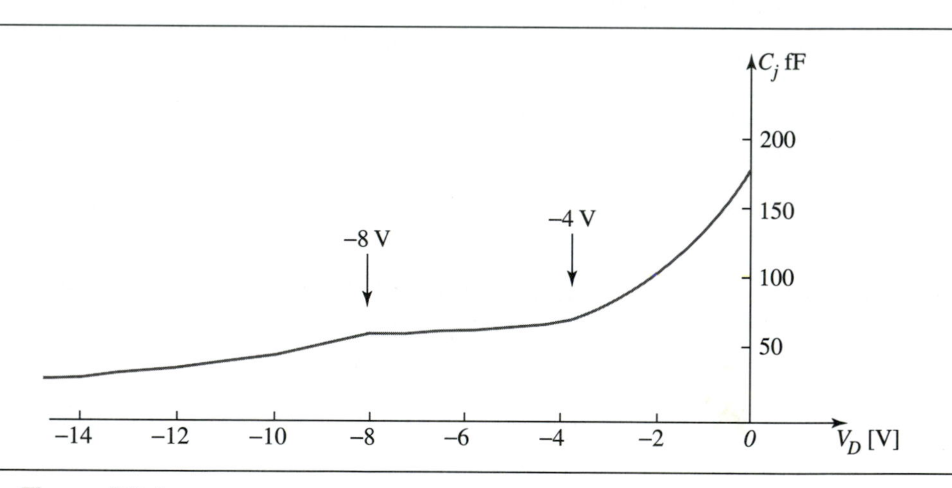

➤ **Figure D3.1**

chapter 4

The MOS Field-Effect Transistor

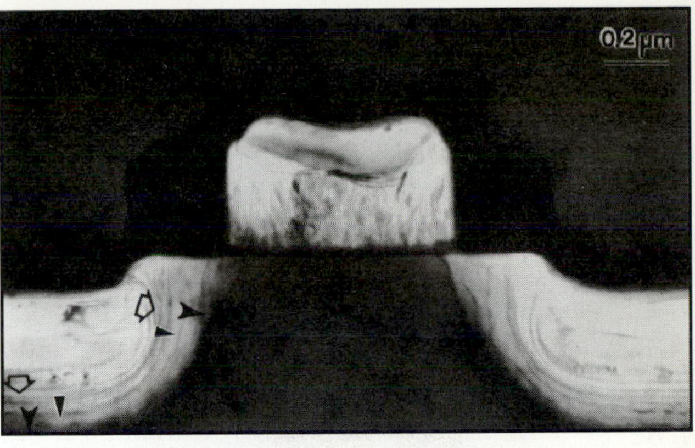

Transmission electron micrograph (TEM) of a 0.6 μm channel length MOSFET. (Courtesy of Prof. R. Gronsky, Dept. of MSME, Univ. of California at Berkeley)

This chapter introduces our first multi-terminal microelectronic device—the MOS field-effect transistor (MOSFET). Your understanding of the MOS capacitor will be the foundation for seeing how MOSFETs work and developing more accurate models. MOSFETs are the dominant transistor type for digital ICs and also have important applications in analog signal processing circuits.

Chapter Objectives

- MOSFET structures, symbols, and current-voltage characteristics.
- Cutoff, triode, and constant current (saturation) regions of operation.
- The gradual channel approximation and its use to model the current-voltage characteristics.
- Large- and small-signal models of n-channel and p-channel MOSFETs for circuit design.
- Level 1 SPICE model for the MOSFET.

4.1 INTRODUCTION

4.1.1 MOSFET Physical Structure

The **MOS field-effect transistor (MOSFET)** is the electronic device behind the explosive growth in digital electronics since 1970. The complexity and capability of MOS integrated circuits have increased many orders of magnitude over the past 25 years. To begin your study of this device, the layout and cross section of a modern n-channel MOSFET is shown in Fig. 4.1. The basic structure is identical to that of the simple MOSFET whose fabrication process was described in detail in Chapter 2. The **gate** of the transistor is a film of n^+ polycrystalline silicon (polysilicon). Underlying the gate is the **gate oxide** which is a layer of thermally grown silicon dioxide (SiO_2).

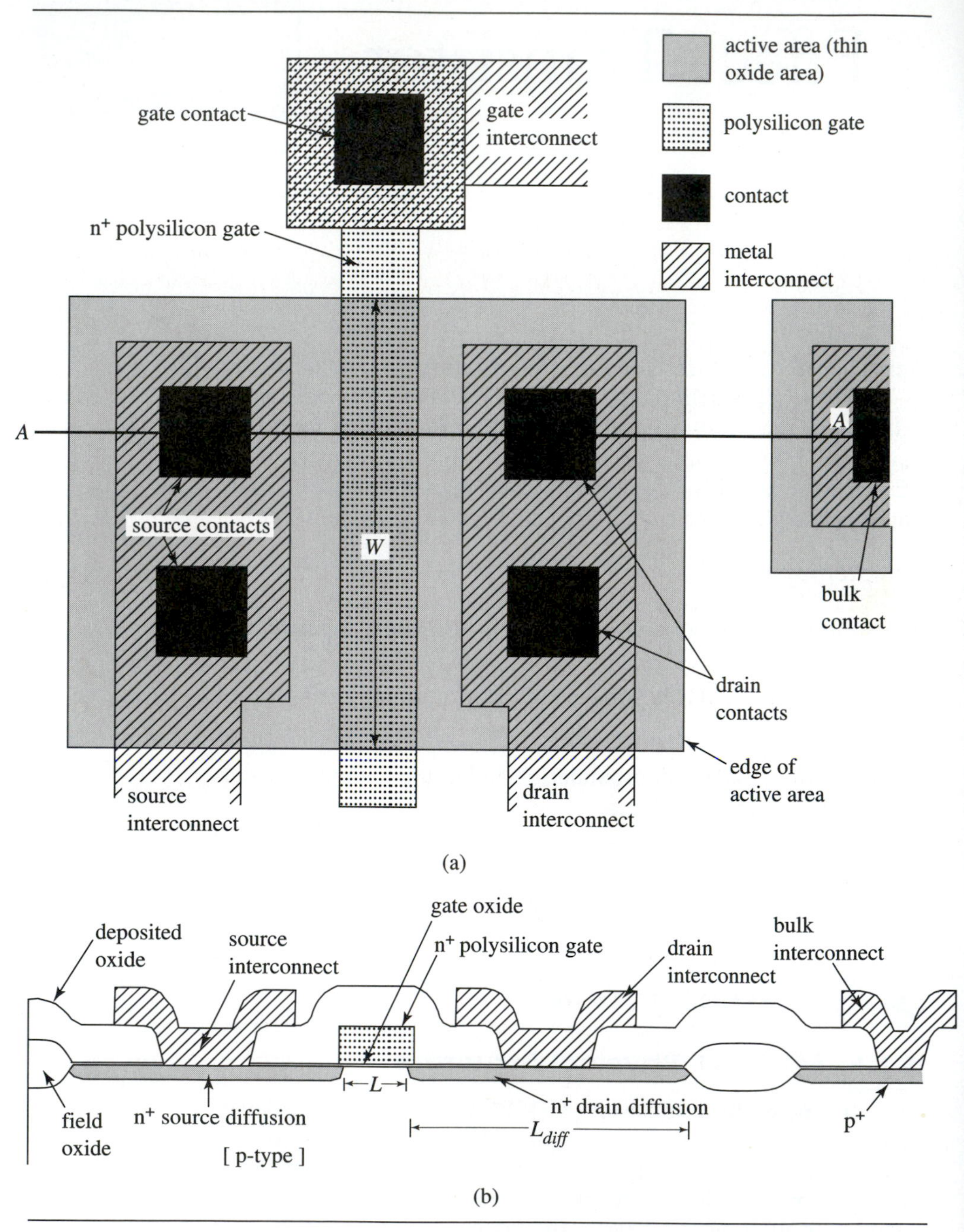

➤ **Figure 4.1** (a) Four-mask layout and (b) cross section of an integrated n-channel MOSFET.

The **channel** is formed under the polysilicon gate between the two n^+ regions as shown in Fig 4.1 (b). The **channel length *L*** of the MOSFET as is defined in Fig. 4.1 (b) as the gap between the n^+ diffusions. On the layout in Fig. 4.1(a), the **channel width *W*** of the MOSFET is defined as the width of the active area. One n^+ diffusion is called the **source**, the other is called the **drain**. The potentials on the diffusions will be the basis for differentiating the source from the drain, as discussed in the next section. The source and drain diffusions have a length L_{diff} and the same width W as the channel. Metal interconnections make ohmic contacts to the gate, source, and drain. In addition, the underlying p-type bulk region (the substrate or a deep p-type diffusion) is contacted with a fourth metal interconnection called the **bulk** or **body**.

4.1.2 MOSFET Circuit Symbol and Terminal Characteristics

Before introducing the current-voltage characteristics of the MOSFET, it is helpful to have a symbol for this device. Unfortunately, there are two common symbols for each MOSFET type. Figure 4.2 shows the circuit symbols and basic structure for an **n-channel MOSFET** (shown in Fig. 4.1) and for the complementary **p-channel MOSFET** that has p^+ source and drain diffusions in an n-type bulk region. In this text we will use the symbols without arrows for digital circuits and the symbols with arrows for analog circuits. The cross sections for the two MOSFETS have been rotated 90° in order to illustrate the close correspondence between the symbols and the physical structure.

The MOSFETs are oriented as they often appear in circuit schematics: the drain for the n-channel MOSFET and the source for the p-channel MOSFET are on top—toward the positive voltage supply. Voltages between terminals are labeled by an ordered pair of subscripts such as $V_{GS} = V_G - V_S$. By IEEE convention, the reference

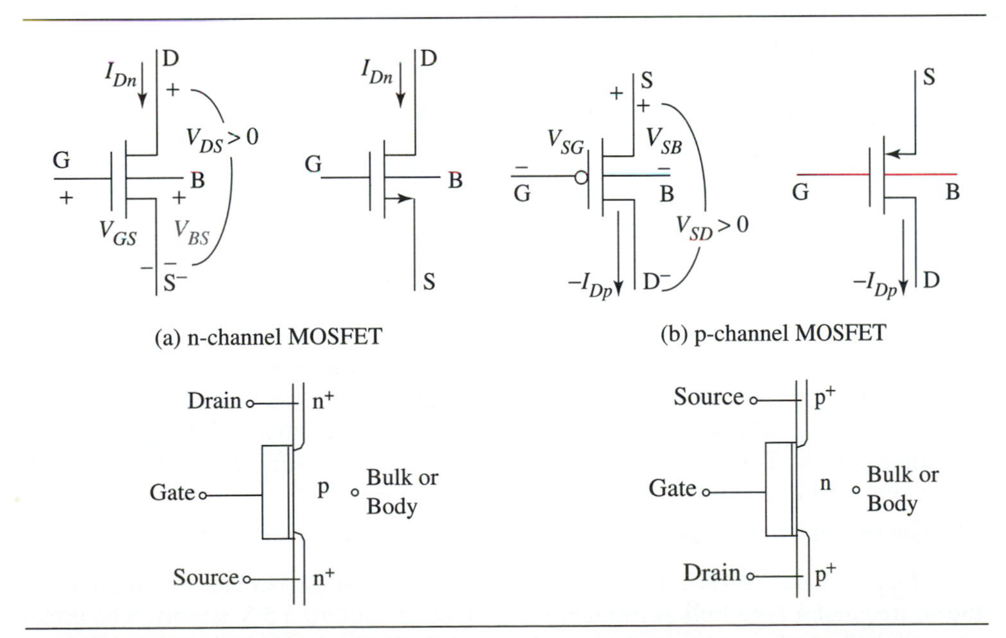

Figure 4.2 (a) n-channel MOSFET symbols and structure and (b) p-channel symbols and structure.

directions for currents are defined as positive *into* the device terminals. Since the drain current for the p-channel device is negative, it is convenient to use $-I_{Dp} > 0$, which is positive leaving the drain terminal. For the n-channel MOSFET in Fig. 4.2(a), the drain and source terminals are selected so that $V_{DS}>0$, which is equivalent to defining the source as the one with the *lower* potential. As shown in Fig. 4.2(b), the labeling convention is opposite for the p-channel MOSFET. Its source and drain are identified so that $V_{SD}>0$, which means its source is defined as the one with the *higher* potential.

The n-channel MOSFET's static (DC) terminal characteristics can be measured with voltage sources and ammeters. Since the MOSFET is a four terminal device, we need to experiment and discover which voltages are most important. For our initial investigation, we short the bulk to the source to set $V_{BS} = 0$ V. Since the voltage *differences* between the drain and source and gate and source are most important, we ground the source ($V_S = 0$ V) for convenience. Since the gate terminal is insulated, the only current of interest is that into the drain terminal I_{Dn}.

Figure 4.3(a) is a schematic of the circuit used to find the drain current's functional dependence on the two voltages under our control: the gate-source voltage V_{GS} and the drain-source voltage V_{DS}. The conventional way to graph the measured data is to plot a family of curves $I_{Dn}(V_{GS}, V_{DS})$, with V_{GS} as a parameter. The drain current versus drain-source voltage curves for a selected set of gate-source voltages are called the MOSFET **drain characteristics**.

Figure 4.3(b) shows measured drain characteristics for an n-channel MOSFET in a digital CMOS process. Although the voltage and current ranges can vary with the technology and device layout, the ranges in Fig. 4.3(b) are typical. In this graph, the drain current I_D is measured as a function of drain-source voltage V_{DS} for eight gate-source voltages

$$I_{Dn}\,(V_{GS} = 0, 0.5, 1, 1.5, 2, 2.5, 3, 3.5\text{ V}, V_{DS}),\ \text{for } 0\text{ V} < V_{DS} < 5\text{V} \tag{4.1}$$

Several observations can be made from studying the drain characteristics in Fig. 4.3(b).

1. The MOSFET is **cutoff** ($I_{Dn} = 0$ A) for gate-source voltages that are less than some critical voltage, which is defined as the **threshold voltage** V_{Tn} of the n-channel device. A typical value for the n-channel threshold voltage is $V_{Tn} = 1$ V. The n^+ source and drain are electrically isolated until sufficient voltage is applied to the gate to create an n-type inversion layer or **channel** between them.
2. The drain current is nearly independent of the drain-source voltage once $V_{DS} > V_{GS} - V_{Tn}$. In this region of operation, called **saturation**, the MOSFET behaves like a current source. Further study reveals that the drain current in saturation is proportional to the square of the gate-source voltage above threshold $(V_{GS} - V_{Tn})^2$. This can be seen from the ratios of drain current for $V_{GS} = 3$ V and for $V_{GS} = 2$ V. This constant-current behavior will be exploited in both analog and digital circuit design.
3. The region where I_{Dn} depends on both V_{GS} and V_{DS} is termed the **triode** region.

The p-channel MOSFET's drain characteristics are shown in Fig. 4.4. In this case, the source and n-type bulk terminals are both connected to a 5 V supply. The effect of varying the drain voltage and gate voltage on the drain current can be investigated. The shape of the drain characteristics are identical to those of the n-channel MOSFET if $-I_{Dp}$ (which is positive) is plotted against the source-drain drop $V_{SD} = 5\text{ V} - V_D$ as a function of the source-gate voltage $V_{SG} = 5\text{ V} - V_G$. The drain current is zero unless the source-gate voltage $V_{SG} > -V_{Tp}$, where V_{Tp} is the p-channel threshold voltage. A typical value for the p-channel threshold voltage is $V_{Tp} = -1\text{ V}$.

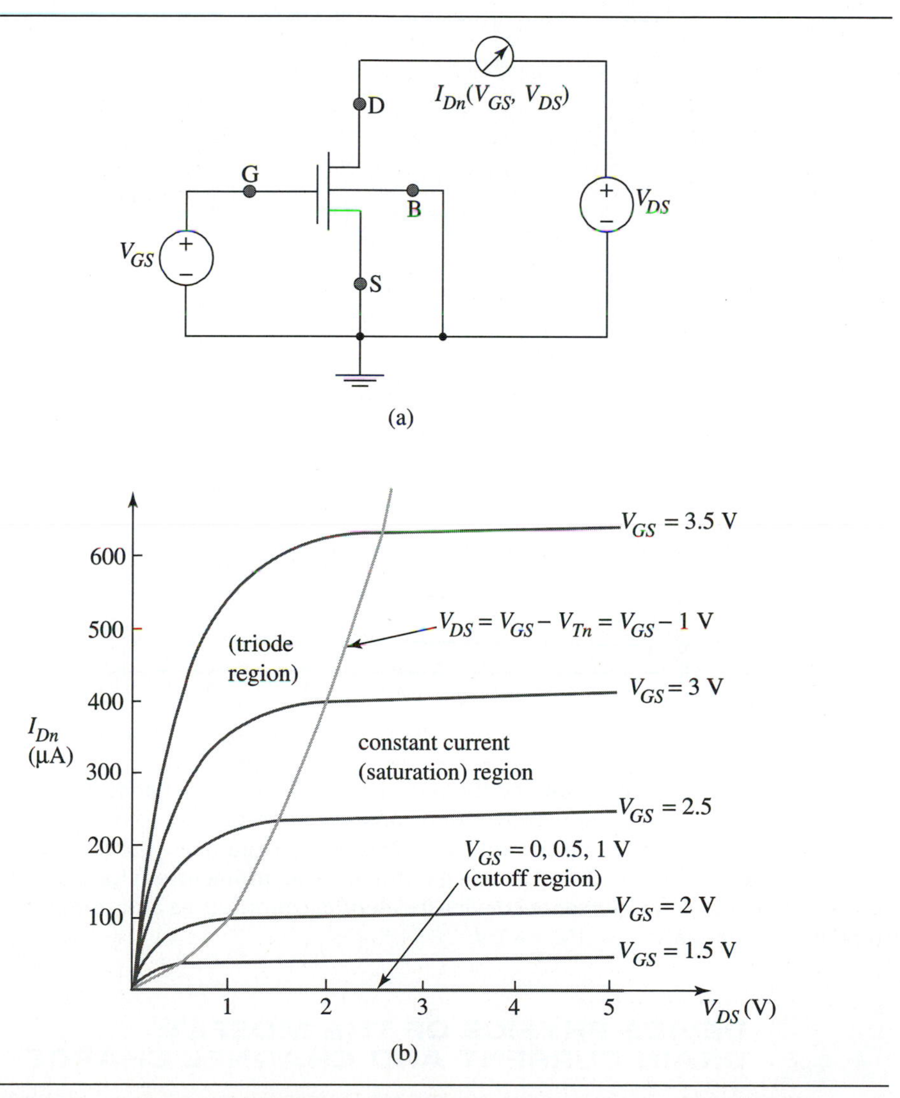

➤ **Figure 4.3** n-channel MOSFET drain characteristics: (a) test circuit and (b) measurements of drain current as function of drain-source voltage, with gate-source voltage as a parameter for a typical integrated device.

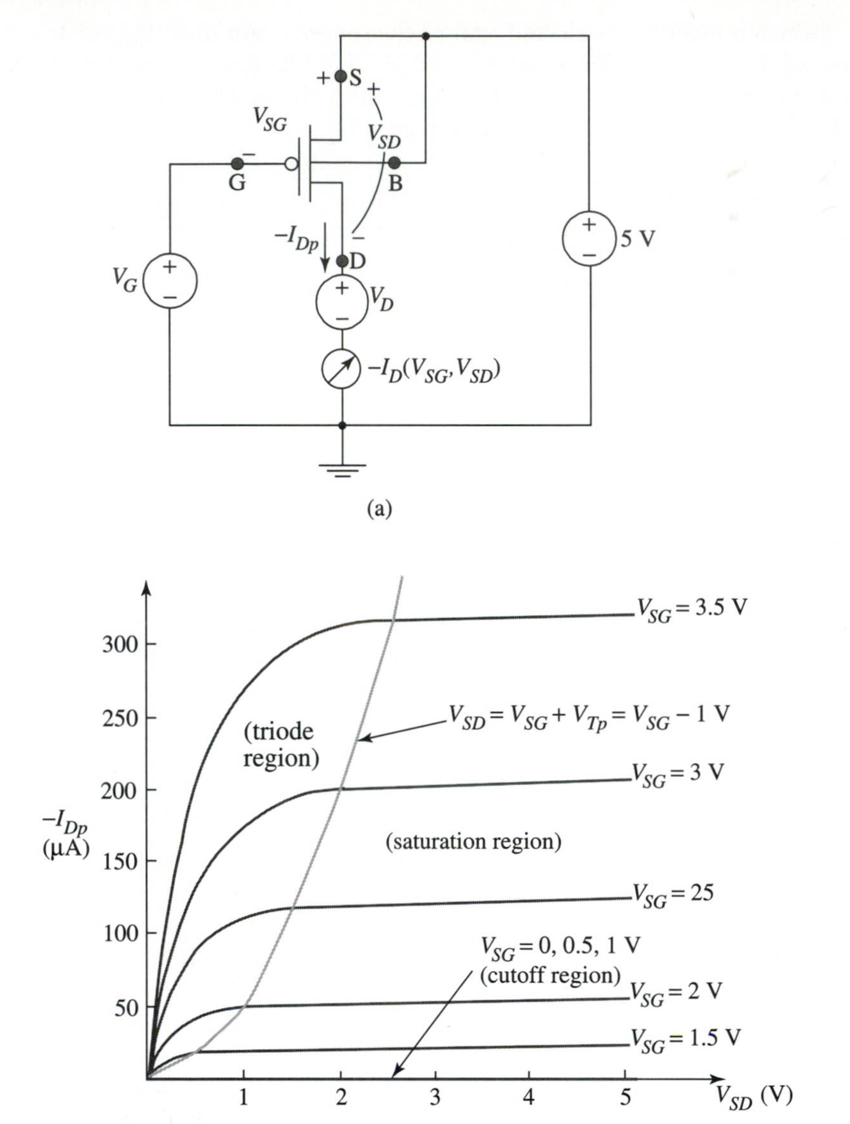

➤ **Figure 4.4** (a) Test circuit and (b) measured drain characteristics of a p-channel MOSFET with the same geometric dimensions as the n-channel device used for Fig. 4.3. The negative of the drain current $-I_{Dp}$ (> 0) is plotted against the source-drain drop V_{SD}, with the source-gate drop V_{SG} as a parameter. The drain current for the p-channel is about half that of the n-channel device for identical geometries and complementary threshold voltages $V_{Tp} = - V_{Tn} = - 1$ V.

➤ 4.2 DEVICE PHYSICS OF THE MOSFET: DRAIN CURRENT AND CHANNEL CHARGE

We are now prepared to quantify how the drain current varies with the terminal voltages for the n-channel MOSFET. As a starting point, we will build upon the electrostatics of the MOS capacitor that were discussed in Chapter 3. Once the channel forms, an inversion layer of electrons exists between source and drain. An applied

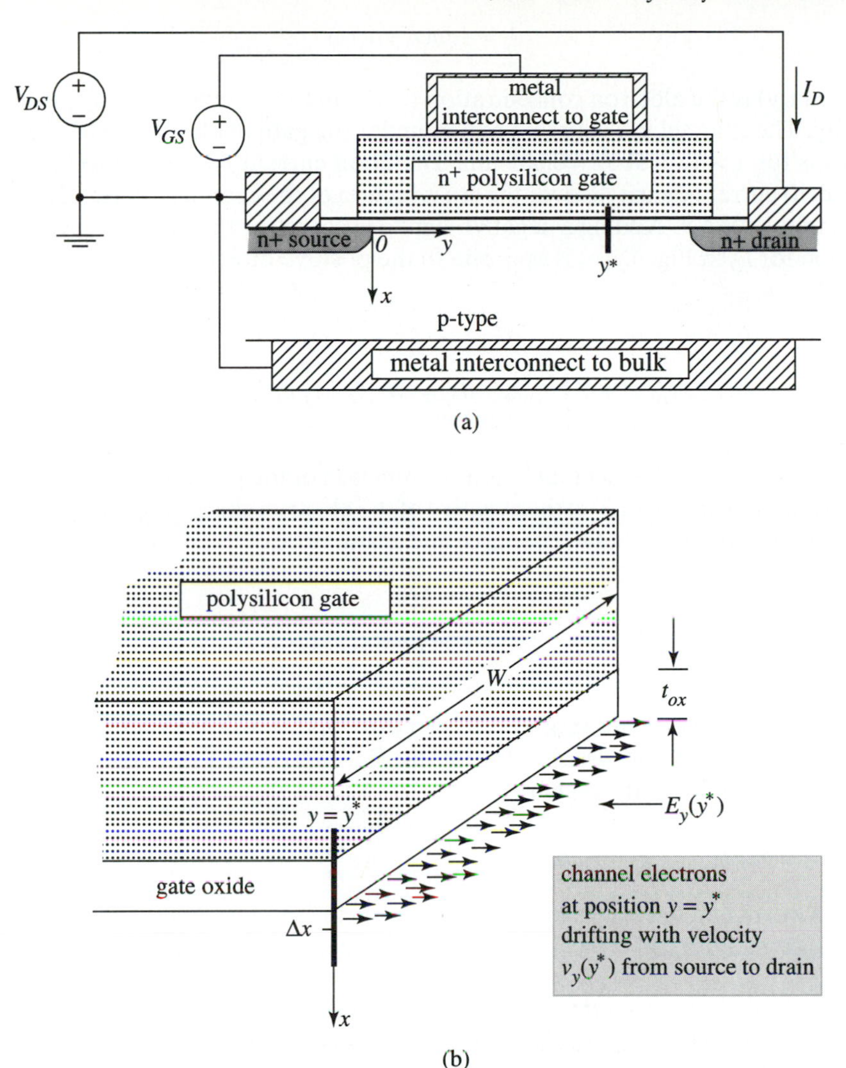

➤ **Figure 4.5** (a) MOSFET biased with $V_{GS}>V_{Tn}$ and $V_{DS}>0$. (b) Cutaway view of drifting electrons in the channel at position y^* (see (a)) between the source ($y=0$) and drain ($y=L$).

voltage between the drain and source will impose an electric field across the channel that results in a drift current density. A good first step is to relate this drift current density, which is the origin of the drain current I_D, to the charge in the inversion layer.

Figure 4.5(a) illustrates an n-channel MOSFET with applied voltages $V_{GS}>V_{Tn}$ and $V_{DS}>0$. For simplicity, we again short the bulk to the source so $V_{BS}=0$. Furthermore, we assume that a channel extends from the source to the drain. The drain current I_D consists of electrons drifting from source to drain along the channel. At any point along the channel, the electron drift current must equal the drain current I_D since there is no current to the substrate or through the insulating oxide to the gate.

Figure 4.5(b) shows a perspective view of a portion of the channel at position y^* between the source and drain. From basic semiconductor physics, the drift current density J_y at any point (x,y) in the channel is given by

$$J_y = -qn(x, y)v_y(y) \tag{4.2}$$

where $n(x,y)$ is the electron concentration (cm^{-3}) in the channel and $v_y(y)$ is the drift velocity. The channel is a very thin region under the gate oxide—there are negligible electrons for $x > \Delta x \approx 50$ Å $= 0.005$ μm. The drain current I_D is equal to the integral of the drift current density J_y over the cross section of the channel in Fig. 4.5(b), which has an area of $W\Delta x$. Adding a negative sign to account for the fact that the reference direction for I_D in Fig. 4.4(a) is opposite to the positive direction of the y coordinate, we find

$$I_D = -W\int_0^{\Delta x} J_y(x, y)dx = Wv_y(y)\left(\int_0^{\Delta x} qn(x, y)dx\right) \tag{4.3}$$

where we have used the fact that v_y is not a function of the position x in the thin channel. The integral in Eq. (4.3) is the negative of the electron charge $Q_N(y)$ (units: C/cm^2) in the channel at point y between source and drain, since

$$Q_N(y) = -\int_0^{\Delta x} qn(x, y)dx \tag{4.4}$$

As we will investigate shortly, the electron charge will vary along the channel between the source and drain. Substituting Eq. (4.4) into Eq. (4.3), we relate the drain current to the channel charge

drain current = product of channel charge, width, and drift velocity

$$\boxed{I_D = -Wv_y(y)Q_N(y)} \tag{4.5}$$

It is worth emphasizing again that the drain current does *not* vary from source to drain, although the drift velocity $v_y(y)$ and channel electron charge $Q_N(y)$ generally do.

The drain current equation in Eq. (4.5) has been derived from basic physics. Next, we must determine how the channel electron charge and drift velocity vary with position in the channel in order to proceed toward finding the functional dependence of drain current on V_{GS} and V_{DS}. Since the full analysis is rather involved, we will make a "first pass" to gain an understanding of the important features of MOSFET operation.

4.3 MOSFET DEVICE PHYSICS: A FIRST PASS

From Chapter 3, we have a quantitative result for the inversion charge under a MOS capacitor

$$Q_N = -C_{ox}(V_{GB} - V_{Tn}) \tag{4.6}$$

In order to apply this one-dimensional result to the two-dimensional channel of a MOSFET, first consider the special case where the drain-source voltage V_{DS} is very small—on the order of 100 mV = 0.1 V or less. As shown in Fig. 4.6, the channel is uniform from source to drain for small drain-source biases. Note that for $V_{DS} = 0$ V,

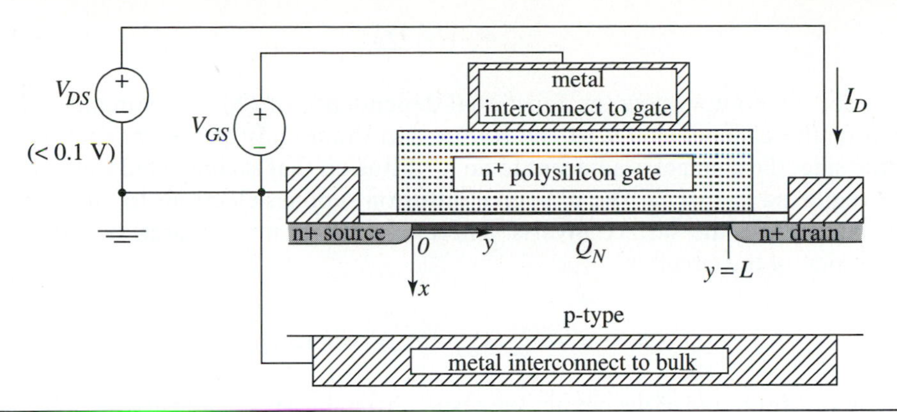

➤ **Figure 4.6** MOSFET biased with $V_{GS} > V_{Tn}$ and very small V_{DS}, so the channel can be assumed to be uniform from source to drain. The channel thickness is greatly exaggerated in order to be visible.

the structure degenerates to a capacitor and Eq. (4.6) is an *exact* expression for the channel charge.

Substituting Eq. (4.6) into the drain current expression from the previous section and noting from Fig. 4.6 that $V_{GB} = V_{GS}$, we find

$$I_D = -Wv_y Q_N = Wv_y C_{ox}(V_{GS} - V_{Tn}), \text{ where } V_{GS} > V_{Tn} \text{ and } V_{DS} < 0.1 \text{ V} \quad (4.7)$$

The remaining unknown in Eq. (4.7) is the drift velocity in the channel $v_y(y)$. Since the channel is nearly uniform, the voltage in the channel must vary linearly from source to drain with a constant electric field in the channel

$$E_y = -\left(\frac{V_{DS}}{L}\right) \quad (4.8)$$

Given the small V_{DS}, the channel electric field is also small in magnitude, making it a good approximation that the drift velocity is proportional to the electric field

$$v_y = -\mu_n E_y = -\mu_n(-V_{DS}/L) = \mu_n(V_{DS}/L) \quad (4.9)$$

Substitution of Eq. (4.9) into (4.7) results in a equation for the drain current in terms of V_{GS} and V_{DS}, for the special case of $V_{DS} < 0.1$ V:

$$I_D = \left(\frac{W}{L}\right)\mu_n C_{ox}(V_{GS} - V_{Tn})V_{DS} \text{ where } V_{GS} > V_{Tn} \text{ and } V_{DS} < 0.1 \text{ V} \quad (4.10)$$

For a fixed gate-source bias and small drain-source bias, Eq. (4.10) indicates that the MOSFET behaves like a voltage-controlled resistor, with resistance given by

$$R = \frac{V_{DS}}{I_D} = \left(\frac{1}{\mu_n C_{ox}(V_{GS} - V_T)}\right)\left(\frac{L}{W}\right) = [R_\square(V_{GS})]\left(\frac{L}{W}\right). \quad (4.11)$$

In order to model the MOSFET for $V_{DS} > 0.1$ V without solving differential equations, first approximate Eq. (4.5) by taking the averages of the channel charge and drift velocity over the channel

$$I_D \approx -W\overline{v_y}\,\overline{Q_N} \tag{4.12}$$

Since we do not know the functional dependence of Q_N, we estimate it as the mean of the channel charge at the source and drain ends of the channel. On the source side, the channel is adjacent to an n^+ diffusion that is connected to the p-type bulk region so $V_{GS} = V_{GB}$. The n^+ source diffusion supplies electrons to the inversion layer as soon as the surface inverts. Therefore, the channel charge stored on the source side of the circuit is

$$Q_N(y=0) = -C_{ox}(V_{GS} - V_{Tn}) \text{ where } V_{GS} > V_{Tn} \tag{4.13}$$

On the drain side of the circuit, the results from the MOS capacitor must be modified since the voltage of the n^+ diffusion adjacent to the drain end of the channel is V_{DS} and the electrostatics of the problem are changed. What is the effect of the higher potential at the drain end? The elevated drain potential lowers the voltage drop from the gate to the channel at the drain end and *reduces* the channel charge. This phenomenon can be approximated by using the gate-drain voltage drop $V_{GD} = V_{GS} - V_{DS}$ rather than the gate-source voltage drop in Eq. (4.13). The channel charge at the drain end is

$$Q_N(y=L) = -C_{ox}(V_{GD} - V_{Tn})\text{ , where } V_{GD} = V_{GS} - V_{DS} > V_{Tn} \tag{4.14}$$

The conditions on V_{GS} and V_{DS} in Eq. (4.14) ensure that the channel charge is greater than zero at the drain end. The reduced charge at the drain end of the channel is shown qualitatively in Fig. 4.7. An estimate of the average channel charge from its values at the ends of the channel is

$$\overline{Q_N} \approx \frac{Q_N(y=0) + Q_N(y=L)}{2} = \frac{-C_{ox}(V_{GS} - V_{Tn}) - C_{ox}(V_{GS} - V_{DS} - V_{Tn})}{2} \tag{4.15}$$

Now that the channel charge varies from source to drain, it is no longer true that the channel voltage will increase linearly along the channel, or equivalently, that the

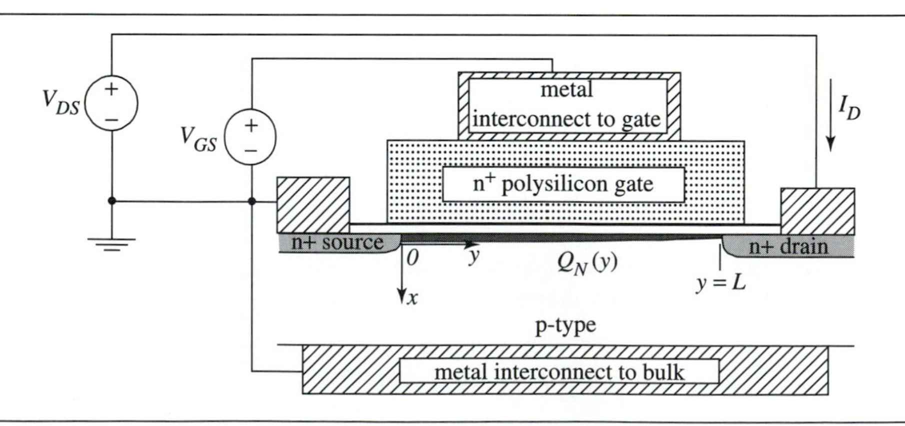

➤ **Figure 4.7** MOSFET biased in the triode region with $V_{GS} - V_{DS} > V_{Tn}$. The channel charge decreases from source to drain. The thickness of the channel is greatly exaggerated.

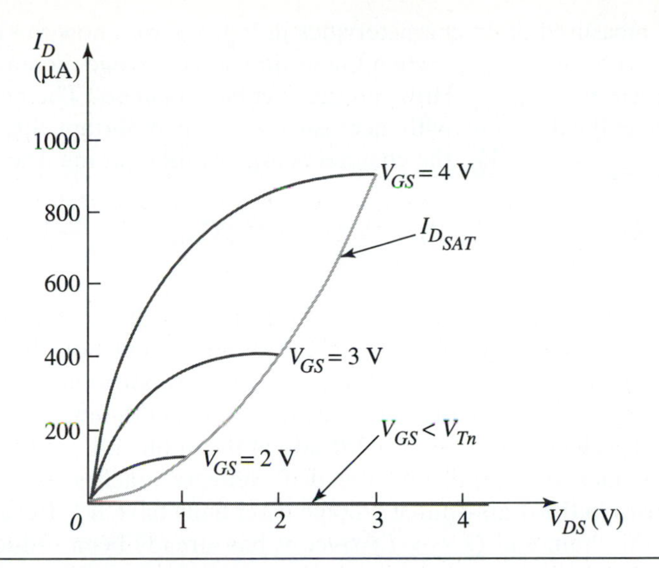

➤ **Figure 4.8** Drain characteristics for $V_{DS} \leq V_{GS} - V_{Tn}$, according to Eq. (4.17), assuming $V_{Tn} = 1$ V, $\mu_n C_{ox} = 50\ \mu\text{A/V}^2$, and $(W/L) = 4$.

channel electric field is constant. However, in the spirit of this approximate analysis, we will estimate the average electric field by

$$\overline{E_y} = -\overline{\left(\frac{dV_C}{dy}\right)} \approx \frac{-V_{DS}}{L} \tag{4.16}$$

where $V_C(y)$ is the channel voltage at position y between source and drain. The average drift velocity is then $\overline{v_y} = -\mu_n \overline{E_y} = \mu_n V_{DS}/L$. Substituting the estimates of average channel charge from Eq. (4.15) and average drift velocity into Eq. (4.12), we find

$$I_D = \left(\frac{W}{L}\right)\mu_n C_{ox}\left[V_{GS} - V_{Tn} - (V_{DS}/2)\right] V_{DS} \quad \text{where } V_{GS} - V_{DS} > V_{Tn} \tag{4.17}$$

As a check, Eq. (4.17) reduces to Eq. (4.10) for small V_{DS}. Figure 4.8 is a plot of Eq. (4.17) for three values of gate-source voltage. The curves have the general shape of the measured MOSFET drain characteristics in the triode region from Fig. 4.3. The inverted parabolas reach their maxima when the drain-source voltage is equal to $V_{DS_{SAT}}$, which is defined by

$$V_{DS_{SAT}} = V_{GS} - V_{Tn}\ . \tag{4.18}$$

This voltage is also the largest drain-source voltage for which Eq. (4.17) is valid. The maximum drain current, called the **saturation current** $I_{D_{SAT}}$, is found by substituting $V_{DS_{SAT}}$ into Eq. (4.17)

$$I_{D_{SAT}} = \left(\frac{W}{2L}\right)\mu_n C_{ox}(V_{GS} - V_{Tn})^2 = \left(\frac{W}{2L}\right)\mu_n C_{ox} V_{DS_{SAT}}^2 \tag{4.19}$$

From the measured drain characteristics in Fig. 4.3, we know the drain current remains nearly constant at $I_{D_{SAT}}$ when the drain-source voltage increases above the saturation voltage $V_{DS} > V_{DS_{SAT}}$. How can this fact be explained? The behavior of the channel charge at the drain end with increasing V_{DS} is an important clue. For the case where $V_{DS} = V_{DS_{SAT}} = V_{GS} - V_{Tn}$, the channel charge vanishes at the drain end

$$Q_N(y=L)\Big|_{V_{DS_{SAT}}} = -C_{ox}(V_{GD} - V_{Tn}) = -C_{ox}(V_{GS} - V_{DS_{SAT}} - V_{Tn}) = 0 \quad (4.20)$$

In Fig. 4.9, the channel charge in a MOSFET biased with $V_{DS} = V_{GS} - V_{Tn}$ is shown qualitatively. How can $Q_N(y = L) = 0$ and the drain current not also vanish? The answer is contained in Eq. (4.5)—the drain current is proportional to the product of the channel charge and drift velocity. For the drain current to remain constant in saturation, the drift velocity must become infinite at the drain end. In reality, the channel charge becomes very small and the drift velocity reaches its limiting value. Further increases in the drain-source voltage essentially have no effect on the channel charge at the drain end $Q_N(y = L)$ since it has already been reduced to nearly zero. As a result, the drain current remains constant, to first order, at its maximum value

$$I_D = I_{D_{SAT}} = \left(\frac{W}{2L}\right)\mu_n C_{ox}(V_{GS} - V_{Tn})^2 \text{ for } V_{DS} \geq V_{DS_{SAT}} \quad (4.21)$$

The very simple models constructed in this section for the triode region (Eq. (4.17)) and the saturation region (Eq. (4.21)) reproduce the essential features of the MOSFET drain characteristics as shown in Fig. 4.10. After some minor refinements in Section 4.6, we will find their simplicity makes them very useful for hand analysis and design of digital and analog circuits.

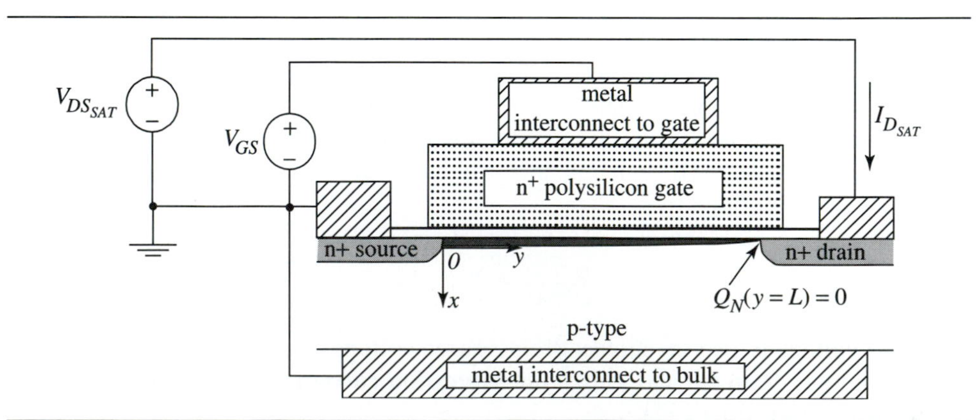

➤ **Figure 4.9** MOSFET biased with $V_{DS} = V_{DS_{SAT}} = V_{GS} - V_{Tn}$ with the channel charge vanishing at the drain end. As in Figs. 4.6 and 4.7, the thickness of the channel is greatly exaggerated in order to be visible.

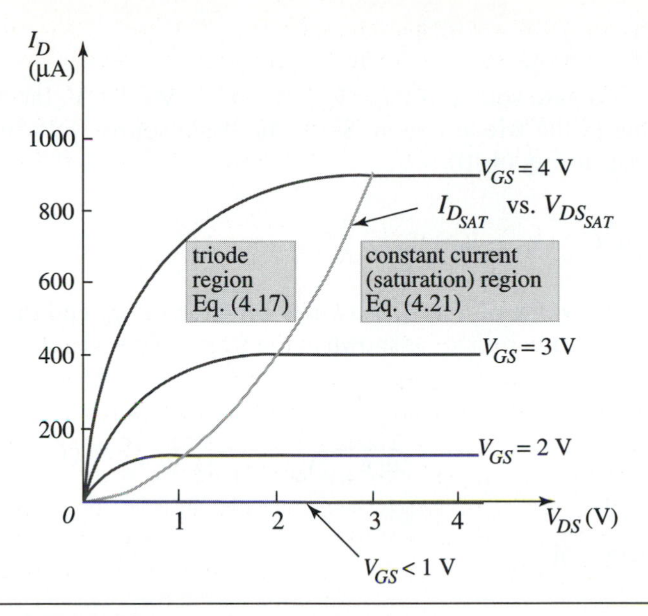

➤ Figure 4.10 MOSFET drain characteristics generated from Eq. (4.17) for the triode region and Eq. (4.21) for the saturation region. Device parameters are the same as for Fig. 4.8.

➤ EXAMPLE 4.1 MOSFET as a Voltage Controlled Resistor

The circuit below shows an n-channel MOSFET that is used as a voltage-controlled resistor. A MOSFET in this configuration is found in circuits such as variable gain amplifiers.

(a) Find the sheet resistance of the MOSFET over the range $V_{GS} = 1.5$ V to $V_{GS} = 4$ V using $\mu_n = 215\ \text{cm}^2\text{V}^{-1}\text{s}^{-1}$, $C_{ox} = 2.3\ \text{fF}/\mu\text{m}^2$ and $V_{Tn} = 1$ V.

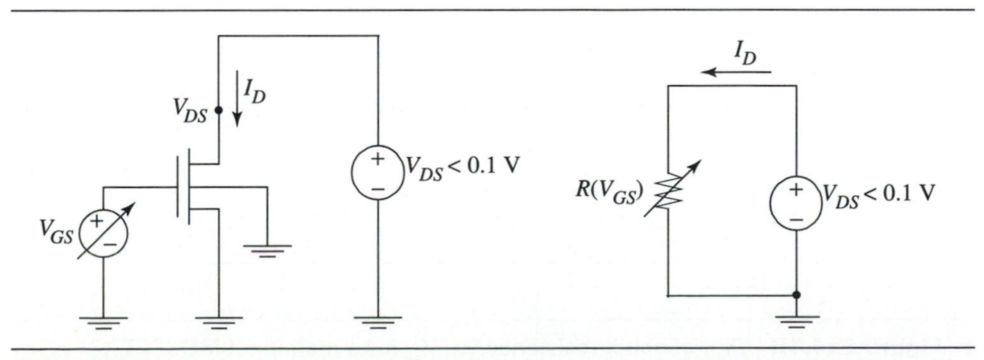

➤ Figure Ex4.1A An n-channel MOSFET used as a voltage controlled resistor.

SOLUTION

Note that for a gate-source voltage $V_{GS} > V_{Tn} + 0.1\text{ V} = 1.1\text{ V}$, the MOSFET will be operating in the **triode** region. Since the drain-source voltage is small, Eq. (4.17) reduces to Eq. (4.10)

$$I_D = \left(\frac{W}{L}\right)\mu_n C_{ox} (V_{GS} - V_{Tn}) V_{DS} = \frac{1}{R} V_{DS}$$

For a particular value of V_{GS}, I_D is a *linear* function of V_{DS} and the circuit model for the MOSFET is a resistor as shown in the figure. Now we relate R to the sheet resistance.

$$R = \frac{1}{\mu_n C_{ox}\left(\frac{W}{L}\right)(V_{GS} - V_{Tn})} = \frac{1}{\mu_n C_{ox} (V_{GS} - V_{Tn})}\left(\frac{L}{W}\right) = R_{\square} (L/W)$$

$R_{\square}$ as a function of V_{GS} is

$$R_{\square} = \frac{1}{(215\text{cm}^2\text{V}^{-1}\text{s}^{-1})\ (2.3 \times 10^{-7}\text{F/cm}^2)}\frac{1}{(V_{GS} - 1V)} = \frac{20\text{ k}\Omega\text{-V}}{(V_{GS} - 1\text{V})}$$

The plot of as $R_{\square}$ as a function of V_{GS} is shown below

(b) For a particular application, we need to control the resistor between 200 Ω and 1 kΩ for $V_{GS} = 1.5$ V to 4 V. How wide should the MOSFET be if the channel length $L = 1.5\ \mu$m?

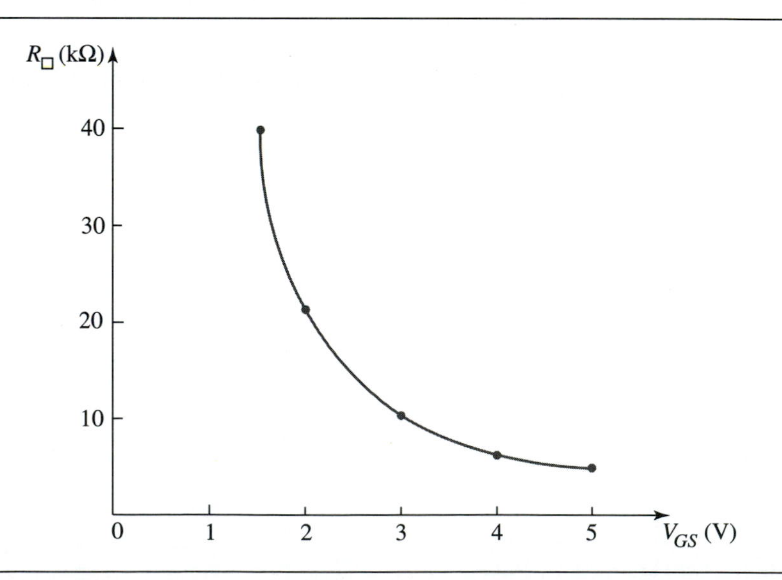

➤ Figure Ex4.1B Plot of sheet resistance vs. V_{GS} for a n-channel MOSFET operating with very small V_{DS}.

SOLUTION

We already solved for the sheet resistance in part (a), so we can find the range of sheet resistances for $V_{GS} = 1.5$ V to 4 V

$$R_{\square min} = \frac{20\ k\Omega\text{-V}}{(4V - 1V)} = 6666.7\Omega \text{ and } R_{\square max} = \frac{20\ k\Omega\text{-V}}{(1.5V - 1V)} = 40k\Omega$$

Solving for (W/L) to obtain $R_{min} = 200\ \Omega$ and $R_{max} = 1\ k\Omega$ yields

$$\left(\frac{W}{L}\right)_{min} = \frac{6666.7\Omega}{200\Omega} = 33.3 \text{ and } \left(\frac{W}{L}\right)_{max} = \frac{40000\Omega}{1000\Omega} = 40$$

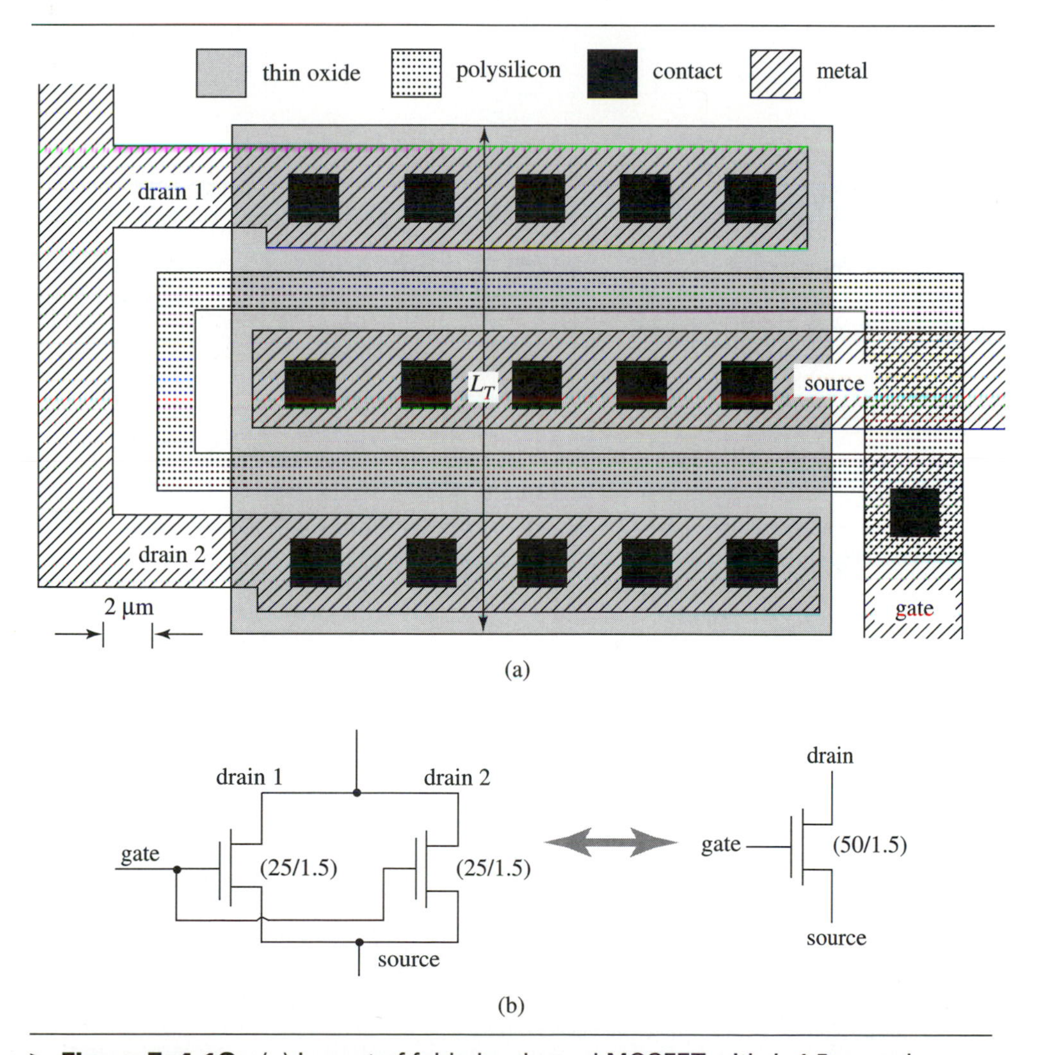

➤ Figure Ex4.1C (a) Layout of folded n-channel MOSFET with L=1.5μm and W= 50 μm and (b) schematic showing the equivalence of the two MOSFETs with W= 25 μm in parallel with a single MOSFET having L = 50 μm

$(W/L) = 33.3$ will satisfy the requirements for both R_{min} and R_{max} so the width of the MOSFET should be

$$W = 1.5\mu\text{m}(33.3) = 50\mu\text{m}$$

(c) Design the layout for this MOS resistor so it occupies a minimum area. The length of the source/drain diffusions is $L_{diff} = 6\ \mu\text{m}$ with contacts that are $2\ \mu\text{m} \times 2\ \mu\text{m}$.

SOLUTION

Given the high ratio of width to length, it is desirable to fold the MOSFET, which is equivalent to putting two MOSFETs of half the width in parallel. Since the diffusions are 6 μm long, the total length is $L_T = 3\ L_{diff} + 2\ L = 3 \times 6\ \mu\text{m} + 2 \times 1.5\ \mu\text{m} = 21\ \mu\text{m}$, which is nearly equal to half the width $W/2 = 25\ \mu\text{m}$. The layout, drawn to scale, and the equivalent circuit are shown in Fig. Ex 4.1C.

➤ EXAMPLE 4.2 MOSFET operating regions

Plot I_D as a function of V_{GS} for V_{GS} ranging from 0 to 5V for the n-channel MOSFET shown in the circuit below with $\mu_n = 215\ \text{cm}^2\text{V}^{-1}\text{s}^{-1}$, $t_{ox} = 150\ \text{Å}$, $V_{Tn} = 1\ \text{V}$, $L = 1.5\ \mu\text{m}$, and $W = 30\mu\text{m}$.

SOLUTION

For $V_{GS} < 1\ \text{V} = V_{Tn}$, the MOSFET is in the *cutoff* region and I_D is zero. Once the gate-source voltage is just high enough to turn on the device, the drain-source voltage $V_{DS} = 2\ \text{V} > V_{GS} - 1$ and the MOSFET is operating in the *saturation* region. The drain current in saturation is given by Eq. (4.21)

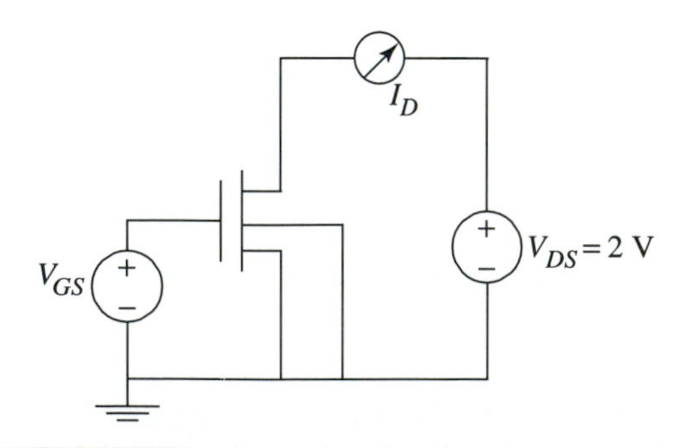

➤ Figure Ex4.2A n-channel MOSFET circuit.

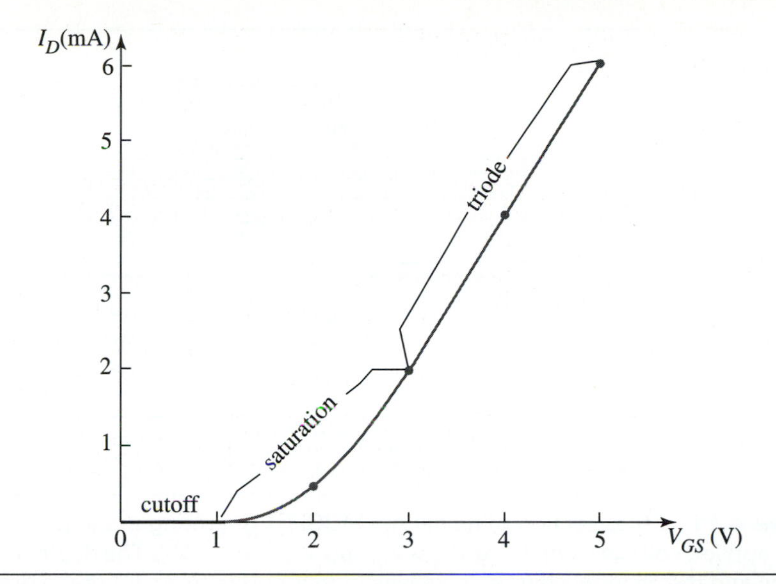

Figure Ex4.2B Plot of the I_D vs. V_{GS} for a n-channel MOSFET with fixed V_{DS} at 2 V.

$$I_D = \left(\frac{20}{2}\right)\left(50\frac{\mu\text{A}}{\text{V}^2}\right)(V_{GS} - 1\text{ V})^2 = 500\frac{\mu\text{A}}{\text{V}^2}(V_{GS} - 1\text{ V})^2$$

Once the gate-source voltage is high enough so $V_{GS} - V_{Tn} = V_{GS} - 1\text{ V} = V_{DS} = 2\text{ V}$, the MOSFET operates in the *triode* region with Eq. (4.17) describing the drain current

$$I_D = (20)\left(50\frac{\mu A}{\text{V}^2}\right)\left(V_{GS} - 1\text{V} - \frac{V_{DS}}{2}\right)V_{DS} = 2\frac{\text{m A}}{\text{V}^2}(V_{GS} - 2\text{V}) \quad \text{for } V_{GS} \geq 3\text{ V}$$

4.4 MOSFET DEVICE PHYSICS: THE GRADUAL CHANNEL APPROXIMATION

The starting point for developing a physical model for the current-voltage characteristics of the n-channel MOSFET is Eq. (4.5), which is repeated here.

$$I_D = -Wv_y(y)Q_N(y) \tag{4.22}$$

Our first step will be to find the functional dependence of the channel electron charge $Q_N(y)$ on the position y between the source and drain. Fig. 4.11 shows the cross section of a MOSFET that has bias voltages V_{GS} and V_{DS}. The bulk terminal is shorted to the source to make $V_{BS} = 0$ to simplify the initial analysis. Later in this section we will consider the case where the bulk is biased at a voltage different from the source.

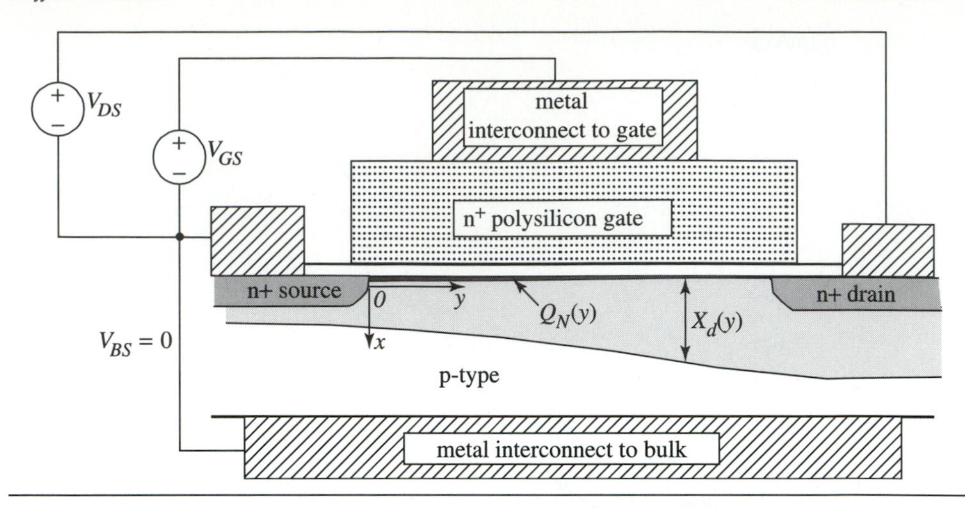

➤ **Figure 4.11** Cross section of n-channel MOSFET operating in the triode region with the applied voltages satisfying $V_{GS} > V_{Tn}$ and $V_{DS} < V_{GS} - V_{Tn}$. The depletion width increases and the channel charge decreases in magnitude from source to drain.

The coordinate system in Fig. 4.11 has its origin at the intersection of the boundary of the n^+ source diffusion with the oxide/silicon interface. In Chapter 3 we saw that the surface potential $\phi_s = \phi(0, y)$ is important in calculating the charge in the bulk depletion region and the charge in the inversion layer (the channel electron charge). Starting with thermal equilibrium, Fig. 4.12(a) plots the surface potential from source to drain. The example structure has a substrate doping of $N_a = 10^{17}$ cm^{-3} and an oxide thickness $t_{ox} = 150$ Å. The potential reference is the internal reference for silicon with intrinsic silicon being defined as zero. Note that the metal source and drain interconnections have the same potential since $V_{DS} = 0$ for equilibrium. The surface potential $\phi(0, y) = 21$ mV as calculated in Ex. 3.8, which corresponds to the surface being very slightly n-type. When a gate-source voltage is applied that is equal to the threshold voltage $V_{GS} = V_{Tn} = 0.6$ V and we maintain $V_{DS} = 0$, the surface potential increases to $\phi(0, y) = -\phi_p = 420$ mV as plotted in Fig. 4.12(b). The surface is inverted and an n-type channel has formed between the source and drain. Further increasing the gate-source voltage to $V_{GS} = V_{Tn} + 0.9$ V $= 1.5$ V and then applying a drain-source voltage of $V_{DS} = 0.5$ V, the sketch in Fig. 4.12(c) shows that the potential $\phi(0, y)$ increases by V_{DS} from the source to the drain. The metal contact to the n^+ source has been kept fixed in Fig. 4.12(c).

In Fig. 4.12(b), note that the potential in the channel is slightly lower than that of the n^+ source and drain regions when the gate is biased at the onset of inversion. For higher gate biases, the surface potential increases somewhat and reaches a maximum value of $\phi_{n+} = 550$ mV, an effect that is ignored in the delta-depletion approximation. In fact, the potential jump at the source becomes negligible for normal gate biases.

gradual channel approximation

We now consider the potential variation perpendicular to the channel (in the x direction in Fig. 4.11). In order to use the one-dimensional analysis from Chapter 3, we assume the lateral electric field (in the y direction) caused by V_{DS} is much smaller in magnitude than the vertical electric field caused by V_{GS}. Overall, this assumption is plausible since the gate oxide thickness (150Å) is much smaller than the channel length (on the order of 1 μm or longer), while the drain and gate voltages are each on

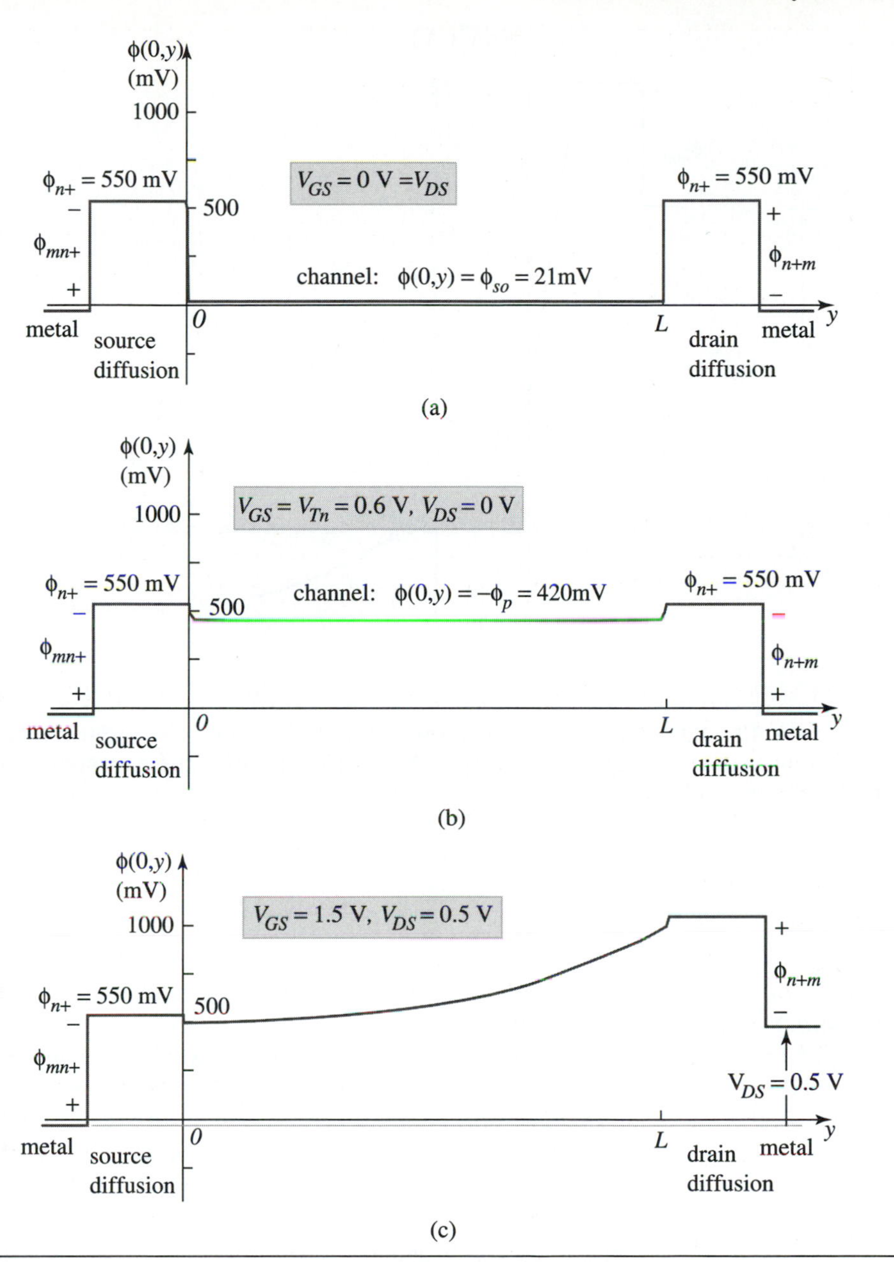

➤ Figure 4.12 Surface potential along the channel of a MOSFET with the substrate doping concentration $N_a = 10^{17}$ cm^{-3} and oxide thickness $t_{ox} = 150$ Å. (a) Thermal equilibrium, (b) gate biased at the threshold voltage with no drain-source bias, and (c) gate bias increased to well above threshold $V_{GS} = 1.5$ V with a drain-source bias $V_{DS} = 0.5$ V that places the MOSFET in the triode region.

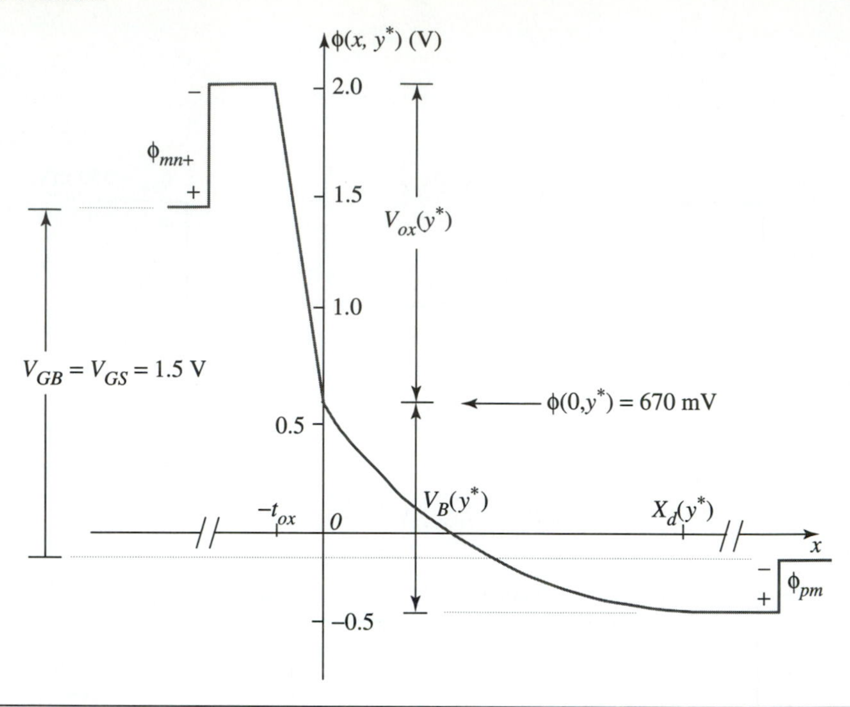

➤ **Figure 4.13** Potential $\phi(x, y^*)$, where the potential at point $y = y^*$ is 250 mV above its potential at the onset of inversion. The MOSFET is biased at the operating point used for Fig. 4.12(c).

the order of a few volts. This standard approach to MOSFET modeling is known as the **gradual channel approximation**.

Applying the one-dimensional results from Chapter 3, we equate the vertical potential drop through the MOSFET at any position in the channel to the sum of the drop across the oxide and the drop across the depletion region. In Fig. 4.13, we plot the potential as a function of x for a point y^* in the channel of a MOSFET biased with $V_{GS} = V_{GB} = 1.5$ V and $V_{DS} = 0.5$ V. In order to plot numerical values, we choose y^* so

$$\phi(0, y^*) = -\phi_p + 250\text{ mV} = 670\text{ mV} \tag{4.23}$$

At a general point y in the channel, we can equate

$$V_{GB} - V_{FB} = V_{GS} - V_{FB} = V_{ox}(y) + V_B(y) \tag{4.24}$$

since $V_{GB} = V_{GS}$ from Fig. 4.10. The drop across the depletion region is

$$V_B(y) = \phi(0, y) - \phi_p \tag{4.25}$$

At the source end of the channel ($y = 0$), the drop across the depletion region is

$$V_B(0) = \phi(0,0) - \phi_p = -2\phi_p \tag{4.26}$$

which is the familiar result for an MOS capacitor from Chapter 3. As shown in Fig. 4.12(c), the applied drain-source bias does not affect the potential at the source end of the channel.

Since we need to find the variation of the channel electron charge $Q_N(y)$, we apply Gauss's Law to relate the oxide field to the charge in the silicon

$$V_{ox}(y) = E_{ox}(y)t_{ox} = \left(\frac{t_{ox}}{\varepsilon_{ox}}\right)(-Q_N(y) - Q_B(y)) = \frac{-Q_N(y) - Q_B(y)}{C_{ox}} \tag{4.27}$$

Substituting for $V_{ox}(y)$ from Eq. (4.24), we can solve for $Q_N(y)$

$$Q_N(y) = -C_{ox}\left(V_{ox}(y) + \frac{Q_B(y)}{C_{ox}}\right) = -C_{ox}\left(V_{GS} - V_{FB} - V_B(y) + \frac{Q_B(y)}{C_{ox}}\right) \tag{4.28}$$

The drop across the depletion region is known in terms of the surface potential from Eq. (4.25), which leads to

$$Q_N(y) = -C_{ox}\left(V_{GS} - V_{FB} - \phi(0,y) + \phi_p + \frac{Q_B(y)}{C_{ox}}\right) \tag{4.29}$$

This result can be compacted by defining the threshold voltage $V_{Tn}(y)$ to account for the variation in the bulk charge $Q_B(y)$ along the channel

$$V_{Tn}(y) = V_{FB} - 2\phi_p - \frac{Q_B(y)}{C_{ox}} = V_{FB} - 2\phi_p + \frac{1}{C_{ox}}\sqrt{2q\varepsilon_s N_a(\phi(0,y) - \phi_p)} \tag{4.30}$$

in which we substitute from Eq. (4.25) for the drop across the depletion region. Note from Eq. (4.30) and Fig. 4.12(c) that the threshold at the drain end is higher than at the source end of the channel because of the larger depletion charge resulting from $V_B(L) > V_B(0) = -2\phi_p$. This varying of threshold voltage along the channel is referred to as the **body effect**. Substituting Eq. (4.30) into Eq. (4.29), we find

$$Q_N(y) = -C_{ox}(V_{GS} - V_{Tn}(y) - \phi(0,y) - \phi_p) \tag{4.31}$$

Another convenient variable is the **channel voltage** $V_C(y)$, which is defined as

$$V_C(y) = \phi(0,y) + \phi_p \tag{4.32}$$

and is the component of the surface potential in the channel due to the drain-source voltage. Note from Fig. 4.12(c) that $V_C(0) = 0$ at the source end and $V_C(L) = V_{DS}$ at the drain end. Substituting for $V_C(y)$ into Eq. (4.31), the electron charge in the channel at point y is given by

$$Q_N(y) = -C_{ox}(V_{GS} - V_{Tn}(y) - V_C(y)) \tag{4.33}$$

in which the threshold voltage in terms of $V_C(y)$ is found by substituting into Eq. (4.30)

$$V_{Tn}(y) = V_{FB} - 2\phi_p + \frac{1}{C_{ox}}\sqrt{2q\varepsilon_s N_a(V_C(y) - 2\phi_p)} \tag{4.34}$$

Having expressed the electron charge in terms of the channel voltage in Eqs. (4.33) and (4.34), we now turn to the drift velocity term in Eq. (4.22). Assuming that the drift velocity is proportional to the electric field component along the channel E_y

$$v_y(y) = -\mu_n E_y(y) = \mu_n \frac{dV_C(y)}{dy} \tag{4.35}$$

Note that this assumption will be invalid for high electric fields, due to velocity saturation. The channel electron mobility in Eq. (4.35) is around one-third the bulk electron mobility from Fig. 2.8 due to excess scattering off the gate oxide in the thin channel. Substituting for the drift velocity and channel electron charge in Eq. (4.22), we find a differential equation in the channel voltage

$$I_D = -WQ_N(y)v_y(y) = WC_{ox}[V_{GS} - V_{Tn}(y) - V_C(y)]\mu_n \frac{dV_C(y)}{dy} \tag{4.36}$$

This equation can be integrated along the channel by separation of variables to eliminate V_C, resulting in an analytical expression for drain current as a function of V_{GS} and V_{DS} in the triode region. The final expression for $I_D(V_{GS}, V_{DS})$ is too complicated to be useful for hand design or analysis. Note that the square-root term in $V_{Tn}(y)$ leads to three-halves powers of the terminal voltages.

To find a simpler expression for the drain current, we neglect the complication of a varying threshold voltage along the channel. We can see the nature of this approximation by first evaluating Eq. (4.34) at the source end of the channel, which is defined as the MOSFET threshold voltage V_{Tn}

n-channel MOSFET threshold voltage

$$V_{Tn} = V_{Tn}(0) = V_{FB} - 2\phi_p + \frac{\sqrt{2q\varepsilon_s N_a(-2\phi_p)}}{C_{ox}} \tag{4.37}$$

Substituting V_{Tn} into Eq. (4.30), the threshold voltage variation $V_{Tn}(y)$ due to the body effect is

$$V_{Tn}(y) = V_{Tn} + \frac{-Q_B(y) + Q_B(0)}{C_{ox}} = V_{Tn} + \frac{qN_a}{C_{ox}}(X_d(y) - X_d(0)) \tag{4.38}$$

Neglecting the body effect, we approximate $V_{Tn}(y) \approx V_{Tn}$ along the channel. The error in Eq. (4.38) due to body effect is small near the source and only depends on the square root of the channel voltage. This implies that we can model the essential features of the channel charge $Q_N(y)$ by approximating Eq. (4.33) with

$$Q_N(y) \approx -C_{ox}(V_{GS} - V_{Tn} - V_C(y)) \tag{4.39}$$

The differential equation for drain current Eq. (4.36) simplifies to

$$I_D = WC_{ox}(V_{GS} - V_{Tn} - V_C(y))\mu_n \frac{dV_C(y)}{dy} \tag{4.40}$$

We can eliminate the channel voltage by separating variables and integrating

$$\int_0^L I_D dy = \int_0^{V_{DS}} W\mu_n C_{ox}(V_{GS} - V_{Tn} - V_C(y))\,dV_C \tag{4.41}$$

which leads to a closed-form expression for the drain current

$$I_D = \left(\frac{W}{L}\right)\mu_n C_{ox}(V_{GS} - V_{Tn} - V_{DS}/2)V_{DS} \tag{4.42}$$

Surprisingly, this expression is identical to Eq. (4.17) that we derived using very crude approximations for the drift velocity and channel electron charge. As mentioned in the previous section, this equation for drain current is valid in the triode region when the inversion layer extends from source to drain. This implies from Eq. (4.39) that

$$V_{GS} \geq V_{Tn} + V_C(L) = V_{Tn} + V_{DS} \tag{4.43}$$

or equivalently, that $V_{DS} \leq V_{GS} - V_{Tn}$

4.4.1 Backgate Effect

In some circuit configurations, the bulk terminal of the MOSFET is biased at a voltage V_{BS} with respect to the source, thereby making it a four-terminal device. The voltage V_{BS} in Fig. 4.14 is called the **backgate bias** since the bulk potential affects the electron charge in the channel and functions as a second gate. Since the source and drain form pn diodes with the bulk terminal, the voltage V_{BS} is negative (V_{SB} positive) in order to ensure that these diodes are reverse biased.

The effect of the backgate bias is to shift the potential of the substrate lower, thereby increasing the potential drop across the depletion region as shown in Fig. 4.15. The MOSFET is under the same gate-source and drain-source bias ($V_{GS} = 1.5$ V and $V_{DS} = 0.5$ V) and the potential is plotted where $\phi(0, y^*) = 670$ mV as in Fig. 4.13.

From Fig. 4.15, the potential drop across the depletion region at a general point y between the source and drain is increased by the backgate bias to

$$V_B(y) = \phi(0, y) - (\phi_p + V_{BS}) = V_C(y) - 2\phi_p - V_{BS} \tag{4.44}$$

in which we substituted the definition of the channel voltage from Eq. (4.32). The larger $V_B(y)$ increases the bulk charge $Q_B(y)$, resulting in a larger threshold voltage. At the source end of the channel, the channel voltage is zero and $V_B(0) = -2\phi_p - V_{BS}$. Substitution of the increased drop at the source end of the channel into Eq. (4.37) yields the dependence of the MOSFET threshold on the backgate bias

$$V_{Tn} = V_{FB} - 2\phi_p + \frac{1}{C_{ox}}\sqrt{(2qN_a\varepsilon_s)(-2\phi_p - V_{BS})} \tag{4.45}$$

Note that both ϕ_p and V_{BS} are negative numbers. The conventional form for Eq. (4.45) is

n-channel MOSFET threshold voltage with backgate bias

$$\boxed{V_{Tn} = V_{TOn} + \gamma_n(\sqrt{-2\phi_p - V_{BS}} - \sqrt{-2\phi_p})} \tag{4.46}$$

where the n-channel MOSFET **backgate effect parameter** is defined as

$$\boxed{\gamma_n = (\sqrt{2qN_a\varepsilon_s})/C_{ox}} \tag{4.47}$$

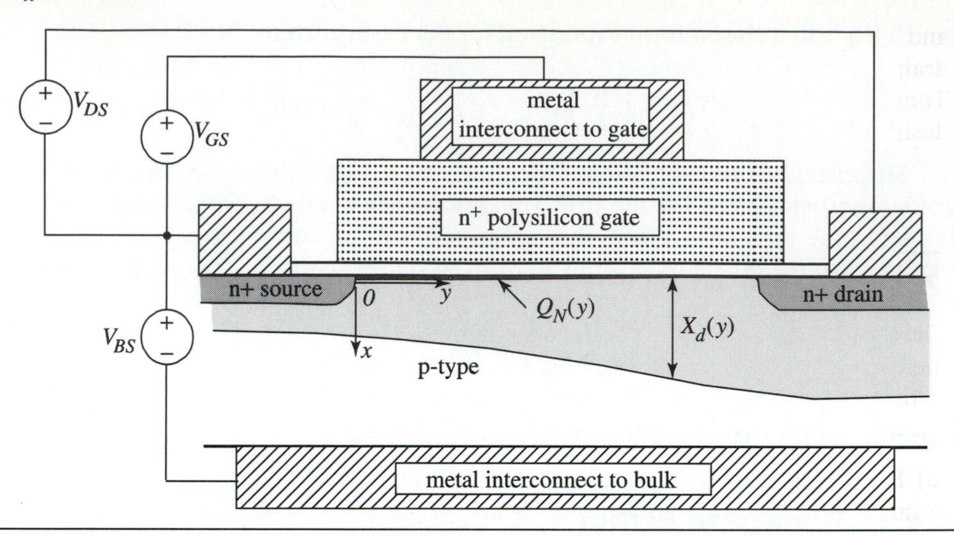

➤ **Figure 4.14** n-channel MOSFET with backgate bias V_{BS}<0. The depletion region has widened in comparison with Fig. 4.11 due to the larger potential drop across it.

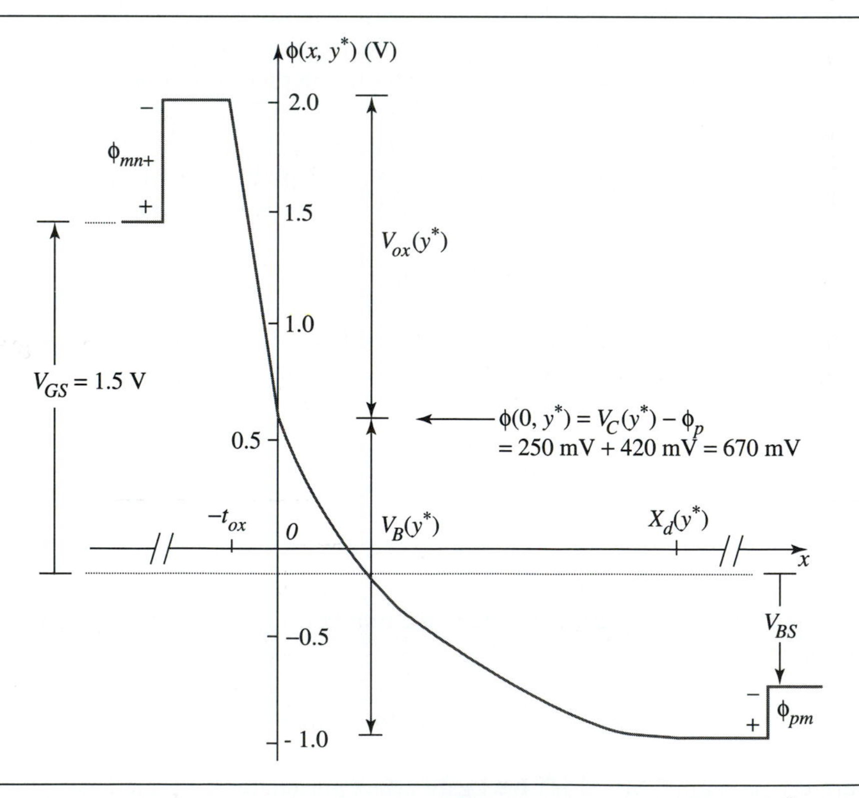

➤ **Figure 4.15** Potential variation in x direction through the MOSFET at y^* where the channel voltage is $V_C(y^*) = 250$ mV. The bias voltages are $V_{GS} = 1.5$ V, $V_{DS} = 0.5$ V, and $V_{BS} = -0.5$ V.

and V_{TOn} is the threshold voltage without backgate effect (Eq. (4.37)). Therefore, the drain current is a function of the backgate bias through its effect on V_{Tn}. In Chapters 8 and 9 we will see that the backgate bias can have significant consequences for circuit design with MOSFETs.

n-channel backgate effect parameter

➤ EXAMPLE 4.3 Measuring the Backgate Effect Parameter

The test circuit from Example 4.1 can be modified to find an experimental value for the backgate parameter, γ_n. Note that a negative voltage V_{BS} is applied from the bulk to the source of the MOSFET. The circuit varies V_{GS} continuously from 0 to 5V, for $V_{BS} = 0$ and $V_{BS} = -5$ V. The drain-source voltage is $V_{DS} = 100$ mV.

(a) From the drain current measurements plotted below, find the backgate effect parameter. Assume that the device parameters are the same as in Example 4.1 with the substrate doping $N_a = 10^{17}$ cm^{-3}.

SOLUTION

Since the drain voltage is only 100 mV, the MOSFET operates in its triode region once V_{GS} exceeds the threshold voltage. The drain current I_D is given by Eq. (4.42)

$$I_D = \left(\frac{W}{L}\right)\mu_n C_{ox} [V_{GS} - V_{Tn}(V_{BS}) - V_{DS}/2] V_{DS}$$

where the functional dependence of the threshold voltage on the backgate bias is indicated explicitly. The triode equation can be approximated with little error by neglecting $V_{DS}/2 = 50$ mV in the parentheses

$$I_D \cong \left(\frac{W}{L}\right)\mu_n C_{ox} [V_{GS} - V_{Tn}(V_{BS})] V_{DS}$$

This equation is linear in V_{GS}, with $V_{Tn}(V_{BS})$ as the intercept. From the graph, we can see $V_{TOn} = 1.0$ V and for $V_{BS} = -5$ V, $V_{Tn} = 2$V.

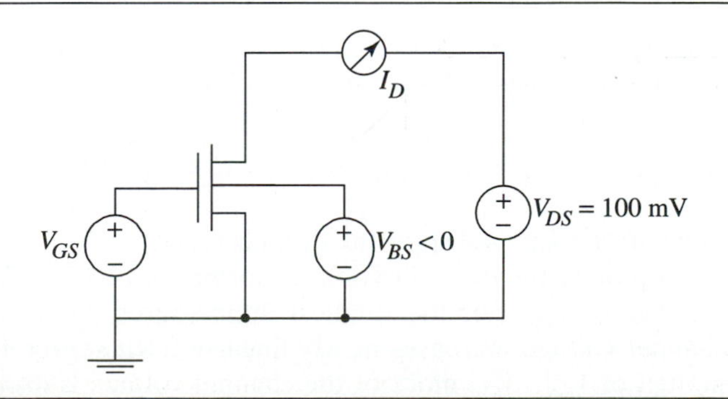

➤ Figure Ex4.3A Circuit to find the backgate effect parameter.

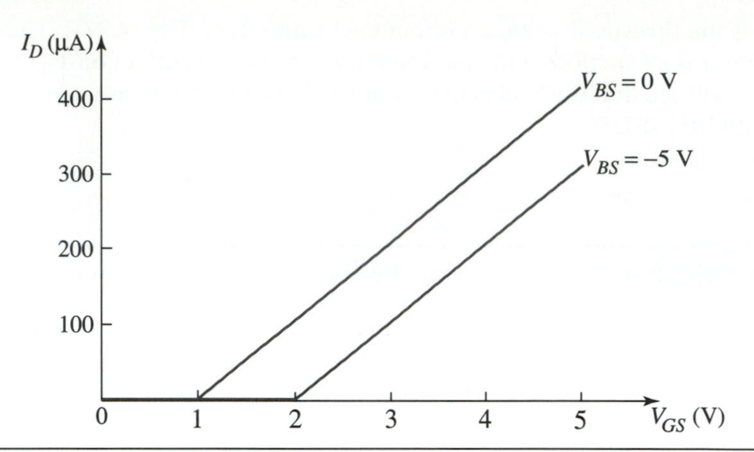

➤ **Figure Ex4.3B** Experimental data for circuit in Example 4.3

Applying Eq. (4.46) with $\phi_p = -0.42$ V, we can extract the backgate effect parameter

$$2\,\mathrm{V} = 1\,\mathrm{V} + \gamma_n\left(\sqrt{(0.84\,\mathrm{V} + 5\,\mathrm{V})} - \sqrt{0.84\,\mathrm{V}}\right) \rightarrow \gamma_n = 0.67\ \mathrm{V}^{1/2}$$

4.4.2 The MOSFET in Saturation

We will now consider the device physics of the MOSFET in saturation where the drain-source voltage increases up to and beyond the point where the channel charge vanishes at the drain end of the channel, according to our simple one-dimensional analysis. The variation of channel voltage $V_C(y)$ from source to drain can be derived from Eqs. (4.40) and provides insight into this phenomenon. By integrating Eq. (4.40) to a point y along the channel, we find that the channel voltage is given by

$$\int_0^y \frac{I_D}{\mu_n W C_{ox}} dy = \int_0^{V_C} (V_{GS} - V_{Tn} - V_C)\, dV_C = (V_{GS} - V_{Tn})\, V_C - V_C^2/2 \qquad (4.48)$$

After integrating the left hand side (remembering that I_D is constant along the channel), a quadratic equation yields the channel voltage

$$V_C(y) = V_{GS} - V_{Tn} - \sqrt{(V_{GS} - V_{Tn})^2 - 2\,(V_{GS} - V_{Tn} - V_{DS}/2)\, V_{DS}\,(y/L)} \qquad (4.49)$$

where we have substituted Eq. (4.42) for the drain current.

Figure 4.16 is a plot of the channel voltage, normalized to V_{DS}, from source to drain for three values of V_{DS}. For the smallest drain-source voltage, one-tenth of $V_{GS} - V_{Tn}$, the channel voltage increases nearly linearly from source to drain. With V_{DS} equal to one-half of $V_{GS} - V_{Tn}$, more of the channel voltage is dropped near the drain end. Finally, for the saturation drain voltage, $V_{DS_{SAT}} = V_{GS} - V_{Tn}$, most of the

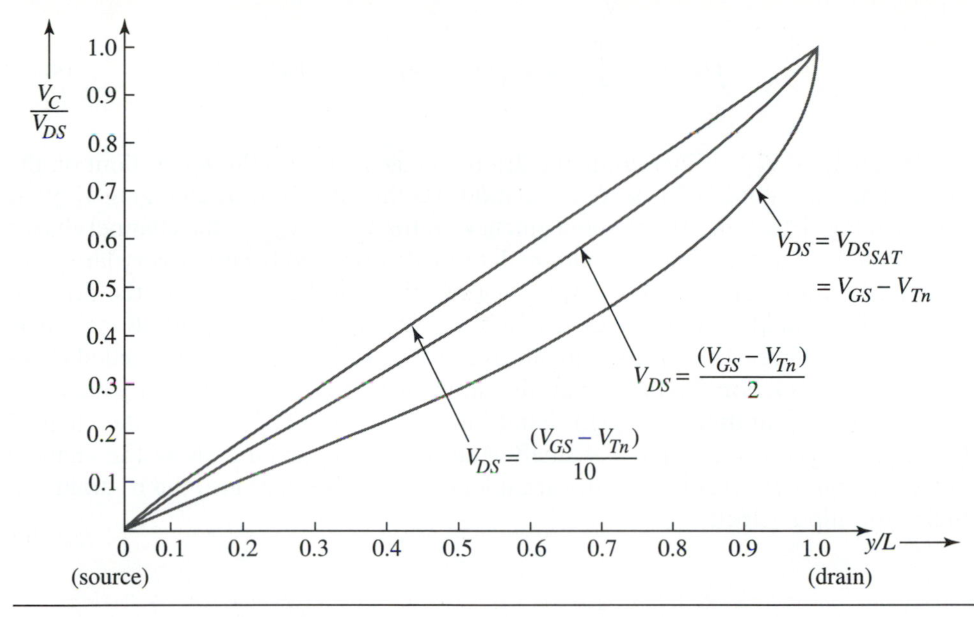

➤ **Figure 4.16** Normalized channel voltage $V_C(y)/V_{DS}$ along the channel for three values of drain-source voltage.

voltage drop occurs at the drain end with the channel voltage becoming vertical at the drain.

The channel electric field E_y is the negative derivative of $V_C(y)$:

$$E_y(y) = -\frac{dV_C}{dy} = \frac{-(V_{GS} - V_{Tn} - V_{DS}/2)\,V_{DS}}{L\sqrt{(V_{GS} - V_{Tn})^2 - 2\,(V_{GS} - V_{Tn} - V_{DS}/2)\,V_{DS}\,(y/L)}} \tag{4.50}$$

which becomes infinite at the drain ($y = L$) when $V_{DS} = V_{DS_{SAT}} = V_{GS} - V_{Tn}$. However, an infinite electric field is not reasonable and we must reexamine the assumptions underlying the gradual channel approximation. The reason why E_y blows up at the drain is seen from the basic equation for drain current, evaluated at $y = L$

$$I_D = -WQ_N(L)v_y(L) = WC_{ox}\,(V_{GS} - V_{Tn} - V_C(L))\,\mu_n\frac{dV_C}{dy}\bigg|_{y=L} \tag{4.51}$$

When the drain voltage is $V_{DS} = V_{GS} - V_{Tn}$, the channel charge vanishes at the drain end of the channel. Consequently, the channel is said to be "pinched off" at the drain. In order for the drain current to remain constant, the drift velocity must be infinite at $y = L$. Since we assumed that the drift velocity is proportional to the electric field, this condition can be satisfied mathematically by the channel voltage having an infinite slope at the drain as shown in Fig. 4.16. In reality, the drift velocity saturates and consequently, the channel electron charge is very small, but does not vanish, at the drain end of the channel.

We know from the measured drain characteristics in Fig. 4.3 that the drain current I_D **saturates** or remains nearly constant as the drain voltage increases beyond $V_{DS_{SAT}}$. The explanation for this phenomenon lies in the drain current equation (Eq. (4.41) that is repeated here for convenience.

$$\int_0^L I_D\,dy = \int_0^{V_{DS}} W\mu_n C_{ox}\left(V_{GS} - V_{Tn} - V_C(y)\right) dV_C \tag{4.52}$$

When $V_{DS} < V_{DS_{SAT}}$, increasing the drain voltage increases the upper limit on the integral on the right-hand side, but also modifies the integrand by changing $V_C(y)$ as shown in Fig. 4.16. Once the channel pinches off for $V_{DS} = V_{DS_{SAT}}$ the channel charge nearly vanishes at the drain end and the lateral electric field becomes very large. An increase in drain voltage beyond $V_{DS_{SAT}}$ causes the high-field region at the drain to widen by a small distance ΔL—enough to accommodate the additional drop $V_{DS} - V_{DS_{SAT}}$. Figure 4.17 shows a qualitative picture of this effect that is called **channel length modulation**. To first order, the drain current I_D is unchanged by increases in V_{DS} above saturation. The right-hand integral in Eq. (4.52) is over the channel (where $0 < V_C(y) < V_{DS_{SAT}}$) and is unaffected. Electrons drifting along the channel enter the high-field region near the drain and are simply swept into the n$^+$ drain—at their saturation velocity.

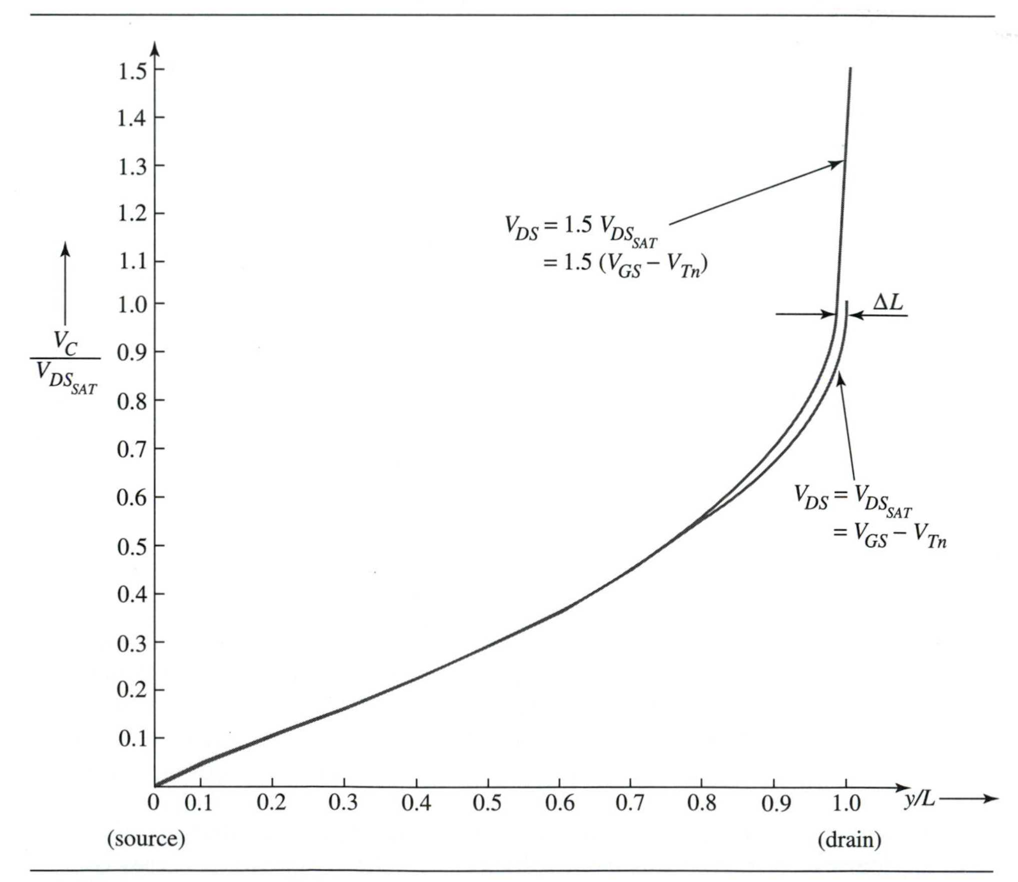

➤ **Figure 4.17** Channel voltage, normalized to the saturation drain-source voltage $V_{DS_{SAT}}$, for $V_{DS} = V_{DS_{SAT}}$ and for $V_{DS} = 1.5\ V_{DS_{SAT}}$.

Upon closer inspection, the drain current is perturbed slightly by an increase in drain voltage above $V_{GS} - V_{Tn}$. The point y in the channel at which $V_C(y) = V_{DS_{SAT}}$ and therefore $Q_N(y) = 0$ moves a small distance ΔL toward the source as the drain voltage increases beyond saturation. The channel length is shortened to $L - \Delta L$, which changes the limit of integration on the left-hand integral in Eq. (4.52)

$$\int_0^{(L-\Delta L)} I_D dy = I_D (L - \Delta L) = \mu_n C_{ox} W \int_0^{V_{DS_{SAT}}} (V_{GS} - V_{Tn} - V_C)\, dV_C \qquad (4.53)$$

where the channel extends over $0 \le y \le L - \Delta L$ and the channel voltage ranges from $0 \le V_C \le V_{DS_{SAT}}$. Carrying out the integration and substituting $V_{DS_{SAT}} = V_{GS} - V_{Tn}$, we find

$$I_{D_{SAT}} = \frac{1}{2}\left(\frac{W}{L - \Delta L}\right)\mu_n C_{ox} (V_{GS} - V_{Tn})^2 \text{ for } V_{DS} \ge V_{DS_{SAT}} = V_{GS} - V_{Tn} \qquad (4.54)$$

An exact calculation of the channel-length modulation ΔL as a function of the terminal voltages involves solving the two-dimensional Poisson's Equation in the drain region. However, a useful approximation can be made by noting that $\Delta L \ll L$

$$\frac{1}{L - \Delta L} \approx \frac{1}{L}\left(1 + \frac{\Delta L}{L}\right) \qquad (4.55)$$

For simplicity, we assume that the fractional change in channel length is proportional to the drain voltage

$$\frac{\Delta L}{L} = \lambda_n V_{DS} \qquad (4.56)$$

where λ_n is the **channel-length modulation parameter**. Device measurements and simulations indicate that λ_n varies roughly with the inverse of channel length. For the MOSFETs in this text we will use

$$\lambda_n = \frac{0.1 \mu\text{mV}^{-1}}{L} \qquad (4.57)$$

where L is in μm. Finally, we substitute Eqs. (4.55) and (4.56) into Eq. (4.54) and find a very useful approximation to the drain current in saturation

$$I_D = \frac{W}{2L}\mu_n C_{ox} (V_{GS} - V_{Tn})^2 (1 + \lambda_n V_{DS}) \qquad (4.58)$$

where $V_{DS} \ge V_{DS_{SAT}} = V_{GS} - V_{Tn}$.

Considering the several simplifying assumptions that have gone into the drain current equations, it is remarkable that they are quite adequate models for hand analysis and design. In practice, the channel electron mobility μ_n is adjusted to fit Eq. (4.42) to the measured drain current. A typical value for channel electron mobility in modern MOSFETs is 200 $\text{cm}^2/(\text{Vs})$, substantially less than the electron mobility in bulk silicon.

➤ EXAMPLE 4.4 **Vertical and lateral electric fields in the channel**

For an n-channel MOSFET with $L = 1.5\ \mu\text{m}$, $t_{ox} = 150\ \text{Å}$, $N_a = 10^{17}\text{cm}^{-3}$, and $V_{Tn} = 0.6\ \text{V}$, evaluate $E_x(0, y)$ and $E_y(0, y)$ to verify the gradual channel approximation, which assumes $|E_y| \ll E_x$.

(a) For $V_{GS} = 1.5\ \text{V}$, $V_{DS} = V_{DS_{SAT}} = 0.9\ \text{V}$, $V_{BS} = 0\ \text{V}$, find the vertical and lateral electric fields at $y = 0.75\ \mu\text{m}$.

SOLUTION

The vertical electric field in the silicon at $x = 0^+$ can be found from the oxide field by applying the boundary condition at the oxide/silicon interface

$$E_x(0^+, y = 0.75\mu\text{m}) = \left(\frac{\varepsilon_{ox}}{\varepsilon_s}\right)\frac{V_{ox}(y = 0.75\mu\text{m})}{t_{ox}}$$

The potential drop across the oxide is given by (see Fig. 4.13)

$$V_{ox}(y) = (V_{GS} + \phi_{n+}) - \phi(0, y) = V_{GS} + \phi_{n+} - V_C(y) + \phi_p$$

where we substituted the channel voltage $V_C(y)$ using Eq. (4.32). We can calculate the channel voltage at $y = 0.75\ \mu\text{m}$ from Eq. (4.49)

$$V_C = (1.5 - 0.6) - \sqrt{(1.5 - 0.6)^2 - 2(1.5 - 0.6 - 0.45)(0.9)(0.75/1.5)} = 0.26\ \text{V}$$

Note that the channel voltage is much less than $V_{DS}/2 = 0.45\ \text{V}$ at the halfway point in the channel, which is consistent with the normalized plots in Fig. 4.16. The vertical electric field at $y = 0.75\ \mu\text{m}$ is

$$E_x(0^+, y = 0.75\mu\text{m}) = \left(\frac{1}{3}\right)\frac{(1.5 + 0.55 - (0.26 - 0.42))}{1.5 \times 10^{-6}\text{cm}} = 3.3 \times 10^5\ \text{V/cm}$$

The lateral electric field at the halfway point in the channel can be found directly from Eq. (4.50)

$$E_y = \frac{-(1.5 - 0.6 - 0.45)(0.9)}{(1.5)\sqrt{(1.5 - 0.6)^2 - 2(1.5 - 0.6 - 0.45)(0.9)(0.75/1.5)}} = -4250\ \text{V/cm}$$

The ratio of the lateral field magnitude to the vertical field is $0.013 \ll 1$. The results of the gradual channel approximation justify the use of one-dimensional results from the MOS capacitor in Chapter 3.

(b) Find the vertical and lateral fields at $y = 1.45\ \mu\text{m}$.

SOLUTION

Applying the same analysis near the drain end of the channel, we find that the channel voltage is $V_C(y = 1.45\ \mu\text{m}) = 0.74\ \text{V}$. The vertical electric field is

$$E_x(0^+, y = 1.4\mu\text{m}) = \left(\frac{1}{3}\right)\frac{(1.5 + 0.55 - (0.74 - 0.42))}{1.5 \times 10^{-6}\text{cm}} = 2.0 \times 10^5\ \text{V/cm}$$

which is lower than at mid-channel because the channel potential is higher near the drain. The lateral electric field magnitude increases dramatically near the drain, as can be seen from Eq. (4.50)

$$E_y = \frac{-(1.5-0.6-0.45)\,(0.9)}{(1.5)\sqrt{(1.5-0.6)^2 - 2\,(1.5-0.6-0.45)\,(0.9)\,(1.45/1.5)}} = -3.3\times 10^4 \text{ V/cm}$$

The ratio $|E_y|/E_x$ at $y = 1.45$ μm is about 0.16—the lateral field is much more significant at the drain end of the channel. In order to model the MOSFET accurately near the drain in saturation, it is necessary to solve numerically the two-dimensional Poisson's Equation

➤ 4.5 MOSFET CIRCUIT MODELS

In the previous section, we developed the groundwork for circuit models for the MOSFET by deriving the steady-state equations for drain current as a function of the terminal voltages. These equations are useful for finding the steady-state behavior of digital circuits and for biasing analog amplifiers. For analog applications, MOSFETs are usually biased in the constant-current (saturation) region and the circuit is designed to process incremental signals. Therefore, we will derive a small-signal model for the n-channel MOSFET in the saturation region. In order to determine the response to time-varying signals, we must include capacitances between the terminals of the saturated MOSFET. We will find that the small-signal analysis of the pn junction and MOS capacitor from Chapter 3 can be applied here to find the capacitances in the MOSFET.

After deriving the n-channel MOSFET circuit models, we can adapt the analysis to the p-channel MOSFET. In a CMOS technology, both n-channel and p-channel MOSFETs are fabricated on the same substrate. One of the transistors is placed in a well of opposite type to the substrate, which affects the device models and the options for the circuit designer.

4.5.1 Steady-State Models

The drain current equations for cutoff, triode, and saturation for an n-channel MOSFET were derived in sections 4.3 and 4.4. The addition of the factor $(1 + \lambda_n V_{DS})$ in Eq. (4.58) to account for channel-length modulation in the saturation region creates a discontinuity in the drain current when $V_{DS} = V_{DS_{SAT}} = V_{GS} - V_{Tn}$. To avoid mathematical difficulties, this factor is added into the current equation for the triode region. The steady-state model for the n-channel MOSFET is

$$I_D = 0 \text{ A} \qquad (V_{GS} \le V_{Tn})$$

$$I_D = (W/L)\mu_n C_{ox}\,[V_{GS} - V_{Tn} - (V_{DS}/2)]\,(1 + \lambda_n V_{DS})\,V_{DS} \qquad (V_{GS} \ge V_{Tn},\ V_{DS} \le V_{GS} - V_{Tn})$$

$$I_{D_{SAT}} = (W/2L)\,\mu_n C_{ox}\,(V_{GS} - V_{Tn})^2\,(1 + \lambda_n V_{DS}) \qquad (V_{GS} \ge V_{Tn},\ V_{DS} \ge V_{GS} - V_{Tn})$$

(4.59)

n-channel MOSFET steady-stae drain current

Channel-length modulation can be neglected in the triode region for hand calculations, since the drain voltage V_{DS} is relatively small and the effect has no practical significance. It is easy to forget that the drain current is also a function of the bulk voltage through the threshold voltage. Therefore, Eq. (4.46) is repeated here

$$\boxed{V_{Tn} = V_{TOn} + \gamma_n\left(\sqrt{-V_{BS} - 2\phi_p} - \sqrt{-2\phi_p}\right)} \tag{4.60}$$

where $\gamma_n = (\sqrt{2q\varepsilon_s N_a})/C_{ox}$ is the backgate effect parameter.

Of the various terms in Eqs. (4.59) and (4.60), the gate oxide capacitance C_{ox} can be found directly from measurements of the gate oxide thickness t_{ox}. The MOSFET threshold voltage expression in Eq. (4.37) is very useful for indicating functional dependencies on device dimensions and parameters, although without additional terms it is not sufficiently accurate to predict V_{Tn}. These additional terms account for the contributions of oxide charges and the threshold-adjustment implant and are discussed in advanced device physics texts. In practice, experimental values for V_{TOn}, μ_n, λ_n, and γ_n are found by "curve-fitting" Eqs. (4.59) and (4.60) to the measured drain characteristics.

At first glance, the ratio W/L in Eqs. (4.59) can be read directly from the device layout. However, the channel length is often designed as the minimum allowed dimension in the fabrication technology. Since the channel length is in the denominator, small errors in L can lead to substantial errors in W/L. Figure 4.18 illustrates the need to be precise in specifying the channel length for the MOSFET. The "as drawn" channel length L_{mask} on the layout becomes the "as etched" length L_{gate} of the polysilicon gate, due to systematic lithography and etching effects. Finally, the ion-implanted source and drain regions diffuse under the polysilicon gate by a dis-

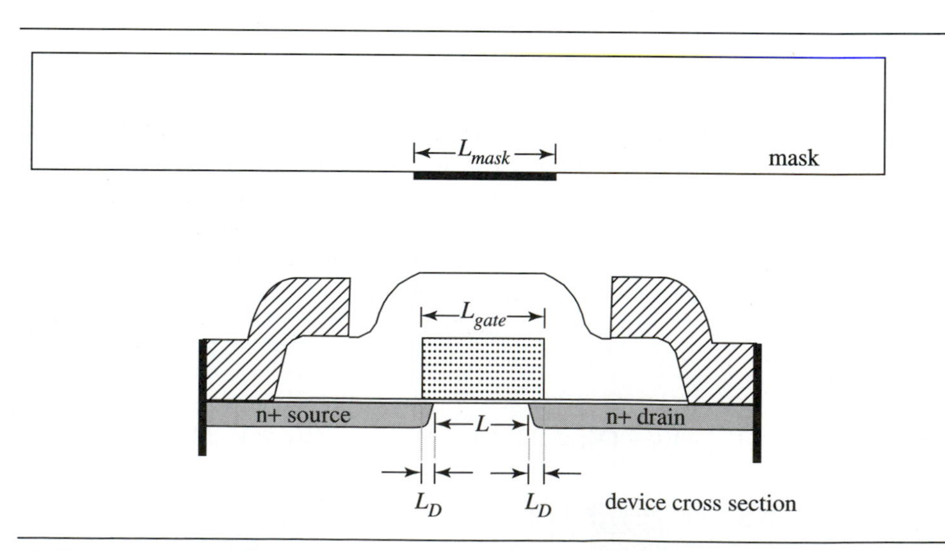

➤ **Figure 4.18** Definitions of the as-drawn gate length L_{mask} on the mask, the "as-etched" polysilicon gate length L_{gate}, and the channel length $L = L_{gate} - 2L_D$.

tance L_D, as shown in Fig. 4.18. The actual channel length, as defined earlier in Fig. 4.1, is given by

$$L = L_{gate} - 2L_D . \tag{4.61}$$

For this reason and because of the limitations of the gradual channel approximation, the ratio W/L is determined by curve-fitting to the measured drain characteristics.

4.5.2 Small-Signal Model Elements

In analog applications, a DC operating point, specified by the bias voltages V_{GS}, V_{DS}, and V_{BS}, is established so $V_{DS} > V_{DS_{SAT}} = V_{GS} - V_{Tn}$. Incremental voltages v_{gs}, v_{ds}, and v_{bs} that are much smaller in magnitude than the bias voltages perturb the operating point. The small-signal model is a circuit representation of the response of the drain current of the MOSFET to these perturbations.

The total drain current $i_D = I_D + i_d$ is a function of the gate, drain, and bulk voltages in saturation. By summing the contributions from the perturbations v_{gs}, v_{bs}, and v_{ds}, we can write a first-order expansion for the drain current

$$I_D + i_d = I_D + \left.\frac{\partial i_D}{\partial v_{GS}}\right|_Q v_{gs} + \left.\frac{\partial i_D}{\partial v_{BS}}\right|_Q v_{bs} + \left.\frac{\partial i_D}{\partial v_{DS}}\right|_Q v_{ds} \tag{4.62}$$

where Q is the DC operating point of the MOSFET, which is defined by V_{GS}, V_{BS}, and V_{DS}. The partial derivatives in Eq. (4.62) represent small-signal circuit elements and are given the following symbols

$$i_d = g_m v_{gs} + g_{mb} v_{bs} + g_o v_{ds} \tag{4.63}$$

where

$$g_m = \left.\frac{\partial i_D}{\partial v_{GS}}\right|_Q , \quad g_{mb} = \left.\frac{\partial i_D}{\partial v_{BS}}\right|_Q , \text{ and } \quad g_o = \left.\frac{\partial i_D}{\partial v_{DS}}\right|_Q \tag{4.64}$$

The MOSFET **transconductance** g_m connects the small-signal gate-source voltage v_{gs} to the resulting incremental drain current. The **backgate transconductance** g_{mb} represents the perturbation of the drain current by an incremental change in the bulk-source voltage. Finally, the **output conductance** g_o relates the incremental change in drain current due to a small-signal, drain-source voltage.

For an operating point Q in the saturation region, Fig. 4.19 is a graphical interpretation of the transconductance. Substituting the saturation drain current expression from Eq. (4.59), we find

$$g_m = \left.\frac{\partial}{\partial v_{GS}}\right|_Q \left\{\left(\frac{W}{2L}\right)\mu_n C_{ox} (v_{GS} - V_{Tn})^2 (1 + \lambda_n v_{DS})\right\} \tag{4.65}$$

Evaluating the partial derivative at the operating point, we find

$$g_m = \left(\frac{W}{L}\right)\mu_n C_{ox} (V_{GS} - V_{Tn}) (1 + \lambda_n V_{DS}) . \tag{4.66}$$

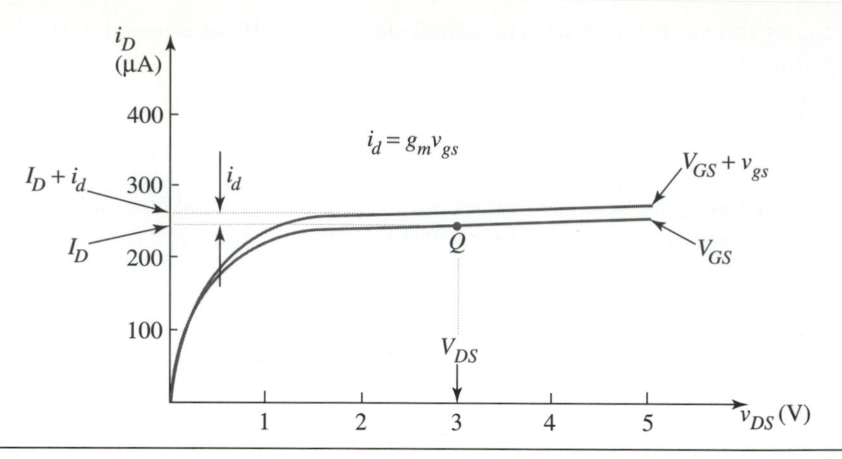

➤ **Figure 4.19** Graphical interpretation of the perturbation of the saturated drain current by the small-signal, gate-source voltage v_{gs} at bias point $Q = (V_{GS}, V_{DS}, V_{BS})$. Note that the drain-source and bulk-source voltages are kept constant so $v_{ds} = 0$ V, $v_{bs} = 0$ V,

A useful approximation to this equation can be made by neglecting the contribution from the channel length modulation, which will be a small error if $\lambda_n V_{DS} \ll 1$

n-channel MOSFET transconductance g_m

$$g_m \cong \left(\frac{W}{L}\right)\mu_n C_{ox}(V_{GS} - V_{Tn}) = \sqrt{2\left(\frac{W}{L}\right)\mu_n C_{ox} I_D} \quad (4.67)$$

These results indicate that g_m is equally sensitive to the bias drain current, the channel width-to-length ratio, and $\mu_n C_{ox}$. In Chapter 7 we will see that the bipolar junction transistor has a transconductance that is linearly related to the bias current, in contrast to the square-root dependence on bias current for the MOSFET. For a typical channel mobility 215 $\text{cm}^2\text{V}^{-1}\text{s}^{-1}$, and oxide capacitance 2.3 $\text{fF}\mu\text{m}^{-2}$, bias drain current $I_D = 100$ μA, and $(W/L) = 10$, we find that the transconductance $g_m = 316$ μS (S = A/V).

Continuing with the backgate transconductance g_{mb}, it is convenient to expand the partial derivative in Eq. (4.64) using the chain rule

$$g_{mb} = \left.\frac{\partial i_D}{\partial v_{BS}}\right|_Q = \left.\frac{\partial i_D}{\partial V_{Tn}}\right|_Q \left.\frac{\partial V_{Tn}}{\partial v_{BS}}\right|_Q \quad (4.68)$$

The first partial derivative can be evaluated by substitution of Eq. (4.59)

$$\left.\frac{\partial i_D}{\partial V_{Tn}}\right|_Q = \frac{\partial}{\partial V_{Tn}}\left(\frac{W}{2L}\right)\mu_n C_{ox}(v_{GS} - V_{Tn})^2(1 + \lambda_n v_{DS})) = -\left.\frac{\partial i_D}{\partial v_{GS}}\right|_Q = -g_m \quad (4.69)$$

where we recognize that the result is the negative of the transconductance. Partial differentiation of the threshold voltage (Eq. (4.60)) yields an expression for the backgate transconductance

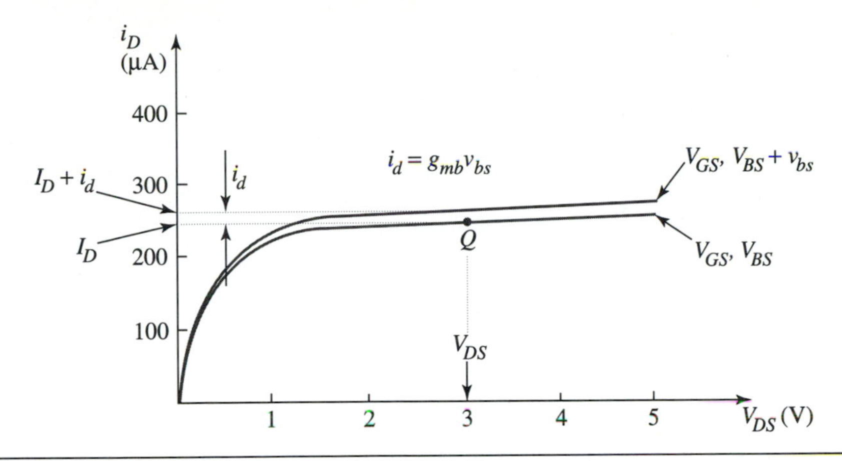

➤ **Figure 4.20** Perturbation of the saturated drain current by the small-signal, bulk-source voltage v_{bs} at bias point $Q = (V_{GS}, V_{DS}, V_{BS})$. Note that the gate-source and drain-source voltages are kept constant so $v_{gs} = 0$ V, $v_{ds} = 0$ V.

$$g_{mb} = (-g_m)\left.\frac{\partial V_{Tn}}{\partial v_{BS}}\right|_Q = (-g_m)\left(\frac{-\gamma_n}{2\sqrt{-2\phi_p - V_{BS}}}\right) = \frac{\gamma_n g_m}{2\sqrt{-2\phi_p - V_{BS}}} \tag{4.70}$$

n-channel MOSFET backgate transconductance g_{mb}

where γ_n is the n-channel backgate effect parameter defined in Eq. (4.47). Figure 4.20 shows how the backgate transconductance can be measured on the drain characteristics.

The physical meaning of the backgate transconductance is that incremental changes in the bulk-source voltage alter the electrostatics of the MOS structure and affect the channel charge exactly as do small-signal gate-source voltages. The bulk thus acts as a **backgate** (actually, "bottom gate" would be more descriptive) that is separated from the channel by a depletion layer. Using Eq. (4.70), we can solve for the ratio of the two transconductances

$$\frac{g_{mb}}{g_m} = \frac{\sqrt{2q\varepsilon_s N_a}}{2C_{ox}\sqrt{-2\phi_p - V_{BS}}} = \frac{1}{C_{ox}}\sqrt{\frac{q\varepsilon_s N_a}{2(-2\phi_p - V_{BS})}} = \frac{C_b(y=0)}{C_{ox}} \tag{4.71}$$

in which we identified the depletion capacitance at the source end of the channel, $C_b(y = 0)$ from

$$C_b(y=0) = \frac{\varepsilon_s}{X_d(y=0)} = \frac{\varepsilon_s}{\sqrt{2\frac{\varepsilon_s V_B(y=0)}{qN_a}}} \text{ and } V_B(y=0) = -2\phi_p - V_{BS}. \tag{4.72}$$

Evaluating Eq. (4.71) with the bulk shorted to the source ($v_{BS} = 0$ V) and $N_a = 10^{17}$ cm^{-3}, the backgate transconductance is about 40% of the "front gate" transconductance g_m. However, the bulk-source signal $v_{bs} = 0$ V for this case, making $g_{mb}v_{bs} = 0$ A. For a MOSFET with $V_{BS} = -1.5$ V, $g_{mb}/g_m \approx 0.25$. In order to minimize

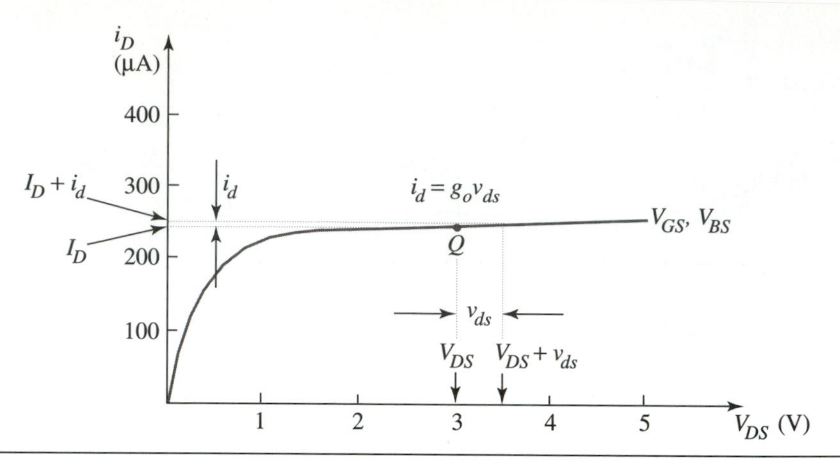

➤ **Figure 4.21** Perturbation of the saturated drain current by the small-signal, drain-source voltage v_{ds} at bias point $Q = (V_{GS}, V_{DS}, V_{BS})$. Note that the gate-source and bulk-source voltages are kept constant, so that $v_{gs} = 0\,\text{V}$, $v_{bs} = 0\,\text{V}$.

the relative magnitude of the backgate transconductance, Eq. (4.71) shows it is necessary to increase the gate capacitance (by reducing the gate oxide) or reduce the depletion capacitance between the channel and the bulk. The latter can be achieved by reducing the bulk doping or increasing the magnitude of the bulk-source bias voltage.

The final small-signal element is the drain conductance g_o that represents the dependence of the drain current on the drain-source voltage

n-channel MOSFET output conductance g_o

$$g_o = \left.\frac{\partial i_D}{\partial v_{DS}}\right|_Q = \left(\frac{W}{2L}\right)\mu_n C_{ox}(V_{GS} - V_{Tn})^2 \lambda_n \cong \lambda_n I_D \tag{4.73}$$

where we again make the approximation that $\lambda_n V_{DS} \ll 1$ at the operating point to obtain a simpler result. The graphical interpretation of the output conductance is shown in Fig. 4.21.

The channel-length modulation parameter λ_n has been found to vary inversely with the channel length

$$\lambda_n = \propto \frac{1}{L}, \tag{4.74}$$

For a MOSFET with $L = 1.5\,\mu\text{m}$, $\lambda_n = 0.06\,\text{V}^{-1}$. At a DC bias current $I_D = 100\,\mu\text{A}$, the inverse of the drain conductance, called the **output resistance**, is

$$(g_o)^{-1} = r_o = 150\,\text{k}\Omega \tag{4.75}$$

4.5.3 Small-Signal Circuit

We can assemble the small-signal model for steady-state perturbations on the operating point by properly connecting these three elements between the four terminals of the MOSFET. Kirchhoff's current law indicates that the three elements are con-

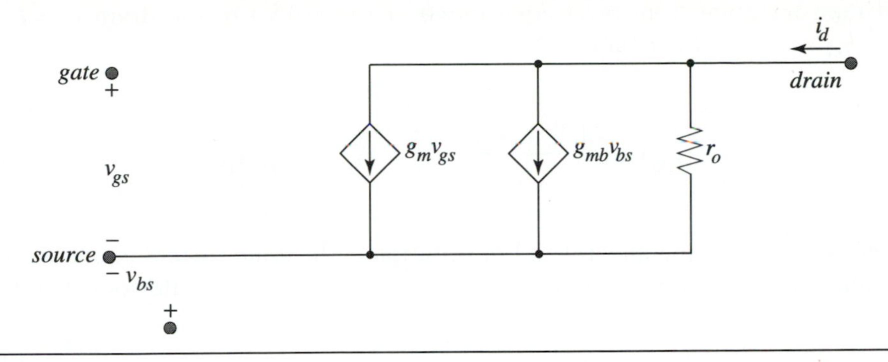

➤ **Figure 4.22** Steady-state small-signal model for the n-channel MOSFET in saturation.

nected in parallel at the drain node as shown in Fig. 4.22, since their contributions sum to give the small-signal drain current

$$i_d = g_m v_{gs} + g_{mb} v_{bs} + g_o v_{ds} \tag{4.76}$$

Note that the small-signal, steady-state gate current of the MOSFET is zero. In addition, the reverse-biased source-bulk junction is an open circuit in steady-state as shown in Fig. 4.22.

4.5.4 MOSFET Capacitances in Saturation

The MOS capacitor formed between the gate and substrate is the central structure of the MOSFET. The inversion-layer electron charge under the gate plays a fundamental role in the MOSFET by enabling current between the drain and source. In order to model the small-signal response to time-varying small signals, we must model the MOS capacitor and other capacitors in the transistor, when the MOSFET is operated in the saturation region.

In saturation, the channel charge is a function only of the gate-source voltage. As discussed in Sections 4.3 and 4.4, changes in the drain-source voltage have negligible effect on the channel for $V_{GS} > V_{DS_{SAT}}$. The total electron charge in the channel in saturation is given by

$$q_N(v_{GS}) = -W\int_0^L C_{ox}(v_{GS} - V_{Tn} - v_C(y))\,dy \tag{4.77}$$

where we integrate the channel charge per unit area from Eq. (4.33) over the area of the channel. Symbols for total channel charge q_N, gate-source voltage v_{GS}, and channel voltage v_C are used since incremental (small) signals are present. We can change variables in the integral using Eq. (4.40)

$$dy = \left(\frac{W\mu_n C_{ox}}{i_D}\right)(v_{GS} - V_{Tn} - v_C)\,dv_C \tag{4.78}$$

and then integrate from the source ($y = 0$ and $v_C = 0$ V) to the drain ($y = L$ and $v_C = v_{DS} = v_{GS} - V_{Tn}$ for saturation)

$$q_N(v_{GS}) = -\left(\frac{W^2 \mu_n C_{ox}^2}{i_D}\right) \int_0^{v_{GS} - V_{Tn}} (v_{GS} - V_{Tn} - v_C^2)\, dv_C \tag{4.79}$$

After evaluation of the integral and substitution of the drain current equation in saturation, we can express the channel charge as a function of the gate-source voltage

$$q_N(v_{GS}) = -\frac{2}{3} WLC_{ox}(v_{GS} - V_{Tn}) \tag{4.80}$$

The gate charge is equal and opposite to the sum of the channel charge and the maximum bulk (depletion region) charge

$$q_G(v_{GS}) = -q_N(v_{GS}) - Q_{B,max} = \frac{2}{3} WLC_{ox}(v_{GS} - V_{Tn}) - Q_{B,max} \tag{4.81}$$

The capacitance in saturation is the derivative of Eq. (4.81)

$$\left.\frac{dq_G}{dv_{GS}}\right|_{V_{GS} \geq V_{DS_{SAT}}} = \frac{2}{3} WLC_{ox} \tag{4.82}$$

where we have used the fact that the bulk charge is constant in saturation.

Figure 4.23 is a cross section of the MOSFET biased in saturation. Due to overlap of the source and drain diffusions and the polysilicon gate and the contribution of fringe electric fields from the gate, there is significant additional capacitance. The overlap capacitance C_{ov} is quantified as a linear capacitance proportional to the gate width, with units of fF/μm. The total gate-source capacitance in saturation is the sum of Eq. (4.82) and the overlap capacitance:

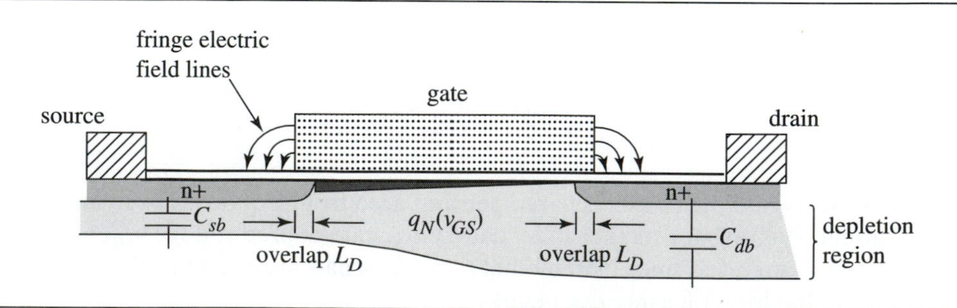

➤ **Figure 4.23** MOSFET cross section in saturation showing the channel charge $q_N(v_{GS})$ and the overlap and fringe contributions to C_{ov}. The source-bulk and drain-bulk depletion capacitances are also shown qualitatively.

$$C_{gs} = \frac{2}{3}WLC_{ox} + WC_{ov} \tag{4.83}$$

gate-source capacitance in saturation

Since the drain has no influence on the channel charge, the only contribution to the gate-drain capacitance is C_{ov}:

$$C_{gd} = WC_{ov} \tag{4.84}$$

gate-drain capacitance in saturation

The drain-bulk and source-bulk capacitances C_{db} and C_{sb} originate from charge storage in the depletion regions between the drain and source n^+ regions and the p-type bulk. Given the area of these reverse-biased junctions $A_{diff} = L_{diff}W$ from the layout in Fig. 4.1, C_{db} and C_{sb} can be found as functions of the junction voltages V_{DB} and V_{SB} using the results from Section 3.6. Finally, the capacitance C_{gb} between gate and bulk is from the overlap of the polysilicon gate onto the field oxide region as shown in Fig. 4.1. This MOS capacitor is in depletion for normal gate voltages and its capacitance can be found from the MOS capacitor analysis in Section 3.9. It is often ignored as negligible, however, and is included only for completeness.

To predict the frequency response of the MOSFET, we add the capacitances between the terminals at the operating point. Figure 4.24 shows the schematic for the small-signal model of the intrinsic n-channel MOSFET with the capacitances added to the steady-state model in Fig. 4.22.

One final, but very important point. We have only modeled the *intrinsic* MOSFET, that is, the device without interconnections. The parasitic capacitances of the interconnections between MOSFETs are often the limiting factor in switching speed and *must* be estimated from the layout and cross section for accurate analysis of a design. Off-chip wiring and package capacitances are also critical for evaluating the performance of any integrated circuit.

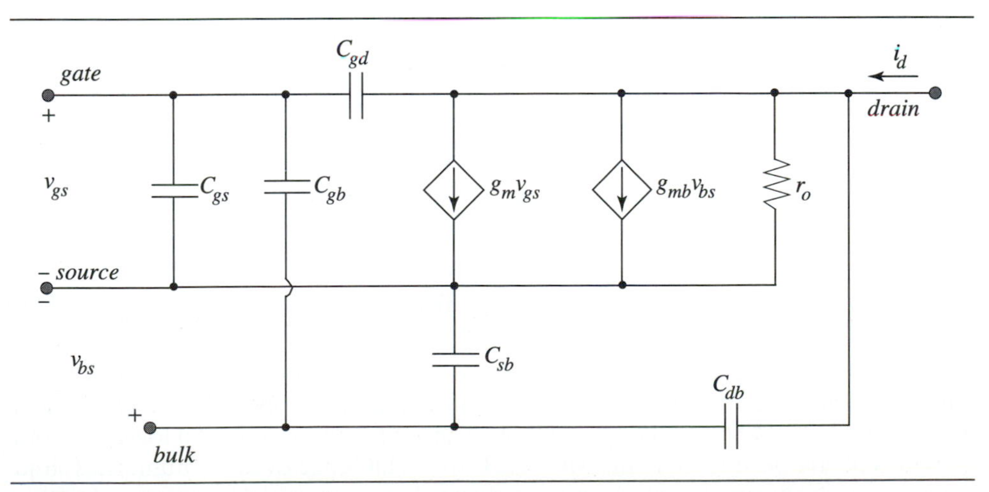

➤ Figure 4.24 Complete small-signal model for the n-channel MOSFET in saturation.

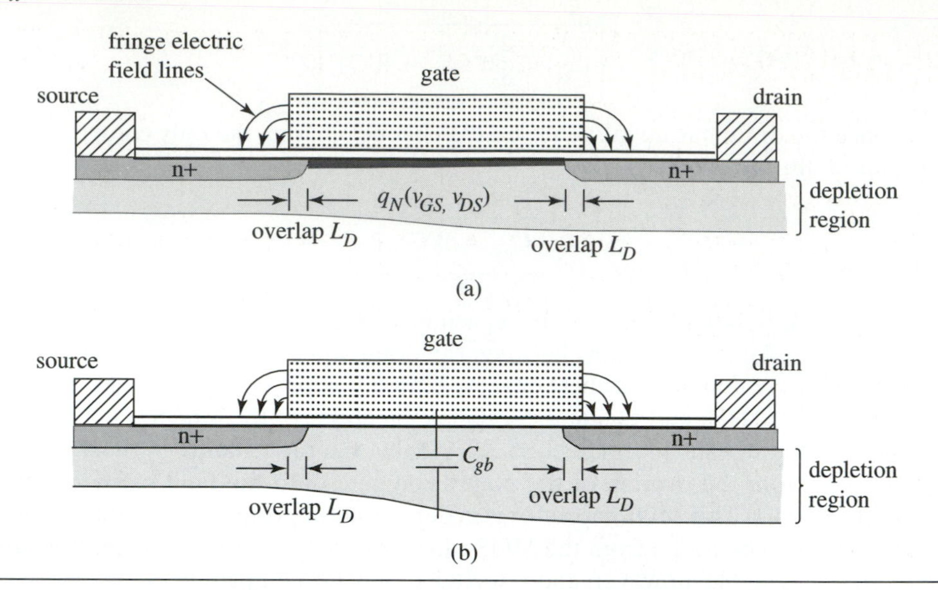

➤ Figure 4.25 MOSFET cross section showing components of C_{gs}, C_{gd}, and C_{gb} in (a) triode and (b) cutoff.

4.5.5 MOSFET Capacitances in Triode and Cutoff

A basic understanding of the terminal capacitances in triode and in cutoff is useful for digital IC design. Figure 4.25 (a) shows the channel charge in triode, which is a function of both v_{GS} and v_{DS}. Rather than derive models for C_{gs} or C_{gd} from $q_N(v_{GS}, v_{DS})$, we split the total channel charge equally between source and drain. After adding in the overlap capacitances, the triode-region capacitances are

C_{gs} and C_{gd} in triode

$$C_{gs} = C_{gd} = \frac{WLC_{ox}}{2} + WC_{ov} = C_{gd} \tag{4.85}$$

In triode, C_{gd} is much larger than its value in saturation, whereas the gate-source capacitance C_{gs} decreases from its value in saturation.

In cutoff, there is no channel present as shown in Fig. 4.25(b). The depletion region under the gate is under gate control and the MOS capacitor theory from Section 3.9 can be applied to find C_{gb}. The gate-bulk capacitance is a sensitive function of the gate-bulk bias voltage, as seen in Fig. 3.35. Since no channel is present in cutoff, the gate-source and gate-drain capacitances are each equal to the overlap capacitance.

C_{gs} and C_{gd} in cutoff

$$C_{gs} = C_{gd} = WC_{ov} \tag{4.86}$$

This result completes our overview of charge storage in the MOSFET. The terminal capacitances are very different in the three regions of operation, as can be seen from Eqs. (4.83)-(4.86). In the case of large transients such as in digital logic, the complexities of charge storage in the MOSFET are a challenge even for numerical simulators. For small-signal analysis in the saturation region, which is the usual case in analog design, the capacitances in Eqs. (4.83) and (4.84) are reasonably accurate for hand analysis.

➤ EXAMPLE 4.5 MOSFET small-signal model

For the n-channel MOSFET in Fig. 4.1, the channel length is $L = 2\ \mu m$, the width is $W = 30\ \mu m$, the length of the source-drain diffusions is $L_{diff} = 6\ \mu m$, the channel mobility is $\mu_n = 215\ cm^2V^{-1}s^{-1}$, the oxide thickness is $t_{ox} = 150$ Å, and the substrate doping is $N_a = 10^{17}\ cm^{-3}$, and the channel length modulation parameter is $\lambda_n = 0.05\ V^{-1}$. The overlap capacitance is $C_{ov} = 0.5$ fF/μm. The threshold voltage without backgate bias is $V_{TOn} = 1$ V. Note that this measured threshold voltage is greater than predicted by the simple MOS theory in Chapter 3 (due to the presence of a shallow threshold-adjustment implant and charge in the oxide film.)

(a) For the DC operating point Q ($V_{GS} = 1.5$ V, $V_{DS} = 1.5$ V, $V_{BS} = 0$ V), find all the small-signal model parameters in Fig. 4.24, except the gate-bulk capacitance C_{gb} that we will consider negligible.

SOLUTION

Saturation drain current:
The first step is to find the DC bias current from Eq. (4.59)

$$I_D = \left(\frac{W}{2L}\right)\mu_n C_{ox}(V_{GS} - V_{Tn})^2(1 + \lambda_n V_{DS})$$

in which
$\mu_n C_{ox} = (215\ cm^2V^{-1}s^{-1})(3.9 \times 8.85 \times 10^{-14}\ F/cm)(1.5 \times 10^{-6}\ cm)^{-1} = 49.5\ \mu AV^{-2}$
The threshold voltage is

$$V_{Tn} = V_{TOn} + \gamma_n(\sqrt{-2\phi_p - V_{BS}} - \sqrt{-2\phi_p}) = V_{TOn} = 1\ V \text{ since } V_{BS} = 0\ V$$

The error in neglecting the channel-length modulation term in finding I_D is $(0.05)(1.5) = 0.075$, less than 10%. Therefore, we neglect it for hand calculations. Substituting $\mu_n C_{ox}$, V_{Tn}, (W/L), and V_{GS} into the saturation drain current, we find

$$I_D = \frac{1}{2}\left(\frac{30}{2}\right)(49.5\mu AV^{-2})\ (1.5V - 1V)^2 = 91\ \mu A$$

Steady-State Parameters:
From Eq. (4.67), the transconductance is

$$g_m = \left(\frac{30}{2}\right)(49.5\mu AV^{-2})\ (1.5V - 1V) = 371\ \mu S$$

The backgate effect parameter is found from Eq. (4.47)

$$\gamma_n = \frac{\sqrt{2 \cdot 1.6 \times 10^{-19} \cdot 11.7 \cdot 8.85 \times 10^{-14} F\ cm^{-1} \cdot 10^{17} cm^{-3}}}{(3.9 \cdot 8.85 \times 10^{-14} F\ cm^{-1})/(1.5 \times 10^{-6} cm)} = 0.79\ V^{1/2}$$

Substitution of γ_n and $\phi_p = -0.42$ V (from the known substrate doping) into Eq. (4.70) yields the backgate transconductance

$$g_{mb} = \frac{(0.79\ V^{1/2}) \cdot (371\mu S)}{2\sqrt{-2(-0.42V) - 0}} = 160\ \mu S$$

The output resistance is

$$r_o = \frac{1}{g_o} = \frac{1}{\frac{1}{2}\left(\frac{30}{2}\right)(49.5\mu\text{AV}^{-2})(1.5\text{V} - 1\text{V})^2(0.05\text{V}^{-1})} = 220\ \text{k}\Omega$$

Capacitances:
The gate oxide capacitance (per unit area) is

$$C_{ox} = \frac{(3.9 \cdot 8.85 \times 10^{-14}\text{F cm}^{-1})}{(1.5 \times 10^{-6}\text{cm})} = 2.3\ \text{fF}/\mu\text{m}^2$$

Substituting C_{ox} and $C_{ov} = 0.5$ fF/μm into Eq. (4.83), together with the MOSFET dimensions, we find

$$C_{gs} = \left(\frac{2}{3}\right)(30 \cdot 2\ \mu\text{m}^2)(2.3\text{fF}\mu\text{m}^{-2}) + (30\ \mu\text{m})(0.5\text{fF}\mu\text{m}^{-1}) = 107\ \text{fF}$$

and the gate-drain capacitance: $C_{gd} = (0.5\ \text{fF}/\mu\text{m})(30\ \mu\text{m}) = 15$ fF.
The remaining capacitances are the pn junction depletion capacitances C_{db} between the n^+ drain and the substrate and C_{sb} between the n^+ source and the substrate. The zero-bias depletion capacitance per unit area for both junctions is

$$C_{jo} = \varepsilon_s \sqrt{\frac{q}{2\varepsilon_s \phi_B}\left(\frac{1}{N_a} + \frac{1}{N_d}\right)^{-1}} \cong \sqrt{\frac{q\varepsilon_s N_a}{2\phi_B}}$$

from Eq. (3.79). Since the source and drain are heavily doped, the junction can be considered one-sided and the contribution of the depletion region on the n-side of the junction can be neglected. The built-in voltage is

$$\phi_B = \phi_{n+} - \phi_p = 0.55\ \text{V} - (-0.42\ \text{V}) = 0.97\ \text{V}$$

The zero-bias junction capacitance is

$$C_{jo} = \sqrt{\frac{(1.6 \times 10^{-19}\text{C})(11.7)(8.85 \times 10^{-14}\text{F/cm})(10^{17}\text{cm}^{-3})}{2(0.97\text{V})}} = 0.92\text{fF}\mu\text{m}^{-2}$$

The source junction has a bias voltage of $V_{BS} = 0$ V, which implies

$$C_{sb} = C_{jo}(WL_{diff}) = (0.92\text{fF}\mu\text{m}^{-2})(30\mu\text{m} \cdot 6\mu\text{m}) = 166\ \text{fF}$$

The drain junction has a bias voltage of $V_{BD} = -V_{DS} = -1.5$ V, so its depletion capacitance is reduced (Eq. 3.80)

$$C_{db} = \frac{(WL_{diff})C_{jo}}{\sqrt{1 - V_{BD}/\phi_B}} = \frac{(30\mu\text{m} \cdot 6\mu\text{m})(0.92\text{fF}\mu\text{m}^{-2})}{\sqrt{1 - (-1.5\text{V})/(0.97\text{V})}} = 104\ \text{fF}$$

We have found the complete small-signal model for the n-channel MOSFET in Fig. 4.1 evaluated at the given operating point.

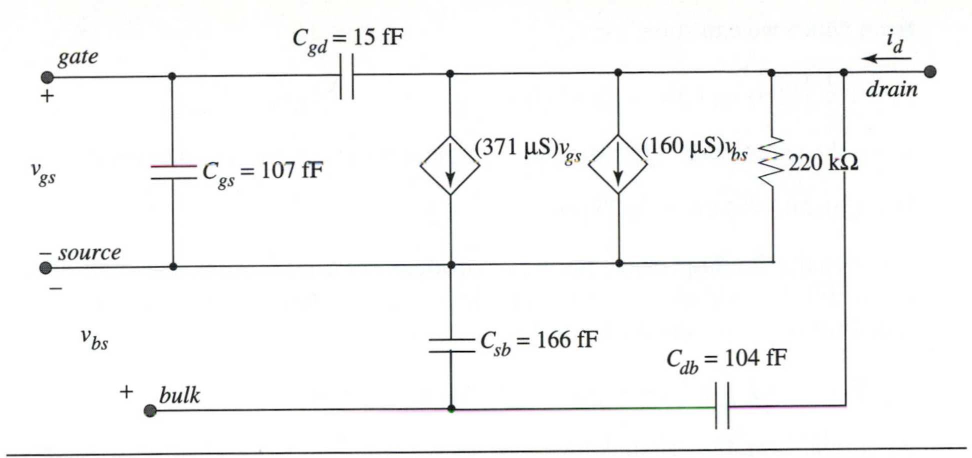

Figure Ex4.5 Small-signal model for the MOSFET in Fig. 4.1 as calculated in Example 4.5.

As we will see in Section 4.6 on the MOSFET SPICE model, the contribution of the perimeters of the source and drain diffusions can be important. The one-dimensional junction capacitance analysis in Chapter 3 is appropriate for the planar bottom of the junction. In order to estimate the perimeter capacitance, a two-dimensional or even three-dimensional analysis is needed.

(b) The product of the transconductance and the output resistance is an important quantity in small-signal amplifiers. For the MOSFET in part (a), this quantity is

$$g_m \cdot r_o = (371 \mu\text{S}) \cdot (220 \text{ k}\Omega) = 82$$

For many applications, $g_m\ r_o$ must be much larger. Given the same operating point as in part (a) and the same width-to-length ratio, redesign the MOSFET so that $g_m\ r_o = 150$. What penalty is paid for improving the product?

SOLUTION

It is worthwhile to examine the functional dependence of $g_m\ r_o$ by multiplying Eq. (4.67) and the inverse of Eq. (4.73)

$$g_m \cdot r_o = \frac{\left(\frac{W}{L}\right)\mu_n C_{ox}(V_{GS}-V_{Tn})}{\left(\frac{W}{2L}\right)\mu_n C_{ox}(V_{GS}-V_{Tn})^2 \lambda_n} = \frac{2}{(V_{GS}-V_{Tn})\lambda_n}$$

Since the operating point is unchanged, the only term that can be adjusted is the channel length modulation parameter λ_n. From Eq. (4.74), λ_n varies inversely with channel length. Given that $\lambda_n = 0.05\ \text{V}^{-1}$ for $L = 2\ \mu\text{m}$, we find the constant of proportionality to be $0.1\ \mu\text{mV}^{-1}$. Solving for λ_n.

$$\lambda_n = \frac{2}{(g_m \cdot r_o)(V_{GS}-V_{Tn})} = \frac{0.1\ \mu\text{mV}^{-1}}{L}.$$

from which we can solve for L

$$L = \left(\frac{1}{2}\right)[150 \cdot (1.5\text{V} - 1\text{V})]\,(0.1\,\mu\text{mV}^{-1}) = 3.75\,\mu\text{m}$$

Since the width-to-length ratio has been held constant, the channel width is

$$W = (30/2)(3.75)\ \mu\text{m} = 56.25\ \mu\text{m}$$

The penalty for improving the $g_m\, r_o$ product is an increase in the area of the MOSFET. In addition, C_{gs} and C_{gd} are increased due to larger W and L. The total area (not counting any backgate contact) is now

$$A = W\,(L + 2\,L_{diff}) = (56.25)(3.75 + 2(6))\ \mu\text{m}^2 = 886\ \mu\text{m}^2$$

By comparison, the original device area was $(30)(2 + 2(6)) = 420\ \mu\text{m}^2$, so the area has more than doubled.

(c) An alternative approach to increasing the $g_m\, r_o$ product is to keep the device geometry the same and modify the operating point. We will also specify that $V_{DS} = 1.5$ V and that $V_{BS} = 0$ V. Find V_{GS} and I_D for $g_m\, r_o = 150$.

SOLUTION

We can solve for the required gate-source bias from

$$g_m \cdot r_o = \frac{2}{(V_{GS} - V_{Tn})\,\lambda_n}$$

The channel length remains $L = 2\ \mu\text{m}$ and so the channel-length modulation parameter remains $\lambda_n = 0.05\ \text{V}^{-1}$. Solving for V_{GS}, we find

$$V_{GS} = V_{Tn} + \frac{2}{\lambda_n\,(g_m \cdot r_o)} = 1\text{V} + \frac{2}{(0.05\text{V}^{-1})\,(150)} = 1.27\ \text{V}$$

The drain current for this gate-source bias is

$$I_D = \left(\frac{30}{2}\right)\left(\frac{49.5\mu\text{AV}^{-2}}{2}\right)(1.27\text{V} - 1\text{V})^2 = 27\mu\text{A}$$

What is the trade-off for lowering the drain current? The transconductance g_m is reduced by more than a factor of two, which degrades the frequency response of the MOSFET, as we will see in Chapter 10. The increased capacitance in (b) also has the same affect, so increasing $g_m r_o$ comes at the cost of diminished bandwidth. This example is an introduction to the trade-offs in device and circuit design in ICs.

4.5.6 p-Channel MOSFET Circuit Models

The drain current equations for a p-channel MOSFET can be derived using exactly the same approach as for the n-channel device since the basic physics is the same. For a p-channel device, the gate must be made negative with respect to the n-type substrate in order to form an inversion layer of holes. Since holes drift across the channel from source to drain in the p-channel MOSFET, the drain voltage must be negative with respect to the source, and the drain current is negative.

A practical problem for the circuit designer is to keep track of the minus signs and negative quantities in the p-channel equations. The common orientation for the p-channel MOSFET on a schematic is with the source "up"—meaning near the positive supply as shown in Fig. 4.2. By taking the source potential as a positive reference, the source-drain potential difference V_{SD}, the source-gate potential V_{SG}, and the bulk-source potential V_{BS} are all positive. By using these quantities in the p-channel drain current equations, only the drain current and the threshold voltage are negative.

p-channel MOSFET steady-state drain current

$$\begin{aligned} -I_D &= 0\text{ A} && (V_{SG} \le -V_{Tp}) \\ -I_D &= (W/L)\mu_p C_{ox}[V_{SG} + V_{Tp} - (V_{SD}/2)](1 + \lambda_p V_{SD})V_{SD} && (V_{SG} \ge -V_{Tp},\ V_{SD} \le V_{SG} + V_{Tp}) \\ -I_D &= (W/2L)(1/2)\mu_p C_{ox}(V_{SG} + V_{Tp})^2(1 + \lambda_p V_{SD}) && (V_{SG} \ge -V_{Tp},\ V_{SD} \ge V_{SG} + V_{Tp}) \end{aligned} \tag{4.87}$$

The p-channel MOSFET threshold voltage is

p-channel MOSFET threshold voltage

$$V_{Tp} = V_{TOp} - \gamma_p(\sqrt{2\phi_n - V_{SB}} - \sqrt{2\phi_n}) \tag{4.88}$$

The small-signal parameters can be found from the drain current in the saturation region. Rather than carry out the derivations again, we adapt the n-channel results. The p-channel MOSFET transconductance is

p-channel MOSFET transconductance

$$g_m = \left(\frac{W}{L}\right)\mu_p C_{ox}(V_{SG} + V_{Tp}) \cong \sqrt{2\left(\frac{W}{L}\right)\mu_p C_{ox}(-I_D)} \tag{4.89}$$

where the transconductance is defined as

$$g_m = \left.\frac{\partial}{\partial v_{SG}}(-i_D)\right|_Q \tag{4.90}$$

The backgate transconductance is

p-channel MOSFET backgate transconductance

$$g_{mb} = g_m \left.\frac{\partial V_{Tp}}{\partial v_{SB}}\right|_Q = \frac{\gamma_p g_m}{2\sqrt{2\phi_n - V_{SB}}} \tag{4.91}$$

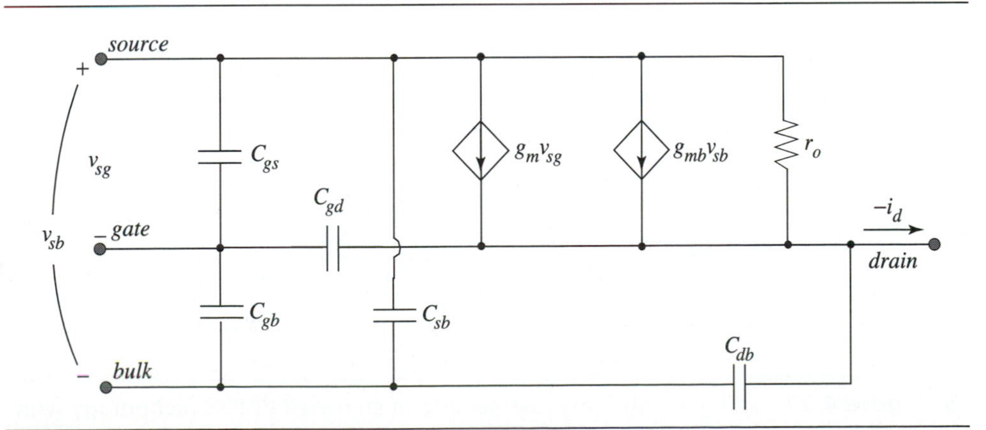

Figure 4.26 Small-signal model for the p-channel MOSFET in saturation.

and the output conductance is

p-channel MOSFET output conductance

$$g_o = \left.\frac{\partial(-i_D)}{\partial v_{SD}}\right|_Q = \left(\frac{W}{2L}\right)\mu_p C_{ox}(V_{SG} + V_{Tp})^2 \lambda_p \cong \lambda_p(-I_D) \qquad (4.92)$$

Finally, the capacitances associated with the p-channel MOSFET are identical to those of its complementary device. Figure 4.26 is the small-signal model for the p-channel MOSFET with the source located at the top, which is its most common orientation.

4.5.7 MOSFET Structures in CMOS Processes

The n- and p-channel MOSFETs can be made together in a Complementary MOS (CMOS) process. Having both types of MOSFETs available has important advantages for both digital and analog circuit design. Since the two devices need a bulk region with the opposite doping type, a CMOS process includes a deep well diffusion. Figure 4.27 shows the layout and a schematic cross section of an n-well CMOS process, in which the p-channel MOSFETs are fabricated in an n-type well in a p-type substrate. The opposite scheme, the p-well CMOS process, is less common.

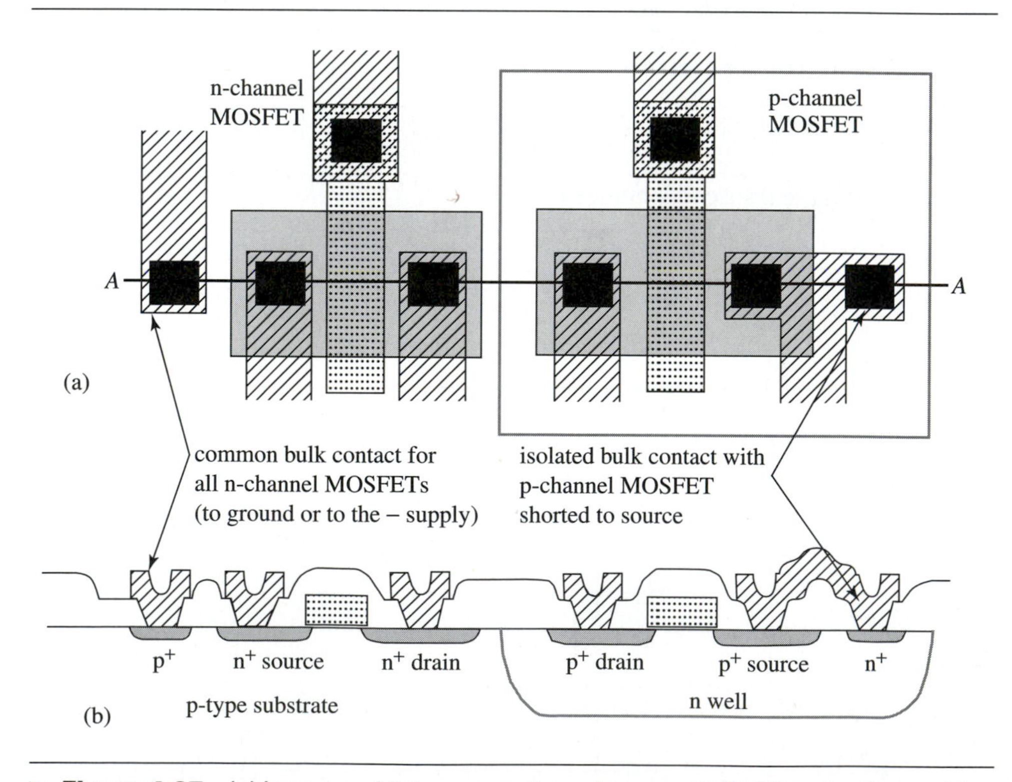

Figure 4.27 (a) layout and (b) cross section of an n-well CMOS technology with isolated bulks regions for p-channel MOSFETs.

All n-channel MOSFETs share a common bulk region (the p-type substrate) in an n-well technology. Multiple contacts are made to the substrate, one of which is shown in Fig. 4.27, to ensure a low-resistance connection to ground. The IC designer has the option of placing a p-channel MOSFET in its own well so that its bulk (the n-well) can be shorted to its source, as shown in Fig. 4.27. Alternatively, a large number of p-channel devices can share a single n-well.

As we will see in Chapter 8, there are circuit performance advantages if the bulk and source contacts can be shorted. For this configuration, the backgate transconductance generator does not add to the small-signal drain current since $v_{BS} = 0$ V. In an n-well process, $V_{BS} = 0$ V only for those n-channel MOSFETs that have their sources grounded, since their shared bulk terminal must be grounded. Other n-channel devices will have their thresholds shifted by the backgate effect through Eq. (4.60) and must have the backgate generator included in their small-signal models. For the p-channel MOSFETs, on the other hand, we can arrange for $v_{SB} = 0$ for a particular device by placing it in its own n-well. There is an area penalty for placing a device in an isolated well, rather than having it share a well with many other p-channel MOSFETs. As a result, an IC designer will only isolate a device when it is necessary for performance reasons.

4.6 SPICE MODELS FOR THE MOSFET

MOSFETs can be simulated using models with varying levels of complexity in SPICE. In this text, we will use the simplest model, which is referred to as the *"Level 1"* model. This model corresponds closely with the large-signal and small-signal models that have been developed in this chapter. In this section, we will introduce the Level 1 MOSFET model and define the notation used in SPICE.

4.6.1 Level 1 DC Model

SPICE models the drain-source current I_{DS} of an n-channel MOSFET in the cutoff, triode, and saturation regions by:

$$I_{DS} = 0 \qquad (V_{GS} \le -V_{TH})$$

$$I_{DS} = \frac{KP}{2}(W/L_{eff})V_{DS}[2(V_{GS}-V_{TH})-V_{DS}](1+LAMBDA\cdot V_{DS}) \qquad (0 \le V_{DS} \le V_{GS}-V_{TH})$$

$$I_{DS} = \frac{KP}{2}(W/L_{eff})(V_{GS}-V_{TH})^2(1+LAMBDA\cdot V_{DS}) \qquad (0 \le V_{GS}-V_{TH} \le V_{DS})$$

$$(4.93)$$

where the threshold voltage is given by

$$V_{TH} = V_{TO} + GAMMA\left(\sqrt{2\cdot PHI - V_{BS}} - \sqrt{2\cdot PHI}\right). \qquad (4.94)$$

The SPICE parameter names correspond closely with the variables used in the analytical model developed in this chapter. In most cases, the SPICE parameter can be deduced from the symbol. However, the SPICE symbols for the channel length unfortunately conflict with the definitions in Fig. 4.18 and Eq. (4.61). The channel length in Eq. (4.93) is called L_{eff} in SPICE and is given by

$$L_{eff} = L - 2\cdot LD, \qquad (4.95)$$

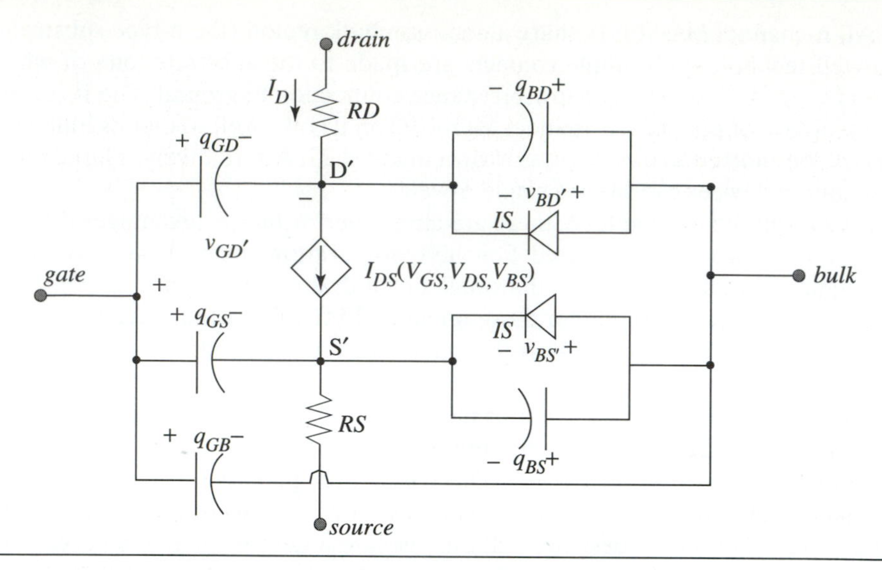

➤ **Figure 4.28** SPICE large-signal model for an n-channel MOSFET, which includes nonlinear charge-storage between the gate and the other terminals and also in the drain and source depletion regions.

where L in SPICE is the length of the polysilicon gate and LD is the gate overlap of the source and drain.

Table 4.1 summarizes the DC SPICE parameter names and their equivalent analytical symbols and units, for the n-channel MOSFET. The elements in the large-signal MOSFET model are shown in Fig. 4.28.

4.6.2 Specifying MOSFET Geometry in SPICE

An n-channel MOSFET is defined by the following statement in SPICE:

M*name D G S B MODname* **L= _ W=_ AD= _ AS=_ PD=_ PS=_ NRD=_ NRS=_**

where *name* identifies the MOSFET with up to seven characters in SPICE2; *D, G, S,* and *B* are the drain, gate, source, and bulk node numbers, *MODname* is the model for the transistor, and the eight geometric parameters are defined in Fig. 4.29. The MOSFET in Fig 4.29 has different diffusion widths for the source and the drain, which though unusual can be modeled in SPICE. The geometry parameters are:

L = Polysilicon gate length,
W = Polysilicon gate width,
AD = Drain area,
AS = Source area,
PD = Perimeter of drain diffusion (not including the edge under the gate),
PS = Perimeter of source diffusion (not including the edge under the gate),
NRD = Equivalent number of squares in the drain diffusion, and
NRS = Equivalent number of squares in the source diffusion.

Table 4.1 n-Channel DC MOSFET SPICE parameters

Parameter Name (SPICE/this text)	SPICE Symbol Eqs. (4.93), (4.94)	Analytical Symbol Eqs. (4.59), (4.60)	Units
Channel length	L_{eff}	L	m
Polysilicon gate length	L	L_{gate}	m
Lateral diffusion/ gate-source overlap	LD	L_D	m
Transconductance parameter	KP	$\mu_n C_{ox}$	A/V^2
Threshold voltage / zero-bias threshold	VTO	V_{TOn}	V
Channel-length modulation parameter	$LAMBDA$	λ_n	V^{-1}
Bulk threshold/ backgate effect parameter	$GAMMA$	γ_n	V$^{1/2}$
Surface potential/ depletion drop in inversion	PHI	$-\phi_p$	V

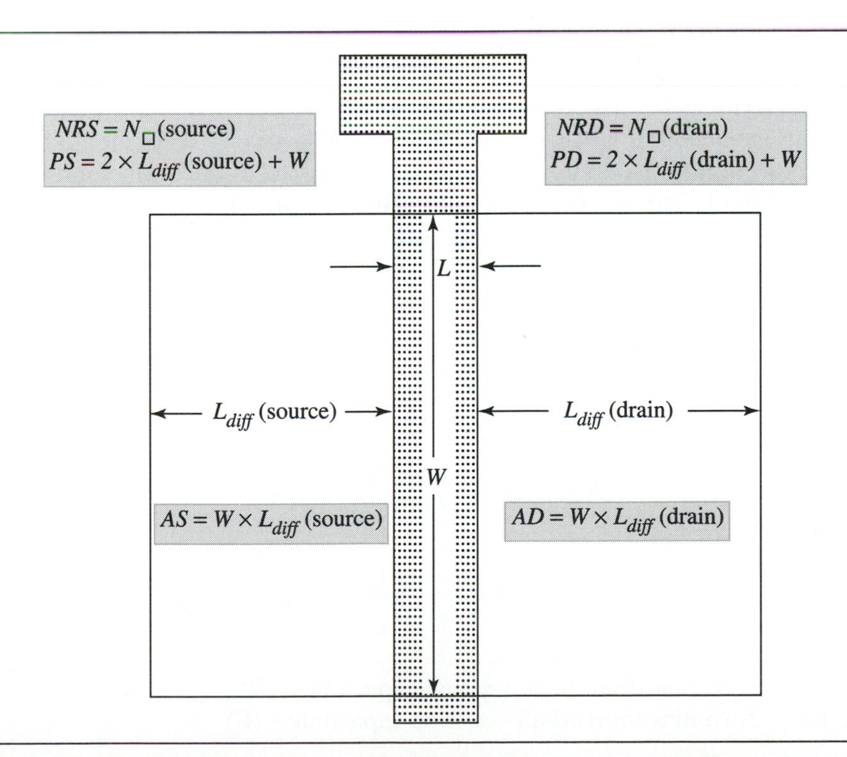

Figure 4.29 SPICE geometry parameters defined on the MOSFET layout.

Two-dimensional modeling (or curve-fitting to measurements) is needed to find the equivalent number of squares for the source and drain diffusions, which often have multiple contacts in order to minimize *NRD* and *NRS* (see the layout in Fig. 4.1, for example).

4.6.3 Level 1 Model Parameters

The MOSFET model statement in SPICE is

.MODEL *MODname* **NMOS/PMOS VTO=_ KP=_ GAMMA=_ PHI=_ LAMBDA=_ RD=_ RS=_ RSH=_ CBD=_ CBS=_CJ=_ MJ=_ CJSW=_ MJSW=_ PB=_ IS= _ CGDO=_ CGSO=_ CGBO=_ TOX=_ LD=_**

The MOSFET model must be specified as n-channel (**NMOS**) or p-channel (**PMOS**). The model parameters that have not been already defined in the DC drain current model Eqs. (4.93) and (4.94) are defined in this section.

The gate oxide thickness t_{ox} = TOX. When this parameter is specified, SPICE models the gate-source and gate-drain charge storage in cutoff, triode, and saturation. A cruder approximation for these capacitances is used if TOX is not specified. The capacitances *CGDO, CGSO*, and *CGBO* describe the overlap capacitances of the gate with the drain, source, and bulk. The units for all three parameters are F/m, with *CGDO* and *CGSO* scaling with the channel width *W*, while *CGBO* scales with the channel length *L*.

The resistance of the source and drain diffusions and their contacts can be described by directly specifying them or by giving their sheet resistance:

RD = Drain resistance (Ω)

RS = Source resistance (Ω)

RSH = Sheet resistance of the drain and source diffusions (*NRD* and *NRS* must be specified on the MOSFET statement) (Ω/□)

SPICE includes a model for the pn diode at both the drain and the source. *IS* is the saturation current of the diode. As shown in the large-signal circuit model in Fig. 4.28, the current into the drain terminal is the difference of the drain-source current I_{DS} and the diode current from the substrate. In normal operation, the diode is reverse biased and $I_D = I_{DS}$.

The final set of parameters concern the modeling of charge storage in the drain-bulk and source-bulk depletion regions. Again, SPICE allows alternative ways to specify the depletion capacitances. By specifying the parameters *CBD, CBS*, *PB*, and *MJ,* SPICE computes the voltage dependences of the drain-bulk and source-bulk capacitances:

$$C_{BD}(V_{BD}) = \frac{CBD}{(1 - V_{BD}/PB)^{MJ}} \tag{4.96}$$

$$C_{BS}(V_{BS}) = \frac{CBS}{(1 - V_{BS}/PB)^{MJ}} \tag{4.97}$$

CBD = Zero-bias drain-bulk junction capacitance (F)

CBS = Zero-bias source-bulk junction capacitance (F)

PB = Built-in potential for the bulk junction (V)
MJ = Bulk junction grading coefficient (dimensionless)

A more accurate simulation of these important capacitances incorporates both the planar junction capacitance and the perimeter, or sidewall, capacitance. If the parameters *CJ*, *CJSW*, *MJ*, *MJSW*, and *PB* are included, SPICE computes the drain-bulk and source-bulk capacitances as follows:

$$C_{BD}(V_{BD}) = \frac{CJ \cdot AD}{(1 - V_{BD}/PB)^{MJ}} + \frac{CJSW \cdot PD}{(1 - V_{BD}/PB)^{MJSW}} \tag{4.98}$$

$$C_{BS}(V_{BS}) = \frac{CJ \cdot AS}{(1 - V_{BS}/PB)^{MJ}} + \frac{CJSW \cdot PS}{(1 - V_{BS}/PB)^{MJSW}} \tag{4.99}$$

CJ = Zero-bias planar bulk junction capacitance (F/m^2)
$CJSW$ = Zero-bias sidewall bulk junction capacitance (F/m)
$MJSW$ = Sidewall junction grading coefficient (dimensionless)

The SPICE small-signal n-channel MOSFET model is identical to Fig. 4.24, except for the addition of the drain and source resistances *RS* and *RD*.

4.6.4 Example Level 1 SPICE Models for an n-well CMOS Technology

It is convenient to define "standard" SPICE models for the n-channel and p-channel MOSFETs in an n-well CMOS technology. These models can then be used throughout the text for simulating both digital and analog circuits. With the exception of the additional capacitance due to the sidewalls of the source and drain diffusions, the device parameters are similar to those derived for the n-channel MOSFET in Ex. 4.5.

For these models, we assume that the lateral diffusion $LD = L_D = 0$ µm, to avoid confusion in specifying the channel length. The Level 1 models are not so accurate that the omission of L_D will make a significant difference, in any case. The length of the source and drain diffusions is $L_{diff} = 6$ µm. Therefore, the drain and source areas and perimeters are given by

$$AD = AS = L_{diff} \cdot W = (6 \cdot W) \times 10^{-12}\ \text{m}^2 \text{ and} \tag{4.100}$$

$$PD = PS = 2L_{diff} + W = (12 + W) \times 10^{-6}\ \text{m}, \tag{4.101}$$

where W is in µm.

The gate-bulk overlap capacitance will be neglected by letting $CGBO = 0$ F/m (the default value in SPICE). Finally, the source and drain resistances are small and will be omitted by letting $RD = RS = 0\ \Omega$.

Table 4.2 describes the MOSFET parameters for the example n-well process.

Table 4.2 n-well CMOS Level 1 SPICE Models.

Level 1 SPICE Parameter	n-channel MOSFET	p-channel MOSFET	Units
Gate oxide thickness *TOX*	150	150	Å
Transconductance parameter *KP*	50×10^{-6}	25×10^{-6}	A/V^2
Threshold voltage *VTO*	1.0	−1.0	V
Channel-length modulation parameter *LAMBDA*	$0.1/L$ L in μm	$0.1/L$ L in μm	V^{-1}
Bulk threshold parameter *GAMMA*	0.6	0.6	$V^{1/2}$
Surface potential *PHI*	0.8	0.8	V
Gate-drain overlap capacitance *CGDO*	5×10^{-10}	5×10^{-10}	F/m
Gate-source overlap capacitance *CGSO*	5×10^{-10}	5×10^{-10}	F/m
Zero-bias planar bulk depletion capacitance *CJ*	10^{-4}	3×10^{-4}	F/m^2
Zero-bias sidewall bulk depletion capacitance *CJSW*	5×10^{-10}	3.5×10^{-10}	F/m
Bulk junction potential *PB*	0.95	0.95	V
Planar bulk junction grading coefficient *MJ*	0.5	0.5	[ ]
Sidewall bulk junction grading coefficient *MJSW*	0.33	0.33	[ ]

For n-channel and p-channel MOSFETs with $L = 3$ μm, the parameters in Table 4.2 lead to the following model statements:

```
.MODEL MODN NMOS LEVEL = 1 VTO = 1 KP = 50U LAMBDA = .033 GAMMA = .6
+ PHI = 0.8 TOX = 1.5E–10 CGDO = 5E–10 CGSO = 5E–10 CJ = 1E–4 CJSW = 5E–10
+ MJ = 0.5 PB = 0.95
.MODEL MODP PMOS LEVEL = 1 VTO = –1 KP = 25U LAMBDA = .033 GAMMA = .6
+ PHI = 0.8 TOX = 1.5E–10 CGDO = 5E–10 CGSO = 5E–10 CJ = 3E–4 CJSW = 3.5E–10
+ MJ = 0.5 PB = 0.95
```

An n-channel MOSFET M_1 with $W = 150$ μm and $L = 3$ μm has source and drain areas and perimeters that are found from Eqs. (4.100) and (4.101)

$AD = AS = 9 \times 10^{-10}\ \text{m}^2$ and $PD = PS = 1.62 \times 10^{-4}\ \text{m}$

Transistor M_1 is specified in SPICE by

```
M1 4 3 2 1 MODN W = 150U L = 3U AD = 9E–10 AS = 9E–10 PD = 1.62E–4 PD = 1.62E–4
```

where the drain is node 4, the gate is node 3, the source is node 2, and the bulk is node 1.

➤ SUMMARY

This chapter has covered the device physics and device modeling of the MOS field-effect transistor, which is by far the most important microelectronic device. It is essential that you grasped the basic device operation and the large- and small-signal models, since the MOSFET will be used in digital and analog circuits throughout this text. Also, a good understanding of the circuit models for the MOSFET will be helpful in learning how to model the other important multi-terminal device—the bipolar junction transistor.

The most important topics in Chapter 4 are:

- The piecewise modeling of the n-channel and p-channel MOSFETs in the three regions of operation: cutoff, triode, and saturation.
- The relationship between drain current, channel charge, and drift velocity.
- The key assumption behind the gradual channel approximation.
- The channel electron charge variation from source to drain in the triode region, neglecting the body effect.
- The concept of saturation.
- The backgate effect on the threshold voltage of the MOSFET.
- MOSFET capacitances in the three regions of operation.
- Small-signal models for the n-channel and the p-channel MOSFETs.
- The Level 1 SPICE model for the MOSFET.

➤ FURTHER READING

MOSFET Device Structures

1. W. Maly, *Atlas of VLSI Technologies*, Addison-Wesley, 1986. Has layouts for simple MOS circuits in several technologies.

MOSFET Device Physics

1. R. F. Pierret, *Field Effect Devices*, Modular Series on Solid State Devices, Vol. IV, R. F. Pierret and G. W. Neudeck, eds., Addison-Wesley, 1983, Chapter 5 has a good discussion of the MOSFET at slightly more advanced level.
2. R. S. Muller and T. I. Kamins, *Device Electronics for Integrated Circuits*, 2nd ed., Wiley, 1986. Chapter 9 has a good description of basic MOSFET operation; derivations use energy bands rather than potential.

MOSFET SPICE Models

1. A. Vladimirescu, *The SPICE Book*, Wiley, 1994. Chapter 3 has a complete description of the Level 1 model; selected examples of the more advanced MOSFET models are detailed in an appendix.

➤ PROBLEMS

EXERCISES

Unless stated otherwise, use the following parameters in the exercises and problems.

n-channel MOSFET	p-channel MOSFET
$\mu_n C_{ox} = 50\ \mu\text{A/V}^2$	$\mu_p C_{ox} = 25\ \mu\text{A/V}^2$
$V_{TOn} = 1.0$ V	$V_{TOp} = -1.0$ V
$\gamma_n = 0.6\ \text{V}^{1/2}$	$\gamma_p = 0.6\ \text{V}^{1/2}$
$\lambda_n = (0.1/L)\ \text{V}^{-1}$, L in μm	$\lambda_p = (0.1/L)\ \text{V}^{-1}$, L in μm
$\phi_p = -0.42$ V	$\phi_n = 0.42$ V

E4.1 For the n-channel MOSFET with terminal voltages in Fig. E4.1, determine (a) the operating region (cutoff, triode, or saturation) and (b) the drain current I_D.

E4.2 Repeat E4.1 for the terminal voltages in Fig. E4.2.

E4.3 Repeat E4.1 for the terminal voltages in Fig. E4.3.

E4.4 For the n-channel MOSFET with terminal voltages in Fig. E4.4, determine (a) the operating region (cutoff, triode, or saturation) and (b) the drain current I_D. Note that the bulk-source voltage V_{BS} is *not* zero.

E4.5 Repeat E4.4 for the terminal voltages in Fig. E4.5.

E4.6 Repeat E4.4 for the terminal voltages in Fig. E4.6.

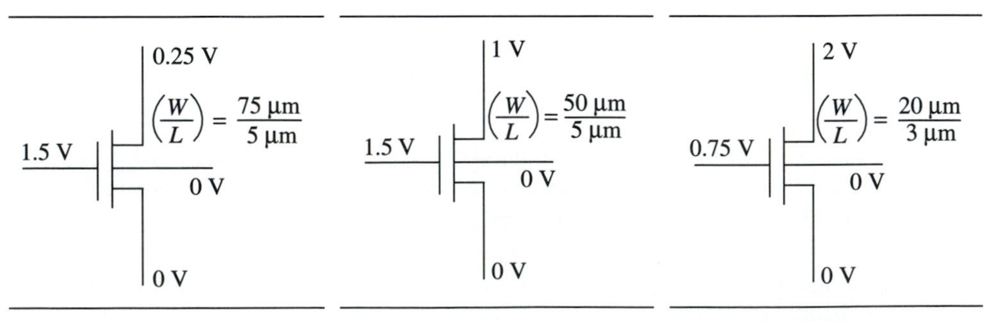

➤ **Figure E4.1** ➤ **Figure E4.2** ➤ **Figure E4.3**

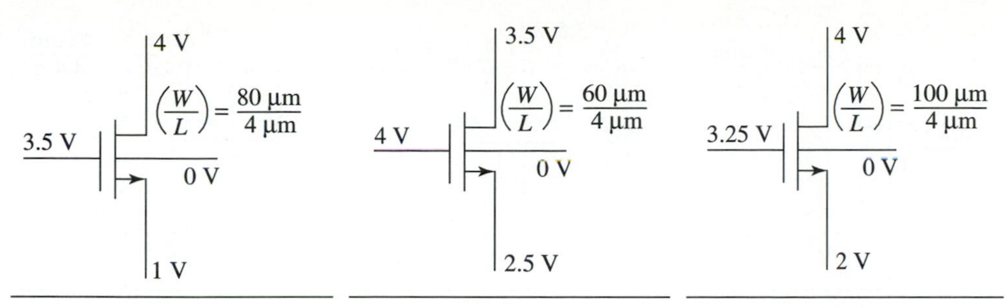

➤ **Figure E4.4** ➤ **Figure E4.5** ➤ **Figure E4.6**

E4.7 For the n-channel MOSFET with terminal voltages in Fig. E4.8, (a) label the source and the drain terminals, (b) determine in which region the transistor is operating (cutoff, triode, or saturation), and (c) find the drain current I_D.

E4.8 Repeat E4.7 for the terminal voltages in Fig. E4.8.

E4.9 For the p-channel MOSFET with terminal voltages in Fig. E4.9, determine (a) the operating region (cutoff, triode, or saturation) and (b) the drain current $-I_D$.

E4.10 Repeat E4.9 for the terminal voltages in Fig. E4.10.

E4.11 Repeat E4.9 for the terminal voltages in Fig. E4.11.

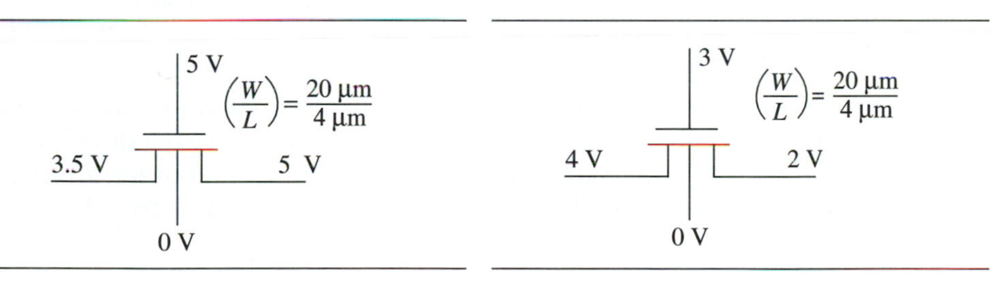

➤ **Figure E4.7** ➤ **Figure E4.8**

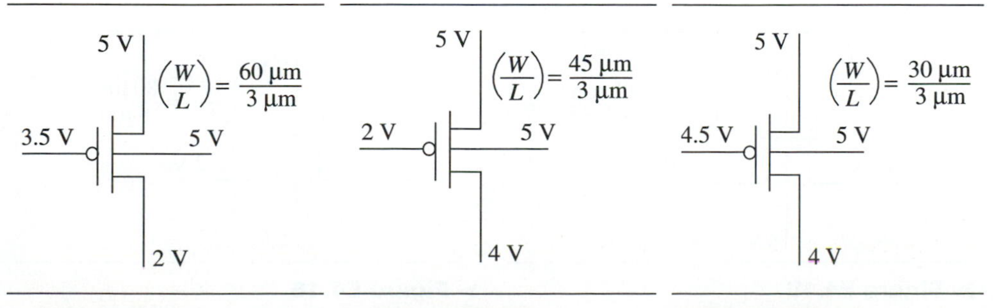

➤ **Figure E4.9** ➤ **Figure E4.10** ➤ **Figure E4.11**

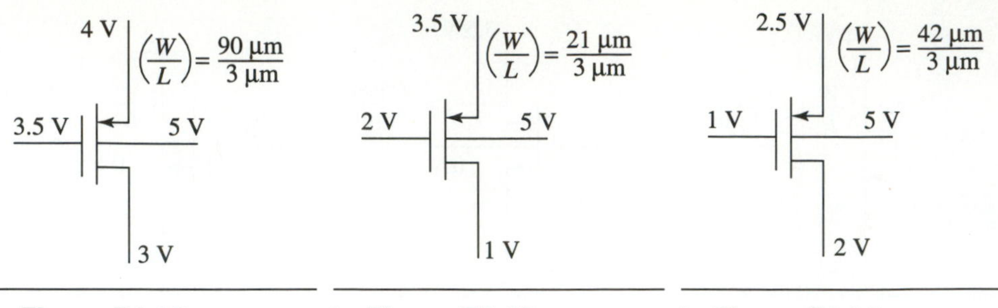

➤ **Figure E4.12** ➤ **Figure E4.13** ➤ **Figure E4.14**

E4.12 For the p-channel MOSFET with terminal voltages in Fig. E4.12, determine (a) the operating region (cutoff, triode, or saturation) and (b) the drain current $-I_D$. Note that the source-bulk voltage V_{SB} is *not* zero.

E4.13 Repeat E4.12 for the terminal voltages in Fig. E4.13.

E4.14 Repeat E4.12 for the terminal voltages in Fig. E4.14.

E4.15 For the p-channel MOSFET with terminal voltages in Fig. E4.15, (a) label the source and the drain terminals, (b) determine in which region the transistor is operating (cutoff, triode, or saturation), and (c) find the drain current I_D.

E4.16 Repeat E4.15 for the terminal voltages in Fig. E4.16.

E4.17 Consider an n-channel MOSFET with $(W/L) = 25\ \mu m/2.5\ \mu m$.

(a) Plot the drain current I_D versus V_{GS} for $V_{DS} = 250$ mV and $V_{BS} = 0$. Note that the threshold voltage can be found from this plot of the MOSFET in the triode region.

(b) Plot the square root of drain current $(I_D)^{1/2}$ versus V_{GS} for $V_{DS} = 5$ V and $V_{BS} = 0$. The threshold voltage can also be extracted from this plot of the MOSFET in the saturation region. Due to second-order effects, the measurement will differ somewhat from the threshold found in the triode region in part (a).

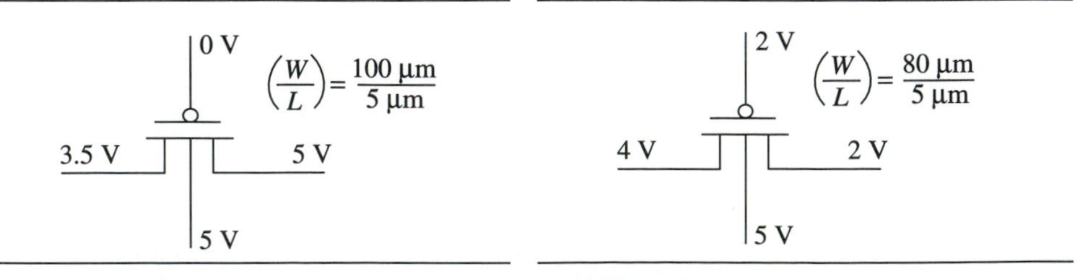

➤ **Figure E4.15** ➤ **Figure E4.16**

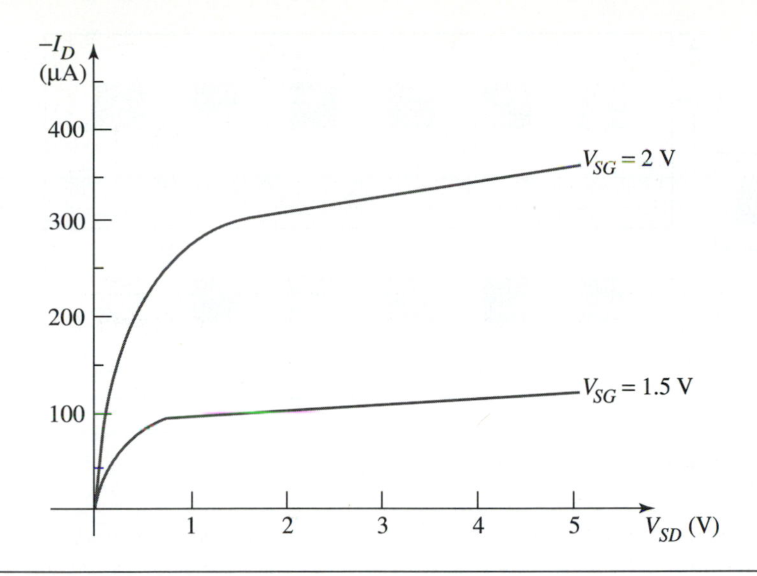

➤ **Figure E4.18**

E4.18 Figure E4.18 is a plot of two drain characteristics of a p-channel MOSFET, for $V_{BS} = 0$.

(a) From these measurements, what is the threshold voltage, V_{Tp}?

(b) Estimate the channel-length modulation parameter, λ_p.

E4.19 The n-channel MOSFET shown in layout in Fig. E4.19 is biased at the operating point: $V_{GS} = 3.5$ V, $V_{DS} = 5$ V, $V_{BS} = -1$ V. For this problem, include $L_D = 0.1$ μm in finding the channel length L from the layout.

(a) Find the small-signal parameters g_m, g_{mb}, and r_o at this operating point.

(b) Find the capacitances C_{gs}, C_{gd}, C_{db}, and C_{sb}. In calculating the overlap capacitances, only consider the under-diffusion of the drain and source

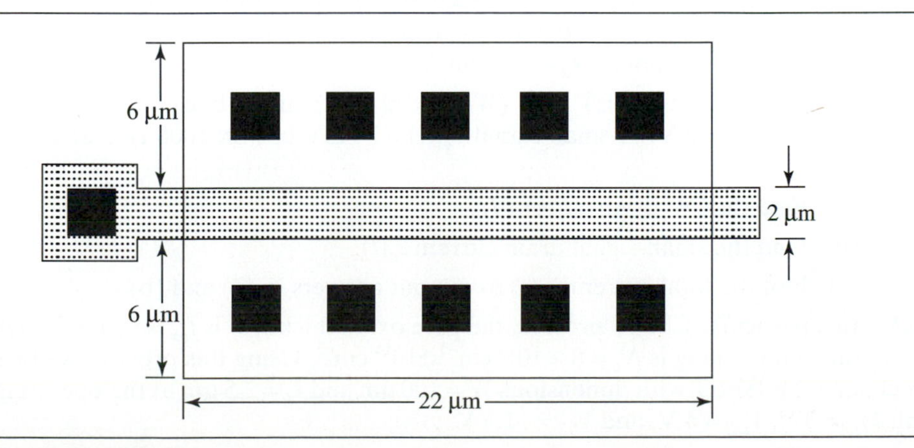

➤ **Figure E4.19**

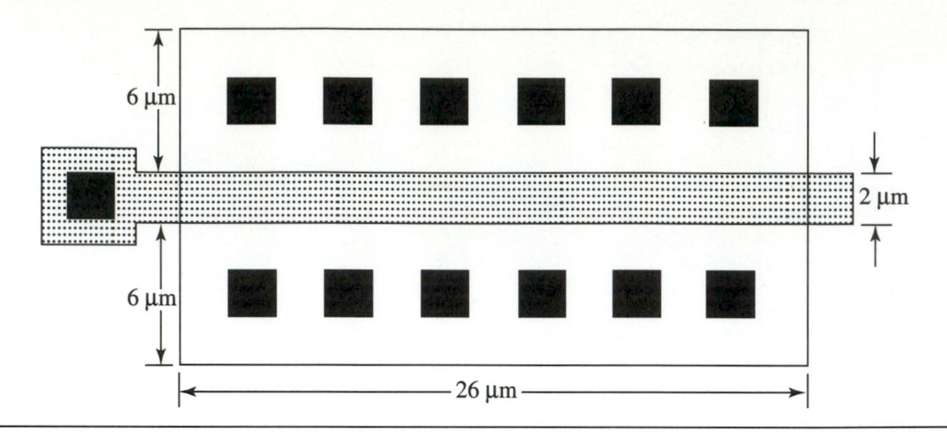

➤ **Figure E4.20**

diffusions by L_D. Also, you can neglect the sidewall capacitance and use a substrate doping $N_a = 10^{17}$ cm^{-3}.

E4.20 The p-channel MOSFET shown in layout in Fig. E4.20 is biased at the operating point: $V_{SG} = 1.5$ V, $V_{SD} = 4$ V, $V_{SB} = -2$ V. For this problem, include $L_D = 0.1$ μm in finding the channel length L from the layout.

(a) Find the small-signal parameters g_m, g_{mb}, and r_o at this operating point.

(b) Find the capacitances C_{gs}, C_{gd}, C_{db}, and C_{sb}. In calculating the overlap capacitances, only consider the under-diffusion of the drain and source diffusions by L_D. Also, you can neglect the sidewall capacitance and use a substrate doping $N_a = 10^{17}$ cm^{-3}.

E4.21 A p-channel MOSFET has $W/L = 500$ μm/2.5 μm. The source and drain diffusions have a length $L_{diff} = 6$ μm. The operating point is: $V_{SG} = 2.5$ V, $V_{SD} = 4$V, $V_{SB} = 0$ V.

(a) Find the small-signal parameters g_m, g_{mb}, and r_o at this operating point.

(b) Find the capacitances C_{gs}, C_{gd}, C_{db}, and C_{sb}. In calculating the overlap capacitances, only consider the under-diffusion of the drain and source diffusions by L_D. Also, you can neglect the sidewall capacitance and use a substrate doping $N_d = 10^{17}$ cm^{-3}.

E4.22 An n-channel MOSFET with $(W/L) = 20$ μm/3 μm is biased at $V_{GS} = 2$ V, $V_{DS} = 3.5$ V, and $V_{BS} = 0$ V. A small signal $v_{gs}(t) = 20$ mV $\cos(2\pi\ 1000\ t)$ is added in series with V_{GS}.

(a) Find the DC drain current I_D.

(b) Find the small-signal drain current $i_d(t)$.

(c) Plot the total current $i_D(t)$ from your answers in (a) and (b).

E4.23 In a particular CMOS process, the gate oxide thickness is $t_{ox} = 150$ Å ±20 Å and the substrate doping is $N_a = 8 \times 10^{16}$ cm^{-3} ±10^{16} cm^{-3}. Using this process, we bias an n-channel MOSFET with dimensions $W = 100$ μm and $L = 2.5$ μm at the operating point: $V_{GS} = 3$ V, $V_{DS} = 4$ V, and $V_{BS} = -1.5$ V.

(a) What is the average threshold voltage and its uncertainty?

(b) What is the DC drain current I_D and its uncertainty at this operating point?

(c) What is the transconductance and its uncertainty at this operating point?

(d) What is the backgate transconductance and its uncertainty at this operating point?

PROBLEMS

P4.1 Figure P4.1 plots the drain current I_D of an n-channel MOSFFET as a function of the gate-source voltage V_{GS}, for a drain-source voltage $V_{DS} = 10$ mV and at three values of backgate bias V_{BS}. From the layout, you also know that $W/L = 10$.

(a) Extract the following device parameters from these measurements: the substrate doping concentration N_a, the oxide thickness t_{ox}, and the channel mobility μ_n.

(b) Estimate the uncertainty in your parameters, given that the current measurements have an uncertainty of ± 1 μA.

P4.2 In a particular CMOS technology, the substrate doping is $N_a = 4 \times 10^{17}$ cm^{-3}, the gate oxide is $t_{ox} = 150$ Å, and n$^+$ polysilicon is used for the gate.

(a) Calculate the surface potential ϕ_{so} in thermal equilibrium.

(b) Calculate the threshold voltage V_{TOn}.

(c) If we require $V_{TOn} = 0.7$ V, what oxide thickness is required?

(d) If the oxide thickness couldn't be changed, what must the substrate doping concentration become in order to achieve $V_{TOn} = 0.7$ V.

P4.3 In a particular CMOS technology, the n-well doping is $N_d = 2.5 \times 10^{17}$ cm^{-3}, the gate oxide is $t_{ox} = 200$ Å, and n$^+$ polysilicon is used for the gate.

(a) Calculate the surface potential ϕ_{so} in thermal equilibrium.

(b) Calculate the threshold voltage V_{TOp}.

(c) If we require $V_{TOp} = -1.5$ V, what oxide thickness is required?

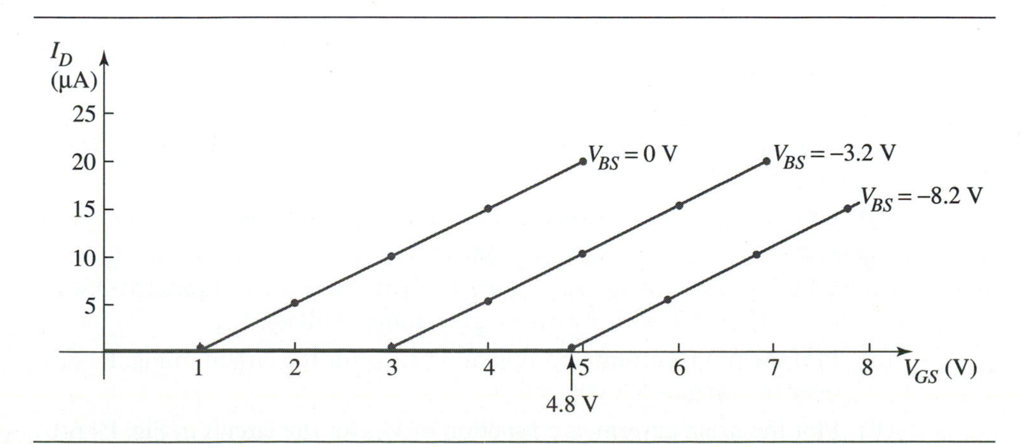

➤ **Figure P4.1**

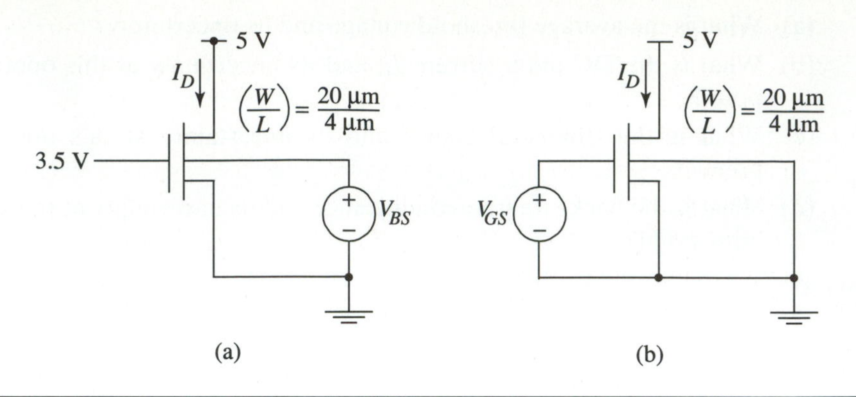

➤ Figure P4.6

(d) If the oxide thickness couldn't be changed, what must the substrate doping concentration become in order to achieve $V_{TOp} = -1.5$ V.

P4.4 In a particular CMOS technology, the n-well doping is $N_d = 2.5 \times 10^{17}$ cm^{-3}, the gate oxide is $t_{ox} = 200$ Å, and p$^+$ polysilicon is used for the gate. The thermal-equilibrium potential of p$^+$ polysilicon is $\phi_{p+} = -550$ mV.

(a) Calculate the surface potential ϕ_{so} in thermal equilibrium.

(b) Calculate the threshold voltage V_{TOp}.

(c) If we require $V_{TOp} = -0.7$ V, what oxide thickness is required?

(d) If the oxide thickness couldn't be changed, what must the substrate doping concentration become in order to achieve $V_{TOp} = 0.7$ V.

P4.5 Silicon nitride (Si_3N_4, $\varepsilon_n = 6.64 \times 10^{13}$ F/cm) is a potential replacement for silicon dioxide as a gate insulator material. Assume the same operating point and transistor dimensions as used in Ex. 4.5.

(a) What thickness of Si_3N_4 is needed for $V_{TOn} = 0.6$ V, if the substrate doping is $N_a = 10^{17}$ cm^{-3}?

(b) Find the backgate effect parameter, γ_n for the silicon nitride gate insulator thickness from (a).

(c) Find the transconductance, backgate transconductance, and output resistance at the operating point and compare with the values for an SiO_2 from Ex. 4.5.

(d) Find the MOSFET capacitances and compare to those for the SiO_2 gate.

P4.6 Figure P4.6(a) shows an n-channel MOSFET with $V_{GS} = 3.5$ V and $V_{DS} = 5$ V with a variable bulk-source voltage V_{BS}. Figure P4.6(b) shows an n-channel MOSFET with $V_{BS} = 0$ V and $V_{DS} = 5$ V with a variable gate-source voltage V_{GS}.

(a) Plot the drain current as a function of V_{BS} for the circuit in Fig. P4.6(a), over the range $-5\text{ V} < V_{BS} < 0$.

(b) Plot the drain current as a function of V_{GS} for the circuit in Fig. P4.6(b), over the range $0\text{ V} < V_{GS} < 5$ V.

P4.7 This problem uses the results of P4.6 to extract small-signal parameters.

(a) From the $I_D(V_{BS})$ plot in P4.6(a), find graphically the backgate transconductance g_{mb} at the operating point $V_{GS} = 3.5$ V, $V_{DS} = 5$ V, $V_{BS} = -2$ V. Compare with the value you calculate from the analytical expression.

(b) From the plot of $I_D(V_{GS})$ in P4.6(b), find graphically the transconductance g_m at the operating point $V_{GS} = 3.5$ V, $V_{DS} = 5$ V, $V_{BS} = 0$ V. Again, compare your result with the analytical expression.

P4.8 A large-value, voltage-controlled resistor is useful in some analog ICs. Using the standard parameters, design a p-channel MOSFET with a resistance of $R = 1$ MΩ for $V_{GS} = 1.5$ V.

(a) What are the required channel width and length?

(b) Find the capacitance between the channel and the gate and between the channel and the substrate. Assume that the source and drain diffusion lengths are $L_{diff} = 6$ μm. Note that the device is operating in the triode region for use as a voltage-controlled resistor.

(c) Model the transistor as a low-pass filter and find the break (–3 dB) frequency.

P4.9 For the case of a p-channel MOSFET, apply the gradual channel approximation.

(a) Find the channel voltage $V_C(y)$ between the source and drain, assuming that $V_{SD} \le V_{SG} + V_{Tp}$.

(b) Find the lateral electric field $E_y(y)$ between the source and drain for $V_{SD} \le V_{SG} + V_{Tp}$.

(c) Find the vertical electric field $E_x(y)$ at the silicon surface ($x = 0$) between the source and drain for $V_{SD} \le V_{SG} + V_{Tp}$.

(d) For $V_{SD} = (3/4)(V_{SG} + V_{Tp})$ plot the ratio E_y/E_x between the source and the drain.

P4.10 In the gradual-channel approximation, we found Eq. (4.36) prior to eliminating the variation of the bulk charge.

(a) Starting with Eq. (4.36), separate variables and find a new triode region equation for the drain current as a function of V_{GS} and V_{DS}. You can assume that $V_{BS} = 0$.

(b) What is the saturation drain current, from the triode equation in (a)?

(c) For $W/L = 25$ μm/2.5 μm, plot the drain characteristics for $V_{GS} = 1.5$ V, 2 V, 3 V, and 4 V over the range $0 < V_{DS} < 5$ V using your results from (a) and (b). In saturation, you can use the factor $(1 + \lambda_n V_{DS})$ in modeling channel-length modulation. On the same graph, plot Eqs. (4.59) and compare.

P4.11 For this problem, use the approach used to derive Eq. (4.80) for the channel-charge in saturation.

(a) Find an expression for the channel charge in the triode region, $q_N(v_{GS}, v_{DS})$.

(b) From your result in (a), find expressions for the gate-source capacitance C_{gs} and the gate-drain capacitance C_{gd} in the triode region. Omit the overlap capacitance.

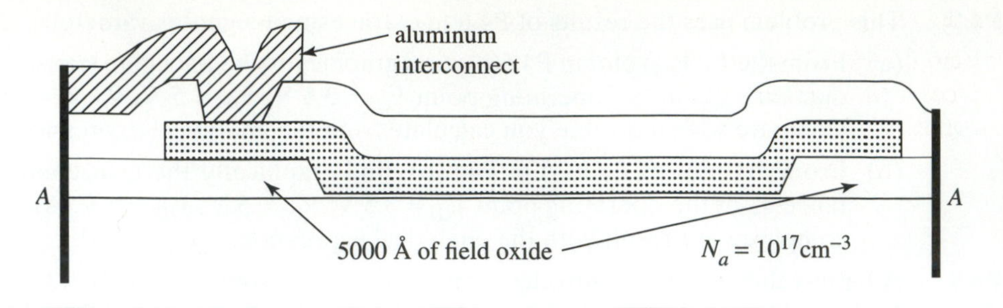

➤ **Figure P4.13**

(c) Plot the gate-source capacitance as a function of V_{DS} over the range 0 to $V_{GS} - V_{Tn}$. Show that the triode result converges with the saturation expression for C_{gs} at $V_{DS} = V_{GS} - V_{Tn}$.

P4.12 This problem concerns itself with the question of how small a "small signal" must be for the linearized model to be adequate. Consider an n-channel MOSFET with width-to-length ratio of 25 μm/5 μm biased at $V_{GS} = 2.5$ V, $V_{DS} = 4$ V, and $V_{BS} = 0$ V.

(a) Find the DC drain current I_D and the transconductance g_m.

(b) Expand the expression for the total drain current in a power series, retaining only the DC, linear, and quadratic terms.

(c) The small-signal model neglects the quadratic term. For a particular application, we can only tolerate a quadratic term that is 5% of the linear term. How large is the "small-signal" V_{gs} that meets this requirement, for the given bias condition?

P4.13 For the layout in Fig. E4.19, the polysilicon gate overlaps thick (field) oxide on both sides of the channel, as shown in cross section *AA* in Fig. P4.13. The field oxide thickness is $t_{fox} = 5000$ Å. The substrate is doped at $N_a = 10^{17}$ cm^{-3} in the field region. The n-channel MOSFET is biased at $V_{GS} = 2.5$ V, $V_{DS} = 4$ V, and $V_{BS} = 0$ V.

(a) What is the threshold voltage of the gate-bulk capacitor?

(b) Find the gate-bulk capacitor C_{gb} at the given operating point. Note that there is an overlap area at both ends of the gate. You will need to apply the MOS capacitor theory from Chapter 3.

(b) Sketch the gate-bulk capacitance as a function of the bulk-source bias V_{BS} over the range $-5 < V_{BS} < 0$.

P4.14 The field oxide thickness in a CMOS process (see Fig. 4.27) must be thick enough so that neither the p substrate nor the n-type well inverts under the extensions of the n$^+$ polysilicon gate. Assume that we have a case of symmetrical doping ($N_a = 10^{17}$ cm^{-3} $= N_d$) in the field regions for the substrate and the well.

(a) What field oxide thickness is needed to have the field threshold voltage equal to $V_{Tnf} = 4$ V for the substrate?

(b) What field oxide thickness is needed to have the field threshold voltage equal to $V_{Tpf} = -4$ V for the n-well?

P4.15 In a CMOS process it is sometimes desirable to match, as much as possible, the circuit models of the n-channel and the p-channel MOSFETs. Assume that both devices have channel lengths of $L_n = L_p = 2$ μm and that the n-channel MOSFET has a channel width $W_n = 20$ μm.

(a) In order to have identical saturation currents for the same gate-source drive ($V_{GS} = V_{SG}$), what is the width of the p-channel transistor?

(b) Given the dimensions from (a), lay out both devices assuming $L_{diff} = 6$ μm.

(c) Derive all the small-signal parameters (including capacitances) for the n-channel and the p-channel transistors.

P4.16 In some cases, it is helpful to have a small-signal model for the MOSFET in the triode region. Using Eq. (4.59) for the triode region.

(a) Find the transconductance g_m as a function of operating point in the triode region.

(b) Find the output resistance r_o as a function of operating point in the triode region.

(c) Plot the output conductance g_o as a function of V_{DS} and show that it approaches its saturation value at $V_{DS} = V_{GS} - V_{Tn}$.

P4.17 Figure P4.17 shows the layout of an n-channel MOSFET, in which the channel width tapers from $W_s = 50$ μm at the source to $W_d = 20$ μm at the drain. The channel length is $L = 10$ μm.

(a) Adapt the gradual channel approximation in Eq. (4.36) to this new geometry and find an expression for the drain current in the triode region.

(b) What is the effective channel width W_{eff} for this MOSFET?

(c) Find the gate-drain and gate-bulk capacitances for this layout and compare to those for a standard MOSFET with $W = W_{eff}$. The overlap capacitance between gate and drain is $C_{gdo} = 0.5$ fF/μm and the operating point is $V_{GS} = 2$ V, $V_{DS} = 4$ V, and $V_{BS} = 0$ V.

P4.18 Figure P4.18 is a square MOSFET layout, which is an approach that is sometimes used to minimize the drain capacitance.

(a) Derive an approximate equation for the saturation drain current for this n-channel MOSFET. What is an estimate of the effective gate width W_{eff}? *Hint*: consider the role of the corners.

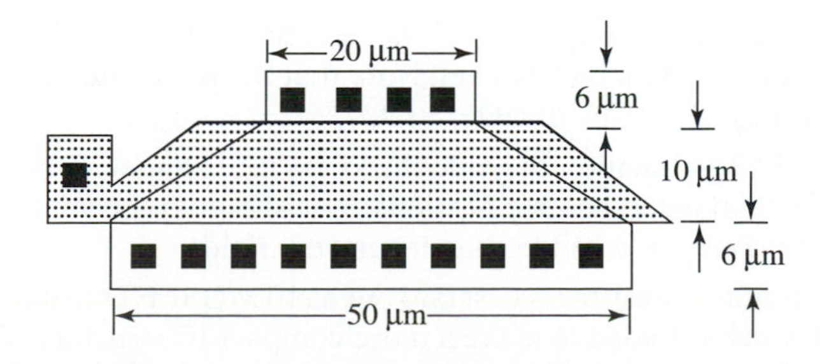

➤ Figure P4.17

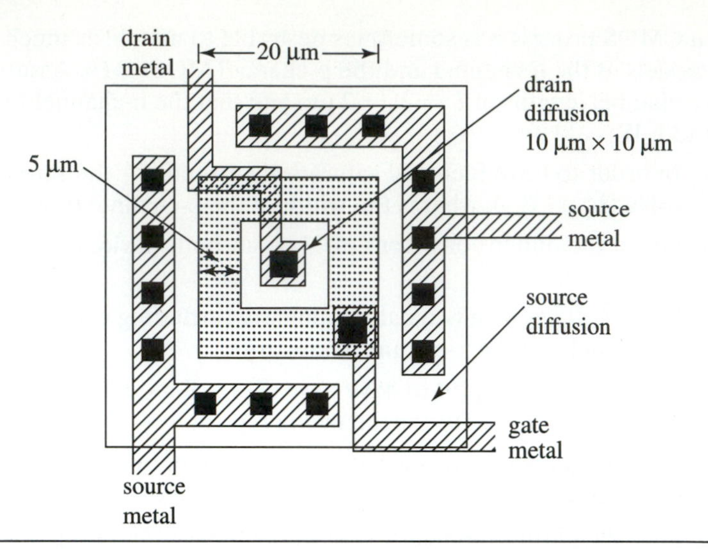

➤ **Figure P4.18**

(b) Find the gate-drain and gate-bulk capacitances for this layout and compare to those for a standard MOSFET with $W = W_{eff}$. The overlap capacitance between gate and drain is $C_{gdo} = 0.5$ fF/μm and the operating point is $V_{GS} = 2$ V, $V_{DS} = 3$ V, and $V_{BS} = 0$ V. The CVD SiO_2 thickness is 5000 Å, which is needed to find C_{gd}.

DESIGN PROBLEMS

D4.1 MOSFETs are useful as switches in power supplies. For this application, the *on resistance* of the transistor is a critical parameter. In this problem, the goal is to design an n-channel MOSFET to satisfy:

$R_{on} = 0.01\ \Omega$ for $V_{GS} = 5$ V

Given that the maximum oxide field is $E_{ox,\max} = 3 \times 10^6$ V/cm and that we limit the average lateral field in the channel to at most (1/200) of the vertical field in the channel. The electron mobility in the channel is a function of the vertical field in the silicon $E_x(x = 0)$:

$$\mu_n = \frac{\mu_{no}}{1 + \Theta E_x(0)} = \frac{215\ \text{cm}^2/\text{Vs}}{1 + (2 \times 10^{-6}\ \text{cm/V})\, E_x(0)}$$

The substrate doping is fixed at $N_a = 10^{16}$ cm^{-3}. The drain current should be linear with the drain-source voltage up to $V_{DS} = 500$ mV, which is why there is a restriction on the lateral electric field in the channel. Note that the peak drain current for linear operation is $V_{DS}/R_{on} = 500$ mV/10 mΩ = 50 A!

(a) Find a combination of oxide thickness t_{ox}, channel length L, and (W/L) ratio that meet the specifications. You should try to minimize L by using the maximum possible channel electric fields.

(b) This huge transistor is best laid out as 10 MOSFETs in parallel, each one of which is folded to make a more compact IC structure. Assuming that the source and drain diffusions have a length $L_{diff} = 6$ μm, what is the area required by each of the 10 MOSFETs? Draw the layout of one of the 10 MOSFETs, using as many 2 μm × 2 μm contacts as you can fit in the source and drain in order to minimize the series resistance. Note that the MOSFET switch may consume most of the chip area in a "smart power" CMOS integrated circuit.

D4.2 In analog design, it is sometimes necessary to have a device with a very large output resistance. Given the standard p-channel MOSFET parameters, we would like to achieve

$$r_o = 650 \text{ k}\Omega \text{ with } I_D = -250 \ \mu\text{A}$$

(a) What are the channel length and channel width to meet these specification?

(b) Draw the layout to scale, using $L_{diff} = 6$ μm and 2 μm × 2 μm contacts.

(c) Find the total capacitance between the drain diffusion and ground, assuming that $V_{SD} = V_{SB} = 2.5$ V. Assume that the gate can be considered at ground. You can assume that the substrate doping is $N_d = 10^{17}$ cm^{-3} and neglect the sidewall capacitance.

(d) What is the product of r_o and the capacitance from part (c)? The time constant that results will determine the break frequency beyond which the output impedance of the MOSFET begins to fall below r_o.

chapter 5

Digital Circuits Using MOS Transistors

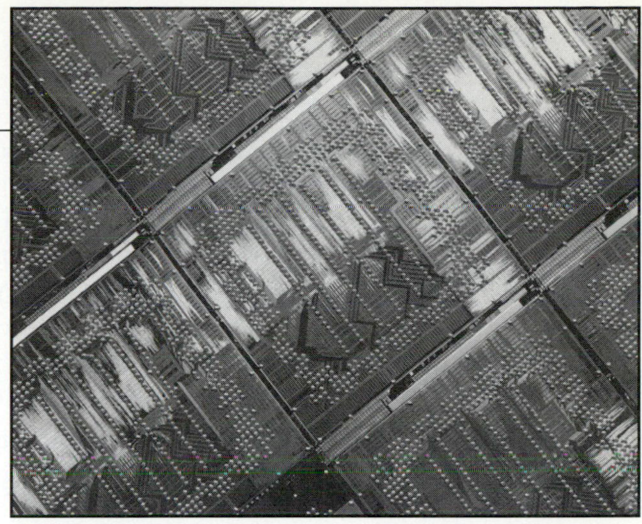

A section of a 200 mm wafer containing Power PC 601 chips. The silver balls are solder bumps used to connect the chip to the package. (Courtesy of International Business Machines Corporation. Unauthorized use not permitted.)

The number of metal-oxide-semiconductor (MOS), transistors used in digital logic and memory circuits has been growing exponentially for the last thirty years. The technology was in its infancy in the late 1960s when semiconductor memory chips with densities on the order of 256 to 1,024 bits were replacing magnetic core memory. By the mid–1970s, digital integrated circuits for electronic watches and calculators became available. The 1980s brought us personal computers and a variety of consumer electronic products. The heart of these products includes microprocessors and memory chips to store information and programs.

In the future, applications will include personal communication systems, high-definition television, and other sophisticated products. Without a doubt, digital integrated circuits based on complementary metal-oxide-semiconductor (CMOS) technology have been the driving force for the information age.

The processing of electronic signals and data in a digital format has the advantage of arbitrary accuracy depending on the complexity of the design. For example, if you want to multiply two numbers with 64-bit accuracy, it is theoretically possible to build a digital processor that can accomplish this goal. The major drawback is the complexity and number of transistors required to perform such an operation.

Because CMOS technology has advanced at an exponential rate over the past thirty years, we currently have integrated circuits with over one million transistors able to perform a wide variety of functions. It is expected that technological advances will continue and push the number of transistors per chip to well over ten million early in the next century.

Chapter Objectives

- Design and analysis of basic logic gates that are the building blocks for sophisticated microprocessors and other digital signal processing systems.
- Basic digital concepts and definitions for the voltage transfer function and transient characteristics of an inverter.
- Analysis of a static CMOS inverter circuit including its voltage transfer function and transient performance.
- Formation of logic gates from the basic static CMOS inverter to perform simple Boolean functions.
- Logic gate design using dynamic logic circuits that save power and area.
- Logic design using area-efficient CMOS transmission gates.

5.1 LOGIC CONCEPTS

Digital electronic systems process information represented by the binary numbers **1** and **0**. **Boolean algebra** is a system of logical operations used to carry out a desired function. Logic gates are used to implement the Boolean algebraic equations. It is assumed that the reader has had some exposure to Boolean algebra and some familiarity with the basic concepts of binary numbers and logic gates. However, a brief review is included on some basic concepts of digital systems and fundamental logic functions.

5.1.1 Inverter

The most basic logic function and the building block for all digital circuits is the inverter. The logical function of the **inverter** is simply to change the state of the input from a **0** to a **1** or a **1** to a **0**. The inverter function is defined in Boolean algebra by

$$Y = \bar{A} \tag{5.1}$$

The logic symbol and the binary truth table of the inverter are shown in Fig. 5.1. In logic diagrams, the circle is often used as a shorthand notation to represent binary inversion.

5.1.2 AND/NAND Gate

An **AND gate** is a logic element that has two or more inputs and a single output. The output of an AND gate is equal to a **1** if all of its inputs are equal to a **1**. If any input to an AND gate is equal to a **0**, the output of the AND gate is a **0**. The logic function for an AND gate is given in

$$Y = ABC \;....\; N \tag{5.2}$$

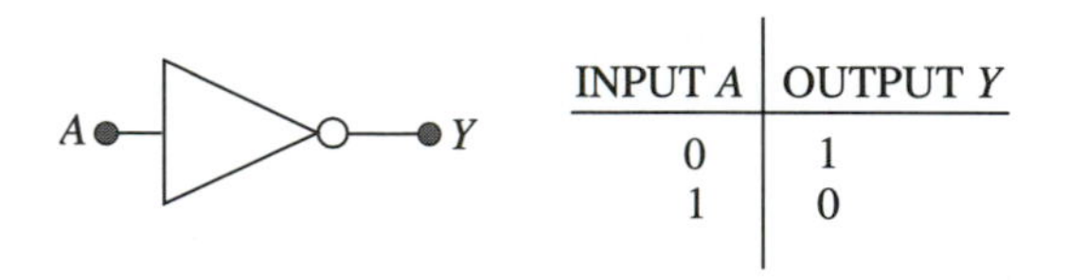

INPUT A	OUTPUT Y
0	1
1	0

Figure 5.1 The inverter logic symbol and truth table.

The **NAND gate** is equivalent to adding an inverter to the output of an AND gate. Its output is the inverse of the AND operation. The logic function for a NAND gate is given in

$$Y = \overline{ABC \ldots N} \tag{5.3}$$

The logic symbols and truth table for AND/NAND gates are given in Fig. 5.2(a). Rather than add an inverter to an AND gate to form a NAND gate, one uses the shorthand notation of the circle at the output of the AND gate to represent the necessary inversion to form a NAND gate.

5.1.3 OR/NOR Gate

An **OR gate** is a logic element that has two or more inputs and a single output. The output of an OR gate is a logic **0** if all of its inputs are logic **0**, otherwise the output is a **1**. Another way to view an OR gate is that its output is a **1** if any of its inputs are a **1**. The **NOR gate** adds an inverter to the output of the OR gate similar to that of the AND/NAND function.The output of the NOR gate is the inverse of an OR gate. The OR and NOR operations are defined for N inputs by the Boolean expressions

$$Y = A + B + C \ldots + N \tag{5.4}$$

$$Y = \overline{A + B + C \ldots + N} \tag{5.5}$$

The logic symbols and truth table for the OR/NOR function are shown in Fig. 5.2(b).

5.1.4 Exclusive OR Gate

Given the ability to have the AND/NAND and OR/NOR functions, as well as inversion, any Boolean algebraic function can be synthesized. To illustrate this fact, we describe an important function in digital systems; namely, the exclusive OR (XOR) gate.

In an **XOR gate**, the output is a logic **1** when the total number of inputs with a logic **1** is odd. If the number of inputs with a logic **1** is even, the output becomes a logic **0**. The logic symbol and truth table for a 2-input XOR gate is shown in Fig. 5.2(c). The XOR function can be synthesized by using the AND/NAND, OR/NOR, and inversion gates described above. The Boolean algebra expressions for the XOR gate are

$$Y = \overline{A}B + A\overline{B} \tag{5.6}$$

$$Y = (A + B)\,\overline{AB} \tag{5.7}$$

These expressions are combinations of the fundamental logic operations previously described. Therefore, in our study of the implementation of logic functions using MOS transistors, we will concentrate only on the fundamental logic operations.

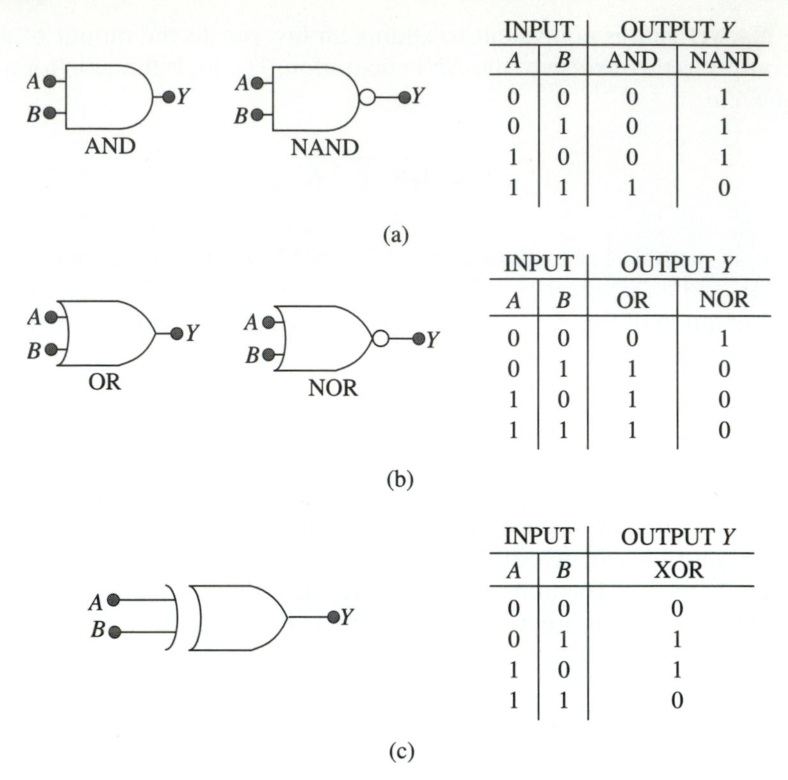

INPUT		OUTPUT Y	
A	B	AND	NAND
0	0	0	1
0	1	0	1
1	0	0	1
1	1	1	0

INPUT		OUTPUT Y	
A	B	OR	NOR
0	0	0	1
0	1	1	0
1	0	1	0
1	1	1	0

INPUT		OUTPUT Y
A	B	XOR
0	0	0
0	1	1
1	0	1
1	1	0

➤ **Figure 5.2** (a) AND/NAND logic symbols and truth table. (b) OR/NOR logic symbols and truth table. (c) XOR logic symbol and truth table.

5.1.5 De Morgan's Laws

Two Boolean algebraic relationships known as **De Morgan's Laws** are often used in digital design. They are

$$\overline{A\,B\,C\,....N} = \overline{A} + \overline{B} + \overline{C} + ... + \overline{N} \tag{5.8}$$

$$\overline{A + B + C + ... + N} = \overline{A}\,\overline{B}\,\overline{C}....\overline{N} \tag{5.9}$$

These laws demonstrate that it is not necessary to use all three fundamental logic operations (OR, AND, and inversion) to perform Boolean logic functions. The OR and inversion function are sufficient because by De Morgan's Law, Eq. (5.9) the AND function can be synthesized from these two. Similarly, the AND and inversion function may be chosen as the fundamental logic operations. The OR function can be synthesized using these gates. Figure 5.3 illustrates De Morgan's Laws and how only two of the three fundamental logic gates are required to synthesize any Boolean algebraic function.

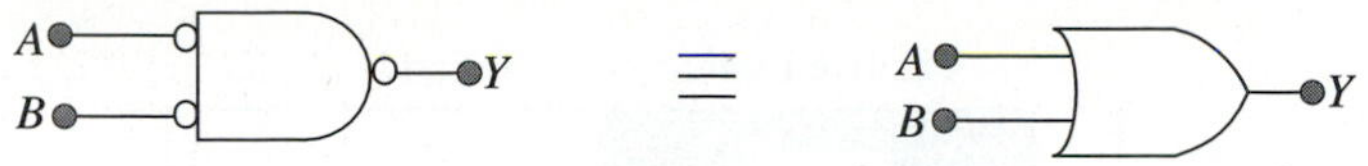

$$Y = \overline{\bar{A}\,\bar{B}} = A + B$$

(a)

$$Y = \overline{\bar{A} + \bar{B}} = A\,B$$

(b)

➤ **Figure 5.3** Illustration of De Morgan's Laws. (a) Inverted inputs to a NAND gate is equivalent to an OR gate. (b) Inverted inputs to a NOR gate is equivalent to an AND gate.

This fact becomes important in the implementation of the logic gates using circuits since either the AND or the OR gate can be easier to implement in a given technology.

5.1.6 Fan-In/Fan Out

Complex digital operations are formed with a variety of gates interconnected to yield the desired logic function. Often, a large number of inputs to a gate are required. In addition, the function may require that a particular gate's output, drive a large number of gates that follow it. The term **fan-in** is defined as the maximum number of logic gates that can be connected at the input of a gate without altering its performance. The term **fan-out** is defined as the maximum number of logic gates that can be connected at the output of a gate without altering its performance.

Limitations on the fan-in and fan-out of a logic gate occur because the gate does not perform to specification: (1) statically, meaning full logic levels are not reached, or (2) dynamically, meaning that the time to reach the full logic levels in the output is longer than the time available between transitions at the input. A typical number for fan-in and fan-out is approximately three. Special circuits that can be designed when large fan-in/fan-out is required will be discussed later.

➤ 5.2 INVERTER CHARACTERISTICS

In digital circuits, two distinct voltage levels can be used to represent the values of the binary numbers **1** and **0**. A voltage has meaning if it falls within one of the two specified ranges as shown in Fig. 5.4.

A digital system uses **positive logic** if a high voltage is defined as a logic **1** and a low voltage is defined as a logic **0**. Similarly, the system uses **negative logic** if the high voltage is defined as a logic **0** and a low voltage is defined as a logic **1**. In this text we will use positive logic and represent a logic **1** with a high voltage. It should be noted that other physical quantities, such as current, flux, charge, etc. could also be used to

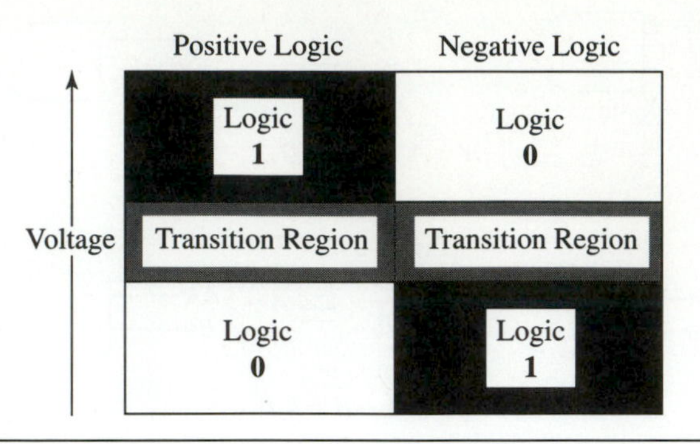

➤ **Figure 5.4** Voltage levels represent logic states. Positive logic represents a **1** with a high voltage and a **0** with a low voltage. Negative logic represents a **0** with a high voltage and a **1** with a low voltage.

represent the logical values. Since modern microelectronic integrated circuits are dominated by the use of MOS transistors, it is most natural to use voltages to represent the logic values.

5.2.1 Inverter Voltage Transfer Characteristic

Since it has been established that we will represent logic **1's** and **0's** as voltages, we will now define the ideal voltage transfer characteristic for the inverter. An input voltage V_{IN} is applied to the input of the inverter in Fig. 5.5(a). The output voltage V_{OUT} is measured at the output of the inverter and a power supply voltage V^+ is used to power the logical element.

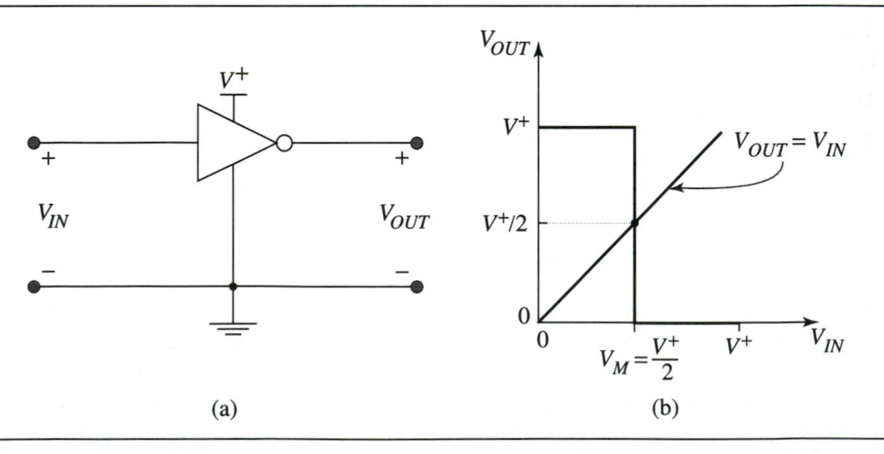

➤ **Figure 5.5** (a) The inverter with an input voltage, V_{IN}, output voltage, V_{OUT}, and power supply voltage V^+. (b) Ideal transfer characteristic for the inverter.

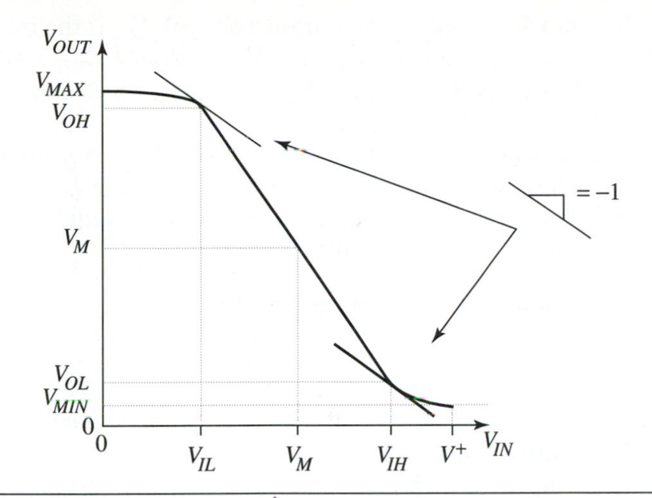

Figure 5.6 Typical inverter transfer characteristic. For $V_{IN} = 0$ V, $V_{OUT} = V_{MAX}$; for $V_{IN} = V_{IL}$, $V_{OUT} = V_{OH}$; for $V_{IN} = V_M$, $V_{OUT} = V_M$; for $V_{IN} = V_{IH}$, $V_{OUT} = V_{OL}$; for $V_{IN} = V^+$, $V_{OUT} = V_{MIN}$.

The ideal voltage transfer characteristic for an inverter is shown in Fig. 5.5(b). Given that the input voltage is small indicating a logic **0**, the output voltage is equal to the power supply voltage. As we increase the input voltage past the point labeled V_M, which in this particular case is equal to $V^+/2$, the inverter switches from a high to a low voltage near 0 V, corresponding to a logical **0**. The **switching point** of the inverter V_M is defined as the voltage at which the input voltage is equal to the output voltage. In the ideal case V_M is equal to one-half of the power supply voltage.

An actual implementation of a digital inverter will only approximate the voltage transfer function given above. A typical inverter transfer characteristic is shown in Fig. 5.6.

We define specific input and output voltages as follows.

V_{IL} **(voltage input low)**—input voltage where the slope of the transfer characteristic is equal to –1

V_{IH} **(voltage input high)**—larger input voltage where the slope is equal to –1

V_{OH} **(voltage output high)**—output voltage given an input voltage of V_{IL}

V_{OL} **(voltage output low)**—output voltage given an input voltage of V_{IH}

V_M **(voltage midpoint)**—input voltage at which the inverter yields an output voltage equal to the input voltage

When the input voltage is equal to 0 V, $V_{OUT} = V_{MAX}$ **(maximum output voltage)**, which in most cases of interest is V^+. For a small input voltage between 0 V and V_{IL}, the output voltage will be between V_{MAX} and V_{OH}. V_{OH} is the minimum output voltage for a valid logic **1**. As we further increase the input voltage, the output voltage rapidly falls through the transition region. When the input voltage is between V_{IL} and V_{IH}, the output voltage is in the transition region where the logic level is undefined. As the input voltage is increased to a value between V_{IH} and V^+, the output voltage is a low value between V_{OL} and $\boldsymbol{V_{MIN}}$ (**minimum output voltage**). V_{OL} is the maximum output voltage for a valid logic **0**. When V_{IN} is equal to V^+, we define $V_{OUT} = V_{MIN}$. In general, V_{MIN} may not be 0 V, however for static CMOS circuits, which you will learn about shortly, $V_{MIN} = 0$ V.

From this transfer characteristic one can see that if $V_{IN} < V_{IL}$ the output voltage is a valid logic **1**. Correspondingly, if $V_{IN} > V_{IH}$ the output voltage is a valid logic **0**. If the input voltage is between V_{IL} and V_{IH} the logic output is undefined. This range of input voltages is referred to as the transition region.

5.2.2 Logic Levels and Noise Margin

As mentioned in the previous section, the inverter characteristic has three distinct regions: the low region where $V_{IN} < V_{IL}$, the high region where $V_{IN} > V_{IH}$, and the transition region where $V_{IL} < V_{IN} < V_{IH}$. In Fig. 5.7. we have one inverter driving a second inverter. The input and output voltage levels corresponding to each of these inverters are indicated. If we look at the region where the first inverter's output is connected to the second inverter's input, we see that $V_{OH1} > V_{IH2}$ and $V_{OL1} < V_{IL2}$.

We can define a **noise margin high**, NM_H which ensures that a logic **1** output from the first inverter is interpreted as a logic **1** input to the second inverter. Similarly, we can define a **noise margin low**, NM_L which ensures that a logic **0** output from the first inverter is interpreted as a logic **0** input to the second inverter. The expressions for both noise margins are given by

noise margins

$$NM_H = V_{OH} - V_{IH} \tag{5.10}$$

$$NM_L = V_{IL} - V_{OL} \tag{5.11}$$

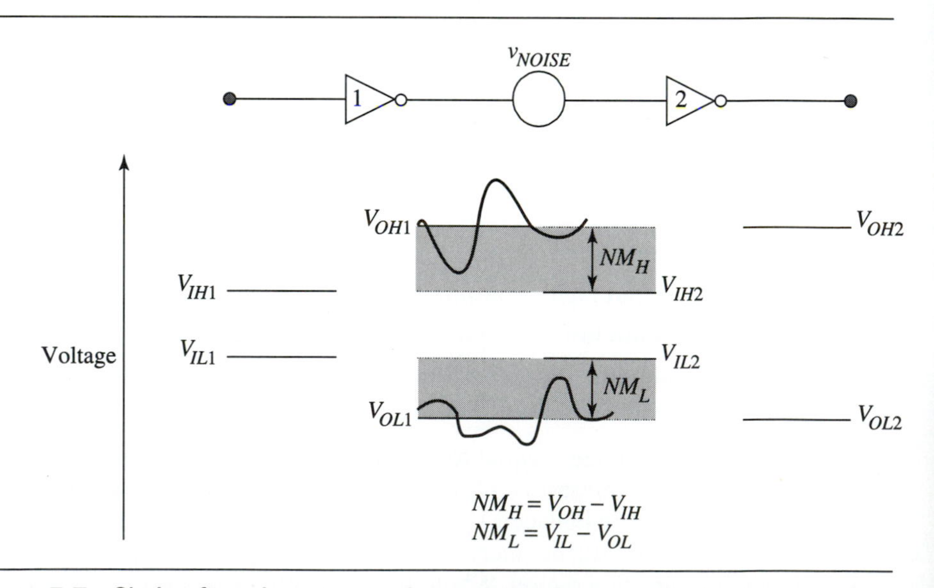

➤ **Figure 5.7** Chain of two inverters to demonstrate the concept of noise margin. A noise source that represents a variety of noise sources such as capacitive coupling is placed between the inverters.

In summary, we have defined several parameters that indicate the voltage transfer characteristics of the inverter. The conventional notation used to represent voltages in digital circuits is upper case variable names and upper case subscripts.

definition of parameters for the voltage transfer characteristic

- V_{OH} - the minimum output voltage from an inverter which indicates a logic **1**
- V_{OL} - the maximum output voltage from an inverter which indicates a logic **0**
- V_{IH} - the minimum input voltage to an inverter to output a logic **0**
- V_{IL} - the maximum input voltage to an inverter to output a logic **1**
- V_M - the voltage at which the input and output voltages are equal

With these values we have defined noise margins to ensure the proper logical output.

5.2.3 Transient Characteristics

In this section we will determine the time required for an output to change state, given that the input changes state. This time is called a **propagation delay**. Figure 5.8 shows a waveform describing a change in input voltage as a function of time. Figure 5.8 also shows how the output voltage of an inverter changes as a function of time. The **rise time** t_R is defined as the time required for the input or output voltage to change from 10% of its high value to 90% of its high value. The **fall time** t_F is defined as the time required for the input or output voltage to change from 90% of its high value to 10% of its high value.

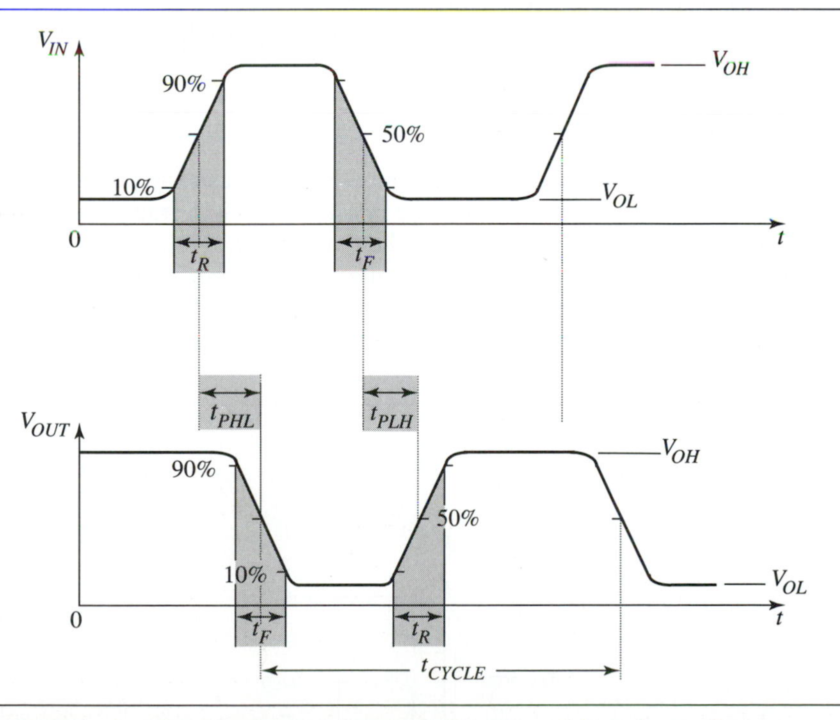

➤ **Figure 5.8** Input and output voltage as a function of time for an inverter. Definitions of t_R, t_F, t_{PHL} and t_{PLH} are depicted graphically.

Referring to the output waveform, we define the **propagation delay from high-to-low** t_{PHL} as the delay between the 50% points of the input and output waveforms. Similarly, we define the **propagation delay from low-to-high** t_{PLH} as the time between the 50% points of the input and output waveforms during this transition of the output.

5.3 MOS INVERTER CIRCUITS

In this section we will examine how to design circuits using MOS transistors to perform the inverter function.

5.3.1 NMOS:Resistor Pull-up

Figure 5.9 shows one circuit implementation for an MOS inverter. The input voltage is applied to the gate of an NMOS transistor and the output voltage is taken from the drain of that transistor. A resistor is placed between a power supply V_{DD} and the drain of the transistor to provide current to charge or "pull-up" the load capacitance C_L. In this case, the load capacitance is simulating the input capacitance to other inverters connected to this particular circuit and any parasitic capacitance.

The backgate is assumed to be shorted to the source so $V_{BS} = 0$ V. For clarity, the backgate terminal is not shown in Fig. 5.9. In this text, when the backgate terminal is not shown, it is assumed to be tied to the source.

To determine the static inverter transfer characteristic of this circuit, we remove capacitor C_L since it is an open circuit for DC signals. The NMOS transistor characteristics are shown in Fig. 5.10. On the horizontal axis we plot V_{DS}, which is equal to the output voltage V_{OUT}. On the vertical axis we plot the drain current I_D. The parameter that is changing is V_{GS} which is equal to the input voltage V_{IN}.

On this same graph we plot the resistor current-voltage characteristic. Equation 5.12 represents this line.

$$V_{OUT} = V_{DD} - I_D R \tag{5.12}$$

The operating points of this circuit can be found by looking at the intersection of the two curves since the DC current through the resistor and MOS transistor must be the same. This graphical analysis demonstrates how the output voltage and corresponding drain current will change with the input voltage. By further examining these characteristics, we see that if the input voltage is less than the NMOS transistor threshold voltage V_{Tn}, the MOSFET is off $I_D = 0$ A and $V_{OUT} = V_{DD}$. As we increase the input

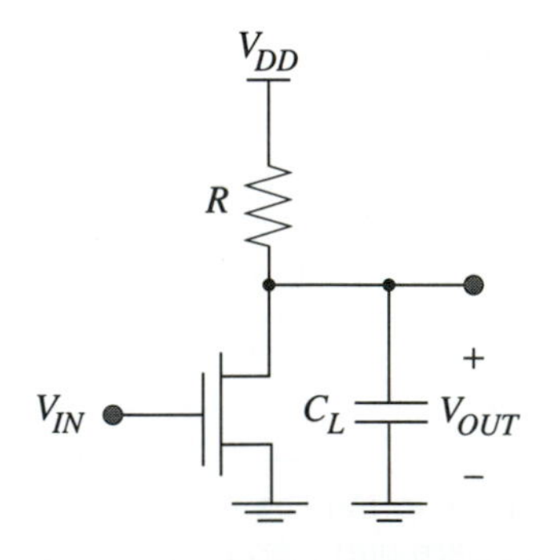

Figure 5.9 NMOS inverter with a resistor pull-up.

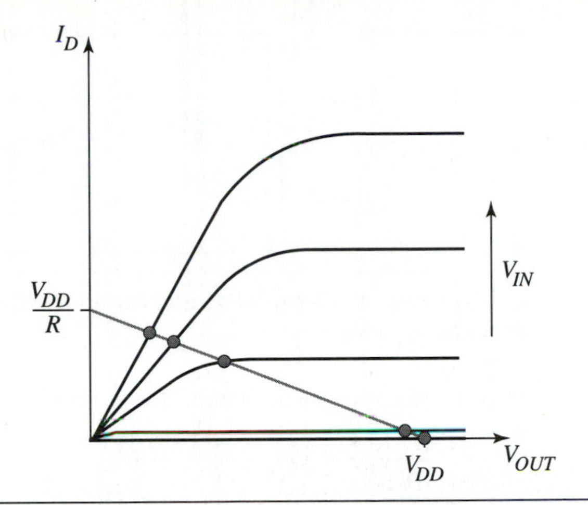

➤ Figure 5.10 NMOS transistor characteristics with resistor $I-V$ characteristic.

voltage, the output voltage decreases by moving along the $I-V$ characteristic of the resistor. Finally, when the input voltage is large, the output voltage becomes small. The voltage transfer characteristic of this inverter is shown in Fig. 5.11.

To improve the noise margin of this inverter, we must reduce the transition region by increasing V_{IL} and decreasing V_{IH}. This change implies increasing the voltage gain of the inverter in the transition region. To evaluate the voltage gain across a small voltage excursion, the small signal model developed in Chapter 4 is used to calculate the voltage gain. In Fig. 5.12 we show a small-signal model for the inverter around the operating point where $V_{IN} = V_M$. At the operating point the drain and gate of the NMOS device are at the same potential and the device is in the constant-current region. Since we are evaluating static characteristics (DC) all capacitors are open circuits. The backgate transconductance generator is an open circuit since $v_{bs} = 0\text{V}$.

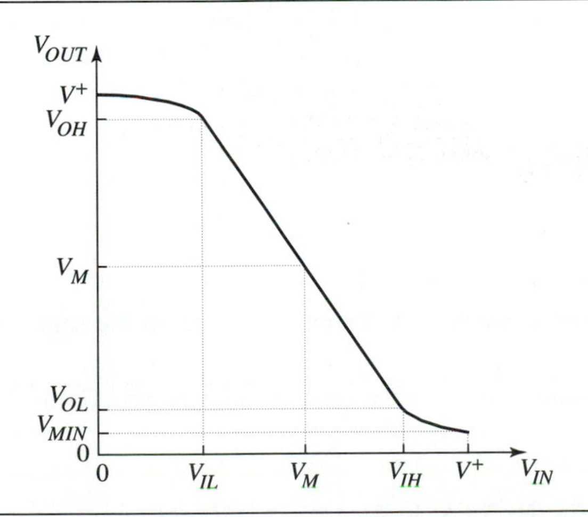

➤ Figure 5.11 Voltage transfer characteristic of an NMOS inverter with a pull-up resistor.

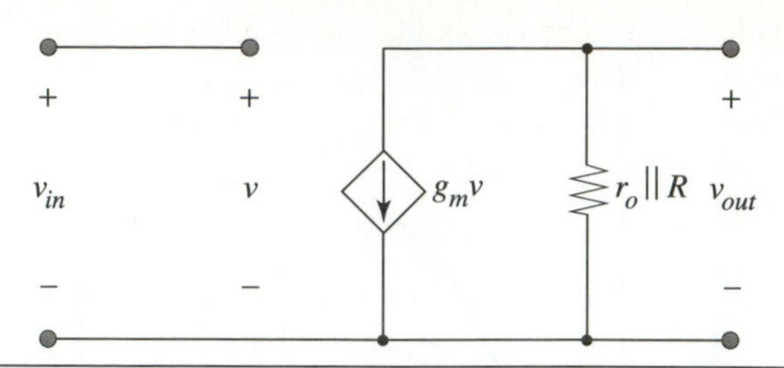

➤ **Figure 5.12** Low frequency small-signal model for an NMOS inverter with a resistor pull-up operating at $V_{IN} = V_M$.

Assuming that the output resistance of the transistor is much larger than resistor R, the voltage gain of the inverter is given by

NMOS inverter with pull-up resistor voltage gain

$$\frac{v_{out}}{v_{in}} = -g_m (r_o \| R) \approx -g_m R \tag{5.13}$$

Increasing the voltage gain will improve our noise margin. This is accomplished by either increasing the transconductance of the MOS transistor or increasing the value of the resistor R. Unfortunately, both of these solutions have a problem. A wider device can be used to increase the transconductance of the MOS transistor. However, this will increase its input capacitance and slow down the transient response at its input. Similarly, if we increase the pull-up resistance, the RC time constant at the output will be increased and a slower transient response at the output will occur. In order to increase the voltage gain without degradation to the transient response, we need a constant current with a high resistance which is a current source. The problem with the resistor is that as V_{OUT} increases, the current through the resistor becomes small. A current source supplies the same current to the load capacitor even when V_{OUT} increases.

➤ EXAMPLE 5.1 Graphical Analysis of Voltage Gain for an NMOS Inverter with Resistor Pull-Up

In this example, we will show by graphical analysis that the voltage gain in the transition region (small signal gain at $V_{IN} = V_M$) can be *increased* by increasing the value of the resistor. Increasing the voltage gain increases the noise margins of the inverter. However, a larger resistor slows down the transient response since the RC time constant needed to charge the load is increased.

Given the NMOS inverter in Fig. 5.9 with $V_{DD} = 5$ V, $(W/L) = 4.5/1.5$, $\mu_n C_{ox} = 50\ \mu A/V^2$, $V_{Tn} = 1$V, and $\lambda_n = 0\ V^{-1}$, sketch:

- I_D vs. V_{DS} for different values of V_{GS}
- $I - V$ characteristics for $R = 5$ kΩ and $R = 25$ kΩ on the same sketch as the device characteristics
- the inverter voltage transfer characteristics for $R = 5$ kΩ and $R = 25$ kΩ

SOLUTION

Refering to Fig. Ex.5.1A, look at $V_{GS} = 0$ V and move along the $R = 5$ kΩ load line. Observe that as V_{GS} increases, V_{DS} is reduced. By using the 25 kΩ resistor,

V_{GS} does not have to increase as high to change V_{DS}, implying a larger voltage gain. The voltage transfer characteristics in Fig. Ex5.1B demonstrate this concept. Note the smaller transition region for the larger resistor.

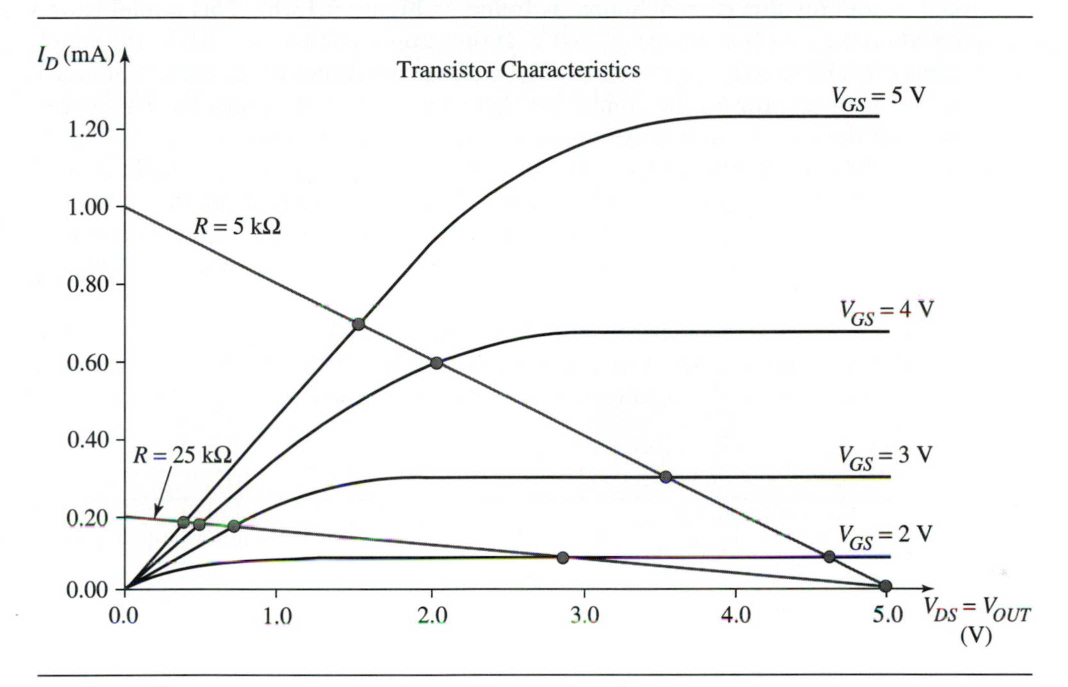

➤ **Figure Ex5.1A**

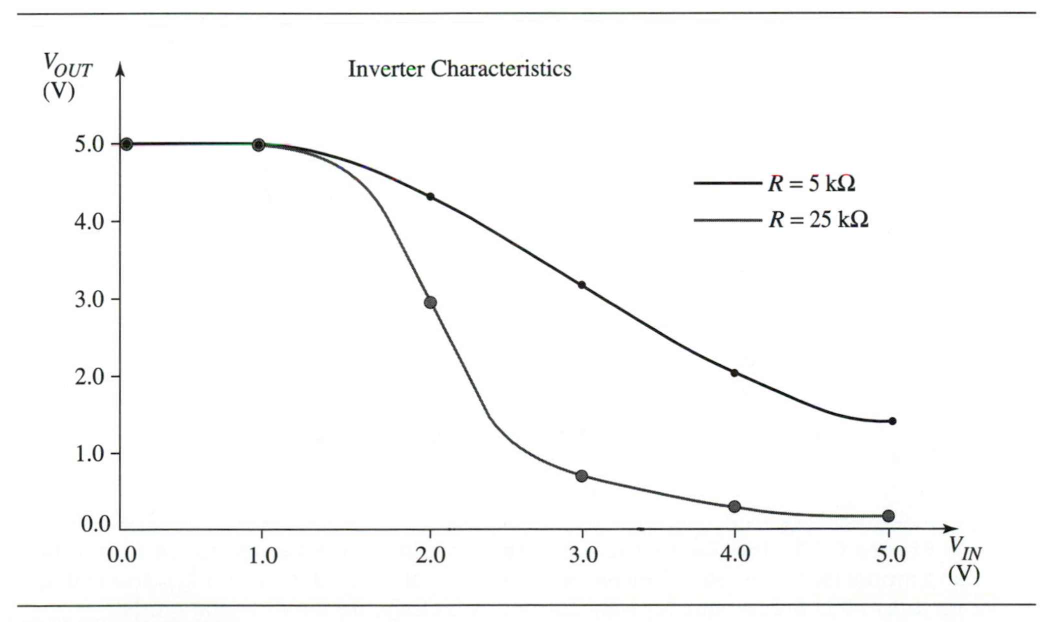

➤ **Figure Ex5.1B**

5.3.2 NMOS-Current Source Pull-up

We will begin by defining an **idealized current source** with internal resistance. The circuit model for this current source is found in Figure 5.13(a). This model is only valid when the supply voltage $v_{SUP} > 0$ V. If the supply voltage $v_{SUP} \leq 0$ V, the supply current is assumed to be equal to 0 A. The **internal resistance** of the current source is modeled with resistor r_{oc}. By combining these two circuit elements, the *I-V* characteristics of the current source are shown in Fig. 5.13(b). The current supplied by the current source has a value of I_{SUP} when the supply voltage v_{SUP} is equal to 0^+. As v_{SUP} increases, the total supply current increases due to the internal resistance that we model with resistor r_{oc}. The small-signal model for this idealized current source is simply the internal resistance r_{oc} since the DC current source is an open circuit (set to zero) for small-signal modeling.

In Fig. 5.14, we replace the resistor used in the inverter circuit in the previous section with the idealized current source. When the input voltage to the NMOS transistor is less than the threshold voltage, no current flows in the cutoff NMOS transistor

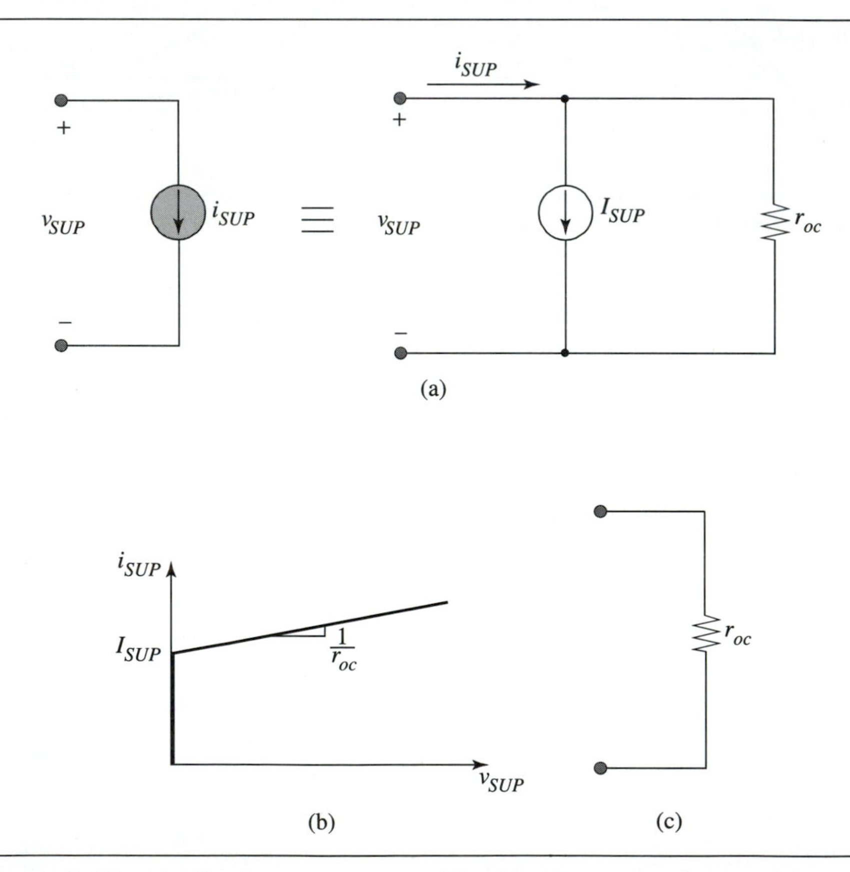

Figure 5.13 Idealized current source with added internal resistance. (a) symbol and model for the idealized current source that includes total current i_{SUP} and voltage v_{SUP}. (b) *I-V* characteristic for total current vs. voltage (c) small-signal model.

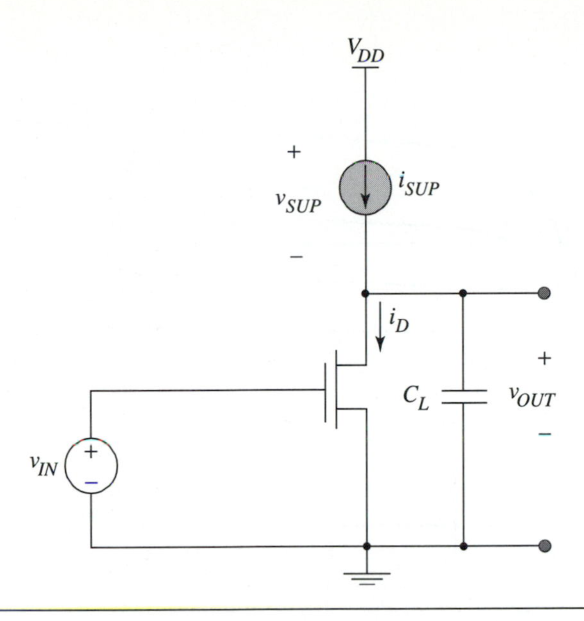

➤ **Figure 5.14** NMOS inverter with idealized current source pull-up.

and the voltage drop across the idealized current source is equal to zero. On the other hand, with a large input voltage, the NMOS transistor goes into the triode region and sinks all the current provided by the idealized current source. Under this condition, the output voltage is equal to V_{MIN}.

To further explain the operation of this NMOS inverter, we will graphically analyze this circuit. We begin by recognizing that under static conditions, the drain current of the NMOS transistor I_D must equal the current supplied by the idealized current source i_{SUP}. The voltage across the idealized current source v_{SUP} is given by

$$v_{SUP} = V_{DD} - v_{OUT} \tag{5.14}$$

Note from this equation that the output voltage v_{OUT} is related to v_{SUP}. This equation says that when v_{SUP} is equal to the power supply voltage V_{DD} the output voltage is equal to zero. Also, when $v_{SUP} = 0\,\text{V}$, the output voltage is equal to V_{DD} and the current supplied to the NMOS transistor is equal to zero. By plotting the *I-V* characteristic of the idealized current source, flipped about the x-axis on the same graph as the NMOS transistor characteristics, we can understand the operation of the NMOS inverter with the current source pull-up. The NMOS characteristics with the idealized current source *I-V* characteristic are shown in Fig. 5.15.

From Fig. 5.15(a), we can qualitatively deduce the voltage transfer characteristic of the NMOS inverter with current source pull-up. Fig. 5.15(b) shows that when the input voltage is less than V_{Tn} of the NMOS transistor, the output voltage is equal to

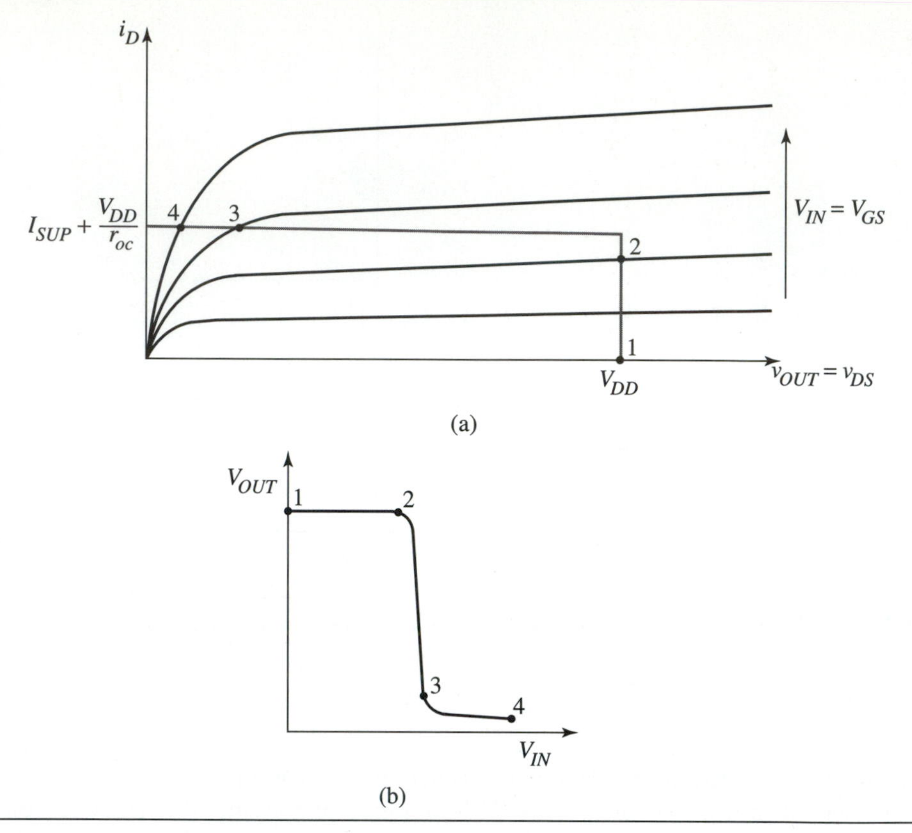

➤ **Figure 5.15** a) NMOS transistor characteristics with the idealized current source *I-V* characteristic. b) Voltage transfer characteristic.

V_{DD} and no current flows in the circuit **(1)**. As we increase the gate voltage to the point where the NMOS transistor current in saturation is equal to the supply current, the output voltage rapidly decreases toward a small value **(2–3)**. As we further increase the gate voltage, the output voltage decreases only slightly since the NMOS transistor is now in its triode region **(4)**. Qualitatively the transfer function for this circuit has a smaller transition region than the NMOS/resistor inverter since the voltage gain is higher.

We can approximate the idealized current source with a p-channel transistor by having its gate tied to a constant bias voltage V_B. Its source is tied to the power supply V_{DD} and its drain is connected to the drain of the NMOS transistor as shown in Fig. 5.16(a). We can plot the supply current from the p-channel transistor as a function of the source-drain voltage for a given bias voltage shown in Fig. 5.16(b). The p-channel transistor approximates our idealized current source when the source-drain voltage is large enough so the transistor enters its constant-current region. The finite output resistance of the transistor is already modeled by resistor r_{oc}. Merging the *I–V* char-

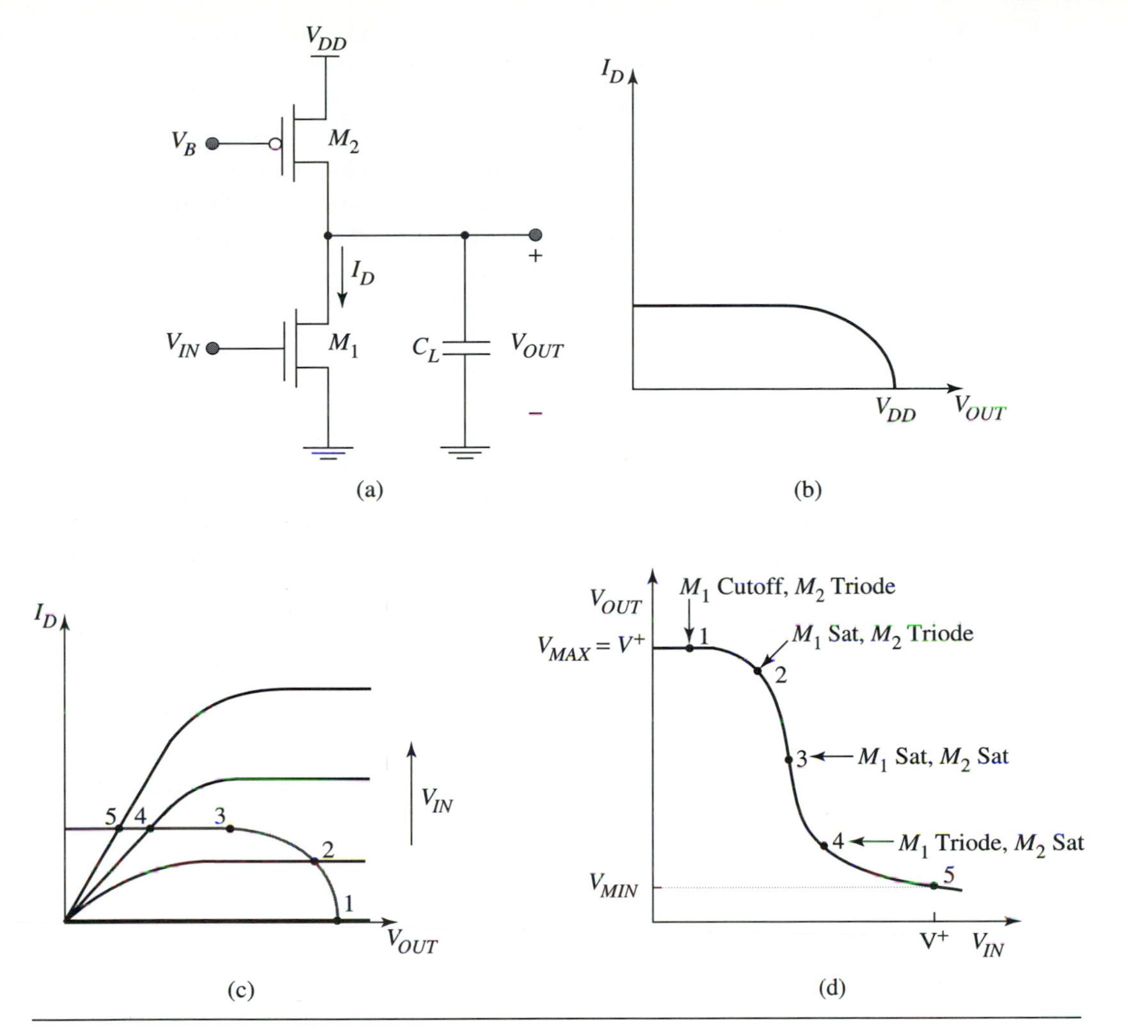

➤ **Figure 5.16** NMOS inverter with p-channel current source pull-up. (a) circuit diagram; (b) p-channel current source *I-V* characteristic; (c) NMOS transistor characteristics with p-channel current source *I-V* characteristic; (d) qualitative voltage transfer characteristic. Numbered points on load line correspond to numbered points on the voltage transfer characteristic.

acteristic from the p-channel device with our n-channel characteristics results in Fig. 5.16(c).

To obtain the transfer characteristic for this inverter, we follow the steps listed below.

1. Start by assuming that the input voltage is less than the threshold voltage of the NMOS transistor. Under this condition the NMOS transistor is cutoff and no current flows through either the n- or p-channel transistor and the output voltage is equal to V_{DD} **(1)**.

2. As we increase the input voltage, the output voltage decreases slightly **(2)**. The NMOS transistor is operating in the constant-current region, while the p-channel transistor is operating in the triode region.
3. By further increasing the input voltage, both the n- and p-channel transistor are operating in the constant-current region and the circuit exhibits a high voltage gain **(3)**. This is the point in the transfer characteristic where the output voltage rapidly changes from a high output voltage to a low output voltage near ground.
4. As the input voltage is raised further, the output voltage moves closer to ground **(4)**. Under this condition the NMOS transistor is in the triode region and the p-channel transistor is in the constant-current region.
5. Finally, when V_{IN} is at its maximum value of V^+, V_{OUT} is equal to V_{MIN} **(5)**.

The qualitative voltage transfer characteristic is shown in Fig. 5.16(d). The slope of the transition region corresponds to the voltage gain of the circuit. Since the output resistance of the p-channel transistor is much higher than the linear resistor used in the previous case, the voltage gain is much larger and the transition region is much smaller for this circuit. To evaluate the voltage gain, we use the small-signal model of the circuit operating at the midpoint where both transistors are saturated. Again, since we are evaluating static characteristics, all capacitors are open circuits. The backgate transconductance generators for both the n- and p-channel transistors are open circuits, since the backgate-source voltage of both devices is shorted ($v_{bs} = 0$ V). The small-signal model for the circuit is shown in Fig. 5.17.

NMOS inverter with p-channel current source pull-up voltage gain

Because a DC voltage has been applied between the source and gate of the p-channel transistor, the transconductance generator of the p-channel transistor is also an open circuit since $v_{sg2} = 0$ V. Only its output resistance affects the voltage gain in this region. The voltage gain of this inverter is

$$\frac{v_{out}}{v_{in}} = -g_{mn}\,(r_{on} \parallel r_{op}) \tag{5.15}$$

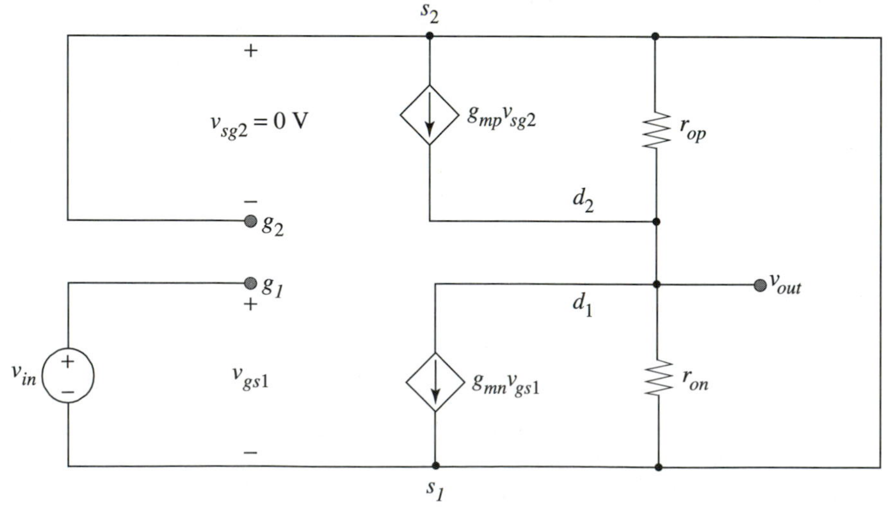

Figure 5.17 Small-signal model for NMOS inverter with p-channel current source pull-up.

➤ EXAMPLE 5.2 NMOS Inverter with P-Channel Current Source Pull-Up

An NMOS inverter with a p-channel current source pull-up and $V_{DD} = 5$ V, as shown in Fig. Ex 5.2, has the following device data:

$\mu_n C_{ox} = 50\ \mu A/V^2$, $\mu_p C_{ox} = 25\ \mu A/V^2$, $V_{Tn} = -V_{Tp} = 1$ V, $L_n = L_p = 1.5\ \mu m$, $\lambda_n = \lambda_p = 0.1 V^{-1}$

(a) When $V_B = 3$ V, find the width of device M_2 so its saturation current is 200 μA.

(b) Calculate the required width of the n-channel device so V_{OUT} is 0.05 V when V_{IN} is 5 V.

(c) Calculate the voltage gain for this inverter in the transition region.

SOLUTIONS

(a) In saturation, $-I_{Dp} = \frac{1}{2}\left(\frac{W}{L}\right)_p \mu_p C_{ox} (V_{DD} - V_B + V_{Tp})^2 \left(1 + \lambda_p \frac{V_{DD}}{2}\right)$

We assume the channel length modulation term is small for hand calculation, so that

$$W_p = \frac{-I_{Dp}}{(V_{DD} - V_B + V_{Tp})^2} \cdot 2\frac{L_p}{\mu_p C_{ox}} = \frac{200\ \mu A \cdot 2 \cdot 1.5\ \mu m}{1V^2 \cdot 25\ \mu A/V^2} = 24\ \mu m$$

(b) Under this condition, the p-channel transistor remains saturated and continues to provide 200 μA to the drain of the n-channel transistor. Since the drain-source voltage of the n-channel transistor, 0.05 V, is much less than $V_{GS} - V_{Tn}$, the n-channel transistor is in the triode region.

In the triode region, $I_{Dn} = \left(\frac{W}{L}\right)_n \mu_n C_{ox}\left(V_{GS} - V_{Tn} - \frac{V_{DS}}{2}\right)V_{DS}$

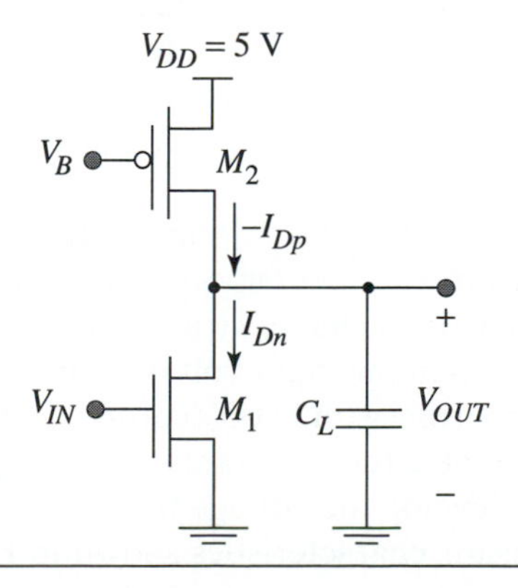

➤ **Figure Ex5.2**

Rearranging,

$$W_n = \frac{I_{Dn}}{\left(V_{IN} - V_{Tn} - \frac{V_{OUT}}{2}\right)V_{OUT}} \cdot \frac{L_n}{\mu_n C_{ox}} = \frac{200\,\mu\text{A} \cdot 1.5\,\mu\text{m}}{(3.95\text{ V})\,(0.05\text{ V}) \cdot 50\,\mu\text{A/V}^2} \approx 30\ \mu\text{m}$$

(c) Recall that in the transition region, both transistors are saturated. The voltage gain is given by

$$A_v = -g_{mn}(r_{on} \| r_{op}),$$

where $g_{mn} = \sqrt{2\left(\frac{W}{L}\right)_n \mu_n C_{ox} I_{Dn}} = 0.63\text{ mS}$

$$r_{on} \| r_{op} = \frac{1}{(\lambda_n + \lambda_p) I_{Dn}} = 25\text{ k}\Omega$$

$$A_v = -15.8$$

This inverter improves the noise margin and maintains a large current to charge up the load capacitance across nearly the full output voltage range. However, this inverter does have a problem with power dissipation. When a high input voltage is applied to the inverter, a constant DC current equal to the current source pull-up flows between the positive power supply and ground. This current is wasted since the output voltage is intended to be in the low state. In fact, what is needed is a switchable current source that is on when the NMOS transistor is off and off when the NMOS transistor is on. This is precisely the operation that we will see in a CMOS inverter.

5.3.3 CMOS Inverter

The concept for a CMOS inverter is demonstrated using switches shown in Fig. 5.18. If the input is high, we turn on the bottom switch to discharge the capacitive load. If the input is low, we turn on the top switch to charge up the capacitive load. At no time (or for a very short time) are both switches on, which prevents DC current from flowing from the positive power supply to ground. A simple implementation of this switching concept is shown by shorting the gates of the p-channel and n-channel devices as shown in Fig. 5.18(c).

Qualitatively, this circuit acts like the switching circuits, since the p-channel transistor has exactly the opposite characteristics of the n-channel transistor. Hence, when the input voltage is high, the p-channel transistor is off (cutoff), and the n-channel transistor is on (triode). When the input voltage is low, the p-channel transistor is on (triode) and the n-channel transistor is off (cutoff). In the transition region, both transistors are saturated and the circuit operates with a large voltage gain.

To understand how we develop the voltage transfer characteristic for the CMOS inverter, consider the transistor characteristics shown in Fig. 5.19 for the n-channel and p-channel transistors.

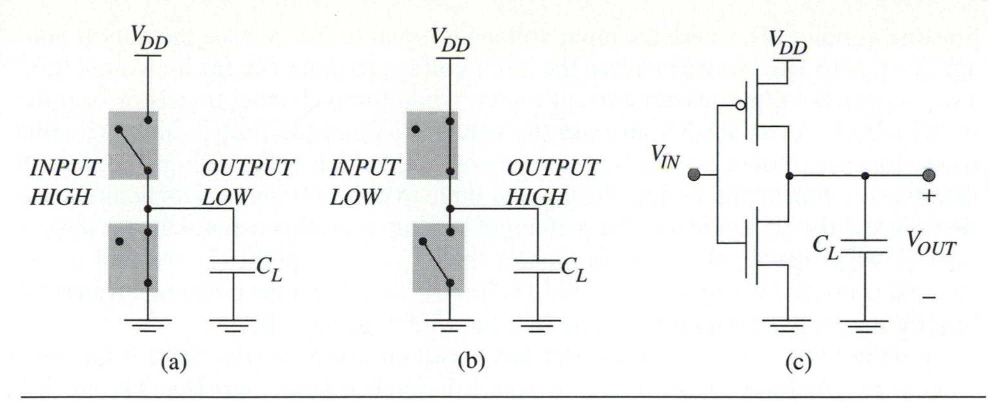

➤ **Figure 5.18** The CMOS inverter. (a) switch level representation V_{IN} = High, V_{OUT} = Low, (b) switch level representation V_{IN} = Low, V_{OUT} = High. (c) circuit diagram for CMOS inverter.

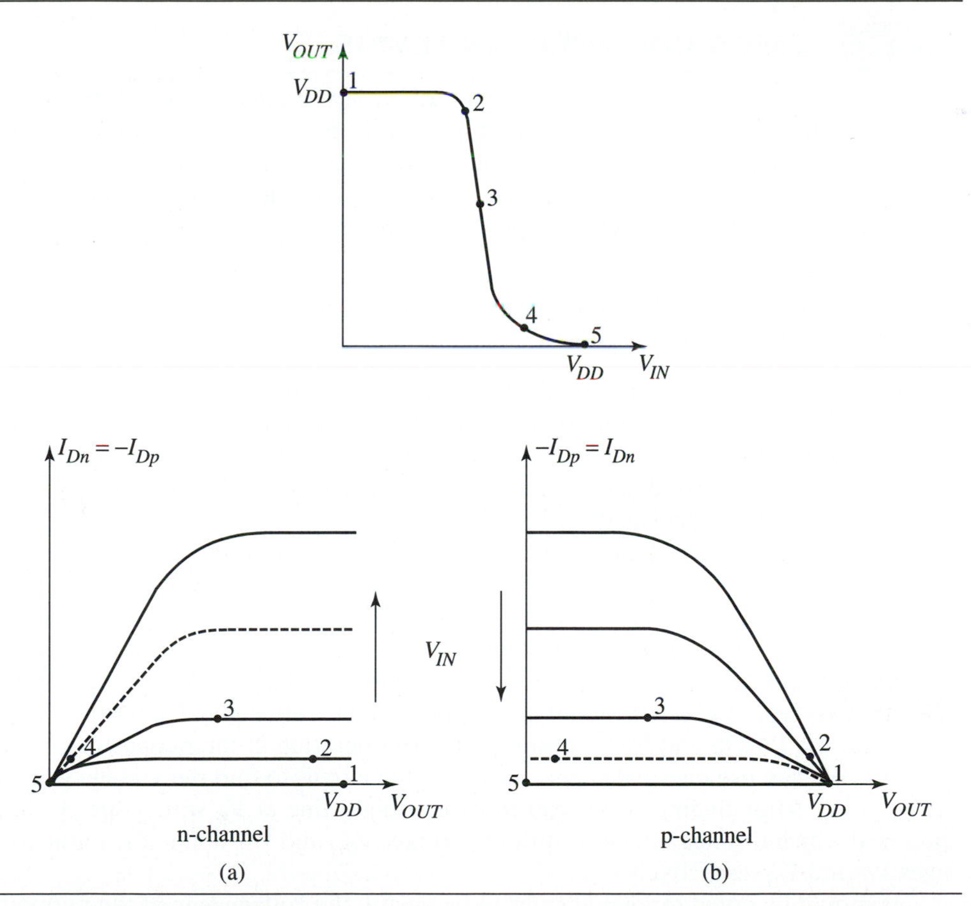

➤ **Figure 5.19** CMOS voltage transfer characteristic shown with (a) NMOS transistor characteristics and (b) PMOS transistor characteristics. Points 1 through 5 correspond to increasing V_{IN} from 0 V to V_{DD}. Note: dashed characteristics on n- and p- channel curves are for equal V_{IN}.

Starting at point **(1)**, where the input voltage is equal to 0 V, we see the output voltage is equal to V_{DD}. As we increase the input voltage to point **(2)**, the n-channel transistor operates in its constant-current region while the p-channel transistor is in the triode region. As we further increase the voltage to point **(3)**, both n- and p-channel transistors are in their constant-current region and we are in the high-gain region of the inverter. Further increasing the input voltage to point **(4)** puts the n-channel transistor in the triode region and the p-channel transistor in its constant-current region. Finally, when the input voltage is greater than $V_{DD} + V_{Tp}$, point **(5)**, we find the p-channel transistor is below its threshold voltage (cutoff) and the n-channel transistor has 0 V across it corresponding to an output voltage equal to 0 V.

Qualitatively, the CMOS inverter has excellent noise margins since it has high gain in the transition region. It has low power dissipation since there is no DC current flowing in either logical state. Finally, the speed of the inverter can be set with a constant current charging and discharging the load capacitor. These three excellent characteristics have made CMOS the technology of choice for complex logic functions, as well as for semiconductor memory.

5.4 CMOS INVERTER ANALYSIS

In the previous section we presented a qualitative view of the CMOS inverter and its operation. In this section, we will perform a static analysis to analyze the inverter transfer characteristic and quantify logic levels and noise margins. In addition, we will perform a transient analysis to calculate the propagation delay for a typical CMOS inverter. Finally, we will calculate its power dissipation.

5.4.1 Simplified Transfer Characteristic for Hand Calculations

Although it is possible to calculate the transfer function and specific voltages V_{IL}, V_{IH}, V_{OL}, V_{OH}, and V_M from the large-signal models for MOS transistors, it is algebraically complex. If we make some simplifications to the typical voltage transfer characteristic for the inverter, we find that the calculations become algebraically simple and reasonably accurate for hand analysis.

The inverter transfer function for hand calculations is shown in Fig. 5.20. For this voltage transfer characteristic we have redefined V_{OH} as V_{MAX} and V_{OL} as V_{MIN}. The error introduced is small and negligible for hand analysis.

The critical points in this transfer function are found by first determining V_{OH} and V_{OL} for the circuit. For CMOS $V_{OH} = V_{MAX} = V^+$ and $V_{OL} = V_{MIN} = 0$ V. V_M is defined as the input voltage at which the input and output voltages are equal. Under this condition both the PMOS and NMOS transistors are operating in their constant-current regions. We can use the small-signal model of the circuit to find the voltage gain A_v at $V_{IN} = V_M$. After finding A_v we can draw a straight line at V_M with slope A_v. The points at which this line intersects output voltages V_{OH} and V_{OL} define the input voltages V_{IL} and V_{IH} respectively.

It should be noted for digital gates to be useful, the voltage gain at the midpoint must be greater than one. In practice, the voltage gain is on the order of ten. Because of the relatively large voltage gain, the error in using these definitions for hand calculations is well within that required for an initial design.

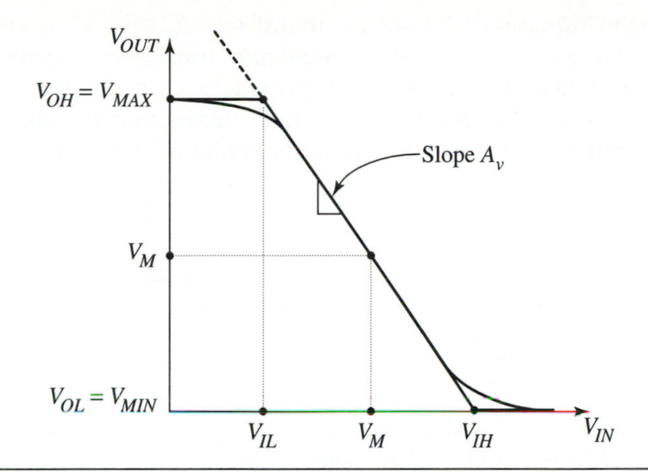

Figure 5.20 Idealized inverter transfer characteristic superimposed on a typical inverter transfer characteristic. Note: The idealized definitions for V_{OH} and V_{OL} are used in this diagram.

5.4.2 Static Analysis of the CMOS Inverter

The goal of this analysis is to determine the logic levels and noise margins for a CMOS inverter. From the previous qualitative description, we know that V_{OH} is equal to V_{DD} and V_{OL} is equal to 0 V. The next step is to find the input voltage V_M when the input and output voltages are equal. Under this condition, both the n- and p-channel devices are operating in their constant-current region. The current for the n-channel device is given by

$$I_{Dn} = \left(\frac{W}{2L}\right)_n \mu_n C_{ox} (V_M - V_{Tn})^2 (1 + \lambda_n V_M) \tag{5.16}$$

and the current for the p-channel device is given by

$$-I_{Dp} = \left(\frac{W}{2L}\right)_p \mu_p C_{ox} (V_{DD} - V_M + V_{Tp})^2 (1 + \lambda_p (V_{DD} - V_M)) \tag{5.17}$$

If we let

$$k_n = \left(\frac{W}{L}\right)_n \mu_n C_{ox} \tag{5.18}$$

and

$$k_p = \left(\frac{W}{L}\right)_p \mu_p C_{ox} \tag{5.19}$$

and set I_{Dn} equal to $-I_{Dp}$, we can solve for V_M. We assume that the channel length modulation terms can be ignored. In fact, for symmetrical transistors ($\lambda_n = \lambda_p$ and $V_M = V_{DD}/2$) the terms precisely cancel. The resulting equation for V_M is

CMOS inverter trip point V_M

$$V_M = \frac{V_{Tn} + \sqrt{\frac{k_p}{k_n}}(V_{DD} + V_{Tp})}{1 + \sqrt{\frac{k_p}{k_n}}} \tag{5.20}$$

Looking at Equation 5.20, we see that if $k_n \gg k_p$ then V_M is approximately V_{Tn}. Physically, a large k_n implies that the n-channel transistor can sink much more current than the p-channel transistor can provide and, hence, the trip point will be reduced. If $k_p \gg k_n$, then the opposite effect occurs and the trip point V_M moves toward the positive power supply, and in the limit becomes equal to $V_{DD} + V_{Tp}$.

➤ EXAMPLE 5.3 CMOS Inverter Static Analysis

You are given a CMOS inverter whose switching point V_M must be reduced from 2.5 V to 2.0 V. Due to layout constraints, the only adjustable parameter is the width of the n-channel transistor W_n. When $V_M = 2.5$ V, $W_n = 5$ µm. Find the new n-channel transistor width given the following data:

$$\mu_n C_{ox} = 2\mu_p C_{ox} = 50\ \mu\text{A/V}^2$$

$$V_{Tn} = -V_{Tp} = 1\text{V},\ L_n = L_p = 1.5\mu\text{m}$$

$$V_{DD} = 5\text{ V}$$

SOLUTION

First find the width of the p-channel transistor. Using Equation 5.20 with $V_M = 2.5$ V, we find $k_p = k_n$.

$$\left(\frac{W}{L}\right)_p = \left(\frac{W}{L}\right)_n \cdot \frac{\mu_n C_{ox}}{\mu_p C_{ox}} = \frac{10}{1.5}$$

so $W_p = 10$ µm.

To find the n-channel width required to lower V_M to 2 V, rearrange Equation 5.20 and use $V_M = 2$V.

$$V_M\left(1 + \sqrt{\frac{k_p}{k_n}}\right) = V_{Tn} + \sqrt{\frac{k_p}{k_n}}(V_{DD} + V_{Tp})$$

$$\sqrt{\frac{k_p}{k_n}} = \frac{V_M - V_{Tn}}{V_{DD} + V_{Tp} - V_M} = \frac{2-1}{5-1-2} = \frac{1}{2}$$

$$\text{so } k_n = 4k_p$$

$$\left(\frac{W}{L}\right)_n = 4\left(\frac{W}{L}\right)_p \frac{\mu_p C_{ox}}{\mu_n C_{ox}} = 4 \cdot \frac{10}{1.5} \cdot \frac{1}{2} = \frac{20}{1.5}$$

Therefore the new n-channel transistor width required is 20 µm.

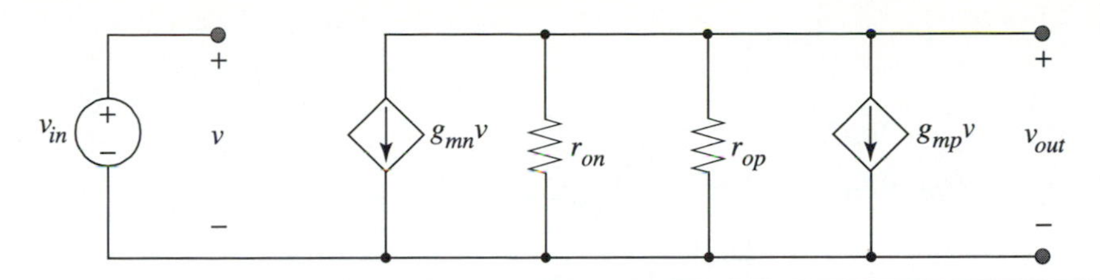

Figure 5.21 Small-signal model of CMOS inverter evaluated at $V_{IN} = V_M$.

The next step is to calculate V_{IL} and V_{IH}. We determine the slope of the transfer characteristic at $V_{IN} = V_M$, (i.e. voltage gain) and use it to project a line to intersect at $V_{OUT} = V_{MIN} = 0$ V to find V_{IH}. Similarly, we project a line to intersect at $V_{OUT} = V_{MAX} = V_{DD}$ to find V_{IL}. To find the voltage gain when the input voltage is equal to V_M, we use the small signal model of both MOS transistors. At that operating point we find the transconductance and the output resistance of the n- and p-channel devices are given by

$$g_{mn} = k_n (V_M - V_{Tn}) \tag{5.21}$$

$$r_{on} = 1/(\lambda_n I_{Dn}) \tag{5.22}$$

$$g_{mp} = k_p (V_{DD} - V_M + V_{Tp}) \tag{5.23}$$

$$r_{op} = 1/(\lambda_p I_{Dp}) \tag{5.24}$$

The small-signal model for the CMOS inverter is shown in Fig. 5.21. All capacitors are open circuits since we are performing a DC analysis. Backgate transconductance generators are open circuits since $v_{bs} = 0$ V. The relationship between the output voltage and the input voltage is given by

CMOS inverter voltage gain:

$$A_v = \left(\frac{v_{out}}{v_{in}}\right) = -(g_{mn} + g_{mp})(r_{on} \| r_{op}) \tag{5.25}$$

In CMOS inverters, one generally uses the shortest channel length transistor that the technology will provide in order to have the largest current possible for a given device size. In Chapter 4 we learned that the output resistance will be small if the channel length is small and, hence, the voltage gain of typical CMOS inverters is on the order of –10.

To find V_{IL} and V_{IH}, we use the known slope, A_v, and find

CMOS inverter V_{IL} and V_{IH}

$$V_{IL} = V_M + \frac{(V_{DD} - V_M)}{A_v} \tag{5.26}$$

$$V_{IH} = V_M - \frac{V_M}{A_v} \tag{5.27}$$

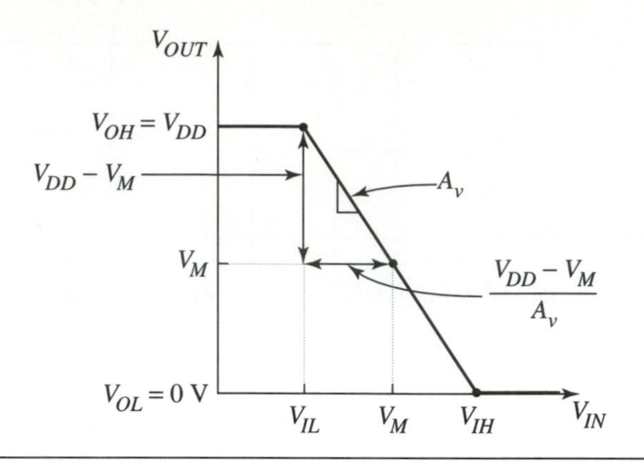

➤ **Figure 5.22** CMOS transfer characteristic illustrating the calculation of V_{IL}, using the hand analysis approximation.

These equations are extremely simple because we made the straight line approximation that the voltage gain was constant between V_{IL} and V_{IH} as shown in Fig. 5.22. Their accuracy is certainly adequate for hand analysis.

The last quantities to calculate for the CMOS inverter static analysis are the noise margins. They are given by

CMOS inverter noise margins

$$NM_L = V_{IL} - V_{OL} = V_M + (V_{DD} - V_M)/A_v \tag{5.28}$$

$$NM_H = V_{OH} - V_{IH} = V_{DD} - V_M + V_M/A_v \tag{5.29}$$

 DESIGN EXAMPLE 5.4 CMOS Static Inverter

In this example, we are going to design a basic CMOS inverter. The inverter's transition voltage must be $V_M = 2.5$ V, the current through the n- and p-channel devices should be 200 μA at $V_{IN} = V_M$, and $NM_L = NM_H \geq 2.25$ V. The power supply voltage V_{DD} is 5 V. Specify to the process engineer the maximum channel length modulation λ that can be tolerated to meet all of the above design specifications. Assume $\lambda_n = \lambda_p$.

MOS Device Data:

$$\mu_n C_{ox} = 2\mu_p C_{ox} = 50\ \mu\text{A/V}^2$$

$$V_{Tn} = -V_{Tp} = 1\ \text{V and } L_n = L_p = 1.5\ \mu\text{m}$$

SOLUTION

To begin this design we need to find the (W/L) ratios of the devices needed to carry 200 μA at $V_{IN} = V_M$. Setting $V_M = 2.5$ V and assuming that λ is small for this calculation,

$$I_{Dn} = \left(\frac{1}{2}\left(\frac{W}{L}\right)_n \mu_n C_{ox} (V_M - V_{Tn})^2\right)$$

$$\left(\frac{W}{L}\right)_n \approx \frac{5.3}{1.5}$$

We will have to make $(W/L)_p$ twice as big as $(W/L)_n$ because $\mu_n = 2 \cdot \mu_p$, so

$$\left(\frac{W}{L}\right)_p \approx \frac{10.6}{1.5}$$

To have the required noise margins of 2.25 V, the minimum voltage gain needed can be found by using Eq. (5.28) and Eq. (5.29).

$$NM_L = V_M + \frac{V_{DD}}{2A_v} \text{ and } NM_H = V_{DD} - V_M + \frac{V_{DD}}{2A_v}.$$

$$\text{So } A_v = \frac{V_{DD}}{2(NM_L - V_M)} = -10$$

Since $(W/L)_p = 2(W/L)_n$ and $\mu_n C_{ox} = 2\mu_p C_{ox}$, $k_n = k_p$. Using Eq. (5.28) and noting $\lambda_n = \lambda_p$, $g_{mn} = g_{mp}$ and $r_{on} = r_{op}$ for $V_M = 2.5$ V. We write

$$A_v = -(g_{mn} + g_{mp})(r_{on} \| r_{op}) = -(2g_{mn})\left(\frac{r_{on}}{2}\right)$$

$$\text{Now } r_{on} = \frac{1}{\lambda_n I_D}$$

Substituting for r_{on} and solving for λ_n, we arrive at $\lambda_n \le \frac{-g_{mn}}{A_v I_D}$

We obtain a value for g_{mn} by

$$g_{mn} = \sqrt{2\left(\frac{W}{L}\right)_n \mu_n C_{ox} I_{Dn}} = 0.27 \text{ mS}$$

Substituting $g_{mn} = 0.27$ mS, $A_v = -10$, and $I_D = 200$ μA, into the expression for λ_n above

$$\lambda_n \le 0.13 \text{ V}^{-1}.$$

5.4.3 Propagation Delay

It is extremely difficult to calculate accurately the voltage waveforms using only hand analysis. Therefore, as we did in the static analysis, we will idealize the transient responses so a simple hand analysis of the propagation delay may be performed. Figure 5.23 shows the idealized transient voltages that we will use for hand analysis.

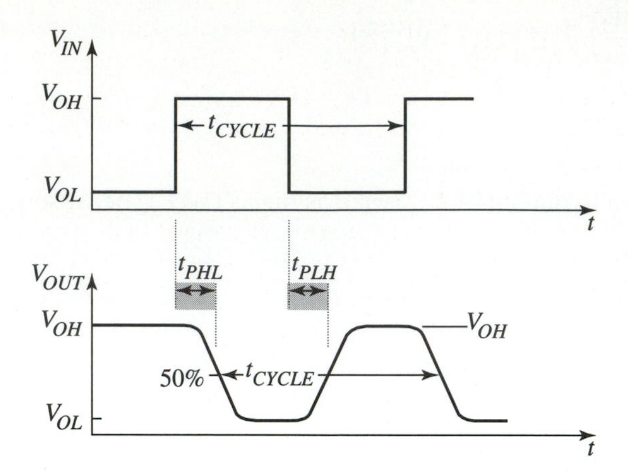

➤ **Figure 5.23** Idealized transient response of an inverter used for hand calculation.

The essential difference between the idealized and real versions is that we assume the input voltage makes its transition infinitely fast. We measure the propagation delay from the edge of the input voltage transition to the 50% point on the output voltage as indicated. The input signal transition is positioned at the 50% point of the actual input voltage waveform. Computer simulation is required for more accurate analysis of the transient response.

A typical CMOS inverter may drive successive gates. The input capacitance of these gates as well as parasitic capacitance due to the depletion capacitance of the drains of the inverters and wiring capacitance are lumped in a load capacitance C_L as shown in Fig. 5.24(a).

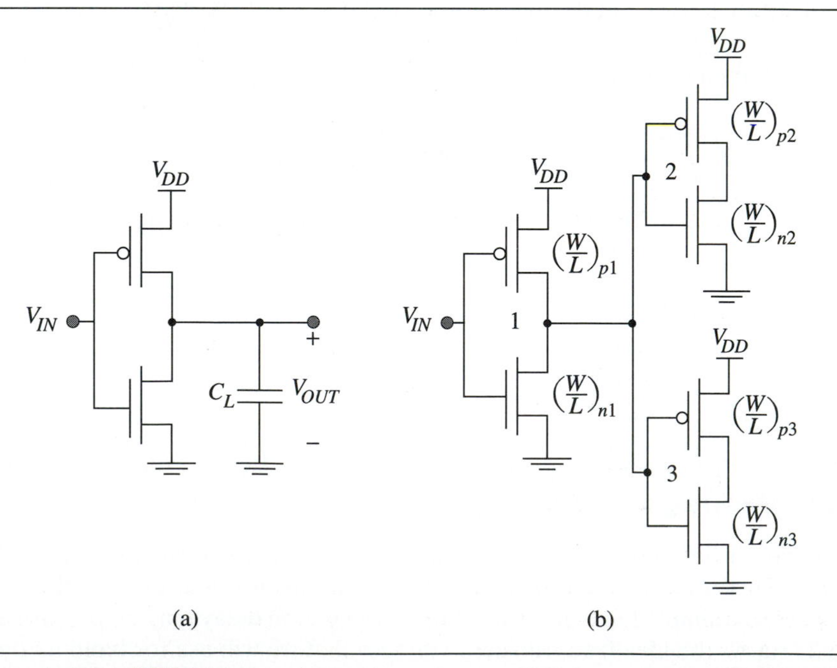

➤ **Figure 5.24** CMOS inverter (a) Driving a lumped load capacitance C_L. (b) driving additional inverters 2 and 3.

Fig. 5.24(b) shows Inverter 1 driving two inverters labeled 2 and 3. In our analysis, we will calculate the propagation delay of Inverter 1. As stated earlier, we assume that the input transition time is infinitely fast. Looking at the high-to-low transition for V_{OUT} for the inverter, we follow the trajectory of the NMOS current-voltage transistor characteristic as shown in Fig. 5.25(a).

At $t = 0^-$ the input voltage is equal to zero and the output voltage is equal to V_{OH}. At $t = 0^+$, we assume the input voltage makes an instantaneous transition to V_{OH}. The output voltage then decreases along the trajectory shown. As discussed in Section 5.2, t_{PHL} is defined as the time it takes the output voltage to discharge from V_{OH} to $V_{OH}/2$. With this in mind we can write an equation for t_{PHL} as

$$t_{PHL} = \frac{C_L \Delta V}{I_D} = \frac{C_L V_{OH}/2}{\frac{k_n}{2}(V_{OH} - V_{Tn})^2} \tag{5.30}$$

We have assumed that the NMOS transistor remains in the constant-current region throughout this transition so the capacitor voltage is linearly discharged by its saturation current as shown in Fig. 5.25(b). Although this may not be strictly true for our definition of $V_{DS_{SAT}}$, it is quite adequate for this simplified hand analysis.

The load capacitance C_L consists of two major components.

1. The total gate capacitance C_G that is the gate capacitance of the transistors being driven by the inverter.
2. A parasitic capacitance C_P that results from the drain diffusions of the inverter being analyzed and the wiring to the gates being driven.

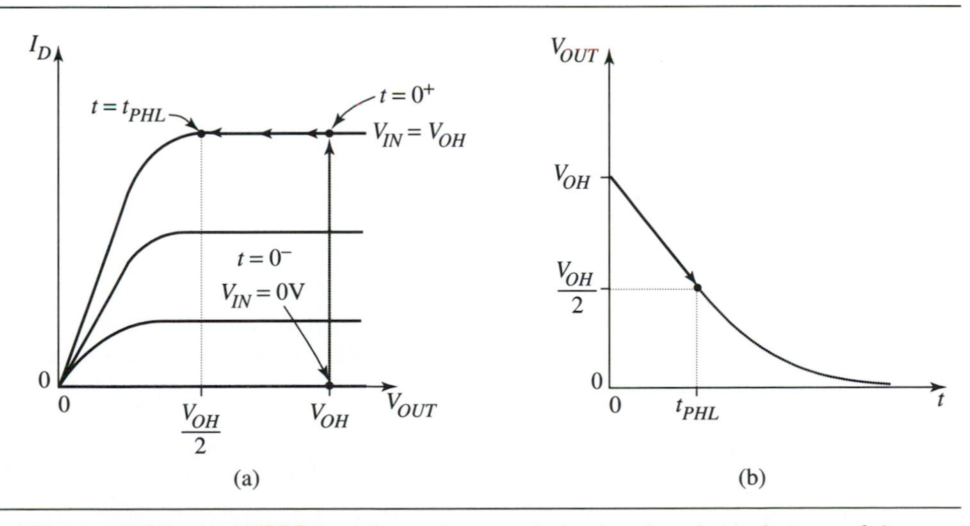

➤ **Figure 5.25** (a) NMOS transistor characteristic showing the trajectory of the current-voltage relationship from $t = 0$ s to $t = t_{PHL}$. (b) Transient waveform for V_{OUT}.

In the following paragraphs we will explore the method to calculate these two components of C_L. Referring to Fig. 5.24, we see that the inverter being analyzed is driving two additional inverters. These two inverters are switching from the triode region through saturation to cutoff during a high-to-low or low-to-high transition. To calculate accurately the total charge supplied to the load inverters, we must perform an analysis with non-linear charge storage elements. However, for a simple hand calculation we can estimate a worst-case capacitance. The total gate capacitance of the load inverters is given by

$$C_G = \sum_{(load)\ inverters} C_{IN} \tag{5.31}$$

where the input gate capacitance of a single inverter is approximated by

$$C_{IN} \approx C_{ox}(W \cdot L)_p + C_{ox}(W \cdot L)_n \tag{5.32}$$

The total gate capacitance C_G for the two inverters in Fig. 5.24(b) can be estimated as

$$C_G = C_{ox}[(WL)_{p2} + (WL)_{n2} + (WL)_{p3} + (WL)_{n3}] \tag{5.33}$$

Refer to the layout and cross-section of an NMOS transistor in Fig. 5.26(a) to calculate the parasitic capacitance due to the drain diffusions of Inverter 1. The n^+ drain diffusion has a depletion region capacitance that is primarily determined by the doping concentration in the p-type region under that diffusion. Along the sides of the n^+ drain diffusion, the doping concentration due to the field implant can be higher than at the bottom of the diffusion. Because of this, we separately account for the perimeter capacitance along the edge of the drain diffusion. Although this capacitance is voltage dependent, for hand analysis we usually use a zero-bias capacitance value since this is a worst-case estimate.

Most MOS process technologies specify to the circuit designer a value for the **bottom junction capacitance per unit area, C_{Jn}**, and a **sidewall junction capacitance per unit length, C_{JSWn}**. Figure 5.26(b) shows a top view of the n-channel MOS transistor. Because there are design rules that limit the minimum size of the contact and the spacing between the contact and polysilicon gate, and the contact and outer region of the active region, there is a **minimum length of the drain diffusion** that we call L_{diff}. This important dimension along with the device width is sketched in Fig. 5.26(c).

The area of the drain diffusion then is $W \times L_{diff}$. The sidewall perimeter of the device is equal to $W + 2L_{diff}$. It should be noted that the diffusion capacitance along the edge of the gate is not included in this calculation. This capacitance is taken into account in the intrinsic MOS transistor model. A similar analysis applies to the PMOS transistors. This discussion shows that the total drain-bulk depletion capacitance C_{DB} for an inverter is given by

$$C_{DB} = W_n L_{diffn}(C_{Jn}) + W_p L_{diffp}(C_{Jp}) \tag{5.34}$$

$$+ (W_n + 2L_{diffn})C_{JSWn} + (W_p + 2L_{diffp})C_{JSWp}$$

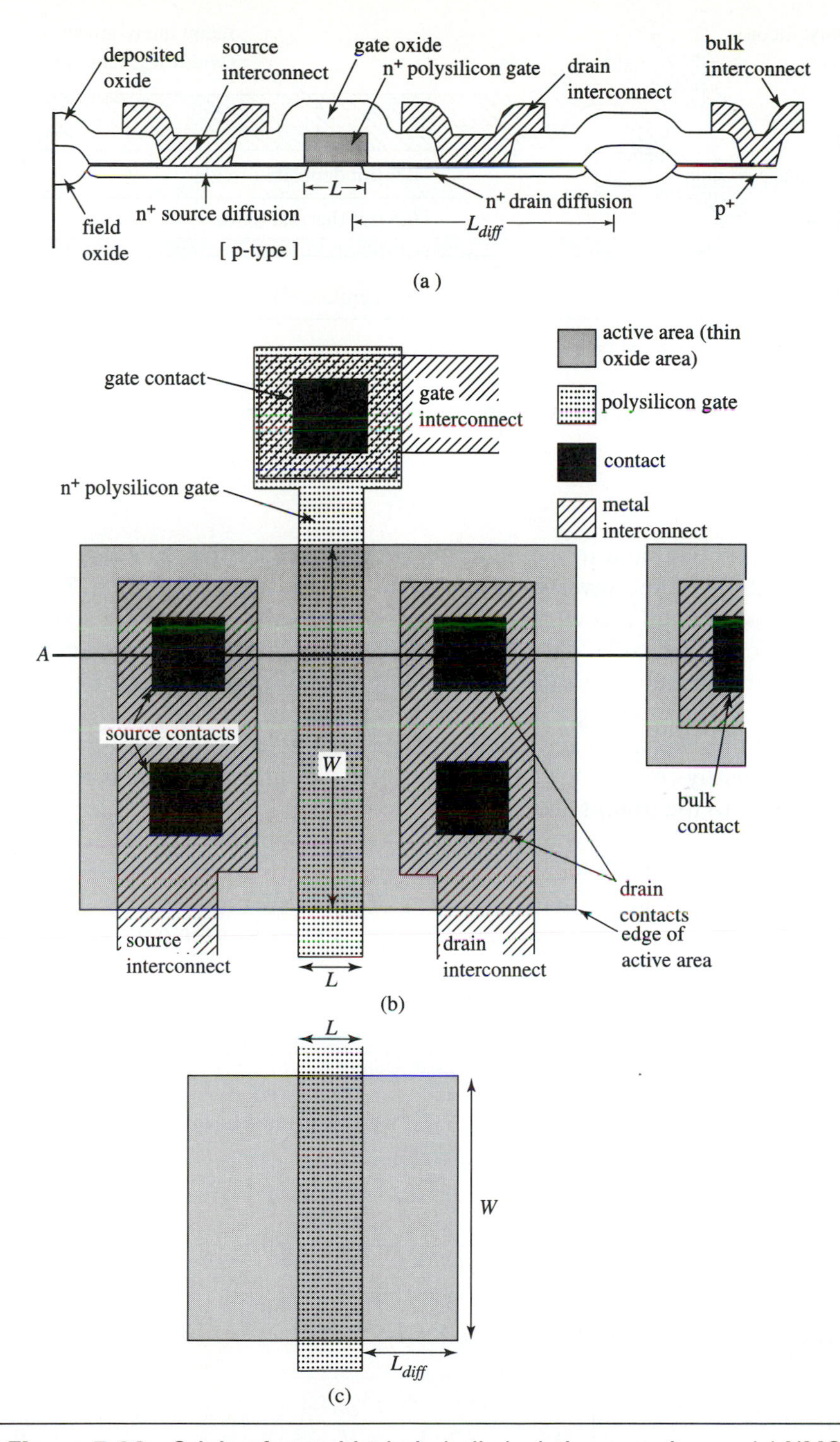

➤ **Figure 5.26** Origin of parasitic drain-bulk depletion capacitance. (a) NMOS cross-section; (b) NMOS top view; (c) definition for dimensions L_{diff} and W used to calculate depletion capacitance.

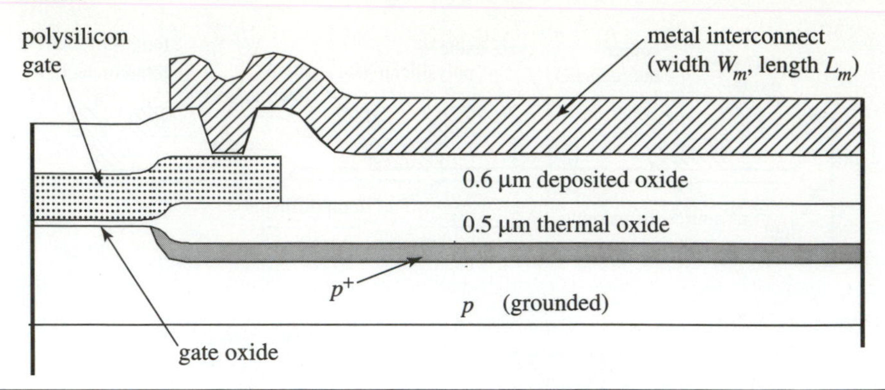

➤ **Figure 5.27** Cross-section showing origin of parasitic wiring capacitance.

Finally, we should add to the parasitic capacitance that which is due to wiring between the drain of the inverter and input of the next gate. Figure 5.27 shows a cross-section of a metal line connecting a polysilicon gate and running across both thermal and deposited oxide. The capacitance of this structure is approximately given by the permittivity of the oxide divided by the total dielectric thickness. To calculate the contribution of the wiring to the parasitic capacitance, multiply this capacitance per unit area times the width and length of the metal interconnect line.

Between tightly coupled logic gates, the metal wiring capacitance is often negligible compared to the drain-bulk depletion capacitance. However, it is important to realize that in large digital systems, we often have metal busses that extend a significant distance across a chip. Under this condition, the wiring capacitance is the dominant capacitance that determines the speed of the transition. In fact, in modern integrated circuits, the ultimate speed of microprocessors is often determined by the delay caused by long wires.

The total parasitic capacitance is given by

$$C_P = C_{DB} + C_{WIRE} \tag{5.35}$$

We add the values of C_G and C_P to obtain the total load capacitance C_L. Substituting $C_G + C_P$ for C_L, we find the value of t_{PHL} to be

CMOS inverter propagation delays t_{PHL} & t_{PLH}

$$\boxed{t_{PHL} = \frac{(C_G + C_P)\,V_{OH}/2}{k_n/2\,(V_{DD} - V_{Tn})^2}} \tag{5.36}$$

Similarly, we can find a value for t_{PLH} to be

$$\boxed{t_{PLH} = \frac{(C_G + C_P)\,V_{OH}/\,2}{k_p/\,2\,(\,V_{DD} + V_{Tp})^2}} \tag{5.37}$$

It is often useful to make the propagation delay symmetrical; namely to make $t_{PHL} = t_{PLH}$. Because the mobility of holes in the p-channel transistor is approximately one-half that of electrons in the n-channel transistor, Eq. (5.36) and (5.37) indicate it is necessary to make $(W/L)_p = 2(W/L)_n$, assuming $V_{Tn} = -V_{Tp}$.

The average propagation delay t_P can be estimated by averaging the transition times as shown in

CMOS inverter average propagation delay

$$\boxed{t_P = (t_{PHL} + t_{PLH})/2} \tag{5.38}$$

In large complex digital systems, many propagation delays are stacked together between system clock cycles. Typical systems have between twenty and fifty propagation delays for every major clock cycle. This implies that a microprocessor with a clock speed of 100 MHz must have gate delays less than 500 picoseconds.

In digital integrated circuit design, it is very important to use computer simulation tools to verify both static and transient analyses. Tools that accurately extract the gate and parasitic capacitance should be used. These capacitance values can be transferred into a circuit simulation program such as SPICE to find more accurate delay times for the particular digital gate being studied. In Design Example 5.6, we will show how to use both hand and computer analysis.

➤ EXAMPLE 5.5 CMOS Inverter Propagation Delay Analysis

A CMOS inverter driving two other identically sized inverters is shown in Fig. Ex5.5. Calculate t_{PHL} and t_{PLH}. Ignore wiring capacitance but include the parasitic drain-bulk depletion capacitance.

MOS Device Data

$\mu_n C_{ox} = 50\ \mu\text{A/V}^2$, $\mu_p C_{ox} = 25\mu\text{A/V}^2$, $V_{Tn} = -V_{Tp} = 1\text{V}$, $C_{ox} = 2.3\ \text{fF}/\mu\text{m}^2$,

$C_{Jn} = 0.1\ \text{fF}/\mu\text{m}^2$, $C_{Jp} = 0.3\ \text{fF}/\mu\text{m}^2$, $C_{JSWn} = 0.5\ \text{fF}/\mu\text{m}$, $C_{JSWp} = 0.35\ \text{fF}/\mu\text{m}$,

$L_{diffn} = L_{diffp} = 6\ \mu\text{m}$

SOLUTION

Begin the solution by finding C_L. Since Inverters 2 and 3 are identically sized, C_G is two times that of Inverter 2. From Eq. (5.33), we find

$$C_G = 2C_{ox}[(WL)_{p2} + (WL)_{n2}]$$

$$C_G = 2\left(2.3\frac{\text{fF}}{\mu\text{m}^2}\right)[(12\ \mu\text{m} \cdot 1.5\mu\text{m}) + (6\ \mu\text{m} \cdot 1.5\mu\text{m})] = 125\ \text{fF}$$

Since the wiring capacitance is neglected, $C_P = C_{DB}$.

$$C_P = C_{DB} = W_{n1}L_{diffn}C_{Jn} + W_{p1}L_{diffp}C_{Jp} + (W_{n1} + 2L_{diffn})\,C_{JSWn}$$
$$+ (W_{p1} + 2L_{diffp})\,C_{JSWp}$$

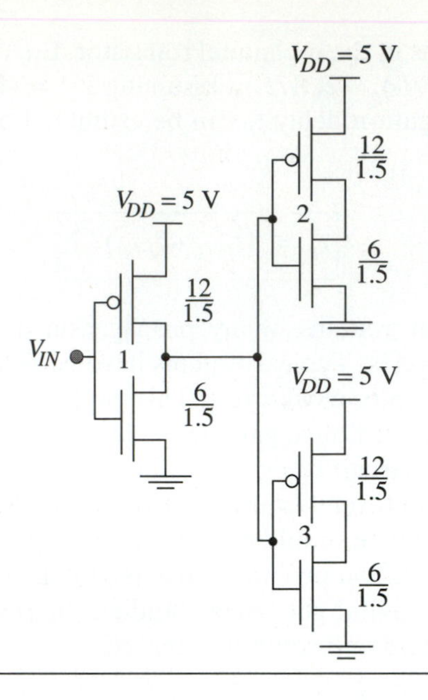

➤ Figure Ex5.5

$$C_{DB} = \left[6\ \mu\text{m} \cdot 6\ \mu\text{m} \cdot 0.1\frac{\text{fF}}{\mu\text{m}^2}\right] + \left[12\ \mu\text{m} \cdot 6\ \mu\text{m} \cdot 0.3\frac{\text{fF}}{\mu\text{m}^2}\right]$$

$$+ \left[(6\ \mu\text{m} + 2 \cdot 6\ \mu\text{m}) \cdot 0.5\frac{\text{fF}}{\mu\text{m}}\right] + \left[(12\,\mu\text{m} + 2 \cdot 6\,\mu\text{m}) \cdot 0.35\frac{\text{fF}}{\mu\text{m}}\right]$$

$$C_{DB} = 3.6\ \text{fF} + 21.6\ \text{fF} + 9\ \text{fF} + 8.4\text{fF} = 42.6\ \text{fF}$$

So $C_L = C_G + C_{DB} = 125\ \text{fF} + 42.6\ \text{fF} = 168\ \text{fF}$.

From Eq. (5.36),

$$t_{PHL} = \frac{(C_G + C_P)\ (V_{DD}/2)}{\frac{k_n}{2}(V_{DD} - V_{Tn})^2} \text{ and } k_n = \left(\frac{6}{1.5}\right)50\ \mu\text{A/V}^2\text{, so } t_{PHL} = 262\ \text{ps}.$$

From Eq. (5.37),

$$t_{PLH} = \frac{(C_G + C_P)\ (V_{DD}/2)}{\frac{k_p}{2}(V_{DD} + V_{Tp})^2} \text{ and } k_p = \left(\frac{12}{1.5}\right) 25\ \mu\text{A/V}^2 \text{ so } t_{PLH} = 262\ \text{ps}.$$

DESIGN EXAMPLE 5.6 CMOS Input Inverter/Buffer

Design a 5V CMOS inverter that is an input to an integrated circuit. The midpoint logic level of the signal driving the CMOS inverter is 1.5 V. The CMOS inverter must be able to drive a 6 μm wire that may traverse the entire 1 cm chip as well as two other inverters sized at $(W/L)_p = 2(W/L)_n = 12/1.5$. The specifications needed for the inverter require t_P to be less than 5 ns and both noise margins to be at least 1.25 V. $\lambda_n = \lambda_p = 0.1\ \text{V}^{-1}$ and the capacitance value of the wire is 0.035 fF/μm². Specify the channel widths needed to build such an inverter. Use the same device data as in Example 5.5. Verify your design in SPICE.

SOLUTION

Start by finding the ratios of the n- and p-channel devices. Rearrange Equation (5.20) to find

$$\sqrt{\frac{k_p}{k_n}} = \frac{V_M - V_{Tn}}{V_{DD} + V_{Tp} - V_M} = \frac{1.5 - 1}{5 - 1 - 1.5} = \frac{1}{5}, \rightarrow k_n = 25\,k_p.$$

Since $k_n \gg k_p$, t_{PHL} will be negligible. This leaves t_{PLH} to be determined by the t_P specification. To find t_{PLH}, first determine the values of C_G and C_P. From Example 5.5, $C_G = 125$ fF. Since the size of the p-channel device is not yet known, we cannot accurately determine C_P. However, we do know it will be dominated by the long wire.

$$C_P \approx C_{WIRE} = 6\,\mu\text{m} \cdot 10,000\,\mu\text{m} \cdot 0.035\ \text{fF}/\mu\text{m}^2 \cdot \frac{1}{1000}\frac{\text{pF}}{\text{fF}} = 2.1\,\text{pF}$$

From Eq. (5.37) we determine $k_p \approx \dfrac{140\ \mu\text{A}}{\text{V}^2}$. This implies that $(W/L)_p = 8.4/1.5$.

We will round up the size of the p-channel device to 10/1.5. To make $k_n = 25k_p$ we need $(W/L)_n = 25/2(W/L)_p$, so $(W/L)_n = 125/1.5$.

Finally, check to see if our inverter will meet the minimum noise margin requirements. To do this, we first find the voltage gain A_v, at $V_{IN} = V_M$.

$$A_v = -(g_{mn} + g_{mp})\,(r_{on} \parallel r_{op})$$

$$g_{mn} = \sqrt{2k_n I_{Dn}}\text{, where } I_{Dn}(V_{IN} = V_M) = \frac{k_n}{2}(V_M - V_{Tn})^2 \approx 435\ \mu\text{A}$$

yielding $g_{mn} = 1.75$ mS and $g_{mp} = \dfrac{g_{mn}}{\sqrt{25}} = 0.35\,\text{mS}$

$$r_{on} = r_{op} = \frac{1}{\lambda I_{Dn}} = 23\ \text{k}\Omega\text{, so } A_v = -24.$$

The noise margins are sufficient, and thus our hand analysis has given us a good approximation. According to SPICE, $A_v = -30$, $t_{PLH} = 4$ ns, and NM_L & NM_H >1.25 V.

5.4.4 Power Dissipation

The increased use of portable electronics such as cellular phones and notebook computers has made power dissipation an important design metric in modern microelectronics. There are two components that make up the power dissipation in CMOS digital gates. The first is **dynamic power** that is needed to charge and discharge the load capacitor. To calculate this power, we begin by noting that the energy coming from the positive power supply to charge the load capacitor is equal to

$$E_{V+} = \int V_{DD} i\,dt = V_{DD} Q \tag{5.39}$$

Since the charge stored in the capacitor is equal to

$$Q = C_L V_{DD} \tag{5.40}$$

the energy from the positive power supply is

$$E_{V+} = C_L V_{DD}^2 \tag{5.41}$$

The energy stored in the capacitor is $(1/2)C_L V_{DD}^2$. Therefore, the energy dissipated by the p-channel transistor in charging the capacitor to V_{DD} is equal to $(1/2)$ $C_L V_{DD}{}^2$. A similar analysis can be performed for discharging the load capacitor. If the inverter is switched on and off f times/second then the total dynamic power P_D dissipated by the CMOS gate is equal to

$$\boxed{P_D = C_L V_{DD}^2 f} \tag{5.42}$$

CMOS inverter dynamic power

The second type of power dissipated in a CMOS circuit happens during the transition time. During this time both the NMOS and PMOS transistors are on, and current can flow from the power supply to ground. It is extremely difficult to quantify this current with hand calculation. However, it can significantly add to the power dissipation and should be taken into account by use of computer simulation.

In general, CMOS power dissipation is quite low when compared to other digital technologies. However, as clock frequencies continually increase, the dynamic power in CMOS digital integrated circuits is becoming quite large. One method to significantly reduce this power is to lower the power supply voltage since this reduces dynamic power dissipation quadratically.

5.5 STATIC CMOS LOGIC GATES

In this section, we will qualitatively and quantitatively describe the operation of the CMOS NAND and NOR gates. As with the CMOS inverter, these circuits should not dissipate any DC power, which means there must be at least one off-transistor between V_{DD} and ground for all possible logic inputs. Current will flow in the circuit during the transitions. Static CMOS logic gates have the property that the output is connected to either the power supply or ground depending on logic state through a low on-resistance transistor.

The CMOS inverter was analyzed in the previous section. To build more complex Boolean functions, additional n- and p-channel transistors need to be combined. We will analyze the static and transient characteristics of static logic gates in a similar manner as for the inverter.

5.5.1 Qualitative Description of CMOS, NAND, and NOR Gates

To understand qualitatively the operation of CMOS logic gates, assume that the transistors behave as perfect switches. The NMOS transistor is on when its gate voltage is high (logic **1**) and off when the gate voltage is low (logic **0**). The p-channel transistor has the opposite characteristic—it is on when the gate voltage is low (logic **0**) and off when the gate voltage is high (logic **1**).

Consider the circuit schematic for a NOR gate in Fig. 5.28(a). If both A and B are a logic **0** or low voltage, transistors M_1 and M_2 are off while transistors M_3 and M_4 are on. Under this condition a low resistance path exists between V_{DD} and the output voltage V_{OUT}. For any other logical input either transistor M_1, transistor M_2, or both

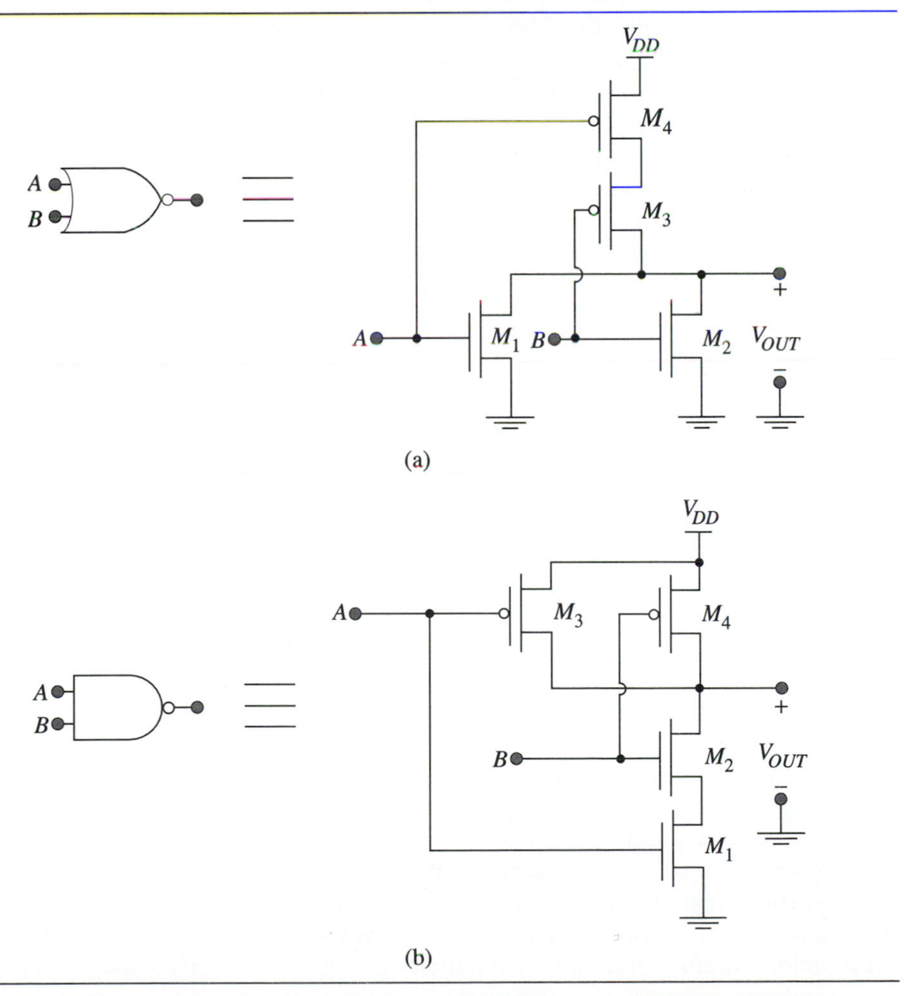

Figure 5.28 Logic gates and CMOS implementation (a) NOR gate (b) NAND gate.

have a low resistance path from the output voltage V_{OUT} to ground while one or both of the p-channel transistors M_3 and M_4 are off. This circuit satisfies our definition of a NOR gate. It has no DC path between the power supply and ground, and the output voltage is always connected to one of the power supplies through a low on-resistance transistor.

Figure 5.28(b) shows a circuit schematic for a NAND gate. If both logic inputs A and B are high, transistors M_1 and M_2 are on while transistor M_3 and M_4 are off. Under this condition V_{OUT} will be equal to 0 V through the low resistance path of transistors M_1 and M_2. With any other logical input A and B, one of the p-channel transistors M_3 or M_4 will be on while one or both of the n-channel transistors M_1 and M_2 will be off. Under this condition the output voltage V_{OUT} will be tied to the power supply V_{DD} through one or both of the p-channel transistors. This circuit performs the logical operation of the NAND function. It also has no DC path between the power supply and ground and the output voltage is connected to either the power supply or ground, through a transistor in the triode region which has a low on-resistance.

As stated in Section 5.1, using DeMorgan's Laws allows us to synthesize any Boolean function with the inverter and either the NOR or NAND gate. The NAND gate is the preferred implementation since the series transistors are n-channel rather than p-channel. Recall in the CMOS inverter that the p-channel transistor had a W/L which was twice that of the n-channel transistor. This provided equal delay times for both the low-to-high and high-to-low transitions. Qualitatively speaking, when transistors are put in series, the W/L ratio must be increased over that of a single transistor. Since the electron mobility is approximately twice that of the hole mobility in MOS transistors, one can place two n-channel transistors in series with approximately the same W/L as the p-channel transistors. When comparing the NOR gate to the NAND gate implementation, we see that the p-channel transistors are in series. Their W/L will have to be nearly four times larger than the n-channel transistors, which increases the overall logic gate area. This qualitative argument is the reason most combinational logic gates in CMOS are composed of NAND and inverter gates to form a variety of Boolean functions.

5.5.2 Static Analysis of CMOS NAND Gate

In this section we will analyze the CMOS NAND gate and determine how the midpoint voltage V_M depends on the transistor characteristics. We have chosen the NAND gate for this analysis. However, a similar strategy can be applied to the NOR gate.

The midpoint voltage for the CMOS NAND gate depends on which input condition is presented to the circuit. First, we will study the case where the input voltage presented to both inputs A and B are simultaneously switched from a logic **0** to a logic **1**. Since we define V_M as the input voltage where the output voltage is equal to V_M, we have set the output voltage to that value. We will use the circuit in Fig. 5.29(a) to calculate V_M.

To begin, we assume that p-channel transistors M_3 and M_4 are in their constant current region. Next, we can guess that transistor M_2 is saturated while transistor M_1 is in the triode region. To test this hypothesis, look at the transistor characteristics for the circuit in Fig. 5.29(b). Transistor M_1 has a gate-source voltage that is equal to V_M. This yields a drain current characteristic larger than device M_2 since its gate-source

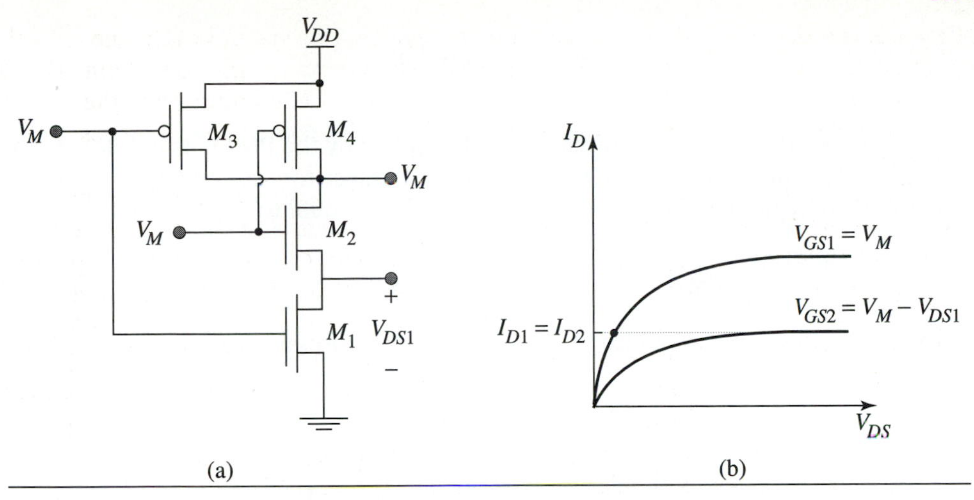

➤ Figure 5.29 (a) CMOS NAND gate circuit for analysis with both inputs simultaneously switching from logic **0** to logic **1**. Input voltage is shown as V_M. (b) NMOS transistor characteristics for M_1 and M_2. Device M_2 is saturated while device M_1 is in the triode region.

voltage is reduced from V_M by the drain-source voltage across device M_1. The mathematical relationship between these voltages is given by

$$V_{GS2} = V_M - V_{DS1} < V_{GS1} = V_M \tag{5.43}$$

Since these transistors are in series, they must have the same current value. Therefore, the current through transistor M_1 and M_2 is equal only when transistor M_1 goes into the triode region of operation. Now that the operating regions of the transistors have been established, we can analyze the circuit to determine a value for V_M. We begin the analysis by writing the drain current for transistor M_2 as

$$I_{D2} = \frac{k_n}{2}(V_M - V_{Tn} - V_{DS1})^2 \tag{5.44}$$

and the drain current for transistor M_1 as

$$I_{D1} = k_n\left(V_M - V_{Tn} - \frac{V_{DS1}}{2}\right)V_{DS1} \tag{5.45}$$

We neglected channel length modulation in both Eq. (5.44) and (5.45). The drain-source voltage across M_1 can be found from Eq. (5.44) yielding

$$V_{DS1} = V_M - V_{Tn} - \sqrt{\frac{2I_{D2}}{k_n}} \tag{5.46}$$

Substituting Eq. (5.46) into Eq. (5.45) and solving for V_M yields

$$V_M = V_{Tn} + 2\sqrt{\frac{I_{D2}}{k_n}} \tag{5.47}$$

We now have V_M in terms of the current supplied by the PMOS transistors to the NMOS transistors. Since we assumed both PMOS transistors are on and in their constant-current regions, we can write that current I_{D2} is equal to

$$I_{D2} = (I_{D3} + I_{D4}) = k_p(V_{DD} - V_M + V_{Tp})^2 \tag{5.48}$$

Substituting this expression into Eq. (5.47) yields an expression for the midpoint voltage given by

static CMOS NAND gate trip point V_M

$$V_M = \frac{V_{Tn} + 2\sqrt{\frac{k_p}{k_n}}(V_{DD} + V_{Tp})}{1 + 2\sqrt{\frac{k_p}{k_n}}} \tag{5.49}$$

It is interesting to compare the equation for V_M for the static CMOS NAND gate to the CMOS inverter given in Eq. (5.20). The only difference is the factor of two multiplying the term $\sqrt{k_p/k_n}$. The value for k_p is two times that for the CMOS

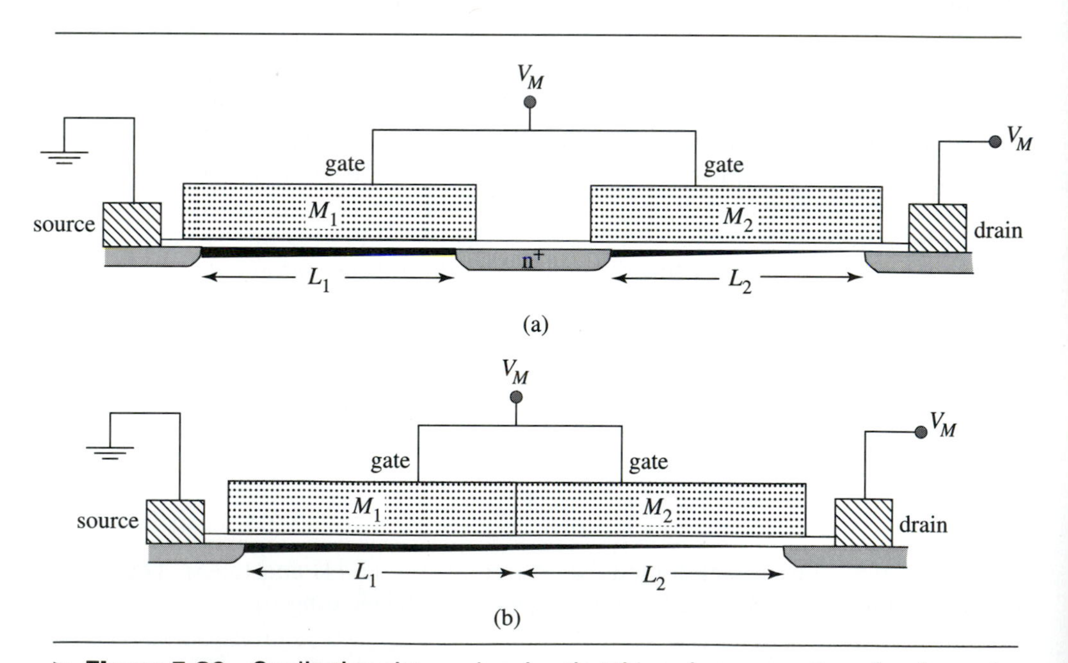

➤ **Figure 5.30** Qualitative picture showing that the series connection of n-channel transistors in the CMOS NAND gate can be thought of as a single n-channel transistor with twice the channel length. (a) Two series transistors with their gate voltage equal to V_M. (b) Equivalent transistor for the two series transistors with the composite length $L = L_1 + L_2$.

inverter since the two p-channel transistors in parallel can be considered as one p-channel transistor with twice the width, increasing k_p by a factor of two.

The series combination of n-channel transistors can be thought of as a single n-channel transistor with twice the length. A qualitative picture in Fig. 5.30 with the saturated transistor M_2 in series with M_1 operating in the triode region demonstrates this fact. The n$^+$ region in between M_1 and M_2 in Fig. 5.30(a) is just connecting the two transistors and from an electrical point of view can be considered negligible since it has a much lower resistance than the channel. An equivalent composite transistor is shown in Fig. 5.30(b) where the channel length is the sum of the two channel lengths $L_1 + L_2$. Since the series combination of n-channel transistors effectively reduces k_n by a factor of 2, the ratio k_p/k_n is increased by a factor of 4.

Given that the threshold voltages of the n- and p-channel transistors are equal to ±1 V respectively and V_{DD} is equal to 5 V, we find V_M is equal to 3 V when k_n equals k_p and is equal to 2.75 V when k_n equals $2k_p$. It appears that we must continue to increase k_n with respect to k_p to drive the midpoint voltage down towards 2.5 V. However, we must consider another switching condition when only one of the two p-channel transistors switches from a high voltage to a low voltage and one of the n-channel transistors does the same thing

Consider Fig. 5.31(a) where only transistors M_2 and M_4 are switching. Under this condition transistor M_1 has a higher gate voltage than in the previous case. It remains in the triode region but the resistance of this device is decreased. Therefore, its drain-source voltage is decreased as shown in Fig. 5.31(b). The current in M_2 is larger since its gate-source voltage will be higher than in the previous case.

Turning our attention to the p-channel transistors, we see that one of the transistors is off. Hence, the p-channel current will be lower than the previous case. Since the n-channel transistor current is higher and the p-channel current is lower than the previous case, the midpoint voltage V_M will be lower than when both inputs are switching on. Using the same typical values we used to evaluate the midpoint voltage with simultaneous switching, we find that V_M is decreased by approximately 0.3 to 0.5 V

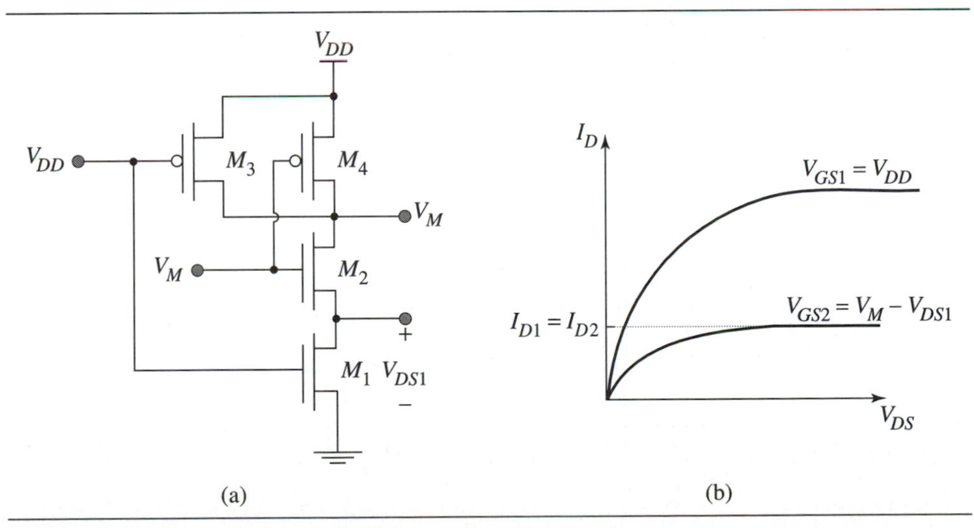

➤ Figure 5.31 (a) CMOS NAND gate circuit for analysis with only one input switching. (b) NMOS transistor characteristics for M_1 and M_2. Device M_2 is saturated while device M_1 is in the triode region.

when only one of the transistors is switched compared to simultaneously switching. Therefore, a value in which k_n is equal to $2k_p$ puts the midpoint switching around 2.5 V.

It should be noted that this hand analysis performed on the NAND gate did not include any backgate effect from transistor M_2. This would further complicate the analysis and make the algebra intractable.

Now that we have calculated V_M, we need to calculate the voltage gain in order to find the noise margin. The voltage gain for the NAND gate is approximately the same as the CMOS inverter, namely, $A_v = -10$. Transistor M_1, which is in the triode region, acts like a resistor between the source of transistor M_2 and ground. As you will see later in Chapter 8, this **degeneration** resistor degrades the transconductance of transistor M_2 by a factor $1+g_m r_{o2}$. However, the output resistance of transistor M_2 is increased by this same factor. Hence, the voltage gain of the composite is approximately the same.

In summary, a CMOS NAND gate has a midpoint voltage approximately centered between V_{DD} and ground when k_n is equal to $2k_p$. The magnitude of the voltage gain of this circuit is usually greater than $|-10|$, which yields sufficient noise margins.

➤ EXAMPLE 5.7 Static CMOS NAND Gate Spice Analysis

Use SPICE to study the transfer functions and noise margins of a CMOS NAND gate (Fig. Ex 5.7A) designed with $k_n = 2k_p$. It should be noted that SPICE will take into account the backgate effect and thus give a view of the accuracy of our hand calculations.

Simulate the transfer function of the NAND gate in Fig. Ex 5.7A with (W/L) of each device equal to 12/1.5. Note that this is the condition for $k_n = 2k_p$. Run all three conditions where only: A switches; only B switches; A and B switch simultaneously. Use $V_{Tn} = 1\text{ V}$, $V_{Tp} = -1\text{ V}$, $\mu_n C_{ox} = 50\ \mu\text{A/V}^2$, $\mu_p C_{ox} = 25\ \mu\text{A/V}^2$, $\gamma_n = \gamma_p = 0.6\text{ V}^{\frac{1}{2}}$, and $\phi_n = -\phi_p = 0.8\text{ V}$. Let $\lambda_n = \lambda_p = 0\text{ V}^{-1}$ so you can accurately see V_M on the simulation.

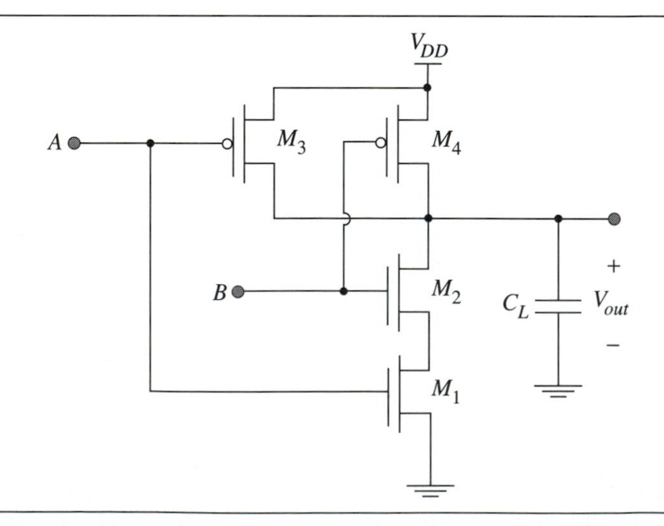

➤ **Figure Ex5.7A**

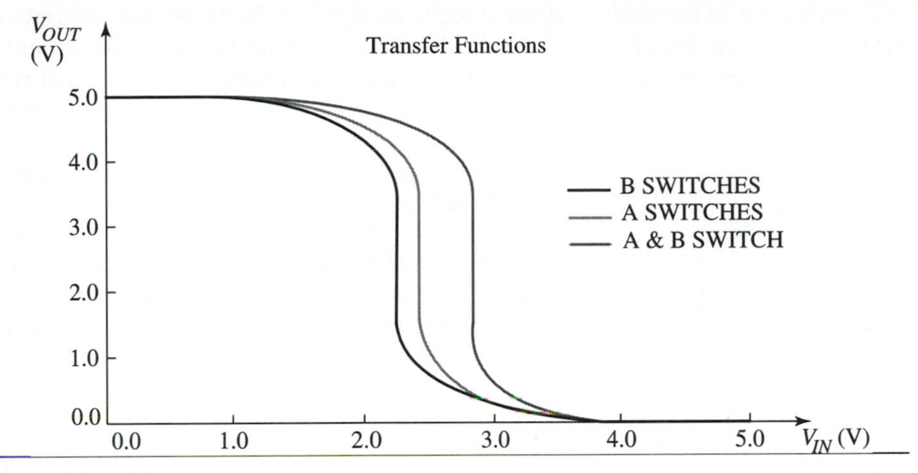

➤ **Figure Ex5.7B**

5.5.3 Transient Analysis of CMOS NAND Gate

The transient analysis for the CMOS NAND gate should be performed under worst-case conditions. The low-to-high transition should be considered when only one of the p-channel transistors is switching. Under this condition the current through the p-channel transistor is equal to

$$-I_{Dp} = \frac{k_p}{2}(V_{DD} + V_{Tp})^2 \tag{5.50}$$

The total capacitance this gate must drive is calculated in the same manner as we presented for the inverter analysis. Calculate the gate capacitance this NAND gate is driving, namely C_G. and the parasitic capacitance, C_p on the output node due to wiring and the drain-bulk depletion regions. As before, we define the transition time from low-to-high to be the time it takes for the output voltage to traverse from 0 to $V_{OH}/2$. With these facts in mind, the low-to-high transition time is given by

$$t_{PLH} = \frac{(C_G + C_P)\,V_{OH}/2}{\frac{k_p}{2}(V_{DD} + V_{Tp})^2} \tag{5.51}$$

To calculate the high-to-low transition time, consider the effect of the series connected NMOS transistors. From our previous discussion, we know that transistor M_2 is in the constant-current region while transistor M_1 is in the triode region. We can write the current equations for devices M_1 and M_2 as

$$I_{D1} = k_n\left(V_{DD} - V_{Tn} - \frac{V_{DS1}}{2}\right)V_{DS1} \tag{5.52}$$

and

$$I_{D2} = \frac{k_n}{2}(V_{DD} - V_{Tn} - V_{DS1})^2 \tag{5.53}$$

As we saw in the static analysis of the CMOS NAND gate, the series combination of n-channel transistors have a current that is equal to that of one composite transistor whose length is twice the length of the individual transistors. Using this simplification we write

$$I_{D2} = \frac{k_n}{4}(V_{DD} - V_{Tn})^2 \tag{5.54}$$

We should try to make the current through the NMOS transistors the same as the current that flowed through the p-channel transistors during the low-to-high transition so that we have symmetrical transition times. By equating Eq. (5.50) and (5.54), we arrive at a relationship between k_n and k_p given in

$$\frac{k_n}{4}(V_{DD} - V_{Tn})^2 = \frac{k_p}{2}(V_{DD} + V_{Tp})^2 \tag{5.55}$$

Assuming the magnitude of the threshold voltage for both the n-channel and p-channel transistors is the same, we find that k_n should be approximately $2k_p$, so that $t_{PHL} = t_{PLH}$.

5.5.4 Transistor Sizing for CMOS Logic Gates

As described in the previous section, the CMOS NAND gate analysis, although algebraically more complex, results in a simple relationship between k_n and k_p in order to have the logic threshold nearly halfway between the power supply and ground and symmetrical transition times. In both cases, k_n should equal $2k_p$. For NAND structures with M inputs, a good first design estimate is that k_n equal $M \times k_p$ in order to have symmetrical transition times and reasonably symmetrical noise margins.

Because the mobility of holes in the PMOS transistors is approximately one-half that of electrons, the W/L of the p-channel transistor must be twice as large as an n-channel transistor in order to have k_n equal k_p. The result is that for an M-input NAND gate one should size the n- and p-channel transistors with the relationship given in

CMOS NAND gate sizing with *M* inputs

$$\boxed{\left(\frac{W}{L}\right)_n = \frac{M}{2}\left(\frac{W}{L}\right)_p} \tag{5.56}$$

A similar analysis can be carried out involving the NOR gate. In general, since the p-channel transistors are in series for the NOR gate, one requires that k_p equal $M \times k_n$ for symmetrical transition times and noise margins. By accounting for the difference in the mobility of holes and electrons, we can write a relationship between the W/L ratio of the p-channel and n-channel devices for an M-input NOR gate as

CMOS NOR gate sizing with *M* inputs

$$\boxed{\left(\frac{W}{L}\right)_p = 2M\left(\frac{W}{L}\right)_n} \tag{5.57}$$

It can be clearly seen that NOR gates require very large p-channel transistors, and hence, the area of a NOR gate is generally much larger than a NAND gate. This fact is the reason that NAND gates are preferred when designing with static CMOS logic.

DESIGN EXAMPLE 5.8 CMOS NAND Gate Design

Design a CMOS NAND gate shown in Fig. Ex5.8 with symmetrical transition times equal to 300 ps. The load capacitance is made up of 50 fF from input gates, 50 fF from input wiring, and the depletion capacitance from the drain-bulk junction of the NAND gate itself. V_{DD}=5 V. Verify your design with SPICE.

MOS Device Data

$\mu_n C_{ox} = 50\ \mu A/V^2$, $\mu_p C_{ox} = 25\ \mu A/V^2$, $V_{Tn} = -V_{Tp} = 1V$, $C_{ox} = 2.3\ fF/\mu m^2$,
$C_{Jn} = 0.1\ fF/\mu m^2$, $C_{Jp} = 0.3\ fF/\mu m^2$, $C_{JSWn} = 0.5\ fF/\mu m$, $C_{JSWp} = 0.35\ fF/\mu m$,
$L_{diffn} = L_{diffp} = 6\ \mu m$

SOLUTION

Begin by finding the (W/L) for the p-channel devices required to guarantee $t_{PLH} = 300$ ps. We need to know the width of the p-channel device to calculate the load capacitance, *but* that is what we are trying to calculate. As with many design problems, an iterative approach is necessary.

Assuming a typical device width of 20 μm, the drain-bulk capacitance of both p-channel devices plus one n-channel device can be estimated to be about 150 fF. (A good initial guess comes from practice and experience.) That makes the total load capacitance, C_L, equal to about 250 fF. From Eq. (5.51),

$$t_{PLH} = \frac{C_L(V_{OH}/2)}{\frac{k_p}{2}(V_{DD} + V_{Tp})^2}$$

yielding $k_p = 313\ \mu A/V^2$, and therefore, $(W/L)_p = 18.75/1.5$. Let's size the p-channel device at 20/1.5. To make the transition times symmetrical we set $k_n = 2k_p$, so the n-channel device is sized at 20/1.5.

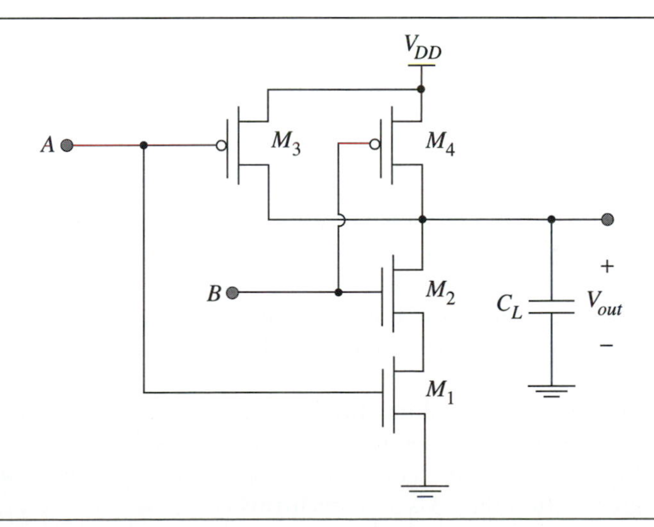

➤ Figure Ex5.8

To calculate the load capacitance we first find C_{DB}.

$$C_{DB} = W_n L_{diffn} C_{Jn} + 2W_p L_{diffp} C_{Jp} + (W_n + 2L_{diffn}) C_{JSWn} + 2(W_p + 2L_{diffp}) C_{JSWp}$$

Plugging in the values we find

$$C_{DB} = 12 \text{ fF} + 72 \text{ fF} + 16 \text{ fF} + 22 \text{ fF} = 122 \text{ fF}.$$

Therefore

$$C_L = C_G + C_{WIRE} + C_{DB} = 50 \text{ fF} + 50 \text{ fF} + 122 \text{ fF} = 222 \text{ fF}.$$

Substituting our value for C_L, we get $t_{PLH} = t_{PHL} = 208$ ps.

Refering to Fig. Ex 5.8, SPICE simulation shows us that our particular sizing of the devices yields a t_{PLH} of 240 ps, if M_2 and M_4 are switched from high-to-low, and the gates of M_1 and M_3 are at V_{DD}. If M_1 and M_2 are switched from high-to-low and the gates of M_2 and M_4 are at V_{DD}, SPICE yields a $t_{PLH} = 340$ ps. Under this condition, device M_2 is on and in the triode region during the whole transition and its total gate capacitance is added to the load capacitance. For $(W/L)_2 = 20/1.5$, $C_G = 69$fF. This increases C_L to 291 fF and consequently increases the t_{PLH} to approximately 275 ps for the hand calculation.

In this example we are seeing a second order effect due to the gate capacitance of the NAND gate transistor which our hand calculation neglected. Furthermore, we see that the effect is data dependent. Further iteration in SPICE is required to meet the specification. Setting $(W/L)_n = (W/L)_p = 40/1.5$ yields a t_{PLH} in SPICE of approximately 300 ps.

It should be noted that during the high-to-low transient either device M_2 or M_3 is on and therefore, t_{PHL} turns out to be approximately 300 ps for both switching conditions.

5.6 DYNAMIC LOGIC

Oftentimes in digital circuit design we require a large fan-in logic gate. These gates are difficult to implement using the static CMOS approach since this approach requires $2M$ devices to implement an M input NOR or NAND gate. In addition, a series chain of M devices will cause the logic gate to operate slowly. In this section we will introduce dynamic logic which is particularly useful for the design of large fan-in gates.

Static CMOS logic gates perform the Boolean logical function with both the n- and p-channel devices.To see this logic redundancy, consider the static CMOS NOR gate shown in Fig. 5.32. If the p-channel transistors were replaced by an appropriate resistor, the logic function that this gate performs is

$$Q = \overline{A + B} \tag{5.58}$$

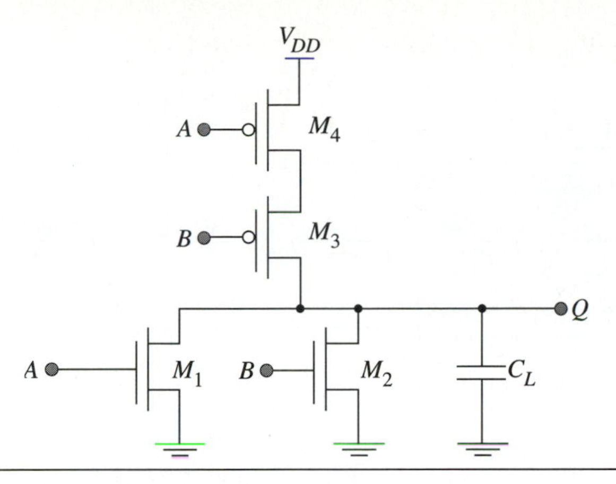

➤ **Figure 5.32** CMOS NOR gate showing redundant logic.

The modified circuit performs this logic, since if the input to either NMOS transistor is a high gate voltage (logic **1**) the output is discharged to a low voltage (logic **0**). If the n-channel transistors were replaced by a resistor and only the p-channel transistors were used to perform the logic, the resulting logic function is

$$Q = \overline{A}\,\overline{B} \tag{5.59}$$

The modified circuit performs this logic since both p-channel devices must have a low gate voltage (logic **0**) to charge the output to a high voltage (logic **1**). By DeMorgan's Laws, these logic functions (Eqs. (5.58) and (5.59)) are equivalent. In other words, the static NOR gate has redundant logic functions. Both the n-channel and p-channel transistors are producing the same Boolean function. In principle, we should be able to eliminate either the n- or p-channel network and perform the desired logic function with considerably fewer transistors.

5.6.1 Dynamic Logic - Basic Operation

We will discuss the basic operation of a dynamic logic gate by using an M input NOR gate as an example. Since the NOR gate has parallel n-channel transistors and series p-channel transistors, we will eliminate the redundant p-channel network to build the dynamic logic gate. The circuit implementation of this gate is shown in Fig. 5.33(a).

We have replaced the p-channel network with a single p-channel transistor whose gate is tied to a clock signal shown in Fig. 5.33(b). The n-channel NOR logic function is implemented with NMOS transistors in parallel. A series n-channel transistor is connected between the logic function and ground, while its gate is connected to the clock signal. The operation of this dynamic NOR gate can be divided into two phases—pre-charge and evaluate.The state of the clock signal ϕ determines whether the circuit is in the pre-charge or evaluate condition.

When the clock signal is at 0 V, the p-channel transistor is on and pre-charges the output load capacitance to the power supply voltage V_{DD}. During this pre-charge condition, the series n-channel transistor M_E is off. Therefore, no DC current can flow

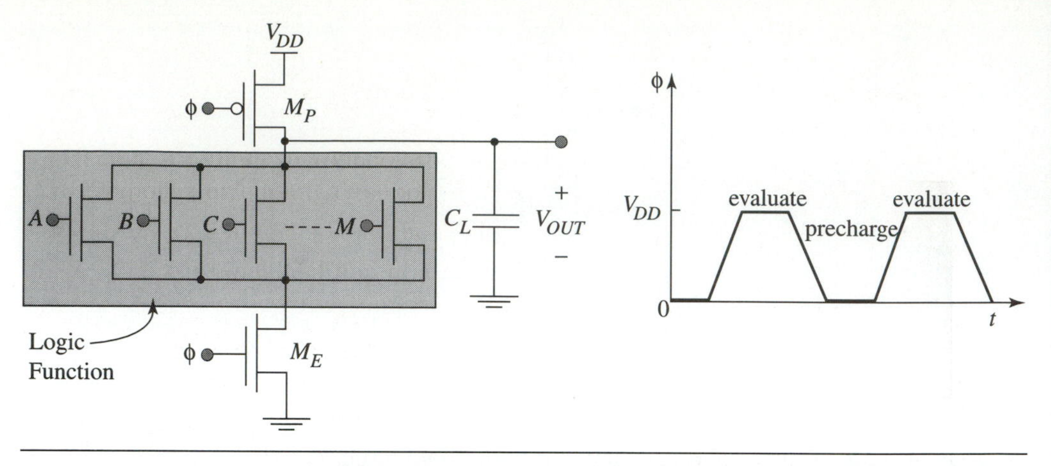

➤ **Figure 5.33** (a) Circuit implementation of an *M*-input NOR gate using dynamic logic. (b) Voltage waveform for clock ϕ. During $\phi = 0$ V, the output is precharged high. During $\phi = V_{DD}$, the logic function is evaluated.

from the power supply to ground, regardless of the logic values presented to the n-channel transistors in the logic network.

During the pre-charge time, valid logic levels for inputs A through M should be established on the gates of the n-channel transistors in the logic network. When the clock signal ϕ goes to V_{DD}, the p-channel transistor, M_P turns off while the series n-channel transistor M_E turns on to begin the evaluation of the logic function. If any of the n-channel transistors in the logic network have a logic **1** input, then the output Q will be discharged to ground (logic **0**). However, if none of the n-channel transistors in the logic network have a valid logic **1** at the inputs, the output will remain pre-charged to the power supply, V_{DD}. Therefore, we have implemented the NOR function in one full clock cycle, which was divided into a pre-charge and evaluation phase.

A number of important statements can be made regarding dynamic logic.

1. N-channel or p-channel transistors can be used to implement the logic function. One should choose between n- or p-channel transistors to minimize the number of series connections in the logic network. For example, if an M input NAND function were desired, a p-channel logic network would have M parallel p-channel transistors and would be preferred over a series connection of M n-channel transistors.
2. The number of transistors to build an M input logic gate is equal to $M + 2$ rather than $2M$ as in the static logic case.
3. As we will see in the following section, the noise margin of dynamic logic does not depend on the transistor ratios W/L.
4. The dynamic logic gate has no DC path to ground and therefore, consumes only dynamic power.

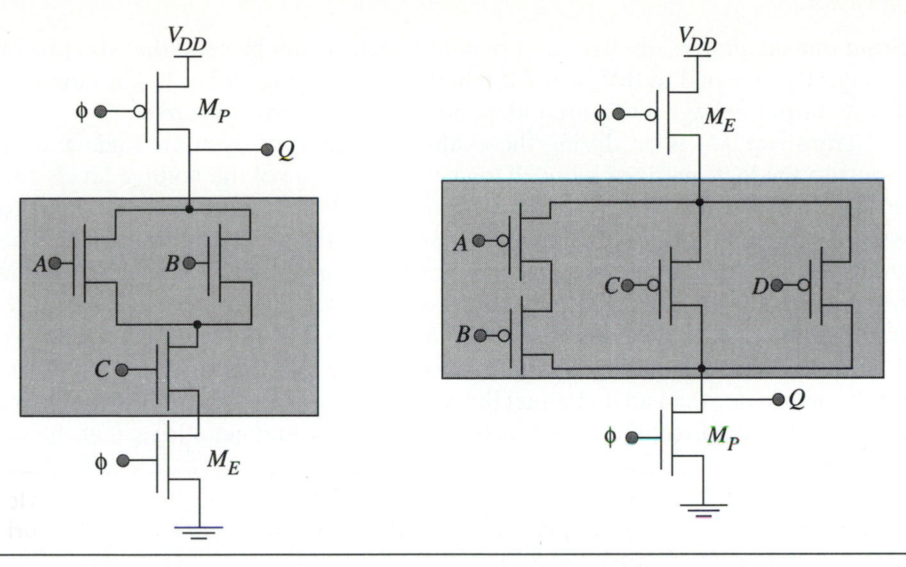

➤ **Figure 5.34** Logic functions implemented using dynamic logic. (a) N-channel logic network $Q = \overline{(A + B)\,C}$. (b) P-channel logic network $Q = \overline{A}\overline{B} + \overline{C} + \overline{D}$.

The logic functions that can be implemented in dynamic logic are not limited to a NAND or NOR gate. Complex logic functions can easily be implemented using n- or p-channel transistors for the logic network. In Fig. 5.34(a), an n-channel logic network is implementing the function given by

$$Q = \overline{(A + B)\,C} \tag{5.60}$$

In Fig. 5.34(b), a p-channel logic network is shown that is implementing the logic function given by

$$Q = \overline{A}\overline{B} + \overline{C} + \overline{D} \tag{5.61}$$

In the p-channel implementation, the precharge and evaluate waveforms are switched so Q is precharged to ground through M_P when ϕ is high and evaluated when ϕ is low.

5.6.2 Dynamic Logic Voltage Transfer Function

In this section we will quantitatively discuss the voltage transfer function of a dynamic CMOS logic gate. We will concentrate on an n-channel network shown in Fig. 5.35(a) for this discussion. A similar treatment can be used when a p-channel logic network is analyzed. The analysis of the voltage transfer function for dynamic logic is quite different than that of static logic since it is a time-dependent transfer function. This time-dependent voltage transfer function is the reason for the name *dynamic logic*.

To begin this discussion, let's assume an ideal network with no leakage currents so the voltage across capacitor C_L does not change unless driven by the transistor.

From our qualitative discussion of dynamic logic, it can be seen that the low output voltage V_{OL} is equal to 0 V, while the high output voltage level V_{OH} is equal to V_{DD}. These output voltage levels are independent of the transistors sizes.

Transistor M_E is on during the evaluation phase of dynamic logic gate. If the inputs to the logic network were all logic **0** and any one of the voltage levels got near a threshold voltage of the n-channel transistor, the charge stored on capacitor C_L would begin to leak through the logic network and the evaluation transistor. This fact shows that if the input voltage gets higher than a threshold voltage on any n-channel transistor in the logic network, the output voltage will incorrectly yield a logic **0**. This statement implies that V_{IL} is equal to the n-channel threshold voltage V_{Tn}.

Rather than calculating V_M and V_{IH}, we can see that if any input to the logic network was greater than an n-channel threshold voltage and we waited enough time for the output to be discharged from its high level, we would get a logic **0** at the output. We can write

$$\boxed{V_{IH} = V_{IL} = V_M = V_{Tn}} \tag{5.62}$$

n-channel dynamic logic noise margin
$NM_H = V_{DD} - V_{Tn}$
$NM_L = V_{Tn}$

The corresponding voltage transfer function is shown in Fig. 5.35(b). This voltage transfer function shows a strong asymmetry in noise margin. NM_L is equal to V_{Tn}, whereas the high level noise margin NM_H is equal to $V_{DD} - V_{Tn}$. However, this asymmetry is fortunate since it is the high value that is most difficult to maintain for dynamic logic. To understand this fact, recall that when a high value is input to the logic network, a low resistance path exists between the output voltage and ground. However, when the inputs to the logic network are a logic **0**, then the output node is floating and its voltage can change by leakage currents or capacitive coupling from other signals.

Leakage currents occur in the reverse-biased drain-well junction of the p-channel transistor and in the drain-bulk junction for the n-channel transistor. These leakage currents cannot be designed to cancel each other precisely. Hence, the charge stored on the output capacitor C_L can drift and cause an error.

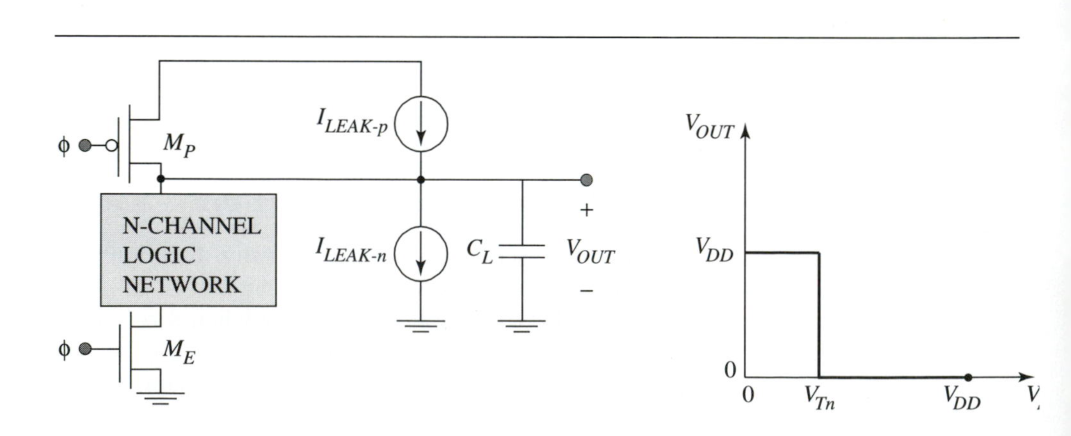

➤ Figure 5.35 Dynamic n-channel network logic gate. (a) Circuit showing possible leakage currents. (b) Voltage transfer function.

5.6.3 Dynamic Logic Transient Analysis

In the following transient analysis of a dynamic logic gate, we will assume an n-channel logic function implementation as shown in Fig. 5.35(a). During the pre-charge phase the output is charged to V_{DD}. This phase is the low-to-high transition time for a dynamic logic gate t_{PLH}. To calculate t_{PLH} recognize that the p-channel device is in the constant-current region and its current is

$$-I_{Dp} = \frac{k_p}{2}(V_{DD} + V_{Tp})^2 \tag{5.63}$$

As with the static inverter, we can assume that the p-channel device will charge the load capacitance with this constant current until the output voltage reaches approximately $V_{OH}/2$.

The load capacitance C_L, in the dynamic logic gate is estimated in the same manner as shown in the static inverter. First calculate the gate capacitance that the logic gate is driving C_G, and add the parasitic capacitance C_P, due to wiring and the diffusions on the output node. All the drain diffusions from the n-channel transistors connected to the output node plus the drain diffusion from the pre-charge device M_P should be included in this parasitic capacitance calculation.The pre-charge time is given by

$$t_{PLH} = \frac{C_L V_{OH}/2}{\frac{k_p}{2}(V_{DD} + V_{Tp})^2} \tag{5.64}$$

dynamic logic propagation delay

The high-to-low transition time for the CMOS dynamic logic gate should be calculated under a worst-case condition as we did with the CMOS static gates. This condition corresponds to the case where only one n-channel transistor in the logic function group is on and is in series with the n-channel transistor M_E. It can be seen that this is the same structure as the CMOS NAND gate analyzed in Section 5.5. Assuming $(W/L)_E = (W/L)_{1,\ldots,n}$ and using Eq. (5.54) to calculate the current I_{Dn} through the two series n-channel transistors, we estimate the high-to-low transition to be equal to

$$t_{PHL} = \frac{(C_L V_{OH}/2)}{\frac{k_n}{4}(V_{DD} - V_{Tn})^2} \tag{5.65}$$

In general, dynamic logic tends to run faster than CMOS static logic since a fewer number of transistors are required to implement a similar function. This fact leads to a smaller parasitic capacitance that needs to be charged and discharged by the transistors. In addition, even though the logic in the dynamic gate cannot be utilized during the pre-charge time, very often the overall digital system can be designed in such a way that the pre-charge time coincides with other system functions.

The major drawback for dynamic logic is that it must operate with a minimum frequency. The **minimum refresh time** is dictated by the amount of leakage current on the output node. Typical numbers for minimum times between precharging are on the order of 1ms. In many digital systems this does not present a problem. However, in a digital system where it is required to stop and save state to inspect internal logic nodes, dynamic logic is unacceptable.

DESIGN EXAMPLE 5.9 4-Input NOR Gate Design with Dynamic Logic

Design a 4-input NOR gate using dynamic logic that drives a static inverter with $(W/L)_p = 12/1.5$ and $(W/L)_n = 6/1$ as shown in Fig. Ex.5.9. The delay time t_{PLH} need not be faster than 500 ps since there is approximately 1ns between precharging pulses. We have chosen the p-channel precharge device size to be the minimum size of (4.5/1.5). t_{PHL} should be as fast as possible but the total active device area should not exceed 1000 μm^2. Use the device data below.

MOS Device Data

$\mu_n C_{ox} = 50\ \mu A/V^2$, $\mu_p C_{ox} = 25\ \mu A/V^2$, $V_{Tn} = -V_{Tp} = 1V$, $C_{ox} = 2.3\ fF/\mu m^2$,

$C_{Jn} = 0.1 fF/\mu m^2$, $C_{Jp} = 0.3\ fF/\mu m^2$, $C_{JSWn} = 0.5\ fF/\mu m$, $C_{JSWp} = 0.35\ fF/\mu m$,

$L_{diffn} = L_{diffp} = 6 \mu m$

SOLUTION

The total active area of each device is estimated as $W(L+2L_{diff})$, thus the p-channel precharge device occupies about 60 μm^2. That leaves 940 μm^2 to be divided equally among the five other n-channel transistors. This allows each device to be about 14 μm wide. In the worst-case scenario, only one n-channel network transistor will be on. It looks like a NAND gate since it is in series with the evaluate transistor M_E. From Eq. (5.54) the current in a series combination of equally sized n-channel devices is

$$I_{Dn} = \frac{k_n}{4}(V_{DD} - V_{Tn})^2 = \frac{1}{4} \cdot \frac{14}{1.5}\left(50\frac{\mu A}{V^2}\right)(5V - 1V)^2 = 1.86\ mA$$

To find t_{PHL}, first calculate C_L. Assuming wiring is negligible,

$$C_P = 4(W_n L_{diffn} C_{Jn} + (W_n + 2L_{diffn})C_{JSWn}) + W_p L_{diffp} C_{Jp} + (W_p + 2L_{diffp})C_{JSWp}$$

$$C_P = 4(8.4fF + 13fF) + 8.1fF + 5.8fF = 99.5fF.$$

$$C_G = C_{IN} = C_{ox}(W \cdot L)_p + C_{ox}(W \cdot L)_n$$

$$C_G = 2.3(12 \times 1.5)\ fF + 2.3(6 \times 1.5)\ fF = 125\ fF$$

$$C_L = C_G + C_P = 125\ fF + 100\ fF = 225\ fF$$

Using Eq. (5.65)

$$t_{PHL} = (C_L V_{OH}/2)/I_{Dn} = (162\ fF \cdot 2.5\ V)/1.86\ mA = 220\ ps$$

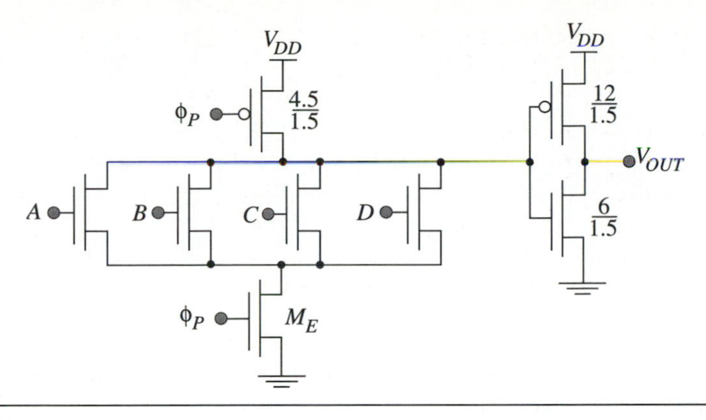

➤ **Figure Ex5.9**

➤ 5.7 PASS TRANSISTOR LOGIC

In this section, we will describe a method to implement logic gates by using a network of switches. MOS transistors make excellent switches, therefore, this logic implementation style is widely used in a variety of VLSI chips.

5.7.1 General Concept

To understand the implementation of logic functions using switches, consider the example of the AND function shown in Fig. 5.36(a). The switch is defined as closed when its logic value is a **1** and open when its logic value is a **0**. Given this definition, the output of the logic gate is only a **1** when both A and B have a logic value of **1**. Since

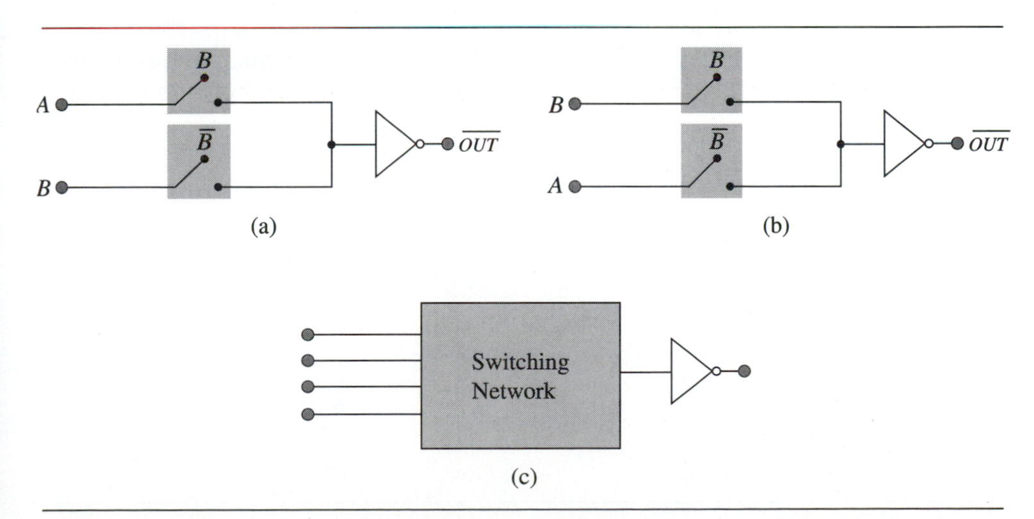

➤ **Figure 5.36** Pass transistor logic: (a) AND gate + inverter; (b) OR gate + inverter; (c) arbitrary switch network + inverter.

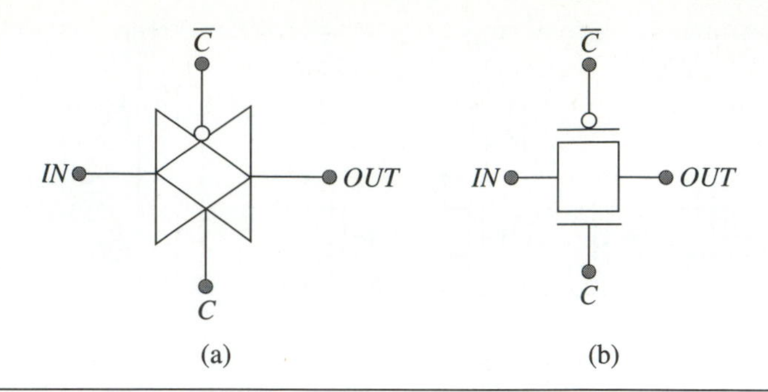

➤ Figure 5.37 (a) CMOS transmission gate. (b) CMOS implementation is a p- and n-channel transistor in parallel.

the switch network is not driven to the supply rails, a static inverter is often inserted to ensure the output signal has a low resistance connection to one of the supplies.

By interchanging inputs *A* and *B*, the same network represents an OR function as shown in Fig. 5.36(b). The output will be equal to a logic **1** in all cases unless both *A* and *B* equal a logic **0**. In general, complex logic functions can be implemented in the switching network with its result buffered by an inverter to drive the next network as shown in Fig. 5.36(c). These complex logic networks are implemented with a minimal number of transistors allowing for fast, area-efficient operation.

5.7.2 Bi-Directional Transmission Gate

To implement a switch with MOS transistors, we use an NMOS and PMOS transistor in a parallel arrangement. The transmission gate acts as a bi-directional switch controlled by the gate signal *C*. When *C* equals logic **1**, the NMOS and PMOS transistors are on allowing the signal to pass through the gate. When C equals logic **0**, both transistors are cutoff creating an open circuit between nodes *A* and *B*. The symbolic representation for a transmission gate is shown in Fig. 5.37(a) and the CMOS implementation is shown in Fig. 5.37(b).

One may ask why is it necessary to place both the n- and p-channel transistors in parallel rather than use a single device. To answer this question, refer to the circuit

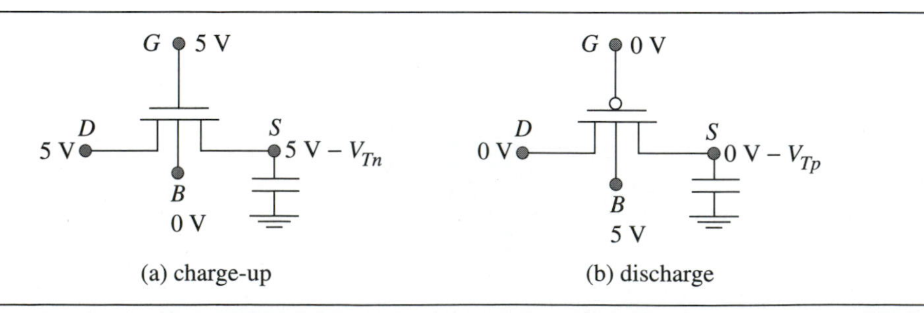

➤ Figure 5.38 Circuit using only one type of transistor as the switch. (a) n-channel charge up circuit. (b) p-channel discharge circuit

in Fig. 5.38(a) where we have a single NMOS transistor attempting to charge up a capacitance to 5 V. With 5 V on the drain and gate, the capacitor will charge to only 5 V minus the n-channel threshold voltage which would be approximately 3.5 V. The additional loss in voltage comes from the increase in threshold voltage from the backgate effect.

The p-channel transistor shown in Fig. 5.38(b) has the opposite effect. If we try to discharge a node by placing 0 V on the gate and 0 V on the drain, we find that the node stops discharging when the source voltage reaches $-V_{Tp}$ which is approximately 1.5 V when the backgate effect is included. The n-channel transistor does a good job of passing low voltages from approximately 0 to 3 V while the p-channel transistor does a good job of passing high voltages from approximately 2 V to 5 V. By placing the two in parallel, we can pass the full range of voltages through the switch.

5.7.3 Static Operation

It is difficult to define noise margin in a similar manner that we defined it for a CMOS static inverter and static logic gates. Instead, we will concentrate on the functionality of the switch. We want the switch to remain off when the voltage on the gate represents a logic **0** and turn on when it represents a logic **1**.

To satisfy this condition the noise margin for the transmission gate is the threshold voltage of the devices. This is a reasonable definition since we want the control voltage to keep the transistors cutoff when we consider the switch open. If the control voltage got slightly higher than the threshold voltage, some current would flow through the transmission gate and possibly upset the node to which it is connected. Specifically,

$$NM_L = V_{Tn} \tag{5.66}$$

$$NM_H = -V_{Tp} \tag{5.67}$$

pass transistor logic noise margin

In general, building complex logic functions with bi-directional transmission gates offers a high speed, area-efficient design with a small penalty in noise margin.

5.7.4 Transient Operation

To understand how to size the n- and p-channel transistors for the transmission gate, we must consider the transient operation of the gate. The discussion begins by considering a single pass gate driven by an inverter and charging a load capacitance as

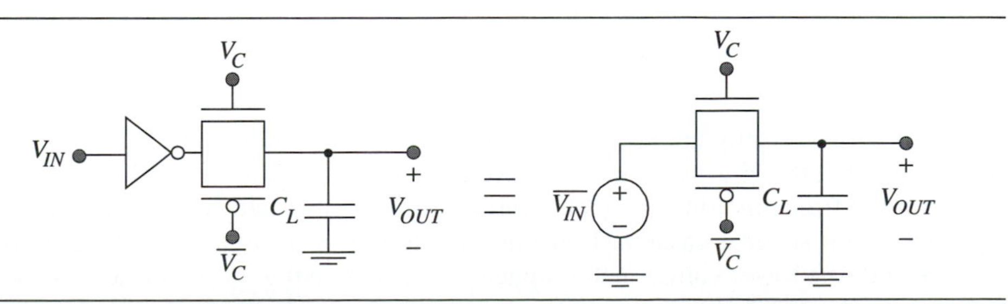

Figure 5.39 Circuit to study transient operation of a CMOS Pass Gate. (a) Inverter driving CMOS Pass Gate. (b) Voltage source modelling output voltage of inverter.

shown in Figure 5.39. We will model the output of the inverter as a voltage source, V_{IN} that has a value of V_{OH} or V_{OL} representing a logic **1** or **0**. The load capacitance C_L is the sum of C_G, the gate capacitances which are being driven, and C_P, the parasitics of the source/drain diffusions and wiring capacitances as explained in the previous section. The goal of this analysis is to find the average current that will charge and discharge the load capacitance and any additional capacitances that are due to the transmission gate.

Consider the time it takes for a low-to-high transition t_{PLH}. For this we consider the input voltage to be equal to V_{OH} which equals the power supply voltage V_{DD}. The output voltage has an initial condition of 0 V, and the control gate is instantaneously turned on by a transition on the n-channel device from 0 V to V_{DD} and on the p-channel device from V_{DD} to 0 V. Since we are interested in a voltage excursion from 0 V to $V_{OH}/2$, we will define the average n-channel current for a low-to-high transition as

$$\overline{I_{DLHn}} = \frac{I_{DLHn}(V_{OUT} = 0\text{ V}) + I_{DLHn}(V_{OUT} = V_{OH}/2)}{2} \tag{5.68}$$

Since the pass transistor has its source and drain interchanged, depending on whether we are considering a high-to-low transition or a low-to-high transition, the backgate of the n-channel transistor cannot be tied to the source and must be connected to the lowest potential, 0 V. During the low-to-high transition the source of the n-channel transistor is the node connected to V_{OUT}. When V_{OUT} is equal to $V_{OH}/2$ the source-to-backgate voltage is $V_{OH}/2$ and the backgate effect will increase the n-channel threshold voltage, V_{Tn}. Assume that the n-channel transistor is saturated for both of these operating points. Hence, the average current is given by

$$\overline{I_{DLHn}} = \frac{\left[\frac{k_n}{2}(V_{DD} - V_{TOn})^2 + \frac{k_n}{2}\left(V_{DD} - \frac{V_{OH}}{2} - V_{Tn}\right)^2\right]}{2} \tag{5.69}$$

where V_{TOn} is the n-channel threshold voltage with $V_{SB} = 0$ V.

Now turning our attention to the p-channel transistor, we see that the backgate must be tied to the highest potential V_{DD}. During the low-to-high transition the source of the p-channel transistor is always at V_{DD}, hence there is no backgate effect. It is assumed that the p-channel transistor is in the constant current region during the transition $V_{OUT} = 0$ V to $V_{OUT} = V_{OH}/2$. Although this is not strictly correct, it is adequate for an initial hand analysis. The p-channel current is approximately constant during the transition and is given by

$$\overline{I_{DLHp}} = \left[\frac{k_p}{2}(V_{DD} + V_{TOp})^2\right] \tag{5.70}$$

Now consider any additional capacitance that should be added to C_L from the transmission gate. We will take into account the self-loading from the gate capacitance of the transmission gate itself, along with the usual load capacitance. Since the n-channel transistor is saturated during the whole transition and the load is on the source end (the lower voltage), the capacitance between the output node and the gate is equal to 2/3 WLC_{ox}. This capacitance appears in parallel with the load capacitor. The p-channel transistor operates in the saturated region and its drain is connected to C_L. The capacitance added to the load from the p-channel device is the overlap capacitance between the drain and gate. The value of this additional capacitance which appears in parallel with the load capacitance is WC_{ov}.

Now that we have calculated the average current through the transmission gate and the capacitance which must be charged through that gate, we can write an approximate expression for the time for a low-to-high transition given by

$$t_{PLH} = (C_{L_{LH}} V_{OH}/2)/(\overline{I_{DLHn}} + \overline{I_{DLHp}}) \quad (5.71)$$

pass transistor logic propagation delay

where

$$C_{L_{LH}} = C_G + C_P + \frac{2}{3}(WL)_n C_{ox} + (W)_p C_{ov} \quad (5.72)$$

A complementary analysis can be performed to find the propagation delay with a high-to-low transition t_{PHL} yielding

$$t_{PHL} = (C_{L_{HL}} V_{OH}/2)/(\overline{I_{DHLn}} + \overline{I_{DHLp}}) \quad (5.73)$$

where

$$I_{DHLn} = \frac{k_n}{2}(V_{DD} - V_{TOn})^2 \quad (5.74)$$

$$I_{DHLp} = \frac{\left[\frac{k_p}{2}(V_{DD} + V_{TOp})^2 + \frac{k_p}{2}\left(V_{DD} - \frac{V_{OH}}{2} + V_{Tp}\right)^2\right]}{2} \quad (5.75)$$

$$C_{L_{HL}} = C_G + C_P + \frac{2}{3}(WL)_p C_{ox} + (W)_n C_{ov} \quad (5.76)$$

➤ EXAMPLE 5.10 Transmission Gate Delay

In this example we will find the propagation delay for a minimum geometry transmission gate driving a single minimum geometry inverter as shown below. Assume the minimum geometry device sizes for the inverter are $(W/L)_n = 3/1.5$ and $(W/L)_p = 6/1.5$. Use these same device sizes for the transmission gate. The MOS device data is shown below.

MOS Device Data

$\mu_n C_{ox} = 50\ \mu A/V^2$, $\mu_p C_{ox} = 25\ \mu A/V^2$, $V_{Tn} = -V_{Tp} = 1$ V, $\gamma_n = \gamma_p = 0.6 V^{\frac{1}{2}}$,

$2\phi_n = -2\phi_p = 0.8V$

$C_{ox} = 2.3$ fF/μm², $C_{Jn} = 0.1$ fF/μm², $C_{Jp} = 0.3$ fF/μm², $C_{JSWn} = 0.5$ fF/μm,

$C_{JSWp} = 0.35$ fF/μm, $C_{ov} = 0.5$ fF/μm, $L_{diffn} = L_{diffp} = 6$ μm

(a) Find t_{PLH} and t_{PHL} for this transmission gate.

SOLUTION

We begin this analysis by solving for t_{PLH} using Eq. (5.71)

$$t_{PLH} = (C_{L_{LH}} V_{OH} / 2) / (\overline{I_{DLHn}} + \overline{I_{DLHp}})$$

From Eq. (5.69) and Eq. (5.70), we can solve for the average current flowing in the n and p-channel transistors of the transmission gate to charge up the input gate capacitance of the inverter.

$$\overline{I_{DLHn}} = \frac{\left[\frac{k_n}{2}(V_{DD} - V_{TOn})^2 + \frac{k_n}{2}\left(V_{DD} - \frac{V_{OH}}{2} - V_{Tn}\right)^2\right]}{2}$$

where

$$V_{Tn} = V_{TOn} + \gamma_n\left(\sqrt{-2\phi_p + \frac{V_{OH}}{2}} - \sqrt{-2\phi_p}\right) = 1.55V$$

and

$$\overline{I_{DLHp}} = \left[\frac{k_p}{2}(V_{DD} + V_{TOp})^2\right]$$

Plugging in our device data we find

$$\overline{I_{DLHn}} = 423\mu A$$

$$\overline{I_{DLHp}} = 800\mu A$$

The load capacitance for the transmission gate is made up of the input gate capacitance of the inverter, C_G, the drain-bulk capacitance of the pass transistors, $C_{DBn} + C_{DBp}$, and the gate capacitance of the pass transistors.

$$C_L = C_G + C_{DBn} + C_{DBp} + \frac{2}{3}(WL)_n C_{ox} + (W)_p C_{ov},$$

$$C_G = (WL)_n C_{ox} + (WL)_p C_{ox} = 10.4\text{fF} + 20.7\text{fF} = 31.1\text{fF}$$

$$C_{DBn} = W_n L_{diffn} C_{Jn} + (W_n + 2L_{diffn}) C_{JSWn} = 1.8\text{fF} + 7.5\text{fF} = 9.3\text{fF}$$

$$C_{DBp} = W_p L_{diffp} C_{Jp} + (W_p + 2L_{diffp}) C_{JSWp} = 10.8\text{fF} + 6.3\text{fF} = 17.1\text{fF}$$

$$C_{L_{LH}} = 31.1\text{fF} + 9.3\text{fF} + 17.1\text{fF} + 6.9\text{fF} + 3.0\text{fF} = 67.4\text{fF}$$

Substituting the value for $C_{L_{LH}}$ and the average currents into the expression for t_{PLH} we find

$$t_{PLH} = 138 \text{ ps}$$

Similarly we repeat this analysis for t_{PHL} using Eq. (5.73) and find

$$t_{PHL} = (C_{L_{HL}} V_{OH}/2) / (\overline{I_{DHLn}} + \overline{I_{DHLp}})$$

From Eq. (5.74) and Eq. (5.75), we can solve for the average current flowing in the n and p-channel transistors of the transmission gate to discharge the input gate capacitance of the inverter.

$$\overline{I_{DHLp}} = \frac{\left[\frac{k_p}{2}(V_{DD} + V_{TOp})^2 + \frac{k_p}{2}\left(V_{DD} - \frac{V_{OH}}{2} + V_{Tp}\right)^2\right]}{2}$$

where

$$V_{Tp} = V_{TOp} + \gamma_p\left(\sqrt{2\phi_n + \frac{V_{OH}}{2}} - \sqrt{2\phi_n}\right) = -1.55 \text{ V}$$

and

$$\overline{I_{DHLn}} = \left[\frac{k_n}{2}(V_{DD} + V_{TOn})^2\right]$$

Plugging in our device data we find

$$\overline{I_{DHLp}} = 423\mu\text{A}$$

$$\overline{I_{DHLn}} = 800\mu\text{A}$$

The load capacitance for the transmission gate for this transition is made up of the *same* input gate capacitance of the inverter, C_G, and drain-bulk capacitance of the pass transistors, $C_{DBn} + C_{DBp}$, added to the gate capacitance of the pass transistors.

$$C_{L_{HL}} = C_G + C_{DBn} + C_{DBp} + \frac{2}{3}(WL)_p C_{ox} + (W)_n C_{ov}$$

The load capacitance for this transition is given by

$$C_{L_{HL}} = 31.1\text{fF} + 9.3\text{fF} + 17.1\text{fF} + 13.8\text{fF} + 1.5\text{fF} = 72.8\text{fF}$$

Substituting the value for C_L and the average currents into the expression for t_{PHL} we find

$$t_{PHL} = 149 \text{ ps}$$

(b) How many minimum geometry inverters can the transmission gate drive and maintain t_{PLH} and $t_{PHL} \leq 250$ ps?

SOLUTION

To find the fan-out for the minimum geometry transmission gate, calculate the additional load capacitance that it can drive while maintaining t_{PHL} and t_{PLH} ≤ 250 ps. Look only at t_{PHL} since it is slightly larger that t_{PLH}. By writing the ratio of the propagation time we calculated, divided by 250 ps and setting that equal to the ratio of the load capacitances, we can determine the additional load capacitance $\Delta C_{L_{HL}}$ that we can drive given $t_{PHL} = 250$ ps.

$$\frac{t_{PHL}}{250\,\text{ps}} = \frac{C_{L_{HL}}}{C_{L_{HL}} + \Delta C_{L_{HL}}}$$

$\dfrac{149\text{ps}}{250\text{ps}} = \dfrac{72.8\text{fF}}{72.8\text{fF} + \Delta C_{L_{HL}}}$. Solving for $\Delta C_{L_{HL}}$ we find that the transmission gate can drive an additional 49 fF. Since each minimum geometry inverter adds an additional gate capacitance of 31.1 fF we can drive one more inverter and the fan-out is two.

➤ SUMMARY

In this chapter we described the analysis and design of simple combinational logic gates implemented with transistors fabricated in a CMOS process. We concentrated on simplified hand analysis of the static and transient characteristics of CMOS inverters and simple logic gates. The most important concepts that you should have mastered are:

- Formal and simplified definitions of logic levels and noise margin from the voltage transfer characteristic.
- Calculation of propagation delay including the effect of load gate capacitance and parasitic wiring and drain/source-bulk capacitances.
- The relationship between device parameters (e.g., V_T, W/L) and the voltage transfer characteristic and propagation delay for the static CMOS inverter.
- Design methods for the implementation of a logic function given specified static and transient requirements.
- Implementation of logic functions using CMOS, static, dynamic, and pass transistor logic.
- Static and transient analysis of logic gates implemented in dynamic or pass transistor logic.

➤ FURTHER READING

1. L. A. Glasser and D. W. Dobberpuhl, *The Design and Analysis of VLSI Circuits*, Addison-Wesley, 1985, Chapter 4. Includes a discussion of the sources of noise in MOS digital circuits.
2. D. A. Hodges and H. G. Jackson, *Analysis and Design of Digital Integrated Circuits 2nd Ed.*, McGraw-Hill, 1988, Chapter 3. Includes a discussion on NMOS logic at about the same level.

3. J. Rabaey, *Digital Integrated Circuits: A Design Perspective*, Prentice Hall, 1996, Chapters 3, 4, and 7. Chapters 3 and 4 have a more in-depth discussion of CMOS static, dynamic, pass transistor logic and several other CMOS logic styles. Chapter 7 points out performance limitations due to interconnect.
4. M. Shoji, *CMOS Digital Circuit Technology*, Prentice Hall, 1988, Chapters 2–5. More advanced treatment of CMOS static and dynamic logic gates.
5. J. P. Uyemura, *Fundamentals of MOS Digital Integrated Circuits*, Addison-Wesley, 1988, Chapters 3–4, 6. Chapters 3 and 4 describe static and transient analysis of an MOS inverter and Chapter 6 analyzes combinational MOS logic circuits at about the same level.
6. N. H. E. Weste and K. Eshraghian, *Principles of CMOS VLSI Design 2nd Ed.*, Addison-Wesley, 1993, Chapters 4–5. More advanced treatment of the practical limitations to circuit performance and a more complete discussion of modern CMOS logic structures.

➤ PROBLEMS

For the following problems use the device data given below unless otherwise specified. Assume $V_{DD} = 5$ V unless otherwise specified.

MOS Device Data

$\mu_n C_{ox} = 50\ \mu A/V^2$, $\mu_p C_{ox} = 25\ \mu A/V^2$, $V_{Tn} = -V_{Tp} = 1$ V,

$-2\phi_p = 2\phi_n = 0.8$ V, $\gamma_n = \gamma_p = 0.6\ V^{-1/2}$, $\lambda_n = \lambda_p = 0.67\ V^{-1}$ @ $L = 1.5\ \mu m$

$C_{ox} = 2.3\ fF/\mu m^2$, $C_{Jn} = 0.1\ fF/\mu m^2$, $C_{Jp} = 0.3\ fF/\mu m^2$,

$C_{JSWn} = 0.5\ fF/\mu m$, $C_{JSWp} = 0.35\ fF/\mu m$, $L_{diffn} = L_{diffp} = 6\ \mu m$

Exercises

E5.1 Draw a logic gate implementation of the XOR function using only NOR gates and inverters.

E5.2 Consider a common two-switch light bulb. When the switches are both OFF or both ON, the light bulb is OFF, and when just one switch is ON the bulb is ON. Assume that when the output of your circuit is a **1**, the light bulb is on.

(a) Write a truth table for the circuit assuming the inputs are the two switches, X and Y.

(b) Draw a logic gate implementation of this control switch using only NAND gates and inverters.

E5.3 One common function that an inverter, acting as an input to a chip, must perform is to translate between the different voltage levels. A typical input to a CMOS chip might be TTL (transistor-transistor logic) levels. Typical TTL levels that will drive the input inverter are $V_{OH} = 2.4$ V and $V_{OL} = 0.8$ V. Choose V_M, V_{IH}, and V_{IL} for the CMOS inverter with $V_{DD} = 5$ V such that the noise margins NM_H, $NM_L \geq 0.5$ V.

E5.4 In Example 5.1 we used graphical analysis to find the voltage transfer function for an NMOS inverter. Repeat this example using a PMOS transistor as shown in Fig. E5.4. Assume $(W/L) = 10/2$, $V_{Tp} = -1$V, $\mu_p C_{ox} = 25\mu A/V^2$ $\lambda_p = 0\ V^{-1}$.

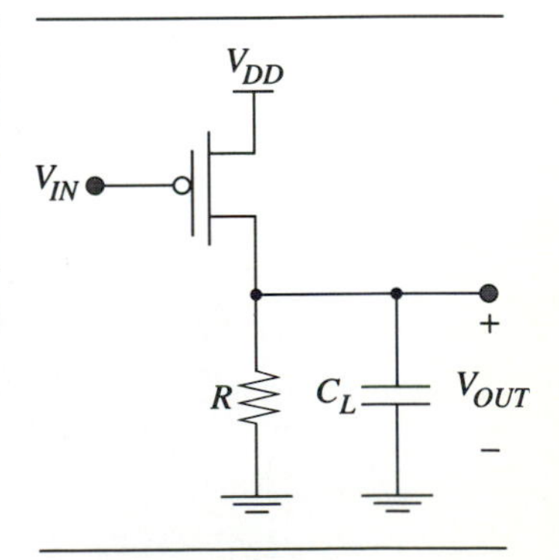

➤ **Figure E5.4**

E5.5 For the PMOS inverter with resistor pull-down in E5.4 assume $\lambda_p = 0\ V^{-1}$.

(a) Find V_{OH}, V_{OL}, and A_V, with $R = 5$ kΩ.

(b) Repeat (a) for $R = 25$ kΩ.

E5.6 Consider a simple NMOS inverter with a resistor pull-up using $V_{DD} = 5$ V.

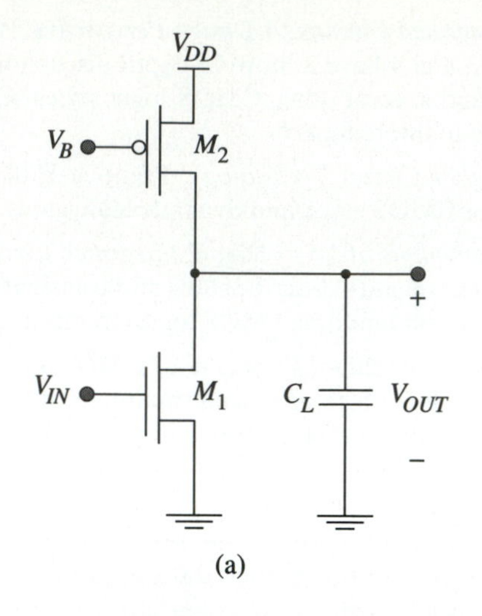

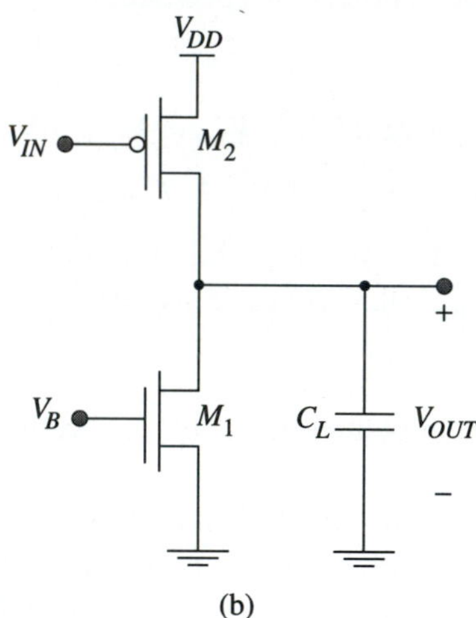

➤ Figure E5.7

(a) What is I_D for $V_{GS} = 4$ V and $V_{DS} = 1$ V and $(W/L) = 3/1.5$?

(b) If the pull-up resistor is 25 kΩ what is the minimum value of W/L to guarantee that $V_{OUT} \leq 1$V with an input of 4 V?

(c) Repeat (b) for $V_{OUT} \leq 0.3$ V.

(d) Repeat (b) for $R = 5$ kΩ.

E5.7 You are given an NMOS inverter with current source pull-up shown in Fig. E5.7(a) with $(W/L)_n = (W/L)_p = 6/1.5$. Use the simplified hand calculation method shown in Section 5.4.1 and neglect channel length modulation for parts (a) and (b).

(a) Calculate V_B such that $V_{MIN} = 0.5$ V

(b) Given the V_B you calculated in part (a), calculate V_M.

(c) Find the small signal gain A_v at $V_{IN} = V_M$.

(d) Find $V_{IL}, V_{IH}, V_{OL}, V_{OH}$.

(e) Find NM_L and NM_H.

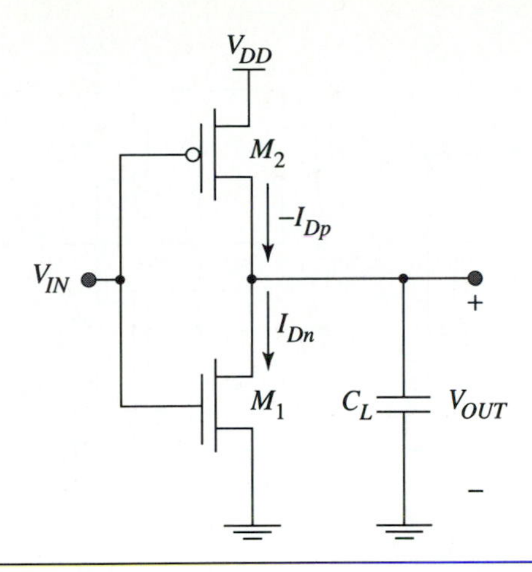

➤ **Figure E5.9**

E5.8 Repeat E5.7 for a PMOS inverter with current source pull-down shown in Fig. E5.7(b) with $(W/L)_n = (W/L)_p = 6/1.5$.

E5.9 For the CMOS inverter in Fig. E5.9 $(W/L)_n = (W/L)_p = 6/2$.

(a) What is V_{IN} and V_{OUT} to make I_{Dn} maximum?

(b) What is V_{IN} and V_{OUT} to make $|I_{Dp}|$ maximum?

(c) Adjust $(W/L)_p$ such that the maximum current for the p-channel and n-channel transistors are the same. Do not neglect channel length modulation for this calculation.

(d) Using $(W/L)_n = (W/L)_p = 6/2$, what is the region of operation for the NMOS and PMOS transistors when $V_{OUT} = V_{IN}$? Find V_M. Neglect channel length modulation for this part.

(e) Draw the small signal model for the CMOS inverter when $V_{IN} = V_M$ and $(W/L)_n = (W/L)_p = 6/2$.

(f) What is the overall g_m and voltage gain using the conditions in part (e)?

(g) Find V_{OH}, V_{OL}, V_{IH}, V_{IL}, NM_H, and NM_L using the conditions in part (e).

E5.10 For the CMOS inverter in Fig. E5.9 with $(W/L)_n = 6/1.5$ and $(W/L)_p = 10/1.5$

(a) Sketch the voltage transfer characteristic and label the important breakpoints. Use the simplified straight line approximation.

(b) Compare your hand analysis to the characteristic generated by SPICE by sweeping the input voltage from 0 V to $V_{DD} = 5$ V.

(c) If C_L is equal to 200 fF, what is t_{PHL} and t_{PLH}. Neglect C_{DB} for this part.

(d) Repeat part (c) with $C_L = 200\text{ fF} + C_{DB}$.

E5.11 To see the effect of device sizing on the voltage transfer characteristic and the propagation delay, repeat E5.10 with

(a) $(W/L)_n = 3/1.5$ and $(W/L)_p = 6/1.5$

(b) $(W/L)_n = 12/1.5$ and $(W/L)_p = 24/1.5$

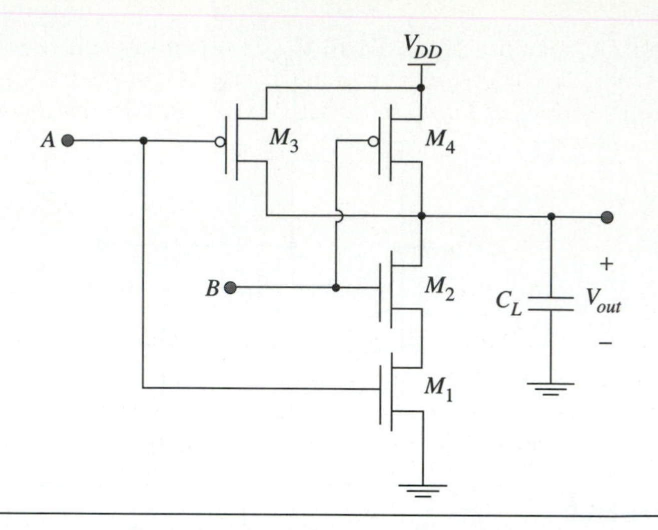

➤ **Figure E5.12**

(c) $(W/L)_n = 12/1.5$ and $(W/L)_p = 6/1.5$

(d) $(W/L)_n = 3/1.5$ and $(W/L)_p = 24/1.5$

E5.12 Given a 2-input NAND gate shown in Fig. E5.12 choose sizes for the p-channel devices such that the propagation delays are approximately symmetrical. Assume $(W/L)_n = 3/1.5$. Repeat this exercise for a 3-input NAND gate.

E5.13 Given a 2-input NOR gate shown in Fig. E5.13, choose sizes for the n-channel devices so the propagation delays are approximately symmetrical. Assume $(W/L)_p = 12/1.5$. Repeat this exercise for a 3-input NOR gate.

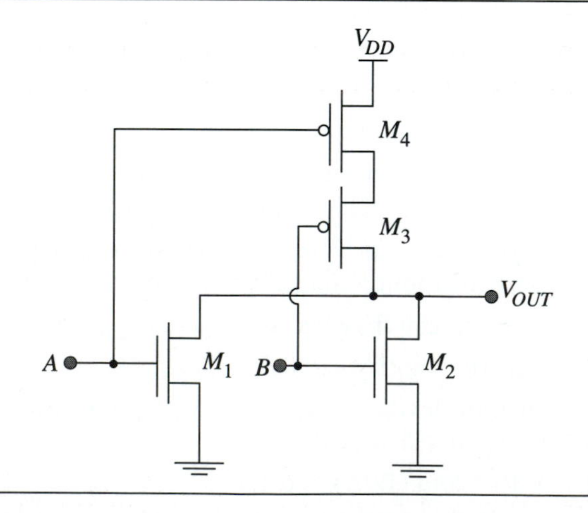

➤ **Figure E5.13**

E5.14 Draw a transistor level circuit diagram that implements the following logic functions using n-channel dynamic logic. Use static inverters to compliment the inputs as necessary. Add a static buffer (1 or 2 inverters depending on logic) at the output.

(a) $Y = ABC$

(b) $Y = \bar{B}C + A\bar{C}$

(c) $Y = (A + \bar{B}) + \overline{(\bar{C} + D)}$

E5.15 Repeat E5.14 using p-channel dynamic logic implementation.

E5.16 You are given both the true and compliment control signals C_0–C_7. Draw a pass transistor logic implementation using n- and p-channel transistors that passes the input logic value Q to one of eight possible outputs assuming only one control signal is a logic **1**.

PROBLEMS

P5.1 The technology you are designing with is optimized for NAND gates and inverters. Draw a logic gate implementation of each of the following functions using only NAND gates and inverters.

(a) $Y = A\bar{B}C$

(b) $Y = \overline{A\bar{B}C} + (D + \bar{E})$

(c) $Y = (A + \bar{C}) + \overline{(\bar{B} + D)}$

P5.2 Noise often comes from capacitive coupling between clock lines and our signal of interest as shown in Fig. P5.2. Given $V_{OH} = 4.5$ V, $V_{OL} = 0.5$ V, $V_{IH} = 3.0$ V and $V_{IL} = 1.5$ V

(a) Calculate NM_H and NM_L.

(b) Find the maximum value for C_1 given the noise margins calculated in (a). Hint: Recall that $\Delta V/V = C_1/(C_1 + C_2)$.

(c) To improve the immunity to capacitive coupling we could increase the size of C_2 by 2X. What is the new value of C_1 which can be tolerated? Note: (This method of reducing coupling also increases propagation delay.)

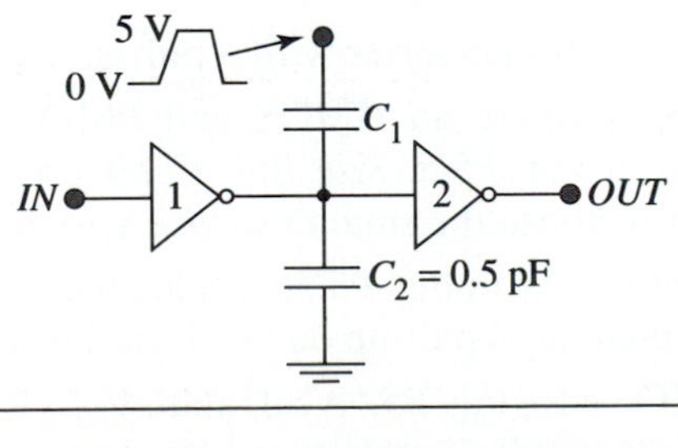

➤ Figure P5.2

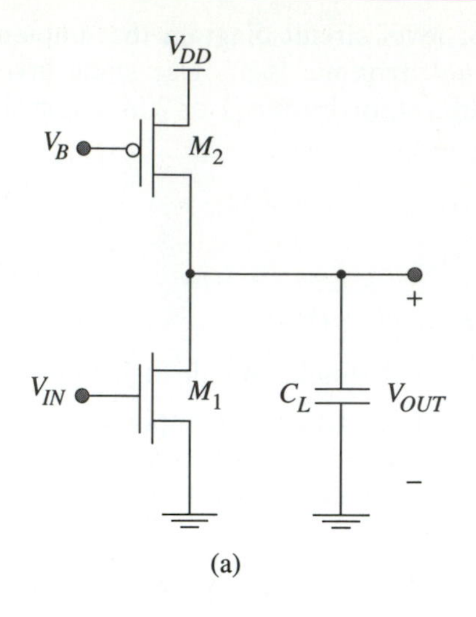

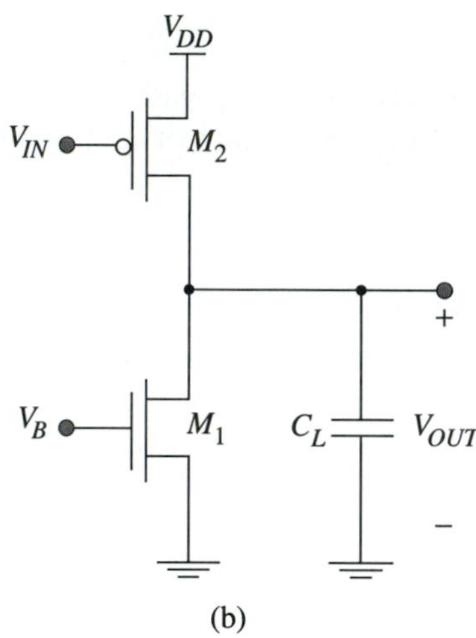

➤ Figure P5.5

P5.3 Given an NMOS inverter with a pull-up resistor of 1kΩ

(a) Find the (W/L) of the NMOS transistor such that $V_{OL} = 0.5$ V

(b) For the NMOS device size found in (a) calculate NM_H and NM_L. Use the simplified hand calculation method shown in Section 5.4.1.

(c) Repeat (a) and (b) for a pull-up resistor of 10 kΩ.

P5.4 Repeat P5.3 for a PMOS inverter with a pull-up resistor of 1 kΩ.

P5.5 In this problem, compare an NMOS and PMOS inverter with a current source pull-up as shown in Fig. P5.5. Size the n and p channel devices such that $V_M = 2.5$ V for both cases. Choose the smaller device size to be 3/1.5.

(a) For the NMOS case in Fig. P5.5(a) calculate V_{OH}, V_{OL}, and A_v given the gate of the p-channel pull-up device is tied to $V_B = 3.0$ V.

(b) For the PMOS case Fig. P5.5b calculate V_{OH}, V_{OL} and A_v given the gate of the n-channel pull-down device is tied to $V_B = 2.0$ V.

P5.6 For the NMOS inverter with current source pull-up shown in Fig. P5.5(a) with $(W/L)_n = (W/L)_p = 6/1.5$ and $V_B = 2.5$ V

(a) Calculate t_{PHL} and t_{PLH} if the inverter is loaded only by an identical inverter. Include C_{DB} for both the NMOS and PMOS transistors.

(b) Calculate t_{PHL} and t_{PLH} if the inverter is loaded with three identical inverters.

(c) Assume that the system must have a clock with period $T \geq 10t_p$, where t_p is the average propagation delay. Find the maximum length of interconnect wires from the output of the inverter to the inputs of the next stages for a fan-out of 3 if we want the system clock to run at 50MHz. Assume the width of the interconnect is 4μm and the capacitance of the wires are 0.05 fF/μm^2.

P5.7 An NMOS inverter with current source pull-up shown in Fig. 5.5(a), has $(W/L)_n = 6/1.5$ and $(W/L)_p = 3/6$ and $V_B = 0$ V. Use the simplified hand calculation method shown in Sections 5.4.1 and 5.4.3. Assume devices are in their constant-current region.

(a) Sketch the voltage transfer characteristic and label V_{IL},V_{IH}, V_M, V_{OH} and V_{OL}.

(b) If a 100 fF load capacitor is connected to the output of the inverter calculate the propagation delay, t_p. Neglect C_{DB} for this part.

(c) Calculate t_p including C_{DB}.

(d) What is the static power consumed by this circuit?

(e) Calculate the device widths such that $C_{DB} = 100$ fF while maintaining the same V_M.

(f) What is t_p for the device sizes calculated in part (e)?

P5.8 To explore how the noise margin can change with device sizing in a CMOS inverter, calculate V_M, NM_H and NM_L when

(a) $(W/L)_n = 1$ and $(W/L)_p = 2$

(b) $(W/L)_n = 10$ and $(W/L)_p = 0.2$

(c) $(W/L)_n = 0.1$ and $(W/L)_p = 20$

P5.9 In this problem we will explore the accuracy of our simplified equation Eq. (5.30), to calculate t_{PHL}, for the CMOS inverter.

(a) Write a differential equation for that is valid when the NMOS transistor is saturated that relates the device current and capacitor current in terms of V_{OUT} *(t)*.

(b) Write another differential equation that is valid when the NMOS transistor is in the triode region that relates the device current and capacitor current.

(c) Solve these equations and write an expression for t_{PHL}, that is the time it takes for the output voltage to drop from V_{OH} to $V_{OH}/2$.

(d) For an NMOS transistor with W/L = 6/1.5 and $C_L = 100$ fF, solve for t_{PHL} using the expression found in (c) and compare it with the value found from Eq. (5.30).

(e) Repeat (d) for W/L = 20/1.5 and $C_L = 500$ fF.

P5.10 In this problem you will size a CMOS inverter with process parameters: $V_{Tn} = 0.7$ V, $V_{Tp} = -0.9$ V, $\mu_n = 500$ cm^2/Vs, $\mu_p = 200$ cm^2/Vs, $t_{ox} = 20$ nm $\lambda_n = \lambda_p = 0.05$ V^{-1}. Assume equal channel lengths, $V_{DD} = 5$ V and all other process parameters are unchanged.

(a) Calculate the ratio W_n/W_p, such that $V_M = 2.5$V

(b) When $V_{IN} = V_M$ we want the current through the inverter to be 1mA. What is W_n and W_p assuming the channel length of both devices is 2μm?

(c) Sketch and label the voltage transfer characteristic.

(d) What are NM_L and NM_H?

P5.11 The CMOS inverter which you have sized in P5.10 must drive 2 identical inverters connected in parallel as shown in Fig. 5.24. Using process parameters from P5.10

(a) What is the component of load capacitance for the drain-bulk capacitance from your inverter?

(b) What is the component of load capacitance from the 2 additional inverters?

(c) Calculate t_{PHL} and t_{PLH}.

P5.12 With a greater emphasis on portable electronics CMOS logic is being designed with lower power supply voltages to reduce power dissipation. In this problem we will investigate the changes in the voltage transfer function and the propagation delay when the power supply is reduced. You are given a CMOS inverter driving 3 identical inverters with $(W/L)_n = 3/1.5$ and $(W/L)_p = 6/1.5$.

(a) Sketch the voltage transfer function for $V_{DD} = 5$ V.

(b) Calculate t_{PHL} and t_{PLH}.

(c) What is the dynamic power dissipation when the inverter is running at 50 MHz.

(d) Repeat a–c with $V_{DD} = 2.5$ V.

P5.13 In P5.12 we saw that the power dissipation was significantly reduced with the reduction of the power supply voltage at the expense of noise margin and propagation delay. Modern processes are using reduced threshold voltages to reduce this problem. Repeat P5.12 with $V_{Tn} = -V_{Tp} = 0.5$ V.

P5.14 As shown in P5.13 the reduction of threshold voltage improves the noise margin and propagation delay but we did not maintain equivalent speed when the power supply was reduced from 5 V to 2.5 V. Another process option to improve the propagation delay is to reduce the channel length.

(a) Find the channel length $L_n = L_p$ such that you achieve the same propagation delay calculated in P5.12(b) with $V_{Tn} = -V_{Tp} = 0.5$ V.

(b) What is the new NM_L and NM_H for the channel lengths calculated in (a). Recall that λ is proportional to $1/L$. Assume $\lambda_n = \lambda_p = 0.1$ V^{-1} at $L_n = L_p = 1.0$ μm.

P5.15 Given a minimum geometry CMOS inverter with $(W/L)_n = 3/1.5$ and $(W/L)_p = 6/1.5$ what is the maximum fan-out possible while keeping t_{PLH} and t_{PHL} ≤ 600 ps with

(a) $V_{DD} = 5$ V

(b) $V_{DD} = 2.5$ V

P5.16 How many identical inverters with (W/L given below) can be driven by a single inverter, if the propagation delay must be less than 10x larger than that of the unloaded single inverter? For the unloaded inverter neglect wiring capacitance but include C_{DB}.

(a) $(W/L)_n = 3/1.5$ and $(W/L)_p = 6/1.5$

(b) $(W/L)_n = 30/1.5$ and $(W/L)_p = 60/1.5$

(c) $(W/L)_n = 100/1.5$ and $(W/L)_p = 200/1.5$

P5.17 For the device data and inverter sizes given in Example 5.5, what is the maximum wire length between the first inverter and inverters 2 and 3 for $t_p \leq 1\text{ns}$? Assume that $C_{WIRE} = 0.1$ fF/μm.

P5.18 A figure of merit for a digital technology is the power-delay product, $P_D t_p$. Plot the power-delay product as a function of V_{DD} for $2.5\text{ V} \leq V_{DD} \leq 5\text{ V}$. For this calculation use a CMOS inverter with $(W/L)_n = 3/1.5$ and $(W/L)_p = 6/1.5$ driving 3 identical inverters (fan-out = 3). Assume a clock frequency of 50 MHz.

P5.19 For a NAND gate using all minimum geometry transistors $(W/L)_n = (W/L)_p = 3/1.5$ and neglecting the backgate effect

(a) Sketch the 3 possible voltage transfer characteristics and label the important breakpoints.

(b) Given the total load capacitance is $C_L = 0.1\text{pF}$ calculate the worst case propagation delays t_{PHL} and t_{PLH}.

P5.20 A 2-input NOR gate implementation requires the p-channel transistors to be 4x the n-channel transistors for symmetrical propagation delays.

(a) Calculate the worst case propagation delay using all minimum geometry transistors $(W/L)_n = (W/L)_p = 3/1.5$. Neglect backgate effect.

(b) Repeat (a) with $(W/L)_n = 3/1.5$ and $(W/L)_p = 12/1.5$

P5.21 Given the n-channel, dynamic 2, input NOR gate in Fig. P5.21

(a) Choose W/L of M_1 M_2 and M_E such that $t_{PHL} = t_{PLH}$. Assume W/L of M_1, M_2 and M_E are the same.

(b) Calculate t_p for this gate assuming $C_L = 0.4$ pF. Neglect C_{DB} and C_{WIRE}.

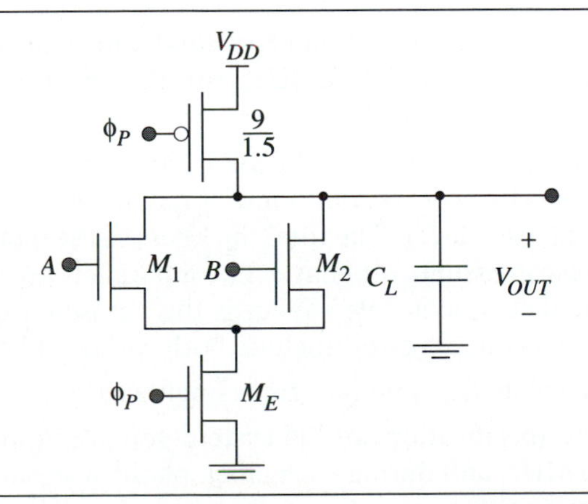

➤ Figure P5.21

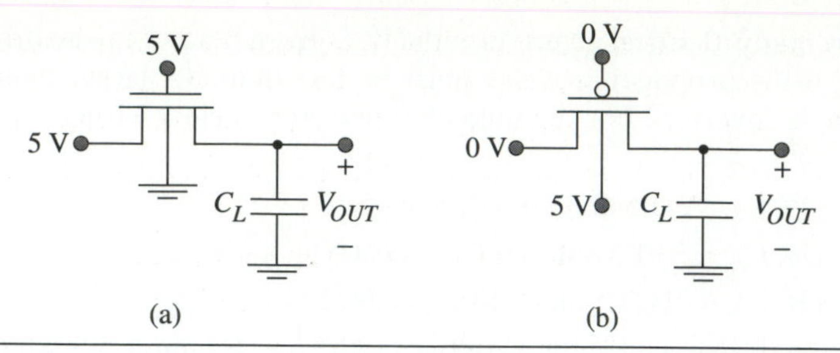

➤ **Figure P5.22**

P5.22 Using a single n-channel transistor pass gate rather than a full CMOS pass gate reduces the voltage swing due to backgate effect. Calculate the maximum voltage on C_L for the voltages shown in Fig. P5.22(a). Assume the initial voltage on C_L was 0 V.

P5.23 Using a single p-channel transistor pass gate rather than a full CMOS pass gate reduces the voltage swing due to backgate effect. Calculate the minimum voltage on C_L for the voltages shown in Fig. P5.22(b). Assume the initial voltage on C_L was 5 V.

Design Problems

For these design problems the technology restricts $W > 3$ µm and $L > 1.5$ µm

D5.1 You are to design an NMOS inverter with resistor pull-up to drive a load capacitance of 0.3 pF. The *RC* time constant must be less than 1ns. Size the NMOS transistor such that $V_M = 2.5$V and NM_L and $NM_H \geq 1.0$ V. Minimize the power dissipation.

(a) For the initial design neglect C_{DB}.

(b) Include C_{DB} for the refined design.

(c) Verify your design in SPICE

D5.2 Repeat the design in D5.1 with PMOS inverter with resistor pull-up.

D5.3 The goal of this problem is to determine the number of identical inverters that a CMOS inverter can drive, given that they are far away from the driving inverter.

(a) Calculate the component of load capacitance from a single inverter load, C_G. The device sizes for the inverter are $(W/L)_n = 6/1.5$ and $(W/L)_p = 12/1.5$

(b) The load inverters are located 1000 µm from the driving inverter. The connecting wire is 2 µm wide aluminum and lies on a deposited glass and field oxide layer. The total thickness of the dielectric layer is 1.0 µm and you can assume it behaves like a parallel capacitor. The permittivity of the dielectric is $3.9\varepsilon_o$. What is the parasitic capacitance resulting from driving one inverter?Include both wiring and drain-bulk capacitance.

(c) Calculate t_{PHL} and t_{PLH} for a single inverter load.

(d) The specifications of the system you are designing has a clock rate of 50 MHz, and during each clock phase, a signal may experience a maximum of 25 propagation delays. Using a 20% safety margin on the propagation delay what is the fan-out of the inverter.

D5.4 Repeat D5.3 with the load inverters 1cm from the driving inverter. Calculate the increase in power dissipation when the larger size inverter is used.

D5.5 In Example 5.6 we designed a CMOS input inverter/buffer. It is often required to have inputs logically ANDED with a select signal. Design a NAND gate which uses the same specifications as Example 5.6.

D5.6 In CMOS technology it is more area-efficient to design with NAND gates rather than NOR gates. Using the specifications of Design Example 5.8, design a 2-input NOR gate which meets the same performance requirements. What is the additional area required by the NOR gate design over the NAND gate design?

D5.7 Design a 2-input NAND gate with $t_{PLH} = t_{PHL} = 2$ ns. Assume C_L is 0.5 pF. Neglect the drain-bulk capacitance for the first iteration. Try to minimize the device sizes used. Verify your design in SPICE.

D5.8 To compare static and dynamic logic, design a 2-input NAND gate using dynamic logic for the same specifications as in D5.7. You can use static inverters if you feel an equivalent logic operation (use DeMorgan's Law) will perform better. After you have completed the design, calculate the total device area $[W \cdot (L + 2L_{diff})]$.

D5.9 Repeat D5.7 and D5.8 for a 2-input NOR gate.

chapter 6

The pn Junction Diode

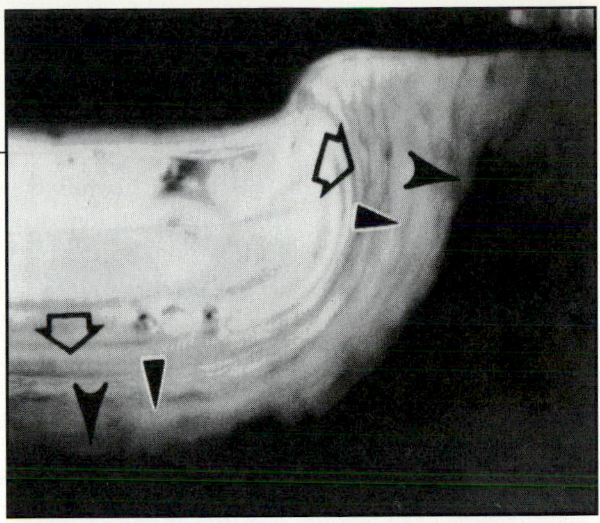

Local variations in arsenic concentration are visible in this TEM of the edge of a source-bulk pn junction. (Courtesy of Prof. R. Gronsky, Dept. of MSME, Univ. of California at Berkeley)

In this chapter, we analyze the internal operation of the pn junction diode and develop large- and small-signal circuit models that describe its behavior as a circuit element. The concepts introduced here are essential in understanding the bipolar junction transistor in Chapter 7.

The current in a pn diode includes drift and diffusion of holes and electrons. Several approximations and considerable analysis are involved in deriving expressions for these current components. Therefore, we will take two passes at the pn diode. The initial goal will be to achieve a basic grasp of the internal operation in order to derive the ideal diode equation and circuit models.

In the second pass, the underlying assumptions and approximations will be considered more carefully. The concept of minority carrier lifetime and the continuity equation are missing pieces of basic semiconductor physics, which are necessary to solve for the minority carrier concentrations in the bulk regions. Finally, we describe two integrated-circuit applications of pn junction diodes.

Chapter Objectives

- The basic operation of the short-base pn diode under applied bias, including the law of the junction and its use to find minority carrier boundary conditions, current components in the short-base diode, and the ideal diode current-voltage equation.
- The piecewise-linear model of the diode and the small-signal model of the forward-biased pn diode, including the diffusion capacitance.
- Minority carrier lifetime and the origin of the continuity equation.
- Minority carrier diffusion length and the general solution of the pn diode, including the long-base approximation.

➤ 6.1 pn DIODE CIRCUIT SYMBOL AND TERMINAL CHARACTERISTICS

The basic structure of a pn diode is shown in Fig. 6.1(a). As discussed in Chapter 3, the p and n regions are contacted by metal layers that form ohmic contacts. The circuit symbol for the pn junction diode and conventional polarity of the diode voltage and current are shown in Fig. 6.1(b).

The p-side of the diode is called the **anode** and the n-side is called the **cathode**. Measurements of the current-voltage characteristics of the pn diode on a linear scale are shown in Fig. 6.2(a). There are three regions of operation:

1. **Forward bias**–diode current $I_D > 0$ with the diode voltage $V_D \approx 0.7$ V
2. **Reverse bias**–diode current $I_D \approx 0$
3. **Breakdown**–diode current $I_D < 0$ and the diode voltage $V_D < V_{BD}$, where V_{BD} is the diode's breakdown voltage

In the forward bias region, the diode approximates a battery with a very small internal resistance. For voltages $V_{BD} < V_D < 0.7$ V, the pn diode approximates an open circuit. The breakdown voltage V_{BD} is typically in the range of –10 V to –25 V. In the breakdown region, it again behaves like a battery with a voltage of V_{BD}. It should be noted that the diode is not necessarily destroyed by operation in the breakdown region. Potentially destructive power dissipation may occur, however, if there is significant reverse current. Specially designed diodes, called Zener diodes, are intended for operation in the breakdown region. Zener diodes have breakdown voltages in the range of –3 to –8 V.

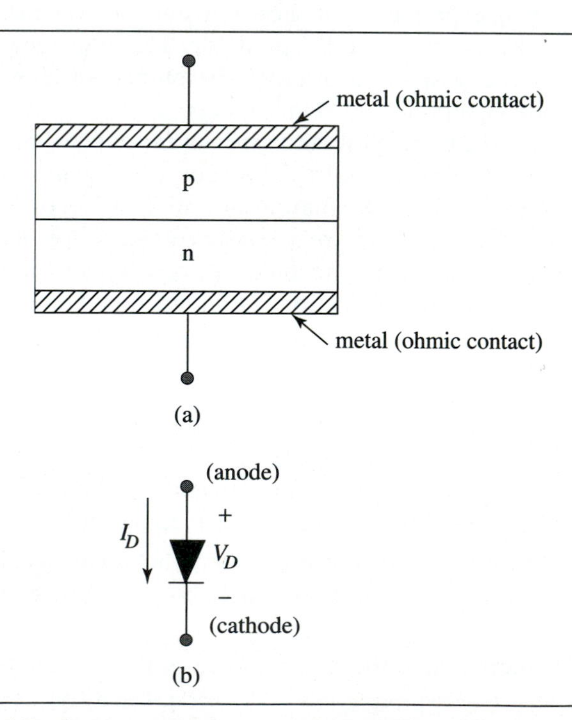

➤ **Figure 6.1** (a) Basic structure of a pn diode and (b) pn diode symbol and current-voltage polarities.

An essential feature of the pn junction diode is revealed by plotting the forward-bias current-voltage characteristics on a log-linear scale in Fig. 6.2(b). Except for deviations at very small and very large currents, Fig. 6.2(b) shows that I_D is proportional to the exponential of the applied voltage over several orders of magnitude of

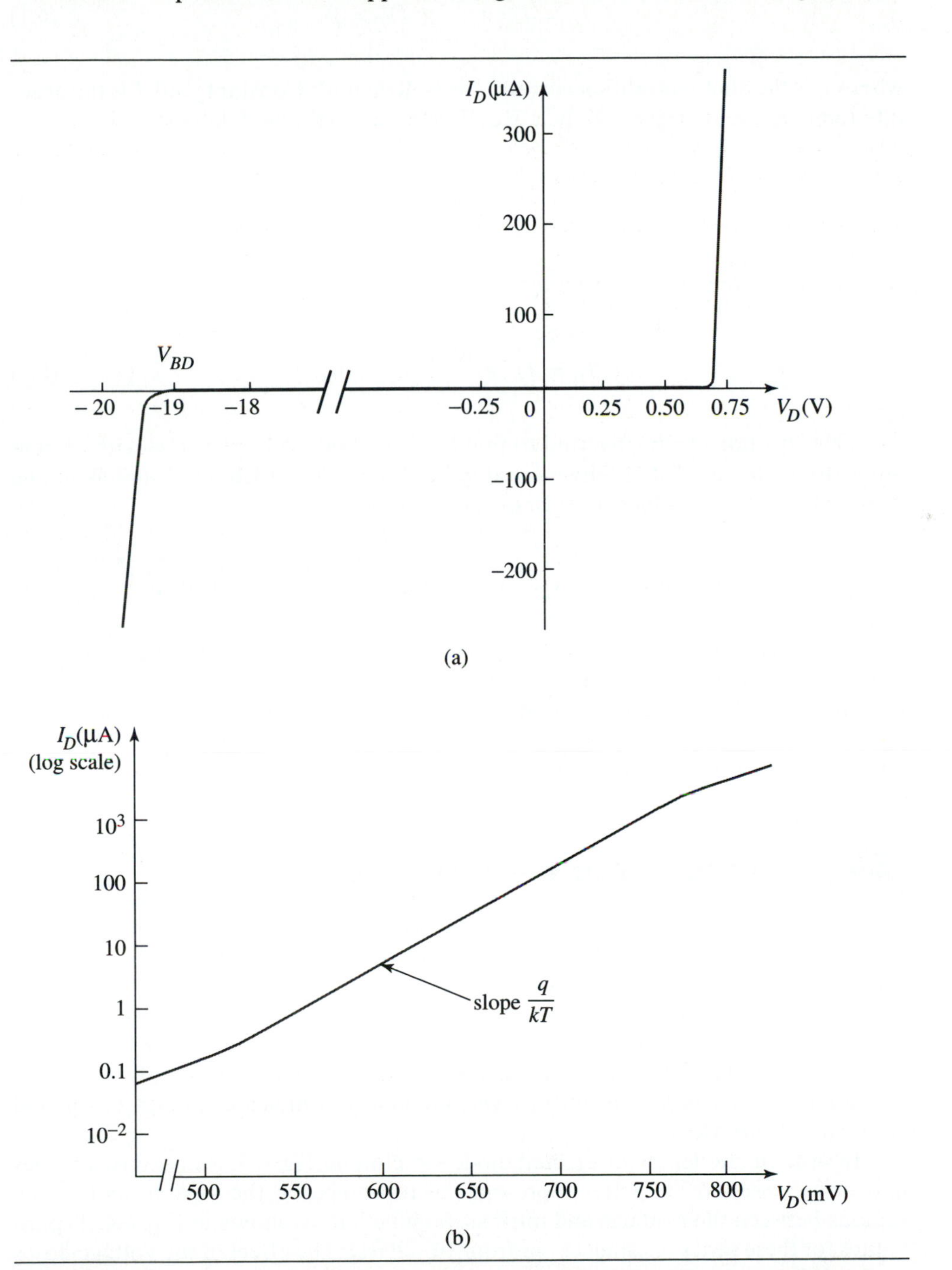

➤ Figure 6.2 (a) Current-voltage characteristics of a typical pn junction diode (note the break in the voltage axis) and (b) log-linear plot of current versus diode voltage for forward bias.

current. Measurements at different temperatures reveal that the current-voltage relationship under forward bias is approximately

$$I_D = I_o \cdot e^{qV_D/kT} \quad (V_D > 0) \tag{6.1}$$

where I_o is the **diode saturation current**, k is Boltzmann's Constant, and T is the absolute temperature in degrees Kelvin. Recalling that the thermal voltage V_{th} is

$$V_{th} = \frac{kT}{q} \tag{6.2}$$

we can express the diode current by

$$I_D = I_o \cdot e^{V_D/V_{th}} \quad (V_D > 0) \tag{6.3}$$

This equation can be inverted to find how the diode voltage varies with current under forward bias. Taking the natural log of both sides of Eq. (6.3) and assuming room temperature at which $V_{th} \approx 26$ mV, we find

$$V_D = V_{th} \cdot \ln\left(\frac{I_D}{I_o}\right) = V_{th} \cdot \ln(10) \cdot \log\left(\frac{I_D}{I_o}\right) = 60\text{mV} \log\left(\frac{I_D}{I_o}\right) \tag{6.4}$$

For every order-of-magnitude increase in the diode current, Eq. (6.4) indicates that the diode voltage increases only 60 mV. Substituting a typical diode saturation current $I_o = 10^{-16}$ A and a representative diode current $I_D = 100$ μA, the diode voltage is $V_D = 720$ mV. As a first approximation, a forward-biased diode is assumed to have a voltage $V_D \approx 700$ mV.

6.2 INTEGRATED CIRCUIT pn DIODES

The pn diode is a ubiquitous structure in integrated circuits. As we saw in Chapter 4, reverse-biased pn junctions are used to isolate wells from the substrate in CMOS technologies. Figure 6.3 shows the cross section and layout of a pn junction diode formed between a diffused p region and an n-type well. Note that this IC structure has two diodes. The diode formed by the n well and substrate maintains its reverse bias since the p-type substrate is connected to the lowest voltage supply—typically ground. It serves to isolate the intrinsic pn junction diode indicated in Figs. 6.3 (a) and (b) from other devices.

In order to model the integrated diode structure in Fig. 6.3, we need two diodes and three resistors. The latter represent the resistances in the bulk p- and n-type regions between the contacts and intrinsic pn junctions as shown in Fig. 6.4. Typical values for these series resistances are from 10 – 250 Ω. The effect of the voltage drops across these resistors is negligible except when the diode is conducting large currents.

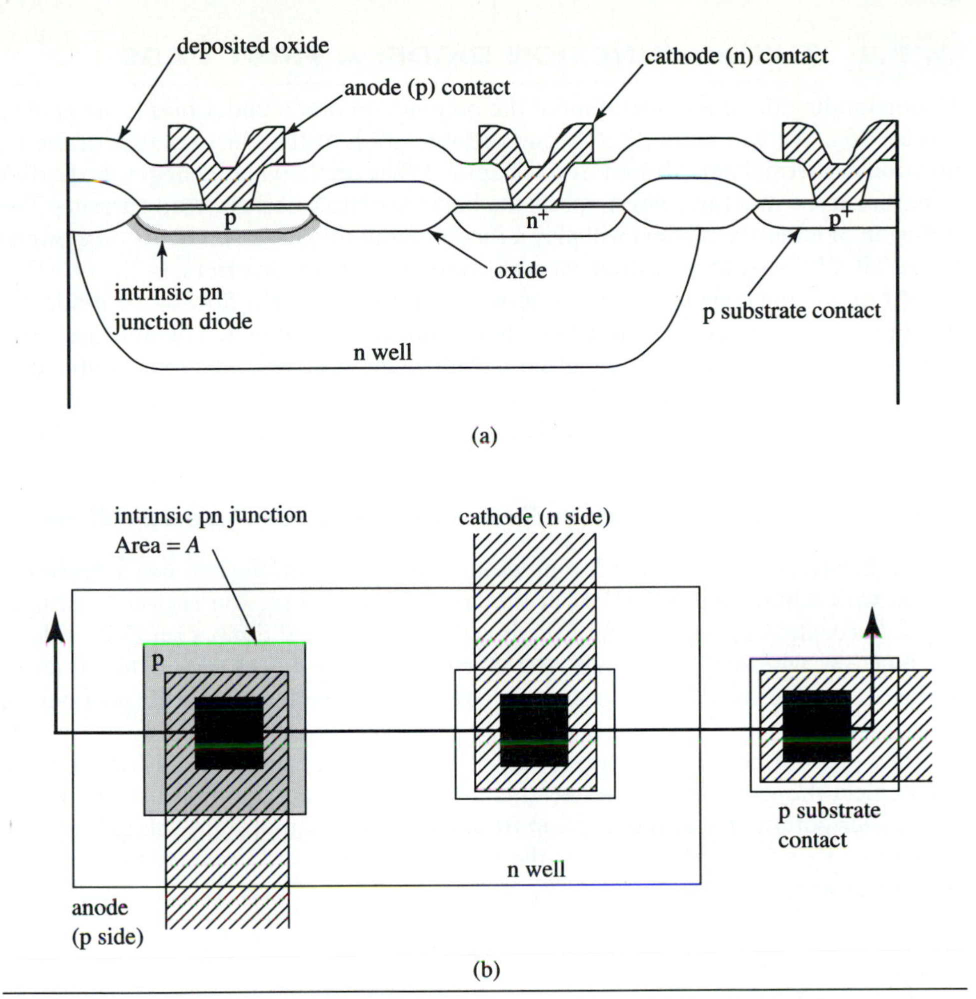

➤ Figure 6.3 pn diode in an n-well CMOS process: (a) cross section and (b) layout. The intrinsic pn junction is indicated in both (a) and (b). The rest of the structure is needed to make electrical contact to the intrinsic junction or for isolation.

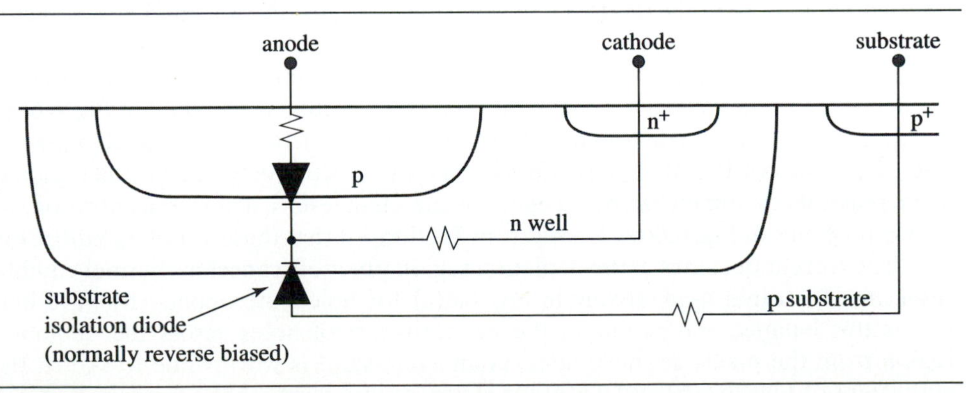

➤ Figure 6.4 Circuit model for IC pn diode superimposed on the cross section in Fig. 6.3(a), including the substrate isolation diode and resistors for the bulk p and n regions.

6.3 THE pn JUNCTION DIODE: A FIRST PASS

Understanding the basic operation of the pn junction diode under bias is the goal of this section. We will build directly on the foundation of the electrostatics of the pn junction in thermal equilibrium from Chapter 3. The applied bias changes the barrier potential at the junction, which upsets the balance of hole and electron currents. The diffusion of minority carriers will play a fundamental role in the pn diode, in contrast to the MOSFET where the drain current consists of drifting carriers.

After gaining insight into the "one-way" nature of current flow in the diode, we model how the carrier concentrations change under applied bias. The hole and electron current components are quantified, from which we derive the exponential current-voltage characteristic of the diode. The complete picture of the current components will be important in modeling charge storage in the diode under forward bias.

6.3.1 Qualitative Carrier Transport under Applied Bias

As we discovered in Section 3.3, a pn diode in thermal equilibrium has a depletion region with a built-in potential ϕ_B at the junction. The depletion region consists of ionized acceptors on the p-side of the junction and ionized donors on the n-side of the junction, with negligible holes and electrons. Figure 6.5(a) shows the pn diode under an applied bias V_D, which we now consider to be either forward (positive) or reverse (negative).

The p and n bulk regions are resistors in the circuit in Fig. 6.5(a); however, we will neglect the voltage drops across them for now. The applied bias acts to change the barrier potential ϕ_j at the junction from its thermal equilibrium value. From the polarity of V_D, a forward bias lowers the barrier potential and a reverse bias raises it. According to Eq. (3.37),

$$\phi_j = \phi_B - V_D \tag{6.5}$$

In Fig. 6.5(b), the potential $\phi(x)$ is sketched for the case of $V_D = 0.7$ V. As an example, the diode is symmetrically doped with $N_a = N_d = 3.2 \times 10^{17}$ cm^{-3} so that the built-in potential is $\phi_B = 0.9$ V. The forward bias raises the potential of the p-side of the junction by 0.7 V, which lowers the barrier potential to $\phi_j = \phi_B - V_D = 0.9 - 0.7 = 0.2$ V. Note that Eqs. (3.23) and (3.25) (the "60 mV rule") do not apply to the diode under applied bias, since the hole and electron currents are not zero.

The lowered barrier at the junction reduces the magnitude of the electric field in the depletion region from its value in thermal equilibrium, as sketched in Fig. 6.6(a). As a result, the drift current densities for holes and electrons are reduced in magnitude. The effect of $V_D > 0$ on the diffusion current densities in the depletion region is to increase them somewhat by steepening the change in carrier concentration, as shown for holes in Fig. 6.6(b). Recall from Section 3.3 that these very large diffusion and drift current densities in the depletion region precisely cancel in thermal equilibrium. As illustrated qualitatively in Fig. 6.6(c) for holes, the applied forward bias upsets this balance and results in the net transport of holes across the depletion region from the p-side to the n-side. As an exercise, it is worthwhile to sketch the equivalent of Figs. 6.6(b) and 6.6(c) for electrons and verify that a forward bias will cause a net transport in the opposite direction (but still a positive current density).

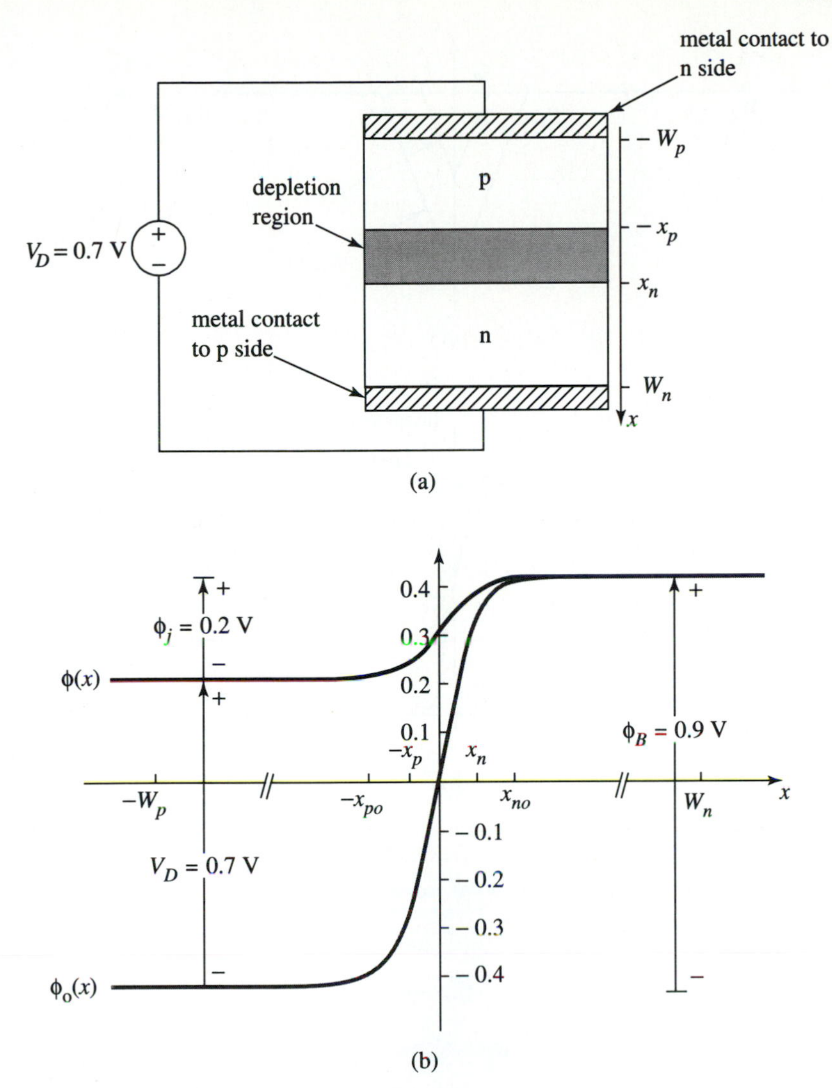

➤ Figure 6.5 One-dimensional diode with $V_D = 0.7$ V. (a) diode structure and coordinates, (b) potential with $V_D = 0.7$ V and for thermal equilibrium, where the built-in potential is $\phi_B = 0.9$ V.

The transport of holes from the p-side across the depletion region into the n-side, where they are minority carriers, under forward bias is called **hole injection**. Similarly, a forward bias causes **electron injection** from the n-side to the p-side of the junction.

From this model for the junction under forward bias, hole and electron injection increase rapidly with increasing forward bias. From Fig. 6.5(b), it appears that we can eliminate the electric field in the depletion region by applying $V_D = \phi_B$. Hole and electron transport across the depletion region would become infinite, according to the simple picture in Fig. 6.6. The resistive voltage drops across the bulk p and n regions tend to prevent this from happening. Second-order effects also contribute to limiting

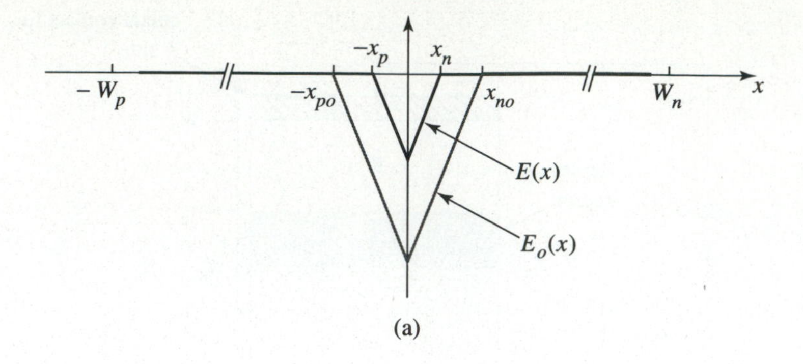

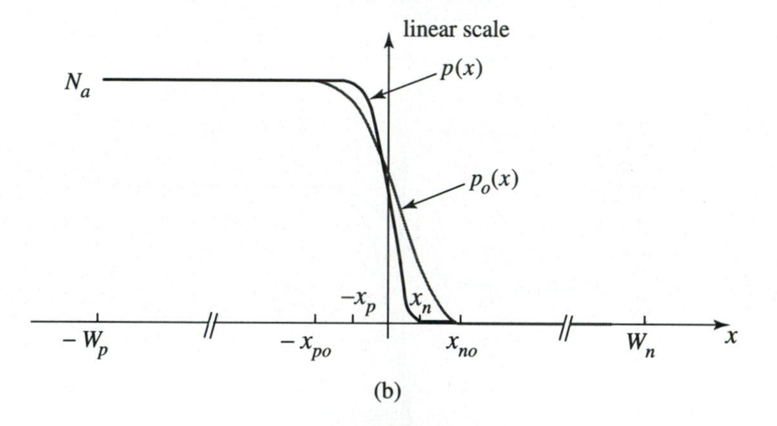

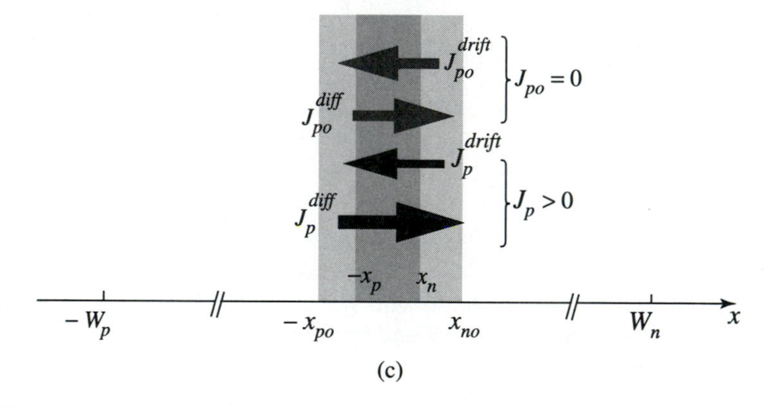

➤ Figure 6.6 Diode from Fig. 6.5 in thermal equilibrium and with V_D= 0.7 V. (a) electric field in depletion region, (b) hole concentration gradient across the depletion region (linear scale), and (c) graphical illustration of the imbalance between hole drift and diffusion caused by the applied bias, which results in hole injection into the n-side of the diode.

the current. Recall from basic electronics and Fig. 6.2 that applying $V_D > 0.7$ V leads to very large diode currents and potential power dissipation problems.

The situation is asymmetrical for $V_D < 0$. From Eq. (6.5), the barrier potential increases under reverse bias, which causes the depletion region to widen. The imbalance of drift and diffusion current densities results in transport of holes from the n-side to the p-side of the junction and of electrons in the opposite direction. These processes are known as **hole extraction** and **electron extraction**. Since the holes are minority carriers on the n-side and the electrons are minority carriers on the p-side of the junction, extraction results in minuscule carrier transport compared with injection.

To complete our intuitive picture of the diode under forward bias, we need to determine how the injected holes and electrons are transported from the depletion region across the n and p bulk regions, in which they are minority carriers. Injection sets up a gradient in the minority carrier concentration across the bulk regions, which strongly suggests that diffusion plays a dominant role. In this section, we will consider the most common (and simplest) situation where the ohmic contacts are very close to the junction. Minority carriers recombine at the ohmic contacts due to the disruption of the silicon lattice at the metal-silicon interface. The electron or hole participating in the recombination event is supplied by the adjacent metal region, with the wire and voltage supply completing the circuit. One of the properties of an ohmic contact is that the minority and majority carrier concentrations are kept at their thermal equilibrium values at the metal-silicon interface, regardless of the current density through it.

We conclude that the steady-state minority-carrier concentrations under forward bias are relatively high at the depletion region edges due to injection and decrease to their low thermal equilibrium values at the ohmic contacts. Rather than proceeding mathematically, it is worthwhile to consider first a simple example that can help to avoid some conceptual difficulties in thinking about steady-state diffusion.

Figure 6.7(a) shows a container filled with water. Adjacent to the left side of the container is a vessel of ink with a fill tube for maintaining its level. The two containers are initially separated by a thin sheet of plastic, so that no diffusion of ink into the water can occur. Next to the right-hand side of the water container, we have a "vacuum" for ink molecules that can maintain the ink concentration at zero on that side of the container of water. At $t = 0$, we remove the plastic sheet from left side and attach the "ink vacuum" to the right side of the water container. The ink diffuses across the water container from left to right; when the ink molecules reach the right-hand side, they are removed from the water container by the vacuum. As the ink level drops in the ink vessel, we add more to maintain the ink concentration at the left-hand side of the container at n_I^*.

Figure 6.7(b) shows the experiment after steady-state is reached (i.e., after the ink concentration $n_I(x)$ in the water is no longer changing). The flux of ink F_I [# $\text{cm}^{-2}\text{s}^{-1}$] is constant at each point across the water container, since otherwise the ink would "pile up" at some point in the container. In other words, each ink molecule that enters from the ink vessel diffuses across the container and is extracted at the right-hand side. From Fick's Law, the ink flux is proportional to the derivative of the ink concentration, as discussed in Chapter 2:

$$F_I = \text{constant} \propto \frac{dn_I}{dx} \rightarrow n_I(x) = a + bx \tag{6.6}$$

where a and b are constants. The ink concentration $n_I(x)$ is linear and is n_I^* on the left side and zero on the right side. The ink concentration plot is shown in Fig. 6.7(c).

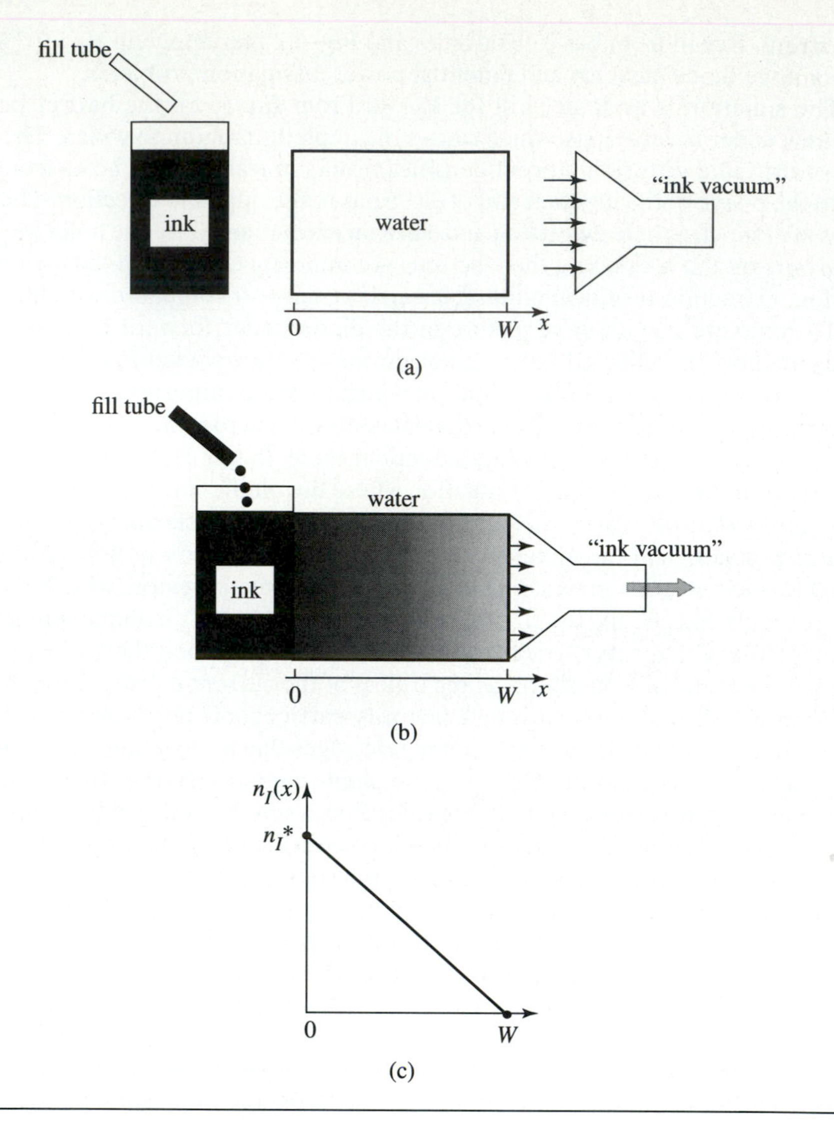

➤ **Figure 6.7** Ink diffusion experiment: (a) set up at $t = 0$, (b) after steady-state diffusion is achieved, and (c) steady-state ink concentration $n_I(x)$.

At first glance, the plot of $n_I(x)$ in Fig. 6.7(c) is puzzling. The ink concentration decreases from left to right, which would seem to indicate that ink molecules are disappearing as they diffuse through the water. The misunderstanding comes from not realizing that the plot is for the *steady-state*. Ink molecules are continuously entering at $x = 0$, diffusing across the water container, and being removed at $x = W$. Since the concentration is higher on the left side than on the right, a linear concentration profile develops in steady state which corresponds to a constant diffusion flux. In this text, we will not consider the more complicated problem of finding *transient* solutions, such as $n_I(x,t)$ for $t > 0$.

The analogy to the pn junction diode is clear: the forward-biased junction injects minority carriers, which diffuse across the bulk region to the ohmic contacts. The bulk regions are short enough that we can neglect the loss of minority carriers due to recombination, which results in a linear decrease in carrier concentration and therefore, a constant diffusion current density. The case where recombination in the bulk regions can be neglected is termed the **short-base diode**. Figure 6.8 plots the minority carrier concentrations on a linear scale under forward bias. We have yet to determine the functional dependence of the minority carrier concentrations at the depletion region edges $p_n(x_n)$ and $n_p(-x_p)$ on the applied bias voltage, where the subscript indicates the type of the silicon region. Although we could formally differentiate $p_n(x)$ and $n_p(x)$ to find the minority carrier diffusion current densities in each region, we will first fill in the missing piece of how $p_n(x_n)$ and $n_p(-x_p)$ depend on the applied bias voltage.

6.3.2 The Law of the Junction

From Eq. (3.34), the potential barrier in thermal equilibrium is given by

$$\phi_B = \phi_n - \phi_p = V_{th}\ln\left(\frac{N_d}{n_i}\right) - \left(-V_{th}\ln\left(\frac{N_a}{n_i}\right)\right) = V_{th}\ln\left(\frac{N_d N_a}{n_i^2}\right) \tag{6.7}$$

This equation can be rearranged, by first labelling the minority hole and majority electron concentrations on the n-side of the junction as:

$$p_{no} = \frac{n_i^2}{N_d} \text{ and } n_{no} = N_d \tag{6.8}$$

where the subscript n refers to the n-side and the o indicates thermal equilibrium. Similarly, the minority electron and majority hole concentrations on the p-side of the junction are:

$$n_{po} = \frac{n_i^2}{N_a} \text{ and } p_{po} = N_a \tag{6.9}$$

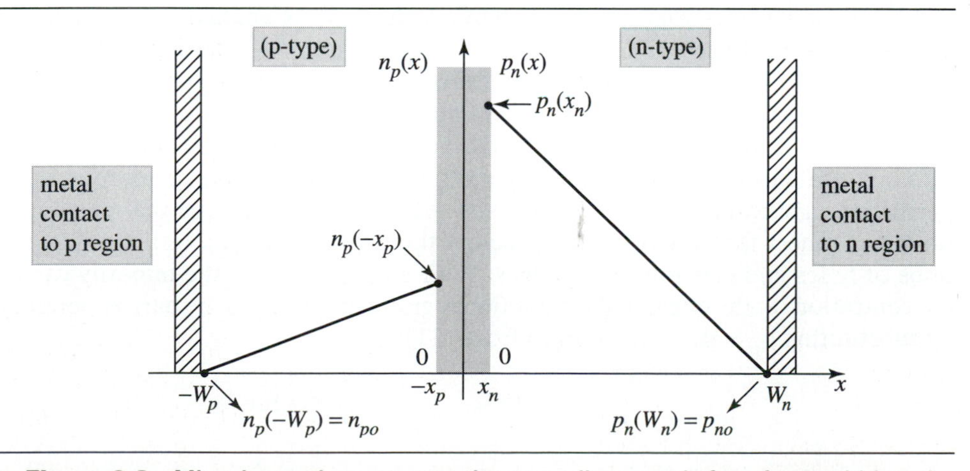

Figure 6.8 Minority carrier concentrations on a linear scale for a forward-biased short-base diode.

Substituting into Eq. (6.7), we find that:

$$\phi_B = V_{th}\ln\left(\frac{p_{po}}{p_{no}}\right) \text{ and } \phi_B = V_{th}\ln\left(\frac{n_{no}}{n_{po}}\right) \quad (6.10)$$

By solving for the minority carrier concentrations, we find that:

$$p_{no} = p_{po}e^{-\phi_B/V_{th}} \text{ and } n_{po} = n_{no}e^{-\phi_B/V_{th}} \quad (6.11)$$

The new perspective in Eq. (6.11) is that the minority carrier concentrations are proportional to the majority carrier concentrations on the *other* side of the junction, multiplied by an exponential factor. At room temperature, the built-in potential can be estimated by looking at the ratio of majority carrier concentration to the minority carrier concentration on the other side of the junction. By converting Eq. (6.10) to base–10 logarithms,

$$\phi_B = (60\text{ mV})\log\left(\frac{p_{po}}{p_{no}}\right) = (60\text{ mV})\log\left(\frac{n_{no}}{n_{po}}\right) \quad (6.12)$$

For example, a difference of 14 orders of magnitude corresponds to a built-in potential $\phi_B = 60\text{ mV} \times 14 = 840\text{ mV}$.

The results in Eq. (6.11) are special cases of Boltzmann statistics, a fundamental physical principle. This principle relates the number of charged particles N_1 at a potential level ϕ_1 to the number N_2 at a different potential level ϕ_2, for the case of thermal equilibrium. Boltzmann's Principle for positively charged particles states that

$$\frac{N_2}{N_1} = e^{-[q\phi_2 - q\phi_1]/kT} = e^{-\Delta\phi/V_{th}} \quad (6.13)$$

This expression corresponds to the hole equation in Eq. (6.11). The built-in potential ϕ_B is the potential difference $\Delta\phi$ across the depletion region, which separates regions where the concentration of holes is p_{no} (the n-side) and p_{po} (the p-side). Equation (6.11) also gives Boltzmann's Principle for negatively charged electrons.

Now we consider what changes when a voltage V_D is applied to the diode, as shown in Fig. 6.5. In general, the answer is a very complicated one, depending on the magnitude of the voltage and the dimensions and the physical parameters of the diode. Fortunately, typical forward biases do not move the semiconductor that far from thermal equilibrium. We assume that the balance between the very large diffusion and drift current densities in the depletion region in thermal equilibrium is not perturbed significantly by the forward bias. As a result, we can use the same equations to connect the potential change across the depletion region to the concentrations of holes and electrons at its edges. With an applied bias, the minority carrier concentrations at the edges of the depletion region are related to the barrier potential by substituting $\phi_j = \phi_B - V_D$ for ϕ_B in Eq. (6.11):

$$p_n(x_n) = p_p(-x_p)e^{-\phi_j/V_{th}} = p_p(-x_p)e^{-(\phi_B - V_D)/V_{th}} \quad (6.14)$$

$$n_p(-x_p) = n_n(x_n)e^{-\phi_j/V_{th}} = n_n(x_n)e^{-(\phi_B - V_D)/V_{th}} \quad (6.15)$$

Note that Eqs. (6.14) and (6.15) only relate the minority and majority carrier concentrations under bias at the edges of the depletion region, $x = x_n$ and $x = -x_p$. We know from Fig. 6.8 that the minority carrier concentrations vary with x under an applied bias.

The majority carrier concentrations are also related to the minority concentrations by the requirement that the bulk regions on either side of the junction remain charge-neutral, even under applied bias. On the p-side of the junction, $\rho(x = -x_p) = 0$ implies that

$$p_p(-x_p) - n_p(-x_p) - N_a = 0 \rightarrow p_p(-x_p) = N_a + n_p(-x_p) \approx p_{po} + n_p(-x_p) \quad (6.16)$$

where we have approximated $p_{po} \approx N_a$. The majority hole concentration on the p-side therefore increases by the injected electron concentration, in order to maintain charge neutrality. Similarly, the majority electron concentration at the edge of the depletion region is given by

$$n_n(x_n) \approx n_{no} + p_n(x_n) \quad (6.17)$$

Although we could substitute into Eqs. (6.16) and (6.17) back into Eqs. (6.14) and (6.15) to solve for the minority concentrations, there is an important simplification that applies to many practical situations. Under conditions of **low-level injection**, we can neglect the slight increase in majority carrier concentration due to minority carrier injection in Eqs. (6.16) and (6.17):

$$\boxed{n_p(-x_p) \ll p_{po} \approx N_a \text{ and } p_n(x_n) \ll n_{no} \approx N_d} \quad (6.18)$$

low-level injection

For low-level injection, the majority carrier concentrations are nearly unchanged from thermal equilibrium

$$p_p(-x_p) \approx p_{po} \approx N_a \text{ and } n_n(x_n) \approx n_{no} \approx N_d \quad (6.19)$$

In this chapter, we will assume low-level injection and furthermore, use the equal sign in Eq. (6.19) in deriving expressions for the minority carrier concentrations. By substituting Eq. (6.19) into Eq. (6.14) we find that the minority hole concentration at the n-side of the depletion region is:

$$p_n(x_n) = N_a e^{-(\phi_B - V_D)/V_{th}} = N_a e^{-\phi_B/V_{th}} \cdot e^{V_D/V_{th}} \quad (6.20)$$

Similarly, substitution of Eq. (6.19) into Eq. (6.15) results in an expression for the minority electron concentration at the p-side of the depletion region:

$$n_p(-x_p) = N_d e^{-(\phi_B - V_D)/V_{th}} = N_d e^{-\phi_B/V_{th}} \cdot e^{V_D/V_{th}} \quad (6.21)$$

The first terms on the right-hand sides of Eqs. (6.20) and (6.21) are the equilibrium minority carrier concentrations, as can be seen from Eq. (6.11).

Substituting, we find the following important results

$$\boxed{p_n(x_n) = p_{no} \cdot e^{V_D/V_{th}} \text{ and } n_p(-x_p) = n_{po} e^{V_D/V_{th}}} \quad (6.22)$$

law of the junction

An applied forward bias $V_D > 0$ exponentially increases the minority carrier concentrations at the edges of the junction above their equilibrium levels. A reverse bias $V_D < 0$ exponentially decreases the minority carrier concentrations below their equilibrium levels, which are small already. The results in Eqs. (6.22) hold for low-level injection and are central to modeling the pn diode; they are known together as the **law of the junction.**

➤ EXAMPLE 6.1 Minority Concentrations under Applied Bias

Given a diode with the p-region doped with $N_a = 10^{18}$ cm^{-3} and the n-region doped with $N_d = 10^{16}$ cm^{-3}, plot the minority carrier concentrations at the edges of the depletion region as a function of the applied bias voltage for -0.4 V $< V_D < 0.8$ V using a log-linear scale.

SOLUTION

We begin by finding the minority carrier concentrations in thermal equilibrium:

$$p_{no} = n_i^2/N_d = 10^{20}/10^{16} = 10^4 \text{ cm}^{-3}$$

$$n_{po} = n_i^2/N_a = 10^{20}/10^{18} = 10^2 \text{ cm}^{-3}$$

Note that the minority carriers are more numerous on the side of the junction that is *less* heavily doped.

It is convenient to express the Law of the Junction as a power of 10:

$$p_n(x_n) = p_{no} \cdot 10^{(\log e) V_D/V_{th}} = p_{no} \cdot 10^{V_D/60\text{mV}} \quad \text{and}$$

$$n_p(-x_p) = n_{po} \cdot 10^{(\log e) V_D/V_{th}} = n_{po} \cdot 10^{V_D/60\text{mV}}$$

Taking the logarithm of both sides, the dependence of the minority carrier concentrations on the applied voltage can be expressed for plotting on a log-linear graph:

$$\log [p_n(x_n)] = 4 + V_D/(60\text{mV})$$

$$\log [n_p(-x_p)] = 2 + V_D/(60\text{mV})$$

These results are plotted on the graph below.

As the plots show, the minority carrier concentrations are negligible for reverse biases of more than a few hundred mV. In the case of the minority hole concentration on the n-side of the junction, forward biases of greater than about $V_D = 0.7$ V result in values that approach the majority electron concentration (10^{16} cm^{-3} for this example), which invalidates the low-level injection assumption underlying the law of the junction.

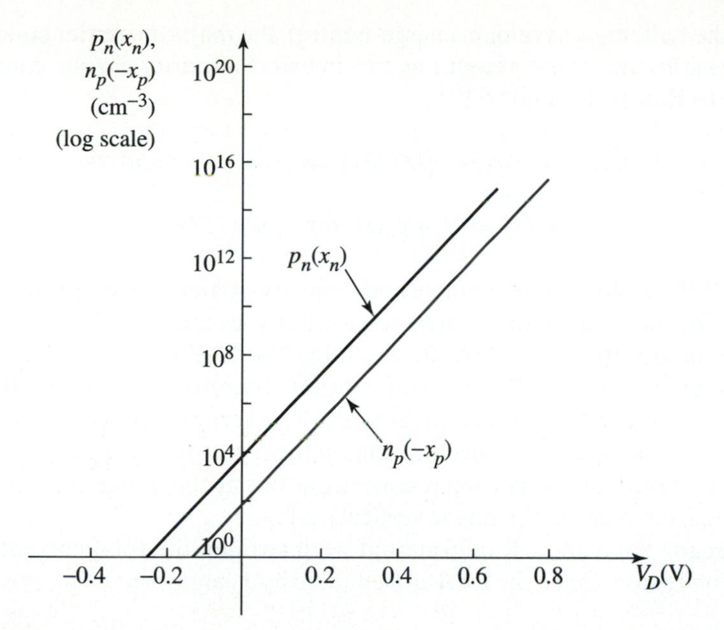

➤ **Figure Ex6.1** Minority carrier concentrations at the edges of the depletion region as functions of the applied bias voltage for a diode with $N_a = 10^{18}$ cm^{-3} and $N_d = 10^{16}$ cm^{-3}.

6.3.3 pn Junction Currents under Forward Bias

The minority carrier distributions are the key to understanding the current components in the diode under applied bias. The law of the junction gives the minority carrier concentrations at the edges of the depletion region $n_p(-x_p)$ and $p_n(x_n)$ as a function of V_D, which completes the graphs of the steady-state concentrations $n_p(x)$ and $p_n(x)$ in the bulk regions in Fig. 6.8. Recall that the linear functions apply to a short-base diode, in which the bulk regions are short enough that the minority carriers diffuse across them without significant loss due to recombination.

In the n-type bulk region, the minority hole concentration in Fig. 6.8 is a straight line between $p_n(x_n)$ at the edge of the depletion region and $p_n(W_n) = p_{no}$ at the ohmic contact:

$$p_n(x) = p_n(x_n) - \left(\frac{p_n(x_n) - p_{no}}{W_n - x_n}\right)(x - x_n) \quad \text{for } x_n \leq x \leq W_n \tag{6.23}$$

Similarly, the minority electron concentration in the p-type bulk region, is given by:

$$n_p(x) = n_p(-x_p) + \left(\frac{n_p(-x_p) - n_{po}}{W_p - x_p}\right)(x + x_p) \quad \text{for } -W_p \leq x \leq -x_p \tag{6.24}$$

Since the bulk regions remain charge-neutral, the majority carrier concentrations must increase by the same amount as the injected minority carrier concentration. According to Eqs. (6.16) and (6.17)

$$p_p(x) = N_a + n_p(x) \quad \text{for } -W_p \le x \le -x_p \text{ and} \tag{6.25}$$

$$n_n(x) = N_d + p_n(x) \quad \text{for } x_n \le x \le W_n \tag{6.26}$$

Figure 6.9 is a plot of the minority and majority carrier concentrations under forward bias. The injected minority carrier concentrations are much less than the majority carrier concentrations, since we are assuming low-level injection. The vertical axis in Fig. 6.9 is broken to show that the majority and minority carrier concentrations are plotted on different ranges of the linear scale. The acceptor concentration on the p-side is higher than the donor concentration on the n-side for the plots in Fig. 6.9. The thermal equilibrium minority carrier concentrations at the ohmic contacts are indistinguishable from zero on the linear vertical scale.

We already have enough information to determine the total current density in the depletion region. Since the total current density J cannot vary with position in the

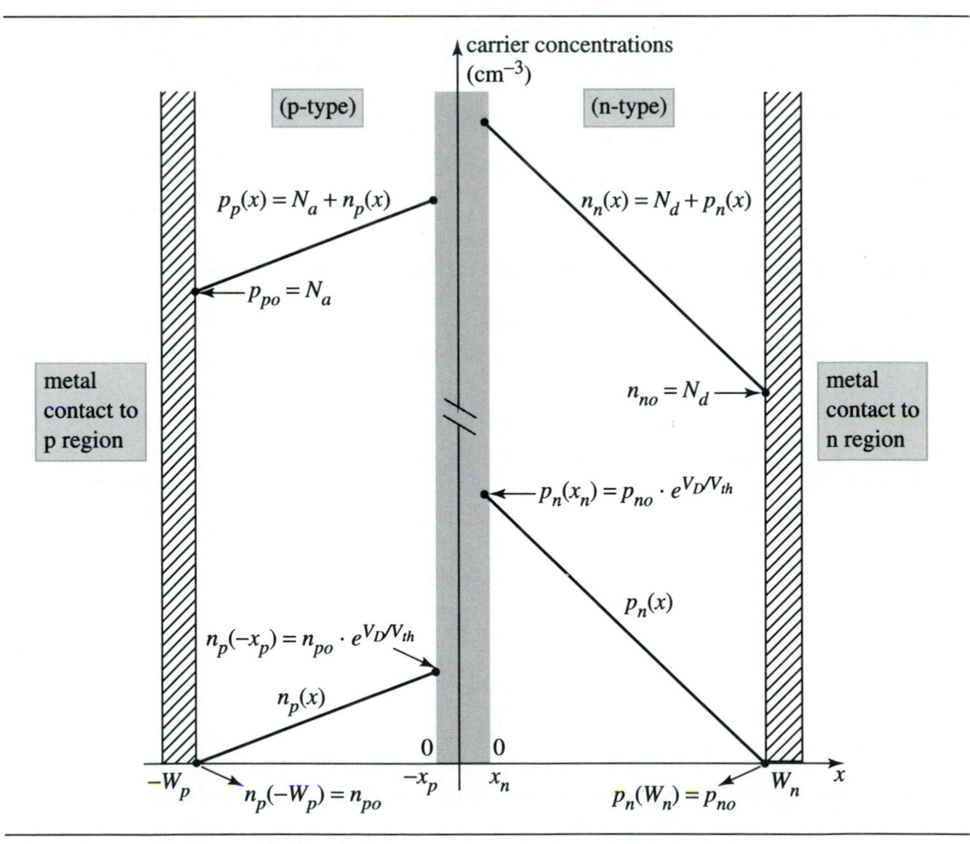

➤ **Figure 6.9** Plots of the minority and majority carrier concentrations in a short-base diode under a forward bias $V_D > 0$ V. The vertical scale is different for the minority and majority carriers, since the former concentrations are much lower under the low-level injection assumption. For this example, we have assumed that $N_a > N_d$.

diode in steady state, we therefore know its value throughout the structure. Minority carrier injection is the net carrier transport process operating across the depletion region. Of course, there are large "built-in" drift and diffusion current densities there, but these nearly cancel. The net current density in the depletion region consists of holes injected from the p-side to the n-side and electrons injected in the opposite direction. The total current density J is therefore the sum of the minority carrier diffusion current densities at the edges of the depletion region:

$$J = J_p^{diff}(x_n) + J_n^{diff}(-x_p) = -qD_p \frac{dp_n}{dx}\bigg|_{x=x_n} + qD_n \frac{dn_p}{dx}\bigg|_{x=-x_p} \tag{6.27}$$

Substituting the minority carrier concentrations from Eqs. (6.23) and (6.24), we can evaluate the derivatives and find that

$$J = -qD_p\left[\frac{p_n(W_n) - p_n(x_n)}{W_n - x_n}\right] + qD_n\left(\frac{n_p(-x_p) - n_p(-W_p)}{W_p - x_p}\right) \tag{6.28}$$

The minority carrier concentrations at the ohmic contacts are their thermal equilibrium values $p_n(W_n) = p_{no}$ and $n_p(-W_p) = n_{po}$. At the depletion region edges, the law of the junction in Eq. (6.22) can be substituted, which leads to

$$J = \left(\frac{-qD_p}{W_n - x_n}\right)(p_{no} - p_{no}e^{V_D/V_{th}}) + \left(\frac{qD_n}{W_p - x_p}\right)(n_{po}e^{V_D/V_{th}} - n_{po}) \tag{6.29}$$

This result can be simplified by neglecting the width of the depletion region edges compared to the bulk regions ($x_n \ll W_n$ and $x_p \ll W_p$), substituting for the thermal equilibrium minority carrier concentrations p_{no} and n_{po} from the mass-action law, and then factoring out the term with the exponential

$$J = qn_i^2\left(\frac{D_p}{N_d W_n} + \frac{D_n}{N_a W_p}\right)(e^{V_D/V_{th}} - 1) \tag{6.30}$$

The final expression for the total current density in the diode can be multiplied by the cross sectional area of the diode A from Fig. 6.3(b) to find the diode current I_D

$$I_D = qn_i^2 A\left(\frac{D_p}{N_d W_n} + \frac{D_n}{N_a W_p}\right)(e^{V_D/V_{th}} - 1) = I_o(e^{V_D/V_{th}} - 1) \tag{6.31}$$

short-base diode $I_D(V_D)$ characteristic

where the diode saturation current is

$$I_o = qn_i^2 A\left(\frac{D_p}{N_d W_n} + \frac{D_n}{N_a W_p}\right) \tag{6.32}$$

short-base diode saturation current

Although we have made several simplifying assumptions, the form of Eq. (6.31) is close to the experimental measurements of $I_D(V_D)$ for IC pn junction diodes.

In order to complete the picture, we now find the components of J in the different regions of the diode under forward bias. The minority carrier diffusion current densities in the bulk regions are given by the two terms in Eq. (6.28), since the derivatives are constant. The holes diffusing across the n bulk region are supplied from the majority holes on the p- side of the junction. Therefore, there must be a majority hole current on the p-side of the junction to supply the injection of holes across the depletion region. Similarly, on the n-side of the junction, a majority electron current flows to supply the injection of electrons across the depletion region to the p-side of the junction. The majority carrier current densities consist of both drift and diffusion components. Fig. 6.10 shows the current components through the pn diode.

6.3.4 pn Junction Diode Currents under Reverse Bias

The previous section presented the short-base diode currents under forward bias; however, the minority carrier concentrations in Eqs. (6.23) and (6.24) are general. Under a reverse bias of more than 0.1 mV, the minority carrier concentrations at the depletion region edges become negligible compared with their thermal equilibrium concentrations

$$p_n(x_n) = p_{no}e^{V_D/V_{th}} \ll p_{no} \text{ and } n_p(-x_p) = n_{po}e^{V_D/V_{th}} \ll n_{po}\ (V_D < -0.1\text{ V}) \quad (6.33)$$

Figure 6.11 plots the minority carrier concentrations in Eqs. (6.23) and (6.24) under reverse bias. Although the majority carrier concentrations are not plotted in Fig. 6.11, they have the same variation with x as the minority carriers, except offset by the orders-of-magnitude larger doping concentrations in the p and n regions as in Fig. 6.9.

The total diode current density is the sum of the minority carriers extracted across the depletion region by diffusion from the opposite bulk region. Evaluating Eq. (6.28) using Eq. (6.33) for the minority carrier concentrations under reverse bias, the total current density is

$$J \approx -qD_p\left[\frac{p_n(W_n) - 0}{W_n - x_n}\right] + qD_n\left(\frac{0 - n_p(-W_p)}{W_p - x_p}\right) \approx -q\left(\frac{D_p p_{no}}{W_n} + \frac{D_n n_{po}}{W_p}\right) \quad (6.34)$$

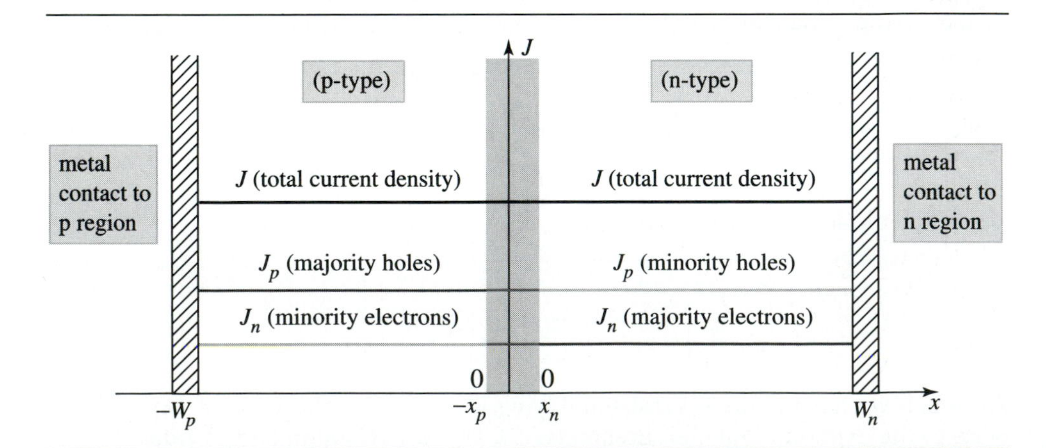

➤ **Figure 6.10** Current-density components in the short-base diode under forward bias.

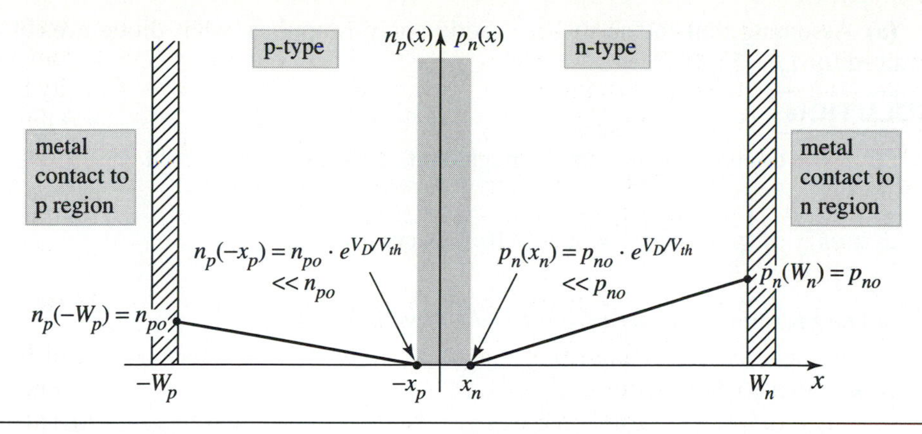

➤ Figure 6.11 Linear plot of the minority carrier concentrations under reverse bias, with $V_D < 4V_{th}$. Note that the scale for this plot is expanded by many orders of magnitude compared to that of Fig. 6.9, where the thermal equilibrium concentrations were negligible.

Note that we could have found I_D directly by substituting $V_D < -0.1$ V into Eq. (6.31). The current $I_D = -I_o$ under reverse bias is given in Eq. (6.32).

The current components in reverse bias are determined using a similar procedure as was used for forward bias in Fig. 6.10. In this case, the minority diffusion currents are extracted and become the majority carrier current on the other side of the junction. The reverse bias characteristic of IC diodes deviates from this model. The reverse current is found to increase with increasing reverse bias and is much larger in magnitude than $-I_o$ for typical biases, although still negligible for most practical applications. The physical origin of the additional components of the reverse current is left for an advanced course.

➤ EXAMPLE 6.2 Short-Base Diode Currents

An IC diode is designed to have a room-temperature saturation current of $I_o = 5 \times 10^{-17}$ A for a particular application. The fabrication process results in the device dimensions and physical parameters listed below. The depletion region width can be neglected in comparison with the bulk region widths. Note that for an IC diode, we estimate the effective width for the side which is contacted lateral to the intrinsic pn junction (the n-side in Fig. 6.3). The *minority* carrier diffusion coefficients are estimated from the plot of mobility as a function of doping concentration in Fig. 2.8 and Einstein's Relation. Since the p-side is doped more heavily, the minority electron mobility and diffusion coefficient are approximately equal to the values for the minority holes on the relatively lightly doped n-side of the junction.

Dimensions:	Doping:	Diffusion Coefficients:
$W_p = 0.5$ μm $W_n = 1.0$ μm	$N_a = 2.5 \times 10^{17}$ cm^{-3} $N_d = 4.0 \times 10^{16}$ cm^{-3}	$D_n = 5$ cm^{2}s^{-1} $D_p = 5$ cm^{2}s^{-1}

(a) Assuming that the short-base diode analysis applies, what diode area A is required for $I_o = 5 \times 10^{-17}$ A?

SOLUTION

Since the depletion width can be neglected, the saturation current is:

$$J_o \cong qn_i^2\left(\frac{D_n}{N_a W_p} + \frac{D_p}{N_d W_n}\right) = 2.64 \times 10^{-11}\ \text{A/cm}^2$$

Solving for the diode area from $I_o = J_o A$, we find that

$$A = 1.89 \times 10^{-6} \text{cm}^2$$

A square layout for the "intrinsic" diode (see Fig. 6.3) with this area would have an edge length of about 14 μm.

(b) Find the current and the minority carrier concentrations at the edges of the depletion region for a forward bias of 720 mV.

SOLUTION

For a forward bias of $V_D = 720$ mV, the diode current is:

$$I_D = I_o(e^{V_D/V_{th}} - 1) \cong (5 \times 10^{-17}\text{A}) \cdot 10^{720/60} = 50\,\mu\text{A}$$

The minority carrier concentrations at the depletion region edges are:

$$n_p(-x_p) = \left(\frac{n_i^2}{N_a}\right)e^{V_D/V_{th}} = \left(\frac{10^{20}}{2.5 \times 10^{17}}\right)10^{720/60} = 4 \times 10^{14}\text{cm}^{-3} \text{ and}$$

$$p_n(x_n) = \left(\frac{n_i^2}{N_d}\right)e^{V_D/V_{th}} = \left(\frac{10^{20}}{4 \times 10^{16}}\right)10^{720/60} = 2.5 \times 10^{15}\text{cm}^{-3}$$

which are much less than the respective majority carrier concentrations. Therefore, the low-level injection assumptions that $n_p(-x_p) \ll N_a$ and $p_n(x_n) \ll N_d$ are valid.

6.4 pn JUNCTION DIODE CIRCUIT MODELS

The basic operation of the short-base pn junction diode was presented in Section 6.3, which contains the essential features needed to develop both large- and small-signal circuit models. Since the diode is a non-linear device, we will find models for each of its operating regions.

6.4.1 Large-Signal Static Model

We begin by developing large-signal models that approximate its static $I_D - V_D$ characteristics. Figure 6.12 shows the current-voltage plot for a typical IC pn diode along with a piecewise-linear approximation. For normal (non-breakdown) operation, the

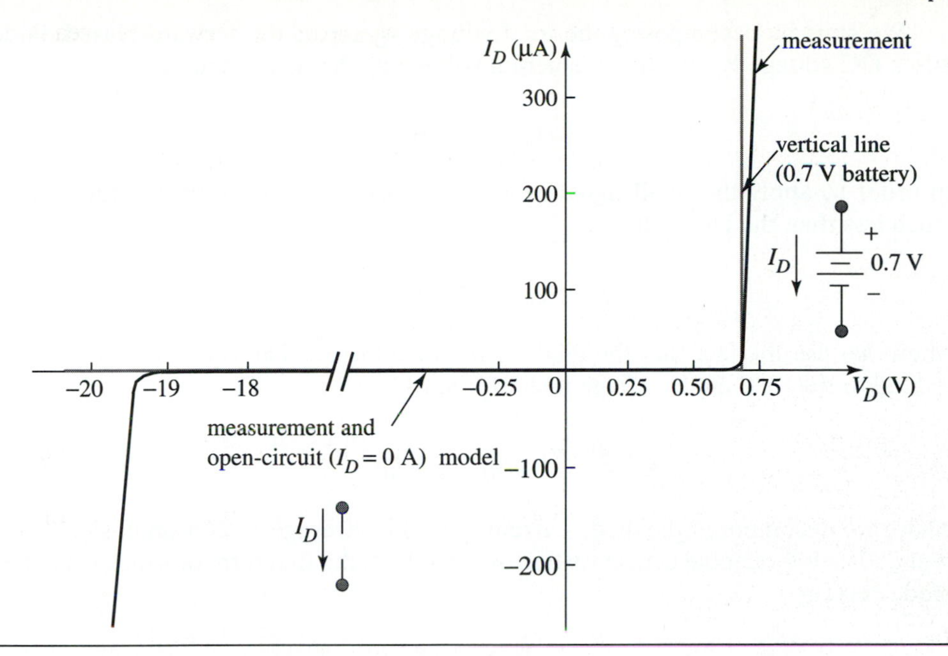

➤ **Figure 6.12** Piecewise-linear approximation to the static pn diode.

static exponential diode characteristic can be represented with adequate accuracy for hand calculations by two straight lines

$$\text{Diode “on”— } V_D = 0.7\text{V} \qquad (I_D > 0) \tag{6.35}$$

$$\text{Diode “off”— } I_D = 0 \qquad (V_D < 0.7\text{V}) \tag{6.36}$$

In contrast to the MOSFET, the diode model is piecewise-linear. Therefore, linear circuit analysis techniques can be applied to diode circuits once the appropriate “on” or “off” model is substituted into the circuit. Later in this chapter, we will apply this model to some typical diode circuits.

This model is useful for hand analysis of large-signal diode circuits such as rectifiers, if the signals vary slowly enough to ignore the switching transients between the *on* and *off* states. Due to the storage of minority carriers in the p and n bulk regions, a significant time delay is needed to switch a diode from forward to reverse bias. This complicated topic is best left for an advanced course in device physics.

6.4.2 Small-Signal Resistance

The diode is a two-terminal element and therefore, controlled sources are not needed in the small-signal model. We can differentiate the current-voltage characteristic, evaluate it at a particular operating point, and relate the small-signal current directly to the small-signal voltage. Considering the rapid increase of the exponential function, it is important to quantify the range of signals for which the small-signal model is a reasonable approximation. In the process, we gain further insight into the basic concept of small-signal modeling.

We start by decomposing the total voltage v_D across the forward-biased diode into a DC voltage V_D and an incremental voltage v_d, that is defined by

$$v_d = v_D - V_D \tag{6.37}$$

In order to apply the small-signal concept, we assume that the magnitude of v_d is much less than the DC voltage

$$|v_d| \ll V_D \approx 0.7 \text{ V} \tag{6.38}$$

where we use the fact that the diode is forward biased. The total diode current is related to the total diode voltage under forward bias by

$$i_D = I_o e^{v_D/V_{th}} \rightarrow I_D + i_d = I_o e^{(V_D + v_d)/V_{th}} \tag{6.39}$$

where we decomposed the diode current i_D into its average I_D and small-signal current i_d. The exponential can be written as a product, the first term of which is the DC diode current

$$I_D + i_d = I_o e^{V_D/V_{th}} e^{v_d/V_{th}} = I_D e^{v_d/V_{th}} \tag{6.40}$$

Expansion of the exponential into a power series is the next step

$$I_D + i_d = I_D\left(1 + \frac{v_d}{V_{th}} + \frac{1}{2}\left(\frac{v_d}{V_{th}}\right)^2 + \frac{1}{6}\left(\frac{v_d}{V_{th}}\right)^3 + \ldots\right) \tag{6.41}$$

Noting that the DC diode current appears on both sides of the equation and can be cancelled, the small-signal diode current is given by

$$i_d = I_D\left(\frac{v_d}{V_{th}}\right) + \frac{I_D}{2}\left(\frac{v_d}{V_{th}}\right)^2 + \frac{I_D}{6}\left(\frac{v_d}{V_{th}}\right)^3 + \ldots \tag{6.42}$$

A linear approximation for i_d can be made by neglecting all but the first term in the power series

$$i_d \cong \left(\frac{I_D}{V_{th}}\right) v_d = g_d v_d \tag{6.43}$$

where g_d is the small-signal diode conductance (units $S = \Omega^{-1}$). The diode's small-signal resistance r_d is the inverse of the conductance

small signal diode resistance

$$\boxed{r_d = \frac{1}{g_d} = \frac{V_{th}}{I_D}} \tag{6.44}$$

For a typical DC diode current $I_D = 100\ \mu A$, the small-signal resistance is $r_d = 260\ \Omega$.

It is worthwhile to estimate the error introduced by neglecting the second (quadratic) term since it is the next largest term in the expansion. A second-order approximation to the small-signal current includes the quadratic term in Eq. (6.42)

$$i_d = g_d v_d\left(1 + \frac{(I_D/(2V_{th}^2))}{I_D/V_{th}} v_d\right) = g_d v_d\left(1 + \frac{v_d}{2V_{th}}\right) = g_d v_d(1+\varepsilon) \qquad (6.45)$$

By neglecting the quadratic term, the linear small-signal conductance underestimates the current by the error

$$\varepsilon = \frac{v_d}{2V_{th}} \qquad (6.46)$$

If we decide to limit this error to 10%, then this condition places an upper bound on the small-signal voltage

$$\varepsilon = \frac{v_d}{2V_{th}} \le 0.1 \rightarrow v_d \le 0.2 \cdot V_{th} = 5\text{mV (at room temperature)} \qquad (6.47)$$

This result gives a measure of what "small" means in modeling the pn diode as a small-signal resistor. As an aside, power series expansions are useful in analyzing distortion in analog ICs. For example, quadratic and cubic terms lead directly to second and third harmonic distortion in an amplifier.

6.4.3 Depletion Capacitance under Forward Bias

We now turn to small-signal charge storage in the pn diode. One source of charge storage in the forward-biased diode is the depletion region. In Section 3.6, we derived Eq. (3.80) for the capacitance of a reverse-biased pn junction ($V_D < 0$)

$$C_j = \frac{C_{jo}}{\sqrt{1 - V_D/\phi_B}} \qquad (6.48)$$

where the zero-bias capacitance C_{jo} given by

$$C_{jo} = A\sqrt{\frac{q\varepsilon_s N_a N_d}{2(N_a + N_d)\phi_B}} \qquad (6.49)$$

Under significant forward bias $V_D > \phi_B / 2$ the approximation that there are no mobile carriers in the depletion region becomes invalid, which is a basic assumption behind Eq. (6.48). A common *ad hoc* approximation to the junction capacitance under forward bias is to evaluate Eq. (6.48) at $V_D = \phi_B / 2$

$$C_j \cong \frac{C_{jo}}{\sqrt{1 - 1/2}} = \sqrt{2}C_{jo} \text{ for forward bias} \qquad (6.50)$$

forward-bias depletion capacitance

6.4.4 Diffusion Capacitance

The depletion capacitance is only part of the small-signal diode capacitance. Under forward bias, there is another component that is often much larger. Its physical origin lies in the storage of charge in the bulk regions due to the injection of minority carriers. As shown in Fig. 6.9, the bulk p and n regions of the diode remain charge-neutral under forward bias, since the majority carrier concentrations increase by the same amount as the injected minority carriers.

A positive increment in the forward bias voltage increases the minority carrier concentrations at the edges of the depletion region. In steady-state, the carrier distributions all increase to higher levels. As a result, additional charge is stored in the majority carrier distributions on each side of the junction—charge that must have been supplied by the external circuit. One way of seeing this unusual charge-storage mechanism is that the equal and opposite increments in minority and majority charge on each side of the junction constitute the virtual "plates" of a capacitor. The capacitors on the p-side and the n-side are in parallel, since the charge storage on both increases with a positive increment in v_D.

Figure 6.13 plots the minority and majority carrier concentrations for an incremental increase v_d in the DC bias voltage V_D. The shaded areas are proportional to the magnitudes of the changes in the charge storage associated with each carrier dis-

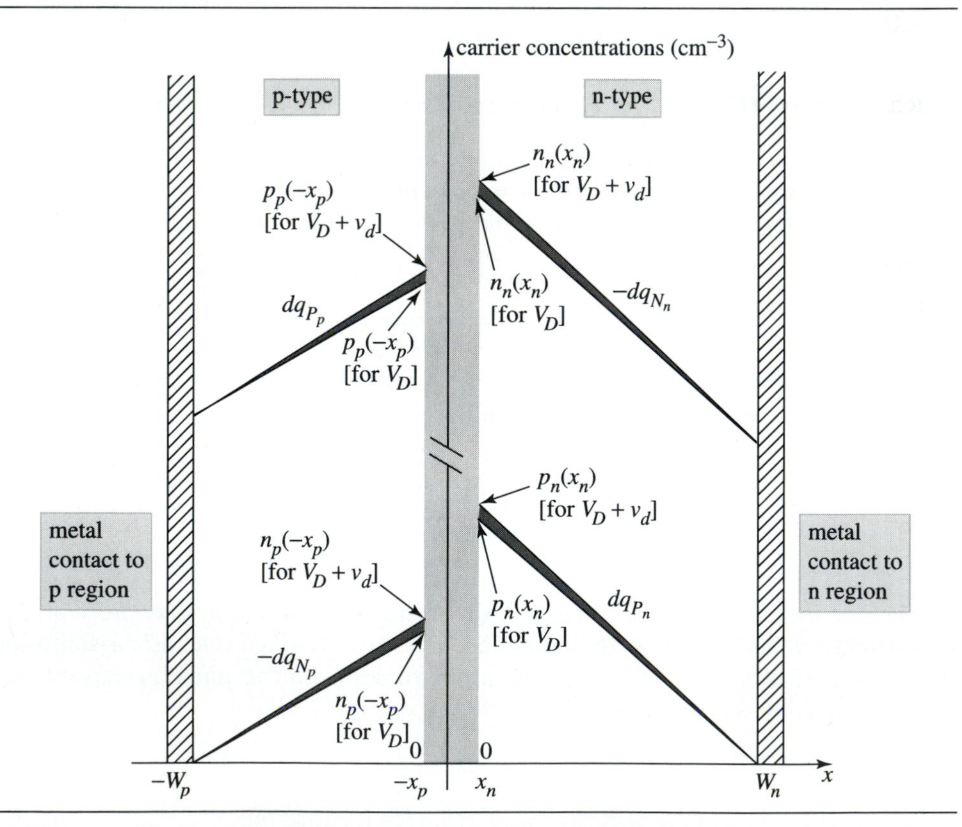

➤ **Figure 6.13** Incremental changes in the short-base minority and majority carrier charge storage due to an incremental increase v_d in the DC diode voltage V_D.

tribution. Since charge neutrality is maintained, the charge increments on each side of the junction cancel

$$dq_{P_p} = -dq_{N_p} \text{ and } dq_{P_n} = -dq_{N_n} \tag{6.51}$$

where the subscript indicates the associated carrier distribution as shown in Fig. 6.13. Note that there is no relationship implied by Eq. (6.51) between the charge increments on opposite sides of the junction and in general $dq_{P_p} \neq dq_{P_n}$ and $dq_{N_p} \neq dq_{N_n}$.

In order to find the capacitance, we must first determine the functional dependence of the charge storage on the total applied bias v_D. There is no stored charge in thermal equilibrium so we must subtract the thermal equilibrium concentrations from the carrier distributions prior to integrating. For convenience, we define the excess minority carrier concentrations by

$$n'_p(x) = n_p(x) - n_{po} \text{ and } p'_n(x) = p_n(x) - p_{no} \tag{6.52}$$

On the p-side of the junction, the stored charge in the minority electron distribution $q_{N_p}(v_D)$ is found by integrating $-qn_p'(x)$. Substituting from Eq. (6.24) and Eq. (6.22), the charge density due to excess minority electrons is

$$-qn_p'(x) = -q\left[n_{po}e^{v_D/V_{th}} + \left(\frac{n_{po}e^{v_D/V_{th}} - n_{po}}{W_p - x_p}\right)(x + x_p) - n_{po}\right] \text{ for } -W_p \leq x \leq -x_p \tag{6.53}$$

which can be rewritten as

$$-qn_p'(x) = -qn_{po}(e^{v_D/V_{th}} - 1)\left[1 + \left(\frac{x + x_p}{W_p - x_p}\right)\right] \text{ for } -W_p \leq x \leq -x_p \tag{6.54}$$

The stored charge in the minority electron distribution is the integral of the excess minority electron charge density:

$$q_{N_p}(v_D) = A\int_{-W_p}^{-x_p} -qn_p'(x)\,dx = -qAn_{po}(e^{v_D/V_{th}} - 1)\int_{-W_p}^{-x_p}\left[1 + \left(\frac{x + x_p}{W_p - x_p}\right)\right]dx \tag{6.55}$$

The integral evaluates to $(1/2)(W_p - x_p)$ and the stored charge is

$$q_{N_p}(v_D) = \frac{-qA}{2}(W_p - x_p)\,n_{po}(e^{v_D/V_{th}} - 1) \tag{6.56}$$

On the n-side of the junction, a similar integration results in an expression for the dependence of the stored minority hole charge as

$$q_{P_n}(v_D) = A\int_{x_n}^{W_n} qp'_n(x)dx = \frac{qA}{2}(W_n - x_n)\,p_{no}(e^{v_D/V_{th}} - 1) \tag{6.57}$$

Electrons and holes must be supplied from the circuit in order to change the charge distributions on each side of the junction. The increment in the hole charge comes from additional injected holes from the p-type bulk (the anode) across the depletion region. The increment in stored electron charge is negative, but results from electrons that are injected from the n-type cathode in the opposite direction. Since the conventional direction for current is from anode to cathode, the charge increments add together. The **diffusion capacitance** C_d therefore is the sum

$$C_d = \left.\frac{-dq_{N_p}}{dv_D}\right|_{V_D} + \left.\frac{dq_{P_n}}{dv_D}\right|_{V_D} \tag{6.58}$$

Substitution of Eqs. (6.56) and (6.57) into Eq. (6.58) and evaluation at the DC diode voltage yields

diode diffusion capacitance

$$C_d = \frac{qA}{2V_{th}}\left((W_p - x_p)\,n_{po} + (W_n - x_n)\,p_{no}\right)e^{V_D/V_{th}} \tag{6.59}$$

Although we have shown the situation for forward bias in Fig. 6.13, Eq. (6.59) is a general result. Under reverse bias, Eq. (6.59) shows that the diffusion capacitance is essentially zero. This result agrees with the reverse-bias carrier distributions in Fig. 6.11, which represent a tiny charge storage that furthermore is not a function of the diode voltage. The diffusion capacitance increases exponentially with the forward bias voltage on the diode and usually C_d is much larger than the forward-bias depletion capacitance in Eq. (6.50).

The physical interpretation of Eq. (6.59) is simpler if we assume a one-sided diode, which is often the case for IC diodes. The diffusion capacitance in this case is dominated by charge storage on the lightly doped side of the junction. For example, if $N_a \gg N_d$, then $n_{po} \ll p_{no}$ and Eq. (6.59) simplifies to

$$C_d \approx \frac{qA}{2V_{th}}(W_n - x_n)\,p_{no}e^{V_D/V_{th}} \tag{6.60}$$

There are several common terms between Eq. (6.60) and the DC bias current for a one-sided diode, which is given by evaluating Eq. (6.31) with $N_a \gg N_d$,

$$I_D \approx \left(\frac{qAD_p p_{no}}{W_n - x_n}\right)e^{V_D/V_{th}} \tag{6.61}$$

After some algebra, the diffusion capacitance for this one-sided diode can be expressed in terms of the DC diode current

$$C_d \approx \frac{I_D}{V_{th}}\left[\frac{(W_n - x_n)^2}{2D_p}\right] \approx g_d\left[\frac{W_n^2}{2D_p}\right] \tag{6.62}$$

in which we approximate $W_n \approx W_n - x_n$. The diffusion capacitance in the one-sided diode is linearly proportional to the DC diode current.

The quantity in brackets has units of time and is called the **transit time** τ_T of the diode

$$\tau_T = \frac{W_n^2}{2D_p} \tag{6.63}$$

for the case where $N_a \gg N_d$. The transit time is the average time required for the minority holes to diffuse across the n-type bulk region. This physical interpretation is consistent with

$$\tau_T = \frac{\text{distance}}{\text{average velocity}} = \frac{W_n}{(2D_p/W_n)} \tag{6.64}$$

Note that the units of the denominator of Eq. (6.64) are (cm^2/s)/cm = cm/s. For diodes that are not one-sided, the transit time τ_T is defined by

$$\tau_T = \frac{C_d}{g_d} = \frac{C_d V_{th}}{I_D} \tag{6.65}$$

where the diode capacitance is found from Eq. (6.59). The complete small-signal model for the forward-biased diode is shown in Fig. 6.14.

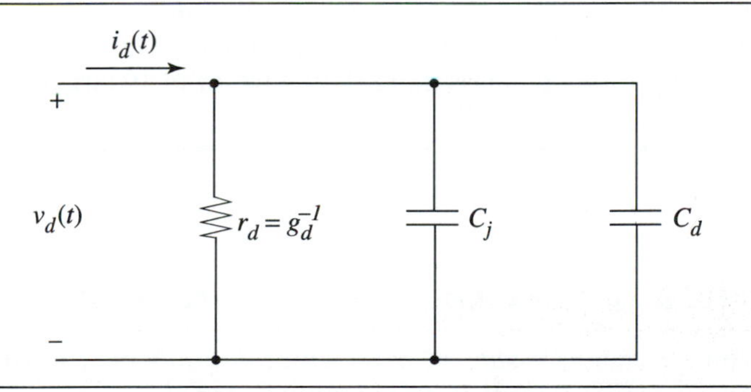

➤ **Figure 6.14** Small-signal model for a forward-biased diode.

➤ EXAMPLE 6.3 Diode Small-Signal Model

Using the dimensions and device parameters in Ex. 6.2, find the numerical values of the small-signal circuit elements for a bias voltage of $V_D = 720$ mV. What is the ratio of the diffusion capacitance to the depletion capacitance?

SOLUTION

From Ex. 6.2, the DC diode current at the operating point is $I_D = 50$ μA. Substitution into Eq. (6.44) yields the small-signal resistance

$$r_d = g_d^{-1} = \frac{V_{th}}{I_D} = \frac{25\text{mV}}{50\ \mu\text{A}} = 500\Omega$$

The depletion capacitance C_j is estimated by substitution into Eqs. (6.49) and (6.50)

$$C_j = 156 \text{ fF}$$

For this pn junction, the depletion region width is $X_d(\phi_B/2) = 124$ nm, which is mostly on the more lightly doped n side. The doping ratio is a factor of 6.25 higher on the p-side, which is not high enough to apply the simple one-sided equation for C_d. Therefore, we substitute the minority carrier concentrations for $V_D = 720$ mV from Ex. 6.2 directly into Eq. (6.59) to find

$$C_d = \left(\frac{(1.6\times10^{-19})(1.89\times10^{-6})}{2\,(0.025)}\right)((0.5\times10^{-4})(4\times10^{14}) + (1\times10^{-4})(2.5\times10^{15}))$$

$$C_d = 1.62 \text{ pF}$$

Note that the depletion region widths $x_n = 0.1$ μm and $x_p = 0.017$ μm are small enough to neglect in finding the diffusion capacitance. The diffusion capacitance under this typical forward bias is more than a factor of 10 larger than the depletion capacitance.

➤ 6.5 SPICE MODEL OF THE pn JUNCTION DIODE

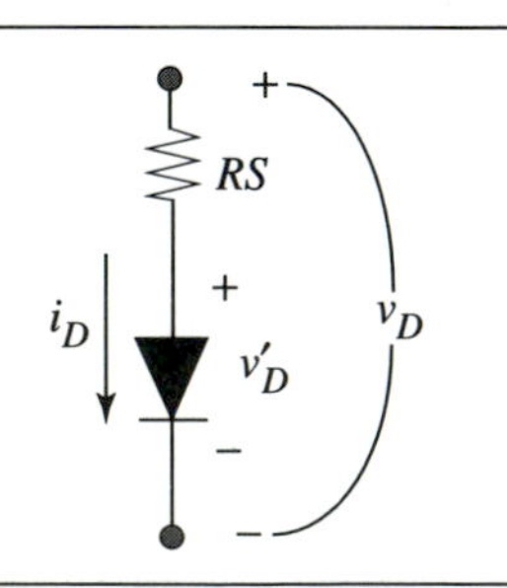

➤ Figure 6.15 DC equivalent circuit used in the SPICE model of the pn diode.

The current-voltage characteristics of a pn diode are modeled in SPICE by two parameters: the saturation current *IS* and ideality factor *N*

$$i_D = IS\,(e^{(v_D')/NV_{th}} - 1) \tag{6.66}$$

It is possible with additional parameters to model the temperature dependence of *IS*. The ideality factor *N* is obtained by curve-fitting the measured characteristic to Eq. (6.66) over the range of currents that the diode will conduct in the particular circuit being simulated. Its value can range from 0.5 – 1, with lower values being needed to model the low-current range as seen in Fig. 6.2(b).

To account for the series resistance in the bulk regions, SPICE includes a parameter RS as shown in Fig. 6.15. The internal diode voltage v_D' in Eq. (6.66) is given by

$$v_D' = v_D - i_D \cdot RS \tag{6.67}$$

Charge storage in the pn junction diode is modeled by the depletion capacitance as described in Chapter 3. The SPICE parameters for modeling the depletion capacitance are CJO, M, and VJ, which are the zero-bias junction capacitance, grading coefficient of the doping profile, and built-in potential. The grading coefficient is 0.5 for the abrupt (step) profiles considered in this book. Minority carrier charge storage is incorporated into SPICE by means of a transit time TT which is extracted from measurements of the diode's small-signal frequency response. The diffusion capacitance is calculated from

$$C_d = \frac{I_D}{V_{th}} \cdot TT \tag{6.68}$$

For one-sided junctions TT can be found from Eq. (6.63).

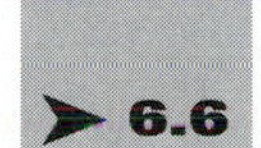

6.6 DEVICE PHYSICS OF THE pn JUNCTION DIODE: NON-EQUILIBRIUM MINORITY CARRIER RECOMBINATION

The intuitive picture of minority carriers diffusing across the narrow bulk regions enabled us to explain the basic operation of a forward-biased diode. However, to develop a solid foundation for modeling minority carrier diffusion, we must first understand generation and recombination of carriers in silicon that is *not* in thermal equilibrium. In developing this topic, we will use an external optical source in order to upset equilibrium and change the carrier densities. The results that we derive are general and will be applied in Section 6.8 to the forward-biased diode.

Suppose an external source of light uniformly illuminates the interior (bulk) of a silicon sample. If the wavelength is shorter than the near-infrared, sufficient energy is absorbed to generate electron-hole pairs. As a result, the concentrations of both types of carriers increase and eventually reach steady-state values n and p in which generation again balances recombination. If the light source is turned off, net recombination of the excess carriers eventually returns the carrier concentrations to their thermal equilibrium values n_o and p_o.

Since we are considering extrinsic (i.e., doped) silicon, one carrier is a majority carrier and has a concentration orders of magnitude greater than that of the minority carrier. It is easy to increase the minority carrier concentration dramatically by means of illumination. However, the majority carrier concentration (which increases by the same amount) is hardly changed at all. In order to make this essential point clear, consider p-type silicon with acceptor concentration $N_a = 10^{15}\ \text{cm}^{-3}$. In thermal equilibrium

$$p_o \cong N_a = 10^{15}\text{cm}^{-3} \qquad n_o = \frac{n_i^2}{p_o} = 10^{5}\text{cm}^{-3} \tag{6.69}$$

If illumination increases the electron concentration to $n = 10^{10}\ \text{cm}^{-3}$ by generation of hole-electron pairs, then the majority hole concentration increases to

$$p = 10^{15}\text{cm}^{-3} + 10^{10}\text{cm}^{-3} \approx 10^{15}\text{cm}^{-3} = p_o \tag{6.70}$$

since holes are created along with electrons by optical absorption. The fact that the majority hole concentration exceeded the minority electron concentration by ten orders of magnitude in equilibrium means that the former essentially remains unchanged, despite a factor of 10^5 increase in the electron concentration. This observation will be used to simplify the mathematics of non-equilibrium recombination later in this section.

In order to understand the steady-state and transient effects of illumination on the minority carrier concentration, start from the balance between generation and recombination in thermal equilibrium that was discussed in Chapter 2. Using p-type silicon for deriving the model, we start from Eq. (2.4) and require that generation and recombination *must* balance in equilibrium. Therefore, the time derivative of the minority electron concentration n_o vanishes

$$\frac{dn_o}{dt} = 0 = G^o - k_o(n_o \cdot p_o) \tag{6.71}$$

where G^o is the sum of all thermal and optical generation processes in thermal equilibrium

$$G^o = G^o_{th} + G^o_{op} \tag{6.72}$$

and the recombination rate is proportional to the product $n_o p_o$. If the sample is illuminated, we are no longer in equilibrium. In this case, we must add the external optical generation term g_l (units: $\text{cm}^{-3}\text{s}^{-1}$) to Eq. (6.71)

$$\frac{dn}{dt} = (G^o + g_l) - k_o(n \cdot p) \tag{6.73}$$

in which we have removed the subscript o indicating thermal equilibrium from the carrier concentrations. The recombination rate constant k_o is assumed to remain at its thermal equilibrium value. We can simplify Eq. (6.73) by writing the minority electron concentration as the sum of the equilibrium concentration and the *excess* concentration n'

$$n = n_o + n' \tag{6.74}$$

Similarly, the majority hole concentration under illumination can be written as

$$p = p_o + p' \tag{6.75}$$

where p' is the excess hole concentration over the equilibrium concentration p_o.

Substituting into Eq. (6.73), we find

$$\frac{dn'}{dt} = (G^o + g_l) - k_o(n_o + n')(p_o + p') \tag{6.76}$$

On the left-hand side of Eq. (6.76), we substituted n' for n in the time derivative since n_o is a constant. Expanding and regrouping the terms, we find

$$\frac{dn'}{dt} = (G^o - k_o n_o \cdot p_o) + g_l - k_o(n'p_o + p'n_o + n'p') \quad (6.77)$$

According to Eq. (6.71), the first term on the right-hand side of Eq. (6.77) sums to zero. This allows us to simplify further

$$\frac{dn'}{dt} = g_l - (k_o p_o)n' - k_o n_o p' - k_o n'p' \quad (6.78)$$

At this point, recall that holes and electrons are generated together by the external source so $n' = p'$. Grouping terms, we find

$$\frac{dn'}{dt} = g_l - k_o(p_o + n_o + n') \cdot n' \cong g_l - k_o(p_o + n') \cdot n' \quad (6.79)$$

since $p_o \gg n_o$ in p-type material.

The final result is a non-linear differential equation in the excess electron concentration n'. In many cases, however, the excess electron concentration over equilibrium is much less than the majority carrier concentration. The case where $n' \ll p_o$ (low-level injection) results in a *linear* differential equation for n'

$$\boxed{\frac{dn'}{dt} = g_l - (k_o p_o) \cdot n' = g_l - \frac{n'}{\tau_n}} \quad (6.80)$$

minority electron lifetime for low-level injection

where we have defined the **minority electron lifetime** $\tau_n = (k_o p_o)^{-1}$. Our analytical models for minority-carrier devices such as the pn junction diode and bipolar transistor assume that low-level injection is valid. The effects of high-level injection (where all terms in Eq. (6.79) must be retained) are significant in minority-carrier device physics and are studied in more advanced courses.

A similar derivation for the case of n-type silicon leads to a first-order differential equation for the excess minority hole concentration p'

$$\boxed{\frac{dp'}{dt} = g_l - (k_o n_o)p' = g_l - \frac{p'}{\tau_p}} \quad (6.81)$$

minority hole lifetime for low-level injection

where the **minority hole lifetime** is defined by $\tau_p = (k_o n_o)^{-1}$. The numerical values for the minority carrier lifetimes depend on the doping level and fabrication sequence. Typical values range from 100 ns to 100 μs for IC processes. For the case of low-level injection, we have shown that the minority carrier concentrations respond to changing illumination levels like a first-order system.

➤ EXAMPLE 6.4 **Minority Electron Concentration under Illumination**

For a p-type silicon sample with doping concentration $N_a = 10^{15}\ \text{cm}^{-3}$ and minority electron lifetime $\tau_n = 1\ \mu\text{s}$, determine the effect of switching on a light source that generates electron-hole pairs in the silicon at a rate of $g_l = 10^{18}\ \text{cm}^{-3}\text{s}^{-1}$.

SOLUTION

We could solve Eq. (6.80) in a formal, mathematical way. However, it is instructive to first find the steady-state excess minority carrier concentration

$$\left.\frac{dn'}{dt}\right|_{t \to \infty} = 0 = g_l - \frac{n'(t \to \infty)}{\tau_n}$$

Solving for the steady-state excess electron concentration, we find

$$n'(t \to \infty) = g_l \tau_n = (10^{18}\text{cm}^{-3}\text{s}^{-1})\,(1\mu\text{s}) = 10^{12}\text{cm}^{-3}$$

The steady-state electron concentration is

$$n(t \to \infty) = n_o + n'(t \to \infty) = 10^5\text{cm}^{-3} + 10^{12}\text{cm}^{-3} \cong 10^{12}\text{cm}^{-3}$$

Since the excess electron concentration is a factor of 10^3 less than the majority carrier concentration, the low-level injection assumption that underlies Eq. (6.80) is valid. The steady-state minority electron concentration has increased a factor of 10^7 over its equilibrium value.

The transient increase in the electron concentration after the light source is switched on can be found from Eq. (6.80), which we recognize as a linear, first-order differential equation. The solution is mathematically the same as for an RC network

$$n'(t) = n'(t \to \infty)\,(1 - e^{-t/\tau_n}) = g_l \tau_n (1 - e^{-t/\tau_n}) \qquad (t \geq 0)$$

Therefore, the minority electron concentration approaches its value for $t \to \infty$ after a few multiples of τ_n, or a few μs for this example.

➤ 6.7 THE CONTINUITY EQUATION

Only one more equation is needed to begin a careful analysis of the pn junction diode under bias. The concept behind the continuity equation is that the holes and electrons entering and leaving a small volume ΔV, taking into account both generation/recombination and drift/ diffusion processes, must "add up."

Figure 6.16 shows a one-dimensional volume $\Delta V = A\ \Delta x$ in which there are $N(x)$ electrons. Defining the change in the number of electrons in the volume in a time interval Δt as $\Delta N(x)$, we add the two terms

$$\Delta N(x) = (G(x) - R(x))\,(A\Delta x)\,\Delta t + \left(\frac{J_n(x)}{-q} - \frac{J_n(x+\Delta x)}{-q}\right)A\Delta t \qquad (6.82)$$

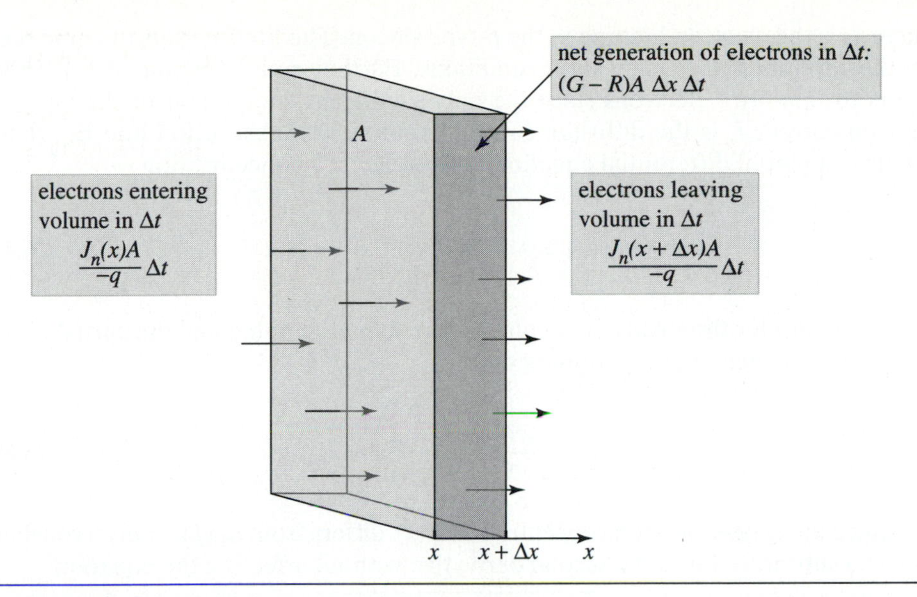

➤ **Figure 6.16** Differential volume used to account for transport and generation/recombination of electrons.

where $(G(x) - R(x))$ is the net generation rate. The current densities are converted into fluxes (units, $cm^{-2}s^{-1}$)) to find the number of electrons entering the volume through the plane at x and the number leaving through the plane at $x + \Delta x$.

Dividing both sides by $\Delta V \Delta t = A \Delta x \Delta t$, we find that

$$\frac{\Delta N(x)}{A\Delta x\Delta t} = G(x) - R(x) + \frac{1}{q}\left(\frac{J_n(x+\Delta x) - J_n(x)}{\Delta x}\right) \tag{6.83}$$

In Eq. (6.83), note that the electron concentration $n(x)$ at x is given by $N(x) / A\Delta x$. If the length Δx and the time increment Δt both approach zero, then we can write Eq. (6.83) in its differential form

$$\boxed{\frac{\partial}{\partial t}n(x, t) = (G - R) + \frac{1}{q}\frac{\partial}{\partial x}J_n(x, t)} \tag{6.84}$$

continuity equation for electrons

Equation (6.84) is referred to as the **continuity equation** for electrons. A similar derivation leads to a corresponding equation for holes

$$\boxed{\frac{\partial}{\partial t}p(x, t) = (G - R) - \frac{1}{q}\frac{\partial}{\partial x}J_p(x, t)} \tag{6.85}$$

continuity equation for holes

Under low-level injection conditions, the last section showed that the net generation rate for minority electrons is

$$G - R = -\frac{n - n_o}{\tau_n} = -\frac{n'}{\tau_n} \tag{6.86}$$

where τ_n is the electron lifetime in the p-type silicon. The lifetime sets the time scale for the relaxation to steady-state conditions. Further simplification of Eq. (6.84) occurs for *minority* electrons since the only significant component of the minority electron current J_n is the diffusion current component. Substitution into Eq. (6.84) results in a partial differential equation for the electron concentration

$$\frac{\partial}{\partial t} n(x, t) = -\frac{n'(x, t)}{\tau_n} + D_n \frac{\partial^2}{\partial x^2} n(x, t) \tag{6.87}$$

If we consider the steady-state, time variation is eliminated and the partial differential equation becomes an ordinary one

$$0 = -\frac{n'(x)}{\tau_n} + D_n \frac{d^2}{dx^2} n(x) \tag{6.88}$$

Since the excess electron concentration $n'(x)$ differs from $n(x)$ by only a constant, it can be substituted into the second derivative without affecting the equation

diffusion equation for minority electrons

$$\frac{d^2}{dx^2} n'(x) = \frac{n'(x)}{D_n \tau_n} = \frac{n'(x)}{L_n^2} \tag{6.89}$$

where $L_n = \sqrt{D_n \tau_n}$ is the **diffusion length** for electrons in the p-type region. This differential equation is known as the **diffusion equation**.

A similar derivation for the case of minority holes in n-type material under low-level injection leads to

diffusion equation for minority holes

$$\frac{d^2}{dx^2} p'(x) = \frac{p'(x)}{D_p \tau_p} = \frac{p'(x)}{L_p^2} \tag{6.90}$$

where $L_p = \sqrt{D_p \tau_p}$ is the diffusion length for holes in the n-type region.

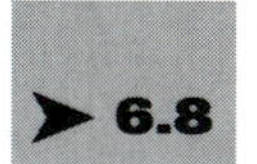

6.8 MINORITY CARRIER DISTRIBUTIONS AND CURRENT COMPONENTS: A SECOND PASS

In this section, we reexamine the pn diode using the diffusion equations to find the steady-state minority carrier concentrations in the bulk regions. We can now quantify the diode current components in the general case and understand the limitations of some of the assumptions made in the "first pass" discussion in Section 6.3. As a starting point, we first collect the five fundamental equations for holes and electrons in silicon.

6.8.1 The Five Semiconductor Equations

The complex behavior exhibited by holes and electrons in silicon can be described by five differential equations. For convenience, the one-dimensional form of these equations are repeated here.

(1) Poisson's Equation: $\frac{d^2\phi}{dx^2} = -\frac{\rho}{\varepsilon_s} = -\frac{q}{\varepsilon_s}(p - n + N_d - N_a)$ (6.91)

(2) Electron current density: $J_n = q\mu_n nE + qD_n\frac{dn}{dx}$ (6.92)

(3) Hole current density: $J_p = q\mu_p pE - qD_p\frac{dp}{dx}$ (6.93)

(4) Continuity of electrons: $\frac{\partial n}{\partial t} = \left(\frac{1}{q}\right)\frac{\partial J_n}{\partial x} + (G_n - R_n)$ (6.94)

(5) Continuity of holes: $\frac{\partial p}{\partial t} = -\left(\frac{1}{q}\right)\frac{\partial J_p}{\partial x} + (G_p - R_p)$ (6.95)

the five semiconductor equations

These coupled, non-linear, partial differential equations can be solved numerically for given initial and boundary conditions to find the time and spatial variation of the electron and hole concentrations: $n(x,t)$ and $p(x,t)$. Two- and even three-dimensional device simulators, based on the 2-D and 3-D versions of these five equations, are routinely used to predict the details of internal device operation, especially under transient conditions. By making judicious approximations to Eqs. (6.91)–(6.95), however, we can find relatively simple analytical solutions that capture the essential features of the internal operation of a device. In this section, we will find that the solution of a practical device problem often requires additional assumptions to find the boundary conditions for the carrier concentrations.

The continuity equations are greatly simplified by considering only steady-state solutions so $\partial n/\partial t$ and $\partial p/\partial t$ are zero. In addition, we assume the special case of thermal generation/recombination processes under low-level injection (Section 6.7) since this case covers the basic operation of non-optical semiconductor devices. The approach is also attractive because it yields analytical solutions. The two continuity equations simplify to

Steady-state, low-level injection, minority electrons: $0 = \left(\frac{1}{q}\right)\frac{dJ_n}{dx} - \frac{(n - n_o)}{\tau_n}$ (6.96)

Steady-state, low-level injection, minority holes: $0 = -\left(\frac{1}{q}\right)\frac{dJ_p}{dx} - \frac{(p - p_o)}{\tau_p}$ (6.97)

The low-level injection assumptions $n \ll p_o = N_a$ in Eq. (6.96) and $p \ll n_o = N_d$ in Eq. (6.97) are necessary for finding linear approximations to the net generation rates.

6.8.2 Internal Potential under an Applied Bias

Our analysis of the diode begins by investigating the potential $\phi(x)$ under an applied bias voltage V_D. As we discovered in Section 6.3, the pn diode is divided into three regions:

- p-type bulk region: $-W_p < x < -x_p$
- depletion region at the junction: $-x_p < x < x_n$
- n-type bulk region: $x_n < x < W_n$

When a voltage V_D is applied across the diode as shown in Fig. 6.17(a), the diode is no longer in thermal equilibrium. Since the metal-silicon contacts are ohmic, their potential drops are unaffected by the diode current. However, there are now potential drops across the bulk regions due to their resistances R_p and R_n. The potential barrier across the pn junction under applied bias is ϕ_j.

Our immediate goal is to relate ϕ_j to the applied bias V_D. Applying KVL around the loop, we find

$$0 = V_D - \phi_{mp} - I_D R_p + \phi_j - I_D R_n - \phi_{nm} \tag{6.98}$$

from which we find the junction barrier potential

$$\phi_j = (\phi_{mp} + \phi_{np}) - V_D + I_D R_p + I_D R_n = \phi_B - V_D + (I_D R_p + I_D R_n) \tag{6.99}$$

In the right-hand equality of Eq. (6.99), the equilibrium barrier height ϕ_B has been substituted for the sum of the metal-silicon contact potentials. To arrive at the basic current-voltage relationship for the diode, we assume that the *I-R* drops in Eq. (6.99) are small enough to be neglected. Except for diodes with long, lightly doped

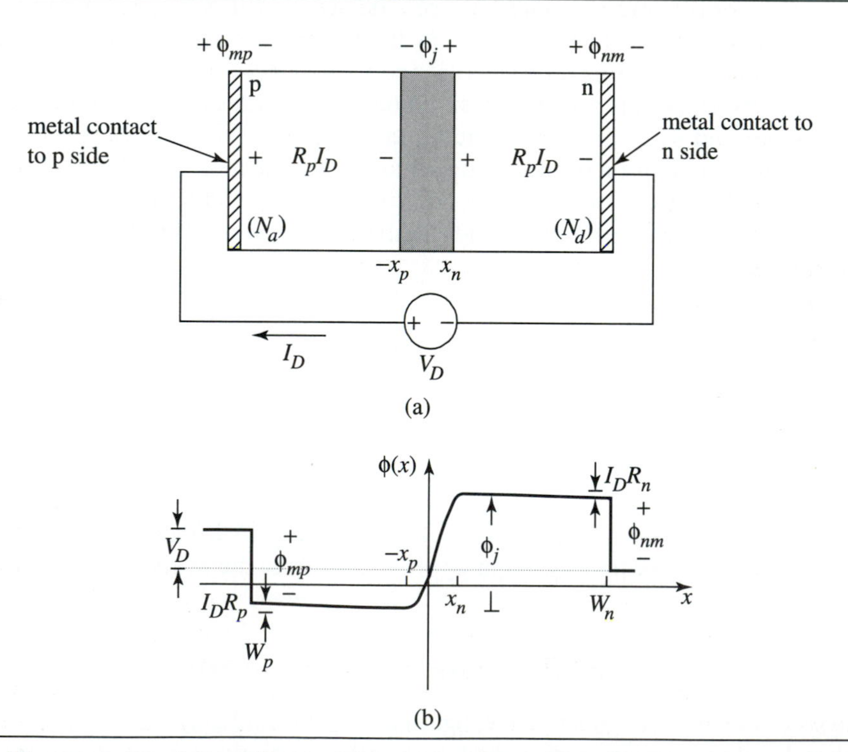

➤ **Figure 6.17** (a) pn diode under applied bias and (b) potential variation through the diode for $V_D > 0$ V.

bulk regions or for the case of very large currents, the error introduced by this approximation is small. With the resistive drops in the bulk regions omitted, Eq. (6.99) simplifies to

$$\phi_j = \phi_B - V_D \tag{6.100}$$

6.8.3 Boundary Conditions for Carrier Concentrations

In order to apply the fundamental semiconductor equations to the diode, we must first determine the carrier concentrations at the boundaries of the p and n bulk regions. Of course, the boundary conditions cannot be found directly from the equations themselves. The ohmic contacts have numerous sites for electron and hole recombination since the silicon lattice is disrupted. As an approximation, we consider that the carrier concentrations are maintained at their thermal equilibrium values at the contacts, regardless of the current density:

$$n_n(x=W_n) = n_{no} = N_d \qquad p_n(x=W_n) = p_{no} = \frac{n_i^2}{N_d} \text{ and} \tag{6.101}$$

$$n_p(x=-W_p) = n_{po} = \frac{n_i^2}{N_a} \qquad p_p(x=-W_p) = p_{po} = N_a \tag{6.102}$$

where the n and p subscripts refer to the n- and p-sides of the diode and the o subscript refers to thermal equilibrium.

We now find the carrier concentrations at the edges of the depletion region $x = -x_p$ and $x = x_n$ as functions of the applied bias. Our approach is to assume that the bias does not perturb thermal equilibrium too much. In particular, we assume that the electron and hole currents in the depletion region do not change appreciably from their thermal equilibrium values, despite the forward bias across the diode. How can this be reasonable?

The answer lies in the balance of very large drift and diffusion current components in the depletion region in thermal equilibrium. For example, electrons go from being minority carriers on the p-side to majority carriers on the n-side over a distance $x_j = x_p + x_n$. A rough estimate of the electron diffusion current density in the depletion region is

$$J_{no,\,diff} = qD_n\frac{dn_o}{dx} \approx qD_n\left(\frac{N_d - n_i^2/N_a}{x_j}\right) = \frac{qD_nN_d}{x_p + x_n} \tag{6.103}$$

For the diode in Ex. 6.2, the n-side doping concentration is $N_d = 4 \times 10^{16}$ cm^{-3} and the total depletion width is $X_d = x_p + x_n = 124$ nm. Equation (6.103) estimates that the equilibrium diffusion current density is about $J_{no,diff} = 12.9$ kA/cm^2 across the depletion region. This huge current density is much larger than the typical forward-bias diode current density of 1–10 A/cm^2. In equilibrium, a balancing drift current density $J_{no,dr}$ is set up by the electric field in the depletion region to cancel precisely $J_{no,diff}$.

As long as the net diode current is a tiny fraction of the originally balanced drift and diffusion currents, the applied bias only slightly perturbs the equilibrium state of the depletion region. Therefore, we still can use equilibrium conditions as good approximations for small diode currents in the depletion region.

$$J_n = q\mu_n nE + qD_n \frac{dn}{dx} \cong 0 \text{ and} \tag{6.104}$$

$$J_p = q\mu_p pE - qD_p \frac{dp}{dx} \cong 0 \tag{6.105}$$

Recall that our goal is to find the carrier concentrations at the edges of the depletion region, therefore, we start integrating Eq. (6.104) from the p-side to the n-side of the depletion region

$$\frac{1}{n}\frac{dn}{dx} = \frac{q}{kT}\frac{d\phi}{dx} \rightarrow \ln\left(\frac{n_n(x_n)}{n_p(-x_p)}\right) = \frac{q}{kT}(\phi(x_n) - \phi(-x_p)) = \frac{q\phi_j}{kT} \tag{6.106}$$

where ϕ_j is the potential difference under forward bias. On the n-side of the junction, assume that the majority (electron) concentration is unaffected by the forward bias (low-level injection), so

$$n_n(x_n) = n_{no} = N_d \tag{6.107}$$

Solving for the minority electron concentration on the p-side, we find

$$n_p(-x_p) = N_d e^{-q\phi_j/kT} = N_d e^{-q(\phi_B - V_D)/kT} = N_d e^{-q\phi_B/kT} e^{qV_D/kT} \tag{6.108}$$

This important result may be simplified by substituting the built-in potential in thermal equilibrium

$$N_d e^{(-q\phi_B)/(kT)} = N_d e^{-\ln\left(\frac{N_a N_d}{n_i^2}\right)} = N_d \frac{n_i^2}{N_a N_d} = \frac{n_i^2}{N_a} = n_{po} \tag{6.109}$$

Substituting Eq. (6.109) into Eq. (6.108), we have rederived the law of the junction that we arrived at in our first pass approach to the diode in Eq. (6.22)

$$n_p(-x_p) = n_{po} e^{qV_D/kT} = n_{po} e^{V_D/V_{th}} \tag{6.110}$$

Similarly for holes, the minority carrier concentration at the edge of the depletion region on the n side is

$$p_n(x_n) = p_{no} e^{V_D/V_{th}} \tag{6.111}$$

The exponential dependence of the boundary values of holes on the n side and electrons on the p side of the junction on the applied voltage strongly suggests that we should focus on the minority carriers as the key to understanding the exponential current-voltage characteristic of the pn diode. According to our assumption of low-level injection, the applied voltage has negligible effect on the majority carrier concentrations. Hence, they play only a supporting role in the operation of the diode.

6.8.4 Minority Carrier Distributions and Currents

Now that the minority carrier concentrations are known at the boundaries of the p and n bulk regions, we can apply the continuity and current equations to find the distribution of minority carriers within these regions. For minority electrons on the p-side, the current is essentially all diffusion

$$J_n = q\mu_n n_p E + qD_n \frac{dn_p}{dx} \cong qD_n \frac{dn_p}{dx} \qquad -W_p \le x \le -x_p \tag{6.112}$$

since both the carrier concentration n_p is relatively small and the electric field E is small in the p-type bulk region. The latter is implicit in the assumption that the $I_D R_p$ drop is negligible. By substituting the current into the steady-state continuity equation (Eq. (6.96)), the result is a second order ordinary differential equation for the minority electron concentration

$$0 = \frac{1}{q}\frac{d}{dx}\left(qD_n \frac{dn_p}{dx}\right) - \frac{n_p - n_{po}}{\tau_n} = D_n \frac{d^2 n_p}{dx^2} - \frac{n'_p}{\tau_n} \tag{6.113}$$

where we substituted the excess minority electron concentration over the thermal equilibrium concentration: $n'_p = n_p - n_{po}$. For a uniformly doped p region, the equilibrium concentration is constant and therefore, it follows that

$$\frac{d^2 n'_p}{dx^2} = \frac{d^2}{dx^2}(n_p - n_{po}) = \frac{d^2 n_p}{dx^2} \tag{6.114}$$

Finally, we use the equality in Eq. (6.114) to simplify Eq. (6.113), which becomes the diffusion equation

$$\frac{d^2 n'_p}{dx^2} = \frac{n'_p}{D_n \tau_n} = \frac{n'_p}{L_n^2} \qquad (-W_p \le x \le -x_p) \tag{6.115}$$

where $L_n = (D_n \tau_n)^{1/2}$ is the diffusion length for electrons in the p-type bulk region. Minority holes in the n-type bulk region are described by a similar equation

$$\frac{d^2 p'_n}{dx^2} = \frac{p'_n}{D_p \tau_p} = \frac{p'_n}{L_p^2} \qquad (x_n \le x \le W_n) \tag{6.116}$$

The diffusion equation for excess electron concentration has the general solution

$$n'_p(x) = Ae^{-x/L_n} + Be^{x/L_n} \tag{6.117}$$

where A and B are determined by the boundary values for $n'_p(x)$ at the ohmic contact and at the edge of the depletion region

$$n'_p(-W_p) = 0 \text{ and} \tag{6.118}$$

$$n'_p(-x_p) = n_{po}e^{V_D/V_{th}} - n_{po} = n_{po}(e^{V_D/V_{th}} - 1) \tag{6.119}$$

Imposing these boundary conditions on the general solution and simplifying, we find

$$n'_p(x) = n_{po}(e^{V_D/V_{th}} - 1)\left[\frac{e^{(x+W_p)/L_n} - e^{-(x+W_p)/L_n}}{e^{(-x_p+W_p)/L_n} - e^{-(-x_p+W_p)/L_n}}\right] \tag{6.120}$$

An alternative form of the solution is possible by using the sum of "sinh" and "cosh" functions in the general solution. On the n-side of the junction, the same approach results in the solution for the excess hole concentration

$$p'_n(x) = p_{no}(e^{V_D/V_{th}} - 1)\left[\frac{e^{-(x-W_n)/L_p} - e^{(x-W_n)/L_p}}{e^{-(x_n-W_n)/L_p} - e^{(x_n-W_n)/L_p}}\right] \tag{6.121}$$

These complicated equations cannot be easily interpreted. Fortunately, we often consider limiting cases. The **short-base diode** solution used in Section 3.3 is valid when the diffusion lengths are much larger than the dimensions of the p or n bulk regions

$$L_n \gg W_p - x_p \quad \text{and} \quad L_p \gg W_n - x_n \text{ (short base limit)} \tag{6.122}$$

For the short-base limit, the exponentials can be expanded using only the first two terms of the power series

$$e^a \cong 1 + a \qquad (a \ll 1) \text{ and therefore} \tag{6.123}$$

$$n'_p(x) = n_{po}(e^{V_D/V_{th}} - 1)\left[\frac{\left(1 + \frac{x + W_p}{L_n}\right) - \left(1 - \frac{x + W_p}{L_n}\right)}{\left(1 + \frac{-x_p + W_p}{L_n}\right) - \left(1 - \frac{-x_p + W_p}{L_n}\right)}\right] \text{ (short-base)} \tag{6.124}$$

Collecting terms and simplifying, we find that the excess electron concentration is linear for the short-base solution

$$n'_p(x) = n_{po}(e^{V_D/V_{th}} - 1)\left[\frac{x + W_p}{-x_p + W_p}\right] \tag{6.125}$$

After some algebra, this result is shown to be identical to Eq. (6.24). For minority holes on the p side, the solution in the short-base limit reduces to

$$p'_n(x) = p_{no}(e^{V_D/V_{th}} - 1)\left[\frac{W_n - x}{W_n - x_n}\right] \tag{6.126}$$

which is consistent with Eq. (6.23).

Since the first-pass analysis of the diode focused on the short-base solution, we will work out the details of the other limiting case—the **long-base diode**. In this limit, the lengths of the p and n bulk regions are much longer than the minority carrier diffusion lengths

$$W_p - x_p \gg L_n \text{ and } W_n - x_n \gg L_p \text{ (long-base limit)} \tag{6.127}$$

In the long-base limit, the solution in Eq. (6.120) reduces to

$$n'_p(x) = n_{po}(e^{V_D/V_{th}} - 1)\left[\frac{e^{(x+W_p)/L_n}}{e^{(-x_p+W_p)/L_n}}\right] = n_{po}(e^{V_D/V_{th}} - 1)e^{(x+x_p)/L_n} \tag{6.128}$$

Therefore, the excess minority electron concentration decays away from the edge of the junction, eventually approaching zero. For the excess hole concentration in the n-side of the diode, the long-base limit for Eq. (6.121) is

$$p'_n(x) = p_{no}(e^{V_D/V_{th}} - 1)\left[\frac{e^{-(x-W_n)/L_p}}{e^{-(x_n-W_n)/L_n}}\right] = p_{no}(e^{V_D/V_{th}} - 1)e^{-(x-x_n)/L_p} \tag{6.129}$$

The long-base solutions for the excess minority carrier concentrations are plotted in Fig. 6.18(a). In contrast to the short-base limit, none of the injected holes or electrons reach the ohmic contacts. Instead, they recombine with the majority carriers in the bulk n or p regions. As a result, the current components in a long-base diode are functions of x and are considerably more complex than for the short-base diode.

Although we have not plotted the majority carrier concentrations in Fig. 6.18(a), they increase by the amount of injected minority carriers in order to maintain charge neutrality:

$$p_p'(x) = n_p'(x) = n_{po}(e^{V_D/V_{th}} - 1)e^{(x+x_p)/L_n} \tag{6.130}$$

$$n_n'(x) = p_n'(x) = p_{no}(e^{V_D/V_{th}} - 1)e^{-(x-x_n)/L_p} \tag{6.131}$$

The minority carrier diffusion currents have the same decaying exponential shape as the carrier concentrations, as shown in Fig. 6.18(b). The reason is that the derivative of an exponential is also an exponential

$$J_n^{diff} = qD_n\frac{dn_p}{dx} = qD_n n_{po}(e^{V_D/V_{th}} - 1)\left(\frac{1}{L_n}\right)e^{(x+x_p)/L_n} \text{ for } -W_p \le x \le -x_p \tag{6.132}$$

Similarly, the minority hole diffusion current on the n-side of the diode is

$$J_p^{diff} = -qD_p\frac{dp_n}{dx} = -qD_p p_{no}(e^{V_D/V_{th}} - 1)\left(\frac{-1}{L_p}\right)e^{-(x-x_n)/L_p} \text{ for } x_n \le x \le W_n \tag{6.133}$$

Note from Eq. (6.133) and Fig. 6.18(a) and (b) that the hole diffusion current density is positive since it is opposite to the gradient in hole concentration.

What is the contribution of the majority carrier currents to the long-base diode? First, the minority-carrier diffusion currents at the edges of the junction are supplied by injection across the depletion region. For example, the holes diffusing into the n-

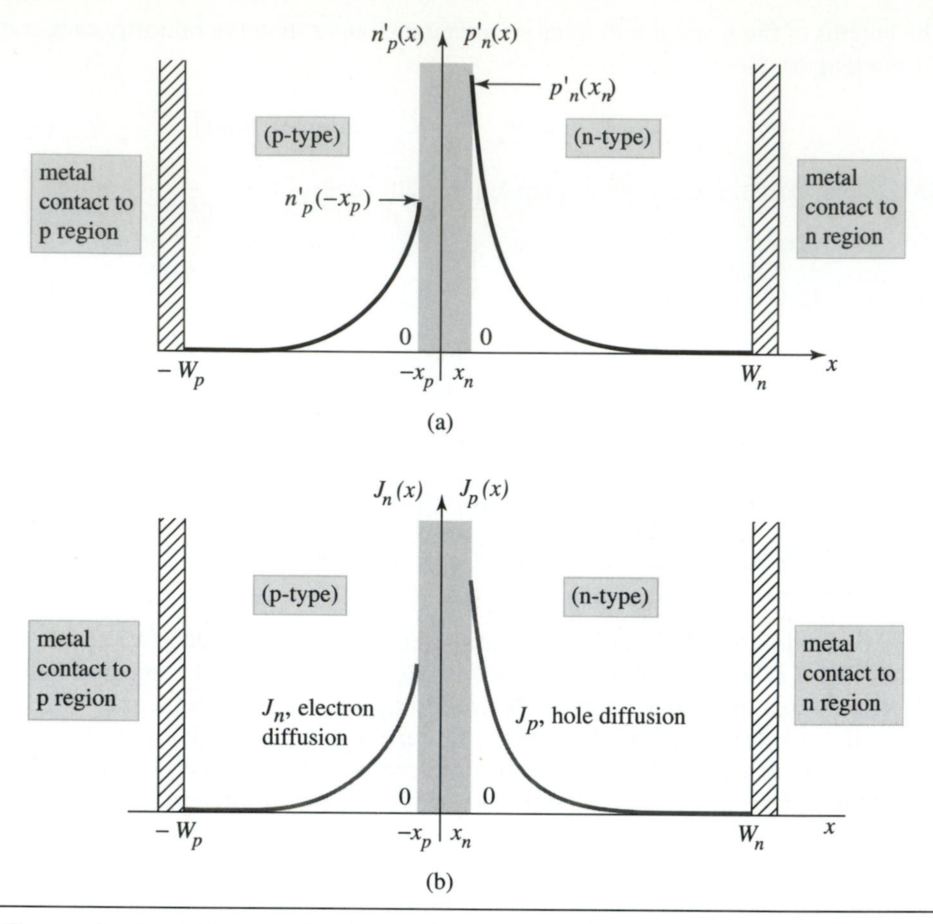

➤ Figure 6.18 a) Excess minority carrier concentrations and (b) minority carrier diffusion currents for the long-base diode.

side of the diode were supplied by a majority hole current on the p-side of the junction. We assume there is no loss of carriers in transport across the depletion region. This is a reasonable assumption in IC diodes except at low applied biases.

A second component of the majority carrier current in a long-base diode is needed to supply carriers for recombination with the injected minority carriers. The key to finding the majority currents is to recognize that the total current density must be a constant throughout the diode in steady-state. Once we determine the total current, we can subtract the minority carrier components. In the depletion region, the only currents are due to injection of carriers. Since we neglect any loss due to recombination in the depletion region, the total current is the sum of the minority carrier currents at the edges of the depletion region

$$J = J_n^{diff}(-x_p) + J_p^{diff}(x_n) = \frac{qD_n n_{po}}{L_n}(e^{V_D/V_{th}} - 1) + \frac{qD_p p_{no}}{L_p}(e^{V_D/V_{th}} - 1) \qquad (6.134)$$

Multiplying by the diode area and putting it into the standard form, we find

$$I_D = A\left(\frac{qD_n n_{po}}{L_n} + \frac{qD_p p_{no}}{L_p}\right)(e^{V_D/V_{th}} - 1) = I_o(e^{V_D/V_{th}} - 1) \quad (6.135)$$

long-base diode current equation

Figure 6.19 sorts out the minority and majority carrier components of the long-base diode current density. The procedure is to use Eq. (6.134) to find that total current density, which must be constant through the device. The total majority carrier current densities are found by graphical subtraction from the known minority carrier diffusion currents.

In this section, we have taken a second look at the diode and found that the steady-state exponential $I_D - V_D$ characteristic is the result of several assumptions, combined with a series of approximations. For forward bias $V_D > 0$ the experimental $I_D - V_D$ curve in Fig. 6.2 shows that Eq. (6.135) (and its counterpart for the short-base diode, Eq. 6.31) are good approximations at intermediate current levels. At high currents, low-level injection is violated and the ohmic drops across the bulk regions become significant. At low currents, a real diode deviates from Eq. (6.135) due to the loss of carriers in the depletion region. Nevertheless, these so-called ideal diode equations contain the essential physics of the device for forward bias in the range of interest for most circuit applications.

In reverse bias, the ideal diode equation predicts that the current will saturate at $-I_o$, a constant current in the range of 10^{-17} to 10^{-15}A. Real diodes show orders-of-magnitude higher reverse currents that are functions of the applied bias, through the contribution of carrier generation in the depletion region. In addition, real diodes will break down at some voltage, which is also not contained in the ideal diode equation.

6.9 DIODE APPLICATIONS

Diodes are very useful as circuit elements since they provide a two-terminal device that has a "one-way" current voltage characteristic. As we found in Section 6.4, when

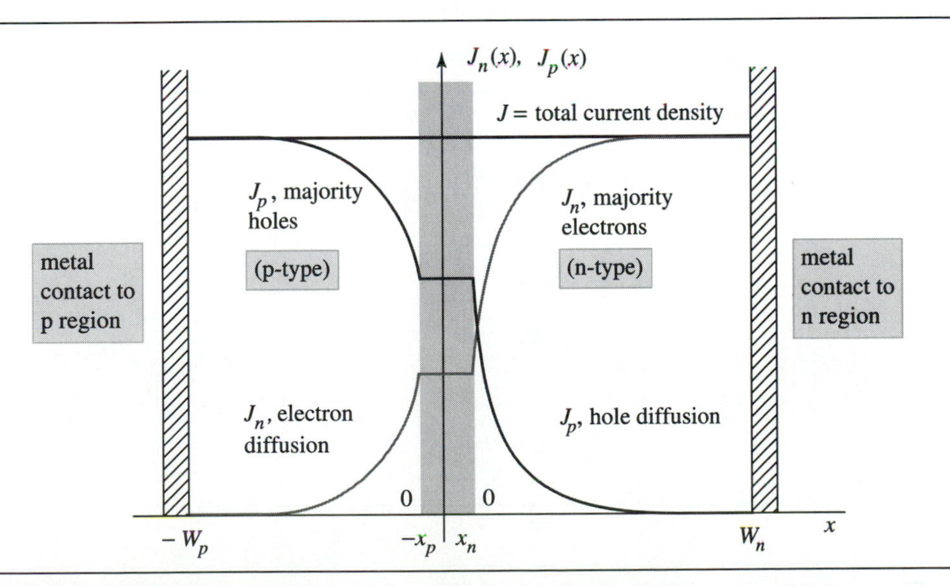

Figure 6.19 Current components in the long-base diode.

forward-biased, a diode is essentially a 0.7 V battery. When reverse-biased, a diode is an open circuit for most practical circuit applications.

In this section we describe two IC applications of pn junction diodes. Figure 6.20 shows a simple circuit which is intended to provide some input protection against damage from *electrostatic discharge.* This phenomenon is a common cause of damage to integrated circuits and is commonly referred to as *static electricity.* The "shocks" that happen when touching a doorknob after walking across a carpet, especially when the air is low in humidity, are from the electrostatic breakdown of air (i.e., arcing).

With a typical breakdown electric field of $E_{BD} = 30$ kV/cm, a spark of length $l = 1$ mm indicates that the potential difference between the person and the (more or less) grounded doorknob is about $V_P = E_{BD}\, l = 3000$ V. Given that a typical person has a capacitance of around $C_P = 1$ pF to ground, the charge stored is $Q = C_P V_P = 3$ nC. This charge is sufficient to zap your finger and any integrated circuit connected to an unprotected pin. In practice, MOS ICs are so sensitive to electrostatic discharge that they are destroyed by insensible levels of discharge. Therefore, they are shipped in special plastic containers with instructions advising customers to wear "ground straps" to drain off any static charge.

The circuit in Fig. 6.20 provides some simple on-chip protection at the input of an IC. We will find the static transfer function relating the input voltage v_I as a function of the off-chip source voltage v_S. Since the diode has a piecewise linear characteristic, we must pick whether it is "on" or "off" before analyzing the circuit and then determine the range of input voltage where the assumed state is valid.

We can use qualitative reasoning to make the analysis more efficient. The two diodes should be open-circuited in normal operation in order to not interfere with the input to the IC. For voltages much higher than the supply V_{DD} or much lower than ground, one of the diodes should turn on in order to prevent an overvoltage at the circuit input. When the input $v_I \gg V_{DD}$, then we assume D_1 is on and D_2 is off. Using the battery model for D_1 and noting the static input current for the MOS IC is zero, we use KVL to find

$$v_S - i_{D1} R_S - v_{D1} - i_{D1} R_{bus} = V_{DD} \tag{6.136}$$

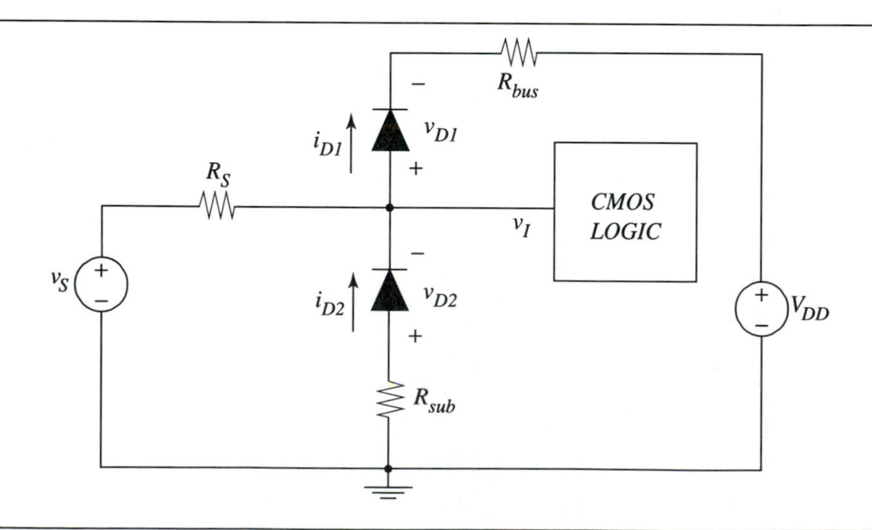

Figure 6.20 Electrostatic discharge (ESD) protection circuit.

where R_S is the source resistance and R_{bus} is the resistance of the interconnect to the supply voltage (including external resistance between the IC and the supply). The diode current is therefore

$$i_{D1} = \frac{v_S - v_{D1} - V_{DD}}{R_S + R_{bus}} \tag{6.137}$$

The current i_{DI} is positive since we have assumed D_I is on, which indicates Eq. (6.137) is valid when $v_S > V_{DD} + v_{DI} = V_{DD} + 0.7$ V. If the bus resistance is small, then the input voltage is prevented from rising much above $V_{DD} + 0.7$ V, as we can show by solving for v_I in terms of the source voltage

$$v_I = v_S - i_{D1}R_S = v_S - \left(\frac{v_S - 0.7\text{ V} - V_{DD}}{R_S + R_{bus}}\right)R_S \tag{6.138}$$

After some algebra, the input voltage can be expressed as

$$v_I = (V_{DD} + 0.7\text{ V}) + R_{bus}\left(\frac{v_S - (V_{DD} + 0.7\text{ V})}{R_S + R_{bus}}\right) \qquad (v_S \geq V_{DD} + 0.7\text{V}) \tag{6.139}$$

For source voltages that are much less than zero, we assume D_2 is on and D_I is off. The current in D_2 is

$$i_{D2} = \frac{-v_{D2} - v_S}{R_S + R_{sub}} = \frac{-0.7\text{ V} - v_S}{R_S + R_{sub}} \tag{6.140}$$

where R_{sub} is the resistance of the ground interconnection. Since i_{D2} is positive by assumption, this equation is valid for $v_S < -0.7$ V. The input voltage v_I is given by

$$v_I = -0.7\text{V} - i_{D2}R_{sub} = -0.7\text{ V} + \left(\frac{R_{sub}}{R_S + R_{sub}}\right)(0.7\text{ V} + v_S) \qquad (v_S \leq -0.7\text{V}) \tag{6.141}$$

Finally, for input voltages in the range $-0.7\text{ V} < v_S < V_{DD} + 0.7$ V, both diodes are open-circuits and the input and source voltages are identical. The transfer curve is plotted in Fig. 6.21, where we assumed that the resistances R_{sub} and R_{bus} are one-tenth

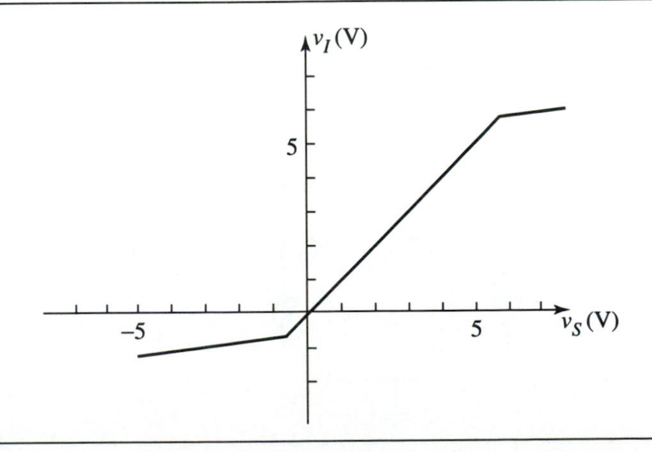

Figure 6.21 Transfer curve for ESD protection circuit in Fig. 6.20.

the source resistance R_S and used $V_{DD} = 5$ V. In circuit jargon, the diodes are said to have "clamped" the voltage at the input and forced it to remain between –0.7 V and 5.7 V.

Although the circuit in Fig. 6.20 effectively protects the IC against large static input voltages, in practice, the rapid rise-time and substantial transient power dissipation which are typical in an electrostatic discharge event often require more elaborate input protection schemes.

A second application of diodes is to construct a temperature sensor. By inverting the forward-bias diode equation, we find

$$V_D = V_{th} \ln\left(\frac{I_D}{I_o}\right) = \left(\frac{kT}{q}\right) \ln\left(\frac{I_D}{I_o}\right) \tag{6.142}$$

For a constant current, the diode voltage has a term proportional to the absolute temperature T. However, the diode saturation current I_o is also a strong function of temperature and therefore, Eq. (6.142) is not suitable for sensing temperature. Two diodes with currents that are in a fixed ratio enables a linear conversion between temperature and voltage. Fig. 6.22 shows the circuit in which the output voltage V_O is the difference between the two diode forward voltage drops.

Assuming the diodes are at the same temperature and that they have identical saturation currents (which are both good assumptions if they are fabricated adjacent to each other on an IC) the output voltage is

$$V_O = V_{D1} - V_{D2} = \left(\frac{kT}{q}\right) \ln\left(\frac{I_1}{I_o}\right) - \left(\frac{kT}{q}\right) \ln\left(\frac{I_2}{I_o}\right) = \left(\frac{kT}{q}\right) \ln\left(\frac{I_1}{I_2}\right) \tag{6.143}$$

As long as the ratio of the current source currents I_1 and I_2 can be kept constant with temperature, then the output voltage will be proportional to the absolute temperature. In fact, there are IC designs for current sources that make the precise control of current ratios easy to achieve. Also, the differential nature of the output voltage is convenient for on-chip amplification. IC temperature sensors based on this concept are commercially available.

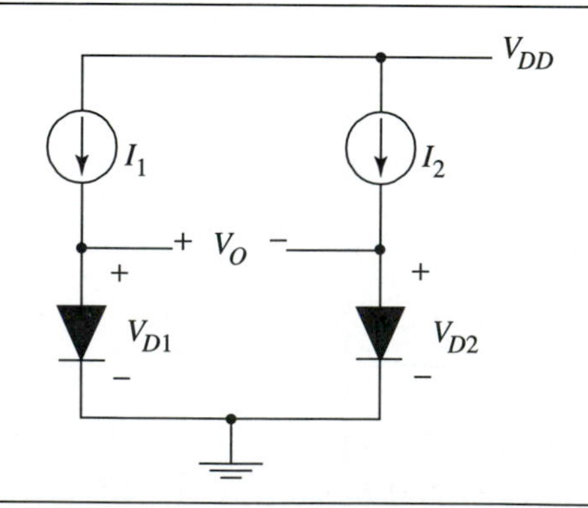

➤ **Figure 6.22** Two-diode temperature sensor.

➤ SUMMARY

This chapter provided an introduction to the internal operation and modeling of the pn junction diode. Key results and concepts that you should understand are:

FIRST-PASS TREATMENT:

- The law of the junction
- Minority carrier concentrations for the short-base diode
- Current components in the short-base diode

CIRCUIT MODELS

- Ideal diode equation (exponential current-voltage relationship)
- Piecewise-linear DC model
- Small-signal steady-state model for the pn diode
- Diffusion capacitance and its physical origin
- SPICE model of the pn diode

SEMICONDUCTOR PHYSICS

- Minority carrier lifetimes and the low-level injection approximation
- Continuity equation
- Steady-state diffusion equation; minority carrier diffusion lengths

SECOND-PASS TREATMENT

- Long-base and short-base approximations
- Minority carrier distributions in the p and n bulk regions
- Current components in the long-base diode

➤ REFERENCES

1. G. W. Neudeck, *The PN Junction Diode*, 2nd ed., Vol. II of the Modular Series on Solid-State Devices, Addison-Wesley, 1989. Comprehensive treatment using energy band theory.
2. R. S. Muller and T. I. Kamins, *Device Electronics for Integrated Circuits*, 2nd ed., Wiley, 1986, Chapter 4 is a good discussion at a more advanced level.
3. C. G. Fonstad, *Microelectronic Devices and Circuits*, McGraw Hill, 1994. Chapter 7 is a parallel treatment at about the same level that includes optical phenomena.

➤ PROBLEMS

EXERCISES

Note: unless given in the problem statement, the diffusion coefficients should be found using the Einstein Relation (Eq. (2.51) and mobilities from Fig. 2.8.

E6.1 The current in a pn junction diode is measured at several forward bias voltages.:

V_D [mV]	I_D [A]
600	3×10^{-6}
660	3×10^{-5}
720	3×10^{-4}
780	3×10^{-3}

(a) Find the saturation current I_o for this diode.

(b) Find an equation relating the diode voltage to the diode current, using a base 10 logarithm.

(c) If the breakdown voltage is $V_{BD} = -18$ V and the diode current is –500 μA for $V_D = -18.5$ V, plot the I_D-V_D characteristic over the range $-18.5 < V_D < 0.8$ V.

E6.2 Given a short-base diode with area 25×25 μm^2 with the following parameters:

$N_a = 10^{17}$ cm^{-3}	$N_d = 10^{17}$ cm^{-3}
$W_p = 2$ μm	$W_n = 2$ μm

(a) Verify that the minority carrier concentrations satisfy low-level injection for $V_D = 700$ mV.

(b) What fraction of the total current is due to minority hole diffusion?

(c) What is the diode current for a forward bias $V_D = 700$ mV? Neglect the width of the depletion region compared to the bulk region widths.

E6.3 Given a short-base diode with area 10×10 μm^2 with the following parameters:

$N_a = 10^{16}$ cm^{-3}	$N_d = 10^{16}$ cm^{-3}
$W_p = 1$ μm	$W_n = 1$ μm

(a) What is the maximum forward bias for which the diode satisfies low-level injection. Consider that the maximum minority carrier concentration for low-level injection is 1/10 the majority carrier concentrations.

(b) What is the saturation current I_o? Neglect the width of the depletion region compared to the bulk region widths.

(c) Sketch the minority carrier concentrations that correspond to a diode current $I_D = 1\ \mu A$.

E6.4 A one-sided, short base diode with area $20 \times 20\ \mu m^2$ has the following parameters.

$N_a = 10^{17}\ cm^{-3}$	$N_d = 10^{19}\ cm^{-3}$
$W_p = 1\ \mu m$	$W_n = 0.25\ \mu m$

(a) What is the built-in voltage ϕ_B?

(b) What is the width of the depletion region on the p-side of the junction in thermal equilibrium and for a forward bias $V_D = 720$ mV? Use the depletion equation with a forward bias of $\phi_B / 2$ to estimate the width for $V_D = 720$ mV.

(c) Sketch the minority electron concentration on the p-side of the junction for a forward bias $V_D = 720$ mV.

E6.5 A one-sided, short base diode with area $20 \times 20\ \mu m^2$ has the following parameters.

$N_a = 10^{18}\ cm^{-3}$	$N_d = 10^{16}\ cm^{-3}$
$W_p = 0.5\ \mu m$	$W_n = 1\ \mu m$

(a) Sketch the minority hole concentration on the n-side for $V_D = -3$ V.

(b) What is the current for $V_D = -3$ V? Be sure to account for the width of the depletion region on the n-side of the diode.

E6.6 For the diode with parameters in E6.2 at a forward bias of $V_D = 700$ mV

(a) Find the small-signal resistance r_d.

(b) Find the depletion capacitance C_j.

(c) Find the diffusion capacitance C_d.

E6.7 For the diode with parameters in E6.2, plot the total capacitance (depletion + diffusion) as a function of V_D over the range $-5\ V < V_D < 0.72\ V$.

E6.8 For the diode parameters in E6.4 at a forward bias of $V_D = 720$ mV.

(a) Find the small-signal resistance r_d.

(b) Find the depletion capacitance C_j.

(c) Find the diffusion capacitance C_d.

E6.9 For the one-sided diode parameters from E6.5 at a forward bias of $V_D = 700$ mV.

(a) Find the small-signal resistance r_d.

(b) Find the transit time τ_T across the n-type bulk region.

(c) Find the average velocity of holes diffusing across the n-type bulk region.

(d) Find the diffusion capacitance C_d.

E6.10 An n-type silicon region is illuminated with an optical source, which generates electron-hole pairs at a rate of $g_l = 10^{20}$ cm^{-3}s^{-1}. The hole lifetime in this sample is $\tau_p = 250$ ns and its doping concentration is $N_d = 10^{17}$ cm^{-3}.

(a) What is the steady-state concentration of holes and electrons, after the sample has been illuminated for a long period of time?

(b) How long is required after the light source is switched on for the hole and electron concentrations to reach 90% of their steady-state values? This interval is called the rise time t_r.

(c) How long is required after the light source is switched off for the hole and electron concentrations to fall to within 10% of their thermal equilibrium vales? This interval is called the fall time t_f.

E6.11 A p-type silicon region has an electron lifetime $\tau_p = 25$ μs and a doping concentration of $N_a = 10^{16}$ cm^{-3}. The sample is illuminated with an optical source.

(a) What is the maximum optical generation rate $g_{l,max}$ (cm^{-3}s^{-1}) so that low-level injection still holds? We will consider that low-level injection is violated when the minority carrier concentration exceeds 1/10 of the thermal equilibrium majority carrier concentration.

(b) Plot the electron concentration as a function of time when a light source with $g_{l,max}$ is switched on.

E6.12 The plot in Fig. E6.12 is for the steady-state minority electron concentration $n_p(x)$ in a silicon sample with $N_a = 10^{17}$ cm^{-3}.

(a) What is the electron diffusion length, L_n?

(b) What is the minority electron lifetime, τ_n?

E6.13 Given a diode with area 20×20 μm^2 with the following parameters:

$N_a = 10^{17}$ cm^{-3}	$N_d = 10^{17}$ cm^{-3}
$W_p = 25$ μm	$W_n = 25$ μm
$\tau_n = 40$ ns	$\tau_p = 40$ ns

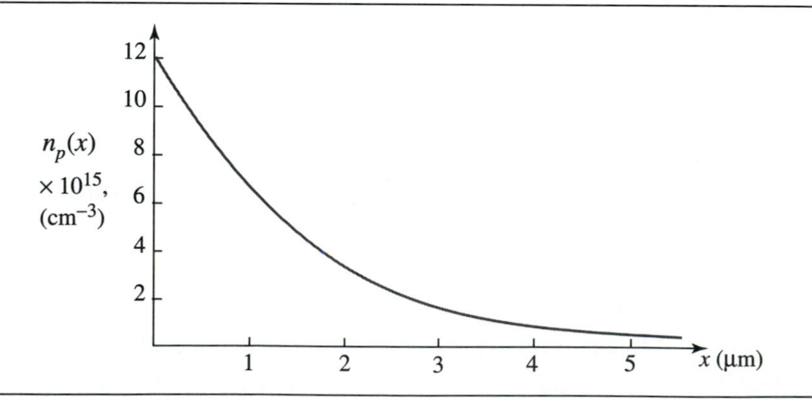

➤ Figure E6.12

(a) Verify that the minority carrier concentrations satisfy low-level injection for $V_D = 700$ mV.

(b) Verify that the widths of the bulk regions are much greater than the minority carrier diffusion lengths.

(c) Sketch the minority carrier concentrations for $V_D = 700$ mV.

E6.14 Given a diode with area $100 \times 100\ \mu m^2$ with the following parameters:

$N_a = 10^{16}$ cm^{-3}	$N_d = 10^{15}$ cm^{-3}
$W_p = 50\ \mu$m	$W_n = 50\ \mu$m
$\tau_n = 100$ ns	$\tau_p = 100$ ns

(a) What is the maximum forward bias for which the diode satisfies low-level injection. Consider that the maximum minority carrier concentration for low-level injection is 1/10 the majority carrier concentrations.

(b) What is the saturation current I_o? If the widths of the p and n regions are doubled, what is the effect on I_o?

E6.15 Given a diode with area $50 \times 50\ \mu m^2$ and the following parameters:

$N_a = 10^{19}$ cm^{-3}	$N_d = 3 \times 10^{16}$ cm^{-3}
$W_p = 4\ \mu$m	$W_n = 20\ \mu$m
$\tau_n = 1$ ns	$\tau_p = 10$ ns

(a) Verify that the widths of the bulk regions are much greater than the minority carrier diffusion lengths.

(b) Sketch the minority carrier concentrations under forward bias.

(c) Find the saturation current I_o for this diode.

(d) What is the fraction of the total current carried by minority electrons diffusing across the p-region?

E6.16 The circuit in Fig. E6.16 is useful as a temperature sensor. The saturation current differs for the two diodes: $I_{o1} = 10^{-17}$ A and $I_{o2} = 10^{-16}$ A.

(a) Assuming that the ideal diode equation is valid, plot the voltage V_O as a function of temperature over the range $-50°\ C < T < 125°\ C$.

(b) Explain how you would design the two diodes to have saturation currents that had a precise ratio of 10:1.

PROBLEMS

P6.1 We define the "on" voltage of a pn diode V_{ON} by the voltage at which the diode current is $I_D = 100\ \mu$A.

(a) Plot I_D as a function of V_D on a linear scale for diodes with these saturation currents: $I_o = 10^{-17}$, 5×10^{-17}, 10^{-16}, 5×10^{-16}, and 10^{-15} A.

(b) Plot V_{ON} as a function of the saturation current I_o.

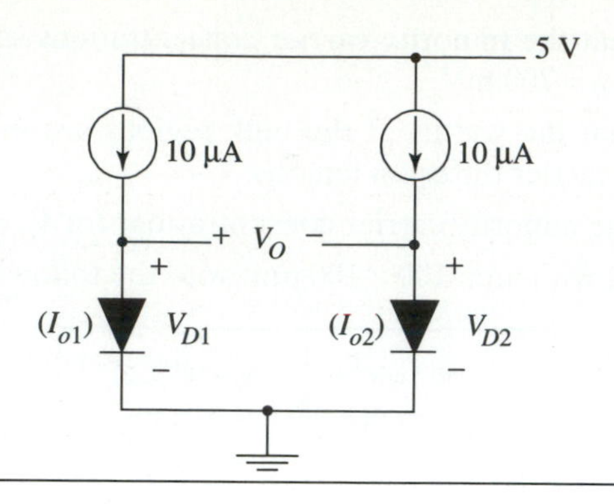

➤ **Figure E6.16**

P6.2 Given a diode with the following parameters:

$N_a = 10^{16}$ cm^{-3}	$N_d = 2 \times 10^{15}$ cm^{-3}
$W_p = 5$ μm	$W_n = 5$ μm
$\tau_n = 500$ ns	$\tau_p = 750$ ns

The diode area is 40×40 μm^2 and the current is $I_D = 10$ μA.

(a) Find the electron diffusion current density at the edge of the depletion region.

(b) What is the forward bias V_D on the diode?

(c) What is the maximum diode current at which low-level injection is still satisfied? We consider low-level injection to be a good approximation when the minority carrier concentrations are at most 10% of the thermal equilibrium majority carrier concentrations.

P6.3 Given a diode with the following parameters:

$N_a = 5 \times 10^{17}$ cm^{-3}	$N_d = 10^{18}$ cm^{-3}
$W_p = 1$ μm	$W_n = 0.5$ μm
$\tau_n = 100$ ns	$\tau_p = 75$ ns

The diode area is 10×10 μm^2 and the forward bias is $V_D = 720$ mV.

(a) Plot the minority carrier concentrations through the diode.

(b) Find the saturation current. If the length of the n-type bulk region is doubled, what is the effect on the saturation current?

P6.4 Given a diode with the following parameters:

$N_a = 5 \times 10^{17}$ cm^{-3}	$N_d = 5 \times 10^{17}$ cm^{-3}
$W_p = 2$ µm	$W_n = 25$ µm
$\tau_n = 100$ ns	$\tau_p = 100$ ns

The diode area is 100×100 µm^2 and the forward bias is $V_D = 700$ mV.

(a) Find the minority carrier diffusion lengths on each side of the diode.

(b) Plot the minority carrier concentrations through the diode. What fraction of the current is due to electron injection into the p-side?

(c) Find the diode current. If the length of the n-type bulk region is doubled, what is the effect on the current?

P6.5 Given a diode with the following parameters:

$N_a = 5 \times 10^{16}$ cm^{-3}	$N_d = 10^{17}$ cm^{-3}
$W_p = 5$ µm	$W_n = 0.5$ µm
$\tau_n = 1$ ns	$\tau_p = 1$ ns

The low minority carrier lifetimes are due to radiation damage. The diode area is 50×50 µm^2 and the diode current is $I_D = 100$ µA.

(a) Find the minority carrier diffusion lengths on each side of the diode.

(b) Plot the minority carrier concentrations through the diode. What fraction of the current is due to hole injection into the n-side?

(c) Find the forward bias on the diode.

P6.6 A diode with very long bulk regions and uniform doping can be formed by wafer bonding. Consider such a diode with the following parameters:

$N_a = 5 \times 10^{14}$ cm^{-3}	$N_d = 5 \times 10^{14}$ cm^{-3}
$W_p = 500$ µm	$W_n = 500$ µm
$\tau_n = 10$ µs	$\tau_p = 10$ µs

The diode area is 1000×1000 µm^2.

(a) Find the minority carrier diffusion lengths on each side of the diode.

(b) Plot the minority carrier concentrations through the diode and find the saturation current I_o.

(c) Find the maximum forward bias on the diode for which the low-level injection approximation is valid. We consider low-level injection to be a good approximation when the minority carrier concentrations are at most 10% of the thermal equilibrium majority carrier concentrations.

(c) Find the diode current for the bias in (c).

P6.7 Given a diode with the following parameters:

$N_a = 10^{17}$ cm^{-3}	$N_d = 10^{17}$ cm^{-3}
$W_p = 1.5$ µm	$W_n = 1.5$ µm
$\tau_n = 0.1$ ns	$\tau_p = 0.1$ ns

The low minority carrier lifetimes are due to radiation damage during plasma etching. The diode area is $25 \times 25 \mu m^2$ and the diode current $I_D = 100$ µA.

(a) Find the minority carrier diffusion lengths on each side of the diode.

(b) Plot the minority carrier concentrations through the diode. In this symmetrically doped diode, what fraction of the current is due to electron injection into the p-side?

(c) Find the forward bias on the diode.

P6.8 Given a diode with the following parameters:

$N_a = 10^{16}$ cm^{-3}	$N_d = 4 \times 10^{16}$ cm^{-3}
$W_p = 20$ µm	$W_n = 10$ µm
$\tau_n = 20$ ns	$\tau_p = 10$ ns

The diode area is 50×50 µm^2 and the diode bias voltage is $V_D = 660$ mV.

(a) Find the minority carrier diffusion lengths on each side of the diode.

(b) Plot the minority carrier concentrations through the diode on a linear scale.

(c) Find the saturation current for the diode.

P6.9 Given a one-sided diode with the following parameters:

$N_a = 10^{18} \pm 10^{17}$ cm^{-3}	$N_d = 10^{16} \pm 10^{15}$ cm^{-3}
$W_p = 0.5 \pm 0.1$ µm	$W_n = 2 \pm 0.1$ µm
$\tau_n = 20 \pm 5$ ns	$\tau_p = 50 \pm 10$ ns

The diode area is 50×50 µm^2.

(a) Assuming that all variations are independent, find the average and variation in the minority carrier diffusion lengths.

(b) Find the average and variation in the saturation current I_o for this diode.

(c) For a diode current $I_D = 50\ \mu A$, find the average and variation in the diode voltage.

P6.10 Given a short-base diode with the following parameters

$N_a = 10^{17}$ cm^{-3}	$N_d = 10^{17}$ cm^{-3}
$W_p = 0.5\ \mu m$	$W_n = 1.5\ \mu m$

(a) Find the DC bias current so that the small-signal resistance is $r_d = 75\ \Omega$.

(b) What is the minimum diode area for the current in (a), such that low-level injection isn't violated at the depletion region edges. The maximum permissible minority carrier concentration is considered to be 1/10 of the thermal equilibrium majority carrier concentration.

(c) What is total small-signal resistance through the diode, including the resistances of the bulk n and p regions?

P6.11 Given that the saturation current of a pn junction diode is $I_o = 5 \times 10^{-17}$A and that the DC bias voltage $V_D = 720$ mV.

(a) If a voltage $v_d = (5\text{ mV})\cos(\omega t)$ is added in series with V_D, plot the small-signal current $i_d(t) = i_D(t) - I_D$.

(b) If a voltage $v_d = (20\text{ mV})\cos(\omega t)$ is added in series with V_D, plot the current $i_d(t) = i_D(t) - I_D$. Note that this voltage is too large for the small-signal model to be accurate.

P6.12 An unusual application of a pn junction is to provide a low capacitance, high small-signal resistance connection to a circuit node. The DC bias is $V_D = 0$ V. The short-base diode has an area of $10 \times 10\ \mu m^2$ with parameters

$N_a = 10^{19}$ cm^{-3}	$N_d = 10^{17}$ cm^{-3}
$W_p = 0.5\ \mu m$	$W_n = 1.5\ \mu m$

(a) Find the small-signal resistance of the diode with $V_D = 0$ V. Note: the answer is not infinity.

(b) Find the depletion capacitance for $V_D = 0$.

(c) Find the diffusion capacitance for $V_D = 0$.

(d) At what frequency does the magnitude of the impedance of the diode capacitance equal the small-signal resistance?

P6.13 The diffusion capacitance was derived for the case of a short-base diode.

(a) Find an expression for the minority carrier charge storage in both bulk regions of a long-base diode.

(b) Find an expression for diffusion capacitance for a long-base diode.

P6.14 A pn junction diode can be used as a tunable capacitor. For the short-base diode from E6.4,

(a) Plot the depletion capacitance as a function of applied bias over the range $-2\ \text{V} < V_D < 0.78\ \text{V}$.

(b) Plot the diffusion capacitance and the total capacitance as a function of applied bias over this range.

P6.15 Using the fundamental results in Eqs. (6.14) – (6.17), the law of the junction can be extended to the case where low-level injection is violated.

(a) Find the minority carrier concentrations at the edges of the depletion region as a function of V_D for the general case, where $n_p(-x_p)$ and $p_n(x_n)$ may exceed the thermal equilibrium majority carrier concentrations.

(b) For $N_a = 10^{16}\ \text{cm}^{-3}$ and $N_d = 2 \times 10^{16}\ \text{cm}^{-3}$, plot on a log-linear scale the minority carrier concentrations $n_p(-x_p)$ and $p_n(x_n)$ as a function of bias over the range $V_D = 600\ \text{mV}\ - 720\ \text{mV}$.

(c) For high-level injection conditions, where the minority carriers greatly exceed N_a and N_d, find an approximate expression for their dependence on V_D.

P6.16 One application of photogeneration is an optical sensor. Consider an ion-implanted resistor with $N_d = 5 \times 10^{16}\ \text{cm}^{-3}$ and a junction depth of $x_j = 1\ \mu\text{m}$. The substrate doping concentration is $N_a = 10^{15}\ \text{cm}^{-3}$. The resistor layout has a total of 125 squares. The hole lifetime in the resistor is $\tau_p = 75$ ns.

(a) What is the resistance without illumination?

(b) What is the steady-state change in resistance when the optical generation rate is $g_l = 5 \times 10^{22}\ \text{cm}^{-3}\text{s}^{-1}$? You should include the effect of illumination on the minority and majority carrier contributions to the conductivity.

(c) For low-level injection to hold, what is the maximum optical generation rate? The maximum minority carrier concentration is considered to be 1/10 of the thermal equilibrium majority carrier concentration for low-level injection. What is the maximum change in resistance?

P6.17 A p-type sample with $N_a = 10^{16}\ \text{cm}^{-3}$ and a minority electron lifetime $\tau_n = 25$ ns is illuminated by a modulated optical source that generates hole-electron pairs at a rate given by

$$g_l(t) = (5 \times 10^{20} \text{cm}^{-3} \text{s}^{-1})(1 + \cos(2\pi 1000 t))$$

(a) Find the average minority electron concentration in the sample.

(b) Plot the minority electron concentration in the sample. Note that the period of the optical source is much longer than the minority carrier lifetime.

P6.18 Given a diode with the following parameters:

$N_a = 10^{17}$ cm^{-3}	$N_d = 10^{17}$ cm^{-3}
$W_p = 1.5$ μm	$W_n = 1.5$ μm
$\tau_n = 0.05$ ns	$\tau_p = 0.05$ ns

The low minority carrier lifetimes are due to contamination by heavy metal ions. The diode area is 50×50 μm^2 and the diode voltage is $V_D = -2$ V.

(a) Plot the minority carrier concentrations on each side of the diode. Note that the solutions in Section 6.8 are valid for reverse bias voltages.

(b) Plot the minority carrier diffusion currents and the majority carrier currents in the diode, using the same procedure as was used in Section 6.8 for forward bias.

(c) Where does the minority carrier diffusion current originate? Hint: look at the net generation rate for minority carriers near the depletion regions.

P6.19 Fig. P6.19 shows a model of a one-sided pn junction diode with a region of damaged silicon (due to a slip dislocation in the crystal) on the n side.

(a) What are the boundary conditions on the minority hole concentration at the boundaries of the damaged region at $x = 1.5$ μm and $x = 2$ μm? Hint: the hole diffusion current should be continuous, as should the hole concentration.

(b) Plot the minority carrier concentration on the n-side of the diode.

(c) Find the saturation current density J_o for this diode.

P6.20 Fig. P6.20 shows a model for a symmetrical diode that has a damaged region near the junction. Both the minority carrier lifetime and the diffusion coefficient are degraded near the junction.

(a) Find the minority carrier diffusion lengths in each region of the diode.

(b) Plot the minority carrier concentrations throughout the diode.

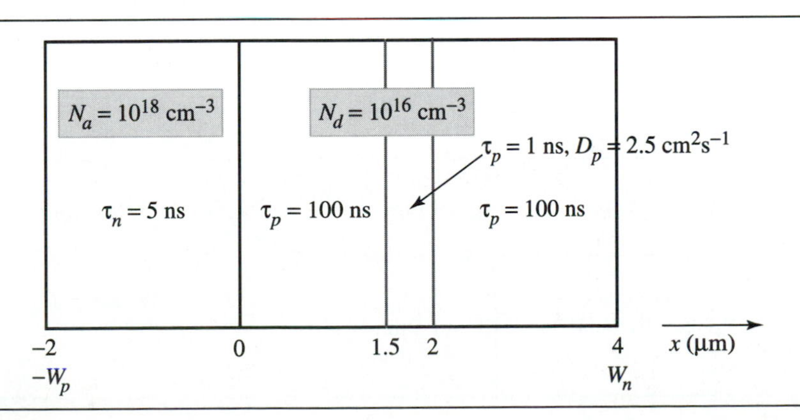

➤ **Figure P6.19**

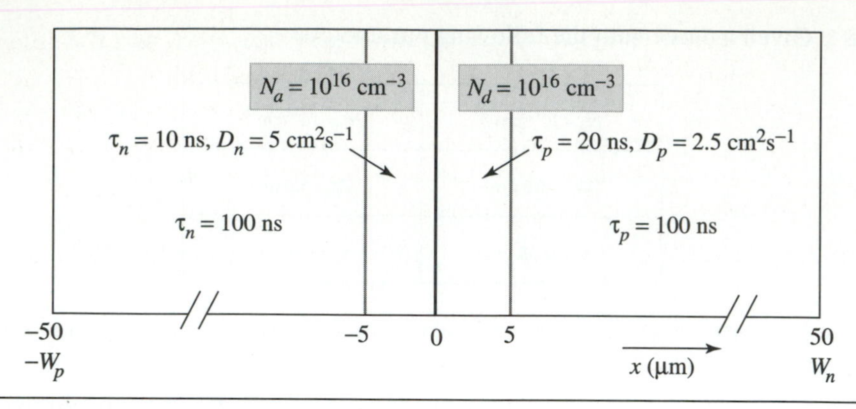

➤ **Figure P6.20**

(c) Find the saturation current density J_o for this diode and compare it with the value for a diode without the damaged region.

P6.21 A simple one-dimensional model for the contribution of the perimeter of an ion-implanted region to the diode current is found in this problem. Figure P6.21 shows the cross section of the structure. The perimeter in cross section is a quarter circle of radius $x_j = 1$ µm. Due to doping gradients in the structure, the perimeter diode has a factor of 5 higher doping concentrations. The width of the p-side is $W_{pe} = x_j = 1$ µm and the width of the n-side is $W_{ne} = 2.5$ µm.

(a) Assuming that it is a short-base diode, find the saturation current I_{oe} of the perimeter diode, per unit length L of perimeter. Use the one-dimensional parameters given in Fig. P6.21.

(b) Find the saturation current I_{ob} per unit area of the planar diode at the bottom of the diffusion.

(c) Plot the ratio of perimeter saturation current to the bottom saturation current as a function of the linear dimension L of the diode's square layout, over the range 2 µm < L < 50 µm.

P6.22 The series resistance of a pn diode can cause a significant deviation from the ideal diode equation in some cases.

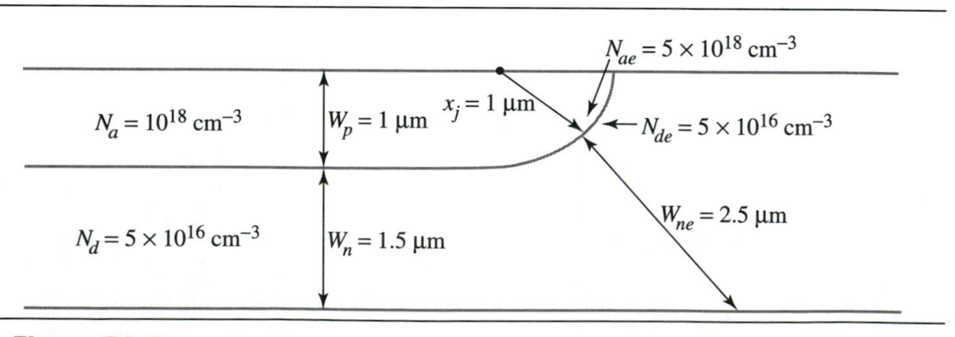

➤ **Figure P6.21**

(a) Derive an implicit equation for the diode current I_D as a function of the diode voltage V_D that includes the series resistance $R_s = R_n + R_p$.

(b) For $I_o = 10^{-15}$ A and $R_s = 1$ kΩ, plot I_D vs. V_D over the range $0 < V_D < 5$ V. Hint: an easy solution is to add graphically the resistive drop and the intrinsic diode drop V_D' since the current I_D is the same.

(c) Repeat (b) for $R_s = 100\ \Omega$ and $R_s = 10$ kΩ.

P6.23 Figure P6.23 shows a a simple rectifier circuit. The source voltage is

$$v_S(t) = (5\text{ V})\cos[(2\pi \cdot 100\text{Hz})t]$$

The frequency of the input signal is low enough that the static large-signal model of the diode is applicable.

(a) Plot the source voltage $v_S(t)$ and the voltage $v_C(t)$ across the capacitor as a function of time on the same graph.

(b) Repeat part (a) if the source voltage amplitude is increased to 10 V.

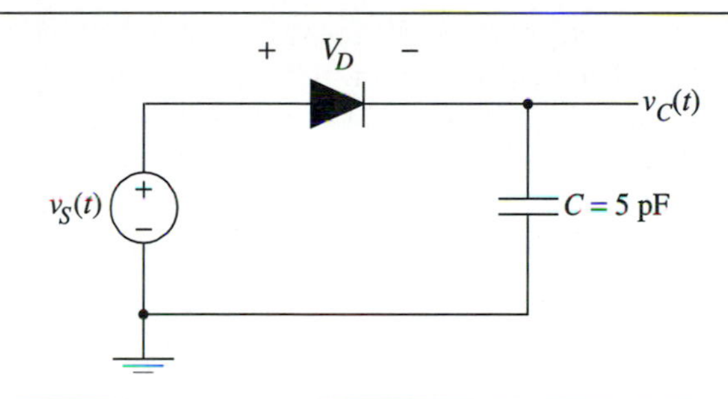

➤ **Figure P6.23**

chapter 7

The Bipolar Junction Transistor

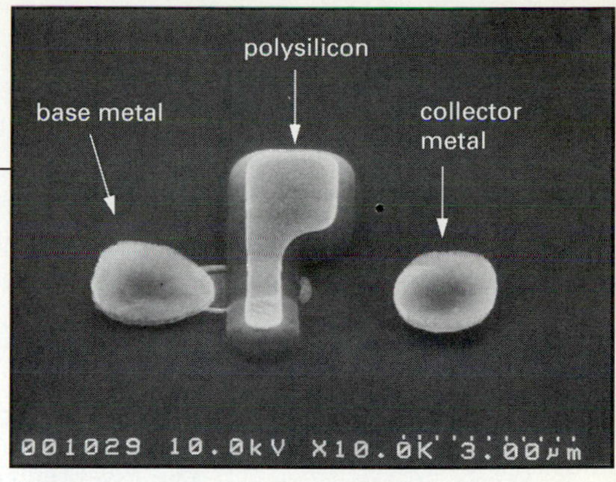

Scanning electron micrograph of a polysilicon-emitter, oxide-isolated npn bipolar transistor, after polysilicon emitter patterning and first-metal contact to the base and collector. The emitter area is $A_E = 0.7\ \mu m \times 1\ \mu m$ and the transition frequency is $f_T = 10$ GHz. QUBiC2 process SEM. (Courtesy of Bill Mack, Philips Semiconductors)

The Bipolar junction transistor (BJT) is the second multi-terminal device we will study. Although its DC terminal characteristics have some broad similarities to those of the MOSFET, the physical principles for its operation are completely different. In subsequent chapters, we will investigate the useful properties of the BJT for integrated-circuit design.

Chapter Objectives

- Bipolar transistor structures, symbols, and current-voltage characteristics.
- The qualitative and quantitative physical basis for BJT operation.
- Ebers-Moll Model for the BJT's terminal characteristics.
- Small-signal hybrid-π models for circuit design.

7.1 INTRODUCTION

Starting in the 1950s, the **bipolar junction transistor** (BJT) was the first solid-state amplifying device to see widespread application in electronics. The BJT was used earlier in commercial applications, even though it was conceived after the field-effect transistor and was based on more complicated physical mechanisms. As you grapple with the intricacies of its internal operation, keep in mind that the three researchers at AT&T Bell Laboratories who fabricated and modeled the first BJT in 1947 were later awarded Nobel Prizes.

The reason for the bipolar transistor's early commercial success was that it can be made using a rather primitive fabrication technology. With the rise of MOS ICs in the 1970s, BJTs were displaced from many digital applications requiring high density or low power. Today, highly integrated digital ICs containing millions of transistors such as microprocessors and memories are generally all-MOS due to the high density of CMOS fabrication processes and the low power dissipation of CMOS logic. However, the bipolar transistor remains the device of choice for maximum performance analog and digital ICs. In the 1990s, the availability of BiCMOS technologies has brought the "best of both worlds" to designers of mixed analog/digital systems such as high-speed modems and analog-to-digital converters. Chapters 8 and 9 will cover the analog circuit properties of BJTs and MOSFETs and develop insights into the advantages and disadvantages of each transistor type.

7.1.1 Bipolar Transistor Physical Structure

The essential core of a BJT is a sandwich of silicon layers of alternating type. The sequence n-p-n results in an ***npn* BJT**, while the sequence p-n-p creates a complementary transistor called the ***pnp* BJT**. These two bipolar transistors are related to each other as circuit elements in a similar manner as n-channel and p-channel MOSFETs. However, unlike the MOSFETs, the npn and pnp transistors in a bipolar IC often have very different physical structures. In this section, we will focus on the npn transistor and defer description of the pnp physical structure until Section 7.7.

Figure 7.1 shows the cross section and layout of an oxide-isolated, polysilicon-emitter npn bipolar junction transistor that is representative of an npn BJT fabricated in modern BiCMOS or high-speed bipolar processes. In contrast to older bipolar processes, a 700 nm-thick field oxide is used instead of a pn junction to provide lateral electrical isolation between adjacent transistors. The silicon substrate is p-type and the pn junction between it and the n-type collector region is reverse-biased in normal operation.

The n-p-n sandwich that is the core of the BJT is highlighted in Fig. 7.1(a). The important features of this device are:

Emitter Region—this very thin, heavily doped n-type layer is electrically contacted by an n^+ polysilicon layer. A metal interconnect makes contact to the polysilicon layer over the field oxide, just as is done for the polysilicon gate of a MOSFET.

Base Region—the p-type layer underneath the emitter. This region must extend laterally to allow contact with the metal interconnect as shown in Fig. 7.1(b).

Collector Region—the n-type layer that lies beneath the p-type base region. This region is contacted through the underlying n^+ layer.

n^+ Buried Layer—the function of this layer is to provide a low-resistance electrical connection to the collector region. A strip of field oxide provides lateral isolation between the p-type base and the n^+ buried layer.

Intrinsic Transistor—the core n^+-p-n sandwich constitutes the intrinsic transistor and is highlighted in the device cross section in Fig. 7.1(a). The area of the sandwich is the same as the emitter area A_E as shown in Fig. 7.1(b). Note that the intrinsic transistor represents a only a small fraction of the total area of the BJT. Most of the area is consumed by oxide isolation regions and by electrical contacts.

7.1.2 BJT Circuit Symbol and Terminal Characteristics

Figure 7.1 illustrated that the bipolar junction transistor has three primary terminals: (1) emitter, (2) base, and (3) collector. The substrate can be considered a fourth terminal, but it does not have a first-order effect on the BJT, in contrast to the MOS-

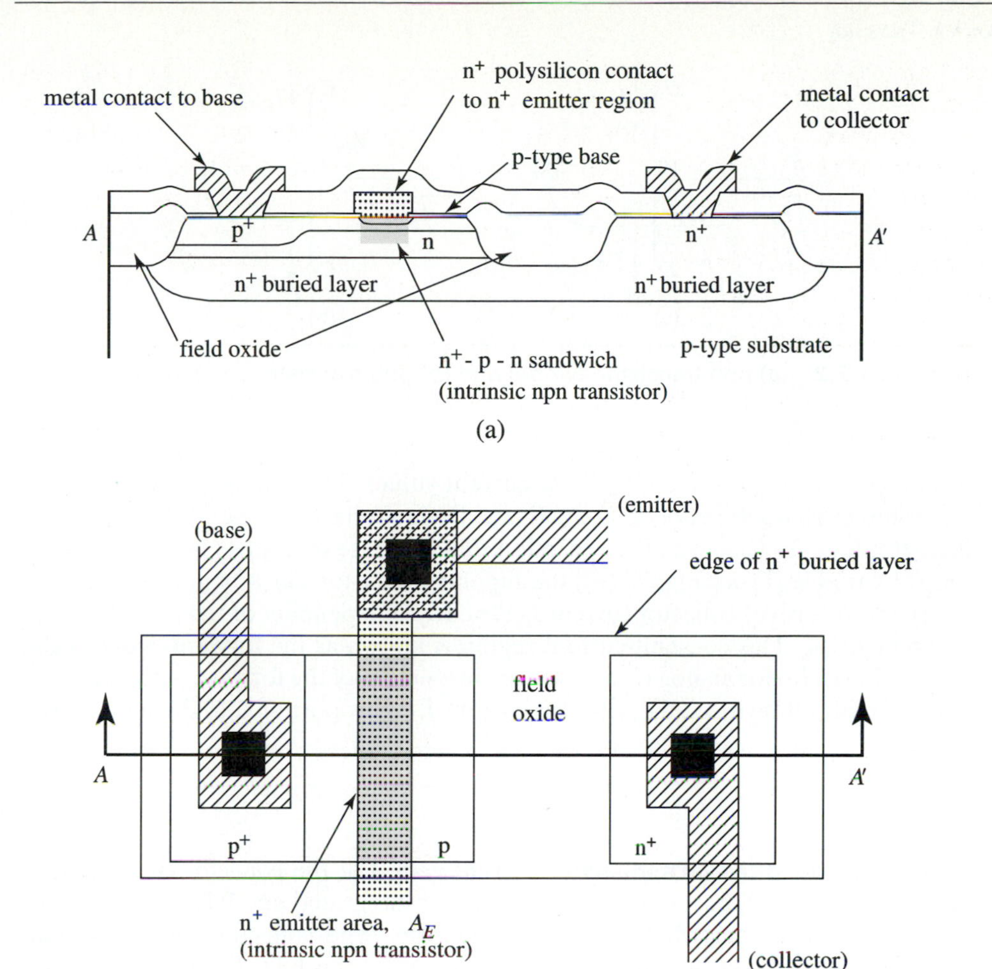

➤ **Figure 7.1** (a) Cross section and (b) layout of a polysilicon-emitter, oxide-isolated npn bipolar junction transistor.

FET. Therefore, we consider the bipolar transistor to be a three-terminal device. The symbols, terminal currents, and voltages for the npn bipolar junction transistor and the pnp BJT are shown in Fig. 7.2. The pnp transistor is shown in its normal configuration on a circuit schematic with the positive supply at the top of the page and the negative supply voltage or ground at the bottom.

By IEEE convention, the reference directions for the base current I_B, the emitter current I_E, and the collector current I_C are defined to be positive *into* the respective terminal. In order to avoid dealing with currents that have negative values, we will use $-I_E$ for the npn transistor and $-I_B$ and $-I_C$ for the pnp transistor as circuit variables. Note in Fig. 7.2 that the reference directions for these currents are reversed and that they are positive in normal operation, e.g., $-I_E > 0$ for the npn transistor.

The terminal characteristics of the bipolar transistor can be measured conveniently using an instrument containing controlled voltage sources, current sources, and ammeters. Figure 7.3(a) shows a schematic of the test circuit for investigating the dependence of the npn BJT's collector current I_C on the base current I_B and the collector-emitter voltage V_{CE}. By convention, Fig. 7.3(b) shows the collector current I_C

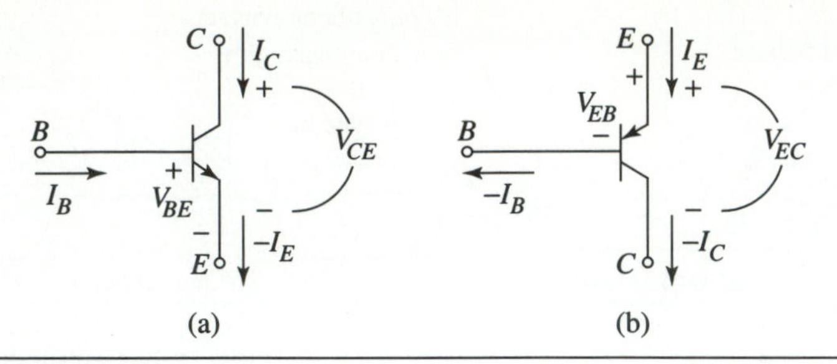

➤ **Figure 7.2** (a) npn transistor symbol and (b) pnp transistor symbol.

as a function of V_{CE} for a range of base current values. This family of curves is called the **collector characteristics** of the BJT. The basic similarity to the MOSFET's drain characteristics is clear, with the exception that the base current I_B is considered the control variable. From Fig. 7.3(b), the bipolar transistor has a region of operation where the (positive) collector current I_C is nearly independent of the collector-emitter voltage V_{CE}. This constant-current region is known as the **forward-active region** and will be useful for analog circuits. Closer inspection of the forward-active collector characteristics shows that the collector current I_C is proportional to the base current I_B with

$$I_C = \beta_F I_B \tag{7.1}$$

where β_F is called the **forward-active current gain**. For the collector characteristics shown in Fig. 7.3(b), $\beta_F = 100$ which is typical for integrated npn BJTs. The collector current increases somewhat as the V_{CE} increases. The collector current deviates strongly from $I_C = \beta_F I_B$ as V_{CE} increases above 5 V for this BJT. If the base current is zero, Eq. (7.1) indicates that the collector current is zero. The transistor is considered **cutoff** in this case.

For negative collector-emitter voltages $V_{CE} < 0$ the collector current also becomes negative as can be seen in the third quadrant of Fig. 7.3(b). The base current, however, is positive. A **reverse-active region** is identified in which the role of the emitter and collector are interchanged. The emitter current is proportional to the base current, with a reverse-active current gain defined by $\beta_R = I_E/I_B$. Noting that the scale is expanded for negative collector currents in Fig. 7.3(b) and that $-I_C = I_E + I_B = (\beta_R + 1)I_B$, the reverse-active current gain is $\beta_R \approx 3$ for this transistor. Because of the very low current gain, IC bipolar transistors are rarely operated in the reverse active region.

Finally, there is a "voltage source" region of operation in which the collector-emitter voltage is about zero—independent of the collector current. This vertical region on the collector characteristics is referred to as the **saturation region**. The terminology is unfortunately inconsistent with that used for MOSFETs. Recall from Chapter 4 that the MOSFET saturation region corresponds to the BJT's forward active region. The origin of the confusion is that the *voltage* is saturated (i.e., not a function of V_{CE}) in the bipolar saturation region, whereas the *current* is saturated (i.e., not a function of V_{DS}) in the MOSFET saturation region. For an IC BJT, a reasonable approximation for the collector-emitter voltage in saturation is $V_{CE_{SAT}} = 100\text{ mV} = 0.1\text{ V}$.

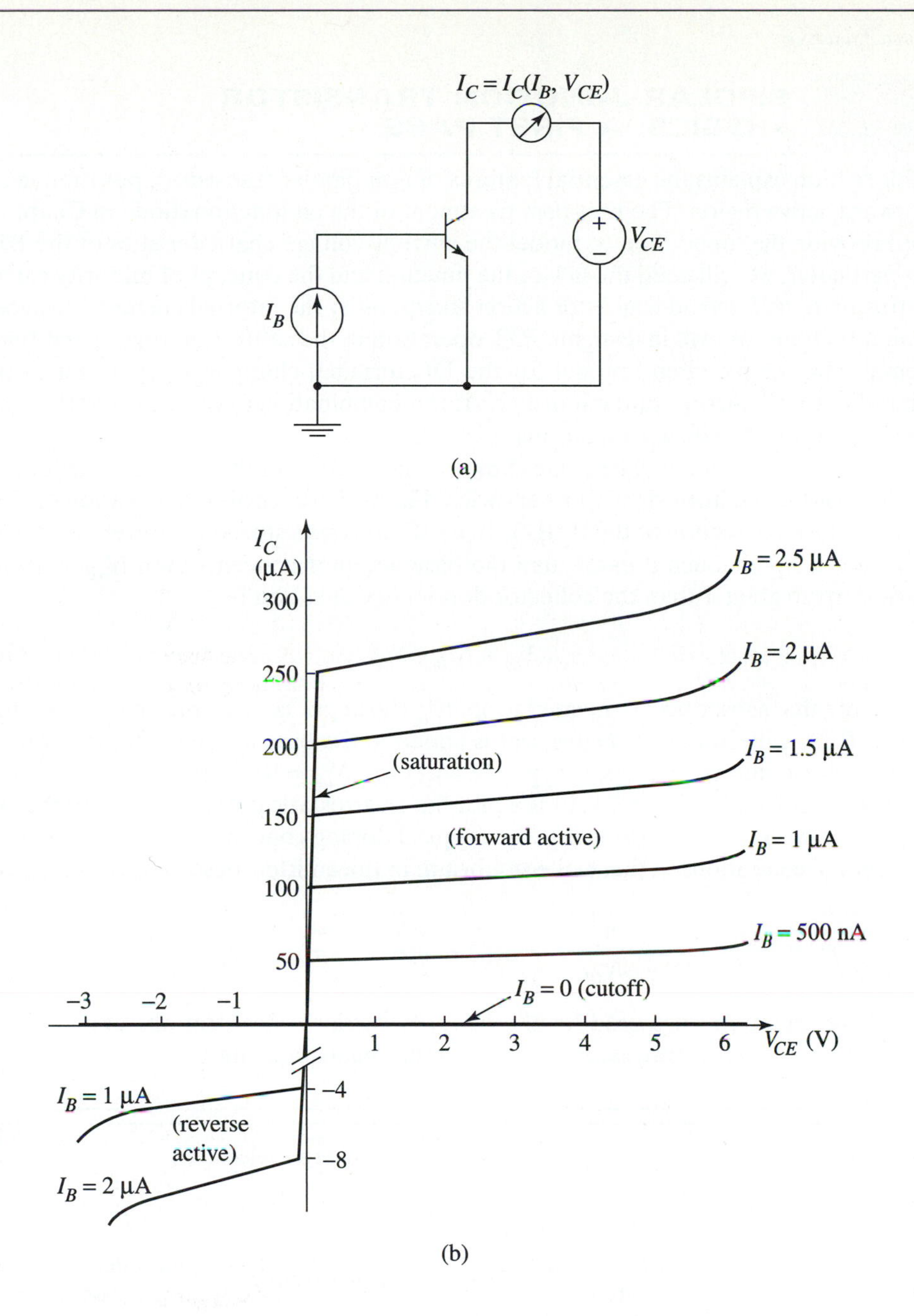

➤ Figure 7.3 npn BJT collector characteristics: (a) test circuit and (b) typical npn I_C (I_B, V_{CE}) measurements. Note the different scale for negative (reverse) collector currents.

➤ 7.2 BIPOLAR JUNCTION TRANSISTOR PHYSICS: A FIRST PASS

This section explains the essential features of npn bipolar transistor operation in the forward active region. The first-pass treatment of the pn junction diode in Chapter 6 will provide the foundation to model the current-voltage characteristics of the BJT. In particular, we will need the law of the junction and the concept of minority carrier diffusion to understand transistor action and quantify the internal currents. In subsequent sections, we will investigate BJT operation in the saturation region and find a remarkably simple circuit model for the DC terminal characteristics. Later in this chapter, we will incorporate minority carrier recombination in the base and the emitter regions into the device equations.

Our starting point will be a one-dimensional analysis of the minority carrier concentrations in the intrinsic n^+-p-n sandwich. Figure 7.4(a) shows the location of the x axis on the cross section of the IC BJT. In an IC npn transistor, the emitter donor concentration N_{dE} is much greater than the base acceptor concentration N_{aB}, which in turn is much greater than the collector donor concentration N_{dC}

$$N_{dE} \gg N_{aB} \gg N_{dC} \tag{7.2}$$

Note that subscripts are needed to specify the donor concentration as E for emitter or C for collector. The B subscript is added to the base acceptor doping concentration, for redundancy's sake. Typical values are $N_{dE} = 10^{19}\ \text{cm}^{-3}$, $N_{aB} = 10^{17}\ \text{cm}^{-3}$, and $N_{dC} = 10^{16}\ \text{cm}^{-3}$. Figure 7.4(b) is a plot the electrostatic potential $\phi_o(x)$ in thermal equilibrium along the x axis, using these typical doping concentrations. The minority carrier concentrations in thermal equilibrium in the emitter, base, and collector are

$$p_{nEo} = \frac{n_i^2}{N_{dE}} \qquad n_{pBo} = \frac{n_i^2}{N_{aB}} \qquad p_{nCo} = \frac{n_i^2}{N_{dC}} \tag{7.3}$$

Three subscripts are used for the minority carrier concentrations to specify the majority carrier type, transistor region, and thermal equilibrium.

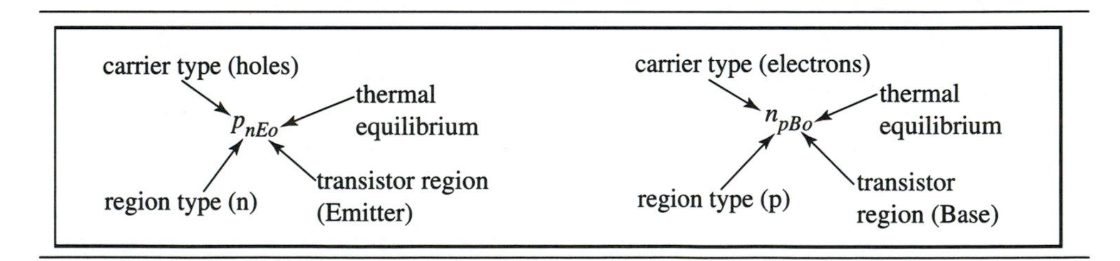

Given the doping concentrations in Eq. (7.2), the relative values for the minority carrier concentrations in thermal equilibrium are

$$p_{nEo} \ll n_{pBo} \ll p_{nCo} \tag{7.4}$$

as shown on a log scale in Fig. 7.4(c). The thermal equilibrium widths of the depletion regions are labeled x_{BEo} and x_{BCo} in Fig. 7.4(c). In addition, the widths of the bulk base and emitter regions in thermal equilibrium are defined as W_{Bo} and W_{Eo}.

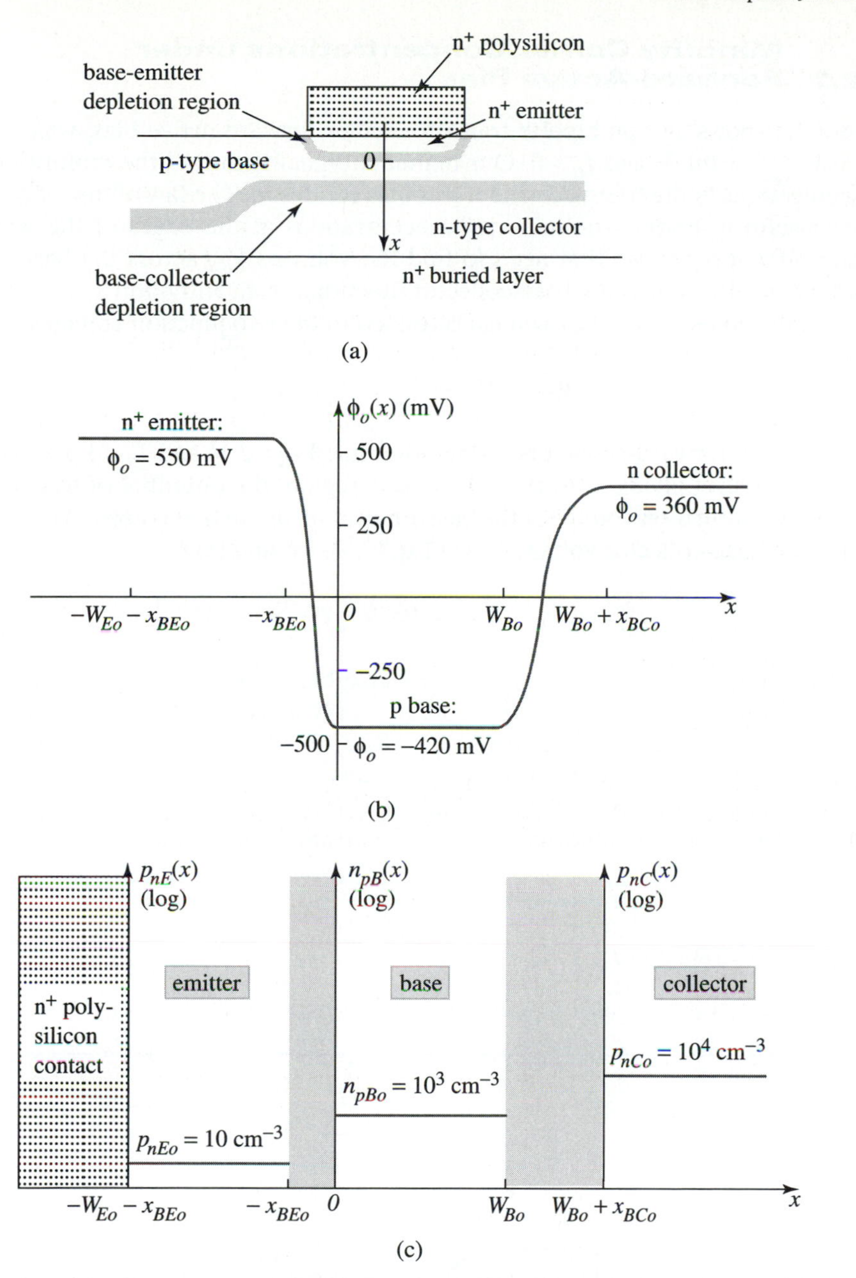

➤ Figure 7.4 (a) Location of the *x* axis in the oxide-isolated npn bipolar junction transistor, (b) electrostatic potential in thermal equilibrium, and (c) one-dimensional log plot of minority carrier concentrations in thermal equilibrium along the *x* axis: p_{nEo} = holes in n$^+$ emitter, n_{pBo} = electrons in p-type base, and p_{nCo} = holes in n-type collector for typical doping concentrations: $N_{dE} = 10^{19}$ cm^{-3}, $N_{aB} = 10^{17}$ cm^{-3}, and $N_{dC} = 10^{16}$ cm^{-3}.

7.2.1 Minority Carrier Concentrations under Forward-Active Bias

Figure 7.5 shows an npn bipolar transistor under forward-active bias, which means $V_{CE} > V_{CE_{SAT}} = 0.1$ V and $I_B > 0$. Our immediate goal is to find the minority carrier concentrations in the transistor under this bias condition. The law of the junction will prove useful in finding the minority concentrations at the edges of the depletion regions. To apply it, we first need to find the voltages V_{BE} across the base-emitter junction and V_{BC} across the base-collector junction. From Kirchhoff's voltage law in Fig. 7.5, the collector-emitter voltage is related to the two junction voltages by

$$V_{CE} = V_{BE} + V_{CB} = V_{BE} - V_{BC} \tag{7.5}$$

A typical forward-active operating point has $V_{CE} = 2$ V and $I_B = 1$ μA. The current source is supplying I_B to the p-type base region, the potential of which cannot rise above the turn-on voltage of the base-emitter pn diode that is around 0.7 V. Solving for the base-collector voltage V_{BC} in Eq. (7.5), we find that

$$V_{BC} = V_{BE} - V_{CE} = 0.7\text{ V} - 2\text{ V} = -1.3\text{ V} \tag{7.6}$$

Thus, the base-collector junction is reverse biased while the base-emitter junction is forward-biased by the current source supplying I_B. In the next section, we will find out how a large collector current $I_C = \beta_F I_B = (100)\ 1\mu\text{A} = 100\ \mu\text{A}$ can be conducted through the sandwich of two pn junctions, one of which is reverse biased.

According to the law of the junction in Eq. (6.22), the two junction voltages V_{BE} and V_{BC} determine the minority carrier concentrations at the edges of the depletion regions

$$\text{B-E junction:} p_{nE}(-x_{BE}) = p_{nEo} e^{V_{BE}/V_{th}} \qquad n_{pB}(0) = n_{pBo} e^{V_{BE}/V_{th}} \tag{7.7}$$

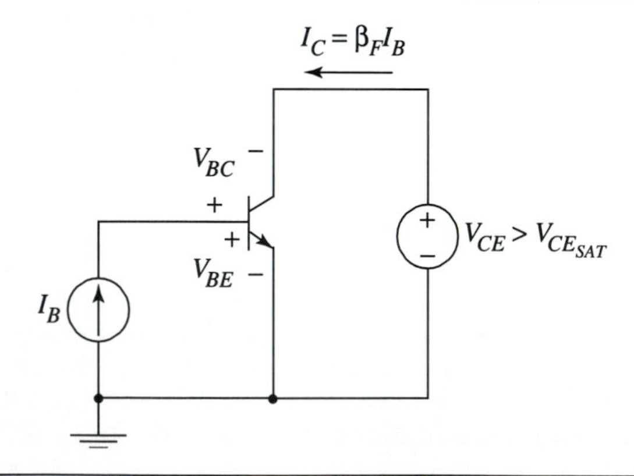

➤ **Figure 7.5** Schematic of an npn transistor biased in the forward-active region by a voltage source V_{CE} and a current source I_B. Using a typical bias point of $V_{CE} = 2$ V and $I_B = 1$ μA, the base-collector junction is reverse biased at $V_{BC} = -1.3$ V, with the base-emitter junction forward-biased at $V_{BE} = 0.7$ V.

$$\text{B-C junction}\ n_{pB}(W_B) = n_{pBo}e^{V_{BC}/V_{th}} \qquad p_{nC}(W_B + x_{BC}) = p_{nCo}e^{V_{BC}/V_{th}} \tag{7.8}$$

There is a potential confusion in the notation for the functional dependence of the minority carrier concentrations on the x coordinate, especially in the case of the electron concentration at $x = 0$

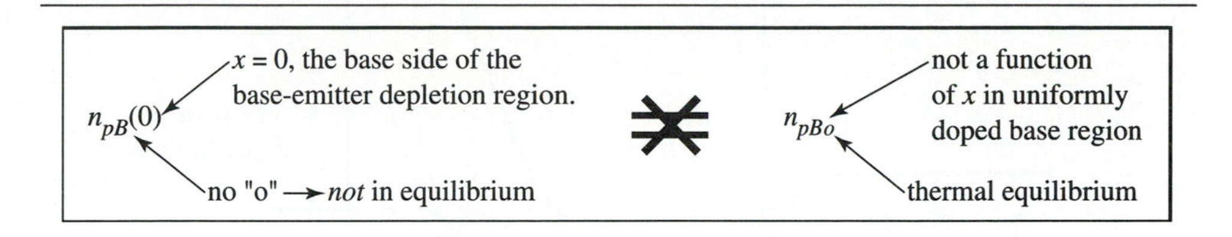

In Eqs. (7.7) and (7.8), the values of the depletion widths and the emitter and base widths have been modified by the bias: $x_{BE} < x_{BEo}$ since the base-emitter junction is forward biased and $x_{BC} > x_{BCo}$ since the base-collector junction is reverse biased. The position $x = 0$ has also shifted slightly since it is located on the base side of the emitter-base depletion region. Assuming the depletion approximation holds, these shifts could be calculated. However, this level of refinement will not be included in our BJT model.

Figure 7.6(a) is a plot of the electrostatic potential for the BJT biased with $V_{BE} = 0.7$ V and $V_{CE} = 2$ V. The minority carrier concentrations at the edges of the base-emitter and base-collector depletion regions are plotted from Eqs. (7.7) and (7.8) in Fig. 7.6(b). According to Eq. (7.8), the plot is necessarily qualitative since the minority carrier concentrations at the base-collector depletion region are essentially zero

$$e^{V_{BC}/V_{th}} = e^{(-1.3\text{V})/(26\text{mV})} = e^{-50} = 2\times10^{-22} \tag{7.9}$$

whereas at the edges of the emitter-base depletion region, the minority carrier concentrations have increased by a factor of $e^{V_{BE}/V_{th}}$. The thermal equilibrium carrier concentrations are included for reference in Fig. 7.6(b).

Although we must precisely know V_{BE} in order to use Eq. (7.7) to find accurate values for the minority carrier concentrations, substitution of $V_{BE} = 0.6$ to 0.7 V shows that $p_{nE}(-x_{BE})$ and $n_{pB}(W_B)$ increase by 10^{10} to 5×10^{11}. As discussed in Chapter 6, this huge increase in minority carrier concentration upon forward bias is referred to as **injection**, whereas the decrease to nearly zero upon reverse bias is referred to as **extraction**.

We assume in this first pass treatment of the BJT that the polysilicon forms an ohmic contact to the n^+ emitter region. The polysilicon/silicon interface has sufficient recombination sites that the carrier concentrations are maintained at their thermal equilibrium values regardless of the current passing through the interface. Figure 7.6(b) shows this assumption for the minority hole concentration at the contact between the n^+ polysilicon and n^+ emitter region under forward-active bias. In our second pass treatment in Section 7.4, we will see this assumption is only approximately true for polysilicon contacts.

It remains to complete the minority carrier distributions in the bulk regions. The n^+ emitter width W_E and p-type base width W_B are both less than 1 μm. Under forward-active bias, Fig. 7.6(b) shows there are huge steady-state gradients in minority

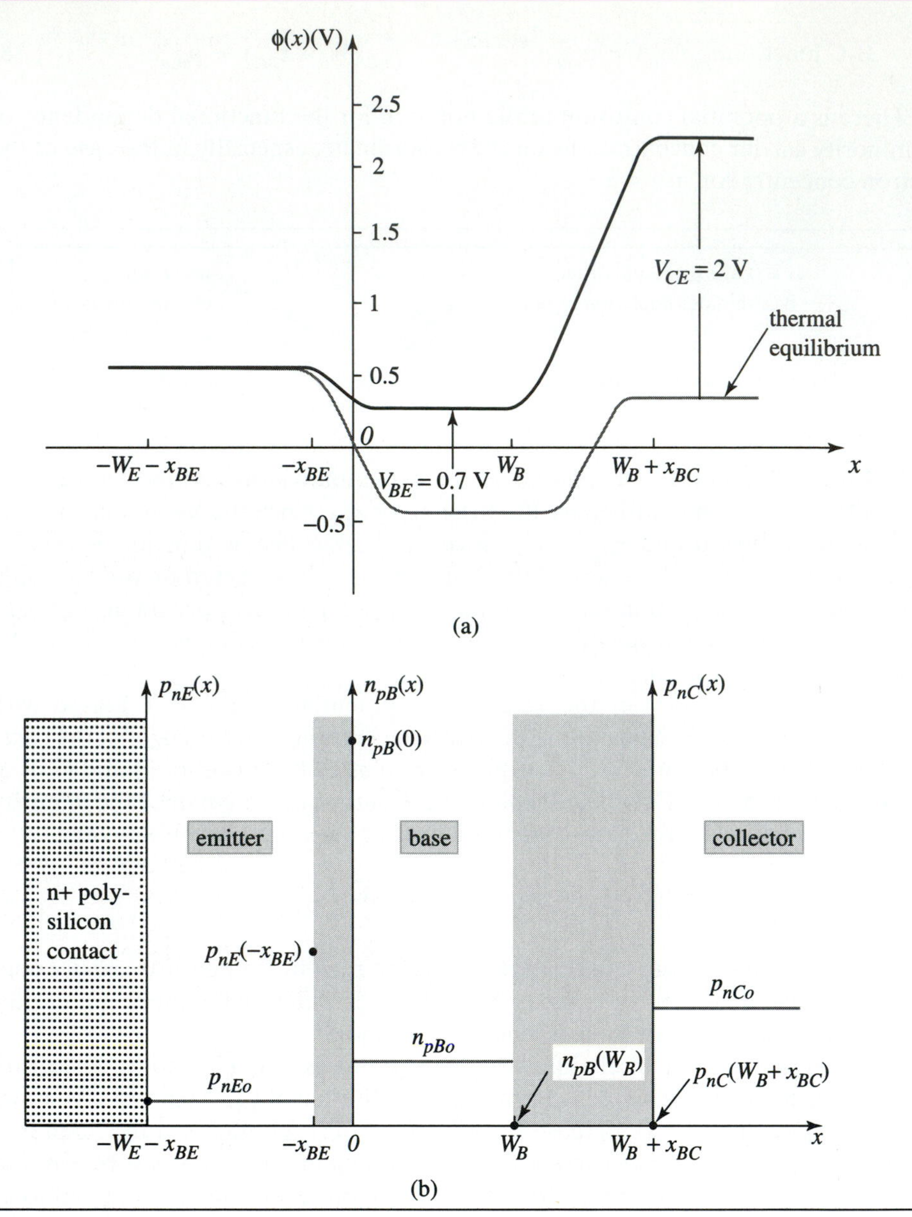

Figure 7.6 (a) The electrostatic potential in an npn BJT biased at $I_B = 1\ \mu A$, $V_{BE} = 0.7$ V, and $V_{CE} = 2$ V. (b) Qualitative plot of the minority carrier concentrations at the edges of the depletion region.

carrier concentrations across the emitter and base bulk regions. These gradients lead to minority carrier diffusion current densities that are the key to understanding the BJT.

Since both regions are narrow, the short-base (straight line) solutions for the steady-state minority carrier concentrations from Chapter 6 apply here. Figure 7.7 is a plot of the hole concentration in the emitter and electron concentration in the base, under forward-active bias.

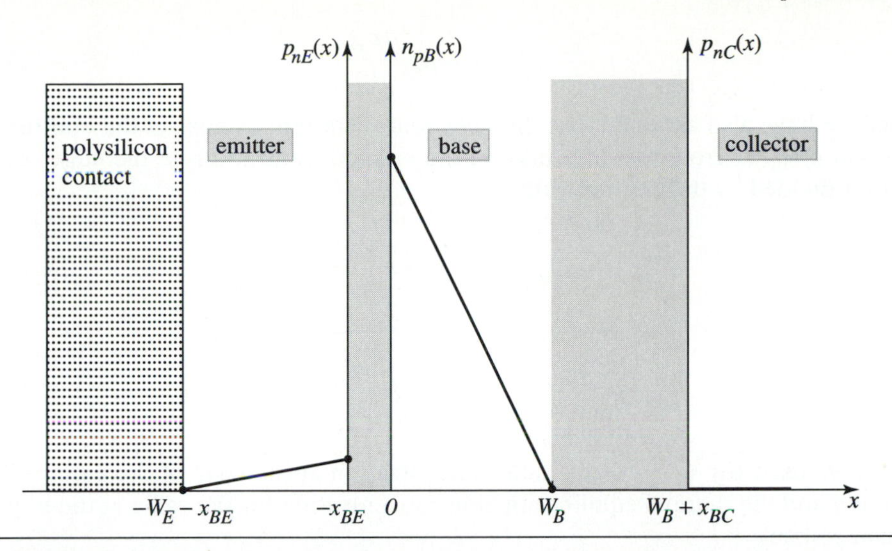

➤ **Figure 7.7** Linear plot of the minority carrier concentrations under forward-active bias. The hole concentration at the edge of the emitter-base depletion region has been exaggerated for clarity. The thermal equilibrium values in the emitter, base, and collector are indistinguishable from zero.

In the base, the minority electrons injected by the forward-biased base-emitter junction diffuse in the $+x$ direction, toward the reverse-biased base-collector junction. The electric field inside the BC junction points in the $-x$ direction, as can be seen in the potential plot in Fig. 7.6(a). Therefore, the diffusing electrons at $x = W_B$ are acted on by an electrostatic force in the $+x$ direction. They are rapidly swept across the base-collector depletion region and enter the n-type collector region where they are majority carriers.

We can quantify the diffusion current densities from Fig 7.7. In the base region, the minority electron diffusion current density J_{nB}^{diff} is given by

$$J_{nB}^{diff} = qD_n \frac{dn_{pB}}{dx} \tag{7.10}$$

The gradient dn_{pB}/dx is the difference in electron concentrations at the edges of the two depletion regions divided by the width of the base

$$J_{nB}^{diff} = qD_n\left(\frac{n_{pB}(W_B) - n_{pB}(0)}{W_B}\right) = qD_n \frac{n_{pBo}(e^{V_{BC}/V_{th}} - e^{V_{BE}/V_{th}})}{W_B} \tag{7.11}$$

where we substituted from Eqs. (7.7) and (7.8). Since the base-collector junction is reverse biased, $n_{pB}(W_B)$ is negligible and Eq. (7.11) can be simplified

$$J_{nB}^{diff} = -\left(\frac{qD_n n_{pBo}}{W_B}\right)e^{V_{BE}/V_{th}} \tag{7.12}$$

electron diffusion current in base

In the emitter region, the minority hole diffusion current density J_{pE}^{diff} is given by

$$J_{pE}^{diff} = -qD_p \frac{dp_{nE}}{dx} \tag{7.13}$$

Since we have also assumed that the hole concentration is linear in the emitter, we can find dp_{nE}/dx from the difference in the hole concentrations at the edges of the emitter divided by the emitter width

$$J_{pE}^{diff} = -qD_p\left(\frac{p_{nE}(-x_{BE}) - p_{nE}(-W_E - x_{BE})}{W_E}\right) \tag{7.14}$$

hole diffusion current in emitter

$$\boxed{J_{pE}^{diff} = -\left(\frac{qD_p\, p_{nEo}}{W_E}\right)\left(e^{V_{BE}/V_{th}} - 1\right)} \tag{7.15}$$

where we used Eq. (7.7) for the hole concentration at the edge of the emitter-base junction and the thermal equilibrium hole concentration for the value at the n^+ polysilicon contact.

7.2.2 Electron and Hole Fluxes under Forward-Active Bias

The minority carrier diffusion current densities under forward-active bias in Eqs. (7.12) and (7.15) are essential to understanding the internal operation of the BJT. However, we need to relate them to the terminal currents I_B, I_C, and I_E.

In order to identify the various components making up the terminal currents, it is helpful to ascertain the *directions* of carrier transport in the actual device structure. Since the electron current density is opposite to the electron transport (due to the negative electron charge), it is convenient to introduce the **electron flux** F_n, where

$$F_n = \frac{J_n}{-q}. \tag{7.16}$$

For holes, the flux F_p is in the same direction as the current density and $F_p = J_p/q$.

Figure 7.8(a) is a cross section of the intrinsic n^+pn sandwich of a forward-active IC bipolar transistor, on which the electron and hole diffusion fluxes in the base and emitter are depicted as "streams" having widths proportional to the magnitude of the flux. In reality, the fluxes are spread over the entire base and emitter regions.

Next we complete the picture by determining where these steady-state fluxes originate and how they are connected to the three terminal currents flowing through the metal contacts. The electrons in the base are injected from the n^+ emitter into the p-type base. Therefore, there must be a majority electron flux in the emitter to supply this injection.

After diffusing across the base, the electrons are swept into the n-type collector region by the base-collector depletion region. It follows that a majority electron flux in the collector, equal in magnitude to the electron diffusion flux in the base, flows out the collector contact and constitutes the collector current I_C. Figure 7.8(b) depicts the electron flux through the emitter, base, and collector regions of the BJT. The majority electron flux in the collector is depicted flowing in the n^+ buried layer since the conductivity is much higher than in the n-type layer. Note in Fig. 7.8(b) that the magnitude of the electron flux is constant from emitter, through the base, and out the collector contact.

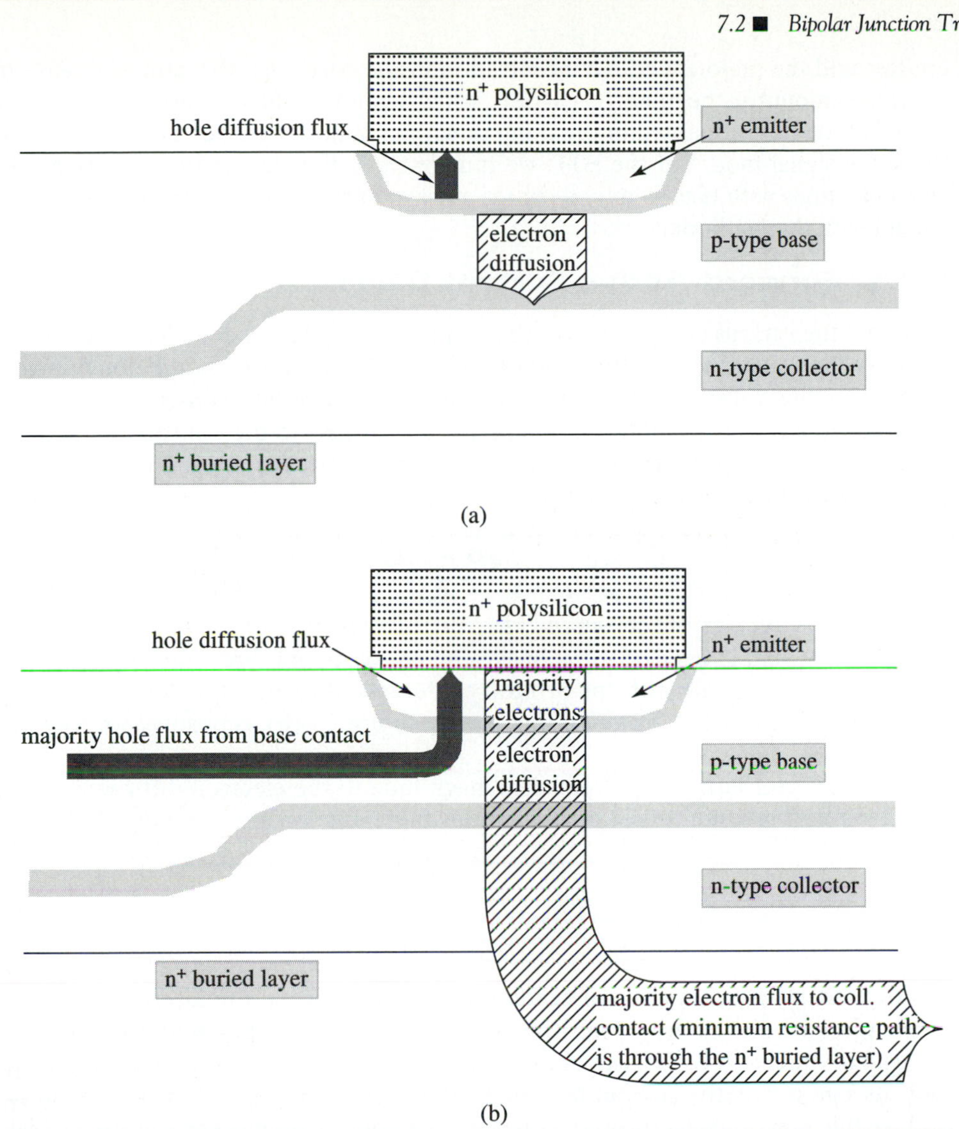

➤ Figure 7.8 (a) Minority carrier diffusion fluxes in the emitter and the base regions, and (b) qualitative picture of the drift and diffusion fluxes throughout the bipolar transistor in forward-active bias, with the width of the stream proportional to the flux magnitude.

Holes are injected from the p-type base into the emitter to supply the hole diffusion flux in the emitter. These holes must be supplied by a flux from the base contact since there is no flux of holes along the x axis in the base. These holes are supplied by a majority carrier flux from the base contact to the emitter-base depletion region as shown in Fig. 7.8(b). In the second-pass treatment of the bipolar junction transistor in Section 7.6, we will investigate what happens to the hole diffusion flux after entering the n^+ polysilicon contact to the emitter.

The majority carrier fluxes in the emitter and base are transported by both diffusion and drift mechanisms. Charge neutrality is maintained in the bulk regions under forward bias, as discussed in Section 6.3. The majority electron concentration in the emitter and the majority hole concentration in the base each therefore increase by the same amount as the minority carriers, which leads to diffusion currents. However, there is little insight gained in breaking down the majority carrier fluxes further. In the small-signal model of the BJT, we must include the change in majority carrier concentrations with bias to model charge storage in the base and emitter regions, as we did with the pn diode.

7.2.3 Forward-Active Current Gains

Now that the various components of the current density have been identified, we can quantify the dependence of the terminal currents I_B, I_C, and I_E on the junction voltages. From Fig. 7.8(b), the base current in forward-active bias consists of holes that are injected across the emitter-base junction and diffuse to the emitter contact. The base current is the product of the emitter-base junction area A_E and the hole diffusion current density from Eq. (7.15)

forward-active base current

$$I_B = -J_{pE}^{diff} A_E = \left(\frac{qD_p p_{nEo} A_E}{W_E}\right)(e^{V_{BE}/V_{th}} - 1) \tag{7.17}$$

The negative sign is needed to reconcile the conventional direction of I_B (positive into the base terminal) with the arbitrary reference direction for the x axis in Fig. 7.4(a). The base current, from the flux picture in Fig. 7.8(b), is positive for forward-active bias.

The collector current I_C is equal in magnitude to the electron diffusion current density in the base, multiplied by the emitter-base junction area (see Fig. 7.8(b))

forward-active collector current

$$I_C = -J_{nB}^{diff} A_E = \left(\frac{qD_n n_{pBo} A_E}{W_B}\right) e^{V_{BE}/V_{th}} \tag{7.18}$$

where the negative sign is again needed to convert between the internal x-axis sign convention and the external "positive in" sign convention. The emitter current I_E is the sum of the electron drift current and the hole diffusion current at the emitter contact. Its sign is negative, as can be seen from the fluxes in Fig. 7.8(b) in the emitter and recalling that the electrons carry negative charge. Dropping the 1 in Eq. (7.15) in comparison with $\exp(V_{BE}/V_{th})$—a negligible error since the base-emitter junction is forward-biased—we find the expression for the emitter current

$$I_E = (J_{nB}^{diff} + J_{pE}^{diff}) A_E = -\left[\left(\frac{qD_p p_{nEo} A_E}{W_E}\right) + \left(\frac{qD_n n_{pBo} A_E}{W_B}\right)\right] e^{V_{BE}/V_{th}} \tag{7.19}$$

This result also follows from applying Kirchhoff's current law to the BJT.

These expressions for the terminal currents are the basis for understanding how the internal dimensions and doping of the BJT influence its operation. The ratio of the collector current to the magnitude of the emitter current is defined as

$$\alpha_F = \frac{I_C}{-I_E} = \frac{\left(\frac{qD_n n_{pBo} A_E}{W_B}\right)}{\left(\frac{qD_p p_{nEo} A_E}{W_E}\right) + \left(\frac{qD_n n_{pBo} A_E}{W_B}\right)} = \frac{1}{1 + \left(\frac{D_p p_{nEo} W_B}{D_n n_{pBo} W_E}\right)} \tag{7.20}$$

"Alpha-F" is always less than one. However, as the term in parentheses in the denominator approaches zero $\alpha_F \rightarrow 1$. Substituting for the minority carrier concentrations in thermal equilibrium, Eq. (7.20) becomes

$$\boxed{\alpha_F = \frac{1}{1 + \left(\frac{D_p N_{aB} W_B}{D_n N_{dE} W_E}\right)}} \quad (7.21)$$

forward-active α_F

As mentioned earlier, a typical emitter doping is $N_{dE} = 10^{19}$ cm^{-3} and a typical base doping is $N_{aB} = 10^{17}$ cm^{-3}. Considering the emitter width to be about equal to the base width and the ratio of hole and electron diffusivities to be around 1/2, a typical α_F is

$$\alpha_F = \frac{1}{1 + \left(\frac{1}{2}\right)\left(\frac{1}{100}\right)(1)} = 0.995 \quad (7.22)$$

The collector characteristics in Fig. 7.3(b) use the base current I_B as the control variable for I_C in the forward-active region of operation. The base current can be found in terms of the collector current using Kirchhoff's current law and Eq. (7.20)

$$I_B = -I_E - I_C = \frac{I_C}{\alpha_F} - I_C = I_C\left(\frac{1 - \alpha_F}{\alpha_F}\right) \quad (7.23)$$

Solving for the collector current, we can express the forward-active current gain β_F in terms of α_F

$$\boxed{\frac{I_C}{I_B} = \left(\frac{\alpha_F}{1 - \alpha_F}\right) = \beta_F} \quad (7.24)$$

forward-active current gain

Substituting the typical value of α_F, we find

$$\beta_F = \frac{0.995}{1 - 0.995} \cong 200 \quad (7.25)$$

Note from Eq. (7.24) that β_F is an extremely sensitive function of the device parameters since α_F is very close to unity. For integrated npn BJTs, the forward-active current gain ranges from 50–250. Special "super-β" transistors can be made with $\beta_F > 1000$. However, other device characteristics are compromised for these very narrow base width devices. In this section, we developed a first-level device model for the npn BJT collector characteristics in the forward-active region. Even with our simple one-dimensional modeling, the expressions for α_F and β_F in Eqs. (7.21) and (7.24) contain the essential parameters that influence the current gain.

➤ EXAMPLE 7.1 npn BJT Device Design

Since the forward-active current gain is an important parameter of the BJT, in this example we examine the degree to which its value can be controlled using IC technology. The transistor's dimensions and parameters are

$$N_{dE} = 7.5 \times 10^{18}\ \text{cm}^{-3},\ N_{aB} = 1 \times 10^{17}\ \text{cm}^{-3},\ N_{dC} = 1.5 \times 10^{16}\ \text{cm}^{-3};$$
$$D_{pE} = 5\ \text{cm}^2/\text{s},\ D_{nB} = 10\ \text{cm}^2/\text{s}.$$

Due variations in the diffusion of impurities during annealing, the emitter junction depth varies from device to device. Assume the base width is $W_B = 300\ \text{nm} \pm 20\ \text{nm}$ and the emitter width is $W_E = 250\ \text{nm} \pm 20\ \text{nm}$ over a wafer lot for a given operating point. These variations are *not* independent—when the emitter width is at its maximum value, the base width is at its minimum value. Therefore, the statistical approach is not appropriate.

Find the minimum and maximum values of the forward-active current gain in this wafer lot.

SOLUTION

We start by substituting the expression for α_F (Eq. (7.21)) into the definition of β_F (Eq. (7.24)) to find the dependence of β_F on the device parameters

$$\beta_F = \frac{\alpha_F}{1-\alpha_F} = \frac{\frac{1}{1+\Delta}}{1-\frac{1}{1+\Delta}} = \frac{1}{\Delta} = \frac{D_{nB} N_{dE} W_E}{D_{pE} N_{aB} W_B}$$

The ratio (W_E/W_B) varies due to the fabrication process

$$\left.\frac{W_E}{W_B}\right|_{max} = \frac{250+20}{300-20} = 0.96 \text{ and } \left.\frac{W_E}{W_B}\right|_{min} = \frac{250-20}{300+20} = 0.72$$

Substituting into the expression for the current gain, we find

$$\beta_F\big|_{max} = \left(\frac{10 \cdot 7.5 \times 10^{18}}{5 \cdot 1 \times 10^{17}}\right)(0.96) = 144 \text{ and}$$

$$\beta_F\big|_{min} = \left(\frac{10 \cdot 7.5 \times 10^{18}}{5 \cdot 1 \times 10^{17}}\right)(0.72) = 108$$

Therefore, the current gain range is 108–144, which is quite wide considering the reasonably tight (7–8%) tolerances on the emitter and base widths. The reason for the large uncertainty is that the emitter and base widths are not independent variables but vary oppositely, which magnifies the spread in β_F. A minimum value for the current gain is usually stated in the device properties section of a design handbook for a bipolar technology. Tight control of β_F is not practical because of effects such as the one in this example.

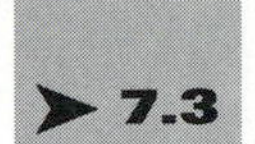

7.3 REVERSE-ACTIVE AND SATURATION OPERATING REGIONS

Although our first-pass investigation of the BJT was developed for the forward-active region, the concept that the electron diffusion current density in the base links the collector current and emitter current is completely general. In this section, we first consider the reverse-active region. We will then superimpose forward and reverse modes to model the transistor when biased in the saturation region.

From Fig. 7.3(b), the reverse active region is where the base current $I_B > 0$ Aand collector-emitter voltage $V_{CE} < -0.1$ V. The operating point $V_{CE} = -2$ V and $V_{BC} \approx 0.7$ V is in the reverse-active region. Therefore, the base-collector junction is forward-biased and the base-emitter junction is reverse-biased at $V_{BE} = -1.3$ V for this operating point. The roles of the two junctions have been reversed. The base-collector junction injects electrons into the base, which diffuse across the base and are collected by the reverse-biased base-emitter junction.

Figure 7.9(a) shows the minority carrier concentrations for reverse-active operation. The collector doping concentration is lower than that of the base. For this reason, the hole concentration on the collector side of the junction is higher than the electron concentration on the base side. The holes injected into the collector diffuse in the $+x$ direction away from the base-collector junction. How far do they diffuse into the n-type underlying layers? The n^+ buried layer has a very high electron concentration that results in a large hole recombination rate. Effectively, the holes do not penetrate far into the buried layer. As a first approximation, we set the hole concentration $p_{nC}(x) = p_{nCo}$, its thermal equilibrium value, at the edge of the buried layer as shown in Fig. 7.9(a).

The flux picture of reverse-active operation is shown in Fig. 7.9(b). The electron diffusion flux in the base links the two pn junctions—it is opposite to that for forward-active bias. There are two components to the majority electron flux in the collector. The first supplies injection into the base, whereas the second supplies recombination in the buried layer of the holes injected from the base into the collector. The flux plot does not reflect the fact that the base-collector junction is much larger in area than the base-emitter junction.

Having clarified the internal operation of the BJT in forward- and reverse-active bias conditions, we are now prepared to understand the more complicated saturation region. In saturation, *both* the base-emitter and base-collector junctions are forward biased. From the law of the junction, electrons are injected into the base from both junctions. In plotting the minority carrier concentrations in Fig. 7.10, we assume that $V_{CE} > 0$ which means that $V_{BE} > V_{BC}$. Therefore, the electron concentration on the base side of the emitter $n_{pB}(0)$ is larger than the concentration at the collector side $n_{pB}(W_B)$. The minority carrier concentration in the base $n_{pB}(x)$ is a straight line connecting these points as shown in Fig. 7.10.

We can gain insight into the current components in saturation by finding the electron diffusion current in the base from the gradient in $n_{pB}(x)$

$$J_{nB}^{diff} = qD_n\left(\frac{n_{pB}(W_B) - n_{pB}(0)}{W_B}\right) = qD_n\frac{n_{pBo}\left(e^{V_{BC}/V_{th}} - e^{V_{BE}/V_{th}}\right)}{W_B} \tag{7.26}$$

electron diffusion current density

By separating the terms due to the base-emitter voltage and base-collector voltage, we can rewrite Eq. (7.26) as

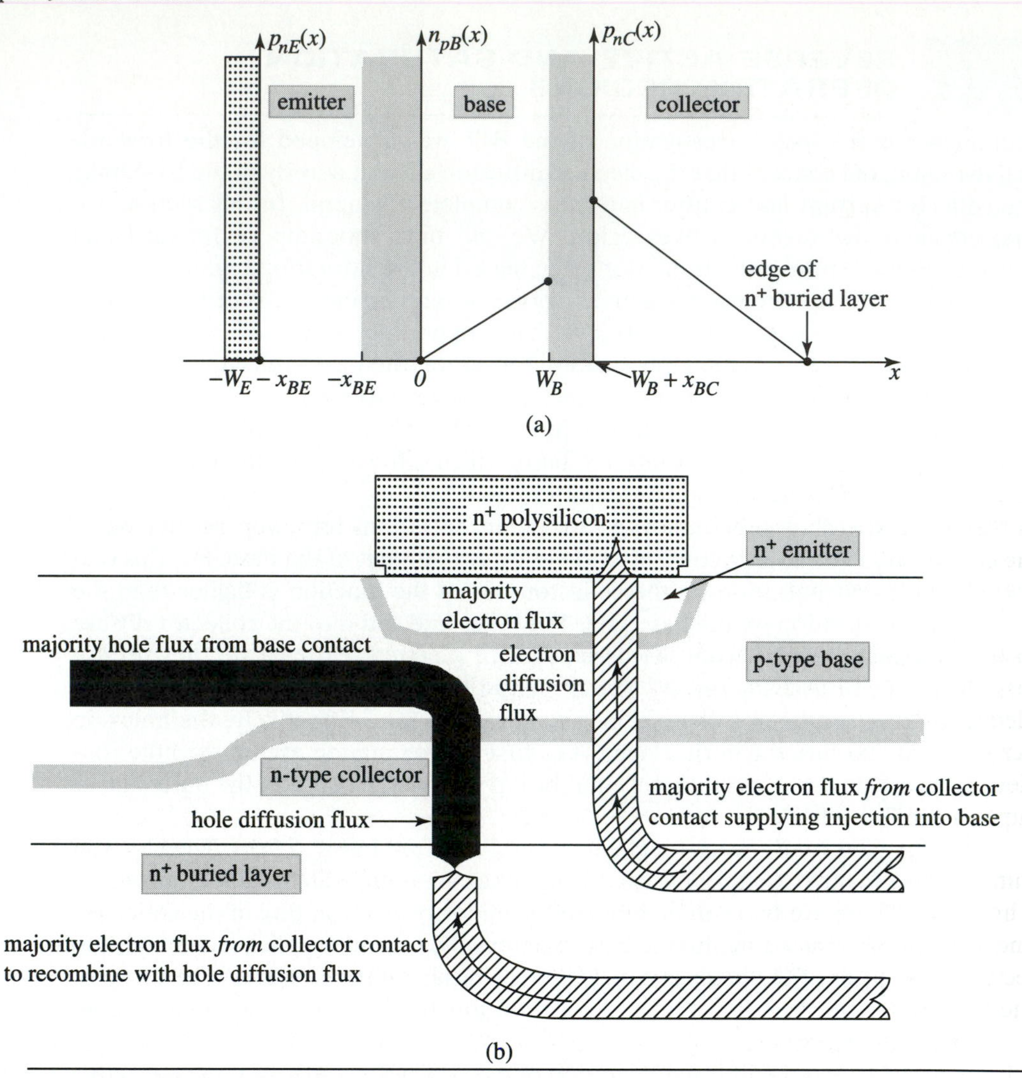

➤ Figure 7.9 (a) Minority carrier concentrations for reverse active bias, and (b) flux picture of hole and electron current components for this bias condition.

$$J_{nB}^{diff} = -qD_n \frac{n_{pBo}e^{V_{BE}/V_{th}}}{W_B} + qD_n \frac{n_{pBo}e^{V_{BC}/V_{th}}}{W_B} = J_{nB}^{diff}\Big|_{\text{forward}} + J_{nB}^{diff}\Big|_{\text{reverse}} \tag{7.27}$$

where the diffusion current densities for forward-active bias (Eq. 7.12) and the similar expression for reverse-active bias have been identified. In Fig. 7.10, we have broken out the minority electron concentration in the base into a forward-active component (injected at the B-E junction and diffusing toward the B-C junction) and a reverse-active electron component (injected at the B-C junction and diffusing in the opposite direction.)

What happens to the diffusing electrons in each component when they reach the forward-biased junction on the other side of the base? Since there is still a potential barrier at the junction, the electric field in each depletion region points toward the

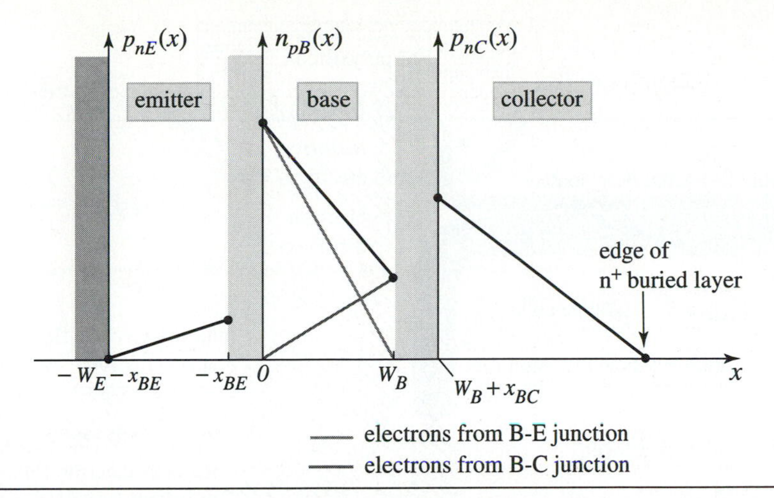

➤ **Figure 7.10** Minority carrier concentrations in the saturation region, where both the B-E and B-C junctions are forward biased. The minority electron concentration in the base is broken into the components injected from the B-E junction and B-C junction.

base side despite the forward bias. In other words, the applied biases V_{BE} and V_{BC} are less than the built-in potentials at the B-E and B-C junctions and the direction of the electric fields in the depletion regions are unchanged. Diffusing electrons are swept across each depletion region into the collector or emitter and result in majority carrier current densities. In summary, the BJT in saturation can be modeled as the superposition of the forward-active and reverse-active regions.

Figure 7.11 shows the carrier flux picture in saturation corresponding to the minority carrier concentrations in Fig. 7.10. Since both junctions are forward-biased, a hole flux from the base contact is needed to supply hole injection into the collector and emitter. As was discussed for reverse-active bias, the holes diffusing in the collector recombine at the n$^+$ buried layer; a component of the majority electron flux in the collector supplies this process.

The two components of the electron diffusion flux in the base are shown in Fig. 7.11. Since $V_{BE} \geq V_{BC}$ the emitter-base junction is injecting more than the base-collector junction for the bias condition in Fig. 7.10, and Eq. (7.26) shows that the net electron diffusion current density is in the $-x$ direction. The flux picture is consistent with Eq. (7.26) since the electron diffusion flux in the base from emitter to collector is greater than the reverse component from collector to emitter.

The transistor bias so that both junctions are forward-biased has $V_{BE} \approx 0.7\text{V}$ and $V_{BC} \approx 0.7\text{V}$. The collector-emitter voltage is therefore nearly zero

$$V_{CE} = V_{BE} - V_{BC} \approx 0\text{V} \tag{7.28}$$

By convention, an npn BJT is considered saturated when $V_{CE} \leq V_{CE_{SAT}} \approx 0.1\text{V}$.

Another way of recognizing that a transistor is in the saturation region is by comparing the ratio of base current to collector current to the forward-active current gain β_F. Comparing Fig. 7.11 with the flux picture for the forward-active region in Fig. 7.8(b), the base current I_B is larger since there is an extra component of hole flux injected into the collector. Also, the collector current I_C is smaller than for forward-active operation, due to both electron injection from the collector into the base and

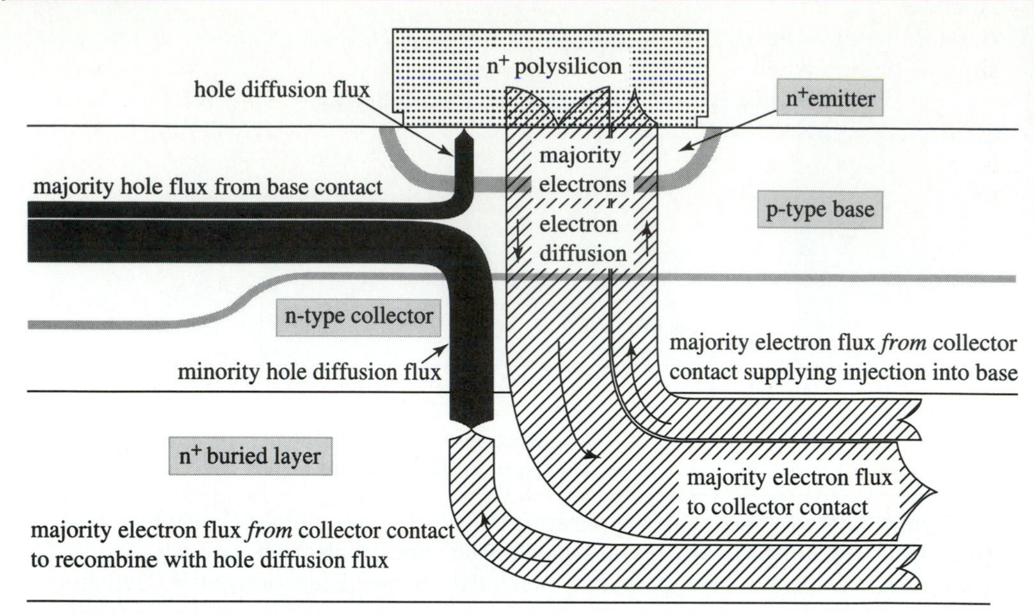

Figure 7.11 Carrier fluxes in the BJT in saturation, which correspond to the minority carrier concentrations plotted in Fig. 7.10.

to recombination with holes diffusing from the base-collector junction. Therefore, the ratio of collector current to base current $I_C/I_B < \beta_F$ in Fig. 7.11, indicating the BJT is saturated.

7.4 THE EBERS-MOLL EQUATIONS

Now that we have a qualitative understanding of the BJT in the forward-active, reverse-active, and saturation regions, we turn to quantifying terminal currents I_C, I_E, and I_B in terms of the applied junction voltages V_{BE} and V_{BC}. Remarkably, the DC characteristics of the BJT can be modeled by a single set of equations that are valid for *all* operating regions. Recall from Chapter 4 that the MOSFET required separate equations for the triode and saturation regions of operation.

Our approach will be to build on the insight that the minority carrier concentrations in saturation can be viewed as the superposition of the forward-active and reverse-active distributions as shown in Fig. 7.10 and 7.11. A good start toward achieving our goal is to write the electron diffusion current density in the base as the sum of forward and reverse currents, each in the form of a diode's current-voltage characteristic. Adding and subtracting one from Eq. (7.26) and then regrouping the exponential terms, we find

$$J_{nB}^{diff} = -qD_n \frac{n_{pBo}(e^{V_{BE}/V_{th}} - 1)}{W_B} + qD_n \frac{n_{pBo}(e^{V_{BC}/V_{th}} - 1)}{W_B} \tag{7.29}$$

Note that the first term in Eq. (7.29) is the diffusion current density for the case when $V_{BC} = 0$ V. This can be verified by substitution and recognizing that the electron concentration at the base-collector junction is $n_{pB}(W_B) = n_{pBo}$ for $V_{BC} = 0$ V from the law of the junction. Similarly, the second term in Eq. (7.29) is the diffusion current density for the case when $V_{BE} = 0$ V.

It remains to connect J_{nB}^{diff} in the base to the terminal currents. Starting with the emitter current, first recall that the multiplication by the emitter area A_E in Fig. 7.1(b) is necessary to convert the current density (units, A/cm^2) into current. It is convenient to collect the pre-exponential terms and define the **transistor saturation current** I_S as

saturation current (npn BJT)

$$I_S = \frac{qD_n n_{pBo} A_E}{W_B} \tag{7.30}$$

which allows us to express the electron diffusion current in the base I_{diff} from Eq. (7.29) by

$$I_{diff} = -I_S(e^{V_{BE}/V_{th}} - 1) + I_S(e^{V_{BC}/V_{th}} - 1) = -I_1 + I_2 \tag{7.31}$$

The currents $-I_1$ and I_2 are known as the forward and reverse diffusion currents. How do these internal currents contribute to the emitter current I_E? The flux picture in Fig. 7.11 shows that the emitter current at the polysilicon contact is the sum of three terms:

1. Forward diffusion current $-I_1$
2. Hole diffusion current in the emitter due to reverse injection I_3
3. Reverse diffusion current I_2 injected at the collector-base junction and collected at the base-emitter depletion region.

The minority hole diffusion current is injected from the base and constitutes the base current in the case of forward-active bias. Since the collector current is equal to the forward diffusion current $-I_1$ in this case, we can apply Eq. (7.24) to relate I_3 to $-I_1$

$$I_3 = I_B\big|_{\text{f.a.}} = \frac{I_C\big|_{\text{f.a.}}}{\beta_F} = \frac{-I_1}{\beta_F} \tag{7.32}$$

Summing the three contributions to the emitter current, we find

$$I_E = -I_1 + I_3 + I_2 = -I_1 - \frac{I_1}{\beta_F} + I_2 = -\left(1 + \frac{1}{\beta_F}\right) I_1 + I_2 \tag{7.33}$$

Note that the positive reference direction for the emitter current is *into* the emitter terminal, which (coincidentally) is the same as the $+x$ direction used for the internal currents. The emitter current can be simplified by substituting for β_F in terms of α_F

$$I_E = -\frac{I_1}{\alpha_F} + I_2 = -\left(\frac{I_S}{\alpha_F}\right)(e^{V_{BE}/V_{th}} - 1) + I_S(e^{V_{BC}/V_{th}} - 1) \tag{7.34}$$

Due to the symmetry of the npn structure, we expect a similar equation for the collector current. The collector current I_C can also be expressed as the sum of three contributions:

1. **Forward diffusion current** $-I_1$ injected at the emitter-base junction and collected at the base-collector depletion region
2. **Hole diffusion current due to reverse injection in the collector** I_4
3. **Reverse diffusion current** I_2.

The hole diffusion current in the collector is the base current in the reverse-active region, where the reverse diffusion current constitutes the emitter current. Therefore, we can relate I_4 to I_2 by

$$I_4 = I_B\big|_{\text{r.a.}} = \frac{I_E\big|_{\text{r.a.}}}{\beta_R} = \frac{I_2}{\beta_R} \tag{7.35}$$

where β_R is the **reverse-active current gain** that relates the emitter and base currents in the reverse-active region. As seen in Fig. 7.3(b), the reverse-active current gain is small in an IC npn BJT because the base doping concentration is larger than that in the collector, which is acting as the emitter in reverse-active operation. The reference direction for the collector current I_C (positive into the terminal) is opposite to the $+x$ direction used as the reference for the internal currents, so the signs of $-I_1$, I_4, and I_2 must be inverted

$$I_C = I_1 - I_2 - I_4 = I_1 - I_2 - \frac{I_2}{\beta_R} = I_1 - \left(1 + \frac{1}{\beta_R}\right)I_2 \tag{7.36}$$

The parameter α_R is defined in analogy to the forward α_F

$$\alpha_R = \frac{\beta_R}{1 + \beta_R} \tag{7.37}$$

Substituting for β_R in Eq. (7.36), we can express the collector current as

$$I_C = I_S(e^{V_{BE}/V_{th}} - 1) - \left(\frac{I_S}{\alpha_R}\right)(e^{V_{BC}/V_{th}} - 1) \tag{7.38}$$

Finally, the base current I_B can be found in terms of I_E and I_C from Kirchhoff's current law

$$I_B = -I_C - I_E \tag{7.39}$$

The emitter and collector current expressions in Eqs. (7.34) and (7.38) are known as the **Ebers-Moll Equations in transport form**. Although they have been derived with the aid of the saturation flux picture in Fig. 7.11, they are valid for forward- or reverse-bias on either junction. For example, Eq. (7.38) under forward-active bias conditions becomes

$$I_C \cong I_S e^{V_{BE}/V_{th}} \tag{7.40}$$

since terms on the order of I_S can be neglected compared with the exponential term. This result is identical to Eq. (7.18).

The standard form of the Ebers-Moll Equations is slightly different from Eqs. (7.34) and (7.38). By defining the emitter saturation current I_{ES} and the collector saturation current I_{CS} by

$$I_{ES} = I_S/\alpha_F \quad \text{and} \quad I_{CS} = I_S/\alpha_R \tag{7.41}$$

we can re-express Eqs. (7.34) and (7.38) in the standard form

$$I_E = -I_{ES}(e^{V_{BE}/V_{th}} - 1) + \alpha_R I_{CS}(e^{V_{BC}/V_{th}} - 1) \tag{7.42}$$

$$I_C = \alpha_F I_{ES}(e^{V_{BE}/V_{th}} - 1) - I_{CS}(e^{V_{BC}/V_{th}} - 1) \tag{7.43}$$

Ebers-Moll Equations (NPN transistor)

There are only three independent parameters in the Ebers-Moll Equations since Eq. (7.41) indicates that

$$\alpha_F I_{ES} = \alpha_R I_{CS}. \tag{7.44}$$

reciprocity relation

This result, known as the **reciprocity relation**, is quite general and applies to transistors with nonuniform doping in the emitter, base, and collector regions.

7.4.1 The Ebers-Moll Circuit Model

The Ebers-Moll Equations, Eqs. (7.42)–(7.44), are remarkable for their simplicity and ability to predict the essential current-voltage behavior of the npn BJT. Since the forward and reverse currents are put into the form of diode currents, it is relatively straightforward to construct the circuit model in Fig. 7.12 that results from these equations. In Fig. 7.12, we define the forward diode current as I_F and the reverse diode current as I_R.

This circuit model can be used in conjunction with network analysis to predict the DC voltages and currents in transistor circuits for all regions of operation. Unfortunately, the nonlinear diode currents make the Ebers-Moll circuit inconvenient for hand analysis. As we have done for the MOSFET and the pn diode, we will find simplified models for the bipolar transistor that are valid in specific operating regions.

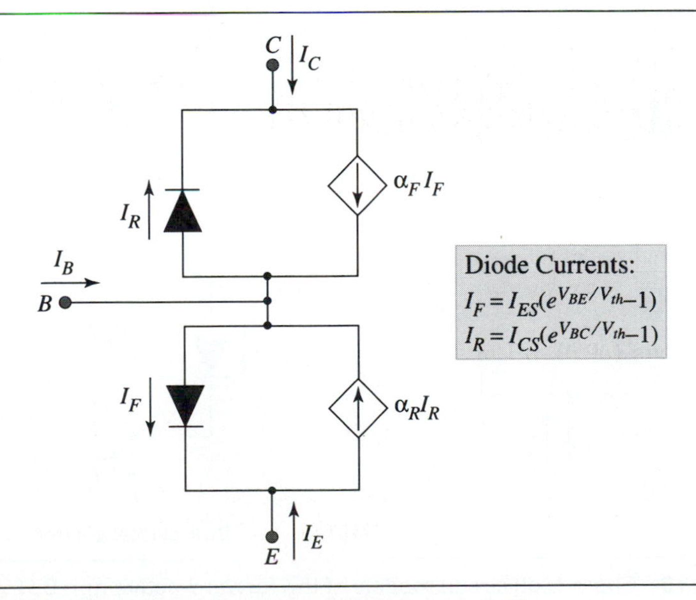

➤ **Figure 7.12** Ebers-Moll circuit model for the npn BJT.

In the forward-active region, the base-emitter junction is forward-biased and the base-collector voltage is *not* sufficiently forward-biased to have I_R be significant. As a result, we can simplify the Ebers-Moll circuit model by omitting the base-collector diode, which also eliminates the current-controlled current source that is dependent on it. Figure 7.13 shows that the remaining two elements are the forward diode and the current-controlled current source that is dependent upon $I_F = -I_E$. It is often more convenient to have the collector current dependent on the base current. We can easily achieve this by changing variables since $I_C = \beta_F I_B = -\alpha_F I_E$ in the forward-active region.

For hand calculation, the forward-biased base-emitter junction diode is replaced in Fig. 7.14(a) by a battery of value $V_{BE} = 0.7$ V. A more convenient form of the model is shown in Fig. 7.14(b), in which the controlled current source is connected between the collector and emitter terminals, which can be done without changing the terminal equations.

Finally, both the forward and the reverse diodes are forward-biased in the case of saturation. The base-emitter junction diode is replaced by a battery of value $V_{BE} = 0.7$ V. The base-collector junction diode is approximated by a battery of value $V_{BC} = 0.6$ V. The controlled current sources in the Ebers-Moll circuit model can be deleted from the circuit since they are in parallel with voltage sources. We have used a slightly lower forward-bias on the base-collector diode in order to model better the measured collector characteristics in Fig. 7.3(b). The collector-emitter voltage in saturation is

$$V_{CE_{SAT}} = V_{BE_{SAT}} - V_{BC_{SAT}} \approx 0.7\,\text{V} - 0.6\,\text{V} = 0.1\,\text{V} \tag{7.45}$$

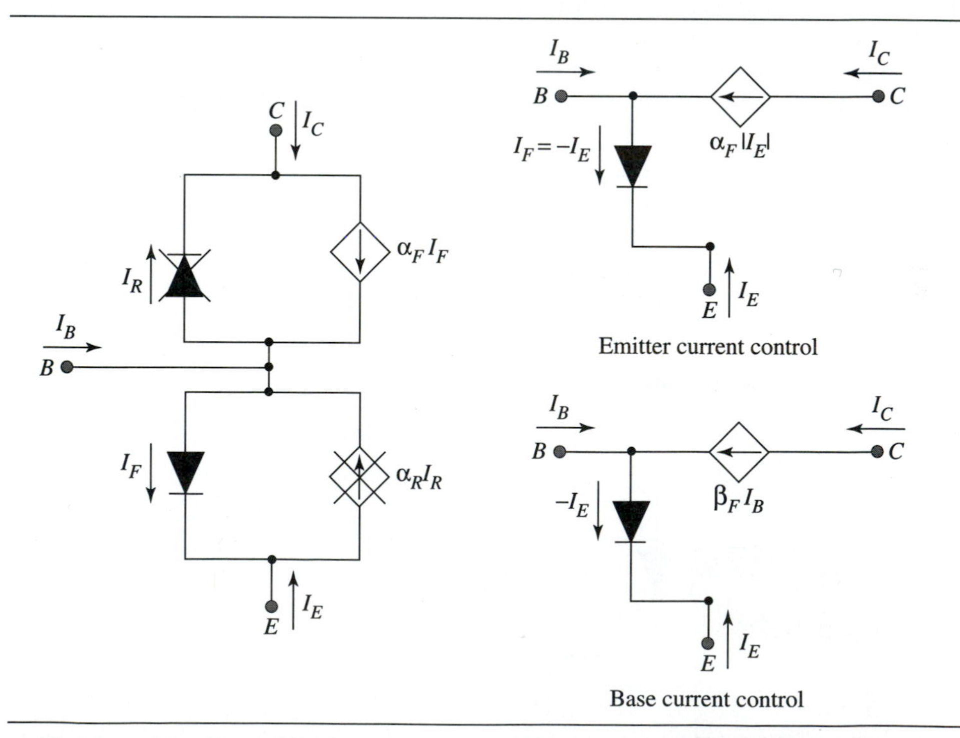

➤ Figure 7.13 Ebers-Moll circuit models of the forward-active npn BJT where I_R is negligible.

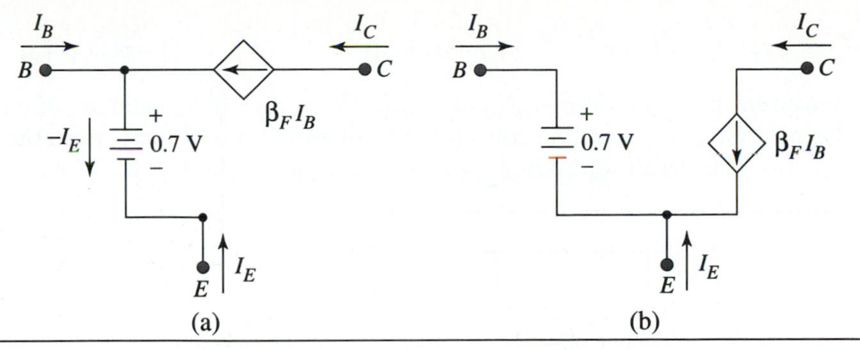

➤ **Figure 7.14** (a) Forward-active model of the npn bipolar transistor for hand calculation, and (b) more convenient form of the model with the controlled source connected between collector and emitter.

The approximate Ebers-Moll circuit model for saturation is shown in Fig. 7.15 below.

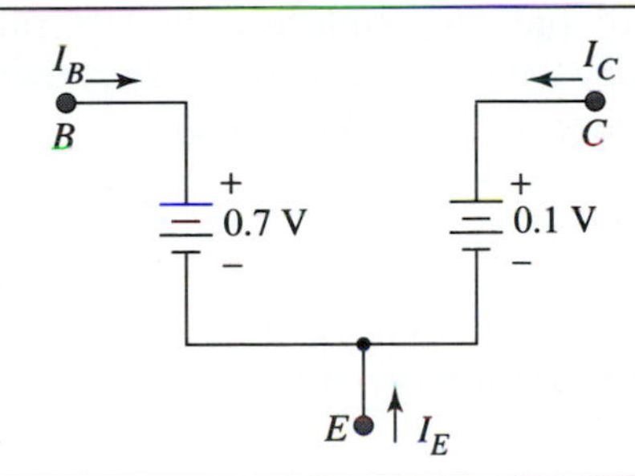

➤ **Figure 7.15** Circuit model of the saturated npn BJT for hand calculation

➤ EXAMPLE 7.2 Measurement of Ebers-Moll Parameters

Devise a way to measure the Ebers-Moll parameters of an npn BJT using simple current-voltage measurements.

SOLUTION

Examining Eq. (7.42), we can separate the forward from the reverse current by setting the base-collector voltage $V_{BC} = 0$ V and measuring I_E as a function of V_{BE}. In addition, we can measure the base current with $V_{BC} = 0$ V and derive the value of α_F. For $V_{BC} = 0$ V, Eq. (7.42) becomes

$$I_E = -I_{ES}(e^{V_{BE}/V_{th}} - 1) \cong -I_{ES} \cdot e^{V_{BE}/V_{th}}$$

for a reasonable forward bias. A log-linear plot of I_E versus V_{BE} will yield an accurate measurement of I_{ES}. The base current for $V_{BC} = 0$ V can be found from Eqs.(7.39), (7.42), and (7.43)

$$I_B = -I_E - I_C = I_{ES}(e^{V_{BE}/V_{th}} - 1) - \alpha_F I_{ES}(e^{V_{BE}/V_{th}} - 1) = (1 - \alpha_F)(-I_E)$$

Measurement of the ratio $-I_E/I_B$ for $V_{BC} = 0$ V, will yield an estimate of α_F. It is also possible to measure the collector current and find α_F directly from the ratio $I_C/(-I_E)$. The circuit for finding I_{ES} and α_F is shown in the Fig. Ex 7.2A.

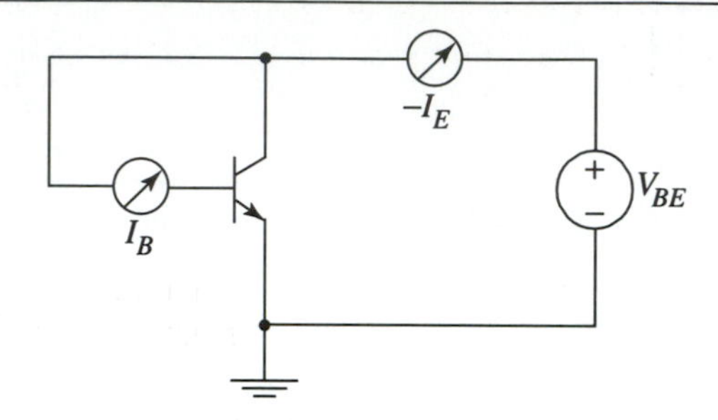

➤ **Figure Ex7.2A** Circuit for measuring the forward saturation current I_{ES} and the forward α_F.

In order to measure the base-collector diode's saturation current, we set $V_{BE} = 0$ V and measure the collector current, as shown in the circuit in Fig. Ex 7.2B. The collector current (Eq. (7.43)) reduces to

$$I_C = -I_{CS}(e^{V_{BC}/V_{th}} - 1) \cong -I_{CS}e^{V_{BC}/V_{th}}$$

for $V_{BE} = 0$ V, so a log-linear plot of I_C versus V_{BC} will allow I_{CS} to be found accurately. The reciprocity relation (Eq. (7.44)) allows α_R to be calculated from the other three parameters

$$\alpha_R = \frac{\alpha_F I_{ES}}{I_{CS}}$$

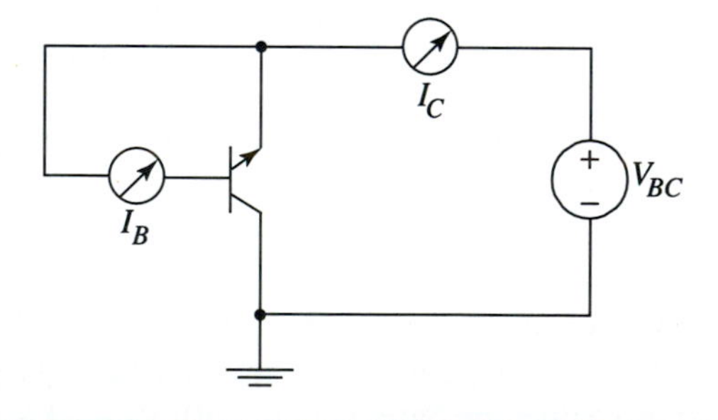

➤ **Figure Ex7.2B** Circuit for measuring I_{CS} and α_R.

By measuring the base current in the above circuit, we can directly measure α_R as a check on the consistency of the other measurements against the reciprocity relation. Since α_F is very close to unity for an IC npn transistor, we must pay careful attention to the measurement uncertainties to obtain an accurate value for this parameter.

➤ 7.5 SMALL-SIGNAL MODEL OF THE npn BJT

In this section, we will develop a small-signal model for the npn bipolar transistor in the forward-active region by following the same general approach used for the MOSFET in Chapter 4. First, the dependence of the collector and base currents on the terminal voltages will lead to resistances and a transconductance in the incremental model. Next, small-signal depletion and diffusion capacitances are added to complete the model. The resulting lumped model is a good representation of the small-signal response of the BJT, up to frequencies of several GHz. At very high frequencies, distributed effects begin to play an important role and a different model is needed.

7.5.1 Transconductance

The incremental change in collector current due to an increment in base-emitter voltage is represented by a voltage-controlled current source. The collector current in the forward active region is given by

$$i_C = I_S e^{(v_{BE}/V_{th})} \tag{7.46}$$

which is repeated here using the symbols for total current i_C and total base-emitter voltage v_{BE}. For small-signal analysis, the total current and total base-emitter voltage are broken into their DC values and an incremental (small-signal) quantity

$$i_C = I_C + i_c \qquad \text{and} \qquad v_{BE} = V_{BE} + v_{be} \tag{7.47}$$

The ratio of i_c to v_{be} at a given DC bias point (V_{BE}, I_C) is defined as the **transconductance** g_m

$$g_m = \left.\frac{i_c}{v_{be}}\right|_{(V_{BE}, I_C)} = \left.\frac{\partial i_C}{\partial v_{BE}}\right|_{(V_{BE}, I_C)} \tag{7.48}$$

Evaluating the partial derivative at the operating point Q (V_{BE}, I_C,), we find

$$g_m = \frac{I_S}{V_{th}} e^{V_{BE}/V_{th}} = \frac{I_C}{V_{th}} = \frac{q}{kT} I_C \tag{7.49}$$

bipolar transistor transconductance

The transconductance connects the base-emitter terminal pair with the collector current and is the central element in the small-signal model. In contrast to the MOSFET transconductance's proportionality to the square root of the DC drain current, the BJT transconductance is directly proportional to the DC collector current. Unlike the MOSFET, the BJT transconductance contains no other parameters under the designer's control. It is also inversely proportional to the absolute temperature.

From a graphical point of view, the transconductance is the slope of the i_C vs. v_{BE} characteristic as shown in Fig. 7.16. The exponential function is so steep that an inset is helpful for clarifying relationships between i_c and v_{be}.

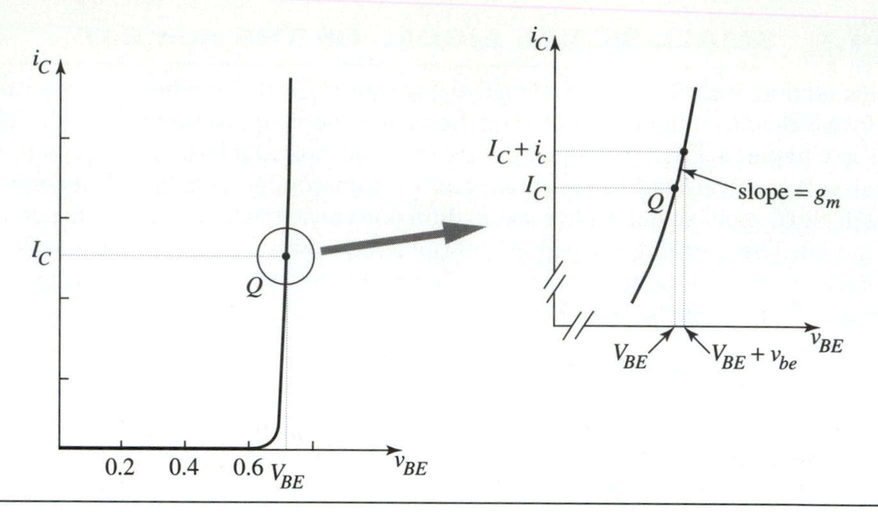

➤ **Figure 7.16** Collector current as a function of base-emitter voltage, showing the definition of the transconductance as the slope of the tangent to the characteristic at the bias point *Q*.

Given the exponential dependence of collector current i_C on the base-emitter voltage v_{BE}, it is worthwhile to quantify how small v_{be} must be for Eq. (7.48) to be an accurate means of finding i_c. As a starting point, we write an exact expression for the collector current as a function of base-emitter voltage

$$i_C = I_C + i_c = I_S e^{(V_{BE} + v_{be})/V_{th}} = I_S e^{V_{BE}/V_{th}} \cdot e^{v_{be}/V_{th}} = I_C e^{v_{be}/V_{th}} \tag{7.50}$$

The right hand side of Eq. (7.50) follows from identifying the bias collector current in the expression for total current. The next step is to expand $\exp(v_{be}/V_{th})$ as a power series in v_{be}/V_{th}

$$I_C + i_c = I_C\left(1 + \left(\frac{v_{be}}{V_{th}}\right) + \frac{1}{2}\left(\frac{v_{be}}{V_{th}}\right)^2 + \ldots\right) \tag{7.51}$$

Multiplying through by the bias current I_C, we find

$$I_C + i_c = I_C + \left(\frac{I_C}{V_{th}}\right)v_{be} + \frac{1}{2}\left(\frac{I_C}{V_{th}^2}\right)v_{be}^2 + \ldots \tag{7.52}$$

Cancelling the bias current from both sides and identifying the multiplier for v_{be} on the right hand side as the transconductance, the increment in collector current i_c is

$$i_c = g_m v_{be} + \frac{1}{2}\left(\frac{g_m}{V_{th}}\right)v_{be}^2 + \ldots \tag{7.53}$$

We only retain the linear term on the right hand side of Eq. (7.53) in the small-signal model. The error ε in neglecting the second order term is given by

$$\varepsilon = \frac{\frac{1}{2}\left(\frac{g_m}{V_{th}}\right)v_{be}^2}{g_m v_{be}} = \frac{1}{2}\frac{v_{be}}{V_{th}} \leq 0.1 \tag{7.54}$$

where we have imposed a error bound of 10% as an example. At room temperature, Eq. (7.54) indicates that the small-signal base-emitter voltage $v_{be} < 5$ mV for the linear small-signal model to be accurate to within 10%. This limitation must be kept in mind in using small-signal analysis for bipolar transistor circuits.

7.5.2 Input Resistance

Unlike the MOSFET, the base current i_B for the bipolar transistor is not zero. Therefore, we would like to find the small-signal resistor that models the incremental base current i_b due to a small-signal base-emitter voltage v_{be}. In the forward-active region, the **small-signal current gain** is defined by

$$\beta_o = \left.\frac{\partial i_C}{\partial i_B}\right|_Q \tag{7.55}$$

The small-signal current gain may vary with operating point and is not necessarily equal to the forward-active current gain β_F. However, this is a second-order consideration and we will consider that $\beta_o = \beta_F$ in this text.

The input resistance r_π is defined by

$$r_\pi^{-1} = \left.\frac{\partial i_B}{\partial v_{BE}}\right|_Q \tag{7.56}$$

Applying the chain rule and substituting the small-signal current gain from Eq. (7.55), we find that

$$r_\pi^{-1} = \left.\frac{\partial i_B}{\partial v_{BE}}\right|_Q = \left.\frac{\partial i_C}{\partial i_B}\right|_Q \left.\frac{\partial i_C}{\partial v_{BE}}\right|_Q = \left(\frac{1}{\beta_o}\right)\left.\frac{\partial i_C}{\partial v_{BE}}\right|_Q = \frac{g_m}{\beta_o} \tag{7.57}$$

where we identified the transconductance g_m. Substituting for the transconductance at the operating point Q, we find the input resistance is

$$\boxed{r_\pi = \frac{\beta_o V_{th}}{I_C} = \frac{\beta_o}{g_m}} \tag{7.58}$$

bipolar transistor input resistance

From Eq. (7.58), the input resistance is inversely proportional to the DC collector current I_C and is proportional to the small-signal current gain β_o.

7.5.3 Output Resistance

The collector current i_C is a weak function of the collector-emitter voltage v_{CE} in the forward-active region as shown in the measured characteristics in Fig. 7.3(b). The physical origin for this effect is called **base-width modulation**. With increasing v_{CE}, the reverse bias v_{CB} on the base-collector junction increases since

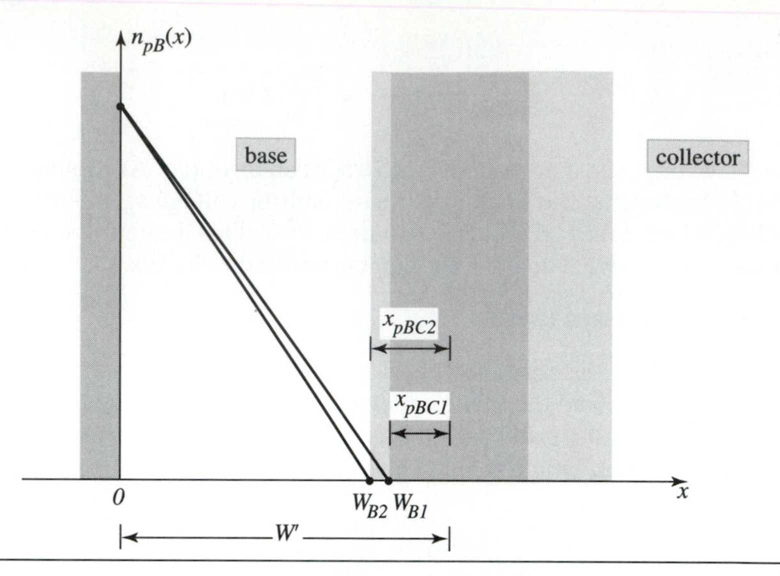

➤ **Figure 7.17** Change in width of p-type base due to changes in the base-collector depletion region width.

$$v_{CB} = v_{CE} - v_{BE} \approx v_{CE} - 0.7\text{ V} \tag{7.59}$$

How does this reverse bias affect the collector current? Figure 7.17 shows the minority electron concentration in the base for two collector-emitter voltages $v_{CE2} > v_{CE1}$. The larger depletion width reduces the width of the p-type base region, which increases the gradient of electron concentration, and therefore the collector current.

To model this phenomenon, first note in Fig. 7.17 that the width of the p-type base region is given by

$$W_B = W' - x_{pBC} \tag{7.60}$$

where W' is the distance between the edge of the emitter-base depletion region and the location of the base/collector junction and x_{pBC} is the depletion width on the p-side of the base-collector junction. The depletion width on the p-side is a function of the base-collector voltage through Eq. (3.61)

$$x_{pBC}(v_{BC}) = \sqrt{\left(\frac{2\varepsilon_s(\phi_B - v_{BC})}{qN_{aB}}\right)\left(\frac{N_{dC}}{N_{dC} + N_{aB}}\right)} \tag{7.61}$$

Substituting for v_{BC} in terms of v_{CE} from Eq. (7.59), we find

$$x_{pBC}(v_{CE}) = \sqrt{\left(\frac{2\varepsilon_s(\phi_B + v_{CE} - 0.7\text{V})}{qN_{aB}}\right)\left(\frac{N_{dC}}{N_{dC} + N_{aB}}\right)} \tag{7.62}$$

We can now find the dependence of the forward-active collector current on v_{CE} by substituting Eqs. (7.60) and (7.62) for the base width into Eq. (7.18)

$$i_C(v_{BE},v_{CE}) = \left(\frac{qA_E D_{nB} n_{pBo}}{W_B}\right) e^{v_{BE}/V_{th}} = \left(\frac{qA_E D_{nB} n_{pBo}}{W' - x_{pBC}(v_{CE})}\right) e^{v_{BE}/V_{th}} \tag{7.63}$$

From Eq. (7.62), we can see why it is desirable to dope the base more heavily than the collector—we minimize the penetration of the depletion region into the base. The increase in collector current seen in Fig. 7.3(b) is modeled quite well by Eq. (7.63). In addition, it indicates that the collector current will increase without bound when the collector-emitter bias is increased to the point where the entire base is depleted, which occurs when $x_{pBC}(V_{CE}) = W'$. **Punchthrough**, the term for this phenomenon, is one form of breakdown that can occur in bipolar transistors. For devices with very thin bases, punchthrough sets an upper bound on the collector-emitter voltage. This phenomenon is evident in the collector characteristics in Fig. 7.3(b).

The output conductance g_o is defined by

$$g_o = \left.\frac{\partial i_C}{\partial v_{CE}}\right|_Q = \frac{1}{r_o} \tag{7.64}$$

where r_o is the output resistance. The expression in Eq. (7.63) can be substituted into Eq. (7.64) and differentiated. However, the result is too cumbersome to be useful in hand calculations. Instead, a purely empirical approximation is used for the dependence of the collector current on the collector-emitter voltage. Figure 7.18 shows how the collector characteristics in the forward-active region can be extrapolated, with their intersection on the negative V_{CE} axis being defined as the negative of the **Early voltage**, V_{An}.

This geometrical construction leads to a modified equation for the forward-active region

$$i_C = I_S e^{v_{BE}/V_{th}}\left(1 + \frac{v_{CE}}{V_{An}}\right) \tag{7.65}$$

Note that the correction term in Eq. (7.65) for the **Early effect**—another term for base-width modulation—is an approximation and extrapolating the measured collector characteristics for a real BJT will lead to a range of intersections with the negative V_{CE} axis. Also, Eq. (7.65) does not model the increase in the slope as punchthrough is approached, as shown in Fig. 7.18.

Substitution of Eq. (7.65) into the definition of the output conductance leads to

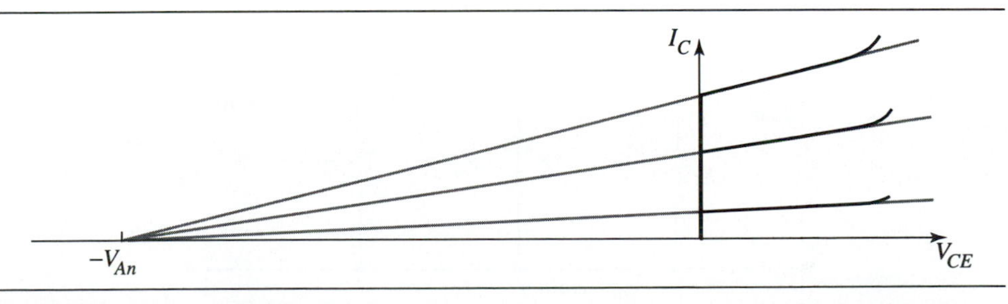

➤ **Figure 7.18** Definition of the Early voltage V_{An}, which is useful for empirically modeling the increase of collector current with collector-emitter voltage in the forward active region.

$$r_o^{-1} = I_S e^{(V_{BE}/V_{th})}\left(\frac{1}{V_{An}}\right) \cong \frac{I_C}{V_{An}} \tag{7.66}$$

On the right-hand side of this equation, we neglected the factor $(1 + V_{CE}/V_{An})$ in order to obtain a simpler result. This is a reasonable approximation if $V_{CE}/V_{An} \ll 1$ at the operating point. Inverting Eq. (7.66), we find that the output resistance is

bipolar transistor output resistance

$$\boxed{r_o = \frac{V_{An}}{I_C}} \tag{7.67}$$

The schematic of the small-signal model of the npn BJT, without charge storage elements, is given in Fig. 7.19. In contrast to the MOSFET's small-signal model, the BJT is a three-terminal device. The relatively small input resistance will have a major impact on circuit design with BJTs, as we will see in Chapters 8 and 9.

➤ EXAMPLE 7.3 Punchthrough Voltage

For an npn transistor, find the punchthrough voltage, which is defined as V_{CE} for which the p-type base is completely depleted. Assume that $W' = 250$ nm, which is the distance from the emitter-base depletion region edge to the base-collector junction. The doping concentrations in the base and collector are: $N_{aB} = 5 \times 10^{16}$ cm^{-3}, $N_{dC} = 2 \times 10^{16}$ cm^{-3}.

SOLUTION

At the point of punchthrough, the depletion width on the base side of the base-collector junction is $x_{pBC} = W' = 250$ nm. Using Eq. (7.62), we can solve for the collector-emitter voltage V_{CE*} at which this occurs

$$x_{pBC}(V_{CE*}) = \sqrt{\left(\frac{2\varepsilon_s(\phi_B + V_{CE*} - 0.7\text{V})}{qN_{aB}}\right)\left(\frac{N_{dC}}{N_{dC} + N_{aB}}\right)} = W'$$

Inverting this equation, the punchthrough voltage is given by

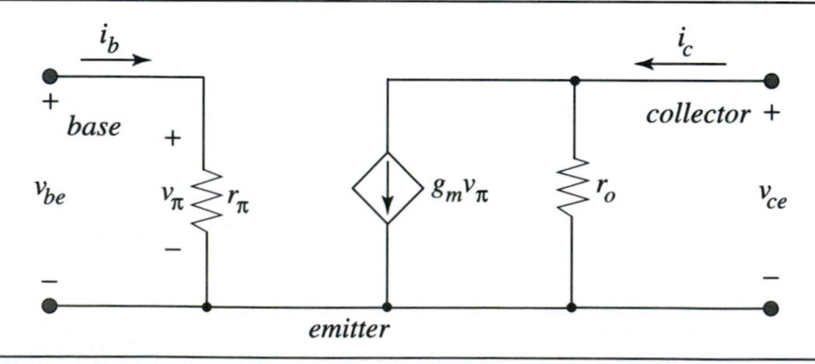

➤ **Figure 7.19** Small-signal model for the npn bipolar junction transistor, omitting the elements that model charge storage.

$$V_{CE^*} = \frac{qN_{aB}(N_{aB}+N_{dC})(W')^2}{2\varepsilon_s N_{dC}} - \phi_B + 0.7\text{V}$$

Substituting in the parameters for this transistor, we find that

$$V_{CE^*} = \frac{1.6\times10^{-19}\cdot 5\times10^{16}\cdot 7\times10^{16}\cdot (2.5\times10^{-5})^2}{2\cdot 1.04\times10^{-12}\cdot 2\times10^{16}} - 0.78 + 0.7 = 8.3\text{ V}$$

This value of the punchthrough voltage is on the same order as that of the transistor whose characteristics are plotted in Fig. 7.3(b).

7.5.4 Base-Charging Capacitance

Under forward-active bias, the steady-state minority electron concentration in the base $n_{pB}(x)$ and the minority hole concentration in the emitter $p_{nE}(x)$ are shown in Fig. 7.20. According to the law of the junction, when the base-emitter voltage is incremented by v_{be}, the minority electron concentration $n_{pE}(x=0)$ increases. When the

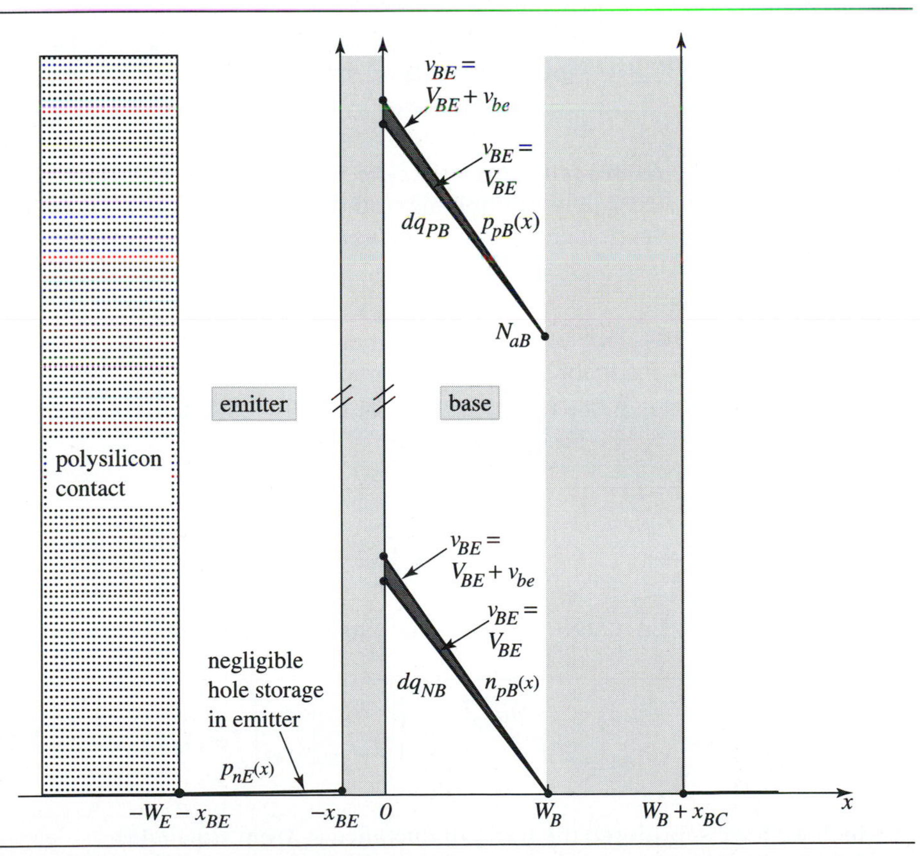

➤ **Figure 7.20** Minority and majority carrier concentrations in the base for $v_{BE}=V_{BE}$ and for $v_{BE}=V_{BE}+v_{be}$, showing the increase in charge storage in the base. The emitter is much more heavily doped than the base: $N_{dE} \gg N_{aB}$, so the reverse-injected hole concentration in the emitter is negligible.

new steady-state condition is reached, the minority electron concentration has increased throughout the base. Since the p-type base remains charge-neutral, the majority hole concentration in the base increases by the same amount, as shown in Fig. 7.20. Charge storage in the n-type emitter is negligible since the emitter doping concentration $N_{dE} \gg N_{aB}$.

The changing charge storage in the base represents a diffusion capacitance, as we saw for the pn diode in Section 6.4. Since the charge stored in the base is under control of the base-emitter voltage, we conclude that a capacitance is needed between base and emitter. We define the **base-charging capacitance** C_b of a BJT as

$$C_b = \left.\frac{\partial q_{PB}}{\partial v_{BE}}\right|_Q \tag{7.68}$$

where q_{PB} is the majority hole charge stored in the base, as shown in Fig. 7.20. For charge neutrality to hold, the hole charge is equal and opposite to the electron charge and $q_{PB} = -q_{NB}$.

It is convenient to calculate the minority electron charge from the triangular-shaped minority electron concentration in the base:

$$q_{NB}(v_{BE}) = -q\int_0^{W_B} n_{pB}(x)A_E\,dx = \frac{-qA_E W_B n_{pBo}}{2}e^{v_{BE}/V_{th}} \tag{7.69}$$

In examining Eq. (7.69), we recognize several common factors with the expression for the collector current i_C. By multiplying and dividing by (D_{nB}/W_B), we can re-express the minority electron charge as a function of the collector current

$$q_{NB}(v_{BE}) = -\frac{1}{2}W_B(W_B/D_{nB})\left(\frac{qA_E D_{nB}}{W_B}\right)n_{pBo}e^{v_{BE}/V_{th}} = -\left(\frac{W_B^2}{2D_{nB}}\right)i_C \tag{7.70}$$

The term in parentheses on the right-hand side of Eq. (7.70) has units of time (charge/current) and is called τ_F, the **base transit time**

base transit time

$$\boxed{\tau_F = \frac{W_B^2}{2D_{nB}}} \tag{7.71}$$

Substituting into Eq. (7.68), the base-charging capacitance is

base charging capacitance

$$\boxed{C_b = \left.\frac{\partial q_{PB}}{\partial v_{BE}}\right|_Q = -\left.\frac{\partial q_{NB}}{\partial v_{BE}}\right|_Q = \left(\frac{W_B^2}{2D_{nB}}\right)\left.\frac{\partial i_C}{\partial v_{BE}}\right|_Q = \tau_F g_m} \tag{7.72}$$

in which we have substituted the transconductance g_m from Eq. (7.48).

The base transit time τ_F is the average time required for the diffusing electrons to transit the p-type base, for forward operation. We can see this important fact from Eqs. (7.70) and (7.71) by writing the electron diffusion current in forward bias I_{diff} as the ratio of the stored minority electron charge and the transit time

$$i_C = I_{diff} = \frac{-q_{NB}}{\tau_F} \quad (7.73)$$

As we will see in Chapter 10, the base transit time sets the maximum speed of the bipolar transistor. A typical value for the base transit time is found by substituting the base width $W_B = 300$ nm and the diffusivity $D_{nB} = 10$ cm^2/s

$$\tau_F = 4.5 \times 10^{-11} \text{ s} = 45 \text{ ps} \quad (7.74)$$

➤ EXAMPLE 7.4 High Speed BJT Design

In a high-performance bipolar technology, a major goal is maximum frequency response. Suppose the transit time must be less than 7.5 ps for a particular communications IC.

(a) What is the required base width? You can assume that the electron diffusivity in the base is $D_{nB} = 10$ cm^2/s.

SOLUTION

The required base width can be found directly from the Eq. (7.71) for the transit time

$$W_B = \sqrt{2D_{nB}\tau_F} = \sqrt{2 \cdot 10 \text{ cm}^2/\text{s} \cdot 7.5 \times 10^{-12} \text{ s}} = 1.2 \times 10^{-5} \text{cm} = 120 \text{ nm}$$

(b) If the collector and base doping concentrations are $N_{dC} = 2 \times 10^{16}$ cm^{-3}and $N_{aB} = 10^{17}$ cm^{-3}and the collector-emitter bias is $V_{CE} = 3$V for part (a), find the distance W' as defined in Fig. 7.17.

SOLUTION

The distance W' between the edge of the base-emitter depletion region and the base-collector junction is

$$W' = W_B + x_{pBC}$$

For $V_{CE} = 3$ V, $V_{BC} = -2.3$V and the penetration of the base-collector depletion region into the base is

$$x_{pBC} = \sqrt{\frac{2\varepsilon_s(\phi_B - V_{BC})}{qN_{aB}}\left(\frac{N_{dC}}{N_{dC} + N_{aB}}\right)} = 82 \text{ nm}$$

from Eq. (3.61). The distance $W' = 120 \text{ nm} + 82 \text{ nm} = 202$ nm.

7.5.5 Depletion Capacitances

The npn BJT structure in Fig. 7.1 has several pn junctions that contribute small-signal depletion capacitances to the small-signal model. Since the base-emitter junction is forward-biased with $V_{BE} = 0.7$ V, the formula for a reverse-biased junction from Chapter 3 cannot be used directly. More accurate models for the forward-biased junction show that the depletion capacitance under forward bias is approximately

$$C_{jE} = \sqrt{2}C_{jEo} \tag{7.75}$$

where C_{jEo} is the zero-bias (thermal equilibrium capacitance) of the emitter-base junction. The sum of C_{jE} and the base-charging capacitance is the input capacitance C_π of the BJT

input capacitance C_π

$$C_\pi = C_b + C_{jE} \tag{7.76}$$

The base-collector junction is reverse-biased; its capacitance is

base-collector capacitance C_μ

$$C_\mu = \frac{C_{\mu o}}{\sqrt{1 + V_{CB}/\phi_{Bc}}} \tag{7.77}$$

where $C_{\mu o}$ is the zero-bias capacitance and ϕ_{Bc} is the built-in voltage of the base-collector junction. Finally, the collector-substrate junction capacitance is

collector-substrate capacitance C_{cs}

$$C_{cs} = \frac{C_{cso}}{\sqrt{1 + V_{CS}/\phi_{Bs}}} \tag{7.78}$$

where C_{cso} is the zero-bias capacitance and ϕ_{Bs} is the built-in voltage of the collector-substrate junction. The substrate is generally connected to the lowest supply voltage (or ground) to avoid forward-biasing the collector-substrate junction. As a result, the substrate terminal is a small-signal ground.

7.5.6 Parasitic Resistances

The base and collector regions in the npn BJT structure in Fig. 7.1 contribute resistances that must be added to the small-signal model. The base resistance r_b is the resistance between the metal interconnect and the emitter-base junction is around 250 Ω for the oxide-isolated npn transistor. An n^+ buried layer is incorporated beneath the core npn sandwich to reduce the resistance between the base-collector junction and the metal contact. The parasitic collector resistance r_c is typically 200 Ω for the npn BJT. Due to the two- and three-dimensional nature of the carrier transport in the base and collector, simulation (or measurement) is necessary in order to find the numerical values for r_b and r_c. Finally, the polysilicon contact introduces an emitter resistance of $r_{ex} = 5\ \Omega$.

For hand calculations, the parasitic resistances are rarely included since they only have a second-order effect on the BJT's behavior. An exception is in modeling the transistor's small-signal behavior at high frequencies, for which the base resistance is an important element.

In Fig. 7.21, we show the complete small-signal model for the npn BJT. This is constructed by adding the capacitors and parasitic resistors to the basic small-signal model in Fig. 7.19. This lumped-parameter circuit is known as the **hybrid-π** model of the BJT. It accurately predicts the small-signal performance of the transistor up to frequencies on the order of the inverse of the base transit time, τ_F. At higher frequen-

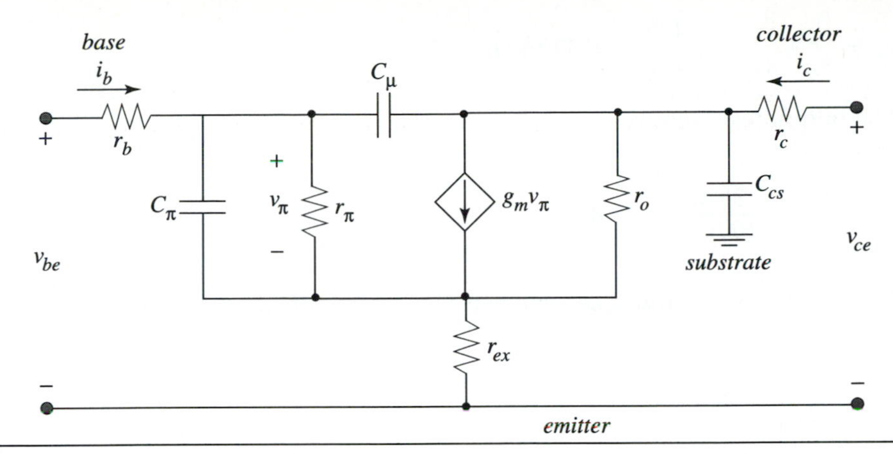

➤ **Figure 7.21** Complete small-signal model for the npn bipolar junction transistor.

cies, a model that takes into account the distributed nature of carrier transport inside the BJT is required.

➤ EXAMPLE 7.5 Small-Signal Parameters for an Oxide-Isolated BJT

For the transistor in Fig. 7.1, with doping levels and dimensions from Ex. 7.1, find the complete small-signal model at the operating point: $I_C = 100\ \mu\text{A}$, $V_{CE} = 2\ \text{V}$, $V_{CS} = 2\ \text{V}$.

GIVEN:

Emitter, collector, and substrate junction areas: $A_E = 25\ \mu\text{m}^2$, $A_C = 100\ \mu\text{m}^2$, and $A_S = 500\ \mu\text{m}^2$;

Base and emitter bulk widths at the operating point: $W_B = 300$ nm and $W_E = 250$ nm
Doping concentrations: $N_{dE} = 7.5 \times 10^{18}\ \text{cm}^{-3}$, $N_{aB} = 10^{17}\ \text{cm}^{-3}$, $N_{dC} = 1.5 \times 10^{16}\ \text{cm}^{-3}$, and $N_{aS} = 5 \times 10^{15}\ \text{cm}^{-3}$.

Diffusion coefficients: $D_{pE} = 5\ \text{cm}^2/\text{s}$, $D_{nB} = 10\ \text{cm}^2/\text{s}$.

Parasitic small-signal resistances: $r_b = 250\ \Omega$, $r_c = 200\ \Omega$, and $r_{ex} = 5\ \Omega$.

Early voltage $V_{An} = 35$ V

SOLUTION

First, we find the forward active current gain using the analysis from Ex. 7.1

$$\beta_F = \frac{D_{nB} N_{dE} W_E}{D_{pE} N_{aB} W_B} = \frac{10\ \text{cm}^2/\text{s} \cdot 7.5 \times 10^{18}\ \text{cm}^{-3} \cdot 250 \times 10^{-7}\ \text{cm}}{5\ \text{cm}^2/\text{s} \cdot 1 \times 10^{17} \cdot 300 \times 10^{-7}\ \text{cm}} = 125$$

The transconductance is

$$g_m = \frac{I_C}{V_{th}} = \frac{100\ \mu\text{A}}{26\ \text{mV}} = 3.85\ \text{mS}$$

In order to find the input resistance, we must assume that $\beta_F \approx \beta_o$:

$$r_\pi = \frac{\beta_o}{g_m} \approx \frac{\beta_F}{g_m} = \frac{125}{3.85\ \text{mS}} = 32.5\ \text{k}\Omega$$

The output resistance is

$$r_o = \frac{V_{An}}{I_C} = \frac{35\ \text{V}}{100\,\mu\text{A}} = 350\text{k}\Omega$$

To find the base-charging capacitance, we compute the base transit time

$$\tau_F = \frac{W_B^2}{2D_{nB}} = \frac{(3 \times 10^{-5})^2}{2 \cdot 10} = 45\ \text{ps}$$

The base-charging capacitance is

$$C_b = g_m \cdot \tau_F = 3.85\text{mS} \cdot 4.5 \times 10^{-11}\text{s} = 173\text{fF}$$

From the doping concentrations, we compute the zero-bias capacitance from Eq. (6.49)

$$C_{jo} = A\sqrt{\frac{q\varepsilon_s N_a N_d}{2\phi_B (N_a + N_d)}}$$

For the three junctions, we have

$$C_{jEo} = 23\text{fF} \qquad C_{\mu o} = 37\text{fF} \qquad C_{cso} = 102\text{fF}$$

where we used the given areas of the B-E, B-C, and C-S junctions. The junction capacitances are as follows, where we used the operating point information to find $V_{CB} = 1.3$ V and $V_{CS} = 2$ V and found $\phi_{Bc} = 0.79\text{V}$ and $\phi_{Bs} = 0.71\text{V}$ from the doping concentrations

$$C_{jE} = \sqrt{2}C_{jEo} = 32\text{fF} \text{ and}$$

$$C_\mu = \frac{37\text{fF}}{\sqrt{1 + 1.5/0.79}} = 22\text{fF} \qquad C_{cs} = \frac{102\text{fF}}{\sqrt{1 + 2/0.71}} = 52\text{fF}$$

Finally, the input capacitance is the sum of the base-charging capacitance and the junction capacitance of the emitter-base junction

$$C_\pi = C_b + C_{jE} = 173 + 32 = 205\text{fF}$$

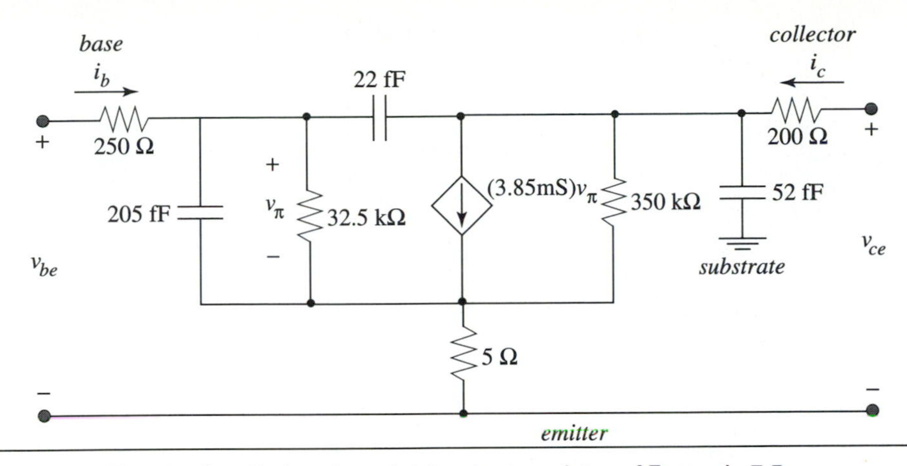

Figure Ex7.5 Small-signal model for the transistor of Example 7.5.

The small-signal model of the npn BJT with numerical values of the components is given in Fig. Ex 7.5.

7.6 BJT DEVICE PHYSICS: A SECOND PASS

In the previous sections, we developed a basic understanding of the operation and circuit modeling for npn bipolar junction transistors. Under forward-active bias, electrons are injected from the emitter into the p-type base, where they diffuse across to the base-collector depletion region and then are swept into the n-type collector. This basic concept of **transistor action** has been explained without solving for the minority carrier concentrations using the continuity equation. In this section, we apply the concepts of minority carrier lifetime and diffusion length from Sections 6.6 and 6.7 to do a more thorough job of modeling the carrier concentrations and the hole and electron fluxes in the polysilicon-emitter npn transistor under forward-active bias.

7.6.1 Hole Concentration in the Emitter

The IC npn transistor shown in Fig. 7.1 has an n^+ polysilicon emitter interconnection. In Section 7.2, we assumed as a first approximation that the polysilicon/emitter interface behaved like an ohmic contact. However, in reality holes diffuse through this interface and penetrate into the polysilicon. Recall that polysilicon consists of small crystallites or grains of silicon, separated by grain boundaries. Due to the effects of grain boundaries, the hole lifetime and hole diffusivity are significantly lower in the n^+ polysilicon than in the adjacent n^+ silicon emitter region. Since the hole diffusion length L_{pP} in the polysilicon is much less than the polysilicon thickness, the hole concentration in this region decays exponentially. For the n^+ silicon emitter region, the hole concentration is linear since the emitter width W_E is much greater than the hole diffusion length L_{pE}. Figure 7.22 shows the minority hole concentration in the n^+ crystalline silicon emitter and the n^+ polysilicon interconnection.

To solve for the interfacial hole concentration p_{nI}, we require that the hole diffusion current density be the same on either side of the interface, which is located at $x_I = -W_E - x_{BE}$

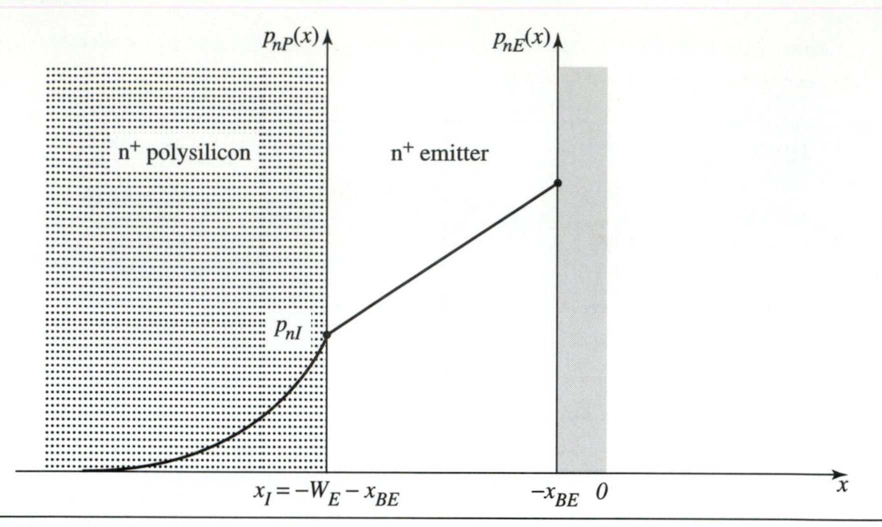

➤ **Figure 7.22** Hole concentration in the polysilicon interconnection and the n⁺ emitter under forward-active bias. The interface is located at $x_I = -W_E - x_{BC}$.

$$J_{pP}^{diff}(x = x_I) = J_{pE}^{diff}(x = x_I) \tag{7.79}$$

The hole diffusion current density on the emitter side of the interface can be found from the (constant) slope in the n⁺ emitter

$$J_{pE}^{diff}(x{=}x_I) = -qD_{pE}\frac{dp_{nE}(x)}{dx}\bigg|_{x = x_I} = -\frac{qD_{pE}[p_{nE}(-x_{BE}) - p_{nI}]}{W_E} \tag{7.80}$$

In the polysilicon, the hole diffusion length is L_{pP} and the excess hole concentration over equilibrium decays exponentially away from the interface

$$p'_{nP}(x) = p_{nI}e^{(x - x_I)/L_{pP}} \tag{7.81}$$

Evaluating the derivative of Eq. (7.81) at the interface, we find that the hole diffusion current density on the polysilicon side of the interface is

$$J_{pP}^{diff}(x{=}x_I) = -qD_{pP}\frac{dp_{nP}(x)}{dx}\bigg|_{x = x_I} = -qD_{pP}\frac{p_{nI}}{L_{pP}}. \tag{7.82}$$

Equating the diffusion current densities from Eq. (7.80) and Eq. (7.82), it follows that

$$-\frac{qD_{pE}[p_{nE}(-x_{BE}) - p_{nI}]}{W_E} = -qD_{pP}\frac{p_{nI}}{L_{pP}} \tag{7.83}$$

Solving for the interfacial hole concentration, we find that

$$p_{nI} = \frac{(D_{pE}/W_E)p_{nE}(-x_{BE})}{(D_{pE}/W_E) + (D_{pP}/L_{pP})} = \frac{p_{nEo}e^{V_{BE}/V_{th}}}{1 + \left(\frac{D_{pP}W_E}{D_{pE}L_{pP}}\right)} \tag{7.84}$$

We can roughly estimate p_{nI} by substituting estimates of the diffusivity D_{nP} and the diffusion length L_{pP} in the n$^+$ polysilicon

$$D_{nP} \approx \frac{D_{nE}}{2} \qquad \text{and} \qquad L_{pP} \approx \frac{W_E}{2} \rightarrow p_{nI} \approx \frac{p_{nE}(-x_{BE})}{2} \tag{7.85}$$

What effect does hole diffusion in the polysilicon have on the BJT? In comparison to the emitter hole concentration for the simple ohmic contact model in Fig. 7.7, the hole concentration in Fig. 7.22 indicates that the hole diffusion current density in the emitter region will be reduced. Substituting Eq. (7.84) into Eq. (7.80), we find that

$$J_{pE}^{diff} = -\left(\frac{qD_{pE}}{W_E}\right) p_{nEo} e^{V_{BE}/V_{th}} \left(1 - \frac{1}{1 + \left(\frac{D_{pP} W_E}{D_{pE} L_{pP}}\right)}\right) \tag{7.86}$$

For the estimated interface concentration $p_{nI} = p_{nE}(-x_{BE})/2$, and the hole diffusion current density in the emitter is reduced by a factor of two. Since these holes are injected from the base, the effect will be to reduce the base current and increase the β_F of the transistor. We will quantify this effect later in this section.

7.6.2 Electron Recombination in the Base

The electrons injected from the emitter diffuse across the narrow p-type base and are swept into the collector by the electric field in the base-collector depletion region. Since the electron concentration far exceeds thermal equilibrium in the base, there is a net recombination rate that is given by

$$R(x) - G(x) = \frac{n_{pB}(x) - n_{pBo}}{\tau_n} \tag{7.87}$$

where we assumed low-level injection ($n_{pB}(x) \ll N_{aB}$). Holes are supplied from the base contact in order to recombine with the excess minority electrons. We quantify this additional component of the base current by integrating the net recombination over the base, as shown in Fig. 7.23.

The hole base-current component I_R needed to supply the recombination process is found by summing up the contribution of each differential volume $A_E\,dx$ in Fig. 7.23

$$I_R = q\int_0^{W_B} (R(x) - G(x))\,A_E\,dx = qA_E\int_0^{W_B}\left(\frac{n_{pB}(x) - n_{pBo}}{\tau_n}\right)dx \tag{7.88}$$

The p-type base is much narrower than an electron diffusion length ($W_B \ll L_{nB}$), making the short-base solution a good approximation for the electron concentration

$$n_{pB}(x) = n_{pB}(x = 0)\left(1 - \frac{x}{W_B}\right) = n_{pBo} e^{V_{BE}/V_{th}}\left(1 - \frac{x}{W_B}\right) \tag{7.89}$$

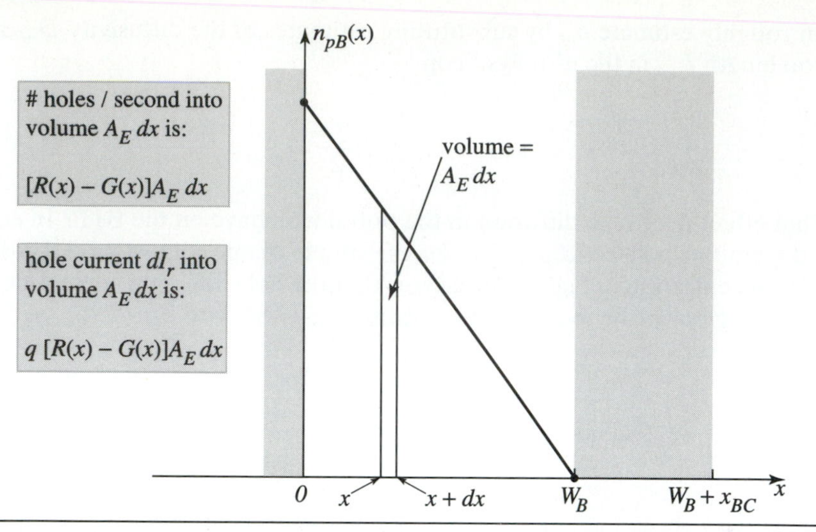

➤ **Figure 7.23** Procedure for finding the hole current due to electron recombination in the base.

Substituting this expression for the electron concentration into Eq. (7.88) for the recombination base current, we find that

recombination base current

$$I_R = \frac{qA_E}{\tau_n} n_{pBo}(e^{V_{BE}/V_{th}} - 1) \int_0^{W_B} \left(1 - \frac{x}{W_B}\right) dx = \frac{qA_E W_B}{2\tau_n} n_{pBo}(e^{V_{BE}/V_{th}} - 1) \quad (7.90)$$

This component of base current does not play a major role in silicon integrated BJT operation since the minority diffusion length is much greater than the base width.

7.6.3 Electron and Hole Fluxes under Forward-Active Bias

We are now able to provide a more complete description of carrier transport in the npn bipolar junction transistor. Holes diffusing into the polysilicon interconnection recombine there with a majority electron flux. The I_R component of base current is a majority hole flux that supplies recombination with electrons diffusing across the base. Figure 7.24 shows a qualitative picture of the fluxes for the case of forward-active bias.

From Fig. 7.24, it is clear that the base current consists of both holes injected from the base into the emitter and holes supplying the net recombination of electrons diffusing across the base. It is helpful to separate the effects of these two components, both of which reduce the ratio of collector current to emitter current. The hole diffusion current in the emitter I_{pE} can be found by multiplying the hole diffusion current density from Eq. (7.86) by the emitter area

$$I_{pE} = J_{pE}^{diff}(x = -x_{BE})A_E = J_{pE}^{diff}(x = x_l)A_E = \frac{-qA_E f' D_{pE}}{W_E} p_{nEo} e^{V_{BE}/V_{th}} \quad (7.91)$$

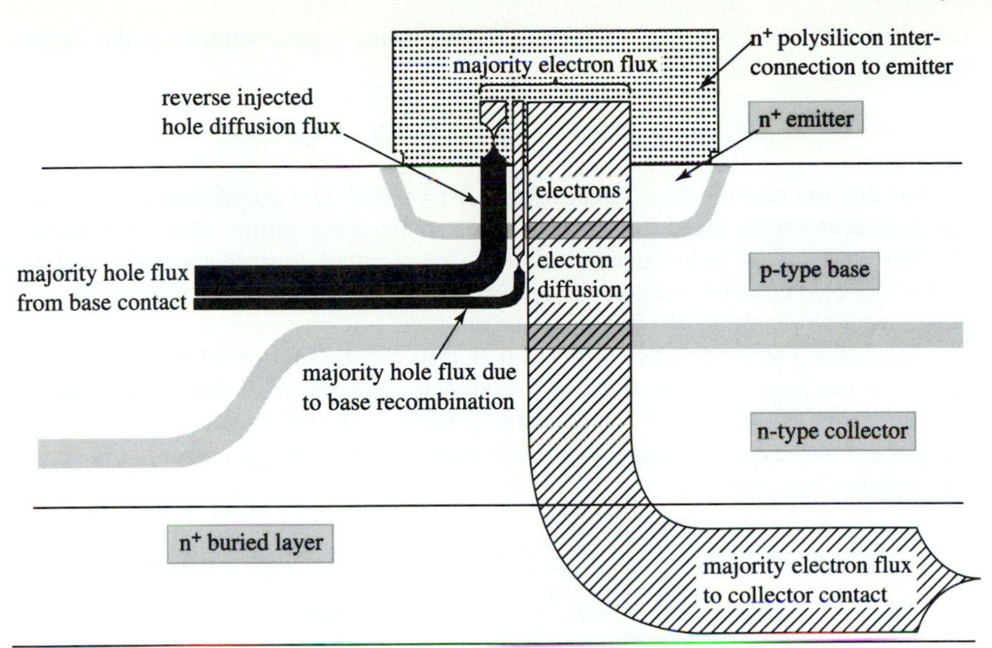

➤ **Figure 7.24** Majority and minority carrier fluxes in the IC npn bipolar transistor for the case of forward-active bias, including the penetration of diffusing holes into the polysilicon interconnection and recombination of electrons in the base region.

where the parameter f' represents the effect of hole diffusion into the polysilicon interconnection and is given by

$$f' = \frac{\dfrac{D_{pP}W_E}{D_{pE}L_{pP}}}{1+\dfrac{D_{pP}W_E}{D_{pE}L_{pP}}} \tag{7.92}$$

The electron component of the emitter current is I_{nE} and is given by

$$I_{nE} = J_{nB}^{diff}(x=0)A_E = \frac{-qA_E D_{nB}}{W_B} n_{pBo} e^{V_{BE}/V_{th}} \tag{7.93}$$

As we can see from the flux picture in Fig. 7.24, the emitter current is the sum of the hole and the electron components

$$I_E = I_{nE} + I_{pE} \tag{7.94}$$

The ratio of the electron component I_{nE} to the total emitter current is defined as γ, the **emitter injection efficiency**

$$\gamma = \frac{I_{nE}}{I_{nE}+I_{pE}} = \frac{\dfrac{D_{nB}n_{pBo}}{W_B}}{\dfrac{D_{nB}n_{pBo}}{W_B} + f'\left(\dfrac{D_{pE}p_{nEo}}{W_E}\right)} = \frac{1}{1+\left(\dfrac{f'D_{pE}N_{aB}W_B}{D_{nB}N_{dE}W_E}\right)} \tag{7.95}$$

emitter injection efficiency γ

Recalling that $f' \approx 0.5$, and substituting the following typical values for the dimensions, doping, and diffusivities

$$D_{pE}/D_{nB} = 0.5,\ W_B/W_E = 2,\ \text{and}\ N_{dE}/N_{aB} = 100 \tag{7.96}$$

we find that the emitter injection efficiency is $\gamma = 0.995$. The polysilicon interconnection increases γ since it reduces the hole component of the emitter current by lessening the gradient in hole concentration in the emitter. For this reason and for processing convenience, the so-called *poly emitter* BJT is commonly used in modern bipolar and bipolar/CMOS (BiCMOS) technologies.

The second source of base current is due to recombination of electrons in the base. In order to quantify its effect on the collector current, we define the **base transport factor** α_T as the ratio of the collector current to the magnitude of the electron component of the emitter current (which is the electron diffusion current at the emitter-base junction edge)

base transport factor α_T

$$\boxed{\alpha_T = \frac{I_C}{-I_{nE}} = \frac{-I_{nE} - I_R}{-I_{nE}} = 1 - \frac{I_R}{-I_{nE}}} \tag{7.97}$$

The flux picture in Fig. 7.24 illustrates that the collector current is from those electrons that diffuse across the base without recombining and therefore, $I_C = -I_{nE} - I_R$. Substituting for the recombination current from (7.90) and the electron component of the emitter current from Eq. (7.93), we find that

$$\alpha_T = 1 - \frac{\left(\dfrac{qA_E W_B n_{pBo}}{2\tau_n}\right)}{\left(\dfrac{qA_E D_{nB} n_{pBo}}{W_B}\right)} = 1 - \frac{W_B^2}{2D_{nB}\tau_n} = 1 - \frac{W_B^2}{2L_{nB}^2} \tag{7.98}$$

where $L_{nB} = (D_{nB}\,\tau_n)^{1/2}$ is the diffusion length for electrons in the p-type base. For a typical base width $W_B = 300$ nm and diffusion length $L_{nB} = 10\ \mu$m, the base transport factor $\alpha_T = 0.9995$.

Having developed a more refined picture of the base and emitter current components, we now find expressions for the current gains in the forward-active region. The ratio of collector current to emitter current is

$$\alpha_F = \frac{I_C}{-I_E} = \left(\frac{I_C}{-I_{nE}}\right)\left(\frac{-I_{nE}}{-I_E}\right) = \alpha_T \cdot \gamma \tag{7.99}$$

Note that the emitter current and its electron component are both negative quantities. Substituting typical values for the emitter injection efficiency and for the base transport factor, we find that

$$\alpha_F = \alpha_T \cdot \gamma = (0.9995) \cdot (0.995) \approx 0.995 = \gamma \tag{7.100}$$

since the base transport factor is nearly equal to 1 in an IC npn BJT. The ratio of collector current to base current is

$$\beta_F = \frac{\alpha_F}{1-\alpha_F} \approx \frac{\gamma}{1-\gamma} \approx 200 \tag{7.101}$$

In concluding this section, we note that a similar approach can be used to model the reverse-active region and to find an expression for β_R. In this region, the base-collector junction is forward-biased and injecting electrons into the base which are collected by the emitter-base junction. The latter junction is much smaller in area as shown in Fig. 7.1. Therefore, two-dimensional diffusion must be incorporated in order to calculate α_R accurately.

7.7 LATERAL pnp BIPOLAR TRANSISTOR

It is not straightforward to fabricate a vertical p^+np sandwich together with the vertical n^+pn structure shown in Fig. 7.1 on the same silicon substrate. However, a modified pnp transistor can be made with little or no modification to the process steps used in making the n^+pn structure. Figure 7.25 shows the cross section and layout of an oxide-isolated **lateral pnp transistor**.

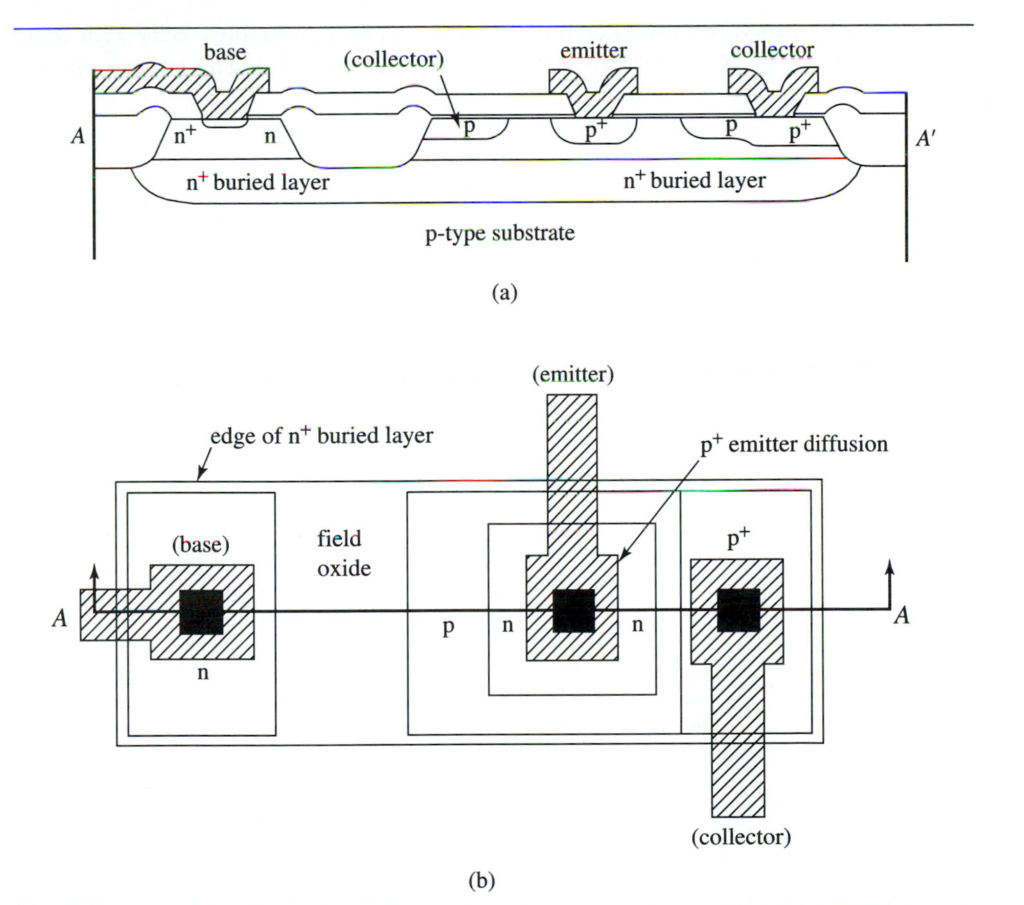

Figure 7.25 (a) Cross section and (b) layout of an oxide-isolated lateral pnp transistor. Note that both p regions that are adjacent to the emitter in (a) are part of the same ring-shaped collector, as shown in (b).

The operation of this transistor is fundamentally the same as the npn, except that the roles of the electrons and holes are reversed. In forward active operation, the emitter-base junction is forward biased by applying $V_{EB} = 0.7$ V (note the reversed polarity compared with the npn BJT) and holes are injected into the n-type base region. Those that diffuse laterally (in the plane of the wafer) cross the base and are swept into the ring-shaped p-type collector by the base-collector depletion region. The holes then are majority carriers and constitute the collector current.

The minority hole flux for the lateral pnp in forward-active operation, where $V_{EB} = 0.7$, $V_{BC} > -0.6$ V, is illustrated in Fig. 7.26. Unlike the vertical npn BJT, the diffusing carriers suffer significant losses due to recombination in the base region. Only those injected laterally have a significant probability of being collected. This problem is not as severe as it appears from Fig. 7.26, since the doping of the emitter region is heaviest near the surface of the device, which results in most of the hole injection occurring there rather than on the relatively lightly doped bottom of the p^+ diffusion. Current gains for the lateral pnp are less than for the vertical npn, with $\beta_F = 25 - 50$ being a typical range for an oxide-isolated lateral pnp BJT.

7.7.1 Ebers-Moll Model for the Lateral pnp BJT

Since the pnp transistor operates on the same physical principles, only with holes replacing electrons, we can adapt our results from Section 7.4. The emitter and the collector currents are

Ebers-Moll Equations (pnp transistor)

$$I_E = I_{ES}(e^{V_{EB}/V_{th}} - 1) - \alpha_R I_{CS}(e^{V_{CB}/V_{th}} - 1) \tag{7.102}$$

$$I_C = -\alpha_F I_{ES}(e^{V_{EB}/V_{th}} - 1) + I_{CS}(e^{V_{CB}/V_{th}} - 1) \tag{7.103}$$

The Ebers-Moll circuit model for the pnp is shown in Fig. 7.27(a) below. Simplifications to the model for hand calculations can be found using the same approach as was done in Section 7.4 for the npn BJT. The forward-active model for hand calcula-

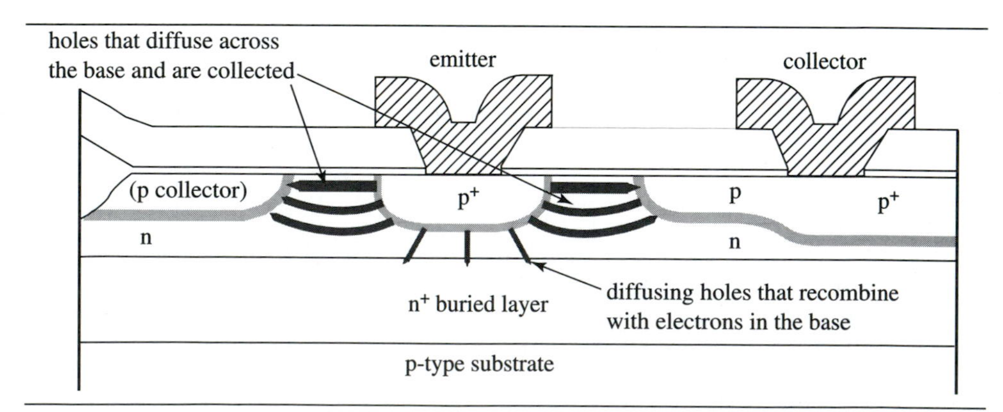

➤ **Figure 7.26** Schematic picture of minority hole diffusion in the n-type base region of the lateral pnp bipolar transistor. Those that are injected down recombine in the base region and fail to reach the collector.

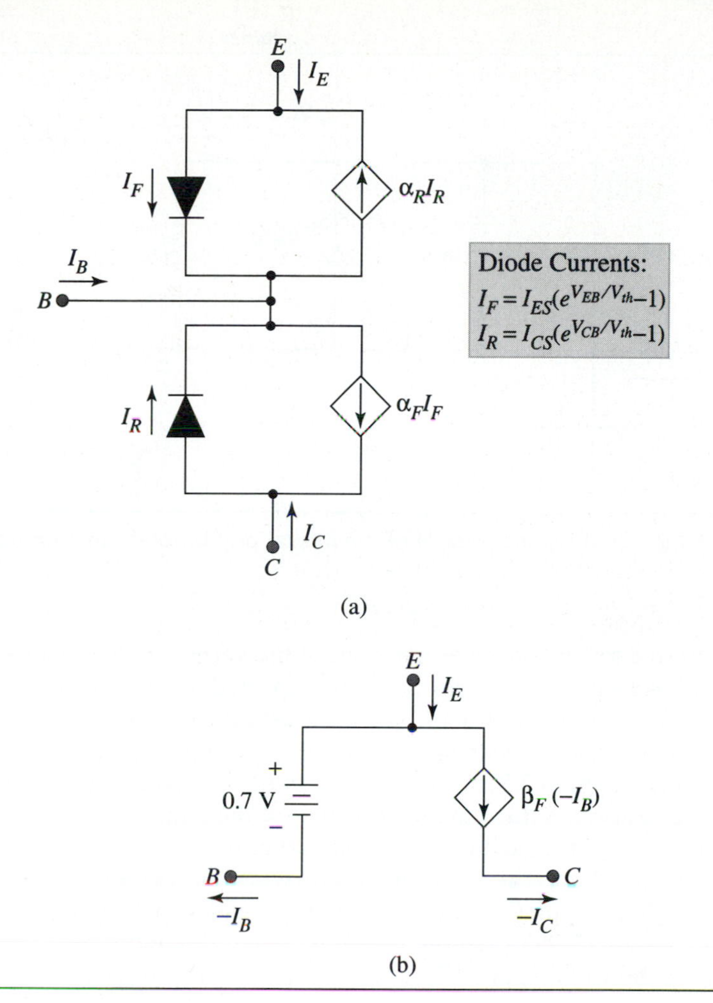

➤ **Figure 7.27** (a) Ebers-Moll model for a pnp transistor and (b) forward-active model for hand calculation.

tion is shown in Fig. 7.27(b). The base current I_B and the collector current I_C are both negative for the forward-active pnp, according to the "positive-in" convention for multi-terminal devices. In order to avoid negative signs, we use $-I_B$ and $-I_C$ as circuit variables.

7.7.2 Small-Signal Model of the Lateral pnp BJT

In the forward-active region, the collector current i_C is a function of the emitter-base voltage v_{EB} and the emitter-collector voltage v_{EC}

$$-i_C = I_S e^{v_{EB}/V_{th}}\left(1 + \frac{v_{EC}}{V_{Ap}}\right) \tag{7.104}$$

where we have incorporated the empirical model for base-width modulation by introducing the Early voltage V_{Ap} for the pnp BJT. In the pnp transistor, the base doping concentration N_{dB} is less than the collector doping concentration N_{aC}, since these lay-

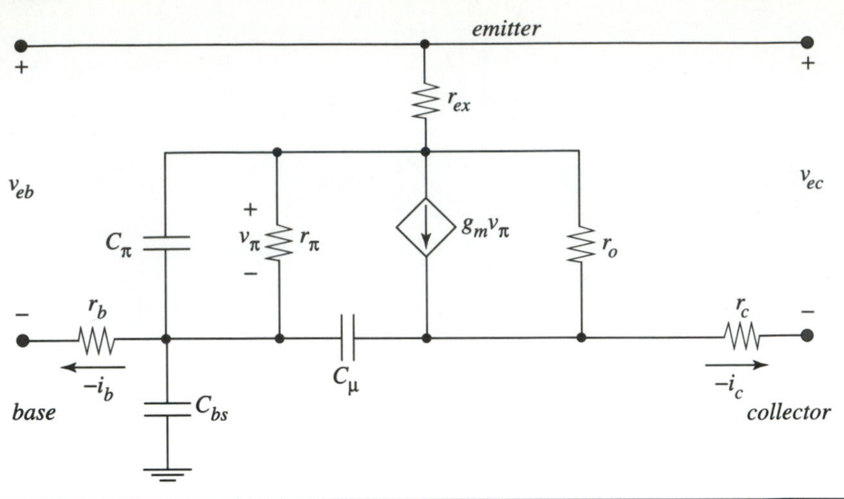

➤ **Figure 7.28** Small-signal model of the lateral pnp bipolar junction transistor.

ers are optimized for the opposite functions in the vertical npn structure. Therefore, the Early voltage V_{Ap} is lower than would be expected for such a wide base width.

Partial differentiation of Eq. (7.104) with respect to v_{EB} and v_{EC} leads directly to the same expressions for the small-signal transconductance g_m and output resistance r_o as we found in Section 7.5 for the npn BJT, with $-I_C$ replacing I_C. The input resistance of the pnp also has the same form as that of the npn.

The base-charging capacitance in the lateral pnp is much higher than for the vertical npn structure, due to the much larger base width W_B. A typical base width for the lateral pnp is about 1 μm. One difference in the small-signal circuit is that there is a depletion capacitor C_{bs} between the n-type base and the underlying substrate in the lateral pnp, as shown in Figure 7.25(a). This capacitor is given by

$$C_{bs} = \frac{C_{bso}}{\sqrt{1 + V_{BS}/\phi_{Bs}}} \tag{7.105}$$

where C_{bso} is the zero-bias capacitance and ϕ_{Bs} is the built-in voltage of the base-substrate junction. Note that there is no capacitor between the collector and the substrate for the lateral pnp. Since the base of the pnp is the same n-type diffused layer as the collector of the npn, C_{bso}/A_S for the pnp and C_{cso}/A_S for the npn have the same numerical value, where A_S is the area of the junction. Figure 7.28 shows the schematic for the small-signal model of the oxide-isolated lateral pnp BJT.

➤ EXAMPLE 7.6 Small-Signal, Diode-Connected pnp BJT

If the base terminal is shorted to the collector, we call the transistor **diode-connected.** For the pnp transistor in Fig. Ex 7.6 , the DC collector current is $I_C = -150$ μA, the current gain is $\beta_F = \beta_o = 30$, and the Early voltage is $V_{Ap} = 45$ V. For low frequencies where the capacitors can be considered open circuits, find the equivalent small-signal resistance between the emitter and collector terminals. All parasitic resistances can be neglected.

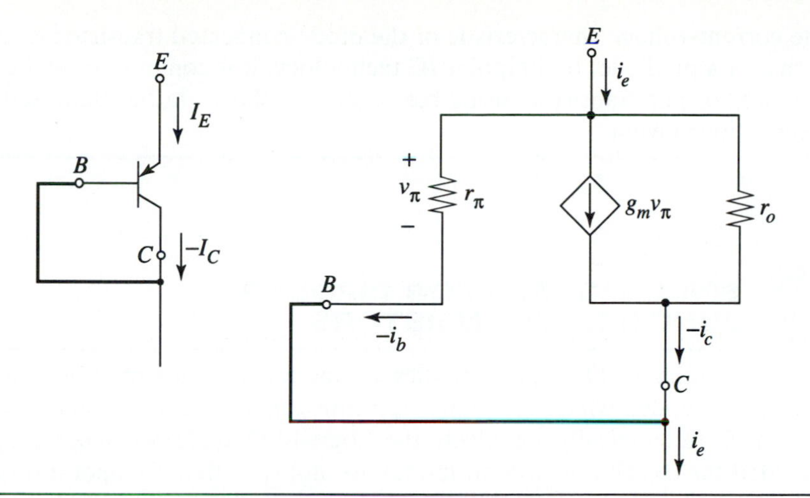

➤ **Figure Ex7.6** Diode-connected pnp transistor and its small-signal model.

SOLUTION

The small-signal model is modified by the short circuit between base and collector, as shown in Fig. Ex 7.6. The small-signal emitter current is the sum of three terms:

$$i_e = \frac{v_\pi}{r_\pi} + g_m v_\pi + \frac{v_{ec}}{r_o} = \frac{v_{ec}}{r_\pi} + g_m v_{ec} + \frac{v_{ec}}{r_o} = v_{ec}\left(\frac{1}{r_\pi} + g_m + \frac{1}{r_o}\right)$$

where we have used the fact that $v_\pi = v_{ec}$ due to the short circuit. The small-signal resistance between the emitter and collector terminals is:

$$\frac{v_{ec}}{i_e} = \left(\frac{1}{r_\pi} + g_m + \frac{1}{r_o}\right)^{-1} = r_\pi \| g_m^{-1} \| r_o$$

For the given DC collector current and transistor parameters, the small-signal elements are:

$$g_m = \frac{-I_C}{V_{th}} = 5.8 \text{ mS} \qquad r_\pi = \frac{\beta_o}{g_m} = 5.2 \text{ k}\Omega \qquad r_o = \frac{V_{Ap}}{-I_C} = 300 \text{ k}\Omega$$

Evaluating the resistance, we find that

$$\frac{v_{ec}}{i_e} = 5.2 \text{ k}\Omega \| 173\ \Omega \| 300 \text{ k}\Omega = 167\ \Omega$$

Since $g_m^{-1} \ll r_\pi \ll r_o$, we could have recognized that the equivalent resistance would have been approximately g_m^{-1} without the need to evaluate r_π or r_o. The origin of the term "diode-connected" is seen in comparing the small-signal resistance to the resistance r_d of a diode carrying the DC current I_E:

$$\frac{v_{ec}}{i_e} \approx g_m^{-1} \approx \frac{V_{th}}{I_E} = r_d$$

The current-voltage characteristic of the diode-connected transistor is identical to that of a pn diode. In a bipolar IC technology, it is common to make diodes from npn or pnp transistors using base-collector shorts, rather than to design a special diode layout.

7.8 SPICE MODELS FOR BIPOLAR JUNCTION TRANSISTORS

The bipolar transistor's DC characteristics are modeled using a modification of the Gummel-Poon model, which is a more accurate generalization of the Ebers-Moll model. SPICE automatically defaults to the Ebers-Moll model when certain parameters needed for the Gummel-Poon model are not specified. Temperature dependence of the saturation current can be specified, as well as a separate temperature dependence of the current gain.

A bipolar transistor is included in a circuit by the following line:

QXXX NC NB NE <NS> MNAME <AREA> <OFF> <IC=VBE,VCE>

XXX is the transistor number
NC NB NE <NS> are the collector, base, emitter, and (optional) substrate node numbers
MNAME is the model name (several BJT models can be specified by this parameter)
AREA is the optional area factor, which specifies the number of parallel devices of the specified model name
OFF is an optional initial condition for DC analysis
IC=VBE,VCE is a user-specified initial condition for transient analysis

The transistor model is specified by a **.MODEL** statement:

.MODEL MNAME TYPE(PNAME1=PVAL1 PNAME2=PVAL2 ...)

TYPE is **NPN** or **PNP**
PNAME1 is model parameter 1, with value **PVAL1**

There are 40 parameters that can be specified in this statement in the basic SPICE model of the BJT. Only those that are commonly specified for a first-order model are listed in the table, along with the corresponding symbols used in this chapter. In SPICE, depletion capacitances are modeled by this general equation:

$$C_{\mu} = \frac{CJC}{\left(1 - \frac{VBC}{VJC}\right)^{MJC}} \qquad (7.106)$$

using the base-collector depletion capacitance as an example. In an IC pn junction, the exponents *MJE*, *MJC*, and *MJS* range from 0.33 to 0.5. An abrupt junction, such as we have assumed in this chapter, leads to an exponent of 0.5.

Table 7.1 npn BJT SPICE Parameters

Name	Parameter Description	Units
IS	Transport saturation current [I_S]	Amps
BF	Ideal maximum forward beta [β_F]	None
VAF	Forward Early voltage [V_{An}]	Volts
BR	Ideal maximum reverse beta [β_R]	None
RB	Base resistance [r_b]	Ohms
RE	Emitter resistance [r_{ex}]	Ohms
RC	Collector resistance [r_c]	Ohms
CJE	B-E zero-bias depletion capacitance [C_{jEo}]	Farads
VJE	B-E built-in potential [ϕ_{Be}]	Volts
MJE	B-E junction exponential factor	None
CJC	B-C zero-bias depletion capacitance [$C_{\mu o}$]	Farads
VJC	B-C built-in potential [ϕ_{Bc}]	Volts
MJC	B-C junction exponential factor	None
CJS	Substrate zero-bias depletion capacitance [C_{cso}]	Farads
VJS	Substrate built-in potential [ϕ_{Bs}]	Volts
MJS	Substrate junction exponential factor	None
TF	Ideal forward transit time [τ_F]	Seconds

For circuit simulations in this text, we need a typical SPICE model for the oxide-isolated npn BJT sketched in Fig. 7.1 This transistor is typical of those available in a modern BiCMOS technology. The SPICE model statement is

```
.MODEL MODQN NPN IS = 1E–17 BF = 100 VAF = 25 TF = 50P CJE = 8E–15 VJE = 0.95
+ MJE = 0.5 CJC = 22E–15 VJC = 0.79 MJC = 0.5 CJS = 41E–15 VJS = 0.71 MJS = 0.5
+ RB = 250 RC = 200 RE = 5
```

➤ SUMMARY

This chapter has built on the models of the pn junction diode in Chapter 6 to construct device and circuit models for the npn and pnp bipolar junction transistors. The key concepts and techniques introduced in this chapter were:

- ◆ Concept of transistor action: injection of minority carriers from a forward-biased pn junction, diffusion across a narrow base region, and collection by a second reverse-biased junction.
- ◆ Physical structure of vertical npn bipolar transistors.
- ◆ Regions of operation of bipolar transistors: forward active, saturation, reverse active, and cutoff.
- ◆ Carrier fluxes in the npn bipolar transistor in the four operating regions.
- ◆ Current gain in forward active operation and its relation to the physical dimensions and parameters of the transistor.

- Ebers-Moll model of the npn bipolar transistor's DC characteristics.
- Hybrid-π small-signal model of the npn and pnp bipolar transistors.
- Base charging capacitance and the base transit time and their physical origin.
- Early voltage and its physical origin in base-width modulation.
- Definition of emitter injection efficiency and base transport factor; the dependence of these parameters on the parameters and dimensions of the transistor.
- Physical structure, Ebers-Moll DC model, and small-signal model of the lateral pnp bipolar transistor.
- SPICE model for the npn bipolar transistor.

REFERENCES

1. G. W. Neudeck, *The Bipolar Junction Transistor*, vol. III of the Modular Series on Solid-State Devices, Addison-Wesley, 1983. This book is a thorough discussion of the BJT at about the level of this book, with the intrinsic pnp structure being used as the example.
2. C. G. Fonstad, *Microelectronic Devices and Circuits*, McGraw Hill, 1994. Chapter 8 discusses the BJT at about the same level, with the addition of optical effects.
3. M. Shur, *Physics of Semiconductor Devices*, Prentice Hall, 1990. Chapter 3 is an in-depth treatment of the BJT, with much more discussion of switching, breakdown, and second-order effects.
4. R. S. Muller and T. I. Kamins, *Device Electronics for Integrated Circuits*, 2nd ed., Wiley, 1986. Chapter 6 is a description of the basic operation, using energy levels. Chapter 7 is a thorough treatment of the limits of the basic models and second-order effects.

PROBLEMS

EXERCISES

For exercises E7.1–E7.10, use the following npn and pnp bipolar transistor parameters:

npn: $\beta_F = 100$, $I_S = 10^{-16}$ A, $V_{CE_{SAT}} = 0.1$ V. pnp: $\beta_F = 50$, $I_S = 10^{-16}$ A, $V_{EC_{SAT}} = 0.1$ V.

E7.1 For the npn transistor biased as shown in Fig. E7.1,

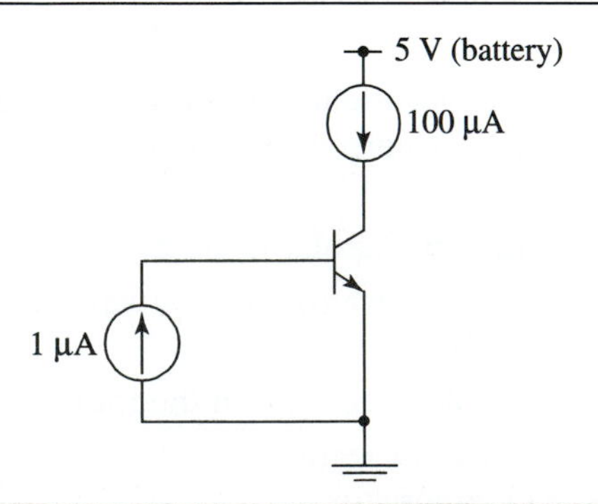

Figure E7.1

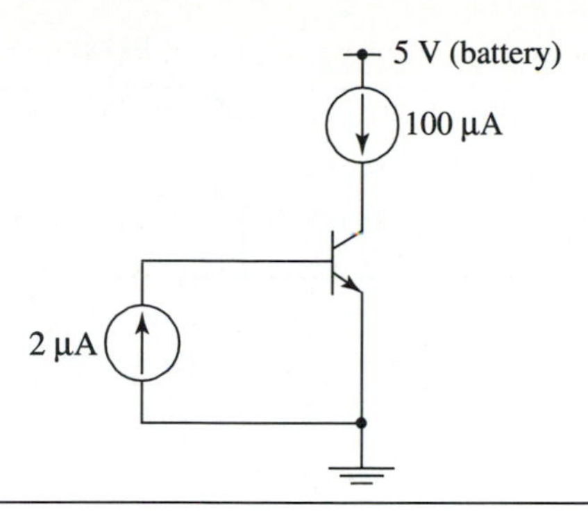

➤ **Figure E7.2**

(a) Identify the operating region (cutoff, forward active, saturation, or reverse active) of the bipolar transistor.

(b) What is the base-emitter voltage V_{BE} for this operating point?

E7.2 For the npn transistor biased as shown in Fig. E7.2,

(a) Identify the operating region (cutoff, forward-active, saturation, or reverse-active) of the bipolar transistor.

(b) What is the approximate value of the collector-emitter voltage V_{CE}?

(c) Locate the operating point on the output characteristics of the BJT. You need sketch only $I_C(I_B, V_{CE})$ for the value $I_B = 2\ \mu A$.

E7.3 For the npn transistor biased as shown in Fig. E7.3,

(a) Identify the operating region (cutoff, forward-active, saturation, or reverse-active) of the bipolar transistor.

(b) Locate the operating point on the transistor's output characteristics.

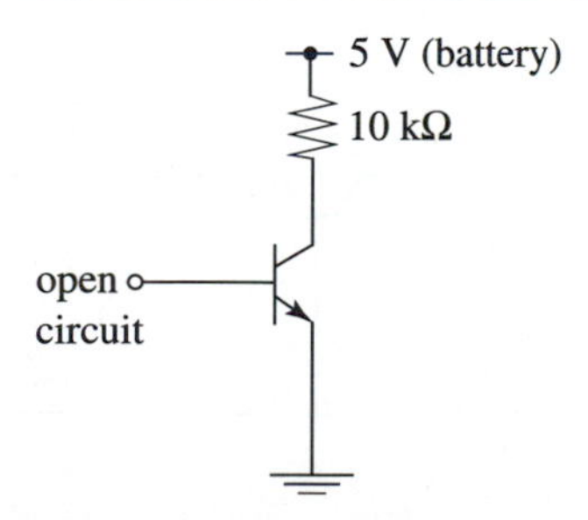

➤ **Figure E7.3**

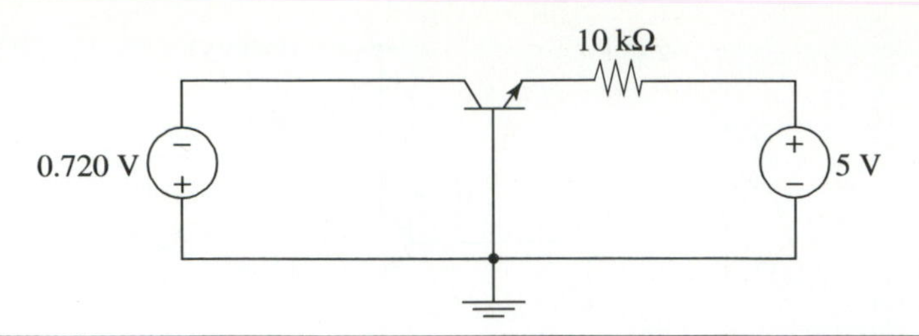

➤ **Figure E7.4**

E7.4 For the npn transistor biased shown in Fig. E7.4,

Identify the operating region (cutoff, forward-active, saturation, or reverse-active) of the bipolar transistor, given that the emitter current is $I_E = 15\ \mu A$.

E7.5 For the npn transistor biased shown in Fig. E7.5,

(a) Identify the operating region (cutoff, forward-active, saturation, or reverse-active) of the bipolar transistor.

(b) Find the collector current I_C.

(c) Find the base current I_B.

(d) Find the collector-emitter voltage V_{CE}.

(e) Locate the operating point on the output characteristics of the BJT; you need only sketch $I_C(I_B, V_{CE})$ for the appropriate value of I_B.

E7.6 For the pnp transistor biased shown in Fig. E7.6,

(a) Identify the operating region (cutoff, forward-active, saturation, or reverse-active) of the bipolar transistor.

(b) What is the emitter-collector voltage V_{EC} for this operating point?

(c) Locate the operating point on the output characteristics of the BJT. You need sketch only $I_C(I_B, V_{EC})$ for the value $I_B = -2.5\ \mu A$.

E7.7 For the pnp transistor biased shown in Fig. E7.7,

(a) Identify the operating region (cutoff, forward-active, saturation, or reverse-active) of the bipolar transistor.

(b) What is the emitter-base voltage V_{EB} for this operating point?

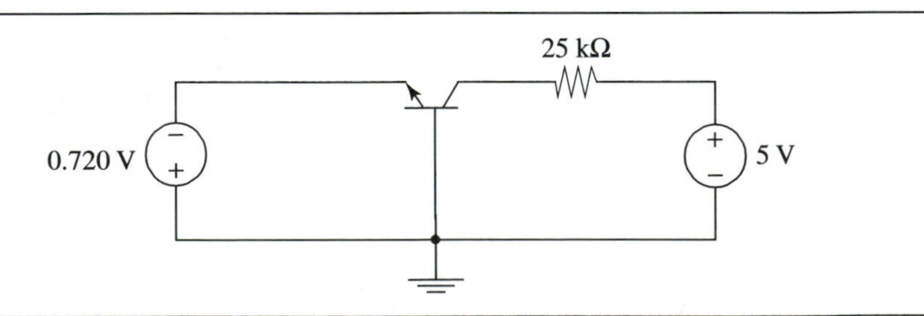

➤ **Figure E7.5**

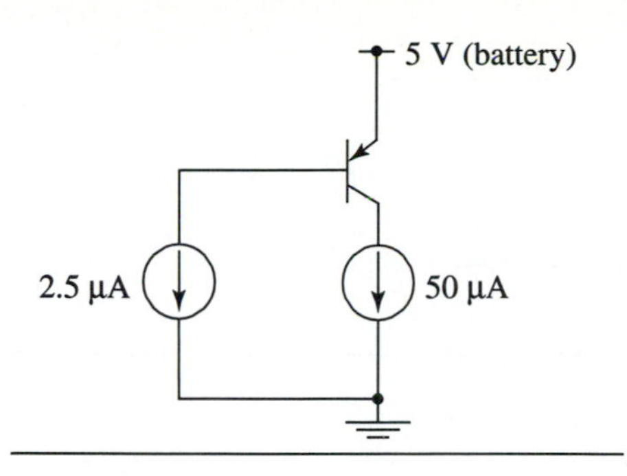

➤ **Figure E7.6**

➤ **Figure E7.7**

E7.8 For the pnp transistor biased shown in Fig. E7.8,

(a) Identify the operating region (cutoff, forward-active, saturation, or reverse-active) of the bipolar transistor.

(b) Locate the operating point on the transistor's output characteristics.

E7.9 For the pnp transistor biased shown in Fig. E7.9, use $I_S = 5 \times 10^{17}$ A.

(a) Identify the operating region (cutoff, forward-active, saturation, or reverse-active) of the bipolar transistor.

(b) Find the collector current I_C.

(c) Find the base current I_B.

(d) Find the emitter-collector voltage V_{EC}.

(e) Locate the operating point on the output characteristics of the BJT; you need only sketch $I_C(I_B, V_{EC})$ for the appropriate value of I_B.

E7.10 For the pnp transistor biased shown in Fig. E7.10,

Identify the operating region (cutoff, forward-active, saturation, or reverse-active) of the bipolar transistor, given that the emitter current is $I_E = -15$ μA.

E7.11 An npn BJT has the doping concentrations: $N_{dE} = 2 \times 10^{17}$ cm^{-3}, $N_{aB} = 10^{16}$ cm^{-3}, and $N_{dC} = 10^{15}$ cm^{-3}. The emitter width is $W_{Eo} = 0.5$ μm and the base width is $W_{Bo} = 0.3$ μm in thermal equilibrium.

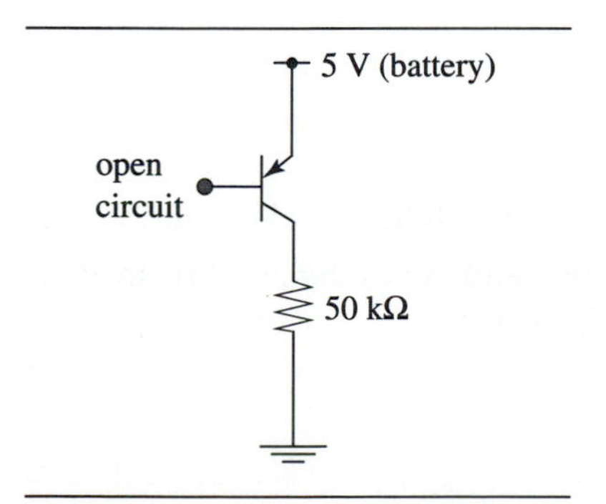

➤ **Figure E7.8**

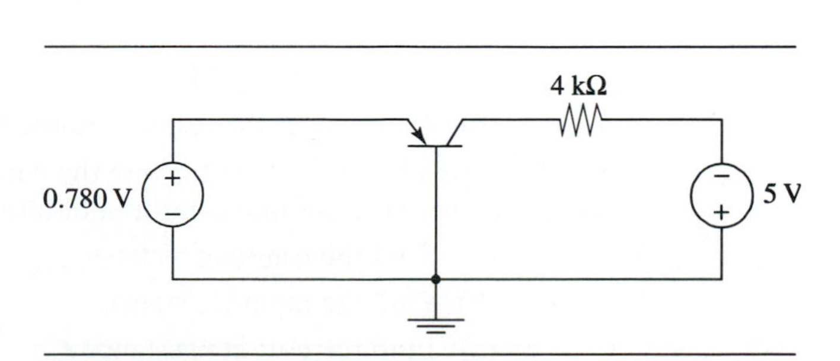

➤ **Figure E7.9**

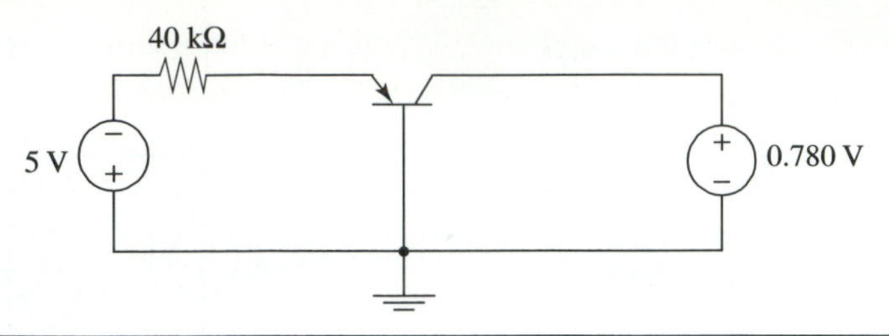

➤ **Figure E7.10**

(a) Plot the charge density $\rho_o(x)$, the electric field $E_o(x)$, and the potential $\phi_o(x)$ through the structure in thermal equilibrium.

(b) Sketch the same quantities with the transistor biased at $V_{BE} = 0.6$ V, $V_{BC} = -1.5$ V.

E7.12 For the transistor in exercise E7.11, sketch the minority carrier concentrations under the operating point defined in E7.11(b). You can assume that the carrier profiles are linear for this transistor.

E7.13 Consider an npn transistor with $W_B = 0.5$ μm and $W_E = 1$ μm. Neglect the effect of varying depletion widths with doping concentrations in this exercise.

(a) For $N_{aB} = 5 \times 10^{16}$ cm^{-3} and $N_{dE} = 5 \times 10^{17}$ cm^{-3}, find the forward current gain β_F. You should use the appropriate minority carrier diffusivities for the given doping concentrations, which can be found from Fig. 2.8 and the Einstein Relation.

(b) Plot the forward current gain as a function of emitter doping concentration over the range $N_{dE} = 5 \times 10^{17}$ cm^{-3} to 10^{19} cm^{-3}, for $N_{aB} = 5 \times 10^{16}$ cm^{-3}.

E7.14 An npn transistor with emitter area $A_E = 10$ μm × 10 μm is biased in the forward active region, with the collector current $I_C = 50$ μA. The emitter and base dimensions and doping are: $N_{dE} = 7.5 \times 10^{18}$ cm^{-3}, $W_E = 0.4$ μm, $N_{aB} = 10^{17}$ cm^{-3}, and $W_B = 0.25$ μm.

(a) Sketch the minority carrier concentrations in the emitter and base.

(b) Find the base-emitter bias V_{BE}.

(c) Find the base current I_B.

E7.15 Given that $I_{ES} = 10^{-17}$ A, $\alpha_F = 0.99$, $I_{CS} = 2 \times 10^{-17}$ A, and $\alpha_R = 0.495$ for an npn bipolar transistor. This exercise requires a math software package.

(a) Use the Ebers-Moll equations to generate the curves $I_C(V_{CE})$ for $V_{CE} > 0$ for $I_B = 0.5$, 1, 2, and 5 μA.

(b) Repeat for $V_{CE} < 0$.

(c) Compare to the results obtained using the SPICE model in Section 7.8.

E7.16 Given the npn transistor with the parameters and operating point from exercise E7.14, with the additional information that $V_{An} = 20$ V.

(a) Find the transconductance g_m.

(b) Find the input resistance r_π.

(c) Find the output resistance r_o.

E7.17 Given the npn transistor with the parameters and operating point from exercise E7.14, with the additional information that the emitter-base depletion width is $x_{BE} = 0.1$ μm.

(a) What is the minority electron storage $Q_{NB}(V_{BE})$ at this operating point?

(b) What is C_π at this operating point?

(c) At what frequency does $|1/j\omega C_\pi| = r_\pi$?

E7.18 The layouts of two adjacent, oxide-isolated npn transistors (see Fig. 7.1 for the cross section) are given in Fig. E7.18. You can assume that the parameters and vertical dimensions of the two transistors are identical.

(a) For identical forward-active biases $V_{BE1} = V_{BE2}$ and $V_{CE1} = V_{CE2} = 2$ V, find the ratio of the collector currents I_{C2}/I_{C1}.

(b) Find the ratio of the base-charging capacitances C_{b2}/C_{b1}.

(c) Find the ratio of the collector-base capacitances $C_{\mu 2}/C_{\mu 1}$.

E7.19 The qualitative effects of parameter and dimensional changes on the model parameters are the subject of this exercise. In Table E7.19, fill in each entry with +, o, or – to represent an increase, no effect, or decrease as a result of an *increase* in the parameter in the left-hand column. The collector width is the distance to the n$^+$ buried layer. The operating point (I_C, V_{CE}) remains fixed.

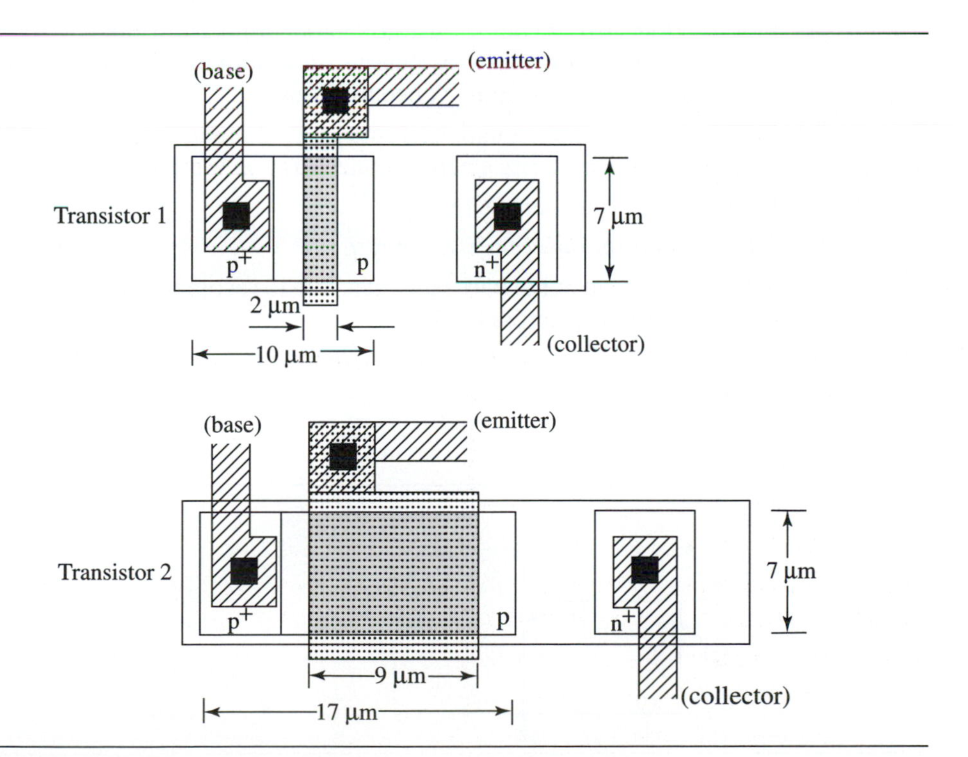

➤ Figure E7.18

Table E7.19 Qualitative Effects of Increase in One Parameter.

Parameter to be increased	Forward gain β_F	Reverse gain β_R	Saturation current I_S	Early voltage V_{An}
N_{dE}				
W_E				
D_{pE}				
N_{aB}				
W_B				
D_{nB}				
N_{dC}				
W_C				
D_{pC}				

E7.20 An npn transistor is biased in the forward-active region by the circuit shown in Fig. E7.20. Identify the direction of the change in the indicated carrier concentration or small-signal model parameter due to an *increase* in the base current I_B in Table E7.20. The carrier concentrations are specified at the edges of the particular junction (see Fig. 7.5)

E7.21 An npn transistor with emitter area $A_E = 10\ \mu m \times 10\ \mu m$ is biased in the forward active region, with the collector current $I_C = 50\ \mu A$. The emitter and base dimensions and doping are: $N_{dE} = 7.5 \times 10^{18}\ cm^{-3}$, $W_E = 0.4\ \mu m$, $N_{aB} = 10^{17}\ cm^{-3}$, and $W_B = 0.25\ \mu m$. The minority carrier lifetimes are $\tau_{pE} = 30$ ns and $\tau_{nB} = 75$ ns.

(a) Find the minority carrier diffusion lengths, using diffusion coefficients calculated from the Einstein Relation and mobilities that are found from Fig. 2.8 and the doping concentrations.

(b) Find the recombination base current I_R by assuming that the short-base solution for the minority electron concentration is sufficiently accurate.

(c) Find the base transport factor α_T.

(d) Find the emitter injection efficiency γ.

(e) What percentage of the base current is due to recombination of injected electrons diffusing across the base?

E7.22 Given that $I_{ES} = 10^{-16}$ A, $\alpha_F = 0.96$, $I_{CS} = 1.76 \times 10^{-16}$ A, and $\alpha_R = 0.545$ for an pnp bipolar transistor.

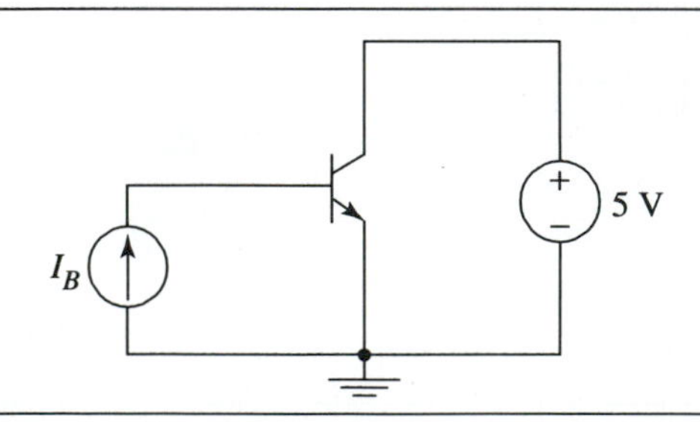

➤ **Figure E7.20**

Table E7.20 Effect of Increase in Base Current.

Carrier concentration or model parameter	Increase	No Change	Decrease
Minority electron concentration at B-E junction $n_p(0)$			
Minority electron concentration at B-C junction $n_p(W_B)$			
Majority hole concentration at B-E junction $p_p(0)$			
Majority hole concentration at B-C junction $p_p(W_B)$			
Minority hole concentration at B-E junction $p_n(-x_{BE})$			
Min. hole concentration at ohmic contact $p_n(-x_{BE}-W_E)$			
Transconductance g_m			
Input resistance r_π			
Output resistance r_o			
Base-charging capacitance C_b			
Base transit time τ_F			

(a) Find the forward current gain β_F and the reverse current gain β_R.

(b) Use a math software package and the Ebers-Moll equations to generate the curves $-I_C(V_{EC})$ for $V_{EC} > 0$ for $I_B = -0.5$, -1, -2, and -5 μA.

(c) Repeat part (b) for $V_{EC} < 0$.

E7.23 A lateral pnp transistor with sidewall emitter area $A_E = 1\ \mu\text{m} \times 50\ \mu\text{m}$ is biased in the forward active region, with the collector current $I_C = -25\ \mu\text{A}$. The emitter and base dimensions and doping are: $N_{aE} = 4 \times 10^{17}\ \text{cm}^{-3}$, $W_E = 1.5\ \mu\text{m}$, $N_{dB} = 1 \times 10^{16}\ \text{cm}^{-3}$, and $W_B = 1.0\ \mu\text{m}$.

(a) Sketch the minority carrier concentrations in the emitter and base.

(b) Find the emitter-base bias V_{EB}.

(c) Find the base current I_B. Neglect hole injection from the bottom of the emitter.

E7.24 Given the lateral pnp transistor with the parameters and operating point from exercise E7.23, with the additional information that $V_{Ap} = 25$ V.

(a) Find the transconductance g_m.

(b) Find the input resistance r_π.

(c) Find the output resistance r_o.

E7.25 Given the lateral pnp transistor with the parameters and operating point from exercise E7.23, with the additional information that the emitter-base depletion width is $x_{BE} = 0.15\ \mu\text{m}$.

(a) What is the minority hole storage $Q_{PB}(V_{EB})$ at this operating point?

(b) What is C_π at this operating point?

(c) At what frequency does $|1/j\omega C_\pi| = r_\pi$?

Problems

P7.1 Consider an npn BJT with the following parameters and dimensions. Recombination in all regions can be neglected.

emitter: $N_{dE} = 9 \times 10^{18}$ cm^{-3}, $W_E = 500$ nm
base: $N_{aB} = 1.5 \times 10^{17}$ cm^{-3}, $W_B = 400$ nm
collector: $N_{dC} = 2 \times 10^{16}$ cm^{-3}, $W_C = 2$ μm
Emitter area = $A_E = 50$ μm^2, collector area = $A_C = 50$ μm^2.

(a) Sketch the minority carrier concentrations in the emitter, base, and collector under forward-active bias, given that $I_C = 25$ μA and $V_{BC} = 0$ V.

(b) Find α_F and β_F for this transistor.

P7.2 For the npn BJT with parameters and dimensions described in P7.1.

(a) Sketch the minority carrier concentrations in the emitter, base, and collector under reverse-active bias, given that $I_C = -5$ μA and $V_{BE} = 0$ V.

(b) Find α_R and β_R for this transistor. Note that $A_E = A_C$ is not possible in a practical transistor.

P7.3 In some special-purpose technologies, a vertical pnp transistor with the same structure an layout as the npn transistor in Fig. 7.1 is available, except that the contact to the emitter is made using aluminum.

(a) Sketch the carrier fluxes under forward-active bias. You can neglect recombination in the emitter and base.

(b) Sketch the carrier fluxes under reverse-active bias. Neglect recombination in the base.

(c) Sketch the carrier fluxes when the pnp transistor is biased in saturation. Again, you can neglect recombination in the base.

P7.4 Given an npn BJT biased with $I_C = (\beta_F/2)I_B$.

(a) What is the approximate value of the collector-emitter voltage V_{CE}?

(b) What is the ratio of $n_p(0)$ to $n_p(W_B)$?

(c) Sketch the minority carrier concentrations in the emitter, base, and collector.

P7.5 A particular bipolar IC process has the following tolerances in the dimensions and parameters of an npn transistor:

emitter: $N_{dE} = 1.2 \times 10^{19} \pm 10^{18}$ cm^{-3}, $W_E = 500 \pm 40$ nm
base: $N_{aB} = 4 \times 10^{17} \pm 2 \times 10^{16}$ cm^{-3}, $W_B = 350 \pm 30$ nm
collector: $N_{dC} = 4 \times 10^{16} \pm 10^{15}$ cm^{-3}, $W_C = 1 \pm 0.05$ μm
The emitter area is $A_E = 100$ μm^2, and the collector area is $A_C = 750$ μm^2.

(a) Assuming that the parameter and dimensional variations are independent, find the average and variation of the saturation current I_S.

(b) For a fixed base-emitter bias $V_{BE} = 720$ mV and base-collector bias $V_{BC} = -2.3$ V, find the average and variation of the collector current I_C.

(c) Assuming that the parameter and dimensional variations are independent, find the average and variation of the forward current gain β_F.

(d) Find the minimum and maximum values of the forward current gain β_F for this process, by using the extreme values of the device parameters and dimension. This approach is illustrated by Ex. 7.1.

P7.6 An npn bipolar transistor has the following parameters:

emitter: $N_{dE} = 1 \times 10^{19}$ cm^{-3}, $W_E = 400$ nm
base: $N_{aB} = 8 \times 10^{16}$ cm^{-3}, $W_B = 500$ nm
collector: $N_{dC} = 8 \times 10^{15}$ cm^{-3}, $W_C = 1.5$ μm
emitter area = $A_E = 2.5$ μm × 2 μm.

Transistors with small emitter areas violate low-level injection at small collector current levels. For this problem we consider the assumption of low-level injection to be invalid with the minority carrier concentration is equal to one-tenth the thermal equilibrium majority carrier concentration.

(a) Find the maximum collector current I_C for which the low-level injection model is valid.

(b) In order to maintain operation in low-level injection, the emitter area of the transistor can be increased. What is the minimum emitter area needed for a transistor with these vertical dimensions and doping levels to satisfy low-level injection at a collector current $I_C = 250$ μA.

P7.7 The layout and cross section (not to scale) of an npn bipolar transistor which lacks an n$^+$ buried layer is shown in Fig. P7.7. The various components are defined in Fig. 7.1. The doping levels for this transistor are

emitter: $N_{dE} = 1 \times 10^{19}$ cm^{-3}, $W_E = 250$ nm
base: $N_{aB} = 2 \times 10^{17}$ cm^{-3}, $W_B = 250$ nm
collector: $N_{dC} = 5 \times 10^{16}$ cm^{-3}.

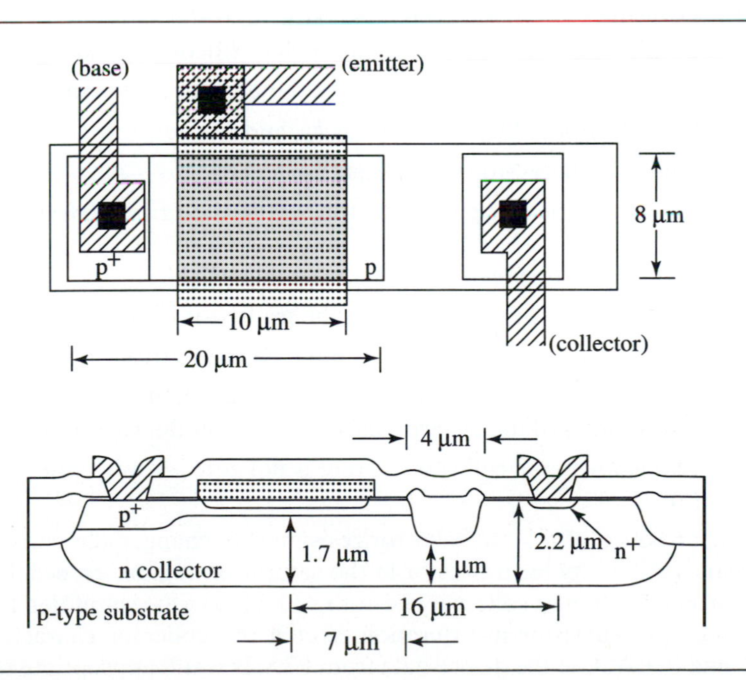

Figure P7.7

(a) From the dimensions given in the layout and cross section in Fig. P7.7, estimate the value of the collector resistor r_c. Hint: the collector region can be modeled as three resistors in series. The contact (with n^+ implant to ensure an ohmic contact) contributes a negligibly small resistance.

(b) What is the forward-active current gain β_F for this transistor?

(c) Model the effect of the high collector resistance by including it in the forward-active simplified Ebers-Moll model. Plot the output characteristics $I_C(V_{CE})$ for $I_B = 2, 5, 10$, and $15\ \mu A$ over the range $0.25\ V < V_{CE} < 5\ V$.

P7.8 In some applications, it is necessary to have a more precise definition of the saturation voltage $V_{CE_{SAT}}$ for a particular current level. As the collector-emitter voltage falls and the transistor enters the saturation region, the collector current starts to decrease abruptly. For this problem, we define $V_{CE_{SAT}}$ as the collector-emitter voltage at which $I_C = (0.9)\beta_F I_B$. In other words, the device enters saturation when the collector current drops 10% below its value in the forward active region. The Ebers-Moll parameters for the npn transistor are: $I_{ES} = 10^{-17}$ A, $\alpha_F = 0.99$, $I_{CS} = 2 \times 10^{-17}$ A, and $\alpha_R = 0.495$.

(a) In saturation, the "1's" can be neglected in the Ebers-Moll equation. Using this approximation, find an expression for V_{CE} in terms of I_C/I_B

(b) Find the numerical value of $V_{CE_{SAT}}$ as defined in the problem statement, using your result from (a).

P7.9 An important development of the late 1980s was Ge-Si (germanium-silicon) alloy emitter BJTs. We will assume for this problem that the intrinsic concentration in the Ge-Si emitter is $n_{iE} = 5 \times 10^9\ cm^{-3}$. Use the transistor doping levels and dimensions from E7.14.

(a) Find α_F and the forward-active current gain β_F.

(b) Determine the base doping that will yield the same value of β_F as the transistor would have if its emitter were silicon instead of Ge-Si.

P7.10 We consider an npn transistor with the dimensions and parameters in P7.1.

(a) Find the Ebers-Moll parameters I_{ES} and α_F.

(b) Find the Ebers-Moll parameters I_{CS} and α_R.

P7.11 For a symmetrical npn transistor with the Ebers-Moll parameters $I_{ES} = 10^{-17}$ A, $\alpha_F = 0.98$, $I_{CS} = 10^{-17}$ A, and $\alpha_R = 0.98$, we apply a voltage $V_B = V_{BE} = V_{BC} = 0.660$ V as shown in the circuit in Fig. P7.11.

(a) What is the operating region of the transistor in Fig. P7.11?

(b) Find the collector current I_C and the emitter current I_E.

(c) Sketch qualitatively the minority carrier concentrations for this operating point. Find the electron diffusion current density in the base.

(d) Explain why the collector current is not zero, despite your results from (c).

P7.12 The presence of a large collector resistance r_c changes the current-voltage characteristics of the bipolar transistor in the saturation region, especially for large collector currents. Adding a collector resistor $r_c = 1.5\ k\Omega$ to the simplified Ebers-Moll model of the npn transistor in saturation, sketch the collector characteristics for $I_B = 0.5, 2$, and $10\ \mu A$. Use the device data from P7.8. Is it still appropriate to consider the transistor to have a single saturation voltage, independent of current?

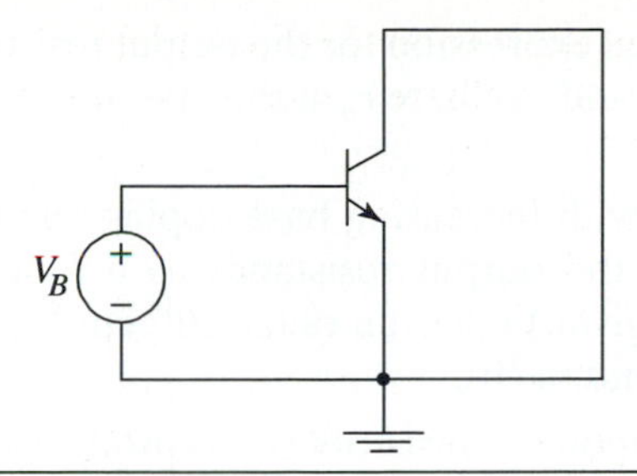

➤ **Figure P7.11**

P7.13 We consider an npn transistor with Ebers-Moll parameters $I_{ES} = 10^{-16}$ A, $\alpha_F = 0.98$, $I_{CS} = 2 \times 10^{-16}$ A, and $\alpha_R = 0.49$.

(a) Plot the collector current I_C as a function of the collector-emitter voltage V_{CE} for $V_{BE} = 660$ mV, 680 mV, 700 mV, and 720 mV. You can assume that the saturation voltage is $V_{CE_{SAT}} = 100$ mV.

(b) From your answer in (a), estimate the transconductance g_m graphically at the operating point $V_{BE} = 700$ mV and $V_{CE} = 2.5$ V. Compare your result with the analytical formula for g_m.

P7.14 Consider an npn transistor with the doping levels and dimensions from P7.1. In this problem, the widths of the regions are interpreted as the distances between the metallurgical junctions. Therefore, the depletion region widths must be calculated to find the widths of the bulk emitter, base, and collector regions. The *total* base-emitter depletion region width is $x_{BE} = 120$ nm for forward-active bias. The transistor is biased in the forward-active region with a base current $I_B = 1.5$ μA.

(a) For a base-collector bias $V_{BC} = -2$ V, find the collector current.

(b) For a base-collector bias $V_{BC} = -2.5$ V, find the collector current.

(c) From your answers to (a) and (b), estimate the Early voltage V_{An} for this transistor at $V_{CE} = 2.7$. V.

P7.15 We can use the approach from P7.14 to find an Early voltage for the reverse active region. Assume that the *total* base-collector depletion width is $x_{BC} = 350$ nm and the that base current is $I_B = 1.5$ μA.

(a) For a base-emitter bias $V_{BE} = -2$ V, find the collector current.

(b) For a base-emitter bias $V_{BE} = -2.5$ V, find the collector current.

(c) From your answers to (a) and (b), estimate the reverse Early voltage V_{An}^R for this transistor at $V_{CE} = -2.7$. V.

P7.16 Consider an npn transistor with the following parameters and region widths, which for this problem mean the widths between the metallurgical junctions.

emitter: $N_{dE} = 1 \times 10^{19}$ cm^{-3}, $W_E = 450$ nm

base: $N_{aB} = 8 \times 10^{16}$ cm^{-3}, $W_B = 550$ nm

collector: $N_{dC} = 5 \times 10^{15}$ cm^{-3}, $W_C = 1.5$ μm

The emitter area is $A_E = 25$ μm^2.

(a) Plot the collector current i_C as a function of v_{CE} for $V_{BE} = 720$ mV using Eq. (7.62) and (7.63), over the range 0.25 V < v_{CE} < 3 V.

(b) Derive the analytical expression for the output resistance of this transistor from Eq. (7.64) and evaluate r_o at the operating point $V_{BE} = 720$ mV, $V_{CE} = 2.5$ V.

(c) The increase in r_o with increasing base doping can be shown from the result in (b). Plot the output resistance as a function of N_{aB} (other parameters are as given) over the range 10^{16} cm^{-3} $< N_{aB} < 2 \times 10^{17}$ cm^{-3} at the operating point in (b).

P7.17 A figure of merit for bipolar transistors is the product of current gain and output resistance. From P7.16(b) we have an expression for r_o as a function of the transistor parameters and dimensions. Use the transistor dimensions from P7.16 and the operating point $V_{BE} = 720$ mV, $V_{CE} = 2.5$ V.

(a) Plot the current gain $\beta_o = \beta_F$ as a function of the base doping over the range 10^{16} cm^{-3} $< N_{aB} < 2 \times 10^{17}$ cm^{-3}.

(b) Plot the product $\beta_o\, r_o$ over the range 10^{16} cm^{-3} $< N_{aB} < 2 \times 10^{17}$ cm^{-3}. What is the optimal base doping?

P7.18 This problem compares the transconductances of the npn BJT and an n-channel MOSFET. The MOSFET has $\mu_n C_{ox} = 100$ μA/V^2, $V_{Tn} = 1.0$ V, and $(W/L) =$ (30 μm/3 μm). The MOSFET is biased in saturation (note that this is equivalent to the BJT being forward active) with $I_D = 125$ μA.

(a) What is the collector current I_C such that the bipolar transistor has the same transconductance as the MOSFET?

(b) If the bipolar transistor is biased with $I_C = 125$ μA, what must the width of the MOSFET become in order to match the transconductance of the BJT? The channel length is maintained at $L = 3$ μm.

(c) Plot the transconductance of the BJT and that of the MOSFET with the original $(W/L) = (30/3)$ versus collector/drain current on a log-log plot. The linear dependence of g_m on collector current is an inherent advantage of the bipolar transistor over the MOSFET.

P7.19 For many signal processing applications, a linear relation between the input and output signals is needed. We found that bipolar transistors have significant second harmonic distortion if v_{be} becomes larger than around 5 mV. Fortunately, we will develop differential amplifiers that cancel the $(v_{be})^2$ term. However, the cubic term cannot be removed using this approach and the resulting third harmonic distortion must be kept as small as possible.

(a) Find an expression for the component of the small-signal collector current i_c that is proportional to $(v_{be})^3$.

(b) For a bias current $I_C = 100$ μA at room temperature, plot on a log-log scale the ratio of the cubic term to the linear term $(g_m\, v_{be})$ in i_c versus v_{be} over the range 1 μV $< v_{be} <$ 50 mV. If this ratio, called the third harmonic distortion, must be less than a 10^{-3}, what is the maximum permissible amplitude of the small-signal base-emitter voltage?

P7.20 Consider the vertical npn transistor with dimensions and doping levels described in P7.1. In this problem, we consider the effect of minority carrier recombination. The hole lifetime in the emitter is $\tau_{pE} = 20$ ns and the electron lifetime in the base is $\tau_{nB} = 50$ ns.

(a) What is the base transport factor α_T for forward-active operation? Show that the short-base solutions for the carrier concentrations are adequate in the base and emitter.

(b) What is the forward-active current gain β_F?

(c) What is the reverse-active current gain β_R? Show that the short-base solutions for the carrier concentrations are adequate in the base and collector.

P7.21 A vertical npn bipolar transistor was initially designed for a forward current gain of $\beta_F = 200$, without considering minority carrier recombination in the base. The doping concentrations and dimensions are

emitter: $N_{dE} = 1 \times 10^{19}$ cm^{-3}, $W_E = 500$ nm, $D_{pE} = 5$ cm^2/s

base: $N_{aB} = 1 \times 10^{17}$ cm^{-3}, $D_{nB} = 10$ cm^2/s

(a) Find W_B so that $\beta_F = 200$. Neglect recombination in the base.

(b) Find the minimum minority electron lifetime τ_{nB} in the base so that the current gain $\beta_F > 100$.

(c) What is the new base width W_B to recover $\beta_F = 200$ using the minimum electron lifetime from (b).

P7.22 The flux picture is helpful in keeping track of minority and majority carrier fluxes, especially when recombination currents are significant.

(a) Draw the flux picture of the vertical npn bipolar transistor in reverse-active operation. Account for the recombination base current, as well as the diffusion of holes into the n$^+$ polysilicon interconnection. (see Fig. 7.24 for the forward-active flux picture).

(b) Draw the flux picture of the vertical npn bipolar transistor in saturation. Account for the recombination base current, as well as the diffusion of holes into the n$^+$ polysilicon interconnection.

P7.23 The general solution of the differential equation for minority electron concentration in the base of an npn transistor leads to an expression similar to Eq. (6.121).

(a) Find the general solution for the minority electron concentration in the base $n_p(x)$ for forward-active bias.

(b) Find the electron diffusion current at the emitter side of the base ($x = 0$) and at the collector side of the base ($x = W_B$).

(c) For the case where $L_{nB} \gg W_B$, the electron concentration is approximately a linear profile. Show that the recombination base current I_R in Eq. (7.90) is the difference between the electron diffusion currents at the two sides of the base from your answers to (b).

P7.24 One way to improve the current gain of a vertical npn transistor is to lengthen the emitter. This problem explores the limits to this approach, for a transistor with the following parameters and dimensions

emitter: $N_{dE} = 1 \times 10^{19}$ cm^{-3}, $D_{pE} = 5$ cm^2/s, $\tau_{pE} = 20$ ns

base: $N_{aB} = 1 \times 10^{17}$ cm^{-3}, $D_{nB} = 10$ cm^2/s, $W_B = 0.3$ μm, $\tau_{nB} = 50$ ns

(a) What is the minority hole diffusion length in the emitter?

(b) Plot the current gain as a function of the emitter length W_E over the range
0.25 μm $< W_E <$ 5 μm.

P7.25 The layout of a vertical npn bipolar transistor with two emitters is shown in Figure P7.25. The area of the first emitter is $A_{E1} = 2.5\ \mu m \times 10\ \mu m$ and the area of the second emitter is $A_{E2} = 5\ \mu m \times 10\ \mu m$. The cross section is similar to Fig. 7.1 but with two emitters.

(a) Find the collector current as a function of V_{BE1} and V_{BE2} for the case where $V_{BC} = -2$ V.

(b) A saturation current I_{S1} and I_{S2} can be found for each emitter. Find the ratio I_{S2}/I_{S1}. Suggest a symbol for this four-terminal device.

P7.26 Although the lateral pnp transistor is fundamentally a two-dimensional device, we can ensure that the "long emitter" is valid by making the emitter region much wider than the electron diffusion length. A lateral pnp structure has the following doping concentrations, dimensions, and parameters:

emitter: $N_{aE} = 1 \times 10^{19}$ cm^{-3}, $W_E = 15\ \mu m$, $D_{nE} = 10$ cm^2/s, $\tau_{nE} = 10$ ns

base: $N_{dB} = 2 \times 10^{17}$ cm^{-3}, $D_{pB} = 5$ cm^2/s, $\tau_{pB} = 25$ ns

collector: $N_{aC} = 1 \times 10^{19}$ cm^{-3}, $W_C = 15\ \mu m$, $D_{nC} = 10$ cm^2/s, $\tau_{nE} = 10$ ns

The sidewall emitter area is $A_E = 50\ \mu m^2$ and the injection from the bottom of the emitter diffusion can be neglected.

(a) In order to achieve a forward-active current gain $\beta_F = 25$, what base width W_B is required?

(b) The Early voltage is lower than would be expected for such a wide base, because of the relatively heavy collector doping, compared with that of the base. Estimate the value of V_{Ap} for the operating point $I_C = -50\ \mu A$, $V_{EC} = 2$ V, using the technique in P7.14.

P7.27 The layout in Fig. P7.27 shows a split-collector lateral pnp transistor, which is a very useful device in analog IC design. There are two C shaped, p-type regions surrounding the p$^+$ emitter diffusion. In order to make good ohmic contacts, p$^+$ areas are formed under the aluminum contacts in the p-type collector regions.

(a) From the layout in Fig. P7.27 and the flux picture for the lateral pnp in Fig. 7.26, what is the ratio of the collector currents I_{C1}/I_{C2} in this split-collector lateral pnp? Suggest a symbol for this four-terminal transistor.

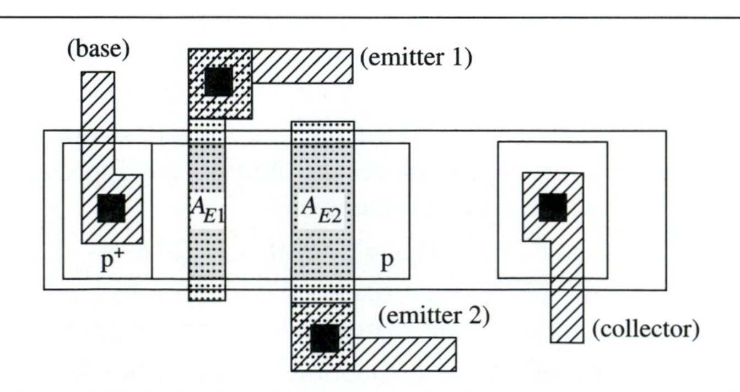

➤ **Figure P7.25**

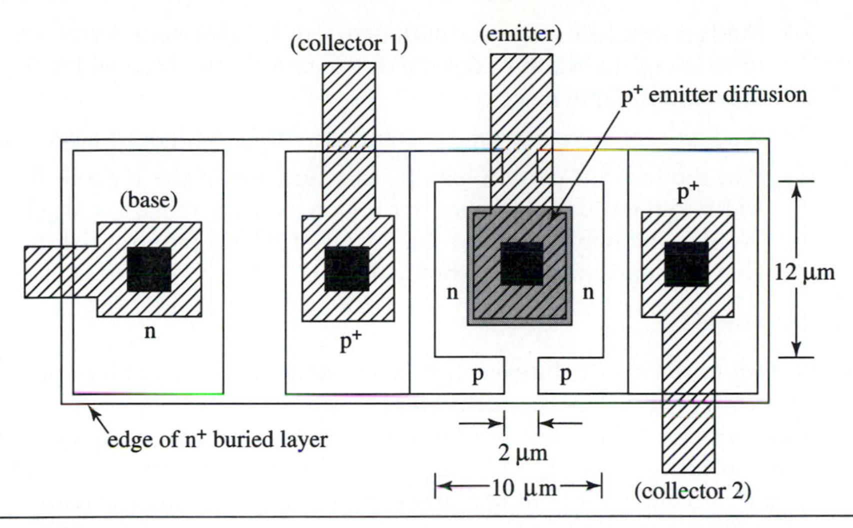

➤ Figure P7.27

(b) There is a penalty for splitting the collector: some carriers are lost by diffusion through the gaps between the "C" shaped collectors. From the layout in Fig. P7.27, estimate the ratio of $I_{C1} + I_{C2}$ to the collector current for the case where a single ring-shaped collector diffusion is used.

P7.28 A simple model for finding the current gain of a lateral bipolar transistor is developed in this problem. Figure P7.28 shows the cross section of the BJT, which is approximated as a rectangular tub with sidewalls of depth $d = (\pi/2)x_j$, where x_j is the junction depth. The emitter doping concentration for the sidewall emitter is taken to be a factor of 10 larger than that of at the bottom of the diffusion. As a further approximation, we neglect base recombination ($\alpha_T = 1$) for the holes diffusing laterally and assume that *all* of the holes injected from the bottom of the emitter diffusion recombine in the base.

The doping concentrations and dimensions are given in Fig. P7.28. For simplicity, we will neglect recombination in the emitter and consider the effective emitter width to be $W_E = W/2$. The electron diffusion coefficient in the emitter is $D_{nE} = 10$ cm²/s and the hole diffusion coefficient in the base is $D_{pB} = 7.5$ cm²/s.

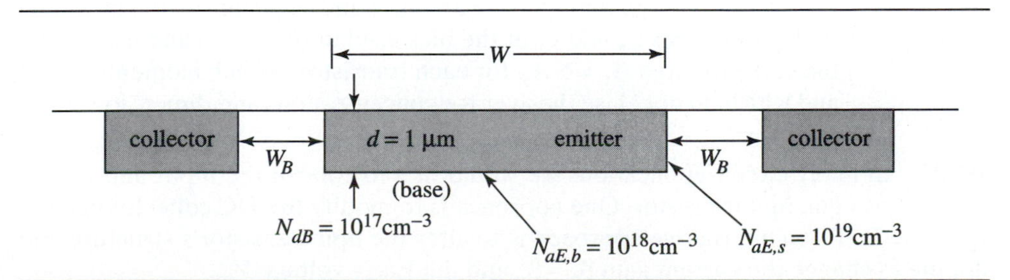

➤ Figure P7.28

(a) Find an expression for the emitter current I_E in forward-active bias, as a function of the width of the emitter region W, the base width W_B, and the emitter depth d.

(b) Find an expression for the collector current I_C in forward-active bias.

(c) Plot the forward current gain β_F as a function of the ratio of the base width to the width of the emitter region over the range $1 < W_B/W < 25$. What is the implication for the layout of lateral pnp transistors, if the goal is to maximize the current gain?

DESIGN PROBLEMS

D7.1 A high-performance bipolar transistor technology for precision analog applications is intended to have perfectly complementary vertical npn and pnp bipolar transistors, with $\beta_{Fn} = \beta_{Fp} = 100$. For this problem, consider that the minority electron diffusion coefficient is $D_n = 10\ \text{cm}^2/\text{s}$ and that the minority hole diffusion coefficient is $D_p = 5\ \text{cm}^2/\text{s}$ for both npn and pnp devices. Also, the collector doping level is $10^{16}\ \text{cm}^{-3}$ for both devices.

The emitter and base doping levels can range from a minimum of $5 \times 10^{16}\ \text{cm}^{-3}$ to a maximum of $2 \times 10^{19}\ \text{cm}^{-3}$. The emitter and base widths can range from a minimum of 250 nm to a maximum of 1 μm.

(a) Find dimensions and doping levels in the emitter and base of the npn and pnp devices that satisfy the requirement that $\beta_{Fn} = \beta_{Fp} = 100$.

(b) If the normalized variation in the emitter and base width is 0.05 and the normalized variation in the doping levels is 0.025, find the variations $\varepsilon_{\beta_{Fn}}$ and $\varepsilon_{\beta_{Fp}}$ in the current gains β_{Fn} and β_{Fp}. You can assume for simplicity that all of the parameter and dimensional variations are independent.

(c) The current gains from the 10,000 chips in a wafer lot are found to be uniformly distributed between $\beta_F(1 + \varepsilon_{\beta_F})$ and $\beta_F(1 - \varepsilon_{\beta_F})$ for both npn and pnp transistors. (This is an artificial distribution that simplifies the math.) A "good" chip must have $95 < \beta_{Fn}, \beta_{Fp} < 105$. From your results in (b), what is the number of good chips from this wafer lot? The yield loss results in an expensive IC that is used only for special-purpose applications.

(d) If the two transistors are biased with $I_{Cn} = 100\ \mu\text{A} = -I_{Cp}$, find the emitter areas of the npn and pnp BJTs, so that the minority carrier concentrations at the edges of the base-emitter depletion region are identical $n_p(0) = p_n(0) = 10^{14}\ \text{cm}^{-3}$.

(e) Find the small-signal models for the npn and pnp transistors, including the capacitances C_π and C_μ at the bias used in (d). You can assume that the collector area $A_C = 5\ A_E$ for each transistor. Which elements match and which do not? Use the average concentrations and dimensions from (a).

D7.2 In many circuit applications we would like to control the input and output resistances of an npn transistor. One approach is to modify the DC collector current I_C of the BJT. An alternative approach is to alter the npn transistor's structure and doping to change the current gain $\beta_F = \beta_o$ and the Early voltage V_{An}.

(a) How should N_{dE} and N_{aB} be changed to increase both the input resistance r_π and the output resistance r_o.

(b) How should N_{dE} and N_{aB} be changed to decrease both the input resistance r_π and the output resistance r_o.

(c) How should N_{dE} and N_{aB} be changed to decrease the input resistance r_π and increase the output resistance r_o.

(d) How should N_{dE} and N_{aB} be changed to increase the input resistance r_π and decrease the output resistance r_o.

(e) If the doping concentrations are fixed, which of these combinations can be achieved by changing the DC collector current I_C?

D7.3 A lateral pnp transistor can be fabricated in an advanced n-well CMOS process, by using the well as the base contact. Holes are injected from the p^+ emitter into the region under the gate oxide and diffuse across to the p^+ collector, as shown in Fig. D7.3. For simplicity, assume that all of the holes injected laterally are collected, while the holes that are injected downward from the bottom of the emitter all recombine in the n-well base. For the device dimensions in Fig. D7.3, the ratio of lateral to downward hole current is 25:1.

emitter (sidewall): $N_{aE} = 1 \times 10^{19}$ cm^{-3}, $W_E = 2$ μm, $\tau_{nE} = 100$ ps

base (n-well): $N_{aB} = 1 \times 10^{17}$ cm^{-3}, $W_B = 500$ nm $= L \ll L_{pB}$

collector (sidewall): $N_{aC} = 1 \times 10^{19}$ cm^{-3}, $W_C = 2$ μm, $\tau_{nC} = 100$ ps

The emitter area $A_E = W \cdot x_j = W \cdot (0.8\ \mu\text{m})$, where W is the width of the emitter diffusion (perpendicular to Fig. D7.3—*not* W_E)

(a) Find the forward-active current gain β_F and the reverse-active current gain β_R.

(b) Design the lateral pnp and sketch the layout for a saturation current $I_S = 5 \times 10^{-17}$ A. You should determine the width W of the emitter diffusion and consider folding the transistor in order to save area.

(c) Find the Ebers-Moll model for the lateral pnp transistor in Fig. D7.3 and sketch the collector characteristics.

(d) Find the small-signal model parameters for this transistor.

(e) The polysilicon gate does not play a fundamental role in the operation of the lateral pnp transistor. However, the gate can be contacted and biased. Suggest the polarity of the bias that would optimize the lateral pnp's performance.

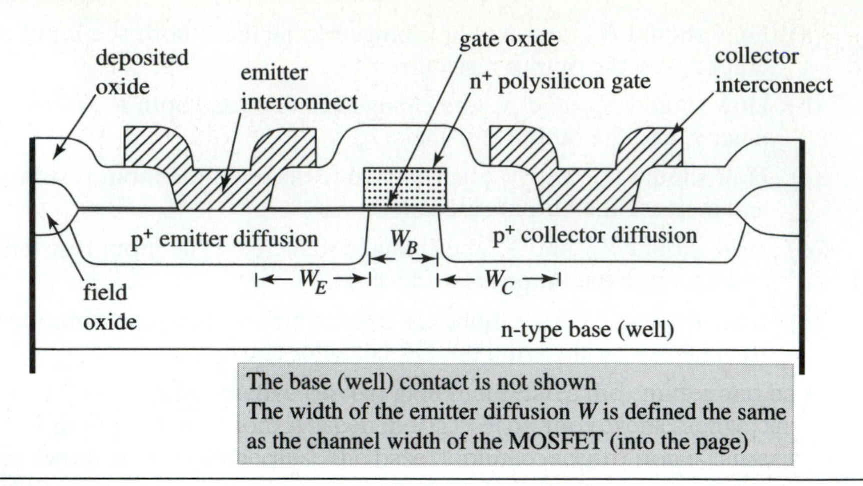
deposited
oxide
emitter
interconnect
gate oxide
n⁺ polysilicon gate
collector
interconnect
p⁺ emitter diffusion
W_B
p⁺ collector diffusion
W_E
W_C
field
oxide
n-type base (well)
The base (well) contact is not shown
The width of the emitter diffusion W is defined the same
as the channel width of the MOSFET (into the page)

➤ Figure D7.3

chapter 8

Single-Stage Bipolar/MOS Transistor Amplifiers

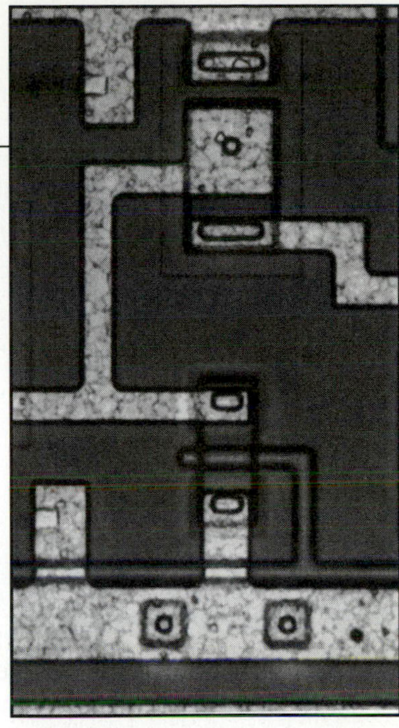
Common-collector amplifier with a MOSFET current-source supply. Designed by Trey A. Roessig, UC Berkeley using the BiMEMS process (Courtesy of Analog Devices Inc.)

The active devices we have studied, namely the MOS and bipolar transistors, can be used in a variety of circuit configurations that have voltage or current as an input and return voltage or current as an output. There are four types of amplifiers that are defined by their input and output parameters.

- Voltage amplifier—both input and output are voltages
- Current amplifier—both input and output are currents
- Transconductance amplifier—input is a voltage and output is current
- Transresistance amplifier—input is a current and output is a voltage

Before analyzing specific amplifier circuits, we will discuss a generalized amplifier structure, a consistent set of notation, and small-signal, two-port amplifier models that will be developed for the four types of amplifiers. We will then study the use of MOS and bipolar transistors to form these amplifier types. Several simplifying assumptions will be made allowing us to see the effects of the device and circuit parameters on the performance characteristics of various amplifiers. It is important to have a thorough understanding of each of these single-stage amplifiers so they can be recognized when embedded in more complex circuits. You should be able to follow the signals through various stages so the overall transfer function of the circuit can be estimated with simple hand analysis. In modern integrated circuit design, detailed analysis is relegated to computer simulation programs such as SPICE. However, an "engineering feel" or intuition of how the circuit and device parameters affect the amplifier performance characteristics is essential in determining the initial circuit design.

Chapter Objectives

- A systematic method for analyzing transistor amplifier circuits.
- Circuit analysis with a current source to supply the current to the amplifier.
- The effect of the internal resistance of the current source on the overall amplifier characteristics.
- Estimating the intrinsic gain and the input/output resistance of the various amplifier configurations.
- The dependence of the input and output resistance and various gain parameters on the supply current and device parameters.

8.1 GENERAL AMPLIFIER CONCEPTS

This section reviews the general concepts regarding amplifier design. The concepts are not specific to any one amplifier topology or use of any specific device. You should be comfortable with these basic concepts since they provide the framework for the rest of the chapter.

A review of the notation used in this text is presented before starting our discussion. Recall that total signals are composed of the sum of DC quantities and small signals. For example, a total input voltage v_{IN} is the sum of a DC input voltage V_{IN} and a small-signal voltage v_{in}. The notation is summarized below.

- Total quantity has a lower case variable name and upper case subscript.
- DC quantity has an upper case variable name and upper case subscript.
- Small-signal quantity has a lower case variable name and lower case subscript.

8.1.1 Generalized Amplifier Structure

Fig. 8.1 shows a generalized amplifier structure. In the center of the diagram, we find an intrinsic amplifier composed of an active device whose device current labeled i_D is controlled by the input signal. Current is supplied to one of the terminals of the active device by a constant DC current called the **supply current** I_{SUP}. Depending on the specific implementation, the active device may be a bipolar or MOS transistor. The supply current may be derived from a resistor tied to the power supply or from a current source.

The input to the amplifier on the left side of the generalized diagram may either be a voltage or a current. If the input is a voltage, then a series combination of voltage sources is the input. The first voltage source is a **DC bias voltage** called V_{BIAS}, that has a value chosen so that the DC current flowing into the active device I_D is precisely the current being supplied from I_{SUP}. When these currents are exactly equal, the DC output current is equal to zero. In series with the DC bias voltage source is a small-signal voltage source v_s and its associated source resistance R_S. The small-signal voltage source perturbs the input voltage to the active device, which in turn perturbs the total device current i_D. Since the supply current is a constant equal to I_D, the output current i_{OUT}, will be equal to the small-signal device current i_d.

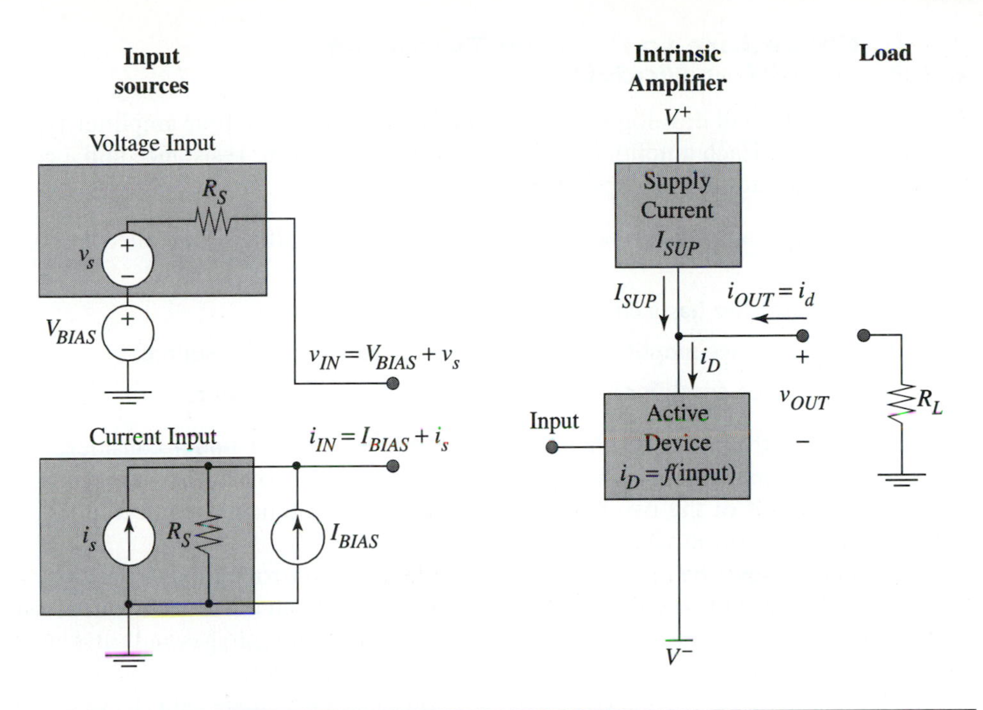

Figure 8.1 General amplifier including input, intrinsic amplifier, and load.

If the input to the general amplifier is a current, a **DC bias current** I_{BIAS} is fed into the input so the DC current flowing into the active device I_D is precisely the current being supplied to the device I_{SUP}. In parallel with the DC bias current source is a small-signal current source and its associated source resistance R_S. The small-signal input current i_s perturbs the total device current i_D. The output current is the small-signal device current i_d.

We attach a **load resistor** R_L at the output of the amplifier. The effect of R_L on the small-signal response of our amplifiers will be studied. Often the important quantity of interest is the small-signal **open-circuit output voltage** that is defined as the small-signal output voltage with the load resistance $R_L \to \infty$. Another important quantity at the output is the small-signal **short-circuit output current**. This small-signal output current flows when the load resistor R_L is set equal to zero.

In many integrated circuit applications it is customary to have only one power supply voltage such as 5 V. Under this condition the positive power supply V^+ is 5 V, the negative power supply V^- is 0 V, and $I_{OUT} = 0$ A when $I_{SUP} = I_D$, as with the dual supply case. The constraint on the choice of voltage for V_{OUT} is the requirement that the active device and current source supply operate in their constant-current regions. Therefore, V_{OUT} should be centered between V^+ and V^- for proper operation. When calculating the quiescent operating point, the input small-signal source and its associated resistance are removed from the circuit. We also neglect the effect of R_L for the DC calculation since under proper bias ($I_D = I_{SUP}$) the current through R_L is zero.

The operation of this generalized amplifier can be summarized as follows. The input bias voltage or current sets the device current to a value that is equal to the supply current. The small-signal input voltage or current source linearly varies the device current and generates an output current and a corresponding output voltage that are proportional to the input voltage or current.

8.1.2 Procedure to Develop Two-port Amplifier Models

In this section we will develop two-port models for each of the four amplifier types shown in Fig. 8.2. Each amplifier type has an input and output resistance and a controlled source to model the amplification.

- In the **voltage amplifier**, the controlled source is a voltage-controlled voltage source.
- A **current amplifier** has a current-controlled current source.
- A **transconductance amplifier** has a voltage-controlled current source.
- A **transresistance amplifier** has a current-controlled voltage source.

To calculate the transfer function of the specific amplifier under consideration the small-signal input voltage or current source, including its source resistance, is connected to the input of the two-port model. The load resistance is connected to the output of the two-port model.

We will now describe the procedure to calculate the controlled sources, as well as the input and output resistances, for the two-port models from the small-signal device models and any other components in the circuit. We apply test voltages and currents to the small-signal model of the circuit to find the two-port parameters.

Referring to Fig. 8.3(a-d), to calculate the **open-circuit voltage gain** A_v, we apply a test voltage at the input with zero source resistance and measure the open-circuit output voltage. To measure the **short-circuit current gain** A_i, we apply a test current source with an infinite source resistance and measure the short-circuit output current. To calculate the **transconductance**, G_m we apply a test voltage source with zero source resistance and measure the short-circuit output current. Finally, to measure the **transresistance**, R_m we apply a test current source with an infinite source resistance and measure the open-circuit output voltage.

To calculate the **input resistance** R_{in} of the amplifier of interest, we apply a test voltage and measure the current coming from the test source, or apply a test current and measure the voltage across the test source. When calculating the input resistance, the load resistance that the amplifier will drive should be connected to the amplifier as shown in Fig. 8.3(e). To calculate the **output resistance** R_{out}, we apply either a test voltage or a test current source and measure the respective current or voltage from the source at the output. When calculating the output resistance, the input source should be set equal to zero. Input voltage sources are shorted; input current sources are open-circuited resulting in only the source resistance being placed across the input terminals as shown in Fig. 8.3(f).

Some amplifier configurations such as the common-emitter and common-source amplifiers are unilateral networks. These amplifier configurations will be discussed later. An amplifier is a **unilateral network** when the input resistance is independent of the load attached to the output and the output resistance is independent of the source resistance. In other configurations the amplifiers may not be unilateral. In those cases, the load resistance can affect the input resistance and the source resistance may affect the output resistance. That is why R_S is left in the circuit to find R_{out} and R_L is left in the circuit to find R_{in}.

The relationships between the various controlled sources and the input/output resistances, with the circuit and device parameters are of primary importance. These

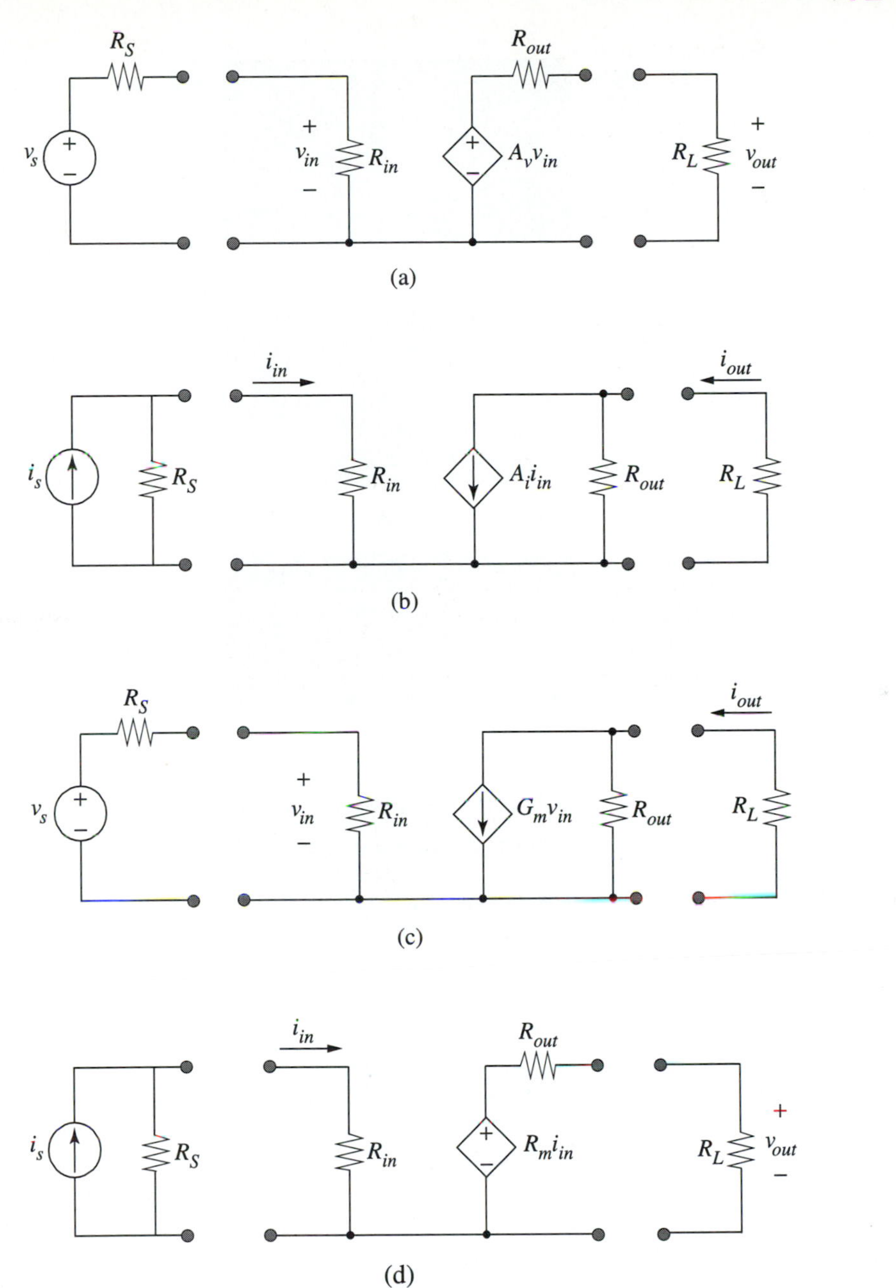

➤ **Figure 8.2** Two port amplifier models with input source and load: (a) voltage amplifier; (b) current amplifier; (c) transconductance amplifier; and (d) transresistance amplifier.

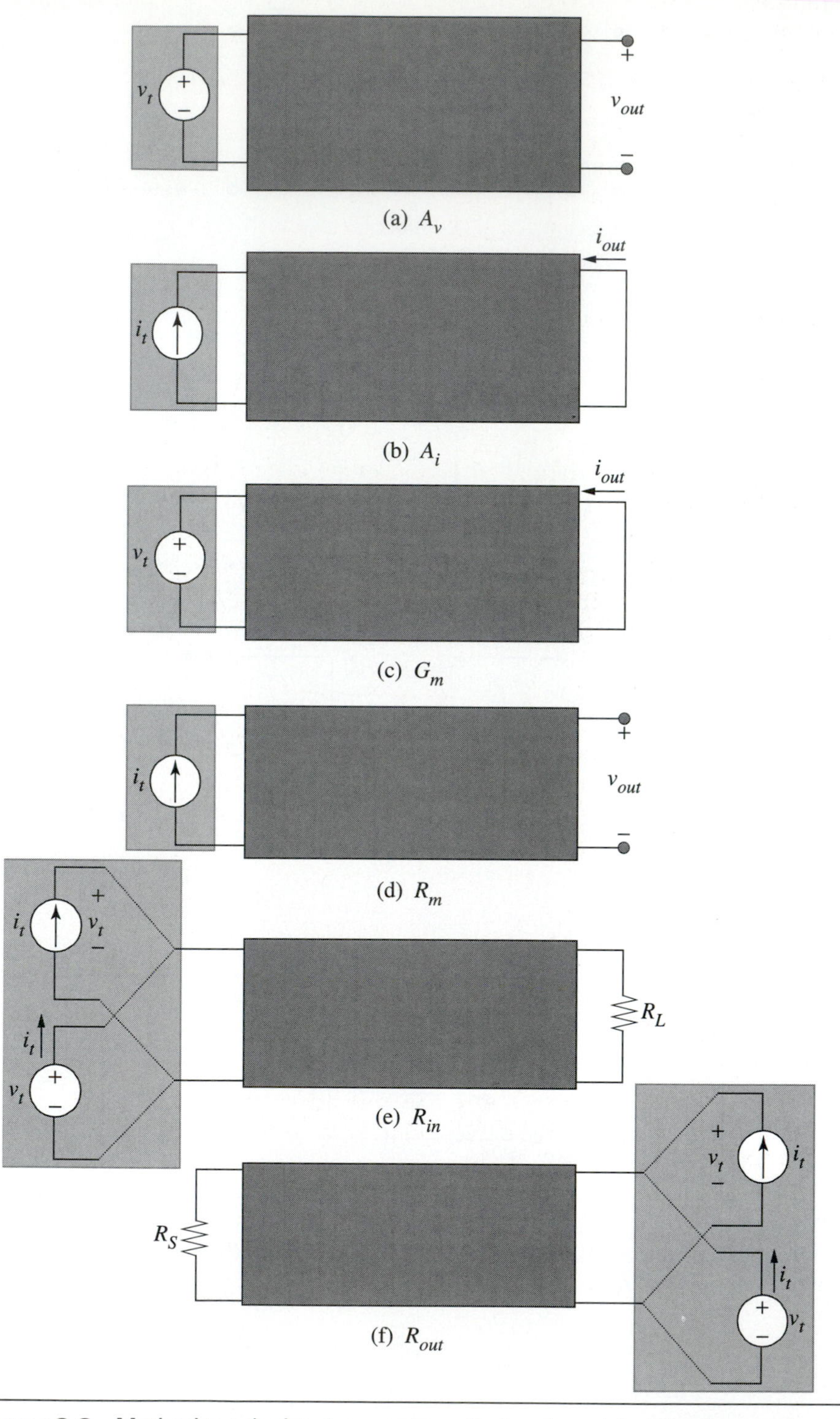

➤ **Figure 8.3** Method to calculate two-port small-signal models: (a) voltage gain A_v, (b) current gain A_i, (c) transconductance G_m, (d) transresistance R_m, (e) input resistance R_{in} , and (f) output resistance R_{out}.

relationships allow the circuit designer to understand how to change either the device or circuit design to improve performance.

8.1.3 Effect Of Source and Load Resistances

Two-port models for amplifiers include an input resistance, an output resistance, and a controlled source depending on the type of amplifier. In the next sections, we will determine their values for various amplifier configurations using the MOS and bipolar devices. We begin by examining how the overall small-signal transfer characteristic of an amplifier is degraded by the addition of source and load resistances.

Fig. 8.4 shows an **intrinsic voltage amplifier** with an input voltage source and associated source resistance R_S driving a load resistor R_L.

The overall voltage gain of this amplifier is shown in

$$\frac{v_{out}}{v_s} = \left(\frac{R_{in}}{R_{in} + R_S}\right) A_v \left(\frac{R_L}{R_L + R_{out}}\right) \tag{8.1}$$

The voltage gain has been degraded by the voltage division between the source resistance and input resistance, and also by the voltage division between the load resistance and output resistance. Equation 8.1 shows that the input resistance should be much greater than the source resistance and the output resistance much less than the load resistance in order to transfer the largest output voltage from the input voltage source.

A transconductance amplifier has a transfer function where current is the output parameter. This is determined by taking the Norton equivalent of the output of the intrinsic voltage amplifier as shown in Fig. 8.5. The overall transconductance of this amplifier is given by

$$\frac{i_{out}}{v_s} = \left(\frac{R_{in}}{R_{in} + R_S}\right) G_m \left(\frac{R_{out}}{R_L + R_{out}}\right) \tag{8.2}$$

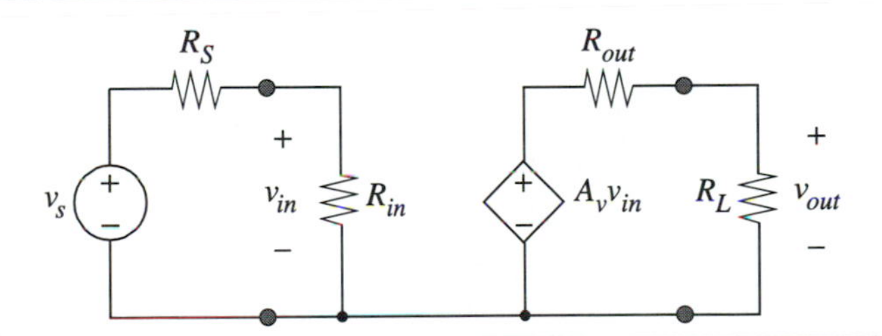

➤ Figure 8.4 Two-port small-signal model of an intrinsic voltage amplifier with a voltage source input and load resistor.

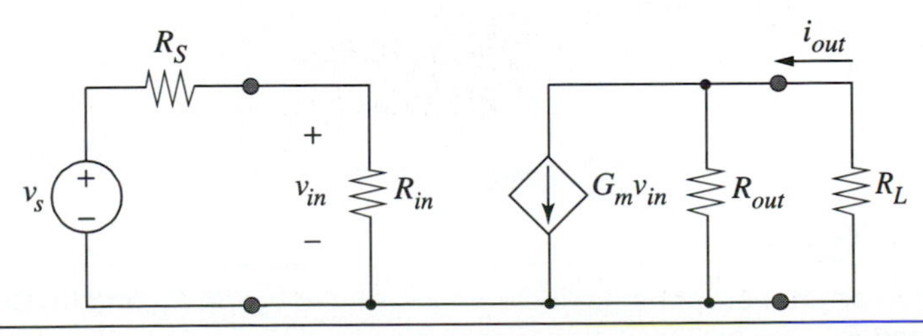

➤ Figure 8.5 Two-port small-signal model of an intrinsic transconductance amplifier with a voltage source input and load resistor.

A large overall transconductance is obtained when the intrinsic amplifier has a large input resistance and a large output resistance.

➤ 8.2 COMMON-EMITTER (CE) AMPLIFIER

The common-emitter amplifier is composed of a bipolar transistor that is biased in the forward-active region. The input signal is driven between the base and emitter of the transistor, and the output signal is taken between the collector and emitter, hence the term **common-emitter amplifier**. As you will see, the common-emitter and common-source amplifiers are natural transconductance amplifiers, meaning their input and output resistances are relatively large. We will analyze these amplifiers with that in mind. One possible circuit implementation of a common-emitter amplifier is shown in Fig. 8.6(a). In this circuit the source resistance is associated with the small-signal input voltage v_s.

8.2.1 Large-Signal Analysis of CE Amplifier

Graphical analysis will be used to understand, qualitatively, the large-signal operation of the common-emitter amplifier. For the large signal analysis, we set our small-signal voltage source and its associated source resistance equal to zero and remove the load resistor as shown in Fig. 8.6(b). Typical bipolar DC transistor characteristics are shown in Fig. 8.7.

On the horizontal axis we plot V_{CE} which is equal to the output voltage V_{OUT}. On the vertical axis we plot the DC collector current I_C. The parameter that is changing is the input bias voltage V_{BIAS}. Because the bipolar transistor collector current is expo-

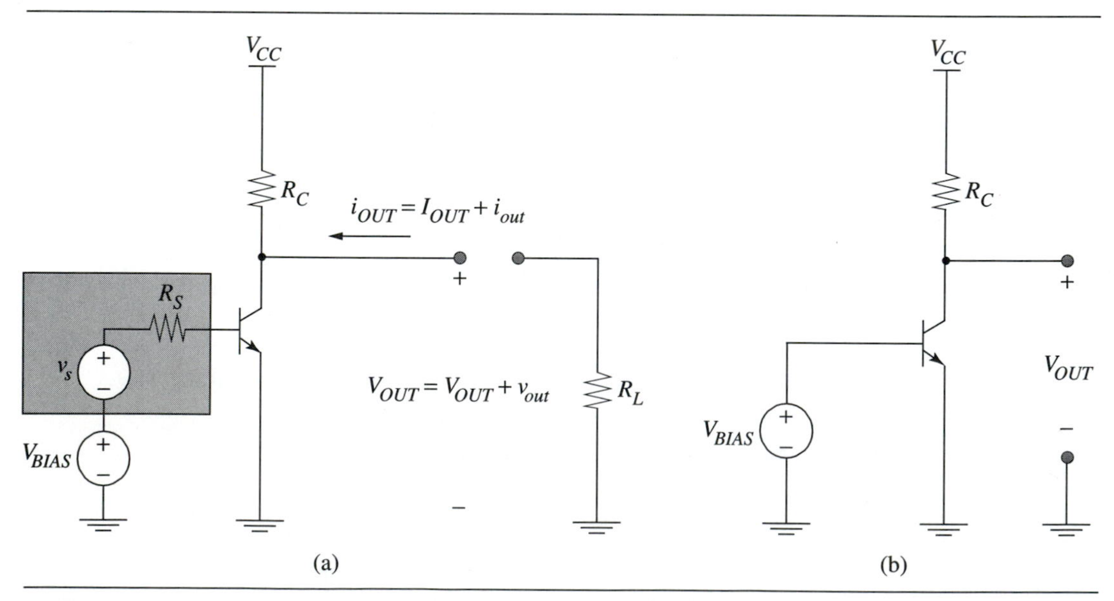

➤ Figure 8.6 Common-emitter amplifier (a) including R_S due to v_s and R_L, and (b) CE amplifier with $v_s = 0$ V, $R_S = 0\ \Omega$ and $R_L \to \infty$ for large-signal analysis.

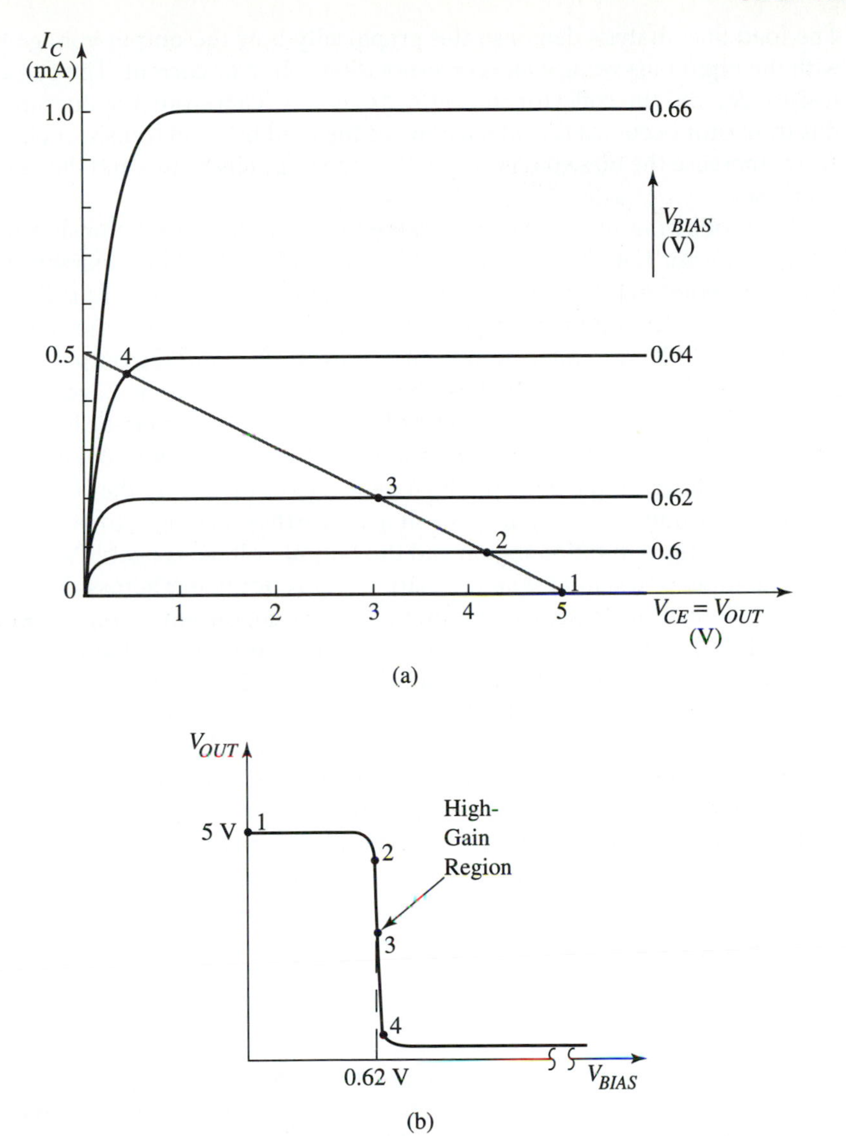

➤ Figure 8.7 (a) Bipolar transistor characteristics for common-emitter configuration with load line. This graph assumes $V_{CC} = 5$ V and $R_C = 10$ kΩ. (b) Common-emitter amplifier transfer characteristic.

nentially dependent on the base-emitter voltage, we see that a very small change in the bias voltage causes a large change in the collector current.

On this same graph we plot a **load line**[1]. This line is a graph of

$$V_{OUT} = V_{CC} - I_C R_C \tag{8.3}$$

1. In our terminology we would call this the *supply current* line. However, the traditional load line terminology is used by most circuit designers. The student should note that resistor R_C is *not* the load.

The load line analysis demonstrates graphically how the output voltage will change with the input bias voltage and corresponding collector current. The current through resistor R_C and the collector current I_C are equal. Therefore, the operating point for this transistor occurs at the intersection of the load line and transistor characteristics. As we increase the bias voltage, the collector current increases and the output voltage decreases.

The following observations will allow you to see the relationship between the voltage transfer function shown in Fig. 8.7(b) and the load line characteristics in Fig. 8.7(a). At point (1) the bipolar transistor is cutoff and no current is flowing. Under this condition the output voltage is equal to V_{CC}. As we increase the bias voltage toward point (2), the collector current begins to flow and the output voltage begins to fall below V_{CC}. At point (3), if we slightly increase the bias voltage the output voltage falls rapidly. This high-gain region of the transfer characteristic is where the bipolar transistor is operating in the constant current region. Our goal is to operate the common-emitter amplifier in the high-gain region by setting the bias voltage so we operate near point (3). As the bias voltage is further increased to point (4), the transistor enters the saturation region and the output voltage is near 0 V.

The input and output voltage relationship is not linear across these large-signal changes. In fact, if we approach either extreme, the output voltage does not appreciably change with input voltage. Therefore, we must ensure that the bias voltage is chosen such that when a small-signal input voltage is added to the bias voltage, the output voltage operates roughly halfway between V_{CC} and ground. Once the DC bias voltage is chosen, the collector current and voltage values are determined. We can then add a small-signal voltage input to vary around this bias point and find how the output current responds to this change. The ratio of the small-signal output current to the small-signal input voltage is the amplifier's transconductance.

To quantify the selection of the bias point we set the input small-signal voltage source equal to 0 Vwhich implies a short circuit. We now use the Ebers Moll Model to model the transistor where

$$I_C = I_S(e^{qV_{BE}/kT} - 1) - \frac{I_S}{\alpha_R}(e^{qV_{BC}/kT} - 1) \tag{8.4}$$

Fig. 8.7 shows that we want to operate the transistor in the forward active region, which implies that the base-emitter junction will be forward biased and the base-collector junction reverse biased. Under this simplification, Eq. (8.4) becomes

$$I_C \approx I_S(e^{qV_{BE}/kT}) = I_S(e^{qV_{BIAS}/kT}) \tag{8.5}$$

We can use Eq. (8.5) to calculate either the collector current or the bias voltage necessary to ensure the amplifier output voltage operates approximately halfway between the power supply V_{CC} and ground.

➤ EXAMPLE 8.1 Common-Emitter Amplifier: Large-Signal Analysis

We would like the DC operating point of the common-emitter amplifier to have an output voltage of 2.5 V which is halfway between the positive power supply of 5 V and ground. Given that the bipolar transistor has $I_S = 10^{-15}$ A and the value of R_C is 10 kΩ, find analytically and graphically the collector current I_C and the bias voltage V_{BIAS}.

SOLUTION

ANALYTICAL METHOD:

$$V_{OUT} = V^{+} - I_C R_C$$

$$I_C = (V^{+} - V_{OUT})/R_C = (5\ \text{V} - 2.5\ \text{V})/10\ \text{k}\Omega = 250\ \mu\text{A}$$

$$I_C = I_S e^{\frac{qV_{BIAS}}{kT}}$$

$$V_{BIAS} = \frac{kT}{q}\ln\frac{I_C}{I_S} = 0.682\ \text{V}$$

GRAPHICAL METHOD:

We use SPICE to plot I_C vs. V_{CE}. Next we find where the load line for resistor R_C intersects I_C at $V_{OUT} = V_{CE} = 2.5$ V. Then we read the values for I_C and V_{BIAS} from the graph. We see from the SPICE output that the optimal setting of V_{BIAS} is 675 mV, which is close to our calculation of 682 mV. The difference is due to the exact temperature chosen, which affects the thermal voltage.

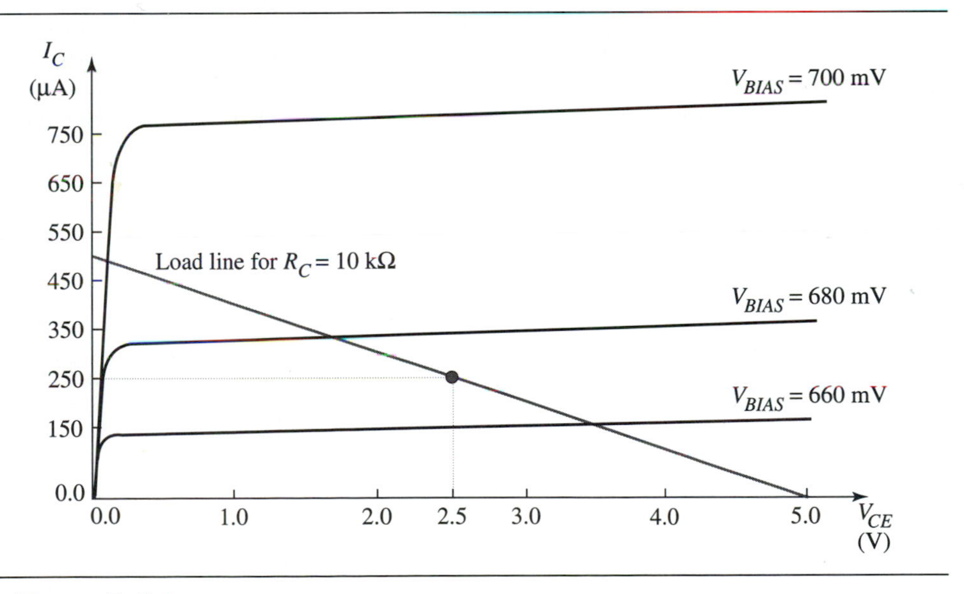

➤ **Figure Ex8.1**

8.2.2 Small-Signal Analysis of CE Amplifier

Now that the DC bias point for the amplifier is determined, we want to analyze the effect of a small-signal input for this amplifier configuration. Recall that for a small-signal model, all independent DC sources are set equal to zero, which implies that voltage sources are shorted and current sources are open-circuited. Using the small-signal model for a bipolar transistor that was derived in Chapter 7, we develop the small-signal model for the intrinsic common-emitter amplifier. The small-signal models for the intrinsic and loaded common-emitter amplifier are shown in Fig. 8.8. The intrinsic amplifier model does not include source and load resistances, while the loaded amplifier model does.

8.2.3 Two-Port Model for CE Amplifier

The procedure outlined in the previous section will be used to obtain a two-port model of the intrinsic common-emitter amplifier shown in Fig. 8.8(a). Our goal is to find the input and output resistance, as well as the transconductance of the intrinsic common-emitter amplifier.

First we find the intinsic transconductance G_m by placing a test voltage source v_t at the input and shorting the output terminals as described in Fig. 8.3(c). The intinsic transconductance is found to be

$$G_m = \frac{i_{out}}{v_t} = g_m = \frac{qI_C}{kT} \tag{8.6}$$

Next we find the input resistance by attaching a test voltage source to the input. We measure the current out of that voltage source with the load resistor R_L attached to the output of the amplifier as shown in Fig. 8.3(e). Since the common-emitter amplifier is a unilateral network, the load resistor has no effect on the input resistance. Observe that the input resistance is simply to r_π. Finally, we calculate the output resistance of this amplifier by attaching a test voltage source at the output and measuring its current while attaching the source resistance to the input as shown in Fig. 8.3(f). Because no voltage appears across v_π, the transconductance generator has a value of zero and appears as an open circuit. We see that the output resistance is the parallel combination $r_o \| R_C$.

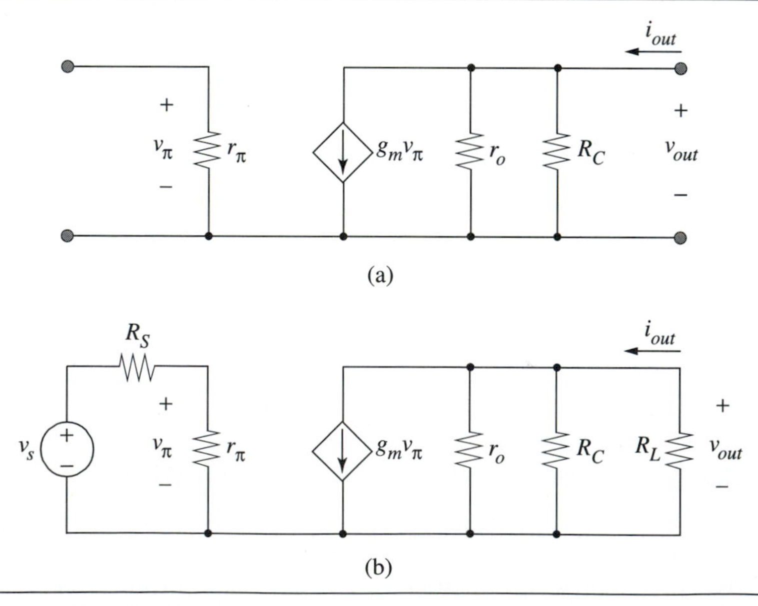

➤ **Figure 8.8** Small-signal model for common-emitter amplifier (a) intrinsic and (b) loaded.

The two-port model for the intrinsic common-emitter amplifier is shown in Fig. 8.9(a). A voltage source with a source resistance can be attached to the input and a load resistor to the output, to calculate how the output voltage and current vary as a function of the input voltage as shown in Fig. 8.9(b).

➤ EXAMPLE 8.2 Common-Emitter Amplifier: Small-Signal Analysis

You are given a common-emitter amplifier with a DC collector current $I_C = 100\ \mu A$ and with $R_C = 25\ k\Omega$. The bipolar transistor has $\beta_o = 100$ and $V_A = 25$ V. Use the two-port models in Fig. 8.9

(a) Find the transconductance of the amplifier when the amplifier is unloaded ($R_S = 0\ \Omega$, $R_L = 0\ \Omega$).

(b) Find the overall transconductance with a source resistance of $25\ k\Omega$.

(c) Find the overall transconductance with a source resistance of $25\ k\Omega$ and a load resistance of $25\ k\Omega$.

SOLUTIONS

(a) $$g_m = \frac{qI_C}{kT} = \frac{100\,\mu A}{25\ mV} = 4\ mS$$

$$r_\pi = \frac{\beta_o}{g_m} = \frac{100}{4\ mS} = 25\ k\Omega$$

$$r_o = \frac{V_A}{I_C} = \frac{25\ V}{100\ \mu A} = 250\ k\Omega$$

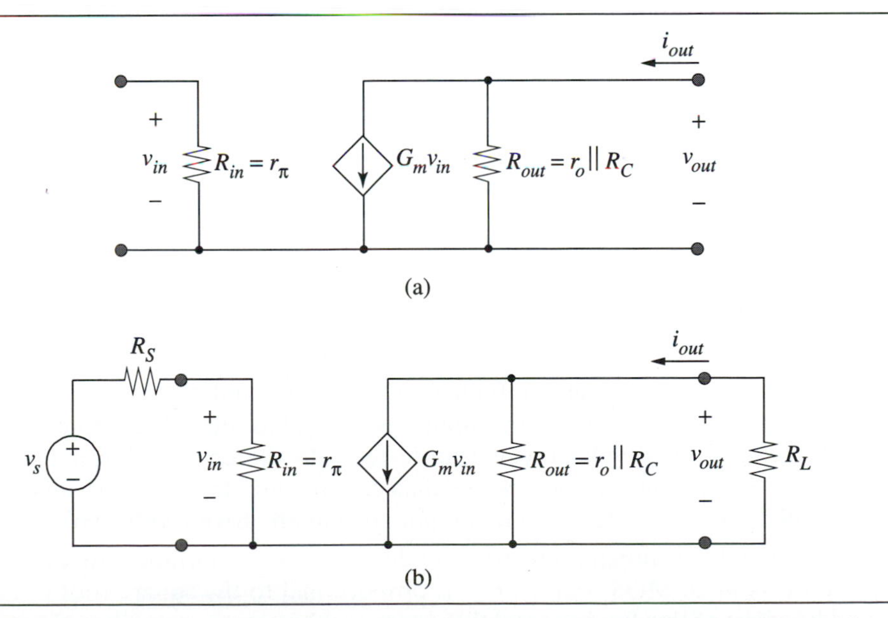

➤ Figure 8.9 Two-port model for the (a) intrinsic common-emitter amplifier and (b) loaded common-emitter amplifier.

Since $R_L = 0\ \Omega$, $i_{out} = g_m v_{in}$, and $v_{in} = v_s$. Thus $\frac{i_{out}}{v_s} = g_m = 4\text{mS}$

(b) With $R_S = 25\ \text{k}\Omega$, the voltage at the source is divided equally between the source resistance and input resistance of the amplifier.

$$v_{in} = \left(\frac{r_\pi}{r_\pi + R_S}\right)v_s = \frac{v_s}{2} \text{ and } i_{out} = g_m v_{in} = \frac{g_m v_s}{2} \text{ so } \frac{i_{out}}{v_s} = \frac{g_m}{2} = 2\text{mS}$$

Half of the transconductance was lost. A large input resistance avoids transconductance degradation for voltage inputs.

(c) Now we will see the effect that a load resistance $R_L = 25\ \text{k}\Omega$ has on the overall transconductance.

$$i_{out} = g_m v_{in} \cdot \left(\frac{R_{out}}{R_{out} + R_L}\right) = g_m v_{in}\left(\frac{(r_o \| R_C)}{(r_0 \| R_C) + R_L}\right)$$

Since $r_o \gg R_C$ and $R_C = R_L$

$$i_{out} \approx \frac{g_m v_{in} R_C}{R_C + R_L} = \frac{g_m v_{in}}{2}$$

From (b) $v_{in} = \frac{v_S}{2}$, so $\frac{i_{out}}{v_s} = \frac{g_m}{4} = 1\ \text{mS}$.

Since our output variable is a current, a larger output resistance avoids degradation. Note that the value of $R_C = 25\text{k}\Omega$ is 10 times less than the output resistance of the bipolar transistor. Later in this chapter we will learn how to avoid the use of a collector resistor and greatly increase the output resistance of the common-emitter amplifier.

8.3 COMMON-SOURCE (CS) AMPLIFIER

Although the MOS transistor is most famous for its use in digital circuits, it can also be used as an amplifying device. In fact, many of the ideas presented in the common-emitter amplifier section can be applied directly to the **common-source amplifier**. There are two major differences between the bipolar and MOS transistor.

1. The MOS transistor has an infinite input resistance and, therefore, is useful only when a voltage is applied at the input. Because it has such a high input resistance, all of the applied voltage gets transferred into the device even if the voltage source has a non-zero source resistance. This infinite input resistance makes MOS devices very attractive for amplifiers that are driven with a voltage source.
2. In a bipolar transistor the transconductance is proportional to the current, whereas in an MOS transistor, it is proportional to the square root of current. The effect of this fundamental difference will be seen in a number of important circuits later in this chapter.

One possible circuit implementation of a common-source amplifier is shown in Fig. 8.10(a). For the large-signal analysis the small-signal input voltage source and its associated source resistance as well as the load resistance are removed as shown in Fig. 8.10(b).

8.3.1 Large-Signal Analysis of CS Amplifier

The qualitative operation of the common-source amplifier can be most easily understood by using the same graphical analysis we performed for the common-emitter amplifier. In Fig. 8.11(a) we show typical NMOS transistor characteristics. On the horizontal axis we plot V_{DS} which is equal to the output voltage V_{OUT}. On the vertical axis we plot the DC drain current I_D and vary the DC input bias voltage as a parameter. The MOS transistor is a **square law** device with a drain current that depends quadratically on the gate-source voltage in the constant-current region. The MOS transistor has a weaker dependence between its output current I_D and its input voltage V_{GS} than the bipolar transistor. In essence, it has a smaller transconductance per unit current.

We plot a load line on the same graph as the MOS transistor characteristics. The equation of this line is given by

$$V_{OUT} = V_{DD} - I_D R_D \tag{8.7}$$

Using the load line analysis, we can graphically determine how the output voltage changes with the input bias voltage and the value of the corresponding drain current. The current through resistor R_D and the drain current I_D are equal. Therefore, the operating point for this circuit occurs at the intersection of the load line and transistor characteristics. The specific operating point is defined by setting either the drain current or the bias voltage.

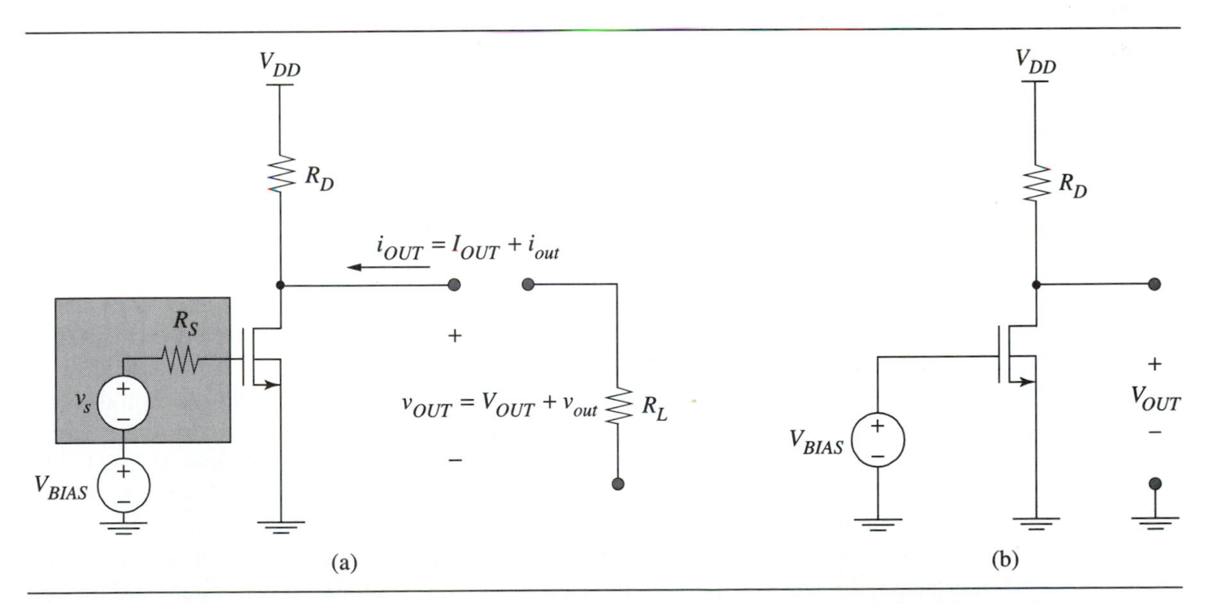

Figure 8.10 Common-source amplifier (a) including R_S due to v_s and R_L (b) CS amplifier with $v_s = 0$ V $R_S = 0\ \Omega$ and $R_L \to \infty$ for large-signal analysis.

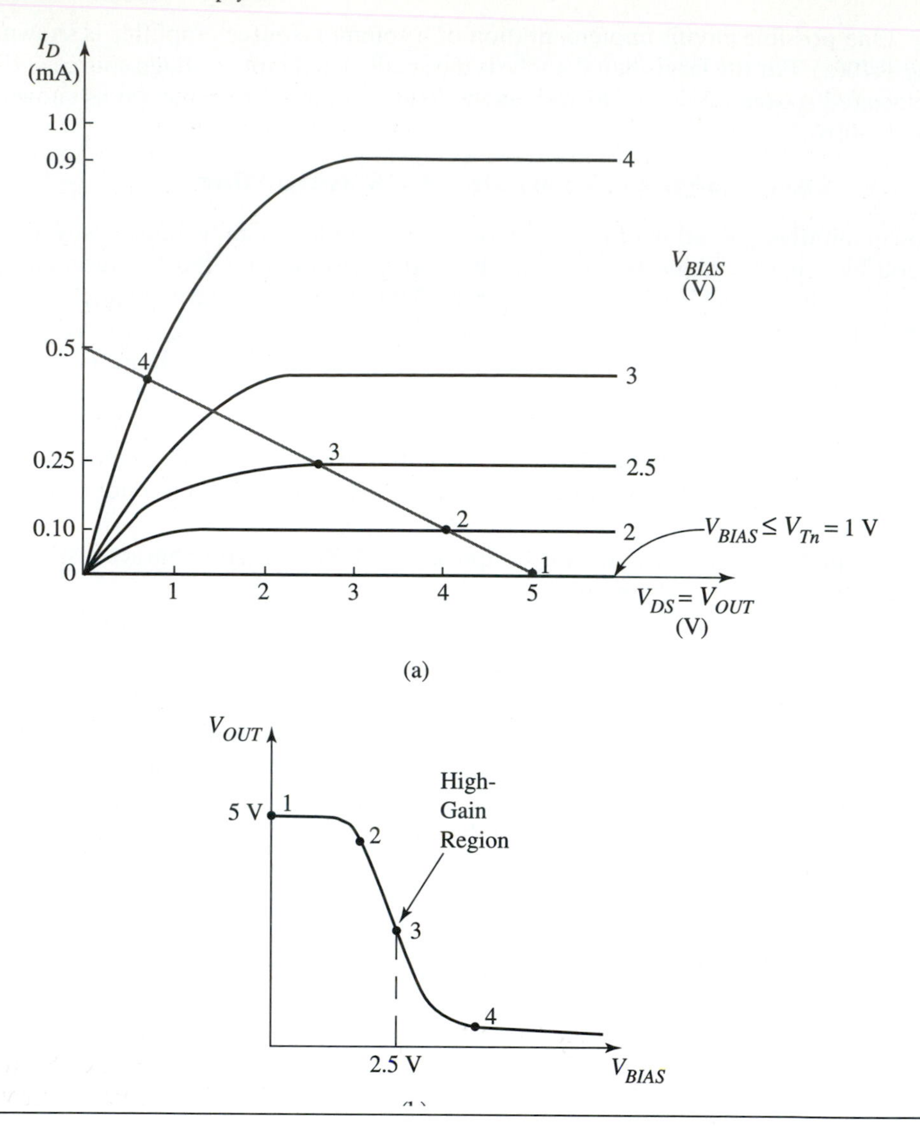

➤ Figure 8.11 (a) NMOS transistor characteristics for the common-source configuration with a load line. This graph assumes $V_{DD} = 5$ V and $R_D = 10$ kΩ. (b) Common-source amplifier voltage transfer function.

To see the relationship between the voltage transfer function shown in Fig. 8.11(b) and the load line analysis, note where the MOS transistor is biased below its threshold voltage (cutoff) and no current is flowing (1). Under this condition the output voltage is equal to V_{DD}. As we increase the bias voltage above the threshold voltage, drain current begins to flow and the output voltage falls below V_{DD} (2). The output voltage rapidly falls toward a low voltage by further increasing the bias voltage. This high-gain region of the transfer characteristic (3) is where the MOS transistor is operating in the constant-current region. As the bias voltage is further increased, the transistor enters the triode region (4). Our goal is to operate this common-source amplifier in the high-gain region by setting the bias voltage so we are operating near point (3).

As the input voltage on the MOS transistor changes, a large change in the output voltage occurs. The relationship is not linear across large-signal changes because the relationship between the drain current and the gate-source voltage is quadratic. In addition, as we approach the extremes in the operation where the MOS transistor is either in cutoff or in the triode region, the output voltage does not appreciably change with the input voltage. We must ensure that the bias voltage is chosen in such a way that the output voltage operates approximately centered in-between the power supply V_{DD} and ground. A small-signal voltage input is applied in series with the input bias voltage to vary the input voltage around the operating point. A corresponding change in output current results from this small-signal input voltage change. The ratio of the small-signal output current to the small-signal input voltage is the amplifier's transconductance.

To quantify the selection of the bias point, we set the input small-signal voltage source equal to zero implying a short circuit. Assuming the MOS transistor is biased in its constant-current region and neglecting channel length modulation ($\lambda_n = 0\ \text{V}^{-1}$), the drain current can be written as

$$I_D = \frac{W}{2L}\mu_n C_{ox}(V_{BIAS} - V_{Tn})^2 \tag{8.8}$$

From this equation we can calculate either the drain current or bias voltage necessary to ensure that the amplifier output voltage operates approximately midway between V_{DD} and ground.

➤ EXAMPLE 8.3 Common-Source Amplifier: Large-Signal Analysis

You are given the following *I-V* characteristics for an MOS transistor. The transistor has $V_{Tn} = 1\text{V}$ and $\mu_n C_{ox} = 50\ \mu\text{A/V}^2$. Assume $V_{DD} = 5$ V.

(a) Draw a load line on this graph for $R_D = 10\ \text{k}\Omega$ and find V_{BIAS} such that $V_{OUT} = 2.5$ V.

(b) Find the *W/L* ratio of the transistor using the *I-V* data from above.

(c) Given that $R_D = 10\ \text{k}\Omega$, analytically find the V_{BIAS} needed to have $V_{OUT} = 2.5$ V. **Note:** The answer to part (c) can be read off the *I-V* characteristics with the load line shown in Fig. Ex 8.3.

SOLUTIONS

(a) To draw the load line, we use Eq. (8.7) to find the drain current when $V_{OUT} = 0$ V and then when $V_{OUT} = 5$ V. Setting $V_{OUT} = 0$ V and $R_D = 10\ \text{k}\Omega$, we find $I_D = 0.5$ mA. When $V_{OUT} = 5$ V, we find that $I_D = 0$ mA. The load line is drawn by connecting these two points on the above graph. We have shown this line on the *I-V* characteristics. By reading off the graph we find that $V_{BIAS} = 2.0$ V and $I_D = 250\ \mu\text{A}$ when $V_{OUT} = 2.5$ V.

(b) To find the *W/L* ratio, we choose any of the above *I-V* lines in the constant-current region and plug into the following equation for the saturation current.

$$I_D = \frac{W}{2L}\mu_n C_{ox}(V_{GS} - V_{Tn})^2$$

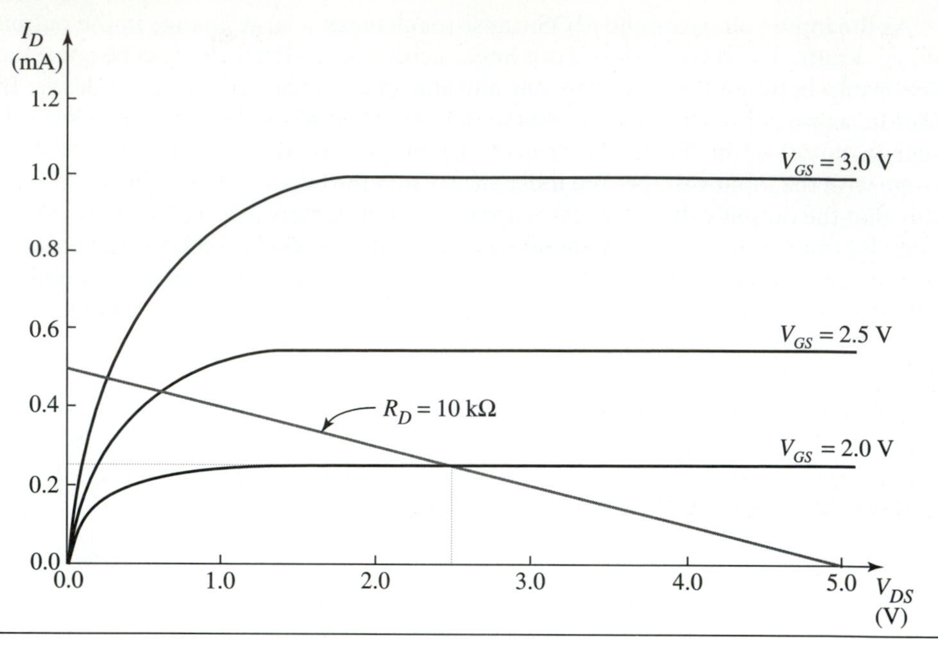

➤ **Figure Ex8.3**

If $V_{GS} = 2.0$ V is chosen when $I_D = 250$ μA we find

$$\frac{W}{L} = \frac{2I_D}{\mu_n C_{ox}(V_{GS} - V_{Tn})^2} = \frac{2 \cdot 250\ \mu\text{A}}{(50\mu\text{A}/\text{V}^2) \cdot (2\text{V} - 1\text{V})^2} = 10$$

(c) To force $V_{OUT} = 2.5$ V we must solve for the drain current required to provide the proper voltage drop across R_D. We begin by writing

$$V_{OUT} = V^+ - I_D R_D$$

Rearranging we can find

$$I_D = \frac{V^+ - V_{OUT}}{R_D} = \frac{5\text{V} - 2.5\text{V}}{10\ \text{k}\Omega} = 250\ \mu\text{A}$$

Finally we can substitute the value for I_D into

$$V_{BIAS} = V_{Tn} + \sqrt{\frac{I_D}{\frac{W}{2L} \cdot \mu_n C_{OX}}} = 1\text{V} + \sqrt{\frac{250\ \mu\text{A}}{\frac{10}{2} \cdot 50\ \frac{\mu\text{A}}{\text{V}^2}}} = 2\text{V}$$

which is the same as we found using the graphical technique.

8.3.2 Small-Signal Analysis of CS Amplifier

Once we have found the DC bias point for the amplifier, the effect of a small-signal voltage input on the small-signal output current can be analyzed. To develop the small-signal model for the common-source amplifier, we recall that all independent DC sources are set equal to zero. This implies that voltage sources are short-circuited while current sources are open-circuited. Applying the small-signal model for an MOS transistor derived in Chapter 4, we find the small-signal model for the intrinsic common-source amplifier shown in Fig. 8.12. The intrinsic amplifier model does not include source and load resistances while the loaded amplifier model does.

8.3.3 Two-Port Model for CS Amplifier

We calculate the two-port model for the common-source amplifier by using the procedure outlined in Section 8.1. Referring to Fig. 8.12(a) we can find the intrinsic transconductance by placing a voltage source v_t at the input and shorting the output terminals as described in Fig. 8.3(c). The intrinsic transconductance is given by

$$G_m = \frac{i_{out}}{v_t} = g_m = \sqrt{2\frac{W}{L}\mu_n C_{ox} I_D} \tag{8.9}$$

Since the input resistance of the MOS transistor is infinite there is no need to use the formal procedure to find R_{in}. Finally, to find the output resistance of this amplifier, we connect a test voltage source at the output and measure its current with the

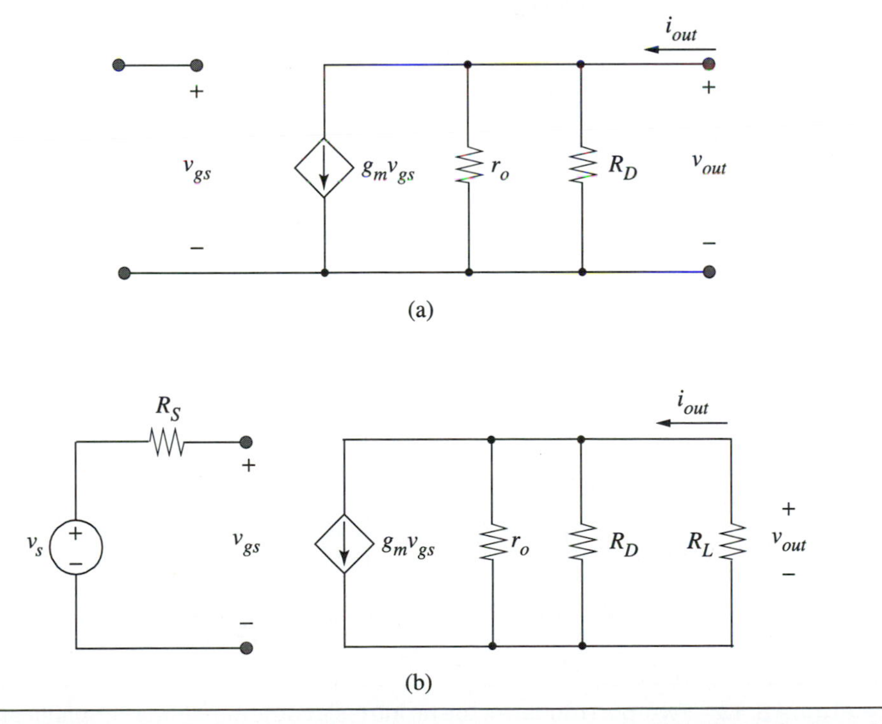

➤ **Figure 8.12** Small-signal model for the common-source amplifier (a) intrinsic and (b) loaded.

source resistance connected to the input as shown in Fig. 8.3(f). Since there is no input voltage, and hence no voltage applied to the transconductance generator, this generator has a value of zero and becomes an open circuit. The output resistance is seen to be the parallel combination $r_o \| R_D$. Since the common-source amplifier is a unilateral network, the output resistance does not depend on the source resistance.

The two-port model for the intrinsic common-source amplifier is shown in Fig. 8.13(a). We can add a voltage source with a source resistor to the input and a load resistor to the output as shown in Fig. 8.13(b) and calculate how the output voltage or current varies as a function of the input voltage.

➤ EXAMPLE 8.4 Common-Source Amplifier: Small-Signal Analysis

In this example we will analyze a common-source amplifier with a DC drain current $I_D = 100\ \mu A$ and $R_D = 25\ k\Omega$.

The MOS device characteristics are $V_{Tn} = 1$ V, $\mu_n C_{ox} = 50\ \mu A/V^2$, $\lambda_n = 0.1\ V^{-1}$, and

$$\frac{W}{L} = \frac{50}{2}.$$

(a) Using the two-port model, find the intrinsic transconductance of the amplifier G_m when it is unloaded ($R_S = 0\ \Omega$, $R_L = 0\ \Omega$).

(b) What is the maximum load resistance R_L this amplifier can drive and still guarantee that the overall transconductance i_{out}/v_s will be greater than 10% of the intrinsic amplifier transconductance?

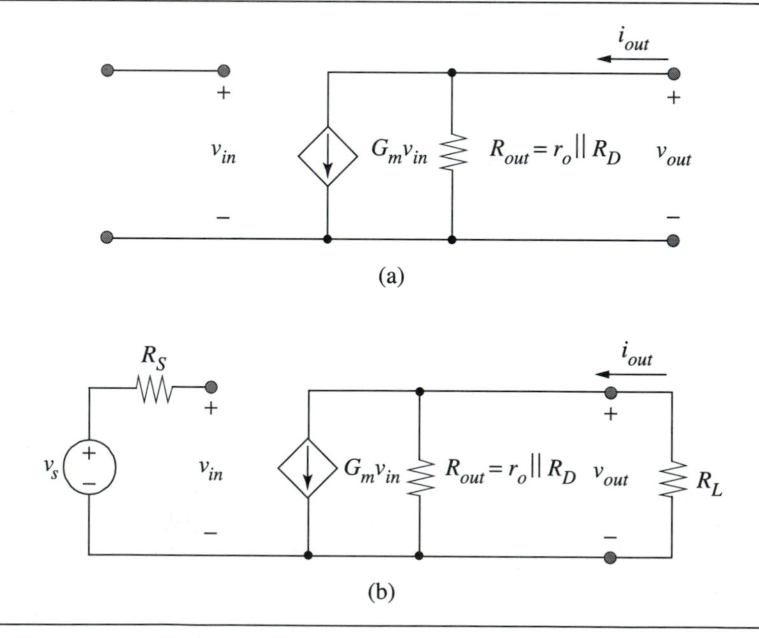

➤ Figure 8.13 Two-port model for the (a) intrinsic common-source amplifier and (b) loaded common-source amplifier.

SOLUTIONS

For the MOS device R_{in} is infinite, so the full voltage input will pass into the amplifier even for a non-zero R_S. A finite load resistance, on the other hand, will degrade the overall transconductance.

(a) When $R_L = 0\ \Omega$ no degradation occurs and the overall transconductance is the same as the intrinsic transconductance and is given by

$$G_m = g_m = \sqrt{2I_D\left(\frac{W}{L}\right)\mu_n C_{ox}} = \sqrt{2 \cdot 100\,\mu\text{A} \cdot \frac{50}{2} \cdot 50\,\mu\text{A/V}^2} = 500\,\mu\text{S}$$

(b) To find the degradation of the overall transconductance due to the load resistor R_L, we begin by writing an expression for R_{out} as $R_{out} = r_o \| R_D$. The transistor output resistance, r_o is given by

$$r_o = \frac{1}{\lambda_n I_D} = \frac{1}{(0.1\text{V}^{-1})\,100\mu\text{A}} = 100\ \text{k}\Omega$$

$$R_{out} = r_o \| R_D = 20\ \text{k}\Omega$$

$$\frac{i_{out}}{v_s} = G_m \cdot \left(\frac{R_{out}}{R_{out} + R_L}\right)$$

Since we require that $\frac{R_{out}}{R_{out} + R_L} > \frac{1}{10}$ to ensure the overall transconductance is greater than 10% of the intrinsic transconductance, we find the maximum R_L to be $R_L < 9R_{out}$, thus $R_L < 180\ \text{k}\Omega$.

8.4 CURRENT SOURCE SUPPLIES

In both the common-emitter and common-source amplifiers, we saw that the output resistance was limited by the resistor supplying the current to the transistor. If we increase the value of the resistor for a given power supply voltage, we lower the current and hence the transconductance. Note that in modern microelectronic systems where low power dissipation is of extreme importance, it is not practical to raise the power supply voltage to increase the current. What we need is an element that can supply a constant current independent of the voltage across it. This requirement is obviously the definition of a current source.

The current sources that will be used to supply the current to transistor amplifiers should have the following ideal characteristics.

- When the voltage across the current source is greater than zero, it should provide a constant current I_{SUP}.
- When the voltage across it is less than or equal to zero the current source should be equal to zero.

We define an **idealized current source** with a finite output resistance with the same characteristics as the one introduced in Chapter 5. The circuit model for this current source is given in Fig. 8.14(a). The model is only valid when the voltage across the current source supply $v_{SUP} > 0$ V. If $v_{SUP} \leq 0$ V, the supply current is set to zero. The internal resistance of the current source is modeled with resistor r_{oc}. By combining these two elements in parallel, we have modeled the *I-V* characteristics of the current source shown in Fig. 8.14(b). The current supplied by this current source is equal to I_{SUP} when v_{SUP} is equal to 0^+ V. As v_{SUP} increases, the total supply current increases due to the internal resistance. The small-signal model shown in Fig. 8.14(c) for this idealized current source is simply the internal resistance r_{oc}, since the DC current source value is set to zero for small-signal modeling.

We will use this idealized current source to supply the current for the amplifiers studied in the rest of this chapter. It allows us to study the effect of its internal resistance on the overall amplifier characteristics. This idealized current source can be approximated using a transistor. We will replace this idealized current source with transistors in the next chapter.

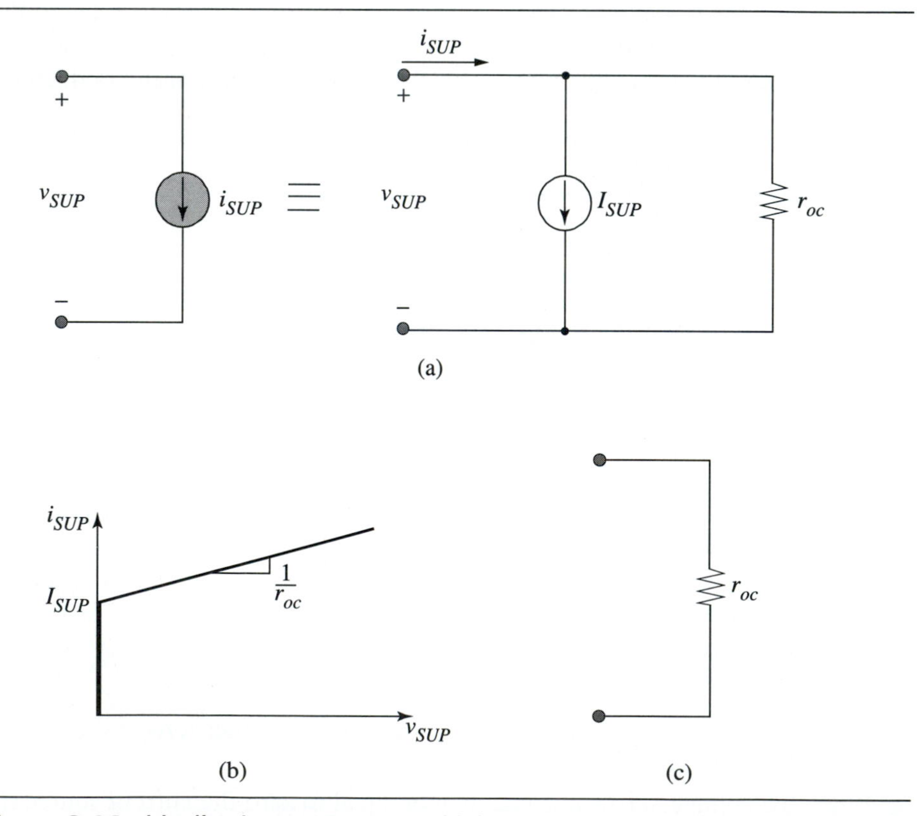

➤ **Figure 8.14** Idealized current source with internal resistance. (a) Symbol and model for the idealized current source that includes the internal resistance, (b) *I-V* characteristic for total current vs. voltage, and (c) small-signal model.

8.5 COMMON-SOURCE (CS) AMPLIFIER WITH CURRENT SOURCE SUPPLY

A common-source amplifier with a current source supply is shown in Fig. 8.15(a). The input signal and bias voltages are connected to the gate. The current source supply is connected to the drain. The source and backgate of the MOS transistor are grounded.

8.5.1 Large-Signal Analysis of CS Amplifier

In order to simplify the large-signal analysis of the amplifier, we assume that the source resistance R_S, associated with the small-signal input voltage source v_s, has a negligible effect on the DC operating point. We neglect the effect of channel length modulation on the drain current ($\lambda_n = 0\ V^{-1}$) and the internal resistance of the current source supply, r_{oc}. We also remove the load resistor since we will bias the transistor such that the DC output current I_{OUT} is equal to zero.The circuit for the large signal analysis is shown in Fig. 8.15(b).

With these simplifications, a graphical analysis can be easily carried out. Fig. 8.16(a) shows the drain current versus output voltage with the input bias voltage as a parameter. The current source supply characteristic is plotted on the same curve. The output voltage and drain current are determined by the intersection of these two characteristic curves. In Fig. 8.16(b) we show how the output voltage changes with input bias voltage.

To see the relationship between the voltage transfer function and the load line characteristics, look at (1) in Fig. 8.16(a) where the MOS transistor is biased below its threshold voltage (cutoff) and no current is flowing. Under this condition the output voltage is equal to V_{DD}. As we increase the bias voltage above the threshold voltage, drain current begins to flow and the output voltage falls slightly below V_{DD} (2). By further increasing the bias voltage, the output voltage will quickly snap from V_{DD}

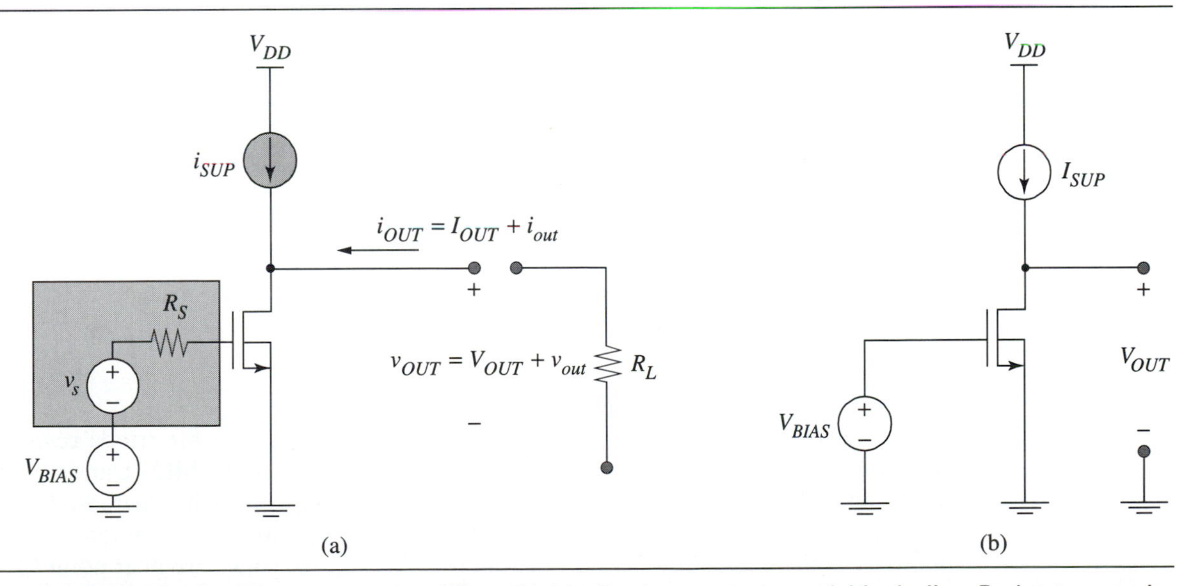

Figure 8.15 Common-source amplifier with idealized current source. (a) including R_S due to v_s and R_L. (b) For large-signal analysis we assume $R_S = 0\ \Omega$, $R_L \to \infty$, and $r_{oc} \to \infty$.

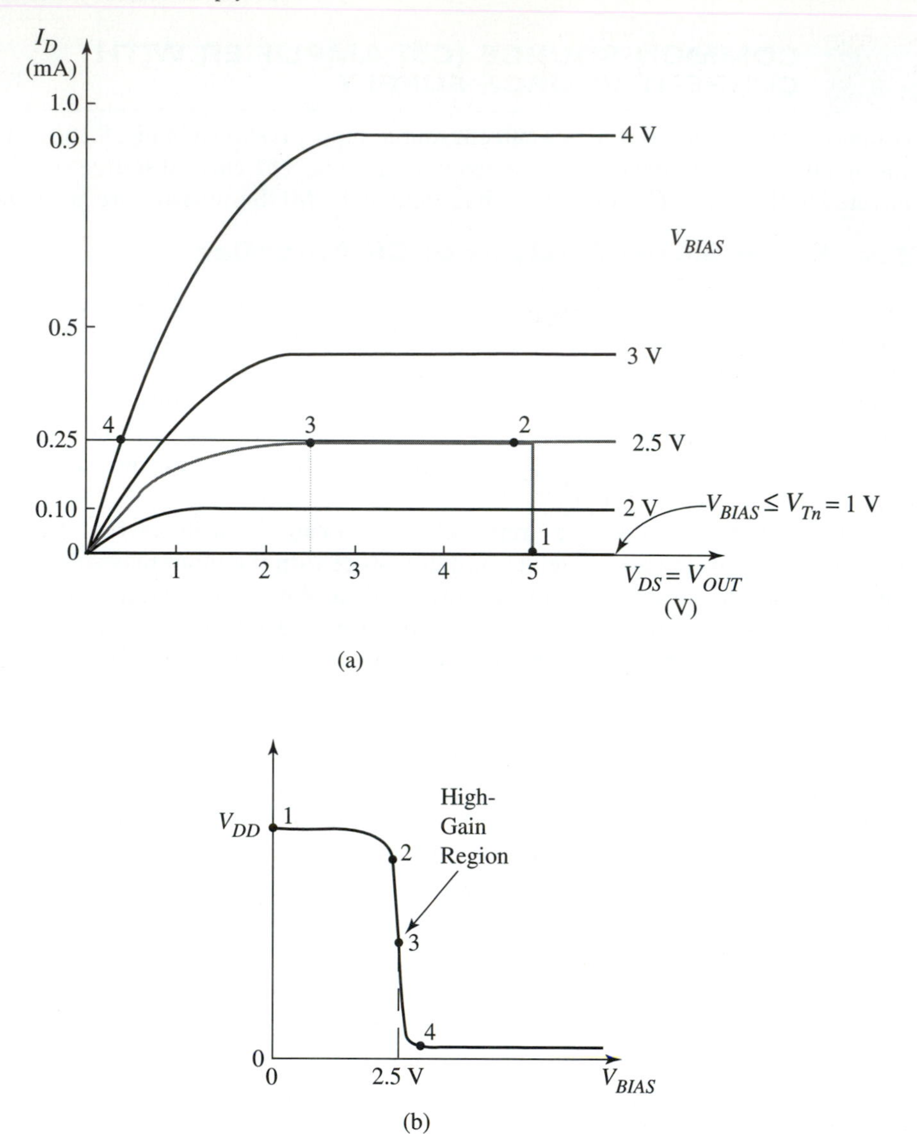

➤ **Figure 8.16** (a) Common-source transistor characteristics with current source load line. (b) Common-source amplifier voltage transfer characteristic.

down to a low voltage where the MOS transistor is operating in the triode region. This high-gain region of the transfer characteristic (3) is where the MOS transistor is operating in the constant-current region. As the bias voltage is further increased, the transistor enters the triode region (4). Our goal is to operate this common-source amplifier in the high-gain region by setting the bias voltage so we are near point (3).

We begin the large-signal analysis by finding the bias voltage and drain current where the common-source amplifier operates in the high-gain region. From Fig. 8.16, the current source supply is shown to be equal to approximately 250 μA. To calculate the bias voltage to operate in the high-gain region, we use the large-signal MOS device model derived in Chapter 4. The drain current is given by

$$I_D = \frac{W}{2L}\mu_n C_{ox}(V_{GS} - V_{Tn})^2(1 + \lambda_n V_{DS}) \quad V_{DS} > V_{DS_{SAT}} \tag{8.10}$$

If we neglect the effect of channel length modulation, or to put it another way, assume the device has an infinite output resistance, we can set $\lambda_n = 0\ \text{V}^{-1}$. This assumption is fine for calculating approximate bias voltages. Setting the supply current equal to the drain current implies that the output current $I_{OUT} = 0$ A. The relationship between the bias voltage and the supply current is

$$I_D = I_{SUP} = \frac{W}{2L}\mu_n C_{ox}(V_{BIAS} - V_{Tn})^2 \tag{8.11}$$

Solving for the bias voltage, we find

$$V_{BIAS} = V_{Tn} + \sqrt{\frac{I_{SUP}}{\frac{W}{2L}\mu_n C_{ox}}} \tag{8.12}$$

From Fig. 8.16(b) we see that if V_{BIAS} is slightly above or below its optimum value (3), the voltage gain is severely degraded. Under this condition the amplifier is said to be saturated or operating outside its high-gain region. In practical circuits, feedback (Chapter 12) will be used so the bias voltage is automatically established. We will explore this point further by using SPICE in Example 8.5.

8.5.2 Small-Signal Analysis of CS Amplifier

Once we have determined the operating point, we can determine the two-port transconductance model of the intrinsic common-source amplifier with the current source supply using the small-signal model in Fig. 8.17.

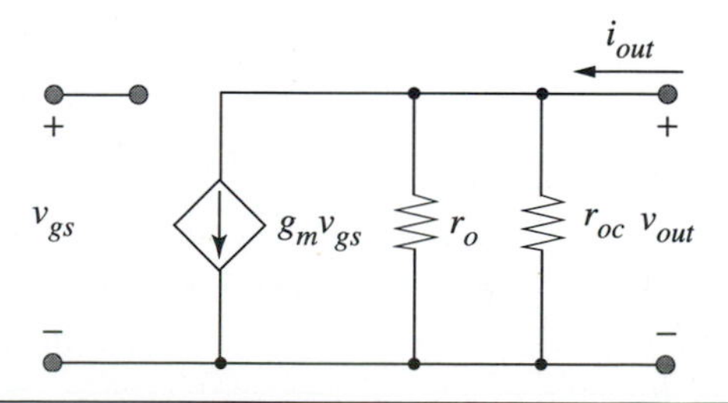

Figure 8.17 Small-signal model of common-source amplifier with current source supply used to find the two-port model.

The intrinsic transconductance is found by placing a test voltage source v_t at the input and shorting the output terminals. By inspection we see

$$G_m = \frac{i_{out}}{v_t} = g_m = \sqrt{2\frac{W}{L}\mu_n C_{ox} I_D} \tag{8.13}$$

Since the MOS transistor has an insulating gate, the input resistance is infinite. To find the output resistance we attach a test voltage source at the output port and measure the current with the source resistance across the input. We find

CS configuration two-port parameters transconductance and output resistance

$$R_{out} = r_o \| r_{oc} = \left(\frac{1}{\lambda_n I_D}\right) \| r_{oc} \tag{8.14}$$

The two-port model for the intrinsic transconductance amplifier using the MOS transistor in the common-source configuration with a current source supply is shown in Fig. 8.18(a).

The value of the input resistance is always infinite, making it an excellent amplifier to accept a voltage source with a non-zero source resistance. The transconductance, G_m of the amplifier goes up with the square root of the supply current, and the device parameters W/L and $\mu_n C_{ox}$. The output resistance increases as we reduce λ_n. It should be noted that λ_n is proportional to $1/L$. This implies that the output resistance can be increased by increasing the channel length.

A useful circuit parameter is the open-circuit voltage gain A_v. By this we mean the voltage gain of the amplifier when $R_L \to \infty$. By using the Thévenin equivalent description of the two-port model in Fig. 8.18(a), the resulting two-port model for the intrinsic voltage amplifier using a common-source configuration is shown in Fig. 8.18(b). Neglecting r_{oc}, the open-circuit voltage gain increases as I_{SUP} is decreased since the output resistance is increasing linearly with decreasing I_{SUP} while the transconductance is only decreasing by $\sqrt{I_{SUP}}$. A similar relationship holds for channel lengths. The output resistance, transconductance, and open-circuit voltage gain dependencies on device parameters are summarized in Table 8.1.

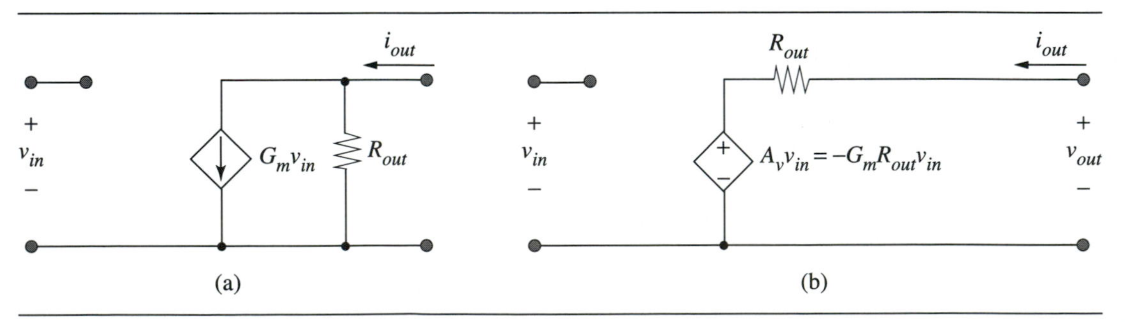

Figure 8.18 Two-port common-source small-signal model (a) intrinsic transconductance amplifier form and (b) intrinsic voltage amplifier form.

Table 8.1 Common-source amplifier circuit parameter dependencies on device parameters and supply current.

Device Parameters	Circuit Parameters			
	A_v	G_m	R_{in}	R_{out}
	$-g_m(r_o \| r_{oc})$	g_m	∞	$r_o \| r_{oc}$
I_{SUP}	↓	↑	–	↓
W	↑	↑	–	–
$\mu_n C_{ox}$	↑	↑	–	–
$L \propto 1/\lambda_n$	↑	↓	–	↑

Arrows indicate which direction to change device parameters to *increase* circuit parameter values.

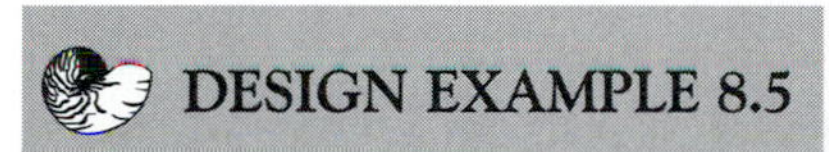

Common-Source Amplifier with Current Source Supply

In this example, we will design a common-source amplifier that drives a load resistor $R_L = 20\ \text{k}\Omega$ as shown in Fig. Ex8.5A. The channel length is set at 2 μm, and we want g_m to be greater than 2 mS (for frequency response reasons you will learn in Chapter 10). Find values for V_{BIAS}, I_{SUP}, and W/L so the overall voltage gain will be greater than |–20|. Verify your result using SPICE.

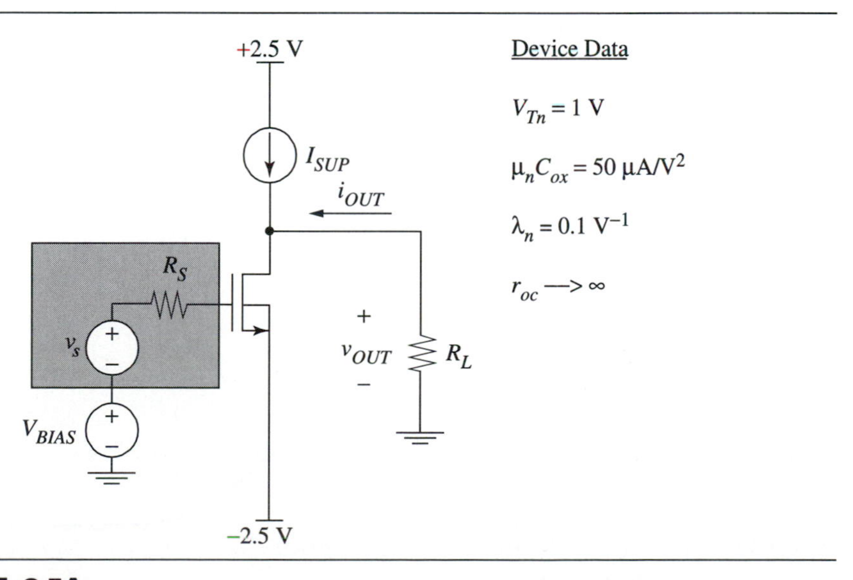

➤ **Figure Ex8.5A**

SOLUTION

Since $g_m \geq 2$ mS, $R_L = 20\,\text{k}\Omega$, and $A_v = g_m(r_o \| R_L) > |-20|$, we find that r_o must be greater than $20\,\text{k}\Omega$. Because R_L is fixed at 20 kΩ, increasing r_o does not increase the gain substantially, so we will leave it at $20\,\text{k}\Omega$. I_{SUP} is determined by the requirement on r_o.

$$I_{SUP} = I_D = \frac{1}{\lambda_n r_o} = \frac{1}{(0.1\,\text{V}^{-1})\,(20\,\text{k}\Omega)} = 500\,\mu\text{A}$$

We can find the transistor W/L by writing

$$g_m = \sqrt{2\frac{W}{L}(\mu_n C_{ox})\,I_D}$$

Solving for W/L above we find $\frac{W}{L} = 80 = \frac{160\,\mu\text{m}}{2\,\mu\text{m}}$

This result leads to finding

$$V_{GS} = V_{Tn} + \sqrt{\frac{I_{SUP}}{\frac{W}{2L}\cdot\mu_n C_{ox}}} = 1.0\,\text{V} + \sqrt{\frac{500\,\mu\text{A}}{\frac{80}{2}\cdot 50\frac{\mu\text{A}}{\text{V}^2}}} = 1.5\,\text{V} = V_G - V_S$$

In this example we are using dual supplies. The source of the MOS transistor is at –2.5 V.

Thus $V_G = V_{GS} + V_S = 1.5\,\text{V} - 2.5\,\text{V} = -1\,\text{V} = V_{BIAS}$

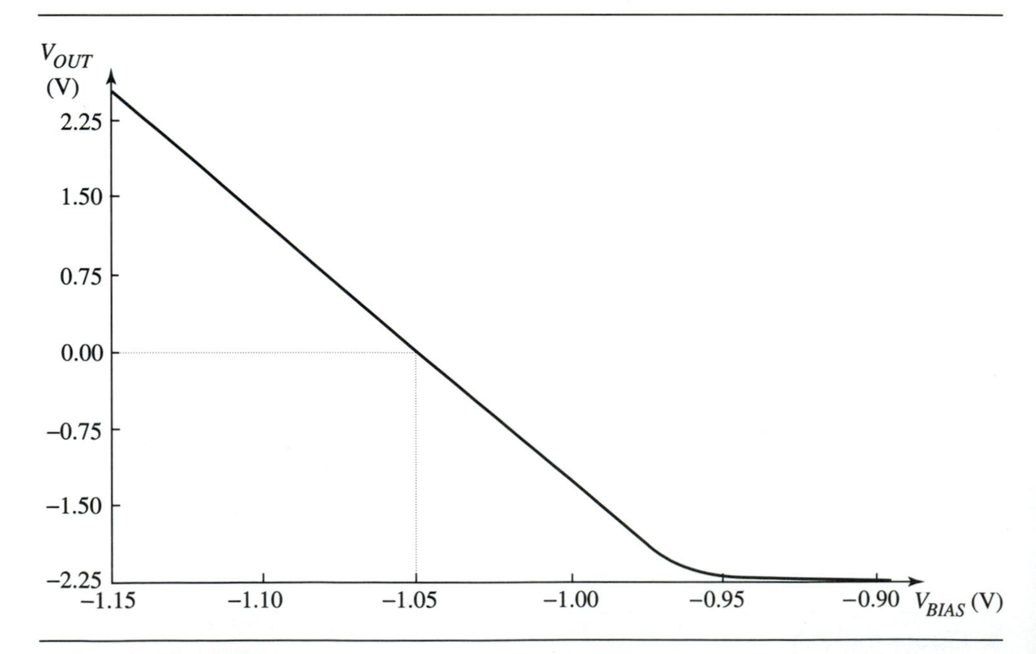

➤ **Figure Ex8.5B**

When running SPICE, you should first get the large-signal transfer function V_{OUT}/V_{IN} by sweeping the bias voltage V_{BIAS} around the –1.0 V bias voltage calculated above. The SPICE output is shown in Fig. Ex8.5B. Now we choose a value for V_{BIAS} so the amplifier is operating in the middle of the high-gain region since SPICE performs a small-signal analysis around the DC operating point it calculates. In this case we choose V_{BIAS} to be –1.05 V. Next, we run a small-signal analysis and find the small-signal voltage gain to be –25.2.

8.5.3 P-Channel Common-Source Amplifier

A p-channel common-source amplifier with a current source supply is shown in Fig. 8.19(a). As with the n-channel common-source amplifier the input signal and bias voltages are connected to the gate of the p-channel transistor and referenced to ground. The current source supply is connected to the drain of the p-channel device. The source and backgate are shorted and connected to the most positive supply, V_{DD}.

The large-signal and small-signal operation of the p-channel common-source amplifier is similar to the n-channel version. However one must take special care to obtain the proper signs. In this section we will show that the small-signal two-port model for the p-channel common-source amplifier is identical to the n-channel version.

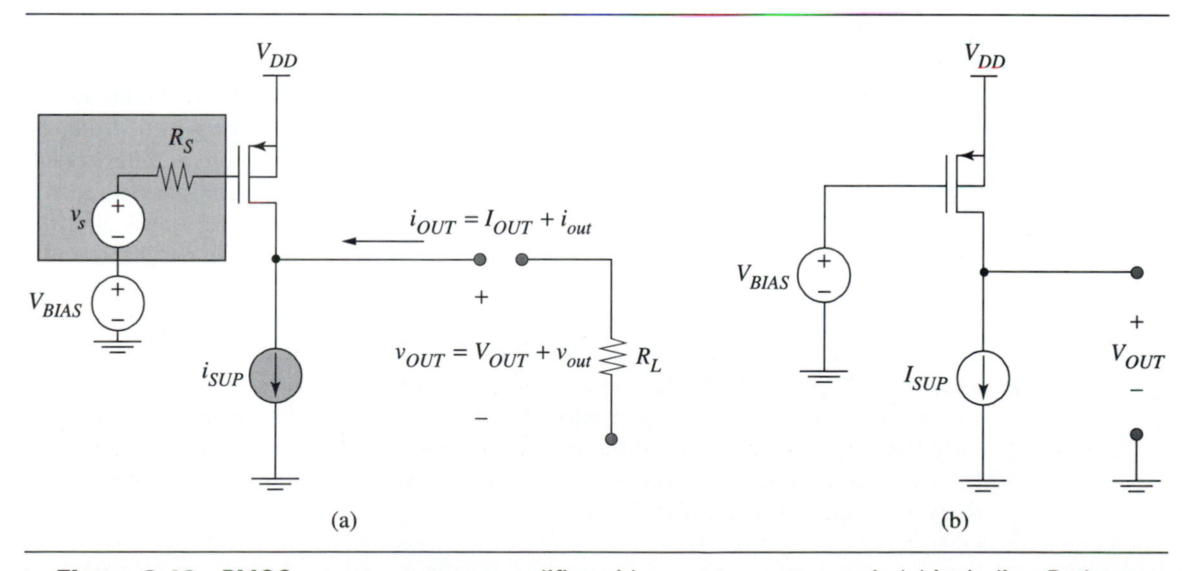

➤ **Figure 8.19** PMOS common-source amplifier with current source supply (a) including R_S due to v_s and R_L. b) For large-signal analysis we assume $R_S = 0\ \Omega$, $R_L \rightarrow \infty$, and $r_{oc} \rightarrow \infty$.

For the large-signal analysis, we assume that the source resistance R_S, that is associated with the small-signal input voltage source v_s, has a negligible effect on the DC operating point. We neglect the effect of channel length modulation on the drain current ($\lambda_p = 0\ V^{-1}$) and the internal resistance of the current source supply, r_{oc}. We also remove the load resistor since we will bias the transistor such that the DC output current I_{OUT} is equal to zero. The circuit for the large-signal analysis is shown in Fig. 8.19(b).

Given the assumptions above and that the PMOS transistor is operating in the constant-current region, the drain current, $-I_D$ is given by

$$-I_D = \frac{W}{2L}\mu_p C_{ox}(V_{SG} + V_{Tp})^2 \tag{8.15}$$

where $V_{SG} = V_{DD} - V_{BIAS}$.
The supply current is set equal to $-I_D$ so that the output current I_{OUT} is zero. Substituting for V_{SG} we find

$$-I_D = I_{SUP} = \frac{W}{2L}\mu_p C_{ox}(V_{DD} - V_{BIAS} + V_{Tp})^2 \tag{8.16}$$

Solving for the bias voltage we obtain

$$V_{BIAS} = V_{DD} + V_{Tp} - \sqrt{\frac{I_{SUP}}{\frac{W}{2L}\mu_p C_{ox}}} \tag{8.17}$$

Comparing the effect of V_{BIAS} on the p-channel vs. n-channel common-source amplifier we find that we must decrease V_{BIAS} to increase V_{SG} and consequently increase the magnitude of the drain current, $|-I_D|$.

Now that we have determined the operating point, we will use the small-signal model of the p-channel transistor developed in Chapter 4 to determine the two-port model for the amplifier. The resulting small-signal model of the amplifier is shown in Fig. 8.20(a).

Although we could solve for the two-port model directly with the circuit shown in Fig. 8.20(a) it is easier to flip it 180° [Fig. 8.20(b)] so that it appears very similar to the n-channel version. Since $v_{sg} = -v_{gs}$ we can change signs at the input and the dependent current source and find that the p-channel common-source amplifier small-signal model is identical to the n-channel version as shown in Fig. 8.20(c). Using the same analysis as with the n-channel version we find that the transconductance and output resistance are the same as given in Eq. (8.13) and Eq. (8.14) except that we replace I_D with $-I_D$. The two-port model is the same as that shown in Fig. 8.18.

The p-channel and n-channel versions of the common-source amplifier have identical small-signal two-port models. This result applies to all the configurations that we will study in this chapter whether MOS or bipolar. Therefore, we will study only the n-type version for all the configurations to avoid further redundancy. The large-signal analysis of p-type amplifiers is sufficiently similar to the n-type that it does not require further discussion.

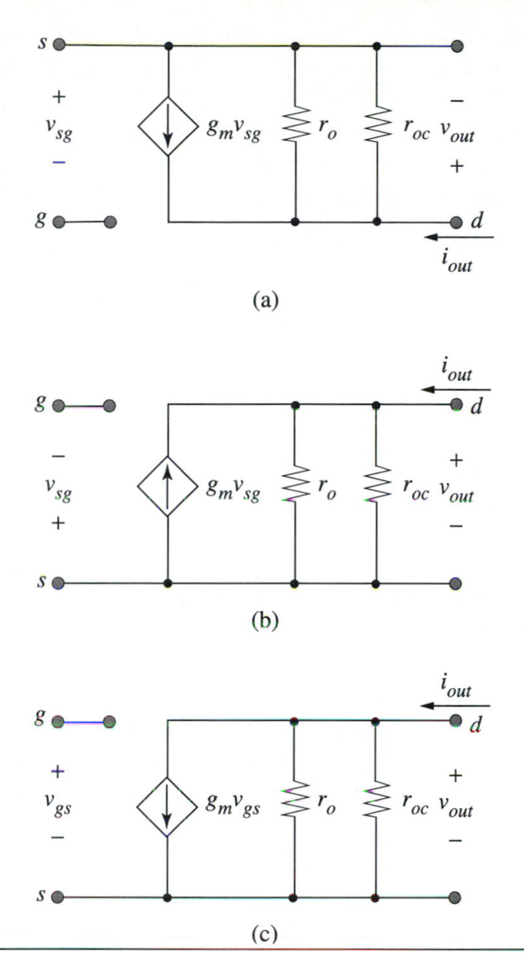

➤ **Figure 8.20** Small-signal model of the p-channel common-source amplifier (a) using the p-channel transistor model directly, (b) resulting from a flipped version of (a), and (c) used to find the two-port model. Note: the model in (c) is identical to the n-channel version in Fig. 8.17.

➤ 8.6 COMMON-EMITTER (CE) AMPLIFIER WITH CURRENT SOURCE SUPPLY

In this section we replace the resistor R_C with an idealized current source in the common-emitter amplifier, perform a large-signal analysis, and develop a two-port model for the small-signal amplifier. The common-emitter amplifier is shown in Fig. 8.21(a).

8.6.1 Large-Signal Analysis of CE Amplifier

In order to simplify the large-signal analysis of the amplifier, we assume that the small-signal resistance R_S associated with the small-signal input voltage source v_s has a negligible effect on the DC operating point. We neglect the Early effect ($V_A \to \infty$), base current ($\beta_o \to \infty$), and the internal resistance of the current source supply, r_{oc}.

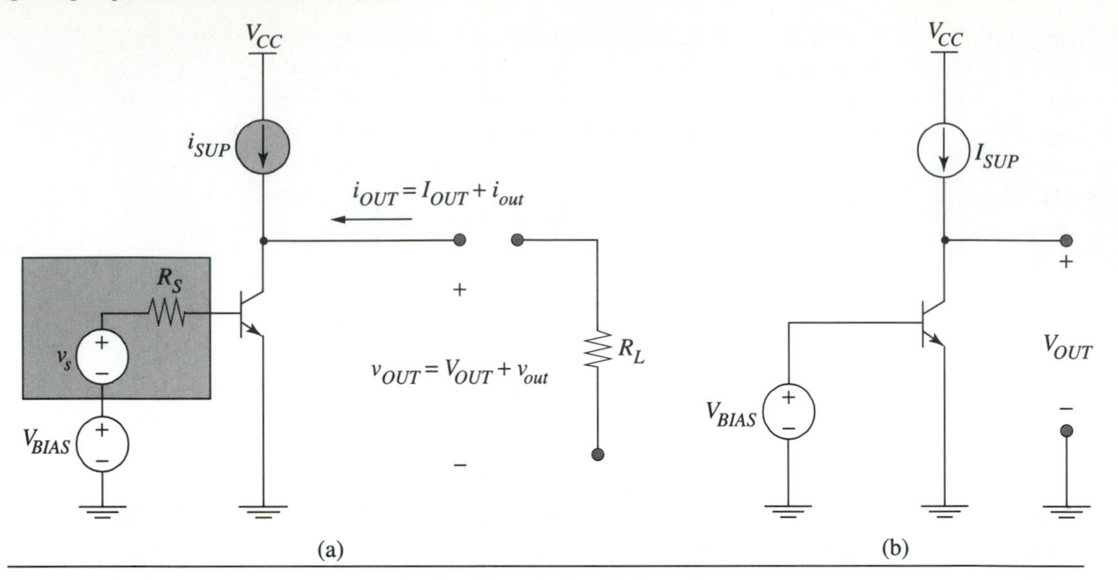

➤ **Figure 8.21** Common-emitter amplifier with idealized current source supply (a) including R_S due to v_s and R_L. (b) For large-signal analysis we assume $R_S = 0\ \Omega$, $R_L \to \infty$, and $r_{oc} \to \infty$.

We also remove the load resistor since we will bias the transistor such that the DC output current I_{OUT} is equal to zero.The circuit for the large-signal analysis is shown in Fig. 8.21(b). With these assumptions we can perform a load line analysis by replacing the collector resistor with the idealized current source as shown in Fig. 8.22(a). As before, the output voltage and collector current can be determined by the intersection of these two characteristics.

Fig. 8.22(b) shows how the DC output voltage changes with the input bias voltage. To understand Fig. 8.22(b), first look at a small bias voltage so the collector cur-

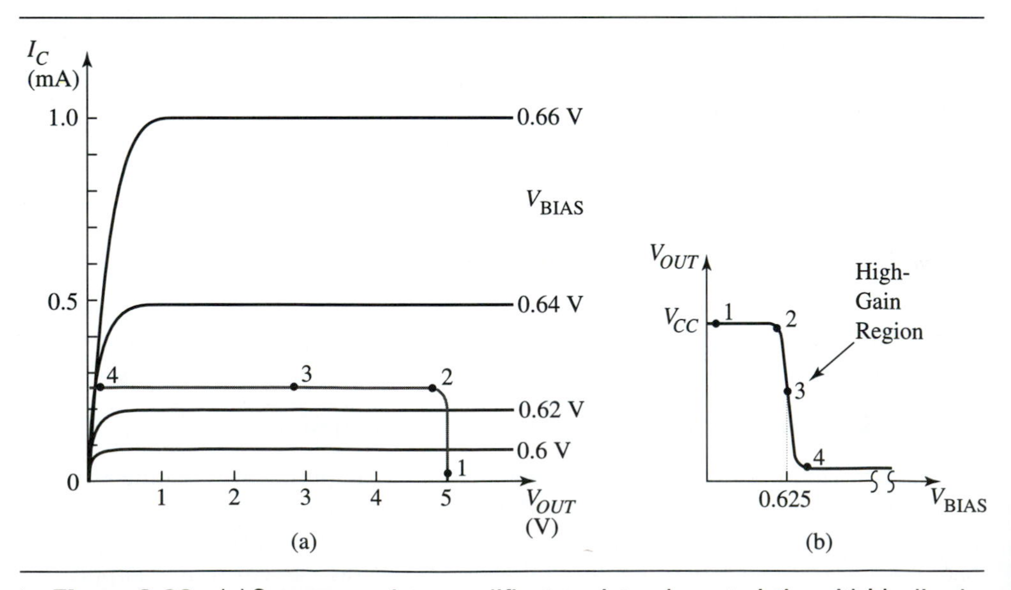

➤ **Figure 8.22** (a) Common emitter amplifier transistor characteristics with idealized current source. (b) Common emitter amplifier DC transfer characteristic.

rent in the bipolar transistor is cutoff (1). Under this condition the output voltage will be equal to V_{CC}. As we increase the bias voltage, the collector current begins to flow and V_{OUT} slightly decreases (2). By slightly increasing the bias voltage the output voltage will quickly snap from V_{CC} down toward the saturation region of the bipolar transistor (3). By further increasing the bias voltage the output voltage becomes $V_{CE_{SAT}}$ (4). We want the amplifier to operate in the region where the bias voltage snaps the output voltage from V_{CC} to a low value. This is the high-gain region of the amplifier. It corresponds to the forward-active region of the bipolar transistor.

We begin the large-signal analysis by finding the bias voltage and collector current where the common-emitter amplifier operates in the high-gain region. Again, assuming forward-active operation we can use Eq. (8.5) for the relationship between the collector current and bias voltage. Assuming that all of the current provided by the current source goes into the collector ($I_{OUT} = 0$ A), we can solve for the bias voltage necessary to maintain the amplifier in the high-gain region.

It should be noted that the bias voltage must be a very precise number or the amplifier will swing to either V_{CC} or $V_{CE_{SAT}}$. This is true because of the very high gain between the input voltage and output voltage. However, it is not necessary to design a bias voltage source that has precision to three decimal places. In Chapter 12 we will introduce feedback which can be used to set the input bias voltage.

8.6.2 Small-Signal Analysis of CE Amplifier

Given that the supply current and bias voltage have been determined, a small-signal two-port transconductance model for the intrinsic common-emitter amplifier with a current source supply can be determined using the circuit in Fig. 8.23.

The intrinsic transconductance G_m for the two-port model is determined by placing a test voltage source v_t at the input and shorting the output terminals. G_m can be written by inspection as

CE configuration two-port parameters transconductance and voltage amps

$$G_m = \frac{i_{out}}{v_t} = g_m = \frac{qI_C}{kT} \tag{8.18}$$

Using the procedure outlined in Section 8.1 to find the input and output resistances, we find that $R_{in} = r_\pi$, and $R_{out} = r_o \| r_{oc}$. The two-port model for the intrinsic transconductance amplifier using a bipolar transistor in the common-emitter configuration is shown in Fig. 8.24(a).

We can calculate the open-circuit output voltage A_v that is defined as the voltage gain with $R_S = 0\ \Omega$ and $R_L \to \infty$. The model in Fig. 8.24(a) is transformed to the Thévenin equivalent as shown in Fig. 8.24(b). In this case A_v is given by

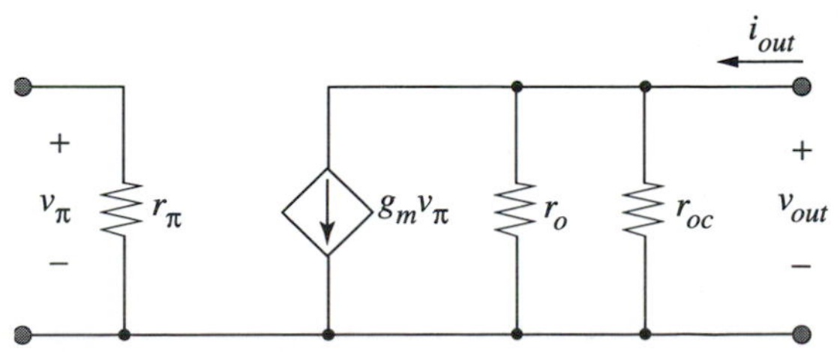

Figure 8.23 Small-signal model of common-emitter amplifier with current source supply used to find the two-port model.

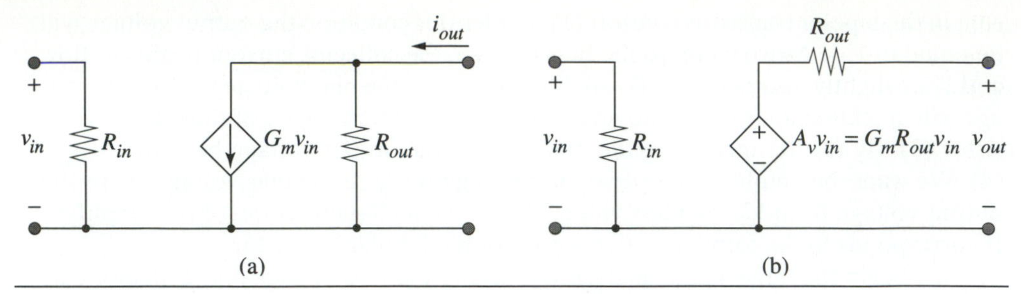

➤ **Figure 8.24** Two-port model for a common-emitter amplifier (a) intrinsic transconductance amplifier and (b) intrinsic voltage amplifier.

$$\boxed{A_v = \frac{v_{out}}{v_{in}} = -G_m R_{out} = -g_m (r_o \| r_{oc})} \tag{8.19}$$

Table 8.2 summarizes the circuit parameter dependencies for the common-emitter amplifier. The transconductance depends on the transconductance of the device and, therefore, can be increased by increasing the supply current. It is independent of the Early voltage and current gain. The input resistance of the common-emitter amplifier is equal to r_π which equals β_o/g_m. To increase the input resistance, we can either reduce the supply current or design the device with increased β_o. The Early voltage has no effect on the input resistance. The output resistance is determined by the Early voltage and supply current, and can be increased by either increasing the Early voltage by device design or reducing the supply current.

Next, we examine the dependencies of the open-circuit voltage gain on the bipolar device parameters. To examine the dependence on only the bipolar transistor, we

■ **Table 8.2** Common-emitter amplifier circuit parameter dependencies on device parameters and supply current.

Device Parameters	**Circuit Parameters**			
	A_v	G_m	R_{in}	R_{out}
	$-g_m(r_o \| r_{oc})$	g_m	r_π	$r_o \| r_{oc}$
I_{SUP}	–	↑	↓	↓
β_o	–	–	↑	–
V_A	↑	–	–	↑

Arrows indicate which direction to change device parameters to *increase* circuit parameter values.

will assume that, $R_S = 0\ \Omega$, and $R_L \to \infty$ and the current source supply's internal resistance is much greater than the output resistance of the transistor, $r_{oc} \gg r_o$. The result is that the open-circuit voltage gain A_v is equal to

$$A_v = -g_m r_o = \frac{-qI_C}{kT}\frac{V_A}{I_C} = \frac{-V_A}{V_{th}} \tag{8.20}$$

The open-circuit voltage gain is dependent on the thermal voltage kT/q, a fundamental physical quantity, and the Early voltage V_A. Therefore, to ensure a large open-circuit voltage gain, the bipolar transistor must be designed with as large an Early voltage as possible. It is also inversely proportional to temperature. In Section 8.7 we will study a configuration that can reduce this strong temperature dependence.

➤ EXAMPLE 8.6 Common-Emitter Amplifier with Current Source Supply

Given the common-emitter amplifier and device data shown in Fig. Ex 8.6:

(a) Calculate the open-circuit voltage gain with $R_S = 0\ \Omega$ and $R_L \to \infty$.

(b) Given that $R_S = 10\ \text{k}\Omega$, find the maximum supply current I_{SUP}, needed to maintain 50% of the open-circuit voltage gain.

(c) From the value of I_{SUP} calculated in (b), find the value for V_{BIAS} so that $V_{OUT} = 0$ V. Use SPICE to verify your results.

SOLUTIONS

(a) Using the small-signal two-port model for the intrinsic voltage amplifier for the common-emitter amplifier shown in Fig. 8.24(b),

$$A_v = -G_m R_{out}$$

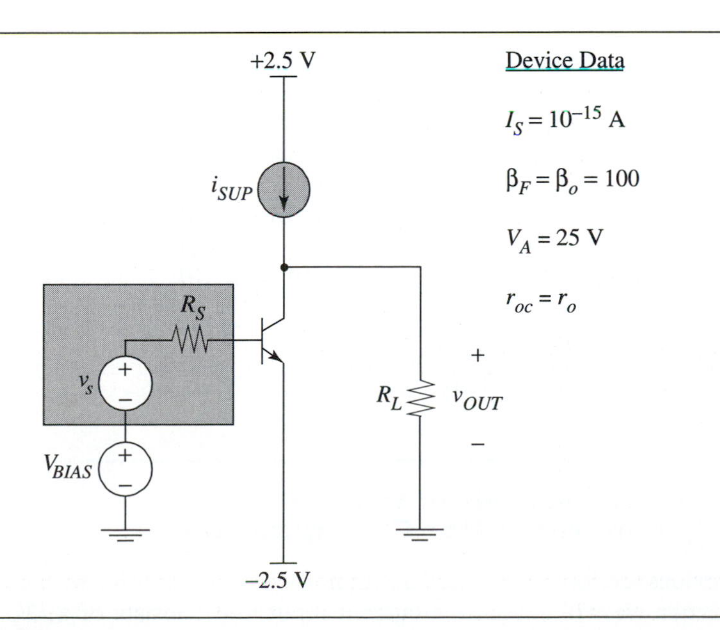

➤ **Figure Ex8.6**

Substituting $G_m = g_m$ and $R_{out} = r_o \parallel r_{oc}$ we get

$$A_v = -g_m(r_o \parallel r_{oc})$$

In this problem $r_o = r_{oc}$ so

$$A_v = -\frac{1}{2} g_m r_o$$

Since $g_m = \frac{qI_C}{kT}$ and $r_o = \frac{V_A}{I_C}$, then $A_v = -\frac{1}{2}\frac{qV_A}{kT} = -500$.

(b) When $R_S = 10\ \text{k}\Omega$, we require $R_{in} \geq 10\ \text{k}\Omega$ so that $\frac{v_{out}}{v_s} \geq \frac{1}{2}A_v$.

For the common-emitter amplifier above $R_{in} = r_\pi = \frac{\beta_o}{g_m} = \frac{\beta_o kT}{I_C q}$

Rearranging,

$$I_C \leq \frac{\beta_o kT}{R_{in} q} = \frac{100 \cdot 25\ \text{mV}}{10\ \text{k}\Omega} = 250\ \mu\text{A}.$$

(c) To find V_{BIAS}, we assume $\beta_o \to \infty$ and $V_A \to \infty$. We want $I_C = I_{SUP} = 250\ \mu\text{A}$.

If we neglect base current $V_{BIAS} \approx V_{BE} - 2.5\ \text{V}$. From $I_C = I_S e^{qV_{BE}/kT}$ we can write

$$V_{BE} = \frac{kT}{q} \ln \frac{I_C}{I_S} = 25\ \text{mV} \cdot \ln\left(\frac{250 \times 10^{-6}}{1 \times 10^{-15}}\right) = 0.656\text{V}$$

$$V_{BIAS} \approx 0.656\ \text{V} - 2.5\ \text{V} = -1.844\ \text{V}$$

Before running a small-signal analysis in SPICE, we must sweep V_{BIAS} around the approximate value we found in part (c) to ensure we are operating the amplifier in the high-gain region. Doing this, in SPICE we find $V_{BIAS} = -1.79$ V and

$$\frac{v_{out}}{v_s} = -256 \text{ with } R_S = 10\ \text{k}\Omega$$

8.6.3 Current Amplifier Using the Common-Emitter Configuration

In the previous sections we studied a common-emitter amplifier with a voltage input. In this section we will examine a current input that consists of a DC bias current source and a small-signal current source in parallel. This common-emitter amplifier with a current source input is shown in Fig. 8.25(a). For the large-signal analysis we will neglect the effect of R_S and R_L. At the input we set $R_S \rightarrow \infty$ and at the output we are interested in the short-circuit current so $R_L = 0\ \Omega$. We also neglect base current, $\beta_F \rightarrow \infty$ and the Early effect, $V_A \rightarrow \infty$. The circuit for the large-signal analysis is shown in Fig. 8.25(b).

As in the previous case, we will use graphical analysis to gain an understanding of the effect of the bias current on the operating point. We plot the collector current versus V_{CE} with the bias current as the parameter in Fig. 8.26(a). The collector current linearly increases with the bias current provided that the device is operated in the forward-active region. In Fig. 8.26(a), we are assuming $V^- = -2.5$ V, so the transistor is biased in the forward-active region. The corresponding current amplifier transfer characteristic is shown in Fig. 8.26(b). To understand the relationship between these two graphs, consider the case when I_{BIAS} is so small that the transistor is cutoff (1). Under this condition, $I_C \approx 0$ A and $I_{OUT} = -I_{SUP}$. As we increase I_{BIAS}, I_C increases and the corresponding output current increases (2). Finally, when I_{BIAS} is further increased, I_{OUT} approaches I_{SUP}.

Now let us quantify the effect of the bias current on the output current. Using the large-signal model for the bipolar transistor and assuming forward-active operation, we know

$$I_C = \beta_F I_B \tag{8.21}$$

In addition, from KCL we know

$$I_{OUT} = \beta_F I_{BIAS} - I_{SUP} \tag{8.22}$$

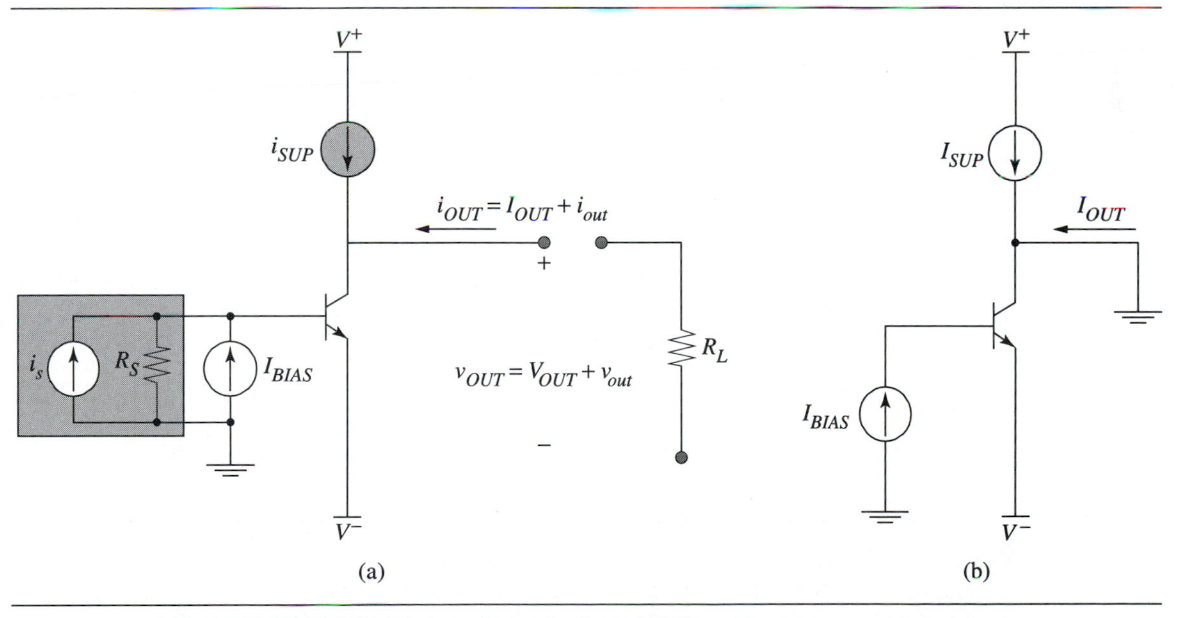

Figure 8.25 Current amplifier using the common-emitter with current source supply configuration (a) including R_S due to i_s and R_L. (b) For large signal analysis $i_s = 0$ A, $R_L = 0\ \Omega$, $R_S \rightarrow \infty$ and $r_{oc} \rightarrow \infty$.

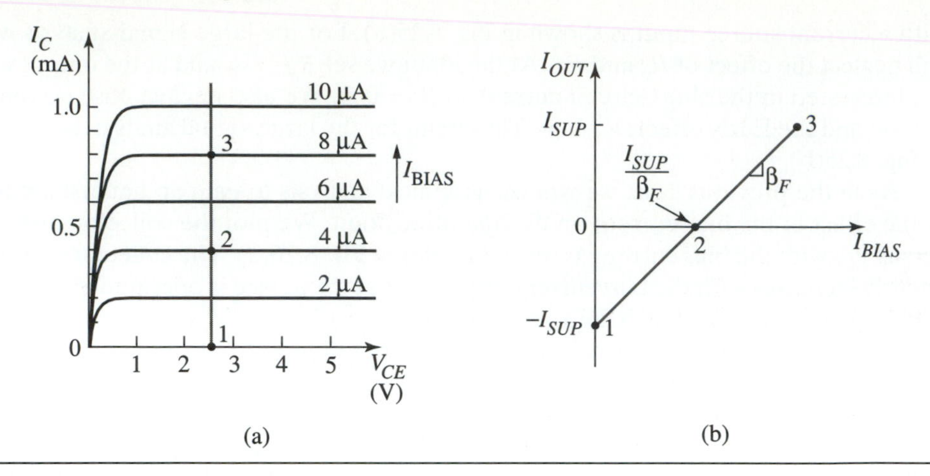

➤ **Figure 8.26** (a) Common-emitter transistor characteristics. (b) Current amplifier transfer characteristic.

Equation 8.22 is the transfer function between the input bias current and short-circuit output current ($R_L = 0\ \Omega$) as shown in Fig. 8.26(b). If we set I_{BIAS} so the collector current I_C is equal to I_{SUP}, I_{OUT} is equal to zero. This bias current yields a symmetrical output current for a small-signal input current and has a value of I_{SUP}/β_F.

Given that the bias current is known, we can perform a small-signal analysis by setting the DC current sources equal to zero. The resulting model is found in Fig. 8.27. To calculate the short-circuit current gain A_i we apply a test current i_t to the input. We begin by noting

CE configuration two-port parameters current gain

$$i_{out} = g_m v_\pi = \beta_o i_b \tag{8.23}$$

$$i_t = \frac{v_\pi}{r_\pi} = i_b \tag{8.24}$$

therefore

$$\boxed{A_i = \frac{i_{out}}{i_t} = g_m r_\pi = \beta_o} \tag{8.25}$$

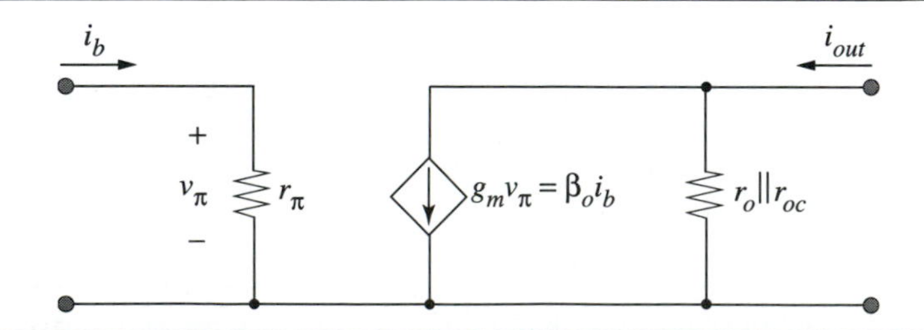

➤ **Figure 8.27** Small-signal model of common-emitter amplifier with current input.

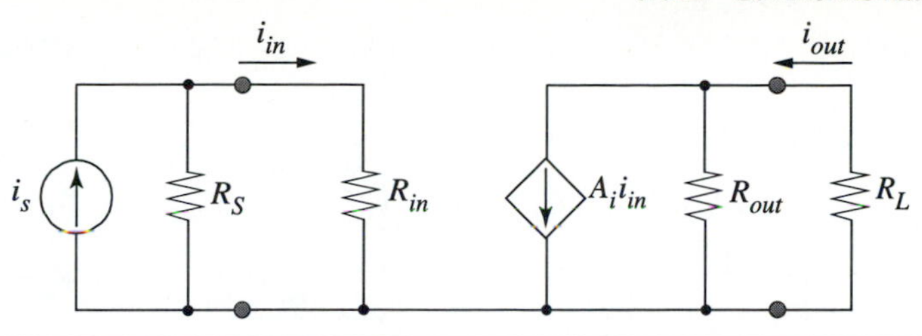

➤ **Figure 8.28** Two-port model of the common-emitter current amplifier with an input current source and associated source resistor and load resistor.

As expected, the current gain is precisely the beta of the transistor. The input and output resistances for the two-port model for an intrinsic current amplifier using a common-emitter configuration are the same as previously calculated. The resulting two-port model is shown in Fig. 8.28.

Now we examine a current amplifier including source and load resistances using the two-port model. The input source is a current source with a source resistance R_S. The output is loaded with a load resistor R_L. The overall current gain is equal to the controlled current source A_i times a resistive divider for the input current to be transferred into the amplifier and the output current to be transferred to the load as shown in

$$\frac{i_{out}}{i_s} = \left(\frac{R_S}{R_S + R_{in}}\right) A_i \left(\frac{R_{out}}{R_{out} + R_L}\right) \tag{8.26}$$

From Eq. (8.26) we see that we require a small input resistance and large output resistance for maximum current gain.

DESIGN EXAMPLE 8.7 Common-Emitter Current Amplifier

Given an input current source with a source resistance of $R_S = 1\text{M}\Omega$, design a common-emitter current amplifier that delivers at least 95 times the input current to a load resistor of $R_L = 1\ \text{k}\Omega$. Design this amplifier with a minimum supply current to meet this requirement. In Fig. Ex 8.7, the supply voltages are ±1.25 V to lower the power dissipation. Verify your design in SPICE.

SOLUTION

To begin this design, we start with the two-port model for the current amplifier using the common-emitter amplifier configuration shown in Fig. 8.28. Using Eq. (8.26), the resulting current gain is

$$\frac{i_{out}}{i_s} = \left(\frac{R_S}{R_S + R_{in}}\right) A_i \left(\frac{R_{out}}{R_{out} + R_L}\right) = \left(\frac{R_S}{R_S + R_{in}}\right) \beta_o \left(\frac{R_{out}}{R_{out} + R_L}\right).$$

To maintain the large current gain, we need R_{in} to be small and R_{out} to be large compared to R_S and R_L, respectively. To meet the current gain of 95 we begin the design by choosing $R_{in} = \frac{R_S}{100}$ and $R_{out} = 100R_L$

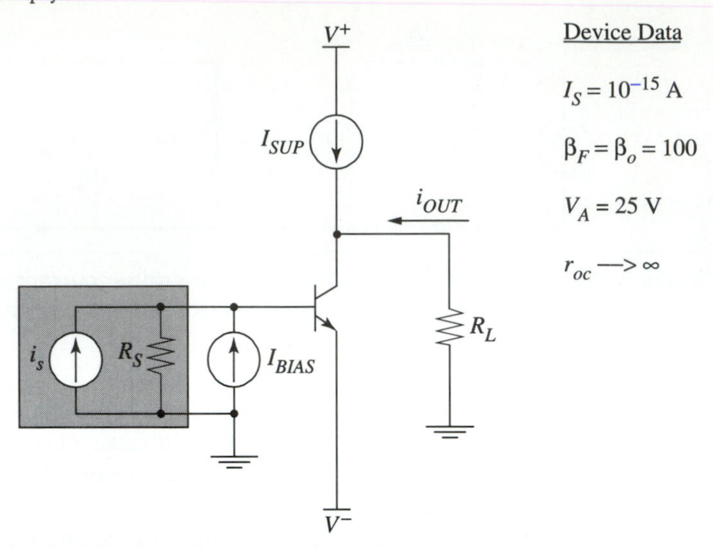

➤ Figure Ex8.7

$$R_{in} = r_\pi = \frac{\beta_o}{g_m} = \beta_o \frac{kT}{qI_C} = \frac{R_S}{100}$$

Let's see how close we are to the specification under this condition. From the equation for R_{in} above

$$I_C = \frac{\beta_o kT}{qR_{in}} = \frac{100 \cdot 25\text{ mV}}{10\text{ k}\Omega} = 250\ \mu\text{A}$$

$$R_{out} = \frac{V_A}{I_C} = \frac{25\text{ V}}{250\ \mu\text{A}} = 100\text{ k}\Omega$$

$$\frac{i_{out}}{i_s} = \left(\frac{1\text{ M}\Omega}{1\text{ M}\Omega + 10\text{ k}\Omega}\right)(100)\left(\frac{100\text{ k}\Omega}{100\text{ k}\Omega + 1\text{ k}\Omega}\right) \approx 98$$

We could further decrease I_C to increase R_{out}, however this action will also increase R_{in}. At this point it is best to use SPICE to iterate to the lowest power solution.

Now we must find the DC bias current I_{BIAS} to set the operating point for the small-signal analysis in SPICE. In this calculation, we neglect the effect of R_S, since it is associated with the small-signal source i_s, which we set to zero for bias calculations. Given that $I_C = \beta_F I_B = \beta_F I_{BIAS}$ and $I_{SUP} = I_C = 250\ \mu\text{A}$, $I_{BIAS} = 2.5\ \mu\text{A}$.

Plugging these circuit parameters into SPICE yields a current gain of 99.6, thus, our design meets the specification. Further iteration is left as an exercise. Before performing the small-signal analysis for this high-gain current amplifier, we had to sweep I_{BIAS} around the approximate value found above. This practice is similar to the voltage amplifiers examined earlier.

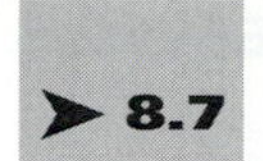

8.7 IMPROVED TRANSCONDUCTANCE AMPLIFIER WITH EMITTER DEGENERATION RESISTOR

The common-emitter and common-source amplifier are natural transconductance amplifiers. However, their input and output resistances can be further increased at the expense of the transconductance by adding a resistor between the emitter/source and ground as shown in Fig. 8.29. Since the MOS input resistance is infinite and its transconductance is generally small, this technique is usually applied only to bipolar transistors.

The emitter resistor causes both the input and output resistance to increase. In addition, the transconductance becomes predominantly dependent on this emitter resistor R_E rather than the intrinsic device g_m. A transconductance amplifier requires high input and output resistance and a well-controlled transconductance value. Integrated circuit technology sometimes has thin film resistors available that are relatively independent of temperature. Recall that the transconductance of a common-emitter amplifier depends on temperature since the transconductance of the device is proportional to $1/T$. The common-emitter amplifier with emitter degeneration resistor is often used as a temperature-stabilized transconductance amplifier with a voltage input and a current output.

Qualitatively, this circuit works as follows. The collector current is governed to first order by the voltage drop across R_E. This is the case because V_{BE} hardly changes as the collector current increases because of the exponential dependence of the collector current on V_{BE}. As the bias voltage increases, we have approximately a 0.7 V drop across the base-emitter junction. The rest of the bias voltage is dropped across the emitter resistance. Therefore, we have approximately a linear relationship between the input voltage and output current, which is dependent on the emitter resistor.

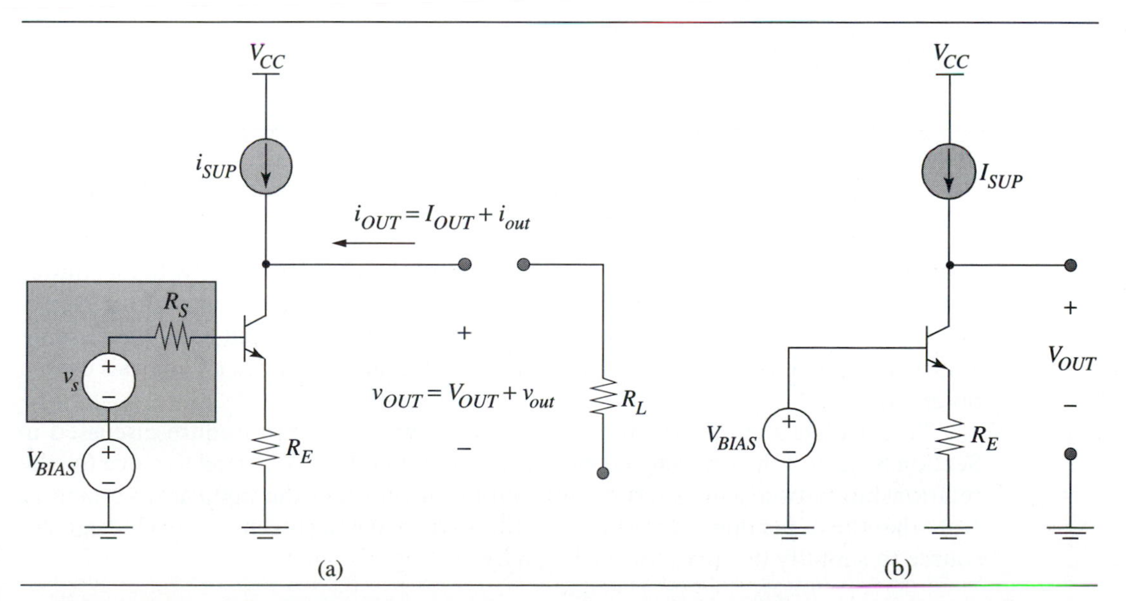

Figure 8.29 Common-emitter amplifier with emitter degeneration resistor (a) including R_S due to v_s and R_L. (b) For large signal analysis $v_s = 0$ V, $R_S = 0\ \Omega$, R_L and $r_{oc} \rightarrow \infty$.

8.7.1 Large-Signal Analysis of CE Amplifier with Emitter Degeneration Resistor

The large-signal analysis of this amplifier is fairly straightforward. The relationship between the bias voltage, base-emitter voltage, and collector current is

$$V_{BIAS} = V_{BE} + \frac{I_C}{\alpha_F}R_E \approx V_{BE} + I_C R_E \tag{8.27}$$

The relation between the base-emitter voltage and collector current, assuming forward-active operation, is

$$V_{BE} = \frac{kT}{q}\ln\left(\frac{I_C}{I_S}\right) \tag{8.28}$$

Given a particular supply current and emitter resistor, we can calculate the necessary bias voltage to ensure that the output current equals zero and that the device is operating in the center of its constant-current region.

➤ EXAMPLE 8.8 CE Amplifier with Emitter Degeneration Resistor: Large-Signal Analysis

Given a bipolar transistor with $\alpha_F = 0.99$ and $I_S = 1 \times 10^{-15}$ A, calculate V_{BIAS} for a supply current of 250 μA and $R_E = 1$ kΩ.

SOLUTION

We want $I_C = I_{SUP}$ to ensure that the output current is equal to zero. From Eq. (8.27) and Eq. (8.28) we can write

$$V_{BIAS} = \frac{kT}{q}\ln\frac{I_{SUP}}{I_S} + \frac{I_{SUP}}{\alpha_F}R_E \approx 0.65\text{ V} + 0.25\text{ V} = 0.9\text{ V}$$

8.7.2 Small-Signal Analysis of CE Amplifier with Emitter Degeneration Resistor

A quantitative small-signal analysis of the common-emitter amplifier with an emitter degeneration resistor can be algebraically complex. To gain some insight in determining the intrinsic transconductance and the input and output resistance, some assumptions will be made so compact relationships suitable for hand analysis can be obtained.

To calculate the intrinsic transconductance, we use the procedure discussed in Section 8.1.2 and the small-signal model shown in Fig. 8.30. We need to solve for the relationship between the short-circuit output current and the input test voltage v_t. Note that the dependent current source will be viewed as a current-controlled current source to simplify the analysis. We begin by writing

$$v_t = i_b r_\pi + (i_{out} + i_b)R_E \tag{8.29}$$

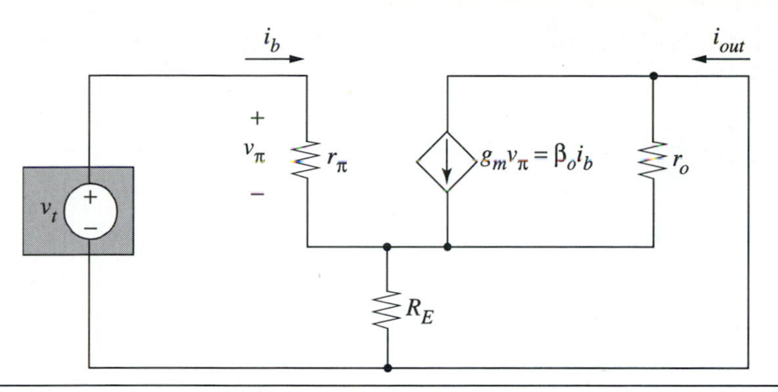

➤ **Figure 8.30** Small-signal model of CE amplifier with emitter degeneration resistor to calculate transconductance.

and

$$i_{out} = \frac{\beta_o i_b r_o}{r_o + R_E} - \frac{i_b R_E}{r_o + R_E} \tag{8.30}$$

Assuming $r_o \gg R_E$ and $\beta_o \gg 1$

$$i_{out} \approx \beta_o i_b \tag{8.31}$$

Substituting Eq. (8.31) into Eq. (8.29)

$$v_t \approx \frac{i_{out} r_\pi}{\beta_o} + i_{out} R_E \tag{8.32}$$

which results in a value for the transconductance

$$\boxed{\frac{i_{out}}{v_t} = G_m = \frac{\beta_o / r_\pi}{1 + (\beta_o / r_\pi) R_E} = \frac{g_m}{1 + g_m R_E}} \tag{8.33}$$

CE amplifier with emitter degeneration resistor two-port parameter: G_m

We see that the transconductance of the amplifier is reduced by the term $(1 + g_m R_E)$. If the product $g_m R_E \gg 1$, the transconductance becomes equal to the reciprocal of R_E. We have built an amplifier with a temperature-independent transconductance if the value of R_E is independent of temperature.

To calculate the input resistance of this amplifier, look at Fig. 8.31 where we are driving the input with a test current source and measuring the voltage across that source. Assuming that $r_o \gg R_E$, all of the current from the dependent current source will flow into resistor R_E. Therefore, we can write

$$v_t = i_t r_\pi + (\beta_o + 1) i_t R_E \tag{8.34}$$

Assuming $\beta_o \gg 1$

$$\boxed{\frac{v_t}{i_t} = R_{in} = r_\pi + \beta_o R_E = r_\pi (1 + g_m R_E)} \tag{8.35}$$

CE amplifier with emitter degeneration resistor two-port parameter: R_{in}

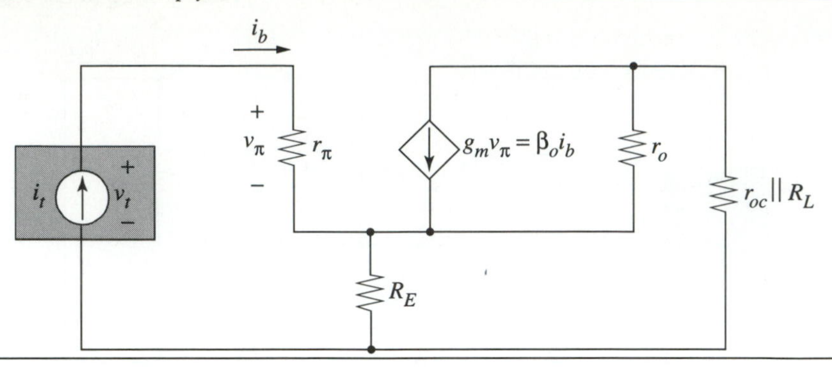

➤ **Figure 8.31** Small-signal model of common-emitter amplifier with emitter degeneration resistor to calculate the input resistance.

Notice that the input resistance is increased by the factor $(1 + g_m R_E)$. This increased input resistance enhances the performance of this amplifier when it is driven with a voltage source.

The output resistance can be calculated using the small-signal model found in Fig. 8.32. We have placed the test current source in between the supply current source's internal resistance r_{oc} and the rest of the amplifier since we know that the output resistance of the remaining circuit will be in parallel with resistor r_{oc}. We begin the analysis by writing

$$i_t = \beta_o i_b + \frac{v_t - v_e}{r_o} \tag{8.36}$$

and the current divider relationship

$$i_b = \frac{-i_t R_E}{r_\pi + R_S + R_E} \tag{8.37}$$

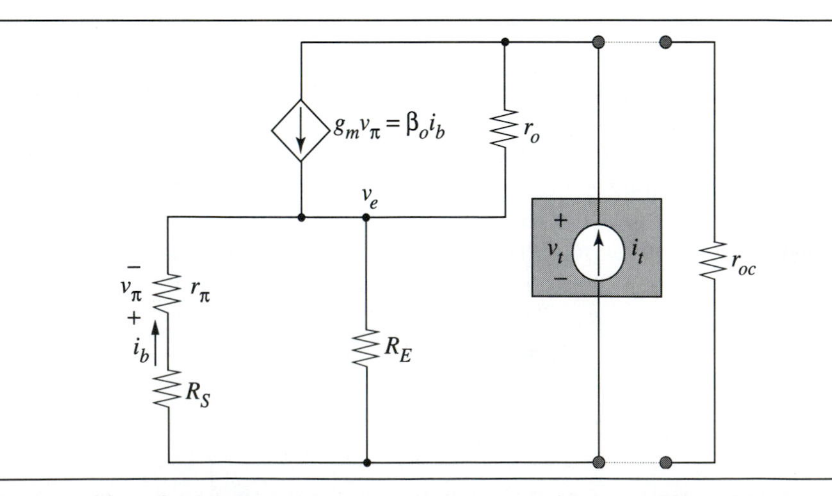

➤ **Figure 8.32** Small-signal model of common-emitter amplifier with emitter degeneration resistor to calculate the output resistance.

Next, we write an expression for the voltage v_e as

$$v_e = i_t[R_E \| (r_\pi + R_S)] \tag{8.38}$$

After some algebraic manipulation we find

$$\frac{v_t}{r_o} = i_t\left(1 + \frac{\beta_o R_E}{r_\pi + R_S + R_E} + \frac{R_E \| (r_\pi + R_S)}{r_o}\right) \tag{8.39}$$

Assuming that $r_o \gg R_E \| (r_\pi + R_S)$

$$\frac{v_t}{i_t} \approx \left(1 + \frac{\beta_o R_E}{r_\pi + R_S + R_E}\right) r_o \tag{8.40}$$

Since the output resistance calculated in Eq. (8.40) is in parallel with the supply current source's internal resistance r_{oc}, we find

$$R_{out} = r_{oc} \| r_o\left(1 + \frac{\beta_o R_E}{r_\pi + R_S + R_E}\right) \tag{8.41}$$

Finally, if we assume $r_\pi \gg R_S + R_E$, we find

$$\boxed{R_{out} \approx r_{oc} \| r_o(1 + g_m R_E)} \tag{8.42}$$

CE amplifier with emitter degeneration resistor two-port parameter: R_{out}

Again, the term $(1 + g_m R_E)$ increases the output resistance of this amplifier, which enhances its performance when current is the output signal. Therefore, the common-emitter amplifier with emitter degeneration resistor performs best as a transconductance amplifier, giving large input and output resistances with a well-controlled transconductance.

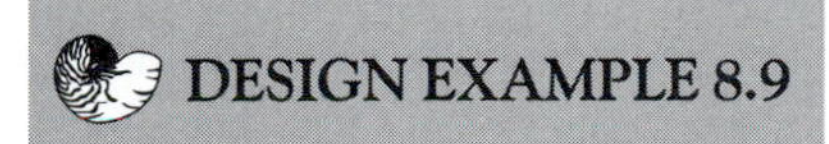

DESIGN EXAMPLE 8.9 CE Amplifier with Reduced Temperature Dependence

Design an amplifier with an overall transconductance that stays within 2% of 1 mS across a temperature range of 0° – 70°C, where $R_S = 1$ kΩ and $R_L = 10$ kΩ. Verify your design in SPICE.

SOLUTION

We start with the two-port model for the common-emitter amplifier with emitter degeneration resistor in Fig. Ex 8.9, where

$$\frac{i_{out}}{v_s} = \frac{R_{in}}{R_{in} + R_S}\left(\frac{g_m}{1 + g_m R_E}\right)\frac{R_{out}}{R_{out} + R_L}$$

We want $R_{in} \gg R_S$, $R_{out} \gg R_L$, and $g_m R_E \gg 1$ in order to reduce the dependence of the overall transconductance i_{out}/v_s on transistor parameters.

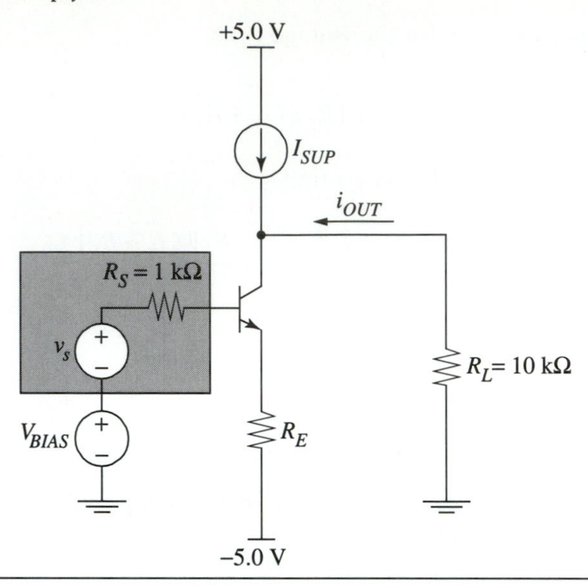

➤ Figure Ex8.9

As a starting point, make $R_{in} \geq 100R_S$, $R_{out} \geq 100R_L$, and $g_mR_E \geq 100$. This ensures that the three terms will contribute less than 1% error. Choose $R_E = 1\text{k}\Omega$ so $G_m \approx 1/R_E$ is approximately 1mS.

Solving for the required g_m of the transistor to make $g_mR_E > 100$ and the corresponding I_C we write

$$g_m = \frac{100}{R_E} = 0.1\ \text{S and } I_C = g_m\frac{kT}{q} = 2.5\text{mA}.$$

From Eq. (8.35) $R_{in} \approx r_\pi(1 + g_mR_E)$

$$\text{Since } r_\pi = \frac{\beta_o}{g_m} = 1\ \text{k}\Omega,$$

$R_{in} \approx 1\ \text{k}\Omega\,(1 + 100) \approx 100\ \text{k}\Omega \geq 100R_S$. So the condition on R_{in} is satisfied.

From Eq. (8.42), $R_{out} \approx r_o(1 + g_mR_E)$

$$\text{Since } r_o = \frac{V_A}{I_C} = 10\ \text{k}\Omega\text{, then } R_{out} \approx 10\ \text{k}\Omega\,(1 + 100) \approx 1\ \text{M}\Omega \geq 100R_L.$$

The condition on R_{out} is satisfied.
Now we find V_{BIAS} to ensure $I_C = 2.5$ mA. Neglecting the voltage drop across R_S, since it is only associated with the small-signal voltage source, we write

$$V_{BIAS} = V_{BE} + \left(\frac{\beta_F + 1}{\beta_F}\right)I_CR_E - V^-$$

$$V_{BE} = \frac{kT}{q}\ln\frac{I_C}{I_S} = 0.714\ \text{V}$$

So $V_{BIAS} = 714\text{ mV} + \frac{101}{100}(2.5\text{ mA} \cdot 1\text{ k}\Omega) - 5.0\text{ V} = 1.761\text{V}$.

Using SPICE we find $G_m = 0.93\text{ mS}$ at 27°C. Since our nominal value is slightly lower than our specification of, $G_m = 1.0\text{ mS}$, we can iterate using SPICE and find that when $R_E = 930\ \Omega$, $G_m = 1.0\text{ mS}$. Running SPICE at 0°C and 70°C, we find that the transductance changes less than 1% from the nominal value.

8.8 COMMON-BASE/GATE AMPLIFIER

In the common-emitter/source amplifier topologies discussed in the previous sections, the input signal was applied to the base/gate and the output signal was taken from the collector/drain. In this section, we will investigate an amplifier topology that will use the emitter/source as the input terminal and the collector/drain as the output terminal. We will find that this transistor amplifier topology has a very low input resistance and a very high output resistance, which is exactly what we want in a current amplifier. Under these conditions, the base/gate of the transistor is common to both the input and output and, hence, this topology is known as the **common-base** or **common-gate amplifier** depending on whether a bipolar or MOS transistor is used.

8.8.1 Common-Base (CB) Amplifier

A circuit diagram for a common-base amplifier is shown in Fig. 8.33(a). The transistor has a current source supply tied to the collector terminal. The input signal enters the emitter and is biased with a DC bias current source I_{BIAS} to ensure the amplifier is operating in its high-gain region. The base of the transistor is tied to an intermediate voltage. In the implementation shown in Fig. 8.33, we use dual power supplies so the base is tied to ground.

As in the previous treatment of amplifiers, we begin analyzing the common-base configuration by performing a large-signal analysis on the circuit in Fig. 8.33(b). We have removed the small-signal source i_s and its associated source resistance R_S, and have set R_L to zero since we will bias the amplifier such that I_{OUT} is zero. By KCL we find

$$I_{OUT} = -\alpha_F I_{BIAS} - I_{SUP} \tag{8.43}$$

For zero bias current, the bipolar transistor will be cutoff and all the supply current will flow out of the output terminal. If the bias current is increased to exactly $-I_{SUP}/\alpha_F$, the collector current will equal the supply current, making the output current equal to zero. This is the point at which we want to bias the transistor.

Once the large-signal collector current is determined, we can analyze the small-signal performance of the common-base amplifier. A small-signal model for this amplifier is shown in Fig. 8.34. We are interested in calculating the short-circuit current gain A_i that is defined as the short-circuit output current for a given small-signal input current. The analysis begins by writing KCL at the top node of the input current source, resulting in

$$i_t = \frac{v_e}{r_\pi} + g_m v_e + \frac{v_e}{r_o} \tag{8.44}$$

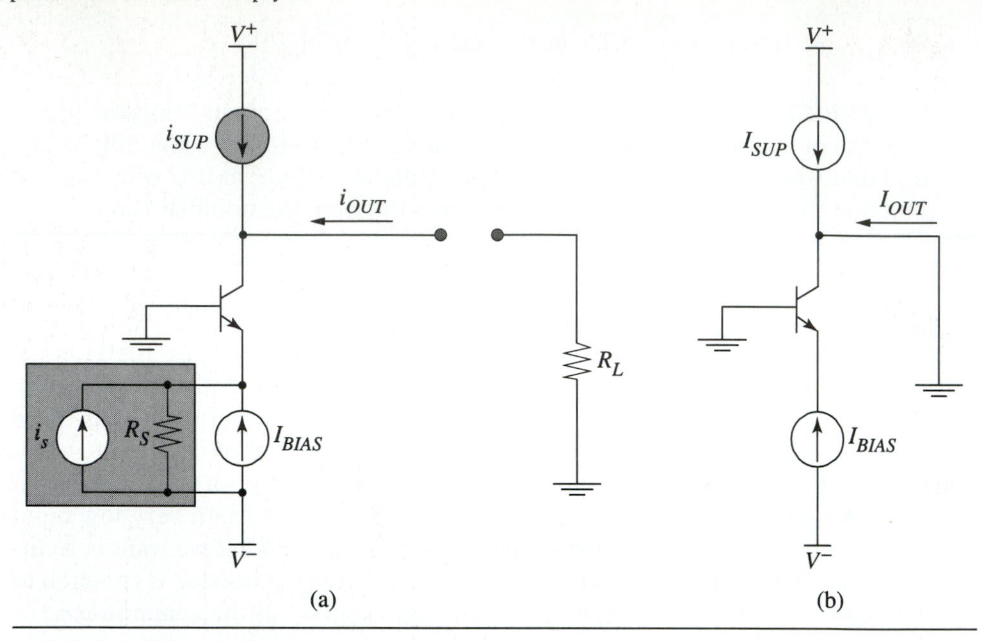

➤ Figure 8.33 Common-base amplifier. (a) including R_S due to i_s and R_L. (b) For large signal analysis $i_s = 0$ A, $R_S \to \infty$, $R_L = 0\ \Omega$ and $r_{oc} \to \infty$.

Also, KCL at the output node results in

$$i_{out} = -g_m v_e - \frac{v_e}{r_o} \tag{8.45}$$

Solving for v_e in Eq. (8.44) results in

$$v_e = \frac{i_t}{1/r_\pi + g_m + 1/r_o} \tag{8.46}$$

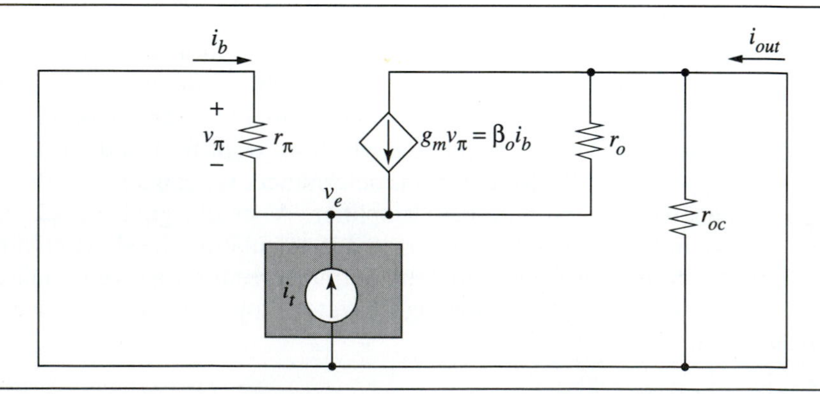

➤ Figure 8.34 Small-signal model of common-base amplifier to calculate current gain.

Substituting Eq. (8.46) into Eq. (8.45), we find the following expression for the current gain

$$\frac{i_{out}}{i_t} = -\frac{g_m r_\pi}{1 + g_m r_\pi + r_\pi / r_o} - \frac{1}{1 + g_m r_o + r_o / r_\pi} \tag{8.47}$$

The second term in Eq. (8.47) is small compared to the first term and $r_\pi / r_o \ll 1$. These simplifications lead to

$$\frac{i_{out}}{i_t} \approx \frac{-g_m r_\pi}{1 + g_m r_\pi} \tag{8.48}$$

Substituting $\beta_o = g_m r_\pi$

$$\boxed{A_i = \frac{-\beta_o}{1 + \beta_o} \approx -1} \tag{8.49}$$

CB amplifier two-port parameter: A_i

for large values of β_o.

The analysis above is algebraically complex. A more intuitive analysis is possible by removing r_o in the small-signal model. This is essentially the same assumption we made in going from Eq. (8.47) to Eq. (8.48). The input current is now equal to the negative of the base current $-i_b$ plus the collector current $-\beta_o i_b$. The output current is equal to $\beta_o i_b$. Hence, the current gain is simply $-\beta_o/(1+\beta_o)$.

In hand circuit analysis it is important to feel comfortable with many approximations. In order to design circuits, very simple models of the transistors and circuits must be used. The assumptions should be verified by computer simulation.

Let us examine the input and output resistance of the common-base amplifier. This amplifier is different from the common-emitter amplifier since the input resistance is a function of the load resistance and the output resistance is a function of the source resistance. This network is non-unilateral because of these interdependencies. In the analysis we will perform, we must take into account the load resistance when looking at the input resistance as shown in Fig. 8.35.

We begin the analysis by writing KCL at the input node, resulting in

$$i_t = \frac{v_t}{r_\pi} + g_m v_t + \frac{v_t(1 - g_m(r_{oc} \| R_L))}{r_o + (r_{oc} \| R_L)} \tag{8.50}$$

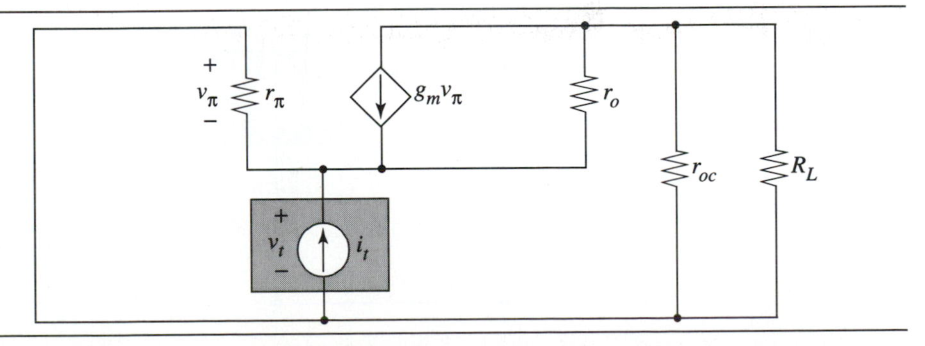

➤ **Figure 8.35** Small-signal model of CB amplifier to calculate input resistance.

Solving for the voltage across our test current source, we find

CB amplifier two-port parameter: R_{in}

$$\frac{v_t}{i_t} = R_{in} = \frac{1}{\frac{1}{r_\pi} + g_m + \frac{1 - g_m(r_{oc} \| R_L)}{r_o + (r_{oc} \| R_L)}} \approx \frac{1}{g_m} \tag{8.51}$$

in which we have used $r_\pi, r_o \gg 1/g_m$, and $r_o, r_{oc} \gg R_L$ which are true in most cases.

To calculate the output resistance, we set the small-signal input source equal to zero (i.e. open circuit) but leave the effect of its source resistance in place as shown in Fig. 8.36. We place a test current source in between the current source supply's internal resistance, r_{oc} and the rest of the circuit, since we know that r_{oc} will end up being in parallel with the resistance that we calculate. This parallel combination will be the total output resistance of the amplifier. The analysis begins by writing

$$i_t = \frac{v_t + v_\pi}{r_o} + g_m v_\pi \tag{8.52}$$

$$v_\pi = -i_t(r_\pi \| R_S) \tag{8.53}$$

Substituting Eq. (8.53) into Eq. (8.52), we get

$$i_t = \frac{\frac{v_t}{r_o}}{1 + \frac{(r_\pi \| R_S)}{r_o} + g_m(r_\pi \| R_S)} \tag{8.54}$$

After some manipulation this becomes

$$\frac{v_t}{i_t} = r_o + (r_\pi \| R_S) + g_m r_o (r_\pi \| R_S) \tag{8.55}$$

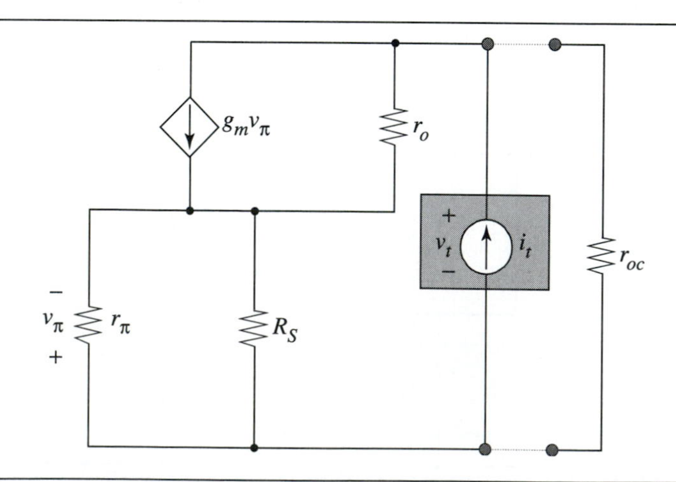

Figure 8.36 Small-signal model of CB amplifier to calculate output resistance.

Placing r_{oc} in parallel we find

$$R_{out} = \left[\left(\frac{1}{r_{oc}} + \frac{1}{r_o + (r_\pi \| R_S) + g_m r_o (r_\pi \| R_S)}\right)\right]^{-1} \tag{8.56}$$

CB amplifier two-port parameter: R_{out}

Assuming $g_m r_o \gg 1$, Eq. (8.56) can be simplified to

$$\boxed{R_{out} \approx r_{oc} \| r_o [1 + g_m (r_\pi \| R_S)]} \tag{8.57}$$

It is useful to examine Eq. (8.57) for the output resistance. Note that it is the parallel combination of the internal resistance of the current source supply r_{oc} with a term $[1+g_m(r_\pi \| R_S)]$ that is multiplied by r_o. If the source resistance is much larger than r_π, the output resistance can be approximated to be $r_{oc} \| \beta_o r_o$. On the other hand, if the source resistance is much smaller than r_π, then the output resistance is approximately $r_{oc} \| g_m r_o R_S$.

Fig. 8.37 shows a two-port diagram of the common-base amplifier. As was previously mentioned, the common-base amplifier is not unilateral so our two-port analysis has some error. (See Appendix A8.1 to quantify the small error introduced by our approach.) Strictly speaking, a two-port cannot be drawn for non-unilateral networks. However, because we are interested in hand analysis and approximate solutions to circuits, the two-port building block approach is extremely valuable to follow the signal flow through complex circuits. Therefore, we have chosen to describe the common-base amplifier as a two-port with the input resistance approximately equal to $1/g_m$, the current controlled current source approximately equal to $-i_{in}$, and the output resistance approximately equal to $r_{oc} \| r_o[1+g_m(r_\pi \| R_S)]$.

The common-base amplifier configuration is an excellent current buffer. We can use a current source with a source resistance that may be only slightly higher than the input resistance of the common-base amplifier, and return nearly the same current with a high resistance at the output of the amplifier. This is precisely the definition of a **current buffer**.

In Chapter 9, which discusses multistage amplifiers, we will see that the common-base amplifier is often used to transform a current source with a medium source resistance into a current source of equal value with a high source resistance, making it a better approximation to an ideal current source.

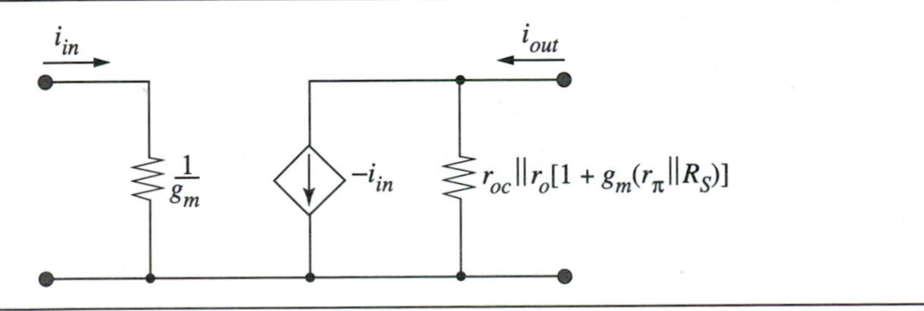

➤ **Figure 8.37** Two-port small-signal model for the common-base amplifier.

Design Example 8.10 Current Buffer

A signal current with a source resistance of 25 kΩ must be increased to have a source resistance of 20 MΩ. Design a common-base amplifier that has an overall short circuit current gain greater than $|0.95|$ and an output resistance greater than 20 MΩ.

SOLUTION

We begin by writing the relationship between the short-circuit output current and i_s using the two-port model given in Fig. 8.37.

$$\frac{-i_{out}|_{R_L = 0\,\Omega}}{i_s} = \frac{-R_S}{R_S + 1/g_m}$$

Make $1/g_m = R_S/25$ so the overall short-circuit gain will be over –0.95. Note that this value yields $A_i = -0.96$. The additional margin will allow for the fact that the intrinsic current gain is not exactly unity, but instead is approximately 0.99 for $\beta_F = 100$.

$$g_m = \frac{25}{R_S} = 1\text{ mS} = \frac{qI_C}{kT}\text{. Thus } I_C = 25\ \mu\text{A}.$$

Let's check R_{out} at this bias condition.

$$r_o = \frac{V_A}{I_C} = 1\text{ M}\Omega \text{ and } r_\pi = \frac{\beta_o}{g_m} = 100\text{ k}\Omega$$

So $R_{out} = r_o(1 + g_m(r_\pi \| R_S)) = 1\text{ M}\Omega(1 + 1\text{ mS}(100\text{ k}\Omega \| 25\text{ k}\Omega)) = 21\text{ M}\Omega$

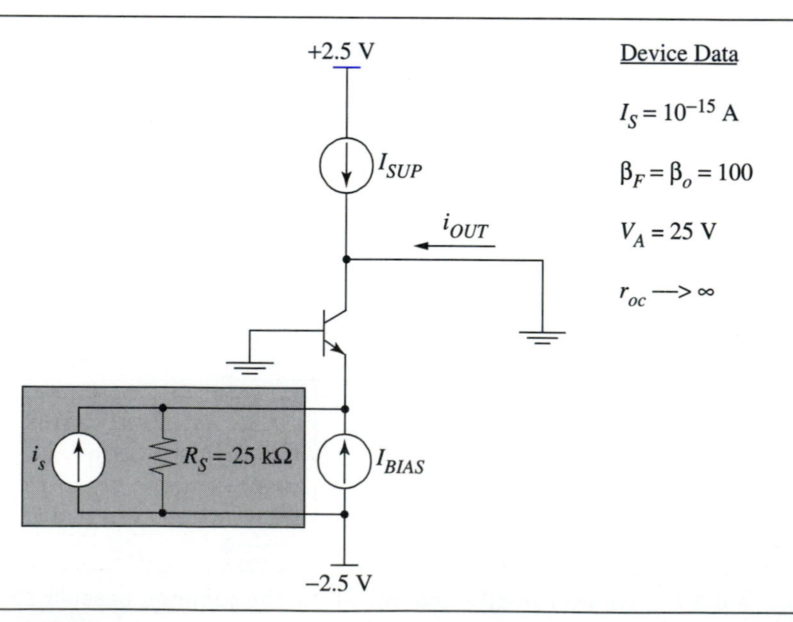

➤ **Figure Ex8.10**

From the DC design requirement, $I_{BIAS} = -\frac{I_C}{\alpha_F} = \frac{-25\ \mu A}{0.99} \approx -25\ \mu A$.

SPICE verifies that our design meets the required specifications $R_{out} = 20.5M\Omega$ and $A_i = -0.953$. When sweeping the DC bias current using SPICE, the program does not remove R_S. Up until this point the voltage drop or current through R_S has been small enough that it had little effect on the value of V_{BIAS} or I_{BIAS}. However, in this example the current through R_S is not negligible and requires $I_{BIAS} = -50\ \mu A$ to ensure the transistor is properly biased and $I_{OUT} = 0$ A.

8.8.2 Common-Gate(CG) Amplifier

The MOS transistor can be configured in a similar manner as the bipolar transistor to form a common-gate amplifier as shown in Fig. 8.38(a). The transistor has a current source supply tied to the drain terminal. The input signal enters the source and is biased with a DC current source I_{BIAS}. The gate of the common-gate amplifier is tied to an intermediate voltage. Since we are using dual power supplies the gate is tied to ground, and the output voltage is referenced to ground.

In Chapter 4 we learned that the MOS transistor is a four-terminal device. In the common-source amplifier, the backgate was tied to the source so it had no effect on the performance of the circuit. However, in the common-gate amplifier, because the source is being used as the input of the device, the backgate may not be able to be tied to the source. Fig. 8.38 shows an n-channel device that has a p-type backgate. If the device is built in a p-well, then the backgate can be tied to the source. If not, the backgate is tied to the most negative supply, which in this circuit is V^-. This configuration results in a voltage between the source and backgate; hence, there is a dependence of the threshold voltage on the source-backgate voltage.

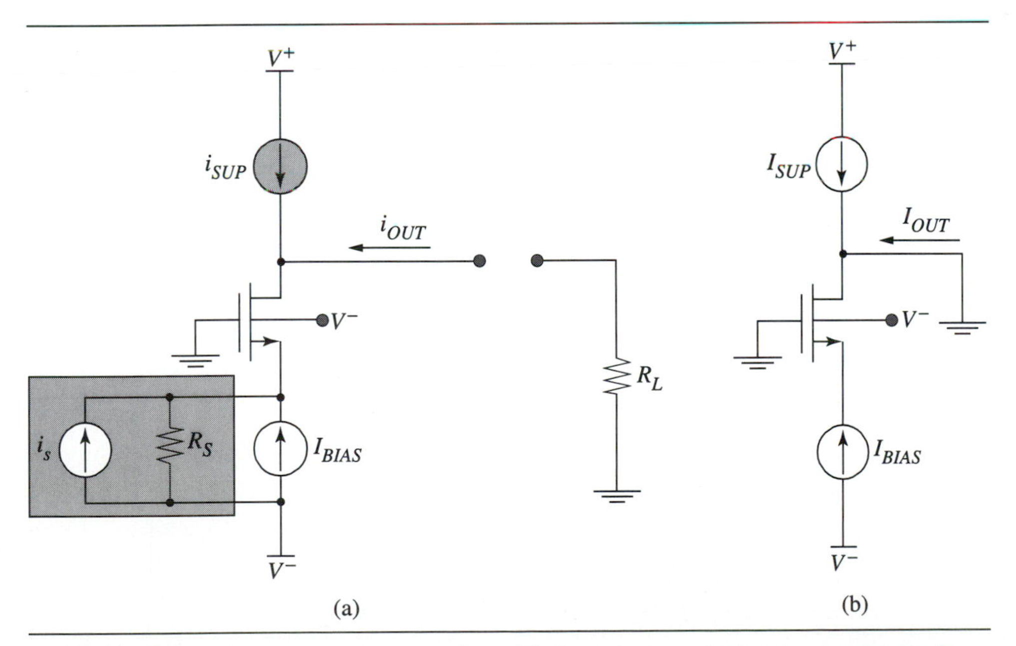

➤ **Figure 8.38** Common-gate amplifier with the substrate tied to the most negative supply. (a) Includes R_S due to i_s and R_L. (b) For large-signal analysis $i_s = 0$ A, $R_S \to \infty$, $R_L = 0\ \Omega$ and $r_{oc} \to \infty$.

To analyze the common-gate amplifier, we begin with the large-signal analysis by removing the small-signal source i_s and its source resistance and setting R_L to zero as shown in Fig. 8.38(b). Performing KCL at the output node results in

$$I_{OUT} = -I_{BIAS} - I_{SUP} \tag{8.58}$$

Note that Eq. (8.58) is exactly the same as that for the bipolar transistor with α_F=1 since there is no base current in an MOS transistor (i.e., the drain current equals the source current). As with the common-base amplifier, we would like to bias this common-gate amplifier with the output current equal to zero. This implies that the bias current should equal the negative of the supply current. Noting that the gate potential is at ground, which is midway between the positive and negative power supplies, we can calculate the source voltage of the MOS transistor necessary to ensure the supply current is equal to $-I_{BIAS}$. Using the large-signal model for the MOS transistor including the backgate bias effect we obtain

$$I_D = I_{SUP} = \frac{W}{2L}\mu_n C_{ox}(V_{GS} - V_{Tn})^2(1 + \lambda_n V_{DS}) \tag{8.59}$$

where

$$V_{Tn} = V_{TOn} + \gamma_n[\sqrt{V_{SB} - 2\phi_p} - \sqrt{-2\phi_p}] \tag{8.60}$$

Neglecting channel length modulation (i.e., $\lambda_n = 0$) and substituting $V_G = 0$ V, we find the equation for the source voltage is

$$V_S = -V_{Tn} - \sqrt{\frac{I_{SUP}}{\frac{W}{2L}\mu_n C_{ox}}} \tag{8.61}$$

If there is a backgate bias $V_S \neq V_B$, V_{Tn} is then a function of V_S and Eq. (8.61) can be solved iteratively with Eq. (8.60).

Given that the bias current is selected such that the output current is equal to zero and the MOS device is in the high-gain region, we can calculate the short-circuit current gain using the small-signal model for the common-gate amplifier as shown in Fig. 8.39.

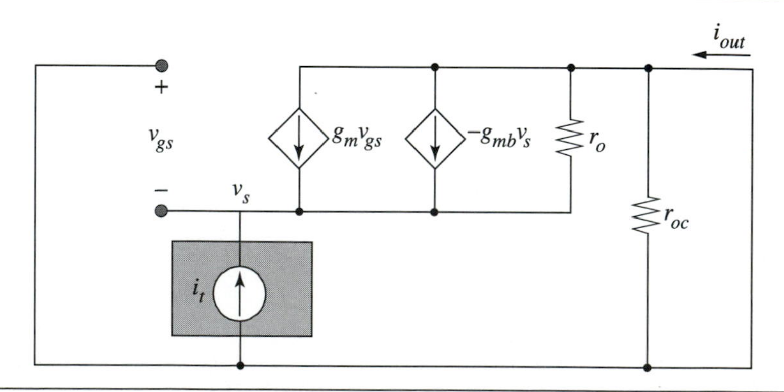

➤ **Figure 8.39** Small-signal model of CG amplifier to calculate short-circuit current gain.

We have drawn both the transconductance and backgate transconductance generator assuming that a voltage between the backgate and source exists—if not, the backgate transconductance generator would be equal to zero. The short-circuit output current can be seen to be equal to the negative of the input current. This implies that the short-circuit current gain A_i of the common-gate amplifier is equal to –1. This result should not be surprising since there is no gate current in the MOS transistor.

CG amplifier two-port parameter: $A_i = -1$

The input resistance of the common-gate amplifier is calculated by using a test current applied at the input and measuring the voltage across that test current source. The circuit to calculate the input resistance is shown in Fig. 8.40.

We begin the analysis by performing KCL at the input node resulting in

$$i_t = g_m v_t + g_{mb} v_t + \frac{v_t - i_t(r_{oc} \| R_L)}{r_o} \tag{8.62}$$

Again, because this is a non-unilateral network, we must include the load resistance when calculating the input resistance of this amplifier. Manipulating Eq. (8.62) results in

$$R_{in} \approx \frac{1 + \dfrac{r_{oc} \| R_L}{r_o}}{g_m + g_{mb} + 1/r_o} \tag{8.63}$$

Assuming $r_o \gg R_L$ and $r_o \gg 1/(g_m + g_{mb})$, the input resistance is approximately

$$\boxed{R_{in} \approx \frac{1}{g_m + g_{mb}}} \tag{8.64}$$

CG amplifier two-port parameter: R_{in}

This is the same result as with the bipolar transistor except for the addition of the backgate transconductance generator due to the fourth terminal in the MOS transistor.

To calculate the output resistance, we set the small-signal input source equal to zero (i.e., open circuit) but leave the effect of its source resistance in place as shown in Fig. 8.41. We place a test current source in between the current source supply's internal resistance, r_{oc} and the rest of the circuit, since we know that r_{oc} will end up

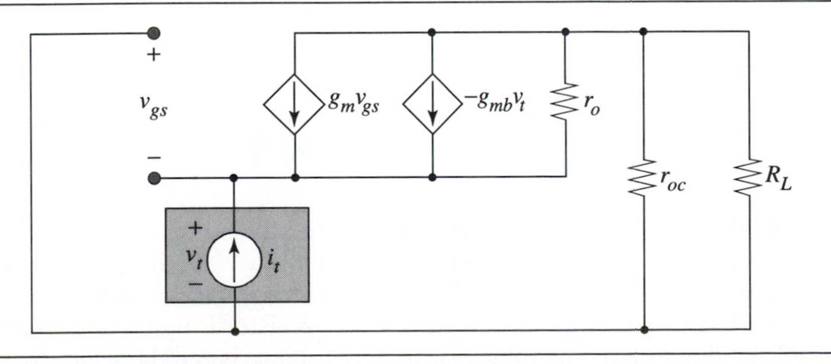

➤ **Figure 8.40** Small-signal model of CG amplifier to calculate input resistance.

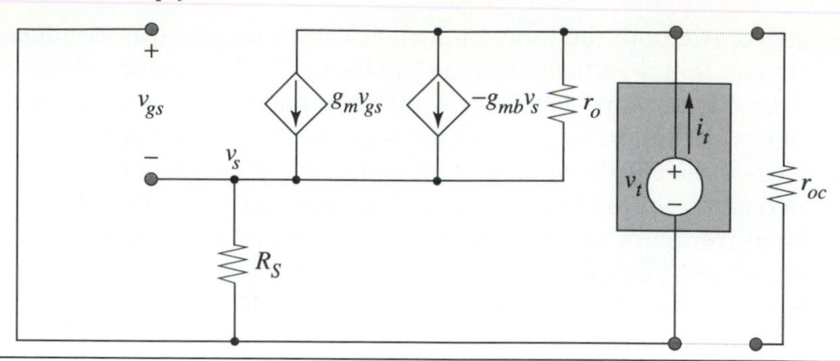

➤ Figure 8.41 Small-signal model of CG amplifier to calculate output resistance.

being in parallel with the resistance that we calculate. This parallel combination will be the total output resistance of the amplifier.

We begin the analysis by noting

$$v_s = i_t R_S \tag{8.65}$$

and writing KCL at the source resistor

$$\frac{v_s}{R_S} = -g_m v_s - g_{mb} v_s + \frac{v_t - v_s}{r_o} \tag{8.66}$$

Solving Eq. (8.66) for v_s we get

$$v_s = \frac{v_t}{r_o(1/R_S + g_m + g_{mb} + 1/r_o)} \tag{8.67}$$

Substituting Eq. (8.65) into Eq. (8.67) results in

$$\frac{v_t}{i_t} = R_S\left(\frac{r_o}{R_S} + g_m r_o + g_{mb} r_o + 1\right) \tag{8.68}$$

Placing r_{oc} in parallel with the resistance calculated in Eq. (8.68) leads to

$$R_{out} = \left[r_{oc} \| \left(\frac{r_o}{R_S} + g_m r_o + g_{mb} r_o + 1\right) R_S\right] \tag{8.69}$$

Assuming $g_m \gg g_{mb}$ and $r_o \gg R_S$ yields

CG amplifier two-port parameter: R_{out}

$$\boxed{R_{out} \approx r_{oc} \| r_o(1 + g_m R_S)} \tag{8.70}$$

Note that the source resistance, R_S is multiplied by the open-circuit voltage gain of the transistor $g_m r_o$ similar to the bipolar common-base amplifier. In the common-base amplifier, we saw from Eq. (8.57) that the source resistance was in parallel with r_π. Since the MOS transistor has no r_π, this source resistor is multiplied directly by its open-circuit voltage gain. The value of output resistance is very large due to this feedback effect on the source resistance. Therefore, the common-gate amplifier also

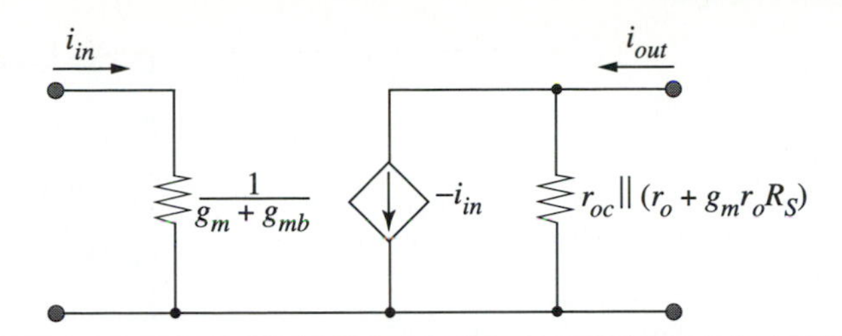

➤ **Figure 8.42** Two-port small-signal model of common-gate amplifier.

operates as an excellent current buffer with a low input resistance, unity current gain, and a high output resistance.

The transconductance should be increased to lower the input resistance. This can be accomplished in an MOS transistor by increasing the supply current or the width-to-length ratio *W/L* of the device. To increase the output resistance, the transconductance can be increased until the limitation becomes the internal resistance of the current source supply r_{oc}.

When making a two-port model of the common-gate amplifier, it should be noted that the network is not unilateral so a slight error in using this two-port for a transfer function will occur. Appendix A8.1 quantifies this small error. However, for hand analysis, the two-port shown in Fig. 8.42 is very useful.

➤ EXAMPLE 8.11 Effect of Varying I_{SUP} on the CG Amplifier Small-Signal Parameters

In this example, we explore the effect of changing the supply current I_{SUP} and *W/L* on the input and output resistance of the common-gate amplifier. Suppose we are given the common-gate amplifier circuit in Fig. Ex 8.11 and are required to have a maximum input resistance of 1kΩ and a minimum output resistance of 1MΩ.

(a) Assume $I_{SUP} = 10\ \mu A$. Calculate the *W/L* ratio necessary to meet the specifications.

(b) If I_{SUP} is increased to 1mA, calculate the new *W/L* required to meet the specifications.

SOLUTIONS

(a) Begin the analysis by calculating the g_m that is required for the input resistance specification. Since $V_{SB} = 0$ V, Eq. (8.64) becomes

$$g_m = \frac{1}{R_{in}} = 1\,\text{mS} = \sqrt{2\frac{W}{L}I_{SUP}\mu_n C_{ox}}$$

$$\frac{W}{L} = \frac{(g_m)^2}{2I_{SUP}\mu_n C_{ox}} = \frac{(1\ \text{mS})^2}{2\,(10\ \mu\text{A})\,(50\ \mu\text{A}/\text{V}^2)} = 1000$$

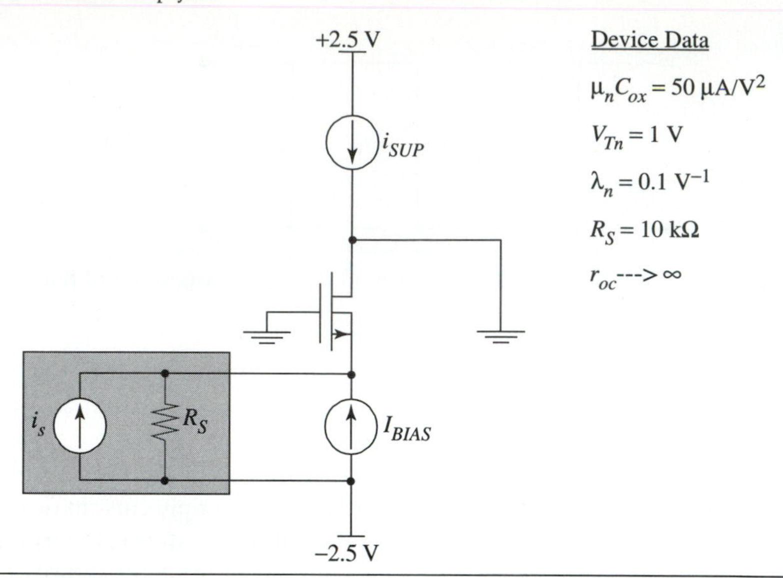

➤ Figure Ex8.11

Since r_{oc} is considered infinite in this example, Eq. (8.70) becomes

$$R_{out} = r_o + g_m r_o R_S$$

where

$$r_o = \frac{1}{\lambda_n I_{SUP}} = \frac{1}{(0.1\text{V}^{-1})(10\mu\text{A})} = 1\ \text{M}\Omega.$$

Thus $g_m r_o = 1000$ and $R_{out} = 1\ \text{M}\Omega + (1000)\ 10\ \text{k}\Omega = 11\ \text{M}\Omega$.

This meets the minimum output resistance specification.

(b) Rather than grind through the equations again, we realize that $g_m \propto \sqrt{I_{SUP}}$ which implies for $W/L = 1000$ that the new $g_m = 10$ mS and $R_{in} = 100\ \Omega$. These values exceed our specifications.

However, $r_o \propto 1/I_{SUP}$. The new $r_o = 10$ kΩ, and $R_{out} = 10\ \text{k}\Omega + (100)10\text{k}\Omega = 1.01$ MΩ which just meets the specifications.

This example demonstrates that for low currents, the transconductance is small, hence, the input resistance for the common-gate amplifier is large. The output resistance of the device is also large, yielding a large overall output resistance for the common-gate amplifier. As the current increases, the input resistance specification is easier to meet, but the output resistance specification becomes more difficult since both r_o and $g_m r_o$ are decreasing as a function of increasing current.

➤ 8.9 COMMON-COLLECTOR/DRAIN AMPLIFIER

In the common-base/gate amplifier we used the emitter/source as an input and the collector/drain as an output. For common-collector/drain amplifier we will use the base/gate as the input and the emitter/source as the output, leaving the collector/

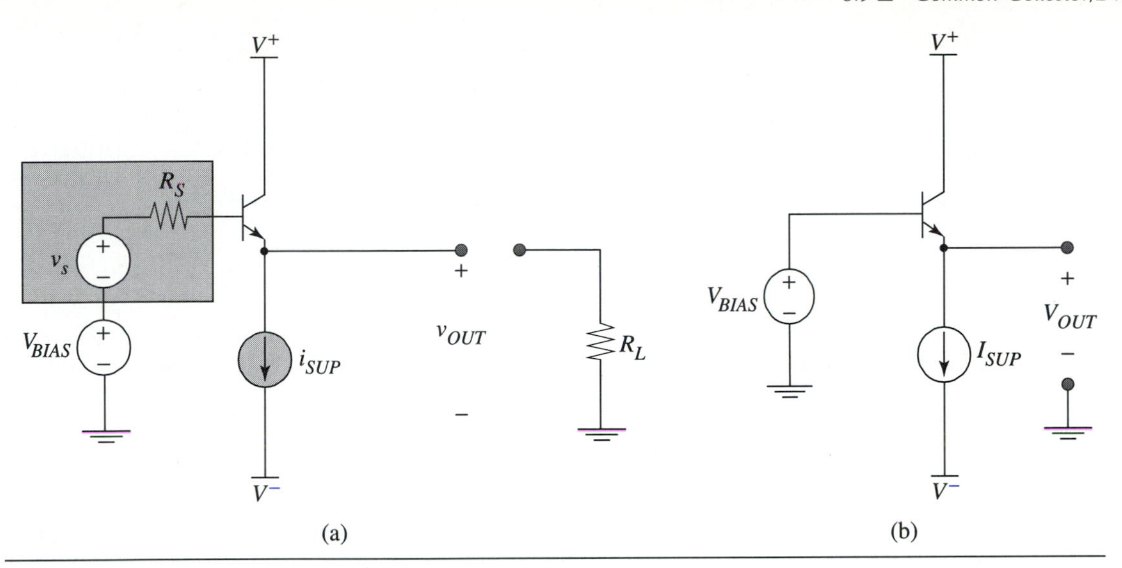

➤ **Figure 8.43** Common collector amplifier (a) Including v_s and its associated R_S and R_L. (b) For large-signal analysis $v_s = 0$ V, $R_S = 0\ \Omega$, $R_L \rightarrow \infty$ and $r_{oc} \rightarrow \infty$.

drain as the common terminal between input and output. This circuit is used as a voltage buffer with a high input resistance and low output resistance. The common-collector/drain amplifier is often called an emitter/source-follower since the output voltage follows the input voltage.

8.9.1 Common-Collector (CC) Amplifier

The circuit for the common-collector amplifier is shown in Fig. 8.43(a). We have a current source supply tied to the emitter where the output voltage is taken. The input bias voltage and signal source are tied to the base. The collector is tied directly to the positive power supply V^+.

Next, we perform a large-signal analysis on this circuit and determine the relationship between the output voltage and input bias voltage. Again we remove v_s, R_S, R_L and neglect r_{oc} for these bias calculations as shown in Fig. 8.43(b). Using KVL we can write

$$V_{BIAS} - V_{BE} = V_{OUT} \tag{8.71}$$

where

$$V_{BE} = \frac{kT}{q} \ln \frac{I_{SUP}}{I_S} \tag{8.72}$$

Note that the output voltage goes from V^- to $(V^+ - V_{BE})$, when the bias voltage moves from V^- to V^+. To bias the transistor in the high-gain region and have a large signal swing, we want $V_{OUT} = 0$ V. From Eq. (8.71), we set $V_{BIAS} = V_{BE}$.

As with the other amplifier configurations, we now perform a small-signal analysis around the DC operating point. The small-signal model to calculate the open-circuit voltage gain for the common-collector amplifier is shown in Fig. 8.44.

To calculate the open-circuit voltage gain A_v between the output and input signal source, we write KCL at the output node as

$$\frac{v_t - v_{out}}{r_\pi} + g_m(v_t - v_{out}) = \frac{v_{out}}{r_o \| r_{oc}} \tag{8.73}$$

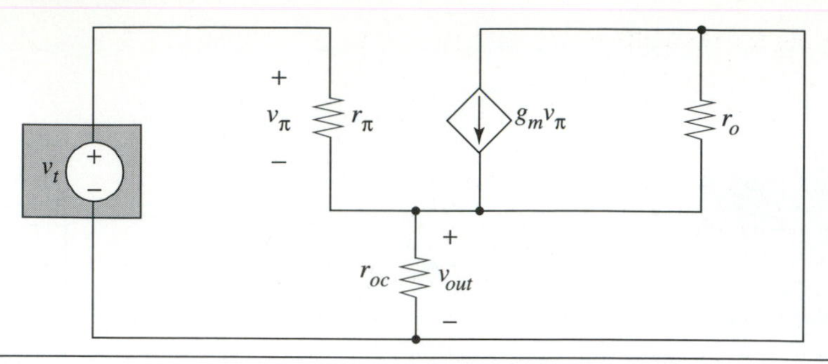

➤ **Figure 8.44** Small-signal model of CC amplifier to calculate the open-circuit voltage gain.

Recognizing that $g_m r_\pi$ is equal to β_o we write

$$v_t(1+\beta_o) - v_{out}(1+\beta_o) = \frac{v_{out} r_\pi}{r_o \| r_{oc}} \tag{8.74}$$

Rearranging terms, we are left with a voltage gain

CC amplifier two-port parameter: A_v

$$\frac{v_{out}}{v_t} = A_v = \frac{1}{1 + \dfrac{r_\pi}{(r_o \| r_{oc})(\beta_o + 1)}} \tag{8.75}$$

For the condition where $\beta_o\ (r_o \| r_{oc})$ is much greater than r_π the voltage gain is approximately equal to 1. This is a good assumption under most conditions, and we will assume that the value for the voltage-controlled voltage source in the common-collector amplifier two-port model is $A_v \approx 1$.

The most important parameters in the common-collector amplifier are the input and output resistances since we will be using it as a voltage buffer. We would like to have a high input resistance and a low output resistance to ensure that most of the voltage is transferred from the input source to the load. The input resistance can be determined from the circuit in Fig. 8.45.

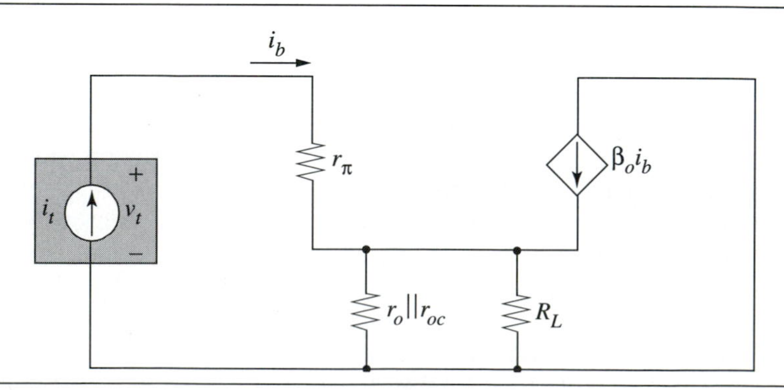

➤ **Figure 8.45** Small-signal model of CC amplifier to calculate input resistance.

Solving for the voltage across the test current source we get

$$v_t = i_t r_\pi + (\beta_o + 1) i_t (r_o \| r_{oc} \| R_L) \tag{8.76}$$

Assuming $\beta_o \gg 1$ we can solve for the input resistance as

$$\boxed{R_{in} \approx r_\pi + \beta_o (r_o \| r_{oc} \| R_L)} \tag{8.77}$$

CC amplifier two-port parameter: R_{in}

Notice that the resistors in the emitter are multiplied by β_o. The input resistance is significantly increased due to the feedback effect of the resistors in the common-collector configuration. We will explore this point further in Chapter 12.

To calculate the output resistance of the common-collector amplifier, we use the circuit in Fig. 8.46. We determine the current out of the test voltage source as

$$i_t = \frac{v_t}{r_\pi + R_S} + \frac{v_t}{r_o \| r_{oc}} - g_m v_\pi \tag{8.78}$$

We write the voltage divider equation for v_π as

$$v_\pi = -\frac{v_t r_\pi}{r_\pi + R_S} \tag{8.79}$$

Substituting Eq. (8.79) into Eq. (8.78) and assuming $\beta_o \gg 1$, we obtain the output resistance as

$$R_{out} = \frac{(r_\pi + R_S)/\beta_o}{\left(1 + \dfrac{r_\pi + R_S}{\beta_o (r_o \| r_{oc})}\right)} \approx \frac{1}{g_m} + \frac{R_S}{\beta_o} \tag{8.80}$$

CC amplifier two-port parameter: R_{out}

Note that the output resistance is equal to $1/g_m$ plus the effect from the source resistance divided by β_o. If the source resistance is larger than $\beta_o/g_m = r_\pi$, this effect must be taken into account. Again, since this circuit is not unilateral, we see that the output resistance can be dependent on the source resistance and the input resistance is dependent on the load resistance. Appendix A8.2 quantifies the error in the two-port model for the common-collector and common-drain amplifiers.

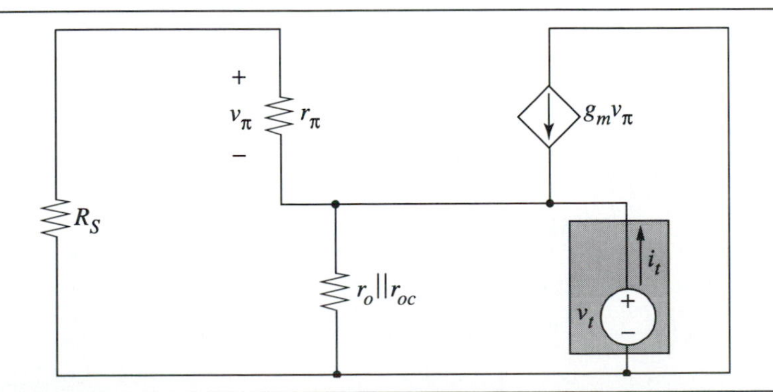

➤ **Figure 8.46** Small-signal model of CC amplifier to calculate output resistance.

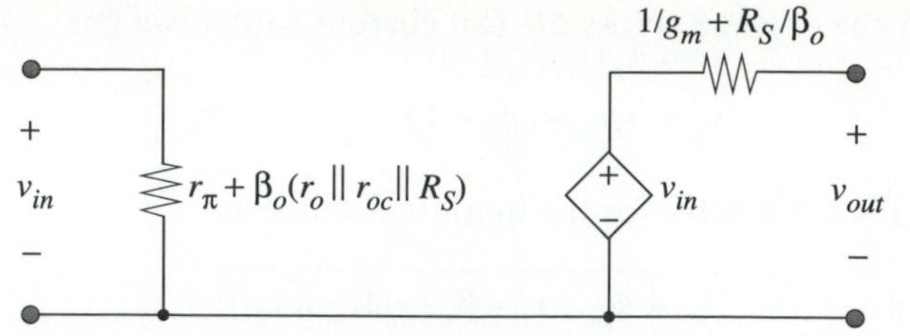

➤ **Figure 8.47** Two-port small-signal model for the common-collector amplifier.

The two-port model for the common-collector amplifier is shown in Fig. 8.47. A large β_o and g_m give the desired result of a high input resistance and a low output resistance.

DESIGN EXAMPLE 8.12 Voltage Buffer Design

You are given a signal voltage source with a source resistance of 1kΩ. Using the circuit and device data in Fig. Ex 8.12A, design a common-collector amplifier setting the value of the current source supply I_{SUP} as small as possible. You need to drive a 50 Ω load with a voltage gain greater than or equal to 0.6. Verify your design in SPICE.

SOLUTION

This design involves finding the minimum I_{SUP} to meet the specification. Begin by using the two-port model in Fig. 8.47 and adding R_S and R_L as shown in Ex 8.12B.

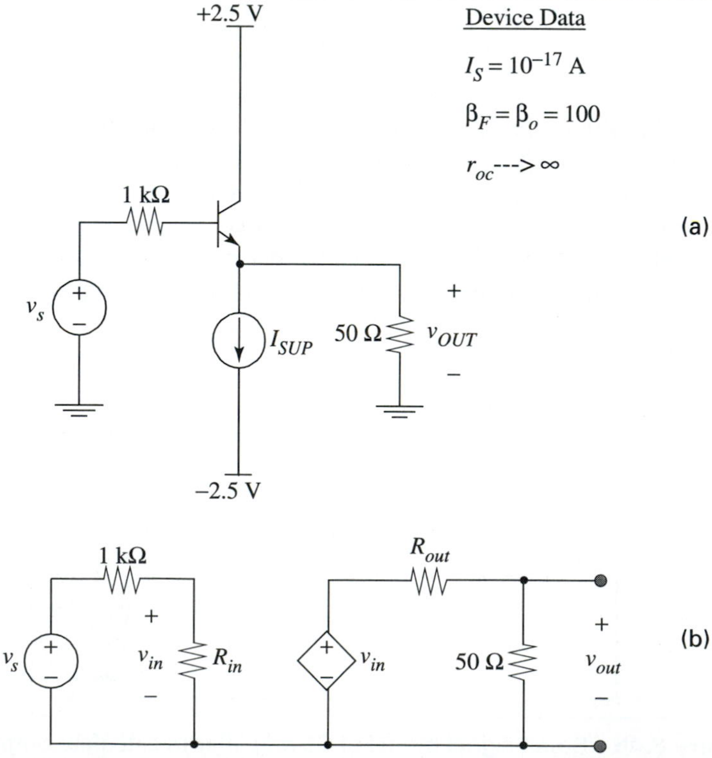

Device Data

$I_S = 10^{-17}$ A

$\beta_F = \beta_o = 100$

r_{oc} ---> ∞

➤ **Figure Ex8.12**

Since our overall gain must be greater than or equal to 0.6, do a first cut design by having the gain $v_{in}/v_s = 0.8$ and $v_{out}/v_{in} = 0.8$. This criterion requires that

$$\frac{R_{in}}{R_{in} + 1\ \text{k}\Omega} \geq 0.8 \ \text{and}\ \frac{50\ \Omega}{50\ \Omega + R_{out}} \geq 0.8$$

Therefore, $R_{in} \geq 4\ \text{k}\Omega$ and $R_{out} \leq 12.5\ \Omega$

From Eq. (8.77) and realizing that r_o, $r_{oc} \gg 50\ \Omega$, we write

$$R_{in} = r_\pi + \beta_o R_L = \frac{\beta_o}{g_m} + \beta_o R_L = \frac{\beta_o}{g_m} + 5\ \text{k}\Omega$$

Thus the R_{in} specification is exceeded no matter what the value of g_m.
From Eq. (8.80) we can write

$$R_{out} = \frac{1}{g_m} + \frac{R_S}{\beta_o}$$

which implies that $g_m = \dfrac{1}{R_{out} - \dfrac{R_S}{\beta_o}} = \dfrac{1}{12.5\Omega - \dfrac{1\,\text{k}\Omega}{100}} = \dfrac{1}{2.5\Omega}$

Since,

$$g_m = \frac{qI_{SUP}}{kT},\ I_{SUP} = 10\ \text{mA}.$$

To further optimize, we can decrease the input resistance and increase the output resistance. On this iteration we try

$\dfrac{R_{in}}{R_{in} + 1\ \text{k}\Omega} \geq 0.9$ and $\dfrac{50\ \Omega}{50\ \Omega + R_{out}} \geq 0.7$. Applying the same analysis as shown above we find $I_{SUP} = 2.17$ mA.

Computing in SPICE, we find the lowest attainable value of I_{SUP} is 1.61mA. With this value of I_{SUP} we find $R_{in} = 8.7\ \text{k}\Omega$ and $R_{out} = 20\ \Omega$ and the overall voltage gain meets the specification.

8.9.2 Common-Drain(CD) Amplifier

The circuit for the common-drain amplifier is shown in Fig. 8.48(a). The drain of the n-channel transistor is tied to the power supply V^+. The input bias and signal voltage sources are applied to the gate and the current source supply is tied to the source of the transistor. The output voltage is taken from the source of the transistor. Because the MOS transistor is a four-terminal device, we must look at the technology to decide if the backgate terminal can be shorted to the source or must be tied to the

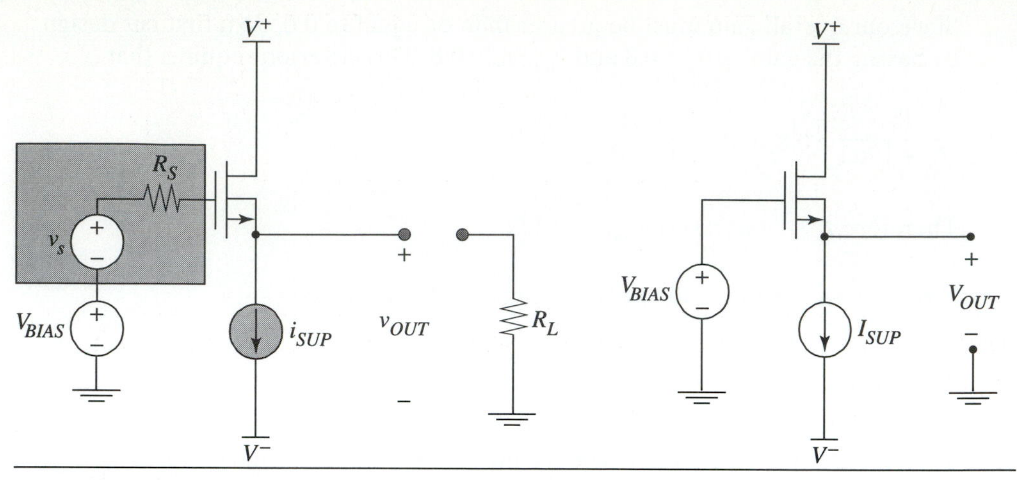

➤ **Figure 8.48** Common drain amplifier.(a) Including v_s and its associated R_S and R_L. (b) For large-signal analysis $v_s = 0$ V, $R_S = 0\ \Omega$, $R_L \rightarrow \infty$ and $r_{oc} \rightarrow \infty$.

most negative supply. The backgate and source can be shorted if the n-channel device is built in an isolated p-well, which removes the dependence of the threshold voltage on the backgate-source voltage. If this is not possible, the threshold voltage will change as the output voltage changes since a backgate-source voltage will be developed.

To calculate a relationship between the input bias voltage and the output voltage, we use the circuit in Fig. 8.48(b) and write

$$V_{BIAS} - V_{GS} = V_{OUT} \tag{8.81}$$

From the large-signal model of the MOS transistor and neglecting channel length modulation, write

$$V_{GS} = V_{Tn} + \sqrt{\frac{I_{SUP}}{\frac{W}{2L}\mu_n C_{OX}}} \tag{8.82}$$

where

$$V_{Tn} = V_{TOn} + \gamma_n [\sqrt{V_{SB} - 2\phi_p} - \sqrt{-2\phi_p}] \tag{8.83}$$

If the backgate and source voltage can be maintained at the same potential, then the threshold voltage reduces to a simple expression and the output voltage is directly proportional to the bias voltage. However, if the backgate is not shorted to the source, a non-linearity is produced between the output voltage and the bias voltage.

As was the case with the common-collector amplifier, we expect the common-drain amplifier to have a voltage gain of approximately one, a large input resistance, and a small output resistance. To calculate these two-port parameters, we begin the small-signal analysis with the circuit in Fig. 8.49. For generality, we include the effect of the backgate transconductance generator.

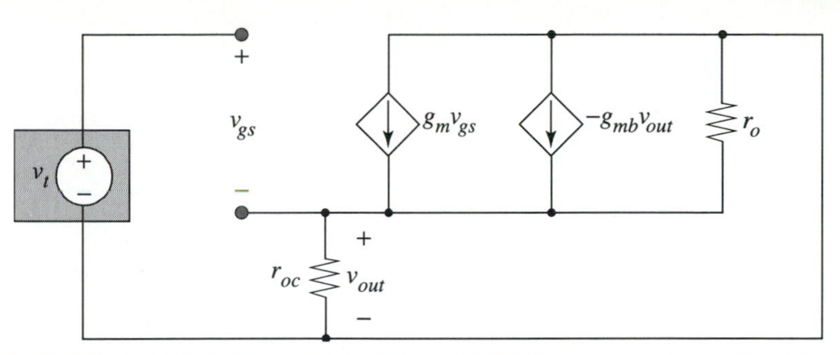

➤ Figure 8.49 Small-signal model of CD amplifier to calculate the open-circuit voltage gain.

To calculate the open-circuit voltage gain, A_v

$$v_{gs} = v_t - v_{out} \quad \text{and} \tag{8.84}$$

$$g_m v_{gs} = v_{out}\left(\frac{1}{r_{oc}} + \frac{1}{r_o} + g_{mb}\right) \tag{8.85}$$

Substituting Eq. (8.84) into Eq. (8.85) we get

$$\boxed{\frac{v_{out}}{v_t} = A_v = \frac{g_m}{\dfrac{1}{(r_o \| r_{oc})} + g_m + g_{mb}} \approx \frac{g_m}{g_m + g_{mb}}} \tag{8.86}$$

CD amplifier two-port parameter: A_v

As expected, the common-drain amplifier has a voltage gain near unity since $r_o \| r_{oc} \gg 1/(g_m + g_{mb})$. It is important to note that if the source and backgate are not tied to the same voltage, the backgate transconductance generator degrades the voltage gain of the common-drain amplifier. A typical gain for a common-drain amplifier is approximately 0.8 if the backgate transconductance generator is allowed to degrade the gain.

CD amplifier two-port parameter: $R_{in} \to \infty$

The input resistance of the common-drain amplifier is infinite because the insulating gate of the MOS transistor passes no current. To calculate the output resistance of the common-drain amplifier, apply a test voltage at the source and measure the current out of that test voltage with the input signal source set equal to zero, as shown Fig. 8.50.

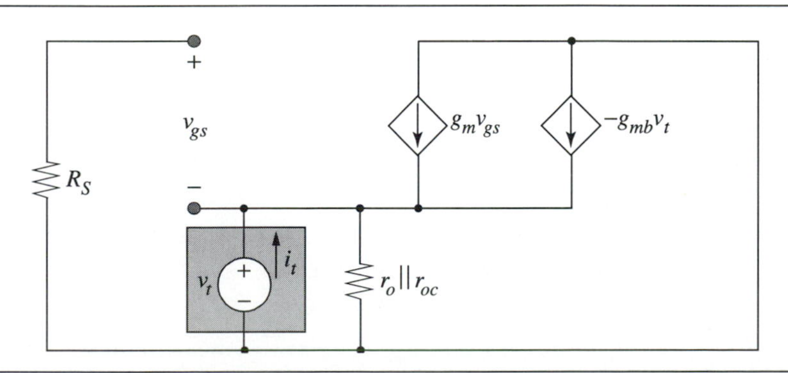

➤ Figure 8.50 Small-signal model of CD amplifier to calculate output resistance.

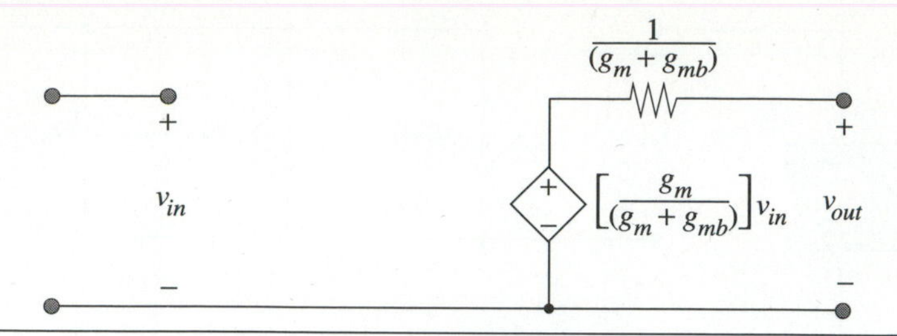

➤ **Figure 8.51** Two-port small-signal model of common-drain amplifier.

We can write the current out of the test source as

$$i_t = g_m v_t + g_{mb} v_t + \frac{v_t}{r_o \parallel r_{oc}} \tag{8.87}$$

Rearranging Eq. (8.87) we get

$$\frac{v_t}{i_t} = \frac{1}{\frac{1}{r_o \parallel r_{oc}} + (g_m + g_{mb})} \tag{8.88}$$

CD amplifier two-port parameter: R_{out}

Finally, by assuming that $r_o \parallel r_{oc} \gg 1/(g_m + g_{mb})$, we find

$$\boxed{R_{out} \approx \frac{1}{g_m + g_{mb}}} \tag{8.89}$$

The two-port model for the common-drain amplifier is shown in Fig. 8.51.

➤ EXAMPLE 8.13 Effect of Backgate Transconductance on CD Amplifier

In this example we will explore the effect of the backgate transconductance on both the voltage gain and output resistance of the common-drain amplifier. Given the common-drain amplifier in Fig. Ex 8.13, calculate the voltage gain and output resistance. Verify in SPICE.

SOLUTION

Begin this analysis by finding the large signal voltage at the MOS transistor's source since it is required to calculate the g_{mb}.

Recall $I_D = \frac{W}{2L}\mu_n C_{ox}(V_{GS} - V_{Tn})^2$ and $V_{Tn} = V_{TOn} + \gamma_n(\sqrt{V_{SB} - 2\phi_p} - \sqrt{-2\phi_p})$

Solving these two equations simultaneously, we find $V_S = V_{OUT} = 0.937\text{ V}$. Given the parameters above, we can solve for

$$g_m = \sqrt{2\frac{W}{L}\mu_n C_{ox} I_D} = 707\ \mu\text{S} \text{ and } g_{mb} = \frac{g_m \gamma_n}{2\sqrt{V_{SB} - 2\phi_p}} = 171\ \mu\text{S}$$

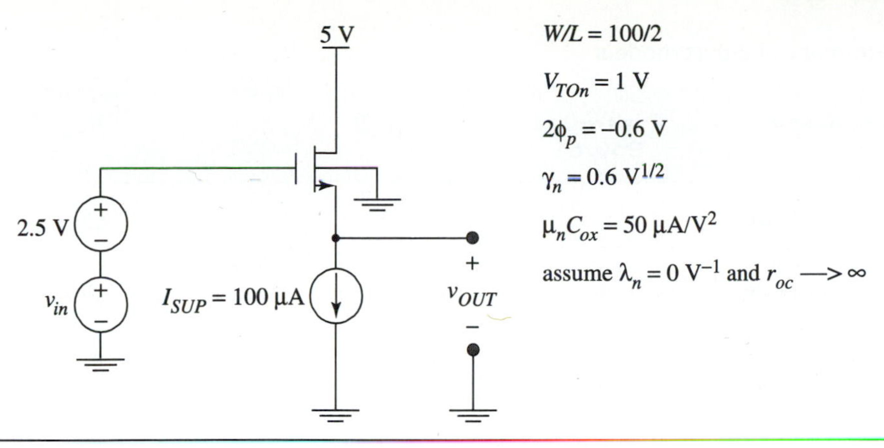

➤ **Figure Ex8.13**

The resulting open-circuit voltage gain is

$$A_v = \frac{g_m}{g_m + g_{mb}} = 0.805$$

and $R_{out} = \dfrac{1}{g_m + g_{mb}} = 1.14\ \text{k}\Omega$.

SPICE shows that the backgate transconductance degrades the voltage gain from nearly unity to 0.805. If at all possible, we should try to use the device type that allows $V_{SB} = 0$ V and avoids this effect. SPICE computes the output resistance of this circuit to be 1.14 kΩ.

➤ SUMMARY

In this chapter we discussed various configurations of amplifiers to provide either gain or a transformation of input and output resistances. In general, the common-emitter or common-source amplifiers are used to provide large transconductance or voltage gain, but often times do not have the proper input or output resistance to match the source or load resistances. The common-base or common-gate amplifiers are used to buffer current sources. The common-collector and common-drain amplifier are used to buffer voltage sources. These various stages give the designer the freedom to properly match the source and load resistances to build an amplifier with the overall small-signal characteristics required for a particular application.

Specifically we showed

- A method to bias amplifiers such that they operate centered between the positive and negative power supplies.
- A methodology to calculate the small-signal two-port parameters for various amplifier configurations.
- How these two-port parameters vary with device parameters and supply current.

Table 8.3 summarizes all the results of the various configurations for both bipolar and MOS transistors. We have assumed a simple form for all of the respective calcu-

Table 8.3 Summary of 2-port models

Type	Application	Controlled Source	R_{in}	R_{out}
CE	Transconductance	g_m	r_π	$r_o \| r_{oc}$
CS	Transconductance	g_m	∞	$r_o \| r_{oc}$
CE_{deg}	Externally controlled transconductance	$\dfrac{g_m}{(1 + g_m R_E)}$	$r_\pi(1 + g_m R_E)$	$r_o\,[1 + g_m R_E] \,\|\, r_{oc}$
CB	Current buffer	–1	$1/g_m$	$r_o[1 + g_m(r_\pi \| R_S)] \,\|\, r_{oc}$
CG	Current buffer	–1	$1/(g_m + g_{mb})$	$[r_o + g_m r_o R_S] \,\|\, r_{oc}$
CC	Voltage buffer	1	$r_\pi + \beta_o(r_o \| r_{oc} \| R_L)$	$\dfrac{1}{g_m} + \dfrac{R_S}{\beta_o}$
CD	Voltage buffer	$\dfrac{g_m}{g_m + g_{mb}}$	∞	$\dfrac{1}{g_m + g_{mb}}$

lations done in this chapter. This table is quite handy when estimating input and output resistances and the various gains when these different stages are cascaded. Table 8.4 shows the circuit diagram for the various configurations studied in this chapter using n- and p-channel MOS transistors and npn and pnp bipolar transistors. This table will help you identify the amplifier type by recognizing its input and output terminals.

➤ FURTHER READING

1. C. G. Fonstad, *Microelectronics Devices and Circuits*, McGraw-Hill, 1994, Chapter 11. Includes discussion of resistive biasing of transistor amplifiers.
2. P. R. Gray and R. G. Meyer, *Analysis and Design of Analog Integrated Circuits 3rd Ed.*, Wiley, 1993, Chapter 3. More advanced treatment of single stage amplifiers.
3. M. N. Horenstein, *Microelectronic Circuits and Devices 2nd Ed.*, Prentice Hall, 1995, Chapters 6, 7 and 8. Chapters 6 and 7 cover large signal and biasing issues in amplifiers. Chapter 8 covers small signal modelling of transistor amplifiers.
4. J. Millman and A. Grabel, *Microelectronics 2nd Ed.*, McGraw-Hill, 1987, Chapter 10. Includes discussion of resistive biasing of transistor amplifiers.
5. A. S. Sedra and K. C. Smith, *Microelectronic Circuits 3rd Ed.*, HRW Saunders, 1991, Chapters 4 and 5. Includes discussion of discrete amplifier design.

■ Table 8.4 Amplifier configurations using n- and p-channel MOS transistors and npn and pnp bipolar transistors. To simplify this diagram all backgates are shown shorted to their respective sources.

Amplifier Type	Transistor Type			
	NMOS	PMOS	npn	pnp
Common Source/ Common Emitter (*CS*/*CE*)	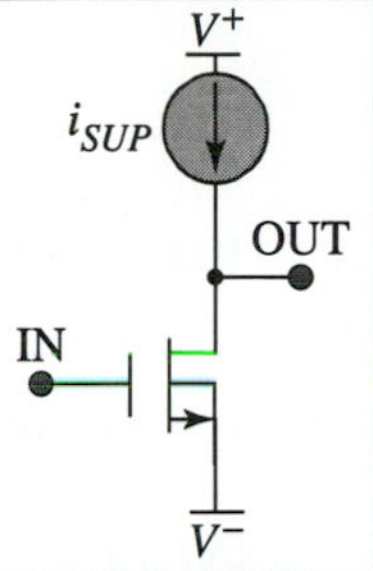 V^+, i_{SUP}, OUT, IN, V^-	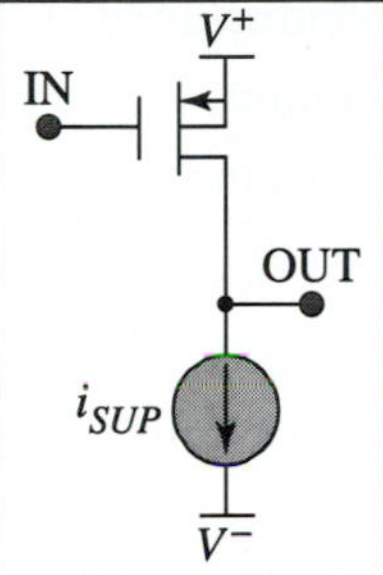 IN, V^+, OUT, i_{SUP}, V^-	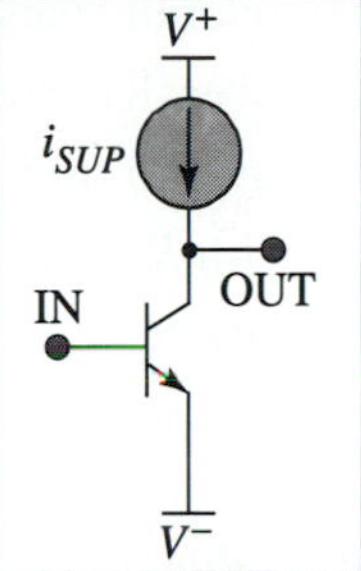 V^+, i_{SUP}, OUT, IN, V^-	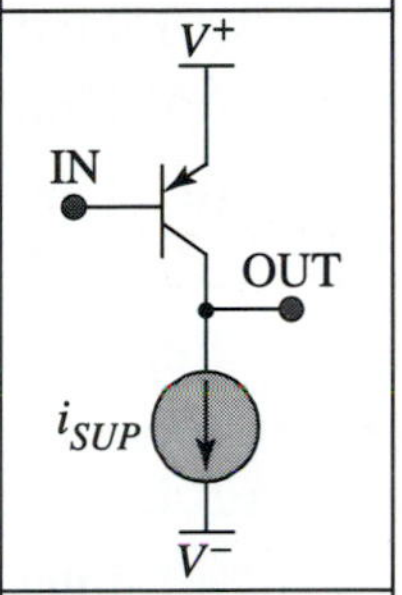 V^+, IN, OUT, i_{SUP}, V^-
Common Gate/ Common Base (*CG*/*CB*)	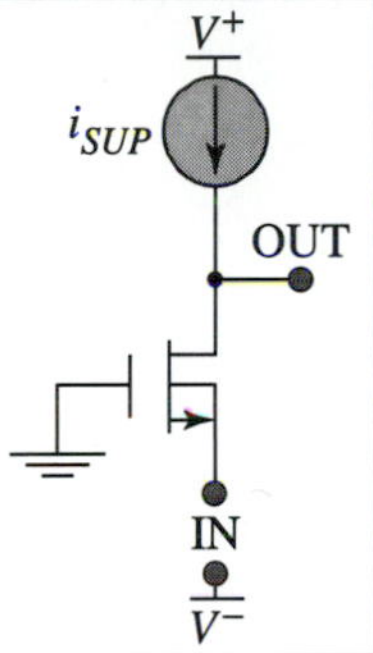 V^+, i_{SUP}, OUT, IN, V^-	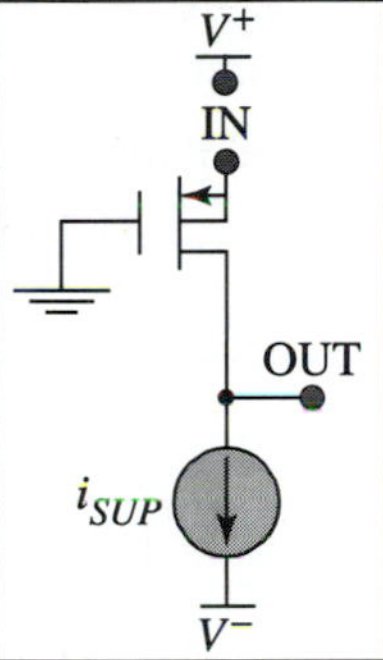 V^+, IN, OUT, i_{SUP}, V^-	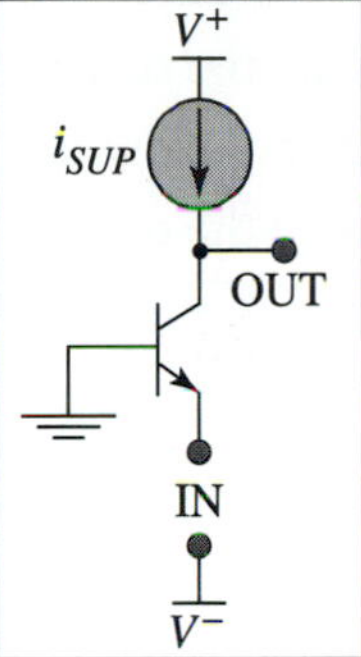 V^+, i_{SUP}, OUT, IN, V^-	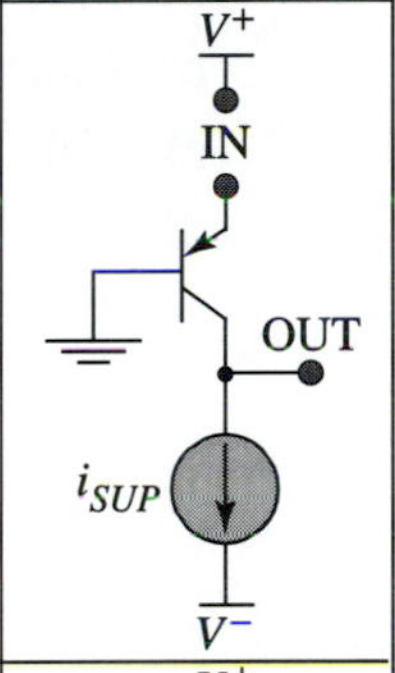 V^+, IN, OUT, i_{SUP}, V^-
Common Drain/ Common Collector (*CD*/*CC*)	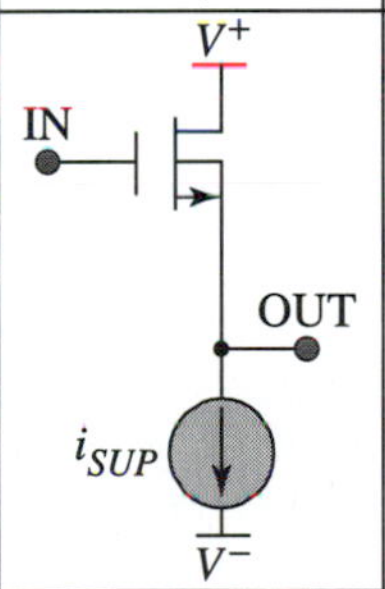 V^+, IN, OUT, i_{SUP}, V^-	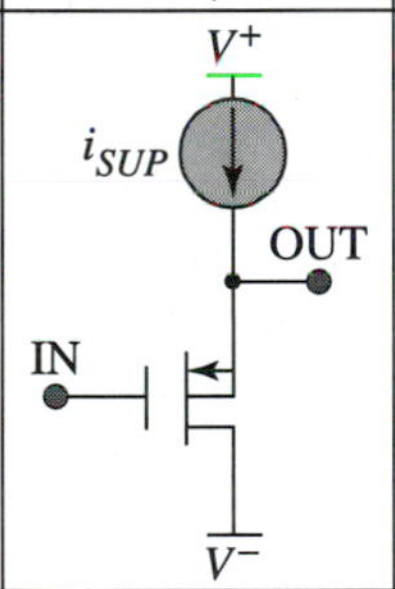 V^+, i_{SUP}, OUT, IN, V^-	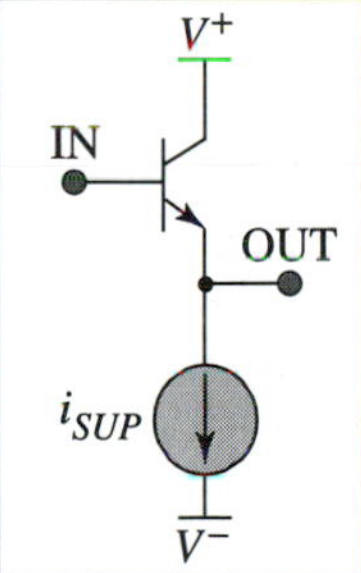 V^+, IN, OUT, i_{SUP}, V^-	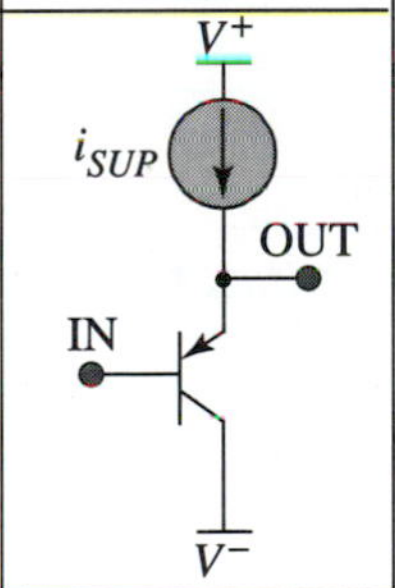 V^+, i_{SUP}, OUT, IN, V^-

➤ PROBLEMS

EXERCISES

E8.1 You are given the two-port model of an amplifier shown below in Fig. E8.1. Given that $R_{in} = 100\ \Omega$, $R_{out} = 10\ \text{k}\Omega$, $R_S = 100\ \Omega$, and $R_L = 10\ \text{k}\Omega$,

(a) Find the overall voltage gain.

(b) Find the overall transconductance.

(c) Find the overall current gain.

(d) Find the overall transresistance.

E8.2 Repeat E8.1 with $R_{in} = 10\ \text{k}\Omega$, $R_{out} = 100\ \Omega$, $R_S = 100\ \Omega$ and $R_L = 10\ \text{k}\Omega$.

E8.3 Repeat E8.1 with $R_{in} = 100\ \Omega$, $R_{out} = 10\ \text{k}\Omega$, $R_S = 10\ \text{k}\Omega$ and $R_L = 100\ \Omega$.

E8.4 You are given an npn common-emitter amplifier with the device data shown in Fig. E8.4.

(a) Find the value for R_C and V_{BIAS} so that $I_C = 500\ \mu\text{A}$ when $V_{OUT} = 0$ V.

(b) Calculate the two-port parameters R_{in}, R_{out}.

(c) Calculate the overall voltage gain.

(d) Calculate the overall transconductance.

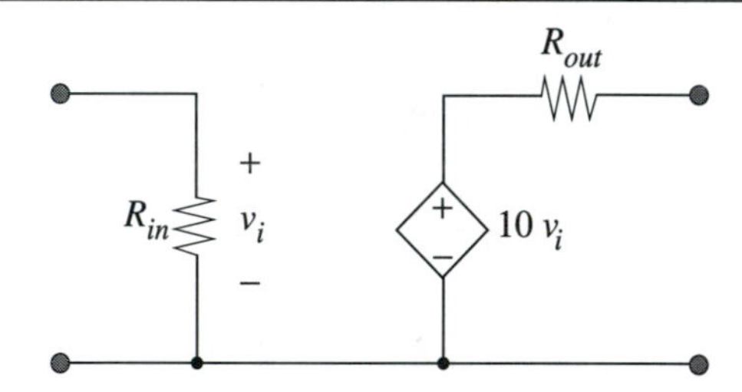

➤ Figure E8.1

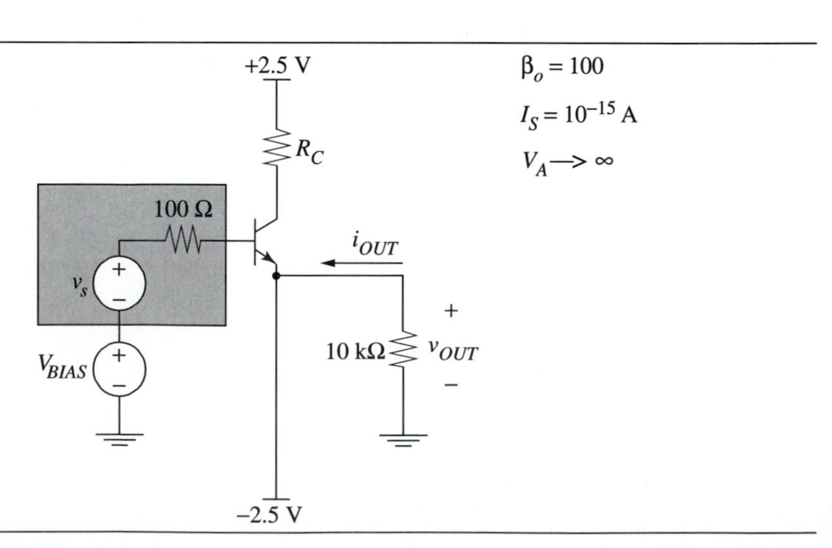

➤ Figure E8.4

E8.5 Repeat E8.4 with $I_C = 50\ \mu A$.

E8.6 You are given a pnp common-emitter amplifier with the device data shown in Fig. E8.6.

(a) Find the value for R_C and V_{BIAS} so that $I_C = 500\ \mu A$ when $V_{OUT} = 0$ V.

(b) Calculate the two-port parameters R_{in}, R_{out}.

(c) Calculate the overall voltage gain.

(d) Calculate the overall transconductance.

E8.7 Repeat E8.6 with $I_C = 50\ \mu A$.

E8.8 You are given an NMOS common-source amplifier with the device data shown in Fig. E8.8.

(a) Find the value for R_D and V_{BIAS} so that $I_D = 500\ \mu A$ when $V_{OUT} = 0$ V.

(b) Calculate the two-port parameters R_{in} and R_{out}.

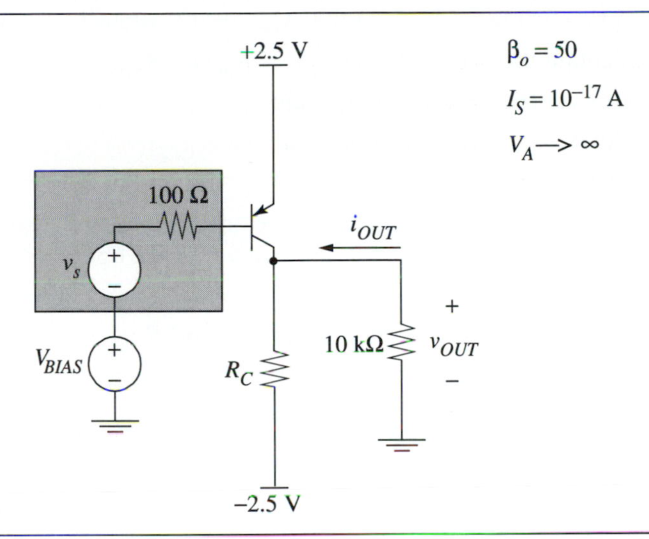

➤ **Figure E8.6**

+2.5 V

R_D

i_{OUT}

1 kΩ

v_s

v_{OUT}

10 kΩ

V_{BIAS}

–2.5 V

$W/L = 30/3$

$V_{Tn} = 1.0$ V

$\mu_n C_{ox} = 50\ \mu A/V^2$

$\lambda_n = 0\ V^{-1}$

➤ **Figure E8.8**

(c) Calculate the overall voltage gain.

(d) Calculate the overall transconductance.

E8.9 Repeat E8.8 with $I_D = 50\ \mu A$.

E8.10 You are given a PMOS common-source amplifier with the device data shown in Fig. E8.10.

(a) Find the value for R_D and V_{BIAS} so that $I_D = 500\ \mu A$ when $V_{OUT} = 0$ V.

(b) Calculate the two-port parameters R_{in} and R_{out}.

(c) Calculate the overall voltage gain.

(d) Calculate the overall transconductance.

E8.11 Repeat E8.10 with $I_D = 50\ \mu A$.

E8.12 You are given an NMOS common-source amplifier with a current source supply as shown in Fig. E8.12. The device data is shown in the figure.

(a) Find the value for V_{BIAS} so that $V_{OUT} = 0$ V when $I_{SUP} = 100\ \mu A$.

(b) Calculate the two-port parameters R_{in} and R_{out}.

(c) Calculate the overall voltage gain.

(d) Calculate the overall transconductance.

E8.13 Repeat E8.12 with $I_{SUP} = 10\ \mu A$.

E8.14 You are given a PMOS common-source amplifier with a current source supply as shown in Fig. E8.14. The device data is shown in the figure.

(a) Find the value for V_{BIAS} so that $V_{OUT} = 0$ V when $I_{SUP} = 100\ \mu A$.

(b) Calculate the two-port parameters R_{in} and R_{out}.

(c) Calculate the overall voltage gain.

(d) Calculate the overall transconductance.

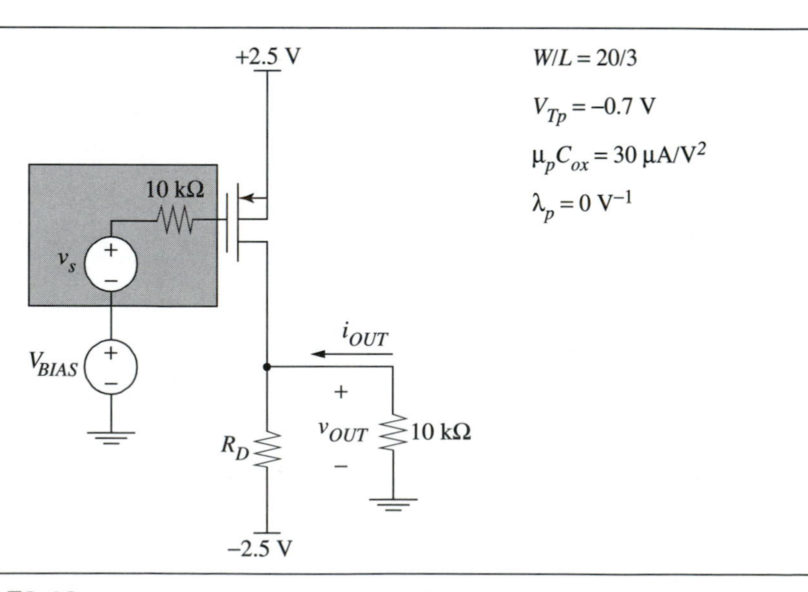

➤ **Figure E8.10**

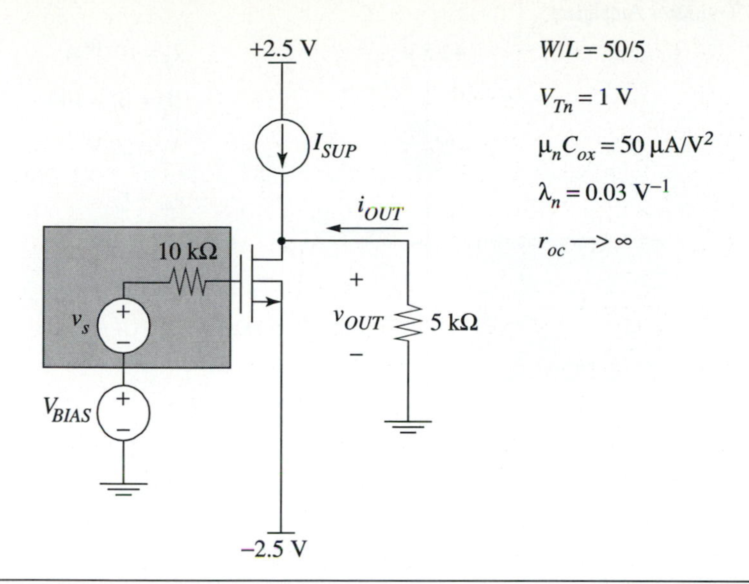

➤ **Figure E8.12**

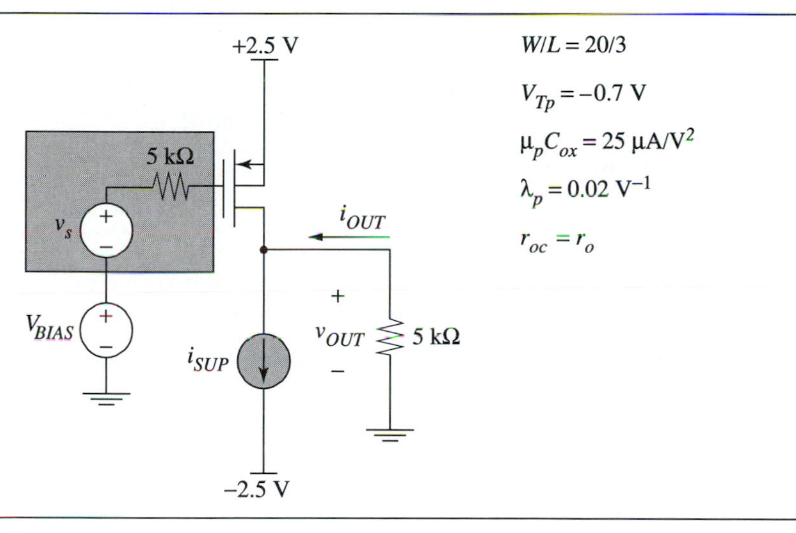

➤ **Figure E8.14**

E8.15 Repeat E8.14 with $I_{SUP} = 10\ \mu A$.

E8.16 You are given an npn common-emitter amplifier with a current source supply as shown in Fig. E8.16. The device data is shown in the figure.

(a) Find the value for V_{BIAS} so that $V_{OUT} = 0$ V when $I_{SUP} = 200\ \mu A$.

(b) Calculate the two-port parameters R_{in} and R_{out}.

(c) Calculate the overall voltage gain.

(d) Calculate the overall transconductance.

E8.17 Repeat E8.16 with $I_{SUP} = 30\ \mu A$.

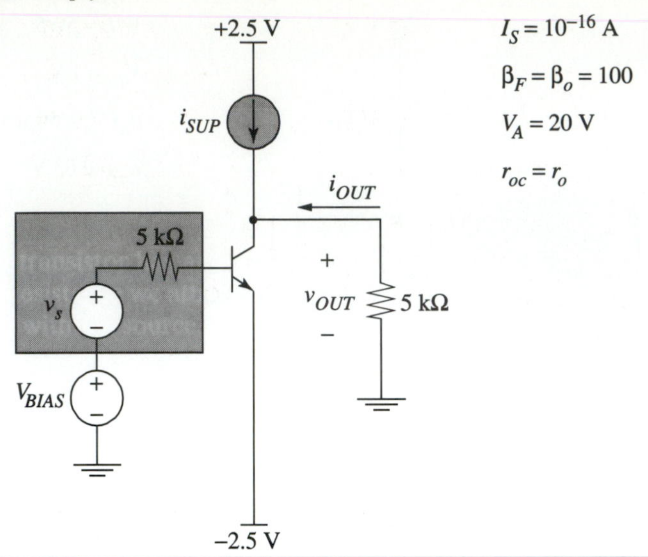

➤ **Figure E8.16**

E8.18 You are given an npn common-emitter amplifier with a current source supply as shown in Fig. E8.18. The input signal source is a current. The device data is shown in the figure.

(a) Find the value for I_{BIAS} so that $V_{OUT} = 0$ V when $I_{SUP} = 200$ μA.

(b) Calculate the two-port parameters R_{in} and R_{out}.

(c) Calculate the overall current gain.

(d) Calculate the overall transresistance.

E8.19 Repeat E8.18 with $I_{SUP} = 20$ μA.

E8.20 You are given an npn common-emitter amplifier with emitter degeneration resistor as shown in Fig. E8.20. The device data is shown in the figure.

(a) Find the value for V_{BIAS} so that $V_{OUT} = 0$ V when $I_{SUP} = 200$ μA.

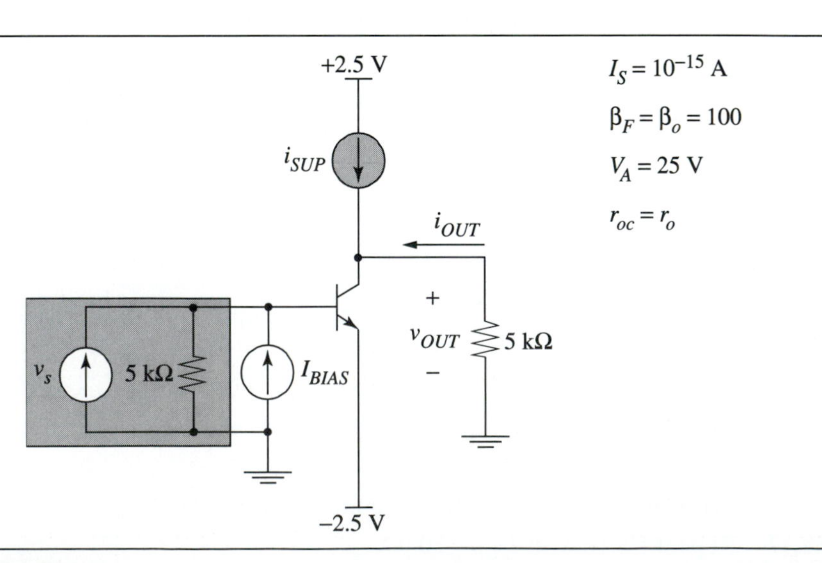

➤ **Figure E8.18**

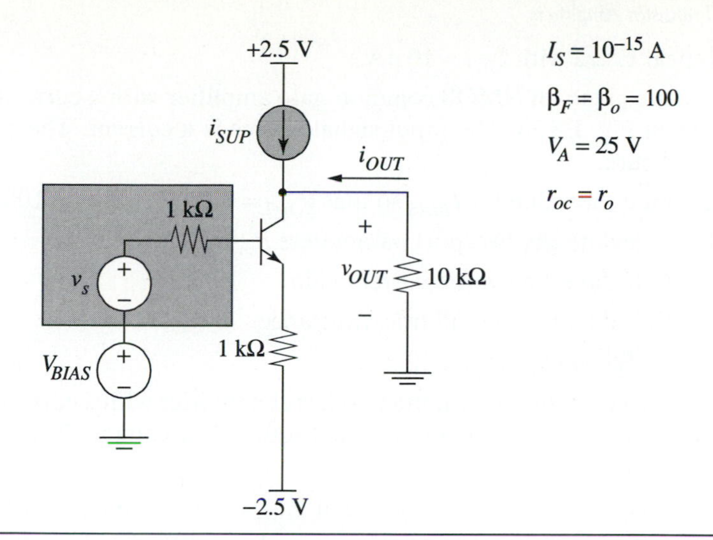

➤ **Figure E8.20**

(b) Calculate the two-port parameters R_{in} and R_{out}.

(c) Calculate the overall voltage gain.

(d) Calculate the overall transconductance.

E8.21 Repeat E8.20 with $I_{SUP} = 20\ \mu$A.

E8.22 You are given an npn common-base amplifier with a current source supply as shown in Fig. E8.22. The input signal source is a current. The device data is shown in the figure.

(a) Find the value for I_{BIAS} so that $V_{OUT} = 0$ V when $I_{SUP} = 100\ \mu$A.

(b) Calculate the two-port parameters R_{in} and R_{out}.

(c) Calculate the overall current gain.

(d) Calculate the overall transresistance.

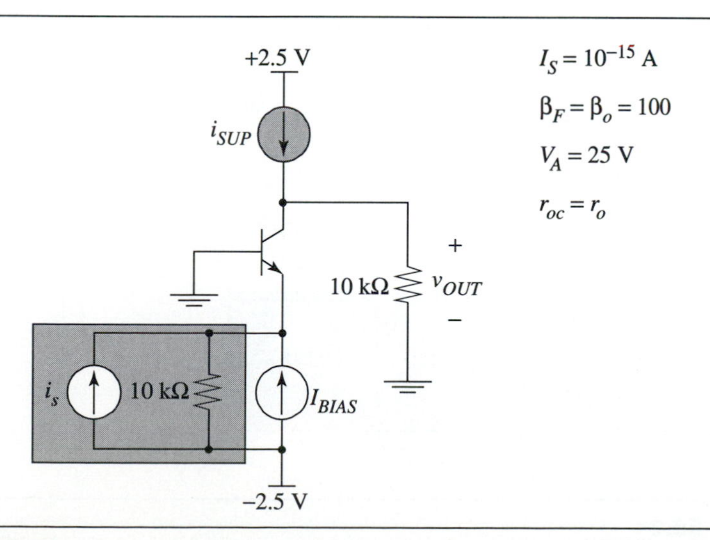

➤ **Figure E8.22**

E8.23 Repeat E8.22 with $I_{SUP} = 10\ \mu A$.

E8.24 You are given an NMOS common-gate amplifier with a current source supply as shown in Fig. E8.24. The input signal source is a current. The device data is shown in the figure.

(a) Find the value for I_{BIAS} so that $V_{OUT} = 0$ V when $I_{SUP} = 100\ \mu A$.

(b) Calculate the two-port parameters R_{in} and R_{out}.

(c) Calculate the overall current gain.

(d) Calculate the overall transresistance.

E8.25 Repeat E8.24 with $I_{SUP} = 10\ \mu A$.

E8.26 You are given an npn common-collector amplifier with a current source supply as shown in Fig. E8.26. The input signal source is a voltage. The device data is shown in the figure.

(a) Find the value for V_{BIAS} so that $V_{OUT} = 0$ V when $I_{SUP} = 100\ \mu A$.

(b) Calculate the two-port parameters R_{in} and R_{out}.

(c) Calculate the overall voltage gain.

(d) Calculate the overall transconductance.

E8.27 Repeat E8.26 with $I_{SUP} = 10\ \mu A$.

E8.28 You are given an NMOS common-drain amplifier with a current source supply as shown in Fig. E8.28. The input signal source is a voltage. The device data is shown in the figure.

(a) Find the value for I_{BIAS} so that $V_{OUT} = 0$ V when $I_{SUP} = 200\ \mu A$.

(b) Calculate the two-port parameters R_{in} and R_{out}.

(c) Calculate the overall voltage gain.

(d) Calculate the overall transconductance.

E8.29 Repeat E8.28 with $I_{SUP} = 20\ \mu A$.

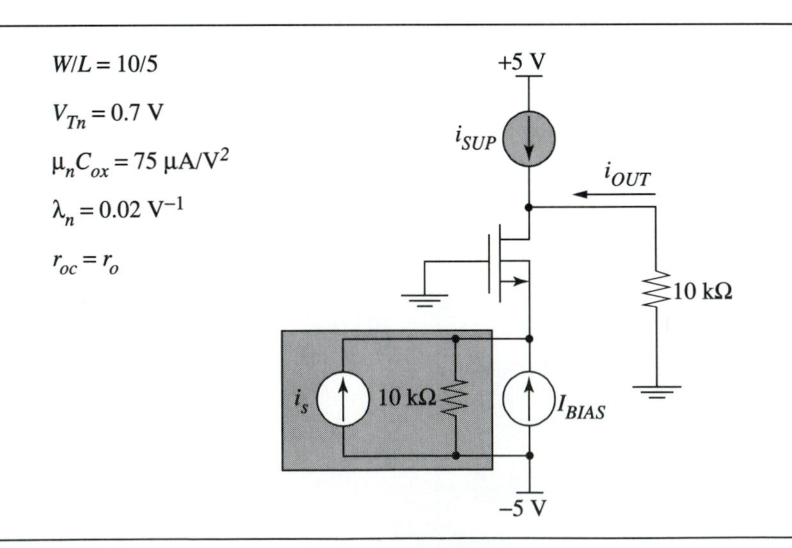

➤ **Figure E8.24**

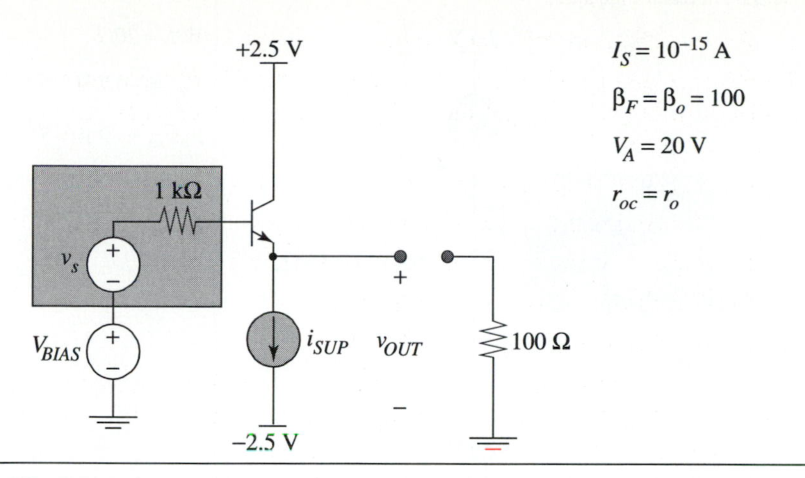

➤ **Figure E8.26**

E8.30 You are given a PMOS common-drain amplifier with a current source supply as shown in Fig. E8.30. The input signal source is a voltage. The device data is shown in the figure.

(a) Find the value for I_{BIAS} so that $V_{OUT} = 0$ V when $I_{SUP} = 200$ μA.

(b) Calculate the two-port parameters R_{in} and R_{out}.

(c) Calculate the overall voltage gain.

(d) Calculate the overall transconductance.

E8.31 Repeat E8.30 with $I_{SUP} = 20$ μA.

+2.5 V
1 kΩ
v_s
V_{BIAS}
i_{SUP}
v_{OUT}
100 Ω
−2.5 V

$W/L = 20/2$

$V_{Tn} = 0.7$ V

$\mu_n C_{ox} = 50$ μA/V^2

$\lambda_n = 0.05$ V^{-1}

$r_{oc} = r_o$

➤ **Figure E8.28**

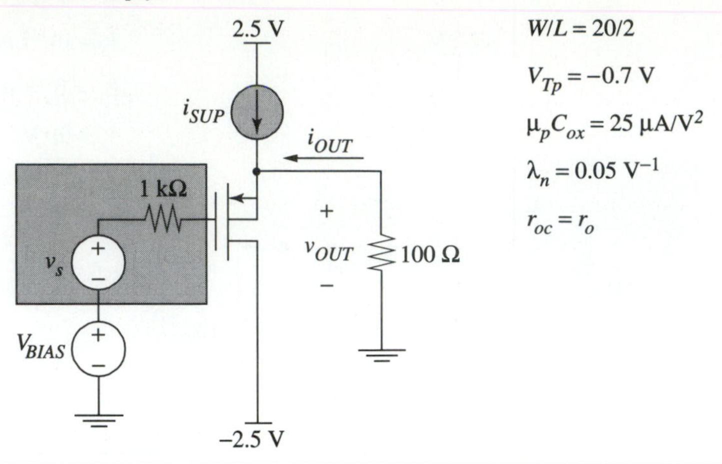

➤ **Figure E8.30**

PROBLEMS

P8.1 Given the small-signal model(Fig. P8.1) for an amplifier circuit

(a) Find the input and output resistance.

(b) Construct a two-port model for a voltage amplifier.

(c) Construct a two-port model for a current amplifier.

(d) Construct a two-port model for a transconductance amplifier.

(e) Construct a two-port model for a transresistance amplifier.

P8.2 You are given an input voltage source with a source resistance, R_S.

(a) Use the two-port model found in P8.1 to find the overall voltage gain when the amplifier is driving a load resistor R_L.

(b) Specify whether the resistances r_b, r_i, r_o, r_c in the small signal model should be increased, decreased or remain the same to improve the overall voltage gain.

(c) Find the overall transconductance of the amplifier.

(d) Repeat (b) for the transconductance amplifier.

P8.3 You are given an input current source with a source resistance, R_S.

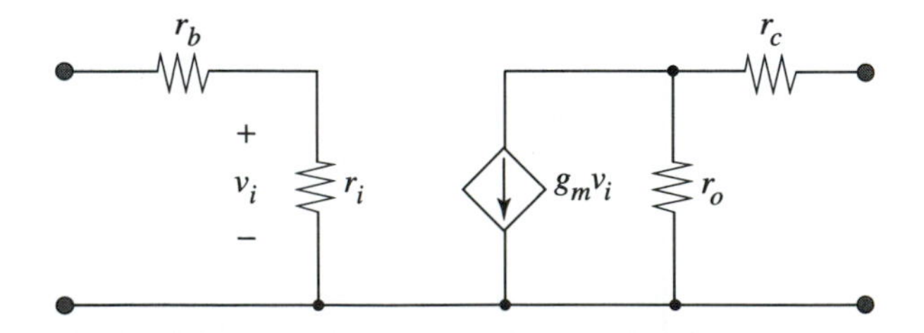

➤ **Figure P8.1**

(a) Use the two-port model found in P8.1 to find the overall current gain when the amplifier is driving a load resistor R_L.

(b) Specify whether the resistances r_b, r_i, r_o, r_c in the small-signal model should be increased, decreased or remain the same to improve the overall current gain.

(c) Find the overall transresistance of the amplifier.

(d) Repeat (b) for the transresistance amplifier.

P8.4 Repeat Problem 8.1 for the small signal model shown in Fig. P8.4.

P8.5 A two-port model of a voltage amplifier is shown in Fig. P8.5.

(a) Draw the two-port model for a transresistance amplifier by converting the model for the voltage amplifier below.

(b) Draw the two-port model for a transconductance amplifier by converting the model for the voltage amplifier below.

(c) Draw the two-port model for a current amplifier by converting the model for the voltage amplifier below.

(d) Based solely on the input/output resistances, what kind of amplifier will have the smallest degradation when 1kΩ source and load resistances are connected?

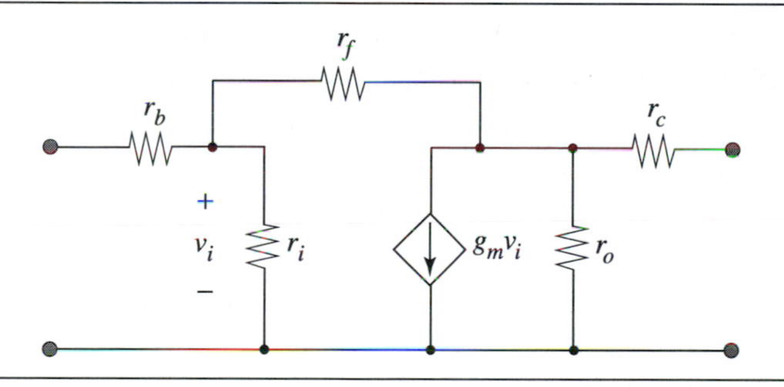

➤ **Figure P8.4**

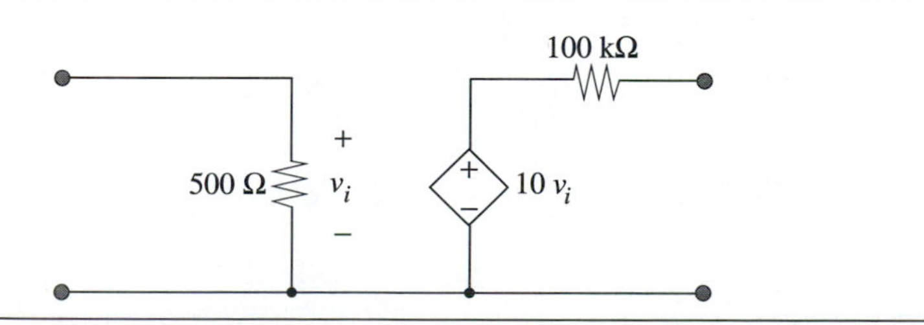

➤ **Figure P8.5**

P8.6 In this problem we will explore the limitation of resistors supplying the current to a transistor amplifier by looking at the intrinsic voltage gain ($R_S = 0\ \Omega$, $R_L \rightarrow \infty$) of the CE amplifier shown in Fig. P8.6.

(a) Calculate I_C and V_{BIAS} such that the DC output voltage is $V^+/2$ in terms of R_C and V^+.

(b) Draw the small signal model of this circuit and calculate g_m and r_π in terms of R_C and V^+.

(c) Show that the voltage gain of this circuit is

$$A_v = \frac{V^+}{2\left(\frac{kT}{q}\right)}$$

Notice that the intrinsic voltage gain can only be increased by increasing the power supply voltage.

P8.7 You are given a pnp common-emitter amplifier shown in Fig. P8.7.

(a) Calculate V_{BIAS} such that $V_{OUT} = 2.5$ V. Neglect R_S and I_B for this part.

(b) Find the small-signal open-circuit voltage gain, $v_{out}/v_{s.}$

(c) What is the minimum value of load resistance that could be driven by the amplifier and lose only a factor of 2 in overall voltage gain?

(d) How high must you increase β_o to increase the open-circuit voltage gain found in (b) by 10%?

P8.8 For the common-emitter amplifier shown in Fig. P8.8

(a) Find V_{BIAS} such that V_{OUT} is at 0 V.

(b) Find the two-port parameters A_v, R_{in}, and R_{out} for a voltage amplifier.

(c) Find the open-circuit voltage gain ($R_L \rightarrow \infty$).

(d) Find the overall voltage gain when $R_L = 1\ \text{k}\Omega$.

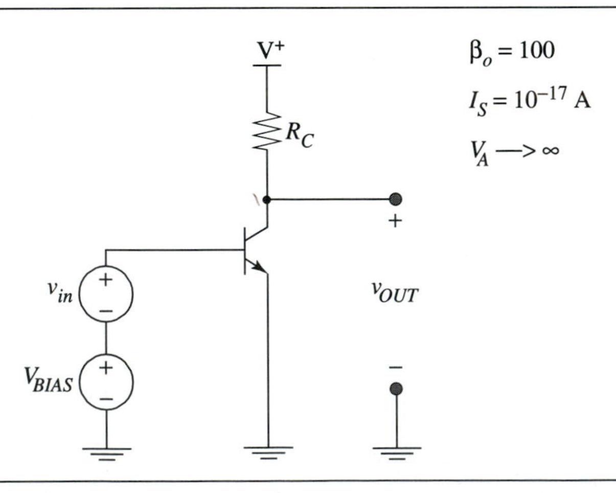

➤ Figure P8.6

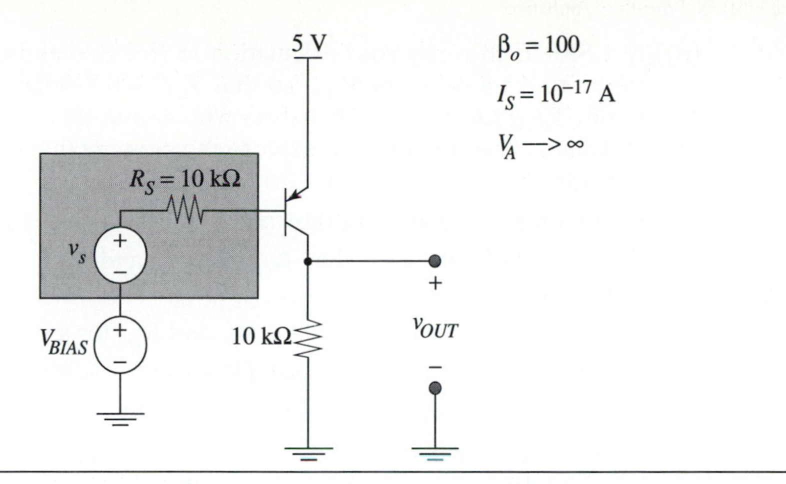

➤ **Figure P8.7**

(e) Verify your calculations in SPICE. You should first perform a DC sweep to determine the V_{BIAS} (close to the value calculated above) that will yield a V_{OUT} near 0 V. Then you can determine the voltage gain using an AC analysis.

P8.9 Using the results from Problem 8.8 we will explore the relationship between the input and output when a low frequency sinusoid is applied. Assume $R_L = 2.5$ kΩ.

(a) If v_s is a sinusoid at a frequency of 1kHz and an amplitude of 2 mV, sketch the total output voltage for one period of the sinusoid.

(b) Calculate the maximum amplitude of the sinusoid so that the output waveform is not clipping.

(c) In part (b) was the limitation due to the transistor going into saturation or cutoff?

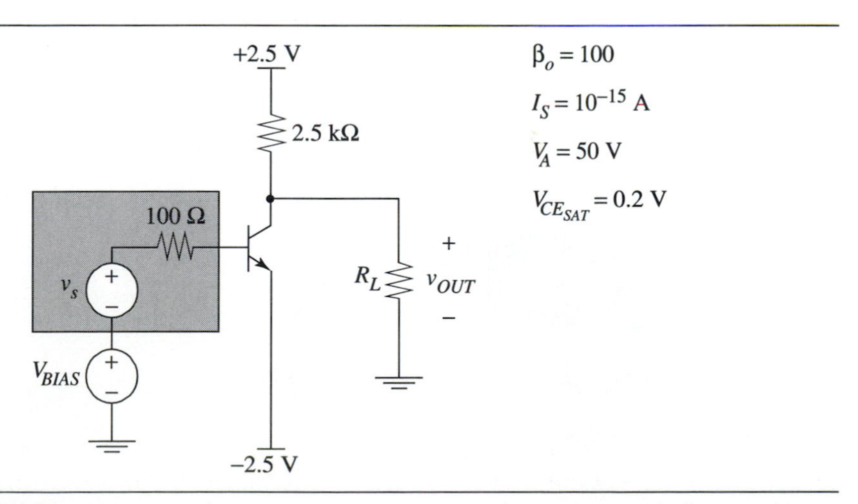

➤ **Figure P8.8**

(d) Use SPICE to verify your calculation in (b). Remember to use a DC sweep to get a value for V_{BIAS} so that $V_{OUT} = 0$ V when $v_s = 0$ V. After setting V_{BIAS} run a transient analysis with a sinusoid using sin(0.002 1k 0 0). Change the amplitude to explore the clipping that occurs when the transistor leaves the constant-current region.

P8.10 For the common-source amplifier shown in Fig. P8.10

(a) Find V_{BIAS} and R_D such that $I_D = 500$ µA and $V_{OUT} = 0$ V given that $W/L = 10$.

(b) Find the two-port parameters A_v, R_{in}, and R_{out} for a voltage amplifier.

(c) Find the open-circuit voltage gain ($R_L \to \infty$).

(d) Find the overall voltage gain when $R_L = 2.5$ kΩ.

(e) Verify your calculations in SPICE. You should first perform a DC sweep to determine the V_{BIAS} (close to the value calculated above) that will yield a V_{OUT} near 0 V. Then you can determine the voltage gain using an AC analysis.

P8.11 Repeat P8.9 for the CS amplifier in P8.10.

P8.12 Repeat P8.10 and 8.11 if W/L is changed to 100.

P8.13 For the p-channel common-source amplifier shown in Fig. P8.13

(a) Given $W/L = 12/2$ and $R_D = 10$ kΩ, calculate V_{BIAS} such that V_{OUT} is 2.5 V.

(b) What is the small-signal voltage gain, $A_v = v_{out}/v_{in}$?

(c) To increase the voltage gain, you increase R_D to 100 kΩ. Calculate the new small-signal voltage gain, A_v. You must re-bias the circuit so that $V_{OUT} = 2.5$ V.

(d) We could also try to increase the voltage gain of the initial circuit by increasing W/L rather than R_D. Calculate the new A_v if $W/L = 120/2$ and $R_D = 10$ kΩ. Be sure to re-bias the circuit so that $V_{OUT} = 2.5$ V.

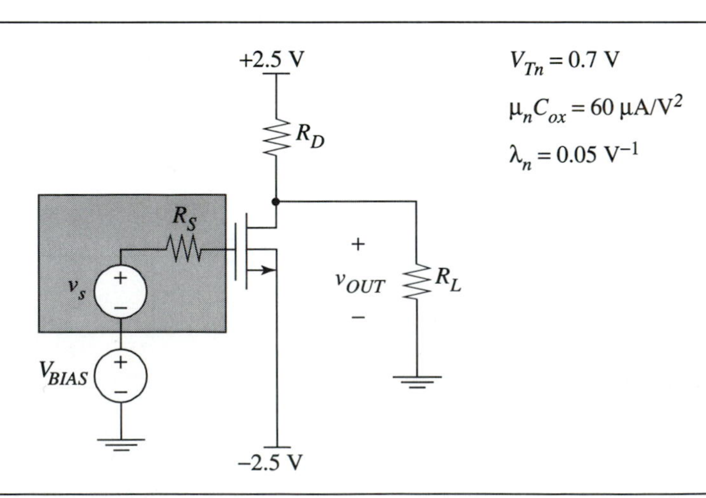

➤ **Figure P8.10**

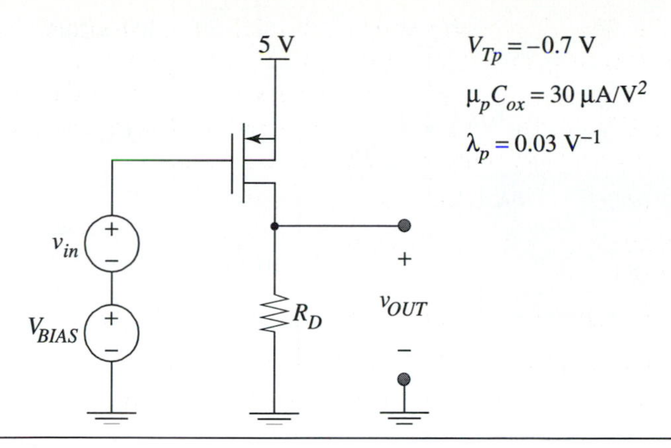

➤ **Figure P8.13**

P8.14 For the same circuit as shown in Fig. P8.13, but with new device parameters $V_{Tp} = -1.0$ V, $\mu_p C_{ox} = 25\ \mu A/V^2$, $\lambda_p = 0\ V^{-1}$, $R_D = 10\ k\Omega$,

(a) Find the W/L of the PMOS device such that V_{OUT} is 2.5 V given that $V_{BIAS} = 1.5$ V.

(b) Find the two-port parameters A_v and R_{out}.

(c) What is the minimum value of load resistance that this amplifier can drive and maintain an overall voltage gain at least 50% of that calculated in (b)?

(d) Resize W/L to increase A_v by a factor of 2 from that calculated in (b). Assume V_{BIAS} is changed such that $V_{OUT} = 2.5$ V.

(e) Calculate the new value of V_{BIAS} to satisfy the assumption $V_{OUT} = 2.5$ V in (d).

(f) Plot $v_{OUT}(t)$ for $v_{IN}(t) = V_{BIAS} + 5\text{mV}\cos(2\pi 1000t)$ for both cases.

P8.15 For the same circuit as shown in Fig. P8.13 but with new device parameters $V_{Tp} = -1.0$ V, $\mu_p C_{ox} = 10\ \mu A/V^2$, $\lambda_p = 0.1\ V^{-1}$, $R_D = 10\ k\Omega$, $W/L = 20/2$.

(a) What V_{BIAS} is required for a DC output voltage of 2.5 V?

(b) For this bias point draw the two-port model for this voltage amplifier and calculate all its parameters.

(c) If the DC output voltage must be increased to 3.5 V, what is the new V_{BIAS}?

(d) Calculate the new two-port parameters for this bias point.

P8.16 In Fig. P8.16 we show an NMOS common-source amplifier with a current source supply, $I_{SUP} = 250\ \mu A$.

(a) Calculate V_{BIAS} such that $V_{OUT} = 0$ V.

(b) Draw the two-port model for the intrinsic voltage amplifier and calculate the parameters.

(c) Calculate the overall voltage gain.

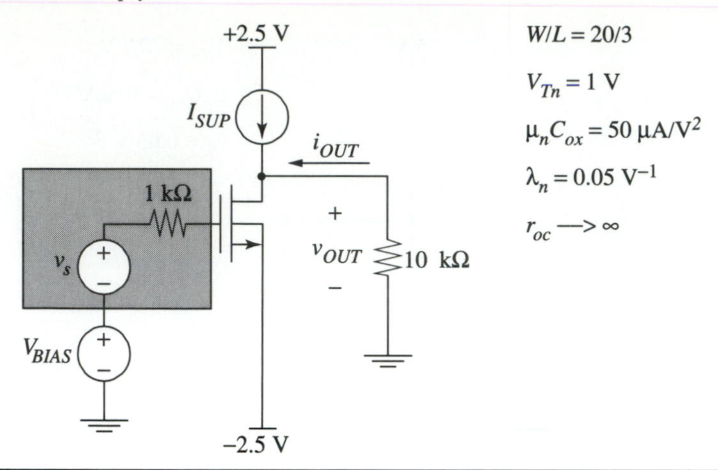

➤ **Figure P8.16**

P8.17 In Table 8.1 we showed how the two-port parameters of the CS amplifier change with device parameters and supply current. Use the parameters calculated or given in P8.16 as an initial starting point.

(a) Change I_{SUP} by $2\times$ and $0.5\times$ and re-calculate V_{BIAS}, the two-port parameters, and the overall voltage gain. Verify your results in SPICE. Be sure to perform a DC sweep in SPICE to find the value of V_{BIAS} for $V_{OUT} = 0$ V.

(b) Repeat (a) by changing W by $2\times$ and $0.5\times$ leaving the other parameters at their nominal values.

(c) Repeat (a) by changing $\mu_n C_{ox}$ by $2\times$ and $0.5\times$ leaving the other parameters at their nominal values.

(d) Repeat (a) by changing L by $2\times$ and $0.5\times$ leaving the other parameters at their nominal values.

P8.18 In Fig. P8.18 we show a PMOS common-source amplifier with a current source supply, $I_{SUP} = 250\ \mu A$.

(a) Calculate V_{BIAS} such that $V_{OUT} = 0$ V.

(b) Draw the two-port model for the intrinsic transconductance amplifier and calculate the parameters.

(c) Calculate the overall transconductance.

P8.19 In Table 8.1 we showed how the two-port parameters of the CS amplifier change with device parameters and supply current. Use the parameters calculated or given in P8.18 as an initial starting point.

(a) Change I_{SUP} by $2\times$ and $0.5\times$ and re-calculate V_{BIAS}, the two-port parameters, and the overall transconductance. Verify your results in SPICE. Be sure to perform a DC sweep in SPICE to find the value of V_{BIAS} for $V_{OUT} = 0$ V.

(b) Repeat (a) by changing W by $2\times$ and $0.5\times$ leaving the other parameters at their nominal values.

(c) Repeat (a) by changing $\mu_p C_{ox}$ by $2\times$ and $0.5\times$ leaving the other parameters at their nominal values.

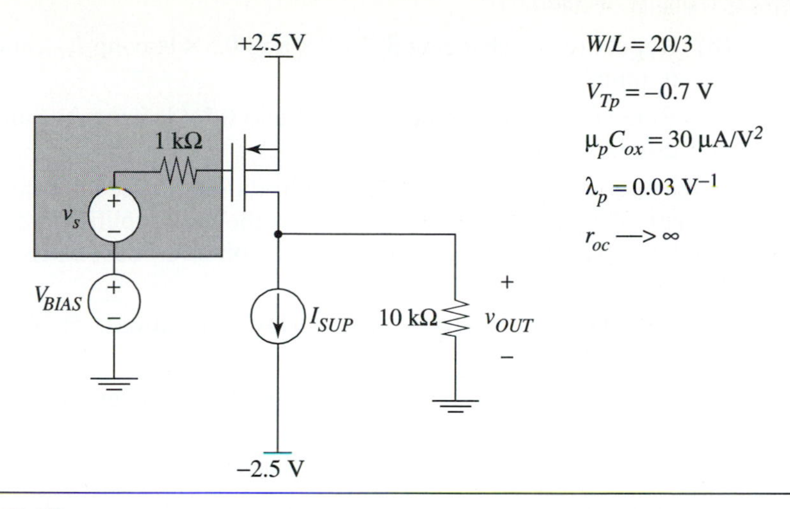

➤ **Figure P8.18**

(d) Repeat (a) by changing L by 2× and 0.5× leaving the other parameters at their nominal values.

P8.20 In Fig. P8.20 we show a CE amplifier with a current source supply.

(a) Calculate V_{BIAS} such that $V_{OUT} = 0$ V.

(b) Draw the two-port model for the intrinsic voltage amplifier and calculate the parameters.

(c) Calculate the overall voltage gain.

P8.21 In Table 8.2 we showed how the two-port parameters of the CE amplifier change with device parameters and supply current. Use the parameters calculated or given in P8.20 as an initial starting point.

(a) Change I_{SUP} by 2× and 0.5× and re-calculate V_{BIAS}, the two-port parameters and the overall voltage gain. Verify your results in SPICE. Be sure to perform a DC sweep in SPICE to find the value of V_{BIAS} for $V_{OUT} = 0$ V.

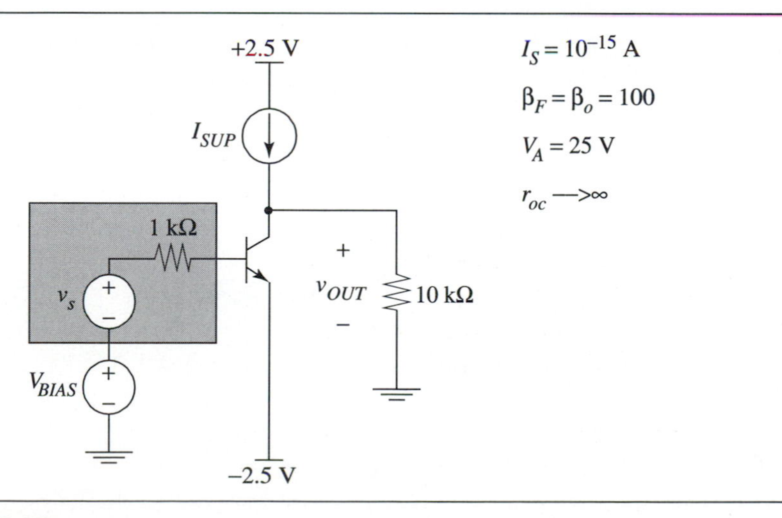

➤ **Figure P8.20**

(b) Repeat (a) by changing β_F by 2× and 0.5× leaving I_{SUP} and V_A at their nominal values.

(c) Repeat (a) by changing V_A by 2× and 0.5× leaving I_{SUP} and β_F at their nominal values.

P8.22 In the circuit shown in Fig. P8.22 the photodiode is used to convert light into a signal current. This signal current is input into the CE amplifier and converted to an output voltage. The source resistance from the photodiode is extremely high and can be neglected.

(a) Calculate I_{SUP} to give a transresistance magnitude of 100 MΩ with $R_L \rightarrow \infty$.

(b) Calculate I_{BIAS} to ensure that the amplifier is operating in its high-gain region.

(c) What is the minimum load resistance that can be driven by the amplifier and still maintain a transresistance magnitude > 20 MΩ?

P8.23 The emitter degeneration resistor in a common-emitter configuration increases the input and output resistance of a transconductance amplifier. Refer to Fig. P8.23 for this problem. Let $R_S = 100\ \Omega$, $R_E = 1\ \text{k}\Omega$, $R_L = 1\ \text{k}\Omega$.

(a) Calculate I_{SUP} such that $V_{OUT} = 0$ V and the voltage drop across R_E is $10kT/q = 250$ mV.

(b) Show that if the voltage drop across R_E is » kT/q that the overall transconductance, G_m of the amplifier is approximately $1/R_E$. Hint: Recall that $G_m = g_m/(1+g_m R_E)$.

(c) Calculate V_{BIAS} such that $V_{OUT} = 0$ V when $v_s = 0$ V. Neglect base current for this calculation.

(d) Calculate R_{in} and R_{out}.

(e) Calculate the overall transconductance.

P8.24 Repeat P8.23 (e) with $R_S = 1\ \text{k}\Omega$ and $R_L = 10\ \text{k}\Omega$.

P8.25 To see how the degeneration resistor R_E has increased R_{in}, R_{out}, repeat P8.23 (e) and P8.24 with $R_E = 0\ \Omega$.

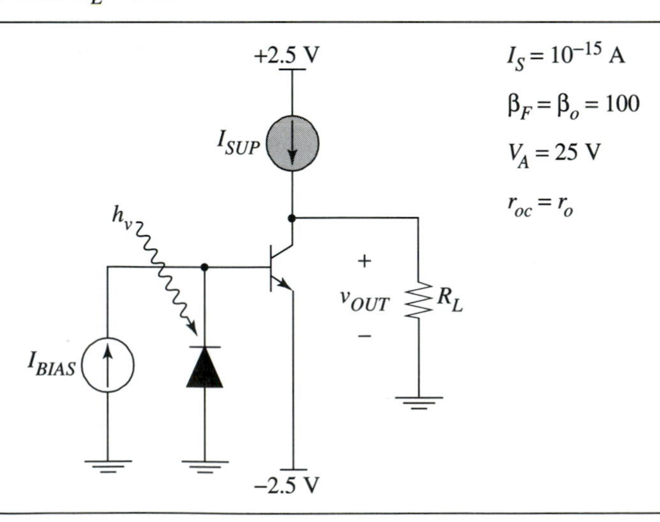

➤ Figure P8.22

$I_S = 10^{-15}$ A

$\beta_F = \beta_o = 100$

$V_A = 25$ V

$r_{oc} = r_o$

➤ **Figure P8.23**

P8.26 Consider an MOS common-source transconductance amplifier with a source degeneration resistor R_{SD}. Derive expressions for the two-port parameters R_{in}, R_{out}, and G_m in terms of I_{SUP}, g_m, r_o, and R_{SD}. Assume $r_{oc} \rightarrow \infty$ and the backgate is shorted to the source. Why is this circuit configuration not very popular?

P8.27 Given the npn common-base amplifier shown in Figure P8.27 with $I_{SUP} = 25\ \mu$A and an input signal $i_s = A\cos\omega t$.

(a) Calculate I_{BIAS} such that $V_{OUT} = 0$ V.

(b) Select V_B such that $v_{out}(t)$ can have the greatest amplitude without any clipping. This requires that the amplifier stay in its high-gain region. Assume I_{SUP} and I_{BIAS} require at least 0.5 V across them.

(c) Given that $R_L = 1\text{M}\Omega$, $R_S = 10\ \text{k}\Omega$, calculate the overall transresistance for this amplifier, v_{out}/i_s.

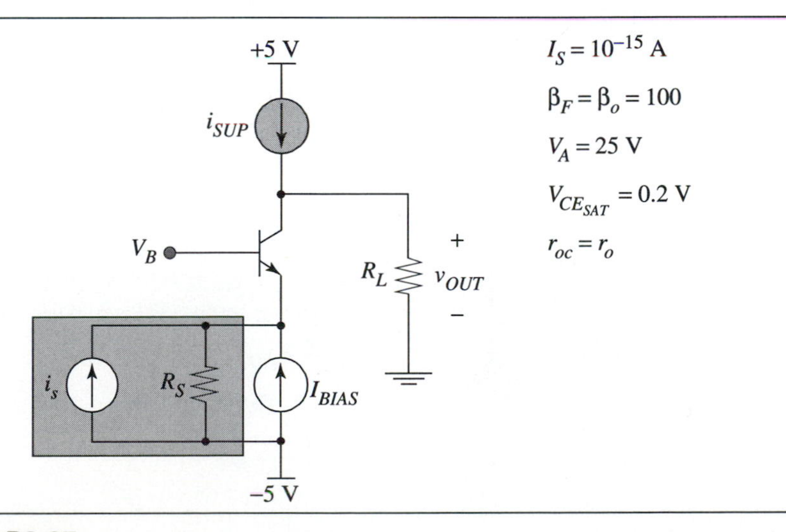

➤ **Figure P8.27**

(d) What is the largest amplitude for i_s that can be input to the amplifier and have no clipping (i.e. the amplifier stays in its high-gain region)?

P8.28 In the circuit in Fig. P8.27 the voltage source V_B which is setting the DC voltage on the base of the transistor is assumed to be perfect, namely that it has 0 Ω internal resistance. In this problem we will explore the effect of a finite internal resistance R_B from the voltage source, on the two-port model parameters for the CB amplifier.

(a) Find an expression for R_{in}

(b) Find an expression for R_{out}

(c) Find an expression for A_i with $R_S \rightarrow \infty$ and $R_L = 0\ \Omega$.

(d) Use these two-port parameters to see how the overall transresistance changes from that calculated in P8.27(c) when $R_B = 100\ \Omega$.

(e) Repeat (d) if I_{SUP} is increased to 1mA.

P8.29 You are given a pnp common-base amplifier with device parameters shown in Fig. P8.29.

(a) For $I_{SUP} = 100\ \mu A$, find I_{BIAS} so that $V_{OUT} = 0$ V.

(b) Calculate R_{in}.

(c) Find the small signal input current to the emitter of the CB amplifier, i_{in}, as a function of R_S.

(d) Calculate R_{out} as a function of R_S.

(e) Find i_{out} as a function of R_S and R_L.

(f) Plot i_{out}/i_s vs. R_S for $R_L = 5\ k\Omega$ over the range $10\ \Omega < R_S < 100\ k\Omega$. Use a log scale.

(g) Plot i_{out}/i_s vs. R_L for $R_S = 20\ k\Omega$ over the range $10\ \Omega < R_L < 10\ M\Omega$.

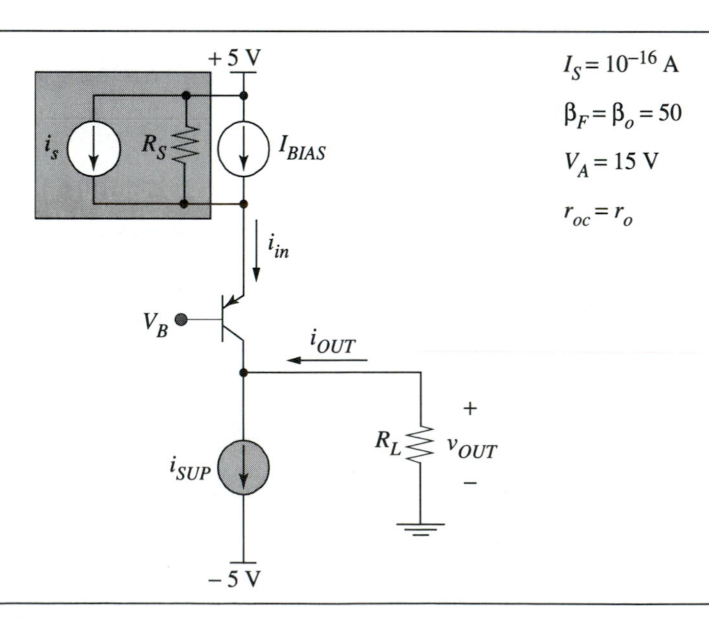

➤ **Figure P8.29**

P8.30 In this problem we will compare the common-base and common-gate configurations shown in Figure P8.30 when they are used as a current buffer. Assume $R_S = 1\ \text{k}\Omega$, $R_L = 10\ \text{k}\Omega$, and $I_{SUP} = 100\ \mu\text{A}$ for both circuits. Essentially we are comparing the use of an MOS vs. bipolar device.

(a) Find the input resistance of both circuits.

(b) Find the output resistance of both circuits.

(c) Find the intrinsic current gain of both circuits.

(d) What is the required W/L of the MOS transistor in the CG amplifier such that the input resistance is the same as the bipolar CB amplifier?

(e) What is the required W/L of the MOS transistor in the CG amplifier such that the output resistance is the same as the bipolar CB amplifier?

P8.31 In this problem we will compute the overall transresistance of the CG amplifier shown in Fig. P8.30(b). The amplifier is driving an MOS CS amplifier with an infinite input resistance, so we consider $R_L \to \infty$. Use the device data given in the figure. The input signal current source has an $R_S = 100\ \text{k}\Omega$.

(a) Given that $I_{SUP} = 80\ \mu\text{A}$, what is the value of I_{BIAS} so that $V_{OUT} = 0$ V?

(b) What is the amplifier's input and output resistance?

(c) What is the overall transresistance?

(d) What is the output voltage when $i_s = 1\text{nA}$?

P8.32 Given the NMOS common-gate amplifier shown in Fig. P8.30(b) except with $V_{SB} = 0$ V, $R_S = 10\ \text{k}\Omega$ and $R_L = 100\ \text{k}\Omega$, $V_{Tn} = 1.0$ V, $\mu_n C_{ox} = 50\ \mu\text{A/V}^2$, and $\lambda_n = 0.05\ \text{V}^{-1}$, at $L = 2\ \mu\text{m}$.

(a) What minimum channel length and width are necessary for $i_{out}/i_s = 0.8$ given $I_{SUP} = 100\ \mu\text{A}$?

(b) If the channel length must be 1 μm, what value of channel width is needed to have $R_{out} = 1\ \text{M}\Omega$ given $I_{SUP} = 100\ \mu\text{A}$?

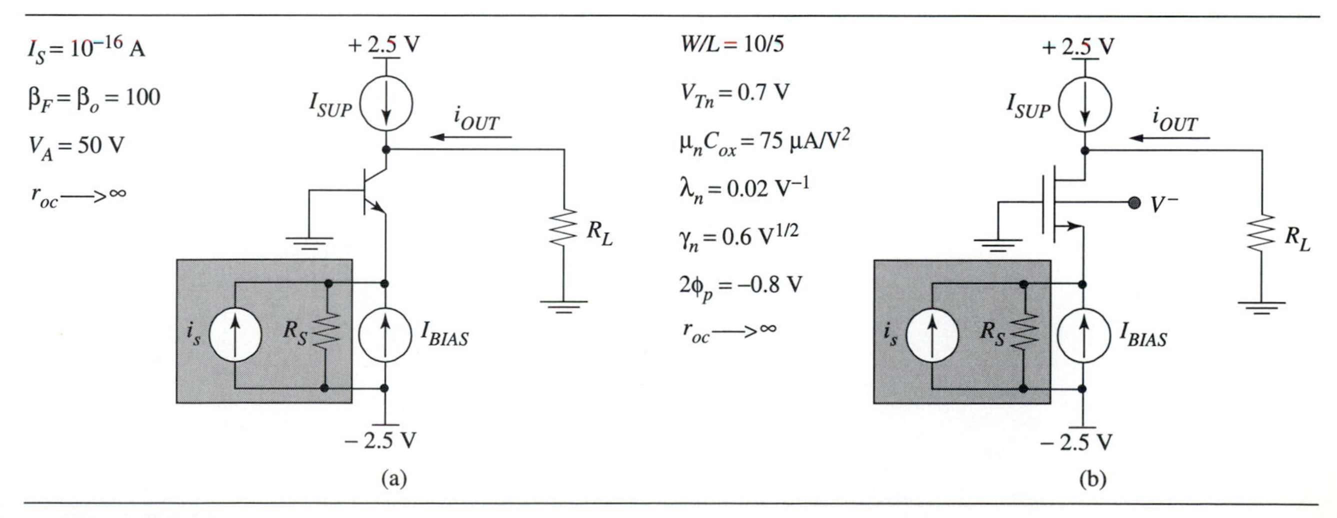

➤ **Figure P8.30**

P8.33 The PMOS common-gate amplifier shown in Figure P8.33 has a power dissipation of 1mW. At the minimum channel length of 2 μm $\lambda_p = 0.2\ V^{-1}$.

(a) What is the W/L required for $i_{out}/i_s = 0.8$?

(b) If the transistor is placed in an n-well so that $V_{SB} = 0$ V, What is the new overall current gain i_{out}/i_s ?

P8.34 The pn junction diode is illuminated with an optical fiber such that its current has a signal component due to photogeneration.The current is input into the common-base amplifier shown in Fig. P8.34.

(a) What is the value of V_G, given $I_{SUP} = 100\ \mu A$ and $W/L = 500$ that will ensure the transistor is operating in its constant-current region? Note this voltage source is providing the bias for the amplifier.

(b) What is the source resistance of the photodiode?

(c) What is i_{out}/i_s?

P8.35 Repeat P8.34 substituting an npn bipolar transistor with $I_S = 10^{-17}$ A, $\beta_o = 100$, and $V_A = 25$ V.

P8.36 The DC voltage supply at the base of the npn transistor has a small-signal component added to it which represents feedthrough from other signal lines as shown in Fig. P8.36. $I_{SUP} = 200\ \mu A$.

(a) Given $R_S = 5\ k\Omega$ and $R_L = 50\ k\Omega$, find i_{out}/i_s for $v_b = 0$ V.

(b) Find i_{out}/v_b for $i_s = 0$ A.

(c) Find the ratio of the output current due to the signal source i_s to that due to v_b.

(d) State whether the value for I_{SUP}, R_S, and R_L should increase, decrease, or stay the same to increase the ratio calculated in (c).

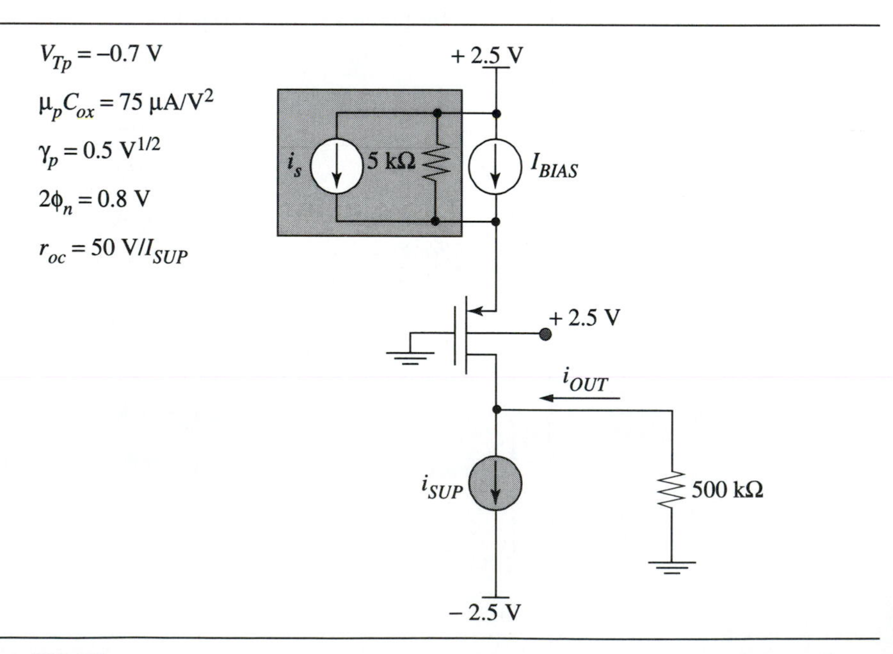

➤ **Figure P8.33**

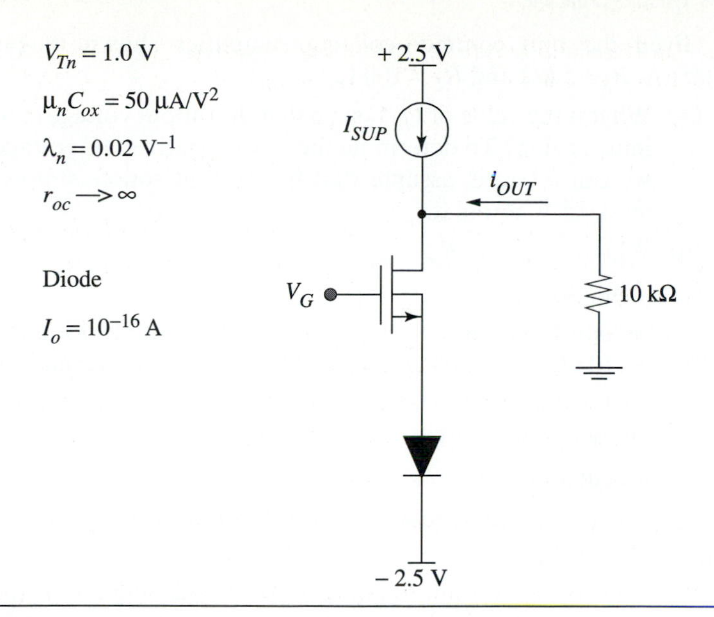

➤ **Figure P8.34**

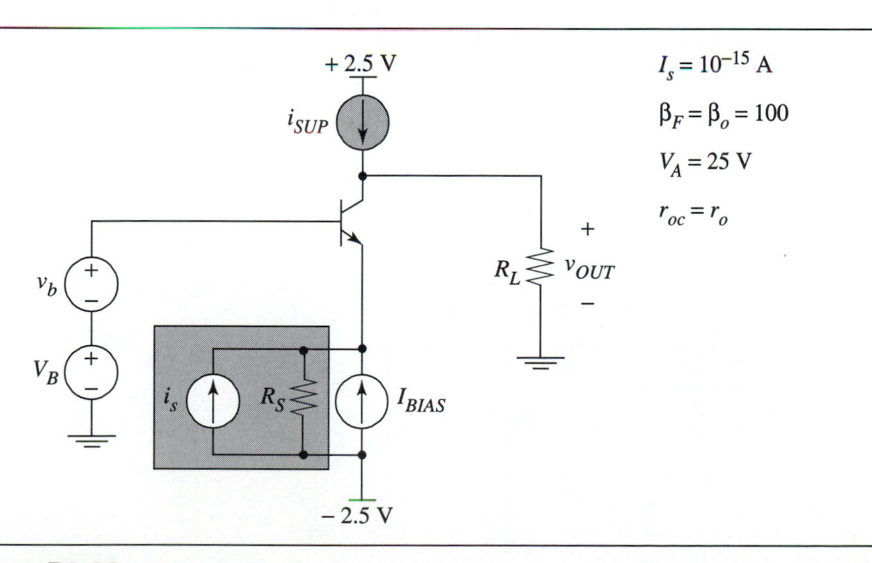

➤ **Figure P8.36**

P8.37 Given the npn common-collector amplifier shown in Fig. P8.37 with $I_{SUP} = 500\ \mu A$, $R_S = 5\ k\Omega$ and $R_L = 100\ \Omega$.

(a) What is the value of V_{BIAS} such that the output voltage can have the maximum swing? To determine the most negative DC voltage at the output we can tolerate, assume that the current source supply must have at least 0.5 V across it.

(b) What is R_{out} and R_{in}?

(c) What is v_{out}/v_s?

P8.38 For the npn common-collector voltage buffer and device data shown in Fig. P8.37 with $R_S = 10\ k\Omega$, $R_L = 500\ \Omega$, $I_{SUP} = 100\ \mu A$ and the internal resistance of the current source supply $r_{oc} = 50\ V/I_{SUP}$.

(a) Draw the two-port model and calculate R_{in}, R_{out}, and A_v.

(b) Calculate the overall voltage gain.

(c) Change I_{SUP} to increase the overall voltage gain by 10% from that found in part (b).

P8.39 In this problem we want to compare the small-signal performance and output voltage swing of an NMOS CD amplifier when the backgate terminal is shorted to the source $V_{BS} = 0$ V and when it is tied to the most negative supply, $V_B = -2.5$ V. You are given an NMOS common-drain amplifier with a current source supply as shown in Fig. P8.39. The device data is also shown in Fig. P8.39.

(a) Find the value for V_{BIAS} so that $I_{SUP} = 500\ \mu A$ when $V_{OUT} = 0$ V and $V_B = -2.5$ V.

(b) What is the maximum and minimum V_{OUT} possible while still keeping the NMOS device in the constant-current region? Assume the current source supply requires at least 0.5 V across it.

(c) What is the overall voltage gain, v_{out}/v_s including the effect of R_L and R_S?

(d) Repeat (a)–(c) for $V_{BS} = 0$ V.

P8.40 In this problem we want to compare the small-signal performance and output voltage swing of a PMOS common-drain amplifier when the backgate terminal is shorted to the source $V_{BS} = 0$ V and when it is tied to the most positive supply,

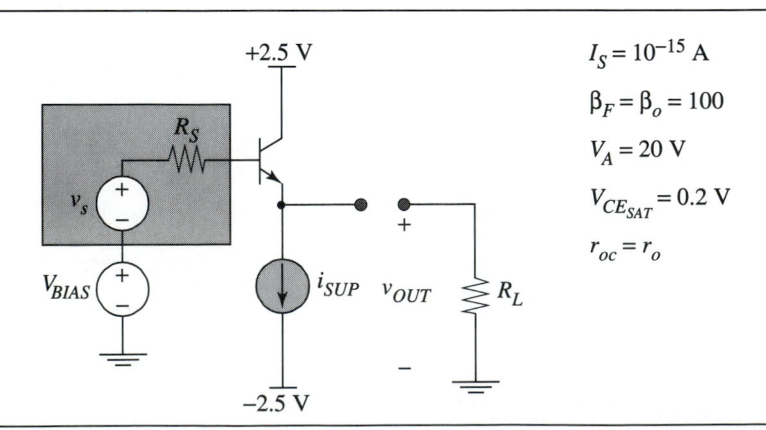

➤ **Figure P8.37**

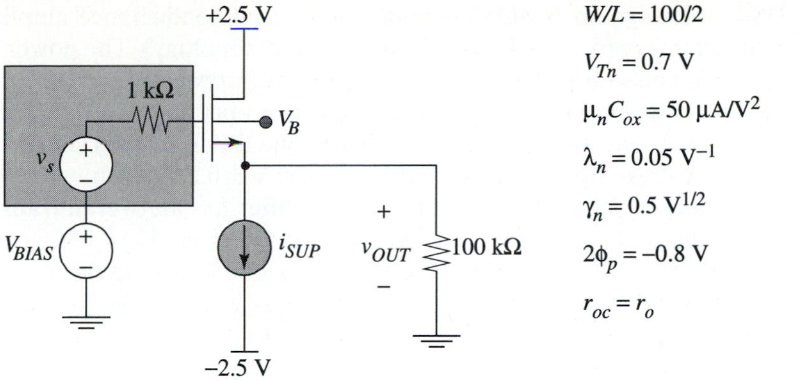

+2.5 V

1 kΩ

v_s

V_B

V_{BIAS}

i_{SUP}

+ v_{OUT} −

100 kΩ

−2.5 V

$W/L = 100/2$

$V_{Tn} = 0.7$ V

$\mu_n C_{ox} = 50\ \mu A/V^2$

$\lambda_n = 0.05\ V^{-1}$

$\gamma_n = 0.5\ V^{1/2}$

$2\phi_p = -0.8$ V

$r_{oc} = r_o$

➤ **Figure P8.39**

$V_B = 2.5$ V. You are given a PMOS common-drain amplifier with a current source supply as shown in Fig. P8.40. The device data is shown in the figure.

(a) Find the value for V_{BIAS} so that $I_{SUP} = 500\ \mu A$ when $V_{OUT} = 0$ V and $V_B = 2.5$ V.

(b) What is the maximum and minimum V_{OUT} possible while still keeping the PMOS device in the constant current region? Assume the current source supply requires at least 0.5 V across it.

(c) What is the overall voltage gain, v_{out}/v_s including the effect of R_L and R_S?

(d) Repeat (a)-(c) for $V_{BS} = 0$ V.

Design Problems

D8.1 Design a PMOS common-source voltage amplifier with a current source supply (See Fig. P8.18 for circuit topology). The bias voltage is set from the previous stage at 1.0 V. The power supply voltages are +2.5 V and –2.5 V. Choose W/L and I_{SUP} so that the amplifier has an overall voltage gain of –50 when $R_L = 50$ kΩ. We require that $V_{OUT} = 0\ V \pm 0.2$ V when $v_s = 0$ V. The PMOS device has $V_{Tp} = -1$ V, $\mu_p C_{ox} = 25\ \mu A/V^2$, $\lambda_p = 0.05\ V^{-1}$ and $r_{oc} = 50$ kΩ. All device parameters can vary ± 10% around the nominal value given.

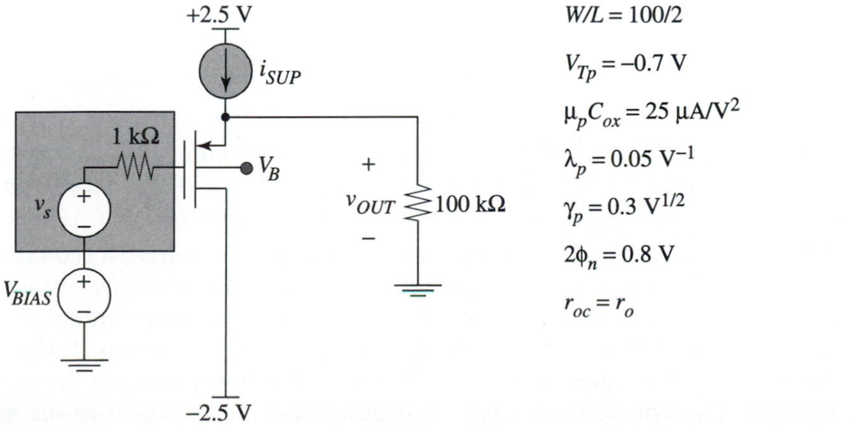

+2.5 V

i_{SUP}

1 kΩ

V_B

v_s

+ v_{OUT} −

100 kΩ

V_{BIAS}

−2.5 V

$W/L = 100/2$

$V_{Tp} = -0.7$ V

$\mu_p C_{ox} = 25\ \mu A/V^2$

$\lambda_p = 0.05\ V^{-1}$

$\gamma_p = 0.3\ V^{1/2}$

$2\phi_n = 0.8$ V

$r_{oc} = r_o$

➤ **Figure P8.40**

D8.2 Design an NMOS common-source transconductance amplifier with a current source supply (See Figure P8.16 for circuit topology). The power supply voltages are 2.5 V and –2.5 V. The value of I_{SUP} must be between 500 μA and 1 mA. The internal resistance of the current source supply is proportional to $1/I_{SUP}$ and is 100 kΩ for $I_{SUP} = 1$mA and 200 kΩ for $I_{SUP} = 500$ μA. The DC input bias voltage is set at –0.5 V. The next stage input requires that $V_{OUT} = 1.0$ V ± 0.2 V when $v_s = 0$ V for this amplifier. Choose W/L and I_{SUP} so that the amplifier has an overall transconductance of 1mS when $R_L = 50$ kΩ. The NMOS device has $V_{Tn} = 1$ V, $\mu_n C_{ox} = 50$ μA/V^2, $\lambda_n = 0.05$ V^{-1}. All device parameters can vary ± 10% around the nominal value given.

D8.3 In Example 8.5 we showed the design of an NMOS common-source voltage amplifier. Repeat the design problem for a PMOS common-source voltage amplifier to achieve the same specification of an overall voltage gain greater than |–20| when driving a load resistor, $R_L = 20$ kΩ. The PMOS device has $V_{Tp} = -1.2$ V, $\mu_p C_{ox} = 25$ μA/V^2, $\lambda_p = 0.05$ V^{-1}.

D8.4 In Example 8.7 we showed the design of an npn common-emitter current amplifier. Repeat the design problem using a pnp device to achieve the same specification of an overall current gain greater than |–50| when driving a load resistor, $R_L = 1$ kΩ. The input current source resistance is 1 MΩ. The pnp device has $I_S = 10^{-14}$A, $\beta_F = \beta_o = 75$, and $V_A = 20$ V. Assume $r_{oc} \to \infty$.

D8.5 In Example 8.9 we showed a design of a common-emitter transconductance amplifier with a reduced temperature dependence over a temperature range of $0° - 70°C$. In some automotive applications temperatures can be as high as $200°C$. Bipolar transistors can have greatly reduced β_F in this temperature range. Design a common-source NMOS transconductance amplifier with source degeneration that meets the specifications of an overall transconductance within 10% of 1mS across a temperature range of $0° - 200°C$, where $R_S = 1$kΩ and $R_L = 10$ kΩ. The NMOS device data at room temperature is $V_{Tn} = 1$ V, $\mu_n C_{ox} = 50$ μA/V^2, $\lambda_n = 0.05$ V^{-1}. Perform the hand design at room temperature and use SPICE to investigate how the temperature dependence of the device parameters affects your design. Recall that the degeneration resistor should desensitize parameter variation on the overall transconductance. The resistors available have a 5% tolerance across $0° - 200°C$. Assume $r_{oc} = r_o$.

D8.6 In Example 8.10 we designed a current buffer with an overall current gain of 0.95 using an npn transistor. You only have a CMOS process available and must achieve the same specifications. Design a current buffer using an NMOS transistor with device parameters $V_{Tn} = 1$ V, $\mu_n C_{ox} = 50$ μA/V^2, $\lambda_n = 0.05$ V^{-1} or a PMOS transistor with device parameters $V_{Tp} = -1.0$ V, $\mu_p C_{ox} = 15$ μA/V^2, $\lambda_p = 0.02$ V^{-1}. Choose the device type based on which design gives the lowest power dissipation.

D8.7 Given $R_S = 20$ kΩ and $R_L = 50$ kΩ, design an npn common-base current buffer such that it minimizes DC power while maintaining an overall current gain $|i_{out}/i_s| \geq 0.9$. The npn transistor has $I_S = 10^{-15}$A, $\beta_F = \beta_o = 100$, and $V_A = 20$ V. The internal resistance of the current source supply $r_{oc} = 50\text{V}/I_{SUP}$.

D8.8 Design a PMOS common-drain voltage buffer to drive a 100 Ω load resistor with an overall voltage gain of 0.5. To begin this design recognize that $1/g_m$ must be < 100 Ω to meet the voltage gain specification. The design trade-off is between area (W/L) and power dissipation (I_{SUP}, supply voltage). In this design we require that the output voltage swing be ≥ 3V. This sets a lower limit on the power supply voltages. Use symmetrical supply voltages. Short the backgate to the source. The

PMOS device data is $V_{Tp} = -1.0$ V, $\mu_p C_{ox} = 25$ μA/V^2, $\lambda_p = 0.05$ V^{-1}. Assume the current source supply requires at least 0.5 V across it.

(a) Design this amplifier to minimize power dissipation by setting 1 μA $\leq I_{SUP} \leq$ 10 μA. We do not allow the current to be any lower due to other design considerations such as frequency response.

(b) Investigate the trade-off between area and I_{SUP} by allowing 100 μA $\leq I_{SUP} \leq$ 1 mA. Find the range of W/L required to meet the design specification over this range of current.

D8.9 In Example 8.12 we showed the design of an npn voltage buffer. Typically pnp devices have a lower β_F compared to npn devices. Design a voltage buffer to meet the same specifications with a pnp transistor having a $\beta_F = 75$. All other device parameters are the same.

APPENDIX

In this appendix we quantify the error which arises by using the two-port parameters calculated for the common-base/gate amplifier used as a current buffer and for the common-collector/drain amplifier used as a voltage buffer. Recall that these amplifiers are not unilateral and hence the method in which we calculated the two-port parameters is not strictly correct. However its simplicity and usefulness in design outweigh this error. As you will see, the error is small under most conditions.

A8.1 Common Base/Gate Amplifier

We begin this error analysis by computing the current gain for the small-signal common-base amplifier shown below. The small-signal current source has a source resistance, R_S and the amplifier is driving a load resistance R_L. We calculate the current gain of the amplifier by finding i_{out}/i_s for the entire circuit. We assume that $r_{oc} \gg R_L$ to make the algebra tractable.

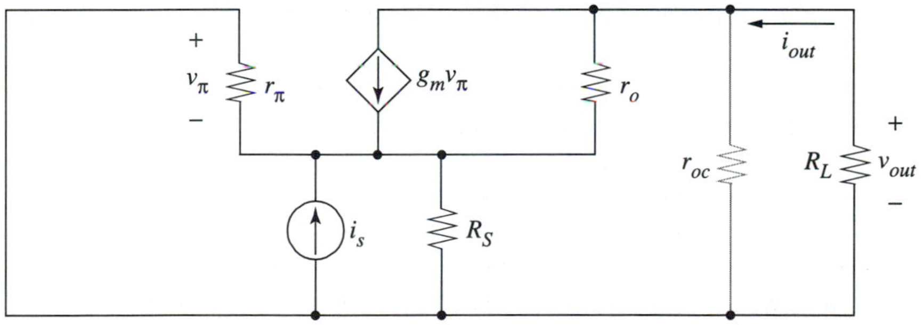

$$i_s = \frac{-v_\pi}{r_\pi \| R_S} - g_m v_\pi - \frac{(v_{out} + v_\pi)}{r_o} \tag{A8.1}$$

$$i_{out} = g_m v_\pi + \frac{(v_{out} + v_\pi)}{r_o} \tag{A8.2}$$

Rearranging we find

$$v_\pi = \frac{i_{out} - v_{out}/r_o}{g_m + 1/r_o} = \frac{i_{out}r_o - v_{out}}{g_m r_o + 1} \tag{A8.3}$$

Substituting Eq. (A8.2) and (A8.3) into Eq. (A8.1) and recognizing that $v_{out} = -i_{out}R_L$

$$i_s = \frac{-i_{out}r_o}{(g_m r_o + 1)(r_\pi \| R_S)} - \frac{i_{out}R_L}{(g_m r_o + 1)(r_\pi \| R_S)} - i_{out} \tag{A8.4}$$

$$\frac{i_{out}}{i_s} = \frac{-(g_m r_o + 1)(r_\pi \| R_S)}{r_o + R_L + (g_m r_o + 1)(r_\pi \| R_S)} \approx \frac{-g_m r_o (r_\pi \| R_S)}{r_o + R_L + g_m r_o (r_\pi \| R_S)} \tag{A8.5}$$

The approximation made in Eq. (A8.5), namely $g_m r_o \gg 1$, is excellent for any reasonably designed transistor and is consistent with the approximation made in the two-port analysis.

Now using the two-port model for the common-base amplifier, we can write an expression for the current gain as

$$\frac{i_{out}}{i_s} = \frac{R_S}{R_S + R_{in}}(-1)\frac{R_{out}}{R_{out} + R_L} \tag{A8.6}$$

From Table 8.3 the expressions for the two-port parameters are $R_{in} = 1/g_m$ and $R_{out} = r_o + g_m r_o (r_\pi \| R_S)$. Substituting these into Eq. (A8.6) we can write

$$\frac{i_{out}}{i_s} = \frac{-g_m R_S}{1 + g_m R_S}\left[\frac{g_m r_o (r_\pi \| R_S)}{r_o + R_L + g_m r_o (r_\pi \| R_S)} + \frac{r_o}{r_o + R_L + g_m r_o (r_\pi \| R_S)}\right] \tag{A8.7}$$

Comparing the expressions Eq. (A8.5) (full analysis) and Eq. (A8.7) (simple two-port) we see that there is both a multiplicative and additive error that results when we apply our two-port method. To quantify these we write

$$Error_{Mult} = 1 - \frac{g_m R_S}{1 + g_m R_S} \tag{A8.8}$$

$$Error_{Add} = \frac{r_o}{r_o + R_L + g_m r_o (r_\pi \| R_S)} \tag{A8.9}$$

The MOS common-gate amplifier current gain can be found by using the same analysis and letting $r_\pi \to \infty$. in Eq. (A8.5). We will assume that the source and backgate are shorted so we can neglect the effect of g_{mb}. The resulting current gain is given by

$$\frac{i_{out}}{i_s} = \frac{-(g_m r_o + 1) R_S}{r_o + R_L + (g_m r_o + 1) R_S} \approx \frac{-g_m r_o R_S}{r_o + R_L + g_m r_o R_S} \tag{A8.10}$$

Now using the two-port model for the common-gate amplifier, and using the expressions for the two-port parameters for the common-gate amplifier, $R_{in} = 1/g_m$ and $R_{out} = r_o + g_m r_o R_S$ we can write

$$\frac{i_{out}}{i_s} = \frac{-g_m R_S}{1 + g_m R_S}\left[\frac{g_m r_o R_S}{r_o + R_L + g_m r_o R_S} + \frac{r_o}{r_o + R_L + g_m r_o R_S}\right] \tag{A8.11}$$

For the MOS common-gate amplifier the multiplicative error is the same as the bipolar common-base amplifier. The additive error is found by letting $r_\pi \to \infty$ resulting in

$$Error_{Add} = \frac{r_o}{r_o + R_L + g_m r_o R_S} \tag{A8.12}$$

These errors are small when $R_S \gg 1/g_m$. Under most conditions when the common-base/gate amplifiers are used they have an input current signal source. Since the input resistance of the amplifier is $1/g_m$, R_S must be much larger than $1/g_m$ or significant degradation in gain will occur. The bottom line is that for most designs the two-port method gives an excellent approximation to the actual circuit performance.

A8.2 Common Collector/Drain Amplifier

We begin this error analysis by computing the voltage gain for the small-signal model of the common-collector amplifier shown below. The small-signal voltage source has a source resistance, R_S and the amplifier is driving a load resistance R_L. We calculate the voltage gain of the amplifier by finding v_{out}/v_s for the entire circuit. We assume that $r_o \| r_{oc} \gg R_L$ to make the algebra tractable.

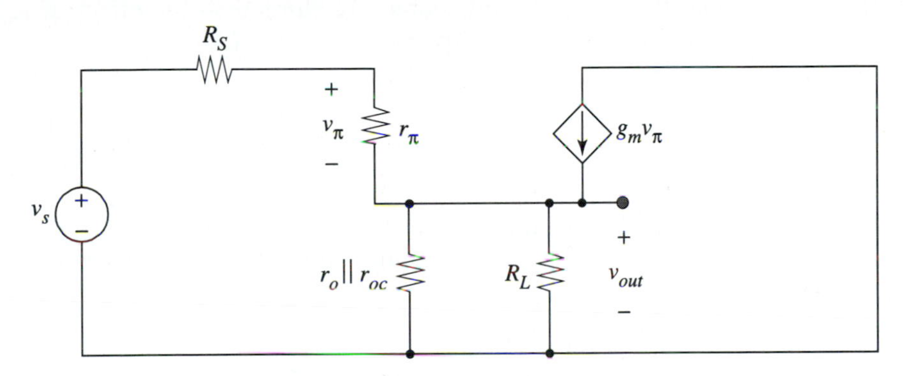

We write KCL at the output node

$$g_m\left[\frac{(v_s - v_{out})}{R_S + r_\pi} r_\pi\right] + \frac{(v_s - v_{out})}{R_S + r_\pi} = \frac{v_{out}}{R_L} \tag{A8.13}$$

Recalling that $g_m r_\pi = \beta_o$ and manipulating Eq. (A8.13) we find

$$\frac{v_{out}}{v_s} = \frac{(\beta_o + 1) R_L}{(\beta_o + 1) R_L + R_S + r_\pi} \tag{A8.14}$$

Assuming that $\beta_o \gg 1$ which is consistent with our two-port analysis we write

$$\frac{v_{out}}{v_s} = \frac{\beta_o R_L}{\beta_o R_L + R_S + r_\pi} \tag{A8.15}$$

Now using the two-port model for the common-collector amplifier, we can write an expression for the voltage gain as

$$\frac{v_{out}}{v_s} = \frac{R_{in}}{R_{in} + R_S}(1)\frac{R_L}{R_L + R_{out}} \tag{A8.16}$$

From Table 8.3 the expressions for the two-port parameters are $R_{in} = r_\pi + \beta_o(r_o \| r_{oc} \| R_L)$ and $R_{out} = \dfrac{R_S}{\beta_o} + \dfrac{1}{g_m}$. Substituting these into Eq. (A8.16) and again assuming $r_o \| r_{oc} \gg R_L$ we can write

$$\frac{v_{out}}{v_s} = \left(\frac{\beta_o R_L + r_\pi}{\beta_o R_L + R_S + r_\pi}\right)\left(\frac{R_L}{R_L + \dfrac{R_S}{\beta_o} + \dfrac{1}{g_m}}\right) = \left(\frac{\beta_o R_L + r_\pi}{\beta_o R_L + R_S + r_\pi}\right)\left(\frac{\beta_o R_L}{\beta_o R_L + R_S + r_\pi}\right) \quad \text{(A.17)}$$

Comparing the expressions Eq. (A8.15) (full analysis) and Eq. (A8.17)(simple two-port) we see that there is only a multiplicative error that results when we apply our two-port method. To quantify the error we write

$$Error_{Mult} = 1 - \left(\frac{\beta_o R_L + r_\pi}{\beta_o R_L + R_S + r_\pi}\right) \quad \text{(A8.18)}$$

From Eq. (A8.18) we see that if $R_S \ll (\beta_o R_L + r_\pi)$ the error is small. For the MOS CD amplifier we assume the backgate and source are shorted so the effect of g_{mb} is neglected. We find the voltage gain by letting $\beta_o \to \infty$ in Eq. (A8.15).

$$\frac{v_{out}}{v_s} = \frac{R_L}{R_L + \cancel{\dfrac{R_S}{\beta_o}} + \dfrac{1}{g_m}} = \frac{g_m R_L}{1 + g_m R_L} \quad \text{(A8.19)}$$

For the MOS common-drain amplifier we use $R_{in} \to \infty$ and $R_{out} = 1/g_m$. Plugging these into Eq. (A8.16) which is also the two-port model for the common-drain amplifier yields

$$\frac{v_{out}}{v_s} = \frac{R_L}{R_L + 1/g_m} = \frac{g_m R_L}{1 + g_m R_L} \quad \text{(A8.20)}$$

When we compare Eq. (A8.19) (full analysis) and Eq. (A8.20) (simple two-port) we have no error. This should not be surprising since the condition required to minimize the error in Eq. (A8.18) is always met for the MOS case since $\beta_o \to \infty$.

The conditions required to minimize the error when using the two-port analysis are summarized below.

CB Amp: $R_S \gg 1/g_m$

CG Amp: $R_S \gg 1/g_m$

CC Amp: $R_S \ll (\beta_o R_L + r_\pi)$

CD Amp: No error

chapter 9

Multistage Amplifiers

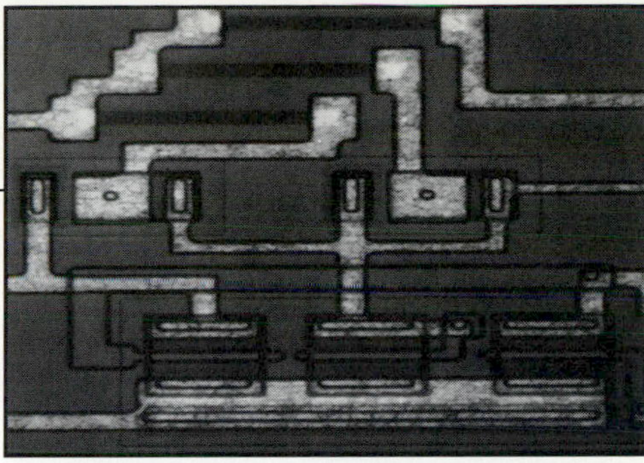

Cascaded common-source stages using degenerated bipolar current-source supplies. Designed by William A. Clark, UC Berkeley using the BiMEMS process. (Courtesy of Analog Devices Inc.)

Multistage amplifiers are used to increase the gain and/or transform input and output resistances to match the source and load resistances for minimum signal attenuation from the input to the output of the entire circuit. The **gain** can be a voltage gain, current gain, transconductance, or transresistance depending on whether the input and output signals are voltages or currents. Two major issues must be solved when combining integrated single-stage transistor amplifiers. First, the input and output resistances of the various stages must be chosen appropriately so maximum signal transfer can occur between stages. Second, DC biasing that sets the quiescent node voltages and currents must be properly chosen so the stages can be directly coupled.

Chapter Objectives

- Design and analyze the gain and input and output resistance of multistage amplifiers based on cascading various single-stage amplifiers.
- Estimate the input/output resistance and the gain of a multistage amplifier.
- Understand when to apply MOS versus bipolar transistor amplifiers.
- Know when to use n-type or p-type transistors for MOS and bipolar implementations.
- Biasing multistage amplifiers with active devices.
- Develop current and voltage sources used to set the quiescent DC operating voltages and currents in the amplifier.
- Develop an intuitive understanding of circuit blocks.
- Trace the signal path through a complicated circuit.

9.1 MOS MULTISTAGE AMPLIFIERS: SMALL-SIGNAL DESCRIPTION

Several examples of cascading the two-port models of various single-stage amplifiers studied in Chapter 8 will be used to help us understand how multistage amplifiers can be used to increase the gain and transform input/output resistances. The desired input and output resistances, as well as high gain, can be achieved with proper selection of the single-stage amplifiers.

9.1.1 Voltage Amplifier

A **voltage amplifier** requires a high input resistance, a low output resistance, and a large voltage gain. Figure 9.1 shows a two-port model of a voltage amplifier including the input source resistance and load resistance. The transfer function between the output voltage delivered to the load resistor and the input signal voltage is given by

$$\frac{v_{out}}{v_s} = \left(\frac{R_{in}}{R_S + R_{in}}\right) A_v \left(\frac{R_L}{R_{out} + R_L}\right) \tag{9.1}$$

This transfer function shows that a high input resistance and low output resistance results in the largest voltage gain between the source and load.

From Chapter 8 we know that a common-source amplifier has an infinite input resistance since the MOS transistor has an insulating gate with no input current. The bipolar transistor counterpart, namely a common-emitter amplifier, has an input resistance equal to r_π. Since the bipolar common-emitter amplifier has a smaller R_{in}, it is less desirable as an input stage for a voltage amplifier.

Now that we know a common-source amplifier is the proper input stage for a voltage amplifier, we can explore cascading two of these stages to increase the voltage gain. The small-signal model of two cascaded common-source amplifiers is given in Fig. 9.2.

two-port parameters CS-CS voltage amplifier

The input resistance of this cascade is infinite and the voltage gain is given by the multiplication of the voltage gain of each stage as shown in

$$A_v = g_{m1}(r_{o1} \| r_{oc1}) g_{m2}(r_{o2} \| r_{oc2}) \tag{9.2}$$

The output resistance of this cascaded common-source voltage amplifier is

$$R_{out} = r_{o2} \| r_{oc2} \tag{9.3}$$

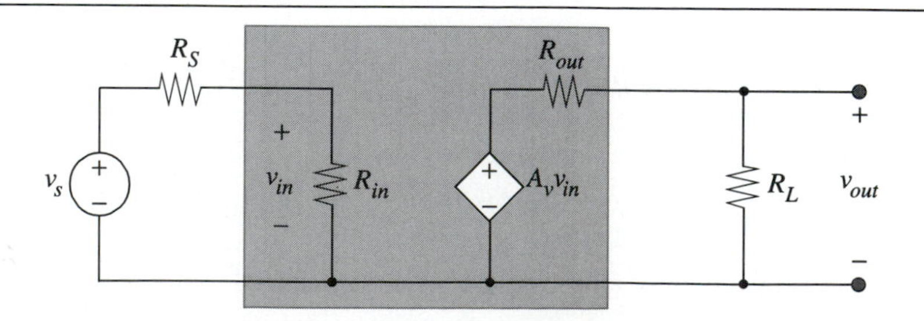

Figure 9.1 Two-port model of a voltage amplifier.

➤ **Figure 9.2** Small-signal model for two cascaded common-source amplifiers.

This cascaded common-source voltage amplifier has two of the three required characteristics, namely a large input resistance and a high voltage gain. However, it still has a very high output resistance that will degrade the voltage transfer from the amplifier to the load resistor.

From Chapter 8 we know that a common-drain amplifier has an infinite input resistance, a low output resistance, and a voltage gain near unity. We can cascade the small-signal, two-port model of a common-drain amplifier with the small-signal model of the common-source cascade described above. This three stage amplifier is shown in Fig. 9.3. In this diagram the cascaded common-source amplifiers are modeled with an infinite input resistance, a voltage gain A_v, and an output resistance given in Eq. (9.2) and Eq. (9.3), respectively.

There is no interstage loss of voltage gain because the common-drain amplifier has an infinite input resistance. In addition, the output resistance is reduced to approximately the reciprocal of the transconductance plus the backgate transconductance of the common-drain amplifier $1/(g_{m3}+g_{mb3})$. This is a significant reduction in output resistance and allows this three-stage voltage amplifier to drive much smaller load resistances while still maintaining a significant transfer of the voltage to the load.

9.1.2 Transconductance Amplifier

A **transconductance amplifier** requires a large input resistance, a large transconductance, and a large output resistance to be able to pass most of its output current into the load. Let us explore using cascaded common-source amplifiers for a transconductance amplifier since common-source amplifiers are good transconductance amplifiers. In Fig. 9.4 we show a small-signal model of two cascaded common-source

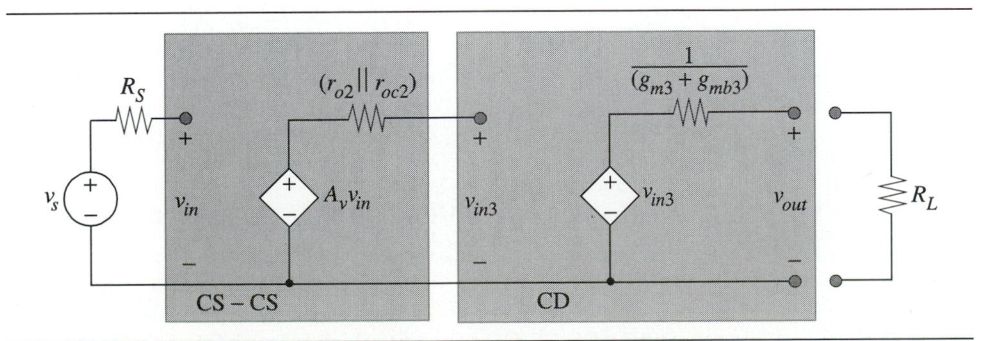

➤ **Figure 9.3** Small-signal model of a CS-CS-CD cascade.

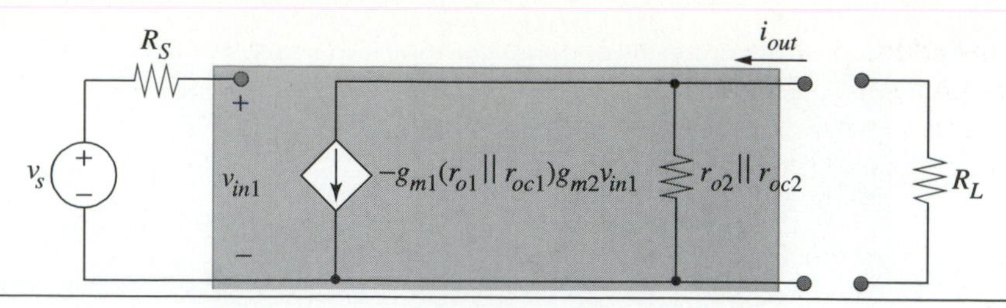

➤ **Figure 9.4** Small-signal model for CS-CS cascade used as a transconductance amplifier.

amplifiers. We use the Norton equivalent output since current is the output variable of interest.

two-port parameters CS-CS transconductance amplifier

The overall transconductance of this common-source cascade is equal to

$$G_m \equiv \frac{i_{out}}{v_{in1}} = -g_{m1}(r_{o1} \| r_{oc1})g_{m2} = A_{v1}g_{m2} \tag{9.4}$$

Notice that the additional common-source stage increased the overall transconductance by the voltage gain of the first stage. There is no interstage loss since the input resistance to the second stage is infinite.

The output resistance of the transconductance amplifier is

$$R_{out} = r_{o2} \| r_{oc2}, \tag{9.5}$$

which is the output resistance of a single common-source amplifier stage.

We can add a current buffer to increase the output resistance of the cascaded common-source amplifier. An **ideal current buffer** is defined as a circuit whose input resistance is very small, output resistance is very large, and has a current gain of unity. The common-gate amplifier studied in Chapter 8 is a good example of a current buffer. A small-signal model of a common-gate amplifier cascaded with two common-source amplifiers is shown in Fig. 9.5.

two-port parameters CS-CS-CG transconductance amplifier

The output resistance of this amplifier is

$$R_{out} \approx g_{m3}r_{o3}(r_{o2} \| r_{oc2}) \| r_{oc3} \tag{9.6}$$

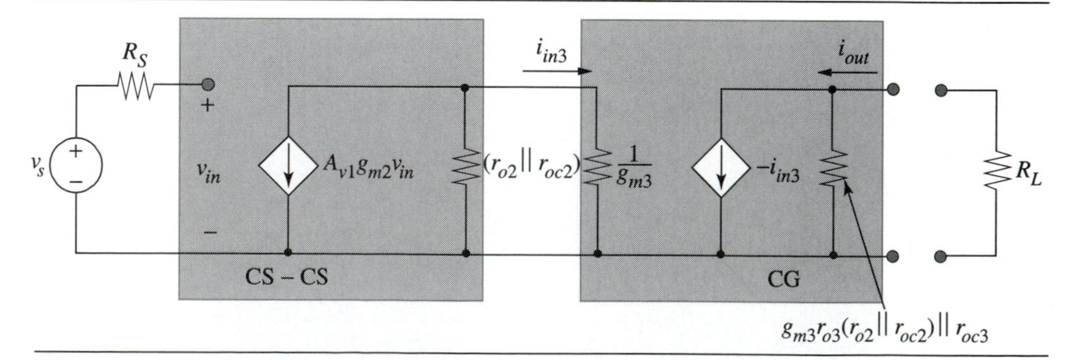

➤ **Figure 9.5** Small-signal model a CS-CS-CG cascade which forms a transconductance amplifier.

With the addition of the common-gate amplifier, R_{out} is increased by its open-circuit voltage gain $g_{m3}r_{o3}$. This increase is only realized if the internal resistance of the current source supply r_{oc3} is large.

The unloaded transconductance is given by

$$G_m \equiv \frac{i_{out}}{v_{in}} = -g_{m1}(r_{o1} \| r_{oc1})\, g_{m2}\left(\frac{r_{o2} \| r_{oc2}}{r_{o2} \| r_{oc2} + 1/g_{m3}}\right) \quad (9.7)$$

The transconductance is slightly degraded due to the parallel combination of the common-source output resistance with the common-gate input resistance. This degradation is negligible since the common-gate input resistance is small compared to the common-source output resistance.

➤ EXAMPLE 9.1 MOS Transresistance Amplifier

This example explores how to cascade two MOS stages to form a transresistance amplifier. Select the stages and calculate R_{in}, R_{out}, and R_m. For this example, assume that the internal resistances of the current source supplies are infinite $r_{oc} \to \infty$. Hint: Recall that a transresistance amplifier requires a low input resistance for the input signal current, a low output resistance for the output signal voltage, and a large transresistance.

SOLUTION

To obtain a low input resistance, we choose a CG amplifier as the first stage. The choice of the second stage depends on the specifications required. Let's try a CS amplifier. The small-signal two-port model for a CG-CS cascade is shown in Fig. Ex 9.1A.

R_{in}, the input resistance, is $1/g_{m1}$. R_{out}, the output resistance, is r_{o2}. We need to find the unloaded ($R_L \to \infty$) transfer function between v_{out} and i_{in1} to calculate R_m. We begin by writing

$$v_{gs2} = i_{in1} r_{o1}(1 + g_{m1}R_S)$$

$$v_{out} = -g_{m2} v_{gs2} r_{o2}$$

$$R_m = v_{out}/i_{in1} = -g_{m2} r_{o1}(1 + g_{m1}R_S) r_{o2}$$

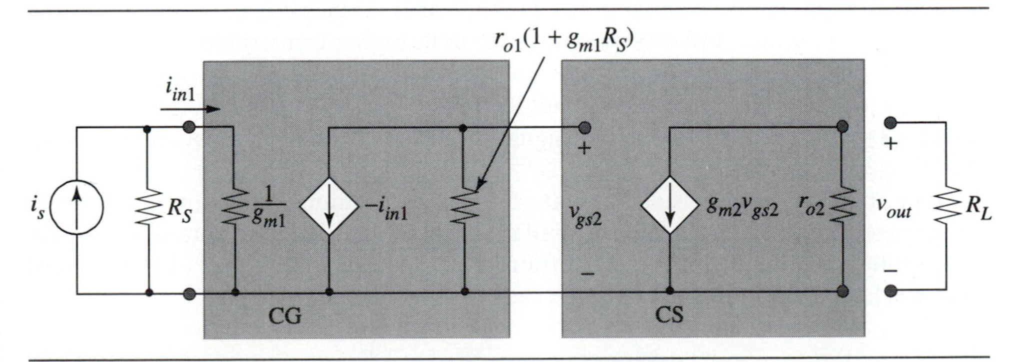

➤ Figure Ex9.1A

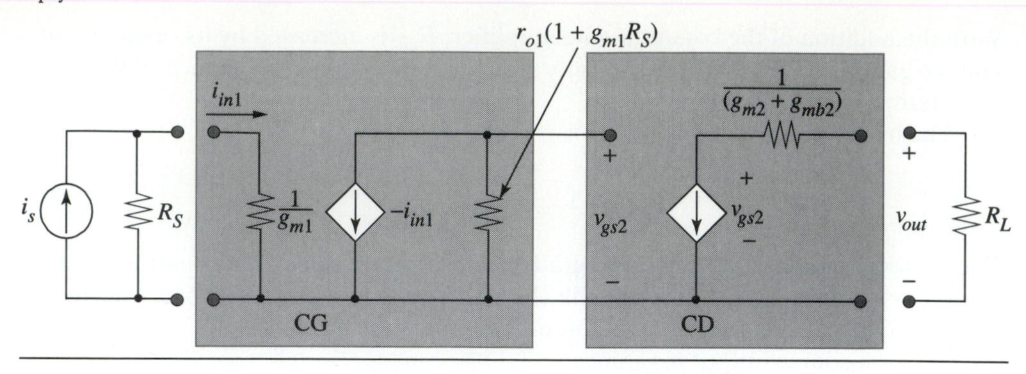

➤ **Figure Ex9.1B**

If we use a CD amplifier instead of a CS for the second stage, we expect a lower output resistance at the expense of transresistance. The small-signal two-port model for a CG-CD amplifier is shown in Fig. Ex 9.1B.

The input resistance is the same for both amplifiers since both use a CG stage as the input. The output resistance is $R_{out} = 1/(g_{m2} + g_{mb2})$ which is approximately $1/g_{m2}$. The transresistance $R_m = v_{out}/i_{in1} = r_{o1}(1 + g_{m1}R_S)$ for the CG-CD configuration. Note that the output resistance and transresistance of the CG-CD configuration are lower than the CG-CS configuration by $g_{m2}r_{o2}$. The proper topological choice depends on the specifications required and the relative value of R_L compared to R_{out}.

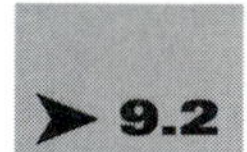

➤ 9.2 BiCMOS MULTISTAGE AMPLIFIERS: SMALL-SIGNAL DESCRIPTION

A bipolar transistor has two major advantages over an MOS transistor for a given current and area: (1) higher transconductance and (2) higher output resistance. This comes at the expense of a non-zero base current that degrades the input resistance of the bipolar transistor when operating in the common-emitter or common-collector configuration. Remember to keep these facts in mind when choosing between using a bipolar or MOS transistor for a specific application.

9.2.1 BiCMOS Voltage Amplifier

We know that a bipolar transistor has a higher open-circuit voltage gain for a given quiescent current than an MOS transistor due to its higher transconductance and output resistance. Therefore, we might attempt to cascade a common-source amplifier with a common-emitter amplifier to yield a larger voltage-gain circuit than cascaded common-source amplifiers. A small-signal model of the common-source/common-emitter cascade is shown in Fig. 9.6.

The input resistance is still infinite since the common-source amplifier is the input stage. The output resistance is similar to that of the MOS configuration, namely the output resistance of the bipolar transistor in parallel with that of the current source is large. The voltage gain of the cascade is given by

$$A_v = A_{v1}\left(\frac{r_{\pi 2}}{r_{\pi 2} + (r_{o1} \| r_{oc1})}\right)A_{v2} \tag{9.8}$$

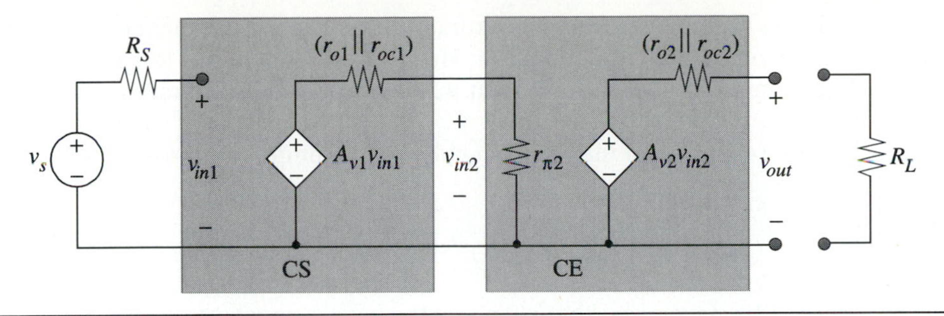

➤ Figure 9.6 Small-signal model for a common-source/common-emitter cascade. $A_{v1} = -g_{m1}(r_{o1} \| r_{oc1})$ and $A_{v2} = -g_{m2}(r_{o2} \| r_{oc2})$.

where $A_{v1} = -g_{m1}(r_{o1} \| r_{oc1})$ and $A_{v2} = -g_{m2}(r_{o2} \| r_{oc2})$.

It can be seen that the overall voltage gain of the CS-CE amplifier is degraded from the ideal cascading of the two voltage gains due to a poor transfer of the voltage from the output of stage-one to the input of stage-two. Since the input resistance of the common-emitter amplifier $r_{\pi 2}$ is much less than the output resistance of the common-source amplifier, the voltage transfer is severely degraded. This degradation in voltage gain can be as large as 100, which wipes out any advantage in voltage gain that the common-emitter amplifier has added when cascaded with the common-source amplifier. Therefore, it is a CS-CS cascade that provides the best small-signal parameters in terms of input resistance and large voltage gain. The problem with the CS-CS cascade is that it still has a large output resistance, and therefore the voltage transfer to a small load resistance is severely degraded.

In the previous section, we added a common-drain amplifier to the CS-CS cascade. We will now explore what happens if we cascade a common-collector amplifier with a two-stage common-source amplifier. A small-signal model of this three-stage cascade is shown in Fig. 9.7.

Because the bipolar transistor has a finite input resistance, there is an interstage voltage gain loss shown in the term in brackets.

$$A_v = A_{v1} A_{v2} \left(\frac{r_{\pi 3} + \beta_{o3}(r_{o3} \| r_{oc3} \| R_L)}{(r_{o2} \| r_{oc2}) + r_{\pi 3} + \beta_{o3}(r_{o3} \| r_{oc3} \| R_L)} \right) \tag{9.9}$$

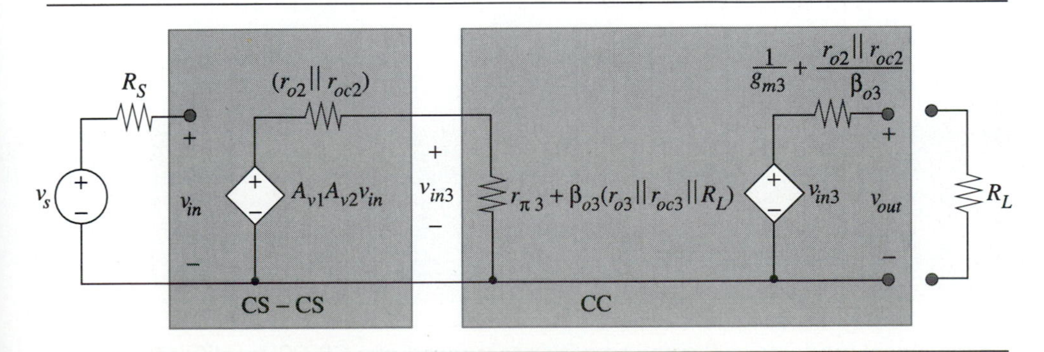

➤ Figure 9.7 Small-signal model for a CS-CS-CC cascade.

two-port parameters CS-CS-CC voltage amplifier

You should note that the common-collector amplifier has a much higher input resistance than the common-emitter amplifier. However, the interstage loss with a common-collector amplifier can be significant when we need to drive reasonably small load resistors.

The output resistance of the common-collector amplifier is approximately

$$R_{out} = \frac{1}{g_{m3}} + \frac{r_{o2} \| r_{oc2}}{\beta_{o3}} \tag{9.10}$$

where it should be recognized that $r_{o2} \| r_{oc2}$ is the source resistance of the common-collector stage of this amplifier. From device physics, we know that the transconductance of a bipolar transistor is generally larger than that of an MOS transistor. This implies that the CS-CS-CC cascade has a lower output resistance compared to a CS-CS-CD cascade provided that the second term in Eq. (9.10) is small compared to $1/g_{m3}$. If this is not the case a CD amplifier could provide a lower output resistance.

The common-collector amplifier has reduced the output resistance of the two-stage CS amplifier. Even though this is a significant reduction of the output resistance compared to that of the pure common-source amplifier cascade, it still does not allow us to drive a very small load resistance. Further reduction of the output resistance is achieved by cascading a common-drain and a common-collector amplifier at the output of the CS-CS stage. A small-signal model of this four-stage voltage amplifier is shown in Fig. 9.8.

two-port parameters CS-CS-CD-CC voltage amplifier

There is no interstage loss between the output of the CS-CS cascade and the CD amplifier. The interstage loss between the common-drain and common-collector amplifier is smaller than the interstage loss of the previous case since the output resistance of the common-drain amplifier is much smaller than that of the common-source amplifier. The interstage loss assuming $R_L \ll (r_o \| r_{oc})$ is shown in brackets as

$$A_v = A_{v1} A_{v2} \left(\frac{r_{\pi 4} + \beta_{o4} R_L}{\frac{1}{g_{m3} + g_{mb3}} + r_{\pi 4} + \beta_{o4} R_L} \right) \tag{9.11}$$

This interstage loss is small since $(r_{\pi 4} + \beta_{o4} R_L) \gg 1/(g_{m3} + g_{mb3})$ and can often be neglected.

➤ **Figure 9.8** Small-signal model for a voltage amplifier composed of a CS-CS amplifier followed by a voltage buffer, which is a CD-CC cascade.

The output resistance of this cascade is given by

$$R_{out} = \frac{1}{g_{m4}} + \frac{1}{\beta_{o4}(g_{m3} + g_{mb3})} \tag{9.12}$$

The CD-CC voltage buffer has an infinite input resistance and a small interstage voltage gain loss, which produces a voltage gain of approximately unity. In addition, it has a very small output resistance dependent on the transconductance of a bipolar transistor. This circuit closely approximates an ideal voltage buffer. By cascading this voltage buffer with the CS-CS amplifier we have a nearly ideal voltage amplifier.

9.2.2 BiCMOS Transconductance Amplifier

We will now explore cascading a common-source amplifier with a common-emitter amplifier and determine its characteristics as a transconductance amplifier. A small-signal model of this cascade is shown in Fig. 9.9

The unloaded transconductance of this two-stage amplifier (neglecting R_S and R_L) is given by the short-circuit output current over the input voltage as shown in

$$\frac{i_{out}}{v_{in}} = G_m = -g_{m1}\left(\frac{(r_{o1} \| r_{oc1})}{r_{\pi 2} + (r_{o1} \| r_{oc1})}\right)g_{m2}r_{\pi 2} \tag{9.13}$$

The resulting transconductance is β_{o2} of the bipolar transistor, $(g_{m2}r_{\pi 2})$ times the transconductance of the MOS transistor g_{m1} with a term that represents the interstage loss. Observe that this loss occurs when the current output of the common-source amplifier is divided between the output resistance of the common-source amplifier and the input resistance of the second-stage common-emitter amplifier, namely $r_{\pi 2}$. In this case, the interstage loss is small since in general $(r_{o1} \| r_{oc1}) \gg r_{\pi 2}$. The overall input resistance of this cascade is infinite and the output resistance is equal to the parallel combination of the bipolar transistor output resistance with the internal resistance of the current source supply $(r_{o2} \| r_{oc2})$.

two-port parameters CS-CE transconductance amplifier

From the previous section, we know the CS-CS cascade is a good transconductance amplifier. It has a transconductance equal to the voltage gain of the first stage times the transconductance of the second stage given in

$$G_m = -g_{m1}(r_{o1} \| r_{oc1})g_{m2} = A_{v1}g_{m2} \tag{9.14}$$

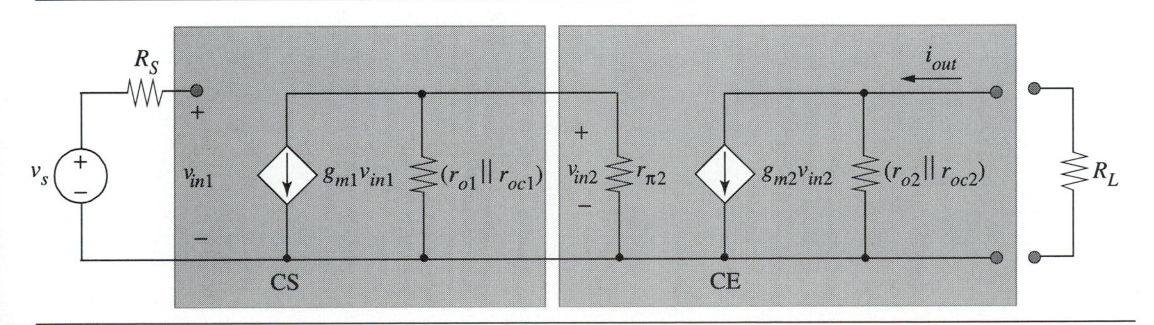

Figure 9.9 Small-signal model of a BiCMOS transconductance amplifier composed of a CS-CE cascade.

The g_m in an MOS transistor is generally smaller than that of a bipolar transistor. However, the transconductance of a CS-CS cascade is on the same order as that of the CS-CE two-stage amplifier since the second stage common-source amplifier has an infinite input resistance and hence no interstage loss. The output resistance of the CS-CS cascade is somewhat lower than the CS-CE since the output resistance of an MOS transistor is generally lower than that of a bipolar transistor. A current buffer would be useful to further increase the output resistance of this cascade.

9.2.3 BiCMOS Current Buffer

An **ideal current buffer** is defined as a circuit that has a very low input resistance, a very high output resistance, and a current gain of unity. From Chapter 8 we know that a common-base stage has an output resistance that is approximately $\beta_o r_o \| r_{oc}$ assuming that $R_S \gg r_\pi$. Since one common-base stage has increased the output resistance by approximately β_o over the common-emitter stage, let us explore what happens if we add another common-base stage to increase the output resistance. A small-signal model of a cascade of two common-base amplifiers is shown in Fig. 9.10.

The output resistance of this cascade is found by using the results from Chapter 8 and realizing that the output resistance of the first CB stage ($\beta_o r_{o1} \| r_{oc1}$) is acting as the source resistance for the second CB stage. R_{out} of the cascade is given by

$$R_{out} = [g_{m2} r_{o2} (r_{\pi 2} \| \beta_{o1} r_{o1} \| r_{oc1})] \| r_{oc2} \quad (9.15)$$

Since $r_{\pi 2} \ll \beta_{o1} r_{o1} \| r_{oc1}$

$$R_{out} \approx [g_{m2} r_{o2} r_{\pi 2}] \| r_{oc2} = (\beta_{o2} r_{o2}) \| r_{oc2} \quad (9.16)$$

Note that the output resistance is still equal to β_o times the output resistance of the bipolar transistor in parallel with the output resistance of the current source. In other words, the second common-base amplifier did not improve the output resistance. The reason for this lack of improvement is the relatively small value for $r_{\pi 2}$. The base current in the common-base amplifier limits the number of cascades of common-base amplifiers to increase the output resistance to only one. Since a common-gate amplifier has no base current, we will look at a cascade of a common-base and common-gate amplifier to increase the output resistance of a current buffer. A small-signal model of this cascade is given in Fig. 9.11.

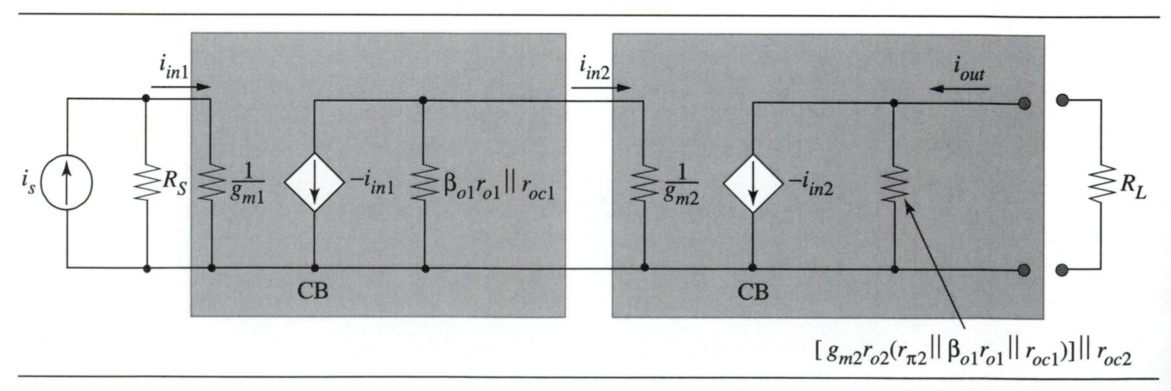

Figure 9.10 Small-signal model of a CB-CB cascade. Note: the increase in the output resistance remains on the order of β_o.

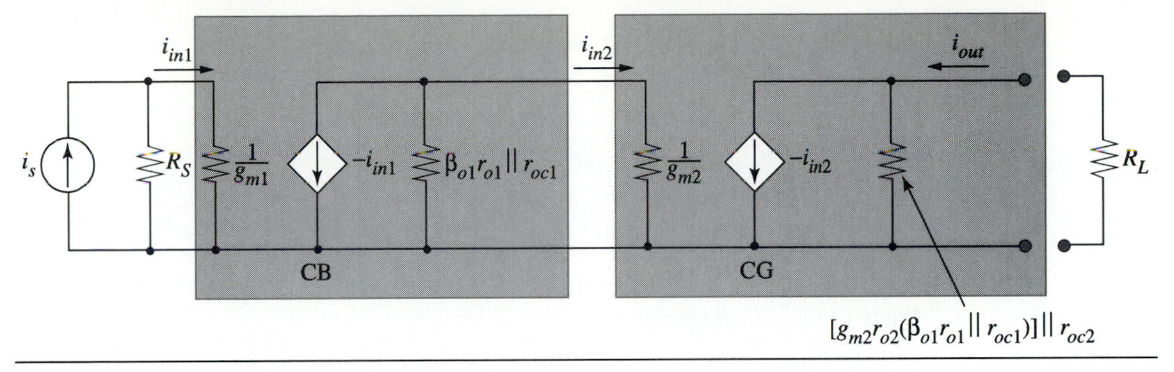

➤ Figure 9.11 Small-signal model of CB-CG cascade to form a current buffer.

Using the results of Chapter 8 and the fact that the output resistance of the CB stage is acting as the source resistance of the CG stage, we find the output resistance of this cascade is given by

$$R_{out} = [g_{m2}r_{o2}(\beta_{o1}r_{o1} \| r_{oc1})] \| r_{oc2} \tag{9.17}$$

Note that the output resistance is now increased by $g_{m2}r_{o2}$ over that of a pure common-base stage. Of course, there is a limit to increasing these output resistances. First, the current source that is supplying the current for the common-gate, common-base cascade, has a finite internal resistance r_{oc2}. This will eventually limit the total output resistance of the amplifier. Second, leakage currents from the pn junctions provide other paths for current limiting the practical value of output resistance. In any case, it is possible to achieve very high output resistances on the order of $10^8\ \Omega$ in modern integrated circuit technology.

➤ EXAMPLE 9.2 Comparison of BiCMOS and CMOS Current Buffers

In this example, we compare a BiCMOS CB-CG topology for a current buffer to an all MOS CG-CG current buffer. As you will see later, these configurations can use the same current source supply. Assume that the current source supply has an infinite internal resistance for this comparison. We want $R_{in} \leq 250\ \Omega$ and $R_{out} \geq 25\ \text{M}\Omega$. Also assume that $R_S = 250\ \Omega$.

MOS Device data

$V_{Tn} = 1\text{ V}$
$\mu_n C_{ox} = 50\ \mu\text{A/V}^2$
$\lambda_n = 0.1\text{V}^{-1}$

Bipolar Device data

$\beta_o = 100$
$V_A = 25\text{ V}$

SOLUTION

Let's begin with the CB-CG configuration.

$$R_{in} = \frac{1}{g_{m1}} = \frac{kT}{qI_C} = 250\ \Omega\text{, thus } I_C = \frac{25\text{ mV}}{250\ \Omega} = 100\ \mu\text{A}.$$

$$\text{At } I_{SUP} = 100\ \mu A,\ r_{o1}(\text{CB}) = \frac{V_A}{I_C} = \frac{25\text{V}}{100\ \mu\text{A}} = 250\ \text{k}\Omega$$

$$r_{o2}(\text{CG}) = \frac{1}{\lambda_n I_D} = \frac{1}{\lambda_n I_C} = \frac{1}{(0.1\text{V}^{-1})(100\ \mu\text{A})} = 100\ \text{k}\Omega$$

$$R_{out}(\text{CB}) = r_{o1} + g_{m1}r_{o1}(r_{\pi 1} \| R_S) \text{ and } R_{out}(\text{CG}) = r_{o2} + g_{m2}r_{o2}R_{out}(\text{CB})$$

For the CB-CG cascade, the output resistance of the CB stage is the source resistance of the CG stage. Therefore,

$$R_{out}(\text{CB}-\text{CG}) = r_{o2} + g_{m2}r_{o2}[r_{o1} + g_{m1}r_{o1}(r_{\pi 1} \| R_S)]$$

$$R_{out}(\text{CB}-\text{CG}) \approx g_{m2}r_{o2}g_{m1}r_{o1}(r_{\pi 1} \| R_S)$$

$$\text{Since } r_{\pi 1} = \beta_{o1}/g_{m1} = \frac{100}{4\text{ mS}} = 25\ \text{k}\Omega,\ r_{\pi 1} = 100R_S,$$

$$R_{out}(\text{CB}-\text{CG}) \approx g_{m2}r_{o2}g_{m1}r_{o1}R_S = g_{m2}(100\ \text{k}\Omega \cdot 4\text{ mS} \cdot 250\ \text{k}\Omega \cdot 250\ \Omega)$$

$$R_{out}(\text{CB}-\text{CG}) \approx g_{m2}(2.5 \cdot 10^{10})\ \Omega$$

To obtain $R_{out} \geq 25\ \text{M}\Omega$, $g_{m2} \geq 1\text{ mS}$.

$$\text{Since } g_{m2} = \sqrt{2\left(\frac{W}{L}\right)_2 \mu_n C_{ox} I_{SUP}},$$

$$\left(\frac{W}{L}\right)_2 = \frac{(g_{m2})^2}{2\mu_n C_{ox} I_{SUP}} = \frac{(1\text{mS})^2}{2 \cdot (50\ \mu\text{A}/\text{V}^2)(100\ \mu\text{A})} = 100.$$

The result is that the MOS device must have a W/L of 100.
Turning our attention to the CG-CG configuration,

$$R_{in} = \frac{1}{g_{m1}} = \frac{1}{\sqrt{2\left(\frac{W}{L}\right)_1 \mu_n C_{ox} I_{SUP}}} = 250\ \Omega.$$

Using the same $I_{SUP} = 100\ \mu\text{A}$, we require

$$\left(\frac{W}{L}\right)_1 = \frac{1}{(250\ \Omega)^2\, 2\left(50\frac{\mu\text{A}}{\text{V}^2}\right)(100\ \mu\text{A})} = 1600.$$

To achieve the same input resistance as the bipolar transistor, we require either a very large MOS transistor or an increase in current. This result is expected since the bipolar transistor has a higher transconductance per unit area and per unit current when compared to an MOS transistor fabricated in equivalent technology. Examining R_{out} of the CG-CG stage, we find

$$r_{o1}(\mathrm{CG}) = r_{o2}(\mathrm{CG}) = \frac{1}{\lambda_n I_D} = \frac{1}{(0.1\mathrm{V}^{-1})(100\ \mu\mathrm{A})} = 100\ \mathrm{k}\Omega \text{ and}$$

$$R_{out}(\mathrm{CG-CG}) \cong g_{m2}\, r_{o2}\, g_{m1}\, r_{o1}\, R_S$$

$$R_{out}(\mathrm{CG-CG}) = g_{m2}(100\ \mathrm{k}\Omega)(1/250\ \Omega)(100\ \mathrm{k}\Omega)(250\ \Omega) = g_{m2}10^{10}\ \Omega$$

To obtain $R_{out} \geq 25\ \mathrm{M}\Omega$, $g_{m2} \geq 2.5\ \mathrm{mS}$.

To find the $(W/L)_2$ to obtain $g_{m2} = 2.5\ \mathrm{mS}$ we write

$$\left(\frac{W}{L}\right)_2 = \frac{(g_{m2})^2}{2\mu_n C_{ox} I_{SUP}} = \frac{(2.5\ \mathrm{mS})^2}{2(50\ \mu\mathrm{A/V}^2)(100\ \mu\mathrm{A})} = 625.$$

To obtain an equivalent output resistance, the MOS implementation requires a larger area. Keep in mind that a BiCMOS process costs more to manufacture than a pure CMOS process for the same geometries. A practical cost trade-off analysis between a BiCMOS and CMOS solution requires business as well as engineering inputs.

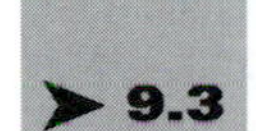

9.3 DIRECT-COUPLED AMPLIFIERS: LARGE-SIGNAL ANALYSIS

In the previous section we have cascaded single-stage amplifiers of various configurations, assuming they were biased at a proper DC operating point such that all devices remain in their constant-current region. In this section we will study how to connect these stages and maintain the proper DC node voltages and branch currents to ensure proper operation of the amplifiers.

9.3.1 Voltage Buffer

In discrete analog circuit design where large value capacitors are readily available, the most common technique to couple two amplifier stages is with a large value capacitor. Each stage is independently designed with the goal of having the output voltage approximately centered between the most positive and negative power supply voltages in the circuit. For the specific circuit shown in Fig. 9.12, we are using a single supply voltage of 5 V. The node voltages of Fig. 9.12 are derived assuming a base-emitter voltage for the bipolar transistor of 0.7 V and a gate-source voltage for the MOS transistor of 1.5 V. Because we are capacitively coupling the two stages with a large value capacitor, the DC voltages (namely the output voltage of the first stage and the input voltage of the second stage) can be different. If the value of the capacitor is large enough, it will approximate a short circuit for AC signals and they can be passed from the first to second stage with little attenuation.

Since large capacitors are not available in integrated circuit technology, DC biasing becomes more difficult. Two stages can be **direct-coupled** as shown in Fig. 9.13. If we center the output voltage of the common-collector stage at 2.5 V, the base of the npn transistor is at 3.2 V, and the input DC voltage of the CD amplifier is at 4.7 V assuming $V_{GS} = 1.5$ V. Note that the input DC voltage is much too high. Therefore,

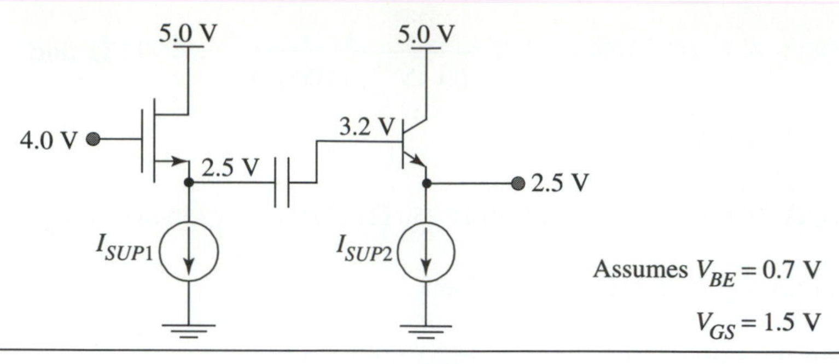

➤ **Figure 9.12** Capacitively-coupled NMOS-CD and npn-CC amplifiers to form a voltage buffer.

we must find another way to ensure that the input voltage is also approximately centered between the positive and negative power supply.

One solution to this problem is to use a PMOS common-drain amplifier at the input as shown in Fig. 9.14. Again, assuming a 1.5 V drop across the gate-source voltage of the PMOS transistor to sink the supply current provided, we see that the input voltage is now 1.7 V, which is much closer to the center of the range.

In setting DC bias voltages with direct-coupled stages, we first ensure that the input and output voltages are near the center of the voltage range. Second, we verify that all the devices are operating in their constant-current region. In common-drain/common-collector circuits when NMOS/npn type transistors are used, the DC input voltage is generally higher than the output. When PMOS/pnp transistors are used, the DC input voltage is lower than the output. Therefore, by using proper combinations of p-type and n-type transistors, one can center the input and output voltage and maintain each device in its constant-current region.

9.3.2 Current Buffer

In the previous section, we showed that the small-signal cascade of a common-base and common-gate amplifier provided a very small input resistance and an extremely large output resistance with a current gain of approximately unity. One can cascade these two stages capacitively as shown in Fig. 9.15.

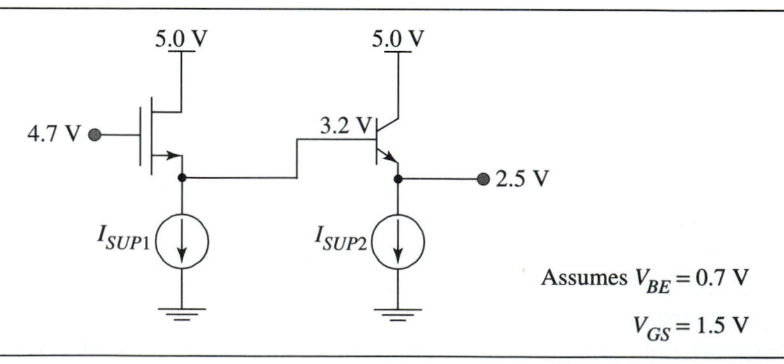

➤ **Figure 9.13** Direct-coupled NMOS-CD and npn-CC amplifier to form a voltage buffer.

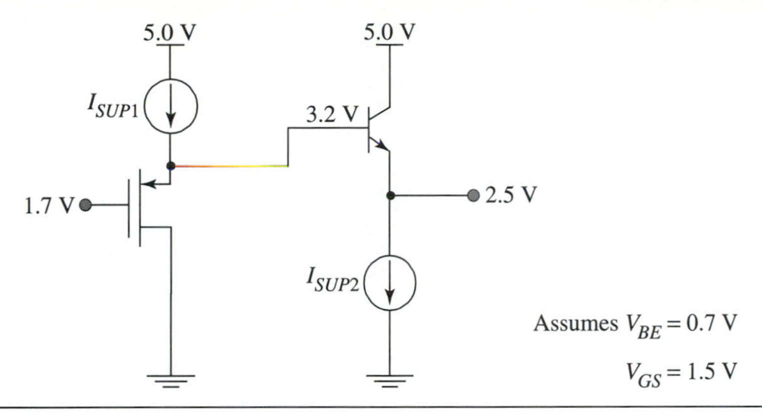

➤ Figure 9.14 Voltage buffer with PMOS CD and npn CC to solve the voltage level shifting problem.

As with the voltage buffer, time-varying signals can be passed from one stage to the next with little attenuation if the value of this capacitor is large enough. However, these large capacitors are not available in integrated circuit technology. Therefore, we need to examine directly coupling the common-base and common-gate stages. Close examination of Fig. 9.15 shows that the current source supplies and bias current sources can be shared as shown in Fig. 9.16. This sharing of DC current sources not only eliminates the need for the capacitor but also reduces the complexity of the circuit.

The design issues to set the DC bias voltages for this current buffer are first to set the supply current equal to the bias current so the output current is equal to zero when the small-signal input current is equal to zero. Second, the base and gate voltages must be set so that all devices operate in their constant-current region. For simplicity, we assume that the current sources and the bipolar and MOS transistors require at least 0.5 V across them (V_{CE}, V_{DS}) to remain in the constant-current region. In addition, assume that the base-emitter voltage is 0.7 V, and the gate-source voltage is 1.5 V to sink the supply current provided. With these assumptions in mind,

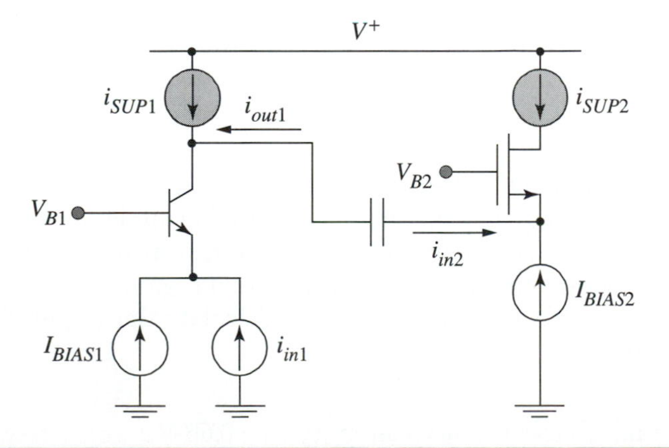

➤ Figure 9.15 Capacitively-coupled CB and CG amplifiers to form a current buffer.

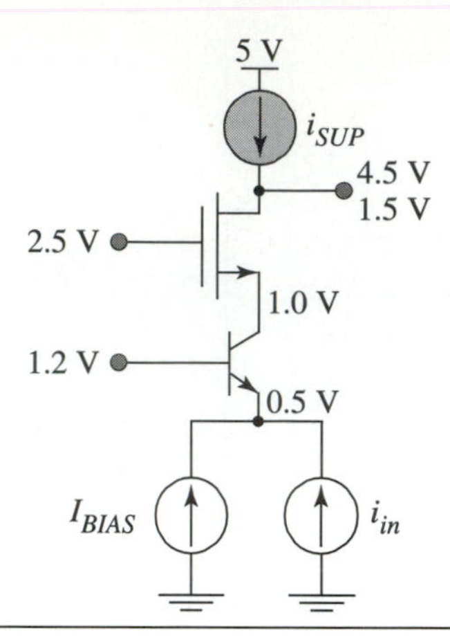

➤ **Figure 9.16** Direct-coupled CB and CG amplifiers to form a current buffer.

we set the base voltage at 1.2 V and the gate voltage at 2.5 V so all devices will be in their constant-current region with an output voltage swing between 1.5 V and 4.5V.

DESIGN EXAMPLE 9.3 Extending the Output Voltage Swing

In this example, we will extend the output voltage swing of the CB-CG amplifier shown in Fig. 9.16. If we continue to assume that each current source requires 0.5 V across it to operate in the constant-current region, we have a voltage swing limit of 0.5–4.5 V. This absolute limit requires $V_{CE_{SAT}}$ of the bipolar transistor and $V_{DS_{SAT}}$ of the MOS transistor to be 0 V—which is impossible. However, there is some room to increase the voltage swing from that shown in Fig. 9.16. Design the W/L of the MOS transistor to achieve a voltage swing of 1.0–4.5 V. Assume the minimum supply current is 100 μA for other design issues such as frequency response. Do *not* assume that $V_{BE} = 0.7$ V and $V_{GS} = 1.5$ V, but *do* assume $\lambda_n = 0$ V^{-1} and V_A is infinite. Verify your design in SPICE.

Device Data

MOS: $V_{Tn} = 1$ V, $\mu_n C_{ox} = 50$ μA/V^2;   Bipolar: $I_S = 10^{-15}$A, $V_{CE_{SAT}} = 0.2$ V

SOLUTION

We begin the design by noting that a smaller current requires a lower V_{GS}, which implies a lower $V_{DS_{SAT}}$. Hence, it is best to use the minimum current to maximize the voltage swing. The emitter of the bipolar transistor must be at 0.5 V for the bias current source to be in its constant-current region. Since $V_{CE_{SAT}} = 0.2$ V, the minimum collector voltage is 0.7 V. To calculate the required voltage on the base, we note that

$$V_{BE} = \frac{kT}{q}\ln\left(\frac{I_C}{I_S}\right) = (26\text{ mV})\ln\left(\frac{10^{-4}\text{A}}{10^{-15}\text{A}}\right) \approx 0.66\text{ V}.$$

$$V_{BE} = V_B - 0.5\text{ V}$$

$$V_B = V_{BE} + 0.5\text{ V} = 1.16\text{ V} \approx 1.2\text{ V}$$

Notice that due to the exponential dependence of I_C on V_{BE}, the voltage required is very close to 0.7 V for a wide range of currents.

With the source voltage of the MOS transistor set at 0.7 V and the minimum output voltage desired to be 1V, we require $V_{DS_{SAT}} < 0.3$ V. We can calculate the *W*/*L* required to have all of the MOS devices operating in their constant-current region.

Recall, $V_{DS_{SAT}} = V_{GS} - V_{Tn}$.

Thus $V_{GS} = V_{Tn} + V_{DS_{SAT}} = V_{Tn} + \sqrt{\dfrac{2I_D}{(W/L)\mu_n C_{ox}}} = 1.3\text{ V}$ and $V_G = V_S + 1.3\text{ V} = 2.0\text{ V}$.

$$V_{DS_{SAT}} = \sqrt{\frac{2I_D}{(W/L)\mu_n C_{ox}}} \le 0.3\text{ V}$$

Using $V_{DS_{SAT}} = 0.3$ V we find

$$\frac{W}{L} = \frac{2I_D}{(V_{DS_{SAT}})^2 \mu_n C_{ox}} = \frac{2(100\ \mu\text{A})}{(0.3\text{ V})^2 (50\ \mu A/\text{V}^2)} = 44.44 \approx 44$$

In SPICE, we find that the devices in our circuit leave the constant-current region when $V_{OUT} < 0.99$ V, which is very close to 1 V. When the devices leave the constant-current region the output resistance of the current buffer is drastically reduced. We have increased the voltage swing from 1.5–4.5 V to 1.0–4.5 V.

9.3.3 Cascode Topology

As with the current buffer, sharing a current source supply among two stages is a very powerful biasing scheme. The **cascode topology** that is made up of a common-source/emitter amplifier followed by a common-base/gate amplifier can be biased with a shared current source supply. An example of the cascode topology with a common-source and common-base stage is shown in Fig. 9.17.

The bias voltages for this circuit are calculated in a similar manner to that of the current buffer. In other words, the input bias voltage applied to M_1 must be set such that M_1 can approximately sink the supply current so the output current is near zero when the small-signal input voltage is equal to zero. The base voltage of Q_2 is set to ensure that the MOS transistor remains in the constant-current region and the output voltage has a large voltage swing centered between the positive and negative power supplies.

The cascode amplifier is an extremely useful circuit that has its origin in vacuum tubes. It has a small-signal two-port model that is the same as a CS-CB cascade. It makes an excellent transconductance amplifier that converts voltage to current with a transconductance equal to the transconductance of the input transistor. In this case the input transistor is an MOS transistor so the two-port parameters are $G_m = g_{m1}$, $R_{in} \to \infty$ and

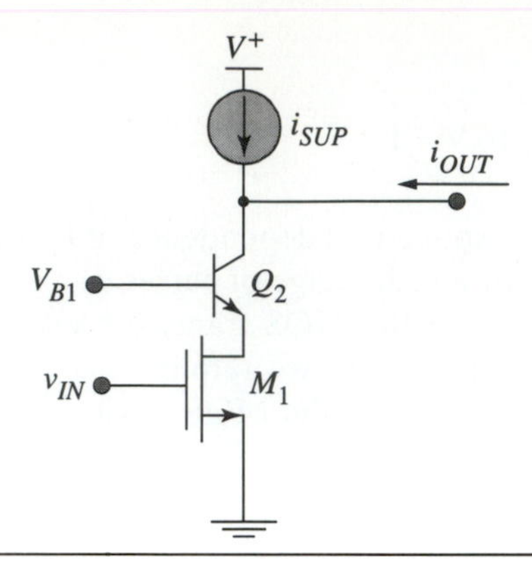

➤ **Figure 9.17** A cascode amplifier formed by directly coupling a CS and CB amplifier. Note the current source supply is shared by both stages.

$$R_{out} \approx \beta_{o2} r_{o2} \| r_{oc} . \tag{9.18}$$

two-port parameters
cascode transconductance amplifier

A cascode amplifier can also be formed by using all bipolar transistors, CE-CB or all MOS transistors, CS-CG. A CE-CB configuration is usually not useful as a transconductance amplifier since its input resistance is small. We will see in Chapter 10 that the frequency response of the cascode amplifier is excellent, which is another reason why it finds its way into many applications.

DESIGN EXAMPLE 9.4 MOS Cascode Amplifier

In this example, we will design the MOS cascode amplifier. The required small-signal specifications are $G_m \geq 1$ mS and $R_{out} \geq 20\ \text{M}\Omega$. We want the amplifier to have an output voltage swing $1\ \text{V} \leq v_{OUT} \leq 4\ \text{V}$. Assume the current source supply has an infinite internal resistance and requires 1.0 V across it to be in its constant-current region. Verify your design in SPICE.

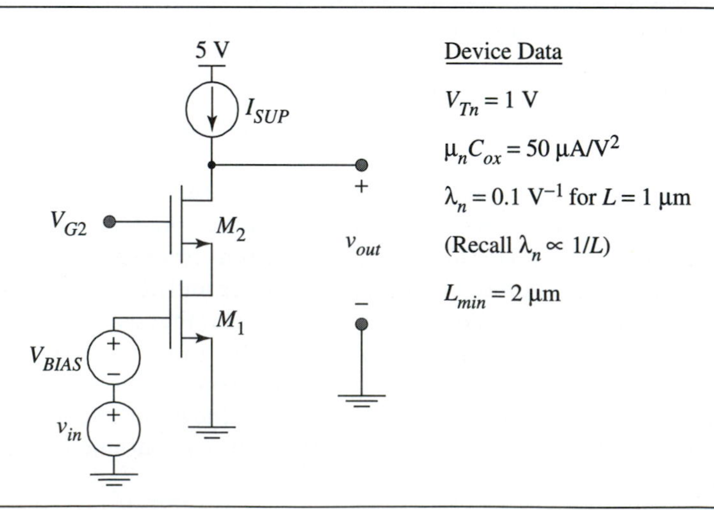

➤ **Figure Ex9.4**

SOLUTION

We approach this design by trying to achieve the small-signal specifications and see if we still meet the large-signal voltage swing specification. The transconductance of this cascode is just the transconductance of M_1. A wide device is needed to achieve the maximum transconductance. As a starting point, let's choose $(W/L)_1$ to be 100/2, and set $g_{m1} = 1$ mS

$$g_{m1} = \sqrt{2\frac{W}{L}\mu_n C_{ox} I_{SUP}} = 1 \text{ mS}\text{, therefore}$$

$$I_{SUP} = \frac{(1 \text{ mS})^2}{2(100/2)(50\mu\text{A}/\text{V}^2)} = 200\ \mu\text{A}$$

The output resistance of this cascode can be found by realizing that the cascode is a CS-CG amplifier. By noting that the output resistance of device M_1 is acting as the source resistance of the common gate amplifier formed by M_2, we can write

$$R_{out} = [r_{o2} + g_{m2}r_{o2}r_{o1}] \approx (g_{m2}r_{o2}r_{o1})$$

$$\lambda_{n1} = \frac{\lambda_n}{2\ \mu m} = 0.05\text{V}^{-1}$$

$$r_{o1} = \frac{1}{\lambda_{n1}I_{SUP}} = \frac{1}{(0.05\text{ V}^{-1})(200\ \mu\text{A})} = 100\text{ k}\Omega$$

If we size M_2 the same as M_1, $g_{m2}r_{o2} = g_{m1}r_{o1} = 100$ and

$$R_{out} \approx (100)\ 100\text{ k}\Omega = 10\text{ M}\Omega$$

We need to double the output resistance to achieve the specification for R_{out}, which can be achieved in a number of ways. However, first we should see how close we are to the voltage swing specification and then decide how to achieve the required R_{out}.

Given $I_{SUP} = 200\mu A$, $(W/L)_1 = (W/L)_2 = 100/2$.

$$V_{BIAS} = V_{Tn} + \sqrt{\frac{2I_{SUP}}{(W/L)_1\mu_n C_{ox}}} = 1\text{V} + 0.4\text{V} = 1.4\text{V}$$

Note to ensure that device M_1 is operating in the constant-current region V_{DS1} must be greater than $V_{DS_{SAT1}} = V_{BIAS} - V_{Tn} = 0.4$ V. Neglecting backgate effect, we find that M_2 will require the same $(V_{GS} - V_{Tn})$, namely 0.4 V.

We can find V_{G2} by

$$V_{G2} = V_{Tn} + \sqrt{\frac{2I_{SUP}}{(W/L)_2\mu_n C_{ox}}} + V_{S2} = 1\text{ V} + 0.4\text{ V} + 0.4\text{ V} = 1.8\text{ V}$$

$$V_{DS_{SAT2}} = V_{GS2} - V_{Tn} = 1.8\text{ V} - 0.4\text{ V} - 1.0\text{ V} = 0.4\text{ V}$$

We can swing the output voltage between $V_{DS_{SAT1}} + V_{DS_{SAT2}}$ and 4 V, namely $0.8\text{V} \le v_{OUT} \le 4.0\text{V}$. This meets the voltage swing specification. The channel length of M_1 can be doubled to make the R_{out} specification. This will require that we double the width of M_1 to maintain the same g_{m1} and V_{BIAS}. Several other options are also possible such as reducing I_{SUP} while increasing W_1 and W_2. Our final design has $(W/L)_1 = 200/4$; $(W/L)_2 = 100/2$; $I_{SUP} = 200\ \mu\text{A}$; $V_{BIAS} = 1.4$ V; and $V_{G2} = 1.8$ V.
When the circuit is simulated in SPICE, the results are

$$G_m = g_{m1} = 1.001\text{mS};\ R_{out} = 24.5\text{M}\Omega,\ V_{DS_{SAT1}} = 0.40\text{ V; and } V_{DS_{SAT2}} = 0.38\text{ V.}$$

Since $V_{DS_{SAT1}} + V_{DS_{SAT2}} \le 1.0\text{ V}$, we have met our voltage swing specification.

9.4 DC VOLTAGE AND CURRENT SOURCES

Until this point, we have used ideal or nearly ideal current sources and voltage sources to set the DC operating point of our single-stage and multistage amplifiers. In this section we will learn how to build voltage and current sources for our amplifiers using integrated circuit technology where "matched" devices with nearly identical characteristics are available.

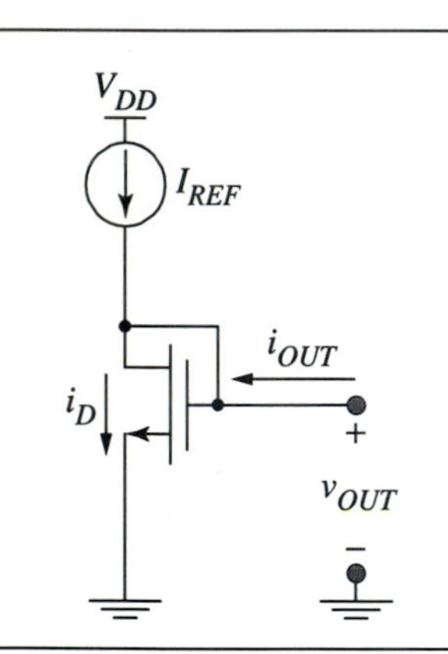

Figure 9.18 Circuit to generate a voltage source.

9.4.1 MOS Voltage Sources

Consider for a moment that we have one **reference current source**, I_{REF}, available to generate all current sources and voltage sources required for the circuit. This reference current source can be formed by a variety of techniques that will be discussed at the end of this section. Consider the circuit shown in Fig. 9.18 where the reference current source is forcing current into a diode connected MOS transistor. The MOS transistor will generate a gate-source voltage sufficient to sink the reference current and any output current. To calculate this voltage, assume that the MOS device is operating in its constant-current region and that the drain current is given by

$$i_D = I_{REF} + i_{OUT} = \frac{W}{2L}\mu_n C_{ox}(v_{OUT} - V_{Tn})^2(1 + \lambda_n v_{OUT}) \quad (9.19)$$

Assuming $\lambda_n v_{OUT} \ll 1$ and solving this equation for the output voltage we find

MOS voltage source *I-V* characteristic

$$v_{OUT} = V_{Tn} + \sqrt{\frac{I_{REF} + i_{OUT}}{\frac{W}{2L}\mu_n C_{ox}}} \quad (9.20)$$

From this equation we can see that the output voltage is controlled by the value of the reference current and the geometry of the device. Qualitative plots of the output current versus the output voltage are shown in Fig. 9.19.

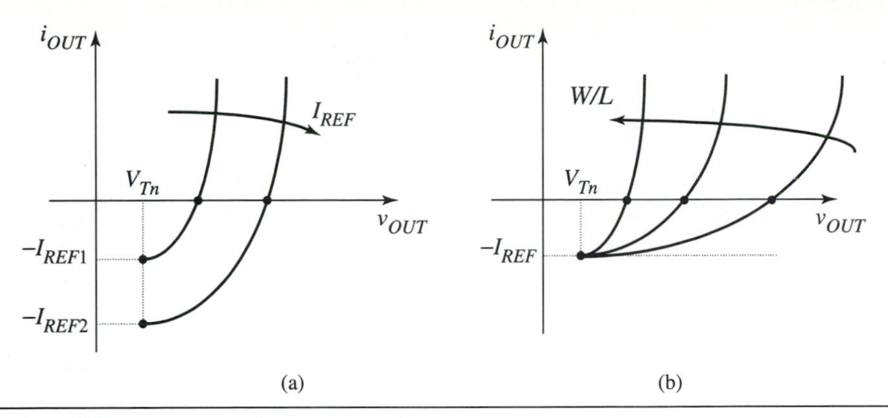

➤ Figure 9.19 i_{OUT} vs. v_{OUT} of MOS diode connected voltage source (a) $I_{REF2} > I_{REF1}$ and W/L is constant. (b) W/L is increasing with I_{REF} constant.

As we increase the reference current for a given W/L, the i_{OUT} vs. v_{OUT} curve is simply shifted down and v_{OUT} at $i_{OUT} = 0$ A is increased. As we increase the W/L ratio for a constant reference current, the voltage source begins to appear more and more ideal, namely a vertical line.

This voltage source can be thought of as having a DC value and an incremental source resistance at the DC operating point. This means we should examine the small-signal characteristics for a given DC voltage about that operating point. Fig. 9.20 shows the small-signal model for the diode connected MOS transistor to calculate the incremental source resistance.

The reference current source of this voltage source becomes an open circuit since it is a DC current source. To find the incremental source resistance, we place a test voltage at the output and measure the current coming from the test voltage source. Solving for the test current we find

$$i_t = \frac{v_t}{r_o} + g_m v_t \tag{9.21}$$

and rearranging terms

$$R_S = \frac{v_t}{i_t} = \frac{1}{\frac{1}{r_o} + g_m} \approx \frac{1}{g_m} \tag{9.22}$$

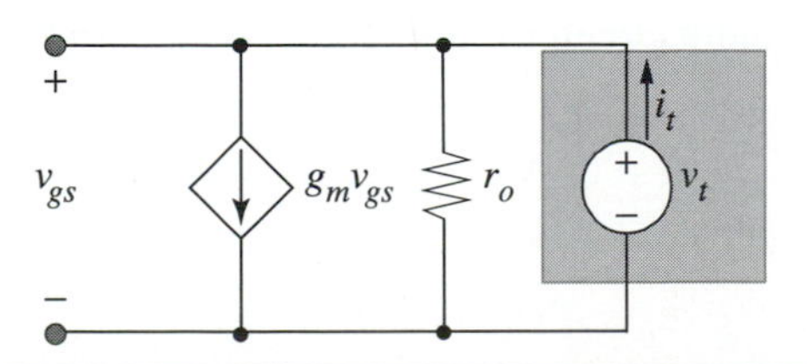

➤ Figure 9.20 Small-signal model for the diode-connected MOS transistor.

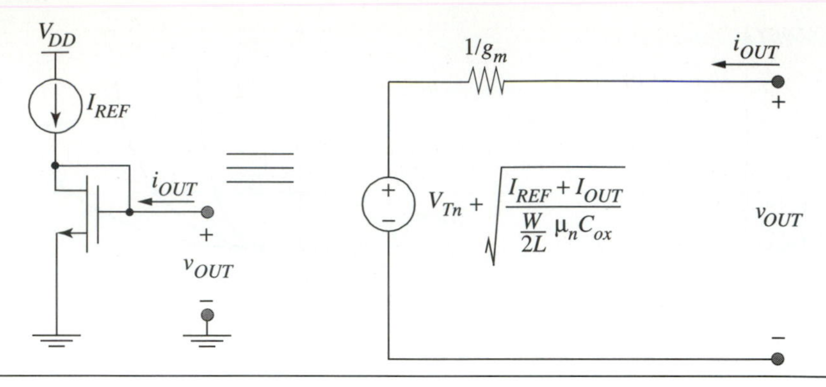

➤ **Figure 9.21** Equivalent circuit for the diode-connected NMOS voltage source.

MOS voltage source incremental source resistance

The incremental source resistance of the voltage source is approximately equal to the reciprocal of the transconductance of the MOS transistor. Fig. 9.21 shows an equivalent circuit model for the voltage source. It has a DC voltage source whose value is set by I_{REF}, W/L, and the DC operating current I_{OUT}. It also includes a small-signal source resistance equal to $1/g_m$ evaluated at the operating point.

NMOS transistors limit the output voltage to values close to the bottom voltage supply. The p-channel version of this circuit as shown in Fig. 9.22 can be used to achieve voltage sources that are closer to the top power supply.

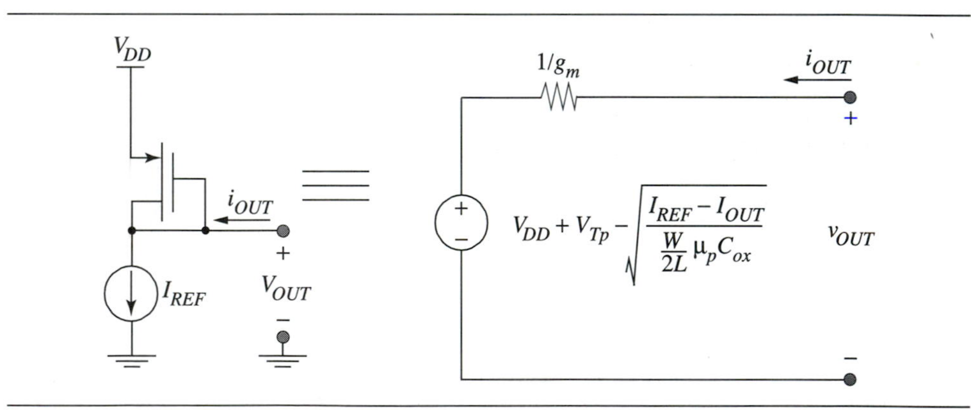

➤ **Figure 9.22** Equivalent circuit for the diode-connected PMOS voltage source.

➤ EXAMPLE 9.5 Minimum Incremental Source Resistance of an MOS Voltage Source

Given a diode-connected PMOS voltage source shown in Fig. Ex 9.5, what is the minimum small-signal source resistance that can be obtained if the maximum reference current is 10 μA and the open-circuit output voltage is 2.5 V?

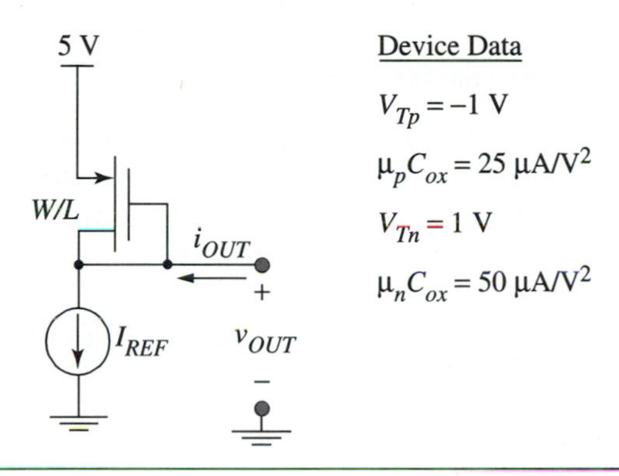

Device Data

$V_{Tp} = -1$ V

$\mu_p C_{ox} = 25\ \mu\text{A/V}^2$

$V_{Tn} = 1$ V

$\mu_n C_{ox} = 50\ \mu\text{A/V}^2$

➤ Figure Ex9.5

SOLUTION

Begin by finding the W/L of the PMOS device which will give us $V_{OUT} = 2.5$ V.

$$V_{OUT} = V_{DD} + V_{Tp} - \sqrt{\frac{I_{REF}}{\frac{W}{2L}\mu_P C_{ox}}}, \text{ so}$$

$$\frac{W}{L} = \frac{2I_{REF}}{\mu_p C_{ox}(V_{DD} + V_{Tp} - V_{OUT})^2} = \frac{20\ \mu\text{A}}{(25\ \mu\text{A/V}^2)(5\text{V} - 1\text{V} - 2.5\text{V})^2} = 0.356$$

To find the incremental source resistance, recall that

$$R_S = 1/g_m = \frac{1}{\sqrt{2I_{REF}\frac{W}{L}\mu_p C_{ox}}} = \frac{1}{\sqrt{2(10\mu\text{A})(0.356)\left(25\frac{\mu\text{A}}{\text{V}^2}\right)}} \approx 75\text{k}\Omega$$

Repeating this exercise for the NMOS case we find

$$V_{OUT} = V_{Tn} + \sqrt{\frac{I_{REF}}{\frac{W}{2L}\mu_n C_{ox}}}$$

$$\frac{W}{L} = \frac{2I_{REF}}{\mu_n C_{ox}(V_{OUT} - V_{Tn})^2} = \frac{20\mu\text{A}}{(50\mu\text{A/V}^2)(2.5\text{V} - 1\text{V})^2} = 0.177$$

$$R_S = 1/g_m = \frac{1}{\sqrt{2I_{REF}\frac{W}{L}\mu_n C_{ox}}} = \frac{1}{\sqrt{2(10\mu\text{A})(0.177)\left(50\frac{\mu\text{A}}{\text{V}^2}\right)}} \approx 75\text{k}\Omega$$

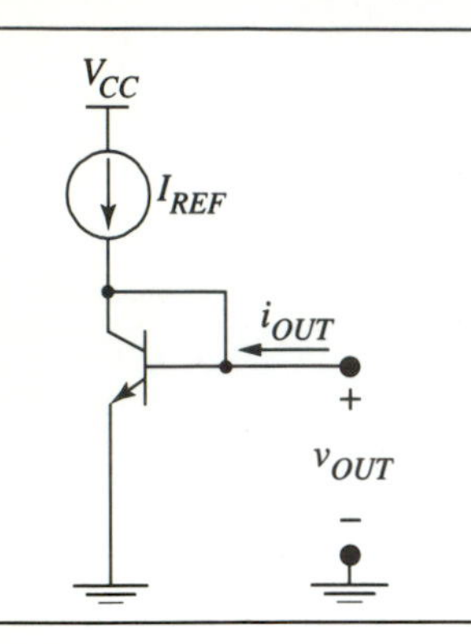

➤ **Figure 9.23** npn bipolar diode-connected voltage source.

The required output voltage and the incremental source resistance are related. You cannot arbitrarily design one quantity independent of the other if the reference current is held constant. To reduce the incremental source resistance, a higher reference current must be used to increase the g_m and a correspondingly larger W/L to obtain the proper V_{OUT}.

9.4.2 Bipolar Voltage Sources

Using a method similar to the one used to generate voltage sources with a diode-connected MOS transistor, one can employ a diode-connected bipolar transistor as shown in Fig. 9.23. Assuming the bipolar device is operating in its constant-current region and neglecting the base current, we can write

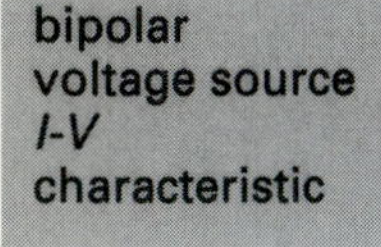

$$i_C = I_{REF} + i_{OUT} = I_S e^{qv_{BE}/kT} = I_S e^{qv_{OUT}/kT} \tag{9.23}$$

and rearranging terms

$$v_{OUT} = \frac{kT}{q} \ln\left(\frac{I_{REF} + i_{OUT}}{I_S}\right) \tag{9.24}$$

The small-signal performance of this bipolar voltage source can be evaluated with the small-signal model shown in Fig. 9.24. By applying a test voltage source at the output and measuring the test current out of this voltage source we find

$$i_t = \frac{v_t}{r_o} + \frac{v_t}{r_\pi} + g_m v_t \tag{9.25}$$

bipolar
voltage source
incremental
source
resistance

Rearranging to solve for the incremental source resistance we find

$$R_S = \frac{1}{\frac{1}{r_\pi} + \frac{1}{r_o} + g_m} \approx \frac{1}{g_m} \tag{9.26}$$

Therefore the incremental source resistance of the bipolar voltage source as with the MOS voltage source is approximately equal to the reciprocal of the transconductance. The equivalent circuit for the bipolar voltage source and the resulting $i_{OUT} - v_{OUT}$ characteristic is shown in Fig. 9.25. Note that in general the bipolar tran-

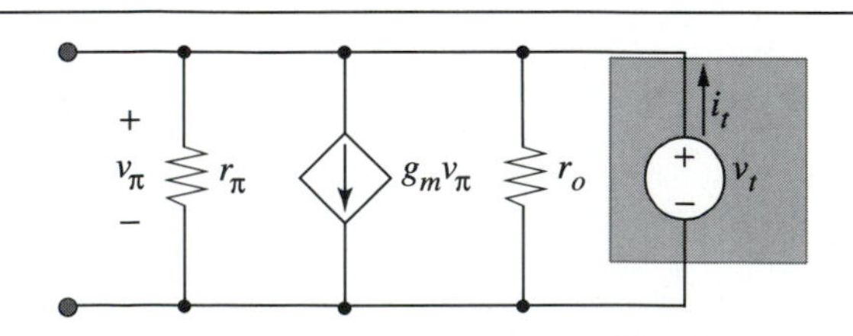

➤ **Figure 9.24** Small-signal model of diode-connected npn voltage source.

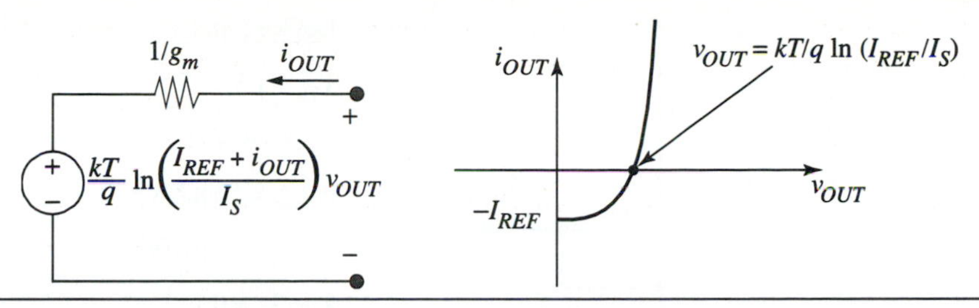

➤ **Figure 9.25** (a) Equivalent circuit for the diode-connected npn bipolar voltage source. (b) i_{OUT} vs. v_{OUT}. Note: The exponential relation between i_{OUT} and v_{OUT} compared to the quadratic relation for the MOS version.

sistor has a larger transconductance than the MOS transistor, which means it makes a better voltage source. However, in order to achieve a reasonable transconductance, the reference current must be on the order of 100 μA. With this value of reference current, the output voltage can only vary by the natural log of the saturation current that can only be changed linearly by scaling the area. For practical geometries, the bipolar voltage source can only be varied approximately ±100 mV around a mean value of approximately 0.7 V. You will see later that this is not a limitation since we will be designing voltage sources that match V_{BE}'s of other bipolar transistors.

These voltage sources find a wide variety of applications. Fig. 9.26 shows a circuit to generate multiple voltage sources. In this circuit we are generating three different voltages that can vary across a wide range. By using a reference current I_{REF} in series with a number of diode-connected devices, a series of bias voltages between the positive and negative supply are derived. We call the circuit shown in Fig. 9.26 a **totem pole voltage source**. V_{OUT1} is approximately 0.7 V whereas V_{OUT2} can be modified by changing the W/L ratio of the MOS transistor to the desired value. V_{OUT3} will then be approximately 0.7 V above V_{OUT2}.

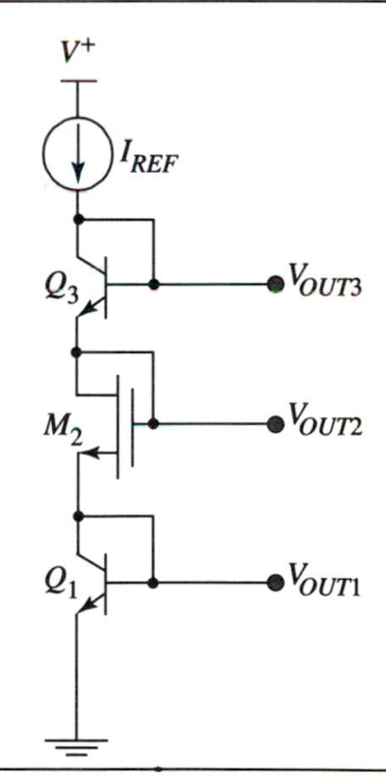

➤ **Figure 9.26** A totem pole voltage source to produce several DC bias voltages.

➤ EXAMPLE 9.6 A CMOS Totem Pole Voltage Source

In this example, we need to find the W/L ratio of the MOS devices in the totem pole voltage source circuit to produce the DC voltages shown in Fig. Ex 9.6. The maximum small-signal source resistance of each voltage source is required to be less than 20kΩ. I_{REF} must have at least 0.5 V across it to be in the constant-current region. Neglect the backgate effect for hand analysis, but run SPICE to see its effect.

SOLUTION

Beginning with device M_3, we want $V_{SG3} = 1.5$ V.

$$V_{SG3} + V_{Tp} = \sqrt{\frac{I_{REF}}{\left(\frac{W}{L}\right)_3 \frac{\mu_p C_{ox}}{2}}} = 1.5\text{V} - 1.0\text{V} = 0.5\text{ V}$$

$$I_{REF} = (V_{SG3} + V_{Tp})^2 \left(\frac{W}{L}\right)_3 \frac{\mu_p C_{ox}}{2} = (0.5)^2 \cdot \left(\frac{W}{L}\right)_3 \frac{\mu_p C_{ox}}{2}$$

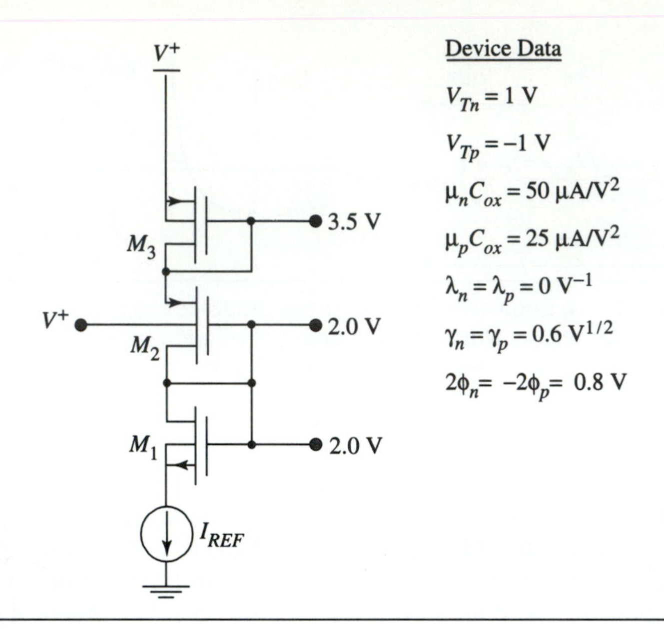

➤ **Figure Ex9.6**

We note the incremental source resistance at the gate/drain of device M_2 is in series with that due to device M_3. We choose each source resistance to be 10 kΩ so the series combination that is seen at the gate/drain of device M_2 is 20 kΩ.

$$\frac{1}{R_{S3}} = g_{m3} = \frac{1}{10\ \mathrm{k\Omega}} = \sqrt{2I_{REF}\left(\frac{W}{L}\right)_3 \mu_p C_{ox}}$$

Substituting our equation for I_{REF}

$$g_{m3} = \sqrt{(0.5V)^2\left(\frac{W}{L}\right)_3^2 (25\ \mu A/V^2)^2} = \frac{1}{10\mathrm{k\Omega}}$$

Rearranging yields $\left(\frac{W}{L}\right)_3 = 8.0$

and thus $I_{REF} = 25\ \mu A$.

Since $V_{SG2} = 1.5$ V, also $\left(\frac{W}{L}\right)_2 = 8.0$.

Since the size of device M_2 is the same size as M_3, $g_{m2} = g_{m3} = 1/10$ kΩ and

$$R_{S2} = \frac{1}{g_{m2}} + \frac{1}{g_{m3}} = 20\ \mathrm{k\Omega}$$

For device M_1, we require that $V_{GS1} \leq 1.5$ V to keep I_{REF} in the constant-current region, namely that it has at least 0.5 V across it. Since I_{REF} is set at 25 µA, we can calculate $(W/L)_1$.

$$\left(\frac{W}{L}\right)_1 = \frac{2I_{REF}}{(V_{GS1} - V_{Tn})^2(\mu_n C_{ox})} = \frac{2 \cdot 25\mu\text{A}}{(0.5)^2 \cdot 50\mu\text{A/V}^2} = 4$$

Since the DC current source is an open circuit under small-signal conditions, the incremental source resistance of device M_1 has no effect on the incremental source resistance at the gate/drain of devices M_2 and M_3.

Without the backgate effect, SPICE yields an incremental source resistance of 10 kΩ at the gate/drain device of M_3 and 20 kΩ at the gate/drain of device M_2. The DC voltages are 3.5 V and 2.0 V at the gates of devices M_3 and M_2, respectively.

When the backgate effect is included for device M_2 the DC voltage at the gate of device M_2 has been reduced to 1.6 V to accomodate the increased threshold voltage magnitude. The DC voltage at the gate of device M_3 remains at 3.5 V. The incremental source resistance at the gate/drain of device M_3 remains at 10 kΩ, while that of M_2 is increased to 22 kΩ due to the backgate transconductance generator.

We can meet the specification for the incremental source resistance by increasing the current. Next, we resize the transistors to obtain the specified DC voltages by taking into account the increased threshold voltage for M_2 due to the source-bulk voltage. This iteration is left to the reader.

9.4.3 MOS Current Sources

An MOS voltage source was developed in the Section 9.4.1. If we attached this DC voltage source to an MOS transistor as shown in Fig. 9.27, we develop a current in device M_2 that can be used as a current source.

To find the value of this output current, assume that transistors M_1 and M_2 are matched. By this we mean that the threshold voltage, mobility, and oxide capacitance are exactly the same for the two devices. This is a reasonable approximation for integrated circuit technology. We also neglect the effects of the output resistance of the

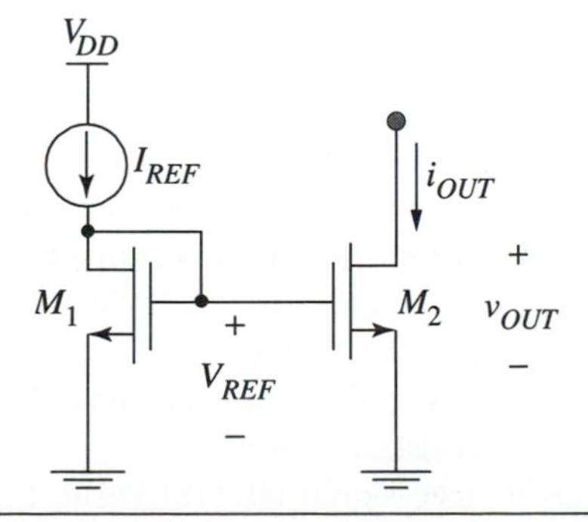

➤ **Figure 9.27** Current source formed by attaching a voltage source to the gate of an MOS transistor.

transistors on this large-signal calculation. Given these assumptions, we find that the gate voltage V_{REF} on devices M_1 and M_2 is given by

$$V_{REF} = V_{Tn} + \sqrt{\frac{I_{REF}}{\left(\frac{W}{L}\right)_1 \frac{\mu_n C_{ox}}{2}}} \tag{9.27}$$

The DC output current is given by

$$I_{OUT} = \left(\frac{W}{L}\right)_2 \frac{\mu_n C_{ox}}{2} \left(V_{Tn} + \sqrt{\frac{I_{REF}}{\left(\frac{W}{L}\right)_1 \frac{\mu_n C_{ox}}{2}}} - V_{Tn} \right)^2 \tag{9.28}$$

Gathering terms we find

$$I_{OUT} = \frac{\left(\frac{W}{L}\right)_2}{\left(\frac{W}{L}\right)_1} I_{REF} \tag{9.29}$$

which shows that the output current is a geometrical ratio times the reference current. The fact that we can build scaled current sources from a reference current is an extremely powerful concept that is used to build all of the current sources necessary in a multistage amplifier.

One can intuitively understand that the output current is ratioed to the reference current as given in Eq. (9.29) with the following argument.

1. The reference current goes into the diode-connected transistor M_1 to generate a reference voltage.
2. The same voltage is used to generate an output current in device M_2.
3. Because the transistors are matched, the output current must be related to the reference current by the geometry of the devices.

If the geometry of M_2 is much larger than the geometry of M_1, then the output current will be much larger than the reference current because the gate-source voltage is the same. It is useful to gain this physical intuition rather than just memorizing the equations.

Next, we will examine the small-signal performance of this current source. A small-signal model for this current source is given in Fig. 9.28. The reference current source is shown as an open circuit. Device M_1 is diode-connected and generates only a DC reference voltage. Hence, its terminals are shorted in the small-signal model. Although its transconductance generator is shown in Fig. 9.28, the value of the current is zero. Therefore, the small-signal input voltage to device M_2 is equal to zero. This in turn sets the transconductance generator of device M_2 equal to zero. Note that the only component in the small-signal model for this calculation of the incremental source resistance is the output resistance of M_2.

The current source has a large-signal current value that is a scaled value of the reference current and a small-signal source resistance equal to the output resistance of device M_2 as shown in Fig. 9.29. The large-signal current source is in parallel with

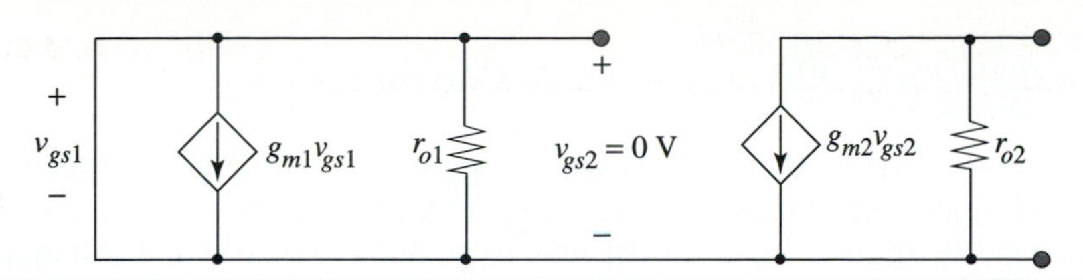

➤ **Figure 9.28** Small-signal model of the current source. The incremental source resistance of the current source is equal to r_{o2}.

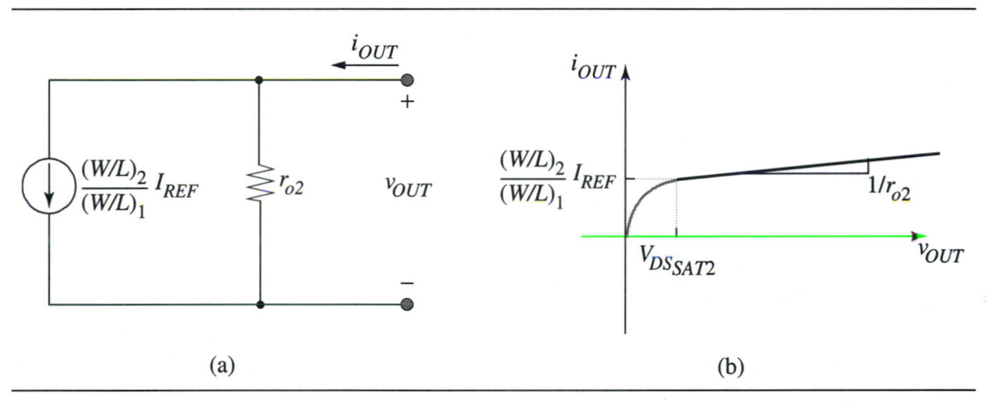

➤ **Figure 9.29** (a) Equivalent circuit for MOS current source, (b) i_{OUT} vs. v_{OUT} characteristic.

the small-signal source resistance. The model is only valid when transistors M_1 and M_2 are in their constant-current region.

To improve the small-signal source resistance of the current source, we can add a cascode device as shown in Figure 9.30. In this particular circuit, the output current is still the ratio of $(W/L)_2/(W/L)_1$ times the reference current, but the incremental source resistance is significantly increased. From previous discussions, we know that device M_4 looks like a common-gate amplifier. A common-gate amplifier has an output resistance that is approximately equal to the open-circuit voltage gain of that device times its source resistance which, in this case, is the output resistance of device M_2. The incremental source resistance of the cascode current source is given by

$$R_S \approx (g_{m4} r_{o4}) r_{o2} \tag{9.30}$$

Current sources can be similarly built with bipolar transistors. However, the finite base current should be taken into account. There are several circuit techniques to help cancel the effect of this base current; however, these techniques are beyond the scope of this text.

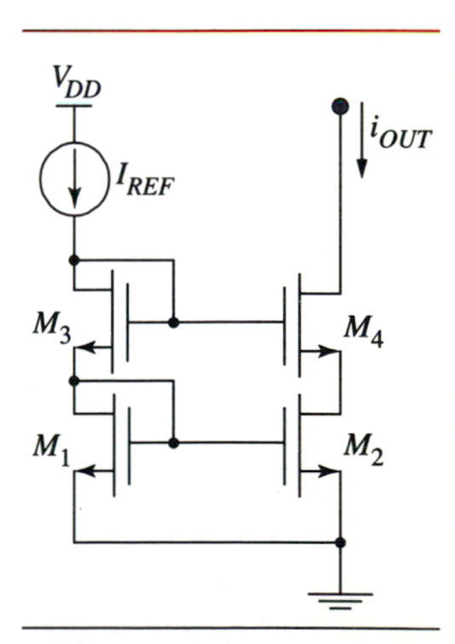

➤ **Figure 9.30** Cascode current source to increase the incremental source resistance.

DESIGN EXAMPLE 9.7 Cascode Current Source

Design a cascode current source, as shown in Figure 9.30, with a DC output current of 10 μA and small-signal output resistance of 100 MΩ. You are given $I_{REF} = 10$ μA. The voltage at the drain of M_4 should be able to go as low as 2.0 V and still maintain the high incremental source resistance. Hint: M_2 and M_4 remain in their constant-current regions. Neglect the backgate effect for this design.

Device data

$V_{Tn} = 1$ V

$\mu_n C_{ox} = 50\ \mu\text{A/V}^2$

$\lambda_n = 0.1\ \text{V}^{-1}$ for $L = 1\ \mu$m

(Recall $\lambda_n \propto 1/L$)

$L_{min} = 2\ \mu$m

SOLUTION

Begin this design by setting the gate voltages of M_1 and M_3 to ensure that M_2 and M_4 are operating in their constant-current region with $V_{D4} = 2.0$ V. As a start, if we choose V_{GS} of M_1, M_2, M_3, and M_4 to be 1.5 V, the gate voltage on M_4 will be 3 V. With $V_{GS4} = 1.5$ V, it implies that the source of M_4, which is the drain of M_2, will be at 1.5 V. Since $V_{DS2} = 1.5$ V and recalling that $V_{GS2} - V_{Tn} = 0.5$ V, $V_{DS2} > V_{DS_{SAT2}}$. Noting that $V_{DS_{SAT4}}$ is also 0.5 V and the source of M_4 is at 1.5 V, the drain voltage of M_4 must be greater than 2.0 V to ensure operation in the constant-current region. Even though this is an arbitrary decision, it is a good starting point.

Next we will set the W/L ratios of M_1 and M_3 to yield $V_{GS} = 1.5$ V with $I_{REF} = 10$ μA. Neglecting backgate effect

$$V_{GS} = V_{Tn} + \sqrt{\frac{I_{REF}}{\frac{W}{L}\frac{\mu_n C_{ox}}{2}}} \Rightarrow \left(\frac{W}{L}\right)_1 = \left(\frac{W}{L}\right)_3 = 1.6$$

Now we match the (W/L) of M_2 and M_4 with M_1 and M_3, respectively, since we want $I_{REF} = I_{OUT}$. With this initial choice of $(W/L)_2 = (W/L)_4 = 1.6$, calculate R_S. We can set the channel length to approximately yield the desired R_S.

$$R_S \approx (g_{m4} r_{o4}) r_{o2}$$

$$\text{For } L = 2\mu\text{m}, \lambda_n = \frac{(0.1)\ V^{-1}\mu\text{m}}{L} = 0.05\ V^{-1} \text{ and}$$

$$r_{o4} = r_{o2} = \frac{1}{\lambda_n I_D} = \frac{1}{(0.05)\cdot 10^{-5}}\Omega = 2\ \text{M}\Omega.$$

$$g_{m4} = \sqrt{2\frac{W}{L}\mu_n C_{ox} I_{REF}} = 40\ \mu\text{S} \text{ and } g_{m4} r_{o4} = 80$$

Thus $R_S = 160$ MΩ

9.4.4 Current Sources and Sinks

In the previous section we used n-channel transistors to form what we called **current sources.** Since we think of current flowing from top to bottom, it may have been more appropriate to call them **current sinks** since current flows into them. P-channel transistors can be used to build current sources in a similar manner as the n-channel current sinks. We also developed a single current sink from a voltage source. It should be noted that if we tie that voltage source to more than one MOS transistor, we can build multiple current sources as shown in Fig. 9.31.

The value of the current source for each of these devices can be varied based on the W/L ratio for each of the particular devices. The value of the output current is given by

$$I_{OUT_n} = \frac{\left(\frac{W}{L}\right)_n}{\left(\frac{W}{L}\right)_R} I_{REF} \tag{9.31}$$

In Fig. 9.31 we showed three current sources, but suppose the circuit also requires current sinks. There is a way to derive a current sink that is ratioed to a current source as shown in Fig. 9.32. The output current from device M_1 is used as a reference current for the voltage source M_3. This voltage source is tied to transistor M_4 to build the current sink.

The value of the DC current I_{OUT1} is equal to

$$I_{OUT1} = \frac{\left(\frac{W}{L}\right)_1}{\left(\frac{W}{L}\right)_R} I_{REF} \tag{9.32}$$

From this current we have derived a current source and current sink with devices M_2 and M_4. The currents are

$$I_{OUT2} = \frac{\left(\frac{W}{L}\right)_2}{\left(\frac{W}{L}\right)_R} I_{REF} \tag{9.33}$$

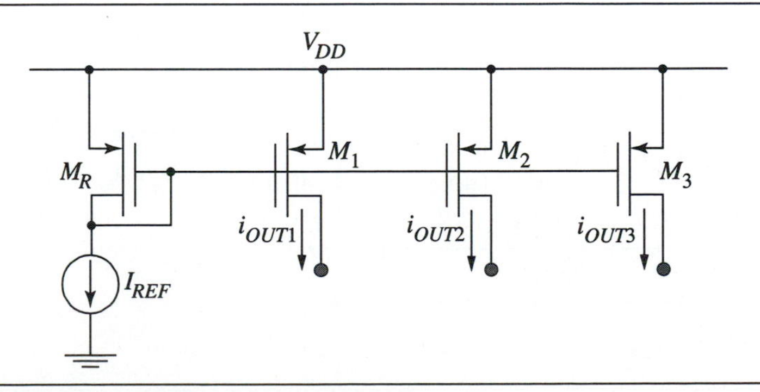

Figure 9.31 Multiple MOS current sources.

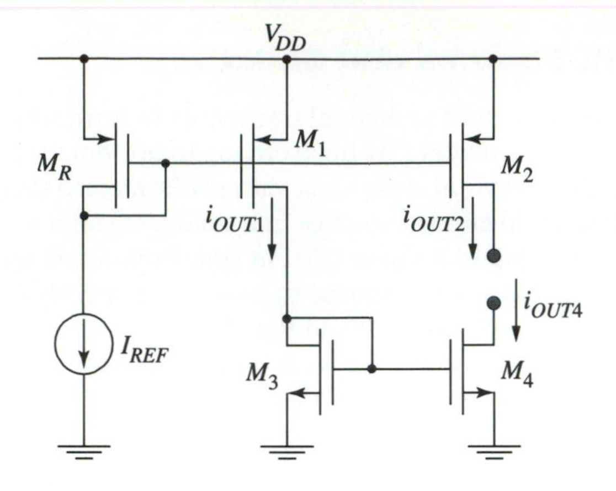

Figure 9.32 Circuit to produce a current source M_2 and current sink M_4.

$$I_{OUT4} = \frac{\left(\frac{W}{L}\right)_4}{\left(\frac{W}{L}\right)_3} I_{OUT1} = \left(\frac{\left(\frac{W}{L}\right)_4}{\left(\frac{W}{L}\right)_3} \cdot \frac{\left(\frac{W}{L}\right)_1}{\left(\frac{W}{L}\right)_R}\right) I_{REF} \tag{9.34}$$

 **DESIGN EXAMPLE 9.8** Current Sources/Sinks

Design current sources with DC current values of 10 μA and 20 μA and current sinks with DC current values of 10 μA and 40 μA. The small-signal source resistance of all current sources and sinks should be larger than 10 MΩ. The $V_{DS_{SAT}}$ of both current sources and sinks must be less than 0.5 V. You are given one reference current source of 10 μA with which you can use to derive the others. Verify your design in SPICE.

SOLUTION

A good topology for this design is shown in Fig. Ex 9.8.

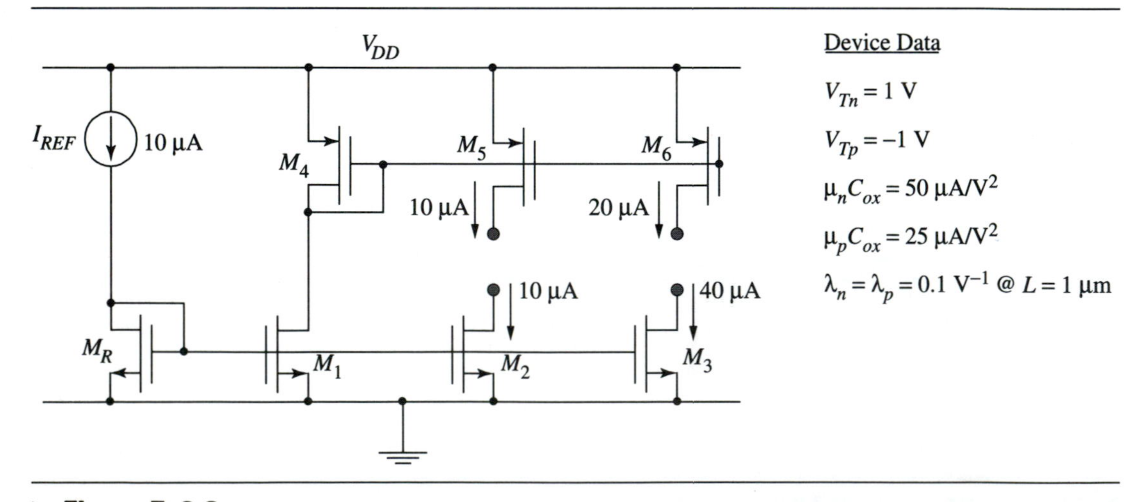

Figure Ex9.8

We begin this design by realizing that in order to meet the $V_{DS_{SAT}}$ requirement $V_{GS} = V_{SG} \le 1.5$ V. This sets a value of $(W/L)_R$.

$$V_{GS} = V_{Tn} + \sqrt{\frac{I_{REF}}{\frac{W}{2L}\mu_n C_{ox}}} \Rightarrow \left(\frac{W}{L}\right)_R = 1.6$$

If we set $(W/L)_1 = (W/L)_2 = 1.6$, $I_{D1} = I_{D2} = 10\ \mu A$. To make $I_{D3} = 40\ \mu A$, let $(W/L)_3 = 4(W/L)_2 = 6.4$. The p-channel devices are sized the same way.

$$V_{SG} = 1.5V = -V_{Tp} + \sqrt{\frac{I_{REF}}{\frac{W}{2L}\mu_p C_{ox}}} \Rightarrow \left(\frac{W}{L}\right)_4 = 3.2$$

To make $I_{D5} = 10\ \mu A$ and $I_{D6} = 20\ \mu A$, $(W/L)_5 = 3.2$ and $(W/L)_6 = 6.4$.
Now we check the small-signal source resistances.
For $I_D = 10\ \mu A$ and $\lambda_n = \lambda_p = 0.1\ V^{-1}$ for $L = 1\mu m$, we need $L = 10\ \mu m$ to get $R_S = 10\ M\Omega$. For $I_D = 20\ \mu A$, we need $L = 20\ \mu m$ and for $I_D = 40\ \mu A$, we need $L = 40\ \mu m$.
Finally: $(W/L)_R = (W/L)_1 = (W/L)_2 = 16/10$, $(W/L)_4 = (W/L)_5 = 32/10$, $(W/L)_3 = 256/40$, and $(W/L)_6 = 128/20$.
SPICE yields: $I_{D1} = 10.20\ \mu A$, $I_{D2} = 9.90\ \mu A$, $I_{D3} = 39.46\ \mu A$, $-I_{D4} = 10.21\ \mu A$, $-I_{D5} = 10.11\ \mu A$, and $-I_{D6} = 20.15\ \mu A$.

SPICE also shows that M_2, and M_3 have small-signal source resistances of 10.2 MΩ, and that M_5 and M_6 have a small-signal source resistance of 9.96 MΩ. These results are very close to our specifications. We could iterate to further increase the small-signal source resistances.

9.4.5 Generating I_{REF}

Students are often puzzled how we get the first current source I_{REF}. There are several techniques to generate this source.

- The current source could come from off-chip.
- One could build a highly sophisticated temperature compensated supply-independent current source. These reference current sources are often based on what is termed as a *bandgap reference*. However the design of these circuits is beyond the scope of this text. See Ref. 2 for further discussion of these references.
- A simple reference current could be generated as shown in Fig. 9.33.

If instead of using a reference current, we tie a resistor between the power supply and the diode-connected MOS transistor, we find the value of the reference current to be

$$I_{REF} = \frac{V_{DD} - V_{OUT}}{R} \tag{9.35}$$

where

$$V_{OUT} = V_{Tn} + \sqrt{\frac{I_{REF}}{\frac{W}{2L}\mu_n C_{ox}}} \tag{9.36}$$

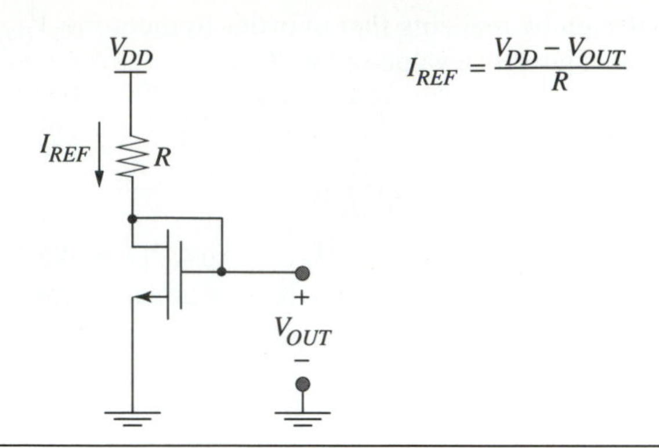

➤ Figure 9.33 Simple circuit to generate a reference current.

If the width-to-length ratio of the MOS transistor is large, the reference current is approximately

$$I_{REF} \approx \frac{V_{DD} - V_{Tn}}{R} \tag{9.37}$$

The value of the reference current is determined by a resistor. One problem with this circuit is that the reference current is directly dependent on the power supply voltage. In sophisticated analog circuit design, a more complex reference current source that is independent of both the power supply value and temperature is usually required.

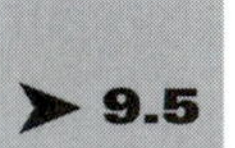

➤ 9.5 A TWO-STAGE TRANSCONDUCTANCE AMPLIFIER DESIGN

This section reviews the concepts discussed in this chapter by showing the design of an MOS transconductance amplifier. Our goal is to construct a transconductance amplifier with a transconductance of 1mS, an output resistance greater than 10 MΩ, and an infinite input resistance. We will use a 5 V power supply and fabricate the circuit in a 2.0 μm CMOS process. The output of the transconductance amplifier drives a capacitive load, i.e., other MOS transistors. Recall that a capacitive load is an open circuit for DC biasing and a short circuit for small-signal design. For frequency response reasons, that you will learn in Chapter 10, we require that minimum channel length transistors of 2.0 μm be used in this design. These n-channel transistors have a $\lambda_n = 0.05\ \text{V}^{-1}$. We also require that all transistors be biased with 100 μA to ensure a reasonable frequency response. We will neglect the backgate effect for this hand design.

From the device data given, we know the output resistance of the NMOS transistor will be equal to 200 kΩ with a bias current of 100 μA. Since a two-stage common-source amplifier has an output resistance on the order of a single device, this topology will not meet our output resistance specification. It is clear that we need a common-gate amplifier as a second stage in order to increase the output resistance to 10 MΩ. Therefore, our topology choice is a cascode amplifier as shown in Fig. 9.34. This stage offers increased output resistance by the common-gate amplifier transistor M_2. However, because we must keep these two transistors plus the current source supply in the constant-current region with a sufficient voltage across them, our voltage signal swing will be limited. In this design we will establish our DC operating point in such a way to maintain a relatively large voltage swing.

9.5.1 Small-Signal Design

It is often useful to begin an amplifier design by establishing the operating point before beginning the small-signal design. However, in this particular problem we have already specified the supply current at 100 μA and our major specifications are small-signal parameters. Our goal in this small-signal design is to establish device sizes for the NMOS transistors M_1 and M_2. We begin by noting that the transconductance of this cascode amplifier is given by

$$G_m = g_{m1} = \sqrt{2I_D\left(\frac{W}{L}\right)_1 \mu_n C_{ox}} \tag{9.38}$$

From the device data shown in Fig. 9.34(b), we find that the width of device M_1 must equal 200 μm to achieve the transconductance specification of 1mS.

Next, we look at the output resistance of this cascode amplifier. By recognizing that the output resistance of device M_1 is the source resistance for the common-gate amplifier formed by device M_2 and using the expression for the output resistance of a common-gate amplifier found in Table 8.3, we can write the approximate output resistance for this cascode circuit as

$$R_{out} \approx (g_{m2} r_{o2}) r_{o1} \tag{9.39}$$

It is important to note that the output resistance of the total amplifier is not only dependent on the output resistance of the cascode amplifier devices (M_1 and M_2), but

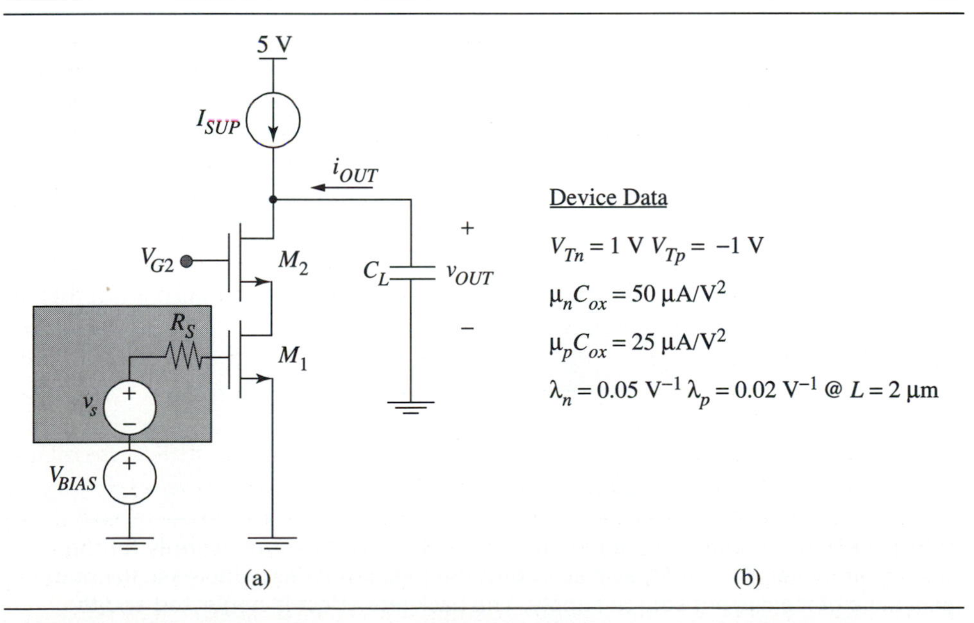

➤ **Figure 9.34** MOS cascode amplifier (a) circuit schematic, (b) device data.

also on the output resistance of the current source supply. If we assume that both of these output resistances are on the same order, we require that each of them be twice the value of our specification, namely 20 MΩ. Since the output resistance of a single transistor is 200 kΩ, we require the open-circuit voltage gain of transistor M_2, $g_{m2}r_{o2}$, be greater than 100. This requires the transconductance of device M_2, g_{m2}, be greater than 0.5 mS. Since we know the current flowing through device M_2 and the transconductance required, we find that $(W/L)_2$ must be greater than 25. Using the minimum channel length device we find $(W/L)_2 = 50/2$.

9.5.2 Current Source Supply Design

Since we know a large output resistance is needed for the current source supply, we will use a p-channel cascode circuit to give a high output resistance. This circuit is shown in Fig. 9.35.

The next step in this design is to establish the gate voltages required for devices M_3 and M_4. Because we lost some voltage swing with the topology choice of a cascode amplifier, we will try to ensure that we have a relatively large voltage swing at the output. We know that V_{SD} must be greater than $V_{SG} + V_{Tp}$ to ensure the device is in the constant-current region. We will choose V_{SG} for our p-channel transistors to be equal to 1.5 V which means our minimum V_{SD} is equal to 0.5 V. Although this choice is arbitrary, if we tried to make V_{SD} smaller, we would require a smaller V_{SG}, and therefore a much larger device size to carry the 100 μA. With the choice of $V_{SG} = 1.5$ V we find

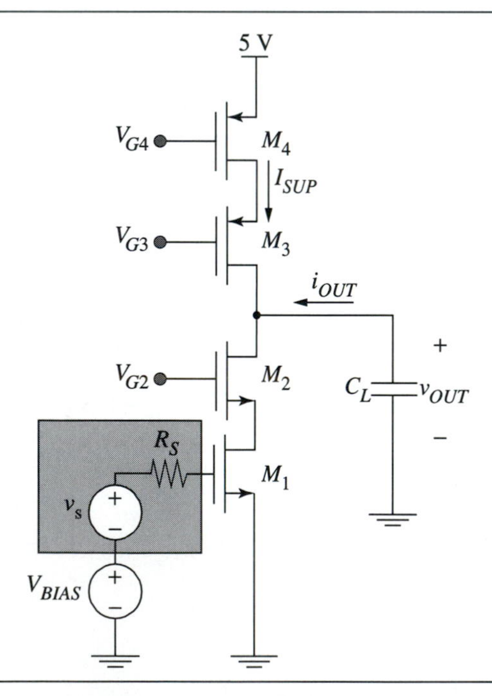

Figure 9.35 P-channel implementation of the current source supply for the cascode amplifier. Device M_3 acts as a common-gate amplifier to increase the output resistance of the current source supply. The backgate effect is neglected and the backgate connection is not shown in this schematic.

the gate voltage, V_{G4} equals 3.5 V and V_{G3} equals 2.0 V. By using the drain current equation for the p-channel devices in the constant-current region

$$-I_{Dp} = \frac{W}{2L}\mu_p C_{ox}(V_{SG} + V_{Tp})^2 \tag{9.40}$$

we calculate the W/L of both device M_3 and M_4 to be equal to 64/2.

The output resistance of the p-channel transistor with 100 μA flowing through it will be 500 kΩ since $\lambda_p = 0.02\ V^{-1}$. Using the W/L of M_3 and M_4 we can calculate the incremental source resistance of the p-channel current source as

$$R_S \approx (g_{m3} r_{o3}) r_{o4} = 0.4\ \text{mS} \cdot 500\ \text{k}\Omega \cdot 500\ \text{k}\Omega = 100\ \text{M}\Omega \tag{9.41}$$

which is sufficiently high.

9.5.3 Voltage References

If we use a totem pole voltage reference structure as shown in Fig. 9.36 to establish V_{G3} and V_{G4}, the easiest way to set the device sizes of the diode-connected transistors is to use the same size devices as M_3 and M_4. Not only does this ease the design, but it helps in matching between the current flowing through the diode-connected transistors and the supply current flowing into devices M_3 and M_4. At this point in our design we have established device sizes for M_1, M_3, M_4, and the diode connected transistors M_{3B} and M_{4B}.

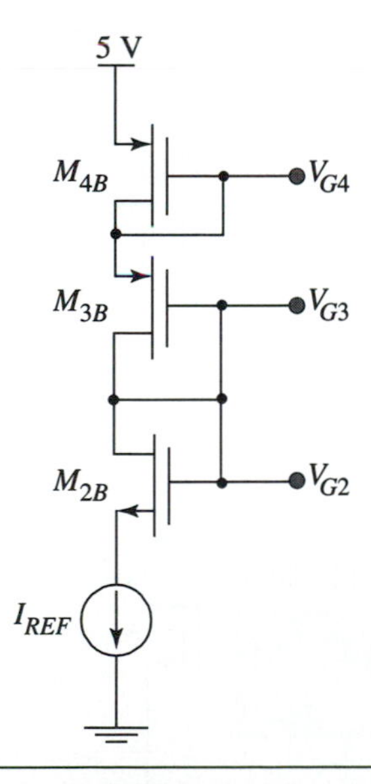

Figure 9.36 Totem pole voltage reference structure used to establish reference voltages V_{G2}, V_{G3}, and V_{G4} for cascode amplifier. Device sizes of M_{2B}, M_{3B}, and M_{4B} should be set equal to M_2, M_3, and M_4 respectively. Backgate connections are neglected.

The next step is to establish a voltage for the gate of device of M_2. One easy way to establish this voltage is with a diode-connected NMOS transistor between the current source I_{REF} and the p-channel transistors. Since we would like the diode-connected NMOS transistor, M_{2B}, and device M_2 to have the same voltage drop from drain-to-source, we size $(W/L)_{2B} = 50/2$. Since we know the drain current and device geometry, we can calculate

$$V_{GS2} = V_{GS2B} = V_{Tn} + \sqrt{\frac{2I_{REF}}{\left(\frac{W}{L}\right)_2 \mu_n C_{ox}}} = 1\text{V} + \sqrt{\frac{2\,(100\mu\text{A})}{\left(\frac{50}{2}\right) 50 \frac{\mu\text{A}}{\text{V}^2}}} = 1.4\text{ V} \quad (9.42)$$

The final design for our two-stage transconductance amplifier is shown in Fig. 9.37. We have specified the DC voltages at the various nodes in the circuit. The reference current source could be replaced by a resistor equal to 6 kΩ to supply the 100 μA to the diode-connected transistors, since the voltage at the source of M_{2B} is 0.6 V. Device M_{2B} is not strictly necessary. However, it provides an additional voltage drop equal to that across device M_2 and helps to match the V_{DS} across all the transistors and the reference current source.

Given the device size for M_1, we can calculate V_{BIAS}, which is the DC voltage to be applied to the gate of M_1 to sink the 100 μA supplied to its drain. We find that $V_{BIAS} = 1.2$ V.

Finally, we want to investigate the output voltage swing of this amplifier. The nominal DC design gave us an output voltage $V_{OUT} = 2.0$ V. The gates of devices M_1–M_4 are set by the totem pole voltage reference and the input bias voltage

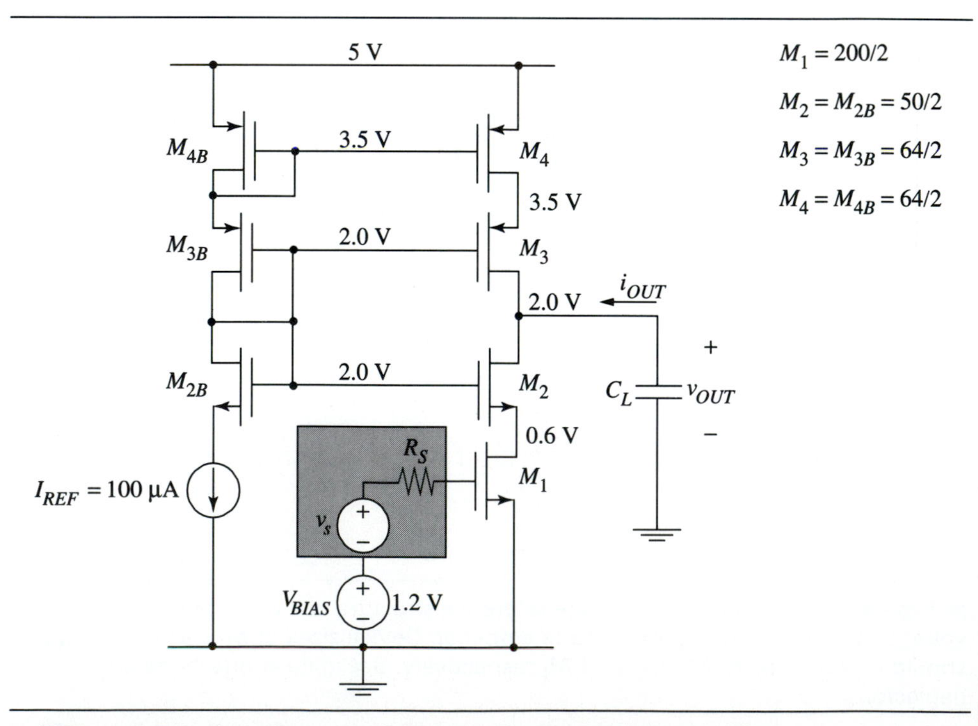

➤ **Figure 9.37** Complete circuit diagram for the cascode amplifier.

V_{BIAS}. The source voltages of the p-channel devices are set since the devices were sized for a $V_{SG} = 1.5$ V. Therefore, we must have $V_{SD} > V_{SG} + V_{Tp} = 0.5$ V to operate the devices in the constant-current region. Devices M_2 and M_{2B} have a $V_{GS} = 1.4$ V, which sets the source voltage of M_2 and M_{2B} at 0.6 V. The drain voltage of device M_2 (output) must be greater than 1.0 V to ensure that the devices are operating in their constant-current region. Device M_3 must have its drain voltage $V_{D3} \leq 3.0$ V so it remains in the constant-current region. Our output voltage swing for this design is $1.0 \text{ V} \leq v_{OUT} \leq 3.0$ V.

At this point the device topology and device sizes should be input to SPICE to verify our hand design. Several variations on topologies or device sizing could have achieved or exceeded our design goal.

9.6 ANALYSIS OF A BiCMOS VOLTAGE AMPLIFIER

This section uses many of the concepts developed in this chapter to show how to estimate the small-signal parameters of an amplifier by following the signal path and identifying the various gain stages. Consider the voltage amplifier shown in Fig. 9.38.

Although, at first, this amplifier looks extremely complex, the best way to attack its analysis is to identify the transistors that make up the signal path. One way to identify these transistors is to look at the output and input, and try to connect these through various gain stages that have been studied in Chapter 8. We start the analysis by finding the input small-signal voltage source at the gate of M_1. We see the drain of M_1 is

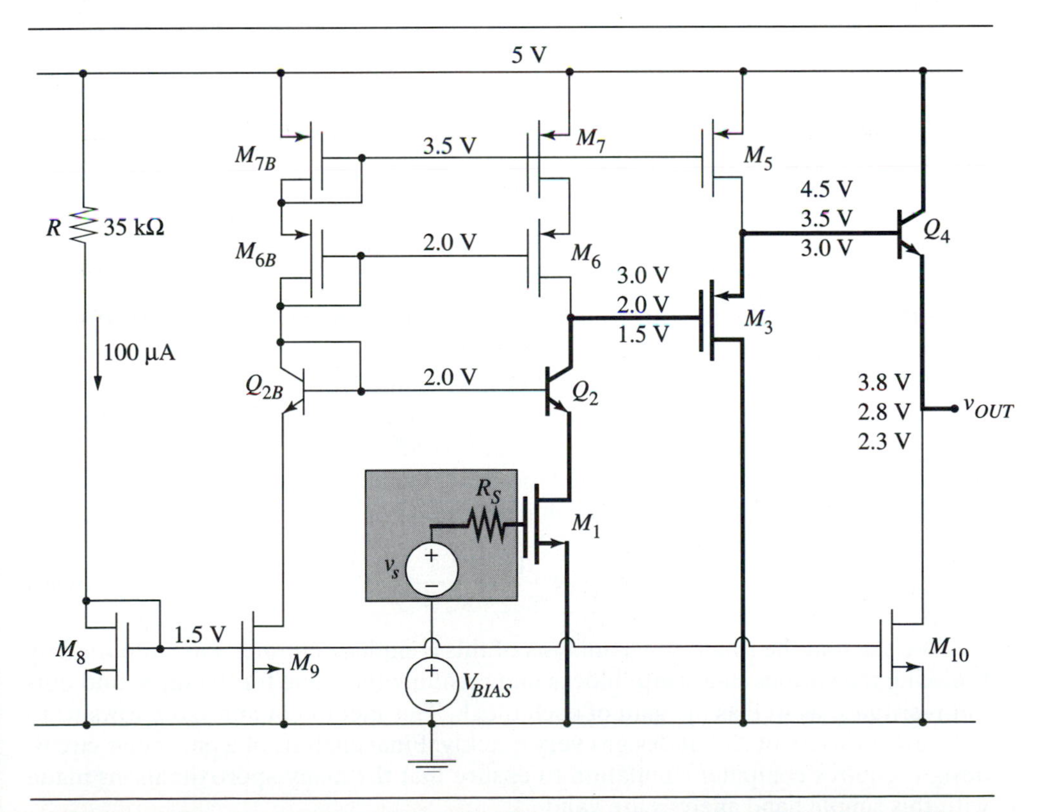

Figure 9.38 A complete voltage amplifier circuit. It contains a cascode amplifier, CS – CB, followed by a CD – CC voltage buffer. The signal path is highlighted.

connected to the emitter of Q_2, indicating a cascode or common-source/common-base connection. The output signal of the cascode is taken at the collector and continues to the gate of M_3 that is a p-channel transistor in the common-drain configuration. The source of M_3 is connected to the base of Q_4 which is a common-collector stage. The combination of devices M_1 and Q_2 form a cascode amplifier and devices M_3 and Q_4 form a voltage buffer. In summary, the signal path of this amplifier is made up of four stages: common-source, common-base, common-drain, and common-collector.

9.6.1 Small-Signal Analysis

The voltage gain of this amplifier can be written by recalling the analyses performed in this chapter. The voltage gain of the cascode amplifier is equal to the transconductance times the output resistance of that cascode amplifier. We define the output of the cascode amplifier, which is the collector of device Q_2, as v_{out2}. The voltage gain from the input to v_{out2} is given by

$$\frac{v_{out2}}{v_s} = -g_{m1}R'_{out} \tag{9.43}$$

The output resistance of the cascode amplifier R'_{out} can be analyzed by looking into the collector of Q_2 and noticing that the resistance seen is approximately equal to β_o times the output resistance of Q_2. This output resistance is in parallel with the cascode current source transistors M_6 and M_7. Recalling that the cascoded current source has an output resistance that is equal to $g_{m6}\, r_{o6}$ times the output resistance of M_7, we find

$$R'_{out} = \beta_{o2} r_{o2} \| g_{m6}\, r_{o6}\, r_{o7} \tag{9.44}$$

Combining these into one equation, we find that the voltage gain from the input of M_1 to the collector of Q_2 is equal to

$$A_v = \frac{v_{out2}}{v_s} = -g_{m1}\left(\beta_{o2}\, r_{o2} \| g_{m6}\, r_{o6} r_{o7}\right) \approx \frac{v_{out}}{v_s} \tag{9.45}$$

Since the voltage gain from M_3 to the output is approximately unity, Eq. (9.45) represents the voltage gain for the entire amplifier.

The input resistance of this amplifier is infinite. The output resistance of this amplifier can be calculated by looking into the emitter of transistor Q_4 and finding the output resistance of the common-collector configuration, which is $1/g_{m4}$ plus the source resistance at the common-collector input divided by β_{o4}. The source resistance of the CC amplifier is equal to $1/g_{m3}$, and we find

$$R_{out} = \frac{1}{g_{m4}} + \frac{1}{g_{m3}\beta_{o4}} \tag{9.46}$$

We see that the small-signal analysis of this complex amplifier was simplified by breaking the various stages into blocks and recalling the value for the input and output resistance as well as the gain of each block. This method of analysis allows us to evaluate a variety of circuit designs very quickly. Final analysis of a particular circuit design requires computer simulation to ensure that the many approximations made with this simple hand analysis are valid.

9.6.2 Large-Signal Analysis

The complexity of this amplifier demonstrates that most of the transistors are used to bias the amplifier rather than handle the signal. This is typical of integrated circuit design where active rather than passive devices are used to bias the amplifier.

To see how we biased this amplifier, we will first point out the function of each bias transistor. Next, we will show how we can estimate the particular DC node voltages. Begin by looking at resistor R. This resistor generates our initial reference current that we assume is approximately 100 μA. The reference current enters the drain of device M_8 generating a voltage which is used for current sinks M_9 and M_{10}. Transistor M_9's current is used to generate a totem pole voltage source consisting of three voltages that bias the base of the common-base amplifier Q_2 and the gates of the cascode current source M_6 and M_7. Finally, the same voltage source generated by M_{7B} is used to bias the current source M_5 for the common-drain amplifier.

To estimate particular node voltages in this amplifier, we assume that all MOS transistors require the magnitude of the gate-source voltage to be approximately 1.5 V to sink or source the current through the devices. Notice that the geometry of the MOS devices, namely the W/L ratio, can be adjusted to significantly change this gate-source voltage. However, since the collector current for bipolar transistors is exponentially dependent on the base-emitter voltage, we find that the base-emitter voltage is approximately 0.7 V regardless of the collector current for practical values of current and device area.

We can now estimate the nominal voltages at each node in the circuit. The gate of M_8 requires 1.5 V to sink the current from the reference resistor. The gate of M_{7B} is at 3.5 V, which also makes the magnitude of its gate-source voltage 1.5 V. Furthermore, the gate of M_{6B} is at 2 V since its gate-source voltage is at 1.5 V.

To determine the other node voltages we need to know the relationship between the currents down the leg of M_7 and M_{7B}. For a moment assume that they are the same and the bias voltage on the gate of M_1 is set to exactly sink that current. If that is the case, by symmetry the collector of Q_2 will be at the identical voltage as the collector of Q_{2B}, namely 2 V. Device M_3 requires $V_{SG3} = 1.5$ V. Its source voltage is approximately 3.5 V. Finally, with $V_{BE4} = 0.7$ V the output is set to approximately 2.8 V which is near the center of the positive power supply and ground. The W/L ratios of various transistors could be adjusted to achieve exactly 2.5 V at the output.

Finally, we should examine the output voltage swing that this amplifier can achieve. To determine the swing we must ensure that all devices remain in their constant-current region. One way to determine the output swing is to start by increasing the output voltage until a device goes out of the constant-current region. We see that if the output voltage is at 3.8 V the base of Q_4 will be at 4.5 V. This means only 0.5 V will be across device M_5, which puts it on the edge of its constant-current region. If the output voltage was any higher than 3.8 V, M_5 would go into the triode region. To find the low value of the output voltage swing, we see that if the output voltage is 2.3 V, the base of Q_4 is at 3 V and the gate of M_3 is at 1.5 V. Since the emitter of Q_2 is at 1.3 V, the collector of Q_2 can go no lower than 1.5 V before it leaves its constant-current region. Therefore, the output voltage swing of this amplifier is between 2.3 V and 3.8 V, or approximately 1.5 V.

In summary, we have shown that the analysis of a complex, multistage amplifier can be significantly simplified by identifying the signal path and estimating the small-signal parameter values. We also showed how to estimate the DC operating point and output voltage swing of this amplifier. These analysis techniques are key to developing skill in the art of analog IC design.

SUMMARY

In this chapter we discussed the analysis and design of multistage amplifiers. The small-signal two-port models were used to investigate the use of various types of amplifiers to transform the input or output resistance as well as increase the gain (voltage, current, transconductance, transresistance). The large-signal analysis of direct-coupled amplifiers and the formation of voltage and current sources were described. We ended the chapter with a design of a cascode amplifier and an analysis of a four-stage voltage amplifier to demonstrate the concepts presented in this chapter.

Specifically we showed

- the power of the two-port models to quickly analyze a complicated circuit and find the signal path from the input to the output.
- how to use the two-port models to find a quick approximation of the overall amplifier performance and any interstage losses.
- when to use MOS or bipolar transistors as well as when n-type or p-type devices are appropriate in a variety of common situations.
- the analysis and design of voltage sources and current sources including their incremental source resistance.

FURTHER READING

1. C. G. Fonstad, *Microelectronics Devices and Circuits*, McGraw-Hill, 1994, Chapter 13. Includes a discussion of capacitively coupled transistor amplifiers.
2. P. R. Gray and R. G. Meyer, *Analysis and Design of Analog Integrated Circuits 3rd Ed.*, Wiley, 1993, Chapter 4. More advanced discussion of current sources.
3. M. N. Horenstein, *Microelectronic Circuits and Devices 2nd Ed.*, Prentice Hall, 1995, Chapters 7 and 12. Chapter 7 covers current sources and large signal issues in amplifiers. Chapter 12 includes a discussion of the limitations of the two-port methodology in the analysis of multistage amplifiers.
4. A. S. Sedra and K. C. Smith, *Microelectronic Circuits 3rd Ed.*, HRW Saunders, 1991, Chapters 5 and 6. Includes an expanded discussion on bipolar and MOS current sources.

PROBLEMS

EXERCISES

E9.1 In Example 9.1 we looked at an MOS transresistance amplifier and compared two topologies. The first was a CG-CS amplifier while the second was a CG-CD amplifier. If you are given that the g_m of the MOS devices is 1 mS and r_o is 100 kΩ, which topology yields the highest overall transresistance, given

(a) $R_S = 1\text{k}\Omega$ and $R_L = 100\ \Omega$

(b) $R_S = 1\text{k}\Omega$ and $R_L = 10\ \text{k}\Omega$

(c) Repeat (a) and (b) when $g_m = 100\ \mu\text{S}$ and $r_o = 10\ \text{M}\Omega$

E9.2 This exercise compares a CS-CS voltage amplifier with a CS-CD voltage amplifier. If you are given that the g_m of the MOS devices is 1mS and r_o is 100 kΩ, which topology yields the highest overall voltage gain, given

(a) $R_S = 1\text{k}\Omega$ and $R_L = 100\ \Omega$

(b) $R_S = 1\text{k}\Omega$ and $R_L = 10\ \text{k}\Omega$

(c) Repeat (a) and (b) when $g_m = 100\ \mu\text{S}$ and $r_o = 10\ \text{M}\Omega$

E9.3 This exercise compares a CS-CS transconductance amplifier with a CS-CG transconductance amplifier. If you are given that the g_m of the MOS devices is 1mS and r_o is 100 kΩ, which topology gives the highest overall transconductance, given

(a) $R_S = 1\text{k}\Omega$ and $R_L = 100\ \Omega$

(b) $R_S = 1\text{k}\Omega$ and $R_L = 10\ \text{k}\Omega$

(c) Repeat (a) and (b) when $g_m = 100\ \mu\text{S}$ and $r_o = 10\ \text{M}\Omega$

E9.4 In a pure CMOS technology where only MOS devices are available there is no single stage which yields current gain. You are given that the g_m of the MOS devices is 1mS and r_o is 100 kΩ.

(a) What two stages would you cascade to form a current amplifier with the highest current gain?

(b) Calculate the overall current gain given $R_S = 1\ \text{k}\Omega$ and $R_L = 100\ \Omega$.

(c) Calculate the overall current gain given $R_S = 1\ \text{k}\Omega$ and $R_L = 10\ \text{k}\Omega$.

E9.5 It is possible to build a current amplifier with a single stage using a CE stage. However it does not have a low input resistance and a high output resistance. Given a BiCMOS technology we will compare a CB-CS current amplifier with a CE-CB current amplifier. If you are given that the g_m of the MOS devices is 1mS and r_o is 100 kΩ, and the g_m of the bipolar devices is 3 mS, $\beta_o = 100$ and r_o is 500 kΩ, which topology gives the highest overall current gain, given

(a) $R_S = 1\ \text{k}\Omega$ and $R_L = 100\ \Omega$

(b) $R_S = 1\ \text{k}\Omega$ and $R_L = 10\ \text{k}\Omega$

E9.6 You are given a BiCMOS technology where the transconductance and output resistance of the bipolar transistors are about 3 times higher than the MOS transistors with equivalent device sizes and power dissipation. The source and load resistances are equal to $1/g_m$ of the bipolar transistor. You are to choose the best topology using 3 stages to build an amplifier with

(a) highest overall voltage gain

(b) highest overall transresistance

E9.7 You are given a BiCMOS technology where the transconductance and output resistance of the bipolar transistors are about 3 times higher then the MOS transistors with equivalent device sizes and power dissipation. The source and load resistances are equal to r_o of the bipolar transistor. You are to choose the best topology using 3 stages to build an amplifier with

(a) highest overall current gain

(b) highest overall transconductance

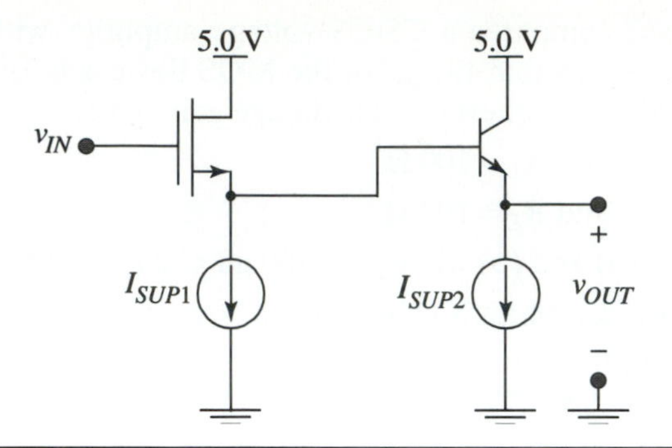

➤ **Figure E9.8**

E9.8 Given the NMOS CD stage cascaded with an npn bipolar CC stage to form a voltage buffer as shown in Fig. E9.8, calculate the input voltage swing and output voltage swing. The devices must remain in their constant-current region. Assume that the current source supplies require 0.5 V across them. Also assume that $V_{BE} = 0.7$ V, $V_{CE_{SAT}} = 0.2$ V, $V_{Tn} = 1$ V, and $V_{GS} = 1.5$ V. Neglect the backgate effect for this exercise.

E9.9 Given the PMOS CD stage cascaded with an npn bipolar CC stage to form a voltage buffer as shown in Fig. E9.9, calculate the input voltage swing and output voltage swing. The devices must remain in their constant-current region. Assume that the current source supplies require 0.5 V across them. Also assume that $V_{BE} = 0.7$ V, $V_{CE_{SAT}} = 0.2$ V, $V_{Tp} = -1$ V, and $V_{SG} = 1.5$ V. Neglect the backgate effect for this exercise.

E9.10 What is the minimum voltage that can be at the output of the current buffer shown in Fig. E9.10 and have all devices operating in their constant-current region? Assume that $V_{BE} = 0.7$ V, $V_{CE_{SAT}} = 0.2$ V, $V_{Tn} = 1$ V, and $V_{GS} = 1.2$ V. Neglect backgate effect for this exercise. Also assume that the current sources all require 0.5 V across them.

E9.11 In Fig. 9.17 we showed a cascode amplifier composed of an NMOS CS amplifier and a npn bipolar CB amplifier. Draw a two-port model and calculate R_{in}, R_{out} and G_m for a cascode amplifier formed with

(a) NMOS CS amplifier and NMOS CG amplifier.

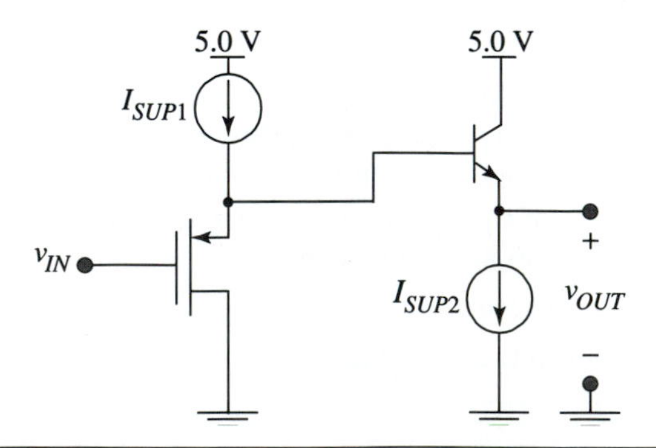

➤ **Figure E9.9**

(b) NMOS CS amplifier and npn bipolar CB amplifier.

(c) npn bipolar CE amplifier and NMOS CG amplifier.

(d) npn bipolar CE amplifier and npn bipolar CB amplifier.

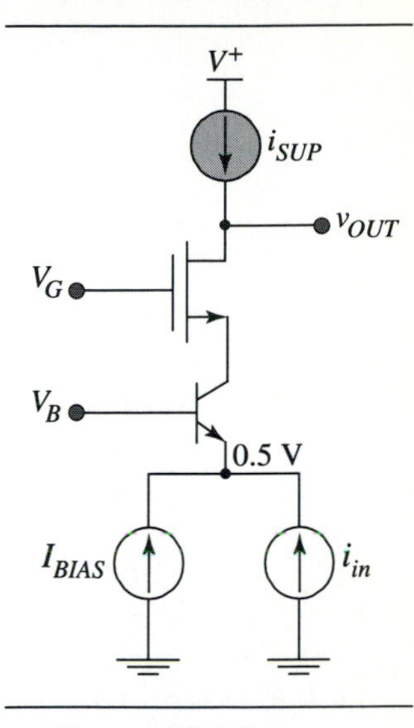

➤ **Figure E9.10**

E9.12 You are given a resistor and diode-connected NMOS transistor as shown in Fig. E9.12 along with its device data.

(a) Find W/L such that $V_{OUT} = 1.5$ V.

(b) Calculate the incremental source resistance of this voltage source.

(c) Calculate W/L to make the incremental source resistance equal 500 Ω.

(d) What is the new V_{OUT}?

(e) If the power supply changes to 4.5 V calculate the new V_{OUT} using the W/L used in part (c).

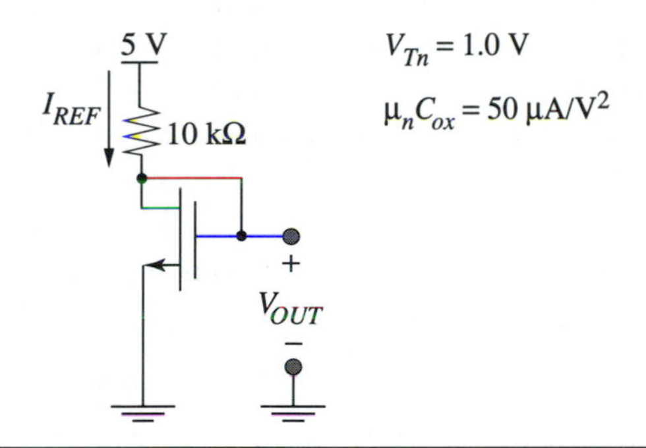

➤ **Figure E9.12**

E9.13 An npn voltage source is shown in Fig. E9.13 where $I_S = 10^{-15}$A. Neglect the base current.

(a) Calculate the open-circuit output voltage.

(b) How much would you have to increase the emitter area of the bipolar transistor to reduce the output voltage calculated in (a) by 120 mV?

(c) Calculate the incremental source resistance for this voltage source.

E9.14 You are given the totem pole voltage source and the device data shown in Fig. E9.14. Neglect base current and backgate effect for this problem. Assume all devices are operating in their constant-current region. $I_{REF} = 200$ μA.

(a) Find V_{OUT1}, V_{OUT2}, and V_{OUT3}.

(b) Keeping V_{OUT1} the same as found in part (a), change the W/L of device M_2 such that $V_{OUT2} = 1.6$ V.

(c) What is the new value of V_{OUT3} when W/L of M_2 is that found in part (b)?

E9.15 Repeat E9.14 given $I_{REF} = 10$ μA.

E9.16 A simple current source along with the device data is shown in Fig. E9.16. Given $I_{REF} = 100$ μA

(a) Find $(W/L)_1$ such that the incremental voltage source resistance of the reference voltage generated by M_1 is equal to 1kΩ.

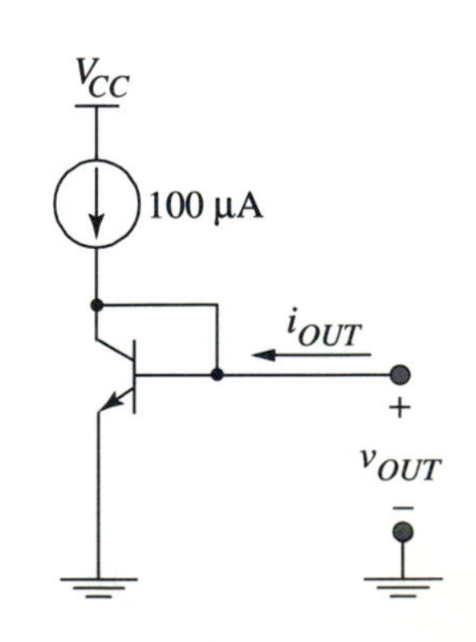

➤ **Figure E9.13**

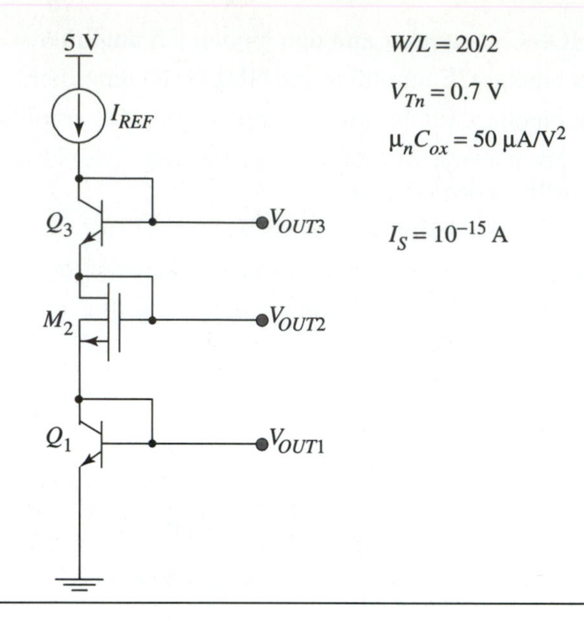

➤ **Figure E9.14**

(b) Find $(W/L)_2$ of M_2 such that $I_{OUT} = 300\ \mu A$.

(c) Find the incremental source resistance of the current source.

E9.17 Draw a p-channel version of the simple current source. Repeat E9.16 with the p-channel device data $V_{Tp} = -0.7$ V, $\mu_p C_{ox} = 20\ \mu A/V^2$, and $\lambda_p = 0.05\ V^{-1}$.

E9.18 In the multiple MOS current sources shown in Fig. E9.18 $I_{REF} = 100\ \mu A$ and the device size of M_R $(W/L)_R = 10/2$. Assume all devices are operating in their constant-current region. Neglect backgate effect.

(a) Find $(W/L)_1$ such that $I_{OUT1} = 40\ \mu A$.

(b) Find $(W/L)_2$ such that $I_{OUT2} = 100\ \mu A$.

(c) Find $(W/L)_3$ such that $I_{OUT3} = 250\ \mu A$.

(d) Repeat parts (a) to (c) with $I_{REF} = 20\ \mu A$.

(e) Repeat parts (a) to (c) with $I_{REF} = 100\ \mu A$ and $(W/L)_R = 30/2$.

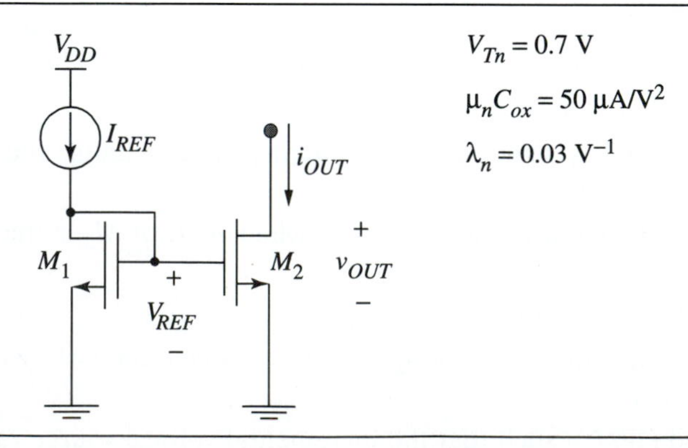

➤ **Figure E9.16**

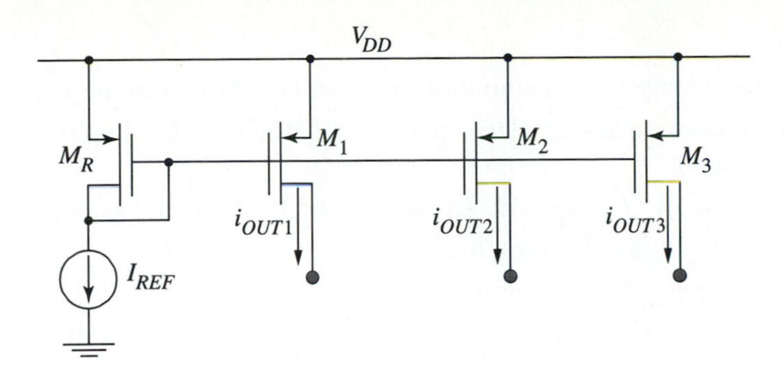

➤ **Figure E9.18**

E9.19 In the multiple MOS current source and sink shown in Fig. E9.19 $I_{REF} = 100\ \mu A$ and the device size of M_R and M_4 is $(W/L)_R = (W/L)_4 = 10/2$. Assume all devices are operating in their constant-current region. Neglect backgate effect.

(a) Find $(W/L)_1$ such that $I_{OUT1} = 40\ \mu A$.

(b) Find $(W/L)_2$ such that $I_{OUT2} = 100\ \mu A$.

(c) Find $(W/L)_3$ such that $I_{OUT3} = 250\ \mu A$.

(d) Repeat parts (a) to (c) with $I_{REF} = 20\ \mu A$.

(e) Repeat parts (a) to (c) with $I_{REF} = 100\ \mu A$ and $(W/L)_R = (W/L)_4 = 30/2$.

E9.20 In the voltage amplifier found in Fig. 9.38 the input CS amplifier was formed with an NMOS device. Redraw the amplifier such that the CS amplifier is a PMOS device, the CB amplifier is a pnp device, the CD amplifier is an NMOS device and the CC amplifier is a pnp device. Draw the complete circuit including all biasing and current sources devices.

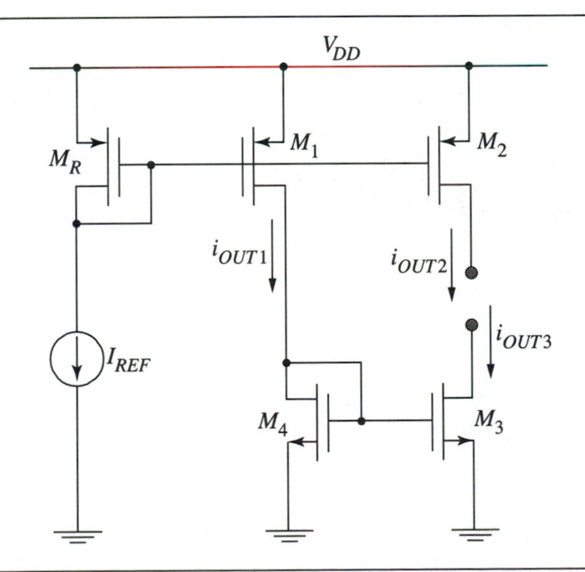

➤ **Figure E9.19**

PROBLEMS

P9.1 Given the 2-stage CS-CS amplifier shown in Fig. P9.1. Assume $V_{GS1} = V_{GS2}$.

(a) Draw a small-signal two-port model of this CS-CS amplifier. Show both stages.

(b) Find $(W/L)_1$ and $(W/L)_2$ such that $I_{D1} = I_{SUP1}$ and $I_{D2} = I_{SUP2}$.

(c) Find R_{in} for the amplifier.

(d) Find R_{out} for the amplifier.

(e) What is the intrinsic voltage gain assuming $R_S = 0\ \Omega$ and $R_L \to \infty$?

(f) What is the voltage gain for $R_S = 25\ \text{k}\Omega$ and $R_L = 10\ \text{k}\Omega$?

P9.2 A 2-stage CMOS transresistance amplifier is shown in Fig. P9.2. The device data is shown in the figure. The circuit is biased such that $V_{OUT} = 2.5$ V. The supply current sources have values of $I_{SUP1} = 200\ \mu\text{A}$, $r_{oc1} = 1\text{M}\Omega$ and $I_{SUP2} = 500\ \mu\text{A}$ and $r_{oc2} = 400\ \text{k}\Omega$. The channel lengths used for both NMOS and PMOS devices are 4 μm.

(a) Calculate the minimum width of M_1 such that its source voltage is at least 0.5 V so that the bias current source remains in its constant-current region.

(b) Calculate $(W/L)_2$ such that its gate voltage is 3 V.

(c) Identify the 2 stages (e.g., CE, CS, etc.).

(d) Calculate the two-port parameters R_{in} and R_{out} for this amplifier.

(e) Calculate the two-port parameter R_m.

(f) Given $R_S = 10\ \text{k}\Omega$ calculate the overall transresistance v_{out}/i_s. Note $R_L \to \infty$.

P9.3 A voltage buffer and the device data is shown in Fig. P9.3. We have assumed that the circuit is fabricated in an n-well CMOS process where we can short the back-gate and source of the p-channel device. Assume $2\phi_n = -2\phi_p = 0.8$ V.

(a) What is V_{OUT} given $V_{IN} = 0$ V?

(b) Find the open-circuit voltage gain ($R_L \to \infty$).

(c) What is the minimum load resistor that the amplifier can drive and still maintain a voltage gain of 0.6?

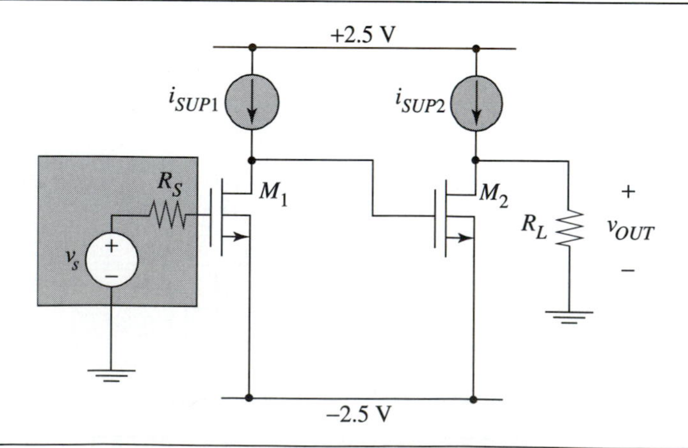

➤ **Figure P9.1**

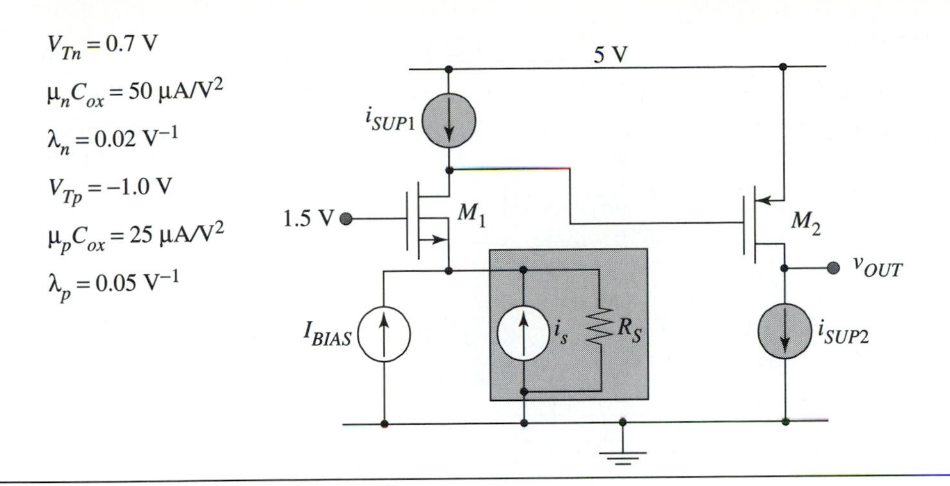

➤ **Figure P9.2**

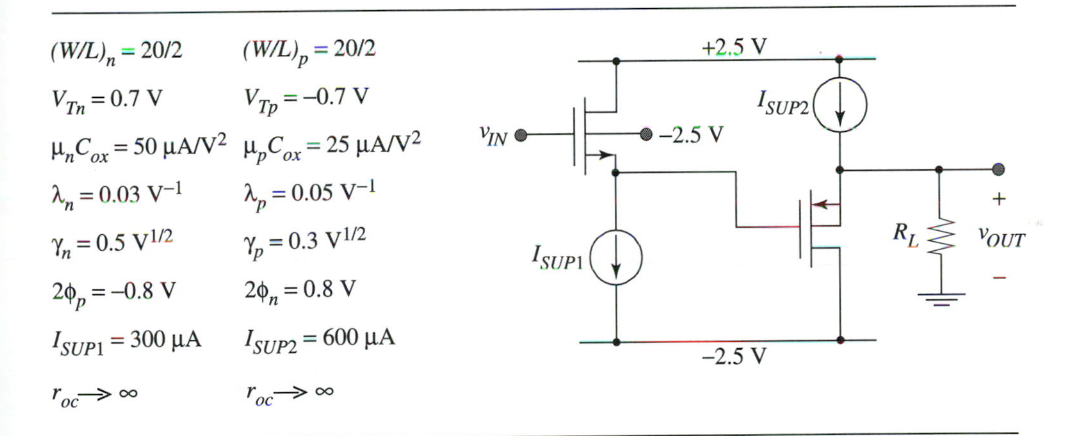

➤ **Figure P9.3**

P9.4 Repeat P9.3 given that the circuit is fabricated in a p-well CMOS process and that the n-channel device has its backgate shorted to the source and the p-channel device has its backgate tied to the positive power supply.

P9.5 A 2-stage BiCMOS transresistance amplifier is shown in Fig. P9.5. The device data is shown in the figure. The circuit is biased such that $V_{OUT} = 0$ V. The supply current sources have values of $I_{SUP1} = 250\ \mu A$, $r_{oc1} = 1 M\Omega$ and $I_{SUP2} = 400\ \mu A$ and $r_{oc2} = 800\ k\Omega$. The length of the n-channel device is 5μm and $R_S = 10\ k\Omega$.

(a) Calculate the width of M_1 such that $V_{GS1} - V_{Tn} = 0.3$ V.

(b) Identify the 2 stages (e.g., CE, CS, etc.).

(c) Calculate the two-port parameters R_{in} and R_{out} for this amplifier.

(d) Calculate the two-port parameter R_m.

(e) Calculate the overall transresistance v_{out}/i_s. Note $R_L \rightarrow \infty$.

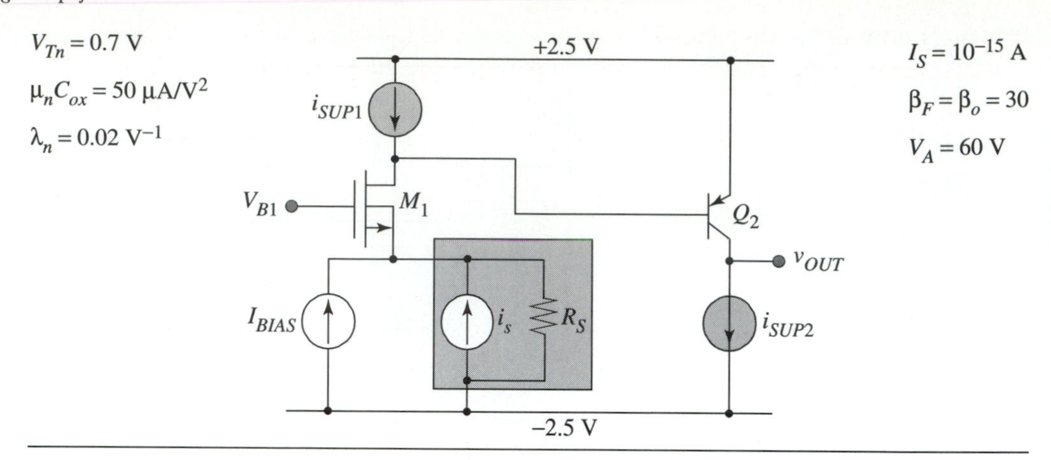

➤ Figure P9.5

P9.6 Repeat problem P9.5 if the backgate of device M_1 is connected to the negative power supply −2.5 V. $2\phi_p = -0.8$ V and $\gamma_n = 0.5$ V$^{-1/2}$

P9.7 You are given a 2-stage voltage buffer shown below in Figure P9.7. The device data is shown in the figure. $R_S = 20$ kΩ and $R_L = 25$ kΩ . Both current source supplies have a value $I_{SUP} = 150\ \mu$A and $r_{oc} = 500$ kΩ. Assume $V_{GS} - V_{Tn} = 0.3$ V and $V_{BE} = 0.7$ V. Ignore I_B for the DC calculations.

(a) Calculate V_{BIAS} such that $V_{OUT} = 0$ V.

(b) What is the voltage at node A?

(c) Draw the cascaded two-port representation including R_S and R_L.

(d) Calculate the overall voltage gain v_{out}/v_s including R_S and R_L.

P9.8 Repeat problem 9.7 but this time switch Q_2 and M_1 (make Q_2 the input).

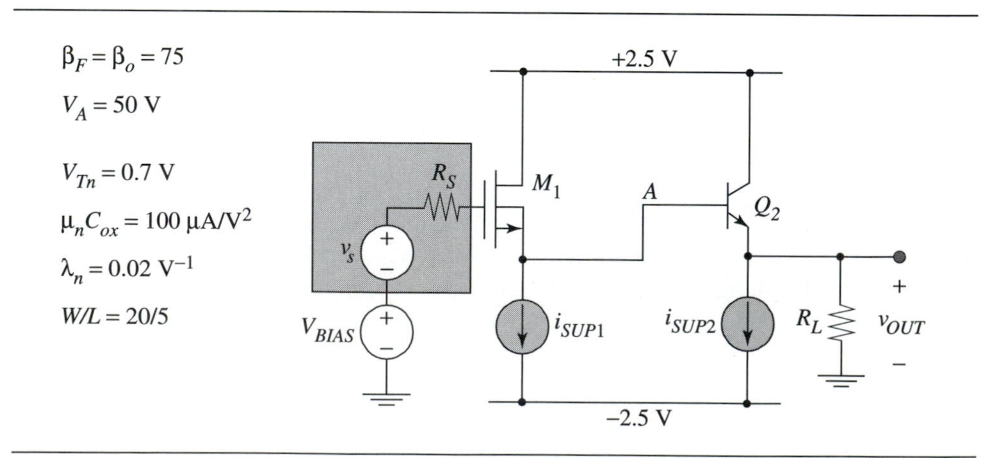

➤ Figure P9.7

P9.9 Given the 2-stage amplifier and the device data shown in Fig. P9.9. Assume V_{BIAS} is set such that both bipolar transistors are operating in their constant-current region.

(a) Draw a small-signal two-port model of this CE-CE amplifier. Show both stages.

(b) Find R_{in} for the amplifier.

(c) Find R_{out} for the amplifier.

(d) What is the intrinsic voltage gain assuming $R_S = 0\ \Omega$ and $R_L \to \infty$?

(e) What is the overall voltage gain for $R_S = 25\ k\Omega$ and $R_L = 10\ k\Omega$?

P9.10 Suppose emitter degeneration resistors are added to a CE-CE amplifier as shown in Fig. P9.10. $V_{BIAS} = 0$ V.

(a) Find R_{E1} such that the current through it is equal to I_{SUP1}. Neglect base current.

(b) Find R_{E2} such that V_{BC} of device Q_1 is equal to 0 V.

(c) Find R_{in} for the amplifier.

(d) Find R_{out} for the amplifier.

(e) What is the intrinsic voltage gain assuming $R_S = 0\ \Omega$ and $R_L \to \infty$?

(f) What is the overall voltage gain for $R_S = 25\ k\Omega$ and $R_L = 10\ k\Omega$?

(g) If you could add one stage of any type to this amplifier to increase the voltage gain what would it be (CB, CS, CC, etc.) and where would you connect it (input or output)?

P9.11 In this problem we compare a bipolar and MOS voltage source as shown in Fig. P9.11. Given the device data shown in the figure and given that $I_{REF} = 100\ \mu A$

(a) Find an expression for V_{OUT} for the bipolar case without neglecting the base current, and find its value.

(b) Find V_{OUT} for the MOS case.

(c) Calculate the incremental source resistance for both sources.

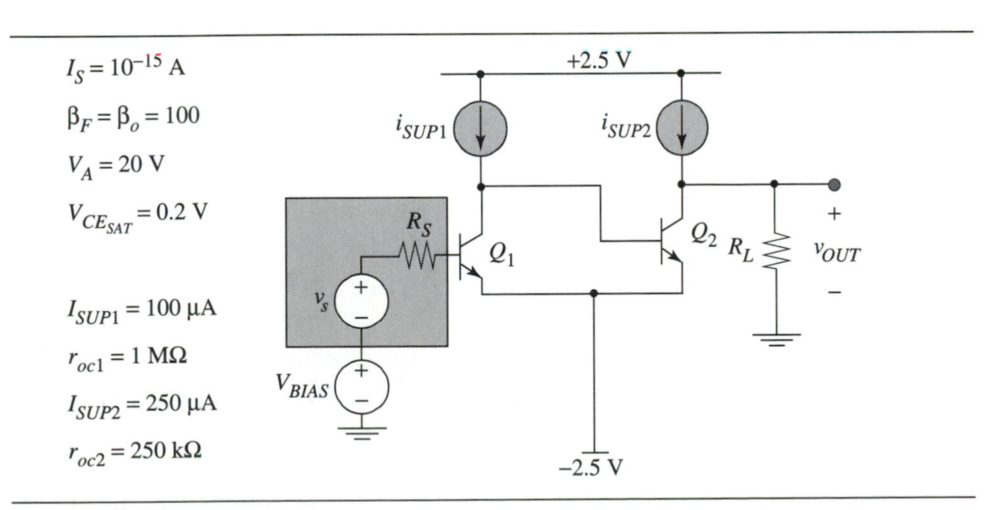

➤ **Figure P9.9**

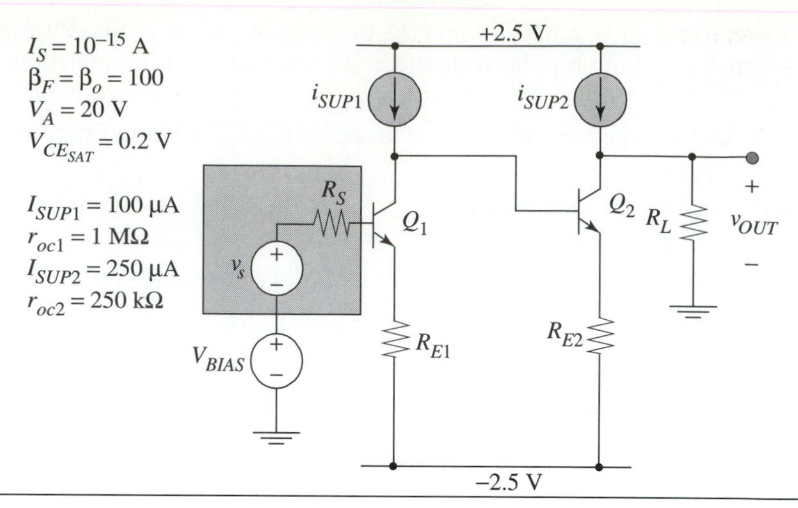

➤ **Figure P9.10**

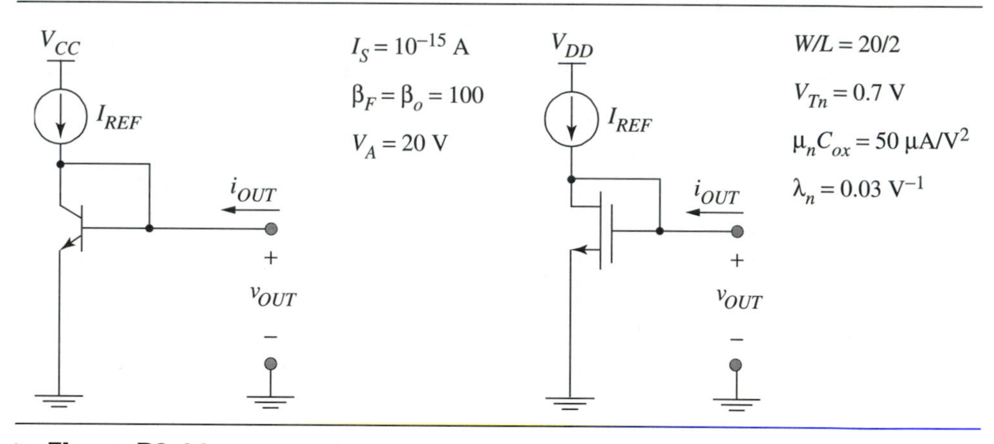

➤ **Figure P9.11**

P9.12 As seen in P9.11 the bipolar source has a smaller incremental source resistance for equivalent device sizes however the bipolar output voltage has a dependence on base current. By adding an NMOS device to "diode-connect" the bipolar device as shown in Fig. P9.12, this dependence is reduced.

(a) Find an expression for V_{OUT} in this new circuit and find its value.

(b) Draw a small-signal model of this circuit to find its incremental source resistance.

(c) Calculate the new incremental source resistance.

P9.13 Given a totem pole voltage source without a reference current we can still generate voltages. Use the circuit and device data shown in Fig. P9.13 and neglect the backgate effect.

(a) If $(W/L)_p = 128$ and $(W/L)_n = 64$ and the drain current of each device is 100 μA, find the voltages V_1, V_2, and V_3.

$W/L = 10/2$

$V_{Tn} = 0.7$ V

$\mu_n C_{ox} = 50\ \mu A/V^2$

$\lambda_n = 0.03\ V^{-1}$

V_{CC}

I_{REF}

i_{OUT}

+

v_{OUT}

–

$I_S = 10^{-15}$ A

$\beta_F = \beta_o = 100$

$V_A = 20$ V

➤ **Figure P9.12**

(b) Given that $(W/L)_p = 2(W/L)_n$ size the devices such that 1mA of current flows through each transistor, while maintaining the same voltages calculated in (a).

P9.14 A cascode current source is shown in Fig. P9.14 along with the p-channel device data. Assume that the backgates are all shorted to their respective sources. Assume the W/L of all devices is 10 and that $I_{OUT} = 100\ \mu A$.

(a) Calculate the incremental source resistance of the current source.

(b) If I_{OUT} is required to be 20 μA, what is the new W/L of devices M_2 and M_{2B} assuming the I_{REF} remains at 100 μA?

(c) What is the new incremental source resistance with the new values calculated in part (b)?

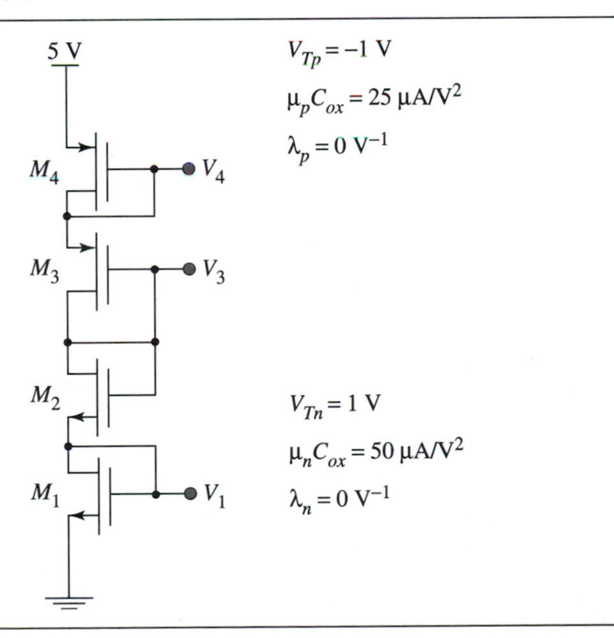

➤ **Figure P9.13**

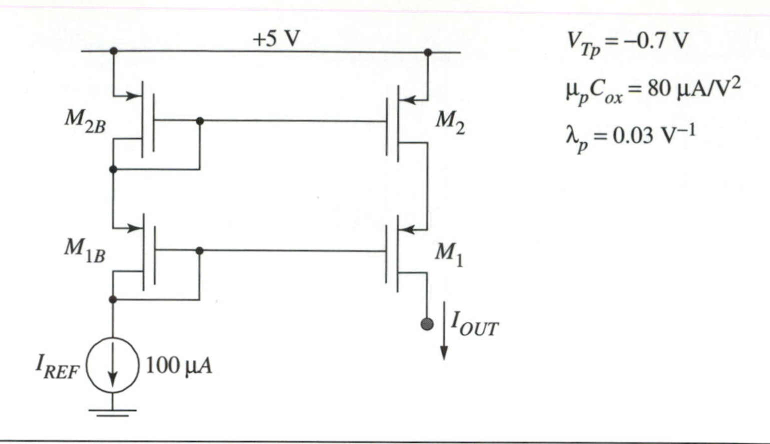

➤ **Figure P9.14**

P9.15 Repeat P9.14 using an n-channel implementation. The NMOS device data is $V_{Tn} = 0.7$ V, $\mu_n C_{ox} = 150\ \mu\text{A/V}^2$ and $\lambda_n = 0.05\ \text{V}^{-1}$.

P9.16 An output stage shown in Fig. P9.16 consists of an amplifying transistor and two transistors as part of a simple current source. The length of all of the transistors is 5 µm. The reference current is 100 µA. $V_{DS_{SAT1}} = 0.5$ V and $V_{DS_{SAT2}} = V_{DS_{SAT3}} = 0.3$ V. The device data is given in the figure.

(a) What are the widths of M_1, M_2 and M_3 and the value of V_{BIAS} such that $V_{OUT} = 0$ V?

(b) What is r_{oc} of the current source?

(c) What is the voltage gain of this output stage?

(d) What is the maximum possible positive and negative swing at the output and still have all devices operating in their constant-current region?

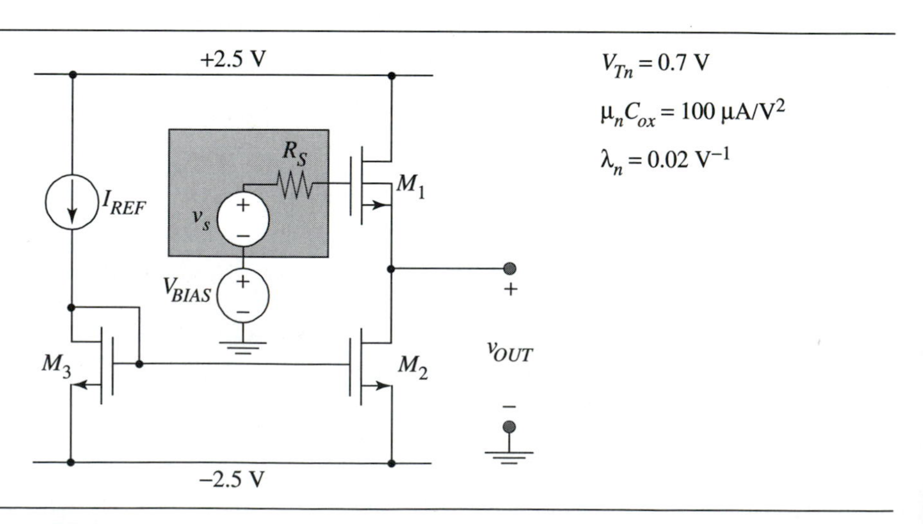

➤ **Figure P9.16**

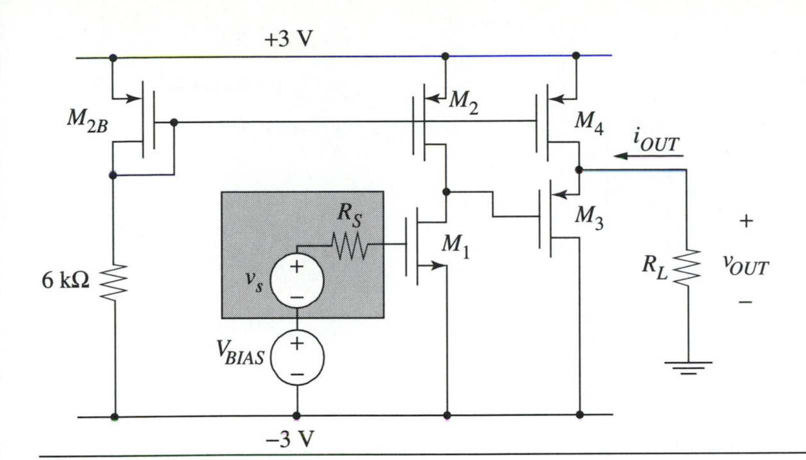

$V_{Tn} = 0.7$ V $V_{Tp} = -0.7$ V

$\mu_n C_{ox} = 50\ \mu A/V^2$ $\mu_p C_{ox} = 25\ \mu A/V^2$

$\lambda_n = 0.03\ V^{-1}$ $\lambda_p = 0.05\ V^{-1}$

➤ **Figure P9.17**

P9.17 Given the voltage amplifier and device data shown in Fig. P9.17 with $(W/L)_1 = 40/2$ and $(W/L)_3 = 100/2$. In this problem we will assume all backgates are shorted to their respective sources.

(a) Find the W/L for M_2, M_{2B}, and M_4 so that each has a drain current of 500 μA.

(b) What is the voltage at the gate of M_3 so that the output level will be 0 V?

(c) Calculate V_{BIAS} so that M_1 sinks the current from M_2.

(d) Draw a two-port model of this CS-CD stage and calculate the parameters.

(e) Calculate the overall voltage gain if $R_S = 10$ kΩ and $R_L = 1$kΩ.

(f) If R_L is reduced to 500 Ω calculate the new device size for M_4 so that the overall voltage gain remains the same as calculated in (e).

P9.18 The circuit in Fig. P9.18 is a standard pnp common-emitter transconductance amplifier. The reference current is 300 μA. The device data for the NMOS and bipolar devices is given in the figure. We require an output voltage swing from +2.25 V to −2.25 V. V_{BIAS} has been set such that V_{OUT} is 0 V. The channel length of the NMOS devices is 5 μm.

(a) Given that the width of M_2 is the same as M_3 calculate the width such that the output voltage swing specification is met.

(b) Calculate the two-port parameters R_{in}, R_{out}, and G_m of this transconductance amplifier.

P9.19 In P9.18 the low input resistance will degrade the transconductance given that $R_S = 5$ kΩ. We can add a common-collector amplifier at the input to increase the amplifier input resistance. The current source supply is formed with two PMOS transistors (M_4 and M_6) of the same size. The PMOS devices have a channel length of 2 μm. Their device data and the circuit are shown in Fig. P9.19. Use the same NMOS device data and bipolar device data as in P9.18. with $I_{REF} = 300$ μA. The NMOS device M_5 is half the width of M_3.

(a) If we want $V_{SD_{SAT}}$ for the p-channel devices to be 0.3 V, calculate the width of devices M_4 and M_6.

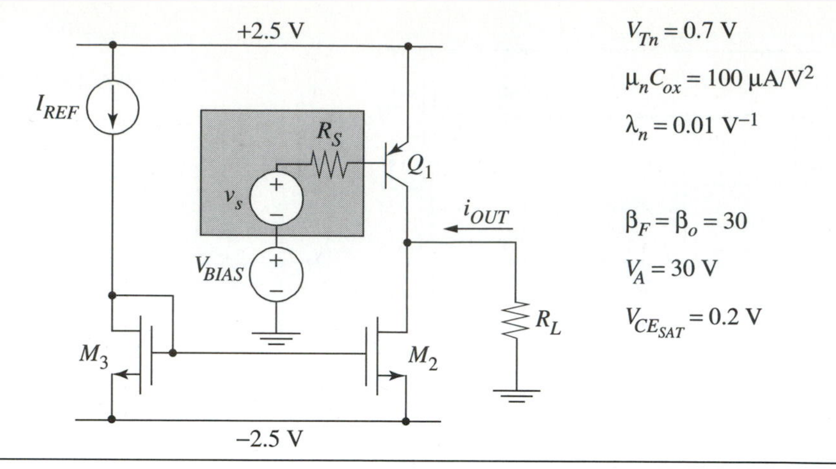

➤ **Figure P9.18**

(b) Calculate the width of the NMOS device M_5.

(c) Calculate R_{in} and R_{out} of both stages.

(d) Calculate the overall transconductance given $R_L = 100\ k\Omega$.

P9.20 A CMOS cascode transconductance amplifier and the device data are shown in Fig. P9.20. Neglect the backgate effect for this problem.

(a) Calculate $(W/L)_1$ of M_1 such that the small-signal transconductance, $i_{out}/v_s = 1\text{mS}$. Assume $R_L = 0\ \Omega$ (short-circuit output current) for this part.

(b) Calculate the value of V_{BIAS} using the $(W/L)_1$ calculated in part (a) such that $I_{OUT} = 0$ A.

(c) Calculate the output resistance of this transconductance amplifier.

(d) What is the maximum value of the load resistor R_L at which the overall transconductance is degraded by 20% from the original value of 1mS?

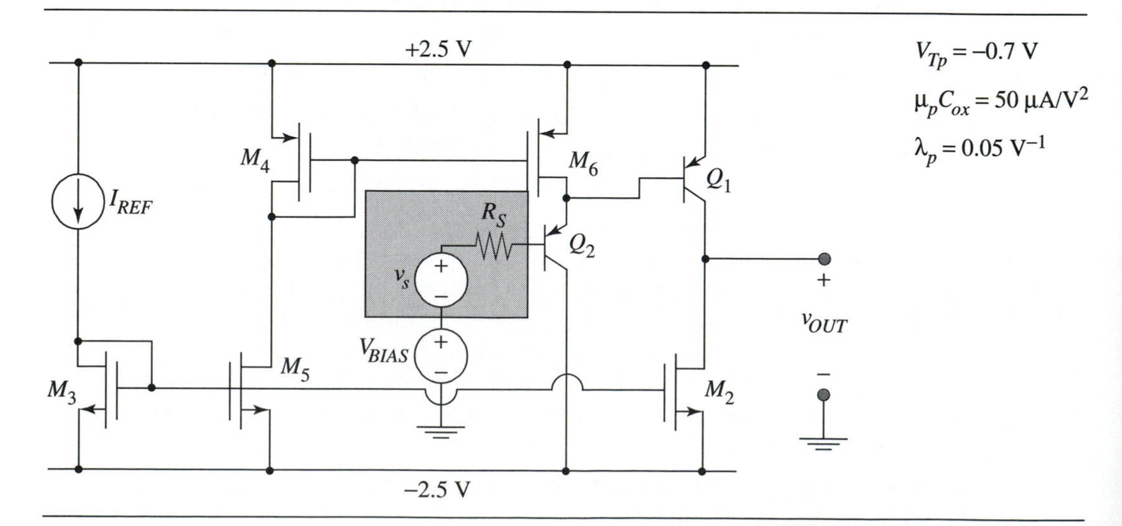

➤ **Figure P9.19**

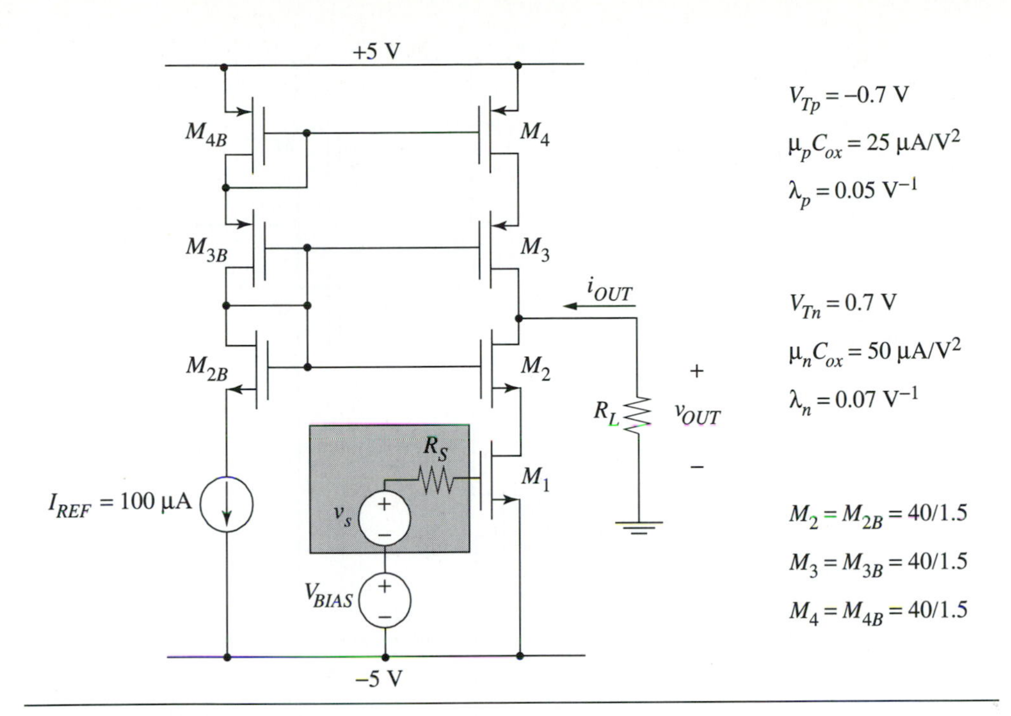

➤ **Figure P9.20**

P9.21 We replace the NMOS devices M_2 and M_{2B} with npn bipolar devices in the transconductance amplifier shown in Fig. P9.20. The bipolar devices have an $I_S = 10^{-17}$ A, $\beta_o = 100$ and $V_A = 20$ V. Neglect the backgate effect for this problem.

(a) Calculate $(W/L)_1$ of M_1 such that the small-signal transconductance, $i_{out}/v_s = 1$mS. Assume $R_L = 0\ \Omega$ (short-circuit output current) for this part.

(b) Calculate the value of V_{BIAS} using the $(W/L)_1$ calculated in part (a) such that $I_{OUT} = 0$ A.

(c) Calculate the output resistance of this transconductance amplifier.

(d) What is the maximum value of the load resistor R_L at which the overall transconductance is degraded by 20% from the original value of 1mS?

P9.22 Given the BiCMOS voltage amplifier and device data shown in Fig. P9.22 with $(W/L)_n = 20/2$ and $(W/L)_p = 40/2$.

(a) What is the current through each of the legs?

(b) Find the value of V_{BIAS} such that device M_1 sinks the current provided by device M_7.

(c) Find the voltage at each node. Assume $V_{BE} = 0.7$ V and $V_{OUT} = 2.5$ V.

(d) What is the voltage swing at the output while keeping all of the devices in their constant-current region of operation?

(e) What is the voltage gain of the cascode stage, v_x/v_s?

(f) What is the voltage gain of the total amplifier, v_{out}/v_s?

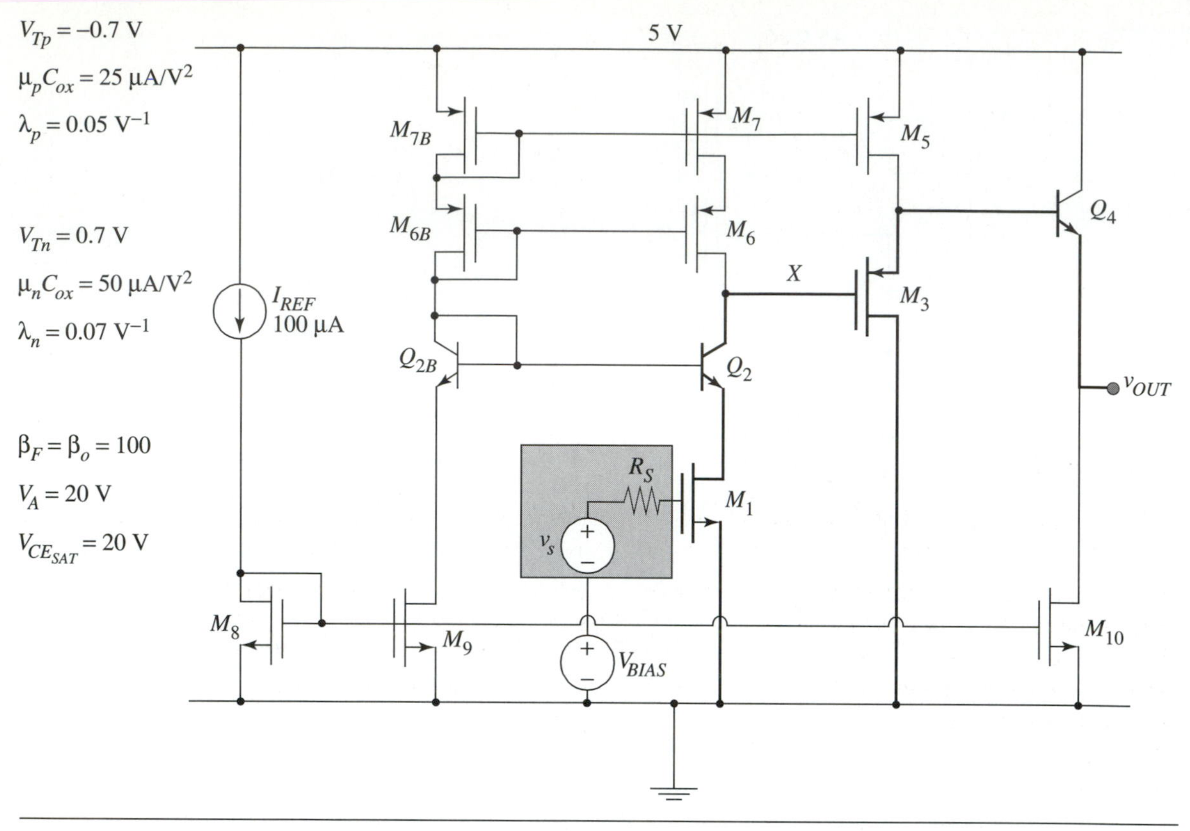

➤ **Figure P9.22**

DESIGN PROBLEMS

D9.1 Design the current source supplies shown in Fig. D9.1 by using 3 PMOS devices and a 50 μA reference current supply tied to the –2.5 V power supply. The PMOS device data is $V_{Tp} = -0.7$ V, $\mu_p C_{ox} = 25\ \mu A/V^2$, and $\lambda_p = 0.05\ V^{-1}$ at a device length of 2 μm. We require $I_{SUP1} = I_{SUP2} = 100$ μA and $r_{oc1} = r_{oc2} = 1$ MΩ.

D9.2 A simple NMOS current source was shown in Fig. 9.27. In this design we will take into account some process variations. The channel lengths have a tolerance of ±0.2 μm and the threshold voltages are equal to 1.0 V ± 0.2 V. However the threshold voltages of the adjacent devices are matched to within 10 mV. Given a reference current of 100 μA design a current source that has an output current between 98 μA and 102 μA. The transconductance of both MOS devices must be at least 0.1mS and you should try to minimize the area.

D9.3 Design a totem pole voltage source so that the output voltages are $V_1 = 1.5$ V $V_2 = 2.7$ V and $V_3 = 3.8$ V. The device data are given below.

(a) Begin the design by neglecting the backgate effect and sizing W/L of each device so that all currents are the same.

(b) What is the incremental source resistance of each voltage source?

(c) Re-size the devices to include the backgate effect.

(d) Re-calculate the output resistance of each voltage source.

(e) Verify your design with SPICE.

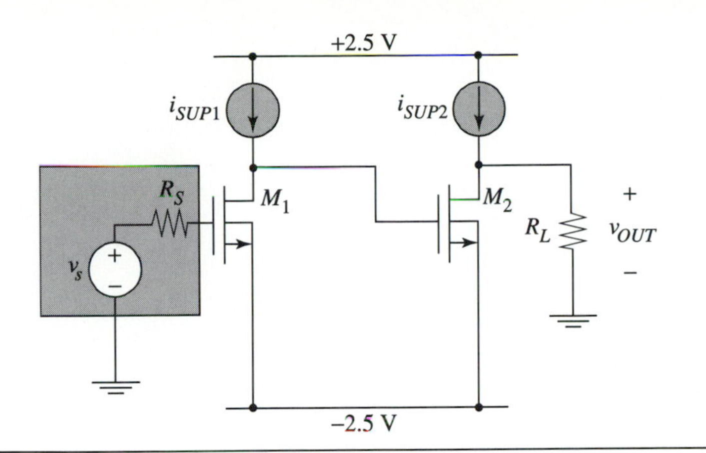

➤ **Figure D9.1**

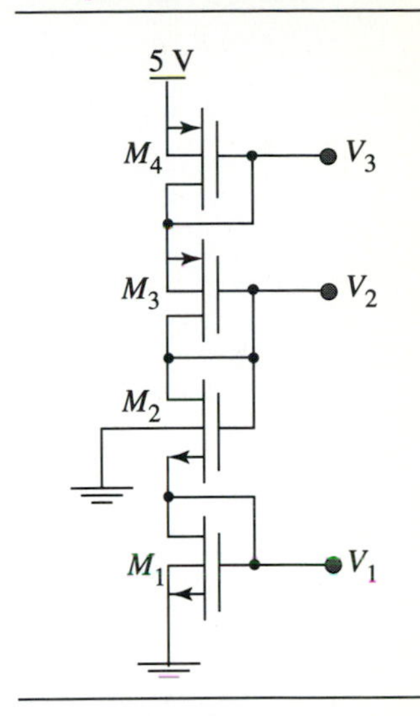

➤ **Figure D9.3**

D9.4 Repeat D9.3 but replace M_1 with a reference current source of 100 μA. Assume that this current source requires at least 0.3 V across it.

D9.5 Design a current source which provides 20 μA and 100 μA and a current sink with a value of 50 μA. Use a reference current source of 100 μA. Since V_{DD} will only be 3 V you cannot use cascodes to increase the incremental source resistance. Design the W/L so that the incremental source resistance of the current sources and sink is greater than 500 kΩ. Use the device data given below. Assume an n-well process so that p-channel devices can have their backgate shorted to their source but the n-channel devices cannot.

D9.6 In Section 9.5 a two-stage transconductance amplifier design was discussed. Applying the same methodology used in that section design a PMOS version that has the same specifications, namely $G_m = 1$ mS and $R_{out} = 10$ MΩ. Use the same device data as shown in Fig. 9.34. Neglect the backgate effect.

D9.7 Using the same design methodology as in Section 9.5, design a double-cascode CMOS transconductance amplifier with an NMOS input. Add a common-gate n-channel cascode transistor and a p-channel cascode transistor to the current source supply to the original circuit. The amplifier should have a transconductance of 2 mS and an output resistance of 500 MΩ with $I_{REF} = 200$ μA. There is a single 5 V power supply voltage. Neglect the backgate effect for the hand design but include it in your SPICE simulation. Assume an n-well process. Use the device data given below.

D9.8 Redraw the BiCMOS voltage amplifier in Fig. 9.38 using the complimentary transistors, namely a PMOS/pnp cascode input and an NMOS/pnp CD-CC output. Given the device data below, design the amplifier to have a voltage gain of > |5000| when driving a 50 Ω load resistor. The minimum current in any one stage must be greater than 100 μA for frequency response considerations that you will learn in Chapter 10.

Device Data

$V_{Tn} = 0.7$ V	$V_{Tp} = -0.7$ V	$I_S = 10^{-16}$A
$\mu_n C_{ox} = 50\ \mu\text{A/V}^2$	$\mu_p C_{ox} = 25\ \mu\text{A/V}^2$	$\beta_F = \beta_o = 50$
$\lambda_n = 0.03\ \text{V}^{-1}$	$\lambda_p = 0.05\ \text{V}^{-1}$	$V_A = 20$ V
$\gamma_n = 0.5\ \text{V}^{1/2}$		
$2\phi_p = -0.8$ V		

chapter 10

Frequency Response

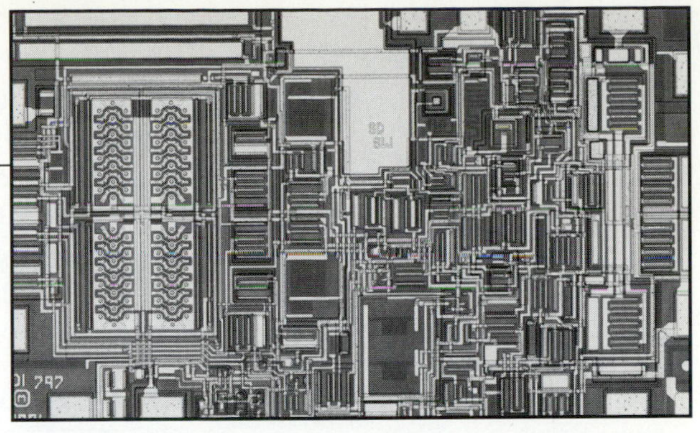

The AD797 low noise operational amplifier has a bandwidth of 8 MHz. It has a settling time on the order of 1 μs to a resolution of 16 bits. (Courtesy of Analog Devices, Inc.)

Microelectronic circuits can be optimized for a variety of frequency bands, depending on the application.

- Audio amplifiers use frequencies from 20 Hz to 20 kHz
- Telecommunication circuits operate at approximately 200 kHz
- Video amplifiers operate at approximately 50 MHz
- Radio frequencies operate at 100 MHz
- Cellular communication circuits operate at 1–2 GHz

The higher frequency signals used for satellite communications cannot be achieved with current state-of-the-art silicon integrated circuit technology.

In this chapter we will consider how the dynamics of an amplifier affect the output signal when the input is a sinusoid. Since we will be investigating only small signals, we can use the powerful analysis techniques derived from linear system theory. Essentially, when the input to a linear amplifier is a sinusoidal signal, the output response can be characterized by the change in amplitude and phase of that sinusoid. This analysis technique is quite general since arbitrary periodic signals can be constructed from a sum of various sinusoids.

Chapter Objectives

- Review the concepts for frequency domain analysis and apply them to single-stage and multistage amplifiers.
- Understand the application of a quasi-static device model for the MOS and bipolar transistors and the frequencies at which these models apply.

- Learn how to analyze the frequency response of the common-emitter/common- source amplifier using the Miller Approximation.
- Learn the method of open-circuit time constants that allows analytical approximation of the frequency response of amplifiers containing multiple capacitors.
- Learn how to apply analysis techniques to other single-stage and multistage amplifiers.

10.1 REVIEW OF FREQUENCY DOMAIN ANALYSIS

In this section, we review linear circuit analysis for simple passive circuits. Keep in mind that the goal of this analysis is to understand the effect of the passive circuit on a steady-state sinusoidal signal's amplitude and phase. It is assumed that you are familiar with the manipulation of complex variables, and phasors. The notation for **phasor quantities** in this text uses an upper case variable name and lower case subscripts such as V_{in} and I_{out}.

In the low-pass filter shown in Fig. 10.1, the voltage is passed from the input to the output with little or no change in amplitude and phase at low ($\omega \ll 1/RC$) frequencies. However, at high frequencies the output voltage amplitude is significantly decreased and the output phase lags the input phase by approximately 90°. The following example demonstrates this point.

EXAMPLE 10.1 Low-Pass Filter

Find the magnitude and phase of the transfer function V_{out}/V_{in} for the low-pass filter in Fig. 10.1.

SOLUTION

By an impedance divider we have $V_{out} = \left(\dfrac{1/j\omega C}{R + 1/j\omega C}\right)V_{in}$

Rearranging,

$$V_{out} = \left(\frac{1}{1 + j\omega RC}\right)V_{in}$$

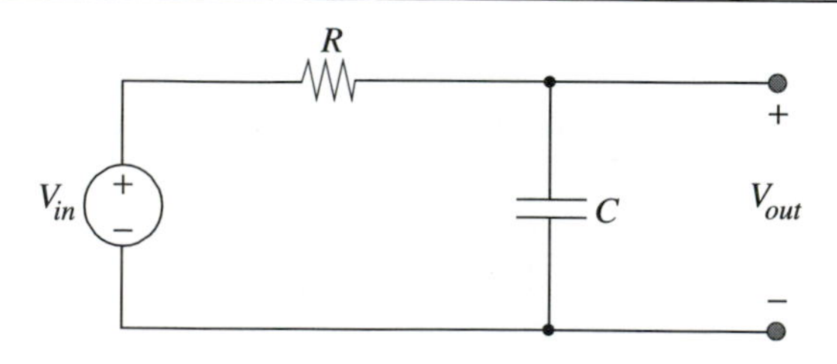

Figure 10.1 A low-pass filter circuit.

The magnitude is $\left|\dfrac{V_{out}}{V_{in}}\right| = \left[\dfrac{1}{1+(\omega RC)^2}\right]^{\frac{1}{2}}$

The phase is $\angle \dfrac{V_{out}}{V_{in}} = \tan^{-1}(-\omega RC)$

For $\omega RC \gg 1$ the sinusoid is attenuated and shifted –90°

$$\left|\frac{V_{out}}{V_{in}}\right| \approx \frac{1}{\omega RC} \qquad \angle \frac{V_{out}}{V_{in}} \approx -90^\circ$$

For $\omega RC \ll 1$, the sinusoid is passed unattenuated and with no phase shift

$$\left|\frac{V_{out}}{V_{in}}\right| \approx 1 \qquad \angle \frac{V_{out}}{V_{in}} \approx 0^\circ$$

10.1.1 Bode Plots

It is essential to recognize that the magnitude and phase are the important parameters for sinusoidal steady-state analysis. In Example 10.1, the transfer function indicated the attenuation of the voltage amplitude and change in its phase at the output, given a particular input voltage at a frequency ω. Usually we want to know the magnitude and phase of transfer functions and other quantities as a function of frequency over many orders of magnitude. A **Bode plot**, which plots the log of the magnitude versus the log of the frequency and the phase angle versus the log of the frequency, describes both magnitude and phase over many orders of magnitude of frequency.

Recall from our example of a low-pass filter that for very low frequencies, where $\omega \ll 1/RC$, the magnitude of the transfer function was equal to approximately 1 and the phase angle was approximately 0°. However, when ω was $\gg 1/RC$, the magnitude and phase were given by

$$\left|\frac{V_{out}}{V_{in}}\right| \approx \frac{1}{\omega RC} \qquad \angle \frac{V_{out}}{V_{in}} \approx -90^o \tag{10.1}$$

Note that the magnitude is decreasing by a factor of 10 for every factor of 10 increase in ω. The interesting point in the Bode plot is where $\omega = 1/RC$. Under this condition the magnitude is given by

$$\left|\frac{V_{out}}{V_{in}}\right| = \left|\frac{1}{1+j\omega RC}\right| = \left|\frac{1}{1+j}\right| = \frac{1}{\sqrt{2}} = 0.707 \tag{10.2}$$

and the phase by

$$\angle\frac{V_{out}}{V_{in}} = \tan^{-1}(-\omega RC) = \tan^{-1}(-1) = -45° \qquad (10.3)$$

The corresponding Bode plot that contains the magnitude and phase information of this transfer function is shown in Fig. 10.2. Both magnitude and frequency are plotted logarithmically, while the phase is plotted on a linear scale. With this plot, one can input any particular sinusoidal frequency into the circuit and see how it affects the magnitude and phase of the signal at the output.

It is customary to express the logarithmic magnitude scale on a Bode plot with a dimensionless unit called a **decibel** (dB). The magnitude of the ratio of voltages and currents in units of dB is:

Ratio of voltages in decibels: $20 \log\left|\frac{V_{out}}{V_{in}}\right|$ dB

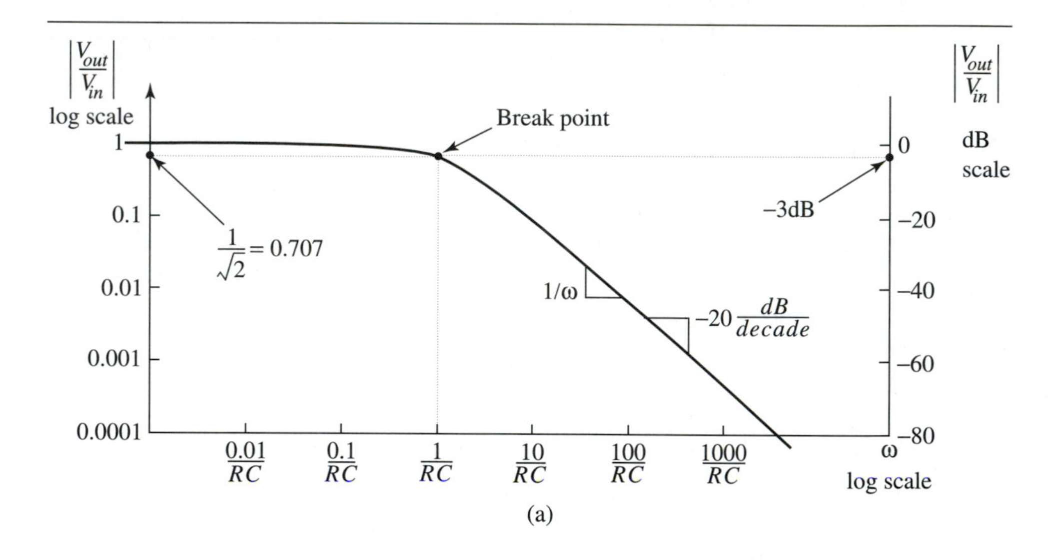

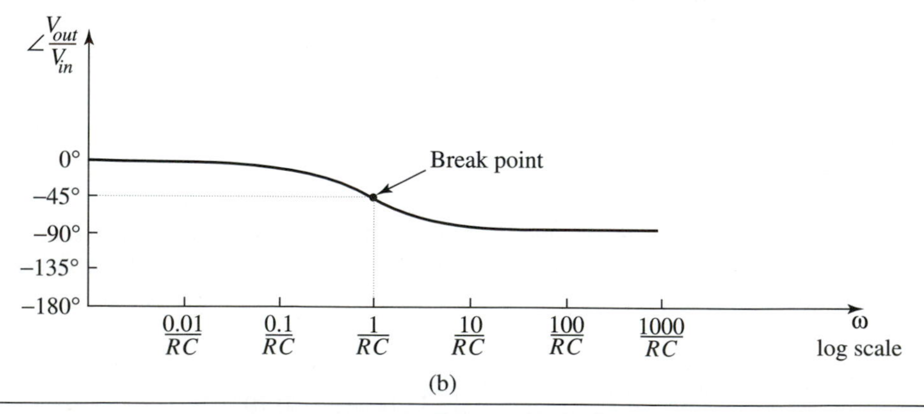

➤ **Figure 10.2** Bode plot for the low-pass filter circuit. (a) Log magnitude vs. log frequency. (b) Phase vs. log frequency.

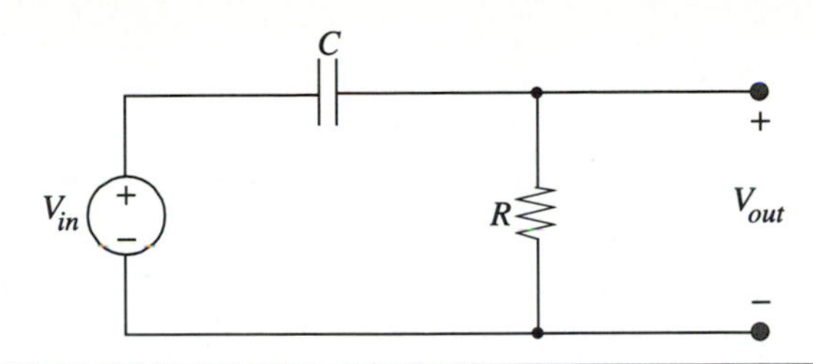

➤ Figure 10.3 A high-pass filter circuit.

Ratio of currents in decibels: $20 \log \left|\frac{I_{out}}{I_{in}}\right|$ dB

Since power is proportional to the voltage squared or current squared, the magnitude of the ratio of powers is expressed in dB as:

Ratio of powers in decibels: $10 \log \left|\frac{P_{out}}{P_{in}}\right|$ dB

The corresponding decibel units with the magnitude units are also shown in Fig. 10.2. The magnitude of the voltage at the breakpoint frequency $\omega = 1/RC$ expressed in decibels is $20 \log(1/\sqrt{2}) = -3$ dB. Above the breakpoint frequency the magnitude falls at −20 dB/decade.

By simply interchanging the resistor and capacitor of the low-pass filter, we obtain a high-pass filter as shown in Fig. 10.3. This implies that the output signal is approximately equal in magnitude and phase to the input signal at high frequencies. However, at very low frequencies ($\omega \ll 1/RC$), the magnitude of the output voltage is greatly attenuated and its phase leads the input voltage by approximately 90°. Example 10.2 shows how to calculate the magnitude and phase of the voltage transfer function for the high-pass filter.

➤ EXAMPLE 10.2 High-Pass Filter

Find the magnitude and phase of the voltage transfer function for the high-pass filter shown in Fig. 10.3 and draw the corresponding Bode plot.

SOLUTION

$$\frac{V_{out}}{V_{in}} = \frac{R}{R + 1/j\omega C} = \frac{j\omega RC}{1 + j\omega RC}$$

$$\left|\frac{V_{out}}{V_{in}}\right| = \left[\frac{(\omega RC)^2}{1 + (\omega RC)^2}\right]^{1/2} \qquad \angle\frac{V_{out}}{V_{in}} = \tan^{-1}\left(\frac{1}{\omega RC}\right)$$

The Bode plot is found in Fig. Ex 10.2.

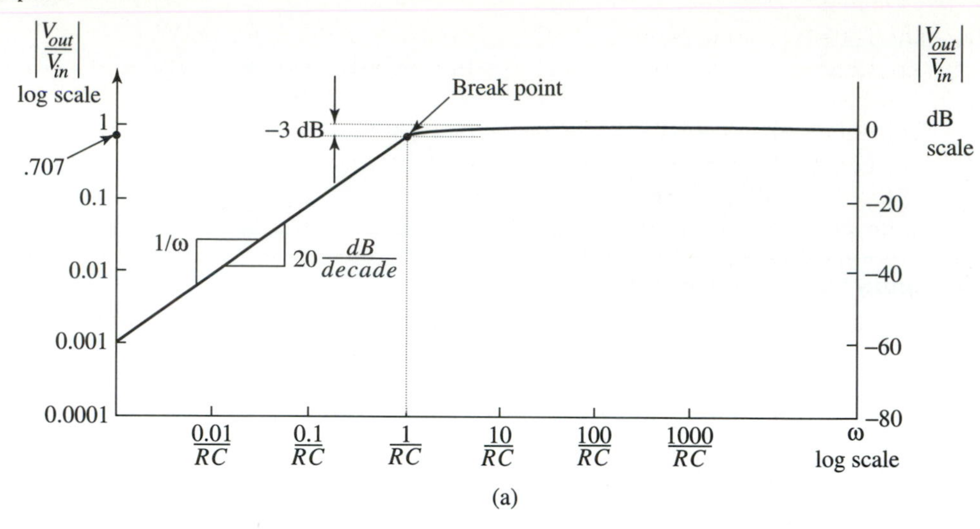

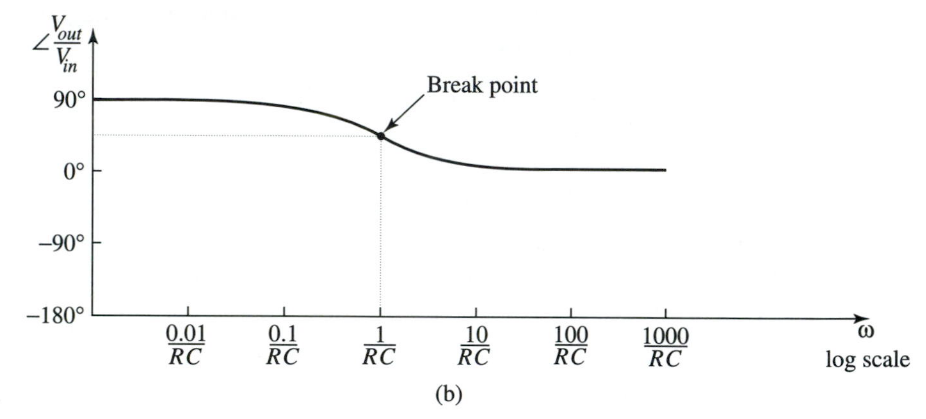

➤ **Figure Ex10.2**

10.1.2 Bode Plot of Arbitrary System Function

As shown in the last section, the Bode plot allows us to see the magnitude and phase of a particular quantity for a given frequency. A Bode plot can be drawn for very complicated transfer functions for many practical situations. In this section, we will outline a number of simple rules that can be used to draw a Bode plot from an arbitrary system function.

A general system function $H(j\omega)$ that may represent a voltage gain, an input impedance, a transconductance, or any other particular quantity is shown in

$$H(j\omega) = \frac{A(j\omega\tau_0)(1+j\omega\tau_2)(1+j\omega\tau_4)\dots(1+j\omega\tau_{2n})}{(1+j\omega\tau_1)(1+j\omega\tau_3)\dots(1+j\omega\tau_{2n-1})} \tag{10.4}$$

In our high-pass filter example, $H(j\omega)$ representing the voltage transfer function is rewritten as

$$H(j\omega) = \frac{j\omega RC}{1+j\omega RC} \tag{10.5}$$

Equation (10.4) can represent Eq. (10.5), when $A = 1$, $\tau_1 = \tau_0 = RC$ and the rest of the time constants $\tau_i = 0$. Our goal is to review plotting Bode plots of this generalized system function.

To begin this task, note that the total number of product terms depend on the quantity being described. The time constants τ_i are the roots of the polynomials in the numerator and denominator and correspond to the breakpoints in the Bode plot. Breakpoints in the numerator are called **zeros**, whereas breakpoints in the denominator are called **poles**. It should be noted that the reciprocals of τ_i are the **breakpoint frequencies**.

To determine the Bode plot from the general system function (Eq. (10.4)), we must assess the effect of each binomial term on the magnitude and phase of the system function. If the frequency is such that $\omega \ll 1/\tau_i$, then the binomial term will have little effect on the magnitude and phase of the system function, as it will simply multiply it by unity. On the other hand, if the frequency is such that $\omega \gg 1/\tau_i$, the system function, magnitude, and phase will be altered. To see its effect, we evaluate the magnitude and phase of $1 + j\omega\tau_i$ for $\omega \gg 1/\tau_i$

$$|1 + j\omega\tau_i| = \sqrt{1 + (\omega\tau_i)^2} \approx \omega\tau_i \tag{10.6}$$

$$\angle(1 + j\omega\tau_i) = \tan^{-1}\omega\tau_i \approx 90° \tag{10.7}$$

For $\omega \gg 1/\tau_i$

- ◆ If the binomial term is in the numerator of the generalized system function, the magnitude will be multiplied by $\omega\tau_i$, and a phase angle of 90° will be added to the total phase.
- ◆ If the binomial term is located in the denominator, the magnitude will be multiplied by $1/\omega\tau_i$ and a phase angle of 90° will be subtracted from the total phase.

When $\omega = 1/\tau_i$, the magnitude and phase are

$$|1 + j\omega\tau_i| = |1 + j| = \sqrt{2} \tag{10.8}$$

$$\angle(1 + j) = 45° \tag{10.9}$$

For $\omega = 1/\tau_i$.

- ◆ If these binomial terms for the breakpoints are located in the numerator, the magnitude of the system function in the numerator is multiplied by $\sqrt{2}$ and a phase of 45° is added to the overall phase.
- ◆ If it is located in the denominator, the magnitude is multiplied by $1/\sqrt{2}$ and a phase of 45° is subtracted from the overall phase of the system function.

Finally, when the numerator of the system function contains the non-binomial term $j\omega\tau_0$, a factor $\omega\tau_0$ will multiply the magnitude and a constant factor of +90° is added to the overall phase angle regardless of frequency.

Given these results, a Bode plot for any linear circuit can be constructed if the system function can be reduced to the form of Eq. (10.4). It is easiest to construct the Bode plot by referring to the following step-by-step procedure.

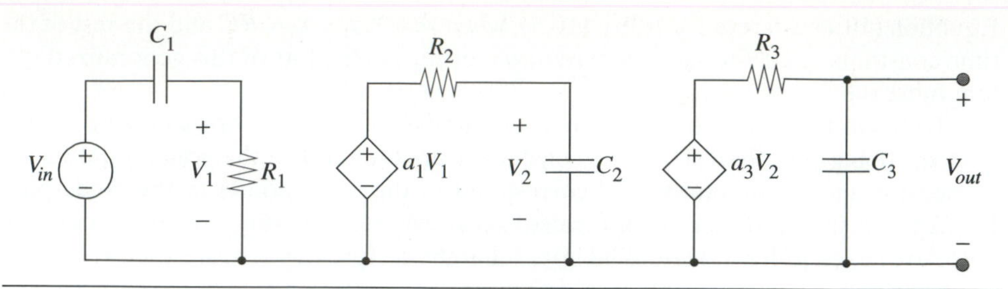

➤ **Figure 10.4** Circuit that contains both high-pass and low-pass filter functions.

- If the term $j\omega\tau_0$ appears in the numerator, the slope is +20 dB at the lowest frequency plotted. It contributes 0 dB to the magnitude at $\omega = 1/\tau_0$ and +90° to the phase.

bode plot construction

- Next find the first breakpoint beginning at the lowest frequency plotted where a binomial term begins to contribute a factor of $\omega\tau_i$ to the magnitude and ±90° to the phase.
- If the binomial term appears in the numerator of the system function, the magnitude slope will be increased by 20 dB/decade, when the frequency is greater than the breakpoint frequency.
- If the binomial term appears in the denominator of the system function, the magnitude of the slope will be reduced by 20 dB/decade when the frequency is greater than the breakpoint frequency.
- To plot the phase, we know that the binomial term will contribute +45° at $\omega = 1/\tau_i$ if it appears in the numerator and −45° if it is in the denominator. We assume that the ±90° phase changes linearly over the interval $0.1/\tau_i < \omega < 10/\tau_i$.

Example 10.3 shows how a Bode plot can be constructed for the transfer function of the circuit shown in Fig. 10.4. This circuit contains both high-pass and low-pass filter functions.

➤ EXAMPLE 10.3 Voltage Transfer Function

Find the voltage transfer function for the circuit in Fig. 10.4. Construct a Bode plot for the circuit showing its magnitude and phase. Label appropriate breakpoints.

The values for the time constants are

$\tau_1 = R_1C_1 = 0.1\text{s}$ $\quad\quad$ $\tau_3 = R_3C_3 = 10^{-7}\text{s}$

$\tau_2 = R_2C_2 = 10^{-4}\text{s}$ $\quad\quad$ $a_1 = a_2 = 1000$

SOLUTION

We can write the transfer function of the circuit by realizing that it is a cascade of three functions. There is a high-pass filter R_1C_1 followed by an amplifier with voltage gain a_1 and a low-pass filter R_2C_2. Finally, there is another amplifier with voltage gain a_2 and a low-pass filter R_3C_3. Solving for the transfer function we find

$$\frac{V_{out}}{V_{in}} = \left(\frac{V_1}{V_{in}}\right)\left(\frac{V_2}{V_1}\right)\left(\frac{V_{out}}{V_2}\right)$$

Substituting each of the individual transfer functions we find

$$\frac{V_{out}}{V_{in}} = \left(\frac{j\omega R_1 C_1}{1 + j\omega R_1 C_1}\right)\left(\frac{a_1}{1 + j\omega R_2 C_2}\right)\left(\frac{a_2}{1 + j\omega R_3 C_3}\right)$$

Rearranging this equation into the form of the general system function leads to

$$\frac{V_{out}}{V_{in}} = a_1 a_2\left(\frac{j\omega\tau_0}{1 + j\omega\tau_1}\right)\left(\frac{1}{1 + j\omega\tau_2}\right)\left(\frac{1}{1 + j\omega\tau_3}\right)$$

First we note that the DC gain $a_1 a_2 = 10^6 = 120$ dB. Next we recognize that a $j\omega\tau_0$ term appears in the transfer function, where $\tau_0 = \tau_1 = R_1 C_1$. This term causes the slope of the magnitude plot to be +20 dB/decade, starting at the lowest frequency plotted. It contributes 0 dB to the magnitude plot at $\omega = 1/\tau_0$. The phase is +90° at the lowest frequency plotted. We also find the first binomial term at $\omega = 1/\tau_1$. Since it appears in the denominator, it reduces the slope of the magnitude plot by 20 dB/decade at $\omega = 1/\tau_1$ and lowers the phase by 90° when $\omega \gg 1/\tau_1$. The next two binomial terms appear in the denominator and also reduce the slope of the magnitude plot by 20 dB/decade at $\omega = 1/\tau_2$ and another 20 dB/decade at $\omega = 1/\tau_3$. The phase is reduced by 90° when $\omega \gg 1/\tau_2$ and an additional 90° when $\omega \gg 1/\tau_3$. Using the rules given above we construct the Bode plot shown in Fig. Ex 10.3.

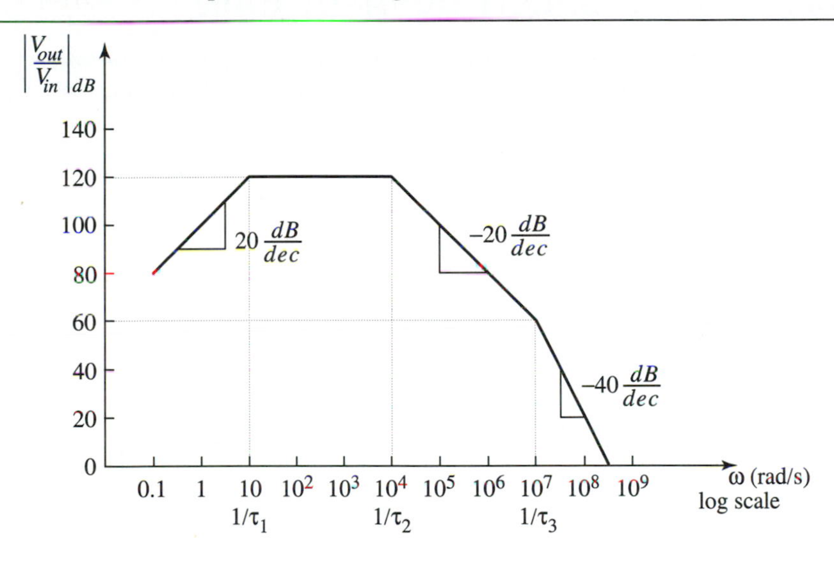

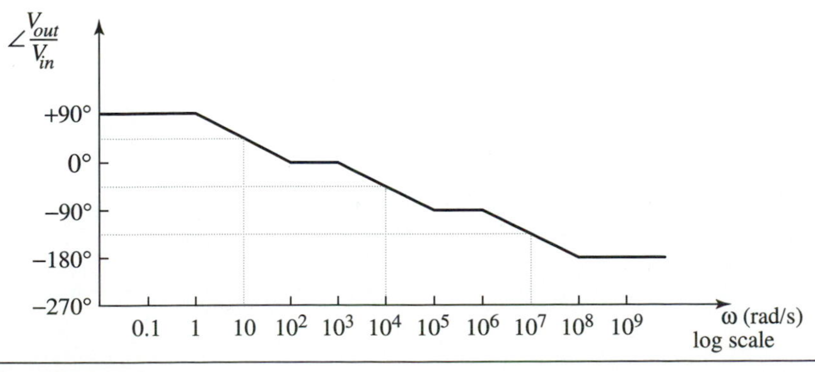

➤ **Figure Ex.10.3**

➤ 10.2 DEVICE MODELS FOR FREQUENCY RESPONSE ANALYSIS

Consider the small-signal models for the bipolar and MOS transistors shown in Fig. 10.5. Because the two models are similar, it is possible to carry out the detailed analysis for one transistor type and then apply the results to the other. In this chapter, we will analyze the frequency response of the bipolar implementation and apply the results to the MOS implementation by substituting C_{gs} for C_π, C_{gd} for C_μ, and letting $r_\pi \rightarrow \infty$.

One should *not* assume that the bipolar transistor is more important than the MOS transistor because of our choice to analyze bipolar implementations first. It is simply a convenience so we do not have to repeat the concepts that apply to both implementations. Furthermore, the difference in the frequency response of bipolar versus MOS implementations is determined by the difference in the capacitance and transconductance values. These values are determined by device parameters and geometry, as well as the current flowing in the transistors. Chapters 4 and 7 covered these dependencies in detail for the MOS and bipolar transistor, respectively.

➤ 10.3 SHORT-CIRCUIT CURRENT GAIN

In this section, we explore the frequency response for a common-emitter/source amplifier, where the output signal is a short-circuit current ($R_L = 0\ \Omega$). We start by inputting an ideal sinusoidal current source to determine the intrinsic frequency response of the active device under consideration.

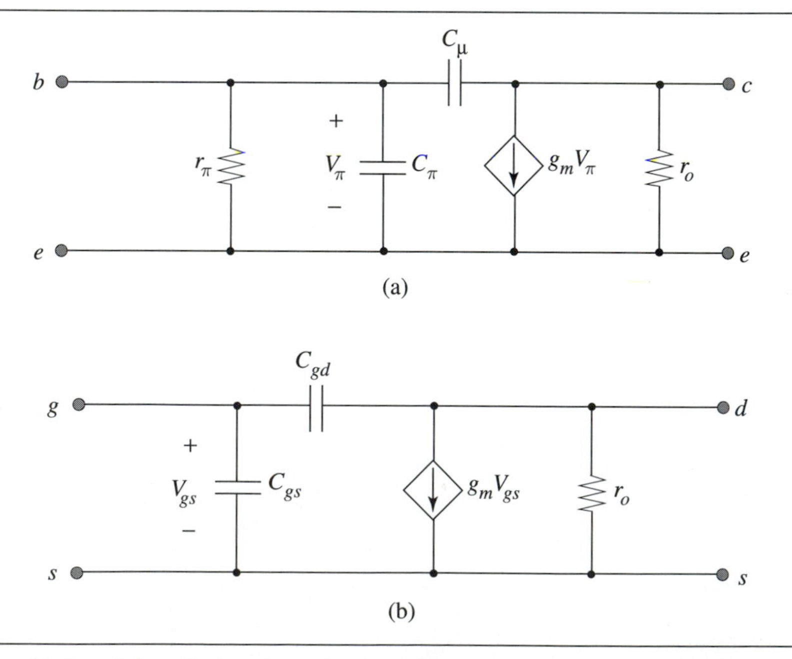

➤ Figure 10.5 a) Small-signal model for (a) bipolar transistor. (b) MOS transistor.

10.3.1 Frequency Response of a Common-Emitter Current Amplifier

Consider a common-emitter amplifier whose input is an ideal sinusoidal current source and output is a short circuit ($R_L = 0\ \Omega$). Assume that the common-emitter amplifier is properly biased in its high-gain region. The resulting small-signal model is shown in Fig. 10.6.

Capacitor C_π, which represents both the base-emitter depletion capacitance as well as the diffusion capacitance from minority carrier charge storage, is connected between the base and emitter of the model. Capacitor C_μ, which represents the depletion capacitance between the collector and the base, is connected between the collector and the base in the small-signal model. The input current source and output current are phasor quantities. In this analysis, we will look at the magnitude of the transfer function representing the current gain as a function of frequency and explore the limits of the quasistatic model.

We begin the analysis by applying KCL to the output node:

$$I_o = g_m V_\pi - V_\pi j\omega C_\mu \tag{10.10}$$

and applying KCL to the input node:

$$I_{in} = \frac{V_\pi}{Z_\pi} + V_\pi j\omega C_\mu \tag{10.11}$$

where $Z_\pi = r_\pi \parallel (1/j\omega C_\pi)$
Solving for V_π in Eq. (10.11) results in

$$V_\pi = \frac{I_{in}}{(1/Z_\pi) + j\omega C_\mu} \tag{10.12}$$

Substituting Eq. (10.12) into Eq. (10.10) results in

$$\frac{I_o}{I_{in}} = \frac{g_m Z_\pi\left(1 - \dfrac{j\omega C_\mu}{g_m}\right)}{1 + j\omega C_\mu Z_\pi} \tag{10.13}$$

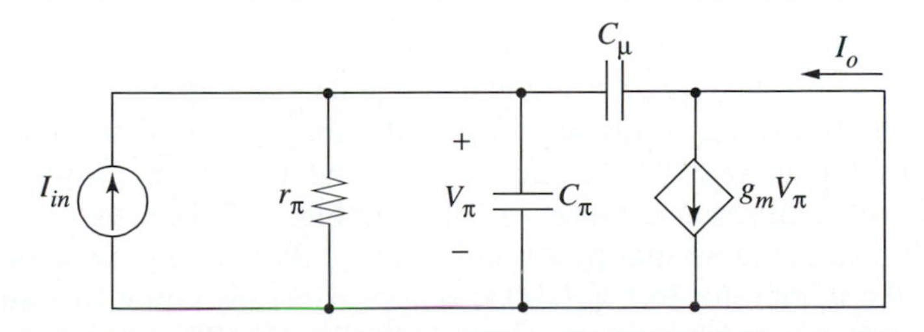

Figure 10.6 Small-signal model for common-emitter amplifier to find the short-circuit current gain frequency response.

By substituting the value for Z_π into Eq. (10.13) and simplifying, we obtain the transfer function for the current gain as

$$\frac{I_o}{I_{in}} = \frac{g_m r_\pi \left(1 - \frac{j\omega C_\mu}{g_m}\right)}{1 + j\omega r_\pi (C_\pi + C_\mu)} \tag{10.14}$$

A Bode plot of Eq. (10.14) is constructed by noting that at $\omega = 0$, the current gain is $g_m r_\pi = \beta_o$. Next, note that there is a pole located at

$$\omega_p = \frac{1}{r_\pi (C_\pi + C_\mu)} \tag{10.15}$$

and a zero located at

$$\omega_z = \frac{g_m}{C_\mu} \tag{10.16}$$

We know that $\omega_z \gg \omega_p$, since $r_\pi \gg 1/g_m$. The Bode plot for the magnitude of the transfer function in Eq. (10.14) is shown in Fig. 10.7(a).

The Bode plot was constructed by noting that the transfer function was equal to β_o at very low frequencies. After we pass the first breakpoint frequency, which is a pole, the slope must decrease by 20 dB per decade. The magnitude will continue to decrease until the zero breakpoint frequency is hit at g_m/C_μ. At this point the zero adds 20 dB per decade to the slope. Since the slope was –20 dB per decade for $\omega < g_m/C_\mu$ the resulting slope for $\omega > g_m/C_\mu$ is zero. The magnitude of the transfer function at frequencies above this point is given by

$$\left|\frac{I_o}{I_{in}}\right|_{\omega \to \infty} = \frac{C_\mu}{C_\pi + C_\mu} \tag{10.17}$$

The Bode plot for the phase is shown in Fig. 10.7(b). The phase is equal to 0° at very low frequencies. At the pole, the phase decreases to –45° and asymptotically approaches –90°. As we increase the frequency, the binomial in the numerator becomes approximately equal to $-j\omega C_\mu/g_m$. The negative sign implies that the zero contributes a negative 90° phase shift. As the frequency increases, the phase asymptotically approaches –180° (–90° from the pole and –90° from the zero).

Although it is theoretically possible to extend the Bode plot to high frequencies, the quasi-static circuit model is not valid at frequencies higher than the point at which the magnitude of the current gain is equal to one. It is extremely important to identify this **transition frequency**, where the magnitude of the current gain is equal to one. Given that $\beta_o > 1$ and $C_\pi > 0$ F the transition frequency ω_T is in between the pole and zero frequency as shown in Fig. 10.7. For reasonably designed transistors, we can assume that $\beta_o \gg 1$ and $C_\pi \gg C_\mu$. By making this assumption, the binomial in the numerator in Eq. (10.14) is approximately equal to 1 and the binomial in the denominator is approximately $j\omega r_\pi (C_\pi + C_\mu)$. This approximation leads to the expression

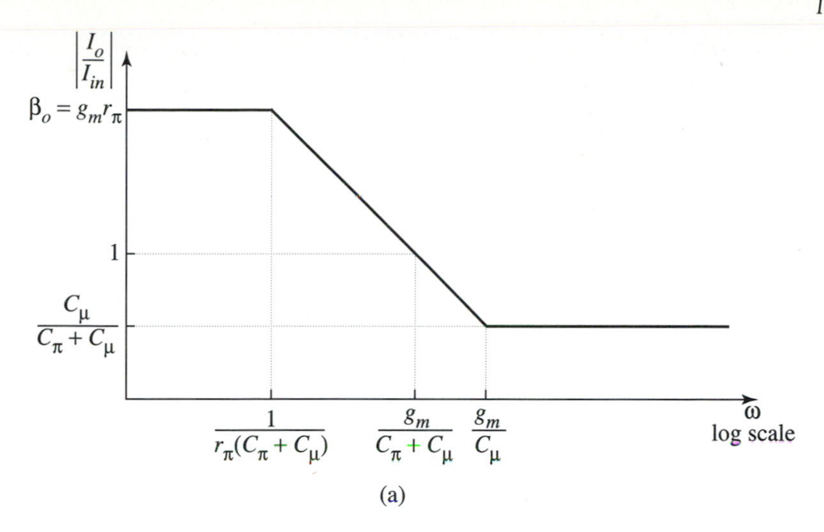

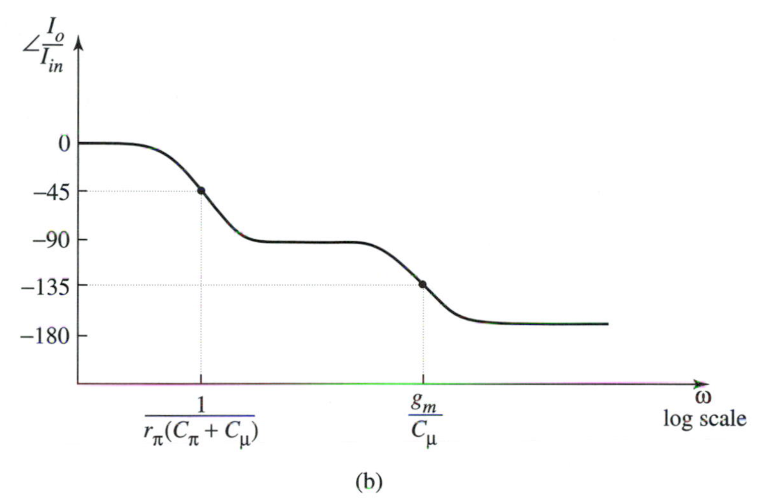

➤ Figure 10.7 Bode plot for the short-circuit current gain of the common-emitter amplifier (a) magnitude; (b) phase.

$$\omega_T = g_m r_\pi \left(\frac{1}{r_\pi (C_\pi + C_\mu)} \right) = \frac{g_m}{C_\pi + C_\mu} \tag{10.18}$$

transition frequency of a CE amplifier where current gain = 1

This very important equation not only quantifies a frequency at which the model for our bipolar transistor breaks down, but also is a figure of merit that can be used to determine the maximum frequency at which circuits containing these bipolar transistors can operate. In practice, signal frequencies must be approximately an order of magnitude lower than the transition frequency in order to obtain reasonable circuit performance.

At this point, it is interesting to examine Eq. (10.18) more closely. The transition frequency is determined by the transconductance and the capacitances C_π and C_μ. Increasing r_π yields an increase in the current gain at DC, but does not affect the transition frequency since the pole due to r_π is reduced by the same amount. This fundamental concept, called **gain-bandwidth product**, is illustrated in Fig. 10.8.

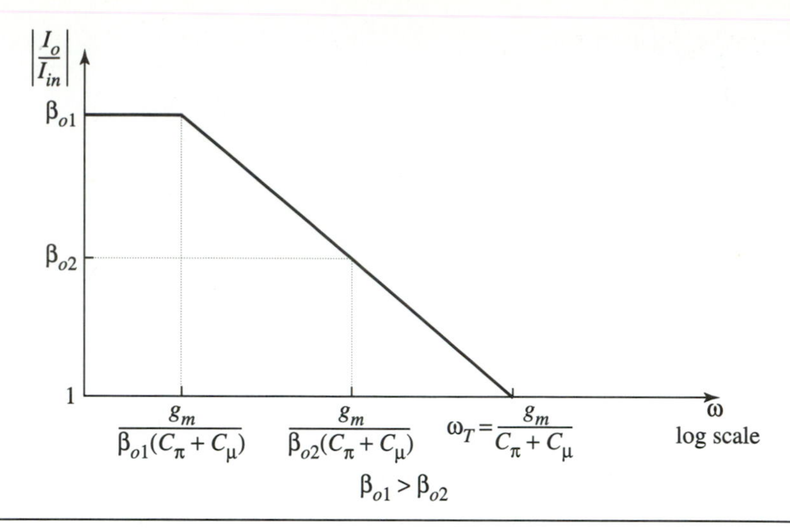

➤ Figure 10.8 Magnitude of the short-circuit current gain demonstrating the principle of gain-bandwidth product.

Now that we know ω_T is a fundamental quantity for the bipolar transistor, we can examine physically how to increase this transition frequency. Recalling the small-signal model for the bipolar transistor presented in Chapter 7, we can write C_π as

$$C_\pi = C_{je} + g_m \tau_F \tag{10.19}$$

where τ_F is the transit time of the minority carriers from the base-emitter junction to the base-collector junction. Note that C_{je} is a capacitance that can be reduced by reducing the emitter area. If C_{je} is made small, C_π will be dominated by $g_m\tau_F$. C_μ can also be reduced by scaling geometries. By increasing g_m or reducing C_{je} and C_μ, the transition frequency can approach its fundamental limit $1/\tau_F$, which is the transit time of carriers across the base.

Finally, let us explore how the transition frequency ω_T varies with the DC collector current flowing in the amplifier. Recalling that the transconductance $g_m = I_C/V_{th}$, we can write

$$\omega_T = \frac{I_C/V_{th}}{(I_C/V_{th})\tau_F + C_{je} + C_\mu} \tag{10.20}$$

The diffusion capacitance component will be small for small collector currents, implying $g_m\tau_F \ll (C_{je} + C_\mu)$. The transition frequency will increase linearly with collector current as shown in Eq. (10.21).

$$\omega_T \approx \frac{I_C}{V_{th}(C_{je} + C_\mu)} \qquad g_m\tau_F \ll (C_{be} + C_\mu) \tag{10.21}$$

However, as collector current increases further, the diffusion capacitance will dominate and the transition frequency will become a constant related to the reciprocal of the transit time of the carriers across the base. At very large collector currents, other physical phenomena beyond the scope of this text cause the ω_T to drop precipitously. A plot of f_T (which is $\omega_T/2\pi$) vs. I_C is shown in Fig. 10.9. The frequency response of the bipolar transistor is limited by a fundamental parameter, the base transit time.

$$\tau_F = \frac{W_B^2}{2D_n} \quad \text{or} \quad \tau_F = \frac{W_B^2}{2D_p} \tag{10.22}$$

In order to reduce this transit time and increase the transition frequency, we use bipolar transistors with small base widths. In addition, npn rather than pnp transistors are preferred since the diffusion coefficient for electrons is larger than the diffusion coefficient for holes, resulting in shorter base transit times.

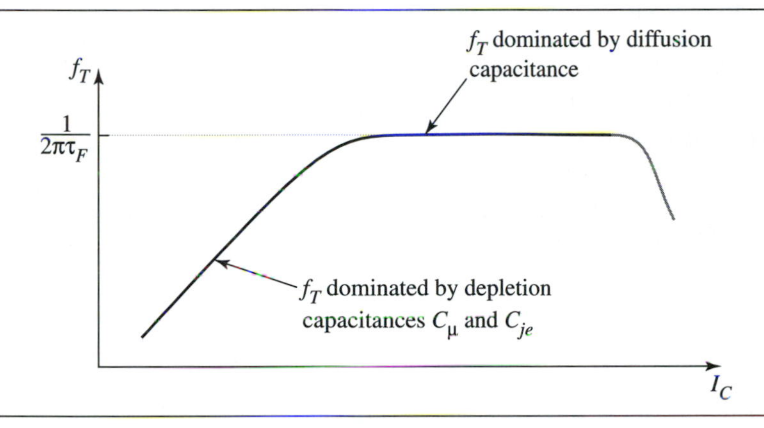

Figure 10.9 Plot of f_T vs. collector current I_C.

EXAMPLE 10.4 Finding f_T vs. I_C

Find the f_T vs. I_C curve using the given typical values for the device parameters. Explore the effect of reducing the base width by a factor of two.

Device Parameters

$D_{nB} = 9\ \text{cm}^2/\text{s}$, $C_{je} = 15\ \text{fF}$ at 0.7 V forward bias, $C_\mu = 9\ \text{fF}$ at 1.8 V reverse bias, $W_B = 300\ \text{nm}$

The circuit to measure f_T has a supply current I_{SUP} that controls the collector current I_C. Assume the bias current I_{BIAS} is chosen to ensure the circuit is operating in the high-gain region. Recall that f_T is defined as the frequency at which the transfer function $|I_{out}/I_{in}| = 1$. In the configuration in Fig. Ex10.4A, $V_{BE} \approx 0.7\ \text{V}$ and $V_{BC} \approx -1.8\ \text{V}$.

a) Sketch a plot of f_T vs. I_C. Label appropriate breakpoints.

b) On the same plot, sketch f_T vs. I_C with the basewidth halved but all other parameters kept the same.

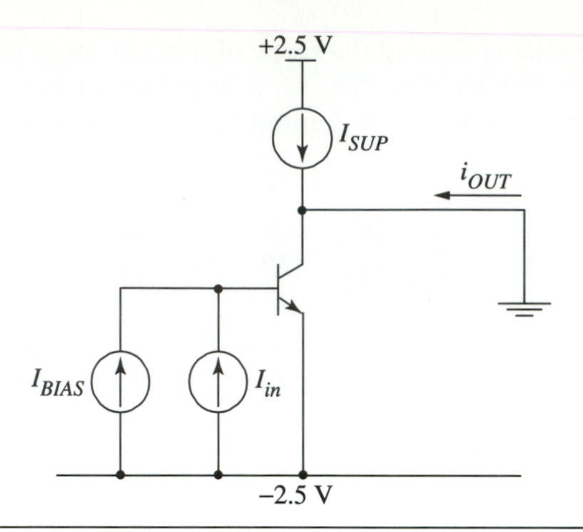

➤ **Figure Ex.10.4A**

SOLUTIONS

a) For high collector currents we define f_{TH} where $g_m \tau_F \gg (C_{je} + C_\mu)$.

$$f_{TH} \approx \frac{1}{2\pi\tau_F} = \frac{2D_{nB}}{2\pi W_B^2} = \frac{2\left(9\ \frac{\text{cm}^2}{\text{sec}}\right)}{2\pi\,(300\text{ nm})^2} = 3.2\text{ GHz}$$

For low collector currents we define (f_{TL}) where $g_m \tau_F \ll (C_{je} + C_\mu)$,

$$f_{TL} \approx \frac{I_C}{2\pi V_{th}\,(C_{je} + C_\mu)}$$

Now we must find the collector current at which the low-current and high-current regions intersect. Setting f_{TH} equal to f_{TL} and solving for I_C, we find

$$I_C = 2\pi V_{th}\,(C_{je} + C_\mu) f_{TH}$$

$$I_C = (6.28)\,(25\text{ mV})\,(15\text{ fF} + 9\text{ fF})\,(3.2\text{ GHz}) = 12\ \mu\text{A}$$

b) Now explore what happens if we halve the base width. Recall $f_T \propto 1/W_B^2$. We define f'_T as the transition frequency with W_B reduced by a factor of two. The new f'_T vs. I_C curve will have the same general shape as the old f_T curve. For high collector current $f'_{TH} \approx f_{TH}/(0.5)^2 = 12.8\text{ GHz}$. At low collector current $f'_{TL} = f_{TL}$ since C_{je} and C_μ are unchanged by reducing the base width (Note, we assumed that the doping was unchanged). The current at which $f'_{TH} = f'_{TL}$ is increased by a factor four to 48 μA. The plot is shown in Fig. Ex 10.4B. For both cases, the difference between the asymptotic plot and the plot using the full equation is largest when $f_{TL} < f_T < f_{TH}$.

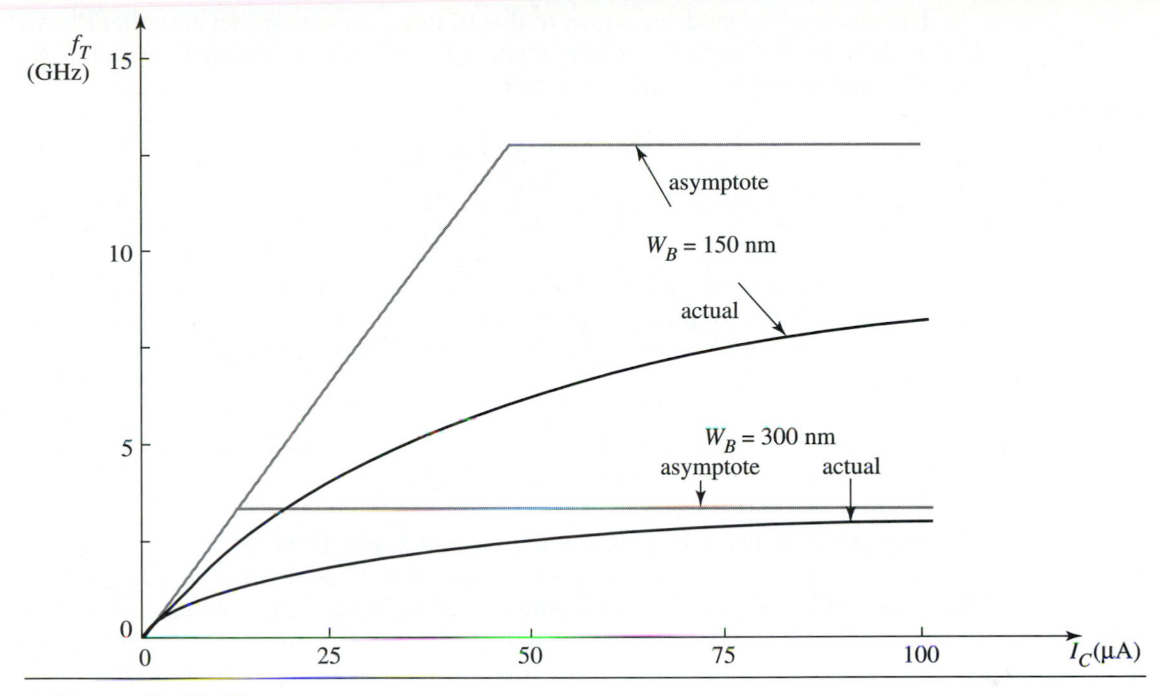

➤ **Figure Ex10.4B**

10.3.2 Frequency Response for a Common-Source Current Amplifier

A similar analysis as presented in the previous section can be conducted for a common- source amplifier. Fig. 10.10 shows a small-signal model for the common-source amplifier, assuming a sinusoidal input current source and measuring the short-circuit output current. We have assumed that the backgate and source are shorted so the backgate transconductance generator is not included in the model.

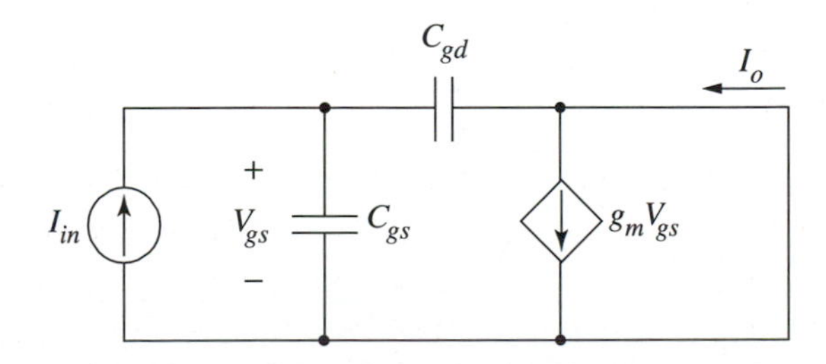

➤ **Figure 10.10** Small-signal model for a common-source amplifier to find the short-circuit current gain frequency response.

This small-signal model is similar to that of the common-emitter amplifier shown in Fig. 10.6. We can apply the results given in Eq. 10.14 by replacing C_π with C_{gs}, C_μ with C_{gd}, and letting $r_\pi \to \infty$. This results in

$$\frac{I_o}{I_{in}} = \frac{g_m\left(1 - \frac{j\omega C_{gd}}{g_m}\right)}{j\omega(C_{gs} + C_{gd})} \tag{10.23}$$

As with the bipolar case, we find that the zero in the transfer function of Eq. (10.23) is at a higher frequency than our transition frequency and, hence, is beyond the validity of the quasi-static model. Neglecting the zero in Eq. (10.23), we find the current gain to be equal to

$$\frac{I_o}{I_{in}} \approx \frac{g_m}{j\omega(C_{gs} + C_{gd})} \tag{10.24}$$

The magnitude of this transfer function is plotted in Fig. 10.11.

Note that the current gain is infinite at DC. This should not be surprising since the MOS transistor has an insulating gate so the DC input current is equal to 0 A. For practical reasons, MOS transistors are not used as current-gain amplifiers at low frequencies since the input signal current source would require an infinite source resistance to drive the MOS transistor. At high frequencies, we find the current gain falls as a function of frequency and has a magnitude equal to 1 at a transition frequency given by

transition frequency of a CS amplifier where current gain = 1

$$\boxed{\omega_T \approx \frac{g_m}{C_{gs} + C_{gd}}} \tag{10.25}$$

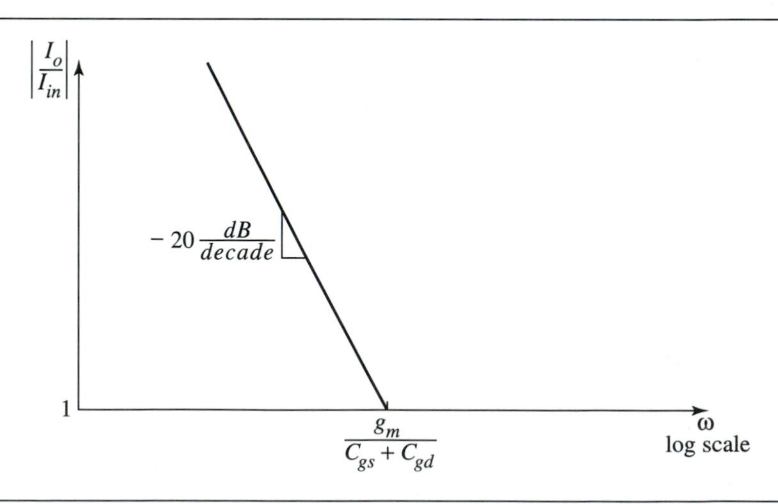

➤ **Figure 10.11** Magnitude of the short-circuit current gain for a common-source amplifier.

As with the bipolar transistor, let us examine the physical limits of this transition frequency. We can write C_{gs} and the transconductance as a function of physical parameters and the DC bias voltage V_{GS} as shown in

$$C_{gs} = \frac{2}{3}WLC_{ox} \tag{10.26}$$

$$g_m = \frac{W}{L}\mu_n C_{ox}(V_{GS} - V_{Tn}) \tag{10.27}$$

For simplicity, we assume that C_{gs} is much greater than C_{gd}. We find that the transition frequency is approximately equal to

$$\omega_T \approx \left(\frac{g_m}{C_{gs}}\right) = \frac{(3/2)\mu_n(V_{GS} - V_{Tn})}{L^2} \tag{10.28}$$

The transit time for the MOS transistor can be written as the reciprocal of the transition frequency ω_T. Since electrons have higher mobility than holes, we use NMOS transistors to reduce this transit time. Most importantly, short channel devices should be used to reduce the transit time. Although Eq. (10.28) shows that the transit time will decrease quadratically with the channel length, another physical mechanism called **velocity saturation** actually reduces this quadratic dependence to a linear dependence for short channel MOS transistors. Regardless, shrinking the channel length will increase the transition frequency of the MOS transistor and circuits that use these devices.

➤ EXAMPLE 10.5 NMOS Transistor f_T

As we determine typical numbers for f_T for an NMOS transistor, we will also observe what the effect of scaling geometries, particularly the channel length, will have on f_T.

a) Given that $C_{ox} = 2.3\ \text{fF}/\mu\text{m}^2$, $C_{ov} = 0.5\ \text{fF}/\mu\text{m}$, $\mu_n = 215\ \text{cm}^2/\text{V-sec}$, $L = 2\ \mu\text{m}$, $V_{Tn} = 1.0$ V, and $V_{GS} = 2.0$ V, find f_T.

b) To see the effect of using a technology with a shorter channel length, find f_T with these new NMOS device parameters: $C_{ox} = 3.0\ \text{fF}/\mu\text{m}^2$, $C_{ov} = 0.5\ \text{fF}/\mu\text{m}$, $\mu_n = 160\ \text{cm}^2/\text{V-sec}$, $L = 1\ \mu\text{m}$, $V_{Tn} = 0.7$ V, and $V_{GS} = 1.5$ V. When the channel length is reduced, typically C_{ox} is increased, μ_n decreases and V_{Tn} is decreased. The applied voltages are also reduced to keep the electric fields approximately constant.

SOLUTIONS

a) From Eqs. (10.25)–(10.27) we can write

$$f_T = \frac{1}{2\pi}\left[\frac{\mu_n C_{ox}(V_{GS} - V_{Tn})/L}{\frac{2}{3}LC_{ox} + C_{ov}}\right] = 1.12\ \text{GHz}$$

Notice that an NMOS transistor has an f_T approximately three times less than a bipolar transistor for equivalent device parameters.

b) Substituting the new device parameters into the formula above, we find that $f_T = 2.45$ GHz. The f_T of this reduced channel length transistor is about two times higher than the transistor in (a). The f_T is not increased by 4× since the mobility of channel electrons is typically lower in the scaled devices, the applied voltages are reduced and C_{ov} does not change substantially with the reduced channel length.

10.4 VOLTAGE GAIN AMPLIFIERS

In this section, we will examine the frequency response of the common-emitter and common-source amplifier including source and load resistances. Since the small-signal model for the MOS transistor is similar to that of the bipolar transistor, we have chosen to analyze more fully the bipolar implementation. This choice is justified since the results for an MOS implementation can be obtained by simply replacing C_π with C_{gs}, C_μ with C_{gd} and letting $r_\pi \rightarrow \infty$. We will obtain the frequency response for these amplifiers using a variety of circuit techniques including the Miller Approximation and the open-circuit time-constant method.

10.4.1 Frequency Response of a Common-Emitter Voltage Amplifier

In the previous section, we studied the frequency response of the intrinsic bipolar transistor with an ideal sinusoidal current source as the input. We also determined that the transition frequency f_T is a fundamental limitation for the bandwidth of amplifiers using bipolar transistors. When a practical voltage source with an associated source resistance and a load resistance are attached to an amplifier, the bandwidth may be lower than f_T.

Consider the common-emitter voltage amplifier shown in Fig. 10.12. We have already determined the voltage gain of this amplifier using the small-signal model

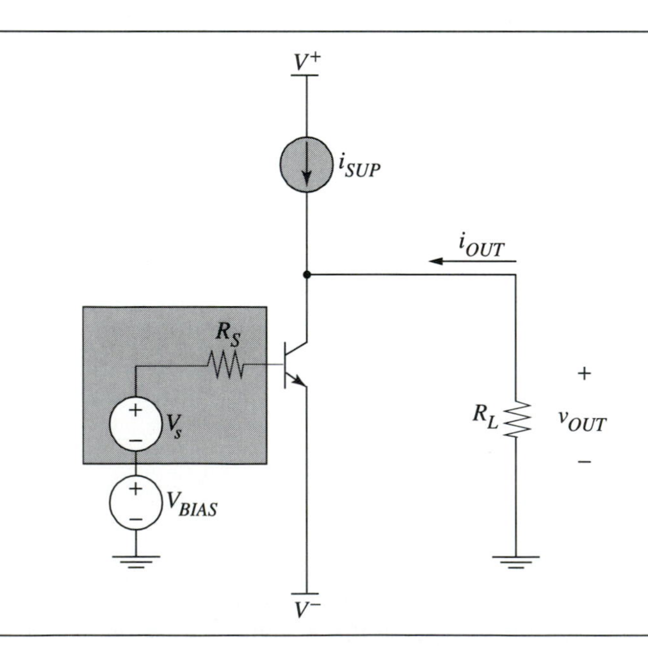

Figure 10.12 Common-emitter voltage amplifier.

in Chapter 8. In this section, we will determine the frequency response of the voltage gain by including the capacitors in the small-signal model. The small-signal model for the bipolar transistor with a sinusoidal voltage source and source resistance R_S driving the transistor, a finite output resistance for the current source supply, and a load resistor R_L is shown in Fig. 10.13.

Capacitor C_π is attached between the base and the emitter, and capacitor C_μ is attached between the base and collector of the bipolar transistor. By considering these capacitors open circuits, we can obtain the voltage gain of this amplifier at DC. As we increase the frequency of the sinusoidal input source, the capacitors affect the dynamics of the amplifier and, hence, the magnitude and phase of the voltage gain. We can solve for the voltage gain of the small-signal circuit shown in Fig. 10.13 by using the impedance transformation for the capacitors and solving KCL and KVL to determine the magnitude and phase of the voltage gain in this amplifier.

10.4.2 Full Analysis of Common-Emitter Voltage Amplifier

To simplify the full analysis of the common-emitter voltage amplifier, we redraw it as shown in Fig. 10.13(b). We have taken the Norton equivalent at the input and combined resistor values at both the input and output to reduce the number of terms carried in the algebra.

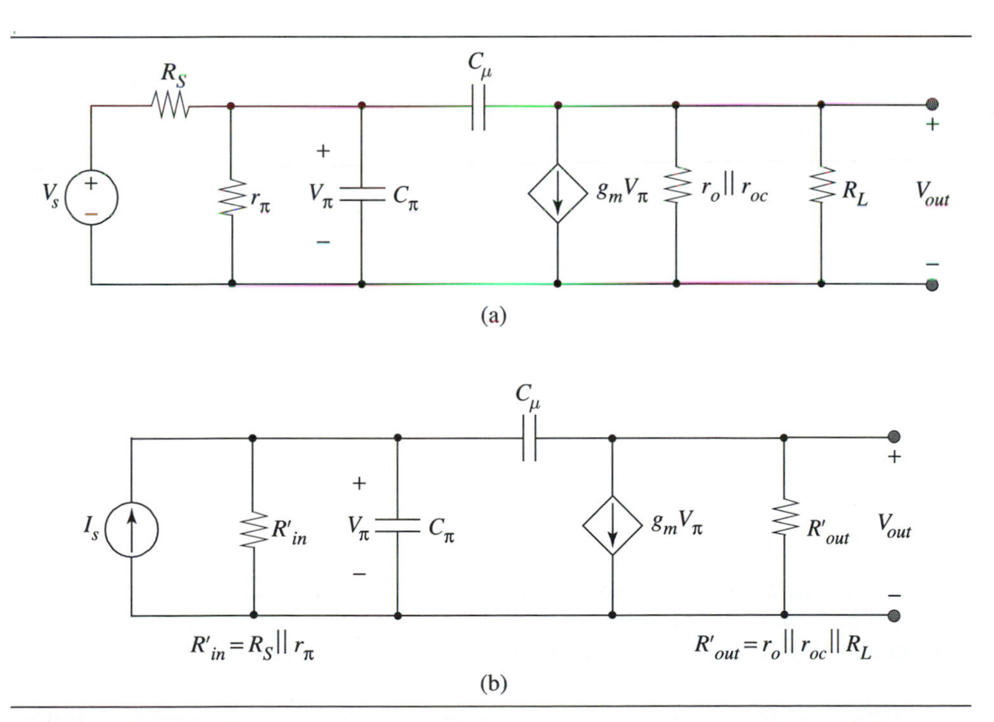

Figure 10.13 Small-signal model for a common-emitter amplifier. (a) Thévenin equivalent input voltage source V_s and source resistance R_S. (b) Norton equivalent input current source I_s.

We begin the analysis by writing a node equation at the left side of C_μ

$$I_s = \frac{V_\pi}{R'_{in}} + V_\pi j\omega C_\pi + (V_\pi - V_{out}) j\omega C_\mu \tag{10.29}$$

We can also write a node equation on the right side of C_μ

$$g_m V_\pi + \frac{V_{out}}{R'_{out}} = (V_\pi - V_{out}) j\omega C_\mu \tag{10.30}$$

Rearranging Eq. (10.30) yields

$$V_\pi = (-V_{out}) \frac{(1/R'_{out}) + j\omega C_\mu}{g_m - j\omega C_\mu} \tag{10.31}$$

Substituting Eq. (10.31) into Eq. (10.29) yields

$$I_s = \left(\frac{1}{R'_{in}} + j\omega C_\pi + j\omega C_\mu\right)\left(-V_{out}\right)\left(\frac{(1/R'_{out}) + j\omega C_\mu}{g_m - j\omega C_\mu}\right) - V_{out} j\omega C_\mu \tag{10.32}$$

Rearranging Eq.(10.32) yields

$$\frac{V_{out}}{I_s} = \frac{-R'_{in} R'_{out}(g_m - j\omega C_\mu)}{1 + j\omega(R'_{out} C_\mu + R'_{in} C_\mu + R'_{in} C_\pi + g_m R'_{out} R'_{in} C_\mu) - \omega^2 R'_{out} R'_{in} C_\mu C_\pi} \tag{10.33}$$

Finally, changing the input current source back to a voltage source by taking the Thévenin equivalent yields

$$\frac{V_{out}}{V_s} = \frac{-g_m R'_{out}\left(\frac{r_\pi}{R_S + r_\pi}\right)\left(1 - \frac{C_\mu}{g_m} j\omega\right)}{1 + j\omega(R'_{out} C_\mu + R'_{in} C_\mu (1 + g_m R'_{out}) + R'_{in} C_\pi) - \omega^2 R'_{out} R'_{in} C_\mu C_\pi} \tag{10.34}$$

Although this analysis is algebraically complex, we can make some observations about the system transfer function of the voltage gain for the common-emitter amplifier. First, we notice at DC, we obtain the same result as setting the capacitors equal to 0 F. In the numerator of the system function we have a zero. The zero implies a frequency at which the voltage gain is equal to 0. This occurs at a frequency given by

$$\omega_z = \frac{g_m}{C_\mu} \tag{10.35}$$

The voltage gain is equal to 0 when all of the current through C_μ is taken up by the transconductance generator and no current is allowed to be delivered to the load. Under this condition the output voltage equals 0 V. Practically speaking the transistor model breaks down before this frequency is attained. Recall that ω_T is given by

$$\omega_T = \frac{g_m}{C_\pi + C_\mu} \tag{10.36}$$

Therefore, ω_z is at a higher frequency than ω_T and the model is not valid.

Let us examine the denominator of the system transfer function of the voltage gain. In general, we can factor the polynomial into binomial terms as shown in

$$\frac{V_{out}}{V_s} = \frac{A_{vo}}{(1 + j\omega\tau_1)(1 + j\omega\tau_2)} = \frac{A_{vo}}{1 + j\omega(\tau_1 + \tau_2) - \omega^2\tau_1\tau_2} \tag{10.37}$$

In this equation we have generalized the DC gain to be equal to A_{vo} and the two time constants as τ_1 and τ_2. A Bode plot of the magnitude of this general second-order system function, assuming $\tau_1 \gg \tau_2$, is shown in Fig. 10.14.

Assuming that time constant τ_1 is larger than the time constant τ_2, the breakpoint frequency ω_1 is at a lower frequency than ω_2. To find τ_1 and τ_2 from our system function shown in Eq.(10.37), we can equate terms that are multiplied by $j\omega$ and $(j\omega)^2$ in Eq.(10.37) and find

$$\tau_1 + \tau_2 = R'_{out}C_\mu + R'_{in}C_\mu(1 + g_m R'_{out}) + R'_{in}C_\pi \tag{10.38}$$

$$\tau_1\tau_2 = R'_{out}R'_{in}C_\mu C_\pi \tag{10.39}$$

Practically speaking, we know that the pole corresponding to τ_2 is at a much higher frequency than τ_1. Therefore, we can make a conservative estimate and assume that τ_2 is much less than τ_1 and Eq.(10.38) can be simplified to

$$\tau_1 \approx R'_{in}[C_\pi + (1 + g_m R'_{out})C_\mu] + R'_{out}C_\mu \tag{10.40}$$

The **3dB frequency**, ω_{3dB}, is the breakpoint frequency at which the magnitude of the voltage gain is reduced by 0.707 and is given in

CE voltage amplifier ω_{3dB} using full analysis

$$\omega_{3dB} = \frac{1}{R'_{in}[C_\pi + (1 + g_m R'_{out})C_\mu] + R'_{out}C_\mu} \tag{10.41}$$

where $R'_{in} = R_S \,||\, r_\pi$ and $R'_{out} = r_o \,||\, r_{oc} \,||\, R_L$.

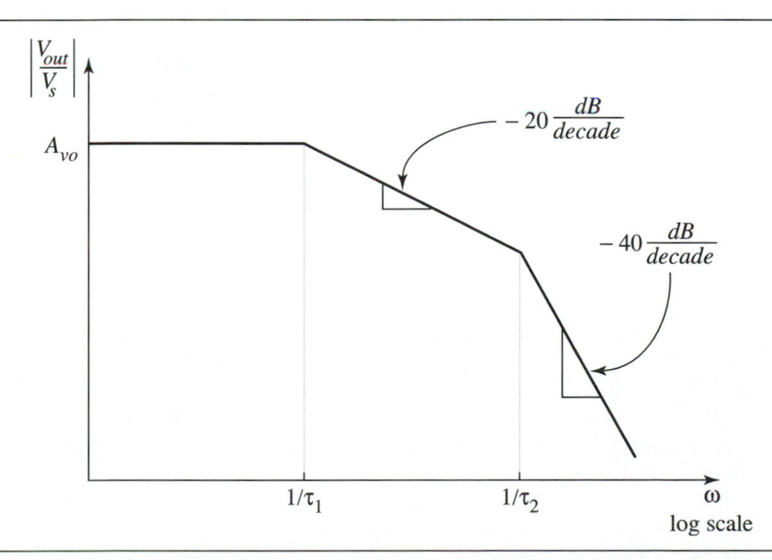

➤ **Figure 10.14** Bode plot of the magnitude of a general 2-pole system assuming $\tau_1 \gg \tau_2$.

10.4.3 The Miller Approximation

Analyzing the small-signal circuit in Fig. 10.13 is complex because of the coupling capacitor C_μ that connects the output to the input. In the following approximation, we will find an effective impedance for C_μ at the input and place it in parallel with C_π.

To begin this analysis, look at the effective impedance at the input of the amplifier caused by C_μ. Using a sinusoidal test voltage source, we measure the current out of the source to determine the effective impedance as shown in Fig. 10.15. We analyze this circuit by calculating the voltage at the output given in

$$V_{out} = -g_m V_t R'_{out} + I_t R'_{out} \quad (10.42)$$

For most practical circuits, the feed-forward current I_t is much less than the current flowing out of the voltage-controlled current source. Mathematically this says

$$I_t \ll |g_m V_t| \quad (10.43)$$

This approximation simplifies Eq. (10.42) to

$$V_{out} \approx -g_m V_t R'_{out} \quad (10.44)$$

We can also write an equation using KVL for the voltage across capacitor C_μ as

$$V_t - V_{out} = \frac{I_t}{j\omega C_\mu} \quad (10.45)$$

By combining Eq.(10.44) and Eq.(10.45), we can solve for the effective impedance as shown in

$$\frac{V_t}{I_t} = Z_{eff} = \frac{1}{j\omega C_\mu (1 + g_m R'_{out})} \quad (10.46)$$

Examining the term in brackets in Eq.(10.46), we see that it is equal to $[1 - A_{vC_\mu}]$, where A_{vC_μ} is the low frequency voltage gain across capacitor C_μ. This is a very

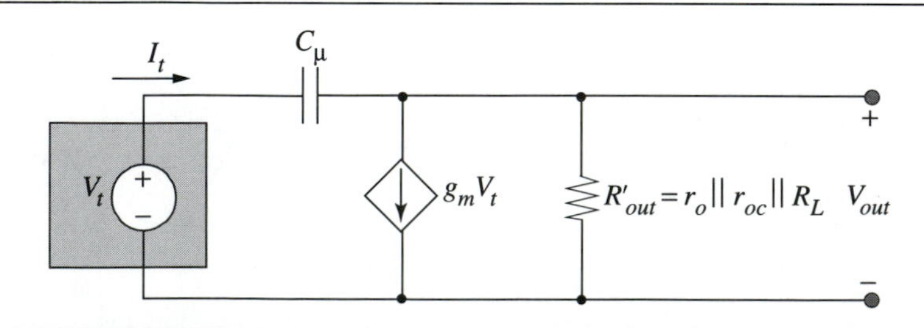

Figure 10.15 Circuit to analyze the effective input impedance caused by capacitor C_μ.

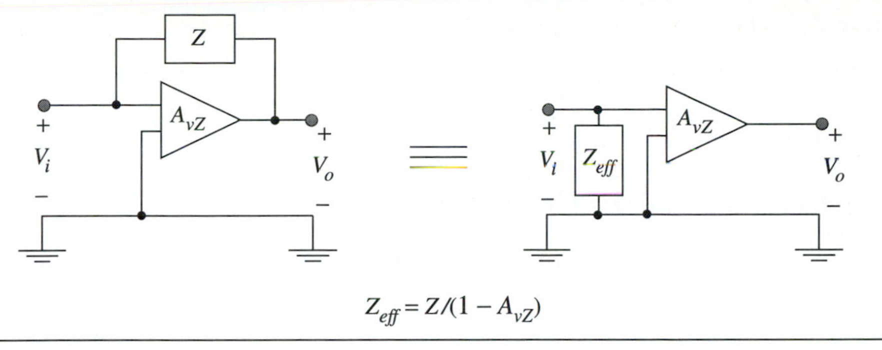

➤ **Figure 10.16** General description of the Miller Approximation.

general result. Any impedance that is connected between the input and output of a voltage amplifier will have its impedance divided by $(1 - A_{vZ})$, as seen between the input and ground. Fig. 10.16 summarizes this general result. This simplification is called the **Miller Approximation**. It is called an approximation since we assumed that the feed-forward current was small. Under most practical situations this is an excellent approximation.

Miller Approximation

Let us physically examine what the Miller Approximation is telling us. Effectively, C_μ is being multiplied by (approximately) the voltage gain of the amplifier. The capacitance has a larger effect because the voltage is being increased at one end of the capacitor. Hence, the charge that needs to be supplied to that capacitor is much larger than had it been simply tied to ground. In the Miller Approximation we take the coupling capacitor and place it between the input and ground and find its effective value. The effective value is much larger. It is increased because of the large voltage swing across that capacitor.

A small-signal model to solve for the voltage-gain system function using the Miller Approximation is shown in Fig. 10.17. Begin analyzing this circuit by looking at the impedance divider at the input and writing

$$V_\pi = \frac{V_s Z_\pi}{R_S + Z_\pi} \tag{10.47}$$

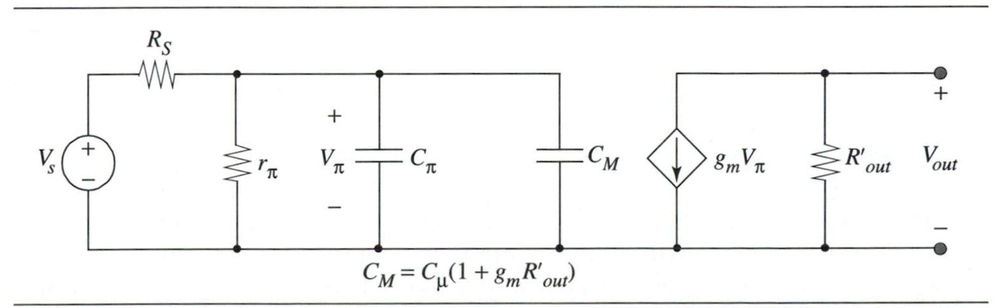

➤ **Figure 10.17** Small-signal model of the common-emitter amplifier using the Miller Approximation.

where Z_π is given by

$$Z_\pi = r_\pi \| \frac{1}{j\omega(C_\pi + C_M)} \tag{10.48}$$

Substituting Eq.(10.48) into Eq.(10.47) results in

$$V_\pi = V_s\left(\frac{r_\pi}{R_S + r_\pi + j\omega(C_\pi + C_M)\, r_\pi R_S}\right) \tag{10.49}$$

We can write an equation for the output voltage as shown in

$$V_{out} = -g_m V_\pi R'_{out} \tag{10.50}$$

By substituting Eq.(10.49) into Eq.(10.50), we find the system transfer function for the voltage gain given in

$$\frac{V_{out}}{V_s} = -g_m\left(\frac{r_\pi}{r_\pi + R_S}\right) R'_{out}\left[\frac{1}{1 + j\omega(C_\mu + C_M)\,(R_S \| r_\pi)}\right] \tag{10.51}$$

We now examine the system function for the voltage gain and draw a Bode plot for the magnitude and phase of this system function. At DC we find the system function to be

$$\frac{v_{out}}{v_s} = -g_m\left(\frac{r_\pi}{r_\pi + R_S}\right) R'_{out} \tag{10.52}$$

This equation gives the same result as if the capacitors were open-circuits. This frequency response analysis has given us one more important term, namely, the frequency at which the magnitude of the voltage gain is reduced by $1/\sqrt{2}$ or 3 dB. This frequency is called ω_{3dB} and is given by

CE voltage amplifier ω_{3dB} using Miller Approximation

$$\boxed{\omega_{3dB} = \frac{1}{(R_S \| r_\pi)\,(C_\pi + C_M)} = \left[\frac{1}{(R_S \| r_\pi)}\right]\left[\frac{1}{C_\pi + (1 + g_m R'_{out})\, C_\mu}\right]} \tag{10.53}$$

For frequencies above this point, the magnitude of the voltage gain decreases a factor of 10 for each factor of 10 the frequency increases or stated another way—the magnitude of the voltage gain will decrease 20 dB/decade. A Bode plot of the magnitude and phase for the common-emitter amplifier using the Miller Approximation is shown in Fig. 10.18.

If we compare the results of the analysis using the Miller Approximation and the full analysis previously performed, we find the full analysis has an additional term $C_\mu R'_{out}$ in its equation for ω_{3dB}. The additional term does not contribute significantly to the value of τ_1. Therefore, the Miller Approximation is an excellent tool for estimating the 3 dB bandwidth of a common-emitter voltage amplifier.

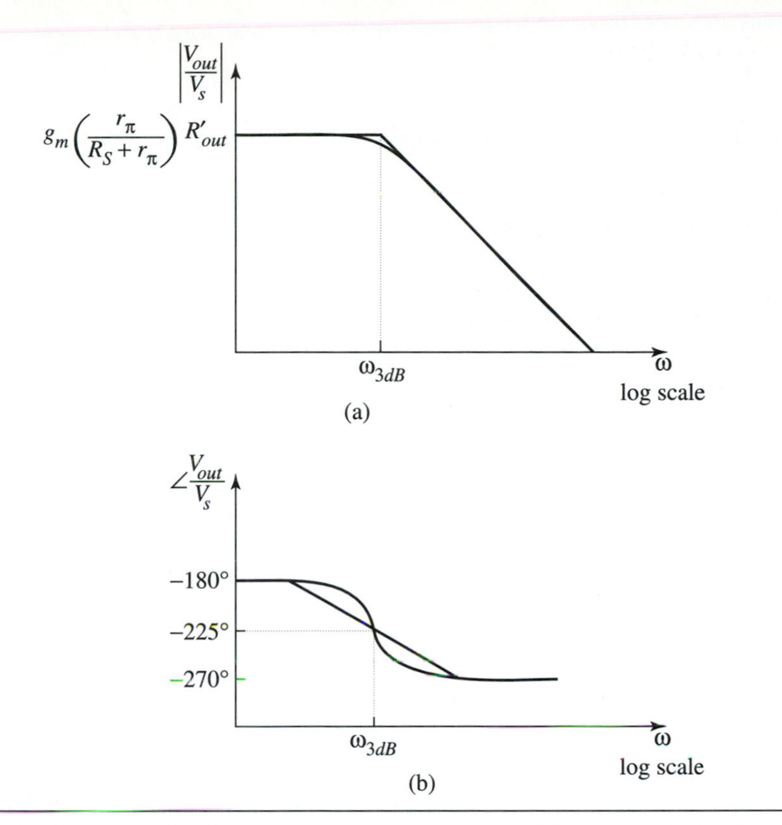

➤ Figure 10.18 Bode plot for the common-emitter amplifier using the Miller Approximation.

➤ EXAMPLE 10.6 Calculating the Common-Emitter Amplifier Frequency Response Using the Miller Approximation

Calculate ω_{3dB} for a common-emitter voltage amplifier first by using the Miller Approximation and then by using the full analysis given

$$I_C = 500\ \mu\text{A} \qquad R_S = 1\ \text{k}\Omega \qquad R_L = 5\ \text{k}\Omega$$

Transistor parameters

$\beta_o = 50$

$r_o, r_{oc} \rightarrow \infty$

$C_\mu = 0.1$ pF

$f_T = 5$ GHz (at $I_C = 500\ \mu$A)

SOLUTION

After determining the transistor small-signal parameters from the data, use Eq. (10.53) to calculate ω_{3dB}.

$$g_m = \frac{qI_C}{kT} = 20\ \text{mS}$$

$$r_\pi = \frac{\beta_o}{g_m} = 2.5\ \text{k}\Omega$$

To determine C_π, recall from Eq.(10.18)

$$f_T = \frac{\omega_T}{2\pi} = \frac{g_m}{2\pi(C_\pi + C_\mu)}$$

$$C_\pi = \frac{g_m}{2\pi f_T} - C_\mu = \frac{20\ \text{mS}}{(6.28)(5\text{GHz})} - (0.1\text{pF}) = 0.537\text{pF}$$

Using these quantities, we find the Miller capacitance C_M is

$$C_M = (1 + g_m R_L)C_\mu = [1 + (20\ \text{mS})(5\ \text{k}\Omega)]0.1\text{pF} = 10.1\text{pF}$$

$$\omega_{3dB} = \frac{1}{(R_S \| r_\pi)(C_\pi + C_M)} = 132\ \text{Mrad/s}$$

Using Eq.(10.41) from the full analysis, we find

$$\omega_{3dB} = \frac{1}{(R_S \| r_\pi)[C_\pi + (1 + g_m R_L)C_\mu] + R_L C_\mu} = 123\ \text{Mrad/s}$$

For this example, the Miller Approximation gave us an estimate of ω_{3dB} within 10% of the full analysis.

10.4.4 Open-Circuit Time-Constant Analysis

The method of open-circuit time constants is a powerful method to determine analytically the bandwidth of an amplifier by finding the location of the lowest frequency pole. The method assumes we have a system transfer function that either has no zeros or that the frequency of the zeros is high enough that they have little or no effect on the transfer function near the lowest frequency pole. In addition, we assume that all the poles in the transfer function are real. These assumptions lead us to the following system transfer function.

$$\frac{V_{out}}{V_s} = \frac{A_{vo}}{1 + b_1 j\omega + b_2 (j\omega)^2 + \ldots + b_n (j\omega)^n} \tag{10.54}$$

It can be shown (see Ref. 2) that the value for b_1 can be found by considering the various capacitors in the network one at a time and setting all other capacitors to an open circuit. The contribution to the total time constant is found by calculating the Thévenin resistance seen by a single capacitor with all other capacitors open-circuited and all independent sources set to zero. This Thévenin resistance R_{Ti} times the capacitance value C_i is the **open-circuit time constant** associated with that capacitor. When all the individual open-circuit time constants are summed together, the result is exactly equal to b_1 in Eq. (10.54)

$$b_1 = \sum_{i=1}^{N} R_{Ti} C_i = \sum_{i=1}^{N} \tau_{C_{io}} \tag{10.55}$$

In principal, we can factor Eq.(10.54) as

$$\frac{V_{out}}{V_s} = \frac{A_{vo}}{(1 + j\omega\tau_1)(1 + j\omega\tau_2)\dots(1 + j\omega\tau_N)} \tag{10.56}$$

By comparing Eq.(10.54) and Eq.(10.56) we see that b_1 is the sum of the time constants representing the pole frequencies. It is important to note that the open-circuit time constants, $\tau_{C_{io}}$ are not the same as the time constants representing the pole frequencies, $\tau_1 \dots \tau_N$. However, their sums are the same.

Our goal is to determine the first breakpoint frequency from the method of open-circuit time constants. If we make the same approximation as we did with the full analysis, namely, that the time constant representing the first pole frequency τ_1 is much larger than $\tau_2 + \tau_3 + \dots\tau_N$, we can write

$$b_1 = \sum_{i=1}^{N} \tau_{C_{io}} \approx \tau_1 \tag{10.57}$$

This analysis is very powerful from a design standpoint since it shows which of the individual open-circuit time constants is contributing most heavily to the lowest frequency pole. To improve the frequency response, we can try to redesign the circuit by lowering the Thévenin resistance or the capacitor value of that dominant time constant.

Consider the common-emitter amplifier as an example to further understand the method of open-circuit time constants. The small-signal model is redrawn in Fig. 10.19. We begin the analysis by setting $C_\mu = 0$ F and the independent sources equal to zero to determine the Thévenin resistance across each of the capacitors in the model. The Thévenin resistance seen by capacitor C_π is $R_S \parallel r_\pi$ and the individual time constant contribution from C_π is

$$\tau_{C_{\pi o}} = R_{T\pi} C_\pi = R'_{in} C_\pi \tag{10.58}$$

where $R'_{in} = (R_S \parallel r_\pi)$.

Next, we determine the individual time constant contribution from capacitor C_μ. To perform this calculation, we open-circuit capacitor C_π and find the Thévenin resistance seen across C_μ by applying a test current source and measuring the test voltage

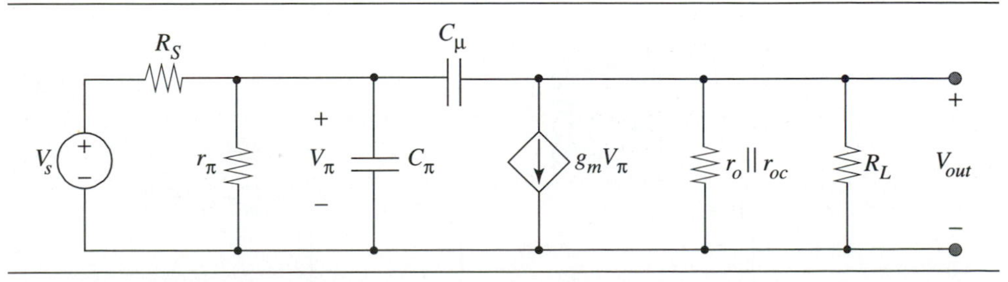

➤ **Figure 10.19.** Common-emitter amplifier small-signal model used for open-circuit time-constant analysis.

as shown in the small-signal model in Fig. 10.20. Writing KCL at the output and input nodes, we find that

$$\frac{v_\pi + v_t}{R'_{out}} + g_m v_\pi = i_t \tag{10.59}$$

$$\frac{v_\pi}{R'_{in}} = -i_t \tag{10.60}$$

Substituting Eq.(10.60) into Eq.(10.59) results in

$$\frac{-i_t R'_{in}}{R'_{out}} + \frac{v_t}{R'_{out}} - g_m i_t R'_{in} = i_t \tag{10.61}$$

Solving for the Thévenin resistance across C_μ we find

$$\frac{v_t}{i_t} = R_{T\mu} = R'_{out} + R'_{in}\ (1 + g_m R'_{out}) \tag{10.62}$$

The individual time constant resulting from C_μ is given in

$$\tau_{C_{\mu o}} = R_{T\mu} C_\mu = [R'_{out} + R'_{in}\ (1 + g_m R'_{out})]\ C_\mu \tag{10.63}$$

Next, we add the individual time constants from Eq.(10.63) and Eq.(10.58), which results in

$$b_1 = R'_{in} C_\pi + R'_{in}\ (1 + g_m R'_{out})\ C_\mu + R'_{out} C_\mu \tag{10.64}$$

Comparing b_1 with the term that multiplied $j\omega$ in the denominator of Eq.(10.34), we find they are exactly the same. This verifies that the method of open-circuit time constants is an exact analysis to determine the factor that multiplies the linear term $j\omega$ in the generalized system function.

Finally, we can estimate the 3 dB breakpoint frequency as

CE voltage amplifier ω_{3dB} using the open-circuit time-constant method

$$\omega_{3dB} \approx \frac{1}{b_1} = \frac{1}{R'_{in}\ C_\pi + R'_{in}\ (1 + g_m R'_{out})\ C_\mu + R'_{out} C_\mu} \tag{10.65}$$

If we compare Eq. (10.65) with the full analysis of the common-emitter voltage amplifier found in Eq. (10.41), we see that the method of open-circuit time constants

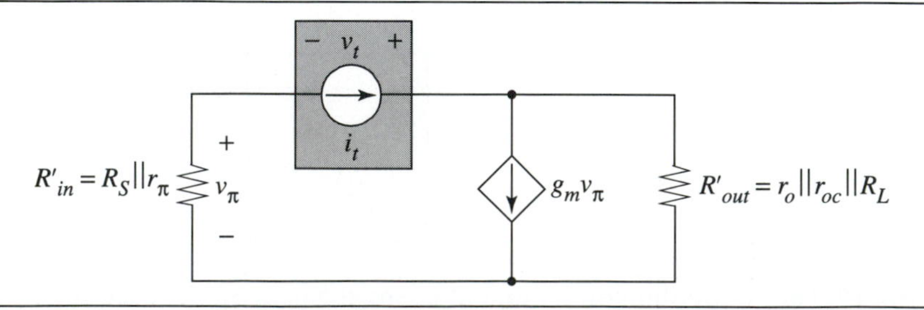

➤ **Figure 10.20** Circuit to determine Thevénin resistance across C_μ.

gives exactly the same result. It is important to note that both the open-circuit time constant method and the full analysis made the approximation that the second pole was at a much higher frequency and could be ignored when determining the first breakpoint frequency ω_{3dB}. Further examination of the terms in the first breakpoint frequency, shows that the dominant term is the Miller-multiplied capacitance of C_μ added to C_π times the input resistance.The additional term of C_μ times the output resistance is usually small compared to the dominant Miller-multiplied term. This fact demonstrates that the Miller Approximation gives a very good estimate of the first breakpoint frequency. In summary, we have analyzed the bandwidth of a common-emitter voltage amplifier using the following three methodologies.

1. The *full analysis* provides an exact answer for the system function. However, we make the approximation that the second pole is at a much higher frequency than the first pole. Therefore, we can assume that the first breakpoint frequency is found by looking only at the term multiplying the linear term $j\omega$ in the denominator of the system function.
2. The method of *open-circuit time constants* also provides an exact answer for the term that multiplies $j\omega$ in the system function. We make the same approximation as the full analysis to determine ω_{3dB}, namely, that the second pole was at a much higher frequency than ω_{3dB}. The method of open-circuit time constants is an excellent design tool since it assists in finding which capacitors and Thévenin resistances are dominating the dynamic performance.
3. The *Miller Approximation* was used to obtain a quick estimate of the 3 dB bandwidth of the common-emitter voltage amplifier. Although it is not an exact calculation, it is very useful for determining an estimate of the bandwidth of the amplifier analytically.

➤ EXAMPLE 10.7 Effect of Parasitic Capacitance C_{cs} on Frequency

In this example, we will explore the effect of a parasitic capacitance between the collector and the substrate on the frequency response of a common-emitter voltage amplifier using the method of open-circuit time constants. The small-signal model with the appropriate device parameters is shown in Fig Ex10.7.

$$R_L \to \infty\,,\ R_S = 25\ \text{k}\Omega\,,\ I_C = 100\ \mu\text{A}\,,\ r_{oc} \to \infty\,,\ C_\pi = 15\ \text{fF} + g_m\tau_F\,,$$

$$\tau_F = 50\ \text{ps}\,,\ C_\mu = 10\ \text{fF}\,,\ V_A = 25\ \text{V},\ \beta_o = 100\,.$$

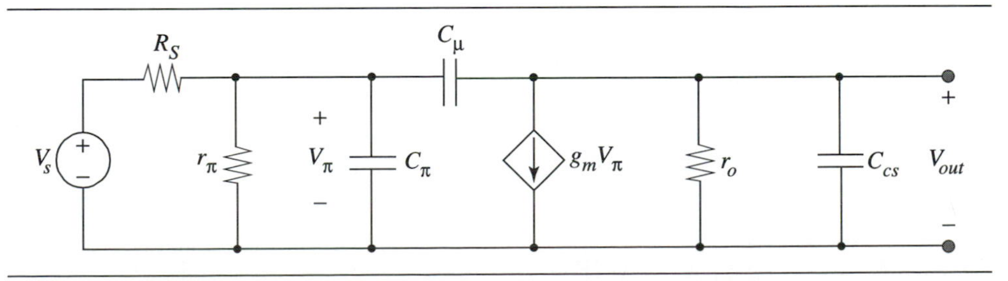

➤ Figure Ex10.7

Use the method of open-circuit time constants to find the 3 dB bandwidth with $C_{cs} = 0$ F. Find the value of C_{cs} which can be tolerated by the circuit and yield only a 10% bandwidth reduction.

SOLUTION

We begin by finding values for g_m, r_π, r_o, and the DC voltage gain.

$$g_m = \frac{qI_C}{kT} = \frac{100\ \mu A}{25\ mV} = 4\ mS\ \text{, and}\ r_\pi = \frac{\beta_o}{g_m} = \frac{100}{4\ mS} = 25\ k\Omega$$

$$r_o = \frac{V_A}{I_C} = \frac{25\ V}{100\ \mu A} = 250\ k\Omega$$

$$A_{vo} = -\frac{r_\pi}{r_\pi + R_S} \cdot g_m r_o = -\frac{1}{2} \cdot (4\ mS)(250\ k\Omega) = -500$$

Now find the value of C_π and its OTC.

$$C_\pi = 15\ fF + g_m \tau_F = 15\ fF + (4\ mS)(50\ ps) = 215\ fF$$

The OTC for C_π is found from Eq.(10.58)

$$\tau_{C_{\pi o}} = (R_S \| r_\pi) C_\pi = (12.5\ k\Omega)(215 fF) = 2.7\ ns$$

The OTC for C_μ is found using Eq.(10.63) with $R'_{out} = r_o$.

$$\tau_{C_{\mu o}} = [r_o + (R_S \| r_\pi)(1 + g_m r_o)] C_\mu = [250\ k\Omega + (12.5\ k\Omega)(1001)] \cdot 10\ fF = 128\ n$$

The sum of the individual OTC's, not including C_{cs}, is $\Sigma\, \tau_{C_{io}} = 131 ns$.

$$\omega_{3dB} = \frac{1}{131 ns} = 7.63\, Mrad/s$$

To determine the value of C_{cs} that can be tolerated and yield only a 10% bandwidth reduction, we first find the OTC for $\tau_{C_{cs}}$. With C_π and C_μ open-circuited and the input voltage signal source set to zero, we find the Thévenin equivalent resistance seen by C_{cs} to be r_o. Therefore,

$$\tau_{C_{cso}} = r_o C_{cs}\ .$$

Reducing the above bandwidth by 10% means we can tolerate ω_{3dB} to be 6.87 Mrad/s.

$$\Sigma\, \tau_{C_{io}} = \frac{1}{6.87\, Mrad/s} = 145\ ns$$

That means $\tau_{C_{cso}} \le 145\ ns - 131\ ns = 14\ ns$. Thus,

$$C_{cs} \le \frac{\tau_{C_{cso}}}{r_o} = \frac{14\ ns}{250\ k\Omega} = 56\ fF$$

A typical collector-substrate capacitance has a value of 0.05 fF/μm^2, and typical dimensions may be 30 μm × 8 μm = 240 μm^2 that results in $C_{cs} \approx 12$ fF. As you can see, C_{cs} is not significant for this example. However, C_{cs} has a significant effect if we are operating at lower currents to save power. For example, for $I_C = 10\ \mu A$,

$g_m = 0.4$ mS, $r_\pi = 250\ k\Omega$, $r_o = 2.5\ M\Omega$,

$$C_\pi = 15\ \text{fF} + g_m \tau_F = 15\ \text{fF} + (0.4\ \text{mS})(50\ \text{pS}) \approx 35\ \text{fF}$$

$$\tau_{C_{\pi o}} = (R_S \| r_\pi) C_\pi = (25\ k\Omega \| 250\ k\Omega) 35\ \text{fF} = 0.80\ \text{ns}$$

$$\tau_{C_{\mu o}} = [r_o + (R_S \| r_\pi)(1 + g_m r_o)] C_\mu = [2.5\ M\Omega + (22.8\ k\Omega)(1001)] \cdot 10\ \text{fF} = 253\ \text{ns}$$

$$\tau_{C_{cso}} = 2.5\ M\Omega \cdot 12\ \text{fF} = 30\ \text{ns}$$

Note that $\tau_{C_{cso}}$ is greater than 10% of the sum of the other time constants. The key point in this analysis is to realize that when considering parasitic capacitances, their effect tends to be most important at *low* bias currents.

10.4.5 Frequency Response of the Common-Source Voltage Amplifier

Because the small-signal models of the common-emitter and common-source amplifier are very similar, all of the analysis techniques shown in the previous section apply

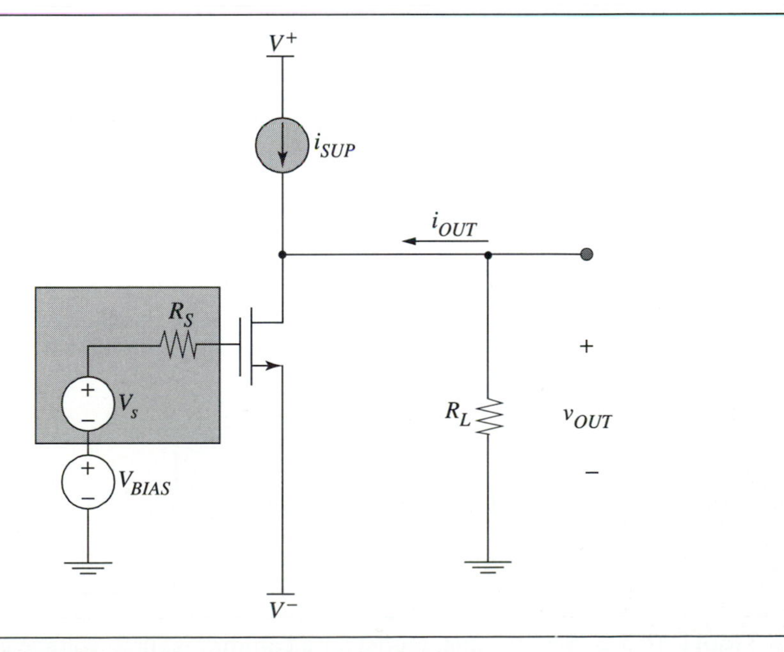

Figure 10.21 Common-source voltage amplifier.

to the MOS common-source voltage amplifier. Let us consider the common-source voltage amplifier shown in Fig. 10.21.

The small-signal model for this circuit including the active device capacitances is shown in Fig. 10.22. Notice that this small-signal model for the common-source voltage amplifier is exactly the same as that shown for the common-emitter voltage amplifier in Fig. 10.19 if we replace C_{gs} for C_π, C_{gd} for C_μ, and let $r_\pi \to \infty$.

At DC, the transfer function between the output voltage and input voltage can be solved as shown in

$$\frac{v_{out}}{v_s} = -g_m R'_{out} \tag{10.66}$$

where $R'_{out} = r_o \parallel r_{oc} \parallel R_L$

We can use the method of open-circuit time constants to determine the 3 dB frequency as with the bipolar case, and determine $\tau_{C_{gso}}$. We open-circuit capacitor C_{gd} and set the independent input voltage source V_s to 0 V. τ_{Cgso} is given by

$$\tau_{C_{gso}} = R_S C_{gs} \tag{10.67}$$

Next, open-circuit C_{gs} and determine the Thévenin resistance across C_{gd}. A small-signal model corresponding to this configuration is shown in Fig. 10.23. The Thévenin resistance can be found by applying a test current source and measuring the voltage across it. It is given by

$$R_{Tgd} = R'_{out} + R_S(1 + g_m R'_{out}) \tag{10.68}$$

By summing the individual time constants we find

$$\Sigma\tau_{C_{io}} = \tau_{C_{gso}} + \tau_{C_{gdo}} = R'_{out} C_{gd} + R_S[C_{gs} + (1 + g_m R'_{out}) C_{gd}] \tag{10.69}$$

where the term involving R_s is exactly the same time constant that would be obtained by using the Miller Approximation. The reciprocal of the sum of the open-circuit time constants yields an approximation to the first breakpoint frequency, and is an excellent analytical approximation of the 3 dB bandwidth of the common-source voltage amplifier. It is given by

CS voltage amplifier ω_{3dB} using the open-circuit time-constant method

$$\boxed{\omega_{3dB} \approx \frac{1}{\Sigma\tau_{C_{io}}} = \frac{1}{R_S C_{gs} + R_S(1 + g_m R'_{out}) C_{gd} + R'_{out} C_{gd}}} \tag{10.70}$$

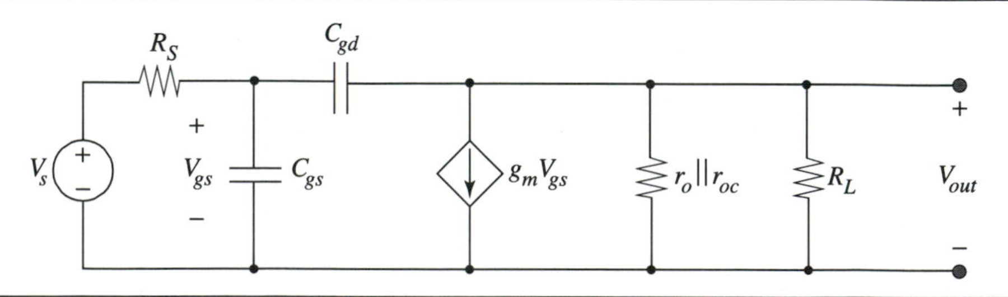

➤ **Figure 10.22** Small-signal model for a common-source voltage amplifier.

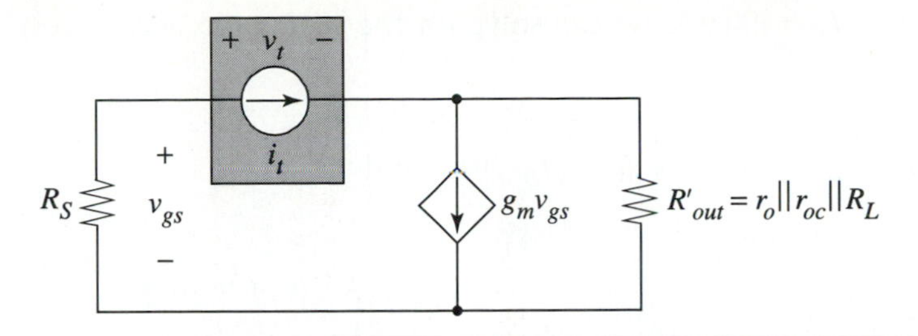

Figure 10.23 Circuit to determine Thévenin resistance across C_{gd}.

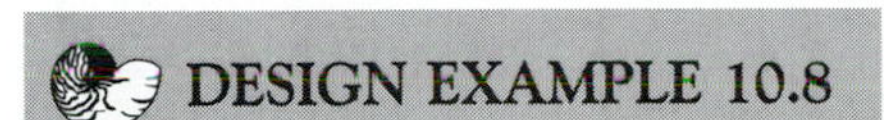

DESIGN EXAMPLE 10.8 Common-Source Voltage Amplifier

Design a common-source voltage amplifier for which the input source has a source resistance of 10 kΩ. The common-source voltage amplifier in Fig. Ex. 10.8 is driving a purely capacitive load of 1pF that includes all parasitics. We need a DC voltage gain of –50 and bandwidth of 1 MHz. Try to minimize the supply current. Verify your design in SPICE.

SOLUTION

Begin by picking a typical g_m of 1mS and see how difficult it will be to make the gain and bandwidth specifications. Because we need a DC gain of 50, we require that

$$r_o = 50/g_m = 50\ \text{k}\Omega$$

Using a 2 μm device, we can find I_D.

$$I_D = \frac{1}{\lambda_n r_o} = 400\ \mu\text{A}$$

$\mu_n C_{ox} = 50\ \mu\text{A/V}^2$

$V_{Tn} = 1$ V

$L_{min} = 2$ μm

$\lambda_n = 0.05\ \text{V}^{-1}$ @ $L = 2$ μm

$C_{ox} = 1\ \text{fF}/\mu\text{m}^2$

$C_{ov} = 0.5\ \text{fF}/\mu\text{m}$

$r_{oc} \rightarrow \infty$

5 V

i_{SUP}

i_{OUT}

$R_S = 10$ kΩ

V_s

V_{BIAS}

1 pF C_L v_{OUT}

Figure Ex10.8

Given that $I_D = 400\ \mu A$, we can solve for the W/L ratio needed to yield a g_m of 1mS.

Rearranging $g_m = \sqrt{2\frac{W}{L}\mu_n C_{ox} I_D}$, we see that

$$\frac{W}{L} = \frac{g_m^2}{2\mu_n C_{ox} I_D} = \frac{(1\text{mS})^2}{2\left(50\frac{\mu A}{V^2}\right)400\ \mu A} = 25$$. So $W/L = 50/2$.

Using the method of open-circuit time constants, we find the time constants due to C_{gs}, C_{gd}, and C_L independently.

$$\tau_{C_{gso}} = R_S C_{gs} = R_S \frac{2}{3} W L C_{ox} = 10\ k\Omega\,(0.67)\,(50 \cdot 2)\,(1.0\ fF) = 0.67\ ns$$

$$\tau_{C_{gdo}} = [r_o + R_S(1 + g_m r_o)]C_{gd} = [50\ k\Omega + 10\ k\Omega\,(51)]\,(0.5\ fF/\mu m \cdot 50\ \mu m) = 14\ ns$$

$$\tau_{C_{Lo}} = r_o C_L = 50\ k\Omega \cdot 1\ pF = 50\ ns$$

$$\Sigma\,\tau_{C_{io}} \approx 65\ ns$$

$$f_{3dB} = \frac{\omega_{3dB}}{2\pi} = \frac{1}{2\pi \cdot \Sigma\tau_{C_{io}}} = \frac{1}{2\pi \cdot 65\ ns} = 2.45\ MHz$$

We see this design has a 3 dB bandwidth that is dominated by C_L. Fortunately, we satisfy the specification required with the value of current we are using. Before beginning to minimize the current, we check the value of V_{BIAS} and simulate this "first cut" circuit in SPICE.

$$V_{BIAS} = V_{Tn} + \sqrt{\frac{I_D}{(W/2L)\mu_n C_{ox}}} = 1.0\ V + 0.8\ V = 1.8\ V$$

SPICE yields $A_{vo} = -60$, $f_{3dB} = 2.2$ MHz using $V_{BIAS} = 1.75$ V.

At this point, we can lower I_D until a bandwidth of 1 MHz is achieved. By cutting the current in half, the new output resistance, r_o' and voltage gain, A'_{vo} are given by

$$r_o' = 2r_o = 100\ k\Omega \text{ and } A'_{vo} = \frac{-g_m}{\sqrt{2}} \cdot 2r_o = -\sqrt{2}g_m r_o \approx -70.$$

$\tau_{C_{gso}}$ is still 0.66 ns, but the new time constant due to C_{gd} is given by

$$\tau'_{C_{gdo}} = [100\ k\Omega + 10\ k\Omega\,(70)]\,(50\ \mu m \cdot 0.5\ fF/\mu m) = 20\ ns$$

$$\tau'_{C_{Lo}} = r_o' C_L = 100\ k\Omega \cdot 1\ pF = 100\ ns$$

$$\Sigma \tau'_{C_{io}} = 120 \text{ ns} \text{ , thus } f'_{3dB} = \frac{\omega'_{3dB}}{2\pi} = \frac{1}{2\pi(120 \text{ ns})} = 1.33 \text{ MHz}$$

Finally, we approximate V_{BIAS} and run SPICE. For $I_D = 200\ \mu\text{A}$ we get

$$V_{BIAS} = V_{Tn} + \sqrt{\frac{I_D}{(W/2L)\,\mu_n C_{ox}}} = 1.0 \text{ V} + 0.57 \text{ V} = 1.57 \text{ V}$$

SPICE yields $A_{vo} = -85$, $f_{3dB} = 1.15$ MHz using $V_{BIAS} = 1.53$ V.
More iterations can be performed to refine this design.

10.5 FREQUENCY RESPONSE OF COMMON-COLLECTOR/DRAIN VOLTAGE BUFFER

In this section, we use the Miller Approximation and open-circuit time-constant method to analyze the frequency response of a voltage buffer composed of either a common-collector or common-drain amplifier. We use the following methodology to attack this problem.

1. Develop a small-signal model for the circuit including the active device capacitances.
2. Open-circuit all of these capacitances to calculate the small-signal DC transfer function.
3. Use either the Miller Approximation or the open-circuit time-constant method to estimate the bandwidth.

If a more detailed analysis is required to find higher order poles or zeros in the transfer function, we must perform a full phasor analysis. In most cases, the approximate solutions are adequate to determine the effect of the various circuit components on the frequency response. However, the analysis should be followed by computer simulation to find the exact frequency response of the circuit being analyzed.

10.5.1 Common-Collector Frequency Response

A common-collector amplifier, including both source and load resistances, is shown in Fig. 10.24. A DC bias voltage source in series with a small-signal sinusoidal source and its associated source resistance is the input to the common-collector amplifier. The output voltage is taken across the load resistor R_L. Our task is to find the low-frequency voltage gain and the bandwidth of this circuit. To accomplish this task, we use the small-signal model of the common-collector amplifier shown in Fig. 10.25.

We have placed capacitor C_μ between the base and collector and capacitor C_π between the base and the emitter. Note for the common-collector amplifier that capacitor C_π is feeding back the output to the input as opposed to C_μ for the common-emitter amplifier. As usual for a voltage output, the Thévenin equivalent for the output port is used.

To determine the low-frequency voltage gain, open-circuit both capacitors and write

$$\frac{v_{out}}{v_s} = \left(\frac{R_{in}}{R_S + R_{in}}\right)(1)\left(\frac{R_L}{R_L + R_{out}}\right) \tag{10.71}$$

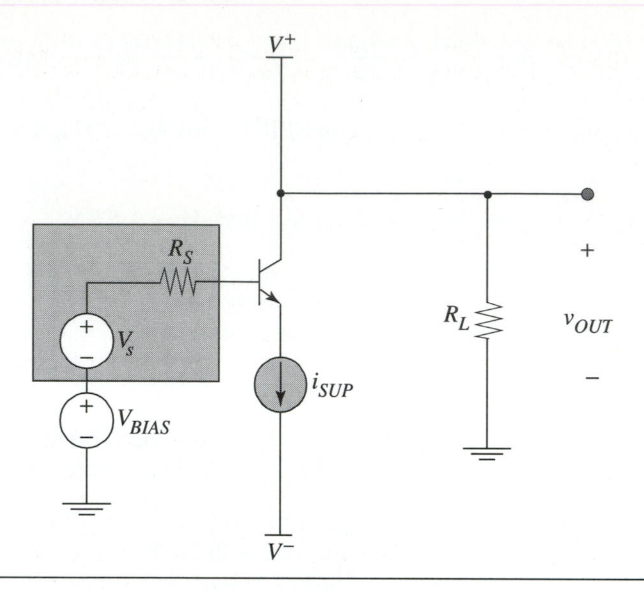

Figure 10.24 Common-collector amplifier.

By substituting the values for the input and output resistances of the common-collector amplifier determined in Chapter 8, we find

$$\frac{v_{out}}{v_s} = \left(\frac{r_\pi + \beta_o R_L}{R_S + r_\pi + \beta_o R_L}\right)(1)\left(\frac{R_L}{R_L + (1/g_m) + (R_S/\beta_o)}\right) \tag{10.72}$$

We can determine the frequency response of the common-collector amplifier by using the Miller Approximation. To find the effective value of C_π when it is placed in parallel with C_μ, we determine the voltage gain across C_π or between the base and emitter of the transistor. Assuming that the intrinsic voltage gain of the amplifier is unity, the voltage gain across C_π, (A_{vC_π}) is given by the resistor series divider of $1/g_m$ and R_L as

$$A_{vC_\pi} = \frac{R_L}{1/g_m + R_L} \tag{10.73}$$

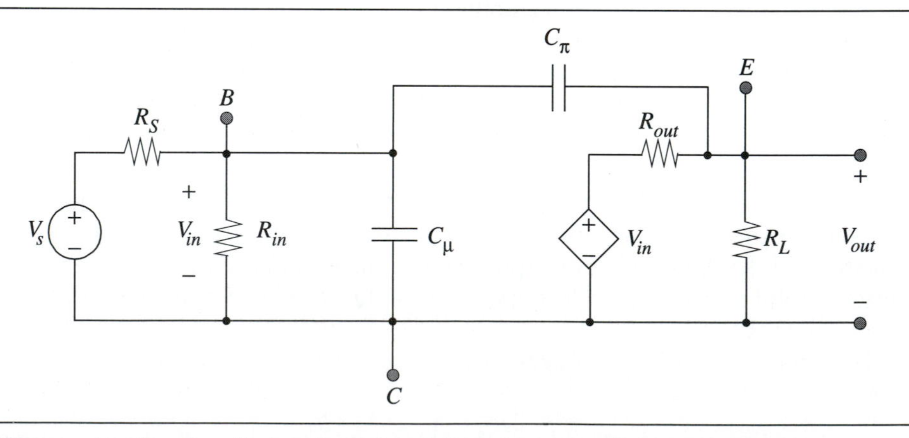

Figure 10.25 Small-signal model of the common-collector amplifier to determine its frequency response. We have added capacitance C_μ and C_π to the two-port model developed in Chapter 8.

Note that A_{vC_π} does not include the term R_S/β_o, since we are only interested in the voltage gain across C_π.The Miller Approximation says that C_π has an effective capacitance across the input when it is multiplied by $(1 - A_{vC_\pi})$. This value is given by

$$(1 - A_{vC_\pi}) = \frac{1/g_m}{1/g_m + R_L} \tag{10.74}$$

The total capacitance across the input C_T is given by

$$C_T = C_\pi\left(\frac{1/g_m}{(1/g_m) + R_L}\right) + C_\mu \tag{10.75}$$

If $1/g_m$ is much less than the load resistance R_L, then the effect of C_π is very small since the voltage gain across it is approximately equal to unity. This result implies that as the voltage on one side of capacitor C_π increases, the voltage on the other side of the capacitor increases by the same amount and the net charge across the capacitor is unchanged. Therefore, C_π has little effect on the frequency response.

To complete the calculation of the bandwidth of the common-collector amplifier, we determine the Thévenin resistance seen by capacitor C_T as

$$R_T = R_S \| R_{in} \tag{10.76}$$

The time constant τ_{C_T} is simply the Thévenin resistance times C_T which is equal to

$$\tau_{C_T} = (R_S \| R_{in})\left[\frac{C_\pi}{1 + g_m R_L} + C_\mu\right] \tag{10.77}$$

Finally, the 3 dB bandwidth is simply the reciprocal of the time constant. By substituting the value for the input resistance in Chapter 8 we find

$$\omega_{3dB} = \frac{1}{\left[(R_S \| (r_\pi + \beta_o R_L))\left(\frac{C_\pi}{1 + g_m R_L} + C_\mu\right)\right]} \tag{10.78}$$

We can also analyze the common-collector amplifier by using the open-circuit time-constant method. To perform this calculation, open-circuit all capacitors and look at the Thévenin resistance at the location of C_μ as shown in Fig. 10.26(a). Note that the input source V_s is set to zero for this calculation. The Thévenin resistance is

$$R_{T\mu} = R_{in} \| R_S \tag{10.79}$$

To find the Thévenin resistance seen by capacitor C_π, we insert a test current source and measure the voltage across that current source in the same manner used previously to calculate input and output resistances. The small-signal circuit model for this calculation is given in Fig. 10.26(b).

To begin the calculation, write KVL across the voltage source as shown in

$$v_t = i_t(R_S \| R_{in}) - v' \tag{10.80}$$

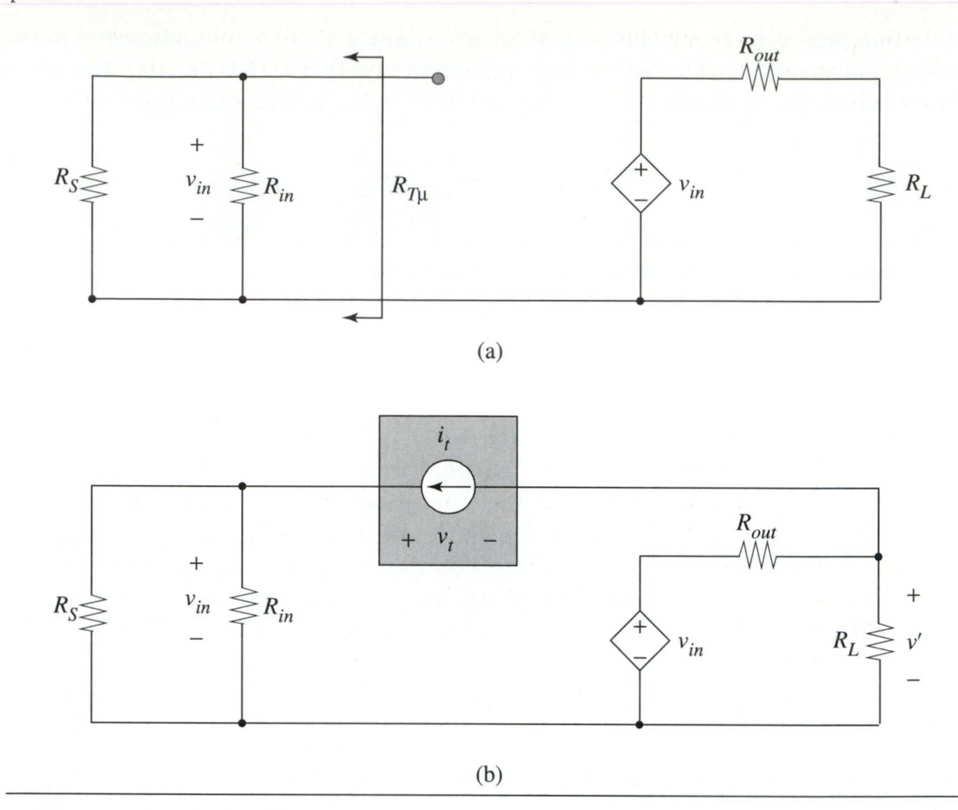

➤ **Figure 10.26** Small-signal circuits to determine open circuit time constants of common-collector amplifier: (a) circuit for $R_{T\mu}$; (b) circuit for $R_{T\pi}$

KCL at the node on the right side of the test current source states that

$$\frac{v'}{R_L} = -i_t + \frac{i_t(R_S \| R_{in}) - v'}{R_{out}} \tag{10.81}$$

Rearranging this equation we get

$$v'\left(\frac{1}{R_L} + \frac{1}{R_{out}}\right) = i_t\left[\frac{R_S \| R_{in}}{R_{out}} - 1\right] \tag{10.82}$$

Substituting Eq.(10.82) into Eq.(10.80) we get

$$v_t = i_t\left[(R_S \| R_{in}) + (R_{out} \| R_L) - \frac{R_L}{R_{out} + R_L}(R_S \| R_{in})\right] \tag{10.83}$$

Solving for the Thévenin resistance seen by capacitor C_π we get

$$R_{T\pi} = (R_S \| R_{in})\left(1 - \frac{R_L}{R_{out} + R_L}\right) + R_{out} \| R_L \tag{10.84}$$

The open-circuit time constant is the sum of the Thévenin resistance seen by C_μ and the Thévenin resistance seen by C_π, multiplied by their respective capacitances as given in

$$\sum \tau_{C_{io}} = R_{T\mu} C_\mu + R_{T\pi} C_\pi \tag{10.85}$$

Substituting Eq.(10.79) and Eq.(10.84) into Eq.(10.85) results in the value of the open-circuit time constant given by

$$\sum \tau_{C_{io}} = (R_S \| R_{in}) \left[C_\mu + C_\pi \left[\left(1 - \frac{R_L}{R_{out} + R_L} \right) + \frac{R_{out} \| R_L}{R_S \| R_{in}} \right] \right] \tag{10.86}$$

Note that the first two terms of Eq. (10.86) are exactly that found by the Miller Approximation. The additional term is due to the effect of C_π and the resistances at the output of the circuit, namely R_{out} and R_L. Assuming R_{out} is much less than R_L, this term is at a higher frequency than ω_T and hence, is not important.

The frequency response of a common-collector amplifier does not have a large Miller-multiplied value of capacitance compared to the common-emitter amplifier. The bandwidth of the common-collector amplifier is primarily determined by the source and load resistances and capacitor C_μ. For values of these resistances frequently encountered ($R_S \approx 1/g_m$ and $R_L > R_{out}$), the frequency response of the common-collector amplifier is near the transition frequency of the active device. When designing multistage amplifiers, the use of a common-collector amplifier stage often has little or no effect on the overall frequency response. As such it is considered to be a **wideband** voltage buffer stage.

10.5.2 Common-Drain Amplifier

A common-drain amplifier is shown in Fig. 10.27. We have replaced the bipolar npn transistor with an n-channel MOS transistor. All the analysis performed in the previous section applies by making the substitutions given in

$$\begin{aligned} C_\pi &\rightarrow C_{gs} \qquad & R_{in} &\rightarrow \infty \\ C_\mu &\rightarrow C_{gd} \qquad & R_{out} &\rightarrow \frac{1}{g_m} \end{aligned} \tag{10.87}$$

As with the bipolar transistor, the MOS common-drain amplifier is a wideband circuit that operates as an excellent voltage buffer up to frequencies near the transition frequency of the active MOS transistor.

10.6 FREQUENCY RESPONSE OF COMMON-BASE/GATE AMPLIFIER CURRENT BUFFER

As with the voltage buffer circuit analyzed in the previous section, we will proceed with the same methodology of developing a small-signal model with the device capacitances included. Next, we will find the low-frequency transfer function and then calculate the bandwidth of the amplifier by using the open-circuit time-constant method.

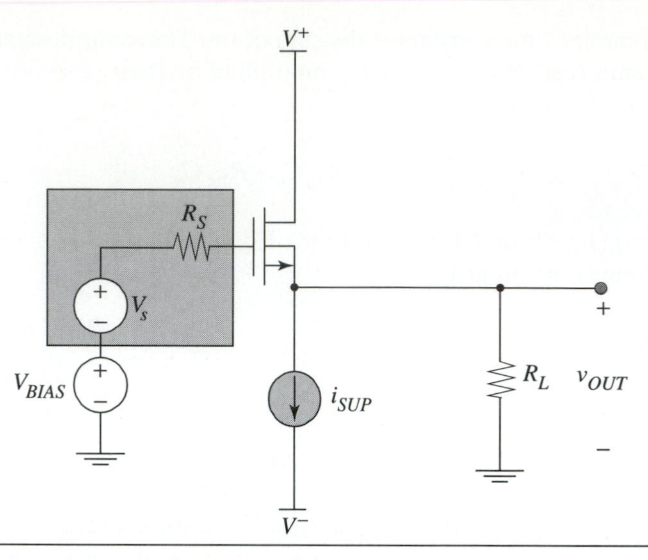

Figure 10.27 Common-drain amplifier.

The common-base amplifier with a sinusoidal input current source, source resistance, and load resistance is shown in Fig. 10.28. The small-signal two-port model for this common-base amplifier with the device capacitances C_π and C_μ included is shown in Fig. 10.29. Notice that C_π is placed between the emitter and base and capacitor C_μ is placed between the collector and base of the transistor.

We begin the analysis by calculating the low-frequency transfer function between the output current and input current with all capacitors open-circuited. The transfer function is given as

$$\frac{i_{out}}{i_s} = -\left(\frac{R_S}{R_{in} + R_S}\right)(1)\left(\frac{R_{out}}{R_L + R_{out}}\right) \tag{10.88}$$

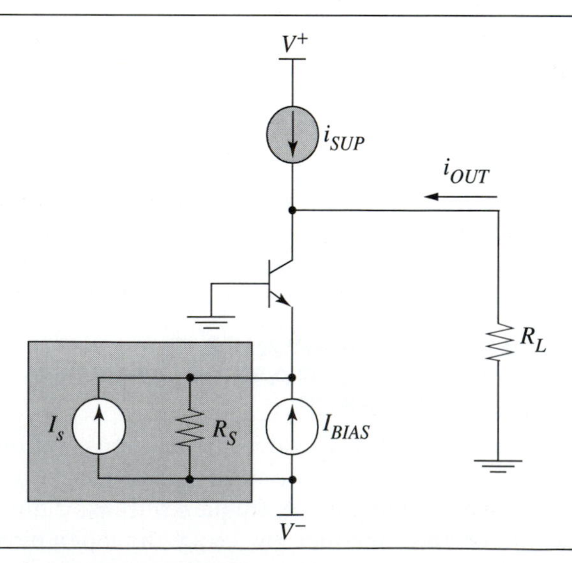

Figure 10.28 Common-base amplifier.

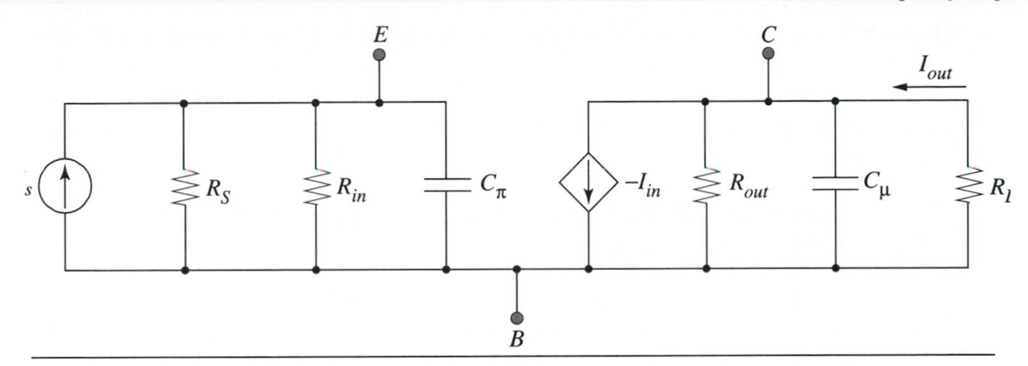

Figure 10.29 Small-signal model to determine frequency response of the common-base amplifier. Capacitors C_π and C_μ were added to the two-port model developed in Chapter 8.

The frequency response of the common-base amplifier can be easily calculated since there are no capacitors connected between the output and the input (i.e., no feedback). The open-circuit time-constant method is used to calculate the bandwidth of this amplifier.

Begin by noting that the Thévenin resistance across C_π is given by

$$R_{T\pi} = R_S \| R_{in} = R_S \| (1/g_m) \approx 1/g_m \tag{10.89}$$

Assuming $R_L \ll \beta_o r_o$, the Thévenin resistance across capacitor C_μ is given by

$$R_{T\mu} = R_{out} \| R_L \approx \beta_o r_o \| R_L \approx R_L \tag{10.90}$$

Summing the open-circuit time constants, we find

$$\sum \tau_{C_{io}} = \frac{C_\pi}{g_m} + R_L C_\mu \tag{10.91}$$

The frequency at which the magnitude of the transfer function in Eq. (10.88) is reduced by $1/\sqrt{2}$ is approximately the reciprocal of the sum of the open-circuit time constants and is given by

$$\omega_{3dB} \approx \frac{1}{(C_\pi/g_m) + R_L C_\mu} \tag{10.92}$$

The common-base amplifier is also a wideband amplifier for low values of the load resistor ($R_L \leq 1/g_m$). When performing a design using a common-base amplifier, we usually assume that its frequency response is near the transition frequency of the active device for hand analysis. The common-gate amplifier is analyzed in a similar manner and also has a wideband frequency response.

10.7 FREQUENCY RESPONSE OF MULTISTAGE AMPLIFIERS

In this section, we explore the frequency response of an important multistage amplifier called the **cascode amplifier**. We examined the DC transfer function of this multistage amplifier in Chapter 9. We will find in this analysis that the cascode amplifier yields high transconductance and voltage gain with a wideband frequency response. Essentially, the cascode circuit eliminates the Miller Effect that severely degraded the frequency response of the common-emitter/common-source amplifier. We will also explore the frequency response of the multistage voltage amplifier circuit that was analyzed in Chapter 9.

10.7.1 Cascode Amplifier Common-Source/ Common-Base

We will analyze the cascode amplifier composed of an MOS common-source amplifier and a bipolar common-base amplifier. This topology was chosen since the MOS transistor has a large input resistance and the bipolar transistor has an excellent frequency response. This device combination takes advantage of the positive aspects of each transistor type. The circuit is shown in Fig. 10.30.

A small-signal model of this cascode amplifier can be obtained by cascading the two-port models for a common-source and common-base amplifier as shown in Fig. 10.31. We assumed the incremental resistance of the current source supply is much larger than the load resistance ($r_{oc} \gg R_L$).

Note that we connect capacitor C_{gs1} between the gate and source of the MOS transistor; capacitor C_{gd1} between the gate and drain of the MOS transistor, capacitor $C_{\pi2}$ between the emitter and base of the bipolar transistor and capacitor $C_{\mu2}$ between the collector and base of the bipolar transistor. We calculate the small-signal voltage gain of this amplifier at DC by open-circuiting the capacitors. This voltage gain is given by

$$A_{vo} = -g_{m1}\left(\frac{r_{o1}}{r_{o1} + (1/g_{m2})}\right)(R_L \| \beta_{o2} r_{o2}) \tag{10.93}$$

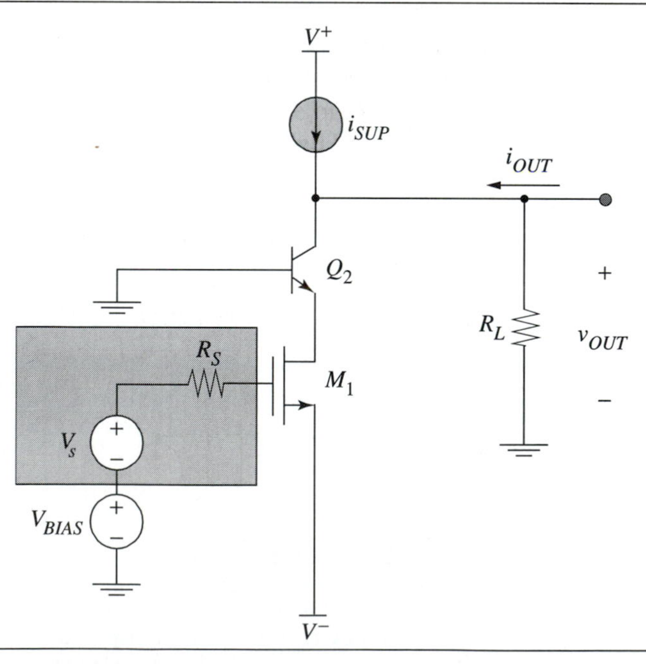

Figure 10.30 BiCMOS cascode amplifier with common-source/common-base stages.

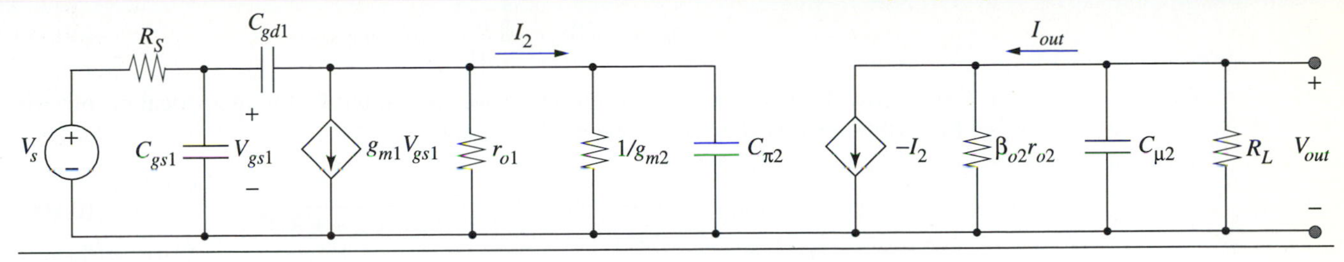

➤ Figure 10.31 Small-signal model for a BiCMOS cascode amplifier.

Assuming the output resistance of the transistors is large, the voltage gain is approximately equal to

$$A_{vo} \approx -g_{m1}R_L \tag{10.94}$$

Note that this is the same as the DC voltage gain that is obtained from a common-source amplifier. We will show that the cascode amplifier possesses a wideband frequency response that makes it advantageous to use over the simple common-source amplifier.

We begin the frequency response analysis of this four-capacitor small-signal circuit by using the Miller Approximation to place C_{gd1} as an effective capacitor in parallel with C_{gs1}. The voltage gain across capacitor C_{gd1}, assuming that $r_{o1} \gg 1/g_{m2}$ is given in

$$A_{vC_{gd1}} \approx \frac{-g_{m1}}{g_{m2}} \tag{10.95}$$

Since the transconductance of the bipolar transistor is usually higher than that of the MOS transistor, the voltage gain will be small. Assuming a worst case condition where $g_{m1} = g_{m2}$ the voltage gain is approximately unity. The total capacitance at the input C_T is

$$C_T \approx C_{gs1} + \left(1 + \frac{g_{m1}}{g_{m2}}\right)C_{gd1} \approx C_{gs1} + 2C_{gd1} \tag{10.96}$$

The Miller Effect is drastically reduced for capacitor C_{gd1} since the voltage gain across it is very small. This reduction occurs because we have loaded the drain of device M_1 with a common-base amplifier whose extremely small input resistance is simply the reciprocal of the transconductance of that bipolar transistor.

Now use the open-circuit time-constant method to estimate the bandwidth of the cascode amplifier. The time constant due to capacitor C_T is given by

$$\tau_{C_{To}} = R_S(C_{gs1} + 2C_{gd1}) \tag{10.97}$$

The open-circuit time constant for capacitors $C_{\pi 2}$ and $C_{\mu 2}$ are

$$\tau_{C_{\pi o}} = C_{\pi 2}\left(\frac{1}{g_{m2}} \parallel r_{o1}\right) \approx \frac{C_{\pi 2}}{g_{m2}} \tag{10.98}$$

$$\tau_{C_{\mu o}} = (\beta_{o2} r_{o2} \| R_L) C_{\mu 2} \approx R_L C_{\mu 2} \tag{10.99}$$

The bandwidth of the cascode amplifier is approximated by the reciprocal of the sum of the open-circuit time constants as shown in

$$\omega_{3dB} \approx \frac{1}{R_S(C_{gs1} + 2C_{gd1}) + (C_{\pi 2}/g_{m2}) + R_L C_{\mu 2}} \tag{10.100}$$

The cascode amplifier has provided a voltage gain comparable to a common-source amplifier with a frequency response comparable to a common-base amplifier. This circuit is often used when a wideband voltage amplifier is required.

DESIGN EXAMPLE 10.9 CMOS Cascode Amplifier

Design a CMOS cascode amplifier using a supply current of 100 μA shown in Fig. Ex 10.9. The source resistance $R_S = 50\ \Omega$ and the load resistance $R_L = 100\ \text{k}\Omega$. We also require that the maximum V_{GS} across the MOS transistors be 1.5 V to ensure a sufficient voltage swing at the output. Design the circuit to have the largest bandwidth possible and still have a voltage gain magnitude greater than 50. Assume a p-well process and that the sources of the MOS transistors are shorted to their respective backgates.

SOLUTION

From the constraint given, we know that $V_{GS} \le 1.5\ \text{V}$. we start with $V_{BIAS} = 1.5\text{V} - 2.5\text{V} = -1\text{V}$.

$$V_{DS_{SAT1}} = V_{GS1} - V_{Tn} = 1.5\ \text{V} - 1.0\ \text{V} = 0.5\ \text{V}\ .$$

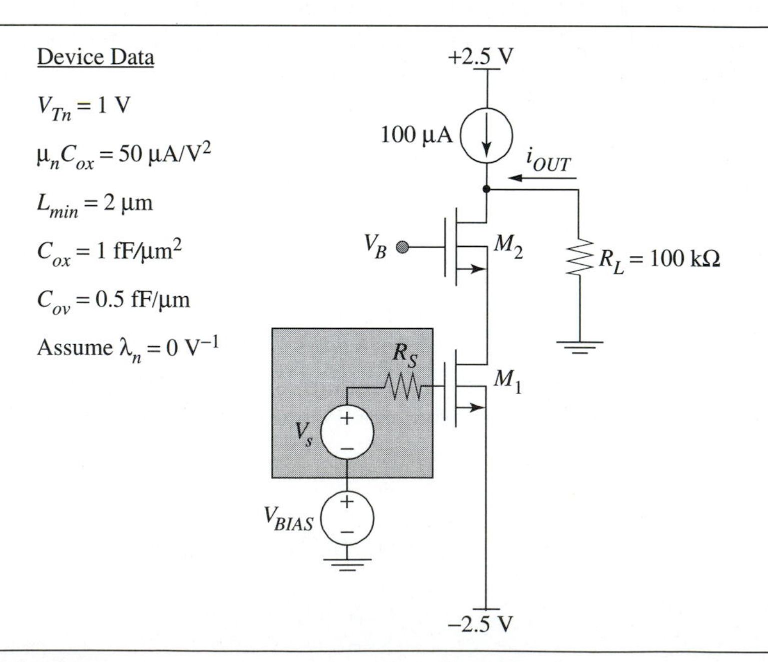

➤ Figure Ex10.9

If we choose $V_B = 0$ V and design $\left(\frac{W}{L}\right)_2$ so that $V_{GS2} = 1.5$ V, we find $V_{DS1} = 1$ V which is sufficient to keep M_1 operating in the constant-current region. Note that we have chosen the largest gate drive possible so we can use the smallest W/L ratio leading to the smallest device capacitances. Now we investigate whether the minimum W/L ratio of device M_1 is set by the voltage gain required or by the biasing constraint. In this CMOS cascode circuit $A_{vo} \approx -g_{m1}R_L$, so

$$g_{m1} = \frac{-A_{vo}}{R_L} = \frac{50}{100\ \text{k}\Omega} = 0.5\text{mS} = \sqrt{2\left(\frac{W}{L}\right)_1 \mu_n C_{ox} I_D} \ .$$

$$\left(\frac{W}{L}\right)_1 = \frac{g_{m1}^2}{2\mu_n C_{ox} I_D} = \frac{(0.5\text{mS})^2}{2\,(50\,\mu\text{A/V}^2)\,(100\ \mu\text{A})} = \frac{50}{2}$$

The DC bias constraint requires $(W/L)_1$ to be set by

$$V_{GS1} = 1.5\ \text{V} = V_{Tn} + \sqrt{\frac{I_D}{\frac{1}{2}\left(\frac{W}{L}\right)_1 \mu_n C_{ox}}}$$

$$\left(\frac{W}{L}\right)_1 = \frac{I_D}{(0.5\text{V})^2 \frac{1}{2}\mu_n C_{ox}} \approx \frac{32}{2}$$

The voltage gain specification sets the value for $(W/L)_1 = 50/2$. We used the minimum channel length since it will give the best frequency response.

The ω_{3dB} frequency of this CMOS cascode amplifier is governed by the same equation as the BiCMOS circuit in Fig. 10.31 except $C_{\pi 2}$ is replaced by C_{gs2} and $C_{\mu 2}$ is replaced by C_{gd2}. Device M_2 should be sized approximately the same as M_1 for the initial design. Therefore $(W/L)_2 = 50/2$ and $g_{m2} = 0.5$ mS.

Next we must find the capacitances.

$$C_{gs1} = C_{gs2} = \frac{2}{3}WLC_{ox} = 67\ \text{fF, and}$$

$$C_{gd1} = C_{gd2} = (0.5\text{fF}/\mu\text{m})\,(50\ \mu\text{m}) = 25\ \text{fF}$$

We can write the expression for ω_{3dB} by using Eq. (10.100) and realize that since we are using an MOS transistor for the cascode transistor, we must substitute C_{gs2} for $C_{\pi 2}$ and C_{gd2} for $C_{\mu 2}$.

$$\omega_{3dB} = \frac{1}{R_S(C_{gs1} + 2C_{gd1}) + C_{gs2}/g_{m2} + R_L C_{gd2}} \approx \frac{1}{C_{gs2}/g_{m2} + R_L C_{gd2}} = 380\text{Mrad/s}$$

$$f_{3dB} = \frac{\omega_{3dB}}{2\pi} = 60.4\ \text{MHz}$$

If we substitute an npn bipolar transistor for M_2 with $\tau_F = 50$ ps, $C_{je} = 15$ fF, and $C_\mu = 10$ fF, we find $g_{m2} = 100\ \mu A/25\ mV = 4$ mS, $C_{\pi 2} = 215$ fF, and $C_{\mu 2} = 10$ fF.

$$\omega_{3dB} \approx \frac{1}{\dfrac{C_{\pi 2}}{g_{m2}} + R_L C_{\mu 2}} = 949\,\text{Mrad/s}\,,\ \text{so}\ f_{3dB} = \frac{\omega_{3dB}}{2\pi} = 151.1\,\text{MHz}\ .$$

When comparing the BiCMOS and CMOS implementations, observe that the BiCMOS bandwidth is approximately 2.5 times higher than for the CMOS implementation. However, the CMOS design could be further improved by setting $(W/L)_2 = 32/2$ so that $V_{GS2} = 1.5$ V. Under this condition C_{gd2} is reduced to 16 fF and ω_{3db} increased approximately 575 Mrad/s.

10.7.2 Voltage Amplifier Circuit

In Chapter 9, we developed a voltage amplifier circuit that combined a cascode amplifier with a two-stage voltage buffer. We have redrawn that circuit in Fig. 10.32. A small-signal model could be drawn for this entire circuit, with all the device capacitors in their proper location. The method of open-circuit time constants could then be used to solve for the bandwidth of the amplifier. However, we can make use of some of the rules-of-thumb that we learned in this chapter concerning the frequency response of single-stage amplifiers. For example, the voltage buffer consisting of a

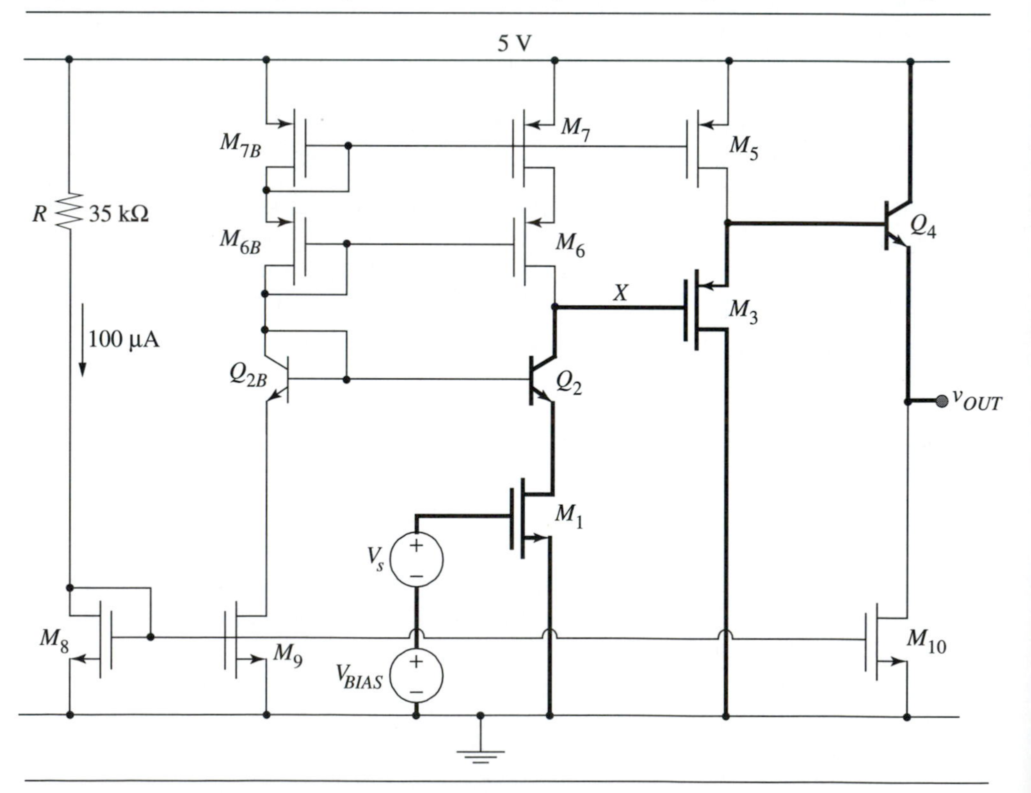

➤ **Figure 10.32** Multistage voltage amplifier circuit using a BiCMOS cascode amplifier cascaded with a voltage buffer.

common-drain and common-collector amplifier will have a frequency response near the transition frequency of the devices. The cascode amplifier consisting of devices M_1 and Q_2 has an extremely large output resistance. Therefore, the ω_{3dB} frequency of this amplifier will be governed by the time constant at the node labeled X in Fig. 10.32.

Let us draw the small-signal model of the voltage amplifier including capacitances and incremental resistances at node X. The small-signal model is shown in Fig. 10.33. Note that the incremental resistance at this node is larger than the output resistance of an individual transistor since the current source supply M_7 is cascoded by M_6 and the input device M_1 is cascoded with device Q_2. The total capacitance at node X, neglecting drain-bulk and collector-substrate capacitances, is the sum of the various gate-drain and collector-base capacitances from devices M_6, M_3, and Q_2. Also we add the gate-source capacitance of the source follower M_3 which is Miller-multiplied by $(1-A_{vC_{gs3}})$. The voltage gain of this source follower will be about 0.8–0.9. Therefore about 10%–20% of the gate-source capacitance of device M_3 will be added to the total capacitance at node X. We can approximate the voltage amplifier's transition bandwidth as

$$\omega_{3dB} = \frac{1}{(\beta_{o2}r_{o2} \| r_{oc})\,[C_{\mu 2} + C_{gd6} + C_{gd3} + (1 - A_{vC_{gs3}})\,C_{gs3}]} \qquad (10.101)$$

The DC voltage gain for this voltage amplifier is approximately

$$A_{vo} \approx -g_{m1}\beta_{o2}r_{o2} \qquad (10.102)$$

Using Eq. (10.101) and Eq. (10.102), we can draw a Bode plot of the magnitude of the voltage gain vs. frequency as shown in Fig. 10.34. Although the 3 dB frequency for this amplifier is at a low frequency, the point where the magnitude of the voltage gain goes to unity is at quite a high frequency. This point is called the **unity-gain bandwidth** of the amplifier. It is calculated by multiplying the low frequency gain times the 3 dB frequency or Eq. (10.101) times Eq. (10.102) yielding

$$\omega_{unity} = \frac{g_{m1}}{C_{\mu 2} + C_{gd6} + C_{gd3} + (1 - A_{vC_{gs3}})\,C_{gs3}} \qquad (10.103)$$

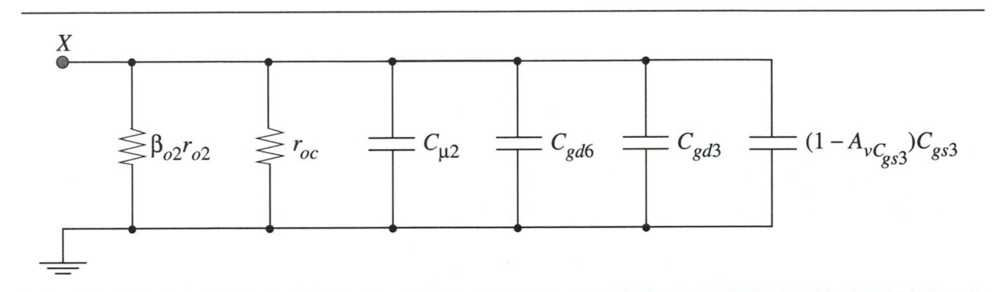

Figure 10.33 Small-signal model of the voltage amplifier including capacitance and incremental resistances at node X, the "high impedance node". We have neglected drain-bulk and collector-substrate capacitances.

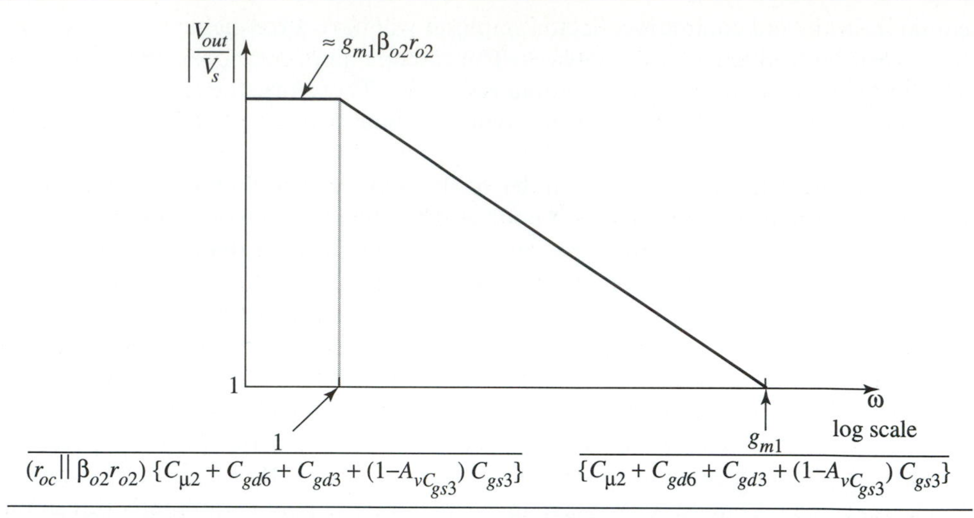

➤ **Figure 10.34** Bode Plot of the magnitude of the voltage gain vs. frequency.

The unity-gain bandwidth is an important figure of merit for amplifiers. We see that the approximate unity-gain bandwidth of this voltage amplifier is on the order of the transition frequency of the active transistors, which is theoretically the best one can expect to achieve for a given technology.

➤ SUMMARY

In this chapter, we have shown several useful analytical techniques to estimate the frequency response of various single-stage and multistage amplifier configurations. We also presented some design techniques to improve the bandwidth of single and multistage amplifiers. Specifically we showed

- A review of the concepts of frequency domain analysis.
- The definition of f_T for both the bipolar and MOS transistors and its relationship to device and technology parameters.
- The analysis of voltage amplifiers using the Miller Approximation.
- The method of open-circuit time constants to estimate the bandwidth of a variety of single and multistage amplifiers.
- Current buffers, using common base/gate amplifiers and voltage buffers using common collector/drain amplifiers, typically, have a frequency response that is near the transition frequency of the active devices.
- The cascode amplifier, which improves the bandwidth of a common-emitter/source amplifier by adding a common-base/gate stage.
- Analysis of the frequency response of multistage amplifiers.

➤ FURTHER READING

1. C. G. Fonstad, *Microelectronics Devices and Circuits*, McGraw-Hill, 1994, Chapter 14. Includes a discussion of capacitively-coupled transistor amplifiers.

2. P. E. Gray and C. L. Searle, *Electronic Principles Physics, Models, and Circuits*, Wiley, 1969, Chapter 15. Excellent explanation of the method of open circuit time constants.
3. P. R. Gray and R. G. Meyer, *Analysis and Design of Analog Integrated Circuits 3rd Ed.*, Wiley, 1993, Chapter 7. More advanced discussion of frequency response.
4. M. N. Horenstein, *Microelectronic Circuits and Devices 2nd Ed.*, Prentice Hall, 1995, Chapter 9. Includes a discussion of discrete transistor amplifiers.
5. A. S. Sedra and K. C. Smith, *Microelectronic Circuits 3rd Ed.*, HRW Saunders, 1991, Chapter 7. Extensive discussion on the frequency response of single and multistage amplifiers.
6. R. D. Thornton, C. L. Searle, D. O. Pederson, R. B. Adler, E. J. Angelo, Jr., *Multistage Amplifier Circuits, Semiconductor Electronics Education Committee, Vol. 5*, Wiley 1965, Chapter 1. Complete derivation of the open-circuit time-constant method.

PROBLEMS

EXERCISES

In the following exercises and problems use the device data given below for the MOS and bipolar transistors unless otherwise specified.

MOS Device Data

$\mu_n C_{ox} = 50\ \mu A/V^2$, $\mu_p C_{ox} = 25\ \mu A/V^2$, $V_{Tn} = -V_{Tp} = 1$ V, $-2\phi_p = 2\phi_n = 0.8$ V

$\gamma_n = \gamma_p = 0.6 V^{-\frac{1}{2}}$, $\lambda_n = \lambda_p = 0.05\ V^{-1}$ @ $L = 2\ \mu m$,

$C_{ox} = 2.3\ fF/\mu m^2$, $C_{jn} = 0.1\ fF/\mu m^2$, $C_{jp} = 0.3\ fF/\mu m^2$, $C_{jswn} = 0.5\ fF/\mu m$,
$C_{jswp} = 0.35\ fF/\mu m$, $C_{ovn} = 0.5\ fF/\mu m$, $C_{ovp} = 0.5\ fF/\mu m$, $L_{diffn} = L_{diffp} = 6\ \mu m$

npn Bipolar Device Data

$I_S = 10^{-17}$A, $\beta_F = \beta_o = 100$, $V_A = 25$ V, $\tau_F = 50$ ps, $C_{je} = 15$ fF @$V_{BE} = 0.7$ V,
$C_\mu = 10$ fF @$V_{BC} = -2.0$ V

E10.1 Sketch the Bode plot for the magnitude, $|I_o/I_s|_{dB}$ and phase $\angle I_o/I_s$ of the circuit shown below in Fig. E10.1, given

(a) $R_1 = 10\ k\Omega$, $R_2 = 100\ k\Omega$, $C = 1$ pF
(b) $R_1 = 0.1\ k\Omega$, $R_2 = 100\ k\Omega$, $C = 1$ pF

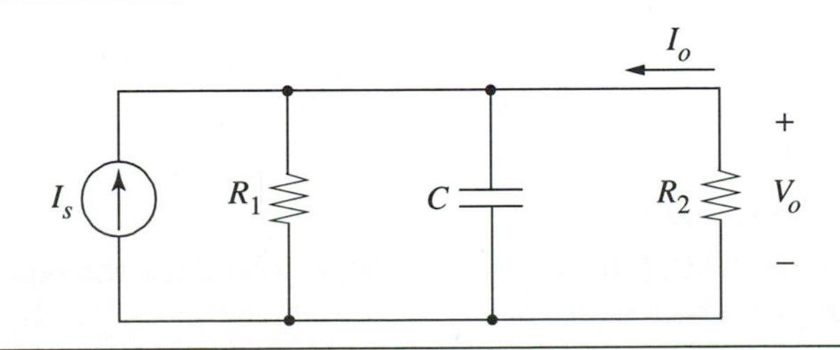

Figure E10.1

(c) $R_1 = 10\ \text{k}\Omega$, $R_2 = 100\ \text{k}\Omega$, $C = 10\ \text{fF}$

E10.2 Repeat E10.1 for the transfer function V_o/I_s.

E10.3 Sketch the Bode plots (magnitude and phase) for the following transfer functions. Assume $R_iC_i \gg R_kC_k$ if $i > k$.

(a) $[1/(1 + j\omega R_1C_1)][(j\omega/(1 + j\omega R_2C_2)]$

(b) $(j\omega R_3C_3)[(1 + j\omega R_4C_4)/(1 + j\omega R_5C_5)]$

(c) $[(1 + j\omega R_6C_6)/(1 + j\omega R_8C_8)][(1 + j\omega R_7C_7)/(1 + j\omega R_9C_9)]$

E10.4 A system has a DC gain of 500, zeros at 10 kHz and 1 MHz and poles at 100 kHz, 10 MHz, and 100 MHz.

(a) Write the transfer function that describes this system.

(b) Make a Bode plot for both the magnitude and phase of this system.

(c) Switch the poles and zeros and repeat parts (a) and (b).

E10.5 For an intrinsic npn common-emitter current amplifier biased such that the device is operating in its constant-current region, calculate the f_T, given

(a) $I_C = 1\ \mu\text{A}$

(b) $I_C = 100\ \mu\text{A}$

(c) $I_C = 10\ \text{mA}$

E10.6 For an intrinsic NMOS common-source current amplifier biased such that the device is operating in its constant-current region, calculate the f_T, given

(a) $W/L = 50/2$ and $I_D = 1\ \mu\text{A}$

(b) $W/L = 50/2$ and $I_D = 100\ \mu\text{A}$

(c) $W/L = 50/2$ and $I_D = 10\ \text{mA}$

(d) $W/L = 25/2$ and $I_D = 100\ \mu\text{A}$

E10.7 You are given an NMOS common-source voltage amplifier with a current source supply with $I_{SUP} = 50\ \mu\text{A}$ and $r_{oc} \to \infty$. The NMOS device has a $W/L = 50/2$. The source resistance, $R_S = 10\ \text{k}\Omega$ and the load resistance $R_L \to \infty$. Assume that all the devices are operating in their constant-current region.

(a) Calculate the open-circuit voltage gain at low frequency.

(b) Calculate ω_{3dB} using the Miller Approximation and considering only C_{gs} and C_{gd} of the NMOS device.

(c) Repeat (b) using the open-circuit time-constant method.

E10.8 Repeat E10.7 for a PMOS device with the same dimensions.

E10.9 Repeat E10.7 for an npn common-emitter voltage amplifier.

E10.10 Given the common-collector amplifier shown in Fig. E10.10, assume that V_{BIAS} is set such that $V_{OUT} = 0$ V and that $r_{oc} \to \infty$. Find the low frequency voltage gain and ω_{3dB} for

(a) $R_S = 1\ \text{k}\Omega$ and $R_L = 500\ \Omega$.

(b) $R_S = 10\ \text{k}\Omega$ and $R_L = 1\ \text{k}\Omega$.

(c) $R_S = 5\ \text{k}\Omega$ and $R_L = 250\ \Omega$.

E10.11 Substitute an NMOS transistor with $W/L = 50/2$ for the bipolar transistor in Fig. E10.10 and repeat Exercise 10.10.

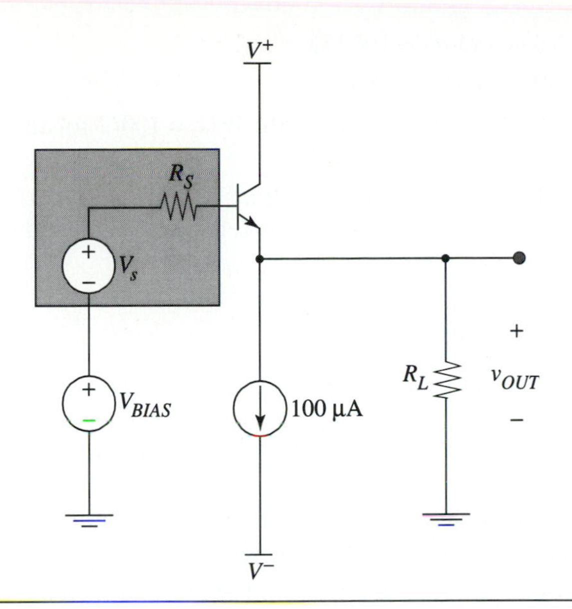

➤ **Figure E10.10**

E10.12 Given a PMOS common-drain amplifier with $I_{SUP} = 100\ \mu A$ and $r_{oc} \to \infty$, find the smallest W/L necessary to have a low frequency |voltage gain| ≥ 0.3 and $\omega_{3dB} \geq 100$ Mrad/s when

(a) $R_S = 1\ k\Omega$ and $R_L = 500\ \Omega$.

(b) $R_S = 10\ k\Omega$ and $R_L = 1\ k\Omega$.

(c) $R_S = 5\ k\Omega$ and $R_L = 250\ \Omega$.

E10.13 Given the common-base amplifier shown in Figure E10.13 assume that I_{BIAS} is set such that $I_{OUT} = 0$ A, $I_{SUP} = 200\ \mu A$ and $r_{oc} \to \infty$. Find the low frequency current gain and ω_{3dB} for

(a) $R_S = 100\ \Omega$ and $R_L = 10\ k\Omega$.

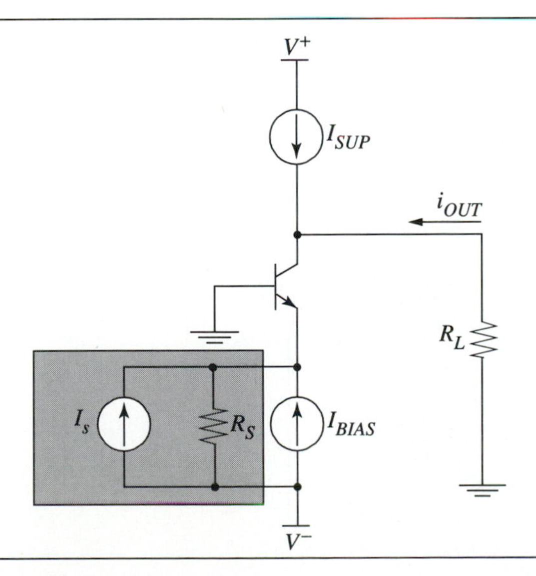

➤ **Figure E10.13**

(b) $R_S = 1\ \text{k}\Omega$ and $R_L = 100\ \text{k}\Omega$.

(c) $R_S = 500\ \Omega$ and $R_L = 5\ \text{k}\Omega$.

E10.14 Substitute an NMOS transistor with $W/L = 100/2$ for the bipolar transistor in Fig. E10.13 and repeat Exercise 10.13.

E10.15 Given a PMOS common-gate amplifier with $I_{SUP} = 200\ \mu\text{A}$, find the smallest W/L necessary to have a low frequency |current gain| ≥ 0.3 and $\omega_{3dB} \geq 100$ Mrad/s when

(a) $R_S = 100\ \Omega$ and $R_L = 10\ \text{k}\Omega$.

(b) $R_S = 1\ \text{k}\Omega$ and $R_L = 100\ \text{k}\Omega$.

(c) $R_S = 500\ \Omega$ and $R_L = 5\ \text{k}\Omega$.

E10.16 A bipolar cascode transconductance amplifier is shown in Figure E10.16. Given $R_S = 1\ \text{k}\Omega$, $R_L = 50\ \text{k}\Omega$, $C_L = 0.1$ pF, $I_{SUP} = 100\ \mu\text{A}$ and $r_{oc} \rightarrow \infty$, find I_{out}/V_s at DC and ω_{3dB}. Assume that V_{BIAS} is set such that all devices are operating in their constant-current region. Include $C_{cs} = 50$ fF in the calculations.

E10.17 Replace Q_1 in E10.16 with an NMOS transistor with $W/L = 100/2$. Use the same component values given in E10.16. Find the transconductance at DC and ω_{3dB}. Neglect backgate effect, but include $C_{cs} = 50$ fF in your calculations.

E10.18 Replace Q_1 and Q_2 in E10.16 with NMOS transistors with $(W/L)_{1,2} = 100/2$. Use the same component values given in E10.16. Find the transconductance at DC and ω_{3dB}. Neglect backgate effect, but include C_{db} of M_2 in your frequency response calculation.

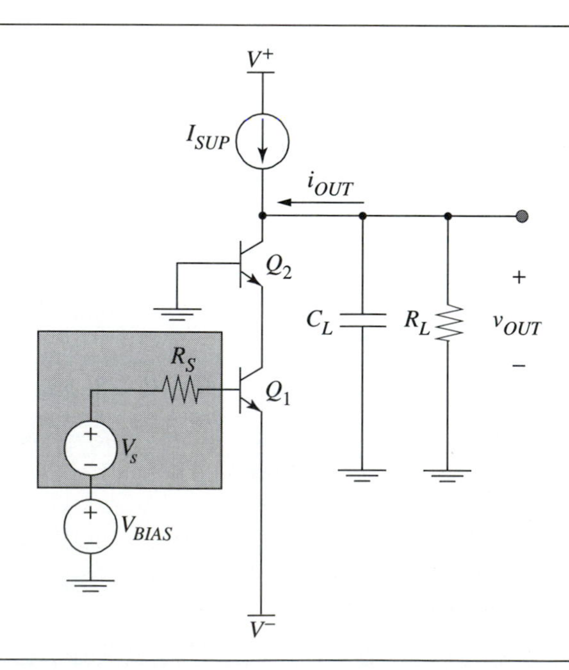

➤ **Figure E10.16**

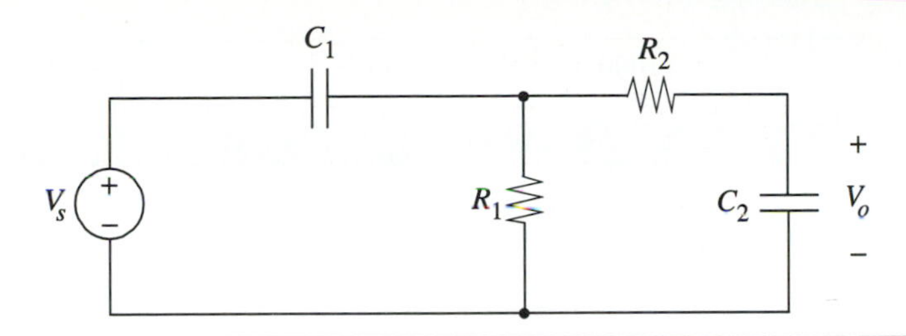

➤ **Figure P10.1**

PROBLEMS

P10.1 Given the circuit shown in Fig. P10.1 with $R_1 = 100\ \text{k}\Omega$, $R_2 = 10\ \text{k}\Omega$, $C_1 = 5\ \text{pF}$, and $C_2 = 1\ \text{pF}$

(a) Write the transfer function V_o / V_s.

(b) Make a Bode plot for both the magnitude and phase of this circuit.

P10.2 Repeat P10.1 given the input sinusoidal source V_s has a source resistance $R_S = 10\ \text{k}\Omega$.

P10.3 Consider a voltage amplifier with 1 zero at 10^3 rad/s and 3 poles at 10^1, 10^4, 10^6 rad/s respectively. The amplifier has a DC voltage gain of 10^5.

(a) Write the transfer function for the voltage gain of this amplifier.

(b) Construct a Bode plot for this amplifier showing both magnitude and phase.

(c) Find the unity-gain frequency.

P10.4 A current amplifier has a gain of 1000 at 10^5 rad/s. It has two zeros at 0 and 10^6 rad/s and three poles at 10^4, 10^7, 10^8 rad/s.

(a) Sketch the Bode plot (magnitude and phase) for this amplifier.

(b) Find the peak current gain of the amplifier.

(c) Find the frequency and phase shift corresponding to the peak current gain found in part (b).

P10.5 A 1mm by 3 μm polysilicon line placed over a 0.5 μm oxide on top of heavily doped silicon is shown in Fig. P10.5. This polysilicon line is often modeled with a number of *RC* segments as shown in the figure. The polysilicon line has a sheet resistance of 50 Ω/□.

(a) By considering this 1mm long line as 10 *RC* segments, calculate the value of the *R* and *C* in the model to approximate the frequency response of the polysilicon line.

(b) A two-element model is the worst case for this situation. Calculate the values of the *R* and *C* for the single *RC* segment.

(c) Construct a Bode plot for the two-element case in part (b).

(d) Using SPICE find V_o / V_s for the model in part (a) and compare it to the analytical result found in part (c).

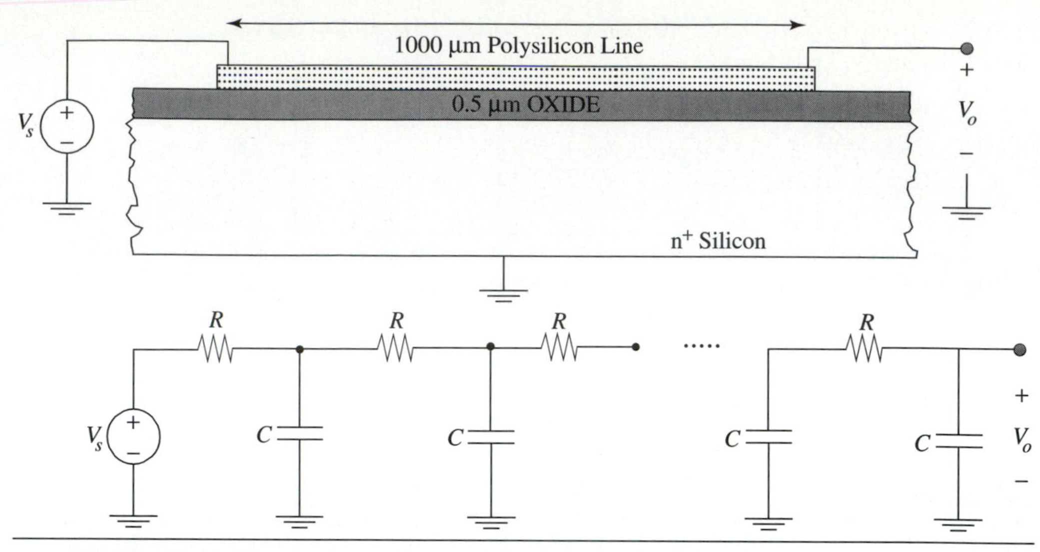

➤ **Figure P10.5**

P10.6 In this problem we want to explore the effect of the collector voltage on the f_T of the device under low current operation. We will use the circuit in Fig. P10.6 assuming $I_{SUP} = 1\mu A$, $r_{oc} \rightarrow \infty$, and I_{BIAS} is set so that $I_{SUP} = I_C$. You are given that $C_\mu = 15$ fF @$V_{CB} = 0$ V, $\phi_{BC} = 0.8$ V and all other device parameters are the nominal values.

(a) For $V_C = 0.6$ V what is f_T?

(b) Given that the current source supply requires 0.5 V across it, the maximum voltage for V_C is 4.5 V. Calculate f_T under this condition.

(c) Calculate the required I_{SUP} to reach the same f_T as part (b) with $V_C = 0.6$ V.

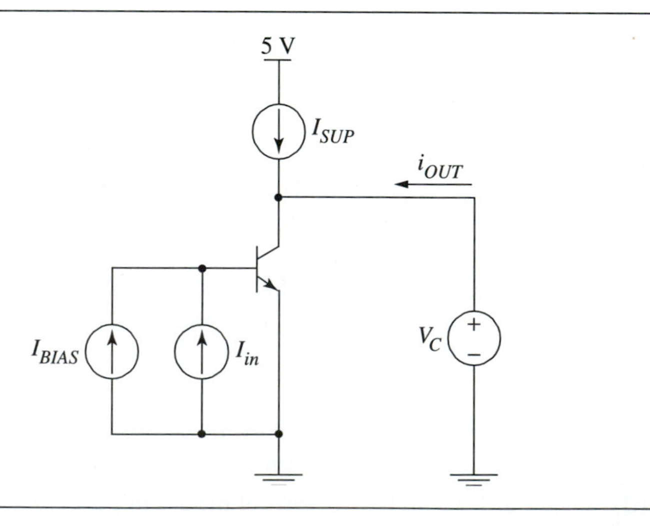

➤ **Figure P10.6**

P10.7 Given a PMOS common-source amplifier with $W/L = 50/2$ as shown in Fig. P10.7. Assume I_{SUP} is equal to $-I_D$ when $I_s = 0$ A.

(a) Find the DC current $(-I_D)$ in the PMOS device.

(b) Plot the Bode plot for $|I_o/I_s|$ for $R_S \rightarrow \infty$.

(c) To see the effect of R_S plot the Bode plot for $|I_i/I_s|$.

(d) What is the value of R_S where the overall unity-gain frequency is reduced by a factor of 10 from that calculated in part (b)?

P10.8 Calculate the f_T of an intrinsic NMOS common-source amplifier ($R_S = 0\ \Omega$ and $R_L \rightarrow \infty$) with a $W/L = 100/2$ given

(a) $I_{SUP} = 1\ \mu A$

(b) $I_{SUP} = 100\ \mu A$

(c) $W/L = 50/1$; $I_{SUP} = 100\ \mu A$

P10.9 In the small-signal model for the common-emitter amplifier in Fig. 10.13 we did not include a load capacitance. Suppose we add a load capacitance C_L. Even though we have added a capacitor the circuit still only has two poles and one zero. Show that the overall transfer function V_{out}/V_s is given by

$$\frac{V_{out}}{V_s} = \frac{-g_m R'_{out}\left(\dfrac{r_\pi}{R_S + r_\pi}\right)\left(1 - \dfrac{C_\mu}{g_m} j\omega\right)}{1 + j\omega[R'_{out}(C_\mu + C_L) + R'_{in}C_\mu(1 + g_m R'_{out}) + R'_{in}C_\pi] - \omega^2 R'_{out}R'_{in}(C_L C_\pi + C_\mu C_L + C_\mu C_\pi)}$$

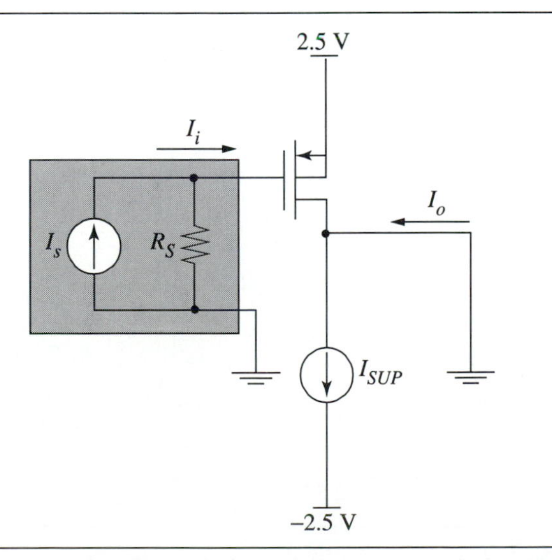

➤ **Figure P10.7**

P10.10 A complete p-channel common-source amplifier with source resistance $R_S = 10\ \text{k}\Omega$ and load resistance $R_L = 1\text{M}\Omega$ is shown in Fig. P10.10. The PMOS transistor has a $(W/L)_1 = 50/2$ and the NMOS transistors also have a $(W/L)_{2,3} = 50/2$. I_{REF} is equal to 100 μA.

(a) Find the value of V_{BIAS} such that the DC output voltage $V_{OUT} = 0$ V.

(b) Draw the small-signal model of this amplifier.

(c) Using the results from P10.9, solve for the complete transfer function V_{out}/V_s in terms of the small-signal parameters. Note: You should include the PMOS transistor device capacitances C_{gs1} and C_{gd1} as well as the capacitance at the load which is the sum of C_{db1}, C_{db2} and C_{gd2}. Even though there are 5 capacitors, this is still a 2nd order system (2 poles).

P10.11 Using the circuit and parameters specified in Problem P10.10

(a) Use the Miller Approximation to find the transfer function V_{out}/V_s.

(b) Find ω_{3dB}.

(c) Compare ω_{3dB} found for using the Miller Approximation to the solution found in P10.10.

P10.12 You are given an NMOS common-source amplifier with an ideal current source supply with $I_{SUP} = 100\ \mu\text{A}$ and $r_{oc} \rightarrow \infty$. The source resistance is 10 kΩ and the load resistance $R_L \rightarrow \infty$. Assume that all the devices are operating in their constant-current region. Use the Miller Approximation for this problem and consider only C_{gs} and C_{gd} of the NMOS device.

(a) Given $W/L = 40/2$ find the open-circuit voltage gain and ω_{3dB}.

(b) From Chapter 8 we know that if we lengthen the device that we will increase the open-circuit voltage gain. Using a $W/L = 80/4$, find the new open-circuit voltage gain and ω_{3dB}.

(c) Compare the product of the gain and bandwidth for parts (a) and (b).

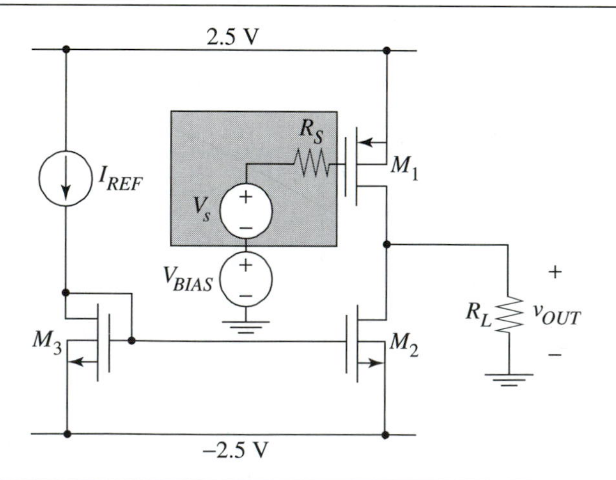

➤ **Figure P10.10**

P10.13 Given the common-collector amplifier shown in Fig. P10.13 assume that V_{BIAS} is set such that $V_{OUT} = 0$ V and that $r_{oc} \rightarrow \infty$. Find the low frequency voltage gain and ω_{3dB} for $R_S = 10$ kΩ, $R_L = 1$ kΩ, and $C_L = 0$ F.

P10.14 Often the common-collector amplifier must drive a load capacitance as well as resistive loads.

(a) Derive an expression for ω_{3dB} using the open-circuit time-constant method including the effect of C_L.

(b) To see the relative effect of C_L repeat P10.13 with $C_L = 100$ fF, 1pF, and 10 pF.

P10.15 You are given an NMOS common-drain amplifier driving a capacitive load with $I_{REF} = 500$ μA as shown in Fig. P10.15. Assume all devices are operating in their constant-current region.

(a) Derive an expression for ω_{3dB} using the method of open-circuit time constants including the effect of C_L. Do not neglect backgate effect.

(b) Calculate ω_{3dB} for $(W/L)_1 = 500/2$ $R_S = 1$kΩ and $C_L = 1$pF. Do not neglect C_{db} and C_{sb}.

(c) Repeat (b) for $(W/L)_1 = 100/2$ and $(W/L)_1 = 1000/2$.

P10.16 In the derivation of the frequency response for the common-base amplifier we assumed that $R_S \gg 1/g_m$ and $R_L \ll \beta_o r_o$. When the amplifier is driving a large load resistance or the supply current is high this assumption may be violated. Given $R_S = 10$ kΩ and $R_L = 10$ MΩ

(a) Find the supply current I_{SUP} at which $\beta_o r_o = R_L$. Assume $r_{oc} \rightarrow \infty$.

(b) Find ω_{3dB} at this operating point.

P10.17 A load capacitance on a common-base amplifier can degrade the frequency response. The load capacitance is made up of parasitic collector-substrate depletion

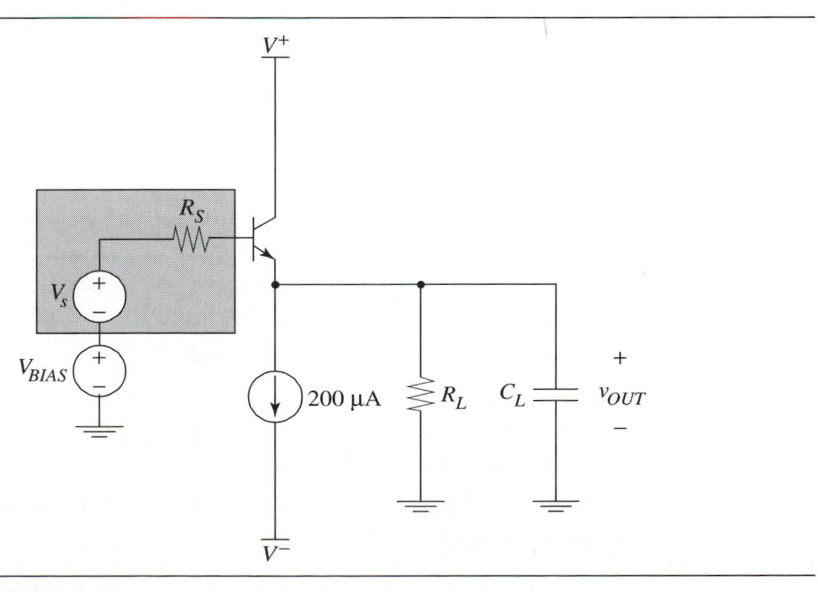

➤ Figure P10.13

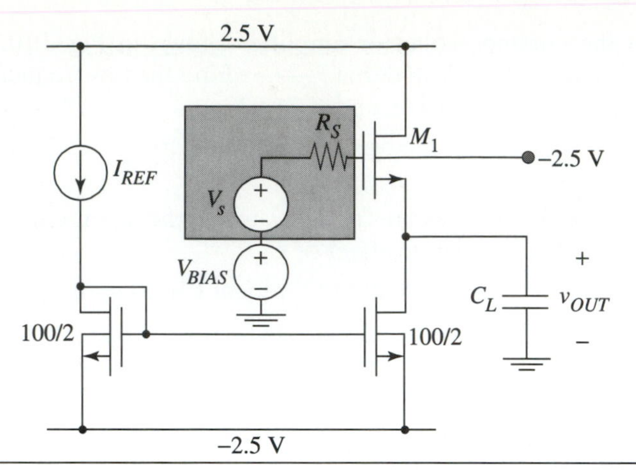

➤ Figure P10.15

capacitance, wiring capacitance, and other elements that the amplifier is driving (i.e. next amplifier).

(a) Derive an expression for ω_{3dB} for the common-base amplifier including C_L.

(b) Find ω_{3dB} given $R_S = 5\ \text{k}\Omega$, $R_L = 1\text{M}\Omega$, $C_L = 100$ fF, and $I_{SUP} = 1$ mA .

(c) Repeat (b) for $C_L = 10$ pF.

P10.18 Derive an expression for ω_{3dB} similar to Eq. (10.92) for a common-gate amplifier.

P10.19 As with the common-base amplifier, the frequency response of the common-gate amplifier depends on C_L. To increase g_m and improve the frequency response one can increase the W of the device. However, as the width of the device is increased C_{db} increases degrading the frequency response. Neglecting the backgate effect and including $C_L = C_{db}$

(a) Derive an expression for ω_{3dB} for the common-gate amplifier including C_L.

(b) Calculate ω_{3dB} with $I_{SUP} = 100\ \mu\text{A}$ and $W/L = 50/2$.

(c) Calculate ω_{3dB} with $I_{SUP} = 100\ \mu\text{A}$ and $W/L = 200/2$.

(d) Calculate ω_{3dB} with $I_{SUP} = 400\ \mu\text{A}$ and $W/L = 50/2$.

P10.20 You are given a 2-stage CS-CS voltage amplifier shown in Fig. P10.20 with $W/L = 100/2$, for both M_1 and M_2 , $R_S = 1\text{k}\Omega$ and $R_L \rightarrow \infty$. Assume V_{BIAS} is set such that all devices are in their constant-current region. Neglect backgate effect.

(a) Draw a small-signal two-port model of this CS-CS amplifier. Show both stages and include all device capacitances.

(b) Find the voltage gain at DC.

(c) Using the Miller approximation and the open-circuit time-constant method, find ω_{3dB} for this circuit.

(d) Use the Miller Approximation across the second transistor only and compare your result with part (c).

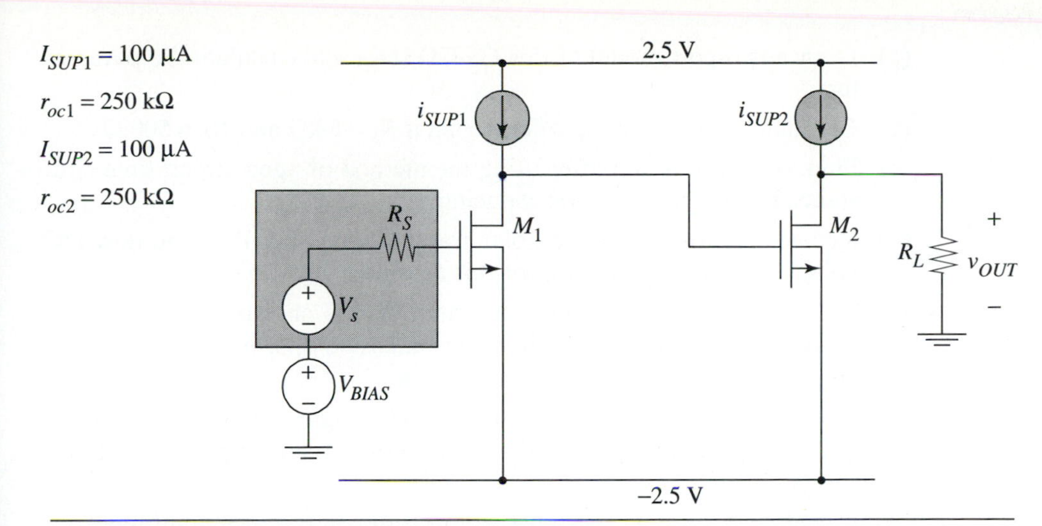

➤ **Figure P10.20**

(e) We add a PMOS common-drain stage in between the common-source stages to improve the bandwidth of the circuit. Assume the W/L of this PMOS device is 100/2 and the current source supply has $I_{SUP} = 100\ \mu A$ and $r_{oc} \to 250\ k\Omega$. Repeat (a)–(d).

P10.21 Repeat P10.20 if the NMOS transistors are replaced with npn bipolar transistors.

P10.22 Given the voltage amplifier shown in Fig. P10.22 with $(W/L)_1 = 50/2$, $(W/L)_3 = 100/2$ and $(W/L)_{2B} = (W/L)_2 = (W/L)_4 = 50/4$. In this problem we will assume all backgates are shorted to their respective sources. Assume that V_{BIAS} is set such that all devices are operating in their constant-current region.

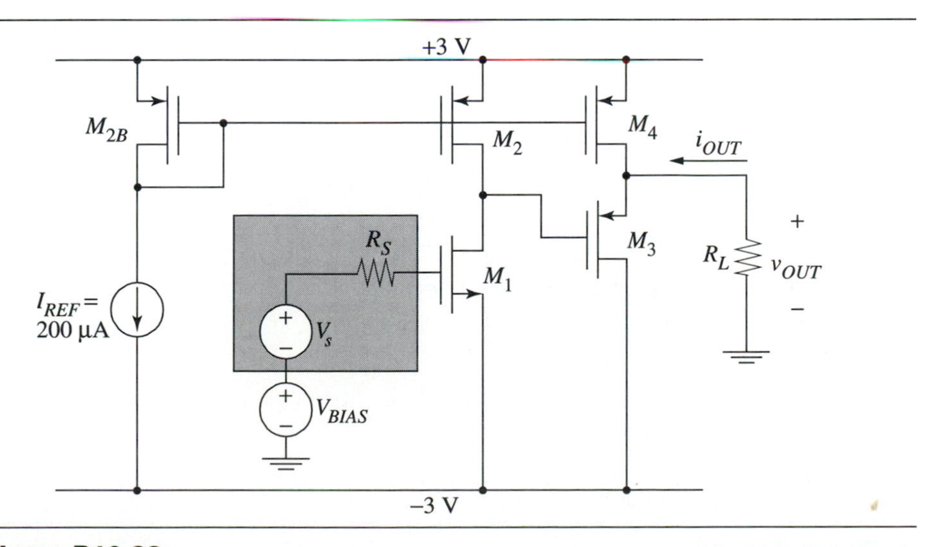

➤ **Figure P10.22**

(a) Draw a two-port model of this CS-CD stage and calculate the parameters.

(b) Calculate the overall DC voltage gain if $R_S = 5\ \text{k}\Omega$ and $R_L = 500\ \Omega$.

(c) Find ω_{3dB} for this amplifier using the method of open-circuit time constants. Include the parasitic capacitances.

(d) Replace M_1 with an npn bipolar transistor and calculate the new DC voltage gain and bandwidth for this amplifier.

P10.23 A CMOS cascode transconductance amplifier is shown in Fig. P10.23. Neglect the backgate effect for this problem. Assume that V_{BIAS} is set such that all devices are operating in their constant current-region.

(a) Calculate $(W/L)_1$ of M_1 such that the small-signal transconductance at DC, $i_{out}/v_s = 2$ mS. Assume $R_L = 0\ \Omega$ (short-circuit output current) for this part.

(b) Calculate the output resistance of this transconductance amplifier.

(c) Calculate the transconductance at DC given that $R_S = 10\ \text{k}\Omega$ and $R_L = 1\text{k}\Omega$.

(d) Estimate the bandwidth of this circuit given that $R_S = 10\ \text{k}\Omega$ and $R_L = 1\text{k}\Omega$.

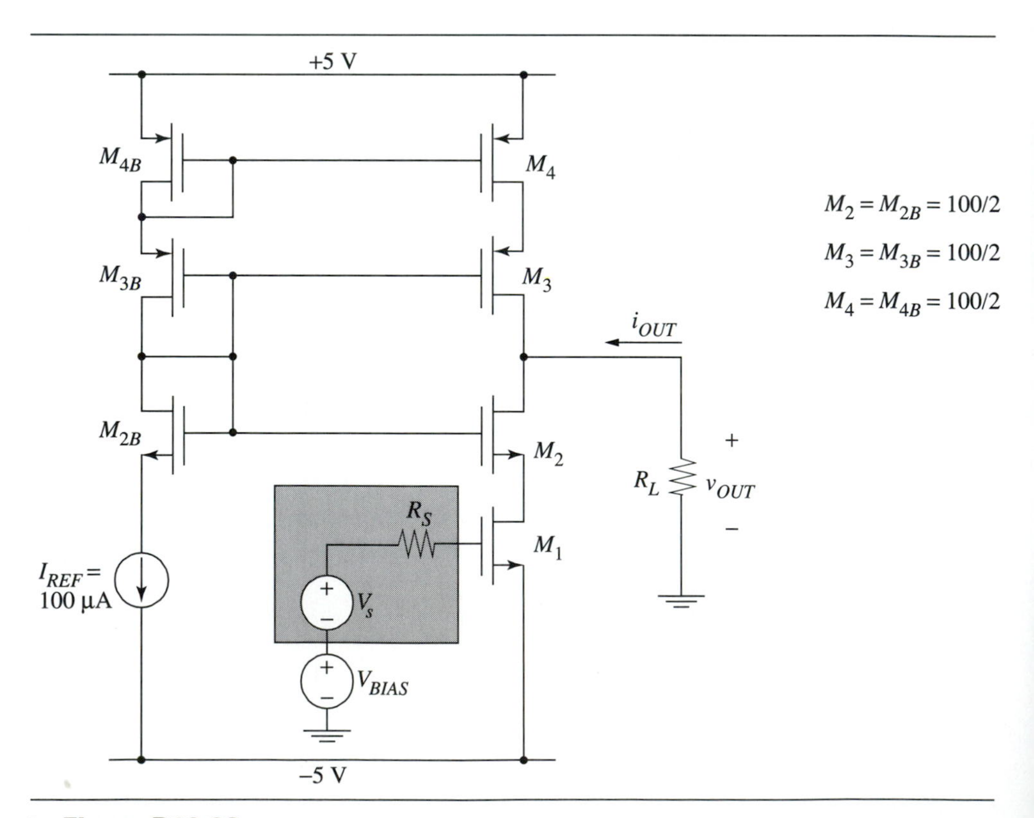

➤ **Figure P10.23**

P10.24 You are given the BiCMOS voltage amplifier shown in Figure P10.24 with $(W/L)_n = 20/2$ and $(W/L)_p = 40/2$. Neglect backgate effect and assume that V_{BIAS} is set such that all devices are operating in their constant-current region.

(a) Draw the two-port model for this voltage amplifier.

(b) Calculate the DC voltage gain.

(c) Estimate the bandwidth of the amplifier using the open-circuit time-constant method. Assume $C_L = 0$ F for this part.

(d) What is the value of C_L that this amplifier can drive and only lose 20% of the bandwidth calculated in part (c)?

DESIGN PROBLEMS

D10.1 A cross section and top-view of two pads at the edge of an integrated circuit are shown in Fig. D10.1. In this problem we want to estimate the coupling between these pads using the equivalent circuit model also given in Fig. D10.1. Assume the p and n regions are both at ground.

(a) Estimate the numerical values of each element in the circuit model.

(b) Find the transfer function I_o/V_s and plot the magnitude Bode plot.

Make reasonable approximations and eliminate some of the circuit elements to make your hand calculations more meaningful.

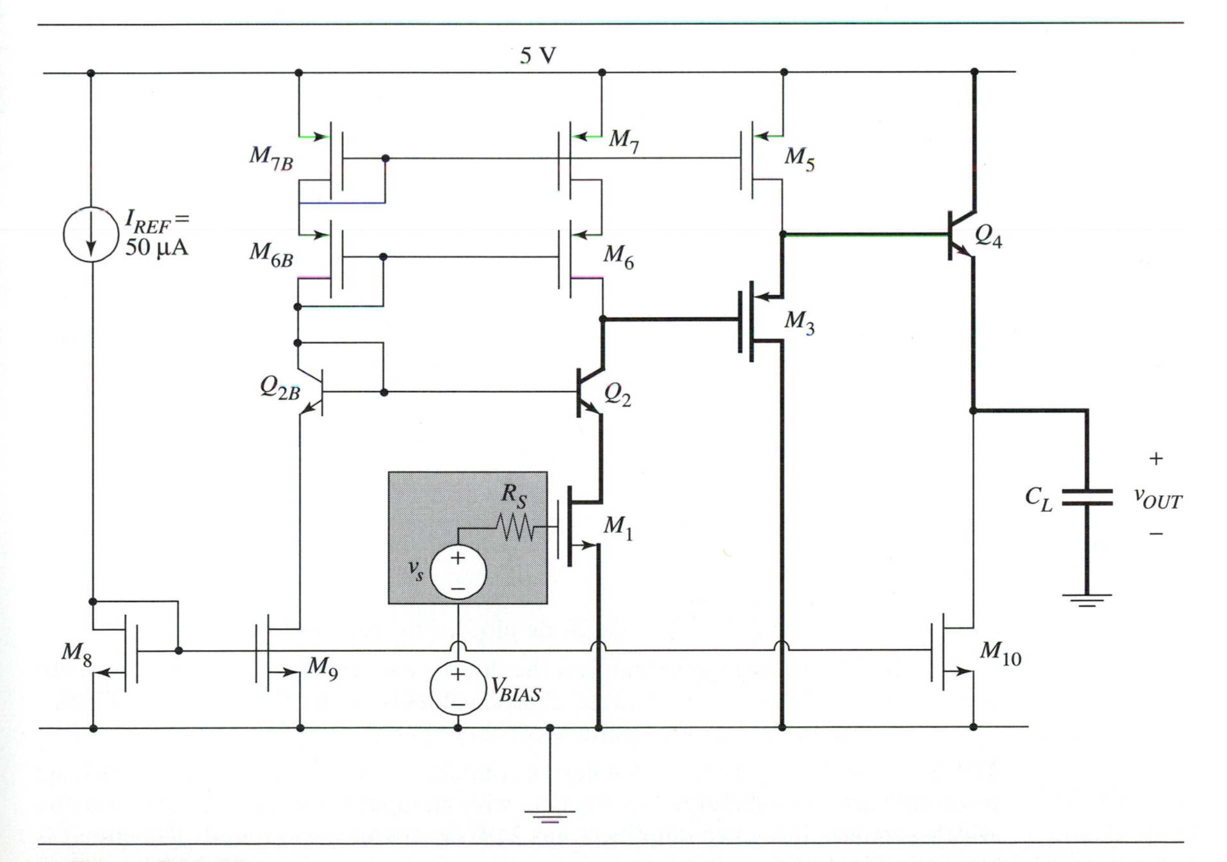

➤ **Figure P10.24**

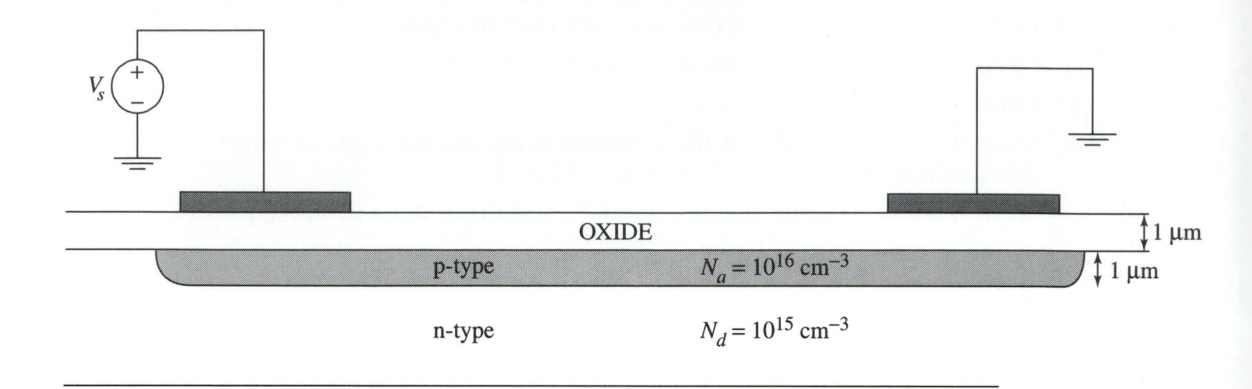

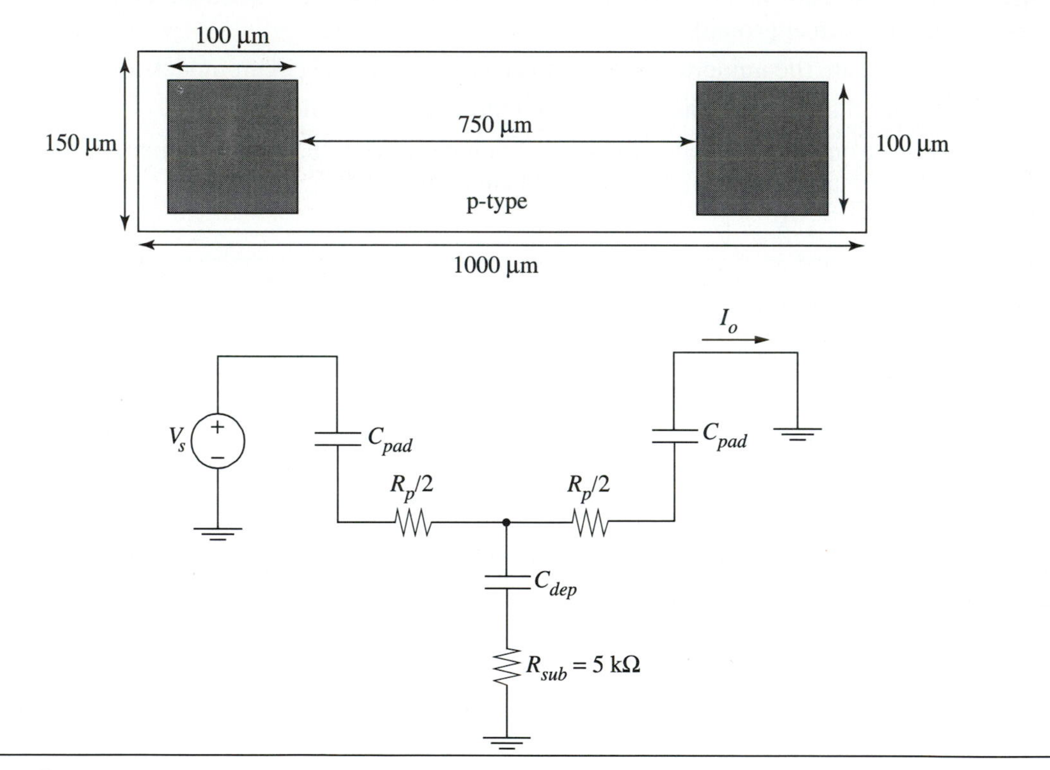

➤ **Figure D10.1**

(c) Use SPICE to find the Bode plot for the full model.

(d) Given that you can adjust the doping concentrations of both the p-well and the n-type substrate, choose values between 10^{15} cm^{-3} and 10^{19} cm^{-3} to minimize the magnitude of the coupling.

D10.2 In this design problem we have a common-emitter current amplifier driving a second current amplifier across the chip with an input resistance of 5 kΩ. The line which connects these two amplifiers has 2 pF capacitance to ground. The circuit is shown in Fig. D10.2. Assume $r_{oc} = r_o$.

(a) Choose I_{SUP} such that the DC current gain $i_{out}/i_s = -50$.

(b) The pole at the output of the amplifier is due to the parasitic line capacitance and the output resistance of the amplifier in parallel with the 5 kΩ load resistance. What is the frequency of this pole in rad/s?

(c) Choose a new value of I_{SUP} to ensure that the overall gain $|I_{out}/I_s| = 35$ at the frequency calculated in part (b). You will need to iterate between part (a) and (c).

(d) Verify and refine your design in SPICE.

D10.3 In Example 10.5 we explored the f_T of an NMOS device. We found that by using a large DC gate-source voltage and a short channel length that we obtained the highest f_T. In modern microelectronics one must try to minimize power dissipation and therefore cannot use a large V_{GS}. Given that the $V_{DS_{SAT}}$ of the NMOS device must be less than 0.2 V to ensure a reasonable voltage swing, find the largest f_T possible for $L_{min} = 1.5$ μm. Does the value of W/L influence your answer?

D10.4 In this problem you are to design a common-collector amplifier with $R_S = 1\text{k}\Omega$, $R_L = 50\ \Omega$ and a load capacitance $C_L = 100$ fF. Refer to Fig. P10.13 for the circuit topology. The amplifier must have a voltage-gain magnitude greater than 0.6 up to 900 MHz. In this design you are to choose values for V_{BIAS} such that $V_{OUT} = 0$ V and I_{SUP} to minimize power dissipation. Assume $r_{oc} \to \infty$.

D10.5 In the design of the common-collector amplifier in D10.4 there is a minimum I_{SUP} required to meet the specification. Suppose you could change one of the device parameters given for the bipolar transistor (i.e. τ_F, β_F, C_{je} etc.). Which device parameter would allow you to maximize the reduction of I_{SUP} and still meet the specification? Describe a technological method to achieve the device parameter change (i.e. change doping, change area, change base width, etc.).

D10.6 In this problem you are to design a common-drain amplifier with $R_S = 1\text{k}\Omega$, $R_L = 5\ \text{k}\Omega$ and a load capacitance $C_L = 100$ fF. In this problem do not neglect backgate effect and account for C_{sb} by adding its capacitance to the load capacitance. The amplifier must have a voltage-gain magnitude greater than 0.6 up to 900 MHz. In this design you are to choose values for V_{BIAS} such that $V_{OUT} = 0$ V and I_{SUP} and W/L to minimize power dissipation. Assume r_{oc} of the current source supply is 25 V/I_{SUP}, the maximum W is 500 μm, and minimum L is 2 μm. The circuit is shown in Fig. D10.6.

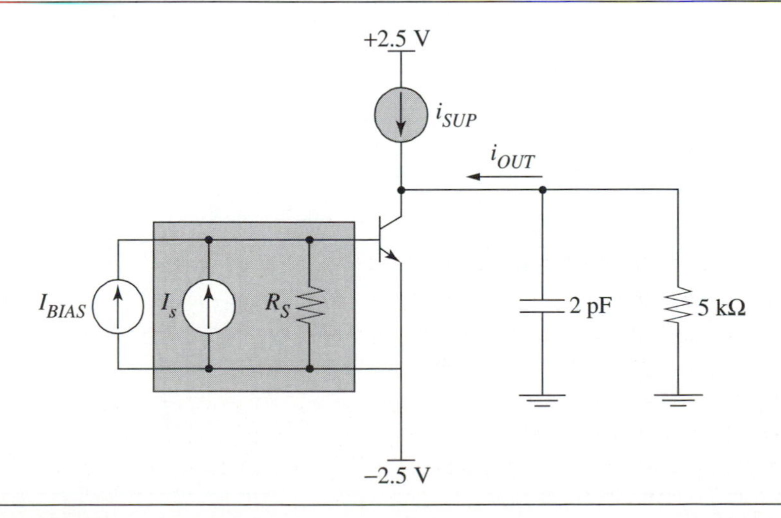

➤ **Figure D10.2**

D10.7 In the design of the common-drain amplifier in D10.6 there is a minimum I_{SUP} required to meet the specification. Suppose you could change one of the device parameters given for the NMOS transistor (i.e. C_{ox}, C_{ov}, V_{Tn} etc.). Which device parameter would allow you to maximize the reduction of I_{SUP} and still meet the specification? Describe a technological method to achieve the device parameter change (i.e. change doping, change area, change gate oxide thickness etc.).

D10.8 A new hot-shot device engineer is assigned to improve a BiCMOS process for your application. You have several common-base and common-gate amplifiers that are suffering from insufficient bandwidth. Make a priority list of the changes in the process that would benefit your design. Justify your list by doing a design of both a common-base and common-gate amplifier with the nominal parameters. Then change the parameters which you have highest on your priority list to quantify the improvement.

D10.9 In Design Example 10.9 a CMOS cascode amplifier was designed to drive a load resistance of 100 kΩ. If we add the capacitance at the load, the frequency response is degraded. Add C_{db} from device M_2, and an additional 50 fF for the wiring and current source supply.

(a) Re-calculate ω_{3dB} to include the effect of the load capacitance.

(b) Re-design the amplifier to achieve the original ω_{3dB} and |voltage gain| ≥ 50. You can increase I_{SUP} if necessary.

D10.10 Repeat D10.9 for the BiCMOS amplifier also designed in Example 10.9.

D10.11 Design a voltage amplifier with the topology given in Fig. 10.32. You need a DC |voltage gain| of 5000 and a unity-gain frequency of 100 MHz. Try to minimize device area and power dissipation. To begin the design, estimate the device sizes required to achieve biasing and the DC gain. Use the results of Example 10.9 to help you find a starting point. Calculate the bandwidth and unity gain frequency. Iterate on the design until you are close to the specifications. Use SPICE to refine the design.

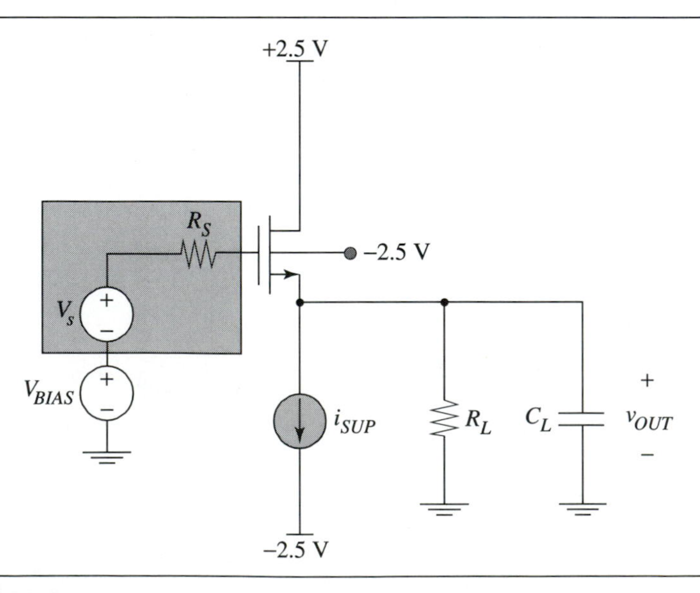

➤ **Figure D10.6**

chapter 11

Differential Amplifiers

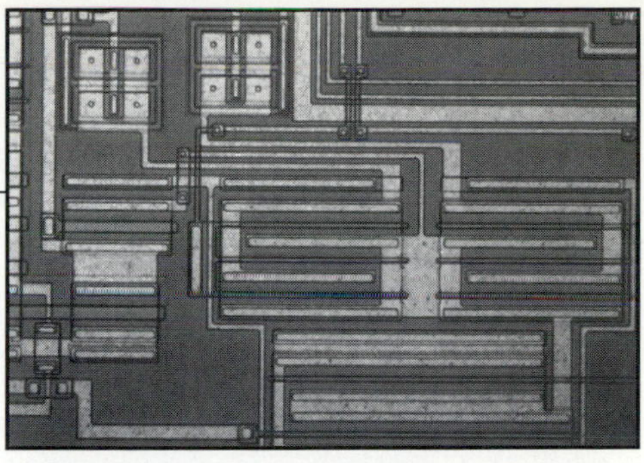

NMOS/npn bipolar cascoded differential amplifier with an NMOS bias current source. Designed by Mark A. Lemkin, UC Berkeley using the BiMEMS process. (Courtesy of Analog Devices, Inc.)

Differential amplifiers are a class of amplifiers that process the difference between two signals rather than the absolute value of those signals. The difference between two signals is often observed in our everyday lives. Consider any sporting event in which team A is leading team B by two points. It is the difference in the score that is important to the listener, especially at the end of the game.

In integrated circuits, differential signals are often used to remove unwanted signals that are common to both sources. For example, if the information of a particular physical quantity is carried by the difference between signal A and signal B, any unwanted signal that adds directly to both input signals will be rejected when the difference of those signals is processed. Examples of unwanted common-mode signals are:

1. Variation in the power supply voltage as a function of time
2. Variation of the substrate voltage of the integrated circuit
3. Variation in the temperature of the integrated circuit

By processing the difference between two signals while rejecting signals that are common to both inputs, the differential amplifier plays an essential role in IC design.

Chapter Objectives

- Differential circuits and how they amplify the difference between two signals independent of their average value.
- Two-port models for differential amplifiers.
- Frequency response of a differential amplifier.
- Analysis of a differential amplifier with a single-ended output.
- Large-signal behavior of MOS and bipolar differential amplifiers.

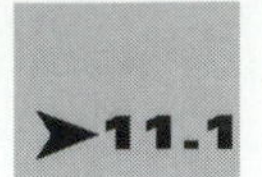

11.1 GENERAL CONCEPTS FOR DIFFERENTIAL AMPLIFIERS

This section examines one advantage of the differential amplifier topology, namely, that it simplifies the DC biasing. In addition, the concepts of differential and common-mode signals are defined.

11.1.1 DC Biasing Considerations

Consider the differential amplifier shown in Fig. 11.1. The components of a common-emitter/source amplifier can be seen in the shaded area on the left side of this figure. The active devices denoted by three-terminal boxes (1, 2) could be a matched pair of bipolar or MOS transistors. The input to each device consists of a DC voltage source V_{I1} in series with a small-signal voltage source v_{i1}. For the purposes of this discussion, we will consider the small-signal input voltage source equal to zero. The current into the collector/drain of the active device I_1 is supplied through resistor R_{C1}. The output voltage of this amplifier is denoted as V_{O1}.

A differential amplifier is formed by placing an identical common-emitter/source amplifier in parallel with the first common-emitter/source amplifier and connecting the emitter/source to a bias current source called I_{BIAS}. For the analysis in this chapter, we will use symmetrical power supply voltages $\pm V$ that are large enough to ensure all active devices are operating in their constant-current region. We will study this assumption and its impact on the input voltage common-mode range in Section 11.6.

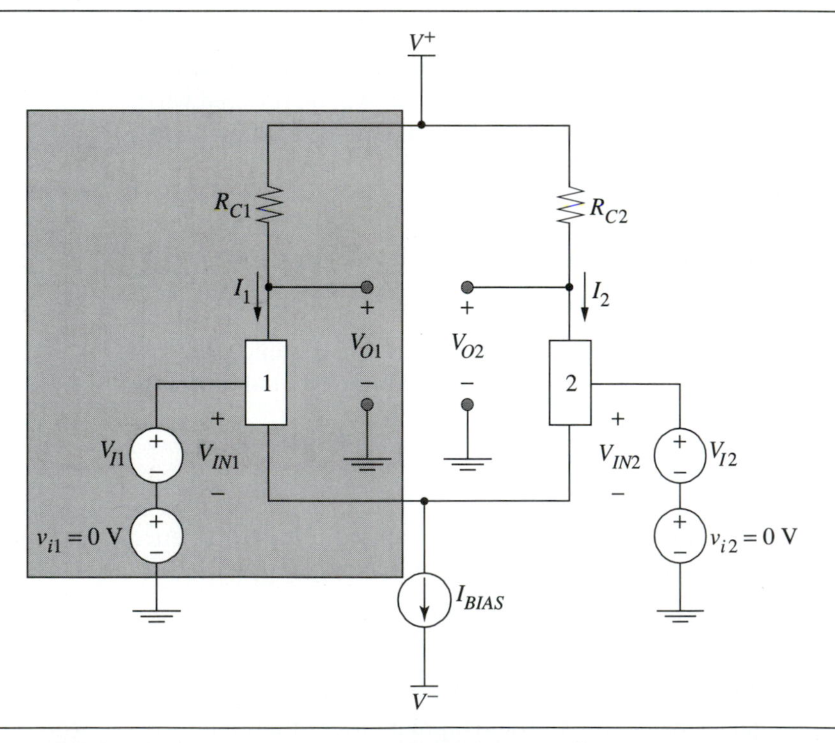

Figure 11.1 Circuit topology for a differential amplifier.

In the following analysis of biasing the differential amplifier, we assume that the input current into the active devices is negligible. The goal of biasing the differential amplifier is to center the output voltage of both amplifiers between V^+ and V^-. This is analogous to the situation when we used a DC bias source at the input of the common-emitter/source amplifier to force the output voltage halfway between the power supply voltages. We want the output voltage to be approximately equal to zero since our power supply voltages are symmetrical in this configuration.

Consider that the large-signal input voltage sources are set equal, namely $V_{I1} = V_{I2}$. Performing KVL around the input loop, we see that the voltages across the input of the active devices must be equal, $V_{IN1} = V_{IN2}$. Since we are assuming that the active devices are perfectly matched, the drain/collector current flowing into these devices must also be equal, namely $I_1 = I_2$.

We begin the analysis by finding device currents I_1 and I_2 and assume the output voltages V_{O1} and V_{O2} are close to 0 V.

$$I_1 = \frac{V^+ - V_{O1}}{R_{C1}} \approx \frac{V^+}{R_{C1}} \tag{11.1}$$

$$I_2 = \frac{V^+ - V_{O2}}{R_{C2}} \approx \frac{V^+}{R_{C2}} \tag{11.2}$$

By KCL, the drain/collector current in the active devices I_1 and I_2 must sum to be equal to the bias current I_{BIAS}, assuming the input current to the active devices is approximately equal to zero. By using this fact and assuming $R_{C1} = R_{C2} = R_C$, we find a value for the bias current source given in

$$\boxed{I_{BIAS} = I_1 + I_2 = \frac{2V^+}{R_C}} \tag{11.3}$$

value of I_{BIAS} to center the output voltage

Since the devices are perfectly matched and $V_{I1} = V_{I2}$, the drain/collector currents will also be identical. The bias current source is the sum of these device currents. By setting the value of the bias current source by the criterion in Eq. (11.3), we ensure that the DC output voltages of each side of the differential amplifier are equal to zero. In Chapter 12, we will describe how to bias practical differential amplifiers where the input devices are not perfectly matched.

It should be noted that the large-signal input voltages could have been equal to zero. This means that small-signal voltage sources can be applied directly to the inputs of the differential amplifier, which greatly simplifies the biasing of the amplifier. Under this condition, the bias current source I_{BIAS} is pulling the proper amount of current through the active devices such that the output voltages are *centered* and approximately equal to zero. The emitter/source of the active devices adjusts itself to the correct potential to ensure that the voltages across the inputs of the active devices are exactly the proper value to carry the collector/drain currents $I_1 = I_2 = I_{BIAS}/2$.

11.1.2 Common-Mode and Differential-Mode Signals

Assume we have two signals x_1 and x_2. These signals could be in the form of voltages or currents depending on the particular circuit being analyzed. The **differential-mode**

signal is defined as the difference between these two signals, whereas the **common-mode signal** is defined as the average of the signals. The definitions are given in

differential and common-mode signal definitions

$$x_{DM} = x_1 - x_2 \tag{11.4}$$

$$x_{CM} = \frac{x_1 + x_2}{2} \tag{11.5}$$

Now we apply the differential and common-mode definitions to our differential circuit. In this analysis, the DC input voltage sources are set equal to zero and the bias current source is set to a proper value so the output voltages are equal to zero. Figure 11.2(a) shows a differential amplifier with small-signal input voltages v_{i1} and v_{i2} as the inputs to the amplifier.

These small-signal input voltages can be decomposed into their purely differential- and common-mode signals as shown in Fig. 11.2(b). By using KVL, we see that the input voltages v_{i1} and v_{i2} are given in

$$v_{i1} = v_{ic} + \frac{v_{id}}{2} \tag{11.6}$$

$$v_{i2} = v_{ic} - \frac{v_{id}}{2} \tag{11.7}$$

Using the definitions of differential and common-mode, we can write their relationship to v_{i1} and v_{i2}. Rearranging Eq. (11.6) and Eq. (11.7) we find

$$v_{ic} = \frac{v_{i1} + v_{i2}}{2} \tag{11.8}$$

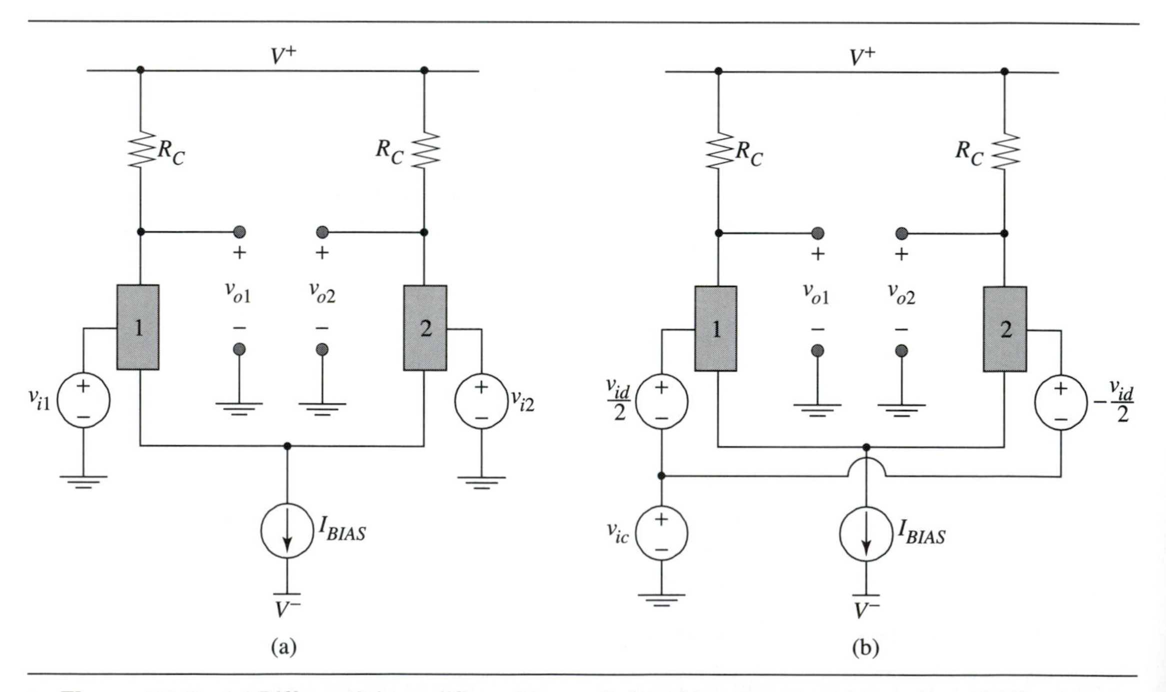

➤ **Figure 11.2** (a) Differential amplifier with small-signal input voltages v_{i1} and v_{i2}. (b) The input voltages have been decomposed into purely common-mode and differential-mode voltages.

$$v_{id} = v_{i1} - v_{i2} \tag{11.9}$$

We find that indeed, v_{id} is the differential-mode signal and v_{ic} is the common-mode signal for the inputs. To demonstrate the utility of such manipulations, consider the circuit shown in Fig. 11.3(a). Assume this circuit is biased properly and that a small-signal input voltage is applied to just one side of the amplifier. We can decompose the input voltage into both a common- and differential-mode signal as shown in Fig. 11.3(b). The differential-mode signal v_{id} is equal to v_{in} and the common-mode signal v_{ic} is equal to $v_{in}/2$.

In summary, we can decompose the input signals into their differential and common-mode parts, then analyze the differential amplifier assuming purely differential and common-mode signals are applied. Because small-signal inputs with linear circuits are being considered, superposition can be used to obtain the differential- and common-mode output responses of the circuit to any combination of differential- and common-mode input signals. These powerful concepts will be used in the next several sections.

11.2 SMALL-SIGNAL ANALYSIS OF DIFFERENTIAL AMPLIFIERS

In this section we analyze the differential amplifier using bipolar transistors as the active devices. A similar analysis can be performed for MOS transistors by letting $r_\pi \to \infty$. The goal of the differential amplifier is to amplify the difference between the input signals and reject any common-mode signal. The differential amplifier

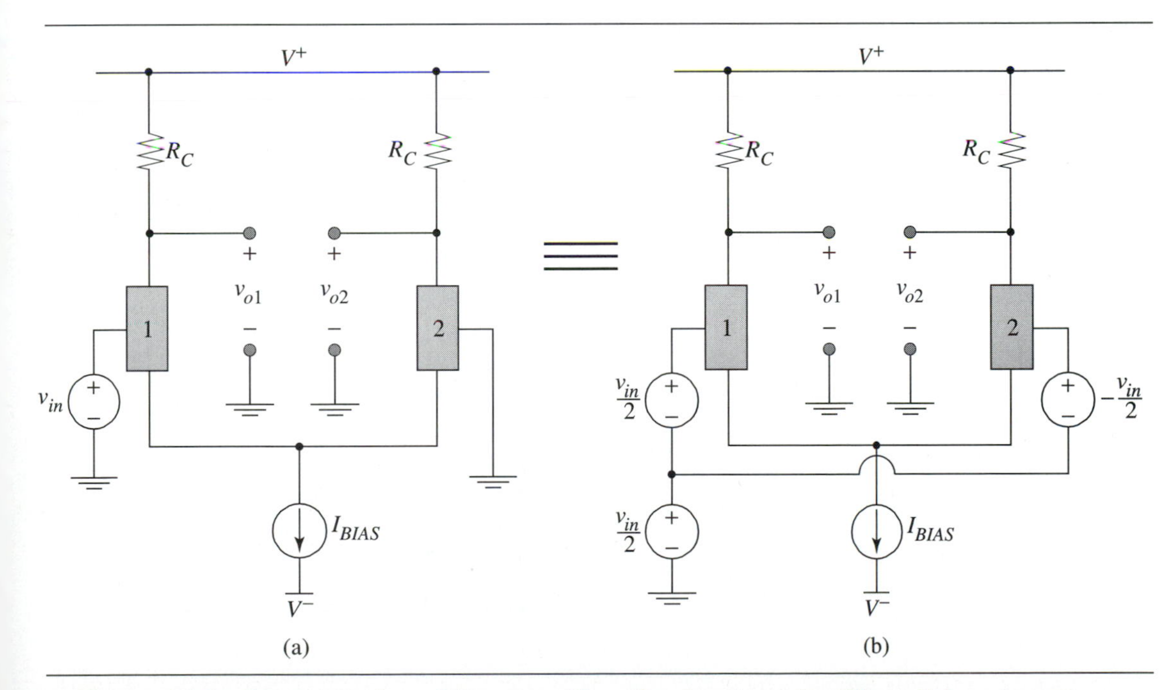

Figure 11.3 Demonstration of decomposition of a signal into its differential- and common-mode parts: (a) original circuit; (b) decomposition of the input voltage v_{in}.

shown in Fig. 11.4 is assumed to have a bias current source such that the output voltages are equal to zero when $v_{I1} = v_{I2} = 0$ V. The small-signal input voltage sources are directly applied to the differential amplifier. For convenience, we have chosen the DC input voltages $V_{I1} = V_{I2} = 0$ V. We account for the finite internal resistance of the bias current source with resistor r_{ob}.

11.2.1 Purely Differential-Mode Input, Small-Signal Model

For this portion of the analysis, we assume that the input voltages are purely differential, meaning that the common-mode voltage is set equal to zero. Under this condition, the input voltages are equal and opposite and have a magnitude $v_{id}/2$. A small-signal model of the differential amplifier is shown in Fig. 11.5. We have assumed that the transistors are matched implying $r_{\pi 1} = r_{\pi 2} = r_\pi$ and $g_{m1} = g_{m2} = g_m$.

The small-signal model for the bipolar transistor was simplified by assuming the output resistance of the transistor can be ignored since it is large compared to R_C. In addition, all DC sources were set equal to zero, namely the bias current source became an open circuit and the power supply voltages were short-circuited.

Note the tremendous amount of symmetry in this circuit before trying to determine the output voltages from this small-signal model. For example, currents i_1 and i_2 are equal and opposite since the transistors are perfectly matched and the input voltages are equal and opposite. Therefore, no current can flow through resistor r_{ob} and voltage v_x is equal to zero.

By redrawing the small-signal circuit as shown in Fig. 11.6, we can solve for the output voltages v_{o1} and v_{o2} as

$$v_{o1} = -g_m R_C \frac{v_{id}}{2} \tag{11.10}$$

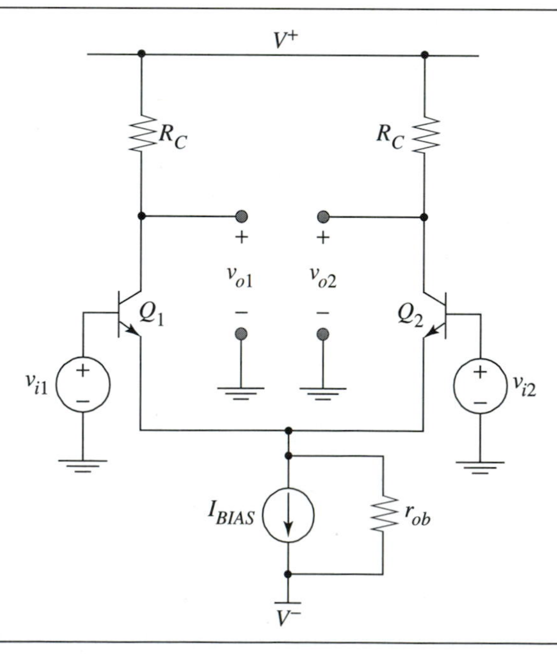

Figure 11.4 Bipolar differential amplifier. Resistor r_{ob} represents the internal resistance of the bias current source.

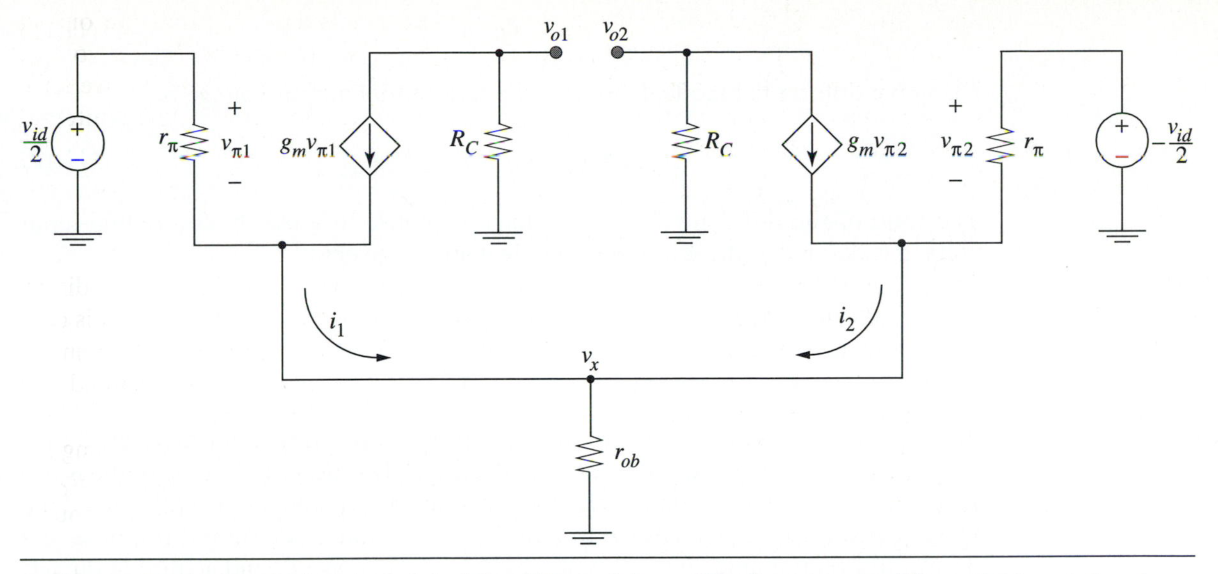

➤ **Figure 11.5** Small-signal model of the differential amplifier for a purely differential-input signal.

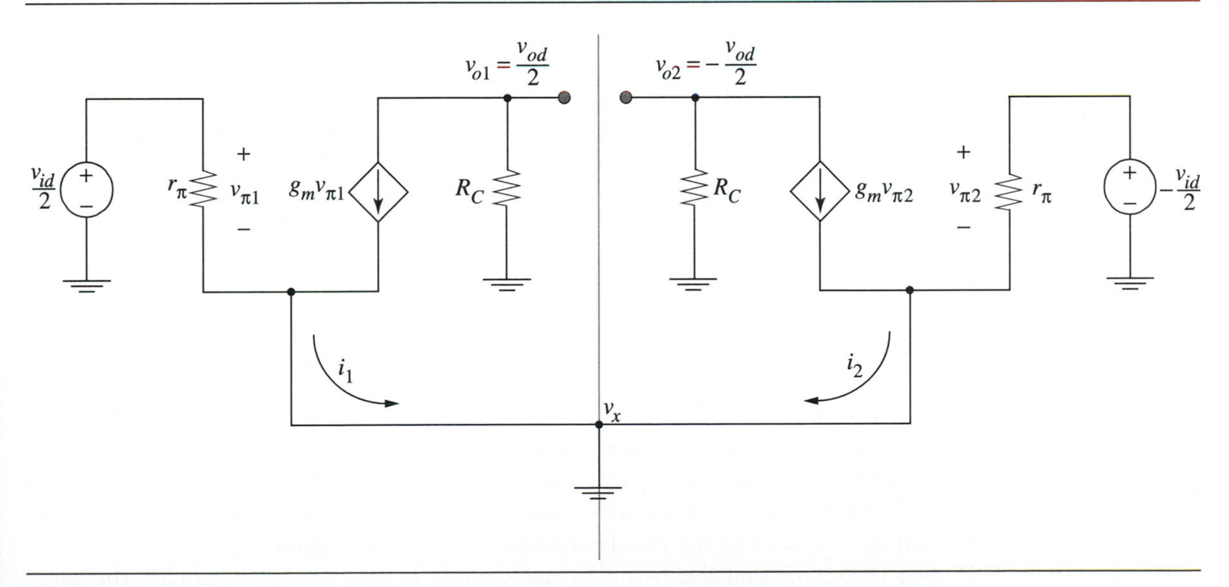

➤ **Figure 11.6** Small-signal model of the differential amplifier for a purely differential-input signal redrawn after realizing v_x is at incremental ground for purely differential signals.

$$v_{o2} = +g_m R_C \frac{v_{id}}{2} \tag{11.11}$$

Using the differential-mode definition, the differential output voltage is

$$v_{od} \equiv v_{o1} - v_{o2} = -g_m R_C v_{id} \tag{11.12}$$

The ratio of the differential output voltage response to a purely differential-input voltage is defined as the **differential-mode gain** a_{dm} given by

$$a_{dm} \equiv \frac{v_{od}}{v_{id}} = -g_m R_C \tag{11.13}$$

It is important to understand how to take advantage of this circuit's symmetry using a powerful concept called the **half-circuit technique**. If we apply a purely differential-mode input voltage and want to find the purely differential-mode output voltage, then the circuits are completely symmetric as shown in Fig. 11.6. Only one of the two circuits needs to be analyzed to find the differential-mode output response of a purely differential-mode input. Using this half-circuit concept and analyzing the left half of the circuit in Fig. 11.6, we find

$$\frac{v_{od}}{2} = -g_m R_C \frac{v_{id}}{2} \tag{11.14}$$

$$\frac{v_{od}}{v_{id}} = -g_m R_C \tag{11.15}$$

which is the same result obtained in Eq. (11.13). The half-circuit concept can *only* be used for purely differential input signals when the circuit is perfectly symmetric and we are interested in calculating the differential output signal.

11.2.2 Purely Common-Mode Input, Small-Signal Model

For this analysis we will assume that the input signal is purely common-mode and the differential-mode input signal is set equal to zero. Under this condition, we have a small-signal common-mode voltage source v_{ic} driving both sides of the differential amplifier. The resulting small-signal model is shown in Fig. 11.7.

Before plowing through KCL and KVL, it is worthwhile to examine carefully the symmetry of this circuit. Because all the components are matched and the input voltage source v_{ic} is the same on both sides, currents i_1 and i_2 must be exactly equal. The value of v_x is given in

$$v_x = (i_1 + i_2)\, r_{ob} = 2 i_1 r_{ob} = 2 i_2 r_{ob} \tag{11.16}$$

We can split the circuit in Fig. 11.7 into two identical circuits as shown in Fig. 11.8. The value of the internal resistance due to the bias current source is $2r_{ob}$ in order to generate the proper voltage v_x with only half of the current that flowed through it in Fig. 11.7. These identical circuits can now be analyzed to find the common-mode output voltage response to the purely common-mode input signal.

We recognize that the two identical circuits in Fig. 11.8 have exactly the same topology as the common-emitter amplifier with an emitter degeneration resistor that we analyzed in Chapter 8. In this configuration, the emitter degeneration resistor is

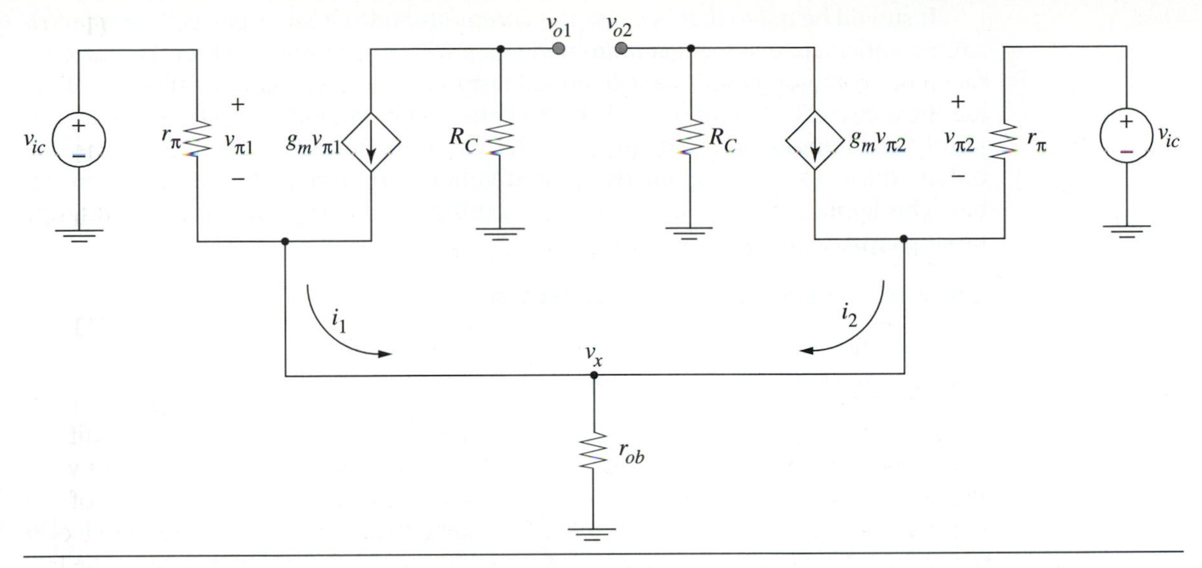

Figure 11.7 Small-signal model of the differential amplifier with a purely common-mode input signal.

equal to $2r_{ob}$. By assuming $\beta_o \gg 1$, the transfer function for the output voltages given the common-mode input voltage is

$$\frac{v_{o1}}{v_{ic}} = \frac{v_{o2}}{v_{ic}} \approx \frac{-g_m R_C}{1 + 2g_m r_{ob}} \tag{11.17}$$

As with the differential-mode response, we define the **common-mode gain** a_{cm} as the ratio of the output common-mode voltage response to a purely common-mode input voltage given by

$$a_{cm} \equiv \frac{v_{oc}}{v_{ic}} = \frac{(v_{o1} + v_{o2})/2}{v_{ic}} = \frac{-g_m R_C}{(1 + 2g_m r_{ob})} \tag{11.18}$$

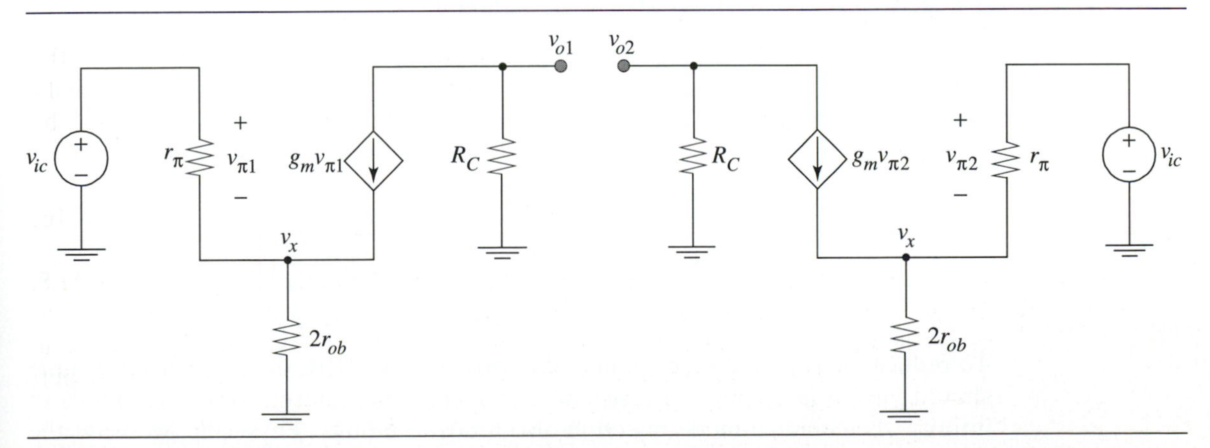

Figure 11.8 Small-signal model of the differential amplifier with a purely common-mode input signal redrawn after realizing $i_1 = i_2$ for purely common-mode signals.

It should be noted that because the common-mode circuit is perfectly symmetrical, we anticipate that the common-mode gain will be the same whether the output is taken in a common-mode or single-ended fashion from either output. Only one of the identical circuits requires analysis to find the common-mode output response to a purely common-mode input. Because the circuit is perfectly symmetrical, zero is obtained when subtracting the two output voltages to form a differential output signal. This implies that the voltage gain for a differential output with a common-mode input is equal to zero.

11.2.3 General Input Signals

Superposition can be used to find the output voltages v_{o1} and v_{o2} when input signals having both common- and differential-mode components are applied. The first step is to write the output voltages as a function of the purely common-mode input voltage v_{ic} and the purely differential-mode input voltage v_{id} as

$$v_{o1} = v_{oc} + \frac{v_{od}}{2} = a_{cm} v_{ic} + a_{dm} \frac{v_{id}}{2} \tag{11.19}$$

$$v_{o2} = v_{oc} - \frac{v_{od}}{2} = a_{cm} v_{ic} - a_{dm} \frac{v_{id}}{2} \tag{11.20}$$

The relationship between the general input voltages v_{i1} and v_{i2} to the purely differential and common-mode input voltages v_{id} and v_{ic} given in Eq. (11.8) and Eq. (11.9), can be substituted into Eq. (11.19) and Eq. (11.20) to yield

relationship between output voltages and arbitrary small-signal input voltages

$$v_{o1} = a_{cm} \frac{(v_{i1} + v_{i2})}{2} + a_{dm} \frac{(v_{i1} - v_{i2})}{2} \tag{11.21}$$

$$v_{o2} = a_{cm} \frac{(v_{i1} + v_{i2})}{2} - a_{dm} \frac{(v_{i1} - v_{i2})}{2} \tag{11.22}$$

By knowing the differential and common-mode gain, we can find the output voltages v_{o1} and v_{o2} for any arbitrary small-signal input voltages.

11.2.4 Common-Mode Rejection

As previously stated, the goal of a differential amplifier is often to amplify the difference signal and reject the common-mode signal. A figure of merit that describes the quality of the differential amplifier is the ratio of the differential-mode gain to the common-mode gain which is called the **common-mode rejection ratio** (*CMRR*). From Eq. (11.15) and Eq. (11.18), this rejection ratio is given by

common-mode rejection ratio definition

$$CMRR \equiv \left| \frac{(v_{od}/v_{id})}{(v_{oc}/v_{ic})} \right| = \frac{a_{dm}}{a_{cm}} = (1 + 2g_m r_{ob}) \tag{11.23}$$

To reduce the common-mode gain and improve the CMRR, the bias current source should have a large internal resistance. In fact, as the internal resistance tends to infinity, the common-mode rejection also tends to infinity. Physically, we want the internal resistance to be large so that as the input voltage changes, the voltage at the common node, namely the emitter/source, follows the input voltage perfectly. Under this condition the device current does not vary and there is no change in the output voltage.

In this section, we have analyzed the common-mode and differential-mode response to purely differential- and common-mode input signals to the differential amplifier. Using superposition, we analyzed the differential amplifier with arbitrary small-signal input voltages. We also performed the analysis using bipolar transistors as our active devices. The results for MOS transistors are identical since we assumed β_o of the bipolar transistors was large.

➤ EXAMPLE 11.1 Analysis of a Differential Amplifier

In this example we will analyze a differential amplifier by looking at its purely differential- and common-mode signals. In the circuit shown in Fig. Ex11.1, which is driven by a single-ended voltage source, the output voltage is taken from the right hand side of the amplifier as a single-ended voltage. A **single-ended** voltage is defined to be a voltage that is referenced to the ground. Assume that all devices are perfectly matched and are operating in their constant-current regions and $R_C = 10\ \text{k}\Omega$, $r_{ob} = 50\ \text{k}\Omega$. Also, assume the input current to be negligible $\beta_o \to \infty$.

SOLUTION

Begin by writing expressions for the purely differential- and common-mode input voltages in terms of the input voltage source v_{in}. Note: $v_{i1} = v_{in}$ and $v_{i2} = 0$ V.

$$v_{id} \equiv v_{i1} - v_{i2} = v_{in} \text{ and } v_{ic} \equiv (v_{i1} + v_{i2})/2 = v_{in}/2$$

Now we write the output voltage v_o in terms of the differential- and common-mode output voltages.

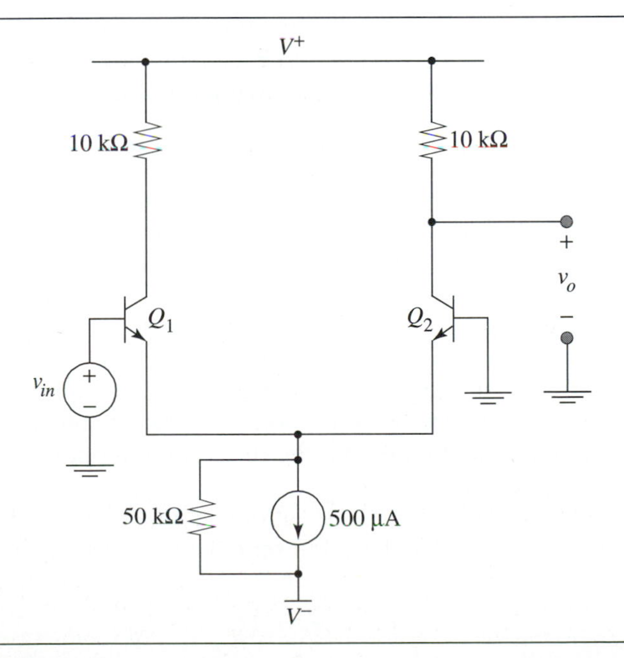

➤ **Figure Ex11.1**

$$v_o = v_{o2} = v_{oc} - v_{od}/2$$

Recognizing that $v_{oc} = a_{cm}v_{ic}$ and $v_{od} = a_{dm}v_{id}$, we can write

$$v_o = a_{cm}v_{ic} - a_{dm}v_{id}/2 = a_{cm}v_{in}/2 - a_{dm}v_{in}/2\text{, leading to } \frac{v_o}{v_{in}} = \frac{a_{cm}}{2} - \frac{a_{dm}}{2}.$$

Now the transconductance of one of the devices is

$$g_m = \left(\frac{q}{kT}\right)\frac{I_{BIAS}}{2} = \frac{1}{25\text{mV}}\frac{500\ \mu\text{A}}{2} = 10\text{ mS}$$

From our previous analysis, $a_{dm} = -g_mR_C = -10\text{ mS}\cdot 10\text{ k}\Omega = -100$

$$a_{cm} = \frac{-g_mR_C}{1 + 2g_mr_{ob}} = -\frac{-100}{1 + 2(10\text{ mS})(50\text{ k}\Omega)} \approx -0.1$$

$$\frac{v_o}{v_{in}} = -0.05 + 50 \approx 50$$

Note that the differential gain was reduced by a factor of 2 because we used a single-ended output. We will learn how to restore that factor of 2 in Section 11.5. In addition, you can see that the common-mode rejection ratio is about 1000.

11.3 TWO-PORT MODEL FOR THE DIFFERENTIAL AMPLIFIER

In this section, we develop a two-port model for the differential amplifier. The model is developed assuming a purely differential-input signal and a differential-output signal. The two-port model developed in this section is not valid for common-mode signals. Since common-mode signals are largely rejected by differential amplifiers, the purely differential two-port model does not have serious limitations.

11.3.1 Differential Amplifier Two-Port Model Parameters

The two-port model for this purely differential voltage amplifier is shown in Fig. 11.9. As with previous two-port models, it contains an input resistance, a gain term, and an output resistance.

The **differential voltage gain** a_{dm}, of this amplifier is defined as the differential output voltage response to a purely differential input voltage which was given in Eq. (11.13) as $-g_mR_C$. Now we need to determine the differential input resistance and differential output resistance for the two-port model. We can use the half-circuit concept developed in the previous section to find the differential input resistance. The half-circuit model is shown in Fig. 11.10.

By applying a test voltage equal to $v_t/2$ at one input and measuring the current i_t out of that voltage generator we find the differential input resistance given in

$$R_{id} = \frac{v_t}{i_t} = 2r_\pi \tag{11.24}$$

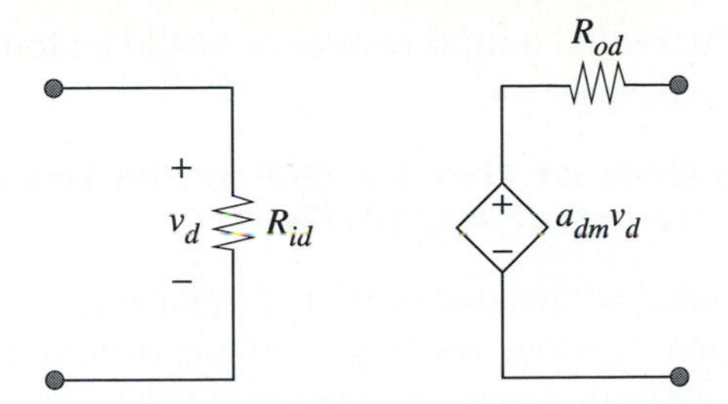

➤ **Figure 11.9** Two-port model for a purely differential voltage amplifier.

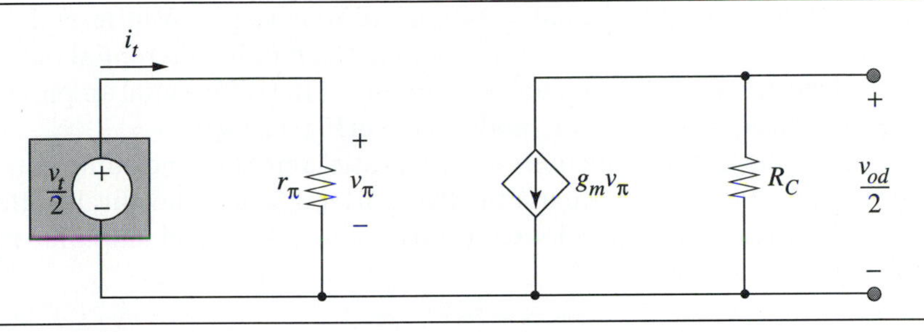

➤ **Figure 11.10** Half-circuit model to determine the differential input resistance for the two-port model.

If an MOS transistor is used as the active device for the differential amplifier, the input resistance is infinity.

To find the differential output resistance, we again use the half-circuit concept and apply a differential test voltage to the output as shown in Fig. 11.11. The input voltage source is set to zero. Under this condition $v_\pi = 0$ V and the differential output resistance is

$$R_{od} = \frac{v_t}{i_t} = 2R_C \tag{11.25}$$

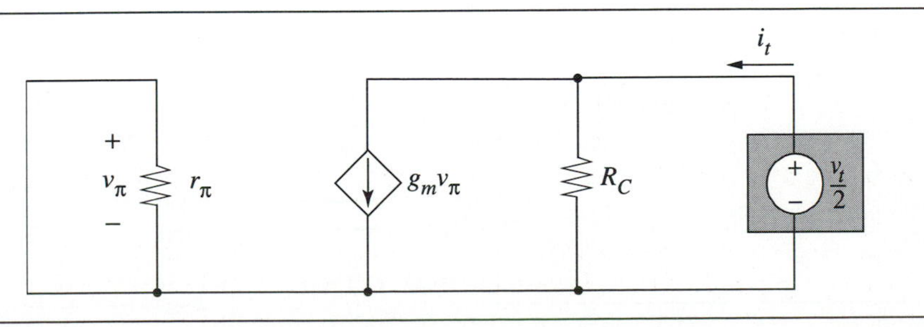

➤ **Figure 11.11** Half-circuit model to determine the differential output resistance for the two-port model.

Note that the same differential output resistance would be found if MOS transistors were used as the active devices.

11.3.2 Application of the Two-Port Model for the Differential Amplifier

As discussed in previous chapters, two-port models for amplifiers are useful to understand the effect of source and load resistances on the overall transfer function of the amplifier. A purely differential input voltage with source resistance R_S is shown in Fig. 11.12(a). The equivalent circuit for this purely differential input is given in Fig. 11.12(b). As expected, we have a purely differential input voltage v_{id} and a source resistance equal to twice the source resistance of the individual differential voltage sources.

At the output of the differential amplifier, we attach a purely differential load as shown in Fig. 11.13(a). Since we are interested in the purely differential output voltage v_{od}, the load resistors attached to both outputs of the differential amplifier add in series, yielding an equivalent circuit model given in Fig. 11.13(b).

To evaluate the transfer function of a differential amplifier, including both source resistance and load resistance, we use the two-port model and the purely differential source and load resistances developed in this section. The total amplifier model is shown in Fig. 11.14.

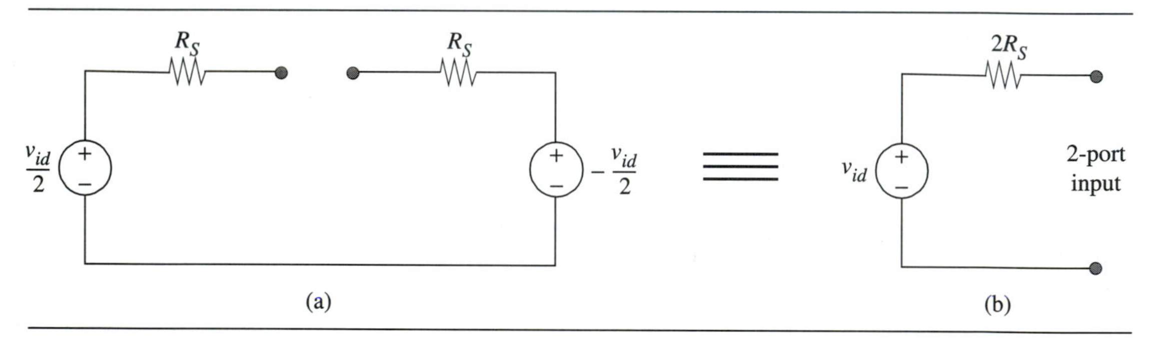

➤ **Figure 11.12** (a) Purely differential input voltage with source resistance R_S; (b) Equivalent circuit.

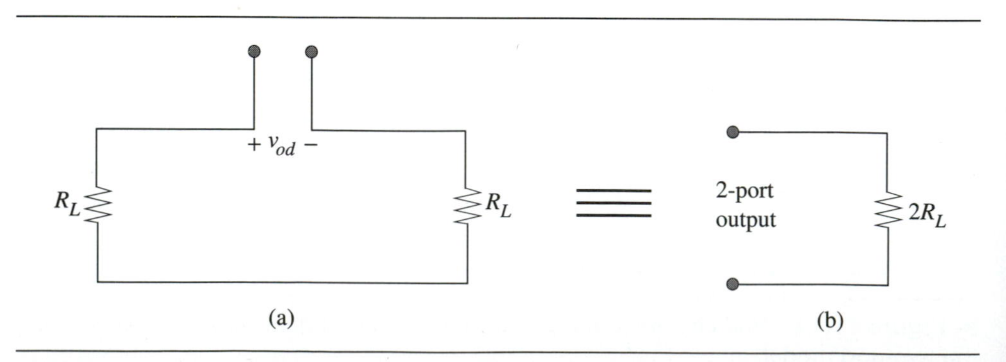

➤ **Figure 11.13** (a) Purely differential output load resistance; (b) Equivalent circuit.

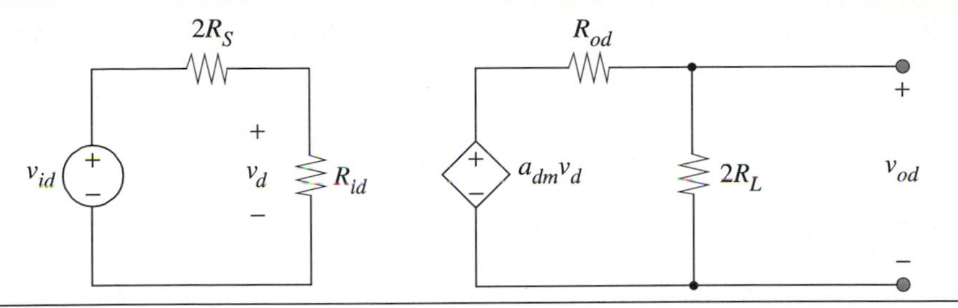

Figure 11.14 Full model for a purely differential amplifier including source and load resistances.

The complete transfer function for the differential voltage gain is given in

$$\frac{v_{od}}{v_{id}} = \left(\frac{R_{id}}{R_{id} + 2R_S}\right)(-g_m R_C)\left(\frac{2R_L}{R_{od} + 2R_L}\right) \tag{11.26}$$

The transfer function for this model has three terms.

1. Gain reduction for the finite input resistance of the amplifier.
2. Intrinsic voltage gain of the differential amplifier.
3. Gain reduction due to the finite output resistance of the amplifier.

Substituting the results we obtained for the differential input and output resistances for the bipolar implementation of the differential amplifier, we obtain

$$\boxed{\frac{v_{od}}{v_{id}} = \left(\frac{r_\pi}{r_\pi + R_S}\right)(-g_m R_C)\left(\frac{R_L}{R_C + R_L}\right)} \tag{11.27}$$

differential voltage gain for bipolar and MOS differential amplifiers including R_S & R_L

Since the MOS transistor has an infinite input resistance, the first term in Eq. (11.27) becomes unity and the differential voltage gain expression is

$$\boxed{\frac{v_{od}}{v_{id}} = (-g_m R_D)\left(\frac{R_L}{R_D + R_L}\right)} \tag{11.28}$$

when MOS transistors are used as the active devices.

In this section we developed a two-port model for the differential amplifier assuming purely differential inputs and outputs. By using this two-port model, we can account for the effect of source and load resistances on the overall transfer function of the differential amplifier.

11.4 FREQUENCY RESPONSE OF DIFFERENTIAL AMPLIFIERS

In this section we analyze the frequency response of the bipolar differential amplifier shown in Fig. 11.15. The same analysis techniques apply to an MOS implementation. First we analyze the frequency response of a differential output voltage given a purely differential input voltage. We then analyze the common-mode output voltage

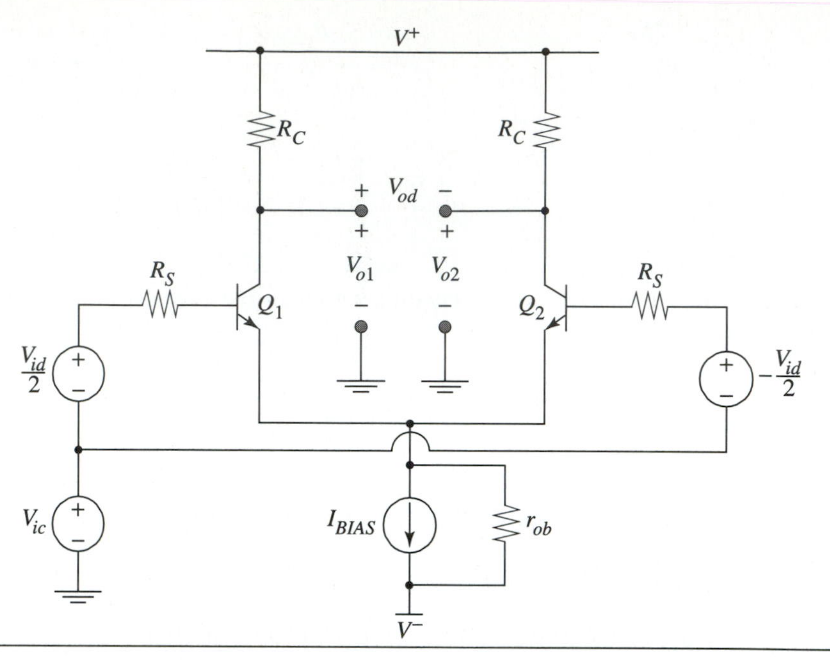

➤ **Figure 11.15** Differential amplifier circuit to be analyzed for frequency response.

given a purely common-mode input voltage. We will apply the half-circuit techniques described in the previous section to simplify the analysis.

11.4.1 Differential-Mode Frequency Response

Since we are analyzing a perfectly symmetrical differential amplifier for its differential-mode response, the half-circuit technique can be used to generate a small-signal model for the amplifier as shown in Fig. 11.16. We begin the analysis of this half-circuit by calculating the low frequency small-signal gain, neglecting the effect of capacitances C_π and C_μ. The ratio of the differential output voltage to the differential input voltage is given by

$$\frac{v_{od}}{v_{id}} = -\left(\frac{r_\pi}{r_\pi + R_S}\right)(g_m R_C) \tag{11.29}$$

To determine the frequency response of this circuit, the Miller Approximation is used to find the effect of capacitor C_μ. Since the voltage gain across capacitor C_μ is equal to $(-g_m R_C)$, the total capacitance, C_T at the input of the intrinsic amplifier is

$$C_T = C_\pi + (1 + g_m R_C)C_\mu \tag{11.30}$$

The Thévenin resistance seen across this capacitance is simply the source resistance R_S in parallel with r_π. The dominant time constant for the differential voltage gain is given by

$$R_T C_T = (R_S \| r_\pi)C_T \tag{11.31}$$

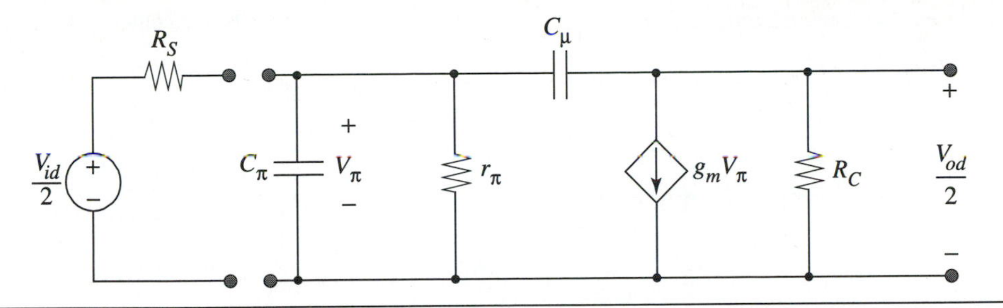

Figure 11.16 Differential mode half-circuit small-signal model used to analyze the differential-mode frequency response of the differential amplifier.

We can write the differential voltage gain as a function of frequency as

$$\frac{V_{od}}{V_{id}} \approx -\left(\frac{r_\pi}{r_\pi + R_S}\right)(g_m R_C)\left[\frac{1}{1 + j\omega(R_S \| r_\pi)[C_\pi + (1 + g_m R_C)C_\mu]}\right] \quad (11.32)$$

differential voltage gain frequency response of bipolar differential amplifiers

Other analysis techniques such as the open-circuit time constant method can be used to solve for the frequency response of the differential-mode voltage gain. The Miller Approximation gives a reasonably accurate result with an extremely simple analysis.

Cascode transistors can be placed in the differential amplifier to increase the bandwidth of the differential-mode response using the same method discussed in Chapter 10. The cascode transistors will keep the collector voltages of the input transistors Q_1 and Q_2 from moving significantly, which minimizes the effect of C_μ on the overall differential voltage gain frequency response.

The differential-mode frequency response for an MOS implementation can be obtained by letting $r_\pi \to \infty$ and replacing R_C with R_D, C_π with C_{gs} and C_μ with C_{gd}. The resulting transfer function is

$$\frac{V_{od}}{V_{id}} \approx -(g_m R_D)\left[\frac{1}{1 + j\omega R_S[C_{gs} + (1 + g_m R_D)C_{gd}]}\right] \quad (11.33)$$

differential voltage gain frequency response of MOS differential amplifiers

EXAMPLE 11.2 Differential-Mode Frequency Response of an MOS Differential Amplifier

The MOS differential amplifier shown in Fig. Ex11.2 has a bias current source of 50 μA and source resistances $R_S = 5$ kΩ. Using this amplifier, specify the W/L of the MOS transistor and the value of R_D needed to achieve a low frequency differential voltage gain of –50. Also, estimate the 3 dB bandwidth of the circuit. For this example, set the DC common-mode output voltage to $V_{O1} = V_{O2} = 0$ V.

SOLUTION

Begin by ensuring the common-mode output voltage is equal to 0 V when $v_{id} = 0$ V.

$$V_{O1} = V_{O2} = 2.5\text{ V} - \frac{I_{BIAS}}{2} R_D = 0\text{V}$$

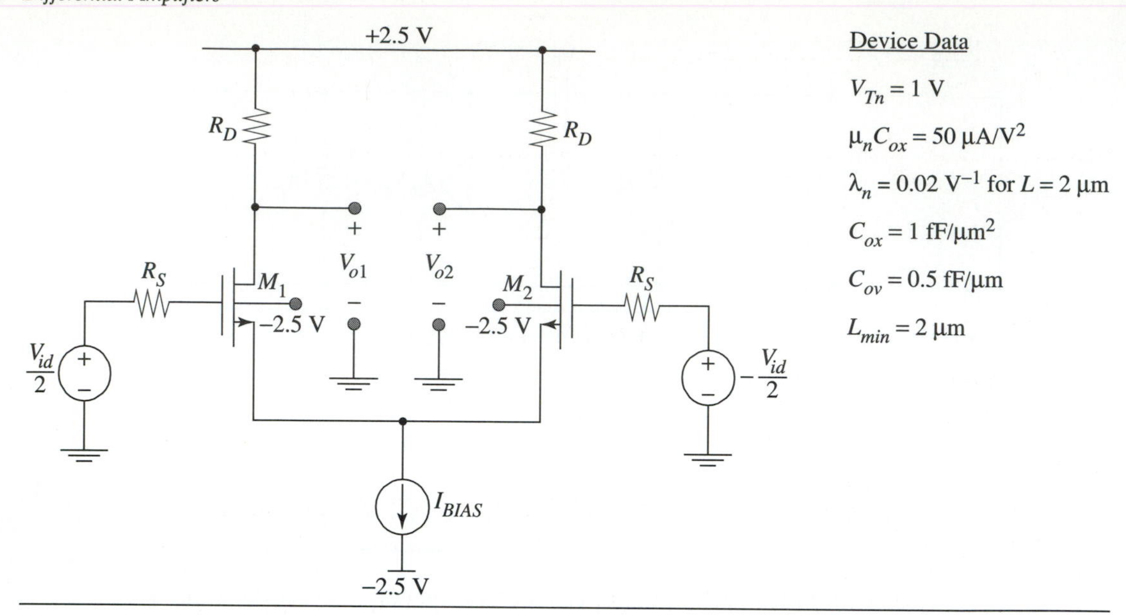

➤ **Figure Ex11.2**

$$\text{So } R_D = \frac{2 \cdot 2.5\ \text{V}}{I_{BIAS}} = \frac{5\ \text{V}}{50\ \mu\text{A}} = 100\ \text{k}\Omega$$

Since we know $R_D = 100$ kΩ, we can find the required g_m needed to achieve a differential-mode gain of −50. From the value of g_m we calculate W/L.

$$a_{dm} = -g_m(R_D \| r_o) \approx -g_m R_D = -50\text{, thus } g_m = -\frac{a_{dm}}{R_D} = \frac{50}{100\ \text{k}\Omega} = 0.5\text{mS}$$

$$g_m = \sqrt{2\left(\frac{W}{L}\right)_{1,2} \mu_n C_{ox} \frac{I_{BIAS}}{2}}\text{, so}$$

$$\left(\frac{W}{L}\right)_{1,2} = \frac{g_m^2}{\mu_n C_{ox} I_{BIAS}} = \frac{(0.5\ \text{mS})^2}{(50\,\mu\text{A/V}^2)\ (50\ \mu\text{A})} = 100$$

We will choose the minimum length of $L = 2\ \mu$m to achieve the best frequency response. Therefore, $(W/L)_{1,2} = 200/2$.
Note: a typical value of λ_n for $L = 2\ \mu$m is 0.02 V^{-1} which yields an output resistance
$r_o = 1/\lambda_n I_{BIAS} = 1$MΩ. Our gain estimate neglected this resistance.

$$C_{gs} = \frac{2}{3} WLC_{ox} = \frac{2}{3} \cdot 200\ \mu\text{m} \cdot 2\ \mu\text{m} \cdot 1\text{fF}/\mu\text{m}^2 = 0.27\text{pF}$$

$$C_{gd} = WC_{ov} = 200\ \mu\text{m}\,(0.5\ \text{fF}/\mu\text{m}) = 0.1\text{pF}$$

The 3 dB frequency can be estimated using the Miller Approximation.

$$C_M \approx g_m R_D C_{gd} = 50\,(0.1\text{pF}) = 5\ \text{pF}$$

$$f_{3dB} = \frac{\omega_{3dB}}{2\pi} = \frac{1}{2\pi R_S (C_{gs} + C_M)} = \frac{1}{2\pi (5\ \text{k}\Omega)\ (5.27\,\text{pF})} = 6.04\ \text{MHz}$$

11.4.2 Common-Mode Frequency Response

In this section we analyze the frequency response of the common-mode output voltage given a purely common-mode input voltage. Since the circuit is perfectly symmetrical, we can again use the half-circuit technique. The half-circuit for the purely common-mode frequency response is shown in Fig. 11.17.

The contribution of the internal resistance of the bias current source to the common-mode half-circuit is twice its actual value. We use a resistor equal to $2r_{ob}$ to account for this fact. A capacitor C_E, in parallel with the bias current source represents the parasitic capacitance at this node. Since this capacitance has the dominant effect on the frequency response of the common-mode voltage gain, we must take this parasitic capacitance into account. The contribution of the impedance of this capacitor to the common-mode half circuit is twice its actual impedance. Therefore, we place a capacitor of value $C_E/2$ in the common-mode half-circuit to account for this impedance.

We begin the analysis by solving for the common-mode voltage gain at DC by open-circuiting all the capacitors. The amplifier looks exactly like the common-emitter amplifier with an emitter degeneration resistor that was analyzed in Chapter 8. The value of the emitter degeneration resistor is $2r_{ob}$. Since the emitter degeneration resistor is large, the input resistance of this amplifier is large and the effect of the

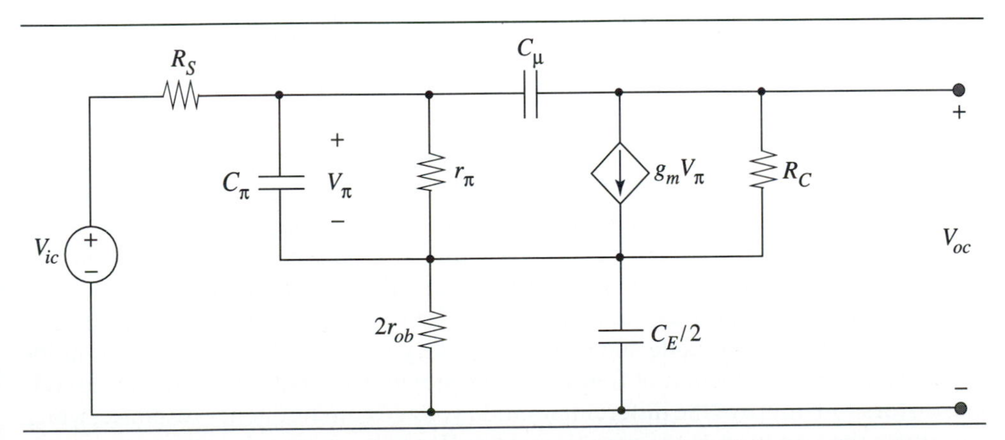

Figure 11.17 Common-mode half-circuit small-signal model to analyze the common-mode frequency response of the differential amplifier.

source resistance R_S can be neglected, provided ($R_S \ll \beta_o r_{ob}$). The low frequency common-mode voltage gain is approximately given by

$$\frac{v_{oc}}{v_{ic}} \approx \frac{-g_m R_C}{1 + g_m 2 r_{ob}} \tag{11.34}$$

Note that the common-mode voltage gain is much less than 1, assuming $R_C \ll r_{ob}$.

Examining the frequency response of this amplifier configuration, we note that since the common-mode voltage gain is very small, the Miller Effect from capacitor C_μ will also be small. In fact, the capacitance that dominates the frequency response of this amplifier is the parasitic capacitance C_E. We perform a first-order analysis of the frequency response of the common-mode voltage gain by considering only capacitor C_E. The impedance of the parallel combination of $C_E/2$ and $2r_{ob}$ is given by

$$Z_e = \frac{2r_{ob}}{1 + j\omega r_{ob} C_E} \tag{11.35}$$

The common mode voltage gain is found by replacing the emitter degeneration resistor with impedance, Z_e yielding

$$\frac{V_{oc}}{V_{ic}} = \frac{-g_m R_C}{1 + g_m Z_e} \tag{11.36}$$

For $|g_m Z_e| \gg 1$which is valid for frequencies $\omega \ll 2g_m/C_E$

common mode voltage gain frequency response of bipolar differential amplifiers

$$\boxed{\frac{V_{oc}}{V_{ic}} \approx \frac{-R_C}{2r_{ob}}(1 + j\omega r_{ob} C_E)} \tag{11.37}$$

Note that the common-mode voltage gain starts at a low value. The voltage gain increases with frequency since the capacitance C_E reduces the degeneration impedance. Hence, the effective transconductance of the amplifier increases. At a higher frequency, capacitors C_μ and C_π decrease the common-mode voltage gain. From an engineering point of view, we are primarily interested in making the common-mode voltage gain as small as possible. Therefore, only capacitor C_E concerns the designer.

The common-mode rejection ratio for the differential amplifier decreases as frequency increases. This reduction occurs since the differential-mode voltage gain falls at high frequency due to the Miller Effect and the common-mode voltage gain increases because of parasitic capacitance C_E. Figure 11.18 shows a plot of the magnitude of the differential-mode and common-mode gain, as well as the common-mode rejection ratio as a function of frequency. As previously mentioned, adding cascode transistors will improve the differential-mode voltage gain frequency response, which improves the common-mode rejection ratio. We will explore the effect of cascode transistors later in Example 11.4.

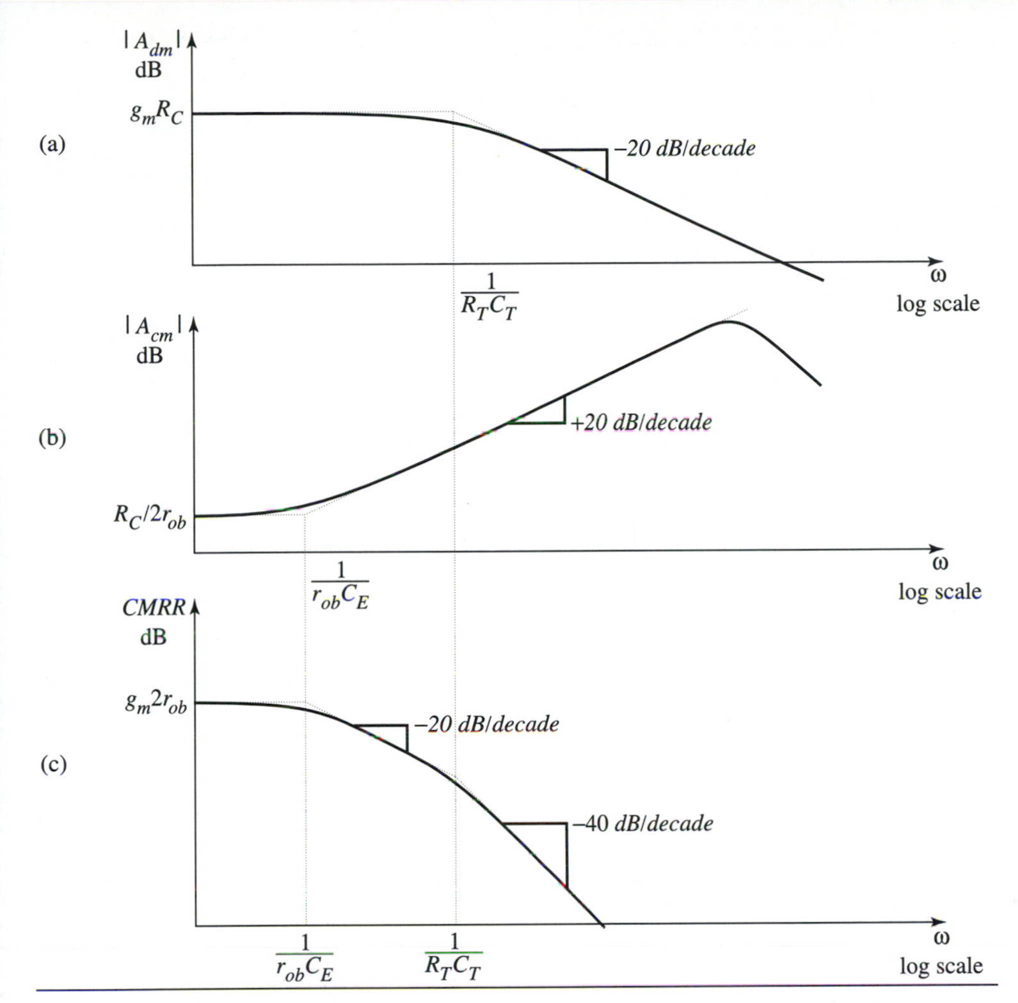

➤ **Figure 11.18** Bode plot of the magnitude of the gain parameters for the differential amplifier in Figure 11.15: (a) differential-mode gain; (b) common-mode gain; (c) common-mode rejection ratio.

DESIGN EXAMPLE 11.3 Common-Mode Rejection Ratio of an MOS Differential Amplifier

In this example we quantitatively investigate the common-mode rejection ratio for an all MOS differential amplifier. The *CMRR* is intimately related to the design of the bias current source. We use the differential amplifier that we analyzed in Example 11.2 and add a transistor-level design of a bias current source. The circuit topology is shown in Fig. Ex 11.3A.

Specify the W/L of M_3, M_4, and M_5, and give a value for R_{REF} so that $I_{BIAS} = 50\ \mu A$ and $CMRR = 2500$ at DC. Recall that $a_{dm} = -50$ from Example 11.2. Analyze the circuit to estimate the frequency at which the *CMRR* is reduced by 3 dB.

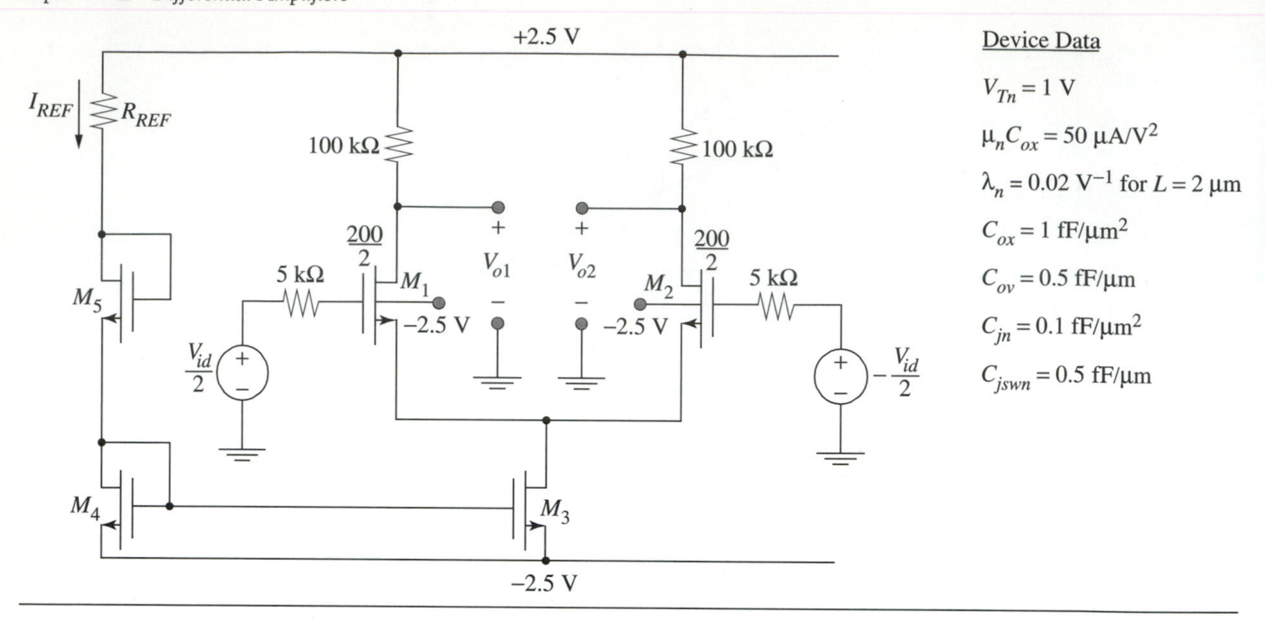

➤ **Figure Ex11.3A**

SOLUTION

We begin this design by reviewing how we picked the circuit topology for this example. R_{REF} is used to generate a reference current. The reference current flows into devices M_5 and M_4. If device M_4 and the bias current source of the differential pair, M_3 have the same size, then the current flowing through the two devices will be equal. Under this condition the bias current is the same as I_{REF}. The best practice for this type of design is to try to use matched device sizes so the voltages across M_5, M_4, and the resistor are close to those in the differential amplifier—namely, $M_{1,2}$, M_3, and the drain resistors respectively.

Since we designed M_3 and M_4 to be identical, 50 μA will flow through R_{REF} and 25 μA will flow through devices M_1 and M_2 and the 100 kΩ drain resistors. Thus there will be 2.5 V across the drain resistors. Since we are trying to match the voltage across R_{REF} with the drain resistors, $R_{REF} = \dfrac{2.5\text{ V}}{50\ \mu\text{A}} = 50\text{ k}\Omega$. In other words, R_{REF} has twice as much current flowing through it than the drain resistors, so it should have half the resistance in order to have the same voltage drop. Similarly, since we want the current flowing through M_5 to be twice that flowing through $M_{1,2}$, we will make the W/L of M_5 twice that of $M_{1,2}$.

$$\left(\frac{W}{L}\right)_5 = 2\left(\frac{W}{L}\right)_{1,2} = \frac{400}{2}$$

To size devices M_3 and M_4 we begin by determining their drain voltages. To do this, we first must determine the voltage at the source of M_5. Recall that the DC gate voltage of M_5 is 0 V. To find its source voltage we write

$$V_{GS5} = -V_{S5} = V_{Tn} + \sqrt{\frac{I_{REF}}{\frac{1}{2}\left(\frac{W}{L}\right)_5 \mu_n C_{ox}}}, \quad \text{so}$$

$$V_{S5} = -\left(1.0\text{ V} + \sqrt{\frac{50\ \mu\text{A}}{100 \cdot 50\ \mu\text{A/V}^2}}\right) = -1.1\text{V}$$

Therefore, devices M_3 and M_4 have $V_{GS3,4} = -1.1\text{ V} - (-2.5\text{ V}) = 1.4\text{ V}$. Now we find the $(W/L)_{3,4}$ ratio of these devices.

$$I_{D3,4} = I_{BIAS} = \frac{1}{2}\left(\frac{W}{L}\right)_{3,4} \mu_n C_{ox} (V_{GS3,4} - V_{Tn})^2$$. Rearranging, we get

$$\left(\frac{W}{L}\right)_{3,4} = \frac{I_{BIAS}}{\frac{1}{2}\mu_n C_{ox} (V_{GS3,4} - V_{Tn})^2} = \frac{50\ \mu\text{A}}{\frac{1}{2} \cdot 50\ \mu\text{A/V}^2 (0.4\text{ V})^2} = 12.5$$

The specification on the *CMRR* will determine the required length for devices M_3 and M_4. Making the devices very long requires a large W, which adds capacitance and degrades the *CMRR* frequency response. To achieve a *CMRR* of 2500 at DC, (recall that $a_{dm} = -50$ from Example 11.2) we need

$$a_{cm} = \frac{a_{dm}}{CMRR} = -0.02 .$$

From Eq. (11.34)

$$a_{cm} = \frac{-g_m R_D}{1 + g_m 2r_{o3}} \approx \frac{-R_D}{2r_{o3}}, \text{ so } r_{o3} \geq \frac{-R_D}{2a_{cm}} = \frac{-100\text{ k}\Omega}{2(-0.02)} = 2.5\text{ M}\Omega$$

We know that $r_{o3} = \dfrac{1}{\lambda_n I_{BIAS}} \rightarrow \lambda_n = \dfrac{1}{r_{o3} I_{BIAS}} = \dfrac{1}{125\text{ V}} = 0.008\text{ V}^{-1}$.

We are given $\lambda_n(L = 2\ \mu\text{m}) = 0.02\text{ V}^{-1}$.

Recalling $\lambda_n \propto \dfrac{1}{L}$, we require $L_{3,4} = 2.5(2\ \mu\text{m}) = 5\ \mu\text{m}$.

$$\left(\frac{W}{L}\right)_{3,4} = \frac{12.5 \cdot 5}{5} = \frac{62.5}{5} \approx \frac{65}{5}$$

To analyze the frequency response, we need to find the capacitance C_E at the drain of M_3. This capacitance consists of $C_{sb1} + C_{sb2} + C_{db3} + C_{gd3}$ where

$$C_{gd3} = W_3 C_{ov} = 65\ \mu\text{m}\,(0.5\text{ fF}/\mu\text{m}) = 32.5\text{ fF}$$

We need to look at the sample layout shown in Fig. Ex 11.3B to find $C_{sb1,2}$ and C_{db3}.

The sum of all three capacitances can be found by looking at the perimeter and area of the gray region. The area of the sources of M_1 and M_2 is approximately ($200\ \mu\text{m} \times 5\ \mu\text{m}$). The perimeter of the source of M_1 and M_2 which is bounded by field oxide is ($5\ \mu\text{m} + 170\ \mu\text{m}$). The area of the drain of M_3 is ($65\ \mu\text{m} \times 2\ \mu\text{m}$). The

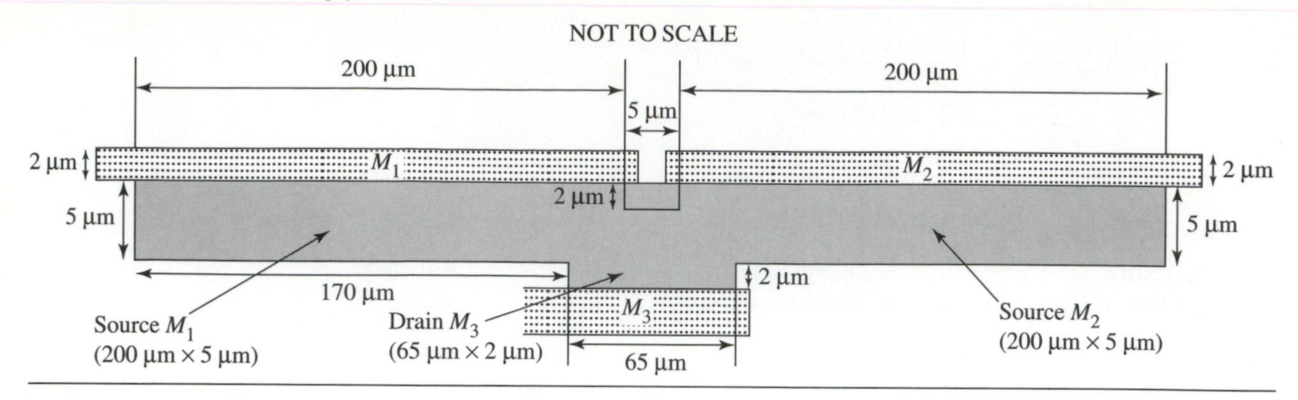

➤ **Figure Ex11.3B**

perimeter is very small since the drain of M_3 is shared with the sources of M_1 and M_2. The perimeter that bounds field oxide is only (2 μm + 2 μm).

$$C_{sb1} = C_{sb2} = (200\,\mu\text{m} \cdot 5\,\mu\text{m})\,C_{jn} + (5\,\mu\text{m} + 170\,\mu\text{m})\,C_{jswn} = 188\text{ fF}$$

$$C_{db3} = (65\,\mu\text{m} \cdot 2\,\mu\text{m})\,C_{jn} + (2\,\mu\text{m} + 2\,\mu\text{m})\,C_{jswn} = 15\text{ fF}$$

So the total capacitance is

$$C_E = C_{gd3} + C_{sb1} + C_{sb2} + C_{db3} \approx 32.5\text{ fF} + 188\text{ fF} + 188\text{ fF} + 15\text{ fF} = 425\text{ fF}.$$

The *CMRR* has a magnitude that is –3 dB below its low frequency value at

$$f_{3dB} \approx \frac{1}{2\pi r_{o3} C_E} = \frac{1}{2\pi\,(2.5\,\text{M}\Omega)\,(425\text{ fF})} = 150\,\text{kHz}$$

➤ 11.5 DIFFERENTIAL AMPLIFIERS WITH SINGLE-ENDED OUTPUTS

It is often important to convert a differential signal into a single-ended signal. A **single-ended signal** is defined as one that is referenced to ground. Many operational amplifiers have this function built into the amplifier as indicated by the symbol for an operational amplifier. The input has two voltage sources that can have both differential and common-mode components, while the output is a single-ended voltage referenced to ground as shown in Fig. 11.19. The easiest way to make a differential-to-single-ended conversion is to take just one output from the differential amplifier analyzed in the previous section, as shown in Fig. 11.20.

The differential- and common-mode gain of this circuit can be analyzed from the results in Section 11.2. Recall in Eq. (11.20) that the expression for v_{o2} was written in terms of the purely differential- and common-mode signals as

$$v_o = v_{o2} = a_{cm} v_{ic} - a_{dm} \frac{v_{id}}{2} \tag{11.38}$$

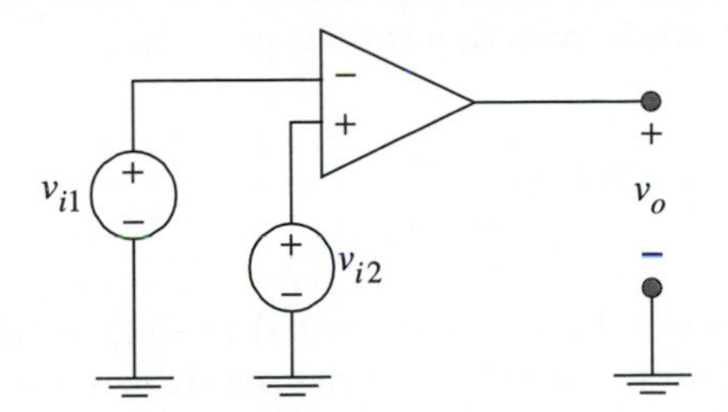

➤ **Figure 11.19** Many operational amplifiers require a differential-to-single-ended conversion.

where $a_{dm} = -g_m R_C$ and $a_{cm} = \dfrac{-g_m R_C}{1 + 2g_m r_{ob}}$.

Using Eq. (11.38) the differential-mode gain for a purely differential input signal for this single-ended amplifier is

$$\frac{v_o}{v_{id}} = -\frac{a_{dm}}{2} = \frac{g_m R_C}{2} \tag{11.39}$$

The common-mode gain for a purely common-mode input signal is

$$\frac{v_o}{v_{ic}} = a_{cm} = \frac{-g_m R_C}{1 + 2g_m r_{ob}} \tag{11.40}$$

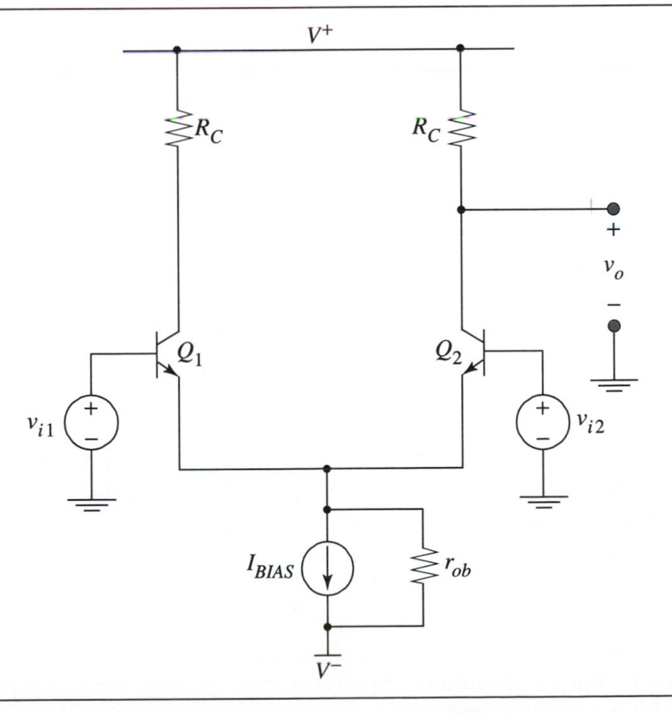

➤ **Figure 11.20** Differential amplifier with single-ended output.

The resulting common-mode rejection ratio is given in

$$CMRR \equiv \left| \frac{(v_o / v_{id})}{(v_o / v_{ic})} \right| = \frac{(1 + 2g_m r_{ob})}{2} \tag{11.41}$$

Although we have accomplished the differential-to-single-ended conversion, the differential gain and *CMRR* are only 50% of the purely symmetrical circuit.

11.5.1 Differential Amplifier with Current Mirror Supply

In this section we develop a circuit topology that accomplishes a differential-to-single-ended conversion while maintaining a gain equal to that of the purely differential amplifier. Consider the differential amplifier shown in Fig. 11.21. MOS transistors are used in this amplifier as the active devices. The resistors that supplied the current in our previous circuits are replaced with current sources to supply the current to the active devices. The value of these current sources is set to precisely half of the bias current source. If we take the output from only one side of this amplifier to perform a differential-to-single-ended conversion, half of the possible voltage gain is lost just as the previous case with resistors supplying the current to the active devices. However, if the current source supplies are replaced with a p-channel current mirror as shown in Fig. 11.22, the current mirror forces the signal current in M_1 to flow in device M_3 and M_4. This additional signal current is added to the output current, which increases the voltage gain by a factor of 2.

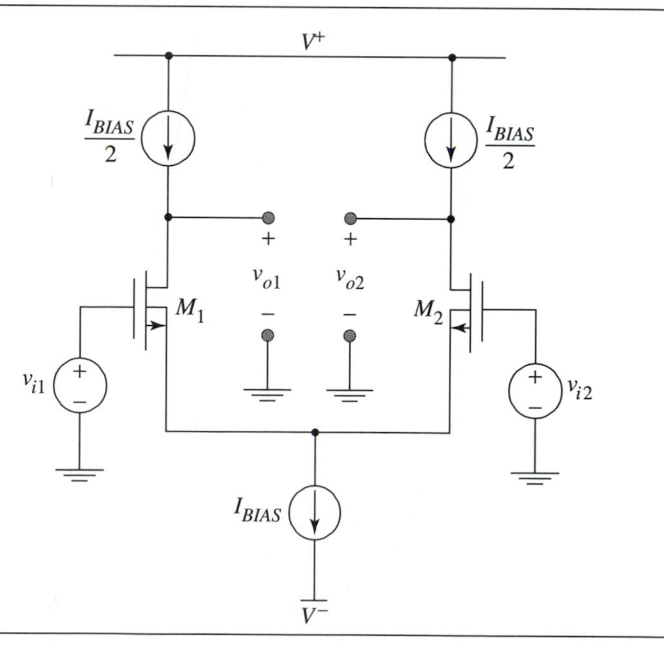

Figure 11.21 Fully differential amplifier with current sources supplying the current to the MOS input devices.

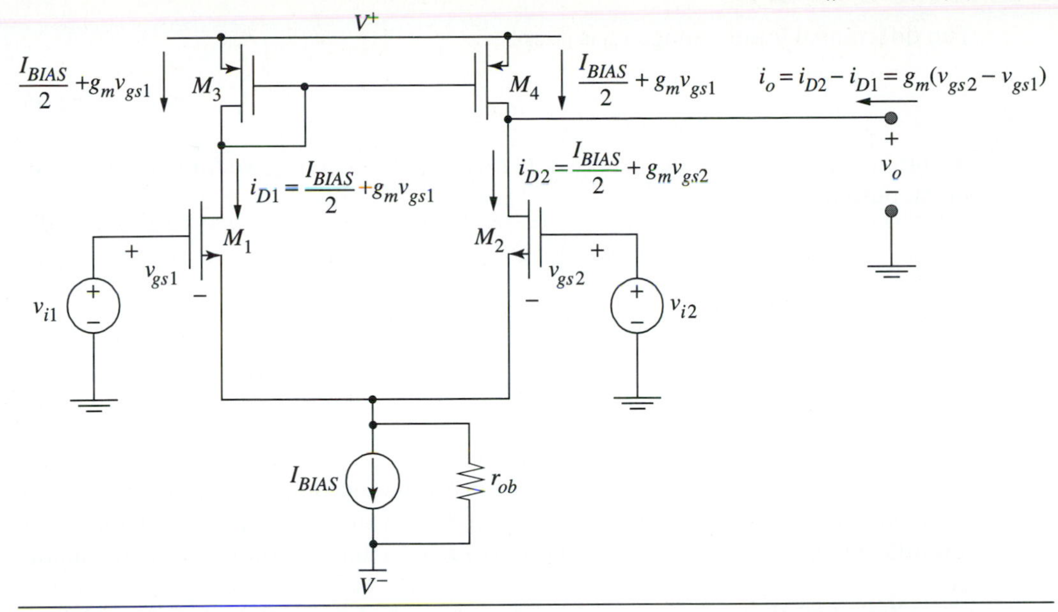

➤ Figure 11.22 Differential amplifier with p-channel current mirror to perform a differential-to-signal-ended conversion. The sources are shorted to the backgates.

Before proceeding with a formal small-signal analysis, it is worthwhile to closely examine Fig. 11.22 to understand intuitively the factor of 2 increase in transconductance and voltage gain. Look at both the large and small-signal quantities flowing through the differential amplifier. We assume that the large-signal voltages at the inputs are equal to zero and, hence, only small signal voltages are applied to these devices which we call v_{i1} and v_{i2}. Starting with transistor M_1, which has a small-signal gate-source voltage v_{gs1}, a small-signal current will be added to half of the bias current flowing into the drain of M_1 so the total current flowing into the drain i_{D1} is given by

$$i_{D1} = I_{D1} + i_{d1} = \frac{I_{BIAS}}{2} + g_{m1} v_{gs1} \tag{11.42}$$

Similarly, the drain current in device M_2 is given in

$$i_{D2} = I_{D2} + i_{d2} = \frac{I_{BIAS}}{2} + g_{m2} v_{gs2} \tag{11.43}$$

By KCL the current in M_3 is equal to the current in M_1. Assuming that the output resistance of devices M_3 and M_4 is large, the p-channel current mirror forces their currents to be identical.

To find the differential transconductance G_{md} of this amplifier, we must find the small-signal output current with the output short-circuited. The small-signal short-circuit output current is the difference between the drain current of M_2 and M_4, which is the same as the difference between the drain current of M_2 and M_1 given by

$$i_o = g_{m2} v_{gs2} - g_{m1} v_{gs1} \tag{11.44}$$

The differential input voltage v_{id} is simply

$$v_{id} = v_{i1} - v_{i2} = v_{gs1} - v_{gs2} \tag{11.45}$$

Assuming $g_{m1} = g_{m2} = g_m$ and using Eq. (11.44) and Eq. (11.45), we find that the small-signal output current is

$$i_o = -g_m v_{id} \tag{11.46}$$

and the differential transconductance G_{md} is given by

$$G_{md} \equiv \frac{i_o}{v_{id}} = -g_m \tag{11.47}$$

This circuit, with a p-channel current mirror, has accomplished a differential-to-single-ended conversion while maintaining 100% of the available gain of a fully differential amplifier. The transconductance is equal to that of a fully differential amplifier.

A full small-signal analysis of the differential amplifier with both differential and common-mode input signals is done by analyzing the small-signal model shown in Fig. 11.23. The DC current flowing in all four devices M_1–M_4 is $I_{BIAS}/2$. This current is set equal to zero for the small-signal analysis.

The p-channel current mirror puts an asymmetry in the circuit, which means that half-circuit techniques are not valid. However, the analysis of the circuit is not difficult if we make a few approximations. We start by writing KCL for current i_{o1} on the left-hand side of the circuit. It is given by

$$i_{o1} = i_{o3} = \frac{v_{sg3}}{r_{o3}} + g_{m3} v_{sg3} \approx g_{m3} v_{sg3} \tag{11.48}$$

In this equation we have assumed that the output resistance of the p-channel transistor is much larger than the reciprocal of the transconductance. We write KCL at the output node and find the short-circuit small-signal output current to be

$$i_o \approx i_{o2} - g_{m4} v_{sg4} \tag{11.49}$$

Since the source-gate voltages of both p-channel transistors are equal, we can substitute Eq. (11.48) into Eq. (11.49) resulting in

$$i_o = i_{o2} - i_{o1} \frac{g_{m4}}{g_{m3}} \tag{11.50}$$

Assuming that the transistors are well matched, their transconductances should be equal. Hence, the output current is given by

$$i_o = i_{o2} - i_{o1} \tag{11.51}$$

To calculate the differential-mode transconductance, we set the common-mode input voltage source equal to zero and write an equation for the short-circuit output current of each active NMOS transistor. Again assuming that $r_{o1,2} \gg 1/g_{m1,2}$ we find

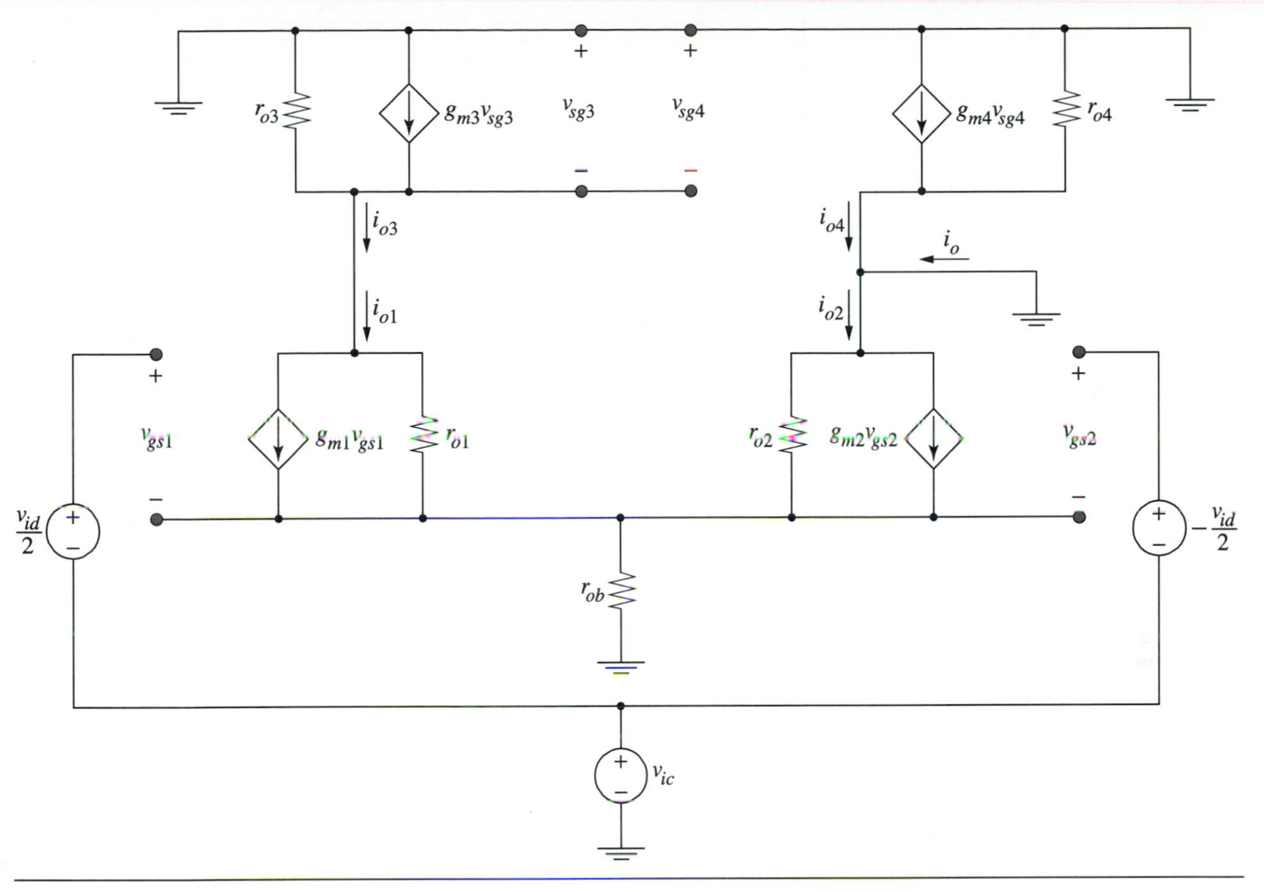

Figure 11.23 Small-signal model of a differential amplifier with a p-channel current mirror.

$$i_{o1} = g_{m1}(v_{id} + v_{gs2}) \quad (11.52)$$

$$i_{o2} = g_{m2}(-v_{id} + v_{gs1}) \quad (11.53)$$

From Eq. (11.45) we know that $v_{id} = v_{gs1} - v_{gs2}$. Substituting these into Eq. (11.51) yields the amplifier's differential transconductance G_{md} to be equal to

$$G_{md} \equiv \frac{i_o}{v_{id}} = -g_{m1} = -g_{m2} \quad (11.54)$$

differential transconductance for differential amplifier with p-channel current mirror

It should be noted that the differential transconductance of the differential pair using the p-channel current mirror is equal to the transconductance of a single MOS transistor.

To obtain the differential two-port model for this amplifier, we recognize that the input resistance is infinite since we are using MOS transistors. The other two-port parameter for this amplifier is its differential output resistance R_{od}. Since its value is not obvious, apply a test voltage source v_t to the output node and calculate the resulting current i_t as shown in Fig. 11.24. The differential input voltage v_{id} is set equal to zero to find the output resistance, which results in the gates of M_1 and M_2 being

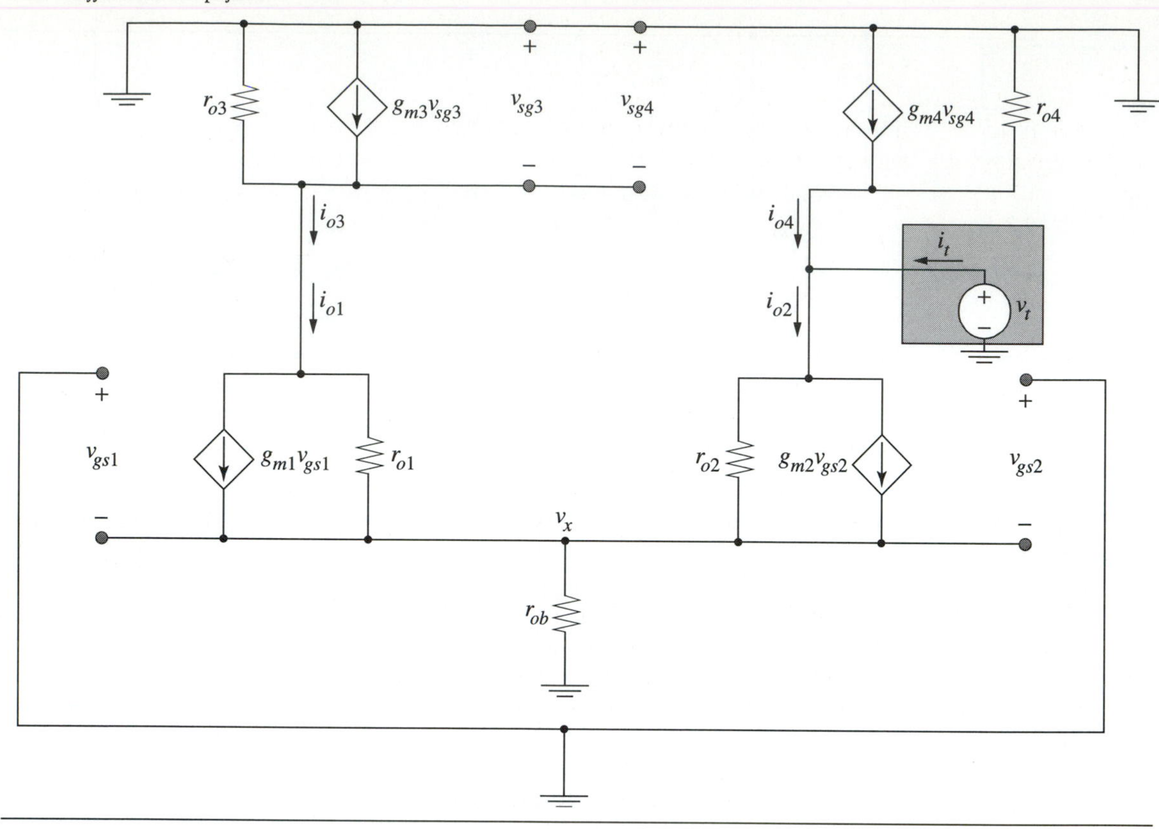

➤ **Figure 11.24** Small-signal model of a differential amplifier with a p-channel current mirror to calculate the differential output resistance.

grounded. From Fig. 11.24, the current i_t is the difference between the drain current of M_2 and the drain current of M_4. By noting that $v_{gs2} = -v_x$ we write

$$i_t = i_{o2} - i_{o4} = \left(-g_{m2}v_x + \frac{v_t - v_x}{r_{o2}}\right) - \left(g_{m4}v_{sg4} - \frac{v_t}{r_{o4}}\right) \tag{11.55}$$

To find an expression for v_{sg4}, note that $v_{sg4} = v_{sg3}$ and write an expression for i_{o1} as

$$i_{o1} = -g_{m1}v_x + \frac{-(v_{sg3} + v_x)}{r_{o1}} = i_{o3} = g_{m3}v_{sg3} + \frac{v_{sg3}}{r_{o3}} \tag{11.56}$$

By noting that $r_{o1}, r_{o3} \gg 1/g_{m1}, 1/g_{m3}$ and neglecting the terms v_{sg3}/r_{o1} and v_{sg3}/r_{o3} in Eq. (11.56) we can solve for v_{sg3}

$$v_{sg3} = -\frac{g_{m1}}{g_{m3}}v_x - \frac{v_x}{g_{m3}r_{o1}} \tag{11.57}$$

Assuming matched devices $g_{m1} = g_{m2}$, $g_{m3} = g_{m4}$, $r_{o2} = r_{o1}$ and $r_{o3} = r_{o4}$ and substituting Eq. (11.57) into Eq. (11.55) we find

$$i_t = v_t\left(\frac{1}{r_{o2}} + \frac{1}{r_{o4}}\right) \tag{11.58}$$

From Eq. (11.58) we find the differential output resistance is the parallel combination of the output resistances of the p-channel and n-channel transistors, M_2 and M_4, given by

differential output resistance for differential amplifier with p-channel current mirror

$$R_{od} = r_{o2} \| r_{o4} \tag{11.59}$$

The resulting two-port model for a purely differential signal is shown in Fig. 11.25(a). The Thévenin equivalent output shown in Fig. 11.25(b) has a differential open-circuit voltage gain a_{vd} that is found by multiplying the differential transconductance of the amplifier times its output resistance given by

differential voltage gain for differential amplifier with p-channel current mirror

$$a_{vd} = -G_{md}R_{od} = g_{m1}(r_{o2} \| r_{o4}) \tag{11.60}$$

Note this circuit produces twice the voltage gain that would be obtained if only a single output was taken from a purely differential amplifier. This fact demonstrates the advantage of using the current mirror to supply current to the differential pair when constructing a differential-to-single-ended converter.

The response to a common-mode input is obtained by setting the differential-mode voltages equal to zero. Referring to Fig. 11.26, the common-mode voltage generator will yield equal values for both output currents i_{o1} and i_{o2}. The circuit forces the drain current of device M_3 to be the same as that of M_1, which implies that $i_{o3} = i_{o1}$. Assuming that $r_{o3,4} \gg 1/g_{m3,4}$, the p-channel current mirror forces the drain current of M_3 and M_4 to be equal. This means that for a common-mode input the circuit behaves perfectly symmetrically, and we can analyze it using the half-circuit technique. The resulting small-signal common-mode half-circuit is shown in Fig. 11.26(a). To simplify the analysis, ignore the output resistances r_{o1} and r_{o3} since they are much larger than $1/g_m$ for both devices. The resulting circuit for the common-mode gain is shown in Fig.

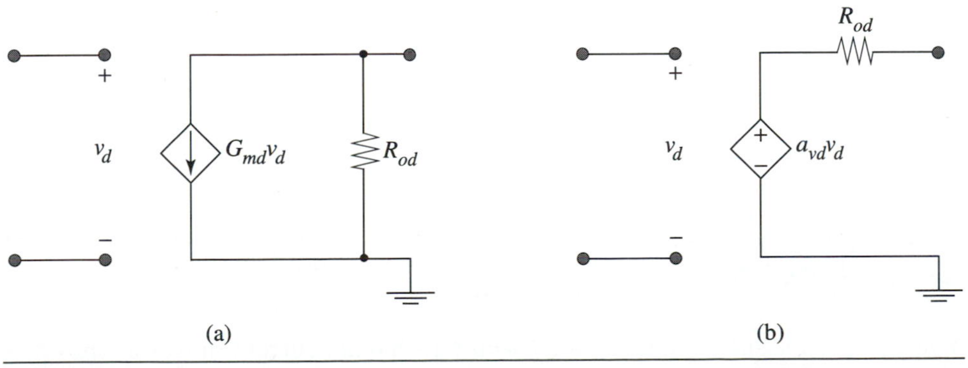

➤ **Figure 11.25** Two-port model for a purely differential voltage applied to a differential amplifier with a p-channel current mirror. (a) Norton equivalent output; (b) Thévenin equivalent output.

11.26(b). By realizing this circuit is a common-source amplifier with a source degeneration resistor equal to $2r_{ob}$, we find the expression for the common-mode gain given in

common-mode voltage gain for differential amplifier with p-channel current mirror

$$a_{vc} \equiv \frac{v_o}{v_{ic}} = \frac{-(g_{m1}/g_{m3})}{1 + 2g_{m1}r_{ob}} \approx \frac{-1}{2g_{m3}r_{ob}} \tag{11.61}$$

In summary, we find that the differential amplifier with a p-channel current mirror provided a differential-to-single-ended conversion while maintaining a large differential voltage gain and good common-mode rejection.

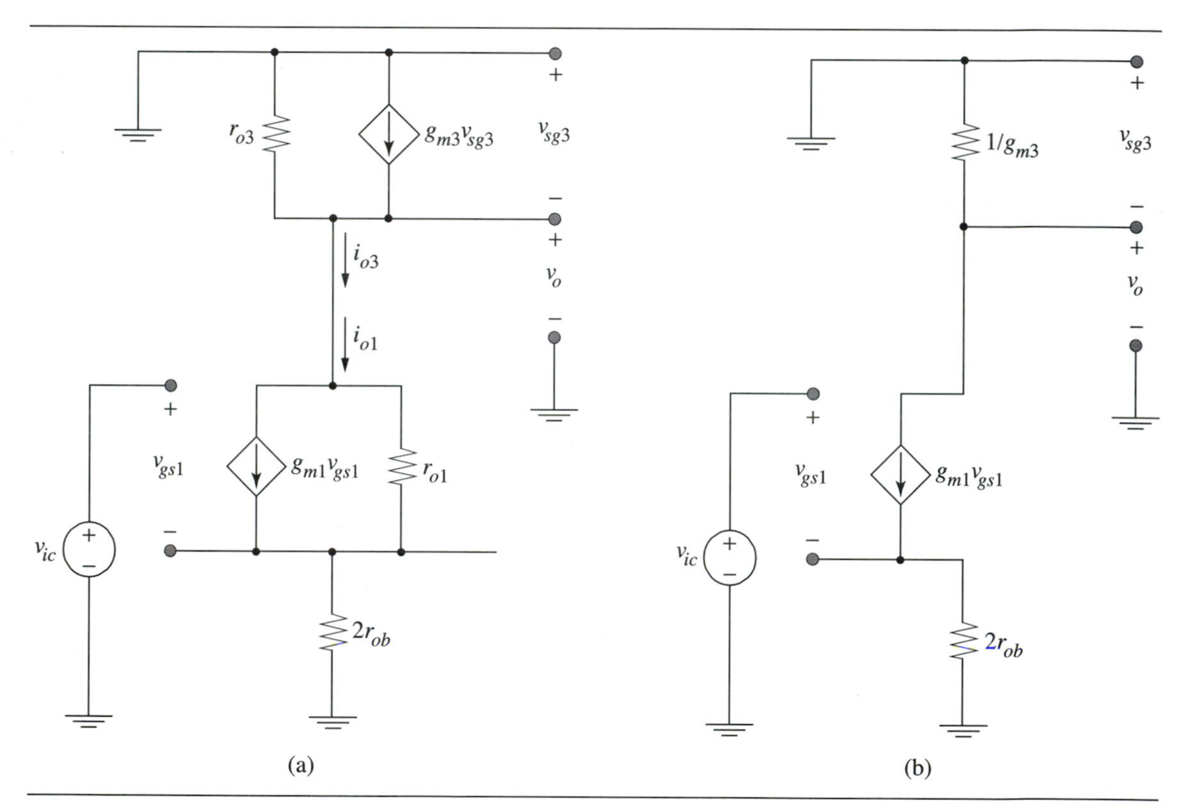

➤ **Figure 11.26** (a) Small-signal common-mode half-circuit model of a differential amplifier with a p-channel current mirror. (b) Simplified model ignoring the effect of r_{o1} and r_{o3}.

➤ EXAMPLE 11.4 Differential Amplifier with p-Channel Current Mirror

You are given a differential amplifier with a p-channel current mirror as shown in Fig. Ex 11.4A. Calculate a_{vd}, a_{vc}, and the *CMRR*. Assume the devices are sized so they are all operating in their constant-current regions with $g_{mn} = g_{mp} = 0.1$ mS and $r_{on} = r_{op} = 1$ MΩ.

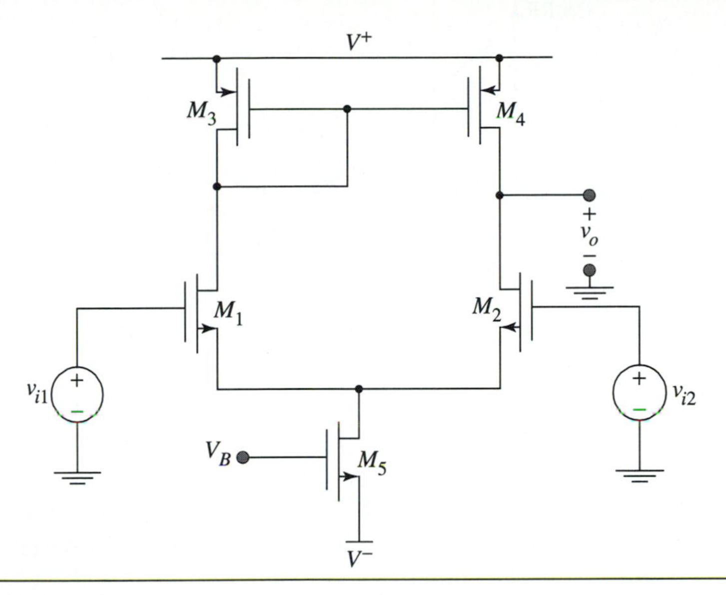

➤ **Figure Ex11.4A**

SOLUTION

To find a_{vd}, we use Eq. (11.60).

$$a_{vd} = g_{m1}(r_{o2} \| r_{o4}) = (0.1\,\text{mS})\,(500\,\text{k}\Omega) = 50$$

We find a_{vc} from Eq. (11.61).

$$a_{vc} = -\frac{1}{2g_{m3}r_{o5}} = -\frac{1}{2\,(0.1\,\text{mS})\,(1\,\text{M}\Omega)} = -0.005$$

So

$$CMRR = \frac{|a_{vd}|}{|a_{vc}|} = \frac{50}{0.005} = 10,000\,.$$

Suppose we want to increase the *CMRR*. We could increase the output resistance of current source device M_5. However, this only increases *CMRR* at low frequencies. Recall that ω_{3dB} for *CMRR* is

$$\omega_{3dB} = \frac{1}{r_{o5}C_E}\,.$$

So by increasing r_{o5}, we increase the *CMRR* at DC but reduce the 3 dB frequency as demonstrated in Fig. Ex 11.4B.

Another way to increase *CMRR* is to increase a_{vd} by adding cascode transistors to our differential amplifier as shown in Fig. Ex 11.4C. We added devices M_6–M_9 to increase the output resistance. We are neglecting the backgate effect to simplify this explanation. Devices M_7 and M_9 increase the output resistance,

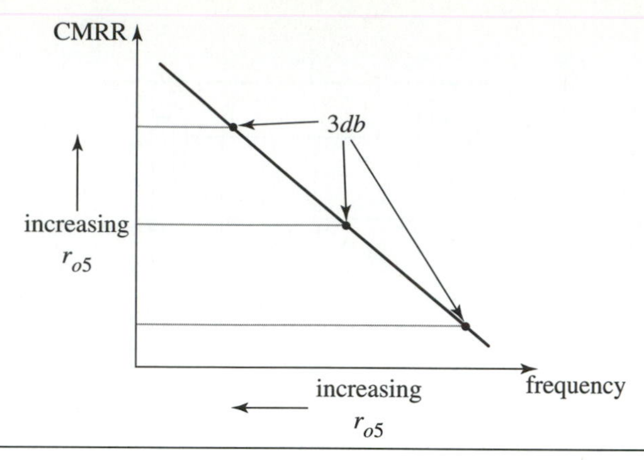

➤ **Figure Ex11.4B**

while M_6 and M_8 provide the proper DC gate voltage. Device M_6 is not strictly necessary. However, it makes the amplifier symmetrical and keeps the node voltages on the left side of the amplifier approximately equal to those on the right side when the differential input voltage is 0 V. The G_{md} of this amplifier is the same as without the cascode devices. We find the output resistance using a similar analysis that led to Eq. (11.59) and adding the effect of devices M_7 and M_9. The result is

$$R_{od} \approx r_{o2}(g_{m7}r_{o7}) \parallel r_{o4}(g_{m9}r_{o9})$$

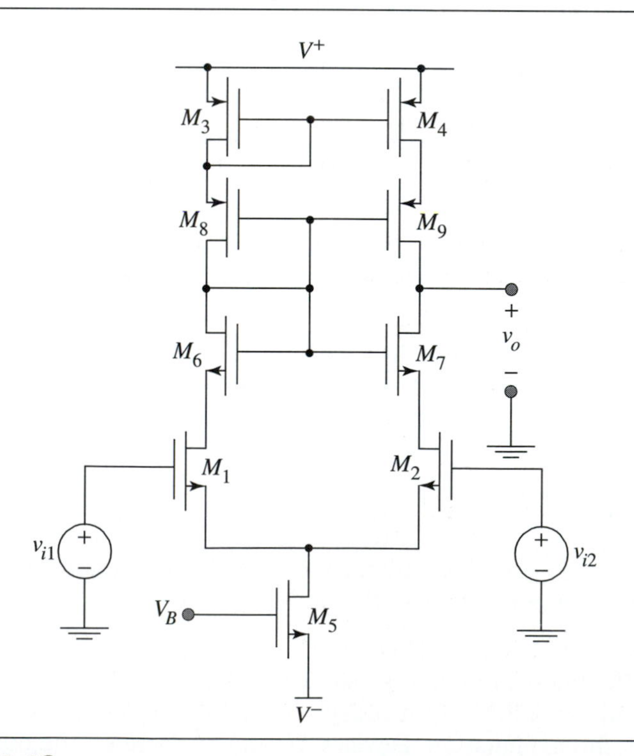

➤ **Figure Ex11.4C**

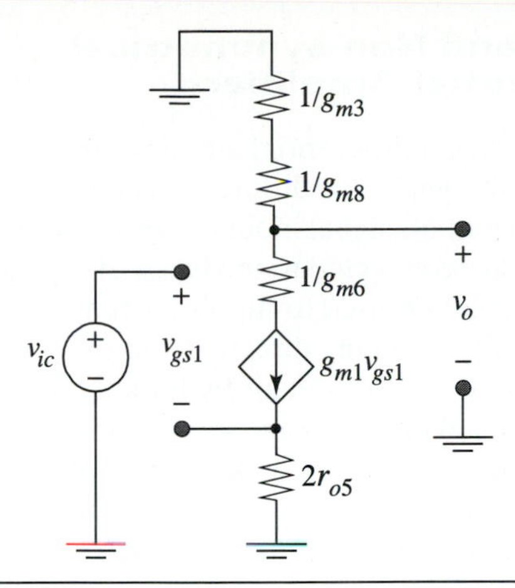

➤ **Figure Ex11.4D**

Recalling that $g_{mn} = g_{mp} = g_m$ and $r_{on} = r_{op} = r_o$

$$a_{vd} = -G_{md} R_{od} = g_m \frac{r_o(g_m r_o)}{2} = 0.1\,\text{mS}\,\frac{1\,\text{M}\Omega\,(100)}{2} = 5000$$

We have increased a_{vd} by a factor equal to the open-circuit voltage gain of one device. To find the new common-mode gain, we employ the same analysis that led to the simplified model in Fig. 11.26(b). A model to analyze the common-mode gain for the cascoded differential amplifier is shown in Fig. Ex 11.4D.

This circuit can be analyzed by recognizing that resistor $1/g_{m6}$ has no effect since it is in series with a current source. To calculate the common mode gain, use Eq. (11.61) and substitute $\frac{1}{g_{m3}} + \frac{1}{g_{m8}}$ for $\frac{1}{g_{m3}}$. The result is

$$a_{vc} = \frac{-g_{m1}\left(\frac{1}{g_{m3}} + \frac{1}{g_{m8}}\right)}{1 + 2g_{m1} r_{o5}} \approx \frac{-2}{2g_m r_o} = -0.01$$

CMRR is now $\frac{|a_{vd}|}{|a_{vc}|} = \frac{5000}{0.01} = 5 \cdot 10^5$, which is a factor of 50 greater than the previous result with the same transitor sizes M_1-M_4 and the same bias current.

11.5.2 Wideband Non-symmetrical Differential Amplifiers

In this section, we analyze a differential amplifier that uses a similar biasing scheme as described for a purely symmetric differential amplifier. This wideband amplifier is shown in Fig. 11.27. The small-signal input to this amplifier is single-ended. Large-signal sources V_{I1} and V_{I2} are connected to the bases of Q_1 and Q_2. The output is also single-ended. Only one resistor is used to supply current to the active device Q_2, whereas the supply current for Q_1 is coming directly from the power supply.

We begin the analysis of this circuit by looking at the large-signal DC voltages and setting the small-signal input voltage source to zero. In an analogous fashion with the differential amplifier, if the input voltage to Q_1 and Q_2 are equal, then the collector currents in each transistor are equal. If the large-signal input voltage sources $V_{I1} = V_{I2} = 0$ V, we should set the bias current source to a value given by

$$I_{BIAS} = \frac{2V^{+}}{R_C} \tag{11.62}$$

With this value for I_{BIAS}, the output voltage will be forced to be near 0 V. Although this amplifier has a single-ended input and output, the biasing approach is similar to that used for the differential amplifier.

To further analyze this amplifier, break the circuit in half as shown in Fig. 11.28. Note that we are not performing a half-circuit analysis since the left and right halves of these circuits are not the same. However, we have broken the circuit in half to show that this non-symmetrical differential amplifier actually looks like a common-collector amplifier cascaded to a common-base amplifier. The signal path is nothing more than the cascade of these two single-stage amplifiers that we have studied in Chapter

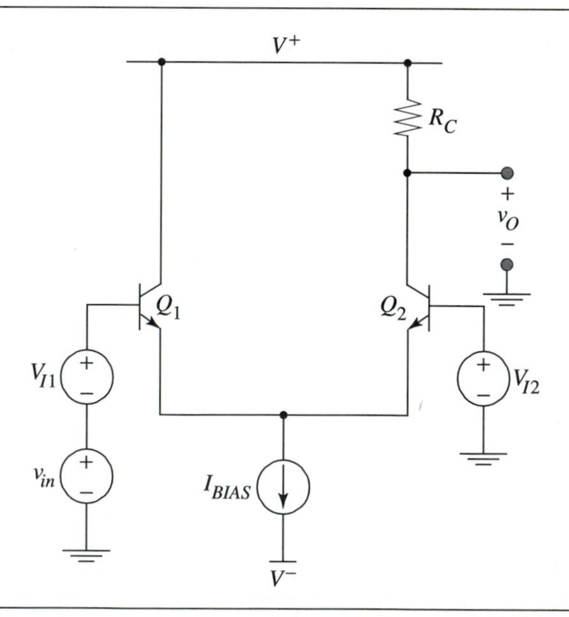

Figure 11.27 Wideband non-symmetrical differential amplifier.

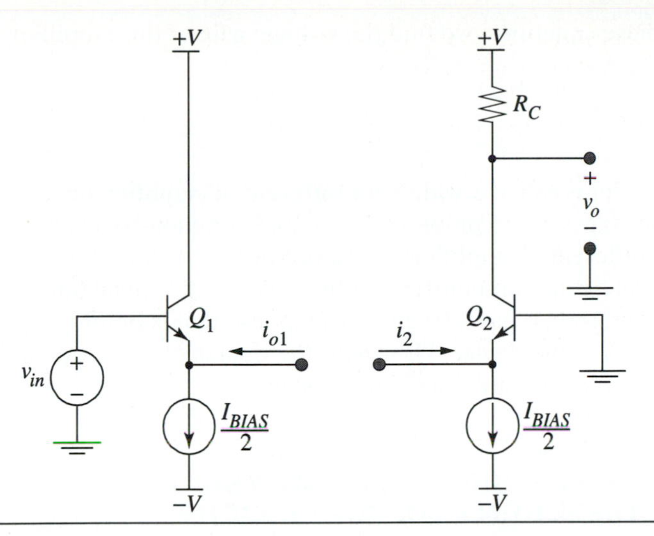

➤ **Figure 11.28** Wideband differential amplifier split into CC and CB stages.

8. We can perform the small-signal analysis using the two-port models from Chapter 8, as shown in Fig. 11.29.

On the left-hand side, we have substituted the two-port model for the common-collector amplifier and on the right-hand side, we have substituted the two-port model for the common-base amplifier. The transconductance for both amplifiers will be equal since the collector current for both transistors Q_1 and Q_2, was set equal by our initial biasing.

We can analyze this model for the transfer function between the output voltage and the input voltage by noting that the output voltage is given by

$$v_o = i_2 R_C \tag{11.63}$$

Recalling that $g_{m1} = g_{m2} = g_m$ the input current to the common-base amplifier i_2 is given by

$$i_2 = \frac{g_m v_1}{2} = \frac{g_m v_{in}}{2} \tag{11.64}$$

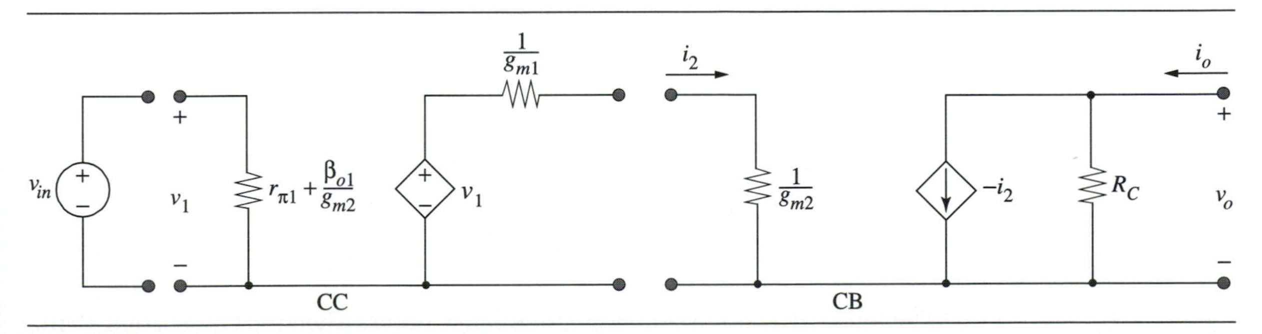

➤ **Figure 11.29** Small-signal model of wideband differential amplifier using a cascade of two-port models for the CC and CB amplifier. Note that we assume $R_C \ll \beta_o r_o$.

Combining these equations, we find the voltage gain of this amplifier is given by

$$\frac{v_o}{v_{in}} = \frac{g_m R_C}{2} \tag{11.65}$$

This amplifier is called a **wideband differential amplifier** since it is composed of two wideband stages—a common-collector and common-base amplifier. It is biased similarly to a differential amplifier. This amplifier has a voltage gain that is only a factor of 2 less than a common-emitter amplifier, yet has no capacitances that are Miller-multiplied. Therefore, it has a frequency response that is potentially near the transition frequency f_T of the device. However, the details of the frequency response will depend strongly on the source and load resistances.

11.6 LARGE-SIGNAL ANALYSIS OF DIFFERENTIAL AMPLIFIERS

In this section we explore the large-signal characteristics of both the MOS and bipolar differential amplifiers. The large-signal common-mode voltage response defines an input common-mode voltage range in which the amplifier operates in its **linear region**. In the linear region, all devices are operating in their constant-current region and the differential amplifier produces useful differential voltage gain with reasonable common-mode rejection.

Another important response to analyze is when a large differential signal is applied to the differential amplifier. We will derive the transfer characteristic for the output voltages of the differential amplifier V_{O1} and V_{O2} as a function of the differential input voltage V_{ID} for both the MOS and bipolar implementations.

11.6.1 Input Common-Mode Range

The **input common-mode range** of a differential amplifier is defined as the range of common-mode voltages V_{IC} over which the amplifier has a linear response to small differential input signals. The differential amplifier will behave linearly provided that the input devices and the device supplying the bias current are all operating in their constant-current region. To investigate the input common-mode range, we apply a purely common-mode large-signal voltage at the input devices, as shown in Fig. 11.30.

The upper limit of the common-mode range is determined when the input devices leave their constant-current region. This occurs when the drain-source or collector-emitter voltage equals its corresponding saturation voltage $V_{DS_{SAT}}$, $V_{CE_{SAT}}$ as given by

$$V_{O1,2} - V_X = V_{DS_{SAT1,2}}, V_{CE_{SAT1,2}} \tag{11.66}$$

Substituting the expression for the output voltage, we find

$$V^+ - \frac{I_{BIAS}}{2} R_C - V_X = \left(V_{DS_{SAT1,2}}, V_{CE_{SAT1,2}} \right) \tag{11.67}$$

The lower limit of the common-mode range occurs when the bias current source does not have sufficient voltage across it to keep it operating in its constant-current region. This condition occurs when

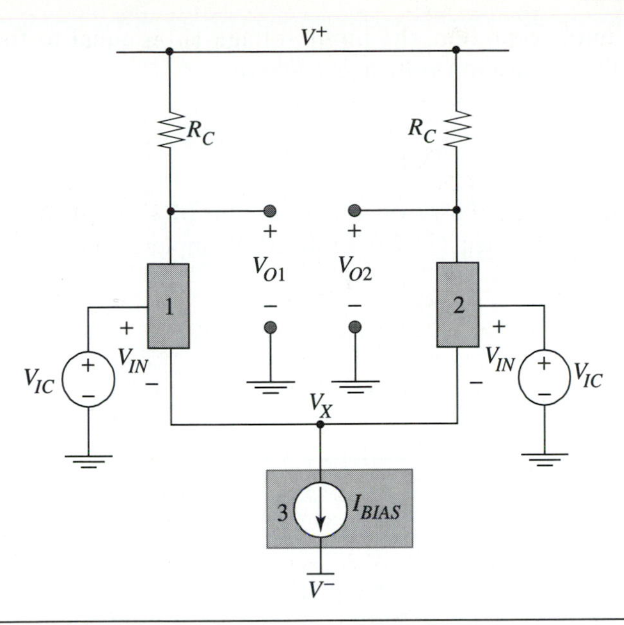

➤ Figure 11.30 Differential amplifier with purely common-mode large-signal input voltage V_{IC}.

$$V_X - V^- = \left(V_{DS_{SAT3}}, V_{CE_{SAT3}}\right) \tag{11.68}$$

Since we are dealing with purely common-mode input signals, the input voltage V_{IN} across the active devices are equal. The relation between V_X and V_{IC} is given by

$$V_X = V_{IC} - V_{IN} \tag{11.69}$$

To determine specifically the input common-mode range, we will consider a particular implementation. For an all-bipolar implementation, both the input devices and the bias current source are formed with bipolar transistors. The input voltage V_{IN} is equal to the base-emitter voltage V_{BE}, the saturation voltage of the input transistors, $Q_{1,2}$ is $V_{CE_{SAT1,2}}$, and the saturation voltage for the bias current source is $V_{CE_{SAT3}}$. We can substitute these quantities and Eq. (11.69) into Eq. (11.67) and Eq. (11.68) to determine the input common-mode range for the bipolar implementation of a differential pair. It is given by

$$V_{IC} \leq V^+ - \frac{I_{BIAS}}{2} R_C + V_{BE1,2} - V_{CE_{SAT1,2}} \tag{11.70}$$

$$V_{IC} \geq V_{CE_{SAT3}} + V_{BE1,2} + V^- \tag{11.71}$$

input common-mode range for the bipolar differential amplifier

where

$$V_{BE1,2} = V_{th} \ln\left(\frac{I_{BIAS}/2}{I_{S1,2}}\right) \approx 0.7\text{V} \tag{11.72}$$

$$V_{CE_{SAT1,2}} = V_{CE_{SAT3}} \tag{11.73}$$

For an all-MOS implementation, the input voltage V_{IN} is equal to the gate-source voltage $V_{GS1,2}$ and the saturation voltage is given by

$$V_{DS_{SAT1,2}} = V_{GS1,2} - V_{Tn} \tag{11.74}$$

To find the input common-mode range for the MOS implementation we can write a similar expression to Eq. (11.70) for the MOS implementation

$$V_{IC} \leq V^+ - \frac{I_{BIAS}}{2} R_D + V_{GS1,2} - V_{DS_{SAT1,2}} \tag{11.75}$$

Substituting Eq. (11.74) into Eq. (11.75)

input common-mode range for the MOS differential amplifier

$$\boxed{V_{IC} \leq V^+ - \frac{I_{BIAS}}{2} R_D + V_{Tn}} \tag{11.76}$$

To find the other extreme, we write an equation similar to Eq. (11.71)

$$\boxed{V_{IC} \geq V_{DS_{SAT3}} + V_{GS1,2} + V^-} \tag{11.77}$$

where

$$V_{GS1,2} = V_{Tn} + \sqrt{\frac{I_{BIAS}/2}{\frac{1}{2}\left(\frac{W}{L}\right)_{1,2}\mu_n C_{ox}}} \tag{11.78}$$

$$V_{DS_{SAT3}} = \sqrt{\frac{I_{BIAS}}{\frac{1}{2}\left(\frac{W}{L}\right)_{3}\mu_n C_{ox}}} \tag{11.79}$$

From the expressions for the input common-mode range, it can be seen that we can operate with a larger range as the power supply voltages increase. However, it is important to note that increasing the power supply voltages also increases the power dissipation of the circuit. In the following example, we will explore some typical numbers for the input common-mode range of an all-MOS differential amplifier.

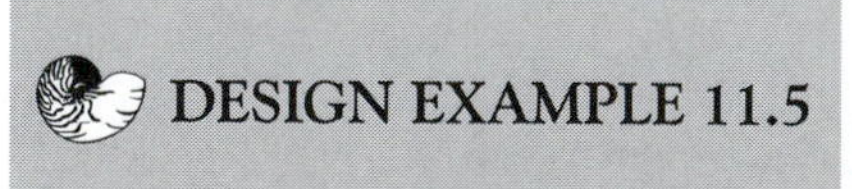

Input Common-Mode Range for an MOS Differential Amplifier with a p-channel Current Mirror

In this design example, we will investigate the input common-mode range of an MOS differential amplifier with a p-channel current mirror shown in Fig. Ex 11.5. Our goal is to achieve an input common-mode range of 1.75 V $\leq V_{IC} \leq$ 4 V with a bias current through device M_5 of 100 μA. We also want to minimize the area of the amplifier. Size the MOS transistors in the amplifier to achieve this specification. Assume that all sources are shorted to their backgates and neglect channel length modulation.

Device Data

$V_{Tn} = 1\ \text{V}$

$\mu_n C_{ox} = 50\ \mu\text{A/V}^2$

$V_{Tp} = -1\ \text{V}$

$\mu_p C_{ox} = 25\ \mu\text{A/V}^2$

$L_{min} = 2\ \mu\text{m}$

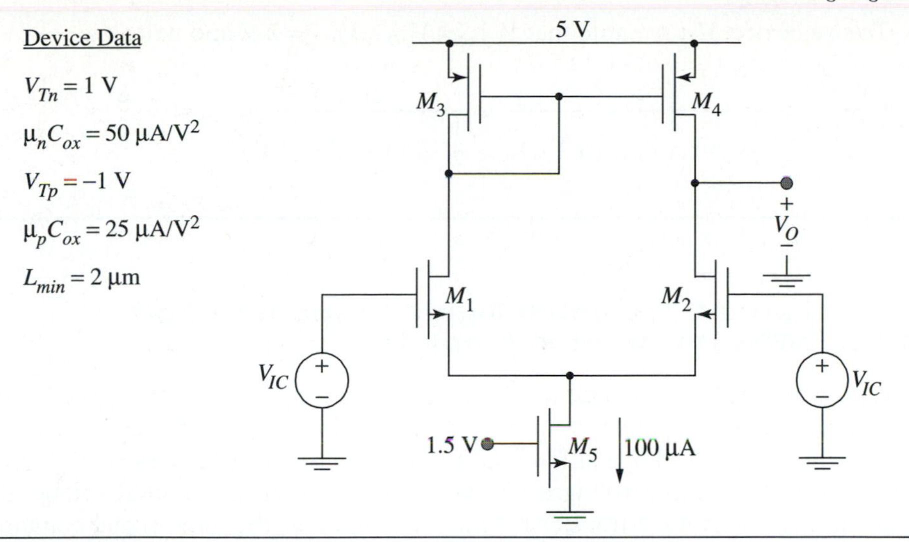

➤ **Figure Ex11.5**

SOLUTION

We begin this design by setting the device size of M_5 to sink 100 μA.

$$\left(\frac{W}{L}\right)_5 = \frac{I_{BIAS}}{\frac{1}{2}\mu_n C_{ox}(V_{GS5} - V_{Tn})^2} = \frac{100\ \mu A}{\frac{1}{2}\cdot 50\mu\text{A/V}^2(0.5\ \text{V})^2} = 16 = \frac{32}{2}.$$

Next we look at the condition for the minimum input common-mode voltage. From Eq. (11.77)

$V_{IC} \geq V_{DS_{SAT5}} + V_{GS1,2} = 1.75\ \text{V}$. Now $V_{DS_{SAT5}} = V_{GS5} - V_{Tn} = 0.5\ \text{V}$ implying that

$V_{GS1,2} = 1.25\ \text{V}$.

Rearranging Eq. (11.78) we find

$$\left(\frac{W}{L}\right)_{1,2} = \frac{I_{BIAS}/2}{\frac{1}{2}\mu_n C_{ox}(V_{GS1,2} - V_{Tn})^2} = \frac{50\ \mu\text{A}}{\frac{1}{2}\cdot 50\,\mu\text{A/V}^2(0.25\ \text{V})^2} = 32 = \frac{64}{2}.$$

To find the maximum V_{IC} we recognize that the size of the p-channel current mirror devices sets the drain voltage of devices $M_{1,2}$, since the gate voltage of $M_{3,4}$ is the same as the drain voltage of $M_{1,2}$. Also recall that $V_{DS} \geq V_{GS} - V_{Tn}$ for an NMOS device to operate in its constant-current region which implies that the condition $V_{GD} \leq V_{Tn}$ must be satisfied. Substituting V_{IC} for the gate voltage of $M_{1,2}$ we find $V_{IC} \leq V_{Tn} + V_{D1,2}$. If we make $V_{D1,2}$ larger then V_{IC} gets larger but, we must make the sizes of the p-channel devices $M_{3,4}$ larger to source their current, since $V_{SG3,4}$ will be reduced (Note: $V_{D1,2} = V_{G3,4}$). Therefore we will set $V_{D1,2} = V_{G3,4}$ at the smallest voltage that will still meet the specification for the maximum required V_{IC}. This will minimize the size of $M_{3,4}$.

$$V_{D1,2} = V_{IC} - V_{Tn} = 4\text{V} - 1\text{V} = 3\ \text{V}.$$

To size device $M_{3,4}$ we note that $V_{SG3,4} = V_{DD} - V_{D1,2} = 2$ V and write

$$\left(\frac{W}{L}\right)_{3,4} = \frac{I_{BIAS}/2}{\frac{1}{2}\mu_p C_{ox}(V_{SG3,4} + V_{Tp})^2} = \frac{50\ \mu\text{A}}{\frac{1}{2} \cdot 25\ \mu\text{A/V}^2(1\text{ V})^2} = 4 = \frac{8}{2}$$

11.6.2 Large Differential Signal Response of an MOS Differential Amplifier

In the previous analysis, we assumed that a small differential signal was applied to the differential amplifiers in order to find the differential-mode voltage gain. In this section, we will explore the large differential signal response to find the transfer function that can relate the output voltages V_{O1} and V_{O2} to the input differential voltage V_{ID}. To determine this voltage transfer function, assume that the large-signal common-mode input voltage is set such that the output voltages of the differential amplifier are 0 V and that all devices are operating in their constant-current region. Before fully analyzing this circuit, determine qualitatively what we expect for the result when we have a large differential input voltage applied, such that $V_{I1} \gg V_{I2}$. Referring to Fig. 11.31 all of the bias current will flow into the drain of device M_1 while device M_2 is cutoff with zero drain current. Under this condition, the output voltages are given by

$$V_{O1} = V^+ - I_{BIAS}R_D \tag{11.80}$$

$$V_{O2} = V^+ \tag{11.81}$$

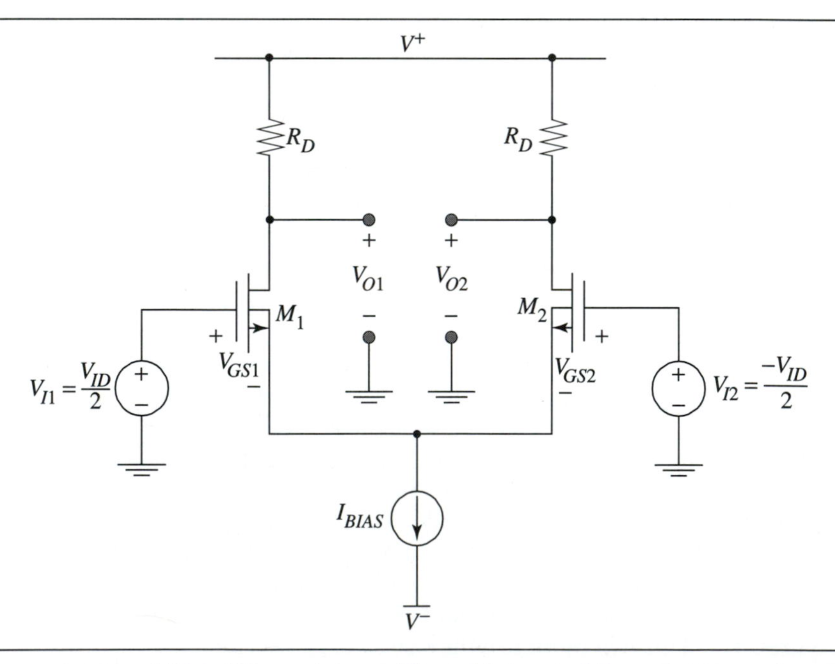

Figure 11.31 MOS differential amplifier with large differential input signals applied.

The minimum differential voltage, required to sink all of the bias current through device M_1 and just begin to cutoff device M_2, occurs when $V_{GS2} = V_{Tn}$. We can determine the differential input voltage when this occurs by writing an expression for the differential voltage in terms of the gate-source voltages of devices M_1 and M_2 as

$$V_{ID} = V_{GS1} - V_{GS2} = V_{Tn} + \sqrt{\frac{I_{BIAS}}{\frac{1}{2}\left(\frac{W}{L}\right)_1 \mu_n C_{ox}}} - V_{Tn} = \sqrt{\frac{I_{BIAS}}{\frac{1}{2}\left(\frac{W}{L}\right)_1 \mu_n C_{ox}}} \quad (11.82)$$

Since the circuit is perfectly symmetrical, the results determined above are similar for those when the differential input voltage V_{ID} is much less than zero.

A plot of the transfer function of the output voltages versus the input differential voltage is shown in Fig. 11.32. It should also be noted that our small-signal analysis is valid around the point where the differential input voltage is nearly equal to 0 V. A simple first order assumption to complete this transfer characteristic would be to draw straight lines and assume linear operation for the entire differential input voltage range. To develop a more rigorous approach to finding the differential voltage transfer function, we will solve for the transfer function between V_{O1} versus V_{ID}.

We begin the analysis by writing an expression for the output voltage V_{O1} and V_{O2} given by

$$V_{O1} = V^+ - I_{D1} R_D \quad (11.83)$$

$$V_{O2} = V^+ - I_{D2} R_D \quad (11.84)$$

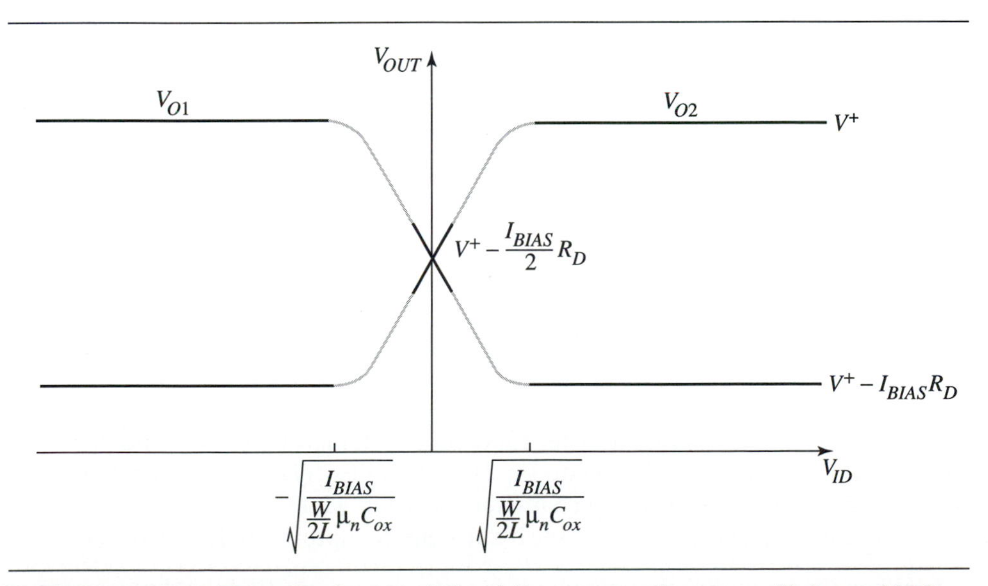

➤ **Figure 11.32** Transfer characteristic of the output voltages vs. V_{ID} for an MOS differential amplifier.

Assuming that both input transistors are operating in their constant-current regions and that we can neglect channel length modulation and the backgate effect, the drain currents in devices M_1 and M_2 are given by

$$I_{D1} = K_n (V_{GS1} - V_{Tn})^2 \tag{11.85}$$

$$I_{D2} = K_n (V_{GS2} - V_{Tn})^2 \tag{11.86}$$

where

$$K_n = \frac{1}{2}\left(\frac{W}{L}\right)_{1,2} \mu_n C_{ox} \tag{11.87}$$

The differential input voltage V_{ID} can be written in terms of the gate-source voltages of the input devices as

$$V_{ID} = V_{GS1} - V_{GS2} \tag{11.88}$$

By manipulating the above equations, we can write

$$\sqrt{I_{D1}} - \sqrt{I_{D2}} = \sqrt{K_n}\ V_{ID} \tag{11.89}$$

By KCL, the sum of the drain currents of devices M_1 and M_2 must be equal to that supplied by the bias current source

$$I_{D1} + I_{D2} = I_{BIAS} \tag{11.90}$$

The equations above can be manipulated to find an expression that relates the drain current in device M_1 to the differential input voltage. The algebraic steps for this analysis are a bit tedious and not informative, so only the result is given as

large-signal transfer function for the MOS differential amplifier

$$I_{D1} = \frac{I_{BIAS}}{2} + \frac{K_n V_{ID}}{2}\sqrt{\frac{2I_{BIAS}}{K_n} - (V_{ID})^2} \quad |V_{ID}| \le \sqrt{\frac{I_{BIAS}}{K_n}} \tag{11.91}$$

A similar analysis can be carried out for the drain current in device M_2. Its current is given by

$$I_{D2} = \frac{I_{BIAS}}{2} - \frac{K_n V_{ID}}{2}\sqrt{\frac{2I_{BIAS}}{K_n} - (V_{ID})^2} \quad |V_{ID}| \le \sqrt{\frac{I_{BIAS}}{K_n}} \tag{11.92}$$

The large-signal voltage transfer function is obtained by substituting the equations for the drain current above into Eq. (11.83) and Eq. (11.84). The resulting equations quantify the dashed transfer characteristic in Fig. 11.32.

Let's investigate Eq. (11.91) to see if it fits our asymptotic analysis given above.

1. When the input differential voltage is equal to zero, we find the drain current in M_1 to be exactly one half of the bias current.

2. When the differential input voltage is equal to $\sqrt{I_{BIAS}/K_n}$, we find the drain current in device M_1 to be equal to the full bias current.
3. When the differential input voltage is equal to $-\sqrt{I_{BIAS}/K_n}$, we find that device M_1 is cutoff and no drain current flows.

A similar result occurs for Eq. (11.92) and both equations check with our simple asymptotic analysis.

It should be noted that the point at which the transfer characteristic saturates depends on the value of K_n. As K_n is made larger corresponding to a larger W/L for the MOS transistor, the differential voltage required to have the full bias current flow through either M_1 or M_2 is decreased. Intuitively, this fact makes sense since with larger device sizes, it takes a smaller gate-source voltage to sink the same amount of current. Since the bias current source is held constant, a smaller gate-source voltage will sink all of the bias current.

11.6.3 Large Differential Signal Response of a Bipolar Differential Amplifier

In this section, we determine the voltage transfer function between the output voltages of the bipolar differential amplifier V_{O1} and V_{O2} and the differential input voltage as shown in Fig. 11.33. The bipolar differential amplifier has a similar large-signal response to that of the MOS differential amplifier—with one important difference. The collector current in a bipolar transistor is exponentially dependent on the base-emitter input voltage. For the MOS transistor, the drain current is related to the square of the input voltage. Therefore, it takes a smaller differential voltage to sink all of the bias current in the bipolar implementation.

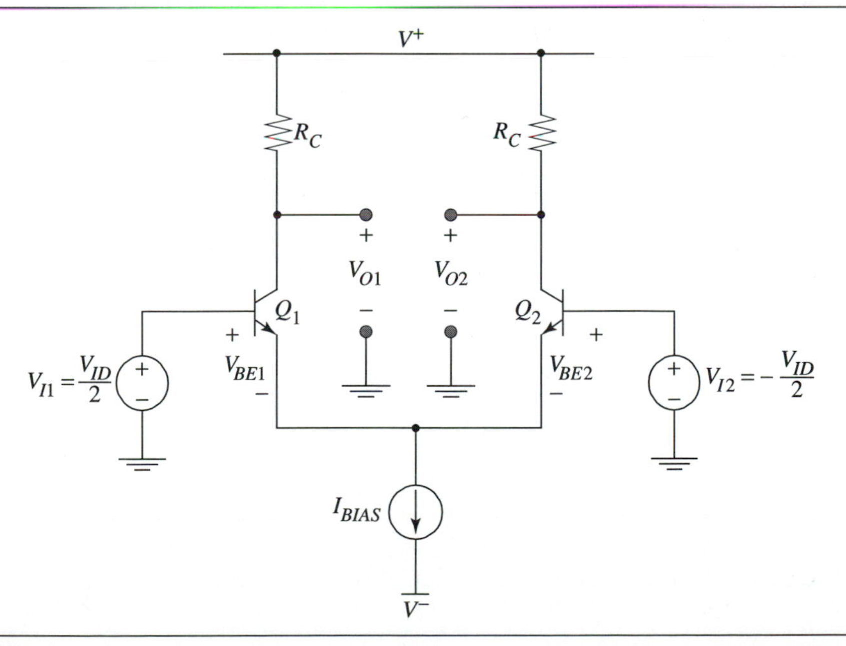

➤ **Figure 11.33** Bipolar differential amplifier with large differential input signals applied.

To determine the approximate differential voltage required to saturate the amplifier, we begin this analysis by writing the differential input voltage as

$$V_{ID} = V_{I1} - V_{I2} = V_{BE1} - V_{BE2} \tag{11.93}$$

Next, we write expressions for the base-emitter voltages of the two input transistors

$$V_{BE1} = V_{th} \ln\left(\frac{I_{C1}}{I_{S1}}\right) \tag{11.94}$$

$$V_{BE2} = V_{th} \ln\left(\frac{I_{C2}}{I_{S2}}\right) \tag{11.95}$$

Using the above expressions and assuming matched devices ($I_{S1} = I_{S2}$), we can relate the differential input voltage to the collector currents as

$$V_{ID} = V_{th} \ln\left(\frac{I_{C1}}{I_{C2}}\right) \tag{11.96}$$

Inverting this equation, we arrive at

$$\frac{I_{C1}}{I_{C2}} = e^{(V_{ID}/V_{th})} \tag{11.97}$$

By recalling the 60 mV/decade rule and examining Eq. (11.97), we see that the collector currents will differ by 10× if the differential input voltage is 60 mV. They will differ by 100× when the differential input voltage is 120 mV and so on. Qualitatively speaking, we can consider the amplifier saturated when the differential input voltage

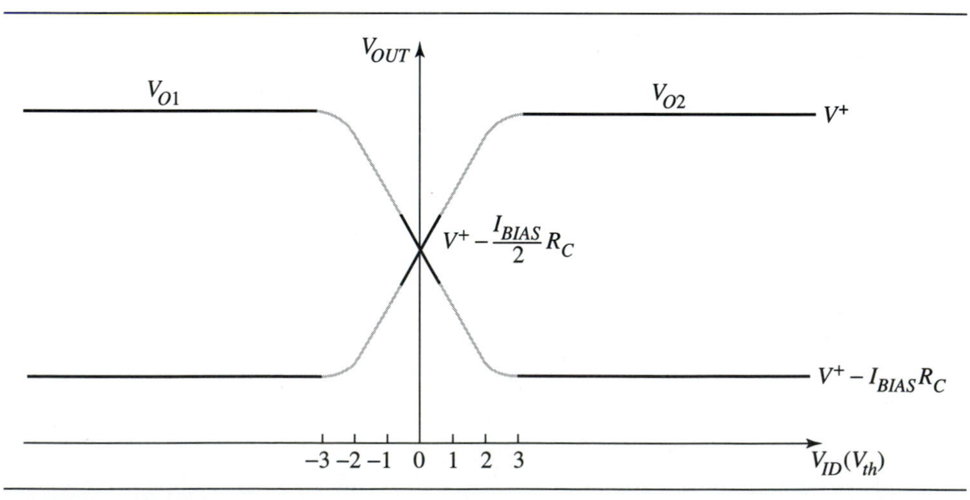

➤ Figure 11.34 Transfer characteristic of the output voltages vs. V_{ID} for a bipolar differential amplifier.

is approximately $\pm 3V_{th}$. Under this condition, nearly all of the bias current will be flowing through either Q_1 or Q_2. The transfer characteristic for the output voltages versus the input differential voltage is shown in Fig. 11.34.

When the differential input voltage is small ($< V_{th}$), we can assume that the output voltages and the differential input voltage are linearly related. In the following analysis we want to connect this small-signal region with the region that is saturated to derive the complete transfer function.

The full analysis for the bipolar differential amplifier is carried out by recognizing that the sum of the collector currents is approximately equal to the bias current source when we neglect the base current

$$I_{C1} + I_{C2} \approx I_{BIAS} \tag{11.98}$$

Solving Eq. (11.97) and Eq. (11.98) simultaneously, we can write expressions for the collector currents of Q_1 and Q_2 given in

$$I_{C1} = \frac{I_{BIAS}}{1 + e^{-(V_{ID}/V_{th})}} \quad I_{C1} < I_{BIAS} \tag{11.99}$$

$$I_{C2} = \frac{I_{BIAS}}{1 + e^{(V_{ID}/V_{th})}} \quad I_{C2} < I_{BIAS} \tag{11.100}$$

large-signal transfer function for the bipolar differential amplifier

The output voltages are given by

$$V_{O1} = V^{+} - I_{C1}R_C \tag{11.101}$$

$$V_{O2} = V^{+} - I_{C2}R_C \tag{11.102}$$

The transfer functions of the output voltages V_{O1} and V_{O2} as a function of the input differential voltage are shown in Fig. 11.34. When comparing the MOS and bipolar implementations of a differential amplifier, we see that the bipolar implementation requires a smaller differential input voltage to swing the output voltages between their two extremes. This fact is related to the exponential dependence of the collector current on the input voltage for the bipolar transistor.

When the MOS transistors are made sufficiently wide, the required differential input voltage to saturate the amplifier is reduced. As the width of the device is continually increased, one might think that the MOS implementation would require a smaller differential voltage than the bipolar implementation to saturate the amplifier. However, it should be noted that as the width of the device is increased further, the gate-source voltage to sink $I_{BIAS}/2$ approaches the threshold voltage ($V_{GS} \approx V_{Tn}$). As a result the models used for the MOS transistor are no longer valid and the device begins to operate in its subthreshold region. In this region, the drain current has an exponential dependence on the gate-source voltage and a similar form for the relation between the output voltages and differential input voltage as the bipolar implementation results.

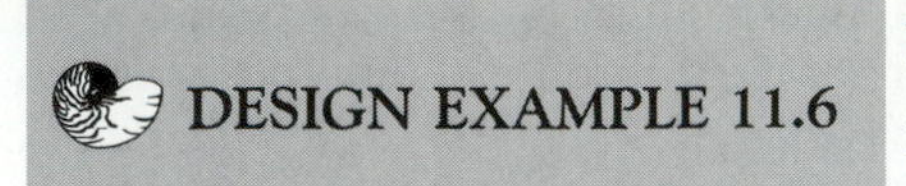

DESIGN EXAMPLE 11.6 Emitter Coupled Logic (ECL): An Application of the Large-Signal Response of a Bipolar Differential Pair

In Chapter 5, we saw that logic gates are made of amplifiers that saturate between two states. The differential amplifier can also be used as a logic gate. The outputs of the differential amplifier swing between V_{CC}, representing a logic **1** and $V_{CC} - I_{BIAS}R_C$, representing a logic **0**. Each output can be used separately. One is the logic output Q and the other is the logical inverse $\overline{Q}$.

A circuit diagram for an emitter coupled logic (ECL) inverter/buffer is shown in Fig. Ex 11.6. The input voltage at transistor Q_1 is compared to the reference voltage at the base of Q_2. If the input voltage is greater than approximately $V_{REF} + 3\ V_{th}$, then the output voltage V_{O1} at the collector of Q_1 will be $V_{CC} - I_{BIAS}R_C$ and the output voltage V_{O2} at the collector of Q_2 will be V_{CC}. From a logic point of view, if the input voltage is greater than V_{REF} by a sufficient noise margin the input logic state is considered a **1**. With a logic **1** input the output logic variable $Q = \mathbf{1}$ and $\overline{Q} = \mathbf{0}$.

In this example, we will design an ECL inverter/buffer. In order to ensure high speed operation, the bipolar transistors must *not* go into saturation implying that $V_{CE} > V_{CE_{SAT}}$. The choice of the reference voltage should be halfway between the two output voltage levels so that the gate can drive other similarly designed ECL gates.

(a) Given that $V_{CC} = 5$ V and $I_{BIAS} = 500\ \mu$A, choose the value of R_C to maximize the output swing while keeping $V_{CE1,2} \geq 0.2$ V.

SOLUTION

Since we know that the differential amplifier will steer the bias current of 500 μA into Q_1 or Q_2 depending on the logic, we can calculate the maximum value of R_C

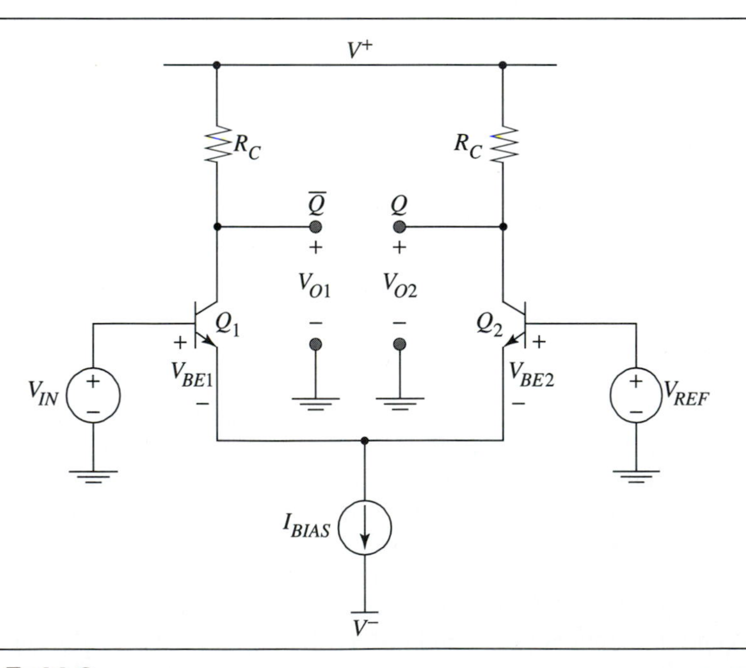

➤ **Figure Ex11.6**

possible to maximize the output voltage swing without saturating Q_1 or Q_2. Assuming 500 μA flows into the collector of Q_1 and neglecting base current we can write

$$V_{O1} = V_{CC} - I_{BIAS} R_C$$

Since the input voltage could be as large as V_{CC} and assuming $V_{BE1} \approx 0.7\text{ V}$, we can write

$$V_{E1} = V_{CC} - V_{BE1} = 5\text{ V} - 0.7\text{ V} = 4.3\text{ V}$$

$$V_{CE1} = V_{O1} - V_{E1} = V_{CC} - I_{BIAS} R_C - V_{CC} + V_{BE1} = V_{BE1} - I_{BIAS} R_C$$

To ensure that the bipolar transistor does not saturate we require

$$V_{CE1} \geq 0.2\text{ V} = 0.7\text{ V} - (500\,\mu\text{A}) R_C$$

Solving for R_C we find

$$R_C = \frac{0.5\text{ V}}{500\,\mu\text{A}} = 1\text{ k}\Omega$$

Given $R_C = 1\text{k}\Omega$ we find $V_{O1} = 4.5$ V and $V_{REF} = 4.75$ V which is halfway between the low and high values of the output voltage.

(b) Analyze the ECL gate to find $V_{IL}, V_{IH}, V_{OL}, V_{OH}, NM_L, NM_H$.

SOLUTION

From part (a) we know that $V_{OH} = 5$ V and $V_{OL} = 4.5$ V. To calculate V_{IL} and V_{IH} we first realize that V_M is the same as V_{REF} in an ECL implementation. Under this condition the differential amplifier is operating in its high-gain region. To calculate the small-signal voltage gain recognize that each output is used individually so that the voltage gain is

$$\frac{v_{o1}}{v_{in}} = \frac{-v_{o2}}{v_{in}} = -\frac{g_{m1,2} R_C}{2}$$

Recall that $g_{m1,2} = \dfrac{qI_{BIAS}/2}{kT} = \dfrac{250\,\mu\text{A}}{25\text{ mV}} = 10\text{ mS}$. With $R_C = 1\text{ k}\Omega$ the magnitude of the small signal voltage gain $|A_v| = 5$.

Recall from Chapter 5 that $V_{IL} = V_M - \dfrac{(V_{OH} - V_M)}{A_v}$ and $V_{IH} = V_M + \dfrac{(V_M - V_{OL})}{A_v}$.

Substituting the values we find $V_{IL} = 4.7$ V and $V_{IH} = 4.8$ V. The noise margins are

$$NM_L = V_{IL} - V_{OL} = 4.7\text{ V} - 4.5\text{ V} = 0.2\text{ V}$$

$$NM_H = V_{OH} - V_{IH} = 5\text{ V} - 4.8\text{ V} = 0.2\text{ V}$$

The small input and output voltage swings are characteristic of ECL and give it a speed advantage over other logic families since any parasitic capacitance does not have to charge over a large voltage range. The bipolar transistor's high g_m allows for small voltage swings at the input to saturate the amplifier. Note: The bipolar transistors are not saturated. The major problem with ECL is that it consumes DC power. ECL is used where speed is the most important specification.

SUMMARY

In this chapter, we developed the general concepts of differential amplifiers and showed the analysis and design of several topologies. In addition we included a discussion on the large-signal characteristics of both the MOS and bipolar differential amplifiers. Specifically we showed

- DC biasing considerations including how to center the DC output voltage.
- General analysis techniques such as half-circuits to simplify hand analysis.
- Small-signal analysis of both bipolar and MOS differential amplifiers including differential- and common-mode gain, common-mode rejection and the frequency response of these parameters.
- The analysis and design of MOS and bipolar differential amplifiers with current mirrors to improve the differential gain.
- Calculation of the DC large-signal input common-mode range for both bipolar and MOS differential amplifiers.
- The derivation of the large-signal response of MOS and bipolar differential amplifiers.

FURTHER READING

1. P. R. Gray and R. G. Meyer, *Analysis and Design of Analog Integrated Circuits 3rd Ed.*, Wiley, 1993, Chapter 3. More advanced discussion on differential amplifiers including an analysis of mismatch effects.
2. M. N. Horenstein, *Microelectronic Circuits and Devices 2nd Ed.*, Prentice Hall, 1995, Chapter 11. A complete description of differential amplifiers.
3. A. S. Sedra and K. C. Smith, *Microelectronic Circuits 3rd Ed.*, HRW Saunders, 1991, Chapter 7. Extensive discussion on differential amplifiers and non-ideal effects.

PROBLEMS

In the following exercises and problems use the device data given below for the MOS and bipolar transistors unless otherwise specified.

MOS Device Data

$\mu_n C_{ox} = 50\ \mu A/V^2$, $\mu_p C_{ox} = 25\ \mu A/V^2$, $V_{Tn} = -V_{Tp} = 1\ V$, $-2\phi_p = 2\phi_n = 0.8\ V$,

$\gamma_n = \gamma_p = 0.6\ V^{-\frac{1}{2}}$ $\lambda_n = \lambda_p = 0.05\ V^{-1}$ @ $L = 2\ \mu m$,

$C_{ox} = 2.3\ fF/\mu m^2$, $C_{jn} = 0.1\ fF/\mu m^2$, $C_{jp} = 0.3\ fF/\mu m^2$,

$C_{jswn} = 0.5\ fF/\mu m$, $C_{jswp} = 0.35\ fF/\mu m$, $C_{ov} = 0.5\ fF/\mu m$, $L_{diffn} = L_{diffp} = 6\ \mu m$

npn Bipolar Device Data

$I_S = 10^{-17} A$, $\beta_F = \beta_o = 100$, $V_A = 25\ V$, $\tau_F = 50\ ps$, $C_{je} = 15\ fF$ @$V_{BE} = 0.7\ V$,
$C_\mu = 10\ fF$ @$V_{BC} = -2.0\ V$

EXERCISES

E11.1 Given the differential amplifier shown in Figure E11.1 with DC bias voltages $V_{I1} = V_{I2} = 0$ V, find the value of I_{BIAS} such that the DC output voltages $V_{O1} = V_{O2}$ are "centered" between the power supplies V^+ and V^- given

(a) $V^+ = 2.5$ V, $V^- = -2.5$ V, $R_C = 10$ kΩ

(b) $V^+ = 5$ V, $V^- = 0$ V, $R_C = 10$ kΩ

(c) $V^+ = 2.5$ V, $V^- = -2.5$ V, $R_C = 5$ kΩ

E11.2 Given the differential amplifier shown in Figure E11.1 with DC bias voltages $V_{I1} = V_{I2} = 0$V, find the value of R_C such that the DC output voltages $V_{O1} = V_{O2}$ are "centered" between the power supplies V^+ and Vy^- given

(a) $V^+ = 2.5$ V, $V^- = -2.5$ V, $I_{BIAS} = 100$ μA

(b) $V^+ = 5$ V, $V^- = 0$ V, $I_{BIAS} = 100$ μA

(c) $V^+ = 2.5$ V, $V^- = -2.5$ V, $I_{BIAS} = 500$ μA

E11.3 In this exercise you are given values for the small-signal input voltages to both sides of a differential amplifier v_{i1} and v_{i2}. Using Fig. 11.2 to help you, find the values of the purely differential and common-mode input voltages, v_{id} and v_{ic} given

(a) $v_{i1} = 1$mV and $v_{i2} = 0.5$ mV

(b) $v_{i1} = -1$mV and $v_{i2} = 1$mV

(c) $v_{i1} = 100.3$ mV and $v_{i2} = 101.3$ mV

(d) $v_{i1} = -100$ mV and $v_{i2} = -110$ mV

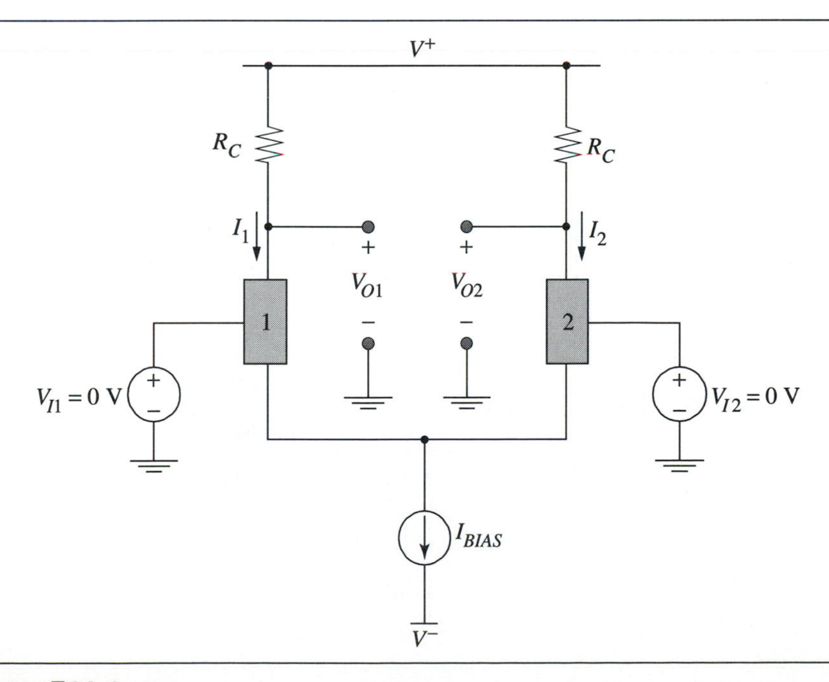

➤ **Figure E11.1** .

E11.4 You are given the differential amplifier shown in Fig. E11.4. The value for r_{ob} is equal to $25\ \mathrm{V}/I_{BIAS}$. Set the value for R_C such that $v_{o1} = v_{o2} = 0\ \mathrm{V}$ when $v_{i1} = v_{i2} = 0\ \mathrm{V}$. Using half-circuit techniques when appropriate find a_{dm}, a_{cm} and $CMRR$ for

(a) $I_{BIAS} = 100\ \mu\mathrm{A}$

(b) $I_{BIAS} = 500\ \mu\mathrm{A}$

E11.5 Repeat E11.4 substituting NMOS transistors with $W/L = 100/2$ for the npn bipolar transistors. Assume that the backgates are shorted to their respective sources.

E11.6 For the differential amplifier shown in Figure E11.6 find the two-port model parameters and the overall voltage gain for a given source and load resistance of $R_S = 1\ \mathrm{k}\Omega$ and $R_L = 50\ \mathrm{k}\Omega$. Use half-circuit techniques when appropriate.

(a) Calculate R_{id}

(b) Calculate R_{od}

(c) Calculate G_{md}

(d) Calculate the overall voltage gain, v_{od}/v_{id}. Note $v_{od} = v_{o1} - v_{o2}$.

E11.7 Repeat E11.6 substituting NMOS transistors for the npn bipolar transistors. Assume that the backgates are shorted to their respective sources. Use

(a) $W/L = 20/2$

(b) $W/L = 200/2$

E11.8 In Example 11.2 the differential mode frequency response was analyzed for an NMOS differential amplifier. Repeat this example using PMOS transistors.

E11.9 In Example 11.2 the differential mode frequency response was analyzed for an NMOS differential amplifier. Repeat this example using npn bipolar transistors. Adjust I_{BIAS} so that the magnitude of $a_{dm} \geq 50$ at DC.

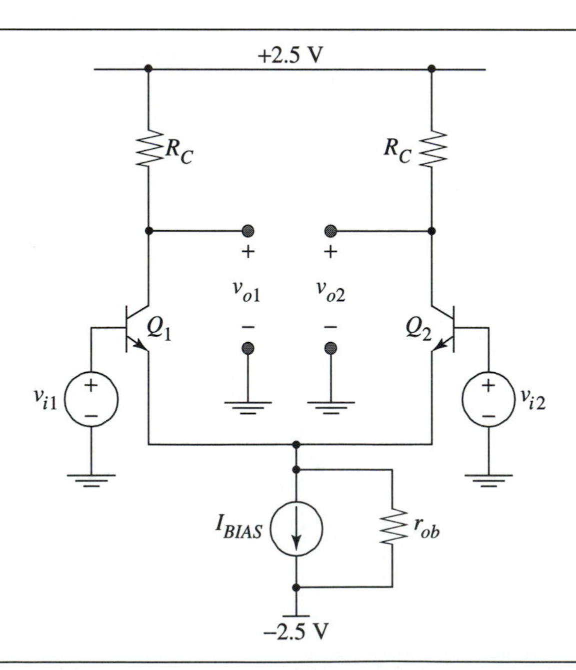

➤ Figure E11.4

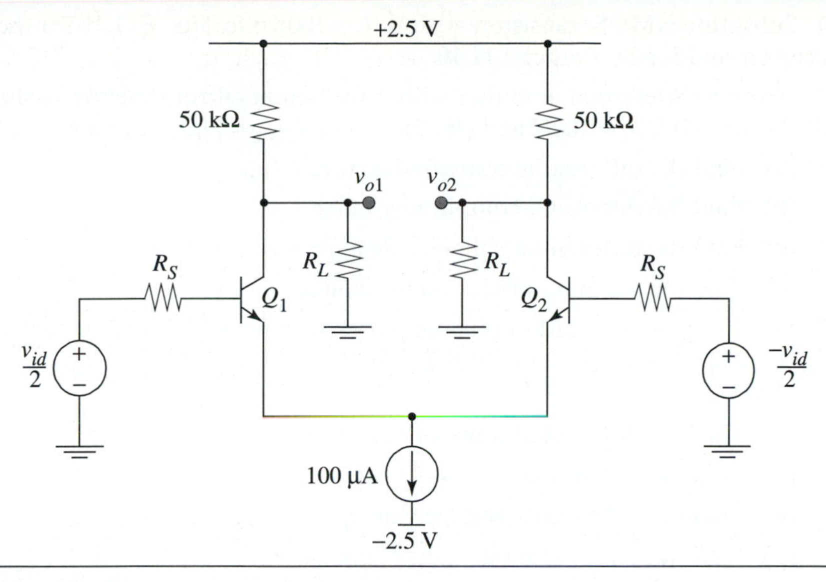

➤ **Figure E11.6**

E11.10 Given the differential amplifier shown in Fig. E11.10 with a single-ended output and $v_{i1} = 10$ mV and $v_{i2} = 12$ mV, $I_{BIAS} = 200$ μA and $r_{ob} = 100$ kΩ,

(a) Find R_C such that the DC output voltages equal 0 V with $v_{i1} = v_{i2} = 0$ V.

(b) Find a_{dm}.

(c) Find a_{cm}.

(d) Using superposition find v_o/v_{id}, where $v_{id} = v_{i1} - v_{i2}$.

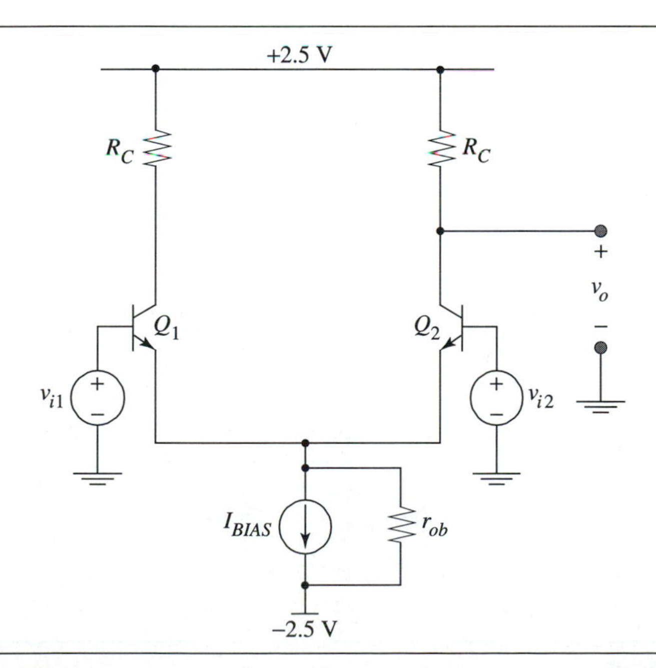

➤ **Figure E11.10**

E11.11 Substitute NMOS transistors with $W/L = 100/2$ into Fig. E11.10 for the bipolar transistors and repeat Exercise 11.10.

E11.12 Given a differential amplifier with a p-channel current mirror as shown in Fig. E11.12 with $(W/L)_n = 100/2$ and $(W/L)_p = 50/4$, $I_{BIAS} = 200$ μA and $r_{ob} = 100$ kΩ,

(a) Find the differential transconductance i_o/v_{id}.
(b) Find the differential output resistance.
(c) Find the differential voltage gain $a_{vd} = v_o/v_{id}$.
(d) Find the common-mode voltage gain $a_{vc} = v_o/v_{ic}$.

E11.13 Draw a p-channel differential amplifier with an n-channel current mirror. Use Fig. E11.12 to help you. Use $(W/L)_p = 100/2$ and $(W/L)_n = 50/4$, $I_{BIAS} = 200$ μA and $r_{ob} = 100$ kΩ.

(a) Find the differential transconductance i_o/v_{id}.
(b) Find the differential output resistance.
(c) Find the differential voltage gain $a_{vd} = v_o/v_{id}$.
(d) Find the common-mode voltage gain $a_{vc} = v_o/v_{ic}$.

E11.14 Draw an npn bipolar differential amplifier with a p-channel current mirror. Use Fig. E11.12 to help you. Use $(W/L)_p = 50/4$, $I_{BIAS} = 200$ μA and $r_{ob} = 100$ kΩ.

(a) Find the differential transconductance i_o/v_{id}.
(b) Find the differential output resistance.
(c) Find the differential voltage gain $a_{vd} = v_o/v_{id}$.
(d) Find the common-mode voltage gain $a_{vc} = v_o/v_{ic}$.

E11.15 Given the wideband differential amplifier shown in Fig. E11.15 with $R_C = 10$ kΩ and $I_{BIAS} = 500$ μA, find the voltage gain.

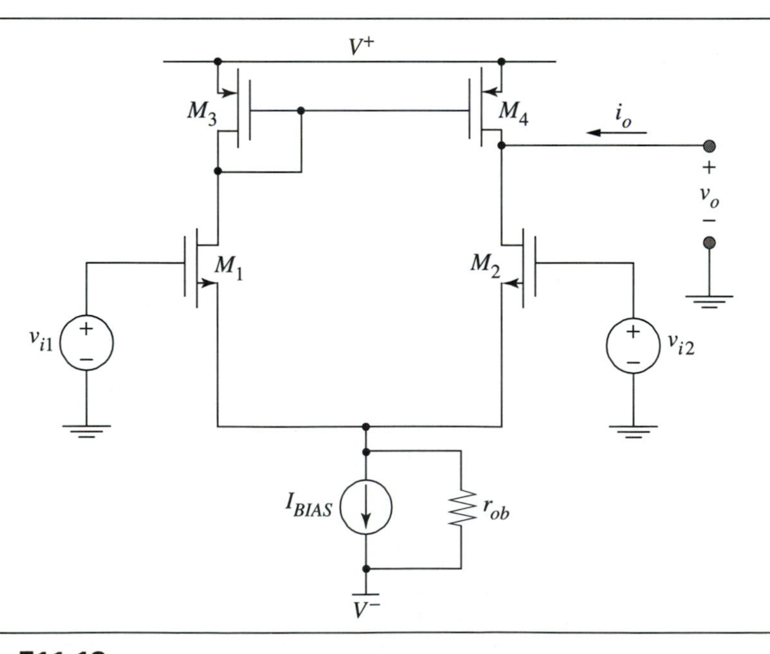

Figure E11.12

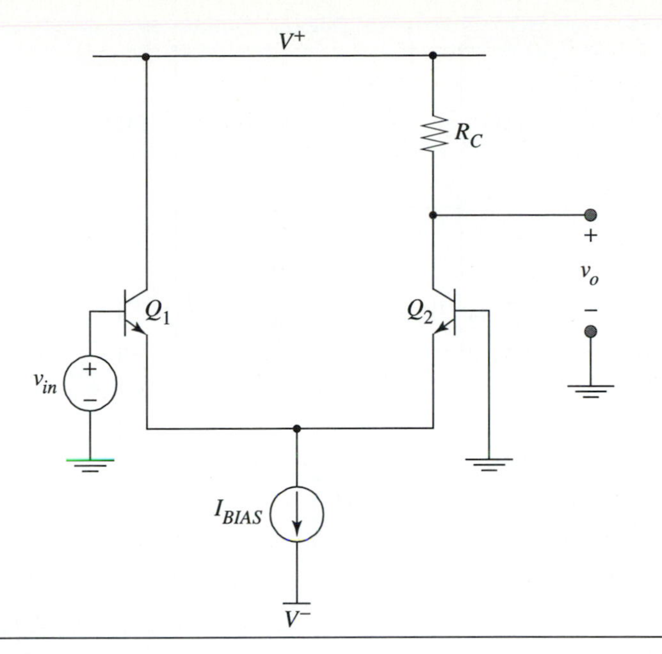

➤ **Figure E11.15**

E11.16 Draw a wideband amplifier configuration using p-channel transistors using Fig. E11.15 as a guide and calculate the voltage gain given $I_{BIAS} = 500\ \mu A$.

E11.17 In the circuit shown in Fig. E11.17 we set $I_{BIAS} = 1$ mA. Assume that $V_{CE_{SAT}} = 0.2$ V and the bias current source requires a voltage across it greater than 0.5 V. Find V_{O1}, V_{O2}, and V_X under the following conditions. State whether the amplifier is operating with all of the devices in their constant-current region.

(a) $V_{I1} = 0$ V, $V_{I2} = 0$ V
(b) $V_{I1} = 2$ V, $V_{I2} = 2$ V
(c) $V_{I1} = 1$ V, $V_{I2} = 0$ V
(d) $V_{I1} = 0$ V, $V_{I2} = -1.5$ V
(e) $V_{I1} = 0$ V, $V_{I2} = 0$ V, with $I_{BIAS} = 500\ \mu A$.

E11.18 Substitute NMOS transistors with a $W/L = 100/2$ for the bipolar transistors in Figure E11.17. Set $I_{BIAS} = 500\ \mu A$. Assume that the bias current source requires a voltage across it is greater than 0.5 V. Find V_{O1}, V_{O2}, and V_X under the following conditions. State whether the amplifier is operating with all of the devices in their constant-current region.

(a) $V_{I1} = 0$ V, $V_{I2} = 0$ V
(b) $V_{I1} = 2$ V, $V_{I2} = 2$ V
(c) $V_{I1} = 1$ V, $V_{I2} = 0$ V
(d) $V_{I1} = 0$ V, $V_{I2} = -1.5$ V
(e) $V_{I1} = 0$ V, $V_{I2} = 0$ V, with $I_{BIAS} = 200\ \mu A$.

E11.19 Repeat E11.18 using NMOS transistors with $W/L = 200/2$.

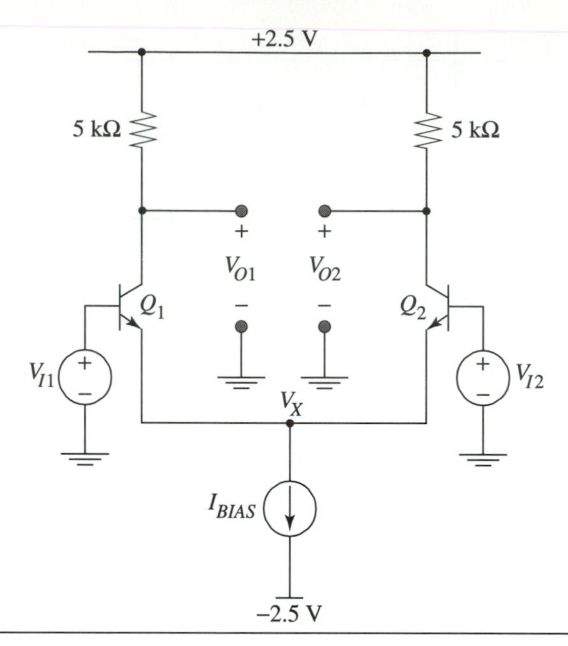

➤ **Figure E11.17**

PROBLEMS

P11.1 You are given a differential amplifier using MOS transistors shown in Fig. P11.1. Assuming $V_{OFF} = 0$ V, $(W/L)_1 = (W/L)_2 = 50/2$, calculate I_{BIAS} such that $V_{O1} - V_{O2} = 0$ V, given

(a) $R_{D1} = R_{D2} = 10$ kΩ

(b) $R_{D1} = R_{D2} = 10$ kΩ nominally, but the absolute value is only controllable to +/− 20%. Using the value for I_{BIAS} calculated in part(a) find the range of $V_{O1} = V_{O2}$.

P11.2 Referring to Fig. P11.1, if there is a mismatch in R_{D1} and R_{D2} such that $R_{D1} = 9.9$ kΩ and $R_{D2} = 10.1$ kΩ and $(W/L)_1 = (W/L)_2 = 50/2$,

(a) Calculate V_{O1} and V_{O2} for $I_{BIAS} = 250$ μA with $V_{OFF} = 0$ V.

(b) By applying an offset voltage, V_{OFF}, at the input we can rebalance the output. Calculate V_{OFF} so that $V_{O1} = V_{O2}$.

(c) What is the value of $V_{O1} = V_{O2}$ found in part (b)?

P11.3 Repeat P11.2 using bipolar transistors in the differential amplifier. Do you see why a smaller offset voltage will rebalance that amplifier?

P11.4 In Fig. P11.4 we are showing a typical input to a differential amplifier from a sensor. This resistor bridge circuit is used to measure a change in the value of resistance of R_X. This change in resistance can be related to a number of physical quantities such as temperature or pressure. The input to the differential amplifier will have a large common-mode voltage which needs rejected and a small differential voltage that needs amplified. In this problem we will calculate the input common-mode and differential-mode voltages for the range of values for R_X.

(a) Find the common-mode voltage if $R_X = 1$kΩ.

(b) Find the common-mode voltage range for 0.9 kΩ ≤ R_X ≤ 1.1 kΩ.

(c) Find the differential-mode voltage if $R_X = 1$ kΩ.

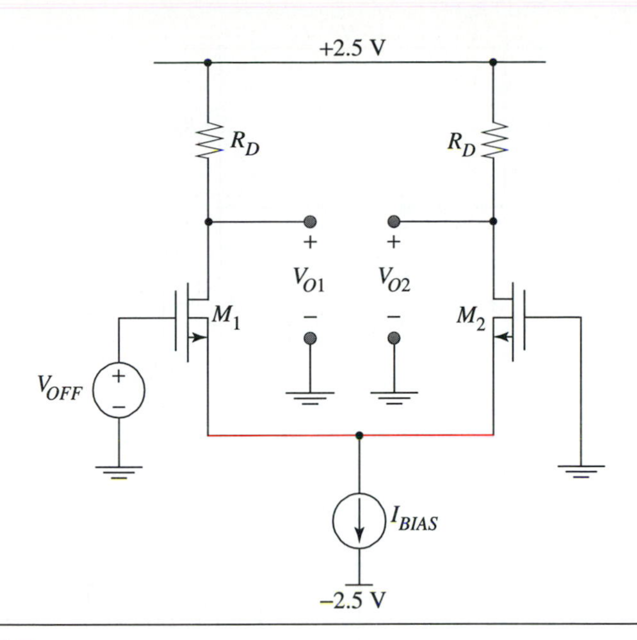

➤ Figure P11.1

(d) Find the differential-mode voltage range for $0.9\ \text{k}\Omega \leq R_X \leq 1.1\text{k}\Omega$.

P11.5 We have added emitter degeneration resistors to the differential amplifier shown in Fig. P11.5. For this problem $R_C = 20\ \text{k}\Omega$, $R_E = 0.5\ \text{k}\Omega$, $I_{BIAS} = 500\ \mu\text{A}$ and $r_{ob} = 50\ \text{k}\Omega$.

(a) Draw the half-circuit for a purely differential-mode input voltage.

(b) Draw the half-circuit for a purely common-mode input voltage.

(c) Find a_{dm}, a_{cm} and $CMRR$ for this amplifier.

(d) Find the differential two-port model parameters R_{id} and R_{od}, and G_{md}.

(e) Draw the two-port model for this amplifier with purely differential input voltages.

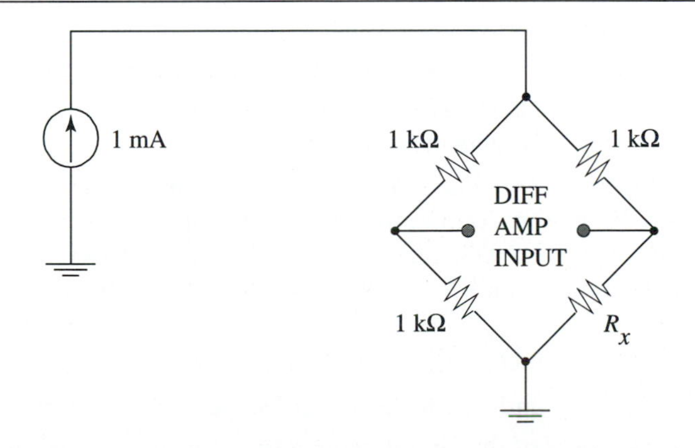

➤ Figure P11.4

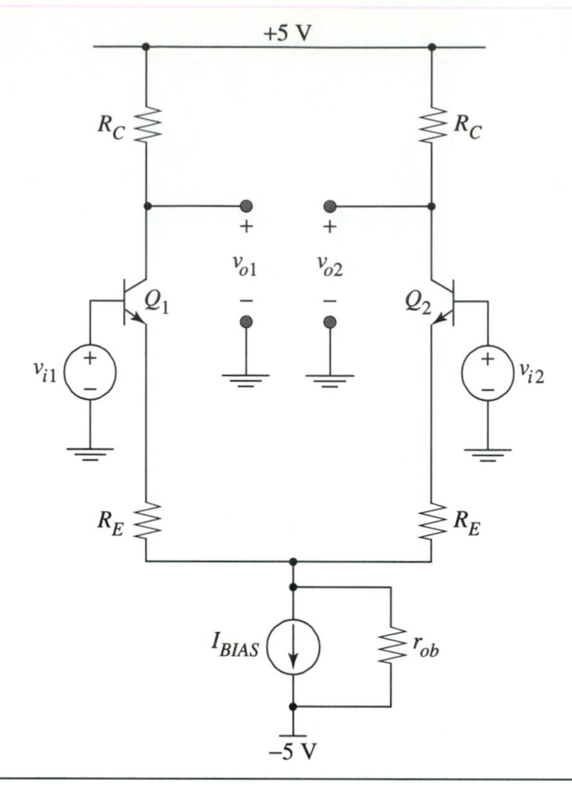

➤ **Figure P11.5**

P11.6 Given the differential amplifier shown in Fig. P11.6 with $I_{REF} = 500\ \mu A$ and the W/L of all the MOS transistors are set equal to 50/2.

(a) Draw the circuit splitting the input into a purely differential and common-mode input.

(b) Draw the half circuits for the purely differential and common-mode input. Note that R_S is not part of the amplifier.

(c) Find a_{dm} and a_{cm}.

(d) Find the small-signal gain from v_{o1}/v_s. You should use superposition. The overall output should be the sum of the outputs when the input is purely differential-mode and when it is purely common-mode.

P11.7 Repeat P11.6 substituting npn bipolar transistors for the NMOS transistors in the differential pair, M_1 and M_2. The current source should still continue to use the NMOS transistors, M_3 and M_4 with $W/L = 50/2$.

P11.8 You are given a two-stage amplifier consisting of an npn differential amplifier followed by a pnp common-emitter amplifier as shown in Fig. P11.8. Both amplifiers have emitter degeneration resistors. $R_C = 2.5\ k\Omega$ and $R_E = 100\ \Omega$ for the differential amplifier and $R_C = 5\ k\Omega$ and $R_E = 500\ \Omega$ for the common-emitter amplifier.

(a) Find R_{in}, R_{out}, and A_v for the first stage of the amplifier.

(b) Find R_{in}, R_{out}, and A_v for the second stage of the amplifier.

(c) Draw a model of the circuit using a two-port model for each stage.

(d) Calculate the overall voltage gain v_{out}/v_{in} for this amplifier.

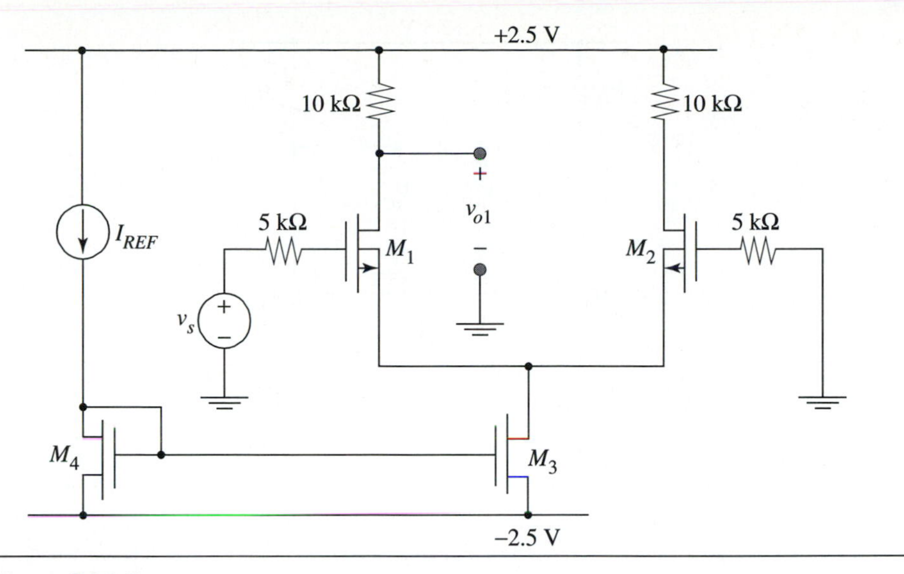

➤ **Figure P11.6**

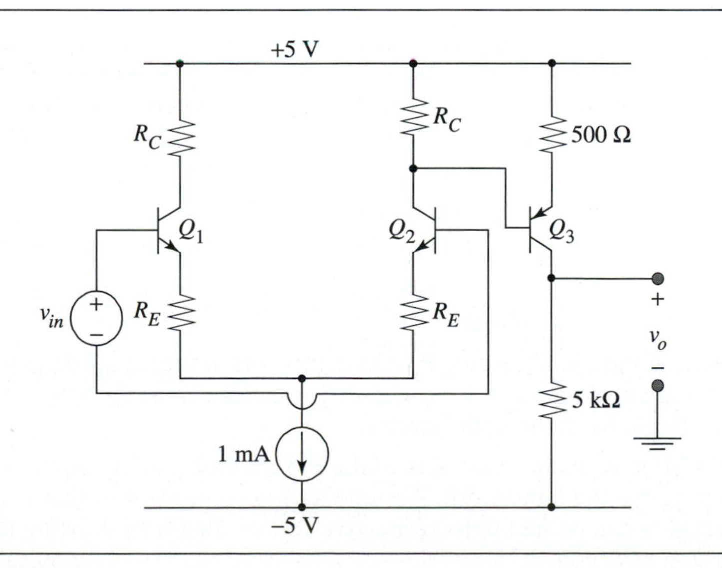

➤ **Figure P11.8**

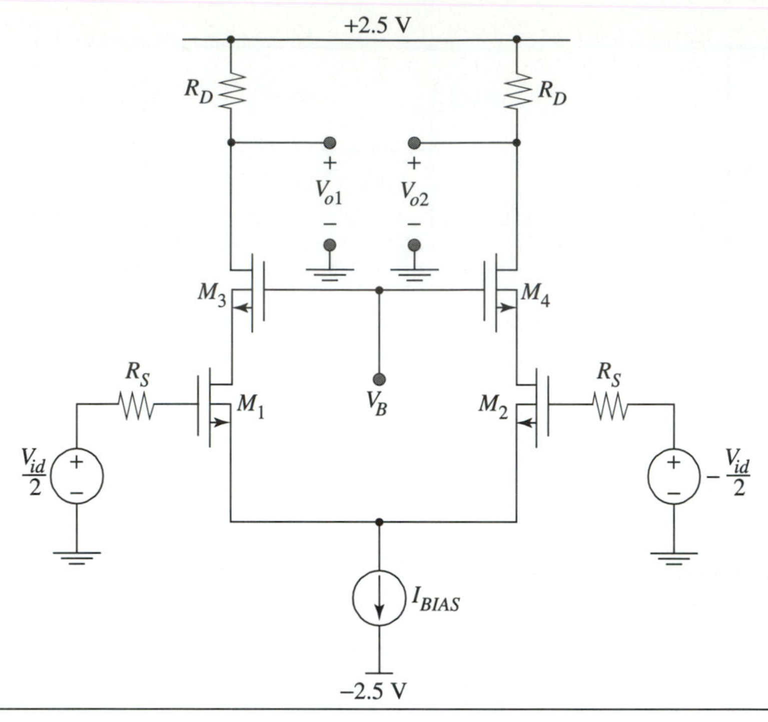

➤ Figure P11.9

P11.9 In Example 11.2, the bandwidth was limited by R_S and the Miller-multiplied C_{gd}. Two methods can be used to increase the bandwidth. (1) Reduce R_S by adding a voltage buffer at the input. (2) Reduce the Miller-multiplied capacitance by adding cascode transistors. The second method is generally better since it requires no extra current. We have redrawn the NMOS differential amplifier with cascode transistors in Figure P11.9. Use the same device data with $R_S = 5\ \text{k}\Omega$, $V_B = 1.5$ V and $I_{BIAS} = 50\ \mu\text{A}$.

(a) Resize resistors R_D such that $V_{O1} = V_{O2} = 1.5$ V when the input voltages are equal to 0 V. (Note: We are raising the DC output voltage to accommodate the additional voltage drop due to the cascode transistors.)

(b) Size the W/L of the NMOS cascode transistors M_3 and M_4 such that their source voltage is equal to 0 V.

(c) Resize the W/L of device M_1 and M_2 such that the magnitude of the differential voltage gain is equal to 50.

(d) Using the method of open-circuit time constants, estimate the bandwidth of the amplifier.

P11.10 Rework Example 11.2 using PMOS transistors at the input. Assume a p-well process so that the backgate of each p-channel transistor must be tied to the +2.5 V power supply. Begin by drawing the circuit.

P11.11 Add PMOS cascode transistors to the PMOS differential amplifier analyzed in P11.10 to improve the bandwidth. Assume an n-well process so that each PMOS transistor backgate can be tied to its respective source. Begin by drawing the circuit. Use Fig. P11.9 to help you.

(a) Resize resistors R_D such that $V_{O1} = V_{O2} = -1.5$ V when the input voltages are equal to 0 V. (Note: We are lowering the DC output voltage to accommodate the additional voltage drop due to the cascode transistors.)

(b) Size the W/L of the PMOS cascode transistors M_3 and M_4 such that their source voltage is equal to 0 V.

(c) Resize the W/L of PMOS devices M_1 and M_2 such that the magnitude of the differential voltage gain is equal to 50.

(d) Using the method of open-circuit time constants, estimate the bandwidth of the amplifier.

P11.12 In the NMOS differential amplifier shown in Fig. P11.12 you are given that $R_S = 1\text{k}\Omega$ and $I_{REF} = 100\ \mu\text{A}$. You are given $(W/L)_{1,2} = 100/2$ and $(W/L)_{3,4} = 50/4$.

(a) Find the value of R_D such that the DC output voltage is equal to 0 V.

(b) Using the half-circuit concept find the differential voltage gain and its ω_{3dB}.

(c) Using the half-circuit concept find the common-mode voltage gain and its ω_{3dB}.

(d) Make a Bode plot of the magnitude of the *CMRR*.

P11.13 Given the PMOS differential amplifier shown in Fig. P11.13 with $(W/L)_{1,2} = 250/2$ and $(W/L)_{3,4} = 50/2$, $(W/L)_5 = 2(W/L)_6 = 25/2$, and $I_{REF} = 100\ \mu\text{A}$,

(a) Find the maximum DC input common-mode voltage.

(b) Find the minimum DC input common-mode voltage.

(c) Find the differential-mode voltage gain.

(d) Find the *CMRR*.

(e) What is the maximum and minimum output voltage?

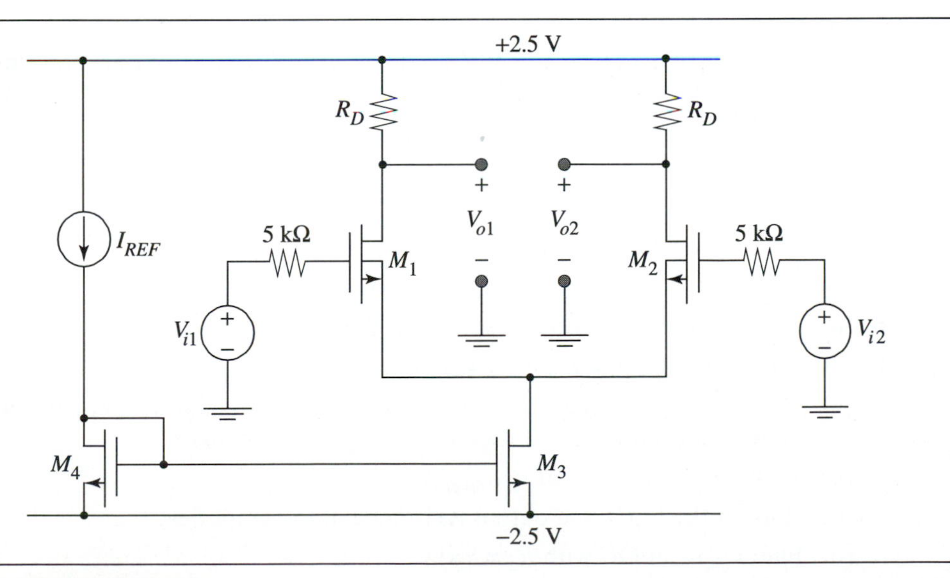

➤ **Figure P11.12**

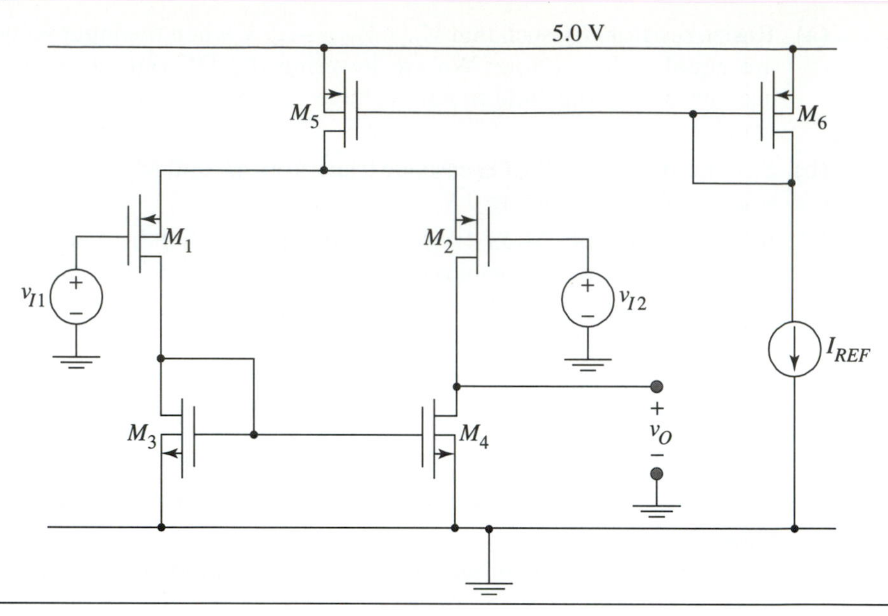

➤ **Figure P11.13**

P11.14 The circuit in Fig. P11.14 is a cascode differential amplifier with n-channel inputs with $I_{REF} = 50\ \mu A$ and $I_{D5} = 100\ \mu A$. If $(W/L)_1 = (W/L)_2 = (W/L)_6 = (W/L)_7 = (W/L)_8 = (W/L)_9 = (W/L)_3 = (W/L)_4$ and we decide to use $L = 2\ \mu m$ for all devices,

(a) What are the transistor sizes including M_5 and M_{10} to have $V_{OUT(max)} = 1.25$ V and $V_{OUT(min)} = -0.75$ V?

(b) What is the differential-mode gain?

(c) What is the common-mode gain?

(d) Estimate the differential-mode bandwidth. Assume $R_S = 1\ k\Omega$.

P11.15 Redraw the circuit in Fig. P11.14 when you add a cascode transistor M_{11} to the bias current source (between M_5 and $M_{1,2}$). $I_{REF} = 50\ \mu A$ and $I_{D5} = 100\ \mu A$. If $(W/L)_1 = (W/L)_2 = (W/L)_6 = (W/L)_7 = (W/L)_8 = (W/L)_9 = (W/L)_3 = (W/L)_4$ and we decide to use $L = 2\mu m$ for all devices,

(a) What are the transistor sizes including M_5, M_{10}, and M_{11} to have $V_{OUT(max)} = 1.0$ V and $V_{OUT(min)} = -1.0$ V?

(b) What is the differential-mode gain?

(c) What is the common-mode gain?

(d) Estimate the differential-mode bandwidth.

P11.16 In the wideband amplifier shown in Fig. P11.16 you are given that $(W/L)_3 = (W/L)_4 = (W/L)_5 = (W/L)_6 = 50/4$, and $I_{REF} = 100\ \mu A$.

(a) Size $(W/L)_7$ such that the amplifier is biased with all devices operating in their constant-current region with $I_{C1} = I_{C2} = 50\ \mu A$.

(b) What are $V_{OUT(max)}$ and $V_{OUT(min)}$?

(c) What is the input and output resistance of the amplifier?

(d) Find v_{out}/v_s at DC with $R_S = 5\ k\Omega$.

(e) Estimate the bandwidth of the amplifier using the method of open-circuit time constants.

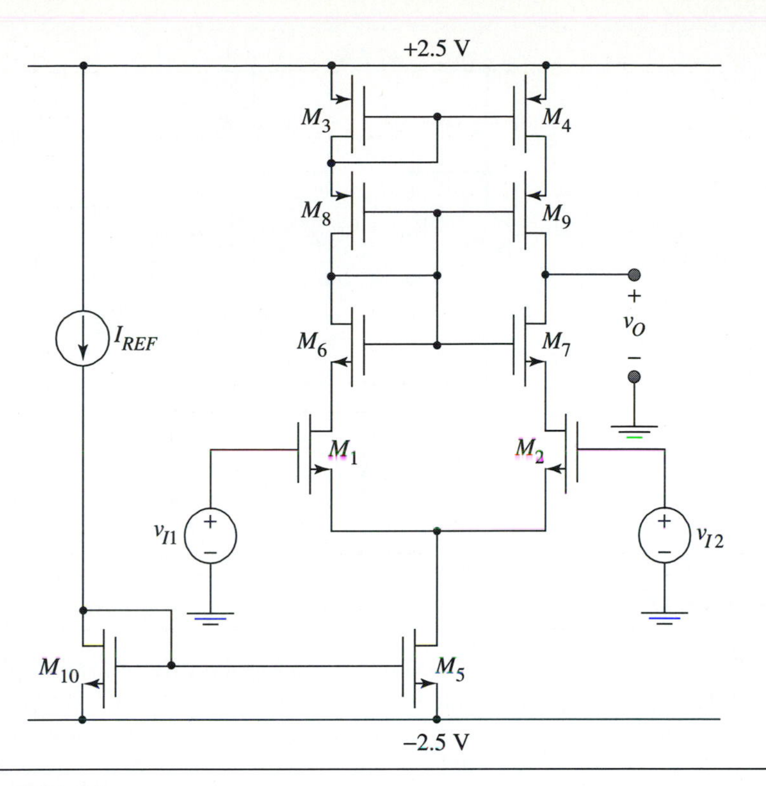

➤ **Figure P11.14**

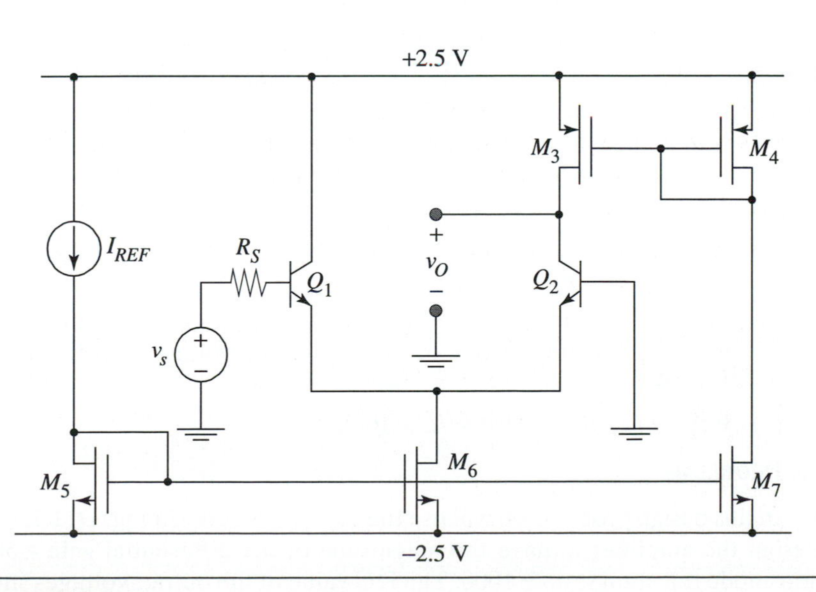

➤ **Figure P11.16**

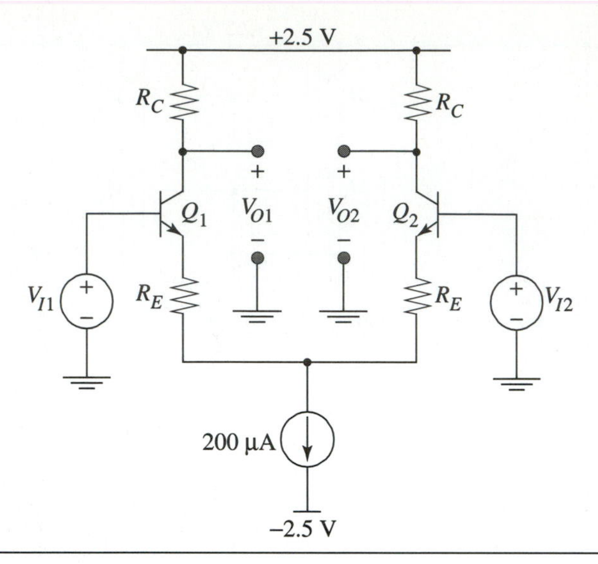

Figure P11.17

P11.17 The large-signal transfer function for the bipolar differential amplifier saturates when $|V_{I1} - V_{I2}|$ is larger than a few thermal voltages (~50 mV). If one adds emitter degeneration resistors to the amplifier we expect the transconductance to be degraded and hence the large signal transfer function to broaden. To see this effect consider the circuit shown in Fig. P11.17. You are given that $R_C = 5\ \text{k}\Omega$.

(a) Sketch the transfer function V_{OD} vs. V_{ID} where $V_{OD} = V_{O1} - V_{O2}$ and $V_{ID} = V_{I1} - V_{I2}$ with $R_E = 200\ \Omega$.

(b) What is the small-signal differential voltage gain of this amplifier. Find it by looking at the slope of the V_{OD} vs. V_{ID} voltage transfer function at $V_{ID} = 0$ V.

(c) Repeat (a) and (b) for $R_E = 500\ \Omega$.

P11.18 In an NMOS differential amplifier we can control the transconductance by the W/L ratio. We expect that the large-signal voltage transfer function should broaden as the W/L ratio is decreased (lower transconductance). Note that decreasing the W/L ratio in the NMOS implementation has the same effect as increasing R_E in the bipolar implementation. To see this effect substitute NMOS transistors for the bipolar transistors in Fig. P11.17.

(a) Sketch the transfer function V_{OD} vs. V_{ID} where $V_{OD} = V_{O1} - V_{O2}$ and $V_{ID} = V_{I1} - V_{I2}$ with $W/L = 100/2$.

(b) What is the small-signal differential voltage gain of this amplifier. Find it by looking at the slope of the V_{OD} vs. V_{ID} voltage transfer function at $V_{ID} = 0$ V.

(c) Repeat (a) and (b) for $W/L = 10/2$.

Design Problems

D11.1 In this design problem we will use the topology shown in Fig. D11.1. Our goal is to design the amplifier to have the magnitude of the differential gain ≥ 50 and a common-mode rejection ratio ≥ 1000. The DC value of the output voltages should be 0 V when $v_{i1} = v_{i2} = 0$ V. Given that you are constrained to have $6\ \mu\text{m} \leq W \leq 100\ \mu\text{m}$,

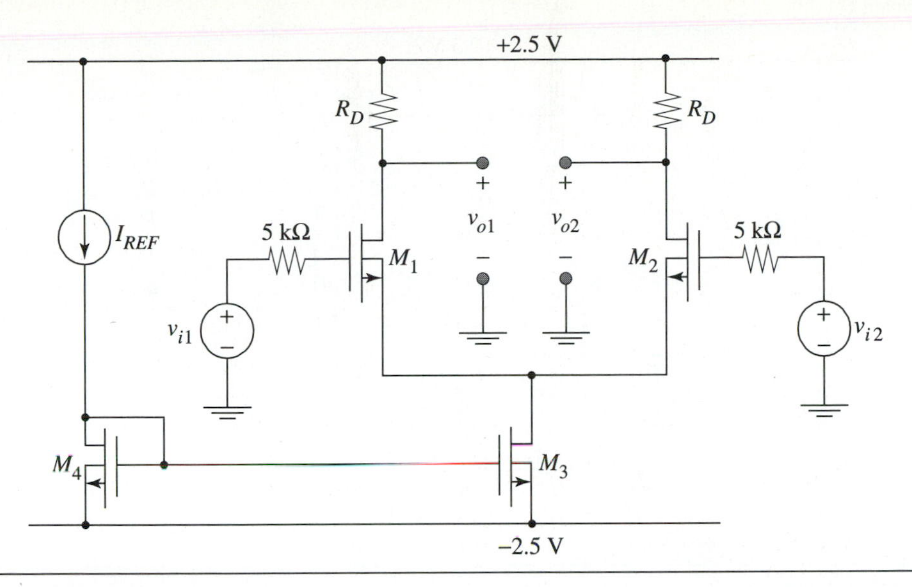

➤ **Figure D11.1**

$2\ \mu m \le L \le 20\ \mu m$, $1\ k\Omega \le R_D \le 10\ k\Omega$, and $100\ \mu A \le I_{REF} \le 300\ \mu A$ try to minimize the area of the circuit. Assume all backgates are shorted to their respective sources.

D11.2 Repeat D11.1 for a PMOS topology. Begin by drawing the circuit and assuming that you have an n-well process so that the backgates of each PMOS transistor can be shorted to its respective source. Use the same constraints and design goals given in D11.1.

D11.3 Repeat D11.1 using npn bipolar transistors in the differential amplifier but continue to use NMOS transistors for the bias current source. Use the same constraints and design goals given in D11.1.

D11.4 In Example 11.3 NMOS transistors were used in the differential amplifier. Repeat the example using a PMOS topology. Begin by drawing the circuit, assuming the p-channel backgates are tied to the positive supply voltage. Use the same constraints and design goals given in Example 11.3.

D11.5 In Example 11.3 NMOS transistors were used in the differential amplifier. Replace M_1, M_2, and M_5 with npn bipolar transistors, but leave devices M_3 and M_4 the same sizes as calculated in the example, namely $(W/L)_{3,4} = 65/5$. You can change R_{REF} to adjust the DC output voltages V_{O1} and V_{O2} to be 0 V. The 5 kΩ source resistance will attenuate the differential- and common-mode voltage gains since the bipolar transistor implementation has a relatively small input resistance.

(a) Adjust the 100 kΩ resistors to achieve $a_{dm} = -50$ at DC.

(b) Calculate the new *CMRR*.

(c) Calculate the frequency where the *CMRR* is reduced by 3 dB.

D11.6 Given an npn differential amplifier shown in Fig. D11.6, the design goal is to have an input common-mode range $-1.5\text{ V} \le V_{IC} \le 1.5\text{ V}$, $a_{dm} = 2000$, and $CMRR = 10000$. Size the four MOS transistors to achieve these specifications.

D11.7 Repeat D11.6 using NMOS transistors in the differential amplifier rather than npn bipolar transistors. Size these devices to meet the same performance specification.

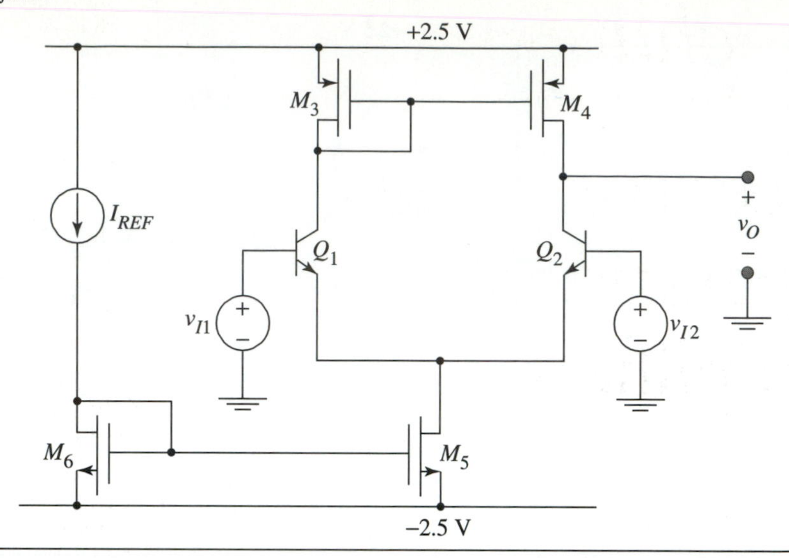

➤ **Figure D11.6**

D11.8 In this design problem we want to compare the frequency response of a differential pair made with npn bipolar transistors as in Fig. D11.6 and one using NMOS transistors. Given that $R_S = 1\text{k}\Omega$, design the amplifier to have a differential gain of 1000 at DC and bandwidth of 60 Mrad/s. In your design try to minimize the area and power dissipation. The *CMRR* should be at least 1000 at DC. How do the two devices compare for this set of specifications?

D11.9 In Example 11.6 an ECL buffer/inverter was designed using a bipolar differential pair. Emitter followers are often placed between the output of the differential amplifier and the next gate as shown in Fig. D11.9. Choose I_{BIAS}, R_C and R_E such that the current through all bipolar transistors is 1mA and that they are not saturated when the input voltage is anywhere between V_{OL} and V_{OH}. Calculate V_{IL}, V_{IH}, NM_L, and NM_H.

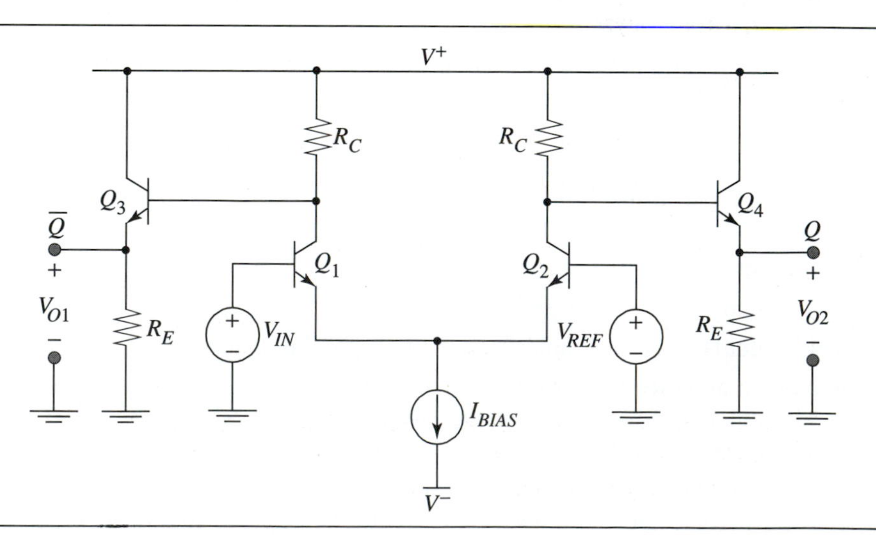

➤ **Figure D11.9**

chapter 12

Feedback and Operational Amplifiers

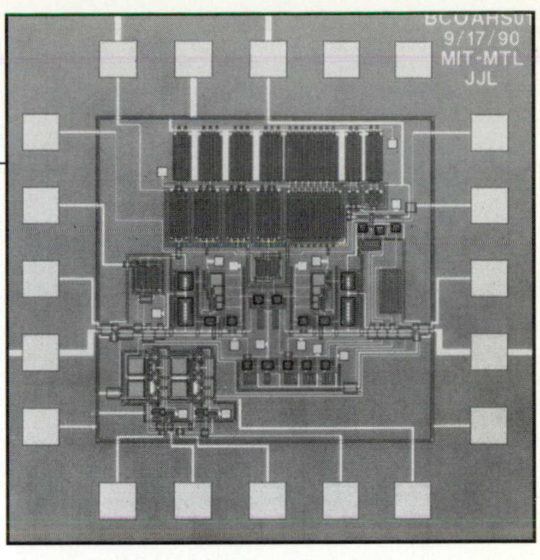

A two stage fully differential BiCMOS operational amplifier designed and fabricated by J. J. Lutsky and K. K. 0 at the MIT Microsystems Technology Laboratories. (Courtesy of J. J. Lutsky)

This chapter introduces the concept of feedback as applied to electronic amplifiers. Several amplifier characteristics can be improved if, at the input, KVL or KCL is used to compare the input signal to a sample of the output signal. We begin by analyzing how ideal feedback networks change the gain and input and output resistances for the four basic amplifier types. After analyzing how feedback alters the amplifier frequency response, we consider practical voltage and transresistance feedback amplifiers that use resistor feedback networks. We turn next to the design of a special-purpose IC building-block for feedback amplifier applications—the operational amplifier (op amp). Using the basic analog IC building blocks from Chapters 8, 9, and 11, we can achieve near-ideal op amp characteristics such as very large voltage gain. The large-signal and small-signal frequency response of basic two-stage CMOS and BiCMOS op amps are analyzed.

Chapter Objectives

- Definitions of feedback factor and loop gain, and the beneficial effects of feedback on the gain, the input and output resistances, and the biasing of amplifiers.
- Frequency response of feedback amplifiers, the concept of stability, and the definition of phase margin.
- Practical voltage and transresistance feedback amplifiers using resistive feedback networks. The effects of loading on the loop gain and input and output resistances.
- Operational amplifiers as an IC building block. Design of basic two-stage CMOS and BiCMOS op amps and the analysis of their DC and small-signal performance.

- Frequency response and compensation of two-stage op amp for unity-gain stability.
- Design philosophy for an improved BiCMOS op amp design and the analysis of its performance.

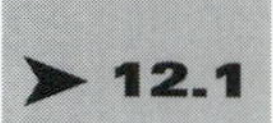

12.1 AMPLIFIER MODELS AND THE FEEDBACK CONCEPT

In order to discuss the concept of feedback as used in electronic amplifier design, we first review two-port small-signal models. As an example, Fig. 12.1 is the voltage form of the small-signal model of a differential amplifier with a "single-ended" output voltage—that is, one referenced to ground. We will use this configuration initially to illustrate how feedback can alter the amplifier properties.

Although in Chapters 8, 9, and 11 we learned the principles for designing amplifiers with very high voltage gains, tight specification of the gain is not possible. The differential voltage gain a_{dm} is a function of transistor small-signal parameters, such as g_m and r_o. Large variations from amplifier to amplifier will occur due to process variations. The gain will typically have a large temperature sensitivity through the dependence of the small-signal parameters on temperature. This results in unacceptably large variations over the temperature ranges demanded by practical applications. The input and output resistances also suffer from a similar lack of control.

How can amplifiers with stable characteristics be designed, given the limitations of IC technology? The answer lies in developing amplifier topologies for which the gain is not a function of device or circuit parameters, but rather is determined solely by ratios of passive circuit elements such as resistors or capacitors. As we learned in Chapter 2, the value of an integrated resistor or capacitor can be specified to only $\pm 10\%$ at best . However, the *ratio* of nearby passive elements on an integrated circuit is set by highly reproducible lithography and etching processes. With careful attention to the layout, the ratios of integrated passive elements can be controlled to within $\pm 0.1\%$ from run-to-run and over wide temperature ranges.

By connecting a passive network between the amplifier output v_o and its inverting input v_{i2} in Fig. 12.1, we can arrange for the inverting input voltage v_{i2} to be proportional to the output voltage

$$v_{i2} = f \cdot v_o \tag{12.1}$$

where $f \leq 1$ is the constant of proportionality. f is determined by ratios of passive elements where possible. This constant is called the **feedback factor**. Applying the

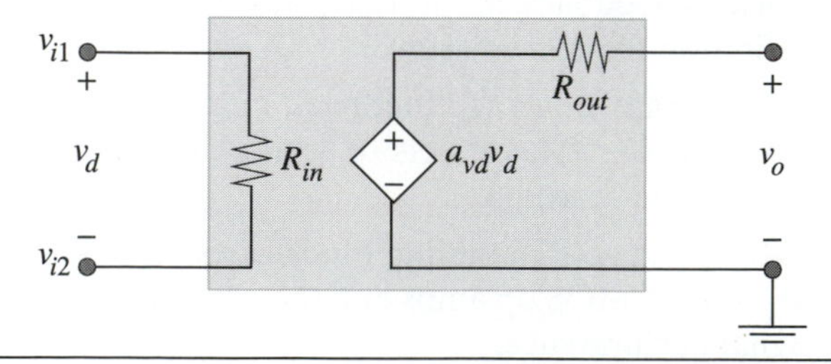

Figure 12.1 Small-signal model of a differential voltage amplifier with single-ended output.

input voltage v_i at the noninverting input v_{i1} in Fig. 12.1, the differential input voltage becomes

$$v_d = v_{i1} - v_{i2} = v_i - fv_o \tag{12.2}$$

From Fig. 12.1, the differential input voltage is related to the output voltage by the inverse of the differential voltage gain

$$v_d = \frac{v_o}{a_{vd}} \tag{12.3}$$

For practical reasons of the output range of the amplifier, the magnitude of the output voltage v_o is limited. If we have a very high differential gain, which is easily designed using current-source supplies, then we can conclude from Eq. (12.3) that v_d is forced to be very small relative to v_i and fv_o. Setting $v_d \approx 0$ in Eq. (12.2) results in the input voltage being related to the output voltage by the feedback factor

$$v_i \approx fv_o \tag{12.4}$$

Solving for the ratio of v_o to the input voltage v_i, which we define as the **closed-loop voltage gain A_{vf}**, we find that

$$A_{vf} \equiv \frac{v_o}{v_i} \approx \frac{v_o}{fv_o} = \frac{1}{f} \tag{12.5}$$

By feeding back the output voltage to the inverting input terminal, we have achieved a stable voltage gain A_{vf} that is set by passive elements independent of the poorly controlled differential gain a_{vd}. All that is required for Eq. (12.5) to hold is for the differential gain to be high enough that v_d can be considered negligible. This promising initial result motivates us to make a systematic exploration of the effects of feedback on the various types of amplifiers.

12.1.1 Voltage Feedback Amplifier

To quantify our understanding of feedback, we introduce the voltage feedback amplifier topology in Fig. 12.2. Rather than using a passive network, we begin by analyzing the feedback amplifier with an idealized feedback network connected in parallel with the differential amplifier. The feedback block feeds the sampled output voltage $f\,v_o$ to the inverting terminal of the differential amplifier. The output voltage is given by

$$v_o = a_{vd}v_d = a_{vd}(v_i - fv_o) \tag{12.6}$$

Solving for the closed-loop voltage gain, we find that

$$\boxed{A_{vf} = \frac{v_o}{v_i} = \frac{a_{vd}}{1 + a_{vd}f}} \tag{12.7}$$

closed-loop voltage gain

In texts on feedback control theory, the denominator of Eq. (12.7) often appears as $1-a_{vd}f$ because the amplifier block has summing rather than differencing inputs. The product of the differential voltage gain a_{vd} and the feedback factor f is defined as the **loop gain** T, and is important in the analysis of feedback amplifiers

loop gain

$$\boxed{T = a_{vd}f} \tag{12.8}$$

For the case where T is much greater than 1, the closed-loop gain is

$$A_{vf} = \frac{a_{vd}}{1+T} \approx \frac{a_{vd}}{T} = \frac{1}{f} \tag{12.9}$$

In Fig. 12.2, we have taken care to input the sampled output voltage fv_o to the inverting terminal of the differential amplifier. Note that both f and a_{vd} are positive numbers for the voltage feedback topology. This connection results in the difference between v_i and fv_o controlling the output voltage. As a result, the output voltage is forced to track the input voltage, creating a **negative feedback** situation. If the terminals at the amplifier input are reversed, then we would be applying **positive feedback**. The effect of positive feedback is to destabilize the amplifier, as can be seen from the closed-loop gain expression for this configuration

$$A_{vf} = \frac{-a_{vd}}{1-a_{vd}f} \text{ (positive feedback)} \tag{12.10}$$

The closed-loop gain blows up if the loop gain $T = 1$. Positive feedback is very useful in some circuits, such as comparators and oscillators, but is to be avoided in designing linear amplifiers. We will see an example of positive feedback in Chapter 13.

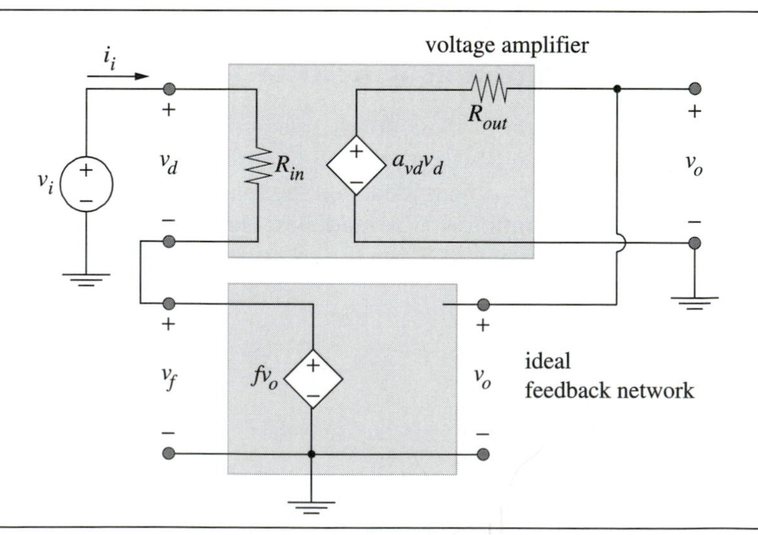

Figure 12.2 Voltage feedback amplifier with ideal feedback network.

The addition of the feedback network greatly changes the input and output resistances. From Fig. 12.2, the input resistance with feedback $R_{in,f}$ is given by

$$R_{in,f} \equiv \frac{v_i}{i_i} = \frac{v_d + v_f}{i_i} = \frac{v_d + f v_o}{i_i} = \frac{v_d(1 + a_{vd}f)}{i_i} \qquad (12.11)$$

Since the ratio $v_d/i_i = R_{in}$, the amplifier's input resistance, then

$$\boxed{R_{in,f} = R_{in}(1 + T) \text{ (voltage amplifier)}} \qquad (12.12)$$

voltage amplifier: input resistance with feedback

Therefore, the input resistance of the feedback voltage amplifier is boosted by the factor $(1 + T)$. For large loop gains, the input resistance can be made quite large.

At the feedback amplifier output, we find the output resistance $R_{out,f}$ by using a test voltage v_t as shown in the circuit in Fig. 12.3, from which

$$i_t = \frac{v_t - a_{vd}v_d}{R_{out}} = \frac{v_t - a_{vd}(-f v_t)}{R_{out}} \qquad (12.13)$$

Solving for the output resistance

$$\boxed{R_{out,f} \equiv \frac{v_t}{i_t} = \frac{R_{out}}{1 + T} \text{ (voltage amplifier)}} \qquad (12.14)$$

voltage amplifier: output resistance with feedback

Feedback *reduces* the output resistance by the same factor of $(1 + T)$. In addition to stabilizing the voltage gain, the addition of feedback moves the input and output resistances in the directions desired for an ideal voltage amplifier. With high loop gains, a voltage amplifier can be designed with extremely large input resistances and negligible output resistances even if the differential amplifier is far from ideal.

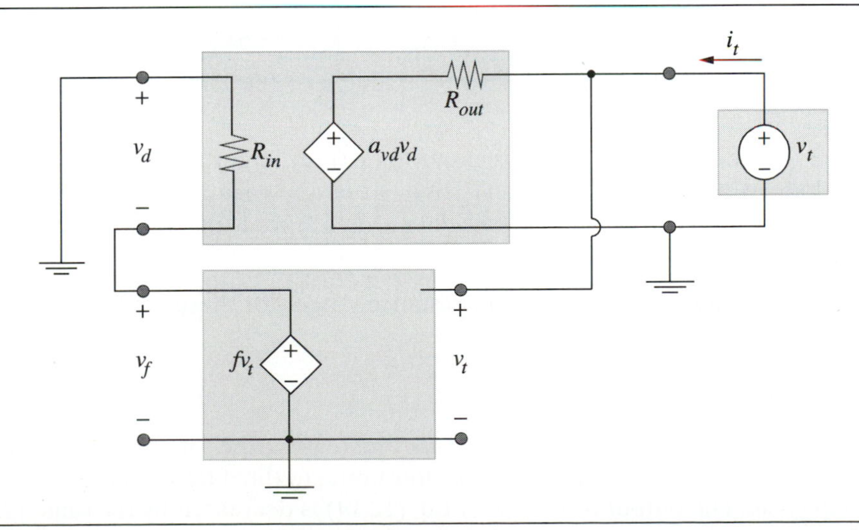

➤ **Figure 12.3** Circuit for finding the output resistance with feedback.

➤ EXAMPLE 12.1 Numerical Example of Voltage Feedback

In this example, we investigate how an ideal negative feedback network around a differential voltage amplifier affects its closed-loop performance. The amplifier has a small-signal differential voltage gain in the range $a_{vd} = 5500 - 7500$, which accounts for fabrication variations and changes over temperature. The input and output resistance are $R_{in} = 500\ \text{k}\Omega$ and $R_{out} = 200\ \text{k}\Omega$.

(a) Find the range in closed-loop gains for $f = 0.01$, 0.05, 0.1, and 1.0.

SOLUTION

Let $a_{vd1} = 5500$ and $a_{vd2} = 7500$. The closed-loop gain can be found for the feedback factors from Eq. (12.7)

$$A_{vf} = \frac{a_{vd}}{1 + a_{vd}f} \quad \rightarrow \quad A_{min} = \frac{a_{vd1}}{1 + a_{vd1}f} \text{ and } A_{max} = \frac{a_{vd2}}{1 + a_{vd2}f}$$

For the four feedback factors, the ranges of closed-loop gains for this amplifier are

Feedback Factor f	A_{min}	A_{max}
0.01	98.2	98.7
0.05	19.93	19.95
0.1	9.98	9.99
1.0	0.9998	0.9999

Note that the closed-loop gain is closer to $(1/f)$ for larger feedback factors since the loop gain $T = a_{vd}f$ is larger and T is a better approximation to $1 + T$. The differential gain has a substantial variation of ±15%, but the effect on the closed-loop gain is small for all cases.

(b) Find the input and the output resistance ranges for these feedback factors.

SOLUTION

According to Eq. (12.12), feedback increases the input resistance by $(1 + T) = (1 + a_{vd}f)$, which is nearly proportional to the differential gain for large loop gains. The output resistance in Eq. (12.14) is decreased by the same factor. The results are summarized in the following two tables.

Feedback Factor f	$R_{in,f}$ minimum	$R_{in,f}$ maximum
0.01	28 MΩ	38 MΩ
0.05	138 MΩ	188 MΩ
0.1	275.5 MΩ	375.5 MΩ
1.0	2.75 GΩ	3.75 GΩ

Feedback Factor f	$R_{out,f}$ minimum	$R_{out,f}$ maximum
0.01	3.57 kΩ	2.63 kΩ
0.05	725 Ω	532 Ω
0.1	363 Ω	266 Ω
1.0	36 Ω	27 Ω

From this example, we conclude that a unity gain buffer with a gain of nearly 1.0, an input resistance of over 2.75 GΩ, and an output resistance under 40 Ω can be made using a differential amplifier having far from ideal characteristics. Finally, note from the above tables that the output and input resistances are not well-controlled by feedback in contrast to the closed-loop gain.

12.1.2 Transresistance Feedback Amplifier

The benefits of feedback have been demonstrated using a voltage amplifier topology. In fact, we can apply feedback to improve the characteristics of all four amplifier types. In the case of a transresistance amplifier, the feedback network must sample the output voltage and return to the input a feedback *current* $i_f = f\,v_o$ that is proportional to the output. The units of the feedback factor in this case are Siemens (1/Ω). For the voltage feedback topology, the differential amplifier inputs were used to subtract the feedback voltage $v_f = f\,v_o$ from the input voltage v_i. For the transresistance feedback amplifier, Kirchhoff's current law is used to subtract i_f from the input current i_i as shown in Fig. 12.4.

The input current i_d is given by

$$i_d = i_i - i_f = i_i - f v_o \tag{12.15}$$

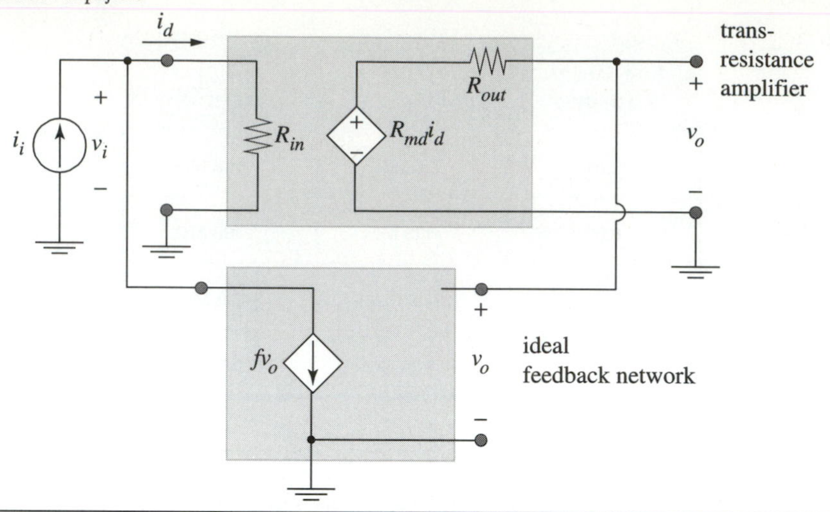

➤ **Figure 12.4** Transresistance feedback amplifier.

The output voltage v_o is the product of the input current and the transresistance R_{md}

$$v_o = R_{md}i_d = R_{md}(i_i - fv_o) \tag{12.16}$$

Solving for the closed-loop transresistance R_{mf}, we find that

closed-loop transresistance

$$R_{mf} \equiv \frac{v_o}{i_i} = \frac{R_{md}}{1 + R_{md}f} \tag{12.17}$$

For negative feedback, we must ensure that the loop gain $T = R_m f$ is positive. If the loop gain $T \gg 1$, then the closed-loop transresistance becomes

$$R_{mf} \approx \frac{1}{f} \tag{12.18}$$

which is the same expression as we found for the closed-loop voltage gain. Note that the units of R_{mf} are Ω (as they must be for a transresistance amplifier) since the units of f are (1/Ω). From Eq. (12.18), we conclude that negative feedback can provide a stable closed-loop transresistance for the case of large loop gain.

The input resistance with feedback $R_{in,f}$ can be found from Fig. 12.4 by finding the ratio of the voltage v_i across the input current source to the input current i_i

$$R_{in,f} = \frac{v_i}{i_i} = R_{in}\left(\frac{i_d}{i_i}\right) = R_{in}\left(\frac{i_i - fv_o}{i_i}\right) = R_{in}\left(1 - f\frac{v_o}{i_i}\right) \tag{12.19}$$

Substituting for the closed-loop gain from Eq. (12.17), we find that

$$R_{in,f} = R_{in}\left(1 - f\left(\frac{R_{md}}{1 + R_{md}f}\right)\right) = R_{in}\left(\frac{1}{1 + R_{md}f}\right) \tag{12.20}$$

Identifying the loop gain $T = R_{md}f$, Eq. (12.20) can be put into the form

$$R_{in,f} = \frac{R_{in}}{1+T} \text{ (transresistance amplifier)} \tag{12.21}$$

transresistance amplifier: input resistance with feedback

The circuit in Fig. 12.5 is used to derive the output resistance. Since the differential input current $i_d = -f\,v_t$ in this case, the test current is given by

$$i_t = \frac{v_t - R_{md}i_d}{R_{out}} = \frac{v_t + R_{md}fv_t}{R_{out}} \tag{12.22}$$

Again, we identify the loop gain $T = R_{md}f$ and express the output resistance as

$$R_{out,f} = \frac{v_t}{i_t} = \frac{R_{out}}{1+T} \text{ (transresistance amplifier)} \tag{12.23}$$

transresistance amplifier: output resistance with feedback

Feedback around the transresistance amplifier reduces both the input and the output resistance by the factor $(1+T)$. Recalling that an ideal transresistance amplifier has input and output resistances that are both zero, feedback again improves the performance. A pattern is emerging from our derivations of the input and output resistances—feedback reduces the resistance by $(1+T)$ when the feedback network is in parallel to the amplifier port and increases the resistance by the same factor when the feedback network is in series with the amplifier port.

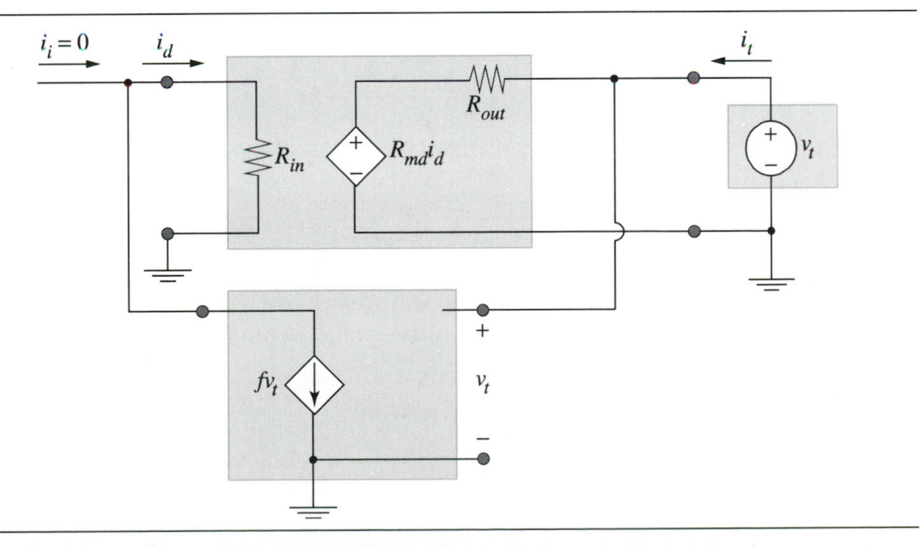

➤ **Figure 12.5** Circuit for finding the output resistance with feedback.

➤ EXAMPLE 12.2 Bipolar Transresistance Feedback Amplifier

In this example, an ideal feedback network is used in conjunction with a bipolar differential amplifier with a p-channel MOSFET current mirror supply. The transistor parameters are: $\beta_o = 100$, $V_{An} = 50$ V and $\mu_p C_{ox} = 25$ μA/V^2, $V_{Tp} = -1$ V , $\lambda_p = 0.02$ V^{-1}, $(W/L)_{3,4} = (50\ \mu\text{m}/3\ \mu\text{m})$.

(a) Find the two-port model parameters of the amplifier block without feedback ($f = 0$ S), using the transresistance form of the two-port model.

SOLUTION

The input resistance of the bipolar differential amplifier is

$$R_{in} = 2r_{\pi 2} = 2\frac{\beta_o}{g_{m2}} = 2\frac{\beta_o V_{th}}{(I_{EE}/2)} = \frac{2\,(100)\,(25\text{ mV})}{50\ \mu\text{A}} = 100\text{ k}\Omega$$

The transresistance R_m of the amplifier is defined as

$$R_{md} = \left.\frac{v_o}{i_i}\right|_{R_L \to \infty} = \left(\frac{v_o}{v_i}\right)\left(\frac{v_i}{i_i}\right) = [-g_{m2}(r_{o2}||r_{o4})]\,(2r_{\pi 2})$$

in which the differential voltage gain $a_{vd} = -g_{m2}(r_{o2} \parallel r_{o4})$ for the current mirror supply (with input on the base of Q_2) has been substituted. The transresistance can be simplified by recognizing the product $g_{m2} r_{\pi 2}$ as the current gain

$$R_{md} = -2g_{m2}r_{\pi 2}(r_{o2}||r_{o4}) = -2\beta_o(r_{o2}||r_{o4}) = -2\beta_o\left(\frac{V_{An}}{I_{C2}}\right)\Bigg|\Bigg|\left(\frac{1}{\lambda_p(-I_{D4})}\right)$$

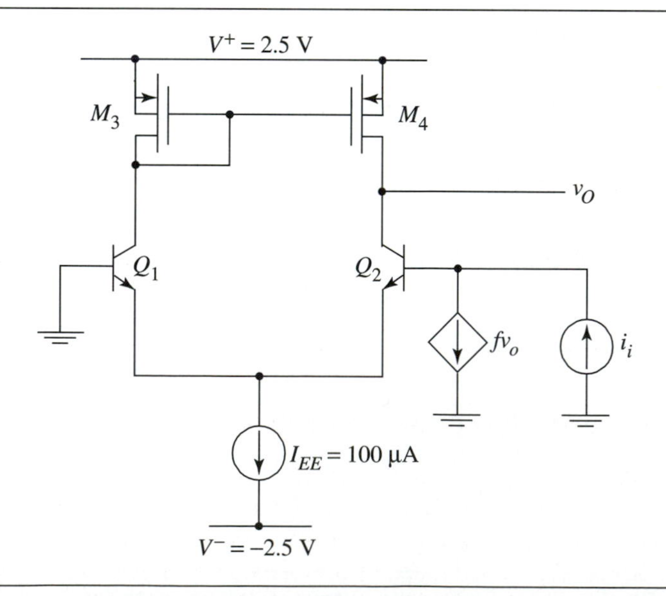

➤ Figure Ex12.2 Transresistance feedback amplifier based on a bipolar differential pair.

Substituting the device parameters, we find that the numerical value of the transresistance is

$$R_{md} = -2\,(100)\left[\left(\frac{50\text{ V}}{50\ \mu\text{A}}\right)\Big\|\left(\frac{1}{(0.02\text{ V}^{-1})\,(50\ \mu\text{A})}\right)\right] = -100\text{ M}\Omega$$

The output resistance of the bipolar differential pair with current-mirror supply is

$$R_{out} = r_{o2}\|r_{o4} = \left(\frac{50\text{ V}}{50\ \mu\text{A}}\right)\Big\|\left(\frac{1}{(0.02\text{ V}^{-1})\,(50\ \mu\text{A})}\right) = 500\text{ k}\Omega$$

(b) Find the two-port parameters of the feedback amplifier for a feedback factor $f = -10\ \mu\text{S}$.

SOLUTION

The loop gain is the product of the amplifier transresistance and the feedback factor:

$$T = R_m f = (-100\text{ M}\Omega)\,(-10\ \mu\text{S}) = 1000$$

Note that the sign of the feedback factor must be negative in order to have $T > 0$ for negative feedback. The closed-loop transresistance is given by Eq. (12.17)

$$R_{mf} = \frac{R_{md}}{1 + R_{md}f} = \frac{-100\text{ M}\Omega}{1 + 1000} \approx -100\text{ k}\Omega$$

Since the loop gain is so high, this result is the inverse of the feedback factor to three significant figures.
According to Eq. (12.21), the input resistance is lowered by feedback to

$$R_{in,f} = R_{in}\left(\frac{1}{1+T}\right) = \frac{100\text{ k}\Omega}{1 + 1000} \approx 100\ \Omega$$

The output resistance with feedback is given by Eq. (12.23)

$$R_{out,f} = \frac{R_{out}}{1+T} = \frac{500\text{ k}\Omega}{1 + 1000} \approx 500\ \Omega$$

Despite the amplifier block being far from ideal as a transresistance amplifier, the high loop gain results in low input and output resistances, as well as a controlled closed-loop transresistance.

12.1.3 Transconductance Feedback Amplifier

Figure 12.6 shows a transconductance feedback amplifier topology, which is our third example of a feedback amplifier. In order to sample the output current, the feedback network is in series with the output. We are not considering loading, so the load resistance $R_L = 0\ \Omega$ in Fig. 12.6. The feedback voltage is generated by a current-controlled voltage source that yields $v_f = f i_o$. The units of the feedback factor are Ohms for this

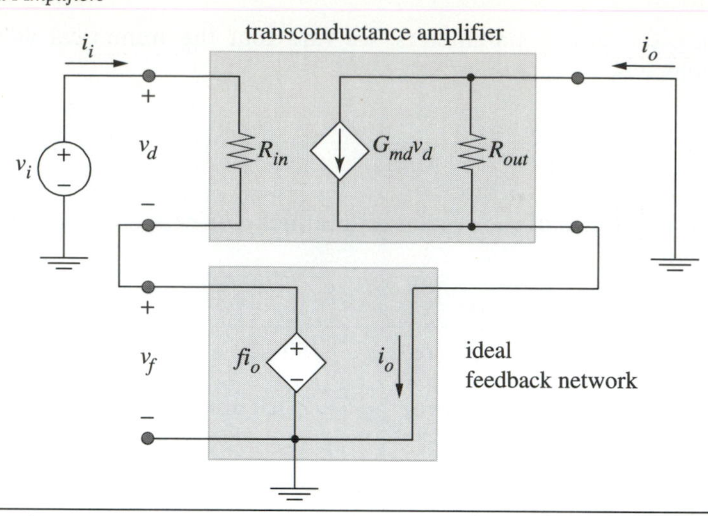

Figure 12.6 Transconductance feedback amplifier with ideal feedback network.

configuration. We can find the closed-loop gain from Fig. 12.6. The output current i_o is

$$i_o = G_{md}v_d = G_{md}(v_i - fi_o) \tag{12.24}$$

The closed-loop transconductance is

closed-loop transconductance

$$G_{mf} \equiv \frac{i_o}{v_i} = \frac{G_{md}}{1 + G_{md}f} \tag{12.25}$$

The input resistance with feedback can be found by calculating the ratio of v_i to i_i from Fig. 12.6. Applying KVL to the input port yields and substituting for the output current $i_o = G_{md}\, v_d = G_{md}\, R_{in}\, i_i$ we find

$$v_i = i_i R_{in} + fi_o = i_{in}(R_{in} + fG_{md}R_{in}) \tag{12.26}$$

Substituting the loop gain $T = G_{md}f$ allows us to put Eq. (12.26) into the standard form

transconductance amplifier: input resistance with feedback

$$R_{in,\,f} = R_{in}(1 + T) \text{ (transconductance amplifier)} \tag{12.27}$$

Figure 12.7 implements the definition of the output resistance for this amplifier topology. Kirchhoff's current law at the output node states

$$i_t = \frac{v_t}{R_{out}} + G_{md}v_d = \frac{v_t}{R_{out}} + G_{md}(-fi_t) \tag{12.28}$$

By definition, the output resistance is the ratio v_t/i_t. From Eq. (12.28), it follows that

transconductance amplifier: output resistance with feedback

$$R_{out,f} = R_{out}(1 + G_{md}f) = R_{out}(1 + T) \text{ (transconductance amplifier)} \tag{12.29}$$

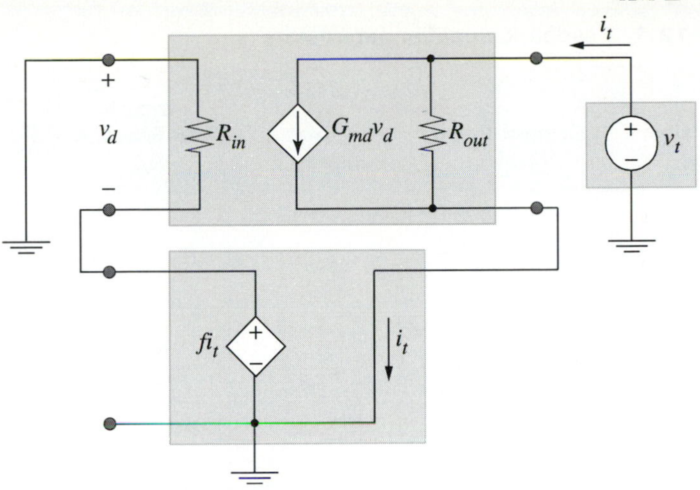

➤ **Figure 12.7** Circuit for finding the output resistance of the transconductance feedback amplifier.

The output resistance is increased by feedback, which is consistent with the feedback network being in series with the amplifier output port.

12.1.4 Current Feedback Amplifier

The final feedback amplifier type is the current amplifier topology, shown in Fig. 12.8. Since the other three types of feedback amplifier have been analyzed, we simply state the results for the closed-loop parameters for the current amplifier in Table 12.1. In all cases, the effect of feedback is to stabilize the gain at approximately $(1/f)$ for cases where the loop gain $T \gg 1$. Feedback raises the input/output resistance by the factor $(1 + T)$ when the feedback network is in series with the amplifier's input or output port. When it is in parallel, the effect of feedback is to reduce the input/output the resistance by the same factor.

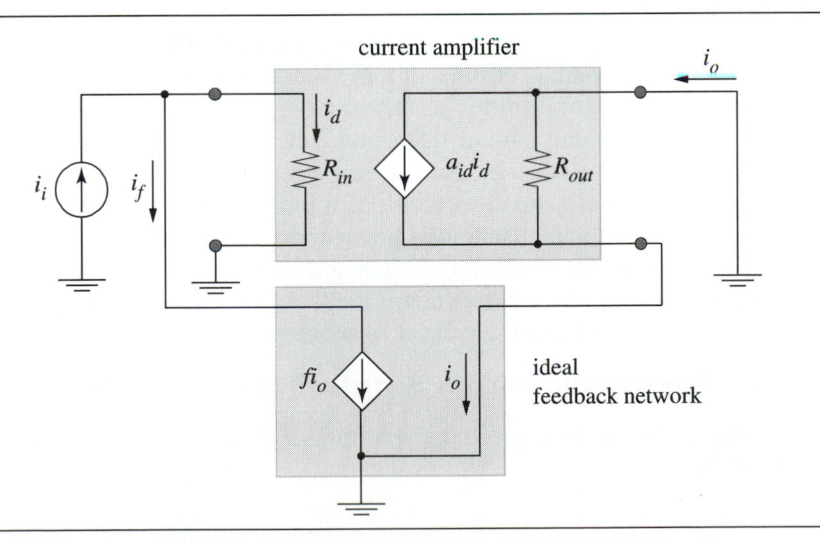

➤ **Figure 12.8** Current feedback amplifier with ideal feedback network.

Table 12.1 Feedback amplifier parameters

Amplifier Type	Closed-Loop Gain	Loop Gain, T	Input Resistance $R_{in,f}$	Output Resistance $R_{out,f}$
Voltage	$A_{vf} = \frac{a_{vd}}{1 + a_{vd}f}$	$a_{vd}f$	$R_{in}(1 + A)$	$\frac{R_{out}}{(1 + T)}$
Trans-resistance	$R_{mf} = \frac{R_{md}}{1 + R_{md}f}$	$R_{md}f$	$\frac{R_{in}}{(1 + T)}$	$\frac{R_{out}}{(1 + T)}$
Trans-conductance	$G_{mf} = \frac{G_{md}}{1 + G_{md}f}$	$G_{md}f$	$R_{in}(1 + T)$	$R_{out}(1 + T)$
Current	$A_{if} = \frac{a_{id}}{1 + a_{id}f}$	$a_{id}f$	$\frac{R_{in}}{(1 + T)}$	$R_{out}(1 + T)$

12.2 FREQUENCY RESPONSE OF FEEDBACK AMPLIFIERS

Negative feedback is useful for achieving stable amplification, along with dramatically improved input and output resistances. We now extend the model of feedback amplifiers to include a frequency-dependent gain in the amplifying block. Feedback can dramatically change the frequency response by increasing the bandwidth of an amplifier. However, there is a potentially serious problem of feedback destabilizing the amplifier. For the voltage feedback amplifier in Fig. 12.2, the differential gain is a negative real number at the frequency where $\angle a_{vd}(j\omega') = -180^{\circ}$ (recall that $e^{-j180^{\circ}} = -1$). Equivalently, we could switch the inverting and noninverting terminals of the amplifier for signals at the frequency ω', which is the configuration for positive feedback. The feedback amplifier could oscillate if it has sufficient gain at this frequency.

In this section, we start by considering voltage feedback around amplifying blocks with a single pole and later consider multipole transfer functions. Feedback allows the designer to make trade-offs between low-frequency gain and bandwidth. Only feedback networks in which the feedback factor f is independent of frequency will be considered. Frequency-dependent feedback factors are quite useful in amplifier design, but we will leave the subject to subsequent courses.

12.2.1 Feedback with a Single Pole

When higher poles can be neglected, the amplifier block's frequency-dependent gain is modeled by

$$a_{vd}(j\omega) = \frac{a_{vdo}}{1 + j\omega/\omega_1} \tag{12.30}$$

Substituting into the closed-loop gain expression in Eq. (12.7) and simplifying, we find

$$A_{vf}(j\omega) = \frac{a_{vd}(j\omega)}{1 + a_{vd}(j\omega)f} = \frac{\dfrac{a_{vdo}}{1 + j\omega/\omega_1}}{1 + \dfrac{a_{vdo}f}{1 + j\omega/\omega_1}} = \frac{a_{vdo}}{1 + a_{vdo}f + j\omega/\omega_1} \tag{12.31}$$

Putting $A_{vf}(j\omega)$ into the standard form for a one-pole transfer function results in

$$A_{vf}(j\omega) = \frac{\dfrac{a_{vdo}}{1 + a_{vdo}f}}{1 + j\dfrac{\omega}{\omega_1(1 + a_{vdo}f)}} = \frac{A_{vfo}}{1 + j\omega/\omega_1'} \tag{12.32}$$

where A_{vfo} is the low-frequency closed-loop gain and the pole with feedback has increased to

$$\omega_1' = \omega_1(1 + a_{vdo}f) \tag{12.33}$$

Bode plots of $a_{vd}(j\omega)$ and $A_{vf}(j\omega)$ are helpful for visualizing Eqs. (12.30) and (12.32). Figure 12.9 presents the magnitude plot of the amplifier gain and the closed-loop gain for the case where $a_{vdo} = 2000$, $\omega_1 = 500$ rad/s, and three values of the feedback factor: $f = 0.01$, $f = 0.1$, and $f = 1.0$. The effect of negative feedback is to reduce the low-frequency gain by the factor $1 + a_{vdo}f = 1 + T_o$ since

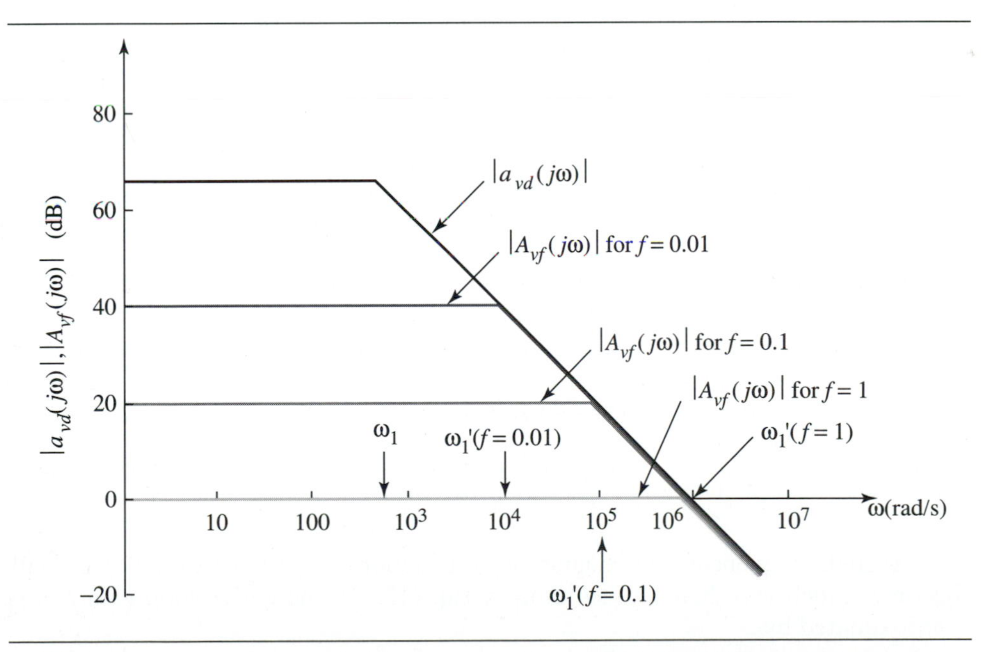

➤ **Figure 12.9** Magnitude Bode plots for the amplifier block gain a_{vd} and closed-loop gain A_{vf}, for the case where $a_{vdo} = 2000$ and $\omega_1 = 500$ rad/s, for $f = 0.01$, 0.1, and 1.0.

$$A_{vfo} = \frac{a_{vdo}}{1 + a_{vdo}f} = \frac{a_{vdo}}{1 + T_o} \tag{12.34}$$

where T_o is the low-frequency loop gain. For large T_o, the closed-loop gain $A_{vfo} \to 1/f$.

Feedback *increases* the –3 dB frequency by the same factor as shown in the Bode plots above and by Eq. (12.33). Another way of expressing this result is that the **gain-bandwidth product**

$$A_{vfo}\omega_1' = \left(\frac{a_{vdo}}{1 + T_o}\right)[\omega_1(1 + T_o)] = a_{vdo}\omega_1 \tag{12.35}$$

is unaffected by feedback. In other words, a designer can directly trade-off the low-frequency closed-loop gain A_{vfo} and the –3 dB closed-loop bandwidth ω_1' by controlling the loop gain T_o through the feedback factor f. Note from the above example that an amplifier bandwidth of 100 krad/s (15.9 kHz) is feasible, starting with a high-gain voltage amplifier with a –3 dB frequency of only 500 rad/s (80 Hz). The penalty is that the gain is reduced from $a_{vdo} = 2000$ to $A_{vfo} = 10$. Of course, the large gain of the amplifier block may not be useful since it is not well-controlled. The broadbanding of amplifiers (increasing the –3 dB bandwidth) by applying negative feedback is a standard analog circuit design technique.

12.2.2 Feedback with Multiple Poles

Amplifying blocks with a single dominant pole as shown in Fig. 12.10 are a special case. What happens when feedback is applied around an amplifier with a multiple-pole transfer function? The amplifier block gain is now

$$a_{vd}(j\omega) = \frac{a_{vdo}}{(1 + j\omega_1)(1 + j\omega_2)(1 + j\omega_2)\ldots(1 + j\omega_n)} \tag{12.36}$$

After substitution into Eq. (12.7)

$$A_{vf}(j\omega) = \frac{a_{vd}(j\omega)}{1 + a_{vd}(j\omega)f} = \frac{a_{vd}(j\omega)}{D(j\omega)} \tag{12.37}$$

Unlike the single-pole case, we cannot factor the denominator $D(j\omega)$ of Eq. (12.37) and derive an analytical result. However, we can investigate $A_{vf}(j\omega)$ in specific frequency ranges where Eq. (12.37) can be simplified. For very low frequencies ($\omega \ll \omega_1$), the differential gain is $a_{vd}(j\omega) \approx a_{vdo}$ according to Eq. (12.36). Substituting into Eq. (12.37), the closed-loop gain at low frequencies is

$$A_{vf}(j\omega) \approx \frac{a_{vdo}}{1 + a_{vdo}f} \approx \frac{1}{f} \quad \text{for } \omega \ll \omega_1 \text{ and } a_{vdo}f \gg 1 \tag{12.38}$$

At high frequencies, the magnitude of the loop gain decreases and eventually becomes much less than 1. According to Eq. (12.37), the closed-loop gain can be approximated by

$$A_{vf}(j\omega) = \frac{a_{vd}(j\omega)}{1 + a_{vd}(j\omega)f} \approx a_{dm}(j\omega) \quad \text{for } \omega \text{ such that } |a_{vd}(j\omega)f| \ll 1 \tag{12.39}$$

Therefore, closed-loop gain asymptotically approaches the differential gain at high frequencies, as shown in the single-pole case in Fig. 12.9.

For the range of frequencies where the loop gain is near unity, it is necessary to examine carefully the magnitude and phase of the loop gain $T(j\omega) = a_{vd}(j\omega)\,f$ to determine the closed-loop response. It is convenient to define the **critical frequency** ω^* as that frequency for which the magnitude of the loop gain is equal to 1

$$\left|a_{vd}(j\omega^*)f\right| = 1 \quad \Leftrightarrow \quad \left|a_{vd}(j\omega^*)f\right|_{\text{dB}} = 0 \text{ dB} \tag{12.40}$$

From this definition, it follows that ω^* is also the frequency at which the magnitude of the amplifier gain is equal to the inverse of the feedback factor

$$\left|a_{vd}(j\omega^*)\right| = |1/f| \quad \Leftrightarrow \quad \left|a_{vd}(j\omega^*)\right|_{\text{dB}} = |1/f|_{\text{dB}} \tag{12.41}$$

The phase of the loop gain at the critical frequency $\angle T(j\omega^*)$ is very important to the closed-loop gain $A_{vf}(j\omega)$. For example, in a case where the phase is

$$\angle T(j\omega^*) = -180^o, \tag{12.42}$$

we can evaluate the closed-loop gain at ω^* using Eq. (12.37)

$$A_{vf}(j\omega^*) = \frac{a_{vd}(j\omega^*)}{1 + a_{vd}(j\omega^*)f} = \frac{a_{vd}(j\omega^*)}{1 + 1e^{-j180^o}} = \frac{a_{vd}(j\omega^*)}{1-1} \to \infty \tag{12.43}$$

Amplifiers with infinite gain at any given frequency are defined as **unstable**. An advanced analysis of feedback amplifiers shows that the condition for instability is

$$\boxed{\angle a_{vd}(j\omega^*)f \le -180 \text{ when } \left|a_{vd}(j\omega^*)f\right| = 1} \tag{12.44}$$

test for feedback amplifier instability

In an unstable amplifier, signal components with frequencies at or near ω^* will grow through positive feedback until limited by non-linearities in the transfer function. No input, other than random noise, is needed to start this positive feedback process.

It is helpful to quantify how far a feedback amplifier is from becoming unstable. To this end, we define the **phase margin** as

$$\boxed{\phi_{PM} = \angle a_{dm}(j\omega^*)f - (-180^o) = \angle a_{dm}(j\omega^*)f + 180^o} \tag{12.45}$$

phase margin of a feedback amplifier

in which ω^* is the critical frequency defined in Eq. (12.40). If the phase margin is small, the amplifier, though stable, may have excessive "peaking" in the closed-loop gain near the critical frequency ω^*. It is common to specify phase margins of 45° to 60° to minimize peaking.

The phase of the loop gain is critically important for assessing the closed-loop stability of the feedback amplifier. In this text, the feedback factor is a constant that is independent of frequency. Hence, its phase is either 0° or 180° depending on the sign of f. After this constant phase is added to the phase of the amplifier gain $a_{vd}(j\omega)$, we can use the latter to find the phase margin. Graphical techniques are convenient for feedback amplifier design, as the next example demonstrates.

➤ EXAMPLE 12.3 **Stability and Phase Margin for Multipole Amplifiers.**

A differential voltage amplifier has a gain given by

$$a_{vd}(j\omega) = \frac{1000}{(1 + j\omega/100)\,(1 + j\omega/10^4)\,(1 + j\omega/10^5)}$$

(a) Plot the magnitude in dB and the phase of $a_{vd}(j\omega)$, using the conventional linear approximations in both plots and find the phase margin in degrees for a feedback factor $f = 0.15$.

SOLUTION

A graphical solution is sufficiently accurate for hand design. From Eq. (12.41), we can find the critical frequency ω^* by intersecting the magnitude Bode plot of $a_{vd}(j\omega)$ with the magnitude of $(1/f)$ in dB as shown in Fig. Ex12.3.

$$\left(\frac{1}{f}\right)_{\text{dB}} = 20 \log\left(\frac{1}{0.15}\right) = 16.5 \text{ dB}$$

From the magnitude Bode plot from part (a), $\omega^* \approx 14$ krad/s . Since the feedback factor has a phase of 0°, the phase of the loop gain is the phase of the differential gain. From the phase Bode plot, the phase at ω^* is approximately –145°. From the definition in Eq. (12.45), the phase margin is

$$\phi_{PM} = \angle a_{dm}(j\omega^*) - (-180^o) = -145^o - (-180^o) = 35^o$$

This phase margin is adequate for stability, but the magnitude of the closed-loop gain will have several dB of peaking near the critical frequency of 14.4 krad/s.

(b) Is the amplifier stable for a unity-gain configuration, when $f = 1$?

SOLUTION

From Eq. (12.41), when $f = 1$ the critical frequency occurs where the magnitude of the amplifier gain intersects 0 dB. This frequency is $\omega^* = 30$ krad/s from the magnitude Bode plot of $a_{vd}(j\omega)$, at which the phase is $\angle a_{vd}(j\omega^*) \approx -180^o$. Therefore, the feedback amplifier is not stable for a unity-gain configuration.

For practical applications, it is convenient to guarantee an adequate phase margin for *all* feedback factors. The critical frequency ω^* is defined in Eq. (12.40) as the frequency where the loop gain has a magnitude of 1, or equivalently, when

$$\left|a_{vd}(j\omega^*)\right| = |1/f| \tag{12.46}$$

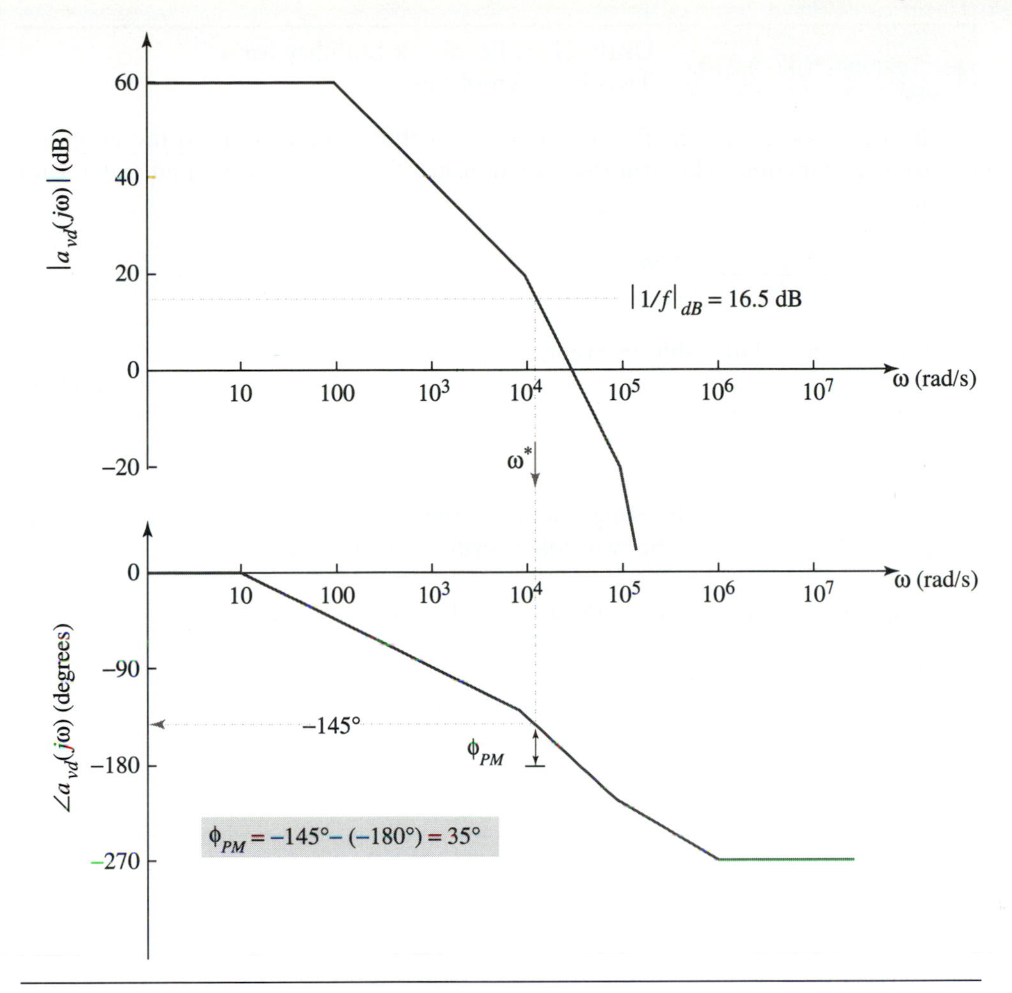

➤ **Figure Ex12.3** Magnitude and phase Bode plots of the amplifier gain, $a_{dm}(j\omega)$, together with the procedure for finding the phase margin for the case where the feedback factor $f = 0.15$.

Since the magnitude of the amplifier gain rolls off with increasing frequency, the critical frequency ω^* is larger for larger feedback factors. The phase of the amplifier gain is decreasing with increasing ω as well, which means that larger feedback factors correspond to lower phase margins and possible instability. The maximum feedback factor is unity, which results in unity closed-loop gain. Therefore, if the amplifier is stable for $f = 1$ with a reasonable phase margin (say, around 45°), then it will also be stable for all feedback factors less than one and have a larger phase margin. The critical frequency for the unity-gain feedback configuration is given by

$$\left|a_{vd}(j\omega^*)\right| = 1 \quad \Leftrightarrow \quad \left|a_{vd}(j\omega^*)\right|_{\text{dB}} = 0 \text{ dB} \ \ (\text{for } f = 1) \tag{12.47}$$

The next example shows how poles of a two-pole amplifier must be located to assure stability.

➤ EXAMPLE 12.4 Unity-Gain Feedback Stability for a Two-Pole Amplifier

In some cases, only the first two poles contribute significantly to the amplifier transfer function. This transfer function has two widely separated poles with $\omega_1 \ll \omega_2$

$$a_{vd}(j\omega) = \frac{a_{vdo}}{(1 + j\omega/\omega_1)(1 + j\omega/\omega_2)}$$

(a) Find the relationship between the positive low-frequency gain a_{vdo} and the locations of the two poles so that the amplifier is stable in a unity-gain feedback configuration with a phase margin of 45°.

SOLUTION

In order to meet the phase margin of 45°, the phase at the critical frequency ω^* is $\angle a_{vd}(j\omega^*) = -135°$. For the two-pole transfer function, the phase is

$$\angle a_{vd}(j\omega) = \angle a_{vdo} - \angle(1 + j\omega/\omega_1) - \angle(1 + j\omega/\omega_2)$$

Since a_{vdo} is given to be a positive real number $\angle a_{vdo} = 0°$. At the second pole, the phase contribution of the first pole is –90° since $\omega_2 \gg \omega_1$

$$\angle a_{vd}(j\omega_2) = 0 - 90^o - \angle(1 + j\omega_2/\omega_2) = 0 - 90^o - 45^o = -135^o$$

Therefore, we conclude that the critical frequency ω^* must equal ω_2, the second pole. In order to meet the condition that the loop gain has unity magnitude at the critical frequency, we require that

$$\left|a_{vd}(j\omega^*)f\right| = \left|a_{vd}(j\omega_2)\right| = \frac{\left|a_{vdo}\right|}{\left|1 + j\omega_2/\omega_1\right|\left|1 + j\omega_2/\omega_2\right|} = 1$$

Since $\omega_2 \gg \omega_1$, the first pole's magnitude is approximately

$$\left|1 + j\omega_2/\omega_1\right| = \sqrt{1 + (\omega_2/\omega_1)^2} \approx \omega_2/\omega_1$$

and $\left|a_{vdo}\right| = a_{vdo}$. The magnitude of the transfer function at the second pole is

$$\left|a_{vd}(j\omega_2)\right| = \frac{a_{vdo}}{(\omega_2/\omega_1)\sqrt{1 + 1}} = \frac{\omega_1 a_{vdo}}{\sqrt{2}\omega_2} = 1$$

Solving for the second pole in terms of the low-frequency gain and the first pole

$$\omega_2 = a_{vdo} \cdot \omega_1/\sqrt{2}$$

For hand calculations, it is convenient to approximate the second pole as

$$\omega_2 \approx a_{vdo} \cdot \omega_1$$

(b) For the special case where $a_{vdo} = 10{,}000$ and the first pole $\omega_1 = 1000$ rad/s, find the location of the second pole for unity-gain stability with a phase margin of 45° and plot the magnitude and phase of the resulting two-pole amplifier gain.

SOLUTION

The second pole is located approximately at

$$\omega_2 \approx a_{vdo} \cdot \omega_1 = (10,000)\,(1000) = 10 \text{ Mrad/s}$$

The magnitude and phase Bode plots (Fig. Ex12.4) are sketched, using the approximate techniques from Chapter 10. In practice, SPICE is essential for

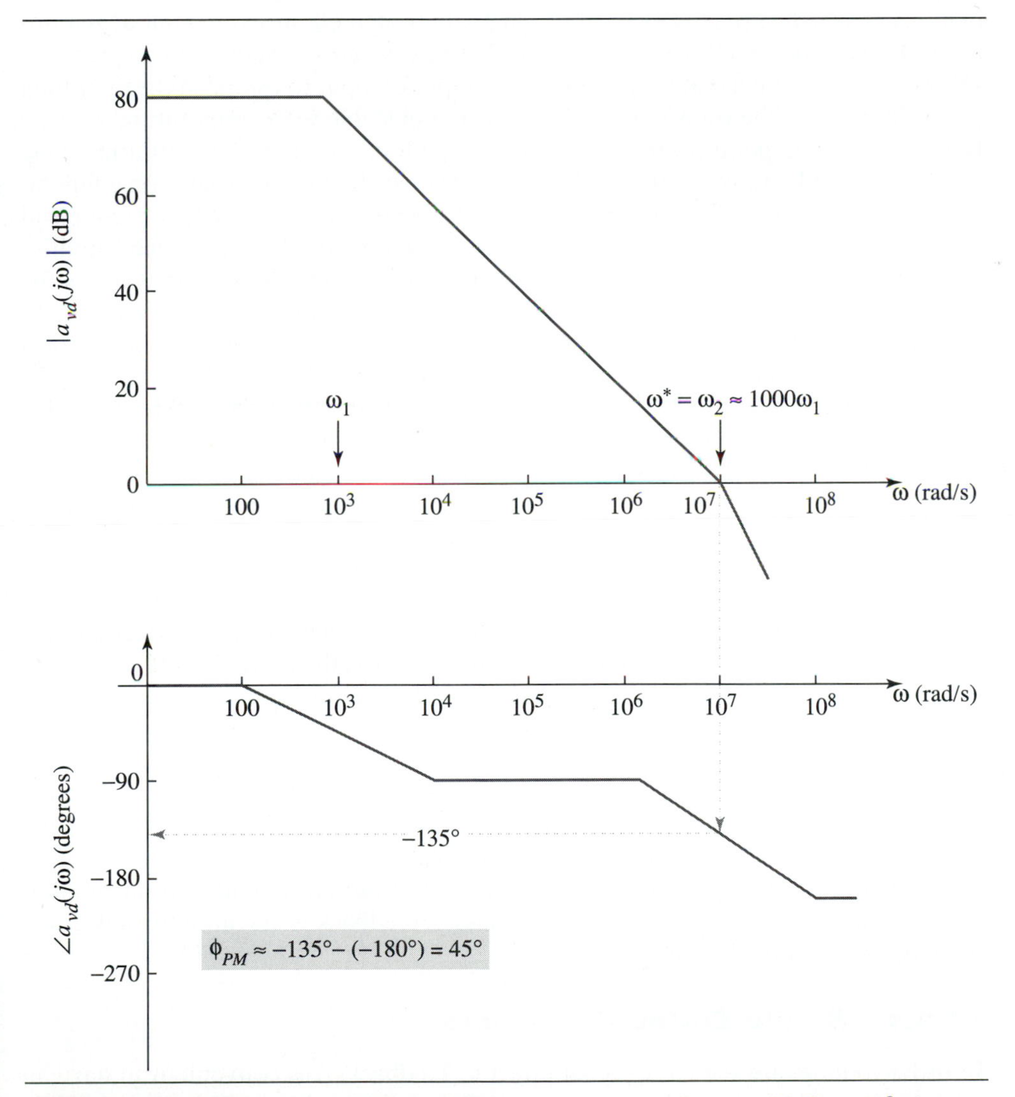

➤ **Figure Ex12.4.** Bode plot of amplifier gain and phase for part (b), for a 45° phase margin in a unity-gain feedback configuration.

refining the amplifier design. The approximate results are adequate as an initial starting point.

12.3 LARGE-SIGNAL BENEFITS OF FEEDBACK

In the first two sections we have seen that feedback enhances the small-signal amplifier performance. With careful attention to the phase of the loop gain, a designer can use feedback to trade-off closed-loop gain for bandwidth. In this section we investigate the important role feedback plays in stabilizing the large-signal transfer function.

In Chapters 8 and 9, we learned that transistor amplifiers using current-source supplies have highly nonlinear transfer functions that are necessary for achieving high small-signal gain. However, this high gain is obtained only for small deviations around a well-chosen DC operating point. Thus far we have assumed that a precision DC bias voltage or current is applied at the amplifier input to establish the operating point. In practice, the transistor parameters are not stable with temperature, causing the DC operating point to drift and possibly result in the amplifier saturating. One solution would be to construct precision DC sources that compensate for temperature drifts in the amplifier's operating point. However, this approach is "easier said than done!" Furthermore, even if we developed such DC bias sources, each amplifier would need a specially adjusted source because of the effect of slight variations in the manufacturing process will have on each one's very nonlinear transfer function. In summary, establishing a stable DC operating point is a "show-stopper" for practical application of the high-gain transistor amplifiers developed thus far in this text.

Fortunately, negative feedback provides an elegant solution to this seemingly intractable problem. Consider the case of the feedback voltage amplifier. As long as the amplifier can control the output voltage, negative feedback acts to minimize the difference between the input voltage v_I and the sampled output voltage fv_O. Therefore, the differential input voltage v_D is maintained as a small-signal difference between the (possibly large) input and feedback voltages: $v_d = v_I - f\,v_O$. Since v_d is negligibly small compared with both v_I and fv_O it follows that $v_O \approx (1/f)\,v_I$. Therefore, the closed-loop transfer function of the amplifier has been linearized and its slope is equal to the small-signal closed-loop gain. Establishing the DC operating point is straightforward since the closed-loop transfer function is linear and is no longer a function of the amplifier parameters. As we saw in Section 12.1, the penalty for using feedback is that the gain is reduced from the amplifier block's gain to $(1/f)$, for the case of high loop gain. Since the extremely high amplifier gain was unstable with temperature and unusable, this gain reduction is a small price to pay for stabilizing the large-signal transfer function of the amplifier.

The effects of feedback on both single-ended amplifiers and differential amplifiers are now considered. In both amplifier classes, feedback is essential for stabilizing the large-signal transfer function.

12.3.1 Single-Ended Amplifiers

In order to illustrate the large-signal effect of feedback, it is convenient to have an analytical model for the transfer function of the amplifying block. This is not feasible for amplifiers with current source supplies, so we employ an example using an amplifier with a resistor current supply. The concepts demonstrated in this example are general and apply to amplifiers with current-source supplies.

➤ EXAMPLE 12.5 Feedback Transconductance Amplifier

For the transconductance amplifier shown in Fig. Ex12.5A, the transistor has $I_S = 10^{-16}$ A at room temperature and $I_S = 2.75 \times 10^{-15}$ A at $T = 75°$ C. Its current gain is $\beta_F = \beta_o = 100$. In order to illustrate the large-signal effect of feedback, we find the functional dependence of the short-circuit output current i_O on the input voltage v_I.

(a) Find the large-signal transfer function $i_O(v_O)$ at room temperature (27° C) and at 75° C.

SOLUTION

Assuming the transistor is forward-active, the short-circuit output current is

$$i_O = -I_{R_C} + i_C = -\frac{5\text{ V}}{10\text{ k}\Omega} + I_S e^{\left(v_I - V^-\right)/V_{th}} = -500\ \mu\text{A} + I_S e^{(v_I + 5\text{ V})/V_{th}}$$

When the input voltage is less than –4.3 V, the transistor is cutoff and the output current is –500 μA. The collector-emitter voltage of the transistor is maintained at 5 V so it can never enter saturation. In the graph below, Fig. Ex12.5B, the output current is plotted as a function of input voltage. A reasonable operating point is to establish zero DC output current. The input bias voltage for setting the output current to zero is $V_I = -4.25$ V at room temperature and $V_I = -4.225$ V for $T = 75°$ C. If we set the bias voltage at room temperature, the DC output current at 75° C drifts to $I_O = -300\ \mu$A, which is hardly an acceptable situation.

(b) We can apply negative feedback to this amplifier by inserting an emitter resistor R_E as shown in Fig. Ex12.5C. For the value $R_E = 500\ \Omega$, find the closed-loop transfer function $i_O(v_O)$ at room temperature and at 75° C.

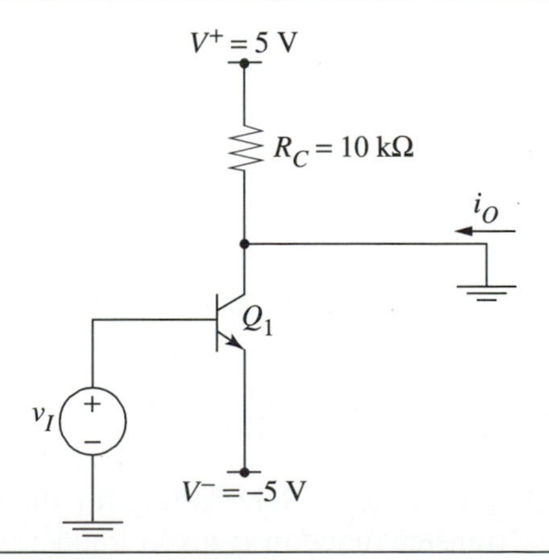

➤ Figure Ex12.5A Single-ended common-emitter transconductance amplifier.

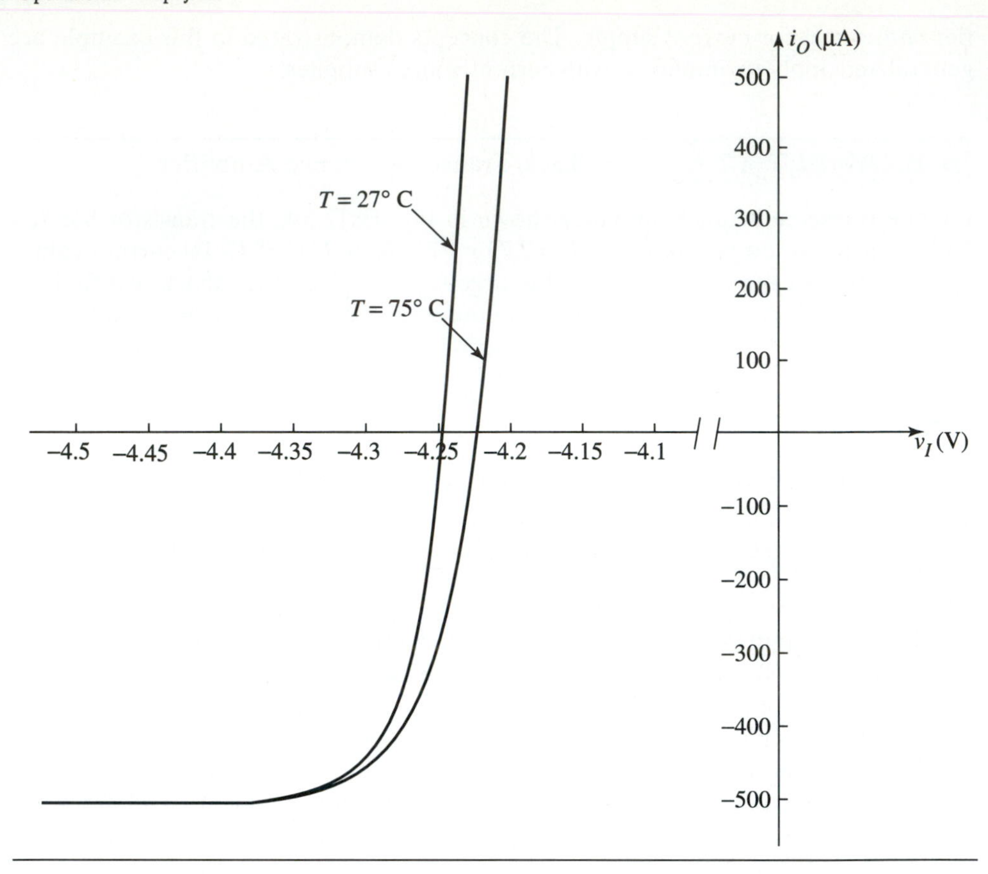

➤ Figure Ex12.5B Transfer characteristics of the transconductance amplifier at room temperature and at 75 °C.

SOLUTION

The resistor R_E samples the collector current of the transistor and returns a feedback voltage $v_F = v_{R_E} \approx i_C R_E$ to the amplifier input. Note that the base-emitter voltage of the transistor is given by $v_{BE} = v_I - v_F$. The output current of the amplifier is not the collector current; however, it is offset from i_C by a constant current

$$i_O = -\frac{V^+}{R_C} + i_C = -500\ \mu\text{A} + i_C$$

We will find that measurement of i_C is sufficient for stabilizing the large-signal transfer function.

The collector current now depends on the feedback voltage, through

$$i_C = I_S e^{\left(v_I - v_F - V^-\right)/V_{th}} = I_S e^{\left(v_I - \alpha_F R_E i_C - V^-\right)/V_{th}}$$

where $\alpha_F = \beta_F / (\beta_F + 1) = 0.99$. Substituting for the output current, we find that the closed-loop transfer function is given implicitly by the transcendental equation

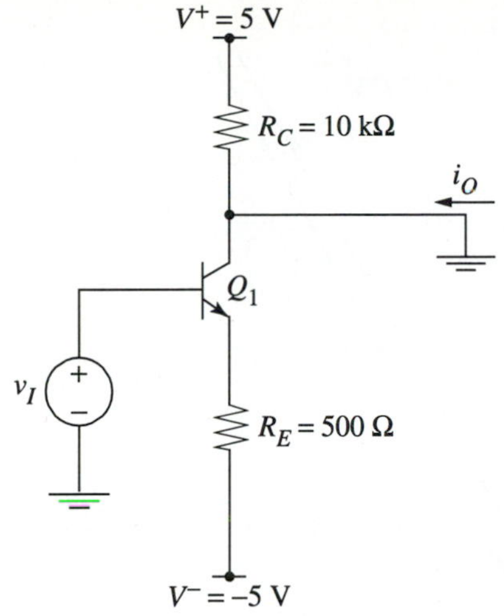

$V^+ = 5$ V
$R_C = 10$ kΩ
i_O
Q_1
v_I
$R_E = 500$ Ω
$V^- = -5$ V

➤ **Figure Ex12.5C** Single-ended feedback transconductance amplifier.

$$i_O = -\frac{V^+}{R_C} + I_S e^{\left[v_I - \alpha_F R_E\left(i_O + V^+/R_C\right) - V^-\right]/V_{th}}$$

For convenience in plotting, we express v_I as a function of i_O

$$v_I = v_F + v_{BE} = \left[\alpha_F R_E i_O + \alpha_F \frac{R_E}{R_C} V^+ + V^-\right] + V_{th} \ln\left[\frac{\left(i_O + \frac{V^+}{R_C}\right)}{I_S}\right]$$

Substituting the numerical values for this design, the inverse transfer function $v_I(i_O)$ is

$$v_I = (495\ \Omega)\, i_O - 4.753\text{ V} + V_{th} \ln\left[\frac{i_O + 500\ \mu\text{A}}{I_S}\right]$$

If the logarithmic term is neglected, the relationship between the input voltage and the output current is linear, which is the result of the feedback provided by R_E. The graph in Fig. Ex12.5D gives the transfer function for room temperature and 75° C, which shows that the DC input voltage can vary over 200 mV around the desired operating point without saturating the amplifier.

It is clear that the effect of feedback is to linearize the transfer function. From Chapter 8, we recognize the feedback transconductance amplifier as a common-emitter amplifier with emitter degeneration. The small-signal transconductance of this stage is

$$G_m = \left.\frac{di_O}{dv_I}\right|_Q = \frac{g_m}{1 + g_m R_E},$$

which is in the familiar form for the closed-loop transconductance. For this case, the feedback factor is $f = R_E$ and the loop gain is $T = g_m R_E$. For an operating

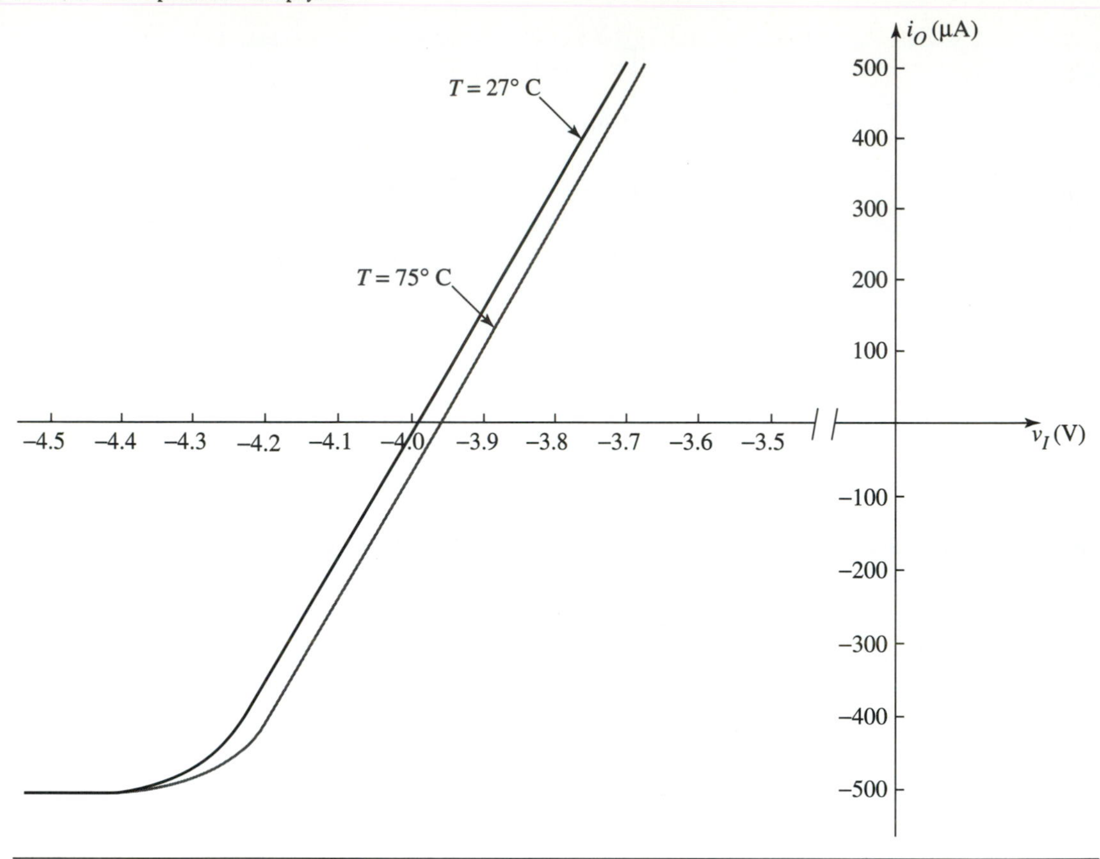

➤ **Figure Ex12.5D** Transfer characteristics of the feedback transconductance amplifier at room temperature and at 75 °C.

point $Q = (V_I, I_O)$ where $g_m R_E$ is much larger than 1, the transconductance is approximately $G_m \approx 1/R_E$. At room temperature and for $i_O = 0$ A, which corresponds to $I_C = 500$ µA, the transconductance is

$$G_m = \frac{(0.5 \text{ mA})/(25 \text{ mV})}{1 + (0.5 \text{ mA})(500\ \Omega)/(25 \text{ mV})} = \frac{0.02}{1 + 10} = 1.82 \text{ mS}$$

Negative feedback causes the slope of the transfer function for $v_I > -4.2$ V to be approximately equal to the small-signal G_m, both at room temperature and at 75° C. Compared with the simple common-emitter stage, the feedback transconductance amplifier is easier to bias. However, since the loop gain is rather small ($T = 10$), the DC operating point for $I_O = 0$ A still drifts significantly due to temperature. For $I_O = 0$ A, the DC input voltage at room temperature is $V_I = -4.0$ V. With this input bias voltage, the DC output current will be about −50 µA at 75° C. Higher loop gains can provide additional stabilization at the cost of reduced closed-loop transconductance.

12.3.2 Differential Amplifiers

At first glance, the differential amplifier topology introduced in Chapter 11 provides a solution to the biasing problem since there is no DC bias voltage applied at either input. In practice, however, asymmetries in the circuit topology or imperfect matching between transistors mean that a differential amplifier is *not* balanced when the two input terminals are shorted together and tied to a constant voltage. It is not uncommon that a high-gain differential amplifier will be so unbalanced in this state that one of the transistors will be operating out of its constant-current region. We must apply a voltage called the **input offset voltage V_{OS}** to a differential amplifier so the DC operating point is established correctly. This voltage will vary both with temperature and with manufacturing variations, which indicates that establishing the DC operating point for differential amplifiers is not an easier problem than for single-ended transistor amplifiers.

With negative feedback, the transfer function is linearized and the biasing problem is simplified. As we will see in the following example, directly connecting the input voltage without considering the amplifier's input offset voltage will provide adequate accuracy for many applications. Again, we will use resistor current supplies to obtain an analytical result for the transfer function.

➤ EXAMPLE 12.6 Effect of Unity-Gain Feedback on a Voltage Amplifier's Transfer Function

For the common-collector, common-base differential amplifier shown in Fig. Ex12.6A, consider the transistors perfectly matched with: $\beta_o = \beta_F = 100$ and $V_A \to \infty$.

(a) Find an analytical expression for the transfer function $v_O(v_{ID})$ and plot over the range $-125\text{ mV} < v_{ID} < +125\text{ mV}$ for the case of room temperature. Repeat for a temperature of 125 °C, at which $V_{th} = 37$ mV. The results from the transfer function of the bipolar differential pair can be used since the Early voltage is assumed to be infinite.

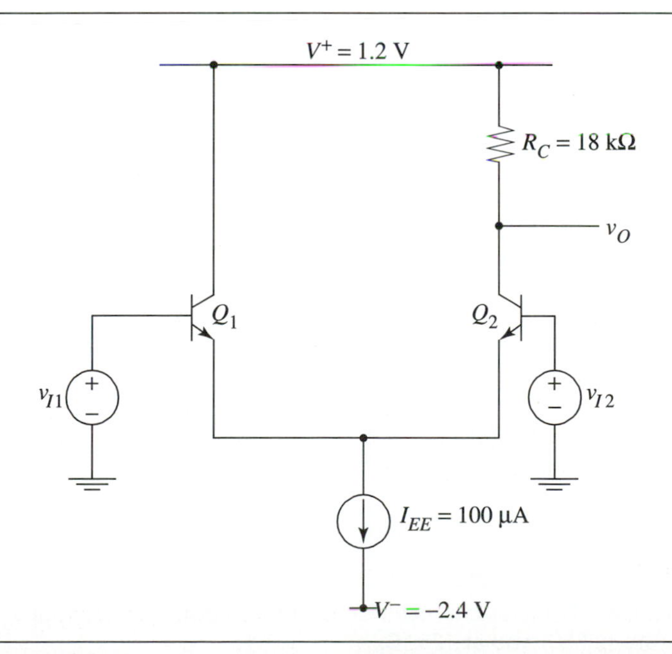

➤ Figure Ex12.6A Common-collector, common-base cascade differential amplifier.

SOLUTION

From Chapter 11, the collector current of transistor Q_2 is given by

$$i_{C2} = \frac{\alpha_F I_{EE}}{1 + e^{v_{ID}/V_{th}}}$$

where $\alpha_F = \beta_F/(1+\beta_F) = 0.99$. The output voltage v_O is given by

$$v_O = V^+ - i_{C2}R_C = V^+ - \frac{\alpha_F R_C I_{EE}}{1 + e^{v_{ID}/V_{th}}} = 1.2\text{ V} - \frac{1.8\text{ V}}{1 + e^{v_{ID}/V_{th}}}$$

The plots of the transfer function $v_O(v_{ID})$ are shown below for room temperature and for $T = 125\text{ °C}$ ($V_{th} = 37$ mV).

We can see from the plot in Fig. Ex12.6B that this amplifier requires careful adjustment of an external offset voltage in order to null the output voltage when $v_{ID} = 0$ V. The input offset voltage, which we consider to be the input voltage needed to set $v_O = 0$ V, shifts from –18 mV at room temperature to –25 mV at 125 °C. In addition, the small-signal gain of the amplifier is a strong function of temperature. Note that we have *not* included mismatches in the characteristics of Q_1

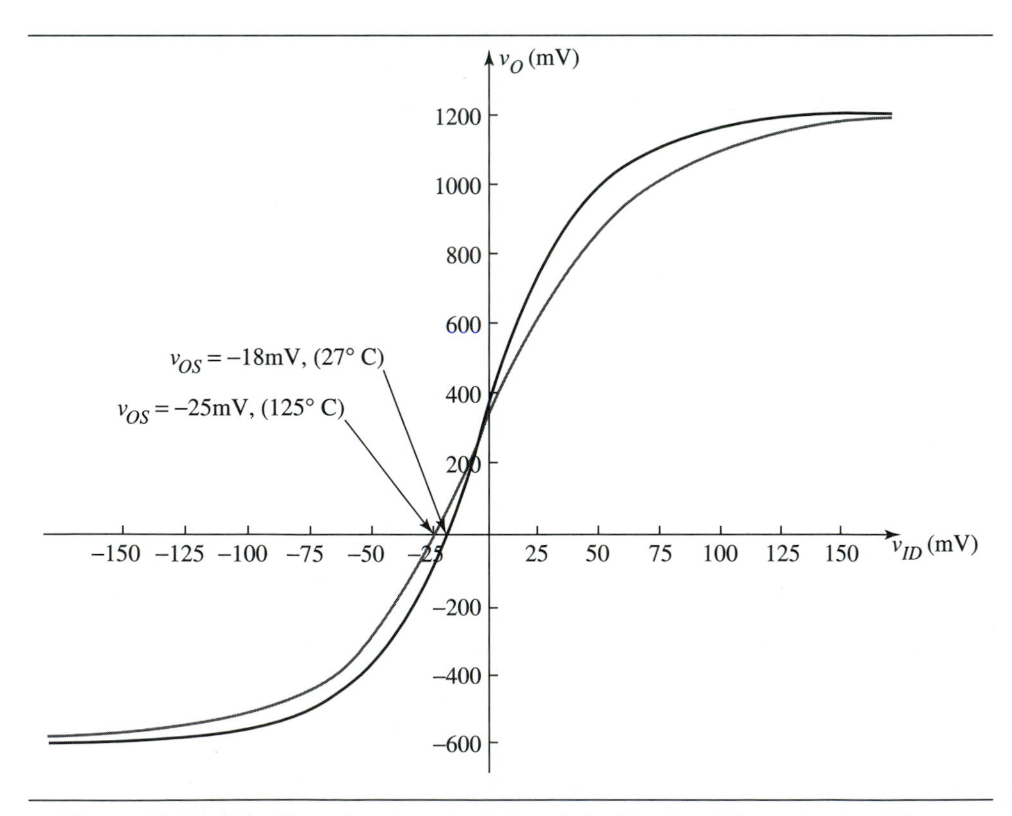

➤ **Figure Ex12.6B** Transfer characteristics of the bipolar differential amplifier at room temperature (27 °C) and at 125 °C.

and Q_2 due to processing variations, which would further compound the range of input offsets and gains for this amplifier topology.

(b) We now apply unity-gain feedback by directly connecting v_O to the inverting terminal v_{I2}, as shown in Fig. Ex12.6C. Using the results from part (a), plot the transfer function $v_O(v_I)$ over the range $-1.2 < v_I < 1.2$ V at room temperature and at $T = 125$ °C.

SOLUTION

The differential input voltage with feedback is $v_{ID} = v_I - v_F = v_I - v_O$. Neglecting the (small) base current into Q_2, we can substitute for v_{ID} in the transfer function from (a)

$$v_O = V^+ - \frac{R_C I_{EE}}{1 + e^{v_{ID}/V_{th}}} = V^+ - \frac{R_C I_{EE}}{1 + e^{(v_I - v_O)/V_{th}}}$$

This result is valid as long as the bipolar transistors and the current supply remain unsaturated, which is the case for the input voltage range considered here. This equation can be inverted to yield v_I as a function of v_O

$$v_I = v_O + V_{th} \ln\left(\frac{v_O - V^+ + R_C I_{EE}}{V^+ - v_O}\right) = v_O + V_{th} \ln\left(\frac{v_O + 0.6\text{ V}}{1.2 - v_O}\right)$$

Note in Fig. Ex12.6D that the unity-gain feedback amplifier is linear nearly to the limits of the amplifier's output voltage. It is also insensitive to temperature, in

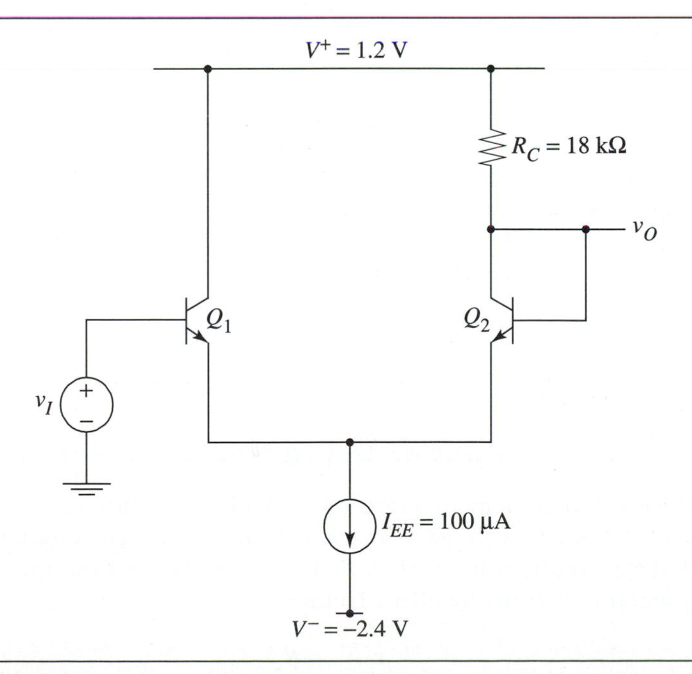

➤ **Figure Ex12.6C.** Unity-gain voltage feedback amplifier, based on the bipolar differential amplifier.

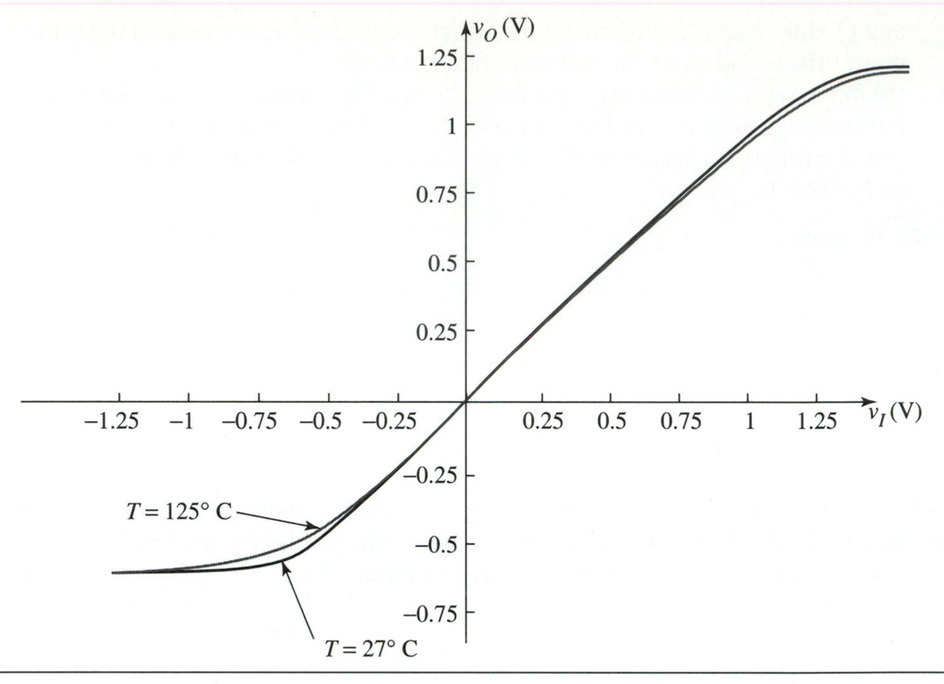

Figure Ex 12.6D Closed-loop transfer function at room temperature and $T = 125$ °C.

marked contrast to the differential amplifier in part (a). In addition, the output voltage for $v_I = 0$ V is only 18 mV, compared to 300 mV from part (a).

12.4 PRACTICAL FEEDBACK AMPLIFIERS

Negative feedback is a powerful tool for enhancing the performance of all types of electronic amplifiers. Thus far, we have assumed an idealized feedback network. In practice, however, feedback is implemented using resistor networks. In this section we analyze how resistive feedback modifies the closed-loop gain and the input and output resistances for the cases of the voltage and transresistance feedback topologies. We will see the benefits that have been demonstrated with ideal networks are realizable in practical feedback amplifiers using resistor networks to couple the output signal back to the input.

12.4.1 Voltage Amplifier Using Resistive Feedback

It is straightforward to implement a resistor network that returns a feedback voltage proportional to the output voltage. A resistor divider is all that is required as shown in Fig. 12.10. If the feedback network is analyzed in isolation from the amplifier, we can see by inspection that the feedback factor is

$$f = \frac{v_f}{v_o} = \frac{R_1}{R_1 + R_2} \tag{12.48}$$

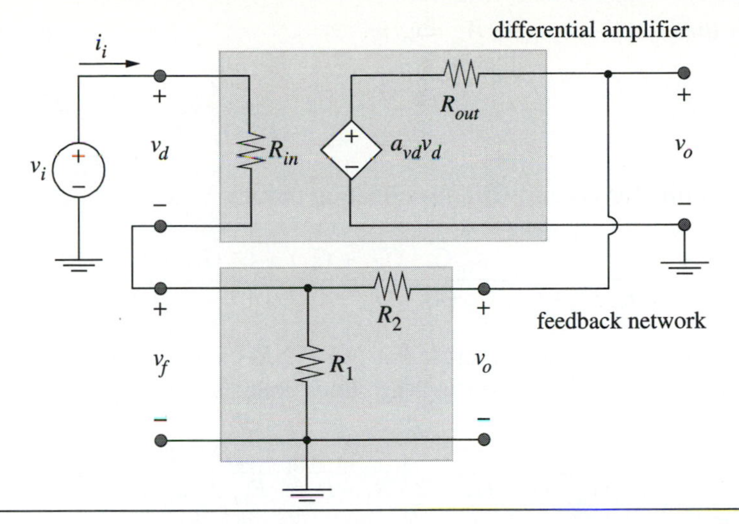

Figure 12.10 Voltage amplifier with resistive divider providing feedback.

As we will see, the circuit analysis becomes surprisingly involved for this simple linear network. It is helpful to define the feedback factor f for this amplifier as the resistor divider expression in Eq. (12.48). With sufficiently large amplifier gain a_{vd}, the closed-loop gain $A_{vf} = v_o/v_i \approx 1/f$, as we found for the idealized case in the previous section.

Applying Kirchhoff's current law at the feedback and the output nodes, we find that

$$G_{in}(v_i - v_f) + G_2(v_o - v_f) = G_1 v_f \tag{12.49}$$

$$G_{out}(a_{vd}(v_i - v_f) - v_o) = G_2(v_o - v_f) \tag{12.50}$$

in which conductances (e.g., $G_2 = 1/R_2$) are used for convenience. To find the closed loop gain v_o/v_i, we solve Eqs. (12.49) and (12.50) for the feedback voltage v_f and then eliminate it

$$v_f = \left(\frac{G_{in}}{G_1 + G_2 + G_{in}}\right)v_i + \left(\frac{G_2}{G_1 + G_2 + G_{in}}\right)v_o = \left(\frac{G_{in}}{G_T}\right)v_i + \left(\frac{G_2}{G_T}\right)v_o \tag{12.51}$$

$$v_f = \left(\frac{-a_{vd}G_{out}}{G_2 - a_{vd}G_{out}}\right)v_i + \left(\frac{G_{out} + G_2}{G_2 - a_{vd}G_{out}}\right)v_o = \left(\frac{-a_{vd}G_{out}}{G'}\right)v_i + \left(\frac{G_{out} + G_2}{G'}\right)v_o \tag{12.52}$$

In Eqs. (12.51) and (12.52), the new symbols $G_T = G_1 + G_2 + G_{in}$ and $G' = G_2 - a_{vd}G_{out}$ have been introduced. Equating Eq. (12.51) and (12.52), we find

$$\left(\frac{G_{in}}{G_T}\right)v_i + \left(\frac{G_2}{G_T}\right)v_o = \left(\frac{-a_{vd}G_{out}}{G'}\right)v_i + \left(\frac{G_{out} + G_2}{G'}\right)v_o \tag{12.53}$$

Solving for the closed-loop gain $A_{vf} = v_o/v_i$

$$A_{vf} = \frac{(-a_{vd}G_{out})G_T - G_{in}G'}{G_2G' - (G_{out} + G_2)G_T} \tag{12.54}$$

After substituting for G_T and G' and collecting terms,

$$A_{vf} = \frac{a_{vd}G_{out}(G_1 + G_2) + G_2G_{in}}{(G_{out}G_1 + G_{out}G_2 + G_{out}G_{in} + G_1G_2 + G_2G_{in}) + a_{vd}G_{out}G_2} \tag{12.55}$$

The second term in the numerator in Eq. (12.55) is negligible compared to the first term, as we can see by converting the conductances into resistances

$$\frac{G_2G_{in}}{a_{vd}G_{out}(G_1 + G_2)} = \left(\frac{R_1}{R_1 + R_2}\right)\left(\frac{R_{out}}{R_{in}}\right)\left(\frac{1}{a_{vd}}\right) \ll 1 \tag{12.56}$$

The first term on the right hand side of Eq. (12.56) is f, the second term is also less than one for reasonable voltage amplifiers, and finally, the inverse of the differential gain is clearly very small. By neglecting this term in the numerator of Eq. (12.55), we can put the closed-loop gain into the form

closed loop voltage gain including loading

$$A_{vf} = \frac{\left(\dfrac{a_{vd}G_{out}(G_1 + G_2)}{G_{out}G_1 + G_{out}G_2 + G_{out}G_{in} + G_1G_2 + G_2G_{in}}\right)}{1 + \left(\dfrac{a_{vd}G_{out}G_2}{G_{out}G_1 + G_{out}G_2 + G_{out}G_{in} + G_1G_2 + G_2G_{in}}\right)} = \frac{a'}{1 + T'} \tag{12.57}$$

This expression will be in the form of Eq. (12.7), if the ratio of T' to the "loaded" open-loop voltage gain a' is the feedback factor. Substituting from Eq. (12.57), we find

$$\frac{T'}{a'} = \frac{a_{vd}G_{out}G_2(G_{out}G_1 + G_{out}G_2 + G_{out}G_{in} + G_1G_2 + G_2G_{in})}{a_{vd}G_{out}(G_1 + G_2)(G_{out}G_1 + G_{out}G_2 + G_{out}G_{in} + G_1G_2 + G_2G_{in})}$$

$$= \frac{G_2}{G_1 + G_2} \tag{12.58}$$

By multiplying Eq. (12.58) by $(R_1 R_2)/(R_1 R_2)$, we find

$$\frac{T'}{a'} = \left(\frac{G_2}{G_1 + G_2}\right)\left(\frac{R_1R_2}{R_1R_2}\right) = \frac{R_1}{R_1 + R_2} = f \tag{12.59}$$

From Eq. (12.57), in cases where $T' \gg 1$, the closed-loop gain $A_{vf} \approx 1/f$. We conclude that the resistive feedback network has stabilized the gain. How does the resistor feedback network affect the closed-loop gain? The loaded open-loop gain can be expressed in terms of resistor ratios

$$a' = a_{vd}\left(\frac{R_{in}(R_1 + R_2)}{R_1R_{in} + R_2R_{in} + R_{out}R_{in} + R_1R_2 + R_1R_{out}}\right) \tag{12.60}$$

From Eq. (12.60), we see that the resistive feedback network, combined with the input and output resistances of the amplifier, reduces or "loads" the open-loop gain. In order to have $A_{vf} \approx 1/f$, the differential gain a_{vd} must be sufficiently high to compensate.

The input resistance of the voltage feedback amplifier in Fig. 12.10 is

$$R_{in,f} = \frac{v_i}{i_i} = \left(\frac{v_d}{i_i}\right)\left(\frac{v_i}{v_d}\right) = R_{in}\left(\frac{v_i}{v_d}\right), \tag{12.61}$$

in which the differential input voltage $v_d = v_i - v_f$. By eliminating v_o from Eqs. (12.51) and (12.52), we can solve for v_f in terms of v_i. The ratio of the input voltage v_i to v_d is

$$\frac{v_i}{v_d} = \frac{G_T(G_{out} + G_2) - G'G_2}{(G_{out} + G_2)(G_T - G_{in}) + G_2(-a_{dm}G_{out} - G')} \tag{12.62}$$

Substituting for G_T and G' in Eqs. (12.62) and (12.61), we find that

$$R_{in,f} = R_{in}\left(\frac{G_{out}G_1 + G_{out}G_2 + G_{out}G_{in} + G_1G_2 + G_2G_{in} + a_{vd}G_{out}G_2}{G_{out}G_1 + G_1G_2 + G_{out}G_2}\right) \tag{12.63}$$

This complicated expression contains terms that are similar to the loop gain T'. By factoring out the first term, we can rewrite Eq. (12.63) as

$$R_{in,f} = R'\left(1 + \frac{a_{vd}G_{out}G_2}{G_{out}G_1 + G_{out}G_2 + G_{out}G_{in} + G_1G_2 + G_2G_{in}}\right) = R'(1 + T') \tag{12.64}$$

in which we identify the loop gain from Eq. (12.57). The resistor R' is given by

$$R' = R_{in}\left(\frac{G_{out}G_1 + G_{out}G_2 + G_{out}G_{in} + G_1G_2 + G_2G_{in}}{G_{out}G_1 + G_1G_2 + G_{out}G_2}\right) \tag{12.65}$$

This expression can be simplified after converting to resistances:

$$R' = \frac{R_{in}(R_1 + R_2) + R_{out}(R_1 + R_{in}) + R_1R_2}{R_1 + R_2 + R_{out}} \tag{12.66}$$

Substituting into Eq. (12.64), the input resistance of the voltage feedback amplifier is

$$\boxed{R_{in,f} = \left(\frac{R_{in}(R_1 + R_2) + R_{out}(R_1 + R_{in}) + R_1R_2}{R_1 + R_2 + R_{out}}\right)(1 + T')} \tag{12.67}$$

input resistance of voltage feedback amp including loading

The input resistance is increased by the same $(1 + T')$ factor that appears in the denominator of the closed-loop gain A_{vf} in Eq. (12.57). The resistive feedback network enters into $R_{in,f}$ through the reduced loop gain, as well as by modifying R'.

We can expect that feedback will also have a dramatic effect on the output resistance of the amplifier in Fig. 12.10. The circuit for finding the output resistance is shown in Fig. 12.11.

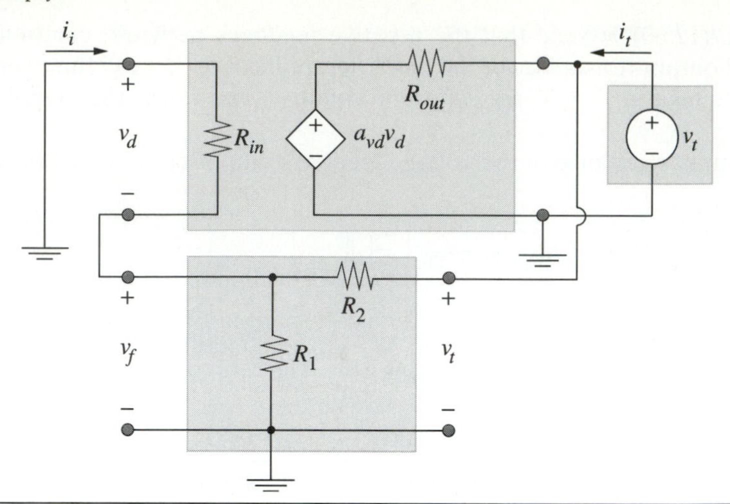

➤ Figure 12.11 Circuit for finding the output resistance of the voltage feedback amplifier.

Kirchhoff's current law at the output node yields

$$i_t = G_{out}(v_t - a_{vd}v_d) + G_2(v_t - v_f) \tag{12.68}$$

Since the input voltage is shorted out for finding $R_{out,f}$, $v_f = -v_d$ and Eq. (12.68) becomes

$$i_t = (G_{out} + G_2)v_t + (a_{vd}G_{out} - G_2)v_f \tag{12.69}$$

The feedback voltage v_f can be expressed in terms of the output voltage by eliminating v_i from Eqs. (12.51) and (12.52). Substituting into Eq. (12.69) and solving for the ratio of v_t to i_t, we find

$$R_{out,f} = \frac{v_t}{i_t} = \frac{a_{vd}G_{out}G_T + G_{in}G'}{a_{vd}G_{out}G_T(G_{out} + G_2) + a_{dm}G_{out}G_2G'} \tag{12.70}$$

After substituting for G_T and G' and further rearrangement, the output resistance can be written as

$$R_{out,f} = \frac{(G_1 + G_2) + [(G_2G_{in})/(a_{vd}G_{out})]}{G_{out}(G_1 + G_2 + G_{in}) + G_1G_2 + G_2G_{in} + a_{vd}G_{out}G_2} \tag{12.71}$$

The term in square brackets in the numerator can be neglected since it is on the order of G_2/a_{vd} or smaller. The denominator can be put into the form $1 + T'$, which leads to

$$R_{out,f} = \left(\frac{G_1 + G_2}{G_{out}G_1 + G_{out}G_2 + G_{out}G_{in} + G_1G_2 + G_2G_{in}}\right)\left(\frac{1}{1 + T'}\right) \tag{12.72}$$

The first term in Eq. (12.72) is easier to interpret after conversion to resistances

output resistance of voltage feedback amp including loading

$$R_{out,f} = \left(\frac{R_{in}R_{out}(R_1 + R_2)}{R_{in}(R_{out} + R_1 + R_2) + [R_1R_2 + R_{out}R_1]}\right)\left(\frac{1}{1 + T'}\right) \tag{12.73}$$

In summary, feedback makes the input and output resistances approach those of an ideal voltage amplifier (infinite input resistance and zero output resistance), even when a resistive feedback network is used. With large values of loop gain, we also achieve a stable closed-loop gain that is largely independent of variations in the differential gain of the amplifying block.

➤ EXAMPLE 12.7 Bipolar Voltage Feedback Amplifier

In this example, a resistive feedback network is used in conjunction with a bipolar differential amplifier with a p-channel MOSFET current mirror supply as shown in Fig. Ex12.7.

Transistor parameters:

$\beta_o = \beta_F = 100$, $V_A = 50$ V and $\mu_p C_{ox} = 25\ \mu\text{A/V}^2$,

$V_{T_p} = -1$ V , $\lambda_p = 0.02\ \text{V}^{-1}$, $(W/L)_{3,4} = (50\ \mu\text{m}/3\ \mu\text{m})$.

(a) Find the two-port model for the differential amplifier; use the voltage form of the model.

SOLUTION

The differential amplifier is identical to that in Example 12.2. We found that $R_{in} = 100$ kΩ and $R_{out} = 500$ kΩ

The differential voltage gain of this amplifier is

$$a_{vd} = g_{m1}R_{out} = \left(\frac{50\ \mu\text{A}}{25\ \text{mV}}\right)(500\ \text{k}\Omega) = 1000$$

(b) For $R_1 = 20$ kΩ, $R_2 = 180$ kΩ, find the feedback factor, the loop gain, the closed-loop gain, and the input and output resistances.

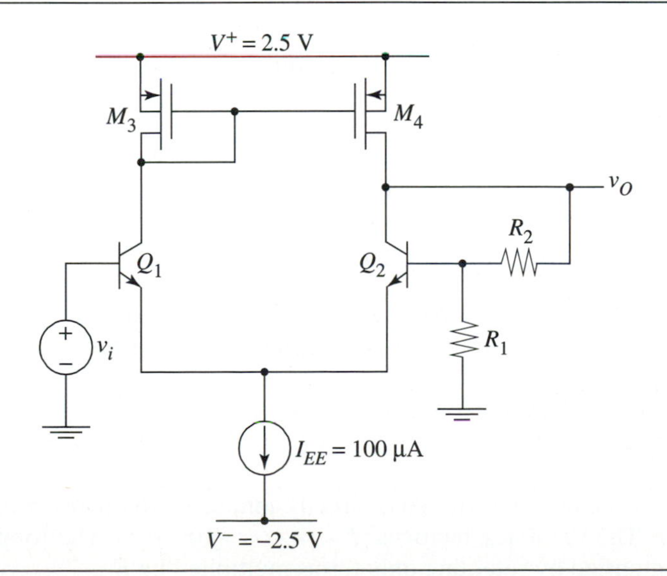

➤ Figure Ex12.7 Voltage feedback amplifier based on a bipolar differential pair.

SOLUTION

From Eq. (12.59), the feedback factor is

$$f = \frac{R_1}{R_1 + R_2} = \frac{20\ \text{k}\Omega}{20\ \text{k}\Omega + 180\ \text{k}\Omega} = 0.1$$

The loop gain is the product of f and the loaded amplifier gain a' from Eq. (12.60)

$$T' = a'f = a_{vd}\left(\frac{R_1 R_{in}}{R_1 R_{in} + R_2 R_{in} + R_{out} R_{in} + R_1 R_2 + R_1 R_{out}}\right)$$

$$T' = (1000)\left(\frac{20 \cdot 100}{20 \cdot 100 + 180 \cdot 100 + 500 \cdot 100 + 20 \cdot 180 + 20 \cdot 500}\right) = 26.8$$

Note that the effect of resistive feedback and the non-ideal differential amplifier has been to reduce the loop gain from $a_{vd}f = (1000)\ (0.1) = 100$ to 27. The closed-loop gain is given by Eq. (12.57)

$$A_{vf} = \frac{a'}{1 + a'f} = \frac{268}{1 + 26.8} = 9.6$$

which is smaller than the limit of $1/f = 10$ because of the relatively low loop gain. The input resistance with feedback is given by Eq. (12.67)

$$R_{in,f} = \left(\frac{R_{in}(R_1 + R_2) + R_{out}(R_1 + R_{in}) + R_1 R_2}{R_1 + R_2 + R_{out}}\right)(1 + T')$$

$$R_{in,f} = \left(\frac{100\,(200) + 500\,(20 + 100) + 20 \cdot 180}{20 + 180 + 500}\right)(1 + 26.8) = 119.4\ \text{k}\Omega \cdot 27.8$$

$$R_{in,f} = (119.4\ \text{k}\Omega) \cdot 27.8 = 3.3\ \text{M}\Omega$$

The output resistance with feedback is found from Eq. (12.73)

$$R_{out,f} = \left(\frac{R_{in} R_{out}(R_1 + R_2)}{R_{in}(R_{out} + R_1 + R_2) + [R_1 R_2 + R_{out} R_1]}\right)\left(\frac{1}{1 + T'}\right)$$

$$R_{out,f} = \left(\frac{100 \cdot 500\,(20 + 180)}{100\,(500 + 20 + 180) + 20 \cdot 180 + 500 \cdot 20}\right)\left(\frac{1}{27.8}\right) = \frac{119.6\ \text{k}\Omega}{27.8} = 4.3\ \text{k}\Omega$$

Even with a relatively small loop gain, the feedback amplifier has much improved input and output resistances compared with those of the differential amplifier.

(c) Repeat part (b) for $R_2 = 0$ and $R_1 \to \infty$.

SOLUTION

In this case, the output terminal is directly connected to the inverting input of the amplifier. The feedback factor is $f = \infty/(\infty + 0) = 1$. The loop gain expression is simplified by retaining only terms multiplied by R_1.

$$T' = a_{vd}\left(\frac{R_{in}}{R_{out} + R_{in}}\right) = (1000)\left(\frac{100}{500 + 100}\right) = 167$$

The closed-loop gain is close to $(1/f) = 1$ because of the higher loop gain

$$A_{vf} = \frac{a'}{1 + a'} = \frac{167}{1 + 167} = 0.99$$

The input resistance with feedback is also increased

$$R_{in,f} = (R_{in} + R_{out})\ (1 + T') = (100 + 500)\ (1 + 167) = 101\ \text{M}\Omega$$

The output resistance with feedback is

$$R_{on,f} = \left(\frac{R_{in}R_{out}}{R_{in} + R_{out}}\right)\left(\frac{1}{1 + T'}\right) = \left(\frac{100 \cdot 500}{100 + 500}\right)\left(\frac{1}{1 + 167}\right) = \frac{83\ \text{k}\Omega}{168} = 496\Omega$$

With unity-gain feedback ($f = 1$), we can design an amplifier with near-ideal voltage buffer characteristics.

12.4.2 Transresistance Amplifier with Feedback

The benefits of negative feedback implemented by a resistor divider have been demonstrated for a voltage amplifier topology. The transresistance amplifier in Fig. 12.4 requires a feedback network that samples the output voltage and delivers a feedback current to the input node of the amplifier block. The simplest solution is to use a single feedback resistor R_F to perform this function, as shown in Fig. 12.12.

The procedure for modeling this feedback amplifier is identical to the one used for the voltage amplifier. Direct analysis of the feedback circuit in Fig. 12.12 leads to the closed-loop gain, which can be placed in the form found for the ideal feedback network case. Rather than present the full details, we present only the key results.

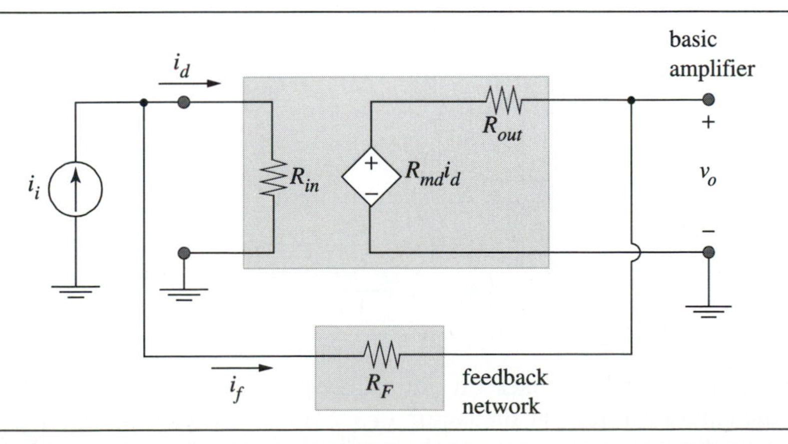

Figure 12.12 Transresistance feedback amplifier topology with feedback resistor R_F.

The closed-loop transresistance is given by:

closed-loop transresistance with loading

$$R_{mf} = \frac{\left(\dfrac{(R_{md}G_{out} - R_{in}G_F)}{(G_{out} + G_F)(1 + R_{in}/R_F)}\right)}{1 + \left(-\dfrac{G_F(R_{md}G_{out} - R_{in}G_F)}{(G_{out} + G_F)(1 + R_{in}/R_F)}\right)} = \frac{R_m'}{1 + R_m'f} \quad (12.74)$$

From Fig. 12.12, the feedback factor is the ratio of the feedback current i_f to the output voltage v_o, for the case where $v_i = 0$

$$f = \left.\frac{i_f}{v_o}\right|_{i_i = 0} = -\frac{1}{R_F} \quad (12.75)$$

For negative feedback, we require that the loop gain $T' = R_m'f > 0$, which implies that the amplifier transresistance is negative. The loop gain can be found directly from Eq. (12.74)

$$T' = R_m'f = \frac{-G_F(R_{md}G_{out} - R_{in}G_F)}{(G_{out} + G_F)(1 + R_{in}R_F)} = \frac{-R_{md}[1 - (R_{in}R_{out})/(R_{md}R_F)]}{(R_{out} + R_F)(1 + R_{in}/R_F)} \quad (12.76)$$

For a reasonable transresistance amplifier $|R_{md}| \gg R_{in}$ and so the second term in the square brackets in the numerator can be neglected

loop gain of transresistance amp including loading

$$T' \approx \frac{-R_{md}}{(R_{out} + R_F)(1 + R_{in}/R_F)} \quad (12.77)$$

From Eq. (12.77), the effect of resistive feedback is to reduce the loop gain compared with the ideal case in Eq. (12.17), where $T = R_{md}f = -R_{md}/R_F$.

The input resistance of the amplifier is modified by feedback

$$R_{inf} = \left(\frac{R_{in}}{1 + R_{in}/R_F}\right)\left(\frac{1}{1 - \dfrac{G_F(R_{md}G_{out} - R_{in}G_F)}{(G_{out} + G_F)(1 + R_{in}/R_F)}}\right) \quad (12.78)$$

From Eq. (12.76), the second term contains the (exact) expression for the loop gain. We can write the input resistance with feedback as

input resistance of transresistance amp including loading

$$R_{inf} = \frac{(R_{in}||R_F)}{1 + T'} \quad (12.79)$$

The effect of resistive feedback on the transresistance topology in Fig. 12.12 is to *reduce* the input resistance by the factor $(1 + T')$, compared to its open-loop value.

Finally, the output resistance of the feedback transresistance amplifier is found by opening the input current source i_i and applying a test voltage v_t. The result is

$$R_{out,f} = \frac{v_t}{i_t} = \frac{1}{G_{out} + \dfrac{(1 - R_{md}G_{out})}{(R_F + R_{in})}} = \frac{(R_{in} + R_F)R_{out}}{R_{in} + R_F + R_{out} - R_{md}} \quad (12.80)$$

For the case where $R_F \gg R_{out}$, the exact result can be rewritten in the standard form

$$R_{out,f} \approx \frac{[R_{out} \| (R_{in} + R_F)]}{1 + T'} \quad (12.81)$$

output resistance of transresistance amp including loading

In summary, the application of resistive feedback around the transresistance amplifier has stabilized its gain and made the input and output resistances closer to the ideal case (zero for both), by reducing them by the factor $(1 + T')$. Resistive loading reduces the loop gain from T to T', as we saw for the voltage amplifier, which makes it necessary to increase the gain of the amplifier block to maintain the same loop gain. As a final observation, the feedback factor is specified by the value of R_F rather than the ratio of resistors as we found for the voltage amplifier. Therefore, the specification on the closed-loop transresistance will be limited by the reproducibility of the integrated resistor and its stability over temperature.

➤ EXAMPLE 12.8 Resistive Feedback Transresistance Amplifier

In this example, we implement a CMOS feedback transresistance amplifier ,using a resistor as the feedback network, as shown in Fig. Ex12.8. The transistor parameters are: $\mu_n C_{ox} = 50\ \mu A/V^2$, $\mu_p C_{ox} = 25\ \mu A/V^2$, $V_{Tn} = 1$ V , $V_{Tp} = -1$ V, $\lambda_n = \lambda_p = 0.033\ V^{-1}$, and $(W/L)_{1,2} = (120\ \mu m/3\ \mu m)$, $(W/L)_{3,4} = (60\ \mu m/3\ \mu m)$.

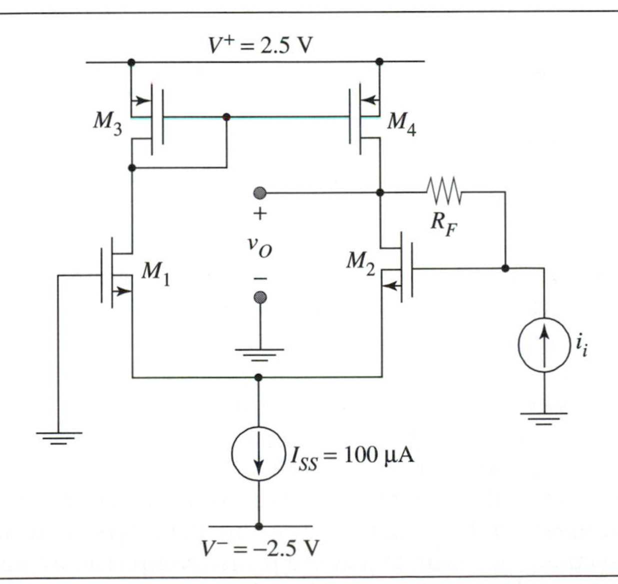

➤ Figure Ex12.8 Transresistance feedback amplifier with resistive feedback.

For $R_F = 100\ \text{k}\Omega$, find the feedback factor, the loop gain, the closed-loop transresistance, and the input and output resistances.

SOLUTION

The feedback factor is

$$f = \frac{-1}{R_F} = -10\ \mu\text{S}$$

Since the MOS differential pair has an input resistance $R_{in} \to \infty$, its differential transresistance $-R_{md} \to \infty$. Equation (12.76) shows that the loop gain $T' \to \infty$ as well. The closed-loop gain is the inverse of the feedback factor for this amplifier

$$R_{mf} = \frac{R_m{}'}{1 + R_m{}'f} = \frac{1}{f} = -100\ \text{k}\Omega$$

The input resistance with feedback is found to be from Eq. (12.79)

$$R_{in,f} = \frac{R_F}{1 + T'} \to 0\ \Omega$$

The output resistance is given by Eq. (12.80), with $-R_{md} \to \infty$

$$R_{out,f} = \frac{(R_{in} + R_F)\,R_{out}}{R_{in} + R_F + R_{out} - R_{md}} \to 0\ \Omega$$

These ideal results are valid only at low frequencies. The input impedance of this amplifier decreases with increasing frequency, as we have seen in Chapter 10, which reduces the loop gain. As a result, the closed-loop input and output impedances increase with increasing frequency.

12.4.3 Techniques for Feedback Amplifier Analysis and Design

The voltage and transresistance feedback amplifier examples have illustrated that resistor feedback networks share the same basic properties with their idealized counterparts. The gain is stabilized with sufficient loop gain and the input and output resistances are modified by the factor $1 + T'$ in the direction desired for an ideal amplifier. The feedback factor f can be found by isolating the feedback network, after identifying the amplifier type. Finally, we have seen that feedback provides overall linearity and solves the problem of setting the DC operating point.

In subsequent analog IC design courses, a procedure is developed for modeling feedback amplifiers by combining the two-port models of the feedback network and amplifying block. An approximate general technique has been developed for finding the loaded loop gain T' without analyzing the closed-loop amplifier directly [ref. 1]. This technique provides the foundation for the efficient analysis and design of feedback amplifiers. However, this is a subject beyond the scope of this text.

From this section, we conclude that the primary requirement for the amplifying block in feedback amplifier applications is very high gain (a_{vd} or R_{md} for the examples here). High gain compensates for the effects of loading from the resistor feedback network and ensures a high loop gain T'. Furthermore, it allows use of smaller feedback factors that correspond to higher closed-loop gains, while still maintaining adequate loop gain.

12.5 INTEGRATED OPERATIONAL AMPLIFIERS

For feedback applications, a differential amplifier with high voltage gain, high input resistance, and low output resistance is extremely useful. These specifications describe an **operational amplifier** (op amp)—a very important IC building block. The generic name for this amplifier class is operational amplifier or op amp. Using the principles of integrated amplifier design from Chapters 8–11, we are well-prepared to tackle the design of this key IC building block.

Figure 12.13 shows the four functional blocks we will consider in designing an op amp. First, there is an input differential gain stage that amplifies the voltage difference between the input terminals, independently of their average or common-mode voltage. Since subsequent stages have inputs that are referenced to ground, a second block converts the differential signal into a single-ended signal. As we will see, the DC level of this signal must be shifted in order to bias the second gain stage properly, a function which is performed in the *level-shift* block. Finally, additional gain is obtained in the second gain stage. This block diagram does not include an output stage, which can be added to provide a low output resistance and the ability to supply and sink large currents. For internal applications in VLSI systems, op amp loads are typically small capacitors and the output stage is unnecessary. Output-stage design is a subject that will be left to a subsequent course.

12.5.1 Two-Stage CMOS Op Amp Topology

We now consider how to implement these functional blocks in CMOS. The input differential amplifier in Fig. 12.13 can be implemented by a differential pair of MOS transistors. The input resistance is infinite and the bias currents into the noninverting (+) and inverting (–) terminals are zero, which are highly desirable features for an op amp. We will use PMOS transistors for the input differential pair, a choice that will be justified after analyzing the op amp's small-signal performance. By using a cur-

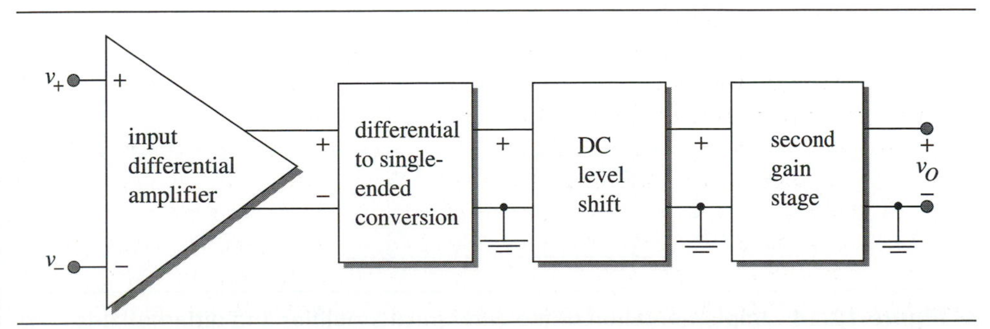

Figure 12.13 Block diagram for an integrated operational amplifier with the output stage omitted.

rent-mirror supply on the differential pair, the first-stage output voltage is referenced to ground without loss of a factor of two in the transconductance. Figure 12.14 shows the input stage. Not counting biasing transistors, we have implemented the first two functional blocks in Fig. 12.13 using only four transistors.

We consider first the DC level-shifting function. Assuming that transistor pairs (M_1, M_2) and (M_3, M_4) are perfectly matched, the drain currents are $-I_{SS}/2$ for M_1 and M_2 and $+I_{SS}/2$ for M_3 and M_4. The drain current of diode-connected M_3 is

$$I_{D3} = \frac{1}{2}\mu_n C_{ox}\left(\frac{W}{L}\right)_3 (V_{GS3} - V_{Tn})^2 (1 + \lambda_n V_{GS3}) = \frac{I_{SS}}{2} \tag{12.82}$$

in which we have substituted $V_{DS3} = V_{GS3}$. The drain current of M_4 is

$$I_{D4} = \frac{1}{2}\mu_n C_{ox}\left(\frac{W}{L}\right)_3 (V_{GS3} - V_{Tn})^2 (1 + \lambda_n V_{DS4}) = \frac{I_{SS}}{2} \tag{12.83}$$

where we substituted $V_{GS4} = V_{GS3}$ from Fig. 12.14 and $(W/L)_4 = (W/L)_3$. Equating Eq. (12.82) and (12.83), we find

$$V_{DS4} = V_{GS3} \tag{12.84}$$

For the case of perfect matching, the current-mirror transistors have the same drain-source voltages, as well as identical gate-source voltages. We can solve for the first-stage output voltage V_{O1} in Fig. 12.14 for $V_{I1} = V_{I2} = 0$ V

$$V_{O1} = V^- + V_{GS3} \approx V^- + V_{Tn} + \sqrt{\frac{I_{SS}}{\mu_n C_{ox}(W/L)_3}} \tag{12.85}$$

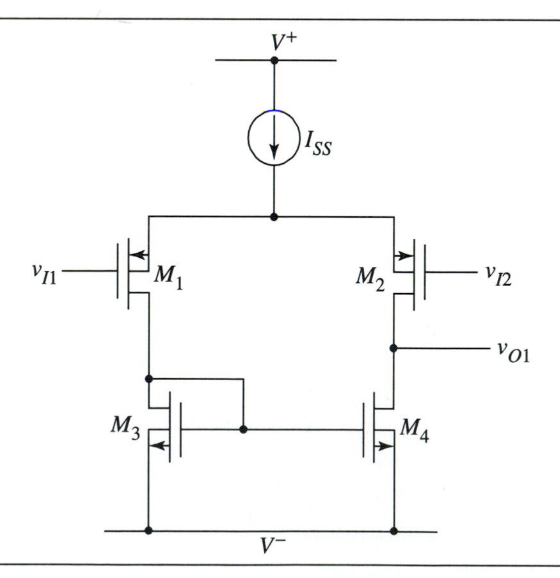

Figure 12.14 Implementation of the differential amplifier and differential-to-single-ended conversion stages using a p-channel differential pair with n-channel current-mirror.

In the right side of Eq. (12.85), we substituted for V_{GS3} from Eq. (12.82) and neglected the channel-length modulation term. Since the first-stage output is one gate-source drop above the negative supply voltage, the next stage must shift the DC level toward the positive supply voltage.

The level-shift function can either be accomplished with a p-channel source follower or by using an n-channel MOSFET connected as a common-source amplifier. The latter choice accomplishes two functions: level-shifting and added gain in a single stage. A p-channel current supply is used to maximize the second-stage gain.

Our basic two-stage CMOS operational amplifier topology is shown in Fig. 12.15. A reference current supply has been added for I_{SS}, consisting of I_{REF} and current mirror M_7 and M_8. The current supply for the second stage is mirrored from I_{REF} by transistor M_6. The load for this op amp is an on-chip capacitor C_L. The capacitor C_c is called the compensation capacitor and is added between the gate and drain of transistor M_5 to ensure that the feedback amplifier is stable with adequate phase margin.

We will analyze the DC bias, small-signal performance, and frequency response of the CMOS two-stage op amp. From the results of this analysis and a consideration of the design constraints, we will specify in a design example the transistor dimensions and bias currents and evaluate the op amp's overall performance.

12.5.2 DC Analysis of the CMOS Two-Stage Op Amp

To find the DC bias point of the op amp, we assume it is placed in a unity-gain feedback configuration, as shown in Fig. 12.16. Within the small offset voltage, the DC levels of the two inputs V_{I1} and V_{I2} and the output V_O are all 0 V. There are only four nodes labeled in Fig. 12.15 for which the DC voltage needs to be calculated.

In addition to the DC bias point, two important DC specifications for an op amp are its **input common-mode voltage range** and **output voltage range**. The former is the range of common-mode DC input voltage for which the op amp maintains its high differential gain and is critical for many op amp applications. The maximum and min-

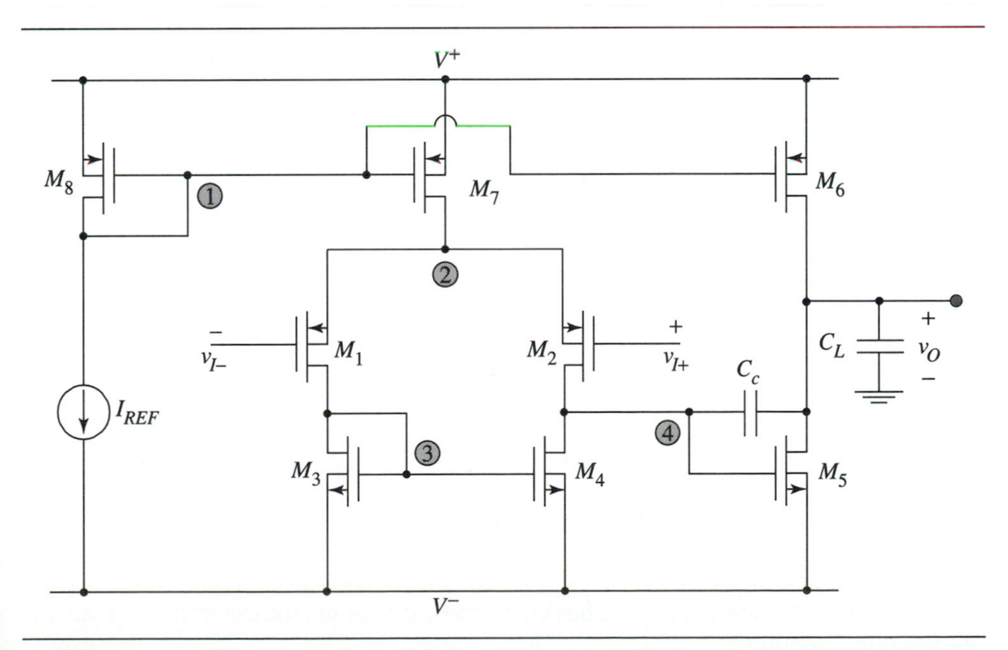

Figure 12.15 Two-stage CMOS operational amplifier topology.

imum output voltages establish the output voltage range for the op amp. We will use judicious approximations in analyzing the DC characteristics of the op amp, such as neglecting the channel-length modulation term in the MOSFET drain current. Our goal is to find simple results that will be useful for developing design guidelines. Circuit simulators such as SPICE are essential for refining the hand calculations of the bias voltages and currents to specify the final design.

To begin the analysis, we assume that all MOSFETs are operating in their constant-current region (saturation). The first step is to calculate the drain currents in the current-supply transistors M_6 and M_7. From Fig. 12.15, the source-gate voltages for transistors M_6, M_7, and M_8 are identical and the drain currents are therefore proportional to their width-to-length ratios:

$$|I_{D8}|:|I_{D7}|:|I_{D6}| = \left(\frac{W}{L}\right)_8:\left(\frac{W}{L}\right)_7:\left(\frac{W}{L}\right)_6 \tag{12.86}$$

Since the drain current in the diode-connected transistor M_8 is identical to the reference current, I_{REF}, we can solve for the bias current I_{SS} of the source-coupled differential pair M_1 and M_2 and the bias current $-I_{D6}$ of the second gain stage

$$I_{SS} = -I_{D7} = I_{REF}\left[\frac{(W/L)_7}{(W/L)_8}\right] \quad \text{and} \tag{12.87}$$

$$-I_{D6} = I_{REF}\left[\frac{(W/L)_6}{(W/L)_8}\right] \tag{12.88}$$

Assuming M_1 and M_2 are identical, the bias currents in the other transistors are

$$-I_{D1} = -I_{D2} = \frac{I_{SS}}{2} \qquad I_{D3} = I_{D4} = \frac{I_{SS}}{2} \qquad I_{D5} = -I_{D6} \tag{12.89}$$

The last equality for I_{D5} follows from the fact that DC load current $I_L = 0$ A in the unity-gain feedback configuration shown in Fig. 12.16.

Node 1: The source-gate voltage of the p-channel current supplies M_6 and M_7 is

$$V_{SG6} = V_{SG7} = V_{SG8} = -V_{Tp} + \sqrt{\frac{2I_{REF}}{\mu_p C_{ox}(W/L)_8}} \tag{12.90}$$

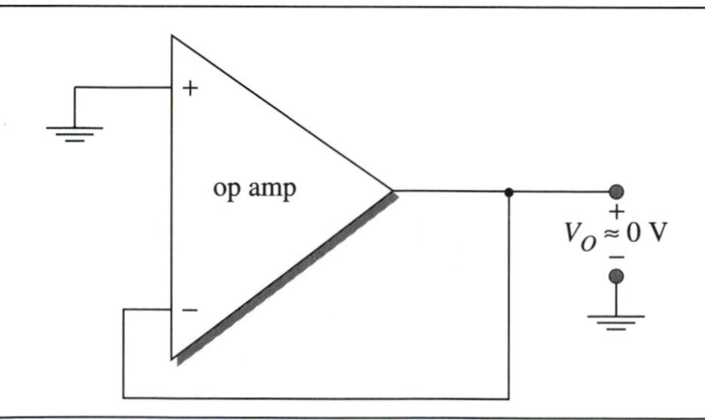

➤ Figure 12.16 Unity-gain feedback configuration for establishing the op amp's DC operating point.

Therefore, the voltage at Node 1 in Fig. 12.15 is

$$V_1 = V^+ + V_{Tp} - \sqrt{\frac{2I_{REF}}{\mu_p C_{ox}(W/L)_8}} \tag{12.91}$$

Node 2: The unity-gain feedback configuration in Fig. 12.16 sets the gate voltages of the input pair M_1 and M_2 equal to zero. The source of these devices is Node 2

$$V_2 = V_{S1,2} = V_{G1} + V_{SG1,2} = 0 - V_{Tp} + \sqrt{\frac{2(-I_{D7}/2)}{\mu_p C_{ox}(W/L)_{1,2}}} \tag{12.92}$$

Node 3: The drain voltage of transistor M_1 is set by the gate-source voltage of diode-connected transistor M_3. Therefore, the voltage at Node 3 is

$$V_3 = V^- + V_{Tn} + \sqrt{\frac{2(-I_{D7}/2)}{\mu_n C_{ox}(W/L)_{3,4}}} \tag{12.93}$$

Node 4: The drain of transistor M_2 is Node 4; its voltage is set by the gate-source voltage of M_5

$$V_4 = V^- + V_{Tn} + \sqrt{\frac{2(-I_{D6})}{\mu_n C_{ox}(W/L)_5}} \tag{12.94}$$

One constraint on the selection of transistor dimensions is to ensure that the drain voltages of M_1 and M_2 be identical, in order to avoid creating a systematic input offset voltage. Therefore, we will impose the constraint that

$$V_3 = V_{D1} = V_{D2} = V_4 \tag{12.95}$$

Equating Eqs. (12.93) and (12.94), Eq. (12.95) requires

$$\frac{-I_{D7}/2}{(W/L)_{3,4}} = \frac{-I_{D6}}{(W/L)_5} \tag{12.96}$$

According to Eq. (12.86), the drain currents in M_6 and M_7 are also related by the ratio of their width-to-length ratios

$$\frac{-I_{D6}}{-I_{D7}} = \frac{(W/L)_6}{(W/L)_7} = \frac{(W/L)_5}{2(W/L)_{3,4}} \tag{12.97}$$

in which the second equality follows from Eq. (12.96). We will delay verifying that all transistors are operating in their constant-current regions until the transistor dimensions and reference current are specified.

The common-mode input range is an important specification for op amp design. For zero differential input voltage (inverting and non-inverting inputs shorted together), it is defined as the highest DC common-mode input voltage $V_{IC,\,max}$ minus the lowest DC common-mode input voltage $V_{IC,min}$ such that *all* devices remain in their constant-current regions.

As shown in Fig. 12.17, as the common-mode input voltage is increased, the source voltage of the two input MOSFETs $V_{S1,2} = V_{D7}$ follows since their source-gate voltages $V_{SG1,2}$ are set by their drain currents (see Eq. (12.92)). Note that we have shorted the bulk to the source for transistors M_1 and M_2, which eliminated the back-gate effect on the threshold voltage. We have implicitly assumed that the op amp is fabricated in an n-well CMOS process, so that these two MOSFETs can be placed in their own wells.

At some point, the current-supply transistor M_7 will enter the triode region as its source-drain voltage drops with increasing common-mode input voltage. From Fig. 12.17, we can write V_{SD7} as

$$V_{SD7} = V^+ - V_{D7} = V^+ - (V_{IC} + V_{SG1,2}) \tag{12.98}$$

The maximum common-mode input voltage is found by evaluating this equation when M_7 is at the edge of the saturation region

$$V_{SD_{SAT,7}} = V_{SG7} + V_{Tp} = V^+ - (V_{IC,max} + V_{SG1,2}) \tag{12.99}$$

from which we find

$$V_{IC,max} = V^+ - V_{Tp} - V_{SG1,2} - V_{SG7} \tag{12.100}$$

According to Eq. (12.100), it is desirable to decrease the source-gate voltages of both the current-source transistor M_7 and the input transistors M_1 and M_2 to maximize $V_{IC,max}$. Increasing the width-to-length ratio is one avenue for accomplishing this goal. In the limit, the two source-gate voltages can be made only slightly larger than the magnitude of the p-channel threshold voltage, which results in a maximum common-mode input voltage approaching $V^+ + V_{T_p}$ for this topology.

The lowest common-mode input voltage is set by the requirement to maintain input devices M_1 and M_2 in saturation. Figure 12.18 is the circuit used to find $V_{IC,min}$.

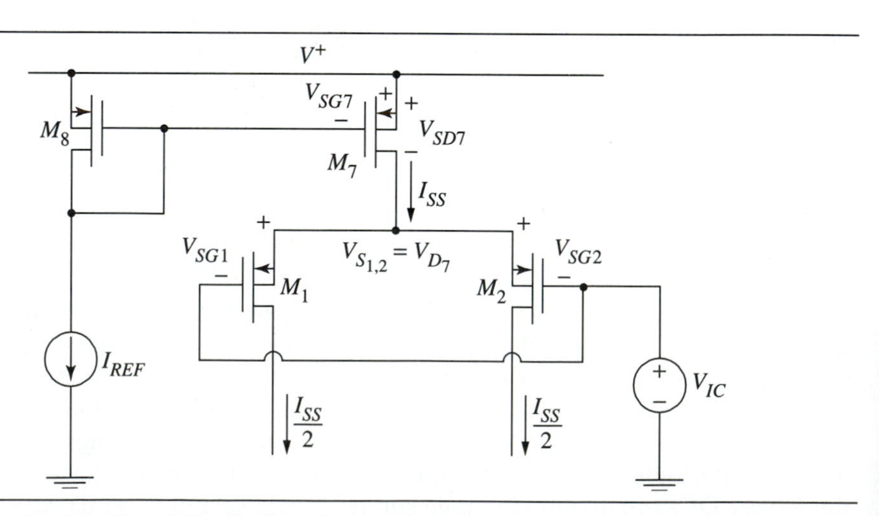

➤ Figure 12.17 Circuit for finding the maximum common-mode input voltage.

The DC drain voltages of the input devices are equal by design. Substituting for the gate-source voltage of M_3, we find

$$V_{D1} = V^- + V_{GS3} = V^- + V_{Tn} + \sqrt{\frac{2(I_{SS}/2)}{\mu_n C_{ox}(W/L)_{3,4}}} \tag{12.101}$$

Since the source voltage of M_1 and M_2 follows the common-mode input voltage, the input transistors may enter the triode region. Applying the test for saturation to M_1

$$V_{SD1} = V_{IC} + V_{SG1} - V_{D1} \geq V_{SG1} + V_{Tp} \tag{12.102}$$

Evaluating this equation when M_1 is at the edge of the triode region, we find

$$V_{IC,min} = V^- + V_{GS3} + V_{Tp} = V^- + V_{Tp} + V_{Tn} + \sqrt{\frac{2(I_{SS}/2)}{\mu_n C_{ox}(W/L)_{3,4}}} \tag{12.103}$$

As we found for $V_{IC,max}$, Eq. (12.103) indicates that increasing the transistor width-to-length ratios (for M_3 and M_4 in this case) will improve (lower) $V_{IC,min}$. In the limit, the minimum common-mode input voltage approaches V^- if the n- and p-channel thresholds are equal and opposite.

For the two-stage CMOS op amp topology in Fig. 12.15, transistors M_6 and M_5 are the obvious ones to limit the output voltage range. The p-channel current source M_6 will remain saturated, provided that

$$V_{SD6} = V^+ - V_O \geq V_{SG6} + V_{Tp} \tag{12.104}$$

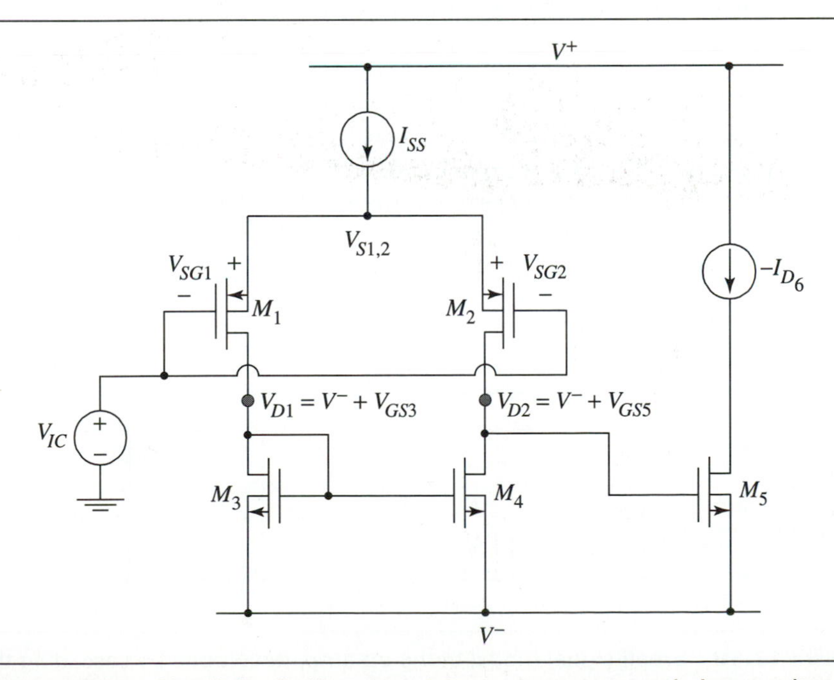

Figure 12.18 Circuit for finding the minimum common-mode input voltage.

Using the equality and solving for the maximum output voltage, we find

$$V_{O,max} = V^+ - V_{Tp} - V_{SG6} = V^+ - \sqrt{\frac{2(-I_{D6})}{\mu_p C_{ox}(W/L)_6}} \tag{12.105}$$

For output voltages that approach the negative supply, M_5 may enter the triode region. The test for saturation for transistor M_5 is

$$V_{DS5} = V_O - V^- \geq V_{DS_{SAT,5}} = V_{GS5} - V_{Tn} \tag{12.106}$$

We can solve for the minimum output voltage by applying the equality in Eq. (12.106)

$$V_{O,min} = V^- + \sqrt{\frac{2(-I_{D6})}{\mu_n C_{ox}(W/L)_5}} \tag{12.107}$$

12.5.3 Small-Signal Analysis of Two-Stage CMOS Op Amp

Figure 12.19 shows the cascade of the two-port models for the input and second gain stages. From Chapter 11, the transconductance and output resistance of the current-mirror differential amplifier are

$$G_{m1} = g_{m1} \text{ and } R_{out1} = r_{o2} \| r_{o4} \tag{12.108}$$

The transconductance and output resistance of the common-source amplifier used as the second gain stage were found in Chapter 8

$$G_{m2} = g_{m5} \text{ and } R_{out2} = r_{o5} \| r_{o6} = R_{out} \tag{12.109}$$

Note that the input terminals in Fig. 12.15 have been labeled so the op amp's differential gain is positive. The gate of transistor M_2 is the non-inverting terminal (v_{I+}) and the gate of transistor M_1 is the inverting terminal (v_{I-}). The labels are opposite the usual convention for differential pairs, which is reflected in a sign change in the first stage gain. The open-circuit differential gain at low frequencies a_{vdo} is given by

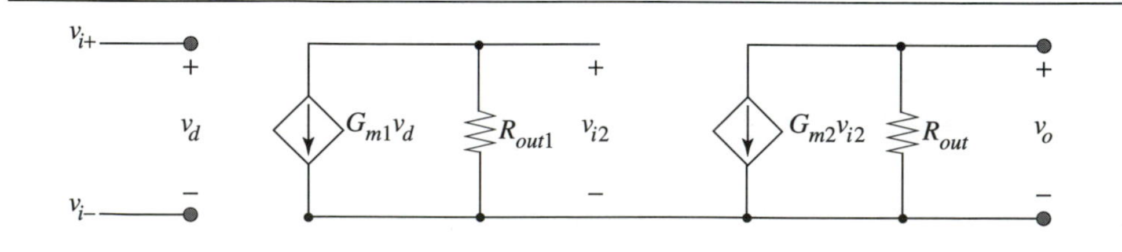

Figure 12.19 Small-signal model of the op amp, developed by cascading the individual two-port models for the input and second stages.

$$a_{vdo} = \frac{v_o}{v_{i+} - v_{i-}} = (-G_{m1}R_{out1})(-G_{m2}R_{out2}) = g_{m1}(r_{o2} \| r_{o4})\, g_{m5}(r_{o5} \| r_{o6}) \quad (12.110)$$

The second gain stage's open-circuit input resistance is advantageous since the full gain of the first stage is preserved without reduction due to loading.

12.5.4 Frequency Response and Compensation of the Two-Stage CMOS Op Amp

We now turn to the frequency response of the two-stage op amp. The function of the compensation capacitor C_c is to introduce a low-frequency dominant pole ω_1 in order to ensure stability. The compensation capacitor is Miller-multiplied by the second-stage voltage gain and appears in parallel with the large output resistance of the first stage. Less obvious, but verified by SPICE, is that the load capacitance C_L at the output node sets the location of the second pole ω_2.

In order to find analytical expressions for the two poles, we find the transfer function of the second gain stage. Since this analysis applies to all two-stage op amp topologies, general symbols are used for the resistances and capacitances. The capacitance C_c' is the total capacitance across the second gain stage, which is equal to $C_c + C_{gd5}$ for the CMOS two-stage topology. The capacitance C_1 is the sum of the output capacitance of the first stage and the input capacitance of the second stage, and C_L' is the total capacitance (load and parasitic) from the output node to ground. The resistor R_1 is the parallel combination of R_{out1} with R_{in2}, the input resistance to the second stage (infinity for the CMOS two-stage op amp.) The resistor R_{out} is the second-stage output resistance, R_{out2}.

The circuit in Fig. 12.20 differs from the common-emitter voltage amplifier in Fig. 10.13 only by the addition of the capacitance C_L'. The analysis is simplified by introducing the impedances

$$Z_1 = R_1 \| \left(\frac{1}{j\omega C_1}\right),\ Z_{C_c}' = \frac{1}{j\omega C_c'},\ \text{and } Z_L' = R_{out} \| \left(\frac{1}{j\omega C_L'}\right) \quad (12.111)$$

Kirchhoff's current law at the input and output nodes states

$$I_s = Y_1 \cdot V_{i2} + Y_{C_c}' \cdot (V_{i2} - V_o) \text{ and} \quad (12.112)$$

$$0 = G_{m2} \cdot V_{i2} + Y_{C_c}' \cdot (V_o - V_{i2}) + Y_L' \cdot V_o \quad (12.113)$$

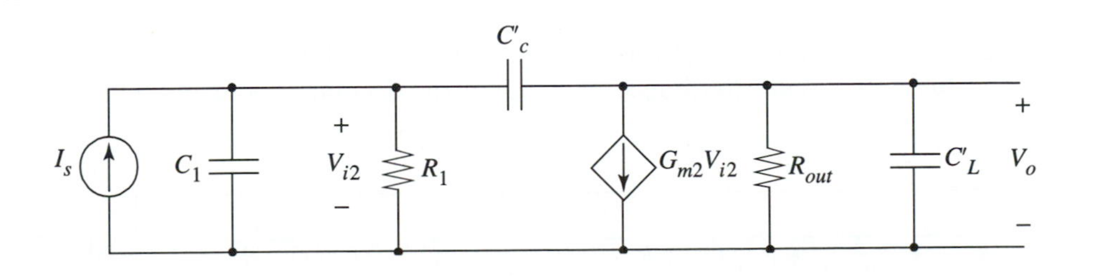

➤ Figure 12.20 Small-signal model of the second gain stage for finding the first two poles of the op amp transfer function. Circuit variables are phasors.

where the admittances Y_1, Y_{C_c}', and Y_L' are the inverses of the impedances that are defined in Eq. (12.111). These two equations can be solved to find the transimpedance $Z_m = V_o/I_s$ of the second gain stage

$$Z_m = \frac{Y_{C_c}' - G_{m2}}{Y_1 Y_L' + Y_{C_c}' Y_L' + Y_1 Y_{C_c}' + G_{m2} Y_{C_c}'} \quad (12.114)$$

Substitution of the admittances that are found by inverting the impedances in Eq. (12.111) and simplifying, we get

$$Z_m = \frac{-G_{m2} R_1 R_{out}(1 - j\omega(C_c'/G_{m2}))}{1 + j\omega A - \omega^2 B} \quad (12.115)$$

where the constants A and B are

$$A = R_1 C_1 + R_{out} C_L' + R_1 C_c' + R_{out} C_c' + G_{m2} R_{out} R_1 C_c' \text{ and} \quad (12.116)$$

$$B = R_1 C_1 R_{out} C_L' + R_1 C_c' R_{out} C_L' + R_1 C_1 R_{out} C_c' \quad (12.117)$$

The denominator polynomial in Eq. (12.115) can be factored as the product of two poles ω_1 and ω_2

$$(1 + j\omega/\omega_1)(1 + j\omega/\omega_2) = 1 + j\omega\left(\frac{1}{\omega_1} + \frac{1}{\omega_2}\right) - \omega^2\left(\frac{1}{\omega_1\omega_2}\right) \quad (12.118)$$

Recall from the analysis of the voltage amplifier in Chapter 10, that the two poles are widely separated, with $\omega_1 = (1/\tau_1) \ll \omega_2 = (1/\tau_2)$. As a result, Eq. (12.118) can be simplified

$$(1 + j\omega/\omega_1)(1 + j\omega/\omega_2) \approx 1 + j\omega\left(\frac{1}{\omega_1}\right) - \omega^2\left(\frac{1}{\omega_1\omega_2}\right) \quad (12.119)$$

since $1/\omega_2 \ll 1/\omega_1$. Using Eq. (12.119), we can find the first pole from the coefficient A in Eqs. (12.116)

$$A = R_1 C_1 + R_{out} C_L' + R_1 C_c' + R_{out} C_c' + G_{m2} R_{out} R_1 C_c' \approx \frac{1}{\omega_1} \quad (12.120)$$

This result is identical to Eq. (10.40) from Chapter 10, with the addition of the term $R_{out} C_L'$ due to the load capacitance. Solving for the dominant pole and rearranging terms

dominant pole of two-stage op amp

$$\boxed{\omega_1 \approx \frac{1}{R_1 C_1 + R_{out} C_c' + R_{out} C_L' + R_1(1 + G_{m2} R_{out}) C_c'} \approx \frac{1}{R_1 (G_{m2} R_{out}) C_c'}} \quad (12.121)$$

The final approximation in Eq. (12.121) takes into account that the second-stage gain $G_{m2} R_{out} \gg 1$. Note that this result is equivalent to estimating the dominant pole from the Miller capacitor $C_M = (1 + G_{m2} R_{out}) C_c'$ at the input to the second stage, which appears in parallel to the resistor R_1.

An expression for the second pole can be found from Eq. (12.119), which indicates that the constant B is the inverse of the product of the two pole frequencies

$$B = \frac{1}{\omega_1 \omega_2} \quad \rightarrow \quad \omega_2 = \frac{1}{\omega_1 B} \tag{12.122}$$

Substituting for B from Eq. (12.117) and approximating ω_1 by Eq. (12.121), we find that

$$\omega_2 \approx \frac{R_1 (G_{m2} R_{out}) C_c'}{R_1 C_1 R_{out} C_L' + R_1 C_c' R_{out} C_L' + R_1 C_1 R_{out} C_c'} \tag{12.123}$$

Cancelling the common terms and factoring out $G_{m2}/(C_L')$, we find

$$\boxed{\omega_2 \approx \left(\frac{G_{m2}}{C_L'}\right)\left(\frac{C_L' C_c'}{C_L' C_c' + C_L' C_1 + C_c' C_1}\right)} \tag{12.124}$$

second pole of two-stage op amp

This result is adequate for hand calculations. For cases where the input capacitance $C_1 \ll C_c', C_L'$, the second pole location can be approximated by

$$\omega_2 \approx \frac{G_{m2}}{C_L'} = \frac{1}{(1/G_{m_2}) C_L'} \text{ for } C_1 \ll C_c', C_L' \tag{12.125}$$

An intuitive interpretation of Eq. (12.125) is that at frequencies near the second pole, the impedance of the compensation capacitor becomes low enough that the gain element in the second stage (transistor M_5) becomes diode-connected. The output impedance of the op amp at the second pole is approximately $1/G_{m2}$ which leads directly to Eq. (12.125).

The first and second poles in $a_{vd}(j\omega)$ are known. The influence of higher poles and zeroes on the phase of $a_{vd}(j\omega)$ will be neglected for hand calculations. We would like this op amp to be stable for unity-gain configurations, with a phase margin of 45°. In Ex 12.4, we found that for a two-pole amplifier, the poles must be related by

$$\omega_2 = \frac{a_{vdo}\omega_1}{\sqrt{2}} \approx a_{vdo}\omega_1 \tag{12.126}$$

in order to meet this requirement. For hand calculations, we will neglect the $\sqrt{2}$ in order to simplify the Bode plots. The process of adjusting the operational amplifier's pole locations to ensure stability is known as **compensation**. It is typical to adjust the compensation capacitor C_c in order to achieve this goal. The load capacitor C_L is usually specified by other constraints and cannot be adjusted.

By substituting into Eq. (12.126) the expressions for the low-frequency gain a_{vdo} from Eq. (12.110) and the two poles from Eq. (12.121) and (12.124), we can find an equation for the compensation capacitor C_c', in terms of the small-signal parameters and the load capacitor:

$$\left(\frac{G_{m2}}{C_L'}\right)\left(\frac{C_L' C_c'}{C_L' C_c' + C_L' C_1 + C_c' C_1}\right) \approx \frac{G_{m_1} R_1 G_{m_2} R_{out}}{R_1 G_{m_2} R_{out} C_c'} = \frac{G_{m1}}{C_c'} \tag{12.127}$$

Applying the quadratic formula to solve the resulting equation for C_c', we find that

$$C_c' = \left(\frac{G_{m1}}{G_{m2}}\right)\left(\frac{C_L'}{2}\right)\left(1 + \frac{C_1}{C_L'}\right)\left[1 + \sqrt{1 + \left(\frac{G_{m2}}{G_{m1}}\right)\left(\frac{4C_1/C_L'}{(1 + C_1/C_L')^2}\right)}\right] \quad (12.128)$$

For the case where $C_1 \ll C_c', C_L'$, the compensation capacitor is directly proportional to the product of the load capacitor and the ratio of the first and second stage transconductances

compensation capacitor for unity gain stability

$$C_c' \approx \left(\frac{G_{m1}}{G_{m2}}\right) C_L' \text{ when } C_1 \ll C_c', C_L' \quad (12.129)$$

To minimize the compensation capacitor and reduce the area of the op amp, Eq. (12.129) indicates that we must maximize the second-stage transconductance. One reason for choosing a PMOS differential pair, followed by an NMOS common-source stage rather than the complementary configuration is the higher transconductance of an NMOS transistor for a given bias current due to its larger mobility.

12.5.5 Two-Stage CMOS Op Amp Design Example

Thus far, we have analyzed the CMOS two-stage op amp topology and developed design equations that describe its DC and small-signal performance, as well as the compensation of its frequency response. In order to specify the bias current and the transistor dimensions, we must place some constraints on the design.

First, the DC supplies are $V^+ = 2.5$ V, $V^- = -2.5$ V and the total DC power dissipation is limited to 1.25 mW for the op amp. As a result, the DC supply currents, together with the reference current, must add to no more than 1.25 mW/(2.5 V − (−2.5 V)) = 250 μA. The second constraint is on the area of the op amp. For reasons of both increased transconductance (improved bandwidth) and lower DC gate-source voltage drop (for higher common-mode input and output ranges), it is desirable to make the width-to-length ratios of the signal-path transistors as large as possible. In order to maintain a reasonably high output resistance for high gain, we will fix the minimum channel length at $L_{min} = 3$ μm. The maximum channel width is limited by area consumption. For this case, we set $W_{max} = 150$ μm. Therefore, the maximum $(W/L) = 150$ μm/3 μm = 50.

Figure 12.21 is a first-cut design for the CMOS two-stage op amp, using the standard MOSFET parameters from Section 4.6. For convenience, the essential device parameters needed for hand calculation and the Level 1 SPICE model are repeated in Fig. 12.21. Transistor M_5, which is critical to the frequency response, is biased at $I_{D5} = 100$ μA and has $(W/L)_5 = (W/L)_{max} = 50$. The input pair is biased at $I_{SS} = -I_{D7} = 100$ μA in order to have adequate first-stage gain. The reference current source consumes the remaining 50 μA allowed in the DC current budget. To avoid a systematic input offset voltage, Eq. (12.97) requires that transistors M_3 and M_4 be dimensioned according to

$$\frac{(W/L)_5}{2(W/L)_{3,4}} = \frac{-I_{D6}}{I_{D7}} = \frac{100\ \mu\text{A}}{100\ \mu\text{A}} = 1 \quad \rightarrow \quad \left(\frac{W}{L}\right)_{3,4} = \frac{1}{2}\left(\frac{W}{L}\right)_5 = 25 \quad (12.130)$$

The drain-bulk capacitance C_{db} for an n-channel MOSFET is given by

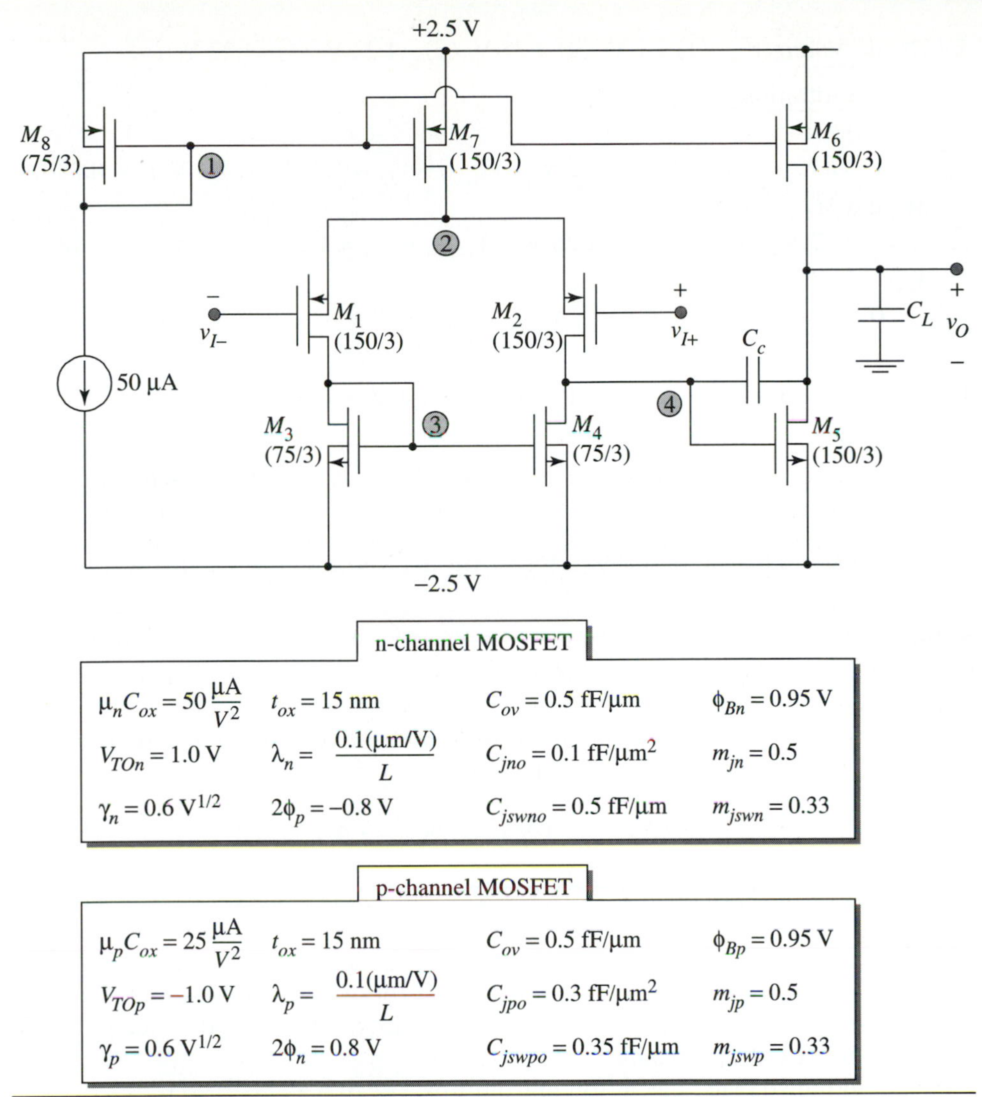

n-channel MOSFET

$\mu_n C_{ox} = 50 \frac{\mu A}{V^2}$	$t_{ox} = 15$ nm	$C_{ov} = 0.5$ fF/μm	$\phi_{Bn} = 0.95$ V
$V_{TOn} = 1.0$ V	$\lambda_n = \frac{0.1(\mu m/V)}{L}$	$C_{jno} = 0.1$ fF/μm^2	$m_{jn} = 0.5$
$\gamma_n = 0.6$ V$^{1/2}$	$2\phi_p = -0.8$ V	$C_{jswno} = 0.5$ fF/μm	$m_{jswn} = 0.33$

p-channel MOSFET

$\mu_p C_{ox} = 25 \frac{\mu A}{V^2}$	$t_{ox} = 15$ nm	$C_{ov} = 0.5$ fF/μm	$\phi_{Bp} = 0.95$ V
$V_{TOp} = -1.0$ V	$\lambda_p = \frac{0.1(\mu m/V)}{L}$	$C_{jpo} = 0.3$ fF/μm^2	$m_{jp} = 0.5$
$\gamma_p = 0.6$ V$^{1/2}$	$2\phi_n = 0.8$ V	$C_{jswpo} = 0.35$ fF/μm	$m_{jswp} = 0.33$

➤ **Figure 12.21** First-cut CMOS two-stage op amp design using MOSFET parameters from Section 4.6. The (W/L) ratios having units of μm/μm.

$$C_{db} = \frac{C_{jno} \cdot (WL_{diff})}{(1 + V_{DB}/\phi_{Bn})^{m_{jn}}} + \frac{C_{jswno} \cdot (W + 2L_{diff})}{(1 + V_{DB}/\phi_{Bn})^{m_{jswn}}} \tag{12.131}$$

in which C_{jno} is the zero-bias junction capacitance per unit area, WL_{diff} is the drain area, C_{jswno} is the zero-bias sidewall capacitance per unit length, and the perimeter of the drain sidewall is $W + 2L_{diff}$. The source/drain diffusion width is $L_{diff} = 6$ μm. The grading coefficients m_j and m_{jsw} for the n-channel and p-channel MOSFETs are given in Fig. 12.21.

Applying the results of Section 12.5.2, we can evaluate the DC node voltages and verify that all transistors are saturated. Figure 12.22 lists the node voltages and applies the test for saturation for the case where $V_O = 0$ V and $V_{I+} = V_{I-} = 0$V.

DC Node Voltages ① 1.1 V ② 1.28 V ③ –1.22 V ④ –1.22 V

Tests for Saturation

M_1 and M_2

$V_{SD1,2} = 1.28\text{ V} - (-1.22\text{ V}) = 2.5\text{ V} > V_{SG1,2} + V_{Tp} = 1.28\text{ V} + (-1\text{ V}) = 0.28$

M_3 and M_4

$V_{DS3,4} = -1.22\text{ V} - (-2.5\text{ V}) = 1.28\text{ V} > V_{GS3,4} - V_{Tn} = 1.28\text{ V} - (1\text{ V}) = 0.28$

M_5

$V_{DS5} = 0\text{ V} - (-2.5\text{ V}) = 2.5\text{ V} > V_{GS5} - V_{Tn} = 1.28\text{ V} - (1\text{ V}) = 0.28\text{ V}$

M_6

$V_{SD6} = 2.5\text{ V} - (0\text{ V}) = 2.5\text{ V} > V_{SG6} + V_{Tp} = 1.4\text{ V} + (-1\text{ V}) = 0.4\text{ V}$

M_7

$V_{SD7} = 2.5\text{ V} - (1.28\text{ V}) = 1.22\text{ V} > V_{SG7} + V_{Tp} = 1.4\text{ V} + (-1\text{ V}) = 0.4\text{ V}$

M_8

$V_{SD8} = 2.5\text{ V} - (1.1\text{ V}) = 1.4\text{ V} > V_{SG8} + V_{Tp} = 1.4\text{ V} + (-1\text{ V}) = 0.4\text{ V}$

➤ **Figure 12.22** DC node voltages and verification of transistor saturation for design in Fig. 12.21 with $V_{I+} = V_{I-} = 0$ V and $V_O = 0$ V.

Now that the DC operating point is established, the common-mode input voltage range can be found from Eqs. (12.100) and (12.103)

$$V_{IC,max} = 2.5\text{V} - (-1\text{ V}) - 1.28\text{ V} - 1.4\text{ V} = 0.82\text{ V} \tag{12.132}$$

$$V_{IC,min} = -2.5\text{V} + 1.28\text{V} + (-1\text{V}) = -2.22\text{V} \tag{12.133}$$

The first-cut op amp design has an asymmetrical common-mode input range. In order to increase $V_{IC,max}$, Eq. (12.100) shows that transistors M_1, M_2, and M_7 could be made wider in order to reduce their source-gate voltage drops. However, this change would increase the area of the op amp.

The output voltage range is found from Eqs. (12.105) and (12.107)

$$V_{O,\,max} = 2.5\text{V} - 0.4\text{ V} = 2.1\text{ V} \tag{12.134}$$

$$V_{O,\,min} = -2.5\text{V} + 0.28\text{V} = -2.22\text{V} \tag{12.135}$$

The small-signal parameters for the transistors in the signal path (M_1–M_6) can be evaluated from the bias currents, node voltages, and transistor parameters

$$g_{m1} = g_{m2} = \frac{2(-I_{D1})}{V_{SG1} + V_{Tp}} = \frac{2(50\ \mu\text{A})}{1.28\text{ V} + (-1\text{ V})} = 357\ \mu\text{S} \tag{12.136}$$

$$g_{m5} = \frac{2I_{D5}}{V_{GS5} - V_{Tn}} = \frac{2(100\ \mu A)}{1.28\ V - 1\ V} = 714\ \mu S \tag{12.137}$$

$$r_{o2} = \frac{1}{\lambda_p(-I_{D2})} = \frac{3\mu m}{(0.1)(V^{-1} \cdot \mu m)(50\ \mu A)} = 600\ k\Omega = r_{o4} \tag{12.138}$$

$$r_{o5} = \frac{1}{\lambda_n I_{D5}} = \frac{3\mu m}{(0.1)(V^{-1} \cdot \mu m)(100\ \mu A)} = 300\ k\Omega = r_{o6} \tag{12.139}$$

The small-signal differential gain at low frequencies is found from Eq. (12.110)

$$a_{vdo} = (357\ \mu S)(600||600\ k\Omega)(714\ \mu S)(300||300\ k\Omega) = 1.15 \times 10^4 \tag{12.140}$$

In decibels, $|a_{vdo}|_{dB} = 81$ dB. Note that the first and second stage gains are the same for this design.

The final design variable is the value of the compensation capacitor needed to ensure unity-gain stability with a phase margin of 45°. We will use a typical load capacitor $C_L = 7.5$ pF for this design. To evaluate Eq. (12.128) to find C_c, we must first find the device capacitances that contribute to C_1 and C_L'. The capacitance between the input of the second stage and ground is

$$C_1 = C_{gs5} + C_{gd4} + C_{db4} + C_{gd2} + C_{db2} \tag{12.141}$$

in which the gates of M_4 and M_2 are considered to be small-signal grounds. Substituting the device parameters from Fig. 12.21 and evaluating the capacitances at the operating point specified in Fig. 12.22, we find

$$C_1 = 765\ fF + 37.5\ fF + 63\ fF + 75\ fF + 179\ fF = 1.12\ pF \tag{12.142}$$

The capacitance between the output node and ground is

$$C_L' = C_L + C_{db5} + C_{db6} + C_{gd6} \tag{12.143}$$

which results in

$$C_L' = 7.5\ pF + 99\ fF + 179\ fF + 75\ fF = 7.85\ pF \tag{12.144}$$

Evaluating Eq. (12.128), we find that the capacitance across the second gain stage is

$$C_c' = 5.32\ pF = C_c + C_{gd5} = C_c + 75\ fF \tag{12.145}$$

Therefore, the compensation capacitor $C_c = 5.25$ pF. If it is implemented using a gate polysilicon-to-polysilicon structure with an oxide thickness of 500 Å, then an area of about $90 \times 90\ \mu m^2$ is required.

Finally, we find the locations of the two poles for this CMOS op amp design. Evaluating Eq. (12.121), the dominant pole is located at

$$\omega_1 = \frac{1}{(600||600\ \text{k}\Omega)\ (1 + 714\mu\text{S}\ (0.3||0.3\ \text{M}\Omega))\ (5.32\ \text{pF})} = 5.8\ \text{krad/s} \qquad (12.146)$$

From Eq. (12.124), the second pole is located

$$\omega_2 = \left(\frac{714\mu\text{S}}{7.85\ \text{pF}}\right)\left(\frac{7.85 \cdot 5.32}{7.85 \cdot 5.32 + 7.85 \cdot 1.12 + 1.12 \cdot 5.32}\right) = 67.2\ \text{Mrad/s} \qquad (12.147)$$

Although there is room for improvement in its DC specifications, the principle deficiency of this op amp is its inadequate bandwidth. In order to increase ω_2, the second-stage transconductance must be increased, as can be seen from Eq. (12.124). By increasing the size or bias current of transistor M_5, we could make some improvement in the bandwidth. However, the MOSFET transconductance depends on the square root of $(W/L)_5$ and I_{D5}, making it difficult to make a significant increase in ω_2. In a BiCMOS technology, bipolar transistors with their large transconductances can be put to good use to improve the basic two-stage op amp's bandwidth.

➤ EXAMPLE 12.9 SPICE Simulation of CMOS Two-Stage Op Amp

In this example, SPICE is used to simulate the DC and small-signal behavior of the two-stage CMOS op amp. Use the standard Level 1 MOSFET models in Section 4.6.

(a) Find the DC operating point and compare with the node voltages in Fig. 12.22.

SOLUTION

We ground the inverting terminal and sweep the non-inverting terminal to find the input offset voltage needed to force $V_O = 0$ V. Surprisingly, this op amp requires essentially no offset voltage to place it in its high gain region with the output voltage near to 0 V. With $V_{I+} = 0$ V, the DC output voltage $V_O = 0.028$ V. Note that SPICE and hand calculations are extremely close for this simple op amp design, as listed in Fig. Ex12.9A.

DC Node Voltages	①	②	③	④	V_O
SPICE	1.105 V	1.276 V	–1.221 V	–1.22 1V	0.028 V
Hand calculation	1.1 V	1.28 V	–1.22 V	–1.22 V	0 V

➤ Figure Ex12.9A DC operating point of the CMOS op according to SPICE and hand calculation. Nodes are referenced in Fig. 12.21.

(b) Find the common-mode input voltage range.

SOLUTION

In using SPICE, it is convenient to find the low-frequency small-signal gain as a function of the DC common-mode input voltage V_{IC}. From the plot in Fig. Ex12.9B, the gain is high for DC common-mode inputs in the range $-2.2\text{ V} < V_{IC} < 0.9\text{ V}$. This result is close to that found by hand calculation of the range over which all MOSFETs remain saturated.

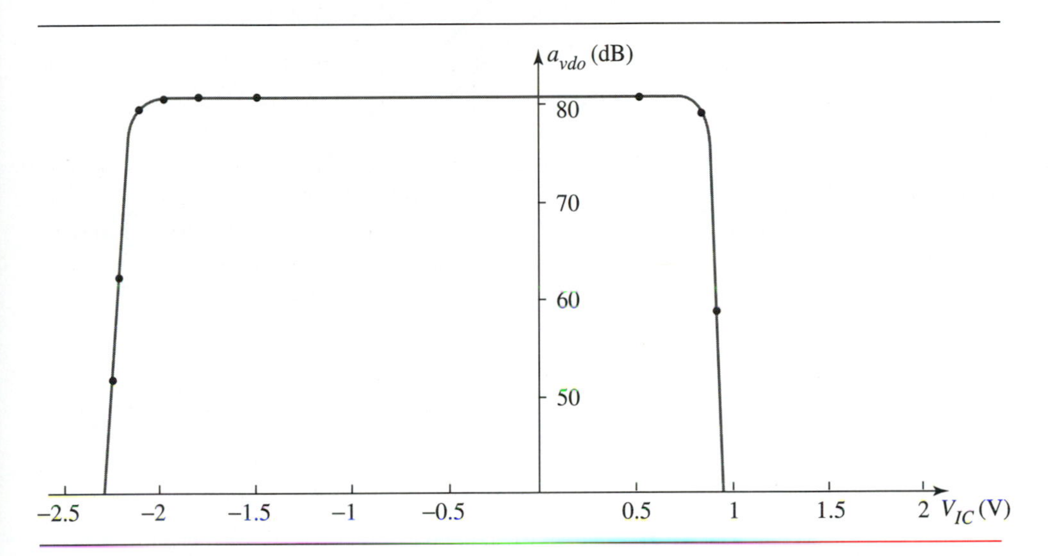

➤ Figure Ex12.9B SPICE simulation of low frequency small-signal gain as a function of DC common-mode input voltage.

(c) Use SPICE to find the frequency response of the op amp and the phase margin for unity-gain feedback applications.

SOLUTION

The Bode plot for $a_{vd}(j\omega)$ is shown in Fig. Ex12.9C. The first pole is at $f_1 = 780$ Hz, which is close to the hand calculation (920 Hz from Eq. (12.146)).
The second pole is at $f_2 = 5.7$ MHz, more than a factor of two lower than the analytical model prediction of 10.7 MHz from Eq. (12.147). The discrepancy indicates that higher poles are contributing excess phase. The frequency at which the gain is zero dB is 8.75 MHz, according to the Bode plot. The phase at this frequency is −153°, which corresponds to a phase margin of only 27°.
To increase the phase margin to 45°, SPICE simulations show that the compensation capacitor must be increased to $C_c \approx 20$ pF. In addition to the area penalty for this large capacitor, the gain-bandwidth product of the op amp is reduced significantly. The first pole is $f_1 = 200$ Hz and the second pole is $f_2 = 3.3$ MHz with $C_c \approx 20$ pF .as shown in Fig. Ex12.9C.

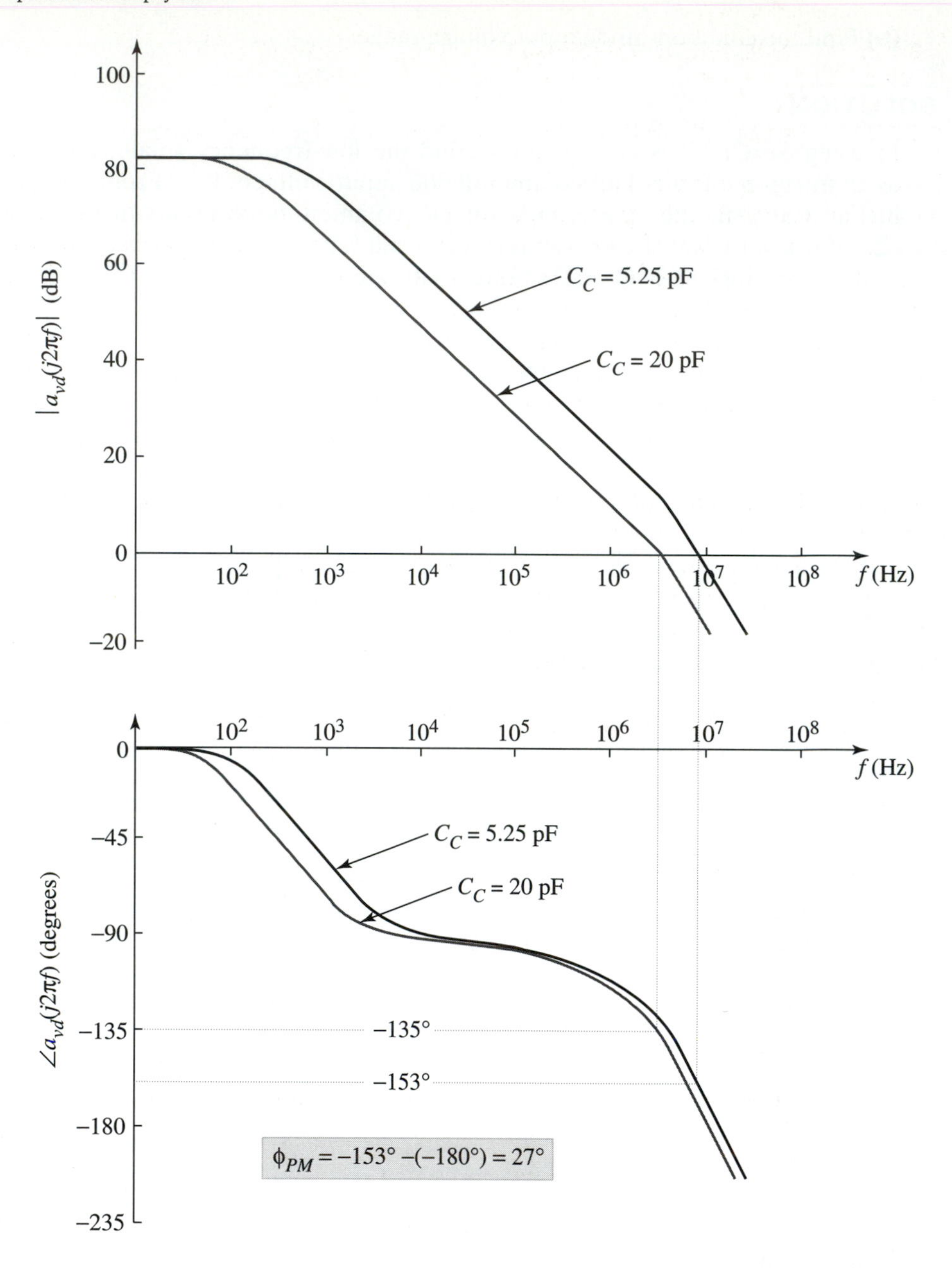

➤ **Figure Ex12.9C** Bode plot for the CMOS two-stage op amp from SPICE. The compensation capacitor is $C_c = 5.25$ pF from hand calculations, which results in a phase margin of only 27°. SPICE simulations show that a compensation capacitor of $C_C = 20$ pF is needed for a 45° phase margin, which reduces the op amp's bandwidth as shown.

12.6 BiCMOS OPERATIONAL AMPLIFIERS

12.6.1 Basic BiCMOS Two-Stage Op Amp Topology

If we simply replace the n-channel MOSFET M_5 with an npn bipolar transistor Q_5 in the second gain stage, the transconductance of the second gain stage can be much higher, for the same bias current. However, several problems are introduced. Substitution of Q_5 for M_5 in the op amp affects the symmetry of the current mirror in the input stage. The voltage $V_{BE5} \cong 700$ mV sets the drain-source voltage of transistor M_4. This voltage should equal the gate-source voltage V_{GS3} of the diode-connected mirror M_3 in order to maintain matched drain currents in the differential pair. Although we could select (W/L) ratios to satisfy this condition, the two voltages are determined by different parameters that will not track with temperature or process variations. Therefore, it is far better to replace the n-channel MOS current mirror with an npn bipolar current mirror, in which the matching and temperature tracking come "for free."

These changes lead to the BiCMOS operational amplifier shown in Fig. 12.23, which has improved bandwidth over the all-MOS version. However, this basic topology suffers from inadequate gain. Figure 12.24 is the small-signal model for this op amp. The two-port parameters for the common-emitter stage were found in Chapter 8

$$R_{in2} = r_{\pi 5},\ G_{m2} = g_{m2},\ \text{and}\ R_{out2} = r_{o5} \| r_{o6} \tag{12.148}$$

The low-frequency differential gain is

$$a_{vdo} = [-G_{m1}(R_{out1} \| R_{in2})](-G_{m2}R_{out2}) \tag{12.149}$$

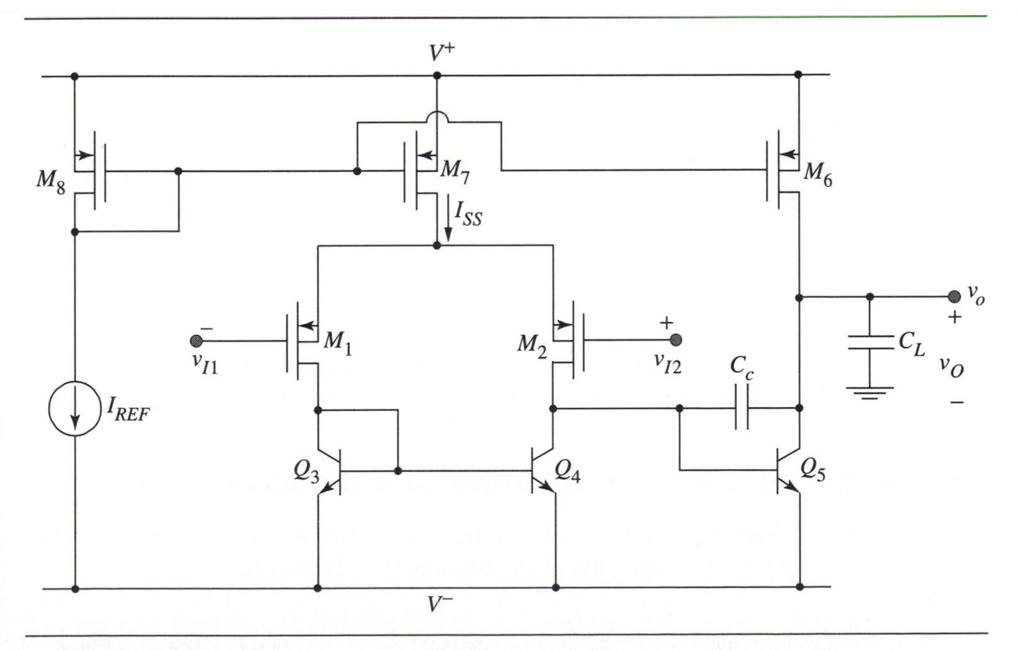

Figure 12.23 Basic two-stage BiCMOS operational amplifier topology.

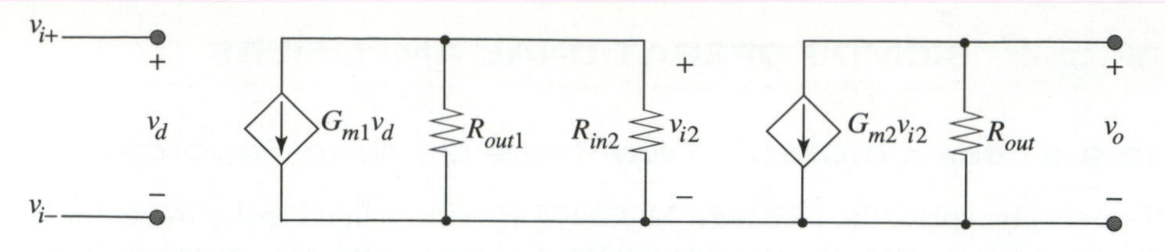

➤ Figure 12.24 Small-signal model of the basic BiCMOS op amp developed by cascading the individual two-port models for the input and second stages.

Substituting the small-signal two-port elements, we find

$$a_{vdo} = g_{m1}(r_{o2} \| r_{o4} \| r_{\pi 5}) g_{m5}(r_{o5} \| r_{o6}) \tag{12.150}$$

From Fig. 12.24 and Eq. (12.150), we see that the output resistance of the first stage appears in parallel with the input resistance of the second stage. Since the common-emitter amplifier used as the second gain stage has a relatively low input resistance, it significantly reduces the low-frequency gain of the first stage.

12.6.2 Improved BiCMOS Op Amp Topology

Although the bandwidth of the basic BiCMOS two-stage op amp is much improved, its low-frequency gain is inadequate. The small-signal analysis indicates that the loss of first-stage gain is caused by the low input resistance of the common-emitter second gain stage. The insertion of a common-collector amplifier is a natural approach to boosting the input resistance of the second gain stage.

Although there are several circuit configurations that will accomplish this goal, the one shown in Fig. 12.25 has the advantage of maintaining a well-balanced differential input stage. The insertion into the signal path of common-collector transistor Q_{11}, with current supplied by M_{12}, increases the input resistance of the second gain stage by a factor of β_o. To supply the bias current to M_{12}, we add a p-channel MOSFET current mirror M_{14}, which supplies current to a current mirror consisting of M_{12} and M_{13}. These devices provide a current supply that sinks current to the negative power supply from the emitter of Q_{11}.

The DC level at the base of Q_{11} is about 1.4 V above the negative voltage supply, which would cause the input stage to have a large systematic offset. Therefore, the op amp in Fig. 12.25 uses second pair of bipolar transistors Q_9 and Q_{10} to implement a bipolar cascoded current mirror supply. For this topology, the drain of M_1 is at $V^- + 2V_{BE}$ which matches the DC level of the drain of M_2, which is set by the two base-emitter drops through Q_{11} and Q_5. As an additional benefit, the output resistance of the input stage is increased somewhat. The compensation capacitor is connected from the output node to the input node of the second gain stage as shown in Fig. 12.25.

12.6.3 DC Analysis of the Improved BiCMOS Op Amp

We will only consider the differences from the CMOS two-stage op amp. Neglecting base currents, the collector currents of the bipolar transistors are

$$I_{C3} = I_{C4} = I_{C9} = I_{C10} = \frac{I_{SS}}{2} = -\frac{I_{D7}}{2} \qquad I_{C11} = I_{D12} \qquad I_{C5} = -I_{D6} \tag{12.151}$$

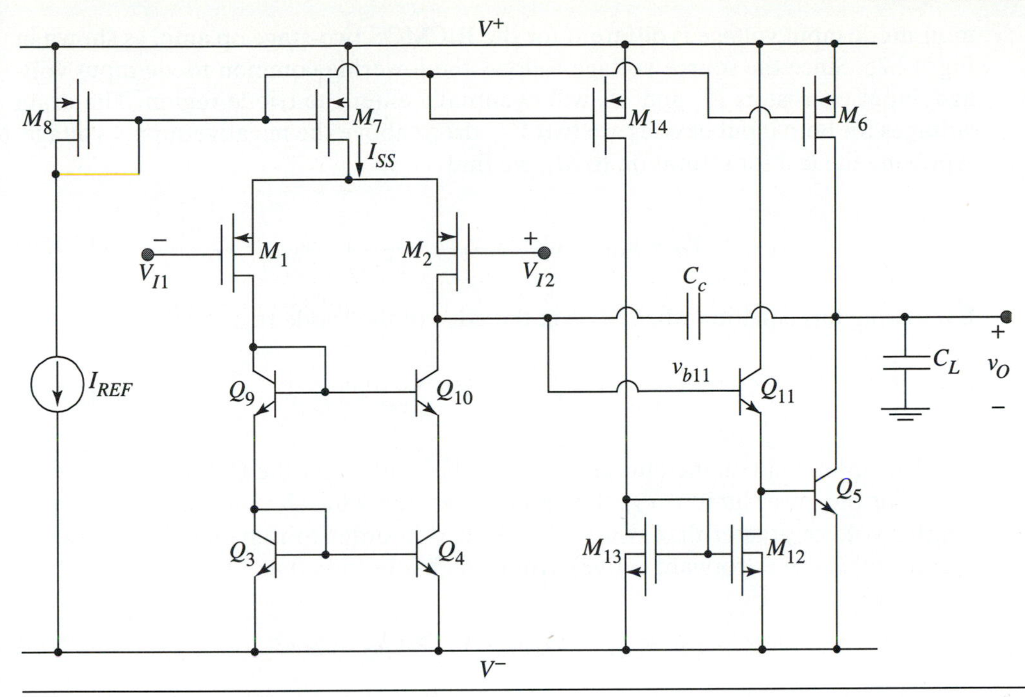

➤ **Figure 12.25** BiCMOS op amp topology with improved low-frequency differential gain.

The drain current in transistor M_{14} is mirrored from the reference current by the ratio

$$-I_{D14} = I_{REF}\frac{(W/L)_{14}}{(W/L)_8} \tag{12.152}$$

Finally, the n-channel MOSFET current mirror formed by M_{12} and M_{13} has drain currents that are determined by the ratio of the transistor dimensions

$$I_{D12} = I_{D13}\frac{(W/L)_{12}}{(W/L)_{13}} = (-I_{D14})\frac{(W/L)_{12}}{(W/L)_{13}} = I_{REF}\frac{(W/L)_{14}}{(W/L)_8}\frac{(W/L)_{12}}{(W/L)_{13}} \tag{12.153}$$

in which we have substituted for $-I_{D14}$ from Eq. (12.152)

The collector of Q_9 is two base-emitter voltage drops above the negative supply

$$V_{C9} = V_{D1} = V^- + V_{BE9} + V_{BE3} \approx V^- + 1.4\text{ V} \tag{12.154}$$

To calculate the base-emitter voltages more accurately, we recall that

$$V_{BE3} = \left(\frac{kT}{q}\right)\ln\left(\frac{I_{C3}}{I_S}\right) = (60\text{ mV})\log\left(\frac{-I_{D7}}{2I_S}\right) \tag{12.155}$$

where I_S is the bipolar transistor saturation current. The bipolar transistors will be forward-active, as long as their collector-emitter voltages are greater than $V_{CE_{SAT}} \approx 100$ mV.

The maximum common-mode input voltage is the same as that of the CMOS op amp and is given by Eq. (12.100). However, the circuit for finding the minimum common-mode input voltage is different for the BiCMOS two-stage op amp, as shown in Fig. 12.26. Since the source voltage follows the lowering common-mode input voltage, input transistors M_1 and M_2 will eventually enter the triode region. The drain voltages for both input devices are two V_{BE} drops above the negative supply voltage. Applying the test for saturation to M_1, we find

$$V_{SD1} = V_{IC} + V_{SG1} - (V^- + V_{BE3} + V_{BE9}) \geq V_{SG1} + V_{Tp} \tag{12.156}$$

Evaluating this equation when M_1 is at the edge of the triode region, we find

$$V_{IC,\,min} = V^- + V_{BE3} + V_{BE9} + V_{Tp} \tag{12.157}$$

The upper limit on the output voltage is the same as for the CMOS two-stage op amp. For the lower limit to V_O, the npn bipolar transistor Q_5 must have a collector-emitter voltage greater than $V_{CE_{SAT}}$ = 100 mV in order to remain in the constant-current regime (i.e., forward active), from which it follows that

$$V_{CE5} = V_{O,\,min} - V^- \geq V_{CE_{SAT}} \tag{12.158}$$

Therefore, the lowest output voltage for constant-current operation is only about 0.1 V above the negative supply.

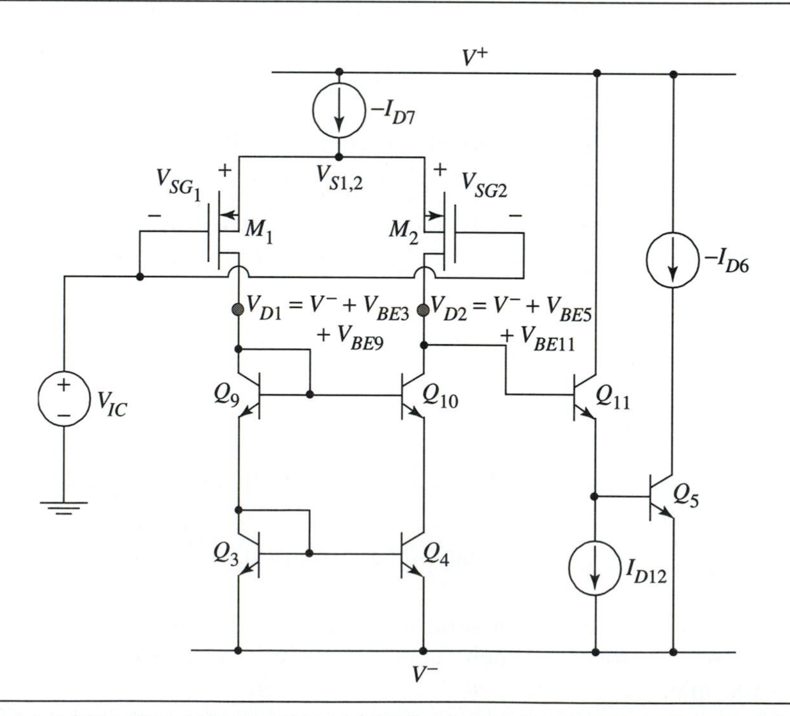

➤ **Figure 12.26** Circuit for finding the minimum common-mode input voltage of the improved two-stage BiCMOS op amp.

12.6.4 Small-Signal Analysis of Improved BiCMOS Op Amp Topology

We now consider the small-signal differences between this op amp topology and the basic two-stage BiCMOS op amp. Transistor Q_{11}, together with its current supply M_{12}, constitute a common-collector amplifier that is driven by the high resistance output of the first stage and loaded by the relatively low resistance seen looking into the base of Q_5.These conditions could lead to significant deviations from the two-port model results. As a check, we will take the trouble to find the voltage gain $A_{v,cc}$ between the controlled source of the first stage and the base of transistor Q_5 by applying the exact analysis from Chapter 8.

Figure 12.27 shows the small-signal model in which we use the Thévenin form of the first stage's output. The voltage gain is

$$A_{v,cc} = \frac{v_{b5}}{A_{dm1}v_d} = \frac{1}{1+\left(\dfrac{r_{\pi 11}+R_{S11}}{(\beta_o+1)R_{L11}}\right)} = \frac{1}{1+\left(\dfrac{r_{\pi 11}+r_{o2}\|\beta_o r_{o10}}{(\beta_o+1)(r_{o11}\|r_{o12}\|r_{\pi 5})}\right)} \tag{12.159}$$

The voltage gain $A_{dm1} = -G_{m1}R_{out1} = -g_{m1}(r_{o2}\|\beta_o r_{o10})$ is the unloaded differential voltage gain of the first stage. The voltage gain $A_{v,ce}$ across the common-emitter stage Q_5 and M_6 is

$$A_{v,ce} = \frac{v_o}{v_{b5}} = -g_{m5}R_{out2} = -g_{m5}(r_{o5}\|r_{o6}) \tag{12.160}$$

Therefore, the overall low-frequency gain of the op amp is

$$a_{vdo} = A_{dm1}A_{v,cc}A_{v,ce} = [-g_{m1}(r_{o2}\|\beta_o r_{o10})]A_{v,cc}[-g_{m5}(r_{o5}\|r_{o6})]$$

$$a_{vdo} = \frac{g_{m1}(r_{o2}\|\beta_o r_{o10})\,g_{m5}(r_{o5}\|r_{o6})}{1+\left(\dfrac{r_{\pi 11}+r_{o2}\|\beta_o r_{o10}}{(\beta_o+1)(r_{o11}\|r_{o12}\|r_{\pi 5})}\right)} \tag{12.161}$$

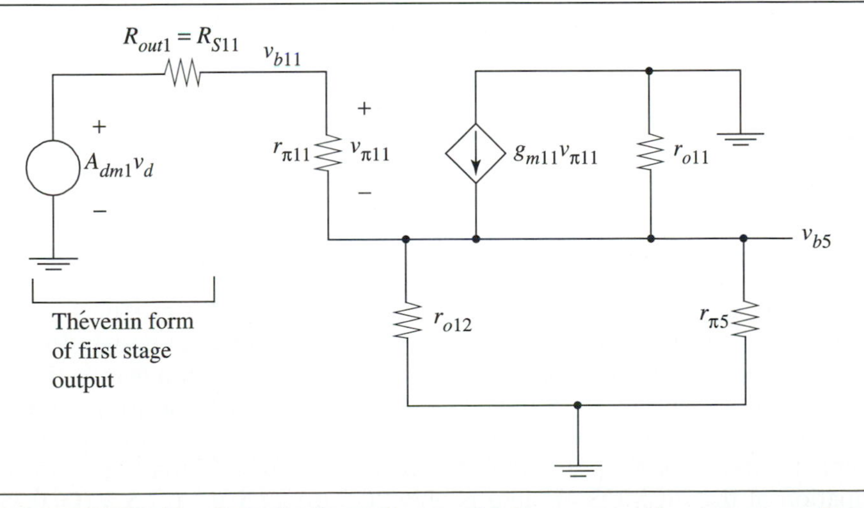

➤ **Figure 12.27** Circuit for finding voltage gain $A_{v,cc}$ across the common-collector stage Q_{11} and M_{12}.

12.6.5 Frequency Response of Improved BiCMOS Op Amp Topology

The analysis in Section 12.5.4 of the pole locations associated with the second gain stage also applies to the improved BiCMOS op amp. The capacitance across the second stage is $C_c' = C_c$ because the base-collector capacitance of transistor Q_{11} is shunted to ground. The Miller capacitor is determined by the gain across the second stage

$$C_M = C_c(1 + G_{m2}R_{out2}) \approx C_c G_{m2} R_{out2} \tag{12.162}$$

The dominant pole is located at

$$\omega_1 = \frac{1}{(R_{out1} \| R_{in2}) C_M} \tag{12.163}$$

where R_{in2} is the input resistance of the composite CC/CE second stage

$$R_{in2} = r_{\pi 11} + (\beta_o + 1)(r_{o11} \| r_{o12} \| r_{\pi 5}) \tag{12.164}$$

The second-stage transconductance G_{m2} is only slightly degraded through the common-collector stage

$$G_{m2} = \left(\frac{v_{b5}}{v_{b11}}\right) g_{m5} \approx g_{m5} \tag{12.165}$$

Substituting into Eq. (12.163) we find that the dominant pole is approximately

$$\omega_1 \approx \frac{1}{(R_{out1} \| R_{in2}) G_{m2} R_{out2} C_c} \tag{12.166}$$

The second pole is a function of the ratio of the first-stage and second-stage transconductances, as well as the capacitances in Fig. 12.20. Equation (12.124) is repeated here for convenience

$$\omega_2 \approx \left(\frac{G_{m2}}{C_L'}\right)\left(\frac{C_L' C_c'}{C_L' C_c' + C_L' C_1 + C_c' C_1}\right) \tag{12.167}$$

12.6.6 Improved BiCMOS Op Amp Design Example

Having developed the design equations for the DC and small-signal performance of the BiCMOS op amp, we now apply them to develop a first-cut design. In order to see the effect of the bipolar transistors on the op amp gain and bandwidth, the dimensions and bias currents of MOSFETs M_1–M_8 are kept the same as for the CMOS op amp design. To provide adequate current to the common-collector stage, the drain currents for transistor M_{14}, M_{13}, and M_{12} are selected to be 25 μA. The static power dissipation of the BiCMOS op amp is only 2(25 μA)(2.5 V – (–2.5 V)) = 0.25 mW higher than for the two-stage CMOS op amp.

From Eq. (12.152), the width-to-length ratio of M_{14} is

$$\left(\frac{W}{L}\right)_{14} = \left(\frac{W}{L}\right)_{8}\left(\frac{25\ \mu\text{A}}{50\ \mu\text{A}}\right) = \left(\frac{75}{3}\right)\left(\frac{1}{2}\right) = 12.5 \qquad (12.168)$$

Keeping the channel length at $L = 3\ \mu\text{m}$, the width of M_{14} is 37.5 µm. Since we have chosen $I_{D13} = I_{D12} = 25\ \mu\text{A}$, $(W/L)_{13} = (W/L)_{12}$. A further constraint on the dimensions of M_{12} is that the transistor must be in its constant-current region with a drain-source voltage $V_{DS12} = V_{BE5}$, which is approximately 0.7 V. This condition will be satisfied with a margin of safety if

$$V_{DS12_{SAT}} = V_{GS12} - V_{Tn} = \sqrt{\frac{2I_{D12}}{\mu_n C_{ox}(W/L)_{12}}} = 0.5\ \text{V} \qquad (12.169)$$

Substituting $I_{D12} = 25\ \mu\text{A}$ in Eq. (12.169), the width-to-length ratio of M_{12} is

$$\left(\frac{W}{L}\right)_{12} = \frac{2I_{D12}}{\mu_n C_{ox}(V_{DS12_{SAT}})^2} = 4 \qquad (12.170)$$

Using $L = 3\ \mu\text{m}$ for M_{12} and M_{13} to ensure a large small-signal resistance of the current supply, Eq. (12.170) indicates that the transistor widths are $W_{12} = W_{13} = 12\ \mu\text{m}$.

Figure 12.28 shows the transistor dimensions for the initial BiCMOS op amp design. The npn bipolar transistors have the model parameters described in Section 7.8, which are presented in Fig. 12.28 for convenience. All of the junctions are assumed to be abrupt, as indicated by the grading coefficients $m_{je} = m_{jc} = m_{js} = 0.5$.

Now that the design is specified, we can verify that all of the MOSFETs and bipolar transistors are operating in their constant-current regions. We will only calculate the node voltages that are different from the two-stage CMOS op amp. The DC levels of Nodes 1 and 2 in Fig. 12.28 are unchanged.

Using Eq. (12.155) and $I_S = 10^{-17}$A from Fig. 12.28, we can calculate more accurate values for the base-emitter voltage drops of Q_3 and Q_9 at room temperature. Neglecting the base currents, we find

$$V_{BE3} = V_{BE9} = (60\ \text{mV})\log\left(\frac{(100/2)\ \mu\text{A}}{10^{-17}\ \text{A}}\right) = 0.76\ \text{V} \qquad (12.171)$$

The DC level of Node 3 is

$$V_3 = V^- + V_{BE3} + V_{BE9} = -2.5 + 0.76 + 0.76 = -0.98\ \text{V} \qquad (12.172)$$

The base-emitter voltage drops of Q_5 and Q_{11} are

$$V_{BE5} = (60\ \text{mV})\log\left(\frac{100\ \mu\text{A}}{10^{-17}\ \text{A}}\right) = 0.78\ \text{V and} \qquad (12.173)$$

$$V_{BE11} = (60\ \text{mV})\log\left(\frac{25\ \mu\text{A}}{10^{-17}\ \text{A}}\right) = 0.74\ \text{V} \qquad (12.174)$$

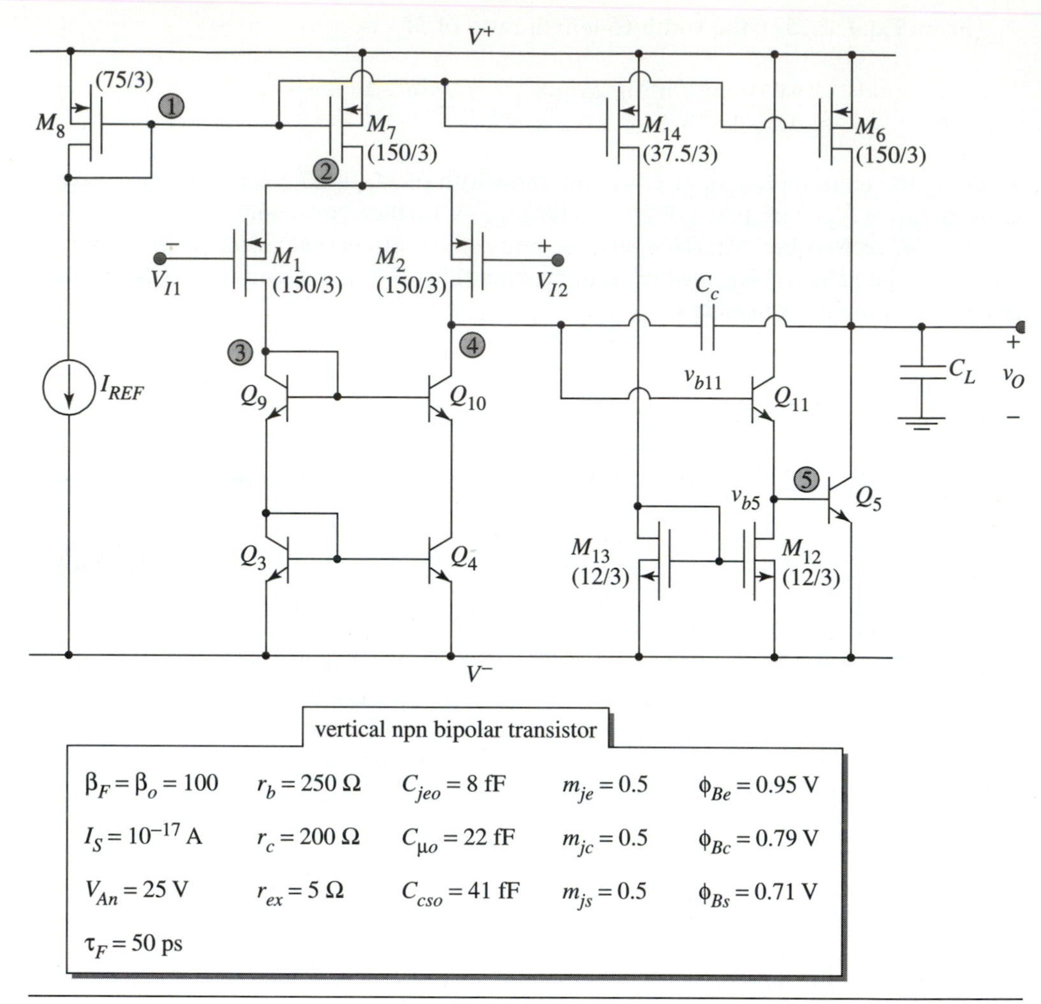

➤ **Figure 12.28** Improved BiCMOS op amp design. The (W/L) ratios have units of (μm/μm)

Node 4 and Node 3 have DC levels that are identical to two significant figures, which will minimize any systematic offset in the input differential pair.

$$V_4 = V^- + V_{BE5} + V_{BE11} = -2.5 + 0.78 + 0.74 = -0.98 \text{ V} \tag{12.175}$$

Finally, Node 5 is one base-emitter drop above the negative supply voltage

$$V_5 = V^- + V_{BE5} = -2.5 + 0.78 = -1.72 \text{ V} \tag{12.176}$$

It is not difficult to verify that all the bipolar transistors and MOSFETs are operating in their constant-current regions.

The maximum common-mode input voltage is unchanged at $V_{IC,max} = 0.82$ V. From Eq. (12.157), the cascoded current mirror supply results in the minimum common-mode input voltage

$$V_{IC,min} = V^- + V_{BE3} + V_{BE9} + V_{Tp} = -2.5 + 0.76 + 0.76 + (-1) = -1.98 \text{ V} \quad (12.177)$$

The maximum output voltage is unchanged from that of the CMOS op amp. According to Eq. (12.158), the minimum output voltage is lowered to

$$V_{O,min} = -2.5 \text{ V} + 0.1 \text{ V} = -2.4 \text{ V} \quad (12.178)$$

Having established the DC characteristics, we turn next to the small-signal performance. To gain insight into the improvement in small-signal gain, it is instructive to find the individual gains in Eq. (12.161) separately. The unloaded differential gain of the input stage is

$$A_{dm1} = -g_{m1}(r_{o2} || \beta_o r_{o10}) = -(357 \ \mu\text{S})(600 \text{ k}\Omega) || (100 \cdot 500 \text{k}\Omega) = -212 \quad (12.179)$$

In Eq. (12.179), we have substituted the output resistance of Q_{10}

$$r_{o10} = \frac{V_{An}}{I_{C10}} = \frac{25 \text{ V}}{50 \ \mu\text{A}} = 500 \text{ k}\Omega \quad (12.180)$$

The common-collector voltage gain from Eq. (12.159) is

$$A_{v,cc} = \frac{1}{1 + \left(\frac{r_{\pi 11} + r_{o2} || \beta_o r_{o10}}{(\beta_o + 1)(r_{o11} || r_{o12} || r_{\pi 5})} \right)}$$

$$= \frac{1}{1 + \frac{0.1 + 0.6 || 50}{101(1 || 1.2 || 0.025)}} = \frac{1}{1 + \frac{0.69}{2.41}} = 0.78 \quad (12.181)$$

where the resistances have units of MΩ. The gain from the base of Q_5 to the output is

$$\frac{v_o}{v_{b_5}} = A_{v,ce} = -g_{m5}(r_{o5} || r_{o6}) = -3.85 \text{ mS} (250 \text{ k}\Omega || 300 \text{ k}\Omega) = -525 \quad (12.182)$$

The low-frequency small-signal gain a_{vdo} is

$$a_{vdo} = (-212)(0.78)(-525) = 86,800 \text{ or } |a_{vdo}|_{dB} = 99 \text{ dB} \quad (12.183)$$

The common-collector amplifier preserves most of the unloaded gain of the input differential amplifier. An analysis based on a cascade of two-port models results in $|a_{vdo}|_{dB} = 94.5$ dB—close enough for hand calculations. In general, SPICE can be relied upon to find unusual situations where the two-port models lead to errors.

In order to find the compensation capacitor needed for unity-gain feedback stability with a 45° phase margin, we must find the capacitance at the input of the second stage C_1 and the capacitance at the output node C_L'. The input capacitance of the common-collector amplifier is

$$C_{in,11} = C_{\pi 11}(1 - v_{b5}/v_{b11}) + C_{\mu 11} \quad (12.184)$$

where $v_{b5}/v_{b11} \approx 1$ is the voltage gain across its base-emitter capacitance $C_{\pi 11}$. Evaluating the capacitances of Q_{11} at the operating point and substituting into Eq. (12.184), we find that

$$C_{in,11} = 67\,\text{fF}\,(1-1) + 9\,\text{fF} = 9\,\text{fF} \tag{12.185}$$

The total capacitance C_1 at the input of the second gain stage is

$$C_1 = C_{in,11} + C_{cs10} + C_{\mu 10} + C_{db2} + C_{gd2} \tag{12.186}$$

Substituting for $C_{in,11}$ and the device capacitances at the operating point, we find

$$C_1 = 9 + 23 + 22 + 185 + 75 = 314\ \text{fF} \tag{12.187}$$

The total capacitance at the output node is

$$C_L' = C_L + C_{gd6} + C_{db6} + C_{cs5} = 7.5\ \text{pF} + (75 + 179 + 19)\ \text{fF} = 7.77\,\text{pF} \tag{12.188}$$

In order to find C_c, we must first evaluate the second-stage transconductance.

$$G_{m2} \approx g_{m5} = 3.85\ \text{mS} \tag{12.189}$$

Substituting Eqs. (12.186), (12.188), and (12.189) into Eq. (12.128), we find that the compensation capacitor for unity-gain stability with 45° phase margin is

$$C_c' = C_c = 980\ \text{fF} \tag{12.190}$$

Using this value for the compensation capacitor in Eq. (12.166), the first pole is

$$\omega_1 = \frac{1}{(R_{out1}||R_{in2})\,G_{m2}R_{out}C_c} = \frac{1}{590\ \text{k}\Omega \cdot 3.85\ \text{mS} \cdot 136\ \text{k}\Omega \cdot 980\ \text{fF}}$$

$$= 3.3\ \text{krad/s} \tag{12.191}$$

The second pole is given by Eq. (12.167)

$$\omega_2 = \left(\frac{3.85\ \text{mS}}{7.77\ \text{pF}}\right)\left(\frac{7.61}{7.61 + 2.44 + 0.308}\right) = 364\ \text{Mrad/s} \tag{12.192}$$

According to the analytical results, the gain-bandwidth product of this op amp is over four times greater than the CMOS two-stage design. Both the low-frequency gain and the location of the second pole have been increased, at the cost of six additional transistors and the need for a BiCMOS technology. Considering the many approximations made in the hand calculations, it is important to simulate the design using SPICE.

➤ EXAMPLE 12.10 Spice Simulation of Improved BiCMOS Two-Stage Op Amp Design

In this example, SPICE is used to simulate the DC and small-signal properties of the improved two-stage BiCMOS op amp design in Fig. 12.28. Use the standard Level 1 MOSFET and the BJT models from Chapters 4 and 7.

(a) Plot the DC transfer function from the non-inverting input to the output, with the inverting input grounded. From this plot, determine the input offset voltage needed for $V_O = 0$ V.

SOLUTION

After some experimentation with range for the input offset voltage, the plot in Fig. Ex12.10 shows the DC transfer function of the op amp over the 2 mV range from 4.6 to 4.8 mV. Note that the output voltage range for high gain is from $V_{O,min} = -2.3$ V to $V_{O,max} = 2.15$ V, which is close to the calculated values.

(b) From the frequency response of this op amp, find the low-frequency gain, the location of the first two poles, and the phase margin for unity-gain feedback. Compare with the results of hand calculations.

SOLUTION

The low-frequency gain $a_{vdo} = 99.9$ dB, within 1 dB of the analytical result of 99 dB in Eq. (12.183). The first two poles are found by finding where the phase is $-45°$ and $-135°$

$f_1 = 560$ Hz compared with 525 Hz from Eq. (12.191)
$f_2 = 37$ MHz compared with 58 MHz from Eq. (12.192)

The analytical model overestimates the second pole location. As a result, the phase margin becomes only about 36°.

(c) According to SPICE, what value of C_c is needed to achieve unity-gain compensation with a phase margin of 45°? What are the pole locations for the final design?

SOLUTION

The compensation capacitor must be increased in order to improve the phase margin. After several simulations, we find that $C_c = 1.3$ pF is adequate for a 45° phase margin. The first two poles for this value of C_c are

$f_1 = 410$ Hz
$f_2 = 38.3$ MHz

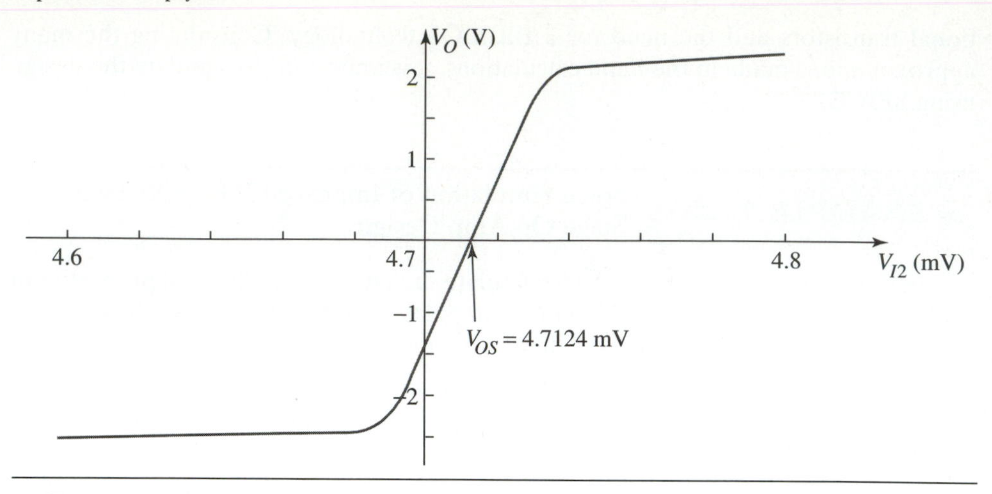

➤ **Figure Ex12.10** SPICE simulation of the DC transfer curve of the improved BiCMOS op amp design.

Note that the second pole increases slightly with the increase in C_c, an effect that is predicted qualitatively by Eq. (12.167). The gain-bandwidth product is more than an order of magnitude greater than that of the CMOS op amp, when both are properly compensated.

12.6.7 Two-Stage Op Amp Design Assessment

To assess the trade-offs between the two op amp designs, it is helpful to look at the Bode plots of the final designs that are verified to be unity-gain compensated by simulation. Fig. 12.29 shows the Bode plots of the differential gain and summarizes their other properties. Depending on the requirements of a particular application and the capabilities of the IC technology, either op amp might prove to be the best choice. The two-stage CMOS op lacks the bandwidth of the BiCMOS designs, but the latter require the high-performance vertical npn bipolar transistors available only in a BiCMOS technology. At the expense of area, the CMOS op amp's performance can be enhanced by widening transistor M_5.

➤ SUMMARY

This chapter introduced negative feedback as a powerful tool in electronic amplifier design. You should be familiar with the following basic results of feedback on the four amplifier topologies

- Closed-loop gain is stabilized as the inverse of the feedback factor for loop gains much greater than unity.
- Input resistance is multiplied by the factor $(1 + T)^{\pm 1}$ depending on whether the feedback network is in parallel or in series with the amplifier input port.

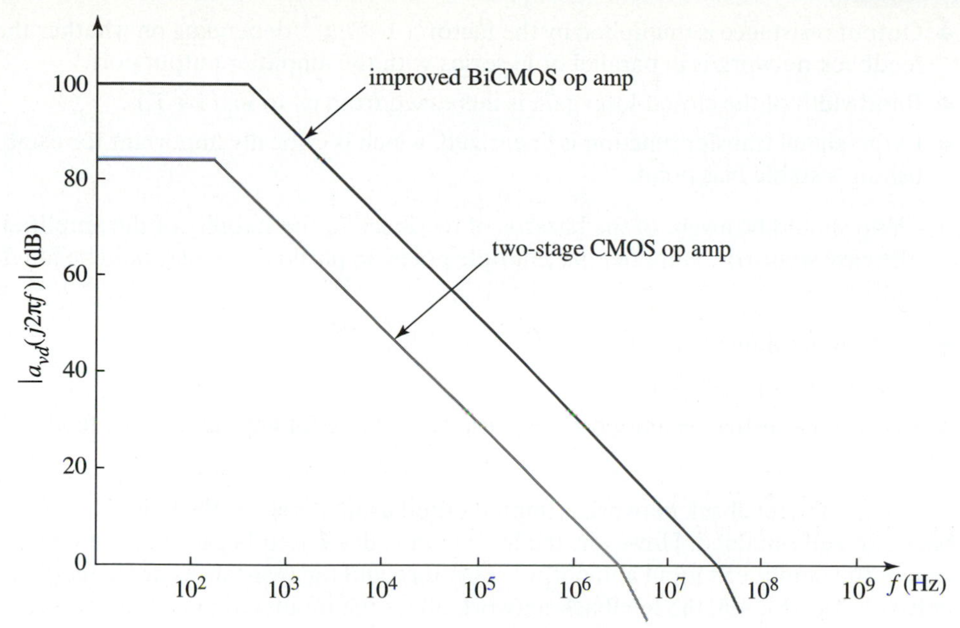

	CMOS	Improved BiCMOS
Transistor Count*	8	14
Static Power	1.25 mW	1.5 mW
Gain a_{vdo}	82.4 dB	99.9 dB
$V_{IC,max}$ $V_{IC,min}$	0.82 V −2.22 V	0.82 V −2.1 V
$V_{O,max}$ $V_{O,min}$	2.1 V −2.22 V	2.1 V −2.4 V
C_c	20 pF	1.3 pF
f_1	202 Hz	410 Hz
f_2	3.3 MHz	38.3 MHz

* not counting the reference current source.

➤ **Figure 12.29** Magnitude Bode plots of op amp gain for the two designs and a summary of performance specifications. The designs are for a load capacitor $C_L = 7.5$ pF and are unity-gain stable with a phase margin of about 45°.

- Output resistance is multiplied by the factor $(1+T)^{\pm 1}$ depending on whether the feedback network is in parallel or in series with the amplifier output port.
- Bandwidth of the closed-loop gain is increased from ω_1 to $\omega_1(1+T)$.
- Large-signal transfer function is linearized, which is critically important for establishing a stable bias point.

You should be aware of the hazards of feedback to the stability of the amplifier, for the case where the amplifier has multiple poles. In particular, you should be familiar with

- Test for amplifier stability
- Definition of phase margin
- Requirements that an amplifier be guaranteed stable for any value of the feedback factor

When the feedback network is implemented using resistors, the benefits of feedback are still obtained. However, the loop gain is decreased because of the interaction of the amplifier's input and output resistance and the resistances in the feedback network. In addition, the feedback network alters the input and output resistance of the feedback amplifiers.

Operational amplifiers (op amps) are an important amplifier class. You should be familiar with the following considerations in designing op amps.

- Basic functional blocks for op amps with the exception of output stages.
- Finding the DC common-mode input voltage range and the output voltage range.
- Finding the frequency response for the basic two-stage op amp topology.
- Compensating the op amp for unity-gain feedback applications.
- Evaluating the advantages and disadvantages of various designs, on the basis of gain-bandwidth product, low-frequency gain, complexity, DC characteristics, and static power consumption.

FURTHER READING

Feedback Amplifier Modelling and Design

1. P. R. Gray and R. G. Meyer, *Analysis and Design of Analog Integrated Circuits,* 3rd ed., Wiley, 1993. Chapter 8 and 9 are treatments of the general techniques for linear feedback amplifier design and compensation.
2. P. J. Hurst, "Exact simulation of feedback circuit parameters," *IEEE Transactions on Circuits and Systems*, **38**, 1382–1389 (1991). Clarifies the approximations involved in different approaches to the modeling and design of feedback amplifiers.

Operational Amplifier Design

1. P. R. Gray and R. G. Meyer, *Analysis and Design of Analog Integrated Circuits,* 3rd ed., Wiley, 1993. Chapter 6 discusses the design of bipolar and CMOS op amps at a more advanced level.
2. J. Dostál, *Operational Amplifiers*, 2nd ed., Butterworth-Heinemann, 1993. Provides a very thorough discussion of second-order effects in op amp applications, with some insights into the origins of nonidealities.

➤ PROBLEMS

EXERCISES

Unless otherwise noted, use the standard MOSFET and bipolar transistor models from Chapters 4 and 7.

E12.1 The amplifying block in Fig. E12.1 has a voltage gain $a_{vd} = 1200$, an input resistance $R_{in} = 250\ \text{k}\Omega$, and an output resistance $R_{out} = 20\ \text{k}\Omega$.

(a) What are the feedback factor f and the loop gain T for this feedback amplifier?

(b) What is the closed-loop gain A_{vf}?

(c) What is the input resistance $R_{in,f}$?

(d) What is the output resistance $R_{out,f}$?

E12.2 The amplifying block in Fig. E12.2 has a transresistance $R_{md} = -1\ \text{M}\Omega$, an input resistance $R_{in} = 50\ \text{k}\Omega$, and an output resistance $R_{out} = 2\ \text{M}\Omega$. The feedback factor is $f = -2 \times 10^{-4}$ S.

(a) What is the closed-loop transresistance R_{mf}?

(b) What is the input resistance $R_{in,f}$ for this feedback amplifier?

(c) What is the output resistance $R_{out,f}$ for this amplifier?

(d) The small-signal model for the amplifying block is valid until its input current i_d reaches 750 nA. What is the maximum input current i_i to the feedback amplifier?

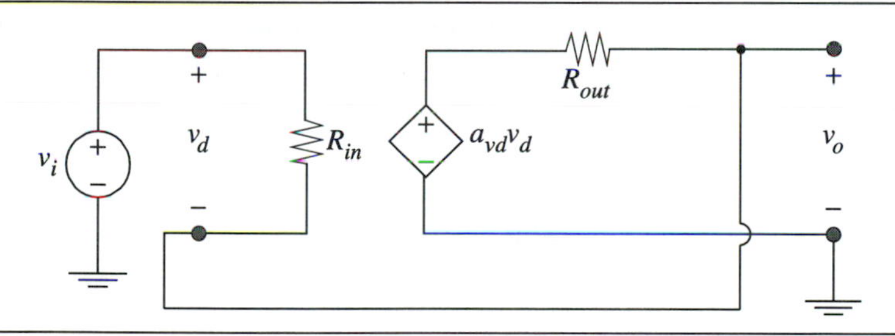

➤ **Figure E12.1**

➤ **Figure E12.2**

E12.3 The feedback transconductance amplifier in Fig. E12.3 has a transconductance $G_{md} = 22$ mS, an input resistance $R_{in} = 500$ kΩ, and an output resistance $R_{out} = 200$ kΩ. The feedback factor is $f = 10$ kΩ.

(a) What is the closed-loop transconductance G_{mf}?

(b) What is the input resistance $R_{in,f}$ for this feedback amplifier?

(c) What is the output resistance $R_{out,f}$ for this amplifier?

(d) If the loop gain is required to be $T \geq 50$, what is the minimum feedback factor f?

E12.4 The current feedback amplifier in Fig. E12.4 has current gain $a_{id} = 350$, input resistance $R_{in} = 25$ kΩ, and output resistance $R_{out} = 2$ MΩ. The feedback factor is selected so that the loop gain $T = 20$.

(a) What is the feedback factor f?

(b) What is the closed-loop current gain, A_{if}?

(c) What is the input resistance $R_{in,f}$?

(d) What is the output resistance $R_{out,f}$?

E12.5 By analyzing the current feedback amplifier circuit in Fig. 12.8,

(a) Show that the closed-loop gain in Table 12.1 is correct.

(b) Show that the input resistance in Table 12.1 is correct.

(c) Show that the output resistance in Table 12.1 is correct.

E12.6 Consider the ideal voltage feedback amplifier in Fig. 12.2.

(a) Solve for the ratio of the differential input voltage v_d to the input voltage v_i.

(b) Solve for the input current i_i into the amplifier in terms of the input voltage v_i. There is no need to derive this result—consider the definition of the input resistance $R_{in,f}$.

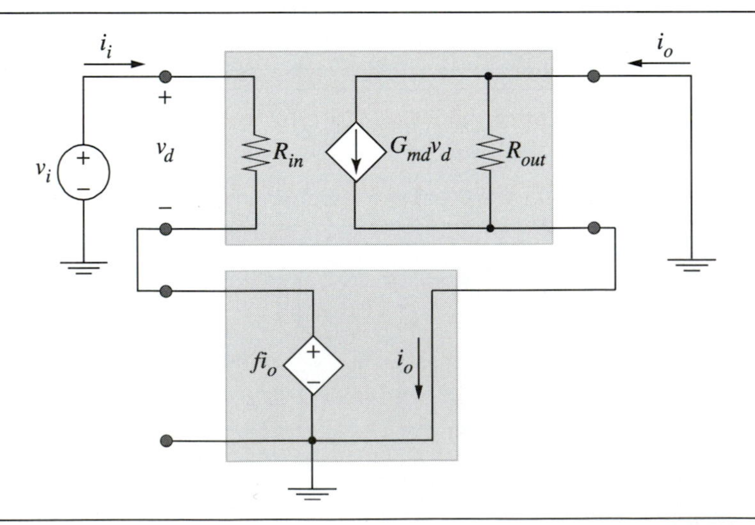

➤ **Figure E12.3**

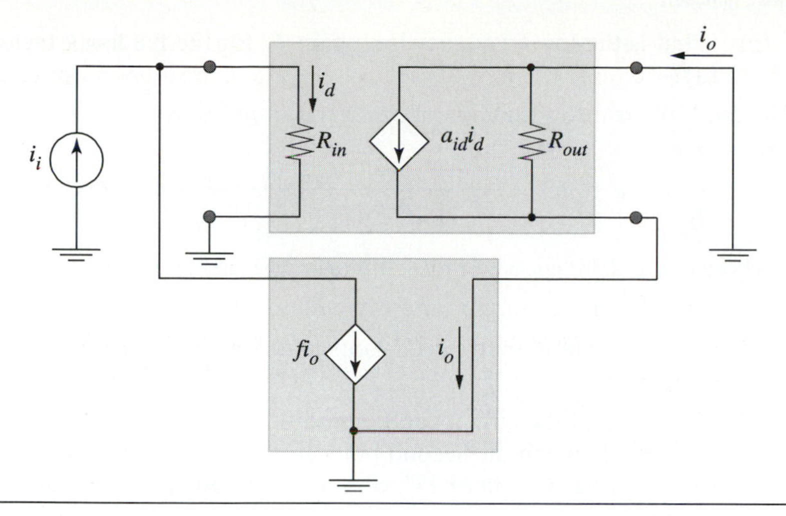

➤ **Figure E12.4**

(c) For a differential amplifier with $R_{in} = 2.8\ \text{M}\Omega$, $R_{out} = 50\ \Omega$, and $a_{vd} = 3 \times 10^5$, we implement voltage feedback with a resistor divider as in Fig. 12.10, with $R_1 = 5\ \text{k}\Omega$ and $R_2 = 45\ \text{k}\Omega$. For this feedback amplifier, find the numerical answers to (a) and (b). Note that these parameters are taken from a commercial bipolar op amp. Your answers should verify the "ideal op amp analysis" taught in basic electronics courses.

E12.7 A transresistance amplifier has a one-pole transfer function given by

$$R_{md}(j\omega) = \frac{-2.5\ \text{M}\Omega}{1 + j(\omega/200)}$$

(a) Graph the magnitude and phase Bode plots of $R_{md}(j\omega)$. Note that the magnitude plot has units of $20 \log_{10}(\Omega)$.

(b) By using feedback, we can modify the transfer function. What is the low-frequency loop gain T_o needed to have the closed-loop pole $\omega_1' = 25$ krad/s?

(c) What is the low-frequency closed-loop transresistance R_{mfo} for the feedback amplifier in (b).

(d) Graph the magnitude and phase Bode plots of the closed-loop transresistance $R_{mf}(j\omega)$. What is the magnitude of the transresistance at 25 krad/s?

E12.8 The one-pole transfer function of a current amplifier is given by

$$a_{id}(j\omega) = \frac{a_{ido}}{1 + j(\omega/\omega_1)}$$

(a) For $a_{ido} = 200$ and $\omega_1 = 2.5$ krad/s, what is the feedback factor needed to achieve a closed-loop pole $\omega_1' = 1$ Mrad/s?

(b) What is the low-frequency closed-loop gain A_{ifo} using the feedback factor in part (a)?

(c) What is the low-frequency loop gain T_o for the feedback factor in part (a)?

E12.9 The two-pole transfer function of a voltage amplifier is

$$a_{vd}(j\omega) = \frac{a_{vdo}}{(1 + j(\omega/\omega_1))(1 + j(\omega/\omega_2))}$$

where $a_{vdo} = 2000$, $\omega_1 = 2$ krad/s, and $\omega_2 = 2$ Mrad/s.

(a) Find the phase margin for unity-gain feedback ($f = 1$).

(b) What is the minimum closed-loop gain, if we require at least a 45° phase margin?

(c) Assume that the circuit can be modified to change the dominant pole, without changing the second pole. In order to be unity-gain stable and have a phase margin of 45°, what is the frequency ω_1 of the new dominant pole?

E12.10 The three-pole transfer function of a transconductance amplifier is

$$G_{md}(j\omega) = \frac{G_{mod}}{(1 + j(\omega/\omega_1))(1 + j(\omega/\omega_2))(1 + j(\omega/\omega_3))}$$

where $G_{mdo} = 250$ mS, $\omega_1 = 10$ krad/s, $\omega_2 = 2.5$ Mrad/s, and $\omega_3 = 20$ Mrad/s.

(a) Graph the magnitude and phase Bode plots of $G_{md}(j\omega)$. Note that the magnitude plot has units of $20 \log_{10}$ (S).

(b) For a feedback factor $f = 500\ \Omega$, what is the critical frequency ω^* at which the magnitude of the loop gain is unity.

(c) Find, using the Bode plots from (a), the maximum feedback factor for a phase margin of 60°.

E12.11 Consider a voltage amplifier with a two-pole transfer function given by

$$a_{vd}(j\omega) = \frac{a_{vdo}}{(1 + j(\omega/\omega_1))(1 + j(\omega/\omega_2))}$$

where $a_{vdo} = 2500$ and $\omega_1 = 5$krad/s. The feedback factor is $f = 0.05$.

(a) What is the location of the second pole such that the phase margin of this feedback amplifier is 30°?

(b) What is the critical frequency ω^* such that $|T(j\omega^*)| = 1$?

E12.12 Differential amplifier asymmetry can be modeled by including an input offset voltage V_{OS} in the two-port amplifier model, as shown in Fig. E12.12.This model is assumed to be valid over the output voltage range $-1\text{ V} < v_O < 1\text{ V}$ (note the uppercase subscript).

(a) For $a_{vd} = 500$ and with no feedback (open loop), plot v_O as a function of the input voltage for $V_{OS} = 2$ mV and $V_{OS} = -2$ mV. Note that the output voltage $|v_O| \leq 1\text{ V}$.

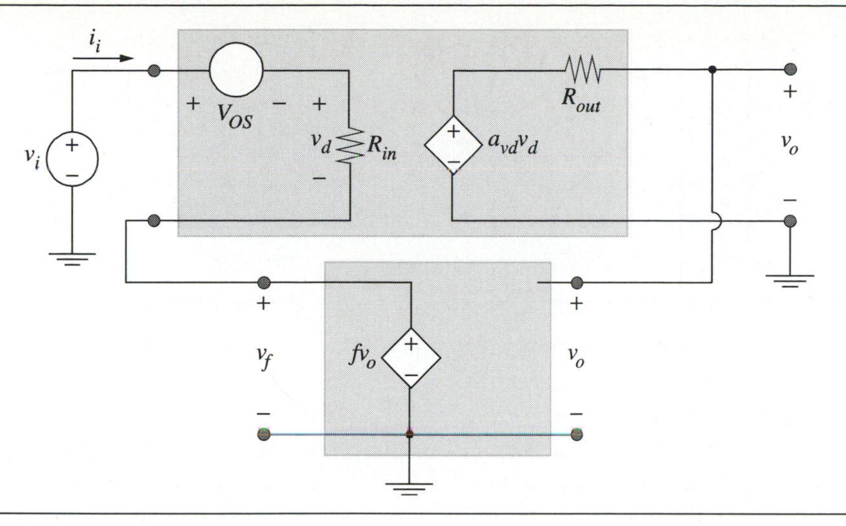

➤ **Figure E12.12**

(b) Now apply feedback with $f = 0.2$. Plot $v_O(v_i)$ for each case of input offset voltage in (a). What is the output voltage for $v_i = 0$ for each case? Does feedback reduce the effect of V_{OS}?

E12.13 Following the same approach as in Ex 12.9, use SPICE to find the DC parameters for the simple BiCMOS operational amplifier design in Fig. 12.23.

(a) Find the maximum and minimum common-mode input voltages; compare with the analytical results.

(b) Find the maximum and minimum output voltages by computing the DC transfer curve $v_O(v_{ID})$ and compare with the analytical results.

E12.14 In an effort to reduce power consumption, the current source in Fig. 12.21 is reduced by a factor of two to 25 μA and the supply voltages are reduced to 2 V and –2 V.

(a) Find the ratio of the static power consumption with the reduced supplies to that of the original op amp in Fig. 12.21.

(b) Find the input common-mode voltage swing.

(c) Find the output voltage swing.

E12.15 An input offset current can be incorporated into the two-port model of a transresistance amplifier, as shown in Fig. E12.15. This model is assumed to be valid over the output voltage range $-1\text{ V} < v_O < 1\text{ V}$ (note the upper-case subscript).

(a) For $R_{md} = -2.5\text{ M}\Omega$ and with no feedback (open loop), plot v_O as a function of the input current for $I_{OS} = 200\text{ nA}$ and $I_{OS} = -200\text{ nA}$. Note that the output voltage $|v_O| \leq 1\text{ V}$.

(b) Now apply feedback with $f = -10^{-5}\text{ S}$. Plot $v_O(i_i)$ for each case of input offset current in part (a). What is the output voltage for $i_i = 0$ for each case? Does feedback reduce the effect of I_{OS}?

E12.16 The VLSI microsystem in which the low-power CMOS op amp in E12.14 will be used requires that it drive a load capacitor $C_L = 5\text{ pF}$ and have a phase margin of 45° for unity-gain feedback.

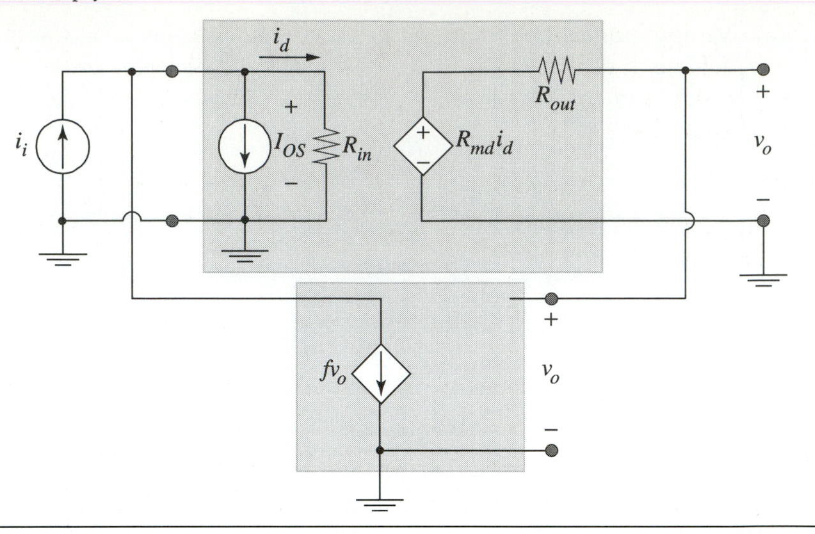

➤ **Figure E12.15**

(a) Find the low-frequency voltage gain a_{vdo} of the op amp.

(b) Find the compensation capacitor C_c needed to meet the phase margin specification.

(c) Verify your calculations using SPICE and if necessary, modify C_c to achieve the phase margin specification. Compare the gain-bandwidth product of the low-power op amp to that of the original op amp in Fig. 12.21.

E12.17 In a particular VLSI chip, the two-stage CMOS op amp in Fig. 12.21 is only used for feedback amplifiers with closed-loop gains $A_{vf} \geq 5$. As usual, we would like these amplifiers to have a phase margin of at least 45°.

(a) State the definition of the critical frequency ω^* for the case of minimum closed-loop gain. Show the location of ω^* on a magnitude Bode plot of the open-loop voltage gain $|a_{vd}(j\omega)|$ in dB.

(b) Find the new value of the compensation capacitor C_c that provides adequate phase margin for the minimum closed-loop gain (or equivalently, the maximum feedback factor.)

(c) Verify your calculations using SPICE and if necessary, modify C_c to achieve the phase margin specification. Compare the gain-bandwidth product of this op amp to that of the original op amp in Fig. 12.21.

E12.18 In a particular VLSI chip, the two-stage BiCMOS op amp in Fig. 12.28 is only used for feedback amplifiers with closed-loop gains $A_{vf} \geq 10$. As usual, we would like all amplifiers to have a phase margin of at least 45°.

(a) State the definition of the critical frequency ω^* for the case of minimum closed-loop gain. Show the location of ω^* on a magnitude Bode plot of the open-loop voltage gain $|a_{vd}(j\omega)|$ in dB.

(b) Find the new value of the compensation capacitor C_c that provides adequate phase margin for the minimum closed-loop gain (or equivalently, the maximum feedback factor.)

(c) Verify your calculations using SPICE and if necessary, modify C_c to achieve the phase margin specification. Compare the gain-bandwidth product of this conditionally stable op amp to that of the original op amp design in Fig. 12.26.

PROBLEMS

P12.1 Using feedback, we would like to design a voltage feedback amplifier with the following properties:

$$A_{vf} > 25,\ R_{in,f} > 2.5\ \text{M}\Omega,\ \text{and}\ R_{out,f} < 250\ \Omega$$

The amplifying block has a voltage gain $a_{vd} = 2500$, an input resistance $R_{in} = 50\ \text{k}\Omega$, and an output resistance $R_{out} = 15\ \text{k}\Omega$.

(a) What is the minimum loop gain required to meet the specifications?

(b) Using your result from (a), what is the feedback factor f?

P12.2 In this problem, we will use the amplifying block from P12.1 in a feedback transresistance amplifier. The specifications are:

$$-R_{mf} > 1.5\ \text{M}\Omega,\ R_{in,f} < 500\ \Omega,\ \text{and}\ R_{out,f} < 250\ \Omega$$

(a) Find the transresistance two-port parameters of the amplifying block.

(b) What is the minimum loop gain required to meet the specifications?

(c) Using your result from (b), what is the feedback factor, f?

(d) Draw the circuit model of the feedback amplifier. Take care that the circuit implements negative—rather than positive—feedback!

P12.3 The two-port parameters of a transconductance amplifier vary over its operating temperature range, so that

$$G_{md} = 70 \pm 10\ \text{mS},\ R_{in} = 200 \pm 50\ \text{k}\Omega,\ \text{and}\ R_{out} = 100 \pm 25\ \text{k}\Omega$$

We desire that the minimum closed-loop transconductance be $G_{mf} = 5$ mS.

(a) What is the maximum loop gain T_o so that the specification on G_{mf} is met over the entire temperature range?

(b) What is the variation of the closed-loop transconductance over the temperature range? Assume that the feedback factor is stable with temperature.

(c) Find the range of closed-loop input resistance $R_{in,f}$ for the feedback factor in part (b).

(d) Find the range of closed-loop output resistance $R_{out,f}$ for the feedback factor in part (b).

P12.4 We would like to design a voltage feedback amplifier with a closed-loop gain that has a normalized run-to-run uncertainty of 0.005 (0.5%), with an average value of 80. The amplifying block has a differential gain $a_{vd} = \overline{a_{vd}}(1 \pm 0.2)$ from run to run.

(a) What is the required feedback factor f? Neglect the uncertainty in the feedback factor.

(b) What is the average differential gain required to achieve the uncertainty specification on the closed-loop gain?

P12.5 A transresistance amplifier is considered to have a one-pole transfer function given by

$$R_{md}(j\omega) = \frac{R_{mdo}}{(1 + j(\omega/\omega_1))}$$

Due to process variations, the low-frequency transresistance is specified as $R_{mdo} = \overline{R_{mdo}}(1 \pm \varepsilon_{R_{mo}}) = -25\ \mathrm{M\Omega}(1 \pm 0.20)$ and the dominant pole is $\omega_1 = \overline{\omega_1}(1 \pm \varepsilon_{\omega_1}) = 1.2\ \mathrm{krad/s}(1 \pm 0.05)$. The uncertainties in transresistance and in the dominant pole are independent.

(a) For a feedback factor $f = -5\ \mu\mathrm{S}$, what is the variation in the low-frequency closed-loop transresistance R_{mof}?

(b) For the same feedback factor as in (a), what is the variation in the pole ω_1' in the closed-loop transresistance?

P12.6 The two-pole transfer function of a transconductance amplifier is given by

$$G_{md}(j\omega) = \frac{G_{mo}}{(1 + j(\omega/\omega_1))(1 + j(\omega/\omega_2))}$$

(a) For $G_{mdo} = 100$ mS, $\omega_1 = 50$krad/s, and $\omega_1 = 2$ Mrad/s, sketch the magnitude Bode plot for this amplifier. Note that the units will be $20\log_{10}$ (S).

(b) For a feedback factor $f = 1\ \mathrm{k\Omega}$, what is the critical frequency ω^* at which the magnitude of the loop gain is unity? What is the phase margin for this feedback factor?

(c) If we would like to have a phase margin of at least 45°, what is the maximum feedback factor? What is the minimum low-frequency closed-loop transconductance for this case?

P12.7 This problem investigates the frequency dependence of the input and output *impedances* of a feedback voltage amplifier. The amplifier block impedances are:

$$Z_{in}(j\omega) = R_{in}\|(1/(j\omega C_{in})) \text{ and } Z_{out}(j\omega) = R_{out}\|(1/(j\omega C_{out}))$$

where $R_{in} = 250\ \mathrm{k\Omega}$, $C_{in} = 1$ pF, $R_{out} = 500\ \mathrm{k\Omega}$, $C_{out} = 2$ pF. The amplifier block's transfer function is

$$a_{vd}(j\omega) = \frac{a_{vdo}}{(1 + j(\omega/\omega_1))(1 + j(\omega/\omega_2))}$$

where $a_{vdo} = 5000$, $\omega_1 = 2$ krad/s, and $\omega_2 = 10$ Mrad/s. The feedback factor is $f = 0.2$.

(a) Plot the magnitude and phase of the closed-loop gain over the frequency range 100 Hz–100 MHz.

(b) For a feedback factor $f = 0.2$, plot the closed-loop input impedance $|Z_{in,f}(j\omega)|$ in units of 20 $\log_{10}$ (Ω) over the frequency range 100 Hz– 100 MHz. Note that the results for voltage feedback based on the resistive two-port model can be applied directly to the case of impedances.

(c) Repeat (b) for $|Z_{out,f}(j\omega)|$.

(d) For a particular application, we require $|Z_{in,f}| > 20$ MΩ and $|Z_{out,f}| < 5$ kΩ. What is the usable bandwidth of the feedback amplifier, using this criterion?

P12.8 This problem investigates peaking in the closed-loop gain. A current feedback amplifier has a two-pole transfer function given by

$$a_{id}(j\omega) = \frac{a_{ido}}{(1 + j(\omega/\omega_1))(1 + j(\omega/\omega_2))}$$

(a) Find analytical expressions for the pole locations of the closed-loop transfer function $A_{if}(j\omega)$ by factoring the denominator polynomial.

(b) Find an analytical expression for the $|A_{if}(j\omega)|$ for the case where the phase margin is less than 45°.

(c) For the case where $a_{ido} = 500$, $\omega_1 = 10$ krad/s, and $\omega_2 = 2$ Mrad/s, graph the magnitude Bode plot $|A_{if}(j\omega)|$ of the closed-loop transfer function for feedback factors $f = 0.25$, 0.5, 0.75, and 1.

P12.9 Figure P12.9 is an NMOS differential amplifier with current-mirror supply, followed by a voltage buffer. The function of the latter is to center the output voltage at 0 V.

(a) Find $(W/L)_5$ so that $V_O = 0$ V if the transistors in the differential amplifier are matched and $v_d = 0$ V.

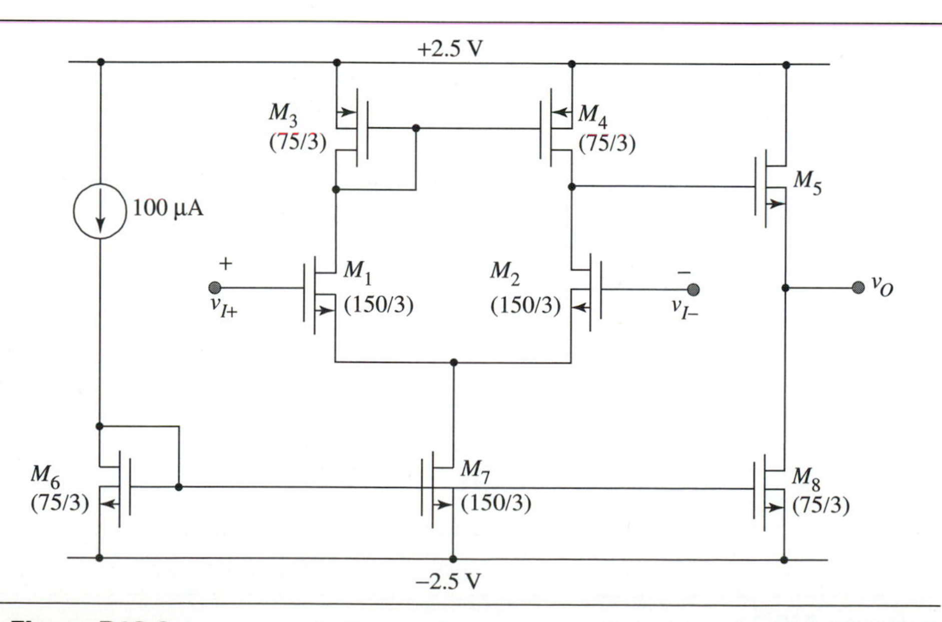

➤ Figure P12.9

(b) Find the differential voltage gain a_{vd} of this amplifier. You can assume that the gain of the common-drain stage is unity.

(c) Find the output voltage swing. You can consider that the inputs are nearly at ground potential for this part.

(d) Sketch the transfer function $v_O(v_D)$ of this amplifier, by extending your result from (b) to the minimum and maximum output voltage from (c).

(e) Compare your approximate transfer function with SPICE.

P12.10 An ideal feedback block is connected between v_O and v_{I-} in the circuit in Fig. P12.9, with a feedback factor is $f = 0.02$.

(a) What is the closed-loop gain A_{vf} of this feedback amplifier?

(b) What is its output resistance $R_{out,f}$?

(c) Sketch the transfer function $v_O(v_{I+})$ assuming that the closed-loop gain is constant until the output voltage reaches its limits (found in P12.9(c)).

(d) Due to mismatches between transistor threshold voltages, an input offset voltage V_{OS} ranging from –2.5 mV to +2.5 mV must be applied at v_{I+} in order to make $v_O = 0$ V. Repeat (b) for the extreme input offsets. The equivalent circuit in E12.12 may be helpful.

P12.11 Figure P12.11 shows a simple feedback transresistance amplifier.

(a) Sketch the open-loop transfer function $v_O(i_I)$ of the amplifier for $i_I \geq 0$ for the extreme values of the forward current gain: $\beta_F = 80$ and 120.

(b) Sketch the closed-loop transfer function with $R_F = 100$ kΩ for $i_I \geq 0$ for the extreme values of the forward current gain: $\beta_F = 80$ and 120. The output voltage $v_O(i_I = 0)$ should be accurate for each plot.

P12.12 A practical feedback transconductance amplifier is shown in Fig. P12.12. The resistor R_E measures the output current, through the emitter current. The voltage v_F across R_E subtracts from the input voltage v_I to yield the base-emitter voltage of Q_1. This amplifier is also known as a degenerated common-emitter amplifier; its two-port parameters are derived in Chapter 8.

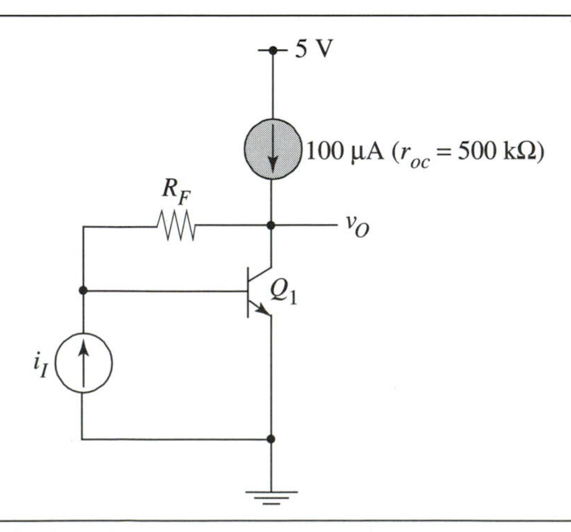

➤ **Figure P12.11**

(a) Using the results from Chapter 8, find the closed-loop transconductance G_{mf} and express it in the standard feedback form. Find the loaded loop gain T'.

(b) Using the results from Chapter 8, find the closed-loop input resistance $R_{in,f}$. Express it in the standard transconductance feedback form $R_{in}(1+T')$.

(c) Find the closed-loop output resistance $R_{out,f}$. from the results in Chapter 8. Under what assumptions can you express it as $R_{out,f} = R_{out}(1+T')$?

P12.13 The circuit in Fig. P12.13 is a variable gain transresistance amplifier that uses MOS transistor M_5 as a voltage-controlled resistor. Note that since the bulk is tied to the source of NMOS transistor M_5, this circuit is fabricated in a p-well CMOS technology.

(a) Determine the (W/L) ratio of M_5 so that the feedback resistor R_F can be varied from 10 kΩ to 1 MΩ by varying V_R from 100 mV over threshold to 2.5 V. The source of M_5 can be considered at 0 V for finding R_F.

(b) Find the two-port parameters of the transresistance amplifier (without feedback). Note that you will have infinite values for more than one parameter.

(c) Find the input resistance with feedback $R_{in,f}$ as a function of the resistor R_F. In evaluating the exact expression for $R_{in,f}$ of the transresistance amplifier, it may be helpful to consider the input resistance R_{in} from (b) as very large, but not infinite. The transresistance R_{md} will then be proportional to R_{in}.

(d) Find the output resistance with feedback $R_{out,f}$ as a function of the resistor R_F. In evaluating the exact expression for $R_{out,f}$ of the transresistance amplifier, it may be helpful to consider the input resistance R_{in} from (b) as very large, but not infinite. The transresistance R_{md} will then be proportional to R_{in}.

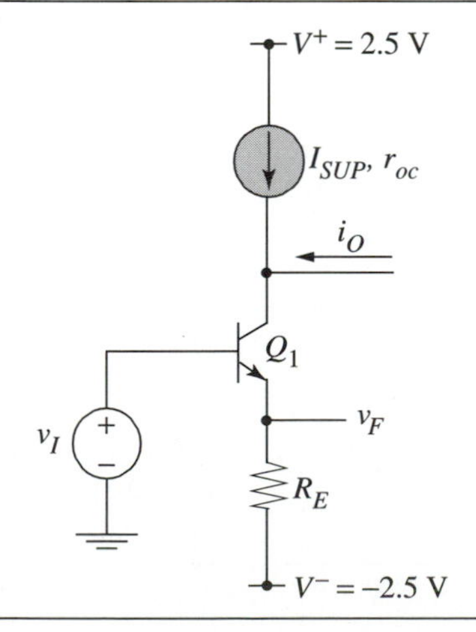

➤ **Figure P12.12**

(e) Plot the closed-loop transresistance $R_{m,f}$ as a function of V_R on a log-linear plot, for $0.1 \text{ V} < V_R < 2.5 \text{ V}$. Since the loaded loop gain T' will vary with the feedback factor, you should verify that there is sufficient loop gain before assuming that $R_{m,f} = -R_F$.

P12.14 Another way to increase the gain of a differential amplifier is to degenerate the bipolar current mirror supply, as shown in Fig. P12.14.

(a) Find R_E so that the resistance $R_{down} = 4\,R_{up}$ in Fig. P12.14.

(b) Find the differential voltage gain a_{vd} of this amplifier in dB for the emitter resistor from part (a).

(c) For the emitter resistance in (a), find the minimum common-mode input voltage $V_{IC,min}$ of this amplifier.

(d) Find the maximum common-mode input voltage $V_{IC,max}$.

P12.15 The basic two-stage CMOS op amp topology in Fig. 12.21 has an asymmetric common-mode input voltage range. In this problem, the goal is to increase $V_{IC,max}$ by making some of the MOSFETs wider.

(a) If the gate widths of M_6, M_7, and M_8 are increased by a factor of 4 (keeping the gate lengths at $L = 3\ \mu\text{m}$), find the new value of $V_{IC,max}$.

(b) In addition to the changes in (a), the channel widths of M_1 and M_2 are increased by a factor of 4, what is the new value of $V_{IC,max}$.

(c) Find the new value of the low-frequency small-signal gain a_{vdo}.

P12.16 In order to improve the gain-bandwidth product of the two-stage CMOS op amp in Fig. 12.21, one approach is to increase the width of transistor M_5 in order to increase the second-stage transconductance.

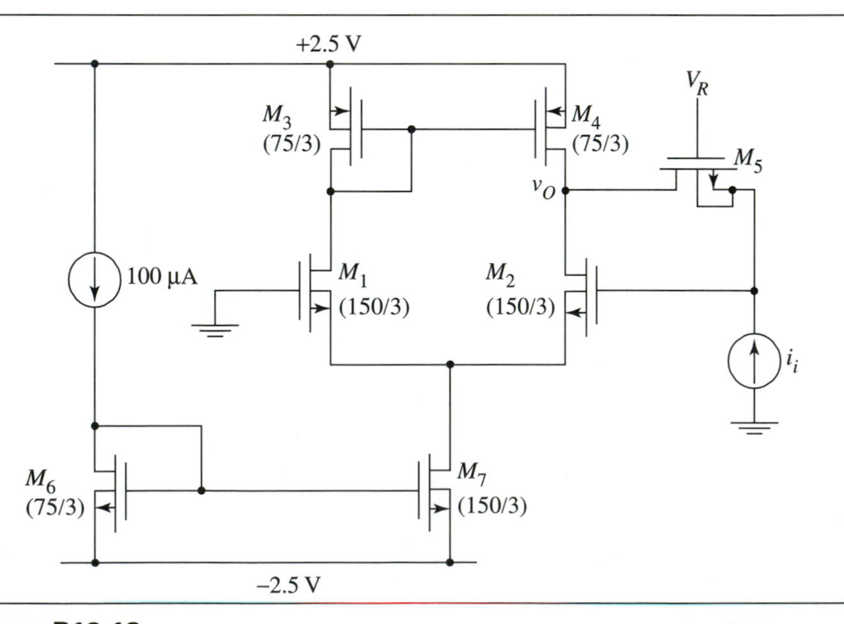

➤ Figure P12.13

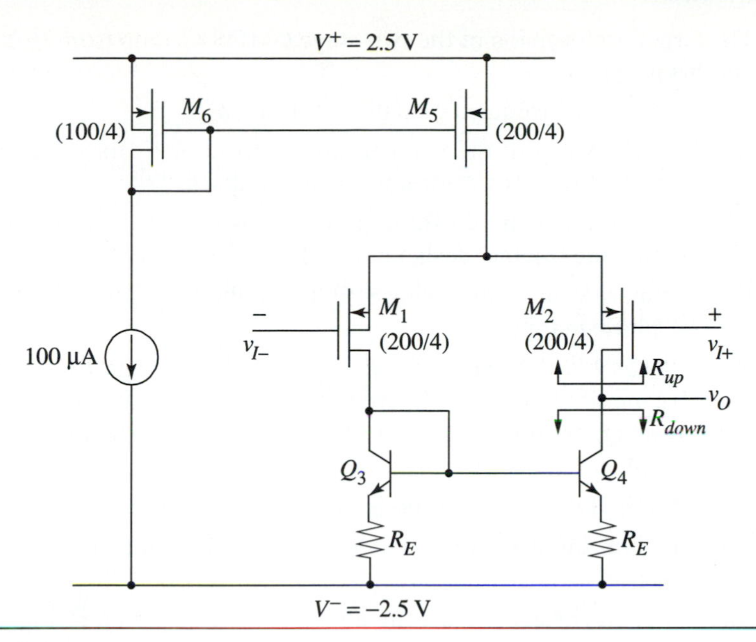

➤ **Figure P12.14**

(a) Find the width of M_5 needed to have the transconductance of the second gain stage equal to that of the simple two-stage BiCMOS op amp. Keep the gate length of M_5 equal to 3 μm.

(b) In addition to needing increased area, another drawback to this approach is creating a systematic input offset voltage V_{OS}. From your answer to (a) and the known differential gain of the first stage, estimate the voltage that needs to be applied in order to have $V_{D4} = V_{G5} = V_{D3}$.

(c) Verify your answer to (b) by using SPICE.

P12.17 In an application where the area of the two-stage CMOS op amp is not a constraint, the gate widths of transistors M_3, M_4, and M_5 can be increased to improve the op amp performance. In order to avoid a systematic input offset voltage, the widths of these transistors should be related by $W_3 = W_4 = W_5/2$. The gate length of these transistors is kept $L = 3$ μm. The other transistors in Fig. 12.21 are unchanged.

(a) Find the widths of M_3, M_4, and M_5 needed to increase the low frequency op amp voltage gain to 90 dB.

(b) A simple estimate of the area required by a MOSFET is

$$A = (L + 2L_{diff}) \cdot W$$

where $L_{diff} = 6$ μm is the length of the source and drain regions. Find the ratio of the total area of the op amp with the new gate widths from (a) to that of the original design in Fig. 12.21.

(c) Find the output voltage swing for the op amp with the new gate widths from part (a).

P12.18 The larger area version of the two-stage CMOS op amp from P12.17 is compensated in this problem.

(a) Find the low-frequency small-signal gain a_{vdo}.

(b) Find the value of the compensation capacitor C_c for ensuring a phase margin of at least 45° for unity-gain feedback.

(c) Graph the magnitude Bode plot for this op amp and compare it with that for the original design in Fig. 12.30.

(d) Compare your results using SPICE and modify C_c in order to achieve a 45° phase margin.

P12.19 Figure P12.19 is the schematic of a modified two-stage BiCMOS op amp using emitter degeneration resistors. In order to avoid a systematic offset, the emitter resistors for Q_3 and Q_4 are double that for Q_5. Note that the current source current is increased to 100 μA in Fig. 12.19.

(a) Find the DC voltages at nodes 1–4.

(b) Find the input common-mode voltage range and the output voltage range.

P12.20 This problem considers the compensation of the op amp in Fig. P12.19 for unity-gain feedback applications, with a phase margin of 45°.

(a) Find the low-frequency voltage gain a_{vdo}.

(b) Select C_c such that the op amp meets the phase margin specification.

(c) Graph the magnitude Bode plot for this amplifier and compare it with the analytical results for the improved BiCMOS two-stage design in Section 12.6.

(d) Simulate the phase margin using SPICE and modify the compensation capacitor if necessary to meet the specification.

DESIGN PROBLEMS

D12.1 The two-stage CMOS op amp in Fig. 12.21 requires too much area for a particular application.

(a) Reduce all MOSFETs to the minimum gate length $L = 1.5$ μm and determine the transistor widths and reference current such that the low-frequency voltage gain is maintained. Reduce the reference current below 50 μA only if necessary.

(b) Find the compensation capacitor C_c so that the op amp is stable for unity-gain feedback, with a phase margin of 45°.

(c) Sketch a rough layout of the op amp with minimum gate lengths and estimate its area. Simple geometric design rules: the source/drain width is $L_{diff} = 6$ μm and separations between source/drain regions within a well are 2 μm, with a 4 μm separation between a source/drain region and the well edge. Metal lines and separations are 2 μm, with 2 μm × 2 μm contacts. MOSFETs with large widths should be folded, as illustrated in Ex 4.1, in order to make them more compact.

D12.2 A one-stage CMOS op amp topology is shown in Fig. D12.2. This problem considers the design of the transistor dimensions and reference current to meet spec-

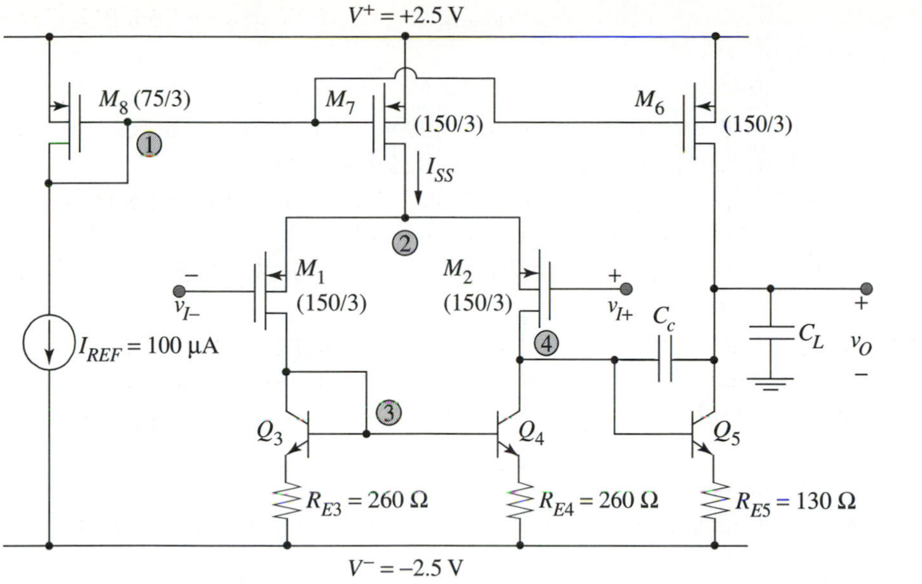

➤ **Figure P12.19**

ifications on the output swing and input common-mode voltage swing. The minimum MOSFET gate length is 1.5 μm for this n-well CMOS technology. Note we are using an n-well CMOS technology and transistors M_5 and M_6 will have their threshold voltages shifted due to the backgate effect.

(a) The output voltage should be $V_O = 0$ V for $V_{I+} = V_{I-} = 0$. The design goal is to achieve an output voltage range of $-1\text{ V} < v_O < 1\text{ V}$. Find a reference current and a set of transistor dimensions that satisfy this specification.

(b) In addition, the range of input common mode voltages should be at least $-1\text{ V} < V_{IC} < +1\text{ V}$. Modify the transistor dimensions and reference current, if necessary, to satisfy this additional specification.

(c) What is the static power required by the op amp (not including the reference current branch)?

(d) Simulate your initial design using SPICE and modify it so that you meet the objectives in (a) and (b).

D12.3 The small-signal analysis and compensation of the one-stage CMOS op amp in Fig. D12.2 are considered in this problem. The low-frequency small-signal gain is targeted at 80 dB. The load capacitor $C_L = 5$ pF. The compensation capacitor C_c is to be chosen such that the op amp is stable for unity-gain feedback with a phase margin of 60°. The higher phase margin corresponds to less ringing in the transient response, which is a primary concern for the particular application.

(a) Evaluate the low-frequency gain and modify the design from D12.2 to achieve 80 dB, if necessary.

(b) From a small-signal model including the device capacitors, C_c, and C_L, find analytical expressions for the dominant pole ω_1 and the second pole ω_2 for this op amp topology. Calculate the dominant pole and adjust C_c to ensure the required phase margin.

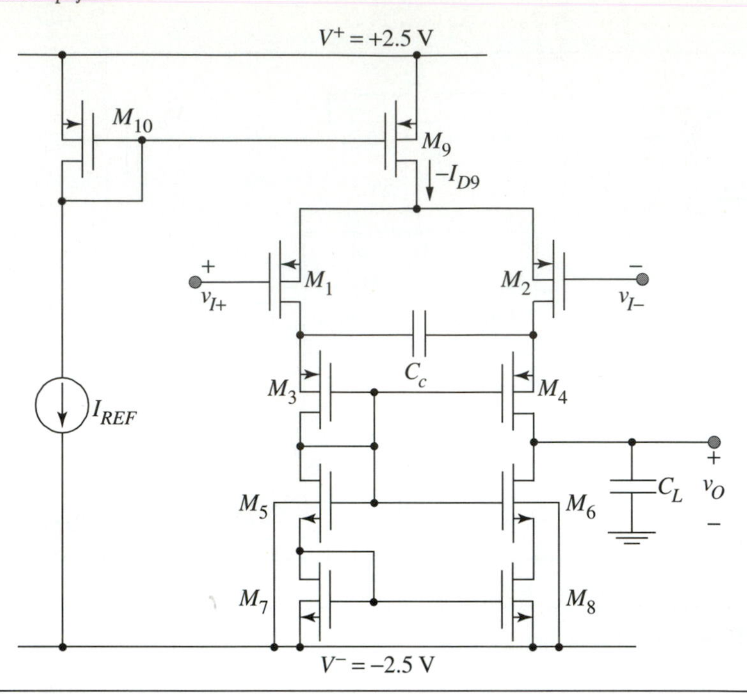

➤ **Figure D12.2**

(c) Check that the op amp meets the specifications using SPICE; if necessary, modify the design.

(d) Compare the one-stage topology to the two-stage topology on the basis of power, area consumption, low-frequency gain, and ease of compensation.

chapter 13

MOS Memories

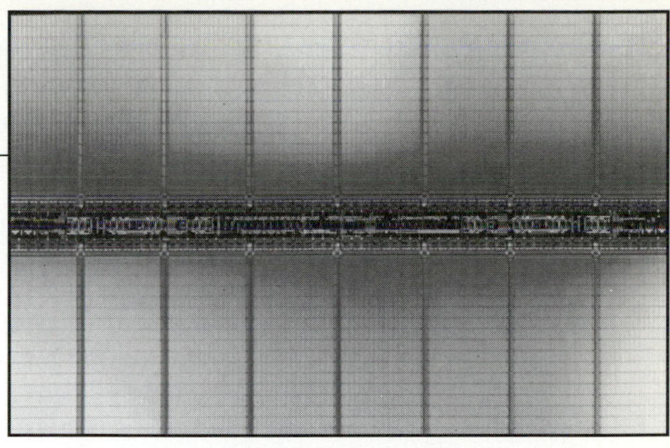

A 256 Mbit dynamic random access memory. Each block contains a 16 Mbit memory. The sensing and other peripheral circuits are located in the center of the die. (Courtesy of International Business Machines Corporation. Unauthorized use not permitted.)

This chapter discusses the analysis and design of three of the most important memories fabricated in MOS technology.

- Read Only Memory, ROM
- Static Random Access Memory, SRAM
- Dynamic Random Access Memory, DRAM

Memory design encompasses the areas of device physics, as well as digital and analog circuit design. Many of the concepts you learned in the preceding chapters will be applied to memory design.

Chapter Objectives

- Overall architecture of semiconductor memories.
- Memory cell structure for ROM, SRAM, and DRAM.
- Sense amplifiers which detect the stored bit from an SRAM or DRAM cell.
- Major peripheral circuitry, including address buffers and decoders to select the specific bit that is being read or written.
- Key points in the design of an embedded static RAM.

13.1 MEMORY CLASSIFICATION

One method of classifying semiconductor memories is to use a hierarchical approach based on the speed and density of the memory. The speed of the memory is usually characterized by the **access time**, which is the time it takes from a request for data to having valid data at the output pins. The **density** of the memory is the number of stored bits per chip. The **cache memory** has the highest speed and lowest density in the hierarchy. The cache memory continuously supplies the processor with data and control information. **Main memory** serves the function of storing programs and data. It has a higher density and lower speed when compared to cache memory. Both cache and main memory are fabricated with silicon integrated circuit technology, predominantly using CMOS or BiCMOS processes. The process technology often has enhancements to reduce memory cell size and increase the storage capability. Higher density, lower speed memories such as hard disks and tape drives complete the hierarchy for a typical digital system.

Memory can also be classified by its function. Memories store programs that contain instructions for processing control and also store the data to be processed. Since many control functions do not change, we can use **read only memory (ROM)** to store the control function. Bits in a ROM are stored either by being "hard wired" during the silicon fabrication process or by a slow electrical programming procedure. ROMs are non-volatile, meaning that the information is maintained even when the power to the digital system is removed.

A second memory function is to store data. This data storage requires both the read and write function since data will change upon computation. The most important memories used for data storage are **random access memories (RAM)** that are fabricated in MOS technology. There are two styles of random access memories.

Static Random Access Memory (SRAM)— stores data as long as the power supply to the digital system is turned on.

Dynamic Random Access Memory (DRAM)—offers approximately a fourfold increase in density over SRAM, but requires the data be refreshed periodically even when the system power is turned on.

Several other types of memories including serial access memories, content addressable memories, and shift registers also have roles in digital system design. A complete survey of all the memory technologies across the memory hierarchy and the various memory functions is beyond the scope of this text.

To appreciate the enormous rate at which memory technology has evolved, Fig. 13.1 shows a graph of DRAM density over the past twenty-five years. The graph shows that density grew at the rate of approximately four times every three to four years. This trend was recognized by Gordon Moore in the early 1970s and has come to be known as **Moore's Law**. This graph also shows the minimum feature size that corresponds to each DRAM generation. You can see that the requirement for increased memory density has driven the minimum feature size down to smaller and smaller geometries.

13.2 MOS MEMORY ARCHITECTURE

Before discussing a typical memory architecture, we will review the major timing definitions for a memory as illustrated in Fig. 13.2.

Read-Access Time—The time it takes from the initiation of a read cycle on the rising edge of the read/write control signal to the time at which the stored data is available at the output of the chip.

Read-Cycle Time—The minimum time between successive read operations. Since most memory architectures require some reset functions after access, the read-cycle time is, in general, longer than the read-access time.

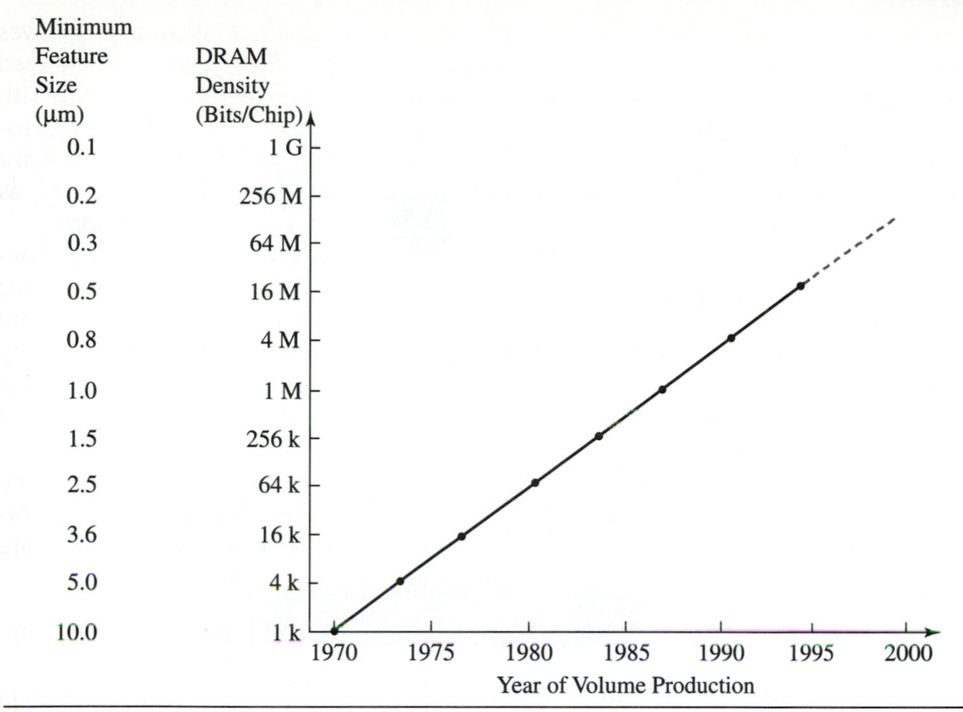

Figure 13.1 DRAM density and corresponding minimum feature size vs. year of volume production.

Write-Cycle Time—The minimum time between successive writes. When writing data into the memory, the data must be valid before the rising edge of the read/write control signal that initiates the write cycle.

Cycle Time—Specified as the larger of the read and write-cycle time.

Most memory is characterized by the read-access time and the cycle time. The engineering trade-off is generally between access and cycle time vs. density of storage.

A typical memory architecture is shown in Fig. 13.3. Memory cells that store one bit of information are arranged in a two-dimensional array. In the memory architecture shown, we are considering a 2^N by 2^M array, meaning that the total number of bits that can be stored is 2^{N+M}. Each memory cell has a **word line** that acts to control

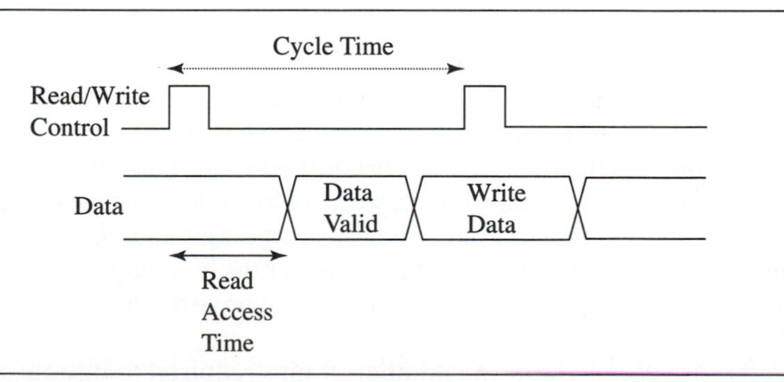

Figure 13.2 Memory timing definitions. The read-access time is the time from the rising edge of Read/Write Control to the time the data is valid. The cycle time is the minimum time between Read/Write Control signals.

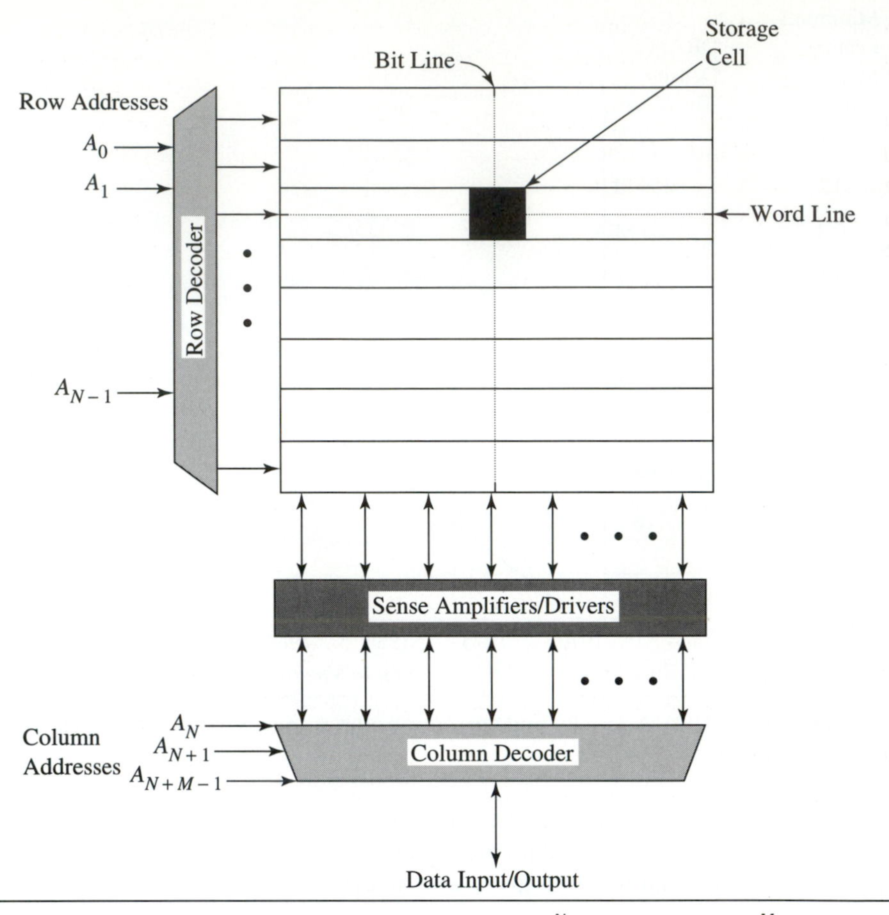

➤ **Figure 13.3** Typical memory architecture with 2^N word lines and 2^M bit lines forming an 2^{N+M} memory.

the cell. The signal that accesses the cell to either read or write data is applied to the word line. Perpendicular to the word lines are **bit lines**. The data that is being written into or read from the memory are found on the bit lines.

A typical memory chip is dominated in size by the area of the memory cells. One way to reduce the size of the memory cells is to reduce the signal swing on the bit lines driven from the cell when being read. By reducing the bit-line signal swing, the size of the individual transistors in the memory cell can be made smaller at the expense of reduced noise margin. Typical bit-line voltage swings are between 100 mV and 1 V depending on the specific memory. To bring the reduced signal-swing levels to the full logic levels required for interfacing with the outside world, a **sense amplifier** is placed at the bottom of each bit line. The sense amplifier's function is to convert the small signal-swing representing the logical value of the bit being read, to full logic levels. These sense amplifiers often double as bit-line drivers when a write operation is required.

Decoders are used to limit the number of pins required to interface with a large memory array. Turning our attention to the horizontal lines, the row decoder decides which word line will be activated and reduces the number of signals required to access a particular row from 2^N to N. A **row decoder** has as its input N addresses and the selection of one word line as its output. Each cell on that word line is connected to a specific bit line that will either access data stored in the cell or write new data into the cell. As previously mentioned, the sense amplifiers/bit-line drivers are at the bottom of each bit line to reduce the requirements on the individual cells in the array. Similar to the row decoder, a **column decoder** is placed below the sense amplifier outputs to decide which bit line to connect to the chip output. This decoder takes 2^M bit lines and selects some smaller number such as 1, 2, or 4 and connects them to the chip output. In our discussion of MOS memories, we will assume that the column decoder selects one-of-M bit lines.

A read operation in this memory architecture begins with a row address being input to the chip. After buffering the address, the row decoder decides which word line to activate. All of the cells connected to that word line output a small voltage-change (~100 mV) to their respective bit lines that represents a stored **0** or **1**. The sense amplifiers amplify the change in bit-line voltage to a full logic level. The column address that was input to the chip is decoded to select a particular bit line. The data from the cell, which is the intersection of the column and row, is buffered and output from the memory chip.

The architecture shown in Fig. 13.3 has a limitation as the amount of stored bits continues to increase. The problem is that the word lines and bit lines will continually become larger and the parasitic capacitance of these lines will slow the memory access time. To solve this problem, an extra dimension to the address space can be used to partition the memory into blocks as shown in Fig. 13.4. The extra address selects one of P blocks to be read or written. This architecture keeps the length of each individual block's word and bit lines smaller, allowing for a faster access time. In addition, only one block needs to be activated at a time, which can reduce the power dissipation since most of the sense amplifiers and decoders can remain inactive.

A great deal of effort is required to find the optimal segmentation of the number of blocks, number of rows, and number of columns in a particular memory. The

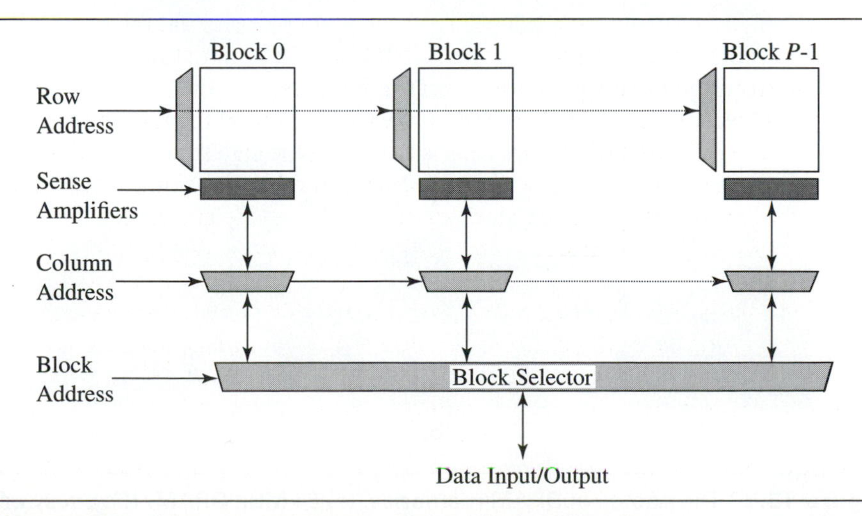

Figure 13.4 Memory architecture using block addressing.

details of a particular technology, as well as the specifications of the memory being designed, are required to optimize this segmentation. In this chapter, we consider only the design of one particular block.

Die photos of memories reflect the memory architecture that has been discussed above. The die photo in Fig. 13.5(a) is a prototype 64 MBit DRAM segmented into blocks. When comparing this chip with the 4 kB DRAM in Fig. 13.5(b), you can see the enormous increase in density in MOS memories over the last twenty years.

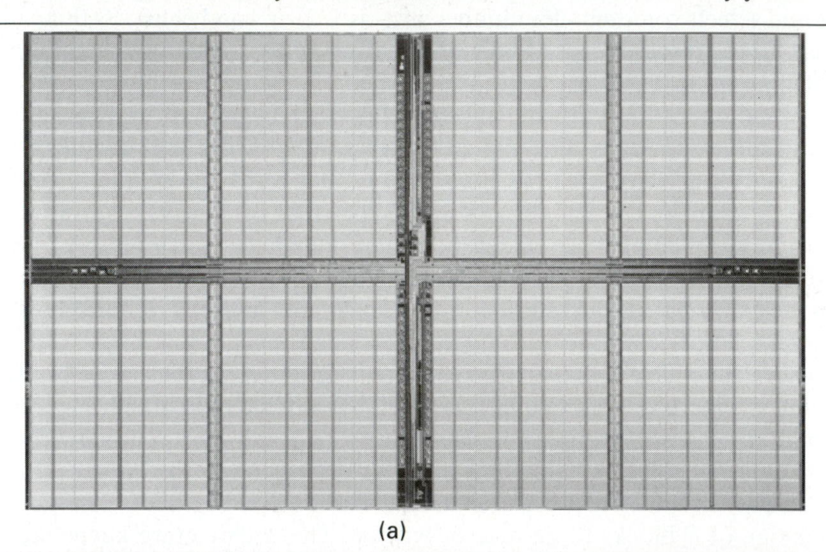

(a)

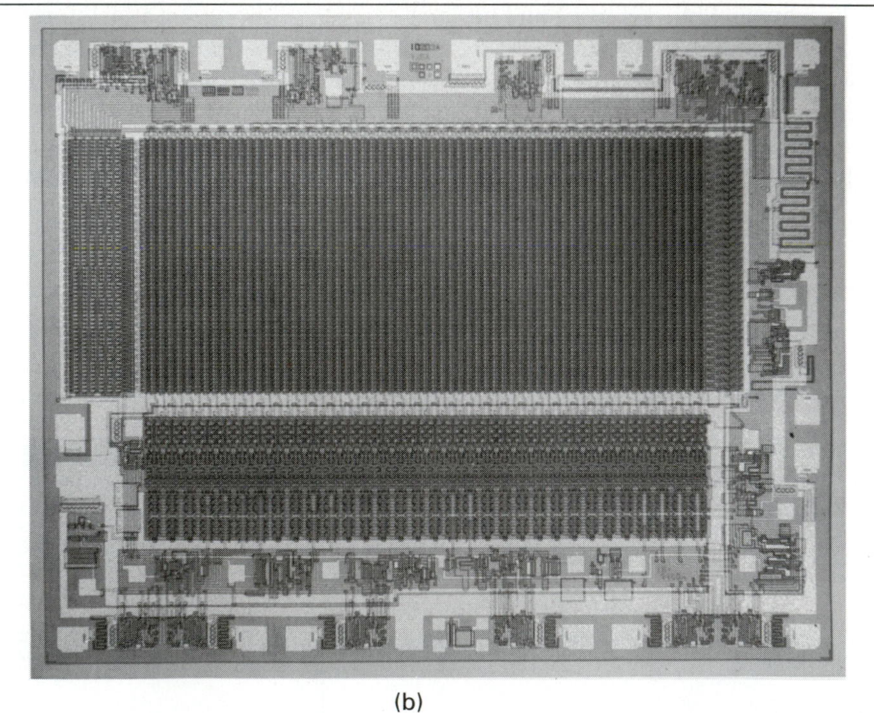

(b)

Figure 13.5 Die photos of DRAM memories. (a) 64 MBit DRAM. *(Courtesy of International Business Machines Corporation. Unauthorized use not permitted.)* (b) 4 kBit DRAM.

13.3 MEMORY CELLS

In this section, we discuss the structure and operation of three of the most important MOS memory cells: (1) read only memory (ROM), (2) static random access memory (SRAM), and (3) dynamic random access memory (DRAM) cells.

13.3.1 ROM Cells

The function of the ROM cell is to present a **0** or **1** to the bit line when the word line is accessed. An NMOS transistor can be used for this function as shown in Fig. 13.6. When the word line is accessed in a cell that contains an NMOS transistor, the bit line voltage will be pulled low. The low voltage level depends on the relative sizing of the precharge pull-up transistor and cell transistor. The opposite state is achieved if the transistor is absent in the cell. Under this condition, the bit-line voltage remains high even when the word line is accessed.

To understand the operation of a typical MOS ROM array, a 4×4 ROM array is shown in Fig. 13.7(a). The absence of a transistor at the intersection of a word and bit line can be considered a stored **1**, while the presence of a transistor at the intersection can be called a stored **0**. To understand the static operation of this ROM array, consider that one word line is activated. The p-channel pull-up devices act as current sources, which keep the bit lines high in the absence of a transistor in the cell. The bit-line voltage will be reduced if a transistor is present and sized appropriately. A sense amplifier can be placed at the bottom of the bit line to sense the small changes in the bit-line voltage. The sense amplifier allows the n-channel transistor sizing in the cell to be greatly reduced since the cell does not have to provide large bit-line voltage swings. Hence, a more dense ROM array is made possible.

The transient operation of this ROM cell is most easily understood by using an equivalent circuit model shown in Fig. 13.7(b). It consists of an inverter with a p-channel current source acting as a pull-up and an n-channel transistor connected between the bit line (output) and ground. The topology is the same as the NMOS inverter with a current-source pull-up studied in Chapter 5. The transient operation of the ROM cell is dominated by the *RC* time constant to charge the word line and the time it takes to discharge the bit line. Accurate modeling of the parasitics associated with the word line and bit line is important since the structure is repeated many times.

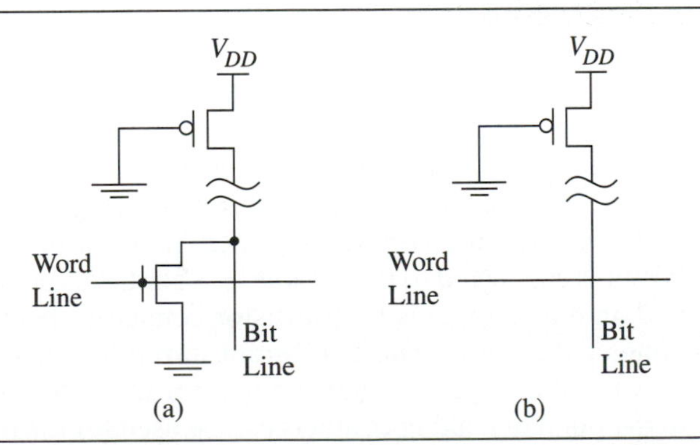

Figure 13.6 MOS ROM cell. (a) Stored **0:** transistor is present. (b) Stored **1:** transistor is absent.

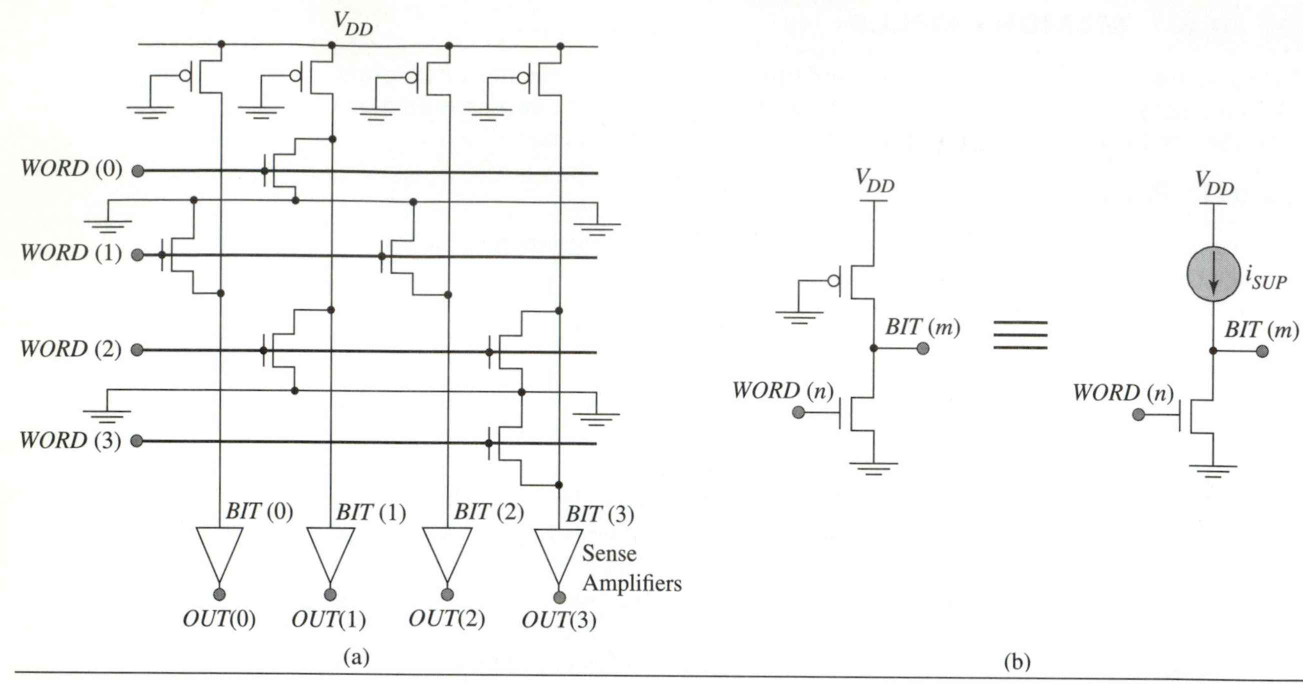

➤ **Figure 13.7** (a) 4 × 4 ROM array. (b) Equivalent circuit models for the ROM cell that contains an NMOS transistor and p-channel pull-up acting as a current source.

The MOS ROM cell described above burns static power since the ROM cell is essentially an NMOS inverter with a PMOS current source. We do not have the option of making each cell a static CMOS inverter since that would require a p-channel transistor in every cell, and hence, significantly increase the memory cell area. However, we can use precharging as outlined in Chapter 5 so the cells do not burn static power. A 4 × 4 array of a precharged MOS ROM is shown in Fig. 13.8(a). The operation of the cell begins with the activation of the precharge clock ϕ_P. Since the precharged devices are p-channel devices, activation occurs when ϕ_P goes from high-to-low. At the time when the memory is accessed, ϕ_P is brought high, turning off the precharge devices. After a sufficient amount of settling, the proper word line is activated and each bit line is either discharged or remains high depending on whether a transistor indicating a stored **0**, or no transistor, indicating a stored **1**, is located in the memory cell. Qualitative waveforms of this operation are shown in Fig. 13.8(b).

One possible layout for a 4 × 4 MOS ROM using a minimum feature size of 1.5 μm is shown in Fig. 13.9. The area of the cell is 6 μm × 9 μm and the ROM cell transistor has a $W/L = 4.5/1.5$. If a contact is placed between the diffusion and bit line, then the transistor is connected to the bit line. If a contact is not placed, the transistor is floating and, in effect, there is no transistor connected between the bit line and ground. By changing the contact mask, different stored patterns can be achieved. Since the contact mask is the only mask that changes for the different data, the same mask designs up to the point of making contacts can be used for the fabrication of the ROM. This ROM is called a **mask programmable memory**. Other ROM cells that are electrically programmable are described in [3].

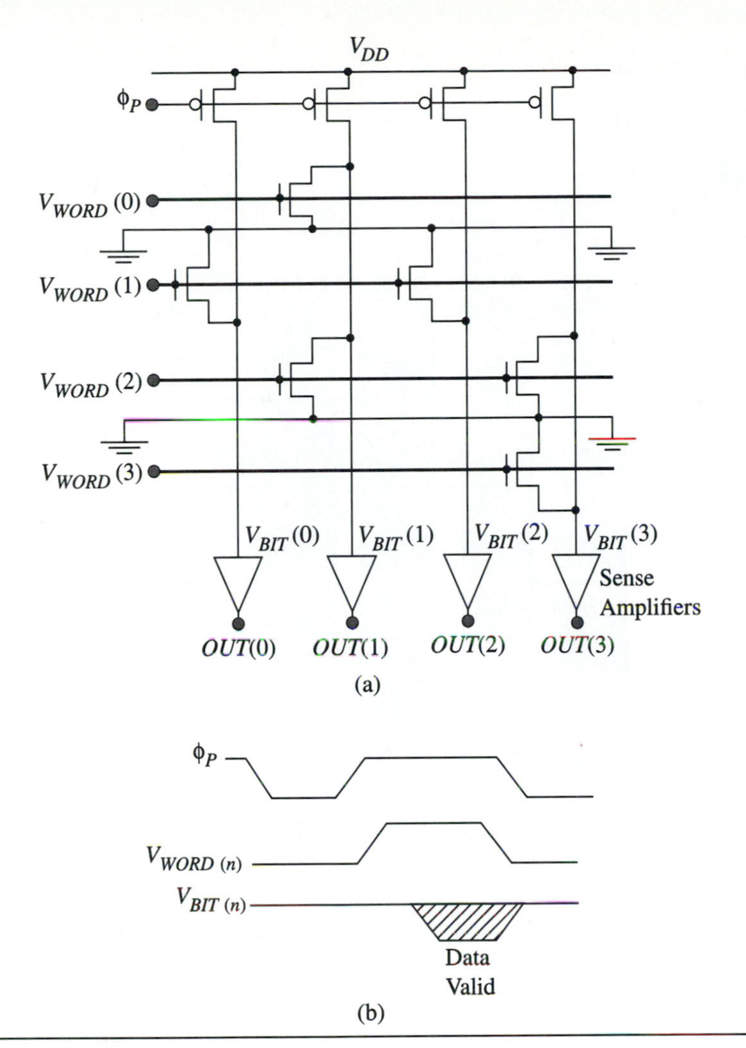

➤ **Figure 13.8** MOS ROM using precharging. (a) 4 × 4 array. (b) Qualitative waveforms showing a read operation.

➤ EXAMPLE 13.1 ROM Array Transient Analysis

In this example, we will explore the transient operation of the ROM cell when it is embedded in a 256 × 256 array and find the access and cycle time given the following device and technology data. The NMOS transistor in the ROM cell has a W/L = 4.5/1.5.

<u>MOS Devices</u>	<u>Parasitic *R* and *C*</u>
$V_{Tn} = 1\text{V}$	$R_{POLY} = 5\ \Omega/\square$
$V_{Tp} = -1\text{V}$	$C_{Jn} = 0.1\ \text{fF}/\mu\text{m}^2$
$\mu_n C_{ox} = 50\ \mu\text{A}/\text{V}^2$	$C_{JSWn} = 0.5\ \text{fF}/\mu\text{m}$
$\mu_p C_{ox} = 25\ \mu\text{A}/\text{V}^2$	$C_{ov} = 0.5\ \text{fF}/\mu\text{m}$
$C_{ox} = 2.3\ \text{fF}/\mu\text{m}^2$	$C_{POLY\text{-}SUB} = 0.06\ \text{fF}/\mu\text{m}^2$
$L_{diffn} = L_{diffp} = 4.5\mu\text{m}$	$C_{METAL\text{-}SUB} = 0.035\ \text{fF}/\mu\text{m}^2$

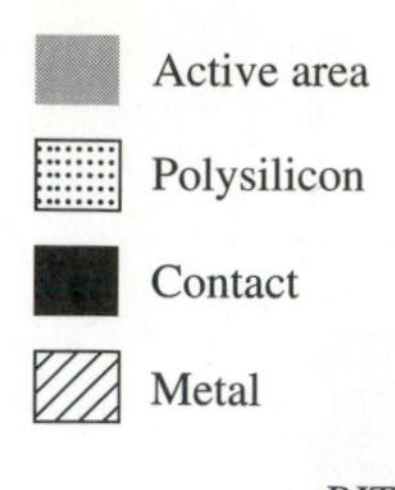

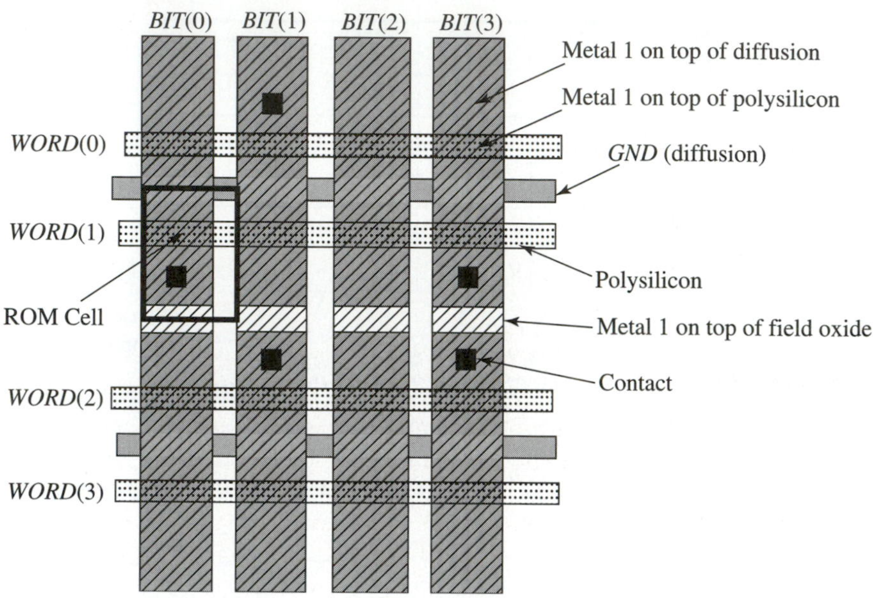

➤ Figure 13.9 Example of a 4 × 4 MOS ROM array layout. The ROM cell dimensions are 6 μm × 9 μm and the ROM cell transistor has $W/L = 4.5/1.5$.

SOLUTION

Begin this analysis by calculating the parasitic resistance and capacitance on the word line and bit line for the ROM cell shown. We will find that the capacitance and parasitic resistance dominate the transient operation. The word line has significant parasitic resistance since it is formed with silicided polysilicon (≈ 5 Ω/□) rather than metal. A metal word line has a significantly reduced resistance at the expense of increased process technology complexity.

Word Line RC

We calculate the resistance of the word line by measuring the number of squares in the ROM cell layout, and find there are four squares per cell. This number comes from the fact that the word line is 6 μm long by 1.5 μm wide in each cell (See Fig. 13.9). The total resistance is given by

$$\#\frac{\square}{\text{cell}} = \frac{6\ \mu\text{m}}{1.5\ \mu\text{m}} = 4\frac{\square}{\text{cell}}$$

$$R_{WORD} = 4\frac{\square}{\text{cell}} \cdot 5\frac{\Omega}{\square} \times 256\ \text{cells} \approx 5\ \text{k}\Omega$$

The capacitance of the word line is made up of the input gate capacitance for the ROM cell transistor and the parasitic capacitance where the polysilicon line runs over the substrate as shown in Fig. Ex 13.1A. For this particular ROM cell, the

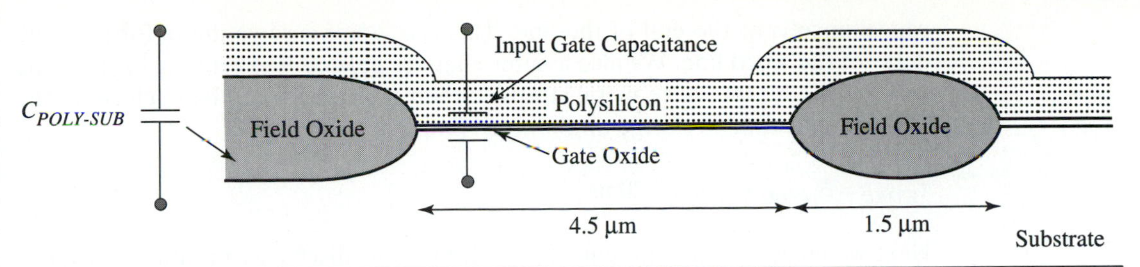

➤ **Figure Ex13.1A**

word-line capacitance is dominated by the gate oxide capacitance compared to the parasitic capacitance as shown by

$$C_{WORD}/\text{cell} = (4.5\ \mu\text{m} \times 1.5\ \mu\text{m})C_{ox} + (1.5\ \mu\text{m} \times 1.5\ \mu\text{m})C_{POLY\text{-}SUB}$$

$$15.5\ \text{fF} + 0.1\ \text{fF} = 15.6\ \text{fF/cell}$$

$$C_{WORD} = 15.6\ \text{fF/cell} \times 256\ \text{cells} \approx 4.0\ \text{pF}$$

Bit-Line RC

Assume the bit line is formed by a metal line, whose resistance is negligible. The capacitance on the bit line is composed of both an area and perimeter component of the junction capacitance from the drain diffusion, as well as the overlap capacitance from each ROM cell transistor in the array. These three components lead to a capacitance given by

$$C_{BIT}/\text{cell} = C_{Jn}(4.5\ \mu\text{m} \cdot 4.5\ \mu\text{m}) + C_{JSWn}(4.5\ \mu\text{m} + 2(4.5\ \mu\text{m})) + C_{ov}(4.5\ \mu\text{m})$$

$$= 2.02\ \text{fF} + 6.75\ \text{fF} + 2.25\ \text{fF} \approx 11.0\ \text{fF/cell}$$

$$C_{BIT} = 11.0\ \text{fF/cell} \cdot 256\ \text{cells} = 2.8\ \text{pF}$$

Transient Operation

Use the circuit model in Fig. Ex 13.1B to estimate the ROM cell transient operation. In this model, we approximate the distributed *RC* time constant of the word line by using a lumped *RC* model that assumes the total resistance and

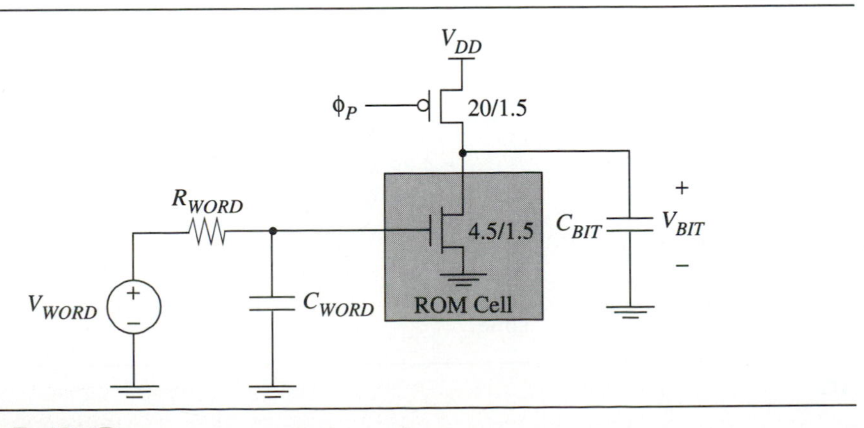

➤ **Figure Ex13.1B**

capacitance is at the end of the line. This is a worst-case estimate for the time delay of the word line. We also assume that the ground line running through the cell and the bit line have metal straps so their resistances can be neglected. The *RC* time constant to charge up the word line is given by

$$\tau_{WORD} = R_{WORD} C_{WORD} = 20 \text{ ns}$$

Next we calculate the amount of time it takes to discharge the bit line by 1 V assuming that the sense amplifier will bring it to full logic levels. We could have chosen a smaller bit-line voltage-swing and require a more stringent sense amplifier design. However, the access time for the ROM is dominated by the *RC* time constant of the word line and not the time it takes to read the cell.

To calculate analytically the amount of time it takes to discharge the bit line, we need to approximate the time-varying gate voltage on the ROM cell transistor with an average voltage as shown in the Fig. Ex 13.1C. A small time-step leads to a better approximation of the waveform. At each time-step, we can calculate the average gate voltage and corresponding device current and the resulting change in bit-line voltage.

A first-order approximation to this method is to assume that the gate voltage is 0 V for $0 \le t \le 0.5\tau$ and is the average voltage of the time-varying waveform between $0.5\tau \le t \le 1.5\tau$. A picture of this first order approximation is also shown in Fig. Ex 13.1C.

We begin calculating the average gate voltage $\overline{V_{WORD}}$ between $0.5\tau \le t \le 1.5\tau$ by writing

$$\overline{V_{WORD}} = \frac{1}{\tau}\int_{0.5\tau}^{1.5\tau} V_{DD}\left(1 - e^{-t/\tau}\right) dt$$

Evaluating this equation with $V_{DD} = 5$ V, we find $\overline{V_{WORD}} = 3.1\text{V}$. This approximation assumes the time to discharge C_{BIT} is less than τ when $\overline{V_{WORD}} = 3.1\text{V}$. It also assumes the value of V_{WORD} at $0 \le t \le 0.5\tau$ is so small that there is very little current. Although the choice of evaluating V_{WORD} between $0.5\tau \le t \le 1.5\tau$ seems arbitrary, the results give a good starting point for your design. The design can be optimized by iterating using SPICE.

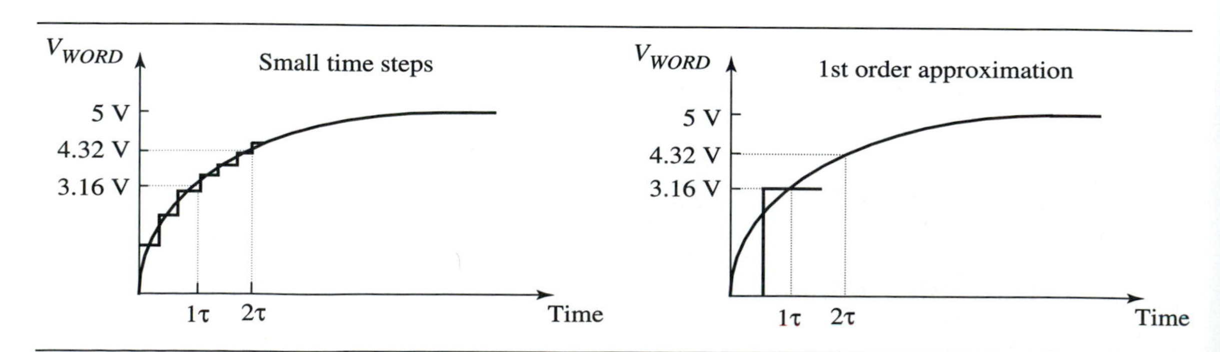

➤ **Figure Ex13.1C**

With this assumption, we can calculate the current that the ROM cell transistor will use to discharge the bit line. We assumed the bit line was precharged to the full V_{DD} value and the ROM cell transistor is operating in the constant-current region. We calculate the current as

$$I_{Dn} = \frac{W}{2L}\mu_n C_{ox}(\overline{V_{word}} - V_{Tn})^2 = \frac{4.5}{3.0}\left(50\,\frac{\mu A}{V^2}\right)(3.1V - 1V)^2 = 330\ \mu A$$

We calculate the time it takes for the bit line to discharge 1 V, $t_{\Delta V_{BIT}}$ as

$$t_{\Delta V_{BIT}} = \frac{C_{BIT}\Delta V_{BIT}}{I_{Dn}} = \frac{(2.8\,\text{pF})\,(1\,\text{V})}{330\ \mu A} = 8.5\ \text{ns}$$

We define the **read time** of this cell as the sum of the word line charging time and bit line discharging time. The read time of this ROM cell is estimated by adding $0.5\tau_{WORD}$ and the time it takes for the bit line to discharge 1 V as given by

$$t_{READ} = 0.5\tau_{WORD} + t_{\Delta V_{BIT}} = 10\ \text{ns} + 8.5\ \text{ns} = 18.5\ \text{ns}$$

This read time is approximately 18.5 ns. The time it takes to charge the word line is a significant portion of the total read time. Note: We have not taken into account the time for addresses to be buffered and decoding operations to take place, which would both be added to the read time to find the access time.
The cycle time is the time to precharge the bit lines added to the access time. In this case it is governed by the time it takes the p-channel device to charge the bit line to V_{DD}. Most of this time is spent with V_{SD} small and the p-channel transistor in its triode region. Given that the p-channel device is sized at 20/1.5 and its $V_{SG} = V_{DD}$ during precharging, we can calculate the time it takes to charge the bit line back to V_{DD} after a read cycle by estimating an average resistance of the p-channel device during charging. We estimate the average resistance by considering the device resistance when $V_{SD} = V_{DD} + V_{Tp}$ and $V_{SD} = 0$ V. These two points are the two extreme points of the triode region and are used to obtain the average resistance of the p-channel device in the triode region. The average p-channel device resistance is given by

$$R_{ON}(V_{SD} = V_{DD} + V_{Tp}) = \frac{1}{\left(\frac{W}{L}\right)_p \mu_p C_{ox}\left(V_{SG} + V_{Tp} - \left(\frac{V_{DD} + V_{Tp}}{2}\right)\right)} = 1500\ \Omega$$

$$R_{ON}(V_{SD} = 0\text{ V}) = \frac{1}{\left(\frac{W}{L}\right)_p \mu_p C_{ox}(V_{SG} + V_{Tp})} = 750\ \Omega$$

$$\overline{R_{ON}} = \frac{R_{ON}(V_{SD} = V_{DD} + V_{Tp}) + R_{ON}(V_{SD} = 0\text{ V})}{2} = 1125\ \Omega$$

The RC time constant to charge the bit line is given by

$$\tau_{BIT} = \overline{R_{ON}}C_{BIT} = 3.15\ \text{ns}$$

Assume it takes approximately three time constants to charge "fully" the bit line. To find the cycle time (not including time for address buffers/decoders), we add the read time and precharge time

$$t_{CYCLE} = t_{READ} + 3\tau_{BIT} = 18.5 \text{ ns} + 3\,(3.15 \text{ ns}) \approx 28 \text{ ns}$$

The most interesting lesson to be learned from this example is that the access and cycle time in memories are dominated by parasitic capacitances and resistances. In this particular ROM cell, if the access time needs to be reduced, a potential technology improvement would be to reduce the resistance of the word line. This reduction could take place by adding an additional level of metal or by reducing the resistance of the word line through process enhancements. The cycle time can be reduced by increasing the size of the p-channel precharge device to overcome the parasitic bit-line capacitance. This example illustrates the close interaction of circuit design, device physics, and process technology in memory design.

13.3.2 Static RAM Cells

As indicated above, we often want the ability to read *and* write data into a memory cell. The first RAM cell that we will describe is a static RAM cell. An SRAM cell stores valid data as long as the power supply is on. To understand the operation of a static RAM cell, we begin by looking at two inverters cascaded in series as shown in Fig. 13.10(a). The transfer function has an output voltage that is low when the input voltage is low, a high-gain region and an output voltage that is high when the input voltage is high. When we connect the output to the input, the operating points can be determined using graphical analysis. We have added a line with $V_{OUT} = V_{IN}$ to the transfer function as shown in Fig. 13.10(b).

There are three operating points for this circuit: (1) at V_{OL}, (2) at V_{OH}, and (3) an unstable operating point in the transition region. To see why the center point is unstable, note that if the input voltage equals the output voltage, and the input increases just a slight amount, the output voltage will also increase a slight amount. Hence, this inverter chain will drive the output voltage to a high level. If the input is decreased from the center point slightly, the inverter chain will drive the output voltage to a low level. Essentially, we have implemented an amplifier with positive feedback. As long as the open-loop gain of the two inverters is greater than one, we will produce two distinct states and regeneration toward one of the two states will occur.

By taking the positive-feedback circuit shown in Fig. 13.10(b) and adding two access transistors, we have the topology of a typical six-transistor CMOS SRAM cell shown in Fig. 13.11. We use n-channel access transistors in order to have high-speed operation between the cross-coupled inverters and bit lines. We use the fully differential signals BIT and $\overline{BIT}$ and store data as Q and $\overline{Q}$ to ensure good common-mode rejection.

Finally, it is interesting to look at the function of p-channel transistors. Recalling the discussion on dynamic logic in Chapter 5, we could store Q and $\overline{Q}$ on the gate capacitance of M_1 and M_3. However, due to the leakage current of the pn junction of the NMOS transistors, the charge would eventually leak to the substrate. This would make both nodes equal to the substrate voltage since no charge is supplied when the word-line voltage is low. This four-transistor circuit is not satisfactory since we want the state of the memory to remain for the whole time the power supply voltage is on.

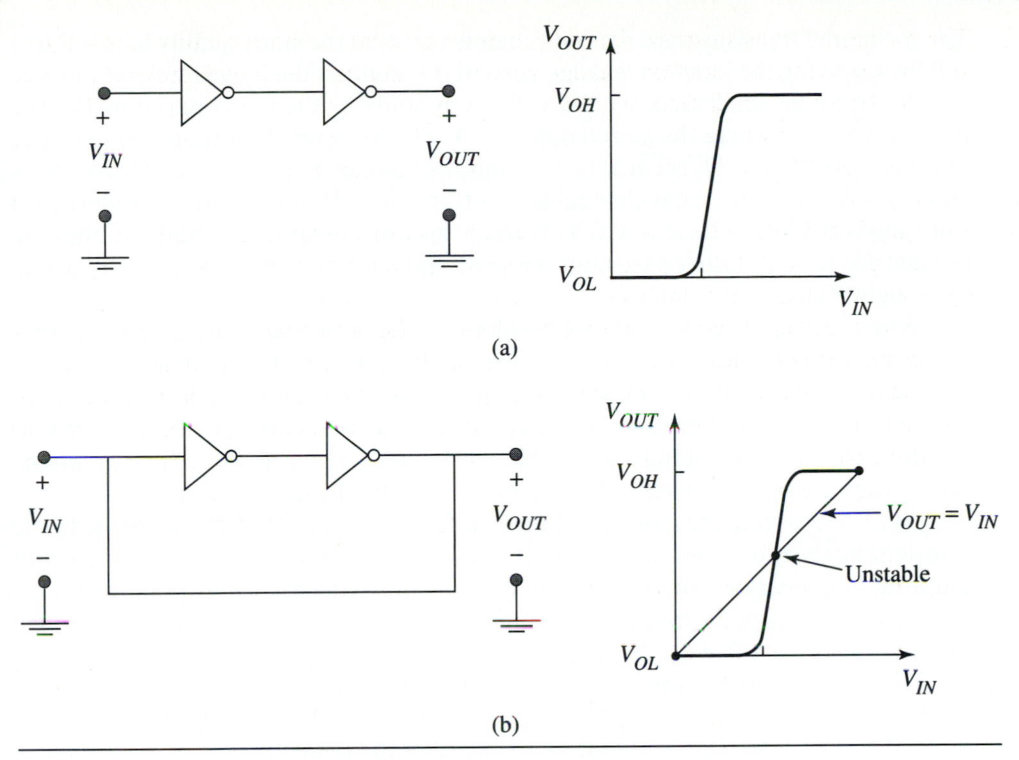

➤ **Figure 13.10** Cascade of two inverters with their voltage transfer characteristic. (a) No feedback. (b) Positive feedback.

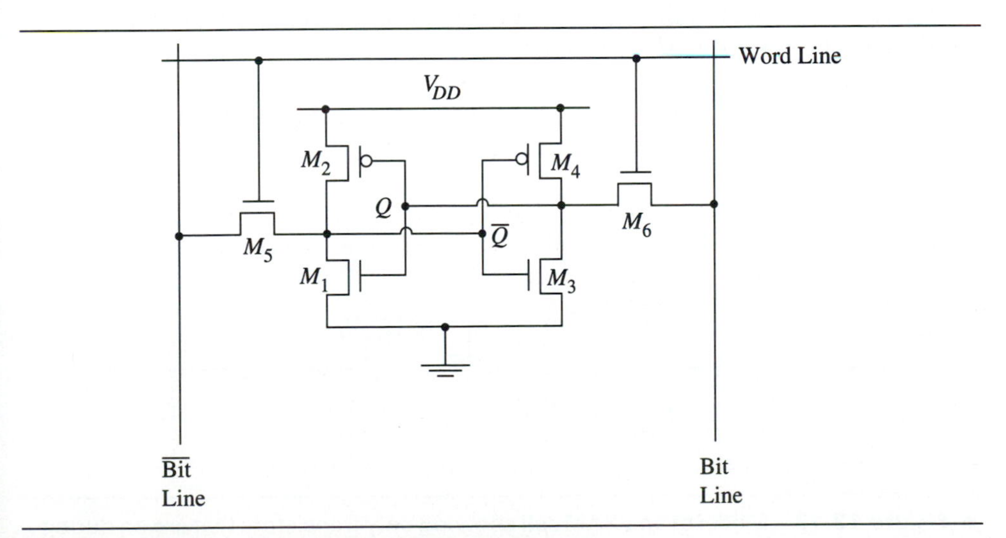

➤ **Figure 13.11** Six-transistor CMOS SRAM cell.

The p-channel transistors are the components that give the static quality to this RAM cell by supplying the junction leakage current to maintain the logical state of the cell.

We begin the analysis of the SRAM cell by studying the read operation. For the read operation, assume the stored data $Q = \mathbf{1}$, which means the voltage on the gates of transistors M_1 and M_2 is equal to V_{DD} and the voltage on the gates of M_3 and M_4 is equal to 0 V. At $t = 0^-$, the n-channel access transistors M_5 and M_6 are considered off since the word-line voltage is at 0 V. During this time, assume that both bit lines are precharged to V_{DD}. Other precharge values could have been chosen, but they would only slightly change the analysis.

With these initial conditions, we break the cell into two parts using only the transistors that are on during the read access as shown in Fig. 13.12. Assume that a sense amplifier is connected as a differential amplifier at the bottom of the two bit lines, and that it requires a difference in voltage ΔV_{BIT} equal to approximately 200 mV to rapidly yield an output of full logic levels. The value chosen for ΔV_{BIT} depends on the sense amplifier design which will be covered in the next section.

At $t = 0^+$, assume the rise time of the word line is infinitely fast as we did with the transient analysis of other digital circuits. We will consider the RC time constant of the word line separately in a later example. The drain current flowing in transistor M_1 and access transistor M_5 will begin to discharge $C_{\overline{BIT}}$ from V_{DD}. Transistors M_4 and M_6 have no current flowing since $V_{SD4} = V_{DS6} = 0$ V.

With these starting conditions, and assumptions, we begin a simple analysis to estimate the time it takes for the SRAM cell to develop a 200 mV differential voltage on the bit lines. Assume capacitance $C_{p\overline{Q}}$, which is the parasitic capacitance between transistors M_1 and M_5, is much smaller than the total bit-line capacitance. This is an excellent assumption for any reasonable sized array. Transistor M_1 can discharge $C_{p\overline{Q}}$ much faster than the discharging of $C_{\overline{BIT}}$. Therefore, transistor M_1 is in the triode region and its current can be estimated as

$$I_{D1} = \left(\frac{W}{L}\right)_1 \mu_n C_{ox}\left((V_{DD} - V_{Tn}) - \frac{V_{DS1}}{2}\right) V_{DS1} \qquad (13.1)$$

Transistor M_5 is in the constant-current region. Its gate voltage is at V_{DD}, while its source voltage is set by the drain-source voltage of M_1. We can write an expression for the current flowing in M_5 as

$$I_{D5} = \left(\frac{W}{L}\right)_5 \frac{\mu_n C_{ox}}{2} (V_{DD} - V_{Tn} - V_{DS1})^2 \qquad (13.2)$$

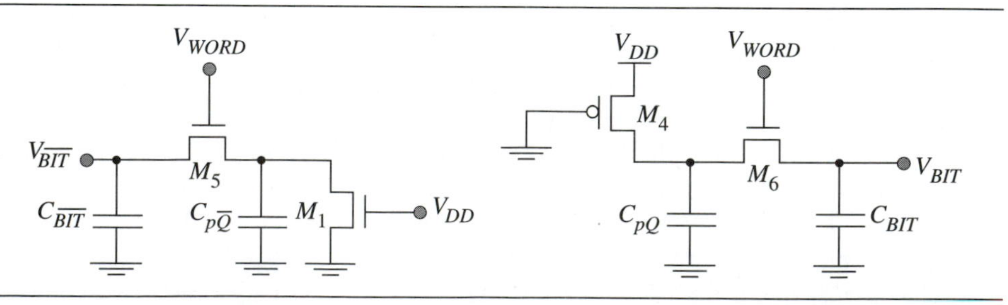

➤ **Figure 13.12** Model of an SRAM cell showing only transistors that are on during a read access with $Q = \mathbf{1}$.

Solving for V_{DS1} in Eq. (13.1) and substituting it into Eq. (13.2), we can solve for I_{D5} after recognizing that $I_{D5} = I_{D1}$

$$I_{D5} = \frac{k_5}{2\left(1 + \frac{k_5}{k_1}\right)} (V_{DD} - V_{Tn})^2 \tag{13.3}$$

$$\text{where } k_1 = \left(\frac{W}{L}\right)_1 \mu_n C_{ox} \quad \text{and} \quad k_5 = \left(\frac{W}{L}\right)_5 \mu_n C_{ox}$$

Since the change on the bit-line voltage is small, we assume the current calculated in Eq. (13.3) does not appreciably change. The time it takes to discharge the bit line ΔV_{BIT} is given by

$$t_{\Delta V_{BIT}} = \frac{C_{BIT} \Delta V_{BIT}}{I_{D5}} \tag{13.4}$$

Now we turn to the write operation for the static RAM cell and first look at its static operation. In order for this SRAM cell to operate regeneratively between the two stored states, we need the voltage gain of the two inverters to be greater than 1. A typical number is on the order of 5. To achieve a maximum density, we should size the two inverters consisting of devices M_1 through M_4 using minimum geometry devices. To analyze the voltage gain of the minimum geometry inverter

1. Find V_M.
2. Find I_D flowing into each inverter.
3. Calculate the transconductance and output resistance at the operating point.
4. Find the voltage gain using

$$A_v = (g_{mn} + g_{mp})(r_{on} \| r_{op}) \tag{13.5}$$

This is exactly the same procedure used in our analysis of static inverters in Chapter 5. In Example 13.2 we will determine a typical value for the voltage gain.

It is difficult to perform an accurate hand analysis of the transient response of the SRAM cell during the write operation. However, with some assumptions, we can gain an excellent intuitive understanding about its operation and then refine the estimate of its performance using SPICE. Assume we have stored a **1** and that we want to write a **0** into the memory cell. To estimate the time it takes to write the SRAM cell, t_{WRITE}, we find the time it takes to get the internal voltages, either V_Q or $V_{\overline{Q}}$, to become V_M and then add the RC delay of the word line. After the internal node voltages reach V_M, regeneration takes place and the cell moves quickly toward the opposite state.

Figure 13.13(a) shows the node voltages just after the word line comes to its full value. We have removed transistors M_2 and M_3 in Fig. 13.13(b) since they are off for a stored **1**. Access-transistor M_5 looks diode-connected and its function is to charge node $\overline{Q}$ to V_M while fighting the pull-down transistor M_1. On the other side of the SRAM cell, transistor M_6 is pulling down node Q from V_{DD} while fighting p-channel transistor M_4. Since the gate drive of M_6 is always large and device M_4 is a minimum geometry p-channel transistor, assume these transistors in the SRAM cell govern the

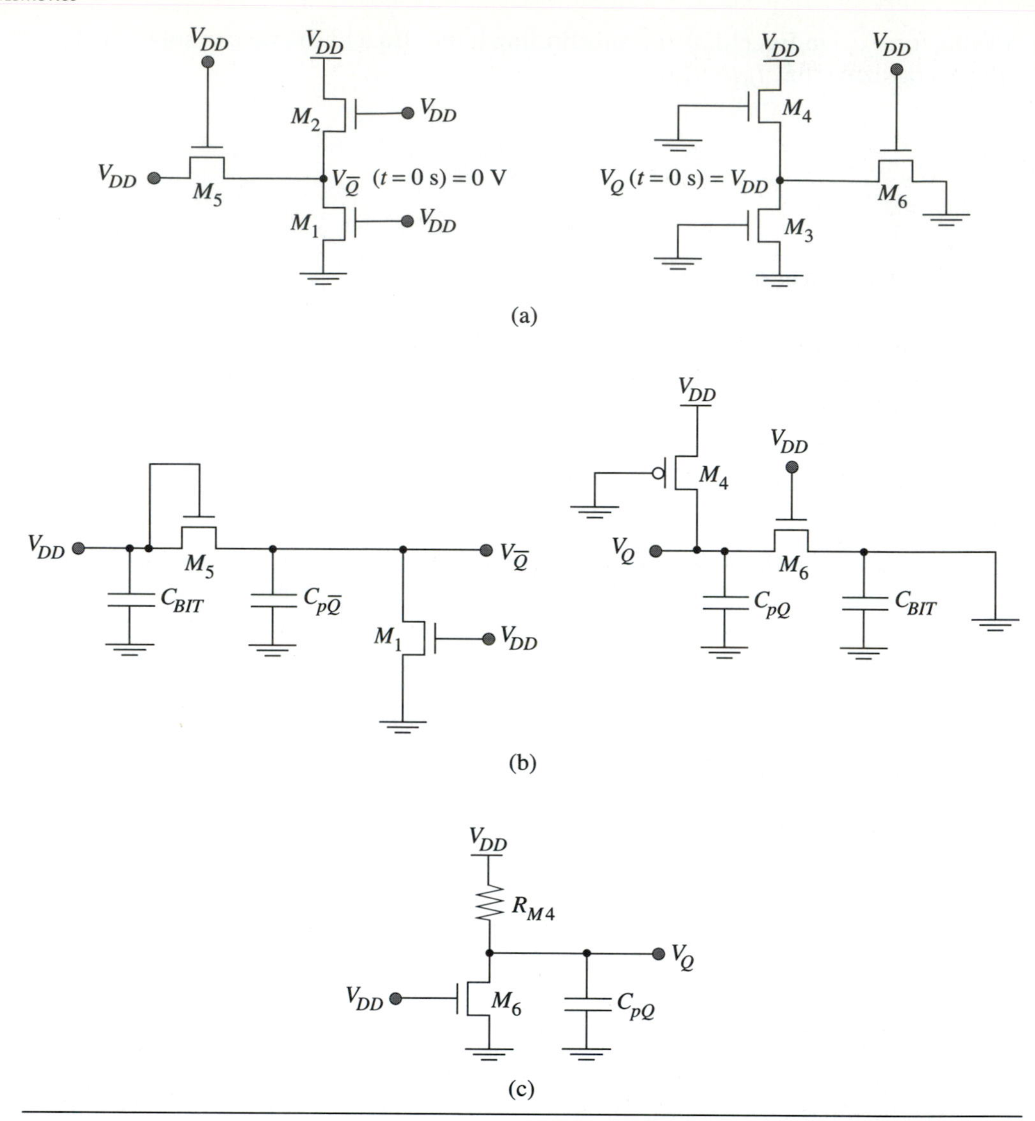

➤ Figure 13.13 Model to analyze SRAM write operation. (a) Node voltages at $t = 0$ S assuming $Q = \mathbf{1}$ and the data to be written is a **0**. (b) Model of an SRAM cell showing only the transistors that are on during a write **0** operation. (c) Simple model of transistor M_6 discharging node Q while fighting the p-channel pull-up M_4 modeled as a resistor during a write **0** operation.

time it takes to move one of the internal nodes to V_M. Figure 13.13(c) shows a simple model of transistors M_4 and M_6 discharging node Q. Since we assume M_4 is in the triode region, V_{SD4} is small and therefore , we model it as a resistor. The change in node voltage V_Q is given by

$$\frac{dV_Q}{dt} = \frac{V_{DD} - V_Q(t)}{R_{M4}C_{pQ}} - \frac{I_{D6}}{C_{pQ}} \tag{13.6}$$

where the on-resistance of the p-channel transistor M_4 is estimated as

$$R_{M4} = \frac{1}{\left(\frac{W}{L}\right)_4 \mu_p C_{ox}\left(V_{DD} + V_{Tp} - \frac{V_{DD} - V_Q(t)}{2}\right)} \tag{13.7}$$

It is clear from both Fig. 13.13(c) and Eq. (13.6) that we require the drain current of the n-channel transistor M_6 to be larger than the current flowing through the p-channel transistor M_4 so the node is discharged towards V_M. Rather than solving the differential equation given in Eq. (13.6) , simply estimate the current in the n-channel transistor M_6 at a worst-case point (smallest current), namely when V_Q is equal to V_M and that it is operating in the triode region. By substituting Eq. (13.7) into Eq. (13.6) and noting $\Delta V_Q = V_{DD} - V_M$, the write time is given by

$$t_{WRITE} = -(V_{DD} - V_M)\left[\frac{V_{DD} - V_M}{C_{pQ}}\left(\frac{W}{L}\right)_4 \mu_p C_{ox}\left(V_{DD} + V_{Tp} - \frac{V_{DD} - V_M}{2}\right)\right.$$
$$\left. - \frac{1}{C_{pQ}}\left(\frac{W}{L}\right)_6 \mu_n C_{ox}\left(V_{DD} - V_{Tn} - \frac{V_M}{2}\right)V_M\right]^{-1} \tag{13.8}$$

We will account for the time to charge the word line in Example 13.2. When numbers are plugged into the above analysis of the static RAM cell, it shows a good first-cut design is to use minimum geometry transistors for both the n and p- channel transistors that form the cross-coupled inverters and to use twice the minimum width over the minimum length for the access transistors to ensure reasonable write times. Since the operation of this circuit was greatly simplified by the hand analysis, SPICE simulations of the full read and write cycle are essential to guarantee proper transient performance. Finally, the access transistor's size is limited, since the extra width adds additional bit line capacitance from each cell in the memory array, and hence, impacts the overall transient performance.

➤ EXAMPLE 13.2 SRAM Cell Read and Write Operation

In this example, we will estimate the time it takes for the SRAM cell read and write operation. The cells are embedded in an array that is 128 words by 32 bits. This notation means each word line has 32 SRAM cells and each bit line has 128 cells. In this example the SRAM cell bit lines and the power supply voltages V_{DD} and V_{SS} are carried using metal; therefore, their resistance can be ignored. However, the word line is formed with a silicided polysilicon line and has a significant resistance, as we saw in the ROM cell example. The technology and device parameters are the same as those used in Example 13.1.

As in most memory arrays, the read and write time will be dominated by the parasitic resistance and capacitance on the word and bit lines. To calculate these parasitics, first design the memory cell itself. Start by assuming that devices M_1–M_4 are minimum geometry devices with $W/L = 3/1.5$, and the access transistors M_5 and M_6 have a $(W/L) = 6/1.5$. These device sizes lead to a layout where the word lines are 27 μm/cell and the bit lines are 54 μm/cell.

SOLUTION

Word-Line Time Constant

The resistance of the word line is estimated by calculating the number of squares per cell and multiplying it by the sheet resistance of the word line as

$$\#\frac{\square}{\text{cell}} = \frac{27\ \mu\text{m}}{1.5\ \mu\text{m}} = 18\frac{\square}{\text{cell}}$$

$$R_{WORD} = 18\frac{\square}{\text{cell}} \cdot 5\frac{\Omega}{\square} \cdot (32\text{ cells}) = 2.88\text{ k}\Omega \approx 3\text{ k}\Omega$$

The parasitic polysilicon-substrate capacitance is a very small percentage of the input transistor's capacitance, as we saw in the ROM cell example. To estimate the capacitance on the word line, we begin by considering the access-transistor gate capacitance during a read operation. Since we assumed in the cell design that bit lines are precharged to V_{DD}, one access transistor will never have a gate-source voltage larger than a threshold voltage, and hence, will be cutoff. Its capacitance is given by

$$C_{ACCESS-OFF} = 2WC_{ov} + WL(C_{ox} \parallel C_b)$$

where $C_{ox} \parallel C_b$ is the gate-bulk capacitance that is composed of the gate dielectric capacitance in series with the depletion capacitance in the semiconductor.
On the other side of the memory cell, the access transistor has its drain connected to V_{DD}, and its source connected to ground. This access transistor is operating in the constant-current region throughout most of the word-line voltage swing, and its capacitance is given by

$$C_{ACCESS-ON} = 2WC_{ov} + \frac{2}{3}WLC_{ox}$$

The capacitance of the on and off access transistor could be summed to find the word-line capacitance. A further complication occurs during the write operation where the capacitance depends on the data to be written.

A simpler first-cut approximation for the hand analysis is to sum the capacitances seen looking into the two access transistors assuming the full gate capacitance for both transistors. Using this assumption, the total word-line capacitance is given by

$$\frac{C_{WORD}}{\text{cell}} = 2WLC_{ox} = 2(6\ \mu\text{m})\,(1.5\ \mu\text{m})\left(2.3\frac{\text{fF}}{\mu\text{m}^2}\right) = 41.4\,\frac{\text{fF}}{\text{cell}}$$

$$C_{WORD} = 41.4\frac{\text{fF}}{\text{cell}} \times 32\text{ cells} = 1.33\text{ pF}$$

and the word line RC time constant is

$$\tau_{WORD} = R_{WORD}C_{WORD} = (3\text{ k}\Omega)\,(1.33\text{ pF}) = 4.0\text{ ns}$$

Bit-Line Capacitance

The bit-line capacitance is dominated by the overlap and drain/source-bulk capacitance of each access transistor. The overlap capacitance given the width of the access transistor is 6 μm is equal to

$$\frac{C_{BIT-ov}}{\text{cell}} = WC_{ov} = 6\ \mu\text{m}\left(0.5\frac{\text{fF}}{\mu\text{m}}\right) = 3\ \text{fF}$$

The depletion capacitance consists of an area and perimeter component and is given by

$$\frac{C_{BIT-diff}}{\text{cell}} = (W \times L_{diff})\,C_{Jn} + [W + 2L_{diff}]\,C_{JSWn}$$

$$\frac{C_{BIT-diff}}{\text{cell}} = (6\ \mu\text{m} \times 4.5\ \mu\text{m})\,C_{Jn} + [6\ \mu\text{m} + 2\,(4.5\ \mu\text{m})]\,C_{JSWn}$$

$$\frac{C_{BIT-diff}}{\text{cell}} = 2.7\ \text{fF} + 7.5\ \text{fF} = 10.2\ \text{fF}$$

Since we are assuming 128 bits on the bit line, the total bit-line capacitance is approximately 1.7 pF.

Read Operation

As in the ROM cell example, we assume it takes $0.5\tau_{WORD}$ to have an average gate voltage on the cell access transistors of 3.1 V. In the analysis of the SRAM cell, we assumed the word line was charged to V_{DD} to calculate the time it took to discharge the bit line. This assumption is too optimistic when dealing with *RC* delays resulting from the large resistance on the word line. We can modify the analysis to include the reduced word-line voltage. Referring to the following figure, we first calculate the effective resistance of transistor M_1.

$$R_{M1} = \frac{1}{\left(\frac{W}{L}\right)_1 \mu_n C_{ox}\,(V_{GS1} - V_{Tn})} = \frac{1}{\left(\frac{3}{1.5}\right)_1 \left(50\frac{\mu\text{A}}{\text{V}^2}\right)(5\ \text{V} - 1\ \text{V})} = 2.5\text{k}\Omega$$

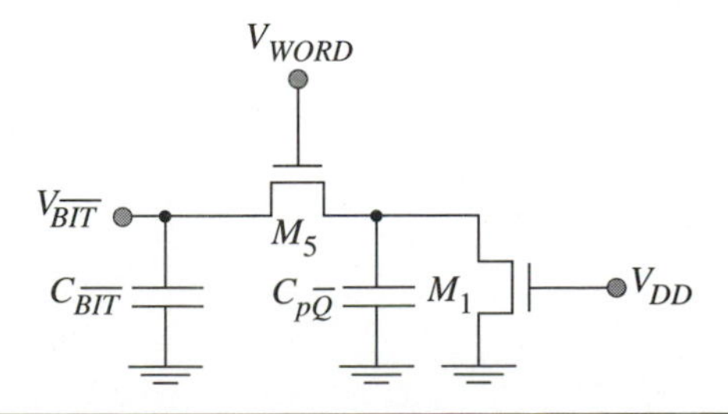

➤ **Figure Ex13.2A**

Next, we calculate the current in access transistor M_5. This transistor is operating in the constant-current region. However, V_{GS5} is reduced by the resistance of device M_1. The current is given by

$$I_{D5} \approx \frac{1}{2}\left(\frac{W}{L}\right)_5 \mu_n C_{ox} (V_{GS5} - V_{Tn})^2$$

where

$$V_{GS5} = \overline{V_{WORD}} - I_{D5} R_{M1}$$

We next solve for the current in transistor M_5 using iteration, starting with an initial value of $V_{GS5} = 2.5\text{V}$. Substituting this value into the expression for the current in M_5, we find $I_{D5} = 225\ \mu\text{A}$. Substituting $I_{D5} = 225\ \mu\text{A}$ and $\overline{V_{WORD}} = 3.1\ \text{V}$ into the expression involving the resistance of device M_1, we find $V_{GS5} = 2.54\ \text{V}$. Continued iteration yields a consistent current $I_{D5} = 231\ \mu\text{A}$ at $V_{GS5} = 2.52\ \text{V}$.

Now that I_{D5} has been calculated, we can estimate the time for the bit line to discharge 200 mV. Assume a sense amplifier is placed at the bottom of the bit lines to drive them to their full logic level. The time to sense is

$$t_{\Delta V_{BIT}} = \frac{C_{BIT} \Delta V_{BIT}}{I_{D5}} = \frac{(1.7\ \text{pF})\ (0.2\text{V})}{(230\ \mu\text{A})} \approx 1.5\ \text{ns}$$

In summary, we have found that the RC time constant of the word line is approximately 4.0 ns and the time to discharge the bit line is approximately 1.5 ns, yielding a total read time given by

$$t_{READ} = 0.5\tau_{WORD} + t_{\Delta V_{BIT}} = 2.0\ \text{ns} + 1.5\ \text{ns} = 3.5\ \text{ns}$$

Write Operation

Before calculating the transient solution to the write operation, we must ensure that the open-circuit voltage gain of each inverter is greater than 1, so we can guarantee regeneration to full logic levels from the positive-feedback circuit. We begin by calculating V_M.

$$V_M = \frac{V_{Tn} + \sqrt{\frac{k_p}{k_n}}(V_{DD} + V_{Tp})}{1 + \sqrt{\frac{k_p}{k_n}}}$$

where

$$\frac{k_p}{k_n} = \frac{\left(\frac{W}{L}\right)_2 \mu_p C_{ox}}{\left(\frac{W}{L}\right)_1 \mu_n C_{ox}} = 1/2$$

$$V_M = \frac{1\text{V} + \sqrt{1/2}\,(4\text{ V})}{1 + \sqrt{1/2}} = 2.25\text{ V}$$

After calculating V_M, the transconductances of devices M_1 and M_2 at V_M are given by

$$g_{mn} = \left(\frac{W}{L}\right)_1 \mu_n C_{ox}(V_M - V_{Tn}) = 125\ \mu\text{S}$$

$$g_{mp} = \left(\frac{W}{L}\right)_2 \mu_p C_{ox}(V_{DD} - V_M + V_{Tp}) = 87.5\ \mu\text{S}$$

When the gate voltage on transistors M_1 and M_2 is equal to V_M, the currents in transistor M_1 and M_2 can be calculated as

$$I_{D1} = I_{D2} = \frac{1}{2}\left(\frac{W}{L}\right)_1 \mu_n C_{ox}(V_M - V_{Tn})^2 \approx 78\ \mu\text{A}$$

Finally, the open-circuit voltage gain is given by

$$A_v = (g_{mn} + g_{mp})\left(\frac{1}{I_{D1}(\lambda_n + \lambda_p)}\right).$$

Given $\lambda_n = \lambda_p = 0.1\text{V}^{-1}$ $A_v = (212.5\ \mu\text{S})(64\text{ k}\Omega) \approx 13.5$

We see that for the design with minimum geometry transistors and the technology parameters given, the open-circuit voltage gain is sufficient to ensure regeneration.

To obtain the time it takes to write an SRAM cell, we use the simple model shown in Fig. Ex 13.2B and calculate the resistance of the p-channel transistor and the current through transistor M_6. The write time will be dominated by the *RC* time constant to charge the word line and turn on the corresponding access transistors. Assume the gate voltage on device M_6 is equal to 3.1 V after one-half time constant. In this example, M_6 is assumed to be in the constant-current region because the word-line voltage is not at the full V_{DD} level. We write its current as

$$I_{D6} = \frac{1}{2}\left(\frac{W}{L}\right)_6 \mu_n C_{ox}(\overline{V_{WORD}} - V_{Tn})^2 = 440\ \mu\text{A}$$

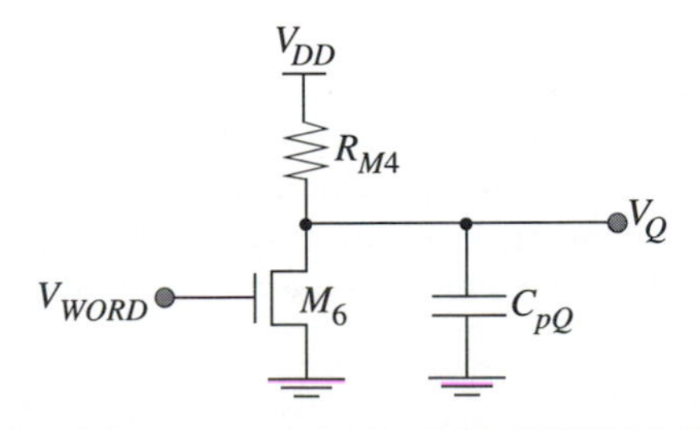

➤ **Figure Ex13.2B**

Using Eq. (13.7) we write the resistance across p-channel transistor M_4 with $V_Q = V_{DD}$ as

$$R_{M4} = \frac{1}{\left(\frac{W}{L}\right)_4 \mu_p C_{ox}\left[(V_{DD} + V_{Tp}) - \left(\frac{V_{DD} - V_M}{2}\right)\right]} = 7.6\ \text{k}\Omega$$

Plugging these values into Eq. (13.6) with $\Delta V_Q = V_{DD} - V_M$ and setting $V_Q = V_M$ yields an estimate of the time it takes to write the SRAM cell in terms of the parasitic capacitance C_{PQ}. At this point, we could attempt to determine accurately capacitance C_{PQ}. However, from previous experience we know that it will be less than 50 fF. Assuming this worst-case value, we find the time it takes to write the SRAM cell is on the order of 1.5 ns. As expected, the *RC* time constant to charge the word line is a significant fraction of the write operation time given by

$$t_{WRITE} \approx 0.5\tau_{WORD} + 1.5\ \text{ns} = 2.0\text{ns} + 1.5\ \text{ns} = 3.5\ \text{ns}$$

In summary, we performed a hand analysis on the SRAM cell and estimated the read and write times to be approximately 3.5 ns for an array of 32 bits by 128 words. SPICE and accurate determination of the parasitic resistances and capacitances are required to find a more accurate solution.

13.3.3 Dynamic RAM Cells

There is a definite trade-off between cell density and the requirement to have full static operation. Dense memories can be achieved if we allow for dynamic operation. We saw this increase in density in Chapter 5 when we looked at dynamic vs. static logic. This increased density is extremely important in memory design.

Although we can use the static RAM cell without the p-channel transistors to yield a fully differential dynamic RAM cell that offers increased density over the SRAM cell, an extremely simple one-transistor cell has become the *de facto* industry standard for DRAM. This cell consists of a single transistor with a capacitor attached to one of its nodes. The other node is attached to the bit line. The word line controls the gate of the MOS transistor. The cell is schematically shown in Fig. 13.14(a). The data is stored as a charge on capacitor C_S. During the write **1** operation, the word line is driven high, the bit line is tied to V_{DD} corresponding to a **1**, and V_{CS} rises toward $V_{DD} - V_{Tn}$. To write a **0** in the cell, the bit line is grounded during the write operation and 0 V is stored on capacitor C_S.

The read operation begins by precharging the bit lines to an intermediate value, namely $V_{DD}/2$. The word line is raised to a high potential and the charge stored on capacitor C_S is shared with that on the bit line. The change in bit-line voltage is given by the change in charge on the bit-line capacitor when the charge stored on capacitor C_S is shared with the bit line. This change in voltage is given by

$$\Delta V_{BIT} = (V_{CS} - V_{DD}/2)\left(\frac{C_S}{C_S + C_{BIT}}\right) \tag{13.9}$$

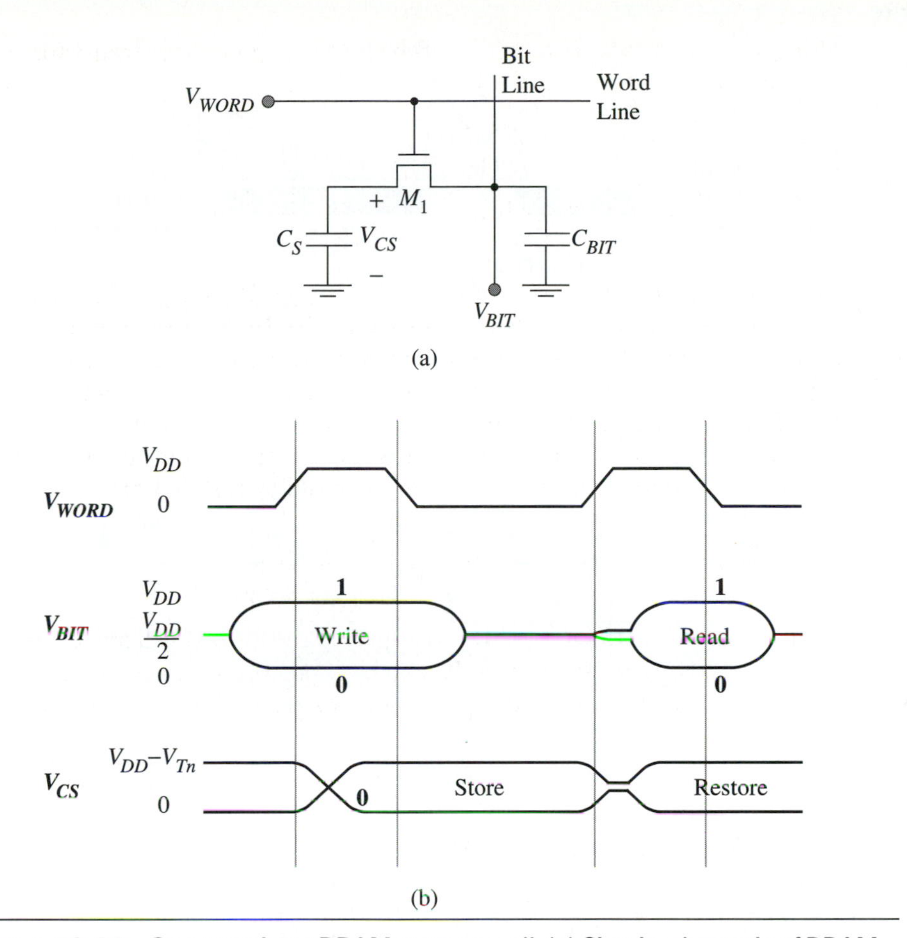

➤ Figure 13.14 One-transistor DRAM memory cell. (a) Circuit schematic of DRAM cell. (b) Signal waveforms describing the read and write operations.

where V_{CS} is the stored voltage on capacitor C_S. The voltage change on the bit line is greatly attenuated by the ratio of the capacitance of the storage cell to that of the bit line. A typical voltage swing between a stored **1** and a stored **0** is on the order of 100 mV. Depending on the exact voltages used in the DRAM design, the capacitance ratio between the storage capacitance and the bit-line capacitance is between 10 and 50. The operation of the cell is described by the signal waveforms shown in Fig. 13.14(b).

It should be noted that the read-out of the one-transistor dynamic RAM cell is destructive. This means that the bit-line voltage needs to be regenerated to a full **1** or **0** level by a sense amplifier in order to restore the full value. This need for restoring the information after the read operation requires that a sense amplifier be used with DRAM cells. With SRAM cells, use of a sense amplifier is only for enhancing the speed of operation and not fundamentally required.

The time it takes to write or read the one-transistor cell is reasonably fast compared to an SRAM cell. The circuit to analyze the transient is a simple *RC* time constant formed by the on-resistance of the single transistor and series combination of the storage capacitor and bit-line capacitor. The series capacitance is on the order of

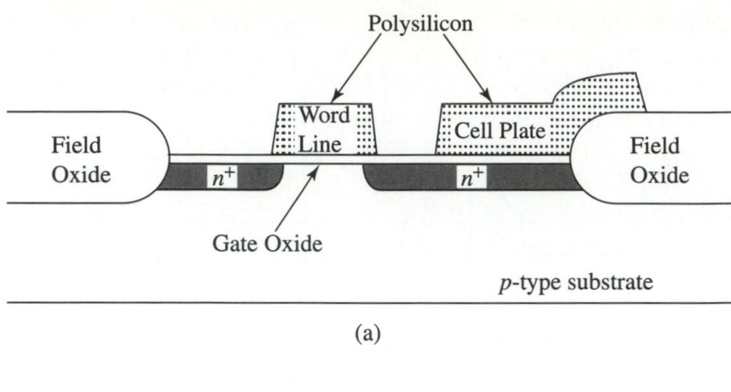

(a)

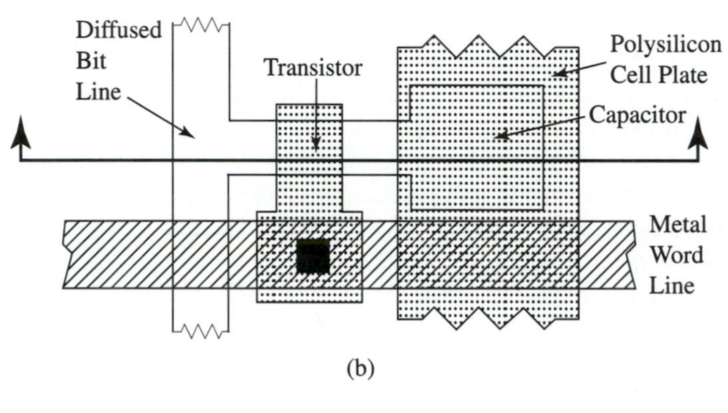

(b)

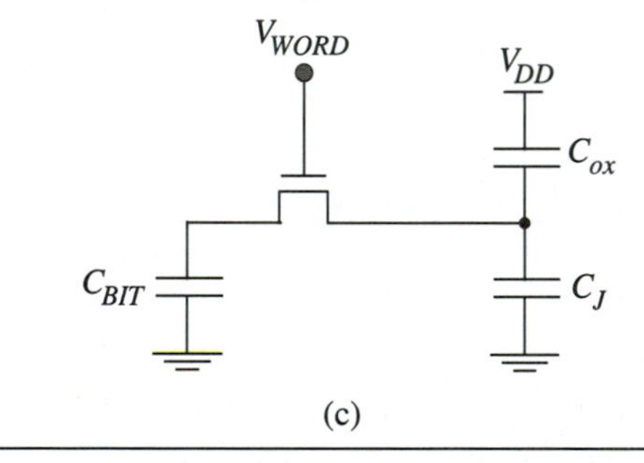

(c)

➤ Figure 13.15 One-transistor memory cell. (a) Cross-section. (b) Layout. (c) Equivalent circuit.

50 fF. The *RC* time constant due to the bit line when performing write and read operations is given by

$$\tau_{BIT} = R_{ON}\left(\frac{C_S C_{BIT}}{C_S + C_{BIT}}\right) \approx R_{ON} C_S \tag{13.10}$$

To achieve the impressive increase in DRAM density that has been seen over the past twenty-five years, the area of the one-transistor and capacitor have continually been decreased. Enhancements in typical MOS technology to reduce the size of the capacitor, as well as a continued reduction in the minimum feature size, have enabled the density of dynamic RAMs to increase.

A tremendous amount of engineering effort has gone into the reduction of DRAM cell area. The following describes some of the major process enhancements to a conventional MOS process that have allowed a dramatic reduction in cell area while maintaining a constant storage capacitance of approximately 30 to 50 fF. The cross-section, layout, and equivalent circuit of the one-transistor DRAM cell are shown in Fig. 13.15. The storage capacitor is made up of the parallel combination of the MOS capacitor and the depletion-region capacitor of the n+ p junction. Although thinning the gate dielectric will increase the capacitance-per-unit area, it is necessary to employ three-dimensional structures for the storage capacitor to have a dramatic reduction in cell area.

One of the first attempts to increase the capacitance-per-unit area of a conventional DRAM cell was to increase the depletion capacitance by locally enhancing the p-type doping concentration under the cell. This increased doping reduces the depletion region width, which increases the capacitance-per-unit area of the pn junction. The total increase in capacitance-per-unit area using this technique was approximately a factor of two.

Clearly, larger increases in capacitance-per-unit area were necessary to continue to increase the density of DRAMs. Truly three-dimensional cells started to appear in production at the 1- and 4-Mbit level. Two methodologies were used: (1) dig a trench to build the capacitor, and (2) build stacks of polysilicon and oxide above the cell. A cross-section of each of these capacitors is shown in Fig. 13.16.

DRAM cells require refresh due to the leakage of charge stored on the capacitor through the reverse-biased pn junction adjacent to the access transistor. For the n-channel implementations with a p-type substrate, the storage node will leak to the substrate potential 0 V, causing a logic **1** to change state to a **0**. Periodicallly refreshing all the memory cells in a DRAM is a requirement. The time interval between refreshing a memory cell is based on the amount of leakage current in the pn junction. Typical refresh times are on the order of tens of milliseconds.

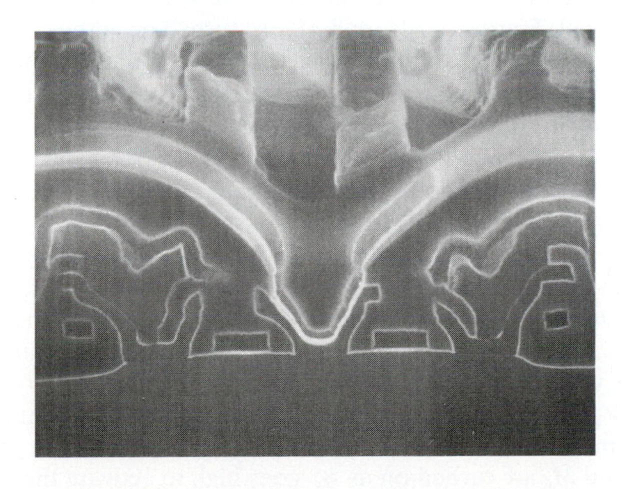

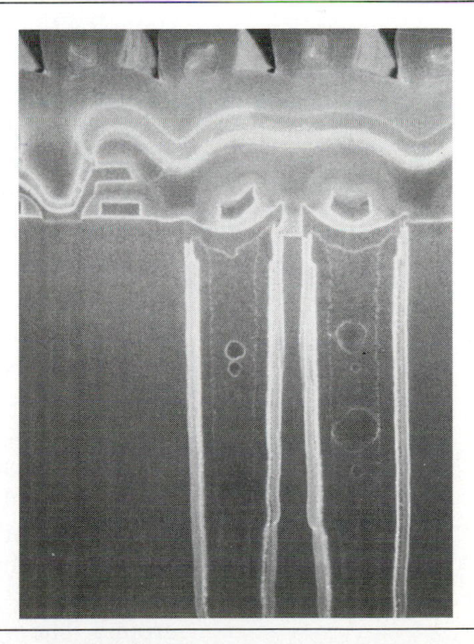

➤ **Figure 13.16** SEM photograph of an early R & D version of advanced DRAMS cells. (a) Stacked capacitor cell. (b) Trench capacitor cell. (Courtesy of Hitachi Corporation)

13.4 SENSE AMPLIFIERS

This section discusses the design and analysis of one of the most common sense amplifier topologies used in dynamic RAM. This sense amplifier also has applications in static RAM. Sense amplifiers are absolutely essential in DRAM and offer performance enhancements in SRAM. Their function in DRAM, is to amplify the signals and to restore the levels on the bit lines to the full logic level since the read operation in a one-transistor cell is destructive. A sense amplifier is not required in an SRAM cell, but can be used to speed up the operation since the bit lines do not have to swing to their full value by discharging and charging through the cell. Instead, the sense amplifier transistors can be made quite large compared to the cell transistors to drive the bit lines to full logic levels.

Sense amplifiers can be built by using a number of topologies. From our work on differential amplifiers, it is known that the common-mode rejection of a differential amplifier allows us to sense smaller signal levels in the presence of common-mode disturbances. We could choose a differential amplifier with a current mirror as described in Chapter 11. However, this amplifier burns static power. Since typical memories have a sense amplifier for each bit line, thousands of sense amplifiers are found on a single chip resulting in huge power dissipation. A linear amplifier is not needed to sense the logic state stored in the memory cell. We only need to compare the state that is stored to some intermediate level and generate a full logic level. This allows us to use positive feedback in the amplifier. In addition, we need to gate the power supplies so there is no direct path between power and ground unless the bit line is being read.

13.4.1 Dynamic-Latch Sense Amplifier: Qualitative Description

The sense amplifier shown in Fig. 13.17 has the attributes to sense both dynamic and static RAM cells. To understand its operation qualitatively, we begin by explaining the function of transistors M_1 through M_4. These transistors form a positive-feedback circuit similar to the SRAM cell. Transistors M_5 and M_6 are turned on at the time the sense amplifier is amplifying and turned off during the rest of the operation, thereby gating the power supply to save power. Transistors M_7, M_8, and M_9 precharge the bit lines to an intermediate value that we choose to be $V_{DD}/2$.

The operation of the sense amplifier can be understood by first recognizing that during the time ϕ_P is high, we are readying the sense amplifier for its next read function. Both bit lines are precharged to $V_{DD}/2$ through transistors M_7 and M_8. During the previous sense cycle, one of the bit lines would be at a full level V_{DD} and the other at 0 V. Transistor M_9 helps share the charge between these bit lines toward $V_{DD}/2$ to speed the precharge time. During this time transistors M_5 and M_6 are off and the positive-feedback circuit is not activated. When ϕ_P goes low, transistor M_7, M_8, and M_9 are disconnected, allowing the bit lines to float. When ϕ_W goes high the word line buffer drives the voltage V_{WORD} toward V_{DD} with a time constant τ_{WORD}, and the input differential signal from the cell is placed on the bit lines as the input to the sense amplifier. Transistors M_5 and M_6 are turned on as ϕ_S goes high to activate the positive-feedback latch forcing the bit lines to swing exponentially toward V_{DD} and ground. The waveforms that qualitatively describe the operation of this sense amplifier are shown in Fig. 13.17(b).

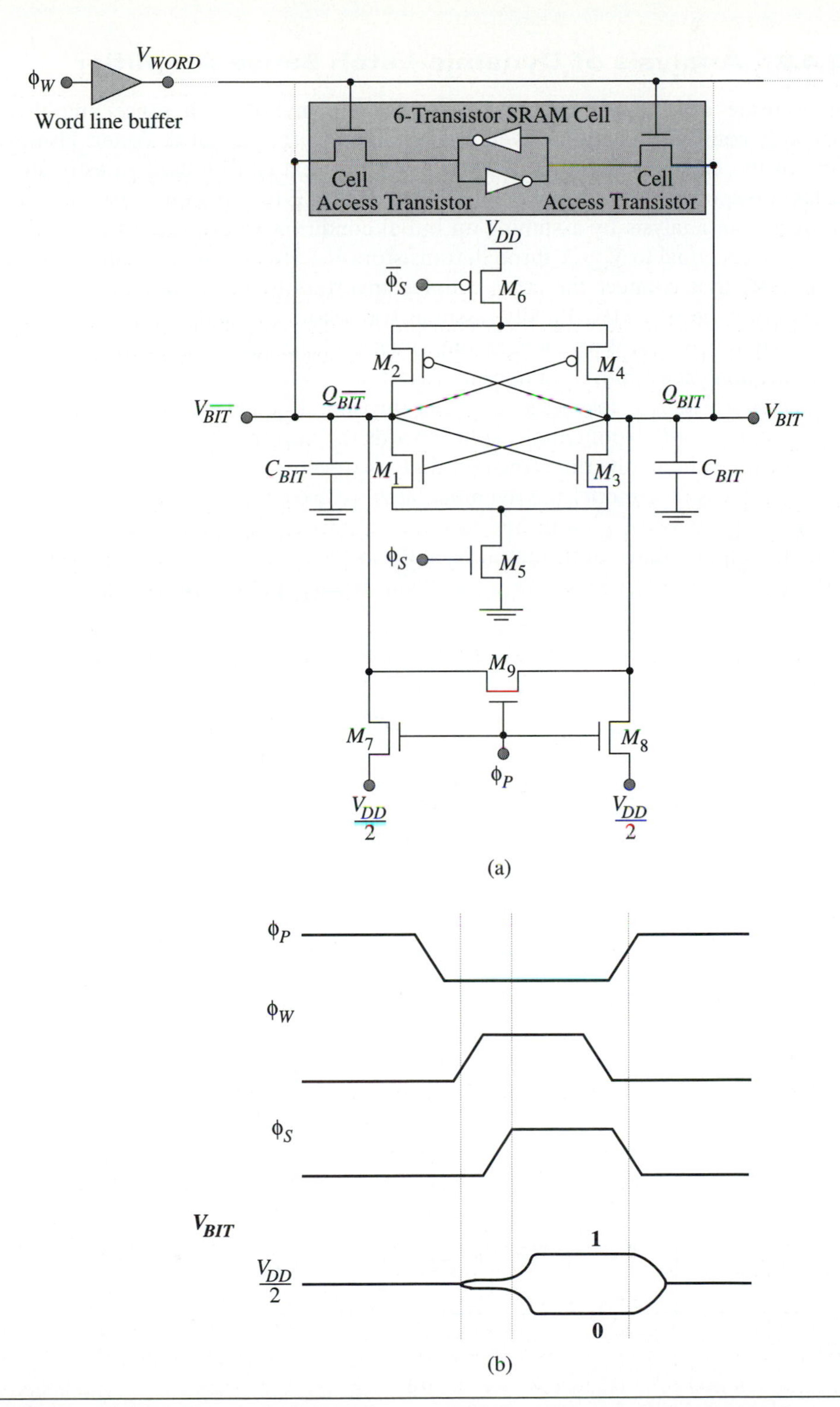

➤ **Figure 13.17** Dynamic-latch sense amplifier. (a) Circuit diagram. (b) Signal waveforms that qualitatively describe the operation. When V_{WORD} goes high the cross-coupled inverters are connected to the bit lines through the access transistors.

13.4.2 Analysis of Dynamic-Latch Sense Amplifier

An accurate large-signal hand analysis of the dynamic-latch sense amplifier is extremely complicated and not very enlightening to the circuit designer. However, many of the concepts we learned throughout this text can be used to estimate the transient response of the amplifier and then use SPICE to give more accurate results. We begin the analysis by assuming an initial condition where the bit-line voltages have been set equal to $V_{DD}/2$ through transistors M_7, M_8, and M_9. Assume transistors M_5 and M_6 that connect the cross-coupled inverters to the power lines, instantaneously turn on at $t = 0^+$. Finally, assume transistors M_5 and M_6 have very small drain-source voltages. With these assumptions in place, we recognize that the amplifier being analyzed is purely differential, and has positive feedback rather than the conventional negative feedback. Because the amplifier has positive feedback, the output voltages will exponentially split towards the power rails.

Now we estimate the time constant that governs the exponential behavior of the output voltages. We model the dynamic latch as two cross-coupled inverters, each having a capacitance C_{BIT} attached to their outputs as shown in Fig. 13.18(a). We define the input voltage of the amplifier ΔV_{IN} as the difference in bit-line voltages at $t = 0$ s, and the output voltage ΔV_{OUT} as the difference in bit-line voltages for $t > 0$ s.

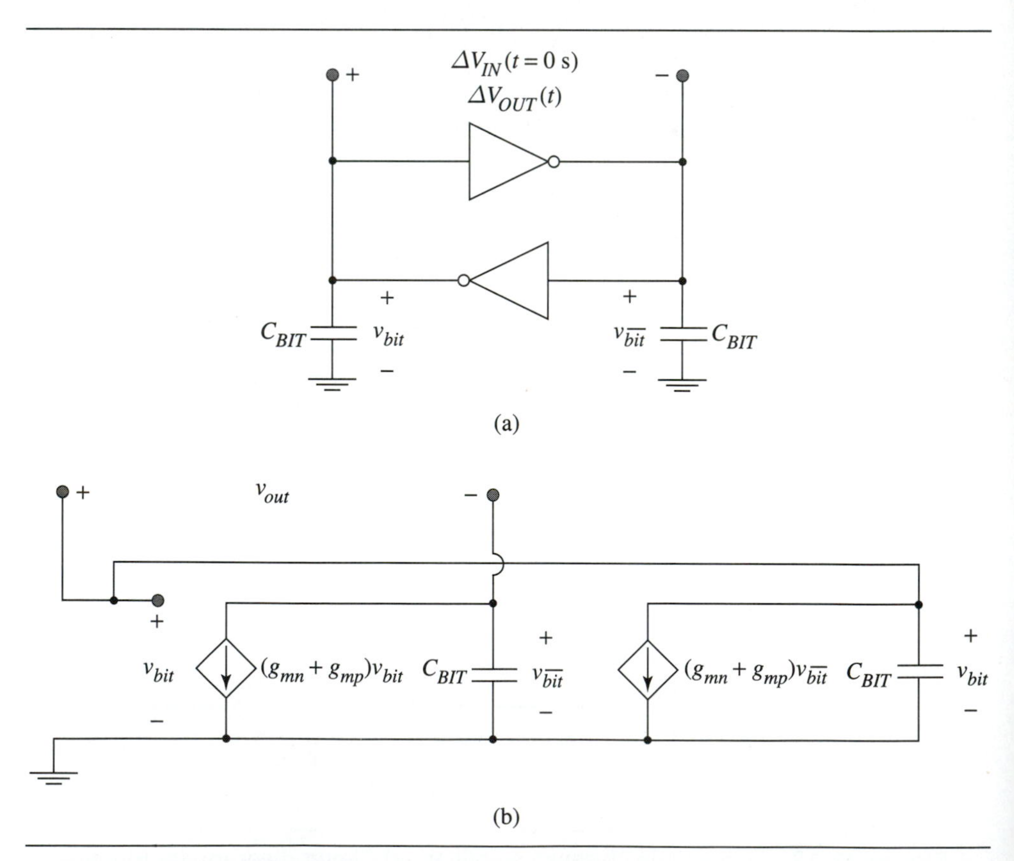

➤ **Figure 13.18** Models to analyze the dynamic latch. (a) Two cross-coupled inverters with C_{BIT} attached to the outputs. (b) Small-signal model to estimate the time response of the bit-line voltages.

A small-signal model of the dynamic latch shown in Fig. 13.18(b) can be used to estimate the time response for the difference in the bit-line voltages. Although the small-signal model is not accurate across the full signal swing, it gives a good estimate of the time constant that governs the exponential time response. In this model, we assume the inverters are identical and neglect the output resistance of the transistors. We start the analysis by writing expressions relating v_{bit} and $v_{\overline{bit}}$ from Fig. 13.18(b):

$$(g_{mn} + g_{mp})\, v_{bit} = -C_{BIT}\frac{dv_{\overline{bit}}}{dt} \tag{13.11}$$

$$(g_{mn} + g_{mp})\, v_{\overline{bit}} = -C_{BIT}\frac{dv_{bit}}{dt} \tag{13.12}$$

Subtracting Eq. (13.12) from Eq. (13.11) we find

$$(g_{mn} + g_{mp})\,(v_{bit} - v_{\overline{bit}}) = -C_{BIT}\left(\frac{dv_{\overline{bit}}}{dt} - \frac{dv_{bit}}{dt}\right) \tag{13.13}$$

We define $v_{out} = v_{bit} - v_{\overline{bit}}$.

Recalling the expression for v_{out} we write

$$\frac{dv_{out}}{dt} = \frac{(g_{mn} + g_{mp})}{C_{BIT}}v_{out} \tag{13.14}$$

Solving Eq. (13.14) we find

$$v_{out} = Ae^{t/\tau} \tag{13.15}$$

where $\tau = \dfrac{C_{BIT}}{(g_{mn} + g_{mp})}$. The constant A is found by noting that at $t = 0$ s, the difference in bit line voltages is ΔV_{IN}. Using these facts and Eq. (13.15) , we write

$$\Delta V_{OUT}(t) = \Delta V_{IN}e^{\frac{(g_{mn} + g_{mp})}{C_{BIT}}t} \tag{13.16}$$

Note that at $t = 0$ s, $\Delta V_{OUT} = \Delta V_{IN}$. As t increases, it will exponentially increase until the amplifier saturates when the output voltage has split toward the power rails. In addition, if the input voltage is equal to 0 V, the output voltage will remain equal to 0 V for all time. This condition, defined as a **metastable state**, is rare.

We can invert Eq. (13.16) to write an equation for the sense time, t_{SENSE} and its relation to the input voltage and final output voltage as

$$t_{SENSE} = \left(\frac{C_{BIT}}{g_{mn} + g_{mp}}\right)\ln\left[\frac{\Delta V_{OUT}(t_{SENSE})}{\Delta V_{IN}}\right] \tag{13.17}$$

We can decrease the sensing time by increasing the transconductance of the input transistors, decreasing the bit-line capacitance, or increasing the initial input voltage.

Finally, note that the input voltage must be large enough to overcome any offsets that might be built into the amplifier due to mismatches in the transistors. In other words, the input signal must be larger than the offset voltage and large enough to ensure a fast sensing time. A reasonable value is approximately 100 mV. In most sense amplifier designs, the channel length of the n and p-channel transistors is not drawn at minimum geometry to help reduce any offsets due to processing mismatches. A typical value for the channel length is approximately 1.5 times the minimum channel length.

13.4.3 DRAM Sense Amplifier

The dynamic-latch sense amplifier described in the previous section is the dominant sense amplifier used in DRAM designs. This sense amplifier is a differential amplifier. However, the signal that comes from a DRAM cell is single ended. We need to construct a reference voltage for one side of the sense amplifier while the other side is connected to the bit line to be sensed. A powerful method to build a reference voltage is to build a replica of the cell that we are sensing. The **replica cell** stores approximately one-half the charge between a **0** and **1**. This cell can be fabricated by using a replica capacitor that is one-half the size of the storage capacitor in a real cell, or by charging the capacitor in the reference cell to one-half the maximum voltage stored in the cell to be sensed. In this discussion, we assume the reference capacitor $C_R = C_S$.

In a typical DRAM, a sense amplifier is placed in the center of two, long bit lines. Each of these bit lines has one replica cell that will serve as the reference. A schematic diagram of this configuration is shown in Fig. 13.19. If a word line accesses a cell on the left side of the sense amplifier, then logic is provided to turn on the replica cell on the right side of the sense amplifier to act as a reference voltage. The input voltage to the sense amplifier for a stored **0** and **1** is given by

$$\Delta V_{IN0} = \left(\frac{C_S}{C_S + C_{BIT}}\right)\left(0 - \frac{V_{DD}}{2}\right) \tag{13.18}$$

$$\Delta V_{IN1} = \left(\frac{C_S}{C_S + C_{BIT}}\right)\left(V_{DD} - V_{Tn} - \frac{V_{DD}}{2}\right) \tag{13.19}$$

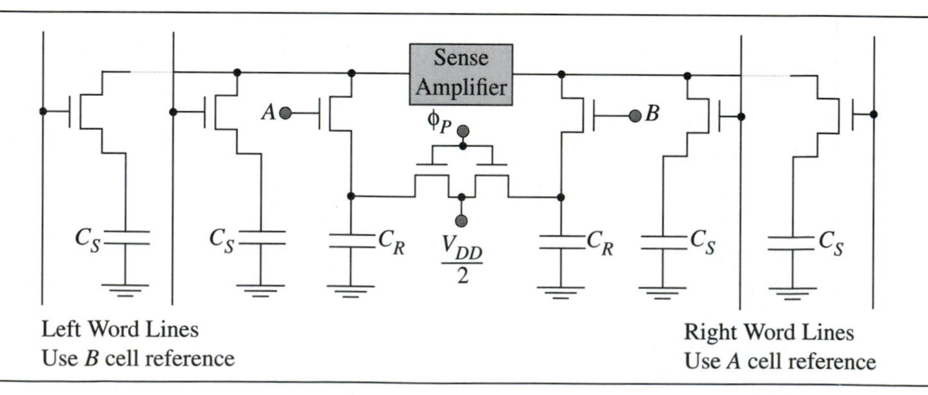

➤ **Figure 13.19** Example of a DRAM sense amplifier between two bit lines. Reference cells are closest to the sense amplifier and have a storage capacitance value $C_R = C_S$.

We assumed that the full storage level is equal to $V_{DD} - V_{Tn}$ and the charge is shared between the storage capacitance and bit-line capacitance. After the charge is placed on the bit line, transistors M_5 and M_6 in the sense amplifier are turned on to activate the sense amplifier and split the bit-line voltages toward the power supply rails.

We have explained how a **1** or **0** is written back into the selected cell after the destructive read. However, the replica cell does not have the proper voltage since its bit line was driven to either V_{DD} or 0 V. After the word lines are turned off, the replica cells are precharged to the value of $V_{DD}/2$. Several enhancements to the circuit shown in Fig. 13.19 can be found in the literature. Most of these are attempting to increase either the accuracy of the sense amplifier by reducing any possible offset voltages or to increase the speed of its operation.

 DESIGN EXAMPLE 13.3 Dynamic-Latch Sense Amplifier

In this example, we design a dynamic-latch sense amplifier for a DRAM using the topology shown in Fig. 13.19. Each DRAM cell has a 50 fF storage capacitor and contributes 25 fF to the total bit-line capacitance. There are 32 bits on each bit line that result in approximately 800 fF of bit-line capacitance. With the addition of replica cells and other parasitics, assume that the bit-line capacitance $C_{BIT} = 1\text{pF}$. The speed of the sense amplifier should be sufficient to drive the difference in bit-line voltages to 2 V in less than 10 ns. Use the device data shown below.

Device Data

$V_{Tn} = 1\text{ V}$ $\quad V_{Tp} = -1\text{ V}$ $\quad \mu_n C_{ox} = 50\ \mu\text{A/V}^2$ $\quad \mu_p C_{ox} = 25\ \mu\text{A/V}^2$

SOLUTION

Begin the design by assuming the bit lines are precharged to 2.5 V. Adjust the W/L of the cross-coupled inverter to ensure $V_M = V_{DD}/2$, implying

$$(W/L)_p = 2(W/L)_n$$

This constraint forces the transconductance of the n and p-channel transistors to be equal.

Next, we determine the level of the input signal supplied by the DRAM cell. We will use a replica cell with $V_{DD}/2$ stored on the reference storage capacitor to provide a reference voltage for the sense amplifier. When connected to the bit line, the reference cell will leave the bit-line voltage at $V_{DD}/2$ since the reference storage capacitor voltage is at the same potential. Ignore any capacitive-coupling effects due to the word-line voltage rising since this common-mode signal will be rejected by the differential sense amplifier. For a stored **0**, the voltage on the cell storage capacitor $V_{CS} = 0\text{ V}$. The change in bit-line voltage is given by

$$\Delta V_{BIT0} = \left(0 - \frac{V_{DD}}{2}\right)\left(\frac{C_S}{C_S + C_{BIT}}\right) = -119\text{ mV}$$

For a stored **1**, the stored voltage on the cell storage capacitor is equal to $V_{DD} - V_{Tn}$. The backgate effect raises the effective threshold voltage of the n-channel transistor when the source voltage is higher than the substrate voltage.

This effect will limit the total voltage stored by the storage capacitor for the **1** state. To quantify this effect, write

$$V_{CS1} = V_{DD} - V_{Tn}$$

where

$$V_{Tn} = V_{TOn} + \gamma_n\left[\sqrt{V_{SB} - 2\phi_p} - \sqrt{-2\phi_p}\right]$$

In the case of the DRAM cell $V_{SB} = V_{CS}$ assuming the substrate is at 0 V.

For $V_{TOn} = 1\text{V}$; $\gamma_n = 0.6\ \text{V}^{\frac{1}{2}}$; and $-2\phi_p = 0.8\text{V}$,

$$V_{Tn} = 1.0\text{V} + (0.6)\left[\sqrt{V_{CS} + 0.8} - \sqrt{0.8}\right]\text{V}$$

By iteratively solving, we find the voltage stored on the storage capacitor for a logic **1** is

$$V_{CS1} = 3.3\text{V}$$

which results in a change in bit-line voltage for a stored **1** as

$$\Delta V_{BIT1} = (3.3\ \text{V} - V_{DD}/2)\left(\frac{C_S}{C_S + C_{BIT}}\right) = 38\ \text{mV}$$

To make the bit-line voltage swings symmetric between a stored **1** and a stored **0**, we could increase the word-line voltage to a sufficiently high potential above V_{DD} so the full V_{DD} value is stored on the storage capacitor. Other techniques to shift the reference voltage are also possible. In this example, we assume our sense amplifier must detect the smallest input voltage, which is approximately 38 mV. The time for the sense amplifier to detect this voltage is given by

$$t_{SENSE} = \left(\frac{C_{BIT}}{g_{mn} + g_{mp}}\right)\ln\left(\frac{\Delta V_{OUT}(t_{SENSE})}{\Delta V_{BIT1}}\right)$$

Now we find the transconductance required to make the output voltage change by 2 V in 10 ns. Sizing the p-channel device width twice the n-channel device width to have $g_{mn} = g_{mp}$ results in

$$t_{SENSE} = 10\ \text{ns} = \frac{1\ \text{pF}}{2g_{mn}}\ln\left(\frac{2.0\ \text{V}}{0.038\ \text{V}}\right)$$, giving a transconductance

$$g_{mn} = 198\ \mu\text{S} \approx 200\ \mu\text{S}.$$

To minimize the offset voltage due to processing variations, do not set the channel length of the inverters to the minimum linewidth but instead, use $L = 3\ \mu\text{m}$. Calculate the required W/L for both the n and p-channel transistors by

$$g_{mn} = \left(\frac{W}{L}\right)_n \mu_n C_{ox}(V_{DD}/2 - V_{Tn}),$$

which yields

$$(W/L)_n = 2.6 \approx 3 = \frac{9}{3}$$

$$(W/L)_p = 18/3$$

The pull-down device M_5 that connects the cross-coupled inverters (M_1–M_4) to ground when the sense amplifier is operating, should have a small voltage drop across it since the voltage directly subtracts from the gate voltage driving the cross-coupled inverters. For the purposes of this design, we choose the voltage drop across M_5 to be less than 100 mV. Its required W/L is found by assuming it is in the triode region and by setting the value of its current to be equal to the sum of the current from both n-channel transistors in the cross-coupled inverters when their gates are at $V_{DD}/2$. Since V_{DS5} is small compared to $V_{DD} - V_{Tn}$, write

$$I_{D5} \approx \left(\frac{W}{L}\right)_5 \mu_n C_{ox}(V_{DD} - V_{Tn})V_{DS5}$$

$$I_{D1} = \frac{1}{2}\left(\frac{W}{L}\right)_1 \mu_n C_{ox}(V_{DD}/2 - V_{Tn})^2 = 170\,\mu\text{A}$$

$$\left(\frac{W}{L}\right)_5 = \frac{2I_{D1}}{\mu_n C_{ox}(V_{DD} - V_{Tn})V_{DS5}} = \frac{340\,\mu\text{A}}{50\frac{\mu\text{A}}{V^2}(5\text{ V} - 1\text{ V})0.1\text{V}} = 17 \approx \frac{25}{1.5}$$

We also need to design a p-channel pull-up M_6 to have a maximum $V_{SD6} = 100$ mV. Since the mobility of holes is half that of electrons, we require the W/L of the p-channel pull-up to be twice that of the n-channel pull-down device. The resulting device sizes for the six transistors in the dynamic-latch sense amplifier are given below.

$M_1 = M_3 = 9/3$; $M_2 = M_4 = 18/3$; $M_5 = 25/1.5$; $M_6 = 50/1.5$

13.5 ADDRESS DECODERS AND BUFFERS

The number of rows and columns in a typical memory can be quite large. For example, a 4 Mbit memory has over 2,000 rows and columns. It is clear that it would be impractical to bring each of these rows and columns off-chip to the outside world. Hence, row and column decoders are required to reduce the number of signals necessary to address particular rows and columns.

In Fig. 13.20, we see that the row decoder must select one out of 2^N rows to drive the word line of choice. We also require the selection of one out of 2^M sense amplifier outputs on the columns to be sent to an output buffer. These functions have an additional constraint in that the decoders must match the **memory cell pitch**. Each memory cell has a particular width and height. If a row decoder should exceed the height of the memory cell, H an extremely difficult wiring problem would exist, causing a large increase in silicon area. There are similar constraints for the column decoder in that it must match the width, W of the memory cell. In this section, we will explore

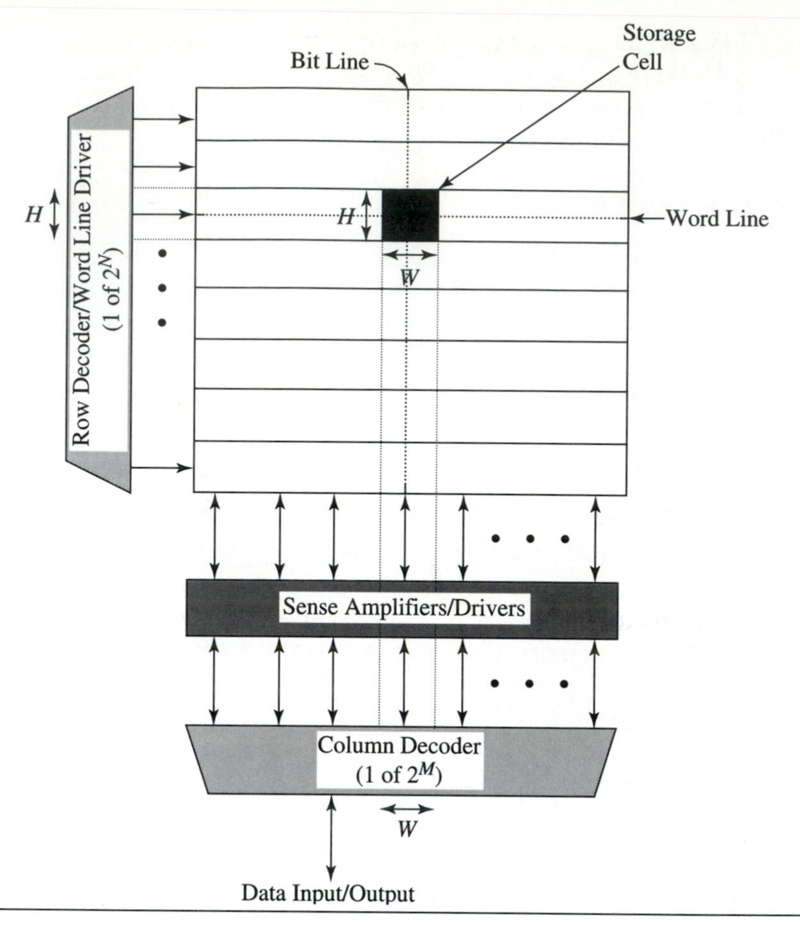

➤ **Figure 13.20** Memory architecture showing pitch-matching requirements for decoders, sense amplifiers and drivers.

typical row and column decoders that are used in memory that satisfy the logical and pitch-matching constraints.

13.5.1 Row Decoders

The function of the row decoder is to choose one out of 2^N word lines to be driven, given a particular address of the word line. The logical function of this decoder is equivalent to 2^N complex logic gates, where the inputs to the logic gates are the true and complement logic values of the addresses for the rows A_0 to A_{N-1}. The logic equations are

$$ROW_0 = \bar{A}_0 \bar{A}_1 \ldots\ldots\ldots \bar{A}_{N-1}$$

$$ROW_1 = A_0 \bar{A}_1 \ldots\ldots\ldots \bar{A}_{N-1}$$

$$ROW_2 = \bar{A}_0 A_1 \ldots\ldots\ldots \bar{A}_{N-1}$$

$$ROW_{2^N-1} = A_0 A_1 \ldots\ldots A_{N-1} \tag{13.20}$$

The N input AND gate can be transformed into an N input NOR gate for each row by using DeMorgan's laws. By using the same structure that was used in the read only memory in Section 13.2 to implement an N input NOR gate, we can greatly reduce the complexity of the decoder so that it can satisfy the cell pitch constraint. A typical implementation of a row decoder using the ROM structure is shown in Fig. 13.21 and its operation is outlined below.

1. Begin the operation of the row decoder by precharging each of the rows to V_{DD} through a p-channel precharge device when ϕ_P is high.
2. The true and complement addresses are input to the decoder during ϕ_A.
3. The n-channel transistors are strategically placed so only the selected row remains charged to V_{DD}. All of the other rows that are unselected by the addresses will be discharged to ground.
4. After an appropriate discharge time, each of the rows is gated into a word line buffer and the selected word line is driven high during ϕ_W.

A qualitative picture of the waveforms in this typical CMOS row decoder is shown in Fig. 13.21(b). The transient operation of the row decoder is determined by the size of the n-channel device used to pull the unselected rows low. One is limited in the size of the n-channel device by the constraint to match the pitch of these rows to the memory cell pitch. The decoding function often takes up to 10 to 20 percent of the total access time in a memory cell operation.

13.5.2 Column Decoders

Unlike the row decoders, the function of the column decoders is to choose one out of 2^M sense amplifier outputs to be driven off-chip. Essentially, this function is a 2^M input multiplexer. There are two common methods to implement this column decoder. The first method uses pass transistors that are driven by a NOR predecoder. The **predecoder**, which selects one out of 2^M inputs, can be the same design as the row decoder discussed above. The line that is driven high then drives a flowtransistor to select the appropriate bit line and pass its voltage on to the output buffer.

The second method that uses less silicon area is a tree decoder shown in Fig. 13.22. The number of devices required for this decoder is drastically reduced from one using the pass transistor method. However, the signals must flow through a series connection of many transistors, and therefore, have a slow response time. In most memory configurations, tree decoders are only used when a slow response time can be tolerated. Combinations of pass-transistor and tree decoder methodologies are sometimes used in block memory architectures. These architectural issues are beyond the scope of this text.

13.5.3 Buffers

In several areas of the memory design, buffers are needed to drive long lines that have large capacitances. For example, the word lines in the memory array can have as much as 1 to 5 pF of capacitance. Clocklines that supply signals to the devices that precharge and discharge specific lines must also be buffered. Signals leaving the chip through the package leads often must drive 25 to 50 pF. Although specific requirements may differ among these various functions, they all have the common requirement to drive a large capacitive load.

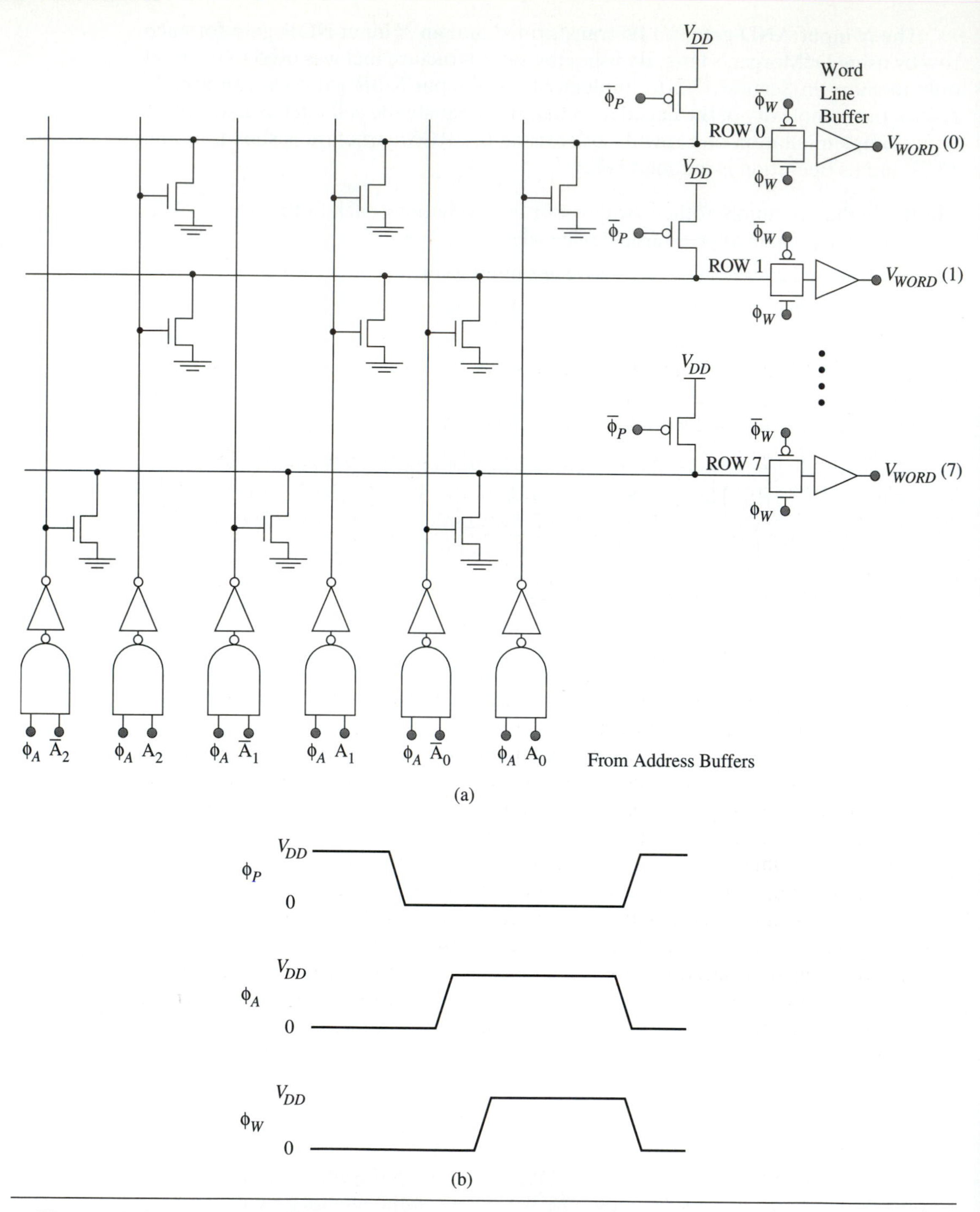

➤ Figure 13.21 Typical CMOS row decoder implementing a 3-input NOR function given address inputs and their compliments. (a) Circuit diagram. (b) Qualitative signal waveforms.

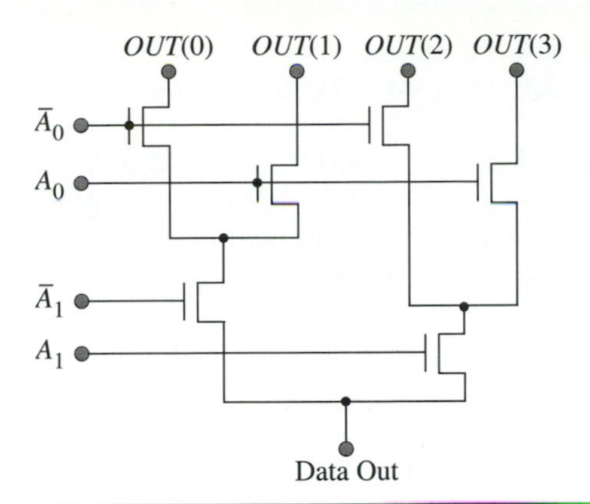

➤ **Figure 13.22** Example of a two-bit tree decoder. It takes four bit-line outputs and selects one.

In Chapter 5, we developed a procedure to design inverters for a specific speed and logic level input. These same constraints apply to building buffers that must drive a large capacitive load. However, it is not optimal to use a single inverter to drive a large capacitive load since its input capacitance is also large. What is needed is a methodology to determine the ratio of the capacitance that can be driven by a particular inverter to its own input capacitance. With that ratio, we can continue to increase the size of the n- and p-channel transistors in the inverter so they are large enough to drive the next successive stage.

Figure 13.23 shows a schematic representation of the increasing size of inverters to drive a large capacitive load. If the opposite logic is required for the final output, we add a minimum-geometry input inverter to minimize area and power dissipation. Although several minimization algorithms have been developed to optimize the capacitance ratio between the load and the input, experienced designers have found that this optimal ratio is somewhere between 3× and 6×. Since other constraints beside the optimal ratio between inverters are often present, we will not spend the considerable amount of time needed to derive exact optimal expressions. Instead, we will make the inverter size a factor of 4 larger than the previous inverter until it is capable of driving the load capacitance in a specified amount of time. Another important point is that by increasing the size of inverters by a constant factor, the rise and fall times of each of the intermediate nodes will be similar. In the design of these buffers, we will assume that the p-channel transistors are twice the size of the n-channel transistors in order to have symmetric rise and fall times, as well as symmetrical noise margins.

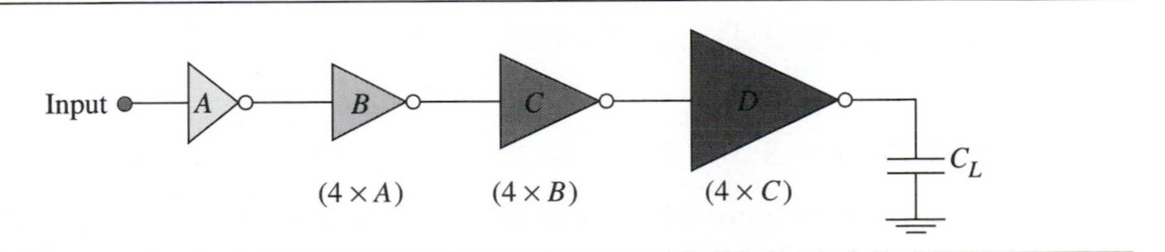

➤ **Figure 13.23** Schematic representation of increasing inverter sizes by 4x to efficiently drive large capacitive loads.

DESIGN EXAMPLE 13.4 Buffer Design

In this example, we will design a buffer that is capable of driving a 10 pF load capacitance. The input signal is a step input into a minimum-geometry inverter $[(W/L)_n = 3/1.5$ and $(W/L)_p = 6/1.5]$. We will design a series of inverters to drive the 10 pF load to a valid output logic level in less than 5 ns. Use the device data given below.

MOS Device Data

$\mu_n C_{ox} = 50\ \mu\text{A/V}^2$, $\mu_p C_{ox} = 25\ \mu\text{A/V}^2$, $V_{Tn} = -V_{Tp} = 1\text{ V}$, $C_{ox} = 2.3\text{ fF}/\mu\text{m}^2$
$C_{Jn} = 0.1\text{ fF}/\mu\text{m}^2$, $C_{Jp} = 0.3\text{ fF}/\mu\text{m}^2$, $C_{JSWn} = 0.5\text{ fF}/\mu\text{m}$, $C_{JSWp} = 0.35\text{ fF}/\mu\text{m}$
$L_{diffn} = L_{diffp} = 4.5\ \mu\text{m}$

SOLUTION

We begin this design by estimating the rise and fall times necessary to meet the delay specification of 5 ns. A simple approximation to this complex problem is to assume the output voltages of the inverters have linear and equal rise and fall times beginning when the input voltage of the inverter is equal to V_M, which we will design to be $V_{DD}/2$. The signal waveforms for a chain of four inverters are shown in Fig. Ex13.4A.

Since the rise and fall times are equal, the delay through two inverters is equal to one rise or fall time as seen in the diagrams. For example, if the rise and fall times are 2.5 ns, then a chain of four inverters could be used to meet the 5 ns delay specification. Begin this design by assuming 2.5 ns rise and fall times. If more than four inverters are needed, then we will reduce the rise and fall time. If less than four inverters are needed, we can relax the rise and fall time requirement.

Starting at the load capacitor, we will find the device sizes for the final inverter necessary to drive the load with a 2.5 ns rise and fall time, given $V_{DD} = 5$ V. This implies $\Delta V/\Delta t = 2\text{ V/ns}$. The gate voltage of this final inverter is also changing with time. Refering to the waveforms shown, the output of each inverter begins to change when the input voltage is $V_{DD}/2$ and completes its transition 2.5 ns later. We assume that the average gate-source (source-gate) voltage is $3V_{DD}/4$ during the transition. Using the expression for current-charging a capacitor from a current source we find

$$\frac{\Delta V}{\Delta t} = 2\text{ V/ns} = \frac{I_{Dn4}}{C_L}$$

Next, calculate I_{Dn4} by assuming the transistor is in the constant-current region with a gate voltage of $3V_{DD}/4$. We find

$$I_{Dn4} = \frac{W}{2L}\mu_n C_{ox}\left(\frac{3V_{DD}}{4} - V_{Tn}\right)^2$$

Substitute the expression above and solve for W/L given $C_L = 10$ pF. The n-channel transistor size is

$$\left(\frac{W}{L}\right)_{n4} = 106 \approx \frac{160}{1.5}$$

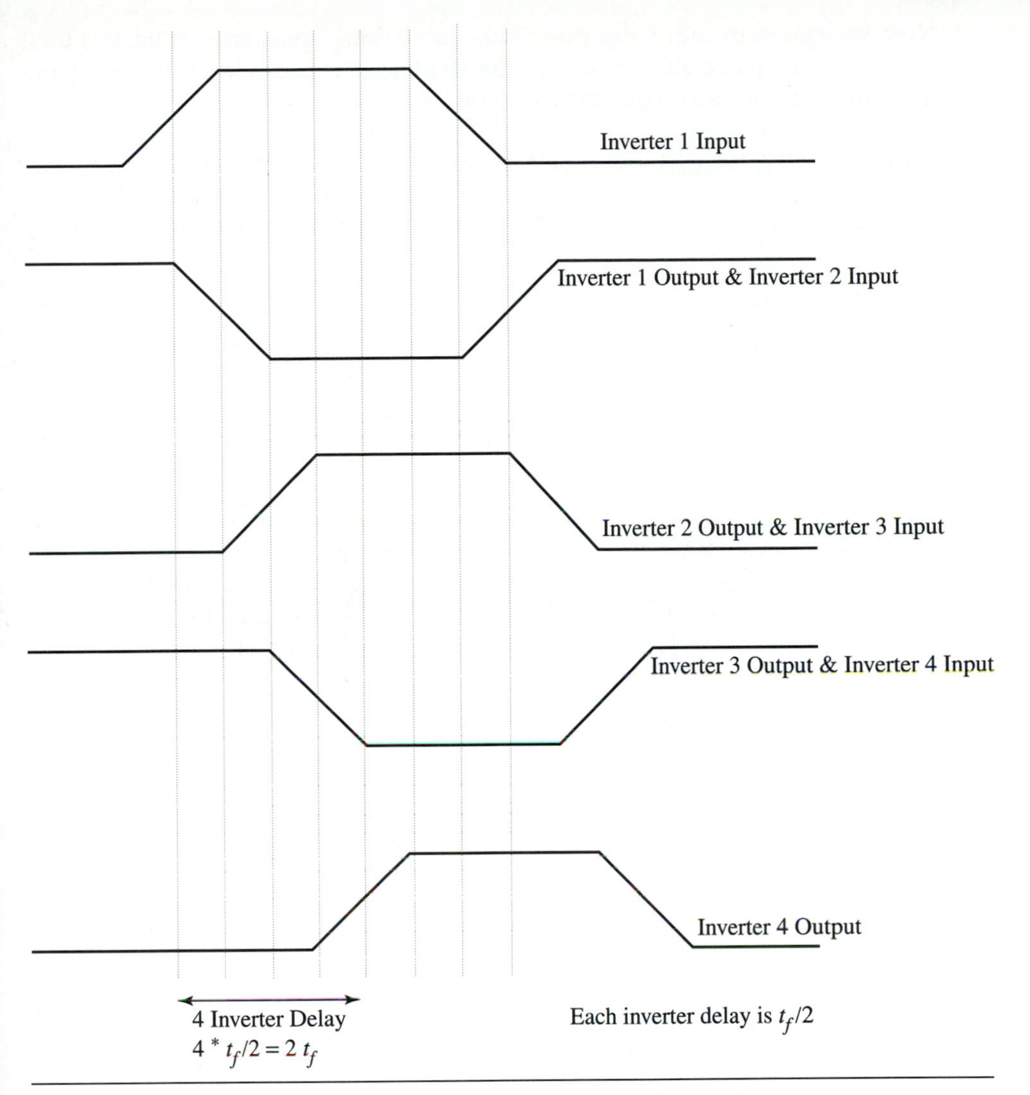

➤ Figure Ex13.4A

To have $V_M = \frac{V_{DD}}{2}$ and symmetrical propagation delays we set

$$\left(\frac{W}{L}\right)_{p4} = 2\left(\frac{W}{L}\right)_{n4} = \frac{320}{1.5}$$

Inverter 3 is sized with $\left(\frac{W}{L}\right)_{n3} = \frac{1}{4}\left(\frac{W}{L}\right)_{n4} = \frac{40}{1.5}$ and $\left(\frac{W}{L}\right)_{p3} = \frac{1}{4}\left(\frac{W}{L}\right)_{p4} = \frac{80}{1.5}$.

Similarly, Inverter 2 has $\left(\frac{W}{L}\right)_{n2} = \frac{10}{1.5}$ and $\left(\frac{W}{L}\right)_{p2} = \frac{20}{1.5}$ and Inverter 1 uses minimum geometry devices, $\left(\frac{W}{L}\right)_{n1} = \frac{3}{1.5}$ and $\left(\frac{W}{L}\right)_{p1} = \frac{6}{1.5}$, which are slightly larger than the 4× scaling requires.

Now we check to see if the drain-bulk depletion capacitance from the final inverter is an appreciable fraction of the 10 pF load capacitance. Calculating the junction and sidewall capacitance, we find

$$C_{DB} = C_{Jn}(W_{n4} \times L_{diffn}) + C_{JSWn}(W_{n4} + 2L_{diffn})$$
$$+ C_{Jp}(W_{p4} \times L_{diffp}) + C_{JSWp}(W_{p4} + 2L_{diffp})$$

$$C_{DB} = (0.1)\frac{\text{fF}}{\mu\text{m}^2}(160\ \mu\text{m} \times 4.5\mu\text{m}) + 0.5\frac{\text{fF}}{\mu\text{m}}(160\ \mu\text{m} + 9\ \mu\text{m})$$

$$+ 0.3\frac{\text{fF}}{\mu\text{m}^2}(320\ \mu\text{m} \times 4.5\ \mu\text{m}) + 0.35\frac{\text{fF}}{\mu\text{m}}(320\ \mu\text{m} + 9\ \mu\text{m})$$

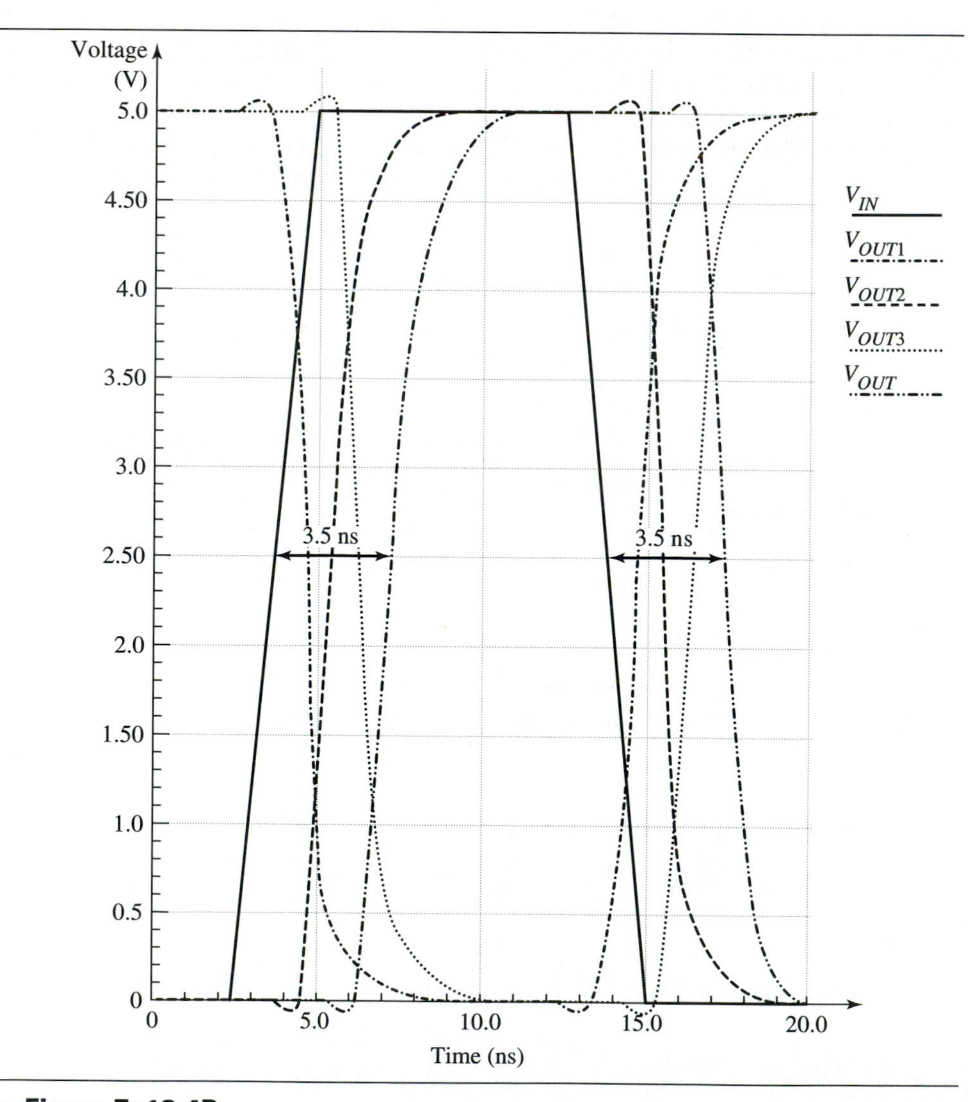

➤ **Figure Ex13.4B**

$$C_{DB} = 72 \text{ fF} + 85 \text{ fF} + 432 \text{ fF} + 115 \text{ fF} = 704 \text{ fF}$$

We could increase the size of the inverter by about 10% to take into account this added depletion capacitance, but the accuracy of this design methodology is not adequate to bother iterating. This methodology to size inverters for a large capacitive load is adequate for a first-cut approximation. Using the initial device sizes based on our hand calculations, SPICE should be run to ensure the device sizing is sufficient to meet the 5 ns delay specification. Some iteration using SPICE is often necessary. The waveforms in Fig. Ex 11.4B show the results of a SPICE simulation. The buffer delay is approximately 3.5 ns.

13.6 SRAM DESIGN EXAMPLE

In this section, we design an SRAM to be embedded in a digital signal processor that requires a read-access time of 15 ns. An architectural schematic of this memory is shown in Fig. 13.24. In this application, the SRAM is required to store 256 words. Each word is comprised of 32 bits. These words represent filter coefficients that the user can program. Since user programming is infrequent, the write time is not a design issue. The word-line selection in the static RAM is decoded in the digital signal processor, and hence, will not be considered part of our memory. Instead we will

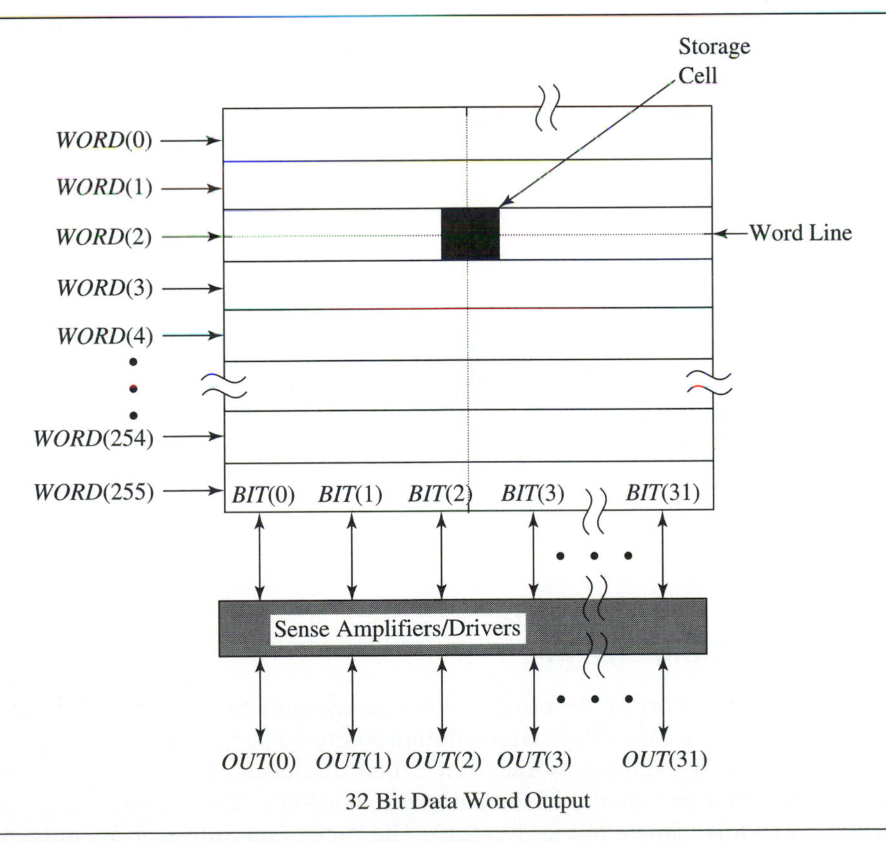

Figure 13.24 Architectural schematic of an embedded SRAM with 256 words × 32 bits.

assume that a step input (logic **1**-to-logic **0**) into a minimum-geometry inverter from the digital signal processor selects the appropriate word line. Bit-line decoders are not needed since all 32 bits in a word are sent to the output. The output of the sense amplifiers will be fed into an output driver capable of driving a 1 pF load that represents the bussing and logic in the digital signal processor. For this example, the read-access time is defined as the time from the step input to a valid output. We will use the same device data found in Example 13.4 and assume $V_{DD} = 5$ V.

We begin with the cell design, followed by the word-line driver design. Based on the results from some of our previous examples, we have a rough idea of the time each of these functions will take. The remaining time is apportioned between the sense amplifier and output driver.

13.6.1 Cell Design

In Section 13.2, we showed a typical static RAM cell uses minimum-geometry transistors for the cross-coupled inverter pair and larger sizes for the access transistors. Applying the results of Example 13.2, the cross-coupled inverter transistors M_1–M_4 have a W/L equal to 3/1.5 and access transistors M_5–M_6 have a W/L of 6/1.5. With these device sizes the SRAM cell area is 27 μm by 54 μm.

From the cell dimensions we can estimate the RC time constant for the word line, as well as the bit-line capacitance. We assume the bit line is fabricated in metal so its resistance is negligible. This SRAM memory has 256, 32-bit words compared with 128, 32-bit words in Example 13.2. We use those results by recognizing that the word line RC time constant is identical and the bit-line capacitance will be 2× in this design.

The following important parameters for this cell design can be calculated in the same manner as Example 13.2. The results are

$$R_{WORD} = 3\ \text{k}\Omega$$
$$C_{WORD} = 1.33\ \text{pF}$$
$$\tau_{WORD} = 4.0\ \text{ns}$$
$$C_{BIT} = 3.4\ \text{pF}$$

The time $t_{\Delta V_{BIT}}$ for this cell to discharge the bit line $\Delta V_{BIT} = 0.2$ V during a read cycle is calculated using the new bit-line capacitance corresponding to 256 bits. Assuming $C_{BIT} = 4$ pF to take the additional capacitive loading from the sense amplifier into account, the sense time is

$$t_{\Delta V_{BIT}} = \frac{C_{BIT}\Delta V_{BIT}}{I_{D5}} = \frac{(4\ \text{pF})\ (0.2\ \text{V})}{(230\ \mu\text{A})} = 3.5\ \text{ns}$$

13.6.2 Word-Line Driver

To design the word-line driver, we use a number of inverters cascaded together. It is our intention to size the final driver large enough so the delay time for the word-line voltage is set by its RC time constant rather than limited by the final driver. To accomplish this, refer to a model of the complete word-line driver (assuming three inverters) and its load shown in Fig. 13.25(a). The p-channel transistor M_3 must be able to source the maximum amount of current that can be driven onto the word line.

This condition occurs when $V' = V_{DD}$ and $V_{WORD} = 0$ V. This maximum current is given by

$$I_{MAX} = V_{DD}/R_{WORD} = (5\text{V})/3\text{k}\Omega = 1.66 \text{ mA}$$

For this design, set the current in M_{p3} equal to 1.66 mA when the average source-gate voltage applied is equal to 3.75 V. Under this condition, the device is in its constant-current region and its current is given by

$$I_{Dp3} = \frac{1}{2}\left(\frac{W}{L}\right)_{p3}\mu_p C_{ox}(3.75\text{ V} + V_{Tp})^2$$

resulting in $(W/L)_{p3} = 26/1.5$
The n-channel transistor in the final word line driver is sized to be half that of the p-channel transistor $(W/L)_{n3} = 13/1.5$ so V_M equals $V_{DD}/2$ and the transient response is symmetrical.

The series inverters are sized by assuming each inverter can drive 4× the capacitive load. This results in Inverter 2 having device sizes

$$(W/L)_{n2} = 4/1.5 \text{ and } (W/L)_{p2} = 7/1.5$$

and Inverter 1 having minimum geometry devices,

$$(W/L)_{n1} = 3/1.5 \text{ and } (W/L)_{p1} = 6/1.5$$

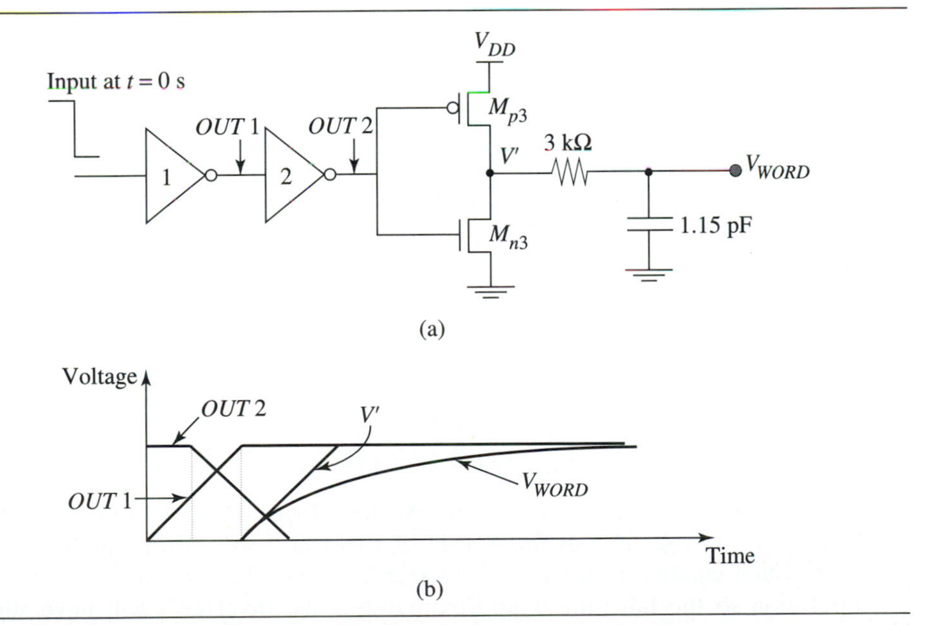

➤ Figure 13.25 Word-line driver model. (a) The word-line driver consists of three inverters. The word line is modeled with a lumped *RC* circuit. (b) Qualitative waveforms.

The minimum-geometry inverter is required to obtain the proper logical output. If our input was a low-to-high transition this inverter could have been eliminated.

Now consider the internal rise and fall times to determine the delay of the word-line driver. This delay is defined as the time from the step input to the minimum-geometry inverter, to the input of the last inverter. Recall that the last inverter's time delay will not be significant since we sized it so the *RC* time constant of the word line is limiting the response. Figure 13.25(b) shows the qualitative waveforms of the output of each inverter and the word-line voltage V_{WORD}. To calculate the internal rise and fall times, assume the p-channel transistor of Inverter 2 has an average gate voltage between $V_M = V_{DD}$ and V_{DD} resulting in $V_{SGp2} = 3V_{DD}/4$. This transistor is operating in the constant-current region, with a current given by

$$I_{D2} = \frac{1}{2}\left(\frac{W}{L}\right)_{p2}\mu_p C_{ox}(3.75 + V_{Tp})^2 = 441\,\mu\text{A}$$

The load capacitance for this p-channel transistor is approximately the total gate capacitance of Inverter 3 and its drain-bulk depletion capacitance, which are

$$C_G = (WL)_{p3}C_{ox} + (WL)_{n3}C_{ox} = (26\,\mu\text{m}\times 1.5\,\mu\text{m})\left(2.3\,\frac{\text{fF}}{\mu\text{m}^2}\right)$$

$$+ (13\,\mu\text{m}\times 1.5\,\mu\text{m})\left(2.3\,\frac{\text{fF}}{\mu\text{m}^2}\right) = 135\text{ fF}$$

and

$$C_{DB} = C_{Jn}(W_{n2}\times L_{diffn}) + C_{JSWn}(W_{n2} + 2L_{diffn})$$

$$+ C_{Jp}(W_{p2}\times L_{diffp}) + C_{JSWp}(W_{p2} + 2L_{diffp})$$

$$C_{DB} = (0.1)\frac{\text{fF}}{\mu\text{m}^2}(4\,\mu\text{m}\times 4.5\,\mu\text{m}) + 0.5\frac{\text{fF}}{\mu\text{m}}(4\,\mu\text{m} + 9\,\mu\text{m})$$

$$+ 0.3\frac{\text{fF}}{\mu\text{m}^2}(7\,\mu\text{m}\times 4.5\,\mu\text{m}) + 0.35\frac{\text{fF}}{\mu\text{m}}(7\,\mu\text{m} + 9\,\mu\text{m})$$

$$C_{DB} = 1.8\text{ fF} + 6.5\text{ fF} + 9.5\text{ fF} + 5.6\text{ fF} = 23.4\text{ fF}$$

Using the current and load capacitance calculated, we find $\Delta V/\Delta t$ is given by

$$\frac{\Delta V}{\Delta t} = \frac{I_{D2}}{C_L} \approx \frac{441\,\mu\text{A}}{135\text{ fF} + 23.4\text{ fF}} = 2.8\text{ V/ns}$$

By sizing the p-channel transistor in Inverter 2 twice the width of the n-channel transistor, the rise time at the output of Inverter 2 will also be approximately 2 ns. The n-channel transistor in Inverter 2 was sized for symmetrical transient operation so the fall time is approximately 2 ns. Inverter 1 will have slightly faster rise and fall times since it is driving less than 4× its input capacitance. As shown in the Section 13.4, we can estimate the delay through two inverters to be equal to the rise or fall time of a single inverter. In this case it is approximately 2 ns.

13.6.3 Sense Amplifier Design

Until this point, we have used approximately 7.5 of the 15 ns for the read-access time. These were comprised of

Word-line driver: 2ns

$t_{READ} = 0.5\tau_{WORD} + t_{\Delta V_{BIT}} = 2.0 + 3.5 = 5.5$ ns

In this sense amplifier design, we assume the bit lines are precharged to V_{DD}. This change from the method used when designing the sense amplifier in the DRAM (Example 13.3) allows us to change the topology of the sense amplifier slightly as shown in Fig. 13.26. When the bit lines are precharged to V_{DD} the p-channel transistors are cutoff. Hence, no p-channel gating transistor is required between V_{DD} and the sources of the p-channel transistors, M_2 and M_4. Only the n-channel transistors will operate until a significant bit-line voltage is established.

As a starting point, we apportion the remaining 7.5 ns of our read-access time as follows.

Sense amplifier: 5 ns

Output driver: 2.5 ns

We determine the transconductance necessary for the n-channel devices M_1 and M_3 so the sense amplifier's time response is 5 ns. We design the sense amplifier to pull

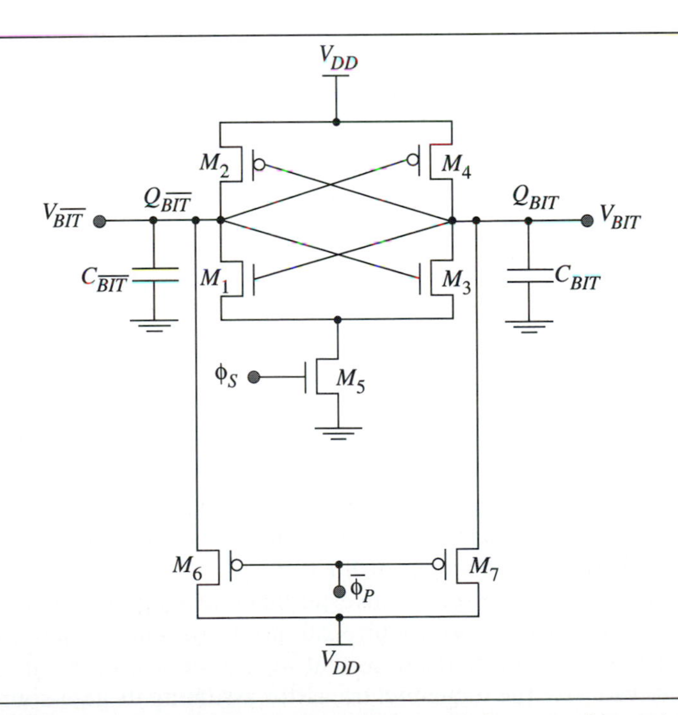

➤ **Figure 13.26** Sense amplifier topology with the bit lines precharged to V_{DD}. There is no need for the p-channel gating device.

the appropriate bit line from 200 mV below V_{DD} to $V_{DD}/2$. The time response for the sense amplifier is given by

$$t_{SENSE} = \frac{C_{BIT}}{g_{mn}} \ln\left(\frac{\Delta V_{OUT}(t_{SENSE})}{\Delta V_{IN}}\right)$$

requiring a transconductance equal to 1.7 mS. To calculate the size of the n-channel transistors for a 1.7 mS transconductance, we begin by assuming the channel length is equal to 3 μm to reduce offset voltages due to lithographic uncertainties. The gate drive on the n-channel transistor (M_1 or M_3) that is pulling the bit line low, is reduced by the voltage drop across the n-channel transistor M_5. If we assume this voltage drop is 300 mV, we can write an equation for the transconductance of the n-channel transistor given by

$$g_{mn} = \left(\frac{W}{L}\right)_{1-3} \mu_n C_{ox} (V_{DD} - 0.3\ \text{V} - V_{Tn})$$

resulting in $(W/L)_{1\text{–}3} \approx 30/3$.
To size the sense amplifier with a midpoint voltage equal to $V_{DD}/2$, we size the p-channel transistors at twice the n-channels so that $(W/L)_{2\text{–}4} = 60/3$.

Next, we must size the pulldown transistor M_5 so its drain-source voltage is less than 300 mV. The current that is flowing through device M_5 can be estimated by assuming both transistors M_1 and M_2 have a gate-source voltage of $(V_{DD} - V_{DS5})$ and are operating in their constant-current region. This assumption yields a total current equal to

$$I_{D1-3} = \frac{1}{2}\left(\frac{W}{L}\right)_{1-3} \mu_n C_{ox} (V_{DD} - V_{DS5} - V_{Tn})^2 = 3.42\ \text{mA}$$

Transistor M_5 is operating in the triode region. Its W/L must be selected to ensure V_{DS5} is less than 0.3 V as shown in

$$I_{D5} = 3.42\ \text{mA} = \left(\frac{W}{L}\right)_5 \mu_n C_{ox} \left(V_{DD} - V_{Tn} - \frac{V_{DS5}}{2}\right) V_{DS5}$$

The resulting device size for M_5 is approximately 90/1.5. The sizing of the p-channel transistors that precharge the bit lines to V_{DD} depends on the cycle time required by the memory.

13.6.4 Output Driver

In this application, the output of the sense amplifier must be buffered to drive a 1 pF load. Since the bit-line capacitance is approximately 5 pF, it can tolerate the additional capacitance from an inverter that is slightly larger than minimum geometry. Since we have allocated 2.5 ns for the output driver, we design the driver with two inverters that have rise and fall times of 2.0 V/ns.

We begin by designing the second inverter that is driving the 1 pF load. To have a rise and fall time of 2.0 V/ns, we require 2.0 mA to be sourced and sinked by the inverter. Using the same approximations that were discussed in the driver section of this chapter, we can size the n-channel transistor assuming its gate-source voltage is $3V_{DD}/4$ and it is operating in the constant-current region. To calculate its W/L, we write

$$I_{Dn2} = \frac{1}{2}\left(\frac{W}{L}\right)_{n2}\mu_n C_{ox}\left(\frac{3V_{DD}}{4} - V_{Tn}\right)^2 = 2.0 \text{ mA}$$

which yields an n-channel transistor with $(W/L)_{n2} = 16/1.5$. For equivalent rise and fall times the p-channel transistor has $(W/L)_{p2} = 32/1.5$.

The inverter that is attached to the bit line should have a midpoint voltage that is higher than $V_{DD}/2$, since the input to this inverter, which is the output of the sense amplifier, only changes between V_{DD} and $V_{DD}/2$. We can accomplish this goal by choosing the n-channel transistor in the first inverter to be four times smaller than its suceeding inverter, yielding a $(W/L)_{n1} = 4/1.5$. To size the p-channel transistor, we select a V_M that is greater than $V_{DD}/2$. Since the threshold voltage of a p-channel transistor is $V_{Tp} = -1$ V, the maximum V_M possible is 4 V. This is an asymptotic maximum and would require $(W/L)_{p1}$ to approach infinity. A more reasonable design is to choose $V_M = 3.0$ V and calculate the required p-channel transistor size

$$V_M = \frac{V_{Tn} + \sqrt{\dfrac{k_p}{k_n}}(V_{DD} + V_{Tp})}{1 + \sqrt{\dfrac{k_p}{k_n}}}$$

We find that $k_p \approx 5k_n$ which implies that $(W/L)_{p1} = 40/1.5$.

Now check that the capacitive loading of this inverter on the bit-line is insignificant compared to its total 5 pF. We calculate the loading of this inverter to be

$$C_G = (WL)_{n1}C_{ox} + (WL)_{p1}C_{ox} = (4\ \mu\text{m} \cdot 1.5\ \mu\text{m})\left(2.3\frac{\text{fF}}{\mu\text{m}^2}\right)$$

$$+ (40\ \mu\text{m} \cdot 1.5\ \mu\text{m})\left(2.3\frac{\text{fF}}{\mu\text{m}^2}\right) = 150 \text{ fF}$$

Since the loading is less than ten percent of the total bit-line capacitance, it is insignificant for hand calculations.

We have completed the initial hand design of our 256 × 32 embedded SRAM. We have apportioned the total 15 ns access time between the various functions, namely the word line driver, cell read time, sense amplifier time, and output driver time. The next step is to simulate the design using SPICE and adjust the device sizes to optimize the design. This exercise will be left to the reader.

➤ SUMMARY

This chapter discussed the analysis and design of three types of MOS memories: ROM, SRAM, and DRAM. Since many of the concepts covered previously are required in memory design, this chapter serves as an excellent review of the entire text. We concentrated on the hand calculations necessary to obtain an initial design. The important new concepts introduced in this chapter were:

- Definition and use of the various elements found in memory architectures including, memory cells, sense amplifiers, row and column decoders, and buffers.
- Basic circuit topology of a ROM, SRAM, and DRAM cell.
- Transient response of most memory cells is limited by the parasitic resistances and capacitances found on the word and bit lines.
- Analysis and design of the dynamic-latch sense amplifier used for the SRAM and DRAM.
- Circuit topologies for row and column decoders.
- A quick method to design buffers assuming the ratio of output-to-input capacitance is approximately four.
- The initial hand design of a memory given a set of specifications for the access and cycle time.

FURTHER READING

1. L. A. Glasser and D. W. Dobberpuhl, *The Design and Analysis of VLSI Circuits*, Addison-Wesley, 1985, Chapters 5 & 7. Chapter 5 has a discussion on sense amplifiers. Chapter 7 discusses several semiconductor memories and other structured arrays.
2. D. A. Hodges and H. G. Jackson, *Analysis and Design of Digital Integrated Circuits 2nd Ed.*, McGraw-Hill, 1988, Chapter 9. Includes an overview of several memory types such as EPROMs and bipolar memories.
3. J. Rabaey, *Digital Integrated Circuits: A Design Perspective*, Prentice Hall, 1996, Chapter 10. More advanced treatment of modern semiconductor memories.
4. N. H. E. Weste and K. Eshraghian, *Principles of CMOS VLSI Design 2nd Ed.*, Addison-Wesley, 1993, Chapter 8. Discusses a variety of memory cells and circuits with example layouts.
5. *IEEE Journal of Solid-State Circuits.* This periodical's coverage includes state-of-the-art memory circuits.
6. *IEEE Transactions on Electron Devices.* This periodical's coverage includes state-of-the-art memory cells.

PROBLEMS

For the following problems use the device data and layout design rules given below unless otherwise specified.

MOS Device Data

$\mu_n C_{ox} = 50\ \mu\text{A/V}^2$, $\mu_p C_{ox} = 25\ \mu\text{A/V}^2$, $V_{Tn} = -V_{Tp} = 1\ \text{V}$, $C_{ox} = 2.3\ \text{fF}/\mu\text{m}^2$, $-2\phi_p = 2\phi_n = 0.8\ \text{V}$, $\gamma_n = \gamma_p = 0.6\ \text{V}^{-1/2}$, $\lambda_n = \lambda_p = 0.066\ \ \text{V}^{-1}$ @ $L = 1.5\ \mu\text{m}$, $C_{Jn} = 0.1\ \text{fF}/\mu\text{m}^2$, $C_{Jp} = 0.3\ \text{fF}/\mu\text{m}^2$, $C_{JSWn} = 0.5\ \text{fF}/\mu\text{m}$, $C_{JSWp} = 0.35\ \text{fF}/\mu\text{m}$, $C_{ov} = 0.5\ \text{fF}/\mu\text{m}$, $L_{diffn} = L_{diffp} = 4.5\ \mu\text{m}$

1.5μm Design Rules

Minimum Active Width: 3.0 μm Minimum Active Space: 1.5 μm
Minimum Poly Width: 1.5 μm Minimum Poly Space: 1.5 μm
Minimum Metal Width: 3.0 μm Minimum Metal Space: 1.5 μm

Contacts: 1.5 μm × 1.5 μm

Contact Nesting: 1.5 μm inside the edges of Active or Polysilicon

Note: 1.0 μm and 0.75 μm design rules can be obtained by scaling the 1.5 μm design rules by 1/1.5 and 0.75/1.5, respectively.

Exercises

E13.1 A 4 Mbit memory architecture has 128 bits on the bit line. Each block of memory should contain 64 kbits.

(a) Calculate the number of word line addresses, N.

(b) Calculate the number of bit line addresses, M.

(c) Calculate the number of block addresses, P.

E13.2 For a 16 Mbit memory architecture, calculate the number of word-line, bit-line, and block addresses such that they are all equal.

E13.3 Suppose the 16 Mbit architecture in E13.2 has 8 data output lines. Re-calculate the number of word-line, bit-line, and block addresses such that they are all equal.

E13.4 In Fig. 13.7(a) an NMOS ROM array is shown. Draw a 4 × 4 PMOS ROM array labelling all lines including V_{DD} and ground.

E13.5 In Fig. 13.8 a precharged NMOS ROM array is shown along with qualitative waveforms showing a read operation.

(a) Draw a 4 × 4 precharged PMOS ROM array labelling all lines including V_{DD} and ground.

(b) Sketch the qualitative waveforms for a read operation.

E13.6 The layout for a 4 × 4 NMOS ROM array is shown in Fig. 13.9 using 1.5 μm design rules.

(a) Layout a 4 × 4 NMOS ROM array using 1.0 μm design rules. Assume the W/L of the ROM cell transistor is 3/1. What is the size of the cell?

(b) Repeat (a) using 0.75 μm design rules. Assume the W/L of the ROM cell transistor is 2.25/0.75.

E13.7 Using the layout for the ROM array with 1.0 μm design rules that you drew in E13.6, calculate the word line time constant, τ_{WORD} and the bit-line capacitance, C_{BIT} for a 256 × 256 array.

E13.8 Calculate τ_{WORD} and C_{BIT} using PMOS transistors in a 256 × 256 array. Each PMOS ROM cell transistor has a $W/L = 4.5/1.5$. Use the layout of the 4 × 4 NMOS ROM array in Fig. 13.9 as a guide.

E13.9 Calculate the read time for the 256 × 256 PMOS ROM array in E13.8.

E13.10 Given an NMOS ROM array with p-channel precharge devices where $C_{BIT} = 5$ pF and $\tau_{WORD} = 5$ ns

(a) Calculate the read time for $(W/L)_n = 4.5/1.5$ and a 0.5 V bit-line voltage swing.

(b) Calculate the cycle time for $(W/L)_p = 15/1.5$.

E13.11 Repeat E13.10 for a PMOS ROM array with an n-channel precharge device. The PMOS ROM cell transistor has a $(W/L)_p = 4.5/1.5$ and the NMOS precharge device has a $(W/L)_n = 15/1.5$.

E13.12 In Example 13.2, the SRAM array analyzed was 128 words × 32 bits. Calculate the word-line time constant, τ_{WORD} and bit-line capacitance, C_{BIT} if the array is

(a) 256 words × 32 bits

(b) 128 words × 64 bits

(c) 256 words × 256 bits

E13.13 In Example 13.2, the SRAM array analyzed was 128 words × 32 bits. Calculate the read time, t_{READ} if the array is

(a) 256 words × 32 bits

(b) 128 words × 64 bits

(c) 256 words × 256 bits

E13.14 In Example 13.2, the SRAM array analyzed was 128 words × 32 bits. Calculate the write time, t_{WRITE} if the array is

(a) 256 words × 32 bits

(b) 128 words × 64 bits

(c) 256 words × 256 bits

E13.15 Given a DRAM cell where a **1** corresponds to a stored voltage on C_S equal to $V_{DD} - V_{Tn}$ and a **0** corresponds to 0 V, calculate ΔV_{BIT} given

(a) $C_S = 50$ fF, $C_{BIT} = 1$ pF, $\gamma_n = 0.6\ \text{V}^{1/2}$

(b) $C_S = 30$ fF, $C_{BIT} = 0.7$ pF, $\gamma_n = 0.6\ \text{V}^{1/2}$

(c) $C_S = 50$ fF, $C_{BIT} = 1$ pF, $\gamma_n = 0.3\ \text{V}^{1/2}$

E13.16 A DRAM array has storage capacitors $C_S = 50$ fF and bit-line capacitance $C_{BIT} = 1$pF. The word-line voltage and W/L of the cell transistor govern the value of the time constant to perform read and write operations. Calculate this time constant for

(a) $W/L = 3/1.5$ $\quad V_{WORD} = 5$ V

(b) $W/L = 3/1.5$ $\quad V_{WORD} = 3$ V

(c) $W/L = 1.5/1.5$ $\quad V_{WORD} = 5$ V

E13.17 The time for the dynamic-latch sense amplifier to amplify the small signal on the bit lines from the cell to a larger signal that can be directly applied to a logic gate is governed by Eq. (13.17). Plot t_{SENSE} vs. $(\Delta V_{OUT})/(\Delta V_{IN})$ given the prefactor $C_{BIT}/(g_{mn} + g_{mp})$ has a value of

(a) 1 ns.

(b) 2 ns.

(c) 5 ns.

E13.18 There are often considerations in DRAM design that require that the array be placed in a well. If the CMOS process is an n-well process then the array transistors must be p-channel. Using Fig. 13.19 as a guide, redraw the circuit interfacing the cells and sense amplifiers using p-channel transistors in the array.

E13.19 In the dynamic-latch sense amplifier shown in Fig. E13.19 you are given that $(W/L)_{1\text{–}4} = 10/1.5$ and $(W/L)_{5\text{–}6} = 20/1.5$. Find t_{SENSE} for $(\Delta V_{OUT})/(\Delta V_{IN}) = 25$.

E13.20 You are given a series of six inverters in which the first inverter uses minimum-geometry devices. Calculate the load capacitance that can be driven assuming that all inverters can drive 4× their input gate capacitance in the specified time.

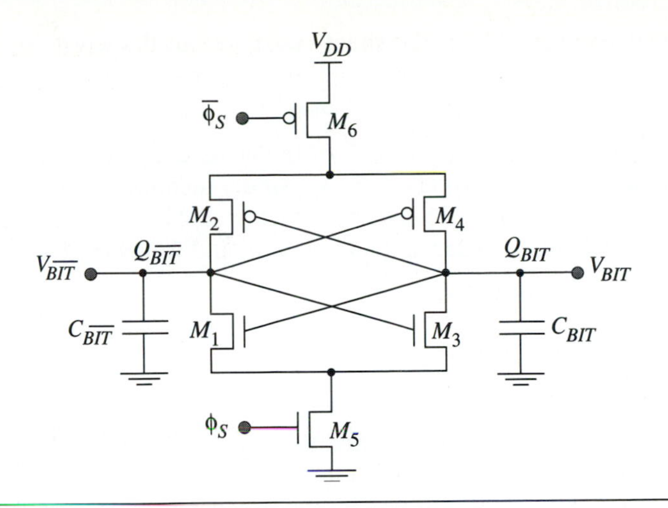

➤ **Figure E13.19**

PROBLEMS

P13.1 Following Example 13.1

(a) Find the read and cycle time for a 512×512 ROM array.

(b) Repeat (a) for a 128×128 ROM array.

P13.2 Following Example 13.1 suppose a process enhancement is made to reduce the word-line resistance to $1\Omega/\square$.

(a) Calculate the new word-line time constant, τ_{WORD}.

(b) Calculate the new read time, t_{READ}.

P13.3 By reducing the width of the ROM cell in Example 13.1 from 4.5/1.5, the word-line time constant, τ_{WORD} will be reduced at the expense of $t_{\Delta V_{BIT}}$.

(a) Find the read time for $(W/L) = 3/1.5$.

(b) Find the read time for $(W/L) = 6/1.5$.

P13.4 In this problem we explore the approximation to the time-varying gate-source voltage on the ROM cell transistor that we used to calculate the read time by using finer time steps.

(a) Find $\overline{V_{WORD}}$ over the intervals $0 \le t \le 0.5\tau$; $0.5\tau \le t \le \tau$; $\tau \le t \le 1.5\tau$; $1.5\tau \le t \le 2\tau$.

(b) Calculate $t_{\Delta V_{BIT}}$ for the $\overline{V_{WORD}}$ calculated in each interval above. Each interval is 0.5τ long.

(c) Add the ΔV_{BIT}'s starting with the first interval until the total ΔV_{BIT} is larger than 1 V.

(d) The total read time is the sum of the time intervals where the sum of each individual $\Delta V_{BIT} \le 1$ V, plus the fraction of the time in the interval required to make the total $\Delta V_{BIT} = 1$ V.

(e) Compare this value to that calculated in Example 13.1.

P13.5 Repeat P13.4 where each interval is 0.25τ.

P13.6 Repeat Example 13.2 for an array size of 256 words $\times$ 64 bits.

P13.7 Repeat Example 13.2 if the sheet resistance of the word line is reduced to $1\Omega/\square$.

P13.8 Repeat Example 13.2 if one uses PMOS access transistors with $W/L = 6/1.5$.

P13.9 In Example 13.2 we set the W/L of the cross-coupled inverter devices to be 3/1.5 and the access devices to be 6/1.5. Assume a technology enhancement allows us to increase the width of all devices by a factor of 2× without any penalty in cell area. Given the cell area remains at 27μm × 54μm and following Example 13.2

(a) Find τ_{WORD} and C_{BIT}

(b) Find t_{READ}

(c) Find t_{WRITE}

P13.10 In Eq. (13.9) ΔV_{BIT} for a DRAM cell is calculated assuming C_S is a constant capacitor (no voltage dependence). Since the capacitance is made up of a parallel combination of C_J and C_{ox} as shown in Fig. 13.15, the constant capacitance assumption is only valid if $C_{ox} \gg C_J$. Recalling that capacitance is defined as dQ/dV we need to integrate C_S across the stored voltage swing to find the change in charge placed on the bit line for a **0** and **1.** To more accurately calculate ΔV_{BIT} we can write

$$\Delta V_{BIT} = \frac{1}{C_{BIT} + C_S(V_{BIT})} \int_0^{(V_{DD} - V_{Tn})} C_S(V)\, dV$$ where V_{BIT} is the precharged voltage on

the bit line and $C_S(V) = C_{ox} + C_J(V)$. Find ΔV_{BIT} for a 5μm × 5μm storage capacitor given

(a) $C_{ox} = 2.3$ fF/μm and $N_a = 10^{15}$ cm^{-3}

(b) $C_{ox} = 2.3$ fF/μm and $N_a = 10^{16}$ cm^{-3}

(c) $C_{ox} = 2.3$ fF/μm and $N_a = 10^{17}$ cm^{-3}

P13.11 Using the top view of a DRAM cell shown in Fig. 13.15 as a guide, carefully layout to scale a DRAM cell with $C_S = 50$ fF using

(a) 1.5 μm design rules.

(b) 1.0 μm design rules.

(c) 0.75 μm design rules.

P13.12 For the layouts drawn in P13.11 calculate the bit-line capacitance C_{BIT} for each cell and the RC time constant to perform write and read operations.

P13.13 In the cross-section of the one-transistor cell shown in Figure P13.13 we see that the cell has a reverse-biased pn junction that can lose stored charge due to leakage current. A stored **1** has $V_{CS1} = V_{DD} - V_{Tn}$ and will leak down toward the substrate potential, 0 V. Assume that $C_S = 50$ fF, $C_{BIT} = 1$pF, the bit line is precharged to $V_{DD}/2$, and that the sense amplifier requires $\Delta V_{IN} \geq 20$ mV to sense the logic **1** in the specified speed. Calculate the maximum leakage current that can be tolerated if V_{DD} and the time between refreshes t_{STORE} are

(a) $V_{DD} = 5$ V and $t_{STORE} = 100$ ms

(b) $V_{DD} = 3$ V and $t_{STORE} = 100$ ms

(c) $V_{DD} = 5$ V and $t_{STORE} = 500$ ms

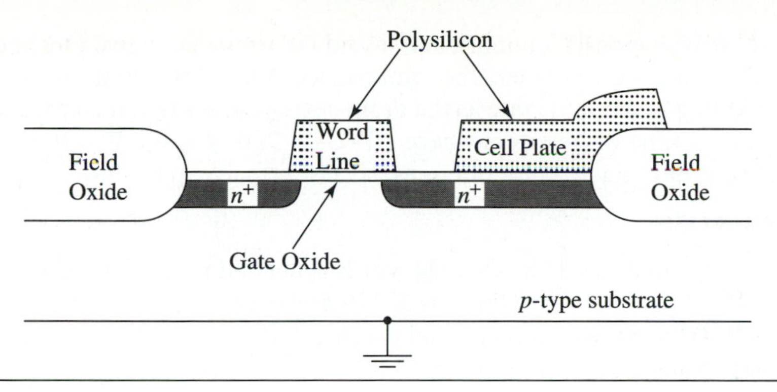

➤ **Figure P13.13**

P13.14 We know from Chapter 4 that when $V_{GS} = V_{Tn}$ a small amount of inversion charge is still present in the channel. In fact the device still passes a small current. Since DRAM cells are sensitive to very small currents this **subthreshold current** is very important in DRAM design. The drain current is exponentially dependent on the gate-source voltage in this region of operation. Assume the transistor passes 1 nA at $V_{GS} = V_{Tn}$ and that the current is reduced by one decade for every 100 mV reduction in gate voltage. If we increase V_{Tn}, the leakage current is reduced at the expense of stored charge since $V_{DD} - V_{Tn}$ will be reduced. Calculate the minimum threshold voltage such that a logic **1** state can be sensed correctly if the storage time between refreshes is 100 ms. Use the same parameters as P13.13 with $V_{DD} = 5$ V.

P13.15 In Example 13.3 the dynamic-latch sense amplifier transistors were sized to meet a 10 ns sense time specification. Suppose we had to minimize the area of the sense amplifier. Using the same parameters as Example 13.3 find the new t_{SENSE} if

(a) $(W/L)_{1\text{–}2} = 3/3$ and $(W/L)_{3\text{–}6} = 6/3$

(b) $(W/L)_{1\text{–}2} = 3/3$ and $(W/L)_{3\text{–}6} = 3/3$

P13.16 In Example 13.3 the replica cell and bit line were precharged to 2.5 V. However, this led to asymmetrical bit-line voltage signals for a logic **1** and **0**. Find the precharge voltage that yields symmetric bit-line voltage swings. Do NOT neglect the backgate effect.

P13.17 Draw a circuit which uses a ROM structure decoder that selects 1 of 8 word lines. Arrange the transistors such that the physical address 000 is

(a) at the top of the array.

(b) the 4th row in the array.

(c) the last row of the array.

P13.18 Draw a circuit which selects 1 bit line out of 8 bit lines using a ROM style predecoder and pass gates.

P13.19 Draw a circuit which selects 1 bit line out of 8 bit lines using a 3-bit tree decoder.

P13.20 Given four inverters in series starting with a minimum-geometry inverter and ending with a 10 pF load, calculate the delay by using inverters which increase by 4×, 5×, and 6×. Compare the results with SPICE.

P13.21 There is obviously a minimum rise and fall time that a given technology can achieve. This limit occurs when the capacitance due to the drain-bulk depletion region is growing at the same rate as the drain current when W is increases. Calculate the rise and fall time of inverters where $(W/L)_p = 2(W/L)_n$ for $W = 10$ μm, 100 μm, 1000 μm, 10000 μm. Make a plot of t_r, t_f vs. W. Use a load capacitance of 1pF.

Design Problems

D13.1 A ROM that has 32 bits × 1024 words is needed to provide instructions to a processor. Note that each word line has 32 bits and each bit line has 1024 bits. Design an NMOS ROM array with a p-channel precharge device that will yield a read time of 10 ns and a cycle time of 15 ns given $\Delta V_{BIT} = 0.5$ V. Use 1.5 μm design rules. The ROM cell transistor should be made just wide enough to meet the specifications while trying to minimize the overall area of the cell.

D13.2 To see the value of a sense amplifier especially when the bit-line capacitance is large repeat D13.1 given $\Delta V_{BIT} = 0.1$ V. Recall that the sense amplifier allows us to reduce the bit-line voltage swing from the cell.

D13.3 Repeat D13.1 using 1.0 μm design rules.

D13.4 In the design of the ROM array in D13.1 several device parameters such as C_{ox}, C_{Jn}, $\mu_n C_{ox}$ etc. determined the transient response. If you could change just one device parameter to reduce the area of the cell while keeping the same performance specifications which one would you choose? Justify your answer.

D13.5 Layout a 6 transistor SRAM cell using 1.5 μm design rules. Assume that $W/L = 3/1.5$ for the cross-coupled inverter n and p-channel transistors and that $W/L = 6/1.5$ for the n-channel access transistors.

D13.6 Design a dynamic-latch sense amplifier for an SRAM with $C_{BIT} = 1$pF. Assume the SRAM bit lines are precharged to 2.5 V and the bit-line voltage swing is ±100 mV. The output of the sense amplifier must be < 1.5 V for a logic **0** and > 3.5 V for a logic **1**. The time to sense t_{SENSE} should be less than 5 ns.

D13.7 Following the design style of Example 13.3 design a sense amplifier for a DRAM requiring 128 bits on each bit line. Assume $C_{BIT} = 3.5$ pF and the rest of the parameters are the same as Example 13.3.

D13.8 In Example 13.3 the replica cell stored an intermediate value of charge since it was precharged to a value of $V_{DD}/2$. Referring to Fig. D13.8 and following the same design style, design the sense amplifier with a replica capacitor value $C_S/2$.

D13.9 In D13.8 we precharged the bit lines to $V_{DD}/2$. In this problem design the sense amplifier using a precharge value for the bit lines that makes the bit-line voltage swing symmetrical for a logic **0** and **1**.

D13.10 Layout a decoder using the ROM array style which selects one word line from 4 address bits. Use 1.5 μm design rules and try to minimize the pitch between word lines.

D13.11 Design a buffer with a series of inverters starting with a minimum-geometry inverter with sizes $(W/L)_n = 3/1.5$ and $(W/L)_p = 6/1.5$. The overall delay from a step input at the minimum geometry gate to the 50 pF load capacitance should be less than 10 ns.

D13.12 Repeat Example 13.4 using inverters that increase by 5× for each stage.

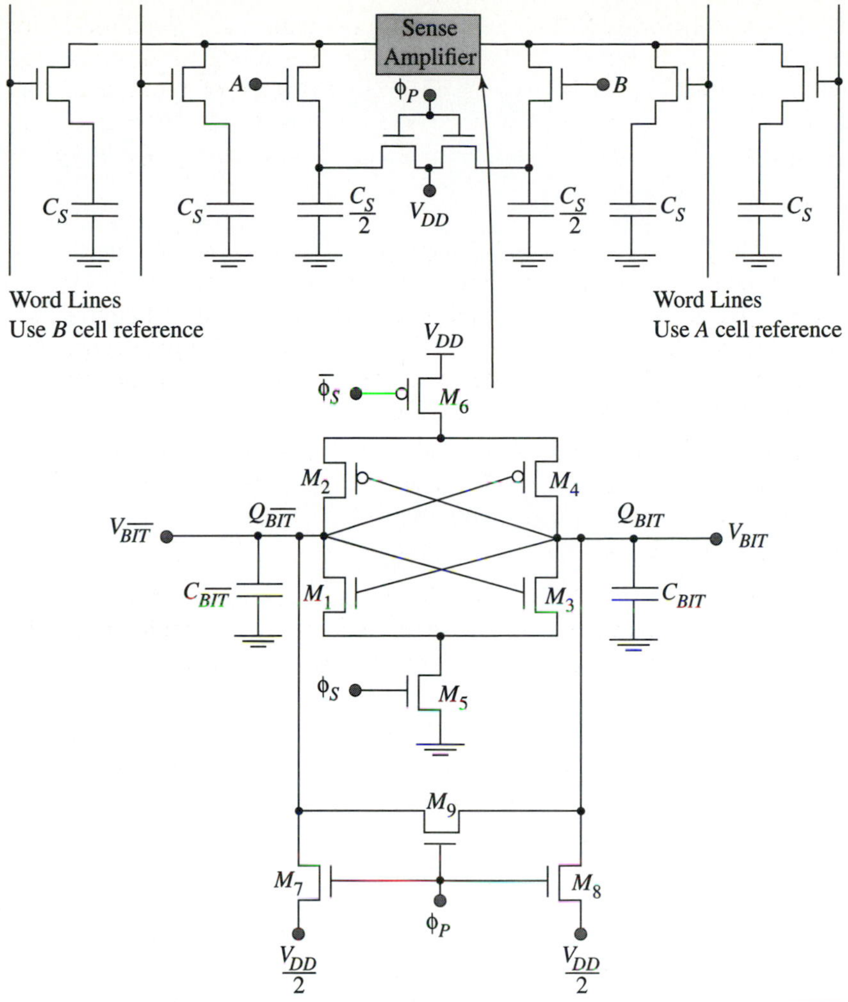

➤ **Figure D13.8**

D13.13 Run SPICE on the SRAM design in Section 13.6. Try to reduce the area of the peripheral circuits (all except the cell) while maintaining the speed specifications.

D13.14 Following the design style performed in Section 13.6, design an SRAM with 256 words × 32 bits with an access time of 20 ns. Use SPICE to optimize the design for minimum area.

D13.15 Following the design style performed in Section 13.6, design an SRAM with 256 words × 32 bits using 1μm design rules with an access time of 15 ns. Use SPICE to optimize the design for minimum area.

Index

D

N